The Concise Oxford Companion to the Theatre

The Concise Oxford Companion

to the

Theatre

Edited by
PHYLLIS HARTNOLL
and
PETER FOUND

Oxford   New York
OXFORD UNIVERSITY PRESS
1992

Oxford University Press, Walton Street, Oxford OX2 6DP

Oxford New York Toronto
Delhi Bombay Calcutta Madras Karachi
Petaling Jaya Singapore Hong Kong Tokyo
Nairobi Dar es Salaam Cape Town
Melbourne Auckland

and associated companies in
Berlin Ibadan

Oxford is a trade mark of Oxford University Press

British Library Cataloguing in Publication Data
Data available

Library of Congress Cataloging in Publication Data
The Concise Oxford companion to the theatre / edited by Phyllis
Hartnoll and Peter Found. —2nd ed.
p.   cm.
1. Theater—Dictionaries.   2. Drama—Dictionaries.   I. Hartnoll,
Phyllis.   II. Found, Peter.   III. Title: Oxford companion to the
theatre.
PN2035.C63   1992   792'.03—dc20   91–23749
ISBN 0–19–866136–3

Typeset by Latimer Trend & Co. Ltd., Plymouth
Printed in Great Britain by
Butler & Tanner Ltd.
Frome, Somerset

Introduction

In the twenty years since the first edition of *The Concise Oxford Companion to the Theatre* was published, the theatrical landscape has changed considerably. In 1972 the Fringe was a comparatively recent phenomenon, and the opening of Britain's first National Theatre building was still four years away; and since then a new generation of actors, playwrights, and directors has now arrived, some of them world famous.

The task of updating and augmenting the material of the first edition, while still not greatly exceeding the original length, presented a formidable challenge, particularly since it was felt that some of the existing entries were too short. Our policy has been to delete some of the less important entries, especially those on minor theatres and biographies of those no longer considered of the first rank, in order to make way for new people and topics and to allow a fuller treatment of others. This runs the inevitable risk of offending some readers by discarding particular favourites, but we believe that a new edition with many more, but shorter, entries would have been more frustrating than useful. In particular, more information has been provided in the entries for playwrights on the subjects they wrote about, and, where it seemed helpful, the careers of actors, directors, and others have been surveyed in greater detail than previously. In the interests of creating space, we have aimed to avoid duplicating information under different headings. This will sometimes necessitate turning to a second, cross-referenced entry to take up the story or to discover more detail. The entry for Peter Brook, for example, ends with the foundation of his International Centre of Theatre Research, the entry on which continues his biography; a similar treatment has been accorded to Tovstonogov, whose entry ends with his move to the Gorky Theatre.

The reader's search for information is aided by plentiful cross-references, indicated by asterisks within the text and small capitals in signpost entries. Names of important people who are mentioned in an article but do not have an entry of their own appear in bold type with dates. English words are used in preference to foreign ones wherever possible, although play titles are normally given in the original language followed by a translation if necessary. Dates following play titles are those of first performance or of revivals, depending on the context; it is indicated that a performance is a revival where this is not obvious. The Royal Shakespeare Company, to which frequent reference is made, is abbreviated to RSC throughout.

As in the first edition, we have aimed at worldwide coverage, though with a discernible bias towards the British and US theatre, which we judge to be of principal interest to most of our readers. The work, while basically mainstream, takes cognizance of the Fringe and the avant-garde. All of the material taken over from the first edition has been looked at afresh, and much of it amended, sometimes drastically. Entries have been written for major new arrivals, and a few unjust omissions from the first edition, such as the Gershwins and Wendy Hiller, have been rectified. We have striven to set the cut-off point for inclusion at a consistent level; in particular, it was decided early on to exclude the careers of those newcomers under the age of forty as a regrettable but necessary means of keeping the work within bounds. Readers searching in vain for rising and risen younger stars are asked to bear this in mind. There will inevitably be disagreement with some of our choices—both inclusions and exclusions—but selection of this kind must ultimately be subjective and in a subject with as long a history and as wide a range of expression as the theatre, the task is particularly daunting.

Our grateful thanks are due to Miss Dorothy Swerdlove, of the Performing Arts Research Center in New York, and to Dr David Gardner, who provided information on the American and Canadian theatre respectively.

Peter Found
Phyllis Hartnoll
June 1991

Illness prevented the Elder Editor from undertaking much of the final editing and from the pleasure of seeing her book through the press, but no better person for such an onerous and in many ways difficult task could have been discovered for this than Peter Found. For my part, I could not have wished for a kinder or more congenial friend as well as associate.

Phyllis Hartnoll
Lyme Regis
1991

A

Abbey, Henry E. (Edwin or Eugene) (1846–96), American theatre manager who, after experience in Akron (his birthplace), Buffalo, and Boston, was encouraged by Charlotte *Crabtree to take over the management of the *Park Theatre 2, New York, in 1876. He was instrumental in attracting to New York many Continental stars, among them *Irving and Ellen *Terry, who with their London company opened Abbey's own theatre (see KNICKER-BOCKER THEATRE) in New York in 1893. Abbey was one of the first managers to provide good plays and companies outside New York, and his early death was a great loss to the burgeoning America-wide theatre.

Abbey Theatre, Dublin, opened in 1904 as the permanent home of the National Theatre Society (see IRISH NATIONAL DRAMATIC SOCIETY). Funds were supplied by Miss *Horniman, who also gave the theatre an annual subsidy. The first directors were Lady *Gregory, *Synge, and *Yeats. From the first the theatre was under pressure from the nationalists, within the company and outside, to conform politically. Yeats, for example, defended Synge's *The Playboy of the Western World* (1907) with a passion equal to that of audiences who condemned it as a betrayal of national ideals. In 1910, however, he refused to close the theatre during the funeral of Edward VII according to Miss Horniman's wishes, and her subsidy was withdrawn. By now the Abbey had achieved an international reputation, chiefly for its naturalistic acting style, largely the work of the *Fay brothers who had been impressed by the *Théâtre Libre in Paris. Although Yeats had hoped to encourage *poetic drama, plays analysing provincial life in the manner of *Ibsen became the staple repertoire, as in the work of Lennox *Robinson and T. C. *Murray.

Foreign tours, organized by Lady Gregory from 1911 to 1914, brought fame if not fortune to the Abbey, though Irish-American audiences took violent exception to several of the plays, and in Philadelphia the entire cast of *The Playboy of the Western World* was summoned on a charge of obscenity. The actors made a considerable impression on discerning playgoers, including the young Eugene *O'Neill. The Abbey's *repertory system influenced the emerging theatres in Europe and the USA. In 1925 an annual subsidy was provided by the newly formed Free State Government. The plays of *O'Casey brought back dwindling audiences helped by the publicity engendered by his treatment of the 1916 rebellion in *The Plough and the Stars* (1926), and new playwrights such as St John *Ervine, George *Shiels, and Brinsley MacNamara came forward with lively comedies. *Shaw's plays were also produced frequently. In 1925 the Peacock Theatre was opened for poetic and experimental productions and was made available to other companies, the *Gate Theatre having its beginnings here in 1928. The late 1920s saw a resurgence at the Abbey, with an excellent company which included F. J. *McCormick, Barry Fitzgerald, Maureen Delaney, and Sara *Allgood in plays characterized by colourful language, exuberant characters, a deft mixture of comedy and tragedy, and a realistic urban or rural kitchen setting.

After the death of Yeats in 1939 a new phase began. The Abbey was managed from 1941 to 1967 by **Ernest Blythe** (1889–1975), who saw its function as being 'to preserve and strengthen Ireland's national individuality'. The cultivation of Gaelic drama became a priority. In 1947 there was a public protest in the theatre over a decline in production standards. The Abbey was destroyed by fire in 1951 and the company moved to the much larger *Queen's, which imposed a mainly commercial quality, though notable plays by Brendan *Behan, Paul Vincent *Carroll, Denis *Johnston, and others were produced. The new Abbey and Peacock Theatres were opened in 1966; the building also contains the Society's fine art collection. New plays by Brian *Friel, John B. Keane, Thomas Kilroy, Tom Murphy, and many others were presented. The work of Samuel *Beckett also enjoyed a special place in the repertoire, and plays by Irish dramatists of earlier centuries, such as *Farquhar, *Goldsmith, *Sheridan, and *Wilde, were more frequently revived, as well as British, European, and American classics. More recently, the work of a younger generation of playwrights has been introduced, and extensive foreign touring has been resumed. At least 10 new plays by Irish authors are produced annually. Since its inception, the Society has presented over 800 new works, probably a world record among state theatres.

Abbot of Misrule, of Unreason, see FEAST OF FOOLS.

Abbott, George Francis (1887–   ), American producer, director, and playwright, originally an actor. He had already had some success in the co-writing and co-directing of many different types of play, among them a melodrama, *Broadway* (1926), and a farce, *Three Men on a Horse* (1935), when he embarked on the long series of musicals for which he is mainly remembered. They include *On Your Toes* (1936), *The Boys from Syracuse* (1938), *Where's Charley?* (1948), based on *Charley's Aunt*, *The Pajama Game* (1954), *Damn Yankees* (1955; London, 1957), and *Fiorello!* (1959), all of which were written or co-authored by himself. In addition, he directed a number of musicals by other writers, among them *Pal Joey* (1940), *On the Town* (1944), *Call Me Madam* (1950), and *A Funny Thing Happened on the Way to the Forum* (1962; London, 1963). Abbott celebrated his 95th birthday by staging a revival of *On Your Toes* (1983; London, 1984), which had previously been revived in 1954. He was honoured in 1966 by having the Fifty-Fourth Street Theatre in New York named after him. It had opened in 1928 as the Craig, in 1934 became the Adelphi, and was pulled down in 1970. Its history was undistinguished, except for 1936–9, when it was taken over by the *Federal Theatre Project.

Abell, Kjeld (1901–61), Danish dramatist and artist, who worked as a stage designer in Paris and with Balanchine at the *Alhambra Theatre, London, in 1931. His first play, *Melodien, der blev væk*, was produced in Copenhagen in 1935 and in London a year later as *The Melody That Got Lost*. None of his other plays has been produced in English, though three have been published in translation: *Anna Sophie Hedvig* (1939), *The Queen on Tour* (1943), a protest against the loss of freedom during the German occupation of Denmark, and *Days on a Cloud* (1947). Abell's work represents a sustained attempt to bring an experimental theatre of dream and vision to Denmark.

Abington, Fanny [*née* Frances Barton] (1737–1815), English actress, wife of a music-master, from whom she soon separated. She made her first appearance on the stage in Mrs *Centlivre's *The Busybody* at the *Haymarket in 1755. On the recommendation of Samuel *Foote she was engaged for *Drury Lane, where she found herself overshadowed by Kitty *Clive. She therefore left London for Dublin, where she remained for five years, returning to Drury Lane at the express invitation of *Garrick, who disliked her but considered her a good actress. During the 18 years that she remained there she played a number of important roles, and was the first Lady Teazle in *Sheridan's *The School for Scandal* (1777). She was also much admired as Miss Prue in *Congreve's *Love for Love*, in which character she was painted by Reynolds. In 1782 she went to *Covent Garden, where she remained until 1790, finally retiring in 1799.

Absurd, Theatre of the, name given by Martin Esslin, in a book of that title published in 1962, to the plays of a group of dramatists, among them *Beckett and *Ionesco and, in England, *Pinter, whose work has in common the basic belief that man's life is essentially without meaning or purpose and that human beings cannot communicate. This led to the abandonment of dramatic form and coherent dialogue, the futility of existence being conveyed by illogical and meaningless speeches and ultimately by complete silence. The first, and perhaps most characteristic, play in this style was Beckett's *Waiting for Godot* (1952), the most extreme—since it has no dialogue at all—his *Breath* (1969). The movement, which now seems to have passed its zenith, nevertheless made a profound and lasting impression on the theatre everywhere.

The English dramatist **N**(orman) **F**(rederick) **Simpson** (1919–) is considered by some to write in this vein; his best-known play, *One-Way Pendulum* (1959), features an attempt to teach 500 weighing machines to sing the Hallelujah Chorus.

Accesi, company of the **commedia dell'arte*, first mentioned in 1590. Ten years later they were under the joint leadership of Pier Maria Cecchini, known as Fritellino, and Tristano *Martinelli, the first to play *Arlecchino, with whom they toured France. With them were Martinelli's brother Drusiano, Flaminio Scala (see CONFIDENTI), and possibly Diana da Ponti, formerly leader of the Desiosi company. On their next visit to France they were without Arlecchino, and soon after Cecchini joined forces with the younger *Andreini; but the quarrels which ensued between their respective wives made it impossible for them to work together and Andreini withdrew. Little is known of the Accesi's later activities, though Silvio *Fiorillo, the first to play *Capitano Matamoros, was with them in 1621 and in 1632, the year of his death.

Achard, Marcel (1899–1974), French dramatist, whose first play *Voulez-vous jouer avec moâ* (1923) was accepted and staged in Paris by *Dullin, for whom a year later Achard wrote an excellent adaptation of *Jonson's *Epicoene* as *La Femme silencieuse*. Other works such as *Jean de la lune* (1929) and *Le Corsaire* (1938) followed, all directed by *Jouvet, attracting attention by their skilful mingling of the fragility and irony of human love, expressed through burlesque which turned unexpectedly to pathos. After the Second World War Achard changed his approach somewhat, becoming more easily accessible and seeking less to 'make an effect', as he had done with his introduction of pantomime creatures into his earlier plays. These later plays proved popular in New York, the first to be seen there being an adaptation of *Auprès de ma blonde* as *I Know My Love* (1949) by S. N. *Behrman.

Achurch, Janet (1864–1916), English actress, who made her first appearance at the *Olympic in 1883. She was one of the first actresses in England to appear in *Ibsen, being seen as Nora in *A Doll's House* in 1889. In 1896 she produced *Little Eyolf* with herself as Rita, Mrs Patrick *Campbell as the Ratwife, and Elizabeth *Robins as Asta. She was also seen as *Shaw's heroine in *Candida* and as Lady Cicely Waynflete in his *Captain Brassbound's Conversion* (both 1900). Shaw called her 'the only tragic actress of genius we now possess', and some excellent descriptions of her acting can be found in his *Our Theatres in the Nineties*. She retired from the stage in 1913.

Ackermann, Konrad Ernst (1712–71), German actor, who in about 1742 joined a travelling company. A handsome man, with a restless vagabond temperament well suited to the life of a strolling player, he played mainly in comedy, and was much admired in such parts as Tellheim in *Lessing's *Minna von Barnhelm*. After a few years he formed his own company, taking with him as his leading lady Sophie *Schröder, whom he married after the death of her husband. Together they toured Europe, being joined eventually by *Ekhof and by Sophie's son F. L. *Schröder, and in 1767 they opened the first and short-lived German National Theatre in *Hamburg, the inspiration behind Lessing's *Hamburgische Dramaturgie*. Schröder left the company, but returned to it shortly before Ackermann's death, after which he took complete control. Ackermann's daughters **Dorothea** (1752–1821) and **Charlotte** (1757–75) both played leading roles in his productions. Dorothea retired on her marriage in 1778, but Charlotte committed suicide at 17, reputedly driven to it by the strain of too many new and taxing roles and constant public appearances, for which Schröder was blamed.

Acoustics. The Greek open-air theatre, built into the hillside, provided a perfect place for sound, as did the Roman, with its towering façade. The medieval cathedral must have presented difficulties not so apparent in smaller churches, though even there some sounds probably got lost in the roof. In the market-place, audibility must have been as chancy as it is today in a flat open space. But the use of rhymed couplets, and a good deal of miming and horseplay, must have helped comprehension. In both the classical and the Elizabethan open-air theatres the use of verse helped to carry the voice over the auditorium, and the actors seem to have had good voices. Acoustics became a problem in the theatre when all plays were given indoors, often in rooms acoustically unsuitable for the purpose. Luckily the development of the Italian opera-house produced in the 18th century buildings which, repeated all over Europe, provided a good place for the sound of music. Because the tiers of boxes were heavily draped, the reverberation was short. The flat ceilings, without domes, and the plentiful use of baroque ornamentation, diffused the sound and so prevented echoes, and the large amount of wood used in the buildings meant that the orchestral tone was adequate, in spite of the size of the auditorium. But for the spoken word these theatres were not so well equipped. The theatre built by Vanbrugh in the Haymarket (see HER MAJESTY'S), with its high vaulted roof, concave in shape, was found, on its opening in 1705, to have sacrificed audibility to architecture. It was said that scarcely one word in ten could be heard distinctly, and that the articulated sounds of the speaking voice were drowned by the hollow reverberation of one word upon another. Luckily the smaller, more intimate English playhouses were better suited to spoken drama, and Vanbrugh's theatre became the first English opera-house, being used for a number of operas by Handel from 1711 onwards. The large theatres built at the beginning of the 19th century, while preserving the horseshoe auditorium which was good for sound, though not always for sight, adopted the domed ceiling, which led to some notable echoes focused from a particular stage position. Even when echoes were not noticeable, curved ceilings gave an unequal distribution of sound, so that some seats were better for hearing than others. But the baroque tradition of ornamentation, stage boxes and heavy drapery was retained, and helped to keep the reverberation short. Later in the century the rise of domestic comedy and drama led to the building of smaller theatres, offering intimate acoustic qualities.

In the 20th century, the design of theatres everywhere was radically changed. Baroque ornamentation was replaced by large continuous surfaces in hard plaster, the stage boxes were removed, the auditorium became fan-shaped, and large areas of sound-absorbing velvet drapery were removed. At first the new fan-shaped plan and splayed proscenium were approved of on the grounds that they provided useful reflecting surfaces, and audibility in the rear seats was improved. But it was then discovered that any return of sound from surfaces at the rear, including balcony fronts, ceiling coves, and balustrades, found its way to the front seats. Complaints of inaudibility now came from the occupants of the expensive stalls. It had been too readily assumed that a powerful sound-absorbing material on the rear wall behind the audience would prevent any return of sound; but in practice, commercial sound-absorbents, often covered with paint, were found to be less effective than modern hard plasters on the reflecting walls and ceiling. Another factor which adversely affected the acoustics was the relatively large area occupied by the rear wall, which was too often given the most dangerous curve possible, struck from a centre near the stage front. The result was not a complete echo but a prolonging of word-endings, likely to obscure rapid speech. It was clear that the fan-shape needed modification, and that the rear wall should not be curved on plan, but straight or polygonal: also that in large theatres it was wise to avoid curved parapets, seat risers, and gallery fronts, and to restore the side-boxes and draped proscenium. A large bare forestage also increases the risk of reverberation in the front of the house. The value of the convex curve, instead of the concave, has now been recognized in the profiling of reflecting canopies and in corrugated ceilings, the latter being also stepped instead of splayed.

The modern demand for the *open stage and particularly for *theatre-in-the-round has brought with it further acoustic problems. The human voice has a direction, and is not equally well heard behind and at the side, particularly since the old-fashioned projected speech has been discarded in favour of an intimate conversational tone. New techniques are being evolved to overcome these problems, and also those of *flexible staging. Where drapes and carpets and upholstered seats are discarded for the sake of easy convertibility, there is all the more need for good distributed sound-absorbents on walls and ceilings in order to reduce reverberation. Electronic amplification has now become commonplace in musicals, though still resisted in the straight theatre. Modern public address systems can compensate for excessive reverberation times and have been used with particular success in large spaces such as cathedrals; these too however are seldom acceptable on orthodox stages.

Act, division of a play, each of which may contain one or more scenes. Greek plays were continuous, the only pauses being marked by the *chorus. Horace was the first to advocate the division of tragedies into five acts, a suggestion followed during the Renaissance by academic dramatists. The first English writer to adopt it was Ben *Jonson. There is no proof that Shakespeare divided his plays thus, and the divisions in the printed copies were probably introduced by the editors in imitation of Jonson. In comedy more licence was allowed to the individual, two or three acts being quite usual, even in *Molière. Modern drama usually keeps to three acts, as being convenient for actors and audience alike, but two acts are sometimes found, and many Shakespeare revivals are performed with only one interval. Division into four acts, found mainly in the 19th century, is now rare.

Act-Drop, name given in the late 18th century to the painted cloth which closed the *proscenium opening between the acts of a play (see CURTAIN).

Acting Company, see CITY CENTER OF MUSIC AND DRAMA.

Actor, Actress, Acting. The need to express emotion, whether in music, dancing, gesture, or speech, seems to be inherent in man, and to have developed originally in connection with religious observances. Nothing is known of the earliest actors, but in classical Greece, where they still took part in a religious festival (see DIONYSUS), they were men of repute, performing excellent tragedies and comedies in large open-air playing-places (see THEATRE BUILDINGS), each with its all-important *chorus. In Rome, where tragedy gave way before low comedy and farce, they were less highly regarded, some being slaves. Although there were no actresses, women appeared on stage as dancers and mimes, sexually provocative. With the spread of Christianity the theatre was proscribed and sank into oblivion, to be rescued, ironically, by those who persecuted it. When the Church decided to educate its illiterate congregations by the 'acting-out' of scenes from the Old Testament, and from the dramatic life of Christ himself (see LITURGICAL DRAMA), the itinerant entertainers who had tried to keep theatrical traditions alive came into their own, as did the *minstrels. Soon all the large towns of Europe had their own *mystery plays, while smaller towns and villages produced their own local plays or pageants.

With Latin no longer a universal language, vernacular drama began to emerge throughout Europe during the 16th century, bringing with it the professional actor and eventually actress. In Italy it first appeared with the *commedia dell'arte, in Spain with the work of Lope de *Rueda, in England with the building of the *Theatre, in France with the establishment of the Hôtel de *Bourgogne. In Germany disunity and internal dissension delayed the rise of theatre companies until the 18th century. In Russia there was no national or professional theatre till the mid-19th century. In England, the arrival of actresses had to wait until the Restoration in 1660 brought to the throne Charles II, who had grown accustomed to women on stage during his exile on the Continent, and demanded the same amenity in his own country. It is interesting to note that in the Far East, where religion maintained its hold on the theatre far longer than in the West, women seem to have retained their original position as singers and dancers in temple ceremonies, but do not appear to have formed part of any static or itinerant group which performed plays in public. Actresses are now beginning to emerge in China and Japan, for example, but mainly in modern plays.

The position of the actor and actress was for a long time precarious throughout Europe. In Catholic countries they were refused the sacraments; legally Shakespeare and his contemporaries were 'rogues and vagabonds' unless under royal or noble protection. It was not until 1895 that the actor in England achieved social credibility, with the knighting by Queen Victoria of Henry *Irving.

Fashions in acting change constantly, one method giving way to another, one convention replacing an earlier one. In Greece the chorus had to be singers and dancers; in tragedy the chief actors were masked, and needed above all a fine voice and a noble presence, as did the tragic actors of 17th-century France. In comedy everywhere the actor needed to be lively, inventive, quick-witted, and something of an acrobat. Some periods imposed their own conditions. When tragedian and comedian were separate employments they seldom crossed each other's boundaries. *Melodrama helped to break down the barriers, while the 'intimate drama' which replaced it gave little scope for ample gestures or raised voices. The modern actor, reared on improvisation and mime, is an 'all-rounder', but increasingly one whose style is tempered by the demands of cinema and television for intimacy and the play of facial expression. As always, the great actor or actress will come at the right moment, equipped to suit the time. Handicaps can be overcome. Something of acting can be taught, but the art of acting is innate. The ideal player is the product of a delicate balance of intuition and hard work, tempered by the fires of experience.

Actors' Company, see McKELLEN.

Actors' Studio, see METHOD and STRASBERG.

Actors Theatre of Louisville, the official State Theatre of Kentucky, founded in 1964. First housed in a loft over a store, it moved to a converted railway station and in 1972 to its present home which consists of two theatres, the Pamela Brown Auditorium, seating 637 round a thrust stage, and the Victor Jory Theatre, seating 161. The former presents seven productions of classical and modern plays in a season which runs Sept.–May, while the latter houses the 'Off-Broadway' Series, a programme of provocative plays that has provided many American and world premières. Both theatres participate in the annual Festival of New American Plays. Plays having their first production there include Tricks (1971; NY, 1973), based on *Molière's Les Fourberies de Scapin; D. L. Coburn's *Pulitzer Prize-winner The Gin Game (1977; NY, 1977; London, 1979), seen on Broadway and in London with Hume *Cronyn and Jessica *Tandy; Marsha Norman's Getting Out (1977); and James McLure's Lone Star (1979), both of which were also seen in New York. The company makes an annual regional tour, and also runs a free children's theatre.

Act-Tunes, musical interludes between the acts of plays. Their mention in a number of stage directions indicates that they were customary in the Elizabethan theatre. In the Restoration theatre the act-tunes became very important, and composers like Purcell were commissioned to write them. The introductory music was sometimes known as the Curtain-Music or Curtain-Tune.

Adam de la Halle (c.1245–c.1288), French trouvère, nicknamed 'le Bossu (Hunchback) d'Arras', and one of the few medieval *minstrels about whom anything is known. He was the author of Le Jeu de la feuillée (c.1276). Bawdy, satirical, and anti-clerical, it marks the beginning of lay, as distinct from ecclesiastical, drama in France. For the Court of Robert II, Count of Artois, Adam wrote also a pastoral, Le Jeu de Robin et de Marion, which, by virtue of its music—for Adam was a composer as well as a poet—is now considered by some the first French light opera. First printed in 1822, it was played in a modernized version in Arras in 1896.

Adamov, Arthur (1908–70), Russian-born French dramatist. His first play, La Parodie, was not produced until 1952, though written in 1947. Two later plays, La Grande et la petite manœuvre and L'Invasion, had been performed in 1950. His early work, including Le Professeur Taranne and Tous contre tous (both 1953), had much in common with the Theatre of the *Absurd, as had Ping-Pong (1955), a satire on the world of commerce and politics. With Paolo Paoli, an exposure of the corruptions of the French social scene which was first produced by *Planchon at Lyons in 1957, Adamov moved towards the *epic theatre of *Brecht, whose influence was even more evident in Le Printemps '71 (1961), dealing with the Paris Commune of 1871, and La Politique des restes (1963). Two of Adamov's later works were inspired by *Gorky—Les Petits Bourgeois (1959) and Les Âmes mortes (1960).

Adams, Edwin (1834–77), American actor, who made his first appearance at Boston in 1853 in Sheridan *Knowles's The Hunchback. For the opening performance of *Booth's Theatre, New York, in 1869, he played Mercutio to the Romeo of Edwin *Booth. His best-known role, however, in which he toured all over the United States, was Enoch Arden in a dramatization of *Tennyson's poem. He made his last appearance in San Francisco, as Iago.

Adams [Kiskadden], **Maude** (1872–1953), American actress, daughter of the leading lady of the Salt Lake City *stock company. As a child she appeared in such parts as Little Eva in one of the many dramatizations of Harriet Beecher Stowe's Uncle Tom's Cabin. In 1888 she made her first appearance in New York, and three years later was engaged to play opposite John *Drew. She first emerged as a star with her performance as Lady Babbie in The Little

Minister (1897), a part which *Barrie rewrote and enlarged specially for her. Her quaint, elfin personality suited his work to perfection, and she appeared successfully in the American productions of his *Quality Street* (1901), *Peter Pan* (1905), *What Every Woman Knows* (1908), *Rosalind* (1914), and *A Kiss for Cinderella* (1916). She was also much admired as the young hero of *Rostand's *L'Aiglon* (1900), and in such Shakespearian parts as Viola, Juliet, and Rosalind. In 1918 she retired, not acting again until 1931, when she appeared on tour in *The Merchant of Venice* as Portia to the Shylock of Otis *Skinner. In 1937 she made her last appearance in Rostand's *Chantecler*, playing the title-role as she had in its first production in 1911.

Addison, Joseph (1672–1719), English politician and man of letters, author of *Cato*, a tragedy on the French classical model seen at *Drury Lane in April 1713. It was supported by the Whigs for political reasons, and by the Tories for effect. Written in unrhymed heroic couplets, it contains some fine poetry, but is not theatrically effective. The part of Cato was originally offered to Colley *Cibber, who declined it, and it was finally played by Barton *Booth, with Anne *Oldfield as Lucia. Addison's only other play was a short comedy, *The Drummer; or, The Haunted House* (1716), also performed at Drury Lane. His dramatic theories and criticisms can be found in several papers of the *Spectator*, which he edited with Richard *Steele, while the *Tatler*, 42 (1709), which he also edited, contains an amusing mock inventory of the properties and furnishings of Drury Lane.

Ade, George (1866–1944), American journalist, humorist, and playwright, famous for his wisecracks, whose plays of contemporary life were full of homely humour and wit. Among the most successful were *The County Chairman* (1903), *College Widow* (1904), which added a new phrase to the American language, *Just Out of College* (1905), and *Father and the Boys* (1908). Ade was also responsible for the books of several musical comedies, among them *The Fair Co-Ed* (1909).

Adelphi Theatre, London, in the Strand, originally the Sans Pareil, was built by a wealthy merchant, to display the talents of his daughter, opening in 1806 with *Miss Scott's Entertainment*. It prospered and after changing hands in 1818 reopened as the Adelphi, playing mainly *melodrama and *burletta. Successful productions were *Moncrieff's *Tom and Jerry; or, Life in London* (1821), and several dramatizations of *Dickens. In 1844 Mme *Céleste and Ben *Webster took over the theatre, making it the home of 'Adelphi drama', mostly written by *Buckstone. A larger theatre was built on the site in 1858, and productions there of *The*

Colleen Bawn (1860) and *The Octoroon* (1861) by *Boucicault were extremely popular. In 1879 the theatre was leased to the Gatti brothers (see GATTI'S). A series of Adelphi melodramas followed, starring William *Terriss, who was assassinated by a madman at the entrance to the theatre in 1897. Rebuilt, the theatre was briefly called the Century, but the old name was restored by popular demand. Under George *Edwardes it housed a series of excellent musical comedies, beginning with *The Quaker Girl* in 1908. In 1930, after further rebuilding, the theatre reopened with Rodgers and *Hart's *Ever Green*, the first of a series of productions by C. B. *Cochran. This was followed by *Knoblock's adaptation of Vicki Baum's novel *Grand Hotel* (1931), *Coward's revue *Words and Music* (1932), and Eric Maschwitz's musical *Balalaika* (1936). The revival of *Novello's *The Dancing Years* (1942) ran for 969 performances. Cochran returned after the Second World War with the musicals *Big Ben* (1946), *Bless the Bride* (1947), and *Tough at the Top* (1949). The first production to reach 1,000 performances was the revue *London Laughs* (1952). Later came *Auntie Mame* (1958) based on Patrick Dennis's novel, with Beatrice *Lillie, and two musicals by Lionel *Bart, *Blitz!* (1962) and *Maggie May* (1964). *Charlie Girl* (1965), another musical, became the longest-running production at the theatre, completing 2,200 performances. Subsequent successes were *Sondheim's *A Little Night Music* (1975) and revivals of the musicals *Irene* (1976), *My Fair Lady* (1979), and *Me and My Girl* (1985).

Adelphi Theatre, New York, see ABBOTT.

Admiral's Men, Elizabethan company which, with Edward *Alleyn as their star actor, was the only real rival of the *Chamberlain's Men. Their patron was Lord Howard, who became an admiral in 1585, and at Christmas in the same year the 'Admiral's players' made their first appearance at Court. In 1590–1 they were housed in the *Theatre with some of *Strange's Men. After a quarrel with Richard *Burbage over finance, most of the two companies moved to the *Rose, under *Henslowe. When the Chamberlain's Men were formed in 1594 some of the Admiral's Men joined them, while the rest formed themselves into an independent company under Alleyn. They had a large repertory of plays, most of which, except for those of *Marlowe, have been lost or forgotten. The retirement of Alleyn in 1597 was a great blow, but in 1600 he returned and the company moved into a new playhouse, the *Fortune. On the death of Elizabeth I they lost Alleyn for good and were renamed Prince Henry's Men. Their young patron died in 1612, and they became the Palsgrave's Men. In 1621 the Fortune was burnt down, and all the wardrobe and playbooks were lost. Two years later a new

Fortune Theatre opened with practically the same company, but the combination of plague and the death of James I proved too much for it. It was disbanded in 1631, and its remnants were probably absorbed into other organizations.

Adrian [Bor], Max (1903–73), Irish-born actor who first came into prominence during a season at the *Westminster Theatre in 1938, when he played Pandarus in a modern-dress *Troilus and Cressida* and Sir Ralph Bloomfield Bonnington in *Shaw's *The Doctor's Dilemma*. He was then with the *Old Vic company, and in 1944 joined *Gielgud's repertory company at the *Haymarket Theatre, playing, among other roles, Tattle in *Congreve's *Love for Love*. He was however an instinctive *revue artist, and appeared successfully in *Light and Shade* (1942), followed by *Tuppence Coloured* (1947), *Oranges and Lemons* (1948), *Penny Plain* (1951), *Airs on a Shoestring* (1953), and *Fresh Airs* (1956). He was seen in *Coward's *Look after Lulu* (1959), and in 1960, as a founder member of Peter *Hall's *RSC, gave outstanding performances as a malevolent Machiavellian Cardinal in *Webster's *The Duchess of Malfi*, and a hauntingly melancholic Feste in *Twelfth Night*. One of his finest characterizations was the tetchy, wily, egotistical Serebryakov in *Chekhov's *Uncle Vanya* at the 1963 *Chichester Festival. Later the same year he joined the *National Theatre company, his roles including the Inquisitor in Shaw's *Saint Joan*. His one-man shows, on Shaw in 1966 and on *Gilbert and Sullivan in 1969, brought him further renown. Adrian was that rare phenomenon in the theatre, a brilliant and fantastic individualist who could nevertheless fit easily into a company.

Advent Play, see LITURGICAL DRAMA.

Advertisement Curtain, see CURTAIN.

Aeschylus (525/4–456 BC), Greek dramatist, born at Eleusis, near Athens, who also won distinction as a soldier in the Persian War. He is said to have written 90 plays, of which the titles of 79 are known, though only seven are extant: the *Suppliant Women* (? c.490 BC), the *Persians* (472), the *Seven against Thebes* (469), the *Prometheus Bound* (? c.460), and the trilogy known as the *Oresteia* (the *Agamemnon*, the *Choephori*, or *Libation-Bearers*, and the *Eumenides*) (458). About a quarter of his plays must have been *satyr-dramas, in which genre he was an acknowledged master. Nothing of these survives except a few fragments.

Aeschylus may reasonably be regarded as the founder of European drama. By reducing the size of the *chorus and introducing a second actor into the play (see AGON), he made the histrionic part as important as the lyric, and so turned oratorio into drama. The transition can be seen in his early plays. In the *Suppliant Women* the chorus is the chief actor; in the *Persians* the chorus still gives the play its formal unity; but the *Seven against Thebes* is clearly dominated by the chief actor. In his later plays Aeschylus used (in a highly individual way) the innovation of the third actor, introduced by *Sophocles.

Dramatists competing at the Athens festival had to present three serious plays and one satyr-play; Aeschylus normally made the three plays a connected 'trilogy' in which each part, though a complete unity, was a coherent part of a larger unity. This gave his drama the amplitude which his vast conceptions needed. The normal scheme may be very baldly summarized as the offence, the counter-offence and the resolution; sin provokes sin, until justice asserts itself. The only complete trilogy which has survived is the *Oresteia*. Of the other plays, the *Suppliant Women* and the *Prometheus Bound* were the first plays of their trilogies, the *Seven against Thebes* the third of its; and, judging by what has been recovered, the scale of these trilogies was hardly less majestic than that of the *Oresteia*.

These conceptions were matched by a bold dramatic technique, an immense concentration, a wonderful sense of structure, and magnificent poetry. Aeschylus made the utmost use of spectacle and colour; and, in virtue of the beauty and strength of his choral odes, he might well be regarded as one of the greatest of lyric poets. By virtue of his many talents, Aeschylus imposed a unity on the theatre which it was soon to lose: he was his own director, chief actor, designer, composer, and choreographer. As a unique honour to him, it was enacted in Athens after his death that his plays might be revived at the festivals, to which normally only new plays were admitted.

Afinogenov, Alexander Nikolaevich (1904–41), Soviet dramatist, who began writing in 1926. His first important play (translated as *Fear* and published in *Six Soviet Plays*, 1936) was performed at the Leningrad Theatre of Drama in 1931. Dealing with the conversion to socialism of a psychologist who has claimed that fear governs the USSR, it was one of the first Soviet plays to combine good technique and dramatic tension with party propaganda. This fusion was even more apparent in a later play, seen at the *Vakhtangov Theatre in 1934, which, as *Distant Point*, was produced at the *Gate Theatre, London, in 1937. The death of Afinogenov, who was killed in an air raid in Nov. 1941, deprived Soviet Russia of one of the country's few early dramatists who might have had a universal appeal.

African Roscius, see ALDRIDGE.

After-Piece, short comedy or farce, performed after a five-act tragedy in London theatres of the 18th century, partly to afford light relief to the spectators already present and partly to attract the middle-class business men and others who found the opening hour of 6 p.m. too early. Half-price was charged for admission. The after-piece was often a full-length comedy cut to one act, but many short plays were specially written for the purpose by *Garrick, *Murphy, *Foote, and others.

Agate, James Evershed (1877–1947), English dramatic critic. From 1923 until his death he was dramatic critic of the *Sunday Times*, succeeding S. W. *Carroll and being succeeded by Harold *Hobson. His weekly articles, many of which were collected and published in book form, were vigorous and outspoken, and always entertaining, in spite of his refusal to admit greatness in any actor later than *Irving.

Agit-Prop, term formed from 'agitation' and 'propaganda', used to describe a movement during the early years of the Russian Revolution which sought to teach Communism and socialism directly through the theatre, mainly by the playing of short, *revue-like sketches, often acted in an apparently impromptu fashion at street corners, factory gates, and the entrances to political meetings. The actors were mainly amateurs, though the best known of the early troupes, the Blue Blouses, were predominantly professional. Agit-Prop was soon overtaken by the new *Socialist Realism but had a great influence on contemporary theatre, particularly in Germany, home of the Red Revels and the Red Rockets.

Agon, Greek word meaning 'contest', used to define the conflict which lies at the heart of Greek *tragedy, transmuted into a clash between two principal characters, sometimes with a hint of physical violence. Normally, however, the *agon* took the form of a debate, where the only weapons were words. For this a second actor, the Deuteragonist, was added to the original *Protagonist of early drama. A third actor, the Tritagonist, was added later, thus widening the scope of the action, all three men being of equal repute and standing.

Ainley, Henry Hinchliffe (1879–1945), English actor, possessor of a remarkably fine voice and great personal beauty and charm. He made his first success in Stephen *Phillips's *Paolo and Francesca* (1902), at the *St James's Theatre under George *Alexander, and soon became known as an excellent romantic actor in such plays as Justin McCarthy's *If I were King* later the same year. In 1912 he made a great impression as Leontes in *The Winter's Tale*, directed by *Granville-Barker, showing his versatility a year later by playing Ilam Carve in Arnold *Bennett's *The Great Adventure*. One of his finest parts was Hassan in James Elroy Flecker's poetic play of that name in 1923, which was staged at His (now *Her) Majesty's

Theatre, one of several London theatres with whose management he was associated. Illness kept him from the stage for many years, but he returned in 1929 to score an instantaneous success as James Fraser in St John *Ervine's *The First Mrs Fraser*, playing opposite Marie *Tempest. A year later he was seen as an excellent Hamlet in a Command Performance; he finally retired in 1932.

Akimov, Nikolai Pavlovich (1901–68), Soviet scene designer and director, who first attracted attention by his designs for *Ivanov's *Armoured Train 14-69* (1929) and *Afinogenov's *Fear* (1931) at the Leningrad Theatre of Comedy. Moving to Moscow, he worked at the *Vakhtangov Theatre and was responsible for the famous *'Formalist' production of *Hamlet* in 1932, in which Hamlet faked the Ghost, and Ophelia, a 'bright young thing', was not mad but drunk: the play was taken off in deference to public opinion. In 1936 Akimov became Art Director of the Leningrad Theatre of Comedy, being responsible for a beautifully staged *Twelfth Night* there. In the 1950s he strove to create an original repertory, and fought for the acceptance of new Soviet comedies. His name is particularly associated with the plays of *Shwartz. From 1955 until his death he was on the staff of the Leningrad Theatrical Institute.

Akins, Zoë (1886–1958), American poet and dramatist, whose first play *Déclassée* (1919) provided an excellent part for Ethel *Barrymore. Her most interesting play was perhaps *First Love* (1926), but all her other plays were overshadowed by the popularity of *The Greeks Had a Word for it* (1930), which was equally successful in London in 1934. In 1935 she was awarded a *Pulitzer Prize for her dramatization of Edith Wharton's novel *The Old Maid*. She dramatized several other American novels, and adapted several plays from the Hungarian and the French, among them Verneuil's *Pile ou face* as *Heads or Tails* (1947). Her last play, produced in 1951, was *The Swallow's Nest*.

Alarcón y Mendoza, see RUIZ DE ALARCÓN Y MENDOZA.

Albee, Edward Franklin (1928–   ), American dramatist, grandson (by adoption) of **Edward Franklin Albee** (1857–1930), who in 1920 owned a circuit of some 70 *vaudeville houses, with an interest in about 300 others. The younger Albee, one of the few major dramatists to emerge in the United States in the 1960s, had his first play, a one-acter entitled *The Zoo Story*, performed in 1959 in Berlin (NY and London, 1960). It was followed by several other short plays, including *The Death of Bessie Smith* (1960) and *The American Dream* (1961). In 1962 he scored a Broadway success with his first full-length play, *Who's Afraid of Virginia Woolf?* (London, 1964), about a night of conflict between an ineffectual professor and his sharp-tongued wife. *The Ballad of the Sad Café* (1963; London, 1969), based on a novel by Carson McCullers, was followed by *Tiny Alice* (1964; *RSC, 1970), which was declared by six New York critics to be incomprehensible, and the *Pulitzer Prize-winner *A Delicate Balance* (1966; RSC, 1969), in which the lives of a middle-aged couple are disrupted by the simultaneous crises of a sister, a daughter, and friends. *All Over* (1971; RSC, 1972), concerning quarrels at a dying man's bedside, was another caustic look at family relationships. *Seascape* (1975), which the author directed, depicts a seashore encounter between two couples undergoing a process of self-assessment—one human, the other belonging to the lizard family. *Seascape* won another Pulitzer Prize, but Albee's later work found little favour. *The Lady from Dubuque* (1980) and *Lolita* (1981), adapted from Nabokov's novel, had only brief runs in New York, as did *The Man Who Had 3 Arms* (1983). The last—in which the protagonist's three arms brought him short-lived fame, which vanished when the third arm withered—was considered an attack on critics who maintained that *Who's Afraid of Virginia Woolf?* was his only notable work.

Albery, Sir Bronson James (1881–1971), English theatre manager. Son of the dramatist James *Albery and the actress Mary Moore (later Lady *Wyndham), he assisted the latter in the management of the *Criterion, *Wyndham's, and New (now *Albery) Theatres until her death, afterwards sharing it with his stepbrother Howard Wyndham. During his long career, and in particular during the 1930s, he was responsible for the introduction to London of many interesting plays. His son **Sir Donald** (1914–), who had been associated with him for many years, successfully continued his policy, becoming Chairman and Managing Director of The Wyndham Theatres Ltd.

Albery, James (1838–89), English dramatist, whose 'decayed gentleman' Digby Grant in *Two Roses* (1870) provided Henry *Irving with one of his first successes in London. Most of Albery's other plays, mainly taken from the French, were intended as vehicles for his wife Mary Moore (later Lady *Wyndham), the best of them being probably *The Pink Dominoes* (1877) and *The Crisis* (1878), the latter based on *Augier's *Les Fourchambaults*.

Albery Theatre, London, in St Martin's Lane, seating 900, built for Charles *Wyndham, who opened it in 1903 as the New Theatre with a revival of *Parker and Carson's *Rosemary*, after which it settled down to a consistently successful career. From 1905 to 1913 Fred *Terry and Julia *Neilson occupied it for a six-month annual season, and many of their most successful plays were seen there, including Baroness Orczy's *The Scarlet Pimpernel* (1905). The theatre

also housed an annual revival of *Barrie's *Peter Pan* for several years. Among outstanding productions have been a dramatization of Louisa M. Alcott's *Little Women* (1919), in which Katharine *Cornell made her only London appearance; A. A. Milne's *Mr Pim Passes By* (1920); and *Shaw's *Saint Joan* with Sybil *Thorndike (1924). A year later came the long run of Margaret Kennedy's *The Constant Nymph*, which saw the first appearance of John *Gielgud at a theatre where he later appeared in Gordon Daviot's *Richard of Bordeaux* (1933), *Hamlet* (1934), *Romeo and Juliet* and Obey's *Noah* (both 1935). Among later productions were *The Taming of the Shrew* (1937), *O'Neill's *Mourning Becomes Electra* (1938) and *Priestley's *Johnson over Jordan* (1939) with Ralph *Richardson. After the bombing of the *Old Vic and *Sadler's Wells, the New became the London headquarters of both companies in 1941. Sadler's Wells withdrew in 1944 and the Old Vic in 1950, in which year T. S. *Eliot's *The Cocktail Party* began a successful run. The theatre housed a series of excellent plays, including Dylan Thomas's *Under Milk Wood* (1956) and Ray *Lawler's *Summer of the Seventeenth Doll* (1957). In 1960 Lionel *Bart's *Oliver!*, a musical based on *Dickens's *Oliver Twist*, began a run of several years. (Another long run began in 1977.) In 1973 the theatre changed its name to honour its former manager, Bronson *Albery. *Shaffer's *Equus* was transferred there from the *National Theatre in 1976. The musical *Pal Joey* (1980) transferred from the *Fringe and Mark Medoff's *Children of a Lesser God* (1981) from the *Mermaid. The musical *Blood Brothers* began a long run in 1988.

Aldridge, Ira Frederick (1804–67), the first great American Negro actor, who in 1863 became a naturalized Englishman. He had already appeared on the New York stage when in 1826, billed as the African Roscius, he made his London début as Othello at the *Royalty Theatre. He was also good as Macbeth and as Mungo in *Bickerstaffe's *The Padlock*. Regarded as one of the outstanding actors of the day, he was the recipient of many honours, amassed a large fortune, and married a white woman. He was last seen in England in 1865, and then returned to the Continent, where he had first toured in 1853. He was immensely popular in Germany, where he played in English with a supporting cast playing in German. His Lear was much admired in Russia, the only country in which he appeared in the part.

Aldwych Theatre, London, three-tier theatre seating 1,100, was built for Seymour *Hicks, who opened it in 1905 with himself and his wife Ellaline *Terriss in a revival of their 'dream fantasy' *Bluebell in Fairyland*. The building was damaged during the First World War but after restoration reopened and in 1923 had its first outstanding success with *Tons of Money*, a farce by Will *Evans and Valentine. In the cast were Ralph *Lynn and Tom *Walls, and both actors stayed on to appear with Robertson *Hare in a succession of 'Aldwych farces' written by Ben *Travers. The series ended in 1933 with *A Bit of a Test*. There were notable productions of American plays: Lillian *Hellman's *Watch on the Rhine* (1943), *Sherwood's *There Shall Be No Night* (1945), with the *Lunts, Tennessee *Williams's *A Streetcar Named Desire* (1949), with Vivien *Leigh, and Maxwell *Anderson's *The Bad Seed* (1955). Edith *Evans starred in *Fry's *The Dark Is Light Enough* (1954).

In 1960, the Aldwych became the London home of the *RSC. Extensive alterations were made to the interior, including the installation of a completely new lighting system and an apron stage, with a proscenium opening 31ft. wide; the seating capacity was slightly reduced, to 1,030. The first season opened with Peggy *Ashcroft in *Webster's *The Duchess of Malfi*, and the theatre subsequently housed new plays and revivals as well as productions transferred from Stratford. From 1964 to 1975 (except for 1974) the Aldwych also housed the annual *World Theatre Season organized by Peter *Daubeny. The first new work presented by the RSC was John *Whiting's *The Devils* (1961), and in the same year *Giraudoux's *Ondine* and *Anouilh's *Becket* represented recent work from abroad, *Brecht's *The Caucasian Chalk Circle* following in 1962. Other Europeans whose work was staged included *Dürren-matt, *Hochhuth, Peter *Weiss, and Marguerite *Duras, and there were two or more works from *Pinter, *Albee, and Peter *Nichols. Notable revivals were *Gogol's *The Govern-ment Inspector* (1965), *Vanbrugh's *The Relapse* (1967), *O'Casey's *The Silver Tassie* (1969), and *Boucicault's *London Assurance* (1970). The virtual 'discovery' of *Gorky in the English theatre was marked by five productions. The year 1980 was particularly successful, with a fine revival of O'Casey's *Juno and the Paycock* and two mammoth productions: *The Greeks*, a three-part adaptation by John *Barton of 10 Greek plays, and David Edgar's eight-hour adaptation of *Dickens's *Nicholas Nickleby*.

In 1982 the RSC transferred its London base to the *Barbican Theatre and The *Pit in the Barbican Centre. The Aldwych's subsequent productions have included Neil *Simon's *Brighton Beach Memoirs* (1986) and Arthur *Miller's *A View from the Bridge* (1987), both transferred from the *National Theatre, and *Stoppard's *Hapgood* (1988).

Aleichem [Rabinovich], **Sholom** (1859–1916), Jewish writer, who in 1888 was owner and editor of a Kiev newspaper. In 1905 he emigrated to the United States, where a number of plays based on his novels and short stories of life in the Jewish communities of the Ukraine were performed in the Yiddish Art Theatres, mainly through the efforts of Maurice *Schwartz. His kindly, simple, but shrewd characters offered considerable scope to such actors as *Mikhoels in the USSR, Zero *Mostel in the USA, and *Topol in London. In 1959 Aleichem's centenary was celebrated by a production at the Grand Palais Theatre—the last surviving Yiddish theatre in London—of his three-act comedy *Hard to be a Jew*. An adaptation of another comedy, *Tevye the Milkman*, as a musical, *Fiddler on the Roof*, was successfully produced on Broadway in 1964 (London, 1967).

Aleotti, Giovanni Battista, see TEATRO FARNESE.

Alexander, Sir George [George Alexander Gibb Samson] (1858–1918), English actor, and manager of the *St James's Theatre from 1891 until his death. He made his first appearance at Nottingham in 1879 and in 1881 was seen in London. In 1890 he entered into management on his own. His tenancy of the St James's was both financially and artistically rewarding. He made his greatest success in the dual role of Rudolf Rassendyll and the King in Anthony Hope's *The Prisoner of Zenda* (1896), and was also much admired as Villon in Justin McCarthy's *If I were King* (1902) and Karl Heinrich in Bleichmann's *Old Heidelberg* (1903). Among his important productions were *Wilde's *Lady Windermere's Fan* (1892) and *The Importance of Being Earnest* (1895), in which he played John Worthing, *Pinero's *The Second Mrs Tanqueray* (1893) and *His House in Order* (1906), Stephen *Phillips's verse-drama *Paolo and Francesca* (1902), and Jerome K. *Jerome's *The Passing of the Third Floor Back* (1908). The last play in which he appeared was Louis N. *Parker's *The Aristocrat* (1917). He was a man of distinguished appearance and great charm.

Alexandra Theatre, Birmingham, see BIRMINGHAM.

Alexandrinsky Theatre, St Petersburg, see PUSHKIN THEATRE, Leningrad.

Alfieri, Vittorio Amedeo (1749–1803), Italian dramatist, chiefly remembered as a writer of austere tragedies in verse. His first, *Cleopatra*, was performed at Turin in 1775 with great success. Of the other 20, of which it has been said that their action 'flies like an arrow to its mark', the best are probably *Saul* (1782) and *Mirra* (1784). A production in 1967 of one of the half-dozen comedies written towards the end of his life, *Il Divorzio* (1802), showed a gift for satiric humour. Alfieri, who was born in Asti of a noble and wealthy family, had an unhappy childhood, and at an early age left home to travel widely in Europe. He became the devoted lover of the Countess of Albany, wife of the Young Pretender, to whom he left all his books and manuscripts.

Alhambra, The, famous London *music-hall, whose ornate Moorish-style architecture dominated the east side of Leicester Square for over 80 years. Opened unsuccessfully in 1854 as an exhibition centre, it was converted into a music-hall seating 3,500 on four tiers. As the Alhambra, a name it retained through successive changes of title (Palace, Music-Hall, Theatre, etc.), it opened in 1860, one of its first successes being the trapeze artist Léotard making his first appearance in London in 1861. In 1870 the theatre lost its licence because its manager presented 'an indecent dance'—the can-can; but a year later a dramatic licence was granted for the first time. In 1882 the theatre was burned down, reopening a year later. Its great days as a music-hall were from 1890 to 1910, when the spectacular ballets rivalled those of the *Empire. In 1912 Charlot inaugurated a series of *revues, and in 1916 came George Grossmith's picture of London life *The Bing Boys Are Here*, with George *Robey, which, like its successor *The Bing Boys on Broadway* (1918), had a long run. In 1923 came the revue *Mr Tower of London*, which made Gracie *Fields a star, and in 1931 *Waltzes from Vienna*, featuring the music of the Strauss family, gave the theatre, whose fortunes were beginning to decline, a long run. There were seasons of Russian ballet in 1933 and 1935, and the building was demolished in 1936, the Odeon cinema being erected on the site.

Alienation, English term for the *Verfremdungseffekt*, aimed at by *Brecht by means of placards, films, strip cartoons, and stylization, all designed to produce in the audience a state of critical detachment from the drama being presented. The actor assists this process by standing outside the character he is portraying, rather than, as in the method employed by *Stanislavsky, identifying with it.

Allen, Chesney, see CRAZY GANG.

Allen, Viola (1869–1948), American actress, who made her first appearance on the stage in 1882 and was from 1891 to 1898 a member of *Frohman's stock company at the *Empire, where she gained a great reputation. After leaving Frohman she played her most famous role, Glory Quayle in Hall Caine's *The Christian* (1898). Other contemporary plays included Clyde *Fitch's *The Toast of the Town* (1905), and she also excelled in Shakespearian roles such as Viola in *Twelfth Night* and Rosalind in *As You Like It*. In 1915–16 she toured with J. K. *Hackett as Lady Macbeth, and she made her last New York appearance in 1916 as Mistress Ford in *The Merry Wives of Windsor*.

Alleyn, Edward (1566–1626), Elizabethan actor and the founder of Dulwich College. He was much admired by *Nashe and *Jonson and

considered the only rival of Richard *Burbage. Plays in which he is known to have appeared include *Marlowe's *Tamburlaine the Great* (c.1587), *The Tragical History of Dr Faustus* (c.1589), and *The Jew of Malta* (c.1590), and *Greene's *Orlando furioso* (c.1591). He was already known as a good actor in 1583 and remained on the stage until the accession of James I in 1603. He married the stepdaughter of *Henslowe, and succeeded to most of his property, and to his papers, which are now at Dulwich. His second wife was the daughter of the poet Donne.

Alley Theatre, Houston, Tex., founded in 1947 by Nina Vance, who led the theatre until her death in 1980. Its first home was a rented dance studio seating 87, which took its name from the narrow passage leading up to it. Forced to move for safety reasons in 1949, the company then converted and occupied a disused fan factory. They created an arena theatre in which their first production was Lillian *Hellman's *The Children's Hour*. By 1957 the resident company had become fully professional, being one of the first three such companies in the United States, the others being the *Arena Stage, Washington, and the *Cleveland Play House. In 1968 it moved into its present permanent home, a striking building whose nine towers give it the appearance of a medieval castle. It has two auditoriums: the Large Stage, seating 824, with a thrust stage which can provide a variety of playing areas, and no seat more than 58ft. from the stage; and the smaller Neuhaus Arena Stage, seating 299 and based on the company's second home, with the same unique alcove playing areas. In a season lasting Oct.–June the theatre presents a balanced repertoire of worldwide drama, including original works and classic revivals, plays for children, and readings of new plays.

Allgood, Sara (1883–1950), Irish actress, sister of Maire *O'Neill. In 1903 she joined the *Irish National Dramatic Society, appearing in the opening productions at the *Abbey Theatre in 1904. In 1907 she created the part of Widow Quin in *Synge's *The Playboy of the Western World*. She was for a time a member of Miss *Horniman's company at Manchester, but returned to the Abbey in 1920 and played Juno Boyle in *Juno and the Paycock* (1924), in which she also appeared in London and America, and Bessie Burgess in *The Plough and the Stars* (1926), both by *O'Casey. In London in 1936 she made a great success in *Bridie's *Storm in a Teacup*. She made her last appearance on the stage in New York in 1940, and then appeared only in films.

All-Russian Theatrical Society, see YABLOCHKINA.

Almeida Theatre, Islington, London, neoclassical 300-seat theatre with an open stage founded in 1981 in premises built in 1836. Its programme features the work of the Almeida Theatre Company, founded in 1979; theatre, music, and dance events produced by visiting companies and artists from Britain and abroad; and the Almeida International Festival, one of the most important festivals of contemporary music. The Almeida Theatre Company has developed a distinctive style of inventive contemporary presentation, combining the work of leading writers, performers, designers, and directors from Britain and abroad. It staged the world premières of *Lyubimov's adaptation of Dostoevsky's *The Possessed* (1985) and Howard *Barker's *The Possibilities* (1988) and the British premières of Botho *Strauss's *The Tourist Guide* and Heiner *Müller's *Hamletmachine* (both 1987). It staged seasons by the *RSC in 1988 and 1989.

Aloni, Nissim, see HEBREW THEATRE.

Alternative Theatre, see COLLECTIVE CREATION, COMMUNITY THEATRE, and FRINGE THEATRE.

Altwiener Volkstheater, see VIENNA.

Álvarez Quintero, Serafín (1871–1938) and **Joaquín** (1873–1944), Spanish dramatists, brothers and collaborators in about 200 light comedies based on the characteristic life and customs of Andalusia and suffused with kindly tolerance and gentle good humour. Several of them were translated into English by Helen and Harley *Granville-Barker, of which *Fortunato* and *The Lady from Alfaqueque* (New York, 1929) were produced in London in 1928 with John *Gielgud in the leading roles. *A Hundred Years Old* (New York, 1939) was also seen in London in 1928, followed by *The Women Have Their Way* (1933; New York, 1939), *Doña Clarines* (1934), and *Don Abel Writes a Tragedy* (1944).

Alvin Theatre, New York, see NEIL SIMON THEATRE.

Amandiers, Théâtre des, see CHÉREAU.

Amateur Theatre. In Britain amateur theatre reached its height in the 19th century, when many of the best-known amateur societies were founded, among them the Old Stagers, who have given performances closely associated with the Canterbury Cricket Festival since 1842; the Manchester Athenaeum Dramatic Society, which has presented plays regularly since 1854; and the student societies of *Cambridge (ADC) founded in 1855 and *Oxford (OUDS) founded in 1885. The dramatic societies of other universities, though formed much later, have also done excellent work, as can be seen by the productions at the annual student drama festival organized by the National Union of Students.

Amateurs include about 5,000 groups attached to organizations such as colleges, youth clubs, women's institutes, and community centres; the dramatic or operatic societies which flourish mainly in large towns and sometimes reach a high standard; and the Little Theatres which, though fewer in number, are important for artistic and aesthetic reasons. They usually own or lease their own theatres and are members of the Little Theatre Guild of Great Britain, founded in 1946. One of the most interesting is the *Maddermarket Theatre at Norwich, which houses the Norwich Players, founded in 1911. The oldest is the Stockport Garrick Society, which dates from 1901.

Although the most significant amateur activity takes place outside London, there are—since the demise of the excellent St Pancras People's Theatre, which lasted from 1926 to 1940—three important centres in the capital—the *Questors Theatre at Ealing, the Tower Theatre in Islington, opened in 1953, and the Mountview Theatre School, Crouch Hill, founded in 1969. The left-wing theatre *Unity was something of a hybrid, having during its career been sometimes amateur and sometimes professional.

Several organizations form a unifying element in British amateur drama. The Central Council for Amateur Theatre, founded in 1977, is the national umbrella organization, with a membership of about 5,000 amateur dramatic and operatic societies. The British Drama League was founded in 1919 by **Geoffrey Whitworth** (1883–1951) to assist and encourage all those interested in the art of the theatre. Later known as the British Theatre Association, it housed Europe's largest collection of theatre books until it was disbanded in 1990. The National Operatic and Dramatic Association was founded in 1899. Scotland has a Scottish Community Drama Association founded in 1926 and Wales a Drama Association of Wales. Ireland also has an Amateur Drama League, founded in 1965, and in 1952 the International Amateur Theatre Association (IATA) was founded. The amateur movement has a natural affinity with the use of drama in education, one outcome of which was the foundation in 1956 by Michael Croft, formerly a teacher, of the *National Youth Theatre. There are also a number of *university departments of drama, not only in Britain but in many European countries. In the USA a strong Little Theatre movement developed in the early 20th century. Over 500 amateur groups were eventually founded, many with their own theatres, and the movement flourished until the 1930s. In 1925 a postgraduate Department of Drama was inaugurated at *Yale under George Pierce *Baker, and many college playhouses opened in the wake of the amateur academic movement. The Little Theatre Movement in Canada also showed how a widespread and flourishing amateur movement could give rise to a soundly based and rapidly growing professional theatre.

Ambassadors Theatre, London, small playhouse in West Street, near St Martin's Lane, seating 460, which opened in 1913. From the following year *Cochran successfully staged a series of intimate revues there, starring Alice *Delysia. From 1919 to 1930 it was leased by **H. M. Harwood** (1874–1959) whose productions in 1920 included his own play *The Grain of Mustard Seed* and Lennox *Robinson's *The White-Headed Boy*, with Sara *Allgood. In 1925 *O'Neill's *Emperor Jones* marked Paul *Robeson's first appearance in London. Sydney *Carroll took over the theatre in 1932, and under him Vivien *Leigh made her successful West End début in Carl *Sternheim's *The Mask of Virtue* (1935). A later success was John Perry and M. J. Farrell's *Spring Meeting* (1938), with the inimitable Margaret *Rutherford. *The Gate Revue*, transferred from the *Gate Theatre early in 1939, had a long run, followed by a sequel, *Swinging the Gate* (1940), and the intimate revues *Sweet and Low* (1943), *Sweeter and Lower* (1944), and *Sweetest and Lowest* (1946). In 1952 the record-breaking run of Agatha *Christie's *The Mousetrap* began. It occupied the theatre until 1973 when it moved to the *St Martin's. Later successes included Hélène Hanff's *84 Charing Cross Road* (1981) and the *RSC's production of *Les Liaisons dangereuses*, adapted by Christopher *Hampton from Laclos's novel of that name, which ran from 1986 until 1990.

American Academy of Dramatic Arts, New York, see SCHOOLS OF DRAMA.

American Actors' Equity Association, see EQUITY.

American Amphitheatre, Boston, see BOSTON.

American Company, small troupe of professional actors which had the elder *Hallam's widow as leading lady and his son Lewis as leading man. The name was first used in a notice of their presence at Charleston in 1763–4. The company played an important part in the development of early American drama, being the first professional group to produce a play by an American—*The Prince of Parthia*, by Thomas *Godfrey, in 1767. It was also responsible for the production of several plays by *Dunlap, who became manager in 1796. The first Joseph *Jefferson joined the company in 1795, moving with it to the new *Park Theatre in 1798. The company's identity was lost when, in 1805, Dunlap went bankrupt and retired, the Park Theatre being taken over by Thomas Abthorpe *Cooper, who had been with the company for some years. The American Company had practically the monopoly of acting in the United States, its only rival being *Wignell's company in Philadelphia.

American Conservatory Theatre (ACT), San Francisco, California, company founded in Pittsburgh in 1965, which was invited to play in San Francisco. It settled in a theatre opened as the Columbia in 1910, one of eight theatres built to replace those destroyed in the earthquake and fire of 1906, and now the only one still in professional full-time operation. Seating 1,456, it was known in the 1920s as the Wilkes, the Lurie, and finally the Geary. Opening with *Molière's *Tartuffe* in 1967, the company became the largest and most active regional theatre in the USA, playing an annual season of 33 weeks in true *repertory, and presenting each season 10 classical and modern plays. It took over a second theatre, the small Marine Memorial, in 1968. In 1972 a programme of Plays in Progress was initiated, offering new plays in the small Playroom, seating 49. From its inception the company attached great importance to theatre training, and since 1978 has offered a master's degree in acting. It also presents guest productions and has been host to important companies from overseas, among them the *RSC and the *National Theatre company from Britain. The company tours in the USA and has visited the USSR and Japan. A financial crisis in the 1980s curtailed its activities, but by the end of the 1980s the situation had improved markedly. A successful revival of Mae *West's *Diamond Lil* was mounted in 1988, and the 1989–90 season included a new version of *Dickens's *A Tale of Two Cities*.

American Laboratory Theatre, training school and production company founded in New York in the 1920s by **Richard Boleslavsky** (1889–1937) and **Maria Ouspenskaya** (1876–1949). They were both former members of the *Moscow Art Theatre, training actors and directors in the techniques of *Stanislavsky, of which they became the leading American exponents. Their students included Lee *Strasberg and Harold *Clurman. After some interesting productions, among which in 1928 were Jean-Jacques *Bernard's *Martine* and Arthur *Schnitzler's *The Bridal Veil*, the company dispersed in 1933.

American Museum, New York, on the southeast corner of Broadway and Ann Street, Broadway showplace opened by **P**(hineas) **T**(aylor) **Barnum** (1810–91) in 1841 which by 1849 had become a theatre with a good stock company and some visiting stars. In 1850 it was enlarged, reopening with an excellent company in *Sedley-Smith's melodrama *The Drunkard*, which had a record run. Barnum sold the theatre in 1855, but in 1860 was able to buy it back. Plays were gradually ousted by freaks, baby-shows, and boxing contests until in 1865 the building was burnt down. On 6 Sept. Barnum opened a New American Museum, also on Broadway, which Van Amburgh took over with his menagerie in 1867. Plays were evidently still being given there, however, as it was during a run of a dramatization of *Uncle Tom's Cabin* that in 1868 the second museum was burnt to the ground; it was never rebuilt.

American National Theatre and Academy (ANTA), organization founded in 1935 under a charter as 'a people's project, organized and conducted in their interest, free from commercialism, but with the firm intent of being as far as possible self-supporting'. The existence of the *Federal Theatre Project and the outbreak of the Second World War made it difficult to raise money privately, but in 1945 the Board of Directors was reorganized to include leading theatre people and the heads of such organizations as American Actors' *Equity. In 1948 ANTA became the US Centre of the *International Theatre Institute and two years later it acquired the Guild Theatre as its headquarters (see VIRGINIA THEATRE). In 1963, pending the completion of the Lincoln Center, ANTA was responsible for the erection of a temporary structure, the *Washington Square Theatre, to house the Lincoln Center repertory company. ANTA joined forces in 1983 with the *John F. Kennedy Center for the Performing Arts in order to establish a national theatre in Washington, DC. Since the move to Washington, the organization has been dormant. The International Theatre Institute/US is now located on West 42nd Street.

American Negro Theatre, see NEGROES IN THE AMERICAN THEATRE.

American Opera House, New York, see CHATHAM THEATRE.

American Place Theatre, New York, in West 46th Street, was founded in 1964 by Wynn Handman and Sidney Lanier, pastor of St Clement's, in which the theatre found its first home. Productions of new and controversial plays, exclusively by American writers, were mounted in the church, services being held in the set of the current production. In 1971 the company moved to a theatre seating 299, with a thrust stage, in a newly built Manhattan skyscraper on the Avenue of the Americas (as permitted under revised building regulations); it incorporated an experimental studio, the Sub-Plot, in the basement. The opening production was a double bill consisting of Ronald Ribman's *Fingernails Blue as Flowers* and Steven Tesich's *Lake of the Woods*; later productions included Sam *Shepard's two one-act plays *Killer's Head* and *Action* (1975). The theatre mounts regular *Off-Broadway subscription seasons, cabaret, and special projects such as the Women's Project and the American Humorists Series.

American Repertory Theatre, Harvard University, see HARVARD UNIVERSITY.

American Repertory Theatre, New York, see CRAWFORD; LE GALLIENNE; and WEBSTER, MARGARET.

American Shakespeare Theatre, Stratford, Conn., until 1972 the American Shakespeare Festival Theatre, was founded through the initiative of Lawrence *Langner. Designed by Edwin Howard and seating 1,534, it stands on the bank of the Housatonic river, its octagonal shape being based on that of Shakespeare's *Globe Theatre in London, and seeks to combine modern and traditional forms in a flexible, functional whole. It opened in 1955 with Raymond *Massey and Christopher *Plummer in repertory in *Julius Caesar* and *The Tempest*. The following year three productions were given, over a longer season, and the nucleus of a permanent company was established. The theatre enhanced its reputation under John *Houseman, its Artistic Director 1956–9, and in 1958 Katharine *Hepburn played Beatrice in *Much Ado about Nothing* in the company's first national tour. The repertory later included classic plays other than those of Shakespeare as well as revivals of modern American plays, and many famous actors were seen there. In 1981 James Earl *Jones and Christopher Plummer played Othello and Iago, but in the same year mounting financial problems forced the theatre to close. It was purchased by the state of Connecticut in 1983, but remained dark until 1989, when it reopened with two productions from the *American Conservatory Theatre.

American Society for Theatre Research, founded in 1956 'to serve the needs of theatre historians in the practice of their profession, and to foster the increase of knowledge of the theatre in America'. Unlike most national theatre societies, it is concerned not only with American theatre but with all aspects of theatre studies. It holds an annual meeting in New York, and publishes an occasional newsletter and an annual volume of essays, *Theatre Survey*.

American Theatre Association, see UNIVERSITY DEPARTMENTS OF DRAMA.

American Theatrical Commonwealth Company, see LUDLOW.

Ames, Winthrop (1871–1937), American director, who used the money inherited from his father, a railroad capitalist, to back non-commercial ventures in the theatre. He built three playhouses in New York: the New Theatre (see CENTURY THEATRE) in 1909; the Little Theatre (see HELEN HAYES THEATRE 2) in 1912; and the *Booth Theatre in 1913. None of these was successful, but in spite of many setbacks he did good work for the American theatre over a long period until his retirement in 1932. Among his productions were a number of plays by Shakespeare and other classics, revivals of *Gilbert and Sullivan, and the first American productions of such modern plays as *Housman and *Granville-Barker's *Prunella* (1913), *Maeterlinck's *The Betrothal* (1918), Clemence

*Dane's *Will Shakespeare* (1923), and *Galsworthy's *Old English* (1924).

Amphitheatre (Amphitheatrum), Roman building of elliptical shape, with tiers of seats enclosing a central arena. It was not intended for dramatic performances, which in ancient theatres were always given in front of a permanent back-scene, but for gladiators and wild beast shows, and mimic sea battles. The first amphitheatre was probably that built by Julius Caesar in 46 BC. The most famous was the Colosseum in Rome, said to be capable of seating 87,000 spectators, which was completed in AD 80 and is still extant.

Anderson, John Murray, see REVUE.

Anderson, Dame Judith [Frances Margaret Anderson-Anderson] (1898–   ), Australian actress of great passion and intensity, who first appeared on the stage in Sydney in 1915. She toured Australia for two years, and then went to the United States, where she adopted her present name and, after some experience in stock companies and on tour, made a great success on Broadway as Elise in Martin Brown's *Cobra* (1924). Later roles included the Unknown One in *Pirandello's *As You Desire Me* (1931). In 1936 she played the Queen to John *Gielgud's Hamlet in New York. A year later she made her first appearance in London, where she gave an outstanding performance as Lady Macbeth with Laurence *Olivier at the *Old Vic; she repeated the role with Maurice *Evans in New York in 1941. She was praised for her Olga in *Chekhov's *Three Sisters* (1942), but her greatest role was probably the name-part in *Euripides' *Medea* (1947), in a new adaptation in which she also appeared in Berlin, Paris, and with the *Elizabethan Theatre Trust. She was seen again at the Old Vic in 1960, as Arkadina in Chekhov's *The Seagull*. She then toured the United States in a recital of scenes from her most famous parts. In 1970 she embarked on a further tour, playing Hamlet at the age of 71 in emulation of Sarah *Bernhardt. In 1982 she played the nurse in a revival of *Medea*.

Anderson, Lindsay Gordon (1923–   ), English director, whose stage work has developed in tandem with a notable film career. He worked predominantly at the *Royal Court, where his productions included Willis *Hall's *The Long and the Short and the Tall* and *Arden's *Serjeant Musgrave's Dance* (both 1959), a musical, *The Lily-White Boys* (1960), *Frisch's *The Fire-Raisers* (1961), and a dramatization of *Gogol's *The Diary of a Madman* (1963), which he helped to adapt. In 1969 he joined William *Gaskill in the running of the theatre, where he directed David *Storey's *The Contractor* and *In Celebration* (both 1969), *Home* (1970), *The Changing Room* (1971), *The Farm* (1973), and *Life Class* (1974), and *Orton's *What the Butler Saw* (1975).

He has also worked for the *National Theatre (Frisch's *Andorra*, 1964, and Storey's *Early Days*, 1980) and the *Chichester Festival Theatre (*Chekhov's *The Cherry Orchard*, 1966). In 1975 he directed Chekhov's *The Seagull* and Ben *Travers's *The Bed before Yesterday* for the company at the *Lyric Theatre. He made rare incursions into the commercial theatre with *Billy Liar* (1960), by Keith Waterhouse and Willis Hall, and William Douglas *Home's *The Kingfisher* (1977; NY, 1978). More recent work includes *In Celebration* in New York (1984), Philip *Barry's *Holiday* at the *Old Vic (1987), and Storey's *The March on Russia* (1989) at the National.

Anderson, Mary (1859–1940), American actress who made her first appearance at the age of 16 at Louisville, playing Shakespeare's Juliet. She toured the United States for some years in a wide variety of parts, being much admired as Julia in Sheridan *Knowles's *The Hunchback*, Pauline in *Bulwer-Lytton's *The Lady of Lyons*, and Parthenia in Mrs Lovell's *Ingomar*, in which she made her first appearance in London, at the *Lyceum in 1883, where in 1887 she played in *The Winter's Tale*, being the first actress to double the parts of Perdita and Hermione. In 1885 she played Rosalind in *As You Like It* at Stratford-upon-Avon, one of her best parts. She also appeared in London in several plays by W. S. *Gilbert, his *Comedy and Tragedy* (1884) being specially written for her. She retired in 1889, and was not seen again on the professional stage. In 1890 she married and settled at Broadway in Worcestershire, where she died. She was part-author with Robert Hichens of *The Garden of Allah* (NY, 1911; London, 1920), a play based on his best-selling novel.

Anderson, (James) **Maxwell** (1888–1959), major and prolific American dramatist who wrote many of his plays in blank verse. His *What Price Glory?* (1924), written in collaboration, was a great popular hit, portraying realistically and sympathetically the American soldier in action during the First World War. Another outstanding popular success was *Saturday's Children* (1927), about the marriage problems of a young couple. A realistic play of modern city life, *Gypsy* (1929), preceded a series of idealistically conceived historical and pseudo-historical plays. The best of these were *Elizabeth the Queen* (1930), the first of his blank-verse plays, *Night over Taos* (1932), *Mary of Scotland* (1933), *Valley Forge* (1934), and *The Wingless Victory* (1936; London, 1943), about a mixed marriage. But Anderson was never content to follow any one dramatic or artistic formula. His realistic satires on political subjects are among his most effective works, among them *Both Your Houses* (1933), a savage attack on political corruption which was awarded the *Pulitzer Prize. In the fantasy *High*

Tor (1937) he combined poetic drama with formal verse, philosophy, and political commentary; and in *Winterset* (1935), based on the Sacco-Vanzetti case, and *Key Largo* (1939), whose action begins in the Spanish Civil War, he sought to make tragic poetry out of the stuff of his own times. Among his later plays were another wartime play, *The Eve of St Mark* (1942; London, 1943), *Joan of Lorraine* (1946), on Joan of Arc, *Anne of the Thousand Days* (1948), on Anne Boleyn, and *The Bad Seed* (1954; London, 1955), a study of inherited homicidal tendencies, based on a novel. He also wrote the book and lyrics for the musicals *Knickerbocker Holiday* (1938) and *Lost in the Stars* (1949).

Anderson, Robert Woodruff (1917–), American dramatist, whose first play, *Come Marching Home* (1945), was awarded first prize in a National Theatre Conference contest. He is best known for *Tea and Sympathy* (NY, 1953; London, 1957); his treatment of the victimization of a schoolboy wrongly accused of homosexuality was seen as an oblique comment on McCarthyism. His later plays include *All Summer Long* (1957); a double bill, *Silent Night/Lonely Night* (1959); *You Know I Can't Hear You when the Water's Running* (1967; London, 1968), four short plays which achieved his longest run; and *I Never Sang for My Father* (1968; London, 1970), about a middle-aged man's fraught relationship with his father. Anderson's plays, though they sometimes verge on sentimentality, deal compassionately with the problems inherent in human relationships, especially within the family.

Andreini, family of Italian actors, outstanding in the annals of the *commedia dell' arte*, of whom the first, **Francesco** (1548–1624), was both actor and author. In 1578 he married **Isabella Canali** (1562–1604), one of the most famous actresses of her time. They were both with the *Gelosi, for whom he at first played opposite her in many lovers' parts; he later abandoned these and achieved success with the role of the *Capitano, which he made a subtle variation on the braggart-soldier type. He took his troupe to France in 1603, playing at the Hôtel de *Bourgogne in Paris and at Fontainebleau, but on the death of his wife disbanded the company and went into retirement, devoting the rest of his life to writing.

Isabella died in childbirth on her way from France to Italy. She seems to have been as highly regarded for her literary works as for her acting. Of their seven children the most famous was the eldest son, **Giovann Battista** (*c.*1579–1654), known as Lelio. He first acted in his parents' company, playing young lovers. Some time before his mother's death he joined the *Fedeli, probably helping in its formation, and provided plays for the troupe. Of these the best known are the *Adamo* (1613) and *La Centaura* (1622), a curious spectacle-play made up of three acts, of which the first is comic, the second pastoral, and the third tragic.

Andrews, Harry Fleetwood (1911–89), English actor gifted with impressive height, strong attack, and a fine, resonant voice. He began his career in 1933 at the *Liverpool Playhouse, and two years later was in London, playing Tybalt in *Gielgud's production of *Romeo and Juliet*. During the next few years he was seen almost entirely in classical parts, notably as Diomedes in a modern-dress *Troilus and Cressida* (1938) and as Laertes in *Hamlet* in Gielgud's company at the *Lyceum (1939). After war service he joined the *Old Vic company in 1945, playing a variety of parts including Creon in *Sophocles' *Oedipus Rex* and Lucifer in *Marlowe's *Dr Faustus*. He was at Stratford-upon-Avon for several years from 1949, appearing with distinction as Brutus in *Julius Caesar*, Benedick in *Much Ado about Nothing*, and the Duke in *Measure for Measure*. He left to make one of his rare appearances in a modern play, as Casanova in Tennessee *Williams's *Camino Real* (1957), but returned to the Old Vic to give an impressive performance in the title-role of *Henry VIII*. Later roles included General Allenby in *Rattigan's *Ross* (1960), the title-role in *Bond's *Lear* (1971), and Serebryakov in *Chekhov's *Uncle Vanya* (1982).

Andrews, Julie, see MUSICAL COMEDY.

Andreyev, Leonid Nikolaivich (1871–1919), Russian dramatist, was encouraged in his early days by *Gorky. He was at first a revolutionary, but after the October Revolution emigrated to Finland, where he died. His plays, permeated with a bitter pessimism, express the despair and desolation of the period between 1905 and 1917. The only one to have survived on the stage is the theatrically effective *He Who Gets Slapped* (1914), an allegorical play set in a circus, first produced in New York in 1922 (London, 1927).

Andronicus, Lucius Livius (*c.*284–*c.*204 BC), Roman dramatist (probably of Greek origin, and a manumitted slave) who in 240 BC produced in Rome the first Latin version of a Greek play. Up to this time the Roman stage seems to have known only a formless medley of dance, song, and buffoonery. The introduction of plays with a regular plot was successful, and Andronicus, who is important as a pioneer though his style was uncouth, continued to translate and produce plays taken from Greek tragedy and *New Comedy until his death. From the fragments of his work that remain it is evident that he introduced into his verse metres which he was, as far as we know, the first to employ, and which later Roman dramatists used for over 200 years.

Angelica, see MARTINELLI.

Angelo, F. d', see PARADISO.

Anglin, Margaret (1876–1958), Canadian-born actress, who made her first professional appearance in 1894 in a revival of Bronson *Howard's *Shenandoah*. Her first outstanding success was as Roxane in *Mansfield's production of *Rostand's *Cyrano de Bergerac* (1898) and she was later the leading lady of *Frohman's *stock company at the *Empire Theatre, New York, where she appeared in a wide variety of parts, being particularly admired in Henry Arthur *Jones's *Mrs Dane's Defence*. Later she was seen in *Camille* by the younger *Dumas, as Antigone, Electra, Iphigenia, and Medea, and in such Shakespearian parts as Viola, Rosalind, and Cleopatra. She was also excellent as Mrs Malaprop in *Sheridan's *The Rivals*, which she first played in 1936.

Animal Impersonation. The representation of birds and beasts by human beings played an important part in early *folk festivals, and may be associated with primitive fertility rites in which live animals were sacrificed, or with the wild-beast skins worn by priests and even participants in ritual religious dances, particularly in hunting communities. In general the short playlets in which these 'animals' appeared followed the main plot of the *mumming play, with the death and resurrection of the chief character. In England the 'animal' was usually a horse, giving rise to the *hobby horse of the mummers. The 'dragon' killed by *St George was a composite creature, with several men inside one skin.

Although *Aristophanes used animal disguises effectively for satirical purposes in his comedies, in the modern theatre they are mainly confined to music-halls and cabarets. They appear occasionally in straight comedies—the animals in *Obey's *Noah*, for instance, or the lion in *Shaw's *Androcles and the Lion*—but otherwise find their chief employment in English *pantomime, Dick Whittington's cat and the Wolf in *Red Riding Hood* being the ones most often seen. In some cases the 'animal' is suggested by one or two costume details—antennae and wings, or heads and tails (as for Bottom the Weaver in *A Midsummer Night's Dream*)—reinforced by appropriate gestures. There is also a form of impersonation limited to sound, where bird-calls and farmyard noises are usually produced by one man. (See also FEMALE IMPERSONATION and MALE IMPERSONATION.)

Animation Culturelle, Centres d', see DÉCENTRALISATION DRAMATIQUE.

Annunzio, Gabriele d', see D'ANNUNZIO.

Anouilh, Jean-Marie-Lucien-Pierre (1910–87), prolific French dramatist, whose plays, in translation, were almost as popular in Britain and America as in his own country. All have in common the theme of the loss of innocence implicit in the struggle for existence in a decadent society. The early ones were divided by Anouilh himself into *les pièces roses*—*Le Bal des voleurs* (*Thieves' Carnival*, 1938), *Léocadia* (*Time Remembered*, 1940), *Le Rendezvous de Senlis* (1941), *Colombe* (1951)—which treated the subject romantically, and *les pièces noires*—*Le Voyageur sans bagage* (*Traveller without Luggage*, 1937), *La Sauvage* (*The Restless Heart*, 1938)—which showed melancholy resignation. Later the mood was transmuted into the glittering wit of *les pièces brillantes*—*L'Invitation au château* (1947), *La Répétition; ou, L'Amour puni* (*The Rehearsal*, 1950)—and the bitter disillusionment of *les pièces grinçantes*—*Ardèle; ou, La Marguérite* (1948), *Le Valse des toréadors* (*The Waltz of the Toreadors*, 1952), *Pauvre Bitos; ou, Le Diner des têtes* (*Poor Bitos*, 1956). Several times Anouilh turned to history for his subjects, as in *L'Alouette* (*The Lark*, 1953), on Joan of Arc, and *Becket; ou, L'Honneur de Dieu* (1959), and to classical themes, as in *Eurydice* (1942), *Antigone* (1944), and *Médée* (1953). *Antigone*, produced in German-occupied Paris, aroused much controversy with its study of personal loyalties in conflict with authority. Admirably written, well constructed, and offering scope for wide-ranging styles of interpretation, Anouilh's plays attracted some of the best talents of his time. In Paris many were directed by *Barsacq. In London his first outstanding success was *L'Invitation au château*, translated as *Ring round the Moon* (1950) by Christopher *Fry, with Paul *Scofield as the twin brothers Hugo and Frédéric. The same translation was used in America, which was not always so with other plays; different titles were sometimes used also: *Le Rendezvous de Senlis* was *Dinner with the Family* in London and *Rendezvous at Senlis* in New York, while *Eurydice* was *Point of Departure* in London but became *Legend of Lovers* in New York. Anouilh's later plays, which were less successful, included *Hurluberlu; ou, Le Réactionnaire amoureux* (1959), a sequel to *Ardèle*, seen as *The Fighting Cock* at *Chichester in 1966; *La Grotte* (*The Cavern*, 1961), *Cher Antoine* (1970), and *Le Directeur de l'Opéra* (1973). His last play *Le Nombril* (1981), depicting a formerly fashionable playwright in old age, was staged in London in 1984 as *Number One* in an adaptation by Michael *Frayn. Anouilh also directed his own and other writers' plays, among the latter being a revival of *Victor; ou, Les Enfants au pouvoir* (1962) by Roger *Vitrac, whom, with *Molière and *Giraudoux, he considered the main influence on his own work.

Ansky [Solomon Rappoport] (1863–1920), Jewish ethnologist and man of letters, whose researches into folklore resulted in his one well-known play, *The Dybbuk; or Between Two Worlds*, a study of demoniac possession and the

Hassidic doctrine of preordained relationship. First seen in Yiddish in a production by the *Vilna Troupe in 1920, shortly after Ansky's death, it was directed two years later in Hebrew by *Vakhtangov for *Habimah, which made it world-famous. It was first seen in New York in Hebrew in 1925 and has since been revived several times, in both Hebrew and Yiddish. In London it was first presented in Hebrew in 1930.

Anspacher Theatre, New York, see PUBLIC THEATRE.

ANTA, see AMERICAN NATIONAL THEATRE AND ACADEMY.

Anta Theatre, New York, see VIRGINIA THEATRE.

Anti-Masque, see MASQUE.

Antoine, André (1858–1943), French actor, director, and theatre manager, one of the outstanding figures in the theatrical reforms of the late 19th century. In 1887 he founded the *Théâtre Libre for productions of the new *naturalistic drama then coming to the fore in Europe. Here in 1890 he staged *Ibsen's *Ghosts*, playing Oswald himself. In 1890 Antoine took over the Théâtre des Menus-Plaisirs, built on what is now the boulevard de Strasbourg in 1866, renaming it the Théâtre Antoine in 1896, and making it a centre for many young dramatists. From 1906 until his retirement in 1916 he was director of the *Odéon.

The Théâtre Antoine had other brilliant periods of management under Firmin *Gémier from 1906, and from 1943 onwards under Simone Berriau, when it was an important focus of Existentialism. She was succeeded on her death in 1984 by her daughter and son-in-law. The theatre has staged the work of authors such as Tennessee *Williams (*Cat on a Hot Tin Roof*), Arthur *Miller (*A View from the Bridge*), and *Stoppard (*Rosencrantz and Guildenstern are Dead*).

Antonelli, Luigi (1882–1942), Italian dramatist, whose works developed further *Chiarelli's Teatro *Grottesco. His *The Man Who Met Himself* (1918) shows that a man is a fool if he hopes to correct the errors of his youth by the wisdom (or experience) of his maturity. Of Antonelli's other plays *The Island of Monkeys* (1922) ironically lays bare the squalor of human civilization; *The Dream Shop* (1927) analyses man's use of dreams; and *Il maestro* (1933) is a neatly turned refutation of *Pirandello's theories, built up on the latter's own foundations. The world of Antonelli's plays is sombre yet visionary, and essentially truthful.

Anvil Productions, see OXFORD PLAYHOUSE.

Anzengruber, Ludwig (1839–89), Austrian dramatist, the first to present realistic peasant life on the modern Austrian stage. Equally successful in tragedy and comedy, he employed a dialect-flavoured dialogue which lent authenticity without being as difficult to understand as the real dialect literature which came later. *Der Pfarrer von Kirchfeld* (1870) is a plea for the exercise of tolerance in religious matters. In *Der Meineidbauer* (1871), where an old farmer who cheats two orphans of their heritage assumes demonic stature, Anzengruber proved that peasant life can furnish matter for true tragedy. A comic battle for matrimonial supremacy is the theme of *Die Kreuzlschreiber* (1872), and *Der Doppelselbstmord* (1876) is a farcical village version of *Romeo and Juliet* which ends happily. Anzengruber's last important play was *Das vierte Gebot* (1877), in which he deserts the country for the town but remains uncompromisingly a realist.

Apollo Theatre, London, in Shaftesbury Avenue, seating 796 in three tiers, opened in 1901 with a musical comedy, *The Belle of Bohemia*. Although it was a failure, later productions had good runs, and the Apollo, while never the permanent home of a great management, became consistently successful. It housed seasons of the *Pélissier Follies*, 1908–12, and the most notable production during the First World War was *Brighouse's *Hobson's Choice* (1916). Immediately after the war Ian Hay's *Tilly of Bloomsbury* (1919) had a long run. *Novello's *Symphony in Two Flats* (1929) and Clemence *Dane's *Wild Decembers* (1933) also played there. During the Munich crisis *Sherwood's *Idiot's Delight* (1938) had a great success, and among wartime productions were Emlyn *Williams's *The Light of Heart* (1940), *Van Druten's *Old Acquaintance* (1941), and *Rattigan's *Flare Path* (1942). *Seagulls over Sorrento* (1950) by Hugh Hastings had a long run, and there were two interesting plays by *Giraudoux: *Tiger at the Gates* (1955) and *Duel of Angels* (1958). In 1960 Marc Camoletti's *Boeing-Boeing*, adapted by Beverley Cross, started a four-year run, and in 1968 *Gielgud appeared in Alan *Bennett's *Forty Years On*, returning in 1970 in David *Storey's *Home*. Later came Peter *Nichols's *Forget-Me-Not Lane* (1971), *Ayckbourn's *Season's Greetings* (1982), Herb Gardner's *I'm Not Rappaport* (1986) with Paul *Scofield, and Hugh Whitemore's *The Best of Friends* (1988) with Gielgud.

Appia, Adolphe (1862–1928), French-speaking Swiss artist whose theories on stage design, and particularly on stage *lighting, had an immense influence on 20th-century methods of play production. He rejected the flat painted scenery of the late 19th century in favour of an environment more suitable for a three-dimensional actor, and employed light (helped by the introduction of electricity) as the visual counterpart of music, enhancing the mood of the play and linking the actor to the setting.

Appia formulated his ideas so clearly in *Die Musik und die Inscenierung* (1899) and his extremely simple but effective designs for plays by *Shaw and *Ibsen that the modern theatre has been able to put them into practice without the need for special apparatus. Mobile lighting, which breaks up and diversifies the direction, intensity, and colour of light, is used today by many who do not realize that the techniques derive from Appia's experiments.

Apron Stage, see FORESTAGE.

Aquarium Theatre, London, see IMPERIAL THEATRE.

Aquatic Drama, spectacular representations of nautical battles, shipwrecks, and storms at sea, which came to London from the circuses of Paris and became exceedingly fashionable in the early 18th century. Many London theatres replaced their stages by large water tanks filled from a neighbouring river and took advantage of the fervour engendered by Nelson's naval victories to put on elaborate reconstructions of the battle of the Nile, the bombardment of Copenhagen, and the battle of Trafalgar. Even *Drury Lane and *Covent Garden fell victims to the popular craze; but the theatre most addicted to aquatic drama was *Sadler's Wells, which changed its name temporarily to the Aquatic Theatre.

Aquatic Theatre, London, see SADLER'S WELLS THEATRE.

Arbuzov, Alexei Nikolayevich (1908–86), Soviet dramatist, one of the few to have found an audience in the West, compared by some English critics to *Chekhov. His first major success came with *Tanya* (1939), about a woman transformed by widowhood. Later plays included a dramatization of *Turgenev's novel *On the Eve* (1948), *City At Dawn* (1957), and the Chekhovian *The Twelfth Hour* (1959), the first to be seen in English, at the *Oxford Playhouse in 1964. *It Happened In Irkutsk* (also 1959), reminiscent in form of *Wilder's *Our Town*, was his greatest success in Russia; it was seen in Sheffield in 1967. By 1963 Arbuzov's plays were running simultaneously at over 70 Russian theatres. *The Promise* (1965) established his reputation abroad, in spite of a sentimental plot covering the lives of a girl and two men from adolescence in 1942 during the siege of Leningrad until 1960. It was produced in London and New York in 1967. *Old World* was produced by the *RSC in 1976, the year after its première in Poland. A two-character play about an autumnal romance between a doctor and his patient, an ex-circus performer, it was staged in Paris (1977) as *Le Bateau pour Lipaïa* and in New York (1978) as *Do You Turn Somersaults?* *Remembrances* (1981) was staged in Watford (1984) as *Chance Visitor*.

Archer, William (1856–1924), Scottish critic and playwright, who had some experience of journalism before migrating to London. From 1894 to 1898 he was dramatic critic of the *World*, reissuing his criticisms in annual volumes. Of his many books on the theatre, the most important was *Masks or Faces* (1888). He was the first to introduce the plays of *Ibsen to the London public with *Quicksands; or, The Pillars of Society* in 1880; *A Doll's House* in 1889; *Ghosts* in 1891; *The Master Builder* in 1893; *Little Eyolf* in 1896; and *John Gabriel Borkman* in 1897. In 1906–8 he published the complete works of Ibsen in English in 11 volumes. Archer, like his friend Bernard *Shaw, was antagonistic to *Irving who, he maintained, had done nothing for the modern British dramatist. He always protested at the critical over-valuation of older plays and the under-valuation of modern ones and upheld the supremacy of the author's script. In 1923 Archer's own play, *The Green Goddess*, an improbable melodrama which had already been seen in New York, was produced in London, running for 416 performances.

Architecture, see THEATRE BUILDINGS.

Arch Street Theatre, Philadelphia, see CHESTNUT STREET THEATRE, DREW, MRS JOHN, and PHILADELPHIA.

Arctic Theatre, see ROYAL ARCTIC THEATRE.

Arden, John (1930–), English dramatist, who has been compared to *Brecht in his preoccupation with moral and social problems and his use of historical themes to illuminate contemporary life. Large sections of his plays are in rhyming verse. Among his early plays, produced at the *Royal Court Theatre, were *Live like Pigs* (1958), on the unsuccessful rehousing of a gipsy family; *Serjeant Musgrave's Dance* (1959; NY, 1966), an anti-military work which was a failure when first produced but is now widely recognized as an important contribution to modern drama; and *The Happy Haven* (1960; NY, 1967), set in an old people's home. In 1963 *The Workhouse Donkey*, a play on English provincial politics, was produced at *Chichester, where *Armstrong's Last Goodnight* (based on a Scottish ballad, and first seen at the *Citizens' Theatre in Glasgow in 1964) was also produced in 1965 with Albert *Finney, subsequently going on to London. In the same year *Left-Handed Liberty*, written to celebrate the 750th anniversary of the sealing of the Magna Carta, was produced at the *Mermaid Theatre. At Christmas 1967 a play for children, *The Royal Pardon; or, The Soldier Who Became an Actor*, a collaboration with his wife Margaretta D'Arcy, was given in London; but the outstanding result of their collaboration has so far been *The Island of the Mighty* (1972), a play based on the Arthurian legends. Its production by the *RSC led to a violent disagreement with the authors, who have since worked only with small, non-professional groups. Their latest works are overtly Marxist and the poetic qualities of the earlier plays are less evident.

Arena Stage, see OPEN STAGE and THEATRE-IN-THE-ROUND.

Arena Stage, Washington, DC, non-profit theatre founded in 1950 by Zelda Fichandler. The group occupied an adapted cinema until 1955, when it moved to a disused brewery, the Old Vat. The first production in the purpose-built arena-stage theatre, which seats 827 on four tiers, was *Brecht's *The Caucasian Chalk Circle*. A second theatre, the Kreeger, opened in 1971 and has a seating capacity of 514 arranged in a fan-shaped auditorium round an end-thrust stage. A rehearsal room in the Kreeger's basement was converted in 1976 into the Old Vat Room, seating 180, with a cabaret-style stage. The Arena Stage company presents a wide range of classic and modern plays, both American and international, during a season which runs from Oct. to June, and has given the first productions of such new American plays as Howard Sackler's *The Great White Hope* (1967) and the reconstruction of the 1925 Marx Brothers' hit *The Cocoanuts* (1988). The Arena Stage also supports Living Stage, a community group which helps the disadvantaged and disabled.

Aretino, Pietro (1492–1556), Italian author and playwright, chiefly remembered for his comedies, which, though written in haste and lacking refinement, are original, amusing, and thoroughly Italian in their realistic and satiric thought and presentation, throwing light on the less creditable aspects of the social life of the day. It has been suggested that Ben *Jonson's *Epicoene* (1609) may owe something to Aretino's comedy *Il marescalco* (*The Sea Captain*, 1533), which was based on *Plautus' *Casina*. Among his other comedies *Lo ipocrito* (1542) may be considered a precursor of *Molière's *Tartuffe*. Aretino's only tragedy *Orazio* (1546) is perhaps the best of those written in his time.

Argent, Théâtre de l'Hôtel d', after the Hôtel de *Bourgogne the second licensed theatre building in Paris. By the end of the 16th century travelling companies were permitted to play at the Paris *fairs, and in 1598 an actor-manager from the French provinces, Pierre Venier, brought a company to the Foire Saint-Germain. From there he went to an improvised theatre in the Hôtel d'Argent in the rue de la Verrerie, where he was allowed to remain for a short time on payment of a tax to the *Confrérie de la Passion. It remained in use intermittently for many years, for in 1610 his daughter Marie *Venier and her husband are found there, having temporarily left the company at the Hôtel de Bourgogne. The assassination of Henri IV in 1610 caused both troupes to leave Paris, and on their return they all went together to the Hôtel de Bourgogne, while a new troupe under *Montdory leased the Hôtel d'Argent. They too later dispersed to other parts of Paris, until in 1634 Montdory opened the Théâtre du *Marais.

Ariosto, Lodovico (1474–1533), Italian poet and playwright, best known for his epic poem, *Orlando furioso*, published in 1532 and adapted for the stage in 1969. He was also one of the first and best writers of early Italian comedy (*commedia erudita*). Though the material of his plays is taken from Renaissance city life, their structure is modelled upon Roman comedy. The first, *La cassaria* (1508) and *I suppositi* (1509), were given at the Court of the d'Este family, Ariosto's patrons, in Ferrara in a theatre built in a classical style under the influence of *Vitruvius, which survived until 1533. The scenery was by Raphael. Both plays were in prose, but were later rewritten in verse, as was *Il negromante* (written 1520, prod. 1530). An English translation of *I suppositi* by George *Gascoigne, was performed at Gray's Inn in London in 1566.

Aristophanes (c.448–c.380 BC), Greek dramatist, author of some 40 comedies, of which 11 are extant (the only Greek comedies to be preserved in their entirety): the *Acharnians* (425), *Knights* (424), *Clouds* (423), *Wasps* (422), *Peace* (421), *Birds* (414), *Lysistrata* (411), *Women at the Festival* (*Thesmophoriazousae*) (410), *Frogs* (405), *Women in Parliament* (*Ecclesiazousae*) (392), and *Plutus* (388). Many of these take their titles from the disguises assumed in them by the *chorus—wasps, clouds, frogs, etc. Aristophanes' direct influence on drama has been slight; the form and spirit of his comedy were so intensely local that they offered no models and little material to comic dramatists of other times and places. On the other hand, his purely literary influence has been great, particularly on Rabelais and *Fielding. The earlier plays have very little plot. Instead a farcical situation, usually having direct reference to some political or social problem of the time, is briefly sketched, and is then exploited in a series of loosely connected scenes. In the *Acharnians*, for example, an Athenian citizen, weary of the war, makes a private treaty with the enemy and consequently enjoys the advantages of trading with them. The iambic scenes develop the ludicrous possibilities of the invention, and enable Aristophanes to hit out at people he dislikes—politicians, busybodies, philosophers. Characters are often burlesques of contemporary Athenians, and even the gods. These earlier plays are an astonishing mixture of fantasy, unsparing (and often violently unfair) satire, brilliant verbal wit, obscenity, literary and musical parody, exquisite lyrics, hard-hitting political propaganda, and uproarious farce. Aristophanes was essentially a popular dramatist, fond of *slapstick and comic business. The *Frogs* marks the transition to a quieter form of comedy in which personal and political invective plays a smaller part and the plot is more elaborate. Some of Aristophanes' plays, notably the *Frogs*, the *Birds*, and *Lysistrata*, have been successfully produced in English translations.

For the 'English Aristophanes', see FOOTE.

Aristotle of Stagira (384–322 BC), Greek philosopher and scientist, whose *Poetics* analyses the function and structural principles of tragedy—a second book on comedy is lost—in reply to the criticisms of Plato and Socrates. To the latter's complaint (in Plato's *Apology*) that poets are unable to give a coherent account of what they do, he opposes a logical theory of poetic composition; and in opposition to Plato's condemnation of poetry and drama because they do not directly seek to inculcate virtue, he defends poetic tragedy because by its representation of a serious action it arouses terror and pity and so leaves the spectator purged and strengthened by catharsis. Within the limits imposed by his concentration on the tragedies of *Sophocles, which he considered representative of the 'mature' form of the art, those of *Aeschylus being the immature and those of *Euripides the enfeebled stage, Aristotle's criticism is penetrating and in many ways final. Although extensively studied and quoted in modern times he has been much misunderstood. The neo-classical critics of the 17th and 18th centuries, especially in France, were anxious to use his authority to support their own doctrines, but of the famous three *unities he mentions only one and a half: he insists on the unity of action, and he remarks, parenthetically, that tragedy 'tries as far as possible to confine itself to 24 hours or thereabouts'. About the unity of place he says nothing, and several extant Greek plays disregarded it. The *Poetics* remains uniquely valuable for its summary of artistic practice appropriate to one time and place, rather than for any attempt to lay down universal laws.

Arlecchino, one of the comic *zanni* or servant roles of the *commedia dell'arte*. Wearing a suit patched with scraps of different colours, with a soft cap on his shaven head, he carried a *slapstick, and was above all a dancer and acrobat. His unique blend of stupidity and shrewdness soon singled him out, and he became an important character in many plays. As Arlequin, still a comic servant, the character appears in later French comedy and in the dumb-shows of the *fairs, while, in the familiar form of Harlequin, it passed into the *harlequinade.

Arliss [Andrews], (Augustus) **George** (1868–1946), English actor, now chiefly remembered for his films, who also had a successful career on the stage in London and New York. He made his first appearance in 1886 and first came into prominence in 1900–1 in revivals of *Pinero's *The Notorious Mrs Ebbsmith* and *The Second Mrs Tanqueray* opposite Mrs Patrick *Campbell, with whose company he subsequently toured the United States. In 1902 he was seen on Broadway in *Belasco's *The Darling of the Gods* with Blanche *Bates, and in *Ibsen's *Hedda Gabler* (1903) and *Rosmersholm* (1907) with Mrs *Fiske. Among his finest parts at this time were the title-roles in *Molnár's *The Devil* (1908) and L. N. *Parker's *Disraeli* (1911), and the Rajah in William *Archer's *The Green Goddess* (1921). In this last part he reappeared in London in 1923, after an absence of over 20 years. He then returned to New York to show his versatility by playing with equal success the elderly gentleman in *Galsworthy's *Old English* (1924) and, his last stage role, Shylock in *The Merchant of Venice* (1928).

Armin, Robert (*c.*1568–*c.*1611), Elizabethan clown, pupil and successor of *Tarleton. He was at the *Curtain and probably the *Globe and appears in the list of actors in Shakespeare's plays, probably playing Dogberry in *Much Ado about Nothing* in succession to William *Kempe. He was also known as a writer. His *Foole upon Foole; or, Six Sortes of Sottes* appeared anonymously in 1600, but his name is found on an enlarged edition published in 1608 as *A Nest of Ninnies*. He was probably the author of *Quips upon Questions* (also 1600), a collection of quatrains on stage 'themes', or improvisations on subjects suggested by the audience, and is credited with the authorship of one play, *The Two Maids of Moreclacke*, produced in 1609.

Armstrong, William (1882–1952), English actor and director, who was for many years connected with the *Liverpool Playhouse where his work was of great value not only to the *repertory movement but to the English theatre as a whole, many of his company becoming leading actors in London. He made his first appearance on the stage at Stratford-upon-Avon in 1908, in Frank *Benson's Shakespeare company, and after touring extensively was with both the *Glasgow and the *Birmingham repertory companies. He went to Liverpool in 1922 as manager and director, remaining there until heavy air raids caused the theatre to close in 1941. After several London productions, including *Van Druten's *Old Acquaintance* (1941), he accepted an invitation from Barry *Jackson in 1945 to go back to Birmingham, where he remained until his death.

Arnaud, Yvonne Germaine (1892–1958), actress who, though born and educated in France, spent her entire professional life in London. Trained as a pianist, she toured Europe as a child prodigy, but at the age of 18, with no previous stage experience, she appeared at the *Adelphi Theatre in the musical comedy *The

Quaker Girl and was an immediate success. She continued to appear both in farce and in musical comedy, her charm and high spirits, added to a musical broken English–French accent, making her a general favourite. Among her outstanding performances were Mrs Pepys in *Fagan's And So to Bed* (1926) and Mrs Frail in *Congreve's Love for Love* (1943). One of her few failures was Mme Alexandra in *Colombe* (1951), where her essential kindliness and good nature was ill-suited to the cruelty and egotism of *Anouilh's ageing actress; but in the following year she had another success as Denise in Alan Melville's *Dear Charles*.

Arne, Susanna, see CIBBER, THEOPHILUS.

Arnold, Matthew (1822–88), English poet, educationist, and critic. His poetic plays, *Empedocles on Etna* (1852) and *Merope* (1858), were for the study rather than the stage; but he did a great service to the theatre by stressing in his essays and lectures its importance as a cultural influence. In 'The French Play in London', an essay published in 1882, he wrote: 'The theatre is irresistible: organize the theatre!'

Aronson, Boris Solomon (1900–80), American scene and costume designer, born in Kiev, where he first worked in the theatre. In 1923 he arrived in New York, where he designed the scenery for Unser Theater in the Bronx and Maurice *Schwartz's Yiddish Art Theatre. His early work was influenced by Chagall's sets in Cubist-fantastic style for the Jewish Theatre in Moscow; but his later work was often more symbolic, restating by form and colour the mood of the play, as in the sets for *Odets's *Awake and Sing* (1935) and *MacLeish's *J.B.* (1959). He also designed the sets for *Coriolanus* at Stratford-upon-Avon in 1959, and for a number of musicals in New York, including *Fiddler on the Roof* (1964), *Cabaret* (1966), and *Sondheim's *A Little Night Music* (1973). He was the inventor of 'projected scenery', a basic permanent set of interrelated abstract shapes made of neutral grey gauze which can easily be 'painted' any desired colour by directing lights upon it through coloured slides.

Arrabal, Fernando (1932–), French dramatist of Spanish extraction, who lived in Paris and wrote in French after 1955. His most characteristic vein derived from his very early plays, written before he left Spain, which were peopled by innocent, childish characters producing ferocious acts in the course of their naïve games. His oppressive upbringing during and after the Spanish Civil War was reflected in the plays that made his reputation, which portrayed a perverse childish revolt against established beliefs. *La Cimetière des voitures* (written in 1957 but not produced until 1966) begins as a satire on suburban community living and ends in a blasphemous crucifixion scene on a bicycle; as *The Car Cemetery* it was seen in London in 1969. In a prolific and varied output Arrabal ventured into formal experiment, as in *Orchestration théâtrale* (1960), a play without actors; social criticism, as in *Pique-Nique en campagne* (written in 1952, produced 1959, seen in London as *Picnic on the Battlefield* in 1964), which juxtaposed banal family life with a military operation; and direct political comment, as in *Et ils passèrent des menottes aux fleurs* (produced and then banned in 1969), an attack on Spanish political prisons, which, translated as *And They Put Handcuffs on the Flowers* and directed by the author, was seen in New York (1972) and London (1973). Arrabal's most successful play, performed in translation in London by the *National Theatre company in 1971, is probably *L'Architecte et l'Empereur d'Assyrie* (1967), a parable of civilization in which two men enter into a role-playing love-hate relationship on a desert island which ends with one eating the other in an attempt to achieve unity. His later plays included *Le Jardin des délices* (1969), *Jeunes Barbares d'aujourd'hui* (1975), a skit on cycle-racing, *Théâtre Bouffe* (1978), and *La Traversée de l'Empire* (1988). He himself has called his work 'Panic Theatre'.

Art, Théâtre d', see LUGNÉ-POË.

Artaud, Antonin (1896–1948), French actor, director, and poet, who is best remembered for his advocacy of the Theatre of Cruelty, which he demonstrated in his production of his own play *Les Cenci* (1935), and defined in a volume of essays, *Le Théâtre et son double*, published in 1938, by which time Artaud, being considered insane, was in an asylum, emerging only two years before his death. Using gesture, movement, sound, and rhythm rather than words, the Theatre of Cruelty as Artaud envisaged it depicted behaviour no longer bound by normal restraints, aiming rather to shock the audience into realizing the underlying ferocity and ruthlessness of human life, and so releasing its own inhibitions. These ideas were adopted by such playwrights as *Arrabal, *Genet, and *Orton, and were apparent in *Barrault's adaptation of Kafka's *The Trial* (1947). Their most successful exposition commercially was seen in Peter *Brook's production of the *Marat/Sade* (1964) by Peter *Weiss. Two other directors much influenced by Artaud's ideas were *Vilar and *Blin. Artaud in his early years was much interested in Surrealist plays, some of which he presented at the Théâtre Alfred *Jarry, of which he was co-founder with Roger *Vitrac.

Arts Council of Great Britain, the main channel for the distribution of public money in subsidies to the British theatre and other arts. It is in essence a continuation of the Council for the Encouragement of Music and the Arts (CEMA), founded in 1940 to bring concerts and plays to the crowded evacuation areas, factory canteens, and other commercially unprofitable centres. The Arts Council was created in 1946, receiving all its funds from the Treasury. Of the sum allotted for drama, a large part goes to the *National Theatre and the *RSC, and a sum is also earmarked for the assistance of touring drama companies. The Arts Council helps to subsidize 12 Regional Arts Associations in England, together with the Scottish and Welsh Arts Councils, that for Northern Ireland being a separate body. These organizations, which receive additional subsidies from sources such as local authorities, give grants to local dramatic organizations, often *civic theatres, and co-ordinate theatre activities within their own areas. The subsidies received by the British theatre are tiny compared to those in countries such as Germany and France.

Arts District Theater, see DALLAS THEATER CENTER.

Arts Theatre, Cambridge, see CAMBRIDGE.

Arts Theatre, London, in Great Newport Street, off St Martin's Lane, opened in 1927 for the production of unlicensed and experimental plays for members only. Seating 347 in an intimate two-tier auditorium, it has a proscenium width of 20ft and stage depth of 18ft. Its first important production, *Young Woodley* (1928) by John *Van Druten, transferred to a commercial theatre, as did several other new plays including Gordon Daviot's *Richard of Bordeaux* (1932), and Norman Ginsbury's *Viceroy Sarah* (1934). In 1942 Alec *Clunes took over, and for ten years made the theatre a vital centre, producing a wide range of plays and winning for it the status of a 'pocket national theatre'. Christopher *Fry's *The Lady's not for Burning* had its first performance here in 1948, with Clunes as Thomas Mendip. The theatre changed hands in 1953, but continued to stage new plays, including *Beckett's *Waiting for Godot* (1955) and *Anouilh's *The Waltz of the Toreadors* (1956). Harold *Pinter's *The Caretaker* received its first performance here in 1960. In 1962 the theatre was leased for six months by the *RSC for a major experimental season, which included new plays and revivals of *Gorky's *The Lower Depths* and *Middleton's *Women Beware Women*. In 1975 Robert Patrick's *Kennedy's Children* came to this theatre from the King's Head, Islington, the first *Fringe theatre production originating in a public house to transfer to the West End. A double bill by *Stoppard, *Dirty Linen* and *New Found Land* (1976), ran for four years. John Godber's *Bouncers* (1986) and *Teechers* (1988) and the musical *A Slice of Saturday Night* (1989) all did well. Since 1967 the theatre has been shared with the Unicorn Theatre for Children, which gives afternoon performances. The evening productions are brought in from outside managements.

Asch, Sholom (1880–1957), Jewish dramatist and novelist, born in Poland, author of several plays in Yiddish, of which the best known is *God of Vengeance*. This was given its first production by *Reinhardt in Berlin in 1907 as *Gott der Rache*. It drew attention to the possibilities of Yiddish drama, and has since been translated and produced in many countries. Some of Asch's Yiddish novels have also been dramatized or adapted for the stage. More than anyone else Asch raised the standard of Yiddish writing and placed it on a literary basis.

Asche, Oscar [John Stanger Heiss] (1871–1936), English actor of Scandinavian descent, born in Australia. He made his first appearance on the stage in 1893, was for a time with *Benson's Shakespeare company, and in 1902 joined *Tree's company at *Her (then His) Majesty's Theatre, taking over the management of the theatre in 1907. There he was seen, in his own productions, as Jaques, Othello, Petruchio (one of his best parts), Shylock, Antony in *Antony and Cleopatra*, and, in 1911, Falstaff in *The Merry Wives of Windsor*. In the same year he made a great success in the part of Hajj in Edward *Knoblock's *Kismet*. He is, however, chiefly remembered for his own oriental fantasy with music *Chu-Chin-Chow* (1916), in which he played Abu Hassan; it ran for five years. Asche's wife, the actress **Lily Brayton** (1876–1953), who made her first appearance in 1896 under Benson, played opposite her husband in most of his productions, being seen as Marsinah in *Kismet* and Sahrat-al-Kulub in *Chu-Chin-Chow*.

Ashcroft, Dame Peggy [Edith Margaret Emily] (1907–91), English actress, one of the foremost of her generation. She made her début in 1926 at the *Birmingham Repertory Theatre, as Margaret in *Barrie's *Dear Brutus*, and first attracted attention in London in 1929, playing Naemi in Feuchtwanger's *Jew Süss* with the simplicity and sense of poetic tragedy that made so many of her later performances remarkable. After enhancing her reputation with an excellent Desdemona to Paul *Robeson's Othello in 1930, she joined the company at the *Old Vic in 1932, playing 10 major roles in eight months, among them Rosalind, Perdita, and Imogen in Shakespeare, of whose women she was always to be the perfect exponent, as well as Cleopatra in *Shaw's *Caesar and Cleopatra*. Much of her best work was done in conjunction with *Gielgud; she played Juliet to his Romeo in 1935, and Ophelia to his Hamlet and Titania to his Oberon in 1944–5. Apart from Shakespeare, she was outstanding as Nina to Gielgud's Trigorin in *Chekhov's *The Seagull* in 1936—and in 1964 played Arkadina with no less mastery of her art—and under his direction played one of her best comedy parts, Cecily Cardew in *Wilde's *The Importance of Being Earnest*, in 1939. In 1950 she was with Gielgud again, playing Beatrice to his Benedict and Cordelia to his Lear at *Stratford-upon-Avon, where she was later outstanding as Cleopatra to Michael *Redgrave's Antony in 1953 and as Katharina in 1960. After the war she gave some outstanding performances in modern plays—Catherine Sloper in *The Heiress* in 1949, based on Henry *James's novel *Washington Square*, Hester Collyer in *Rattigan's *The Deep Blue Sea* in 1952, Miss Madrigal in Enid *Bagnold's *The Chalk Garden* in 1956, and, with amazing versatility, the dual roles of the prostitute and her male cousin in the *Royal Court production of *Brecht's *The Good Woman of Setzuan*, also in 1956. Her Hedda Gabler in *Ibsen's play in 1954 gained her the King's Medal from King Haakon of Norway. In 1961 Dame Peggy became a member of the *RSC (of which she was later made a director), playing not only Margaret of Anjou from youth to old age in *The Wars of the Roses* in 1963 (see BARTON), but appearing in such new works as Marguerite *Duras's *Days in the Trees* (1966) and *Albee's *A Delicate Balance* (1969). In 1976, after being seen at the *National Theatre in Ibsen's *John Gabriel Borkman* and *Beckett's *Happy Days*, she returned to the RSC in *Arbuzov's *Old World*. Her last stage role was the Countess of Rousillon in *All's Well That Ends Well* in 1981.

Ashwell [Pocock], **Lena** (1872–1957), English actress-manager who made her first appearance in 1891, and scored a great success in 1900 in Henry Arthur *Jones's *Mrs Dane's Defence*, in which she was also seen in the USA. On her return to England she took over the *Kingsway Theatre, remaining in management until 1915, when she left to organize entertainment for the troops in France and Germany, for which she was awarded the OBE. She was active in the foundation of the British Drama League (see AMATEUR THEATRE) and in 1925 took over the Bijou Theatre, in Bayswater, renamed it the Century, and produced there several new plays, including her own adaptations of Dostoevsky's *Crime and Punishment* and Stevenson's *Dr Jekyll and Mr Hyde* (both 1927).

Asolo Theatre Company, Sarasota, since 1965 a state theatre of Florida (a title given to only three other theatres in the state), which became fully professional in 1966. It evolved from the summer festival of 18th- and 19th-century comedies first staged in 1960 by the Division of Theatre of Florida State University. The Asolo was originally housed in an 18th-century Italian court theatre, erected in 1798 in Asolo, Italy, and reassembled in Sarasota by the Ringling Museums in 1951. In 1989, the Asolo Theatre Company moved to its own theatre in the new Asolo Center for the Performing Arts, which houses the Asolo Mainstage, the Asolo Second Stage, and the Florida State University/ Asolo Conservatory of Professional Actor Training. Plasterwork wall panels and friezes, cornice work, and carved box fronts from Scotland's Dunfermline Opera House have been reassembled and restored to provide the interior for the Asolo Mainstage. The Asolo Theatre Company is one of the few American companies to play in true *repertory, its season running from early Dec. to July. As many as five matinées a week are often given, with different plays at matinée and evening performances. Productions range from classics to modern plays and musicals as well as premières of new plays. Since 1966 the theatre has been committed to educational work. The Asolo Touring Theatre, founded in 1975, takes its three professional companies to schools throughout the state with a repertory designed for different age groups. Since 1981 Asolo On Tour has annually toured a mainstage work throughout the country.

Asphaleian System, one of the earliest methods of controlling the stage floor; in it the whole stage was divided into individual platforms resting on hydraulic pistons, each of which could be separately raised, lowered, or tipped. The system was first used in the Budapest Opera House in 1884, and was named after the Austrian company which evolved it.

Association of Producing Artists, see PHOENIX THEATRE, New York.

Astley, Philip, see ASTLEY'S AMPHITHEATRE.

Astley's Amphitheatre, London, on the south bank of the Thames, where in 1951 a plaque was unveiled to its memory at 225 Westminster Bridge Road. This hybrid building, remembered mainly for its *equestrian drama, was immortalized by *Dickens in *Sketches by Boz* (1837), and one of its chief actors, **Edward Alexander Gomersal** (1788–1862), is described in Thackeray's novel *The Newcomes* (1855). Its history began when in 1784 **Philip Astley** (1742–1814), a cavalryman and horse trainer, who was responsible for many such 'amphitheatres' in Britain, France, and Ireland, erected a wooden building with a stage for displays of horsemanship on the site of an open circus ring. Burned down in 1794, rebuilt and again destroyed by fire in 1803, it became famous for its 'equestrian spectacles', which continued after Astley's death, the theatre then being renamed Davis's Amphitheatre. One of its great attractions was **Andrew Ducrow** (1793–1848), who because of illiteracy seldom played a speaking part, but was unrivalled in equestrianship. The building was again destroyed by fire, in 1830 and in 1841, after which William Batty (1801–68) rebuilt it and gave it his own name. His successor, William Cooke, is memorable for having turned Shakespeare's *Richard III* into an equestrian drama, giving Richard's horse, White Surrey, a leading role. In 1862 Dion *Boucicault made a disastrous attempt to run the theatre, renamed the Theatre Royal, Westminster. His successor reverted to the old name of Astley's, and drew large audiences across the river to see Adah Isaacs *Menken in a drama based on Byron's poem *Mazeppa*. In 1871 the control of the theatre passed to the circus proprietor 'Lord' George Sanger. In 1893 the building was declared unsafe and closed, being finally demolished by 1895. No trace of it remains.

Aston, Tony [Anthony] (*fl.* 1700–50), Irish actor, known also as Matt Medley, to whom, in default of more precise information, goes the honour of having been the first professional actor to appear in the New World. In the preface to his play *The Fool's Opera; or, The Taste of the Time*, printed in 1731, he refers to his appearances in 'New York, East and West Jersey, Maryland, Virginia (on both sides Cheesapeek), North and South Carolina, South Florida, the Bahamas, Jamaica, and Hispaniola'. He is known to have appeared in Charleston in 1703, and later in the same year in New York, but it is not known in what roles he appeared, or whether he was alone or with a company. He may have given a sort of variety entertainment like the 'medley' in which he appeared in the English provinces in 1717. In 1735 he protested against the proposed bill for regulating the stage, and he was still alive in 1749, when he was spoken of as 'travelling still and as well known as the posthorse that carries the mail'. He was the author of several other plays including *Love in a Hurry* (1709) and *The Coy Shepherdess* (1712).

Astor Place Opera House, New York, one block east of Broadway, was built for Italian opera and opened in 1847. In 1848 the English actor *Macready appeared there, but owing to the jealousy of Edwin *Forrest his visit terminated in 1849 with the anti-British Astor Place riot, in which 22 people were killed and 36 wounded by shots fired by the militia. The theatre then closed for repairs, and reopened in 1849. Actors who came between opera seasons included Charlotte *Cushman, Julia *Dean, and George *Vandenhoff, but the theatre never recovered from its reputation as the 'Massacre Opera House' and closed in 1850.

Astracanadas, see GÉNERO CHICO.

Atelier, Théâtre de l', see DULLIN.

Atellan Farce, see FABULA 1: ATELLANA.

Athénée Louis-Jouvet, Théâtre de l', see JOUVET.

Atkins, Eileen (1934–), English actress, who first appeared in London at the *Open Air Theatre in Regent's Park in 1953, and was at *Stratford-upon-Avon, 1957–9. Her performance as Childie, the child-like 40-year-old lesbian in Frank Marcus's *The Killing of Sister*

George (1965; NY, 1966), first brought her into prominence. She followed it with a portrayal of a neglected, stonily observant wife in David *Storey's *The Restoration of Arnold Middleton* (1967). In New York, also in 1967, she appeared in *Arbuzov's *The Promise*, and back in London was seen as Celia Coplestone in T. S. *Eliot's *The Cocktail Party* (1968). She again proved her versatility and emotional directness as Elizabeth I in Robert *Bolt's *Vivat! Vivat Regina!* (1970; NY, 1972). In 1973, after waging a two-year campaign on its behalf, she brought Marguerite *Duras's *Suzanna Andler* to London, giving a superb performance herself in the title-role. She was seen as Rosalind in *As You Like It* for the *RSC (1973); as Hesione Hushabye in *Shaw's *Heartbreak House* for the *National Theatre (1975); and for *Prospect as Shaw's St Joan (1977) and as Viola in *Twelfth Night* (1978). In 1981 she returned to the RSC as Billie *Whitelaw's *alter ego* in Peter *Nichols's *Passion Play*. She played *Euripides' Medea at the *Young Vic (1986), and Paulina in *The Winter's Tale* and the Queen in *Cymbeline* for the National Theatre (1988).

Atkins, Robert (1886–1972), English actor and director, whose career was devoted mainly to Shakespeare. He made his first appearance at *Her (then His) Majesty's Theatre in 1906 in *Henry IV, Part I*, and in 1915 joined the *Old Vic company, where he returned in 1920 after war service. He remained there until 1925, directing and acting in a number of plays, including *Ibsen's *Peer Gynt* (1922) and the rarely seen *Titus Andronicus* and *Troilus and Cressida* (both 1923). He then toured with his own company. In 1936 he founded the Bankside Players, which he directed in Shakespeare at the *Ring and the *Open Air Theatre, of which he was Director of Productions from its opening until 1960, and manager, 1938–60. He was also Director of the *Shakespeare Memorial Theatre, 1944–5. During 1940, at the height of the bombing of London, he continued to present Shakespeare at the *Vaudeville Theatre. He was an excellent actor, among his best parts being Sir Toby Belch in *Twelfth Night*, Touchstone in *As You Like It*, Caliban in *The Tempest*, Bottom in *A Midsummer Night's Dream*, Sir Giles Overreach in *Massinger's *A New Way to Pay Old Debts*, and James Telfer in *Pinero's *Trelawny of the 'Wells'*.

Atkinson, (Justin) **Brooks** (1894–1984), American dramatic critic. From 1926 to 1960 (except during the war years) he was dramatic critic of the *New York Times*—an unrivalled tenure. His literate, scrupulously honest, and acute writing won a tremendous following. He had great influence: his approval guaranteed a respectable run for a play irrespective of the opinion of other critics. On his retirement the Mansfield

Theatre in New York was renamed the *Brooks Atkinson in his honour.

Auckland Theatre Trust, see MERCURY THEATRE, Auckland.

Auden, W(ystan) **H**(ugh) (1907–73), English poet (later an American citizen) and in his early days a playwright. His first play, *The Dance of Death* (1935), was produced by the *Group Theatre in London in a double bill with T. S. *Eliot's *Sweeney Agonistes*. The same company produced the three satirical left-wing verse plays which Auden then wrote in collaboration with the novelist Christopher Isherwood: *The Dog beneath the Skin* (1935), which has musical choruses; *The Ascent of F6* (1936), which describes the symbolic climbing of a mountain in a mixture of prose and poetry, including pop songs; and *On the Frontier* (1938), with music by Benjamin Britten. Auden also provided the libretti for several operas, and in 1958 prepared a modern verse adaptation of the medieval music-drama *The Play of Daniel*.

Audiberti, Jacques (1899–1965), French dramatist whose Surrealist world is close to that of *Artaud. His first play, *Quoat-Quoat* (1946), in which a French ship bound for Mexico becomes a symbol for organized society at the mercy of atavistic forces, was followed by the commercially successful *Le Mal court* (1947), a tale of innocence corrupted by experience set in the 18th century, and by *Les Femmes du bœuf* (1948), which was produced at the *Comédie-Française, as was *Fourmi dans le corps* (1962). Among his other plays are *La Fête noire* (1948), *Pucelle* (1950), a somewhat irreverent treatment of Joan of Arc's life, *La Hobereaute* (1956), set in medieval Burgundy, *L'Effet Glapion* (1959), and his last work, *L'Opéra du monde* (1965). Audiberti belonged to no particular school of dramatists, but created his own world of fantasy and farce in which his eccentricities found full scope, though his work was often rendered unintelligible by an excess of verbal dexterity.

Audience on the Stage. The practice, common in the early English and Continental theatres, of allowing members of the audience, usually chattering fops in fine clothes, to buy seats on the stage was a fertile source of disorder and of acute irritation to both actor and dramatist. It is first mentioned in England in 1596 and in France not until 1649, but by both dates the tradition was well established. *Voltaire was responsible for the removal of spectators in 1759 from the stage of the *Comédie-Française. When *Garrick took over *Drury Lane he too tried to remove them, but had to wait until the theatre was enlarged in 1762. In other London theatres the practice lingered on into the 19th century, and was remarked on at *Grimaldi's farewell performance at Drury Lane in 1828,

and again in 1838 and 1840, when Queen Victoria visited *Covent Garden.

Audition, trial run by an actor seeking employment, either to display his talents in general or to demonstrate his fitness for a particular role by reading some part of it to the *director and his associates. In earlier times actors were often engaged for a company merely on hearsay, by a letter of recommendation, or after a private audition by the manager. In modern times leading actors may be specifically engaged for a part on their reputation alone, while the supporting cast is chosen only after a number of auditions have been held.

Auditorium, that part of a *theatre building designed or intended for the accommodation of the people witnessing the play—the audience. The word in its present usage dates from about 1727, though Auditory and Spectatory are also to be found. The auditorium can vary considerably in size and shape, placing the audience in front of, part way round, or entirely round the acting area. It can also be in one entity, or divided, with galleries above the main area and boxes at the sides or all round.

Augier, (Guillaume Victor) **Émile** (1820–89), French dramatist, one of the first to revolt against the excesses of the Romantics. After a few plays in verse, he found his true vocation in writing domestic prose-dramas dealing with social questions of the moment, of which *Le Gendre de Monsieur Poirier* (1854) is the best known. Among others successful in their day were *Le Mariage d'Olympe* (1855), which paints the courtesan as she is and not as idealized by the younger *Dumas; *Les Lionnes pauvres* (1858), which shows the disruption of home life consequent on the wife's adultery; and two political comedies, *Les Effrontés* (1861) and *Le Fils de Giboyer* (1862). Under the influence of the new currents of thought running across Europe, he ended his career with two problem plays, *Mme Caverlet* (1876) and *Les Fourchambault* (1878), which as *The Crisis*, in a translation by James *Albery, was produced in London in 1878.

Augustus Druriolanus, see HARRIS, SIR AUGUSTUS.

Aukin, David (1942–   ), English theatre manager and director, originally a solicitor. He began his theatrical career on the *Fringe, being associated with the Joint Stock and Foco Novo companies, both of which he helped to found. From 1970 to 1973 he was literary adviser to the *Traverse Theatre. After a year as Administrator at the *Oxford Playhouse he moved to the *Hampstead Theatre, where he was responsible for some highly successful productions, among them James *Saunders's *Bodies* (1978) and Brian *Friel's *Translations* (1981), which were transferred to London. From 1984 to 1986 he was at the *Haymarket Theatre,

Leicester, where he launched the revival of the musical *Me and My Girl*, which became a hit in both London and New York. He was then appointed Executive Director at the *National Theatre, remaining there until 1990 and sharing the running of it from 1988 with Richard *Eyre.

Aulaeum, see CURTAIN.

Aunt Edna, see RATTIGAN.

Ausoult, Jeanne, see BARON.

Author's Night, see ROYALTY.

Auto Sacramental, Spanish term for the religious play in the vernacular which derived from the Latin *liturgical drama. Although its development followed in general that of the *mystery play, it had by the end of the 16th century come to be recognized as a dramatic restatement of the tenets of the Catholic faith which embodied the preoccupations and ideals of the Counter-Reformation, as the later mystery plays did in Germany. In this form it provided one of the main elements in the celebrations on Corpus Christi (the Thursday after Trinity Sunday), an important feast-day on the Continent. The most distinctive feature of the *auto sacramental* was its use of elaborate allegory. The outstanding writer of such plays was *Calderón, who from 1651 to his death in 1681 was responsible for all those staged in Madrid. Even after his death his plays were frequently revived for the Corpus Christi festival in Madrid, where they continued to be performed long after they had disappeared in France and England. They were finally prohibited by a royal edict of 1765.

Avancini, Nikolaus, see JESUIT DRAMA.

Avenue Theatre, London, see PLAYHOUSE.

Avignon Festival, founded in July 1947 by Jean *Vilar with the first production in French of Shakespeare's *Richard II* in which Vilar himself played the title-role. Productions were presented in the inner court of the papal palace, whose floodlit façade provided an imposing setting. The new festival began modestly, but when in 1951 Vilar was appointed head of the *Théâtre National Populaire he was able to use its resources, including its admirable company, and in the 1950s Avignon sprang into prominence with memorable performances by Gérard *Philipe in *Corneille's *Le Cid* and *Kleist's *Prinz Friedrich von Homburg*. By 1967 the number of spectators had risen to 150,000 and the 17 official performances were regularly sold out. After his resignation from the TNP in 1963 Vilar continued his association with Avignon until his death in 1971, widening its cultural basis, increasing the official theatres to four, and organizing a number of *Fringe activities. There have been many visits from the *Comédie-Française, other notable productions including Peter *Brook's *Mahabharata* (1985), *Vitez's

12-hour production of *Claudel's *Le Soulier de satin* (1987), and Patrice *Chéreau's *Hamlet* (1988).

Avon Theatre, see STRATFORD (ONTARIO) FESTIVAL.

Axer, Erwin (1917–83), Polish director, pupil and successor of Leon *Schiller. In 1945 he established a company in Łódź, and four years later he founded the Contemporary Theatre in Warsaw. From 1955 to 1957 he also directed at the National Theatre, and he worked in many theatres outside Poland. He was responsible for the first production of many Polish plays, including Kruczkowski's *The Germans* (1949), *Witkiewicz's *The Mother* (1970), and *Mrożek's *Tango* (1965), *The Happy Arrival* (1973), and *The Tailor* (1979). He also added a number of foreign classics and new plays to his repertory, among them *Brecht's *Arturo Ui*, which he also staged in Leningrad (1962); *Chekhov's *Three Sisters* (1963); *Pinter's *Old Times* and *Ionesco's *Macbett* (both 1972); and *Bond's *Lear* (1974). In 1980 he staged the world première of *Frisch's *Triptychon*. The company from the Contemporary Theatre was seen in the *World Theatre Season in 1964.

Ayckbourn, Alan (1939–), English playwright and director, who in 1970 was appointed director of productions at Scarborough (see STEPHEN JOSEPH THEATRE IN THE ROUND). Most of his plays were written for this company and then transferred to London, where the first to be outstandingly successful was *Relatively Speaking* (1967). This was followed by *How the Other Half Loves* (London, 1970; NY, 1971) and *Time and Time Again* (1972). *Absurd Person Singular* (1973; NY, 1974) depicts three couples on three successive Christmas Eves, each act being set in the kitchen of one of them. Ayckbourn's next work, *The Norman Conquests* (1974; NY, 1975), was a trilogy of full-length plays, *Table Manners*, *Living Together*, and *Round and Round the Garden*, showing simultaneous events in the dining-room, living-room, and garden of the same house during the same week-end. A musical, *Jeeves* (1975), based on P. G. Wodehouse, for which Andrew Lloyd Webber wrote the music, was not a success, but in the same year *Absent Friends* had a long run. In *Bedroom Farce* (*National Theatre, 1977; NY, 1979), the action passes between the bedrooms of three married couples, which form a *multiple setting. *Just between Ourselves* (also 1977) and *Joking Apart* (1979) were more serious and had comparatively short runs. In *Sisterly Feelings* (NT, 1980) the middle two of the four scenes have alternative versions. *Taking Steps* and *Season's Greetings*, about a family reunion at Christmas, were also seen in London in 1980. Ayckbourn's plays of the 1980s, while still described as comedies, are more persistently disturbing. In *Way Upstream* (NT, 1982) a boating holiday turns into a nightmare; in *Woman in Mind* (1986; NY, 1988), the leading character (embedded, like so many Ayckbourn women, in a profoundly unsatisfactory marriage) goes out of her mind; *A Small Family Business* (NT, 1987) takes a sharp look at business morality; and *Henceforward . . .* (1988) envisages a robotized future. *A Chorus of Disapproval*, however, centring on an amateur production of *Gay's *The Beggar's Opera* (NT, 1985), provides comparatively unclouded pleasure, as does *Man of the Moment* (1990), about a television encounter between a reformed robber and a man who once tried to intercept him. Ayckbourn depicts with humour, accuracy, and an occasional note of cruelty the sexual and other stresses of English middle-class life, and his plays are ingeniously constructed. They have been performed all over the world, and at one time five of them were running in London simultaneously.

Aylmer [Aylmer-Jones], **Sir Felix Edward** (1889–1979), English actor. He began his career in London in 1911 under Seymour *Hicks, and in 1913 joined the newly-formed *Birmingham Repertory Theatre company, playing Orsino in the opening play, *Twelfth Night*. After serving in the RNVR in the First World War he returned to the stage, appearing in several plays by *Shaw at the *Everyman Theatre. From then on he worked almost continuously in London and New York, being seen in many distinguished plays including *Drinkwater's *Abraham Lincoln* (1921), *Robert E. Lee* (1923), and *Bird in Hand* (1928), and making his New York début in *Galsworthy's *Loyalties* (1922).

He returned to Shaw in 1924 to play LordSummerhayes in *Misalliance* and Dr Paramore in *The Philanderer*, adding Sir Colenso Ridgeon in *The Doctor's Dilemma* in 1926. Other roles were in Feuchtwanger's *Jew Süss* (1929), *Sherriff's *Badger's Green* (1930), Galsworthy's *Strife* (1933), and *Granville-Barker's *The Voysey Inheritance* (1934) and *Waste* (1936). He made a reputation as a dependable and immaculate actor with beautiful diction, and was much in demand for parts needing patrician authority and sardonic humour, such as the Earl of Warwick in Shaw's *Saint Joan* (1934). One of his finest performances was as the First Lord of the Admiralty in Charles *Morgan's *The Flashing Stream* (1938; NY, 1939), and the full maturity of his gifts was seen in his portrayals of Sir Joseph Pitts in *Bridie's *Daphne Laureola* (1949) and the Judge in Enid *Bagnold's *The Chalk Garden* (1956).

Ayrer, Jakob (1543–1605), early German dramatist, successor to Hans *Sachs, and like him a prolific author of *Fastnachtsspiele, of which about 60 were published in 1618. Ayrer, who spent most of his life in Nuremberg, and was probably a member of the Guild of Mastersingers there, was much influenced by the *English Comedians, from whom he took the character of the *Fool, recognizable through his many metamorphoses by his first name of Jan. Although Ayrer's actors were amateurs, he aimed at big spectacular effects, and was one of the first German dramatists to write extensive stage directions. Three of his plays were taken from the same sources as those used by Shakespeare for *The Comedy of Errors*, *Much Ado about Nothing*, and *The Tempest*.

B

Babo, Joseph Marius, see RITTERDRAMA.

Bacchus, see DIONYSUS.

Backcloth, flat painted canvas, or a plain surface on which light can be projected. Hanging at the back of the stage, suspended from the *grid, it was used in combination with *wings to form a wing-and-backcloth scene. It was replaced by the *box-set and is now used only in ballet, opera, and pantomime. It is known in America as a backdrop.

Back Stage, term originally applied to a recess in the back wall of the stage, used for the last pieces of scenery in a deep vista, and at other times for storage. By extension the word was applied to all parts of the theatre behind the stage, including the actors' dressing-rooms. The expression 'to go (or work) backstage' is of recent origin, the first recorded use of it to indicate a visit behind the scenes dating only from 1923.

Bacon, Frank (1864–1921), American actor and playwright, best remembered for his playing of Bill Jones in his own play *Lightnin'*. His portrait of a lovable old rogue, who paid his way through life with exciting apocryphal tales of his exploits during the old pioneering days and the war between the states, was based on his memories of his Uncle Morris, known as Lightnin' in recognition of his lackadaisical approach to everything, including work. *Lightnin'* opened on Broadway in 1918 and ran for three years and a day, a record for the time. It was during its equally successful run in Chicago that Bacon died.

Baddeley, (Madeleine) **Angela Clinton-** (1904–76), English actress. She was on the stage as a child, making her adult début as Jenny Diver in Nigel *Playfair's production of *Gay's *The Beggar's Opera* (1920). She was with her sister Hermione (below) in Zoë *Akins's *The Greeks Had a Word for it* (1934), which was followed by Emlyn *Williams's *Night Must Fall* (1935; NY, 1936) and Natasha in *Chekhov's *Three Sisters* (1938) in *Gielgud's company. New plays in which she was seen included Dodie Smith's *Dear Octopus* (1938), Emlyn Williams's *The Light of Heart* (1940) and *The Morning Star* (1941), *Rattigan's *The Winslow Boy* (1946), and *Giraudoux's *The Madwoman of Chaillot* (1951). She was an excellent Miss Prue in Gielgud's revival of *Congreve's *Love for Love* in 1943, and appeared with the *Old Vic and *Stratford-upon-Avon companies in a

variety of parts by Shakespeare under the direction of her second husband Glen *Byam Shaw. In 1961 she was seen in an American musical, *Bye Bye Birdie*, and at the time of her death was appearing in another musical, *Sondheim's *A Little Night Music* (1975).

Baddeley, Hermione Clinton- (1906–86), English actress, sister of Angela (above), who was on the stage as a child, made her adult début in 1923, and soon achieved an enviable reputation in *revue, appearing with *The Co-Optimists* at the *Palace in 1924 and then in *Cochran's *On with the Dance* (1925) and its sequel *Still Dancing* (1926). In 1927 she made a great success as Ninetta, The Infant Phenomenon, in Nigel *Playfair's *When Crummles Played*, an extravaganza based on the theatrical chapters in *Dickens's *Nicholas Nickleby*. She was also much admired as Polaire in Zoë *Akins's *The Greeks Had a Word for it* (1934), in which her sister also appeared. She returned to revue with *Nine Sharp* (1938) and *The Little Revue* (1939) and made her one excursion into classic comedy as Margery Pinchwife in *Wycherley's *The Country Wife* (1940). After two revues in which she was partnered by Hermione Gingold, *Rise above it* (1941) and *Sky High* (1942), she returned to straight plays as Ida Arnold in a stage version of Graham *Greene's *Brighton Rock* (1943), then rejoined Gingold in the revue *Slings and Arrows* (1948) and *Coward's *Fumed Oak* and *Fallen Angels* (1949). She was also in the revues *À La Carte* (1948) and *At the Lyric* (1953) and was excellent as Mrs Pooter in Grossmith's *Diary of a Nobody* (1955). She made her New York début in Shelagh Delaney's *A Taste of Honey* (1961), and was later seen there in Tennessee *Williams's *The Milk Train Doesn't Stop Here Anymore* (1963). Her last appearance in the West End was as Mrs Peachum in *Brecht and Weill's *The Threepenny Opera* in 1972.

Baddeley, Robert (1732–94), English actor, who was first seen on the stage in 1760 at *Drury Lane, where he created the part of Moses in *Sheridan's *The School for Scandal* (1777). It was always his most popular role, and he was dressing to play it when he was taken ill and died. He was excellent in such parts as Fluellen in *Henry V* and Brainworm in *Jonson's *Every Man in His Humour*. Having in his youth been a pastrycook, he left a sum of money for a cake and wine to be partaken of by the Drury Lane company on Twelfth Night, a custom which is still observed. He was married briefly and unhappily to the actress **Sophia Snow** (1745–86), over whom he fought a duel with the brother of David *Garrick. The daughter of the trumpeter Valentine Snow, she was painted by Zoffany as Fanny in *Colman and Garrick's *The Clandestine Marriage*.

Bagley Wright Theatre, see SEATTLE REPERTORY THEATRE.

Bagnold, Enid [Lady Roderick Jones] (1889–1981), English author, who was already well known as a novelist when her *Serena Blandish* (1924) was dramatized by S. N. *Behrman and produced in New York in 1929. It was seen in London in 1938 with Vivien *Leigh in the title-role. Enid Bagnold herself then dramatized two of her novels, *Lottie Dundass* (1943) and *National Velvet* (1946), but her most successful play was *The Chalk Garden* (NY, 1955; London, 1956). It contained two major female roles: an elderly woman (Gladys *Cooper in New York, Edith *Evans in London) whose failure to control her chalk garden symbolized her failure with her daughter and granddaughter; and the companion (Siobhán *McKenna in New York, Peggy *Ashcroft in London), formerly convicted for murder, who helps to solve both problems. Three later plays, *The Last Joke* (1960), *The Chinese Prime Minister* (NY, 1964; London, 1965), and *Call Me Jacky* (1967), were unsuccessful in spite of excellent casts. The last was produced in New York in 1976 as *A Matter of Gravity*, with Katharine *Hepburn.

Baird, Dorothea, see IRVING, HENRY BRODRIBB.

Baker, Benjamin A. (1818–90), American actor-manager and playwright, a well-loved figure in the American theatre, familiarly known as 'Uncle Ben Baker'. He made his first appearance in New York in 1839, and had already written a number of plays, mainly burlettas, when in 1848 he produced, for his own benefit night, *A Glance at New York in 1848*, in which Frank *Chanfrau made a great success as the hero Mose, a New York volunteer fireman. Other 'Mose' plays followed, including *Three Years After* (1849) and *Mose in China* (1850). Their success gave rise to a type of *melodrama, with strong action set against a background of local conditions. Baker was manager successively of several theatres, and in 1856 of Edwin *Booth's company. He continued to work as manager and theatrical agent until his death.

Baker, George Pierce (1866–1935), one of the most vital influences in the formation of modern American dramatic literature and theatre, and the first Professor of Dramatic Literature at *Harvard, where he was educated and where, in 1905, he instituted a course in practical playwriting. This in turn led to the foundation of his famous '47 Workshop' for the staging of plays written under his tuition. Among the playwrights who attended his special courses were Edward *Sheldon, Eugene *O'Neill, Sidney *Howard, and George *Abbott. In 1925, by which time he had seen his pioneer work bearing fruit in many centres of learning, often under his own pupils, Baker left Harvard for *Yale, remaining there as director of the postgraduate Department of Drama until his retirement in 1933. Combining in a rare degree the attributes of the scholar and the practical man of the theatre, he had an immense influence not only in his own country but throughout Europe.

Baker, Josephine [Freda Josephine McDonald] (1906–75), American music-hall artist, who rivalled the legendary *Mistinguett in popularity. Black, illegitimate, and ill educated, she grew up in poverty, and after two brief marriages—taking her surname from her second husband—she went to Paris in 1925 in an American all-black show. An excellent singer and dancer—revealing almost all her beautiful body—she soon starred at the *Folies-Bergère. Idolized by Parisians of both sexes, she seldom left her adopted city. She retired from the stage several times, but always returned, and died while one of her come-back shows was still running.

Baker, Sarah (1736/7–1816), English theatre manageress, whose activities in Kent covered more than 50 years. The daughter of an acrobatic dancer, she married an actor in her mother's company in about 1761, and was widowed in 1769. Left with three small children to support, she went into management on her own account, and from 1772 to 1777 managed her mother's company. The latter having retired, Mrs Baker formed a new company which visited 10 towns in the country and had an ambitious repertory including Shakespeare and *Sheridan. At first a portable theatre or any suitable building was used, but from about 1789 she built her own theatres—10 in all. Edmund *Kean appeared with her early in his career.

Balieff, Nikita (1877–1936), Russian theatre director, deviser and compère of a Russian *cabaret, La Chauve-Souris, seen in Paris soon after the First World War. It was brought to London in 1921 by Charles *Cochran, and after a slow start became part of the London theatrical scene. In New York its success was instant and unflagging. In both places Balieff, a big burly man with a vast genial moon-face, who must have weighed at least 16 stone and who eked out his slender store of English with most expressive shrugs and gestures, gained immense personal popularity. His 'turns', costumed and set with a richness reminiscent of the Russian ballet, consisted of short burlesques and small, often mimed, sketches based on old ballads, folk-songs, prints, engravings, the woodenness of a toy soldier, or the delicacy of a china shepherdess, rendered amusing by very slight and subtle guying of the material.

Ballad Opera, name given in England to a play of topical interest, with spoken dialogue and a large number of lyrics set to existing tunes. The most famous example is *Gay's The Beggar's Opera* (1728). Such was the vogue for ballad operas that in the next 10 years almost 100 were presented in London. In 1743 Charles Coffey's *The Devil to Pay* (1731) was seen in Germany as *Der Teufel ist los*, with such success that a German form of ballad opera became popular there, under the name of *Singspiele*. These undoubtedly influenced the development of the great German operas with spoken dialogue, as exemplified by Mozart's *Die Entführung aus dem Serail* (1782) and Beethoven's *Fidelio* (1805). A second wave of popular plays with musical numbers, in this case the music being specially written, was initiated in London by Isaac *Bickerstaffe's *Love in a Village* (1762), which had music by Thomas Arne. This was less a true ballad opera than a forerunner of English operetta and *musical comedy.

Ballard, Sarah, see TERRY, BENJAMIN.

Banbury, Frith (1912–   ), English actor, manager, and director, who made his first appearance on the stage in 1933, and played a wide variety of parts, both in straight plays and revues, before directing *Dark Summer* (1947) by Wynyard Browne, whose comedy *The Holly and the Ivy* (1950) he also directed. He was then responsible for a number of interesting productions, including N. C. *Hunter's *Waters of the Moon* (1951); *Rattigan's *The Deep Blue Sea* (1952); John *Whiting's *Marching Song* (1954); and *Bolt's *Flowering Cherry* (1957) and *The Tiger and the Horse* (1960). He also directed plays in New York, and Peter *Nichols's *A Day in the Death of Joe Egg* (1968) for the *Cameri Theatre in Tel Aviv. After 1970 he was mainly involved in revivals, including *Sherwood's *Reunion in Vienna* (1971) at *Chichester and Michael *Redgrave's adaptation of Henry *James's *The Aspern Papers* (1984).

Bancroft, Sir Squire (1841–1926), English actor-manager, who, with his wife **Marie Effie** (*née* **Wilton**) (1839–1921), introduced a number of reforms on the British stage and started the vogue for drawing-room comedy and drama in place of *melodrama. Marie Wilton, the daughter of provincial actors, was on the stage from early childhood. She first appeared in London in 1856, scoring a great success as Perdita in Brough's *burlesque on *The Winter's Tale*. She continued to play in burlesque, notably at the Strand in H. K. *Byron's plays, until she decided to go into management on her own account. On a borrowed capital of £1,000, of which little remained when the curtain went up, she opened an old and dilapidated theatre

nicknamed the 'Dust Hole'—later the *Scala Theatre. Renamed the Prince of Wales's, charmingly decorated, and excellently run, it opened in 1865. In the company was Squire Bancroft, who had made his first appearance in Birmingham in 1861, and had played with Marie Wilton in Liverpool. The new venture was a success, and the despised 'Dust Hole' became one of the most popular theatres in London, where the Bancrofts (who had married in 1867) presented and played in the plays of Tom *Robertson, Bancroft giving one of his finest performances in *Caste* (1867). The Bancrofts did much to raise the economic status of actors, paying higher salaries than elsewhere and providing the actresses' wardrobes. Among other innovations, they adopted Mme *Vestris's idea of practicable scenery (see BOX-SET). In 1880 they moved to the *Haymarket and continued their successful career, retiring in 1885.

Bankhead, Tallulah Brockman (1903–68), American actress, a strikingly attractive woman with a deep husky voice, whose off-stage notoriety for some time prevented her from being taken seriously as an actress. She made her first appearance in New York in 1918, and in 1923 was seen in London playing opposite Gerald *Du Maurier. In 1925 she achieved a great success as Julia Sterroll in *Coward's *Fallen Angels*. She remained in London until 1930, when she was seen in the younger *Dumas's *The Lady of the Camellias*, and in 1933 returned to New York, having in the meantime appeared in a number of films. Among her later parts were Sadie Thompson in *Maugham's *Rain* (1935), Regina Giddens in Lillian *Hellman's *The Little Foxes* (1939), and Sabina in *Wilder's *The Skin of Our Teeth* (1942). She also played the Queen in *Cocteau's *The Eagle Has Two Heads* (1947). Before and after the last, however, she was seen in her most popular role, Amanda Prynne in Coward's *Private Lives* (1946–50). She also appeared in revivals of two plays by Tennessee *Williams: as Blanche Du Bois in *A Streetcar Named Desire* (1956) and as Mrs Goforth in *The Milk Train Doesn't Stop Here Anymore* (1964).

Banks, Leslie James (1890–1952), English actor, who made his first appearance in London in 1914. After army service during the First World War he joined the *Birmingham Repertory Company and returned to the West End in 1921, where in a long series of successful productions he established a solid reputation as a player of power and restraint. Among his best parts were Petruchio in *The Taming of the Shrew* (1937), with Edith *Evans as Katharina, the schoolmaster in James Hilton's *Goodbye, Mr Chips* (1938), John Thackeray in Barré Lyndon's *The Man in Half Moon Street* (1939), and the Duke in Patrick Hamilton's *The Duke in Darkness* (1942). He appeared many times in

New York, notably as Captain Hook in *Barrie's *Peter Pan* (1924). His versatility was shown in *Gielgud's repertory season at the *Haymarket Theatre in 1944–5, when he played Lord Porteous in *Maugham's *The Circle*, Tattle in *Congreve's *Love for Love*, Bottom in *A Midsummer Night's Dream*, and Antonio Bologna in *Webster's *The Duchess of Malfi*. Later roles included Father in Howard *Lindsay and Russel Crouse's *Life with Father* (1947) and Aubrey in *Pinero's *The Second Mrs Tanqueray* (1950). He was also an excellent director, usually of plays in which he was appearing.

Bankside Players, see ATKINS, ROBERT.

Bannister, John (1760–1836), English actor, son of the comedian **Charles Bannister** (1741–1804), whose modest reputation was made mainly at the *Haymarket with *Foote. John made his first appearance, also at the Haymarket, in 1778, and was then engaged for *Drury Lane, where he spent most of his career. Overshadowed by John Philip *Kemble in tragedy, he found his true bent in comedy. He was the first to play Don Ferolo Whiskerandos in *Sheridan's *The Critic* (1779), and was also good in such parts as Sir Anthony Absolute in Sheridan's *The Rivals*, Tony Lumpkin in *Goldsmith's *She Stoops to Conquer*, Scrub in *Farquhar's *The Beaux' Stratagem*, and Doctor Pangloss in the younger *Colman's *The Heir-at-Law*.

Banvard's Museum, New York, see DALY'S THEATRE.

Baptiste, see DEBURAU.

Baraka, Imamu Amiri, see NEGROES IN THE AMERICAN THEATRE.

Barbican Theatre, in the Barbican Centre in the City of London, is the London home of the *RSC, which moved there on its completion in 1982. Seating 1,160, it was specially built, with financial help from the Corporation of the City of London, the architects and the artistic directors working closely together. Structurally it is an adaptation of a traditional theatre, with shallow balconies thrusting downward round the stalls audience. The raked stage has a *forestage backed by a *proscenium 44 ft. wide. No spectator is more than 65 ft. from a point centre stage 8 ft. inside the forestage. The theatre presents productions from *Stratford-upon-Avon and original productions of other classical plays, usually non-Shakespearian. It opened with *Henry IV, Parts One and Two, A Midsummer Night's Dream*, and *All's Well That Ends Well*, and its own productions included *Rostand's *Cyrano de Bergerac* (1983; NY, 1984), with Derek *Jacobi, and *Chekhov's *Three Sisters* (1988). It also presented musicals: Peter *Nichols's *Poppy* (1982; *Adelphi, 1983), the enormously successful *Les Misérables* (1985; *Palace, 1986; NY, 1987), and *The Wizard of Oz*

(Christmas 1987 and 1988). The Barbican Centre also contains the RSC's second London theatre, The *Pit.

Barbieri, Niccolò (1576–1641), actor and dramatist of the *commedia dell'arte, who played under the name of Beltrame and is credited with the invention of the mask of *Scapino. He is first found with the *Gelosi in Paris in 1600–4 and then joined the *Fedeli, finally being with the *Confidenti. One of his plays, *L'inavertito* (1629), originally only a *scenario for improvisation, was later published with dialogue in full, and was made use of by *Molière for *L'Étourdi*.

Bard, Wilkie, see MUSIC-HALL.

Barker, character of the fairground or itinerant theatre company, who stood at the door of the booth and by his vociferous and spellbinding patter induced the audience to enter. He is probably as old as the theatre itself, and was known to the ancient world. He reappeared in 16th-century France and in the later fairs and showgrounds of Europe and America.

Barker, Harley Granville-, see GRANVILLE-BARKER.

Barker, Howard (1946–), English playwright, one of the most important of a group whose plays deal with major contemporary issues, usually from a socialist point of view. His first play *Cheek* (1970) was followed by such plays as *Alpha Alpha* (1972), about twin gangsters, *Claw* (1975; NY, 1976), about a pimp, and *Stripwell* (also 1975), in which Michael *Hordern played a judge whose house is invaded by a criminal he has sentenced. *That Good between Us* (1977) is set in a totalitarian Britain of the future; in *The Hang of the Gaol* (1978) a prison governor becomes an arsonist; and the background of *The Love of a Good Man* (1980) is a First World War battlefield in 1920. Barker is a prolific dramatist of great passion, whose complex and often confusing plays convey great moral outrage which occasions both his outlandish plots and his frequently excellent dialogue. His work in the 1980s included *No End of Blame* (London and NY, 1981), portraying the pressures on a Hungarian cartoonist before and after his emigration to England; *Victory* (1983) and *The Castle* (1985), both using historical settings for allegorical purposes. The year 1988 saw the production of *The Possibilities*, 10 short plays; *The Last Supper*, a 'speculation on the life of Christ', with music; and his most demanding play, the 4½-hour *The Bite of the Night*. Barker's adaptation of *Middleton's *Women Beware Women* was seen in 1986 (NY, 1987). Most of his plays have been presented either at the *Royal Court or by the *RSC.

Barker, James Nelson (1784–1858), American dramatist, author of the first American play on an Indian theme, *The Indian Princess; or, La Belle*

Sauvage (1808), which was also the first play from America to be performed in England (as *Pocahontas; or, The Indian Princess,* at *Drury Lane in 1820). Of Barker's other plays, only three have survived: *Tears and Smiles,* a comedy of manners played in Philadelphia in 1807; a dramatization of Scott's *Marmion* (1812), which held the stage for many years, being last revived in 1848; and *Superstition* (1824), the story of a Puritan refugee from England who leads his village against the Indians.

Barnes, Charlotte Mary Sanford (1818–63), American actress and dramatist, the daughter of **John Barnes** (1761–1841), who went with his wife Mary from *Drury Lane to the *Park Theatre, New York, where they remained for many years. Charlotte first appeared on the stage at the age of 4 in 'Monk' *Lewis's *The Castle Spectre,* and in 1834 made her adult début in the same play. She was admired as *Rowe's *Jane Shore,* as well as in such roles as Desdemona and Lady Teazle in *Sheridan's *The School for Scandal.* In 1842 she appeared in London, playing Hamlet among other parts, in which she was well received, though she was never accounted as good an actress in male parts as her mother, who was an excellent Romeo to her daughter's Juliet. Four years later Charlotte married an actor-manager, Edmond S. Connor, and became his leading lady, being associated with him in the management of the Arch Street Theatre, *Philadelphia. Her first play, *Octavia Bragaldi,* written at the age of 18, was produced at the National Theatre, New York, with herself in the leading part. Of her other plays the only one to survive was *The Forest Princess* (1844), based on the story of Pocahontas.

Barnes, Clive Alexander (1927–), English dance and drama critic, who after leaving Oxford became a freelance journalist, writing mainly on the performing arts until in 1961 he became dance critic of the London *Times* and editor of *Dance and Dancers.* In 1965 he became dance critic of the *New York Times,* being appointed drama critic also two years later. In 1977 he moved to the *New York Post,* where he holds the same positions, as well as being Associate Editor. One of the most important and powerful critics in America, he has been attacked both by directors and by other critics for a certain lack of appreciation of American theatrical sensibility, perhaps because he has on occasion supported *Off-Broadway and international theatre at the expense of commercial Broadway productions; his judgements have on the whole been justified by time.

Barnes, Sir Kenneth, see VANBRUGH, IRENE.

Barnes Theatre, London, in Church Road, Barnes, two-tier hall built about 1905 which opened as a small theatre in 1925. After several short runs of new plays came the much-publicized stage version of *Tess of the D'Urbervilles* by Thomas Hardy himself, with Gwen *Ffrangcon-Davies as Tess. *Chekhov's *Uncle Vanya,* which opened in Jan. 1926 with Jean *Forbes-Robertson as Sonia, heralded a remarkable run of Russian plays—Chekhov's *Three Sisters* with *Gielgud as Tusenbach, *Andreyev's *Katerina,* again with Gielgud, *Gogol's *The Government Inspector* with Charles *Laughton—all directed by *Komisarjevsky. A dramatization of Dostoevsky's *The Idiot* was followed by John *Drinkwater's adaptation of Hardy's novel *The Mayor of Casterbridge,* which he also directed. Chekhov returned with *The Cherry Orchard* and a revival of *Three Sisters,* and this interesting and extremely fruitful venture ended in Nov. 1926 after a revival of Drinkwater's *Abraham Lincoln.*

Barnstormers, name given in the late 19th century to the early itinerant companies whose stages were often set up in large barns, and whose work was characterized by ranting and shouting and general violence in speech and gesture.

Barnum, Phineas Taylor, see AMERICAN MUSEUM.

Baron, Michel (1653–1729), French actor, son and grandson of strolling players. His father, **André Baron** [Boyron] (*c.*1601–55), was with *Montdory at the *Marais and later at the Hôtel de *Bourgogne, and is said to have died of a wound in the foot self-inflicted during a spirited performance as Don Diègue in *Corneille's *Le Cid.* He played kings and noblemen in tragedy and rustic characters in comedy. André's wife **Jeanne Ausoult** (1625–62) was much admired in breeches parts. Her early death was much regretted by Corneille, who said he had written a part for her in his next play, probably that of Sophonisbe. Michel, orphaned before he was 10, was acting with the juvenile Troupe du Dauphin when *Molière saw him and took him into his own company, giving him small parts to play. Unfortunately Molière's wife took a dislike to the boy, and on one occasion slapped his face, whereupon he ran away and rejoined his former companions. He remained with them until 1670, when Molière asked him to return and play Domitian in Corneille's *Tite et Bérénice.* On Molière's death Baron went to the Hôtel de Bourgogne to play the young tragic heroes of *Racine's plays. He later became the chief actor of the newly formed *Comédie-Française, being endowed with a fine presence, a deep voice, amplitude of gesture, and a quick intelligence. He was the author of several comedies, of which the best is *L'Homme à bonne fortune* (1686). He retired in 1691, at the height of his powers, but returned to the stage in 1720, remaining until his death. He was the last actor who had known Molière to appear at the

Comédie-Française. His son **Étienne** (1676–1711) was also a member of the Comédie-Française, as were Étienne's son and one of his nieces.

Barraca, La, Spanish theatre company, more formally known as the Teatro Universitario, formed after the creation of the Spanish Republic in 1931 to tour plays from the classical repertoire around rural areas. Headed by *García Lorca and Eduardo Ungarte, the company, consisting largely of students, met with an enthusiastic response and influenced the form of the three great tragedies of Lorca's maturity. It disappeared after the Spanish Civil War.

Barrault, Jean-Louis (1910–), French actor, director, and producer, who made his first appearance on the stage on his 21st birthday, as one of the servants in Jules *Romains's adaptation of *Jonson's *Volpone*. He then began the study of *mime, which was to be so important in his career, and his first independent production was a mime-play adapted with *Camus from Faulkner's novel *As I Lay Dying*. In 1940 he was engaged by *Copeau for the *Comédie-Française, making his début in *Corneille's *Le Cid*, though his most important work was done in his productions of *Racine's *Phèdre*, Shakespeare's *Antony and Cleopatra*, and *Claudel's *Le Soulier de satin*. In 1946 he and his wife Madeleine *Renaud, who was already an established star at the Comédie-Française when he joined it, left to found their own company. Its opening production, at the *Théâtre Marigny, was *Hamlet* (1946). The company remained at the Marigny until 1956, appearing in a mixed repertory of classic and modern works, among them the plays of Claudel—*Partage de midi* (1948), *L'Échange*, *Christophe Colomb* (both 1951), and *Tête-d'or* (1959). The inclusion of *Marivaux plays in the company's repertory can be attributed partly to the outstanding acting of Madeleine Renaud in this author's works. The company also revived several farces by *Feydeau, and gave the first performance of such modern plays as *Ionesco's *Rhinocéros* (1960). In 1956 Barrault left the Marigny for the Théâtre Sarah-Bernhardt, and later occupied the *Palais-Royal. In 1959 he was appointed director of the *Odéon, where his work with Madeleine Renaud included *Beckett's *Oh! les beaux jours* (1960), and where he later transferred the *Théâtre des Nations; but he was summarily dismissed in 1968, after the theatre had been occupied by student demonstrators. A year later he was at the Élysées-Montmartre near Clichy, formerly an indoor wrestling stadium, where he directed his own adaptations of Rabelais and of *Ubu sur la butte*, based on *Jarry's plays. *Rabelais* was seen in London, in both French and English, and toured Europe and America. On returning to Paris in 1972 Barrault set up a circus tent inside the disused hall of the Gare d'Orsay, which was taken from him in 1980 to become a museum. He was then given the Palais des Glaces, an old skating rink to which he transported the staging from the Gare d'Orsay, making a large auditorium seating 920 spectators and a smaller one for 200. In 1985 he again appeared in *Le Cid*.

As an actor Barrault, elegant and outwardly nonchalant, gives the impression of strong passions firmly controlled by disciplined intelligence and physically sustained by the strenuous exercise of mime. As a mime-actor he is best known through his appearance as *Deburau in the film *Les Enfants du paradis* (1945); but the rigour of his training can most effectively be judged on stage, where his slightest gesture speaks volumes.

Barrel, counterweighted tube of metal (or, rarely, wood) $1\frac{1}{2}$–2 in. in diameter, known in America as a pipe batten. It is hung on wire-rope or hemp lines from the *Grid, and is used to attach scenery or lighting equipment, being referred to in the latter case as a spot bar, or, in America, a light pipe.

Barrett, Lawrence (1838–91), American actor and producer, who, as a boy actor, travelled the United States with many outstanding companies, including that of Julia *Dean. In 1857 he was seen in New York in Sheridan *Knowles's *The Hunchback* and other plays, and in 1858 he was a member of the Boston Museum company. He frequently played Cassius in *Julius Caesar*, the part in which he is best remembered, to the Brutus of Edwin *Booth, and in 1871 was manager of *Booth's Theatre. While *Irving was in America in 1884, Barrett took over the *Lyceum in London, and though his visit was not a success financially, he was fêted on all sides. Tall, with classic features, and dark, deeply sunken eyes, he was probably at his best in Shakespeare, to whose interpretation he brought dignity, a dominant personality, and intellectual powers somewhat exceptional in an actor at that date. All the standard classics were in his repertory, however, and he did his best to encourage American playwrights.

Barrett, Wilson (1846–1904), English actor-manager, who had few equals in the melodramatic plays fashionable in his time, being strikingly handsome and having a resonant voice, though not tall. His finest parts were Wilfred Denver in Henry Arthur *Jones and Herman's *The Silver King*, which he produced in 1882, and Marcus Superbus in his own play *The Sign of the Cross*. This melodramatic story of a Roman patrician in love with a beautiful young Christian convert, with whom he goes into the arena to face the lions, was first produced in the USA in 1895, when Barrett was on tour and badly in need of money. It brought him a fortune. It was first seen in England at the Grand Theatre in Leeds later in

1895 and in London a year later, when it created a sensation. It was several times revived and perennially successful on tour. Barrett was less successful in Shakespearian roles, only his Mercutio in *Romeo and Juliet* being thought passable. He had a brother and nephew in his company, and his grandson was also on the stage.

Barrie, Sir James Matthew (1860–1937), Scottish novelist and dramatist, whose whimsical children's play *Peter Pan*, based on one of his own novels, was once a feature of the London theatrical scene at Christmas. It was first produced in 1904, with Nina *Boucicault as Peter. Barrie's early plays were unsuccessful, and he first came into prominence with a farce, *Walker, London* (1892), in which *Toole appeared. The success of *The Professor's Love Story* (1894) and *The Little Minister* (1897) established him as a promising playwright, and he consolidated his position with the romantic costume play *Quality Street* (1902) and the social comedies *The Admirable Crichton* (also 1902) and *What Every Woman Knows* (1908). The mixture of fantasy and sentimentality which runs through much of Barrie's work was uppermost in *Dear Brutus* (1917) and *Mary Rose* (1920), both of which invoke supernatural elements in a realistic setting. Among his shorter plays, the one-act *The Old Lady Shows her Medals* (1917) and *Shall We Join the Ladies?* (1922) have been popular with amateur companies. His last play, *The Boy David* (1936), was specially written for Elisabeth *Bergner, who played the biblical David. It was not a success, and he is more likely to be remembered for *The Admirable Crichton*, whose title has passed into everyday speech to describe a man of all-round excellence, and *Peter Pan*.

Barry, Elizabeth (1658–1713), the first outstanding English actress, who made her first successful appearance, after several failures, in a revival of *Orrery's *Mustapha* (1675). She played opposite *Betterton for many years and is credited with the creation of over 100 roles, including Monimia in *Otway's *The Orphan* (1680) and Belvidera in his *Venice Preserv'd* (1682), and Zara in *Congreve's *The Mourning Bride* (1697). Though good in comedy, she was at her best in tragedy, in which she displayed great power and dignity. Many scandals are attached to her name, some perhaps undeservedly. She retired in 1710 and, as far as is known, never married.

Barry, Philip [Jerome Quinn] (1896–1949), American dramatist. He achieved some success with his first professional production *You and I* (1923), but is best remembered for his light comedies. *Paris Bound* (1927) looks at upper-crust infidelity; *Holiday* (1928) deals with the revolt of youth against parental snobbery; *The Animal Kingdom* (1932) reverses the roles of wife and mistress by making the latter the loyal companion and therefore the true wife; and *The

Philadelphia Story (1939), his most popular play, is a deft comedy of manners and character which was a great triumph for Katharine *Hepburn. His more serious work, including *Hotel Universe* (1930), a probing psychological drama, and *Here Come the Clowns* (1938), a mystifying but provocative allegory of good and evil, had less appeal. He left unfinished a comedy, *Second Threshold*, which was completed by his friend Robert *Sherwood and presented in New York in 1951.

Barry, Spranger (1719–77), Irish actor, first seen on the stage in Dublin in 1744. In 1746 he was at *Drury Lane, where he appeared as Othello to the Iago of *Macklin. When *Garrick took over in 1747 he remained in the company, playing leading roles, and alternating with Garrick in such parts as Jaffier and Pierre in *Venice Preserv'd*, Chamont and Castalio in *The Orphan* (both by *Otway), Hastings and Dumont in *Jane Shore*, and Lothario and Horatio in *The Fair Penitent* (both by *Rowe). In 1750 Barry went to *Covent Garden and set himself up as a rival to Garrick in roles such as Romeo, Lear, and Richard III, not always successfully, though he was an excellent Young Norval in *Home's *Douglas*. Later, having ruined himself by the speculative building of a theatre in *Dublin, he returned to Drury Lane. Shortly afterwards his wife died and he married **Ann Dancer** [*née* **Street**] (1734–1801), an actress who excelled in high comedy roles such as Millamant and Angelica in *Congreve's *The Way of the World* and *Love for Love*, and Mrs Sullen in *Farquhar's *The Beaux' Stratagem*. After the death of Barry, who was buried in Westminster Abbey, she married a fellow actor and retired in 1798.

Barrymore, Ethel (1879–1959), one of the leading actresses of the American stage, granddaughter of Mrs John *Drew and sister of John and Lionel *Barrymore. She made her first appearance in New York in 1894, and in 1898 was in London, where she appeared with Henry *Irving in Laurence *Irving's *Peter the Great*. In New York she scored her first outstanding success as Madame Trentoni in Clyde *Fitch's *Captain Jinks of the Horse Marines* (1901). She played the title-roles in *Maugham's *Lady Frederick* (1908) and *The Constant Wife* (1926), and Rose in *Pinero's *Trelawny of the 'Wells'* (1911); but she was also successful in more serious parts, such as Marguerite in the younger *Dumas's *The Lady of the Camellias* (1917) and Paula Tanqueray in Pinero's *The Second Mrs Tanqueray* (1924). In 1928 she opened the *Ethel Barrymore Theatre with *Martínez Sierra's *The Kingdom of God*, returning there in 1931 as Lady Teazle in *Sheridan's *The School for Scandal*. She was outstanding in two English parts—as Gran in Mazo de la Roche's *Whiteoaks* (1938) and as Miss Moffat in Emlyn *Williams's *The Corn is

Green (1940), probably her best role. She also played in *vaudeville, notably in *Barrie's *The Twelve-Pound Look*.

Barrymore, John (1882–1942), outstanding but very uneven American actor, grandson of Mrs John *Drew, brother of Ethel and Lionel *Barrymore, who became a popular matinée idol and a good light comedian. He proved himself a serious actor also by his performances as Falder in *Galsworthy's *Justice* (1916), in the title-role of the adaptation of George Du Maurier's novel *Peter Ibbetson* (1917), and in *The Living Corpse* (1918), based on *Tolstoy's *Redemption*. His career reached its peak with his Richard III (1920) and Hamlet (1922; London, 1925), which had a New York run of 101 consecutive performances, breaking by one performance the record set by Edwin *Booth. He gave a meticulous and scholarly reading of the part but eventually failed to fulfil his promise, largely because of alcoholism.

Barrymore, Lionel (1878–1954), American actor, brother of Ethel and John *Barrymore and grandson of Mrs John *Drew, under whom he made his first appearance on the stage. After achieving some success, notably in the title-role of *Barrie's *Pantaloon* (1905), he went to Paris to study art. Returning to the United States in 1909, he reappeared on the stage in Conan Doyle's *The Fires of Fate*, and soon became one of New York's leading actors, being particularly admired in dramatizations of George Du Maurier's *Peter Ibbetson* (1917) and Augustus *Thomas's *The Copperhead* (1918). After an unsuccessful Macbeth (1921) and other failures he left the theatre for films in 1925.

Barrymore, Maurice [Herbert Blythe] (1847–1905), English actor, who took his stage name from an old playbill hanging in the *Haymarket Theatre in London. He first appeared on the stage in 1875, and then went to New York, where he played opposite many leading actresses, including *Modjeska, for whom he wrote *Nadjezda* (1886). His best role was probably Rawdon Crawley in *Becky Sharp* (1899), an adaptation of Thackeray's *Vanity Fair*, in which he played opposite Mrs *Fiske. Towards the end of his career he went into *vaudeville. He also wrote several plays. He married in 1876 Georgiana *Drew, actress daughter of Mrs John *Drew, under whom she first appeared on the stage at the age of 16. After her marriage she appeared with her husband in a number of plays, but her career was much hampered by illness and she died young. Their three children, Lionel, Ethel, and John (above), were all on the stage.

Barrymore, Richard Barry, Earl of, see PRIVATE THEATRES IN ENGLAND.

Barsacq, André (1909–73), French director and manager of Russian descent, who was studying design in Paris when a visit to the Atelier, then under the direction of *Dullin, turned his thoughts to the theatre. He worked with Dullin on several plays and films, including the 1928 adaptation of *Jonson's *Volpone* by Jules *Romains, and in 1937 co-founded the Compagnie des Quatre Saisons, a travelling repertory company which was seen in the same year at the International Exhibition and then spent two years in the French Theatre in New York. Barsacq produced *Anouilh's *Le Bal des voleurs* (1938) and almost all his later plays until 1953, including the subversive *Antigone* (1944). After the war Barsacq succeeded Dullin at the Atelier and remained there until his death, in spite of illness and financial difficulties, and introducing to Parisian audiences *Betti, Félicien *Marceau, and *Dürrenmatt. He also directed several successful adaptations of *Chekhov, Dostoevsky, and *Turgenev.

Bart, Lionel (1930–), English lyricist and composer who helped to revive English *musical comedy after a long period of American domination. His first work, *Fings Ain't Wot They Used T' Be* (1959), was produced by *Theatre Workshop and then ran in the West End for two years. He also wrote the lyrics for *Lock up Your Daughters* (1959), a play with music based by Bernard *Miles on *Fielding's *Rape upon Rape*. His greatest success was *Oliver!* (1960; NY, 1963), based on *Dickens's *Oliver Twist*. It had a long run when first produced and was revived in 1967 and 1977. Two other productions, *Blitz!* (1962) and *Maggie May* (1964), about a Liverpool prostitute between the wars, were also successful, but a Robin Hood musical *Twang!* (1965) was a disastrous failure, as was *La Strada* (1969), a musical based on Fellini's film, which had only one performance in New York.

Barter Theatre, Abingdon, Va., proscenium-arch theatre seating 380. Founded during the Depression, it opened in 1933, admission being by barter, each member of the audience paying in kind for the privilege of seeing a play. In this way the actors, though always short of money, at least had enough to eat. The theatre prospered, and in 1946 was declared the official State Theatre of Virginia. In 1953 it was refurbished with some of the furnishings from the old *Empire Theatre, New York, and in 1971 a second theatre, the Barter Playhouse, seating 100 round a three-quarters arena stage, was added. Tours are undertaken by the company outside the resident season of classic and contemporary plays, which runs from Apr. to Oct., and special plays for children are presented in the Playhouse.

Barton, John Bernard Adie (1928–), English director, who has been connected with the *RSC since its foundation, and associate director since 1964, having previously directed plays in *Cambridge for the Marlowe Society and the

ADC. His first production at Stratford was *The Taming of the Shrew*. He devised the recitals *The Hollow Crown* (1961), *The Art of Seduction*, and *The Vagaries of Love* (both 1962), in all of which he took part. He also adapted *Henry VI* and *Richard III* as *The Wars of the Roses* (1963). In 1969 he prepared for Theatregoround, the RSC schools programme, a shortened version of *Henry IV* as *When Thou Art King*, concentrating on the relationship between Prince Hal and Falstaff. He also made a shortened version of *Henry V* as *The Battle of Agincourt*, and in 1970 added to it two more scenes from the history cycle as *The Battle of Shrewsbury* and *The Rejection of Falstaff*. His non-Shakespearian productions included his own version of *Marlowe's *Dr Faustus* (1974); *The Greeks* (1980) based on 10 plays, mainly by *Euripides; *Schnitzler's *La Ronde* (1982); and Aphra *Behn's *The Rover* (1986).

Bassermann, Albert (1867–1952), German actor, nephew of August *Bassermann, who made his first appearance at *Mannheim under his uncle's management and was with the *Meiningen company from 1890 to 1895. Under Otto *Brahm, whom he joined in 1899, he became the outstanding interpreter of *Ibsen in Germany. He was also effective in plays by *Hauptmann and *Schnitzler. From 1909 to 1914 he was with *Reinhardt, and he later extended his range to include Shakespeare's Shylock and Lear, *Goethe's Mephistopheles (in *Faust*), and other great roles. The rise of the Nazi party sent him into exile, and in 1944 he appeared on Broadway in his first English-speaking part, the Pope in *Embezzled Heaven*, based on a novel by *Werfel. He died in Zurich, having returned to Europe in 1945.

Bassermann, August (1848–1931), German actor, uncle of the above, who made his début at *Dresden in 1873 and later appeared at the *Vienna Stadttheater in such parts as Rolla in *Sheridan's *Pizarro* and Karl Moor in *Schiller's *Die Räuber*. In 1886 he became manager of the *Mannheim theatre, and from 1904 until his death he was director of a theatre at Karlsruhe. He went to New York several times, and played in the German Theatre there, being much admired in classic and heroic parts.

Bateman, H(ezekiah) **L**(inthicum) (1812–75), American impresario, who gave up acting to further the careers of his four daughters by his wife **Sidney Cowell** (1823–81), daughter of Joseph *Cowell and herself an actress and playwright. All four were on the stage from early childhood: **Ellen** (1844–1936) retired from the theatre on marriage; **Virginia** (1855–1940) married the actor Edward *Compton and continued her career under his management. Bateman's eldest daughter **Kate** (1842–1917), as an adult actress, made a great success in the title-role of *Daly's *Leah the Forsaken* (1863), with which

she became particularly associated. She retired in 1892. In 1871 Bateman, anxious to launch his third daughter **Isabel** (1854–1934), on her adult career in London, leased the *Lyceum Theatre and engaged Henry *Irving as his leading man, allowing him, after some demur, to put on Lewis's *The Bells*, which was an immediate success. Isabel, though disliking the theatre, was an excellent actress, playing Ophelia to Irving's Hamlet, Desdemona to his Othello, and Queen Elizabeth I in *Tennyson's *Queen Mary*, with her sister Kate as Queen Mary. After her husband's death Mrs Bateman continued to lease the Lyceum until Irving took over in 1878. She then went to *Sadler's Wells with Isabel, who in 1898 left the stage to become a nun.

Bates, Alan Arthur (1934–), English actor, who made his first appearance in London at the *Royal Court in 1956, where later the same year he attracted favourable notice as Cliff Lewis in *Osborne's *Look Back in Anger* (NY, 1957). He then played Edmund Tyrone in *O'Neill's *Long Day's Journey into Night* (1958). It was, however, as Mick in *Pinter's *The Caretaker* (1960; NY, 1961) that he made his first out-standing success. He also appeared as Richard III at the *Stratford (Ontario) Festival in 1967. In 1969 he returned to the Royal Court in David *Storey's *In Celebration* and in 1970 (London, 1971) he was seen as Hamlet at the *Nottingham Playhouse. By now one of the country's leading actors, he embarked on an interesting series of classical and modern roles: from the title-role in Simon *Gray's *Butley* (1971; NY, 1972) to an 'effervescently likeable' Petruchio in *The Taming of the Shrew* at the *Royal Shakespeare Theatre to new plays by Storey (*Life Class*, 1974) and Gray (*Otherwise Engaged*, 1975). In 1976 he played Trigorin in *Chekhov's *The Seagull*, and he was seen in two more of Gray's plays, *Stage Struck* (1979) and *Melon* (1987). He also excelled as Colonel Redl, the Jewish homosexual tricked into treachery, in Osborne's *A Patriot for Me* (*Chichester and London, 1983); as Edgar in *Strindberg's *The Dance of Death* (1985); and in the title-role of the voyeuristic narrator in *Shaffer's *Yonadab* (*National Theatre, 1986). In 1989 he played Chekhov's Ivanov and Benedick in *Much Ado about Nothing* in repertory.

Bates, Blanche (1873–1941), American actress, whose parents were both in the theatre. She first appeared on tour in California, from 1894 to 1898, and then made a brief appearance in New York under the management of Augustin *Daly. It was, however, as the leading lady in plays by *Belasco—*Madam Butterfly* (1900), *The Darling of the Gods* (1902), and *The Girl of the Golden West* (1905)—that she made her reputation. She was also much admired as Cigarette in a dramatization of Ouida's *Under Two Flags* (1901). After leaving Belasco she continued to act, under her own and other managements, but less successfully, until her retirement in 1927.

Bath. The first theatre in Bath was built in 1705. On the passing of the Licensing Act in 1737 it was demolished, and the company moved to a room under Lady Hawley's Assembly Rooms. In 1750 a second theatre was erected in Orchard Street, the shell of which still exists. The two theatres were rivals until 1756, when the companies amalgamated under John Palmer and settled in Orchard Street. The theatre was reconstructed in 1767 and again in 1774; in 1768 it became a *Theatre Royal under a patent from George III, the first provincial theatre in England to be so honoured. From 1778 to 1782 Mrs *Siddons was a member of its *stock company. A new and larger theatre was erected in Beaufort Square in 1805; it flourished for a time but by the 1820s it was in decline and its interior was destroyed by fire in 1862. Less than 12 months later the present Theatre Royal, seating 615, with a further 250 in the gallery, opened on the same site with a production by the *Bristol stock company of *A Midsummer Night's Dream* in which Ellen *Terry, aged 15, played Titania. Since 1867 it has been used by touring companies, Henry *Irving's being the first to visit it. In 1979 the theatre was taken over by a Trust and in 1982 it was refurbished. It is now widely visited by major touring companies and is a prime venue for pre-London tours.

Batten, length of timber used to stiffen a surface of canvas or boards, as by 'sandwich-battening' a cloth (fixing the upper and lower edges between pairs of battens screwed together) or by 'battening-out' a section of boards or a run of *flats with crossbars. A light batten, known in America as a strip light or border light, is a horizontal trough containing a row of lights, often divided into compartments (and then known as a compartment batten), each containing a lamp, reflector, and coloured filter.

Baty, Gaston (1885–1952), French director, who in 1930 opened the Théâtre Montparnasse under his own name. Here he put on an imposing series of old and new plays, many of them foreign classics, and several dramatizations of novels, of which the best were probably his own version of Dostoevsky's *Crime and Punishment* and his own play *Dulcinée* (1938), based on an episode from *Don Quixote*. Baty was sometimes accused of subordinating the text of the play to the décor, which led him to substitute pictorial groupings for action; but this resulted in some fine work, most noticeable in his productions of Gantillon's *Maya*, *Lenormand's *Simoun*, and J.-J. *Bernard's *Martine*. In 1936 Baty was appointed a director of the *Comédie-Française, where his erudition and impeccable theatrical technique helped to re-animate the classical repertory. In the year of

his death he was instrumental in founding the Centre Dramatique du Sud-Est.

Bauer, Wolfgang (1941–   ), Austrian dramatist, one of a group of dialect playwrights who emerged from Graz at the end of the 1960s. His plays are in an Austrian tradition which embraces the realistic dialogue of Horváth (see HAMPTON), *Schnitzler's *ennui*, and *Nestroy's clockwork satire. Like Schnitzler in particular, Bauer presents the comedy inherent in a decadent boredom so severe that in attempting to alleviate it a character may meet his death, as in *Sylvester oder das Massaker im Hotel Sacher* (1971). As in other plays, including the shorter ones such as *Film und Frau* (also 1971), briefly seen in London in 1972 as *Shakespeare the Sadist*, the setting is a jaded milieu of pop, pot, and casual sex, among talentless would-be avant-gardists. His control of pace and mood, his convincing theatrical climaxes, even his liking for fast changes of role and clothes, give his plays the impact of farce; but they are true comedies of modern manners and have enjoyed great success in Germany as well as in Austria.

Bäuerle, Adolf, see STABERL.

Bayes, Nora, see VAUDEVILLE, AMERICAN.

Baylis, Lilian Mary (1874–1937), English theatre manageress, and one of the outstanding women of the British theatre, being only the second woman outside the university to be given an Hon. MA at Oxford (in 1924, for her services to Shakespeare). Born in London, of a musical family, she accompanied her parents and sister to South Africa, with the intention of establishing herself there as a music teacher. Under parental pressure, however, she returned to London in 1895 to help her aunt, Emma Cons, run the old Victoria Theatre as a temperance hall. It later became the *Old Vic. In 1912, on her aunt's death, she took over the management of the theatre, and devoted the rest of her life to it. When the popularity of Shakespeare threatened to engulf that of opera, she leased and renovated *Sadler's Wells in 1931, hoping to run the two theatres in tandem. When this proved impossible, owing to the rapid advance in popularity of the ballet company for which Lilian Baylis, with her usual acumen, had chosen Ninette de Valois as ballet mistress at the Vic in 1928, the two houses divided, the Old Vic thriving on drama, Sadler's Wells on ballet, with opera once again ousted from its original pre-eminence. The history of both theatres bears witness to the strength and tenacity of Lilian Baylis's vision of places where impecunious Londoners could find 'good theatre at reasonable prices', and drama and ballet profited by what opera had lost.

Bayreuth, see THEATRE BUILDINGS.

Bear Gardens Museum, see WANAMAKER.

Beaton, Sir Cecil Walter Hardy (1904–80), English designer and photographer. His first work for the stage was the scenery and costumes for the revue *Follow the Sun* (1936). His first venture into the non-musical theatre was the costumes and scenery for *Gielgud's production of *Wilde's *Lady Windermere's Fan* (1945). A year later he made his first appearance as an actor in an American production of this play, again designing the costumes and scenery. He was associated with revivals of *Pinero's *The Second Mrs Tanqueray* (1950), *Lonsdale's *Aren't We All?* (1953), and *Love's Labour's Lost* (*Old Vic, 1954). He was also responsible for the décor and costumes of *Coward's *Quadrille* (1952; NY, 1954). His later work—including Enid *Bagnold's *The Chalk Garden* (1955), Coward's *Look after Lulu* (1959), and the musical *Coco* (1969)—was seen mostly in New York. His best-known work was the lavish and witty costume design for the musical *My Fair Lady* (1956; London, 1958).

Beaumarchais, Pierre-Augustin Caron de (1732–99), French dramatist, whose first play, *Eugénie* (1767), had a moderate success (after rewriting) at the *Comédie-Française, as did *Les Deux Amis* (1770). His first important play, *Le Barbier de Séville*, originally intended as a play with music, and later used as the basis of an opera by Rossini, was refused by the *Comédie-Italienne, which thought its hero a caricature of their leading actor, formerly a barber. It was, however, accepted by the Comédie-Française soon after its completion in 1772, but constantly delayed and not acted until 1775, before an audience which still felt secure from the Revolution which it presaged, and which forced Beaumarchais himself into temporary exile. Nine years later, after an even harder struggle with the censorship because of its provocative political satire, *Le Mariage de Figaro* (1784)—on which Mozart based his opera—was finally seen by a public which was beginning to realize the dangers that lay before it. Figaro, older and wiser, no longer criticizes an individual, as in the earlier play, but society as a whole.

These two plays sum up Beaumarchais's whole life and character. He is himself the precocious page, the handsome Almaviva—he was three times married—and above all Figaro, the jack-of-all-trades. His later dramatic works are less interesting, though he completed the Figaro trilogy with *La Mère coupable* (1795). More important was the part he played in breaking the stranglehold of the actors on their authors, and in instituting, through the Société des Auteurs, of which he was the founder, the system of payment for plays by means of a fixed percentage on takings (see ROYALTY).

Beaumont, Sir Francis (1584–1616), English dramatist, whose name is so closely associated with that of John *Fletcher that they are usually spoken of in one breath, while scholars are still disentangling their separate contributions, and those of most of their playwright contemporaries, from the more than 50 plays which once passed under their joint names. Even Beaumont's first play, *The Woman Hater* (c.1605), once believed to be his own, is now assigned partly to Fletcher. It is however believed that *The Knight of the Burning Pestle* (1607), a lively comedy which satirizes the romantic-historical plays popular at the time, and gives amusing glimpses of a contemporary audience in action, is by Beaumont alone. He ceased to write for the stage on his marriage in 1613, though he still contributed occasionally to entertainments seen at Court.

Beaumont, Hugh ('Binkie') [Hughes Griffiths Morgan] (1908–73), Welsh theatre manager, who after experience in several small London theatres was appointed business manager of H. M. Tennent, for nearly 30 years the leading theatrical producers of London. He initiated some 400 productions, both of new plays and of revivals, often featuring the outstanding actors and actresses of the day. In 1939, for instance, he was responsible for *Gielgud's production of *Wilde's *The Importance of Being Earnest* and for his *Hamlet*, and in 1944–5 he encouraged this great actor in his repertory season at the *Haymarket. The contemporary theatre owed much to his commercial acumen and notably high standards, such modern playwrights as *Bolt, *Anouilh, *Whiting, and *Fry being presented in London under his aegis. He also encouraged such solo performers as Beatrice *Lillie, Ruth *Draper, and Joyce *Grenfell, and was responsible for a number of musicals, including *My Fair Lady* and *Oklahoma!* Beaumont was elected to the governing body of the *Shakespeare Memorial Theatre in 1950, and to the *National Theatre Board in 1962. H. M. Tennent was taken over in 1981 but some productions still appear under its name.

Beauval [Jean Pitel] (c.1635–1709), French actor who, with his wife **Jeanne Olivier de Bourguignon** (c.1648–1720), was in several provincial troupes before joining *Molière's company in 1670. Beauval, though in general a mediocre actor, was apparently excellent as Thomas Diafoirus in Molière's *Le Malade imaginaire* (1673), in which his wife played Toinette and his daughter Louise the child Louison. Mme Beauval was an excellent comic actress, given to uncontrollable fits of laughter, which Molière incorporated into her part of Nicole in *Le Bourgeois gentilhomme* (1671). After Molière's death the Beauvals joined the company at the Hôtel de *Bourgogne, and became members of the *Comédie-Française on its foundation in 1704.

Beazley, Samuel (1786–1851), English theatre architect, designer of the *Lyceum, the *St James's, the *City of London, that part of the *Adelphi fronting on the Strand, and the colonnade of *Drury Lane. His buildings, though plain and somewhat uninteresting, were good and well adapted for their purpose. He was a prolific dramatist, mainly of ephemeral farces and short comedies.

Beck, Julian, see LIVING THEATRE.

Beck, Martin, see PALACE THEATRE, New York.

Beckett, Samuel (1906–89), Irishman resident in Paris from 1938, who wrote mainly in French and in 1969 was awarded the Nobel Prize for Literature. He was already well known as a novelist when his first play, *En attendant Godot*, was produced by Roger *Blin in 1953. Two years later it was seen in London, in the author's own translation, as *Waiting for Godot* (NY, with Bert *Lahr, 1956). In it Beckett abandoned conventional structure and development both in plot and language, creating a situation in which two tramps, indecisive and incapable of action, suffer and wait hopefully for help which never comes. The play is considered one of the masterpieces of the Theatre of the *Absurd. In 1957 came *Fin de partie*, in which a blind man sits in an empty room flanked by two old people in dustbins; it had its first performance in London. An English version, as *Endgame*, was produced in London in 1958 (NY, 1958), accompanied by *Krapp's Last Tape* (NY, 1960), a monologue originally written in English, in which an old man listens uncomprehendingly to recordings he made as a youth; it was produced in Paris as *La Dernière Bande* (1960). *Oh! les beaux jours* (*Happy Days*), virtually a monologue for an actress progressively buried in the earth until only her head remains visible, had its first production in New York in 1961. In 1962 it was seen in London, where it was also played in French by Madeleine *Renaud (1965). *Comédie* (*Play*), first produced in Germany in 1963 (Paris, London, and NY, 1964), had three characters, again only heads protruding from urns, and their short dialogue was repeated da capo. In this and in his later plays, including *Va et vient* (*Come and Go*, 1966; London, 1970; NY, 1975), *Breath* (1969), which lasts about 30 seconds, beginning with the cry of a newborn child and ending with the last gasp of a dying man, and *Not I* (1971; NY, 1972; London, 1973), featuring only a mouth, Beckett moved further and further away from conventional theatre. The trend continued in his brief final works such as *Footfalls* (1975), about a sleepless woman, and *Rockaby* (1981), about an old one, both written for Billie *Whitelaw, his leading British interpreter. In

any critical assessment of Beckett's work his radio plays—*All that Fall, Embers, Words and Music*, and *Cascando*—and particularly his novels, must also be taken into account.

Beckett Theatre, Melbourne, see MELBOURNE THEATRE COMPANY.

Becque, Henry [also Henri] **François** (1837–99), French dramatist, an outstanding exponent of naturalistic drama in the style of *Zola. His first important play was *Michel Pauper* (1870), which was underrated on its first production and appreciated only in a revival of 1886. Meanwhile Becque had written the two plays usually associated with him, *Les Corbeaux* (1882) and *La Parisienne* (1885), naturalistic dramas of great force and uncompromising honesty which present rapacious or amoral characters who seem unaware of their own degradation. These *comédies rosses*, as they were called, were not wholly successful until Antoine's *Théâtre Libre provided a stage suitable for them.

Bedford Music-Hall, London, in Camden High Street. Built on part of the tea garden belonging to the Bedford Arms, this opened in 1861 and had a most successful career, its interior, which then had a seating capacity of 1,168 on three tiers, being the subject in 1890 of a painting by Walter Sickert. Among the variety stars who appeared there were George *Leybourne, *Little Tich, and Marie *Lloyd. In 1949 it became a theatre for six months, housing revivals of such old favourites as Miss Braddon's *Lady Audley's Secret*, *Jerrold's *Black-Ey'd Susan*, Leopold Lewis's *The Bells*, and Mrs Henry Wood's *East Lynne*. It then closed and was demolished in 1969.

Beerbohm, Sir Max [Henry Maximilian] (1872–1956), English author and critic, half-brother of the actor-manager *Tree. From 1898 to 1910 he was dramatic critic of the *Saturday Review* in succession to *Shaw, who introduced him to his readers as 'the incomparable Max'. In 1908 he married the American actress **Florence Kahn** (1877–1951) soon after he had seen her playing Rebecca West in *Ibsen's *Rosmersholm*. She had been *Mansfield's leading lady in the USA, and was already well known for her playing of Ibsen's heroines. She retired from the stage on her marriage, but made a few guest appearances from time to time, notably as Aase in the *Old Vic production of Ibsen's *Peer Gynt* in 1935. Beerbohm was the author of a one-act play, *A Social Success*, in which George *Alexander made his first appearance on the music-halls in 1913, but is best remembered for *The Happy Hypocrite* (1900), which he based on one of his own short stories; it was produced at the *Royalty Theatre by Mrs Patrick *Campbell. In 1936 a three-act version by Clemence *Dane starring Ivor *Novello was also a success.

Beeston, Christopher (?1570–1638), important figure of the Jacobean and Caroline stage. He began his career as an actor with *Strange's Men, and in 1602 was with Worcester's Men at the *Rose Theatre, remaining with them when they became *Queen Anne's Men in 1603 on the accession of James I. In 1616–17 he took over the *Cockpit and ceased to act. In 1625 it became the home of *Queen Henrietta's Men, with Beeston as manager, and when they disbanded in 1636 he trained a group called Beeston's Boys, some of whom later made their name in the Restoration theatre. On his death control of the company and the theatre passed to his son William (below). His eldest daughter Anne married the actor Theophilus *Bird.

Beeston, William (c.1606–82), English theatre manager, son of Christopher *Beeston. On his father's death he took over Beeston's Boys and appeared with them at the *Cockpit. In 1647 he also acquired *Salisbury Court, which was wrecked by Commonwealth soldiers in 1649 after unauthorized productions during the *Puritan Interregnum. Beeston, who was himself imprisoned, is suspected of being the 'ill Beest' who in 1652 betrayed to the authorities the actors who were appearing secretly. When the theatres reopened in 1660 Beeston refurbished Salisbury Court and leased it to several companies, running it in 1663–4 with a company which he trained himself. He probably passed on some of the traditions of Elizabethan stage business which he had learned from his father. The granting of royal *patents to *Davenant and *Killigrew put him out of business.

Behan, Brendan (1923–64), Irish dramatist, whose early experiences of prison as a member of the IRA gave him the material for his first play, *The Quare Fellow*, first seen in Gaelic and produced in Dublin in 1954 and in London, by *Theatre Workshop, in 1956 (NY, 1958). It was followed in 1957 by *The Hostage*, also first written and produced in Gaelic. This tragi-comic account of IRA activities in a seedy brothel ends with the shooting of an English soldier (the hostage of the title). It reached London in 1959, again at Theatre Workshop (NY, 1961). In its construction, with interspersed songs and dances and direct addresses to the audience, it showed the influence of *Brecht, which can be seen even more clearly in Behan's last play, *Richard's Cork Leg*, left unfinished at his death. Adapted and enlarged for production at the *Abbey Theatre's experimental Peacock Theatre in 1972, this was seen at the *Royal Court a few months later. A dramatization of Behan's autobiography *Borstal Boy* was staged in New York in 1970 (Abbey, 1972). Behan took a Rabelaisian delight in the idiosyncrasies of his eloquent characters, and this small body of original work created a considerable stir.

Behn, Aphra (1640–89), playwright and novelist, and one of the first Englishwomen to make a living (albeit precarious) by her pen. Little is known of her early life, but she was evidently at some time in Surinam, the scene of her novel *Oroonoko* (c.1688), dramatized by *Southerne in 1695. On her return to England in 1664 she married a merchant of Dutch extraction, but was left a widow after two years. She is known to have spied on the Continent for Charles II during the Dutch War, some of her dispatches being still extant. Her first play, a tragi-comedy entitled *The Forced Marriage; or, The Jealous Bridegroom* (1670), was presented at *Lincoln's Inn Fields Theatre, and well received; but it was in comedies of intrigue, often with some reference to a topical scandal, that she had her greatest successes, particularly with *The Rover; or, The Banished Cavalier* (1677). Her later plays were less successful. *The Emperor in the Moon* (1687), a pantomime-farce based on a *commedia dell'arte* scenario recently played in Paris, is historically interesting as being the forerunner of the many *harlequinades which later led to the English *pantomime.

Behrman, S(amuel) **N**(athaniel) (1893–1973), American dramatist, best remembered for the deft characterization and sparkling dialogue of his comedies. His first solo work was *The Second Man* (1927), and in 1929 he made an excellent dramatization of Enid *Bagnold's *Serena Blandish* (London, 1938). It was followed by *Meteor* (also 1929), a study of an egotist in the world of business; *Brief Moment* (1931), which dramatized the misalliance of a socially prominent young man; *Biography* (1932), which dealt with the contrasting temperaments of a carefree woman portrait painter and a crusading journalist; *Rain from Heaven* (1934), which used an English houseparty to establish the untenability of civilized detachment in a strife-torn world; and *End of Summer* (1936), showing the bankruptcy of the idle rich. He then made for the *Lunts an excellent adaptation of *Giraudoux's *Amphitryon 38* (1937; London, 1938) and in his next play, *No Time for Comedy* (1939; London, 1941), dealt with what was possibly his own dilemma, that of a writer who is anxious to come to grips with serious contemporary problems but has a great talent for light comedy. He then embarked on a series of adaptations, two of which, *The Pirate* (1942) from a play by Ludwig Fulda and *I Know My Love* (1949) based on Marcel *Achard's *Auprès de ma blonde*, again provided excellent vehicles for the Lunts; Franz *Werfel's *Jacobowsky and the Colonel* (1944; London, 1945) showed a Jew and an anti-Semitic Pole sharing a common peril; *Jane* (1952; London, 1947) was a straight comedy based on a story by Somerset *Maugham; and *Fanny* (1945) a musical based on Pagnol's 'Marius' film trilogy. *The Cold*

Wind and the Warm (1958) was partly auto-biographical, and his last play was *But for Whom Charlie* (1964).

Béjart, French family of actors intimately connected with *Molière. The eldest daughter **Madeleine** (1618–72) was already well known as an actress when Molière met her and joined the company of which she was a member. In her early years she played the heroines of classical tragedy, and in later life created a number of Molière's witty maids. After touring the provinces for some years she returned with him to Paris in 1658, as did her elder brother **Joseph** (1616–59), her sister **Geneviève** (1624–75), who was better in tragedy than in comedy, and her younger brother **Louis** (1630–78), known as L'Éguisé on account of his sharp tongue. He was slightly lame, a trait which Molière incorporated into his part of La Flèche in *L'Avare* (1668), where it has remained traditional. Joseph, in spite of a slight stutter, was a useful member of the company, and his death so soon after settling in Paris was a great blow.

The youngest and most important member of the family was **Armande-Grésinde-Claire-Élisabeth** (1642–1700), whom Molière married in 1662. She made her first appearance on the stage as Élise in his *Critique de L'École des femmes* (1663), and after 1664 played most of Molière's heroines, which he wrote with her in mind. She was an excellent actress who owed all her training to her husband, but their marriage was not a success. After Molière's death she kept the company together until 1680, when it was merged in the newly founded *Comédie-Française.

Belasco [Valasco], **David** (1859–1931), American actor-manager and playwright, for many years one of the outstanding personalities of the American theatre. On stage as a child, he continued to act as an adult, and also made a success of dramatizing novels, adapting old plays, and devising spectacular melodramas. In 1882 Daniel *Frohman appointed him stage manager of the *Madison Square Theatre and in 1886 he became Steele *Mackaye's stage manager at the *Lyceum in New York where he remained until 1890, continuing to stage a number of his own plays, mainly collaborations. Success came with *The Girl I Left behind Me* (1893), *The Heart of Maryland* (1895), with Maurice *Barrymore, *Zaza* (1899), starring Mrs Leslie *Carter, and *Madam Butterfly* (1900), with Blanche *Bates. This dramatization of a story by John Luther Long was used by Puccini as the basis of an opera, as was *The Girl of the Golden West* (1905). Meanwhile Belasco had begun his association with the actor David *Warfield in *Klein's *The Auctioneer* (1901). In the following year he opened the first Belasco Theatre, where several of his most famous plays were first seen (see REPUBLIC). Its success enabled

Belasco to build a new theatre which opened in 1907 as the Stuyvesant, becoming the *Belasco in 1910. Here Belasco remained until his death, his last production being *Mima* (1928), his own adaptation of *Molnár's *The Red Mill*. Vain and posturing, he contributed little that was original to the American stage, and in no way encouraged national American drama; but he belongs to the great age of American stagecraft. His great contribution to the American scene lay in his elaborate décors and the passion for realism which led him, in his melodrama *The Governor's Lady* (1912), to place an exact replica of a Child's restaurant on the stage. He was a meticulous stage director, and made good use of the mechanical inventions of his time, as well as interesting and far-reaching experiments in the use of light. His long fight against the stranglehold of the *Theatrical Syndicate involved the freedom of the American theatre.

Belasco Theatre, New York. The first theatre of this name opened as the *Republic. The second, on West 44th Street, between Broadway and 6th Avenue, opened as the Stuyvesant in 1907 with *Belasco's *A Grand Army Man*, and was renamed the Belasco in 1910, opening with *The Lily*, an adaptation by Belasco of Leroux's *Le Lys* (1908). Belasco's productions included his own *The Return of Peter Grimm* (1911), with *Warfield as the old man who returns after death to rectify the errors of his life, and *Laugh, Clown, Laugh* (1923), based on *Martini's *Ridi, pagliaccio* (1919). The theatre was at one time leased by Katharine *Cornell, who produced there a translation of *Obey's *Le Viol de Lucrèce* (1931) as *Lucrece* (1932), and Sidney *Howard's *Alien Corn* (1933). Two plays by Elmer *Rice were produced in 1934, *Judgement Day* and *Between Two Worlds*, and from 1935 to 1941 the *Group Theatre had their headquarters there, producing several plays by Clifford *Odets. From 1949 to 1953 the theatre was used for broadcasting, but then housed Teichmann and *Kaufman's *The Solid Gold Cadillac*. In 1964 the National Repertory Theatre, with Eva *Le Gallienne, played a season and two years later came Frank Marcus's *The Killing of Sister George*. In 1975 the interior was converted to cabaret-style seating for *The Rocky Horror Show*, later reverting to conventional seating. The theatre, which is now owned by the *Shubert organization, was used for 'Shakespeare on Broadway for Schools', presented by *Papp, in 1986.

Belfast Civic Arts Theatre, organization which corresponds to the Dublin *Gate Theatre. Founded (as the Mask Theatre) in 1944, it opened in a converted loft with Charles *Morgan's *The Flashing Stream*. Three years later, after many successful productions of European classics, it was renamed. Two important American plays, *Kingsley's *Darkness at Noon* and *Van Druten's *I am a Camera*, had their European premières there, and *Anouilh's *Waltz of the Toreadors* its first production in English. In 1961 the company, which was eventually forced to lower its standards by the demands of its audiences for comedies and thrillers, moved to a new playhouse in Botanic Avenue, which opened with Tennessee *Williams's *Orpheus Descending*. Owing to civil disturbances the theatre was forced to close in 1971, but it was reopened in 1976 by the Ulster Actors Company, formed a year earlier. The enterprise achieved striking success, gaining a large following for children's shows, rock musicals, revivals of plays from London and New York, and productions of work by Irish authors. The company is aided by grants from the Belfast City Council and the Arts Council of Northern Ireland. The comfortable auditorium seats 500, and there are good front-of-house and backstage facilities.

(See also LYRIC PLAYERS, Belfast, and GROUP THEATRE, Belfast.)

Bel Geddes, Norman [Norman Melancton Geddes] (1893–1958), American scenic designer and a pioneer of décor in the American theatre. As early as 1915 he had the idea of a theatre without a proscenium, and in 1923 he won instant recognition with his magnificent designs for *Reinhardt's American production of Volmöller's *The Miracle*. In 1931 he designed a complex of steps and rostrums for a pioneering production of *Hamlet*. Another of his successful experiments was the multiple setting for *Kingsley's *Dead End* (1935). He devised art deco sets for such musical productions as *Lady, Be Good!* (1924) and *Ziegfeld Follies of 1925*. His plan for a monumental production of Dante's *Divine Comedy* at Madison Square Garden, for which he designed an immense circular stage, was unfortunately never carried out; nor was his scheme for a *theatre-in-the-round in 1930. Nevertheless these and other seminal ideas had a great influence on the development of modern American stage design. His activity in the theatre was at its height during the 1930s, after which he concentrated mainly upon industrial design.

His daughter **Barbara** (1922–   ) made a great success in Arnaud d'Usseau and James Gow's *Deep are the Roots* (1945). Among her later parts were Rose Pemberton in Graham *Greene's *The Living Room* (1954), Margaret in Tennessee *Williams's *Cat on a Hot Tin Roof* (1955), and the title-role in Jean Kerr's *Mary, Mary* (1961).

Belgrade Theatre, Coventry, built in 1958, the first new theatre to be built in England since the *Oxford Playhouse in 1938 and also the first *civic theatre, receiving subsidies from local funds and from the *Arts Council. It was named in recognition of a gift of timber used in its construction from the city of Belgrade in Yugoslavia. The auditorium is on two levels, the stalls and circle, and seats 866. The spacious foyer is large enough to house displays by local artists and other exhibitions. The theatre opened with *Half in Earnest*, a musical version of *Wilde's *The Importance of Being Earnest*. The Belgrade is a producing house, also receiving two or three touring productions a year. The Youth and Community Department organizes various festivals and community theatre weeks, including the Coventry Festival held during the summer. The department also runs the Belgrade Youth Theatre, which presents plays and projects. The Belgrade Theatre-in-Education company, founded in 1965, was the first of its kind, and a studio theatre built in 1977 provides a venue for new and experimental plays, local bands, community projects, and other events. A mixed programme is staged in the main house, from musicals, dramas, and thrillers to farces and comedies. The Belgrade is also responsible for the triennial performances of the Coventry *Mystery Plays in the ruins of the bombed cathedral.

Bell, Marie [Marie-Jeanne Bellon] (1900–85), outstanding French actress, who appeared from 1921 at the *Comédie-Française, notably in a number of classical tragic roles including *Racine's *Phèdre*, *Bérénice*, and Agrippine (in *Britannicus*), and in *Cocteau's *Renaud et Armide* (1943). She was also seen at other theatres in Paris in modern plays including *Barrault's production of *Claudel's *Le Soulier de satin* (1943). Having left the Comédie-Française in 1953, she managed the Théâtre du *Gymnase from 1959 until her death, both acting and directing there. She was seen in the first productions of Félicien Marceau's *La Bonne Soupe* (1959) and *Genet's *Le Balcon* (1960), and in 1973 in a version of William Douglas *Home's *Lloyd George Knew My Father* as *Ne coupez pas mes arbres*. She also staged there such works as Herb Gardner's *A Thousand Clowns* and Arthur *Miller's *After the Fall*. In the 1930s she had a distinguished career in films.

Bellamy, George Anne (c.1727–88), English actress, who received her first names from a mishearing of Georgiana at her christening. She is believed to have made her first appearance on the stage at *Covent Garden in 1744 in *Otway's *The Orphan*, but may have been seen there two years earlier as Miss Prue in *Congreve's *Love for Love*. She was at her best in romantic and tragic parts, being an admirable Juliet to the Romeo of *Garrick, who made her one of his leading ladies at *Drury Lane, though she had little professional integrity and much of her success was due to her youth and beauty. Arrogant and extravagant, she was twice married, once bigamously, and because of the scandals in which she was involved, and the loss of her looks through dissipation, managers became chary of engaging her. Though held in little esteem by the public or her fellow-actors, she continued to appear intermittently until her retirement in 1785. A six-volume *Apology* for her life (1785), edited by another hand, is readable rather than reliable.

Belleroche, see POISSON.

Belleville, see TURLUPIN.

Bellwood, Bessie, see MUSIC-HALL.

Beltrame, see BARBIERI.

Benavente y Martínez, Jacinto (1866–1954), the most successful Spanish playwright of his day, awarded the Nobel Prize for Literature in 1922. He was the author of over 150 plays and of translations of Shakespeare and *Molière. The best known of his works outside Spain are *Los intereses creados* (1907), which, as *The Bonds of Interest*, was the first play presented in New York by the *Theatre Guild in 1919 (London, 1920); and *La malquerida* (1913), which, as *The Passion Flower*, was produced in New York in 1920 (London, 1926). The first makes use of the traditional masks of the *commedia dell'arte* to depict modern society as a puppet show in which men are moved by the strings of passion, selfishness, and ambition. The second is a rustic tragedy in which a peasant girl succumbs to evil. Benavente's plays mark the transition from the comedies of intrigue made popular by *Echegaray to the drama of social criticism implicit in such works as *Gente conocida* (*Important People*, 1896), *La noche del sábato* (*Witches' Sabbath*, 1903), *Los malhechores del bien* (*The Evil Doers of Good*, 1905), a plea for tolerance which attacks misguided charity, and *Más fuerte que el amor* (*Stronger than Love*, 1906). Benavente's reputation has declined considerably since his death.

Benefit, special performance, common in the 18th and early 19th centuries, of which the financial proceeds, after deduction of expenses, were given to a member of the company, who was allowed to choose the play for the evening. In the days of the *sharing system and the *stock company an actor might rely on his benefit to provide him with ready money, his weekly 'share' barely paying current expenses; or it could be for a player who was ill or his dependants. Before the introduction of the *royalty system a performance was sometimes given for an author. According to Colley *Cibber in his *Apology* (1740), the first benefit performance in England was given in about 1686 for Mrs Elizabeth *Barry, but it did not become common practice until much later. There is an interesting account of the way it worked in *Dickens's *Nicholas Nickleby* (1838), and the abuses of the system in the early American theatre can be studied in Odell's *Annals of the New York Stage* (1925–49). It was an unsatisfactory arrangement, which exposed the actor to petty humiliations and kept him in a constant state of financial uncertainty, and it gradually died out between 1840 and 1870. A comprehensive account can be found in Sir St Vincent Troubridge's *The Benefit System in the British Theatre* (1967). A slightly more dignified method of making money was the Bespeak Performance, whereby a wealthy patron or local authority would buy most of the tickets for one evening and sell or give them away, choosing the play from the company's current repertory. In that case the proceeds were divided among the company and not given to an individual.

Bennett, Alan (1934–), English dramatist and actor, who first became known through the epoch-making *revue *Beyond the Fringe* (1960), of which he was part-author and in which he also appeared. His first play, *Forty Years On* (1968), starred John *Gielgud as the retiring headmaster of a minor public school in whose honour the boys enact a savagely satiric pageant of recent British history. It was followed by *Getting On* (1971), in which a disillusioned middle-aged Labour MP longs for earlier certainties; a farce, *Habeas Corpus* (1973; NY, 1975), with Alec *Guinness in London and Donald *Sinden in New York; and *The Old Country* (1977), again with Guinness as a British traitor in exile in Soviet Russia. Among Bennett's later plays, *Enjoy* (1980) had a disappointingly short run, despite fine performances by Joan *Plowright and Colin *Blakely as a working-class married couple in old age. *Kafka's Dick* (*Royal Court, 1986) was a rather aimless play featuring Kafka and an English insurance agent who studies him. *Single Spies* (*National Theatre, 1988; *Queen's, 1989) consisted of *An Englishman Abroad*, about Guy Burgess, already seen on television, and *A Question of Attribution*, about Anthony Blunt (played by Bennett). His work for films and television is also important, the latter being especially suited to his minute observation of human foibles.

Bennett, (Enoch) **Arnold** (1867–1931), English novelist and dramatist, whose plays enjoyed considerable success in their day. His plots were ingeniously constructed, and his character-drawing had vitality and was mellowed by a homely humour. They are now forgotten, except for *Milestones* (1912), written in collaboration with Edward *Knoblock; and *The Great Adventure* (1913), based on his own successful novel *Buried Alive* (1908), which owed much to the acting of Henry *Ainley. Bennett's later plays were less successful, the best of his post-war works being *London Life* (1924), again with Ainley, and *Mr Prohack* (1927), in which Charles *Laughton played the title-role. In both Bennett again collaborated with Knoblock.

Bennett, Jill, see OSBORNE.

Bennett, Richard (1872–1944), American actor, whose handsome appearance and magnificent speaking voice made him a matinée idol, though he also had great acting talent. Unpredictable both on and off the stage, he would sometimes interrupt a play to criticize the audience. His début in Chicago in 1891 as Tombstone Jake in Elmer E. Vance's *The Limited Mail* was followed by his first New York appearance in the same role later in the year. Other roles were Father Anselm in Robert Marshall's *A Royal Family* (1900), Hector Malone in *Shaw's *Man and Superman* (1905), Jefferson Ryder in Charles *Klein's *The Lion and the Mouse* (also 1905) in which he made his London début in 1906, and John Shand in *Barrie's *What Every Woman Knows* (1908). He played George Dupont in *Damaged Goods* (1913), an adaptation of *Brieux's notorious study of venereal disease, and important later parts included Robert Mayo in *O'Neill's *Beyond the Horizon* (1920), He in *Andreyev's *He Who Gets Slapped* (1922), and Tony in Sidney *Howard's *They Knew What They Wanted* (1924). One of his most famous roles came at the end of his career—that of Judge Gaunt in Maxwell *Anderson's *Winterset* (1935). He was the father of the film stars Joan, Constance, and Barbara Bennett.

Benson, Sir Frank Robert (1858–1939), English actor-manager, best remembered for the Shakespearian company with which he visited *Stratford-upon-Avon annually from 1886 to 1916, appearing in a repertory of seven or eight plays during the summer festival there, and at other times travelling all over the provinces, thus keeping the plays always before the public and providing a good training-school for a number of young players. While at *Oxford he had been a prominent member of the OUDS. He made his first appearance on the professional stage in 1882, at the *Lyceum under *Irving, playing Paris in *Romeo and Juliet*. In the following year he formed his own company, and in due course he directed it in all Shakespeare's plays with the exception of *Titus Andronicus* and *Troilus and Cressida*. He was a capable actor, with handsome aquiline features and much of the air of an 'antique Roman'. He is the only actor to be knighted actually in a theatre.

Bentley, Eric Russell (1916–), English-born critic, director, and playwright, best known for his promotion of the work of *Brecht in the English-speaking world; having begun to translate Brecht's plays in the 1940s he collaborated with the author on a production of *Mutter Courage und ihre Kinder* in Munich in 1950. His work as a director also includes *Him* by e. e. cummings, with Kenneth *Tynan in the lead, in Salzburg and *García Lorca's *The House of Bernarda Alba* at the *Abbey Theatre,

Dublin (both 1950). In the same year he co-directed *O'Neill's *The Iceman Cometh*, in a German translation, in Zurich. He was drama critic of the *New Republic* from 1952 to 1956. His own plays include *Are You Now or Have You Ever Been . . .?* (1972), *The Recantation of Galileo Galilei* (1973), and *Expletive Deleted* (1974); he has adapted and translated plays by *Pirandello and *Schnitzler as well as Brecht and written a number of books on the theatre.

Beolco, Angelo (c.1502–42), one of the earliest Italian actors and dramatists, connected with the origins of the *commedia dell'arte. He was a gifted amateur who acted in Venice, Ferrara, and Padua during the carnival season from 1520 onwards under the name of Il Ruzzante, 'the gossip', a shrewd peasant who indulges in long and amusing soliloquies. His plays, which are fully written out and not scenarios for improvisation, are in the dialect of Padua, his birthplace. They show a close and sympathetic observation of country life and provide plenty of opportunities for pantomime and comic turns—*burle and *lazzi. They are highly thought of today and a number have been performed on the Italian stage, notably *La moschetta* (*The Coquette*, 1528). They were translated into French in 1925–6 by Beolco's biographer Mortier, and an English version of *Il reduce* (c.1528), as *Ruzzante Returns from the Wars*, is included in vol. i of Eric *Bentley's *The Classic Theatre* (1958).

Bérain, Jean (1637–1711), French theatrical designer, who replaced *Vigarani at the *Salle des Machines. In 1674 he became designer to the King, and the costumes and decorations which he prepared for Court spectacles greatly influenced all forms of contemporary art. Their most striking characteristic is the complete synthesis of fantasy and contemporary taste. There was no attempt at realism or archaeological reconstruction, and even in costumes for Romans, Turks, or mythological characters, the exotic elements were absorbed into an intensely personal style which left its impress on contemporaries and successors alike.

Bérard, Christian (1902–49), French artist and theatre designer whose work, which was characterized by a wonderful feeling for the visual aspects of the theatre and great skill in the use of colour, had a great influence on European stage design. After designing for most of the great choreographers of his time, he began in 1934 a fruitful collaboration with *Jouvet, starting with *Cocteau's *La Machine infernale* and continuing through such important and diverse productions as *Beaumarchais's *Le Mariage de Figaro* (1939), *Giraudoux's *La Folle de Chaillot* (1945), *Genet's *Les Bonnes*, and *Molière's *Don Juan* (both 1947). He also designed a number of productions for other theatres, including the *Comédie-Française, and, for *Barrault's company at the *Marigny, Molière's *Amphitryon* (1946) and *Les Fourberies de Scapin* (1949). He died while supervising the lighting of this last play on the night before its production.

Bergelson, David, see MOSCOW STATE JEWISH THEATRE.

Bergerac, Savinien Cyrano de, see CYRANO DE BERGERAC.

Bergman, Hjalmar Frederik (1883–1931), Swedish dramatist and novelist, one of the most influential in the Swedish theatre after the death of *Strindberg, who, with *Maeterlinck and *Ibsen, had a great influence on Bergman's early work. He first came into prominence with two one-act 'Marionette Plays' produced in 1917—*Dödens Arlekin* and the exquisite psychological tragedy *Herr Sleeman kommen*. Other plays included *Ett Experiment* (1918) and *Vavaren i Bagdad* (*The Weaver of Bagdad*), published in 1923, which he wrote after completing a translation of Sir Richard Burton's *One Thousand and One Nights*. He also adapted two of his own regional novels. Of his later plays, in which comedy replaced the tragic mood of his earlier work, the best known are *Swedenhjelms* (1925), a realistic comedy about the eccentric family of a Nobel Prize-winner, seen at the *Birmingham Repertory Theatre in 1960 as *The Family First*, and *Patrasket* (1928), a Jewish folk comedy.

Bergman, Ingmar (1918–), Swedish director, who at the age of 20 was directing with an amateur company. In 1942 he wrote a play, *Kaspers död* (*Kasper's Death*), which he also directed. He was head of the municipal theatre in Hälsingborg, 1944–6, and of Gothenburg's city theatre, 1946–9, where his opening production was *Camus's *Caligula*, with Anders *Ek. From this time he worked increasingly in the cinema, developing a distinguished international reputation; but he concurrently held appointments at the Malmö city theatre, 1952–60, and at the *Royal Dramatic Theatre, Stockholm, 1963–6, where he introduced a number of reforms and attracted to the company some of Sweden's most distinguished players. His intense, often morbid, concentration on psychological interpretation in the tradition of *Strindberg (at whose plays he excelled) was brought to bear on other authors ranging from *Chekhov via *Pirandello and *Brecht to *Albee. In 1970 he directed *Ibsen's *Hedda Gabler*, with Maggie *Smith, for the *National Theatre company, and his production of Strindberg's *The Dream Play* was seen in London during the *World Theatre Season in 1971. In 1976, after being wrongly charged with tax irregularities, he left Sweden, working at the Residenztheater in *Munich, where his productions included his own *Scenes from a Marriage*

(1981), and not returning permanently to the Royal Dramatic Theatre until 1985. He directed his first *Hamlet* in 1986, and in 1987 his productions of this play and Strindberg's *Miss Julie* were staged by the RDT at the National Theatre in London.

Bergner, Elisabeth (1900–86), Austrian-born actress who was in *Berlin in the 1920s, where, under Max *Reinhardt, she had her first resounding success as *Shaw's St Joan in 1924. Her boyish figure and vivacious temperament were admirably suited to the part, and made her an excellent interpreter of Shakespeare's heroines. She was also good in modern roles, among them *Strindberg's Miss Julie and Queen Christina, Nina in *O'Neill's *Strange Interlude*, and Alcmene in *Giraudoux's *Amphitryon 38*. In 1933 she moved to London, where she made her English-speaking début as Gemma Jones in Margaret Kennedy's *Escape Me Never* (1933; NY, 1935). She also played the title-role in *Barrie's *The Boy David* (1936) and was seen again as St Joan at the *Malvern Festival in 1938. She played in New York during the 1940s, her roles including the Duchess in *Webster's *The Duchess of Malfi* (1946). She returned to London in 1951 in *The Gay Invalid*, adapted from *Molière, and toured Germany and Austria in *Rattigan's *The Deep Blue Sea* (1954), O'Neill's *Long Day's Journey into Night* (1957), and Jerome Kilty's *Dear Liar* (1959). Later she played the Countess Aurelia in Giraudoux's *The Madwoman of Chaillot* in both Germany (1964) and England (1967).

Bergstrøm, Hjalmar (1868–1914), Danish dramatist, one of the best known of his period outside his own country. Showing the influence of *Ibsen, his plays deal with such social problems as the struggle between the classes, as in *Lynggaard & Co.* (1905), and feminine emancipation, as in *Karen Bornemann* (1907). *Dame-Te* (*Ladies' Tea*, 1910) is an anecdotal analysis of spinsterhood.

Berlin. Until the actor-manager **Karl Döbbelin** (1727–93) brought his company to settle in Berlin in 1767, there had been no strong theatrical tradition in the city. Döbbelin's efforts to establish one resulted in the founding in 1787 of a National Theatre, led from 1796 by *Iffland, Döbbelin remaining manager until 1789. After Iffland's death the great actor Ludwig *Devrient led the company, making it notable for its fine productions of classical and modern plays. After Devrient's death the theatre, which had been destroyed by fire in 1817 and rebuilt in a neo-classical style, entered on a period of triviality which was to last for some 50 years.

A new phase in the theatrical life of Berlin began with the founding in 1883 of the *Deutsches Theater followed in 1889 by that of the *Freie Bühne. Both became noted for their productions of realistic drama. In 1905 *Reinhardt took over the Deutsches Theater, and made Berlin one of the outstanding theatrical centres in Europe. The *Volksbühne, founded in 1890, opened its own theatre in 1914, and in 1919 Reinhardt opened the *Großes Schauspielhaus. Between the two world wars *Piscator was in Berlin as director of the Zentral Theater, moving to the Volksbühne in 1924, and to his own theatre from 1927 to 1929. After he, *Jessner, Reinhardt, and many more had left Germany, the Berlin theatre again sank to a low level, though *Gründgens at the Schauspielhaus and Hilpert at the Deutsches Theater maintained a classical repertory until in 1944 all surviving theatres were closed.

The return of *Brecht to East Berlin, in 1949 at the Deutsches Theater and from 1954 at the Theater am Schiffbauerdamm, once again brought Berlin back into the theatrical limelight; while under *Besson the Volksbühne became known for its artistically enterprising programmes. In West Berlin the Freie Volksbühne opened in 1949, and at the rebuilt Schillertheater a good selection of classical revivals was staged from 1951. Meanwhile the Schaubühne, founded in 1962, housed distinguished productions under the direction of Peter *Stein from 1970 to 1985; in 1981 the company moved into a new theatre. It is to be hoped that the reunification of Germany in 1990 will lead to reciprocity and the sharing of mutual interests between the theatres of East and West Berlin.

Berlin, Irving [Israel Baline] (1888–1989), American lyricist and composer, born in Russia to Jewish parents who emigrated to the USA when he was 5. In his teens he was a singing waiter and song plugger, and soon began to write songs himself, the first for which he wrote both words and music being performed in 1908. He became famous in 1911 with 'Alexander's Ragtime Band', and in 1914 composed his first complete score, the revue-cum-musical comedy *Watch Your Step*. His first genuine musical comedy was *Stop! Look! Listen!* (1915), but revues were often his chosen medium, including *The Ziegfeld Follies of 1919*, which featured 'A Pretty Girl Is like a Melody'; the four *Music Box Revues*, 1921–4, which contained 'What'll I Do?' and 'All Alone' and were staged in the new *Music Box which Berlin built with Sam H. Harris (see SAM H. HARRIS THEATRE); *As Thousands Cheer* (1933), which had 'Easter Parade'; and the wartime all-soldier *This is the Army* (1942; London, 1943). Among his musical comedies were *Face the Music* (1932), *Louisiana Purchase* (1940), and two enormous successes: *Annie Get Your Gun* (1946; London, 1947), whose most enduring song is 'There's no Business like Show Business', and *Call Me Madam* (1950; London, 1952). Despite his lack of musical education Berlin had a supreme gift for melody

which made him the most popular composer of his day. His work in films is equally important.

Berliner Ensemble, theatre company founded in 1949 in East *Berlin by *Brecht and his wife Helene *Weigel. It was first housed at the *Deutsches Theater and in 1954 moved to the Theater am Schiffbauerdamm. It achieved pre-eminence in Germany and world-wide acclaim with its productions of Brecht's plays, directed by himself and after his death, when his widow took over, by *Palitzsch, *Besson, and others. By 1974, its 25th anniversary, the company had staged a total of 54 productions and developed a very personal style of acting and presentation designed to prevent the suspension of disbelief and encourage in the audience, as was Brecht's desire, a critical appraisal of the play. To avoid the danger, apparent since the mid-1960s, of turning itself into a Brecht museum, the company has extended its repertoire to include *Büchner, *Shaw, *Wedekind, and Heiner *Müller, as well as continuing to present selected plays from the entire Brecht canon. The company was received with great enthusiasm in Paris in 1954 and 1955 and visited London in 1956 (just after Brecht died) and 1965. **Manfred Wekwerth** (1929–) became its manager in 1977, after an association dating from 1951. His production of Brecht's adaptation of *Coriolanus* was seen at the *Old Vic in 1965 during the company's visit, and in 1971 he directed Shakespeare's original play for the *National Theatre.

Bernard, Jean-Jacques (1888–1972), French dramatist, son of the playwright **Tristan Bernard** (1866–1947), author of a number of amusing but ephemeral light comedies. The son's work is altogether different, since it deals with the tragedy of unrequited or unacknowledged love, and derives from the *école intimiste* (see SILENCE, THEATRE OF) founded by *Maeterlinck. Several of his plays were seen in London in translation, among them *The Sulky Fire* (*Le Feu qui reprend mal*, 1921) in 1926, *The Unquiet Spirit* (*L'Âme en peine*, 1926) in 1928, *Martine* (1922) in 1929 (NY, 1928), *The Springtime of Others* (*Le Printemps des autres*, 1924) in 1934, and *Invitation to a Voyage* (*L'Invitation au voyage*, 1924) in 1937.

Bernhardt, Sarah (-Marie-Henriette) (1845–1923), French actress, known as the 'divine Sarah', a consummate artist whose voice was likened to a 'golden bell' or the 'silver sound of running water'. In 1862 she made her début at the *Comédie-Française, where she was seen only intermittently, not accommodating herself easily to the traditions of that venerable institution, but where in 1872 she achieved a double triumph as Cordelia in *King Lear* and the Queen in *Hugo's *Ruy Blas*. She consolidated her position at the head of her profession by outstanding performances as *Racine's Phèdre

(1874) and as Doña Sol in Hugo's *Hernani* (1877), one of her finest parts. She made her first appearance in London in 1879 in *Phèdre*, and in New York in 1880 in *Scribe and Legouvé's *Adrienne Lecouvreur*, scoring an immediate triumph in both capitals, and returning many times. In Paris, after her final departure from the Comédie-Française in 1880, she managed a number of theatres, including the Ambigu and the *Porte-Saint-Martin, before opening the present Théâtre de la Ville as the Théâtre Sarah-Bernhardt in 1899. Her greatest successes included the younger *Dumas's *La Dame aux camélias*, which she performed all over the world, *Sardou's *Fédora* (1882) and *La Tosca* (1887), and *Rostand's *L'Aiglon* (1900), with which her name is always associated, and in which she played Napoleon's son. She was also seen on several occasions as Shakespeare's Hamlet. She was an accomplished painter and sculptress and wrote poetry and plays, appearing in some of the latter herself. Her career survived the amputation of a leg in 1915.

Bernini, Giovanni Lorenzo (1598–1680), Italian architect, artist, sculptor, and scene and machine designer, some of whose effects were seen in Rome by Richard Lascelles, contemporary of Inigo *Jones, and described by him in his *Italian Voyage* (1670). He was responsible for the décor of the theatre built by the Barberini family, and for some of the stage sets. He excelled in water pieces with boats floating on them, floods, storms, men and chariots flying through the air, houses falling in ruins, the sudden appearance of temples, forests, and populous city streets, serpents, dragons, and articulated monsters, and *transformation scenes of all kinds.

Bernstein [*née* Frankau], **Aline** (1881–1955), American costume and scene designer. Her first work for the stage was for the Indian play *The Little Clay Cart* (1924) produced by the *Neighborhood Playhouse. Her designs were seen in a number of important productions, including *Sherwood's *Reunion in Vienna* (1931) for the *Theatre Guild; *Chekhov's *The Cherry Orchard* (1928) and *The Seagull* (1929), *Romeo and Juliet* (1930), and *Molnár's *Liliom* (1932) for the Civic Repertory Theatre under Eva *Le Gallienne; and, for other managements, *Barry's *The Animal Kingdom* and Sidney *Howard's *The Late Christopher Bean* (both 1932); Elmer *Rice's *Judgement Day* (1934); and Lillian *Hellman's *The Children's Hour* (1934) and *The Little Foxes* (1939). The designs for *Cocteau's *The Eagle Has Two Heads* (1949) were also much admired.

Bertinazzi, Carlo (1713–83), the last of the great Arlequins of the *Comédie-Italienne, which he joined in 1741. Known as Carlin, he was much admired by *Garrick, who said his back wore the expression his face would have shown had the mask not covered it.

Bespeak Performance, see BENEFIT.

Besson, Benno (1922–), Swiss director, who after working in Paris joined the *Berliner Ensemble in 1949. He proved himself the best and most imaginative of *Brecht's company, helping him to adapt *Molière's *Don Juan* (1954) and *Farquhar's *The Recruiting Officer* (as *Pauken und Trompeten*, 1955). He left the company in 1958 and in 1961 joined the *Deutsches Theater, being responsible for the production of adaptations of *Aristophanes' *Peace* (1962) and Offenbach's *La Belle Hélène* (1964), and of *Shwartz's modern fairy-tale *The Dragon* (1965). In 1969 he became director of the *Volksbühne, which he revitalized, making it an entertaining and cosmopolitan 'people's theatre'. In 1973 and 1974 he staged *Spectacles I and II*, mini-festivals involving as many as 12 different performances under one roof—backstage, on stage, in the foyer, etc.—and in 1975 confirmed the pre-eminence of his theatre in East Berlin with an outstanding production of *Die Schlacht* (*The Battle*) by Heiner *Müller. His productions outside Germany included *As You Like It*, *Hamlet*, and Brecht's *The Caucasian Chalk Circle*, seen at the *Avignon Festival in 1976, 1977, and 1978 respectively. He left the Volksbühne in 1978 and in 1982 became Director of the Théâtre de la Comédie in Geneva.

Betterton, Thomas (?1635–1710), English actor, the greatest figure of the Restoration stage. He was in the company which reopened the *Cockpit in 1660, and soon after joined *Davenant's company at *Lincoln's Inn Fields Theatre. After Davenant's death in 1671, the company, of which Betterton was by then joint manager with Henry *Harris, moved to a new theatre in *Dorset Garden, and remained there until it was amalgamated with the company at the Theatre Royal in 1682. In 1695 Betterton broke with the management of the Theatre Royal and successfully reopened the theatre in Lincoln's Inn Fields with the first performance of *Congreve's *Love for Love*, moving 10 years later to *Vanbrugh's new theatre in the Haymarket (see HER MAJESTY'S THEATRE). He was a good manager and an excellent actor, both in comedy and tragedy, his Hamlet and Sir Toby Belch (in *Twelfth Night*) being equally admired. He adapted a number of Shakespeare's plays to suit the taste of the time, and in 1690 turned *Fletcher's *The Prophetess* (1622) into an opera with music by Purcell. Though not so well suited to *Restoration comedy he excelled in *heroic drama, and created many leading roles by contemporary dramatists. Betterton's wife was **Mary Saunderson** (?–1712), one of the first English actresses, whom *Pepys in his Diary always refers to as Ianthe, from her excellent playing of that part in Davenant's *The Siege of Rhodes.*

Betti, Ugo (1892–1953), Italian playwright, much of whose work shows the influence of *Pirandello. A lawyer by profession, he gave many of his plays a legal setting, as in his most important work, *Corruzione al Palazzo di Giustizia* (1949), seen in New York, as *Corruption in the Palace of Justice*, in 1963. His other plays— he wrote nearly 30—include light comedies such as *Una bella domenica di settembre* (*A Fine Sunday in September*, 1937) and more serious works such as *Frano allo scala Nord* (*Landslide on the North Quay*, 1936); *Delitto all'isola delle capre* (1948), seen in Oxford in 1957 as *Crime on Goat Island; Lotta fino all'alba* (*The Struggle Ends at Dawn*, 1949); and *Il giocatore* (1951), seen in New York in 1952 as *The Gambler*. In 1955 three of Betti's most important plays were seen in London in translations by Henry Reed—*Il paese delle vacanze* (1942) as *Summertime; La regina e gli insorti* (1951) as *The Queen and the Rebels* with Irene *Worth; and *L'aiuola bruciata* (1952) as *The Burnt Flower Bed. The Queen and the Rebels* was seen in New York in 1982, with Colleen *Dewhurst. Betti has been called 'the Kafka of drama'. For him the world is on trial and his characters are haunted by visions of a lost Earthly Paradise. Yet in spite of his realistic portraits of degradation, Betti was an optimist, and man's journey led ultimately to the discovery of Christ.

Betty, William Henry West (1791–1874), Ulster-born child prodigy who as the Young Roscius took London by storm in 1804–5. He was first seen at *Covent Garden as Achmet in John Brown's tragedy *Barbarossa*, and there and at *Drury Lane he played such parts as Hamlet, Romeo, Richard III, Osman in *Voltaire's *Zaïre*, Rolla in *Sheridan's *Pizarro*, Frederick in Mrs *Inchbald's *Lovers' Vows*, and Young Norval in *Home's *Douglas*, in which character he was painted by Northcote and Opie. After a brief and hectic success, during which he was more popular than Mrs *Siddons or John Philip *Kemble, opinion turned against him, and he was hissed off the stage when he attempted Richard III. After three years at *Cambridge he returned to the theatre, but without success. He was ignored, his father squandered his money, and the rest of his life was passed in obscurity.

Beverley, William Roxby (c.1814–89), English scene painter, who first worked for the Theatre Royal, Manchester. Later he did good work for Mme *Vestris at the *Lyceum, achieving his greatest success in *Planché's extravaganza *The Island of Jewels* (1849). Beverley's long and fruitful association with *Drury Lane, where his best work was done for the annual *pantomime, began in 1854 and lasted through successive managements until 1885. He also worked intermittently elsewhere, and for Charles *Kean at the *Princess's designed the sets for *King John, Henry IV, Part One*, and *Macbeth*, and for an elaborate production of Milton's *Comus*. Next to *Stanfield, Beverley, a one-surface painter firmly opposed to *'built stuff', was the most distinguished English scene painter of the 19th century.

Biancolelli, Giuseppe Domenico (c.1637–88), actor and playwright of the *commedia dell'arte* who became famous as Dominique, playing the part of Arlequin. Son and grandson of actors, he was invited to join the Italian company in Paris in 1654. It was through Dominique that the Italians, who were restricted to acting in their own language, obtained permission from Louis XIV for them to play in French, in spite of the determined opposition of the *Comédie-Française. Dominique later became a naturalized Frenchman and married an actress. Two of his daughters were on the stage, one marrying an actor at the Comédie-Française. His youngest son **Pietro Francesco** (1680–1734) acted in Paris both as Arlequin and as Pierrot, until the Italian players were banished in 1697. As Dominique le Jeune, he was a member of the company which Lélio (see RICCOBONI) took back to Paris in 1716, and remained with it until his death, playing in the first productions of some of the plays of *Marivaux.

Bibiena [Bibbiena, da Bibbiena], family of scenic artists and architects, originally from Florence, whose work, in pure baroque style, is found all over Europe, though Parma and Vienna probably saw their greatest achievements. The family name was Galli, and Bibiena was added later, from the birthplace of **Giovanni Maria Galli** (1625–65), father of **Ferdinando** (1657–1743) and **Francesco** (1659–1739), who together founded the family fortune and renown. While still a young man, Ferdinando worked in the beautiful *Teatro Farnese built by Aleotti, which he left to go to Vienna. There, with the help of his brother and his sons, he was responsible for the decorations of many Court fêtes and theatrical performances. His eldest son **Alessandro** (1686–1748) became an architect, but the three younger ones, **Giuseppe** (1695–1756), **Antonio** (1697–c.1774), and **Giovanni Maria** (1700–74), worked in the theatre, Giuseppe being probably the first designer to use transparent scenery lighted from behind. His son **Carlo** (1721–87) was associated with his father in the building and decoration of the opera-house at Bayreuth. The family made its home in Bologna, but its members can be traced all over Europe, working so harmoniously that it is sometimes impossible to apportion their work individually. Among their many innovations in scene design was the perspective scene, or *scena d'angolo*, which inaugurated a new era in stage setting and made possible the elaborate architectural settings characteristic of the family style.

Bible-History, see MYSTERY PLAY.

Bickerstaffe, Isaac (1735–1812), English dramatist, considered in his day the equal of John *Gay as a writer of *ballad operas. The first of these, *Thomas and Sally; or, The Sailor's Return* (1760), was given at *Covent Garden. It was followed by *Love in a Village* (1762) and *The Maid of the Mill* (1765), based on Richardson's novel *Pamela*, which held the stage for many years, and is found in the repertory of the *toy theatre. Of Bickerstaffe's later productions, many of them written in collaboration with Samuel *Foote and Charles *Dibdin, the best was *Lionel and Clarissa* (1768). In 1772 Bickerstaffe was suspected of capital crime, and fled to the Continent, where he died in poverty.

Bidermann, Jakob (1578–1639), Jesuit priest, and an outstanding writer of plays in Latin for collegiate production (see JESUIT DRAMA). The best of those which have survived is *Cenodoxus*, the story of a pious hypocrite in Paris whose soul, after death, is tried and cast into Hell. The play ends with the founding by St Bruno, who has witnessed the condemnation, of the religious order of Carthusians. *Cenodoxus* was first performed in Latin in 1602 in Augsburg, and in Munich in 1609. In 1958 it was produced at the Residenztheater in Munich as part of the city's celebration of its 800th anniversary, in an abridged German text which in 1975 was published in an English translation.

Bill-Belotserkovsky, Vladimir Naumovich (1884–1970), Soviet dramatist, who added Bill, his nickname in the USA (where he worked from 1911 to 1917), to his family name. His first play *Echo* (1924) depicts United States dockers refusing to load arms for anti-Soviet use. *Hurricane* (or *Storm*), produced in 1925 at the *Mossoviet Theatre, is a stirring piece of propaganda dealing with the struggles of a Revolutionary leader in a small village during the Russian Revolution. The first realistic play about the new Soviet state, it was produced all over Eastern Europe with immense success. Later plays included *Life is Calling* (1934), also known as *Life Goes Forward* (published in an English translation in 1938), in which socialist duty conflicts with personal happiness.

Billetdoux, François (1927–91), French journalist and dramatist, who studied under *Dullin and achieved success in 1959 with *Tchin-Tchin*, a play considered at the time a sign of renewed vigour in the Parisian avant-garde theatre. As *Chin-Chin* it was staged in London (1960) with Celia *Johnson and New York (1962) with Margaret *Leighton. Billetdoux created unusual situations in order to question the whole social fabric. Thus *Tchin-Tchin* shows a drunken complicity developing between a bricklayer and the wife of a doctor who is having an affair with the bricklayer's wife; *A la nuit la nuit* (1960) revolves on the mysterious identity of a man who visits a prostitute; *Le Comportement des époux Bredburry* (also 1960) is about a woman who tries to sell her husband through the small ads; and *Va donc chez Törpe* (1961) is set in an inn where all the guests have gone to commit suicide. Like *Giraudoux and *Audiberti, Billetdoux writes highly coloured prose which constantly runs the risk of sounding mannered. Among his later plays was *Il faut passer par les nuages* (1964), in which a successful middle-class woman is led by a personal crisis to allow her family to disintegrate, and *Wake up! Philadelphie!* (1988).

Billy Rose Theatre, see NEDERLANDER THEATRE.

Biltmore Theatre, New York, on West 47th Street between Broadway and 8th Avenue. Seating 994, it opened in 1925 and had an undistinguished career until 1928, when *Pleasure Man* by Mae *West was closed by the police after three performances. In 1936 the *Federal Theatre Project presented its experimental *Living Newspaper, *Triple-A Plowed Under*. Warner Brothers then bought the theatre, and George *Abbott staged for them a number of successful shows. Later productions included *My Sister Eileen* (1940), based on stories by Ruth McKenny. The theatre was used for broadcasting, 1952–62, but in 1963 Neil *Simon's *Barefoot in the Park* began a long run, and in 1968 came the epoch-making hippie musical *Hair*, revived in 1977. William Douglas *Home's *The Kingfisher*, starring Rex *Harrison, was staged in 1978. The theatre has been dark since 1987.

Bio-Mechanics, name given to a method of play production devised by *Meyerhold. Reducing the actor to the status of a puppet, it calls for the complete elimination of the player's personality, the stripping of the stage to the bare boards, and the willing co-operation of the audience in supplying, by an effort of the imagination, all that is lacking, thus having much in common with the Japanese *kabuki. A cerebral rather than an emotional approach to the theatre, it served its purpose in the early days of the Russian Revolution by clearing away the inessentials which cluttered up the old Imperial stages; but it was too abstract for mass audiences and was succeeded by *Socialist Realism.

Bird, Robert Montgomery (1806–54), American playwright, by profession a doctor. He first came into prominence when in 1831 Edwin *Forrest appeared as Spartacus in his romantic tragedy *The Gladiator*. It was an immediate success, and Forrest selected it for his first appearance at *Drury Lane in 1836, continuing to act in it until his retirement in 1872. It was seen on the stage as late as 1892. Bird wrote two more plays for Forrest, *Oralloossa* (1832) and a domestic drama, *The Broker of Bogota* (1834), and revised for him Stone's *Metamora*. All three were popular and frequently revived, but owing to the chaotic state of the *copyright laws Bird made no money from them, and withdrew from the theatre to seek a livelihood elsewhere.

Bird, Theophilus (1608–64), English actor, who bridged the gap between the Caroline and Restoration theatres. He was the son of a minor actor who appears frequently in *Henslowe's diary as a member of the *Admiral's Men. Theophilus is first found playing female parts with *Queen Henrietta's Men, but by 1635 he had graduated to male parts. Shortly afterwards he married Anne, eldest daughter of Christopher *Beeston. In 1637 he joined the *King's Men, remaining with them until the closing of the theatres in 1642. He is named first in *Downes's list of actors who appeared immediately on the reopening of the theatres in 1660, and *Pepys notes a rumour that he had broken his leg while fencing in a play in 1662.

Birmingham, England's second largest city, has a long theatre history, typical in its early days of many English provincial towns. Two theatrical *booths existed before 1730, but the first permanent building was erected in Moor Street about 1740. A theatre in King Street opened in 1751, another in New Street in 1774. The last was twice rebuilt after fires, in 1792 and 1820. The father of *Macready became its manager in 1795. *Touring companies were first introduced in 1849, but the resident *stock company lingered on until 1878. The theatre closed in 1901, was rebuilt and reopened in 1904, and finally demolished in 1957. Another theatre, the Prince of Wales's, was destroyed by enemy action during the Second World War. The *Birmingham Repertory Theatre opened in 1913.

The Alexandra, built in 1901 and for its first year known as the Lyceum, staged in its heyday mainly *melodrama, *pantomime, *music-hall, and *revue. In 1927 it became a *repertory theatre, and though rebuilt in 1935 its policy remained unchanged until 1946, when regular spring and autumn visits from touring companies were introduced. In 1964 this theatre housed the first performance by the *National Theatre company of *Othello*, with Laurence *Olivier in the title-role. The building was bought by the Birmingham Corporation in 1968, and it became a 'touring only' theatre in 1976. Extensively renovated in 1979, it now has a seating capacity of 1,562.

The Birmingham Hippodrome, which opened in 1900 as the Tivoli, taking its present name in 1903, was for many years a *variety theatre. It had a somewhat chequered career and in 1968 was in danger of demolition. It is now a major lyric theatre housing opera, ballet, and other large-scale productions, being used extensively by the Welsh National Opera company. It was bought by the local authority in 1979 and is now run by a Trust, having been extensively altered before its reopening in 1981. In 1990 it became the permanent home of Sadler's Wells Royal Ballet.

Birmingham Repertory Theatre, one of the most significant enterprises launched in the English theatre during the first half of the 20th century, began with private theatricals in the home of Barry *Jackson. From these emerged in 1907 the Pilgrim Players, an amateur company which put on at local halls plays unlikely to be seen in the commercial theatre. Inspired by the opening of a repertory theatre in *Manchester in 1908, and of the *Liverpool Playhouse (originally the Liverpool Repertory Theatre) in 1911, Jackson, who was a wealthy man, built and equipped a theatre in Station Street to house a professional company, which opened in 1913 in *Twelfth Night*. During the next 10 years a wide variety of uncommercial plays was produced, including *Drinkwater's *Abraham Lincoln* (1918) and the first British production of *Shaw's *Back to Methuselah* (1923), one of many Shaw plays produced by Jackson. In 1924, disheartened by lack of civic support, Jackson closed the theatre, but the Birmingham Civic Society guaranteed a sufficient number of season-ticket holders to induce him to reopen it. Such plays as *Phillpotts's *The Farmer's Wife* (1924), Drinkwater's *Bird in Hand* (1927), and Besier's *The Barretts of Wimpole Street* (1930) were transferred to London, and there were controversial modern-dress productions of *Hamlet* (1925) and *Macbeth* (1928). From 1929 to 1938 the company also provided the nucleus of that which appeared at the *Malvern Festival. In 1935 Jackson transferred the Birmingham theatre to a Board of Trustees, but remained its director until his death in 1961. It became known as a fine training ground for young actors and actresses, including Laurence *Olivier, Paul *Scofield, Margaret *Leighton, Albert *Finney, and Derek *Jacobi. In the late 1960s the Birmingham City Council donated the site for a new and larger theatre, seating 900, which opened in 1971 in the heart of the City Centre, financed by the Council, the *Arts Council, and public subscription. The size of the auditorium, twice that of the old theatre, has intensified the normal conflict in a subsidized theatre between artistic and commercial considerations; but a mixture of ancient and modern drama is presented in both the main auditorium and the studio theatre, one of the most flexible in the country, seating 140. The theatres have staged the premières of such plays as *Wesker's *The Merchant* (1978) and *Annie Wobbler* (1983) and *Ustinov's *Beethoven's Tenth* (1983).

Bjørnson, Bjørnstjerne (1832–1910), Norwegian novelist, poet, and dramatist, whose work, though perhaps less powerful, was in its day more popular than that of his contemporary Henrik *Ibsen because of his reputation as a novelist and song-writer, his influence in contemporary politics, and his more sanguine tone. After spending the years 1857–9 as Director of the Norske Teatret in Bergen (see NATIONALE SCENE, DEN), Bjørnson embarked on a series of dramas on historical and patriotic themes: *Between the Battles* (1858), *Lame Hulda* (1858), *King Sverre* (1861), the trilogy *Sigurd Slembe* (1862), and *Maria Stuart i Skotland* (1864); he was to return to such historical subjects with *Sigurd Jorsalfar* and *King Eystejn* in 1872. Following his appointment as artistic director of the *Kristiania Theater (1865–7), his plays show an increasing interest in social issues: *The Newly-Marrieds* (1865) had already dealt with a matrimonial problem in a contemporary setting. *The Editor* (1874), *The Bankrupt* (1875), *The King* (1877), and *The New System* (1879) all debate contemporary themes, especially the urban and commercial concerns of the time.

The plays of Bjørnson's maturity fall into two groups: private and largely individual problems are dealt with in *A Gauntlet* (1883), *Love and Geography* (1885), *Laboremus* (1901), *Daglannet* (1904), and the sparkling comedy *When the New Wine Blooms* (1909); the individual's relations with established authority are considered in *Over Ævne I* (1883), best known in English as *Pastor Sang* and incomparably Bjørnson's most powerful work, *Paul Lange og Tora Parsberg* (1893), and *Over Ævne II* (1895, seen in London in 1901 as *Beyond Human Power*). Bjørnson's treatment of social themes has tended to become dated. In Norway, however, he is honoured as a public figure and patriot, as well as a writer of the first rank.

His son **Bjørn** (1859–1942) was a well-known actor and theatre manager who joined the *Meininger company in 1880. He became a director of the Kristiania Theater, 1885–93, and in 1899 took over the newly established *Nationaltheatret. During the next 40 years he served the Norwegian theatre in many different capacities.

Black, George, see LONDON PALLADIUM and REVUE.

Blackface, see MINSTREL SHOW.

Blackfriars Theatre, London. Two Blackfriars theatres were built within the boundaries of the old Blackfriars monastery. The first was used by the Children of the Chapel Royal and of Windsor Chapel under their Master, Richard Farrant, from 1576 until his death in 1581. From 1583 to 1584 it was used by a *boy company. It then lapsed and was let out as lodgings. In 1596 James *Burbage bought another part for use by the *Chamberlain's Men, but was opposed by the authorities, who did not wish to see another playhouse established. In 1600 his son Cuthbert *Burbage leased it to the Children of the Chapel, who played there until 1608, when the Burbages' company, now known as the *King's Men, appeared there, remaining in possession until the closing of the theatres in 1642. The second Blackfriars, certainly bigger than the first, was 46 ft. by 66 ft. with galleries on three sides and the stage on the fourth. Its prices were higher than those of the open-air or 'public' theatres. There was no standing room, the pit being filled with benches. Music was a great feature of the house: musicians paid to play there in the hope of attracting patronage from the nobility in the audience. It may have been the first English playhouse to use scenery, as it is referred to in Suckling's *Aglaura* (1637), presented by the King's Men. An attempt to close the theatre in 1631 was unsuccessful, and it was one of the most popular of the pre-Restoration playhouses. It fell into disrepair after 1642 and was demolished in 1655.

Blagrove, Thomas, see MASTER OF THE REVELS.

Blakely, Colin George Edward (1930–87), rugged Irish-born actor, whose early death robbed the British stage of a powerful talent. He appeared first in Belfast in 1958, making his London début at the *Royal Court Theatre in *O'Casey's *Cock-a-Doodle-Dandy* (1959). In 1963 he joined the newly founded *National Theatre company, where he soon made an outstanding success as Pizarro in *Shaffer's *The Royal Hunt of the Sun* (1964). His career with the National provided many other notable roles, including the title-roles in *Brighouse's *Hobson's Choice* (1965) and *Jonson's *Volpone* (1968), John Proctor in *Miller's *The Crucible* (1965) and Captain Shotover in *Shaw's *Heartbreak House* (1975). He had a season with the *RSC (*Pinter's *Old Times,* 1971, *Titus Andronicus,* 1972), and was seen in the West End in *Ayckbourn's *Just between Ourselves,* *De Filippo's *Filumena* (both 1977), Alan *Bennett's *Enjoy* (1980), and a notable revival of Miller's *All My Sons* (1981). He also starred in the West End productions of two National Theatre plays, Shaffer's *Equus* (1976) and Ayckbourn's *A Chorus of Disapproval* (1986), the latter being his last appearance on the stage.

Blakemore, Michael Howell (1928–), Australian director. He went to England in 1950 and was originally an actor, beginning to direct while at the *Citizens' Theatre (1966–8), his successes there including Peter *Nichols's *A Day in the Death of Joe Egg* (London, 1967; NY, 1968) and the British première of *Brecht's *Arturo Ui* (London, 1969). He continued his association with Nichols in *The National Health* (*National Theatre, 1969), *Forget-Me-Not Lane* (1971), and *Privates on Parade* (*RSC, 1977). During a period as associate director at the National Theatre, 1971–6, he was responsible for such memorable productions as *O'Neill's *Long Day's Journey into Night* (1971), with *Olivier, Hecht and MacArthur's *The Front Page* (1972), and Ben *Travers's *Plunder* (1976). He also directed David *Williamson's *Don's Party* (1975) at the *Royal Court and his *Travelling North* (1980) while Resident Director at the *Lyric, Hammersmith. His work in the commercial theatre included David *Hare's *Knuckle* (1974); Peter *Shaffer's *Lettice and Lovage* (1987; NY, 1990); and Michael *Frayn's *Make and Break* (1980), *Noises Off* (1982; NY, 1983), and *Benefactors* (1984; NY, 1985), as well as Frayn's translation of *Chekhov's *Uncle Vanya* (1988). In 1989 he directed the musical *City of Angels* in New York.

Bland, George, see JORDAN.

Blin, Roger (1907–84), French actor and director, who made his first appearance in 1935 in an adaptation by *Artaud of *Shelley's *The Cenci*. He studied mime with *Barrault, with whom he appeared many times during the late 1930s, and in 1949 began a successful career as a director with a production of *Strindberg's *The Ghost Sonata*. He is best remembered for his long advocacy of *Beckett's *En attendant Godot*, of which in 1953 he staged the first production; he later directed all Beckett's plays up to and including *Oh! les beaux jours* (1965). Blin was also instrumental in the discovery of *Adamov and in spite of political opposition to *Genet helped to popularize his plays, giving them successfully flamboyant productions. In general he was a discreet director, allowing due prominence to text and actors without obtruding himself on either.

Blondin, Charles, see MUSIC-HALL.

Blood-Tub, see GAFF.

Bloom, Claire (1931–), English actress. She first attracted attention in *Fry's *The Lady's not for Burning* (1949) and *Anouilh's *Ring round the Moon* (1950). She then joined the *Old Vic company, and for the next six years played leading roles in Shakespeare. After appearing as Cordelia to *Gielgud's King Lear in the West End in 1955 she returned to the Old Vic, touring America with the company and making her début in New York as Juliet, a role in which she was especially admired. In 1958 she starred in London with Vivien *Leigh in *Giraudoux's *Duel of Angels* and in 1959 she returned to New York, where she appeared in Fay and Michael Kanin's *Rashomon*, adapted from Akutagawa. She reappeared in London in 1961 in *Sartre's *Altona*, and in 1965 played in *Chekhov's *Ivanov* with Gielgud. In New York in 1971 she played

two famous *Ibsen heroines, Nora in *A Doll's House* and Hedda Gabler (repeating the former later in London), and she also gave a fine performance there as Mary, Queen of Scots, in Robert *Bolt's *Vivat! Vivat Regina!* (1972). In London she was seen as Blanche Du Bois in Tennessee *Williams's *A Streetcar Named Desire* (1974) and as Rebecca West in Ibsen's *Rosmersholm* (1977). Although strikingly beautiful, she relies not on her looks but on the depth and intensity of her interpretations of arduous and challenging roles. After her Ranevskaya in Chekhov's *The Cherry Orchard* at *Chichester (1981) she concentrated on film and television.

Blythe, Ernest, see ABBEY THEATRE, Dublin.

Boards, component parts of the stage floor, which run up- and down-stage supported on joists running crossways; used also as a phrase indicating the acting profession, to be 'on the boards' or to 'tread the boards' signifying 'to be an actor'. In modern theatres the boards form an unbroken expanse, but in earlier times they were removable, to facilitate the working of *traps and *machinery.

Boat Truck, platform, running on castors, on which scenes, or sections of scenes, can be moved on and off stage. Two such trucks, each pivoted at the down-stage and side-stage corners, so as to swing in and out over the acting area, are known as a scissor stage. A further variant is the waggon stage run on a system of rails and lifts. As many as five waggon stages can be found in some large theatres, each capable of moving aside from the acting area, or of rising and sinking in the cellar, with a full load of scenery.

Bobèche [Antoine Mandelot] (1791–c.1840), French farce-player, well known on the *boulevard du Temple where, with red jacket and grey tricorne hat with butterfly antennae, he amused the holiday crowds with his *parades in company with **Galimafré** [Auguste Guérin] (1790–1870). Both became extremely popular, and were invited to private houses to entertain the guests, but the topical jokes of Bobèche offended Napoleon and he was banished. He returned under the Restoration and was again successful, but later went into management in the provinces and was not heard of again. Galimafré, who refused to act after the fall of Napoleon in 1814, joined the stage staff of the *Gaîté and elsewhere. A play by the brothers Cogniard based on the lives of these two comedians was produced at the *Palais-Royal in 1837.

Bobo, see GRACIOSO.

Bodel, Jean (c.1165–1210), French poet and playwright, whose *Jeu de Saint-Nicolas* is the earliest surviving *miracle play in any vernacular language. Its theme, the conversion to Christianity of a Saracen king whose stolen treasure is miraculously restored to him through the intervention of the saint, reflects contemporary enthusiasm for the Fourth Crusade, and it was probably first performed in 1200 in the Guildhall at Arras on 5 Dec., the eve of St Nicholas' Day. Its double plot, in which scenes of dramatic intensity are interspersed with comic episodes, and its racy dialogue are unlike any of the extant *liturgical dramas which preceded it; the realistic details of contemporary life are found again in later dramatists of 13th-century Arras, notably *Adam de la Halle. The play was still being performed by students in northern France two centuries after the death of Bodel, who in about 1202 contracted leprosy and retired to a lazar-house in Arras where he died.

Bogdanov, Michael (1938–), director of Welsh/Russian parentage. He was Associate Director of the Tyneside Theatre Company, 1971–3 (see NEWCASTLE PLAYHOUSE), before moving to Leicester as Associate Director of the new *Haymarket Theatre and Director of the Phoenix Theatre, 1973–7. After a spell at the *Young Vic, 1978–80, he became an associate director of the *National Theatre in 1980, soon being involved, as the play's director, in the controversy over *Brenton's The Romans in Britain. His other productions at the National included *Calderón's The Mayor of Zalamea (1981), *Kyd's The Spanish Tragedy (1982), and *Kaufman and Hart's You Can't Take it with You (1983). He worked for the *RSC on such plays as The Taming of the Shrew (1978) and *O'Casey's Shadow of a Gunman (1980), and directed the musical Mutiny (1985).

In 1986 he co-founded and became Joint Artistic Director of the English Shakespeare Company, designed to bring high-quality productions of Shakespeare's plays to the provinces. His populist productions of the plays in the 'Wars of the Roses' cycle featured 20th-century clothes and other anachronisms. The company visited several theatres abroad, amongst them the *Tokyo Globe which it opened in 1988, and the Deutsches Schauspielhaus in *Hamburg, to which Bogdanov returned as Artistic Director, 1989–91.

Bogusławski, Wojciech (1757–1829), Polish actor, director, and playwright, the virtual founder of the theatre in Poland. He joined the company of the National Theatre in 1778, 13 years after its foundation, being its director for 30 years from 1783 and encouraging new Polish playwrights. Of his own plays the most successful, which are still in the repertory, were Henry IV Goes Hunting (1792) and Cracovians and Highlanders (1794), the latter substantially a *ballad opera. He was a fine actor and the first Pole to play Hamlet, appearing in 1797 in his own version of Shakespeare's play based on the German text by F. L. *Schröder. He established theatres throughout Poland—including Vilna

(1785) and Lwów (1794)—and in 1901 a theatre named after him opened in Warsaw.

Boileau-Despréaux, Nicolas (1636–1711), French critic whose writings brought a new and invigorating atmosphere into the literary debates of Paris and exercised a great influence on French literature and drama. He was not a poet but a good writer of verse. Though without imagination or warmth he had plenty of common sense and an uncanny flair for the best in art. He was the friend of many great writers of his time, including *Racine—whom he taught to write verse—and especially *Molière, whose satirical genius complemented his own.

Boker, George Henry (1823–90), American dramatist and a poet of some standing, one of the few successful writers in modern times of poetic tragedy. His romantic dramas based on historical incidents included Calaynos (1849) and Leonor di Guzman (1853), both with Spanish settings. The most popular was Francesca da Rimini, first produced for only eight performances in New York in 1855, but revived in 1883 by Lawrence *Barrett with himself as Lanciotto, Francesca's husband, who in Boker's version is the chief character. It remained in Barrett's repertory for many years. Boker also wrote two comedies in verse, The Betrothal (1850) and The Widow's Marriage (1852), and one in prose, The World a Mask (1851), a melodrama The Bankrupt (1855), and several unproduced plays. He was American envoy to Turkey, 1871–5, and to Russia, 1875–8.

Boleslavsky, Richard, see AMERICAN LABORATORY THEATRE.

Bolshoi Dramatic Theatre, Leningrad, see GORKY THEATRE, GRAND.

Bolt, Robert Oxton (1924–), English dramatist. He first attracted attention with Flowering Cherry (1957; NY, 1959), in which Ralph *Richardson played a man with a pipe-dream of owning an apple orchard. It was followed by The Tiger and the Horse (1960), with Michael *Redgrave and his daughter Vanessa, and A Man for All Seasons (also 1960; NY, 1961), an outstanding portrait of Sir Thomas More, played by Paul *Scofield. Gentle Jack (1963), starring Edith *Evans, had only a short run, but after a play for children, The Thwarting of Baron Bolligrew (1965), and Brother and Sister (1967), which was seen only in the provinces, Bolt had another success with Vivat! Vivat Regina! (1970), about the relationship between Elizabeth I and Mary, Queen of Scots, first seen at *Chichester and transferred to London (1970; NY, 1972). Another historical play, State of Revolution (1977), featuring Trotsky and Lenin, was produced at the *National Theatre. He also wrote several screenplays.

Bolton, Lancashire, see OCTAGON THEATRE.

Bolton, Guy, see MUSICAL COMEDY.

Bolton Theatre, see CLEVELAND PLAY HOUSE.

Bond, Edward (1935–), English dramatist, whose plays have aroused fierce controversy. The first, *The Pope's Wedding* (1962), like much of his later work, was staged at the *Royal Court. *Saved* (1965; NY, 1970) achieved some notoriety, mainly because of a scene in which a baby was stoned to death in its pram. Both these plays were set in contemporary England and their characters were almost inarticulate; *Early Morning* (1968), originally banned, was a Surrealist work in which one of the characters was Queen Victoria, and *Narrow Road to the Deep North* (also 1968; NY, 1972) used Japanese history to attack imperialism. *Black Mass* (1970) and *Passion* (1971) were anti-apartheid and *Lear* (also 1971) was a rewriting of Shakespeare's play in an attempt to give it contemporary relevance. *The Sea* (1973; NY, 1975) centred on a drowning and *Bingo* (1974; NY, 1976) was a portrait of Shakespeare (played in London by John *Gielgud) which condemned him for lack of political commitment. Bond's adaptation of *Wedekind's *Frühlings Erwachen* as *Spring Awakening* (also 1974) was seen at the *National Theatre (NY, 1978). *The Fool* (1976) was based on the life of the poet John Clare. *The Bundle* (*RSC, 1978) was again set in historical Japan, while the characters in *The Woman* (NT, also 1978) were taken from Greek tragedy. *Restoration* (1981), one of his best plays, was a black comedy with a historical setting, which used the dramatic tradition of *Restoration comedy to attack current values. *Summer* (1982; NY, 1983) was a confrontation between two women set in Eastern Europe; and *The War Plays* (1985) a trilogy set after an atomic disaster.

Bonehill, Bessie, see MUSIC-HALL.

Bonstelle, Jessie [Laura Justine Bonesteele] (1872–1932), American actress and theatre manager, nicknamed 'the Maker of Stars'. She began her career in 1890 with a touring company under Augustin *Daly, and a year later was managing the *Shuberts' theatre in Syracuse. After similar ventures in other towns, including Toronto, she leased the Garrick Theatre in Detroit in 1910, remaining there until 1922. She then bought the Playhouse in the same city, opening it in 1925; in 1928, having aroused the interest of the townsfolk, she was able to make it one of America's first civic playhouses, run on the lines of the *Theatre Guild. Under her control it flourished until her death. She encouraged much native American talent, among her discoveries being Katharine *Cornell.

Bontempelli, Massimo (1878–1960), Italian dramatist, akin to Rosso di San Secondo (see GROTTESCO) in his musicality and his recognition of man's creative need for fantasy, and to *Pirandello in his vision of a world turned upside down. In *La guardia della luna* (*The Guardian of the Moon*, 1920) he traces with genuine pathos the madness and eventual death of a woman after the loss of her infant daughter, and in *Minnie la candida* (1929) he portrays a woman driven to suicide by her belief that human beings are literally puppets, and that she too may be one. The intuitive sympathy shown in these plays is contradicted by *Nostra Dea* (1925), which is based on the satirical concept that woman can acquire a personality only from her clothes.

Bon Ton Theatre, New York, see KOSTER AND BIAL'S MUSIC HALL.

Book, spoken dialogue of a *musical comedy or musical play. (The word libretto is sometimes used, but more commonly denotes the text of an opera.) The book and the lyrics of the songs are often by the same author, but composers such as Irving *Berlin and Cole *Porter wrote their own lyrics, as does Stephen *Sondheim, while authors such as Guy Bolton were book rather than lyric writers. The rare feat of writing book, lyrics, and music was achieved by Ivor *Novello and Noël *Coward.

Book-Holder, Elizabethan name for the prompter, not to be confused with the *book-keeper. In addition to prompting the actors, the book-holder was responsible for seeing that they were ready to enter on cue (a function performed later by the *call-boy) and for handing them their *props, of which he was in sole charge.

Book-Keeper, important member of an Elizabethan company, responsible for the custody of the play scripts belonging to it and for the copying and distributing of individual parts, which were supposed to be returned to him when not in use. They were probably taken home by him when the theatre closed, or the company migrated to another theatre, or went on tour, and the low survival rate of Elizabethan play texts may in part be accounted for by their concentration in so few hands, with the subsequent risk of loss by fire, theft, or carelessness.

Book Wings, or Revolving Wings, method of changing the side pieces in a scenic set used in some early Victorian theatres, among them the Theatre Royal at Ipswich where they are known to have been installed before 1857. There were usually four book wings at each wing position, each quartet being fastened like pages in a book to its own central upright spindle which passed down through a hole in the stage; at its lower end was a grooved wheel. By means of a connecting rope passing over these wheels, all the spindles could be rotated and all the *wings of a scene changed simultaneously by means of one master handle.

Booth, Agnes [*née* Marian Agnes Land Rookes] (1846–1910), Australian-born actress who became well known in America, making her New York début in Tom *Taylor's *Our American Cousin* (1865), followed soon after by *Bulwer-Lytton's *Richelieu* with Edwin *Forrest. In 1874 she was seen as Pauline in the same author's *The Lady of Lyons*, and she showed a gift for comedy in *Gilbert's *Engaged* (1879). She also played Juliet (1874), Cleopatra (1877), and Audrey (in *As You Like It*, 1891), the last for A. M. *Palmer's company which she had joined in 1885. Her emotional acting was greatly admired by some, though it ultimately became outmoded and her career declined.

She married **Junius Brutus Booth** junior (1821–83) in 1867. Though a lesser actor than his father, he was for many years a useful member of the company at the *Bowery Theatre.

Booth, Barton (1681–1733), English actor, who while still at Westminster School showed great aptitude as an actor in a schoolboy production of *Terence's *Andria*. In 1700, after gaining experience in the provinces and in Dublin, he was engaged by *Betterton for *Lincoln's Inn Fields Theatre. He was slow to establish himself, mainly owing to the jealousy of *Wilks, but played a wide range of tragic parts. The most striking feature of his acting was his adoption of 'attitudes'—the pose, for instance, in which as Othello he listened with appropriate gestures to Emilia's address to the dying Desdemona: even those who were not moved by his acting conceded his effectiveness at such moments. Though not tall he had a dignified appearance and a richly resonant voice, and made a great success as Pyrrhus in *Philips's *The Distrest Mother* (1712) and in the title-role of *Addison's *Cato* (1713). He became manager of *Drury Lane with *Cibber and Wilks after *Doggett retired.

Booth, Edwin Thomas (1833–93), outstanding tragedian, and the first American actor to achieve a European reputation. Son of Junius Brutus *Booth senior, he made his first appearance at 16 in his father's company, playing Richard III at 18. Although at his best in tragedy he was, like his father, much admired as Sir Giles Overreach in *Massinger's *A New Way to Pay Old Debts*, which he played on his first visit to London in 1861. He was manager of the Winter Garden, 1863–7 (see METROPOLITAN THEATRE), where in 1864 he played 100 consecutive performances of *Hamlet*, a record unbroken until John *Barrymore's 101 in 1922. The same year saw a performance of *Julius Caesar*, in which Edwin and his two brothers, John Wilkes and Junius Brutus junior, appeared together for the only time. After the destruction

by fire of the Winter Garden in 1867 he built his own theatre (see BOOTH's), which opened in 1869 with *Romeo and Juliet*. The venture was not a success and Booth went bankrupt in 1873, though returning successfully to tour in the USA, England, and Germany. In 1881 he appeared at the *Lyceum in London by invitation of Henry *Irving, alternating with him the roles of Othello and Iago. In 1888 he presented his house in Gramercy Park to the newly founded *Players' Club, of which he became the first president. He was an unhappy man, his habitual melancholia aggravated by the madness of his father, second wife, and younger brother, the assassin of Abraham Lincoln.

Booth, John Wilkes (1839–65), American actor, younger brother of Edwin *Booth, who at 10.22 p.m. on 14 Apr. 1865 assassinated Abraham Lincoln in *Ford's Theatre, Washington, during a performance of Tom *Taylor's *Our American Cousin*. There are two theories as to the reason for this act, one asserting that Booth, an indifferent actor, was jealous of his elder brother's success and sought notoriety through crime, the other ascribing his action to mistaken patriotism. The whole question is systematically surveyed in *The Mad Booths of Maryland* (1940) by Stanley Kimmel.

Booth, Junius Brutus, senior (1796–1852), English-born American actor, eldest of a notable theatre family. Trained for the law, he went on the stage at 17 and after touring for some years appeared at *Covent Garden as Richard III, almost immediately entering into rivalry with Edmund *Kean. He was also seen as Sir Giles Overreach in *Massinger's *A New Way to Pay Old Debts*, Posthumus Leonatus in *Cymbeline*, and in 1818 as Shylock. Two years later he went to *Drury Lane, where he played Iago to Kean's Othello, Edgar to his Lear, and Pierre to his Jaffier in *Otway's *Venice Preserv'd*. In 1821 he went to America, and from then until his death he was constantly seen there, making his first appearance as Richard III, and becoming manager of several theatres. Among his other achievements was the playing in French of Oreste in *Racine's *Andromaque*, and he is said to have played Shylock in Hebrew. He was a rough and unpolished actor but full of grandeur and eloquence, with a resonant voice and ample gestures. There was a streak of madness in him, aggravated by intemperance. His sons Edwin and John Wilkes (above) were on the stage, as was Junius Brutus junior (see BOOTH, AGNES).

Booths, portable theatres which provided travelling companies with an adequate stage, supplanting the adapted innyards (see INNYARDS USED AS THEATRES), barns, or makeshift rooms which were at first the only places available for theatrical productions. Booths were probably first erected in the grounds which accommodated *fairs in Britain and on the Continent, and consisted originally of tents housing a small stage on which were presented short sketches and *drolls interspersed with juggling and rope dancing. Portable buildings, easily dismantled and re-erected, later made their appearance. In the 19th century the stage itself was solidly constructed, sometimes on the carts that carried the show around, and had a few simple backcloths and properties. The auditorium consisted of a canvas tent that could easily be rolled up and transported on the wagons, the seats being almost always plain wooden planks. Sometimes there was a platform outside from which the performers could tempt the audience in. This was a particular feature of the French fairground theatres which evolved from the early booths, and the first permanent fairground theatres in Paris—which had no counterpart in Britain— had large balconies for the performance of introductory *parades. Within the booths performances were gone through as quickly as possible so as to clear the seats for another audience, often as many as 10 or 12 a day. The plays were necessarily simple, farces or strong drama based on popular legends, stories from the classics, the Bible, or history. There were seldom any written scripts, though sometimes a popular play would be adapted for a local audience. The classic example is *Maria Marten; or, The Murder in the Red Barn*, an anonymous dramatization of a murder which took place in 1828. The booths lasted longest for the accommodation of *puppet shows, of which the portable stage of the *Punch and Judy show is a vestigial remnant.

Booth's Theatre, New York, on the southwest corner of the Avenue of the Americas and 23rd Street. Built for Edwin *Booth, it had several unusual features. The stage had no *rake and there was a tall 'stage house', believed to be the first in New York, with large, hydraulically powered elevator trays for lowering three-dimensional scenery to the cellar, though during Booth's own management *box-sets were often used. The venture began auspiciously in 1869 with *Romeo and Juliet*, starring Booth, followed by *Othello*. During subsequent seasons Booth was seen in many of his finest parts, supported by such outstanding players as Lawrence *Barrett and Charlotte *Cushman. But Booth was not able to make a success of his venture and in 1873 went bankrupt, leaving the management of the theatre to his elder brother Junius Brutus. A year later it passed out of the control of the Booth family and in 1879 the lease was taken over by *Boucicault, under whom many stars including Rose *Coghlan and John *Brougham made successful appearances. The last production, *Romeo and Juliet* with *Modjeska and Maurice *Barrymore, closed on 30 Apr. 1883,

after which the building was pulled down and a large department store was erected on the site.

Booth Theatre, New York, on West 45th Street. Built by Winthrop *Ames in association with Lee *Shubert, this small playhouse seating 766 opened in 1913 with *The Great Adventure* by Arnold *Bennett. In 1925 there was a production of *Hamlet* in modern dress. *Kaufman and *Hart's *You Can't Take it with You* (1936) and *Saroyan's *The Time of Your Life* (1939) both won the *Pulitzer Prize. Later productions included *Inge's *Come Back, Little Sheba* (1950), the *revue *At the Drop of Another Hat* (1966), Jason Miller's *That Championship Season* (1972), and *Sondheim's *Sunday in the Park with George* (1984), the last two being also Pulitzer Prize-winners.

Boquet, Louis-René, see COSTUME.

Borchert, Wolfgang (1921–47), German dramatist, whose only play, *Draussen vor der Tür*, was performed on 21 Nov. 1947, the day after the author's death. A passionate protest against contemporary corruption and decadence, it made a deep impact on war-torn Germany, and has been frequently revived, being considered a link between the plays of *Expressionism and the later Theatre of the *Absurd. Its central character was a soldier who returns from a prisoner-of-war camp (of which Borchert had personal experience after being captured on the Russian front) and was unable to adapt himself to civilian life. Through him Borchert voiced the nihilism and despair of Germany after the Second World War. As *The Man Outside*, it was published in 1952 in translation.

Border, narrow strip of painted cloth, battened at the top edge only, used to hide the top of the stage from the audience. If the lower edge is cut to represent foliage it is known as a Cloud or Tree Border. The use of clouds for masking the top of almost any scene was formerly common (see CLOUDINGS). In Victorian times plain sky borders were sometimes known as Blues. Tails, or Legs, on a border are long vertically hanging extensions at each end, forming with the border an arch over the scene.

Boston, Mass. Early efforts to introduce plays into Boston met with fierce puritanical opposition, and *Othello*, on its first showing, had to be presented as a 'Moral Lecture against the Sin of Jealousy'. Plays continued to be disguised in this way even after the building of the first permanent playhouse in 1792. At least two more theatres were built before the end of the century, the Federal Street in 1794 and the Haymarket in 1796; the former, a brick building, was destroyed by fire in 1798, immediately rebuilt, and then survived until 1852. Its greatest rival for many years was the Tremont Street Theatre, which opened in 1827 and was destroyed by fire in 1852, having ceased to be used as a theatre in 1843. In 1832 the American Amphitheatre was opened, but it was also burned down in 1852. The best known and most successful theatre in the town was the Boston Museum. The first playhouse of this name opened in 1841, and proved so popular that in 1846 a much larger theatre was erected. It underwent major renovation in 1872, and survived until 1893, reaching the peak of its prosperity between 1873 and 1883, when it housed an excellent resident *stock company with good visiting stars. In 1846 J. H. *Hackett opened the Howard Athenaeum, and in 1854 the Boston Theatre was built to replace the Federal Street Theatre. With a seating capacity of 3,140 and an almost circular auditorium, it was one of the most advanced theatre buildings of its time. The centenary of the Federal Street Theatre saw the opening of the Castle Square Theatre. Situated at the junction of Tremont, Chandler, and Ferdinand Streets, it held 1,800. Boston's pre-eminence in theatre, however, has now waned, and it no longer ranks as one of the outstanding theatrical cities of the United States.

Boucicault (**Bourcicault, Boursiquot**), **Dion**(-ysius) **Lardner** (1820–90), Irish playwright, actor, director, and theatre manager, born in Dublin of Huguenot extraction. He began his career as an actor, his first success as a playwright being *London Assurance* (1841), produced in London by Mme *Vestris with herself and her husband Charles James *Mathews in the cast, and constantly revived during the 19th century. Boucicault collaborated with Ben *Webster on two plays at the *Adelphi in 1844, and while resident dramatist at the *Princess's Theatre made adaptations from the French of *The Corsican Brothers* (1852) and *Louis XI* (1855). His successful career in the USA dates from *The Poor of New York* (1857), later known, according to the locality in which it was being produced, as *The Poor* (or *The Streets*) *of Liverpool*, etc. It was followed by *Jessie Brown; or, The Relief of Lucknow* (1858), in which the part of Jessie was played by Boucicault's common-law wife **Agnes Robertson** (1833–1916), who thereafter appeared in many of his plays including the spectacular *The Octoroon; or, Life in Louisiana* (1859), the first play to treat the American Negro seriously. He made a successful dramatization of *Dickens's *The Cricket on the Hearth* as Dot (also 1859). In 1860 Boucicault returned to England, after producing for Laura *Keene the first of his highly successful Irish plays, *The Colleen Bawn; or, The Brides of Garryowen*. Other plays of 'authentic' Irish life and character, but still cast in the melodramatic mould of the day, were *Arrah-na-Pogue; or, The Wicklow Wedding* (1864) and *The Shaughraun* (1874). Among Boucicault's later plays were a new version of Washington *Irving's *Rip Van Winkle* (1865); *Hunted Down* (1866), originally known as *The Two Lives of Mary Leigh*; and *The Flying Scud; or, Four-Legged Fortune* (also 1866), one of the first of the popular horse-racing melodramas. In 1872 he returned to America, where he remained, with occasional visits to England, until his death, and re-entered the New York theatre with *Mimi* (1872) and *Belle Lamar* (1874), one of the first plays to depict the American Civil War.

During his lifetime Boucicault was probably more esteemed as an actor than as a dramatist, in spite of his authorship of about 250 plays. He was highly praised for his skill in characterization, his timing, and his technical perfection. His facility and inventiveness as a director and his innovations in theatre management made him one of the outstanding personalities of the 19th-century theatre. He also helped to get the first American dramatic *copyright law passed in 1856. His demands for a more equitable deal for playwrights from managements materially assisted the establishment of the *royalty system; he was the first dramatist in England to receive a royalty. His four children by Agnes Robertson were on the stage, the best-known being 'Dot' and Nina (below).

Boucicault, 'Dot' [Darley George] (1859–1929), English actor and dramatist, elder son of Dion *Boucicault and Agnes Robertson. He made his first appearance in his father's company in New York in 1879, and was then seen in London and toured Australia. From 1901 to 1915 he directed the plays given at the *Duke of York's Theatre under the management of Charles *Frohman. He was an excellent actor, particularly in later life, two of his outstanding parts being Sir William Gower in *Pinero's *Trelawny of the 'Wells'* (1898) and Carraway Pim in A. A. Milne's *Mr Pim Passes By* (1901). He married in 1901 the English actress Irene *Vanbrugh, who appeared in many of his productions.

Boucicault, Nina (1867–1950), English actress, daughter of Dion *Boucicault and Agnes Robertson. Born in London, she first appeared in her father's company in America, playing Eily O'Connor in a revival of his *The Colleen Bawn* in 1885. She then toured Australia with him, and was in London in 1892 when she appeared as Kitty Verdun in the first London production of Brandon *Thomas's *Charley's Aunt*. In 1903 she was much admired as Bessie Broke in a dramatization of Kipling's *The Light that Failed*, and thereafter had a long and successful career, making her last appearance in 1936 as the Countess Mortimer in the first public performance of *Granville-Barker's *Waste*. She is, however, chiefly remembered as the first actress to play Peter in *Barrie's *Peter Pan* (1904).

Bouffes du Nord, Théâtre des, Paris, see INTERNATIONAL CENTRE OF THEATRE RESEARCH.

Boulevard du Temple, fairground in Paris which became a centre of entertainment, with circuses, *booths (in which *Bobèche and Galimafré revived memories of the earlier *parades), children's theatres, and puppet-shows. During the Revolution a number of small permanent theatres were built there (see FUNAMBULES and GAÎTÉ) in which actors such as *Deburau and *Frédérick appeared in plays, often by the elder *Dumas and later by *Pixérécourt. From the latter's works it got its nickname of 'the Boulevard of Crime'. The whole picturesque scene, with its sideshows, waxworks, fireworks, museums, cafés, concerts, and perambulating ballad-singers, was swept away in Haussmann's rebuilding of Paris in 1862, and the boulevard Voltaire now occupies most of the site.

Boulevard Plays (*Théâtre de Boulevard*), term used to describe the type of play usually found in the commercial theatres of Paris. When in the 1850s the old moats round the city were filled in to provide the foundations for the new boulevards, many of the unlicensed theatres of Paris, which had developed outside the walls from the old *booths of the fairgrounds, found themselves advantageously placed to cater for holiday audiences which came only to be entertained. Their staple fare was farce, comedy, melodrama, and romance, and substantial rewards were offered to those who could satisfy their demands. The apotheosis of the boulevard play came during 'la belle époque'—roughly 1880–1914—and changes in public taste meant that by 1970 they were to be found at only about 20 theatres in Paris, among them the Théâtre du *Palais-Royal and the *Comédie des Champs-Élysées. Most of these rely for financial support on foreign and provincial visitors and provincial tours, and their repertory includes many adaptations of West End and Broadway successes, though their favourite themes—witty topical satire and light-hearted attacks on sexual morality—are considered typically Parisian.

Bourchier, Arthur (1863–1927), English actor-manager, husband of Violet *Vanbrugh whom he married in 1894. He was one of the founders of the Oxford University Dramatic Society (OUDS) (see OXFORD). On leaving the university he joined Lillie *Langtry's company and after his marriage he toured with his wife, who played the leading parts in his productions, many of them comedies and farces adapted from the French. From 1900 to 1906 he was in management at the *Garrick Theatre, where he appeared as Macbeth, Shylock in *The Merchant of Venice*, and in a number of modern plays, among them *Pinero's *Iris* (1901). In 1910 he joined *Tree's company at *Her (then His)

Majesty's Theatre and was again seen in Shakespeare, being particularly admired as Henry VIII. He was at his best in truculent, fiery, or hearty parts, but had little subtlety and resented criticism, spoiling much of his best work by impatience.

Bourgeois Drama, see DRAME BOURGEOIS.

Bourgeois Tragedy, see BÜRGERLICHES TRAUERSPIEL.

Bourgogne, Théâtre de l'Hôtel de, the first theatre in Paris, built in 1548 in the ruins of the palace of the Dukes of Burgundy in the rue Mauconseil for occupation by the *Confrérie de la Passion. The theatre was long and narrow, some 102 ft. long and about 42 ft. wide. The total stage depth from front edge to back wall would have been approximately 43 ft. and the acting area even more restricted. The greater part of the auditorium was occupied by a pit in which the spectators stood; at the back, on a base of only 10 ft., were sharply rising tiers of benches. There were two rows of boxes, seven down each side and five along the back in each row—38 in all. Both stage and auditorium were lit by candles which had to be snuffed frequently during the performance. The medieval open-air *multiple setting was still in vogue, in a cramped and curved indoor version which forced the actors to declaim downstage.

As soon as the new theatre was ready for occupation the Confrérie were forbidden to appear in religious plays, on the grounds that the mixture of sacred and profane elements which had by now become general in the *mystery plays was bringing religion into disrepute. The company struggled along with farces and secular plays but gradually lost their audiences and towards the end of the 16th century were forced to hire out their hall to travelling companies from the provinces. As early as 1578 Agnan *Sarat appeared there, and an English company is believed to have been there 20 years later. The first permanent company to occupy the theatre was that of *Valleran-Lecomte, known as the King's Players; it reigned supreme in Paris until in 1634 the Théâtre du *Marais opened under *Montdory. His early retirement again left the Hôtel de Bourgogne, under Belleville's successors *Floridor and *Montfleury, in an unchallenged position until the arrival of *Molière in Paris in 1658. Many of the outstanding plays of the 17th century were first seen there, until in 1680 the company was finally merged with the other actors in Paris to form the *Comédie-Française. The stage of the Hôtel de Bourgogne was then occupied intermittently by the *Comédie-Italienne until 1783.

Bourguignon, Jeanne Olivier de, see BEAUVAL.

Bouschet, Jan, see ENGLISH COMEDIANS.

Bowery Theatre, New York. **1.** The first Bowery Theatre opened in 1826 with *Holcroft's *The Road to Ruin*. The theatre was lit by gas enclosed in glass shades, not, as elsewhere, with naked jets. Edwin *Forrest made many of his early successes at the Bowery, first appearing there as Othello. In 1828 the theatre was burnt down, and the actors had to migrate to the Sans Souci (see NIBLO'S GARDEN). A second Bowery opened in the same year, again with Forrest as its star, and became the first theatre in New York to have continuous runs. In 1836 the theatre was burnt down, as was a third Bowery in 1838 and a fourth in 1845, rebuilt in the same year. In 1851 an actor who was to be the idol of the Bowery audiences, **Edward Eddy** (1822–75), made his appearance in a series of strong parts. He took over in 1857 but closed it after one season. In 1858 the theatre reopened under George L. *Fox and James W. Lingard with plays and *pantomimes. When they left for the New Bowery (below), the old one, by now the oldest playhouse in New York, was subjected to a series of incompetent managers. During the Civil War it was occupied by the military, and it then became a circus. Fox reopened it as Fox's Old Bowery, with *melodramas as the staple fare, though old-fashioned *farce continued to flourish there long after it had vanished from New York's new playhouses. The audiences were notoriously uncouth. The theatre then fell a victim to the prevalent craze for *burlesque, and closed in 1878, reopening a year later as the Thalia for plays in German. It was destroyed by fire in 1923 and again, for the sixth and last time, in 1929.

2. The New Bowery opened in 1859, under Fox and Lingard, with a good company. It had a short and undistinguished career, enlivened only by visits from guest stars and the inevitable *Uncle Tom's Cabin*. It was destroyed by fire in 1866.

Boxes, small compartments in the *auditorium which formerly ran right round the *pit, each holding up to 20 people. This arrangement can still be seen in some opera-houses, but elsewhere they are situated only on each side of the stage and hold 4 to 6 people. Though often the most expensive seats in the theatre they were also the most frustrating, since they gave a distorted sideways view of the stage; they were usually occupied by those who wished to be seen more than to see. In theatres that retain them they are now seldom used, except to house lighting or television equipment. In many Court and Continental theatres a large Royal or King's Box was placed in the centre of the first tier over the entrance to the pit, directly facing the stage, a position it still retains even in the absence of royalty, except in Britain where in Georgian times it was shifted to one side.

Box-Office, that part of a theatre devoted to the selling of seats. The name dates from the time when the majority of seats in the house, except in the *gallery, were in *boxes. The term pay-box, used in the 18th century for the recesses at which entrance money was paid, is now generally reserved for those windows which collect money at the door. The term 'box-office' is used metaphorically to indicate the appeal of a play or player to the public.

Box-Set, scene representing the three walls and ceiling of a room, not by means of perspective painting on *wings, *backcloth, and *borders, as in early scenery, but by an arrangement of *flats which form continuous walls, with practicable doors and windows, completely covered in by a ceiling-*cloth. In London the box-set was first used in Mme *Vestris's production of W. B. Bernard's *The Conquering Game* in 1832, and brought to perfection in her production of *Boucicault's *London Assurance* at *Covent Garden in 1841. It was in general use for more than a century, especially for contemporary drama; but since it can only function behind a *proscenium arch, the widespread acceptance of various kinds of *open stage as well as the cost of changing the scene for each act, has almost led to its disappearance even from traditional theatres.

Boy Actors, troupes of choirboys who performed in London in the 16th and early 17th centuries, and were an important element in its theatrical life. The Children of the Chapel (based on London's Chapel Royal) played regularly at Court from about 1517. The Children of Paul's (based on St Paul's Cathedral) played both at Court and in the cathedral courtyard and their school hall. Their repertory eventually included most contemporary dramatists. The first group to play in a theatre (the *Blackfriars) was that from the Chapel Royal at Windsor in 1576. The Children of the Chapel, virtually professional, achieved great popularity in the second Blackfriars theatre in the early 1600s, giving the first performance of plays by *Jonson, *Chapman, and others. For several years the boys drew audiences away from the adult companies, and some of them—including Ezekiel *Fenn and Nathan *Field—became well known as adult actors. The last group to survive, they were disbanded in about 1615.

Individual boy actors played women's roles until after the Restoration, when women were finally allowed on the British stage.

Boy Bishop, see FEAST OF FOOLS.

Boyle, Roger, see ORRERY, LORD.

Boyle, William (1853–1923), Irish dramatist, whose first play *The Building Fund* (1905) was produced at the newly established *Abbey

Theatre. With this and with his later plays, *The Eloquent Dempsey* and *The Mineral Workers* (both 1906), Boyle helped to establish a new type of 'realistic' Irish play which materially altered the character of the Abbey Theatre as *Yeats and Lady *Gregory had originally envisaged it.

Bracco, Roberto (1862–1943), Italian playwright who achieved a solid success in the theatre during the 1890s and up to 1914. His opposition to Fascism drove his plays from the stage, though they later became known again. Like *Giacosa, he came under the influence of both the younger *Dumas, as in *A Woman* (1892) and *Masks* (1893), and of *Ibsen's 'theatre of ideas', as can be seen in such plays as *The Faithless One* (1894), *Trionfo* (1895), and the later *Ghosts* (1906). Though a brilliant writer of comedy, as he showed in *One of the Honest Men* (1900) and *Il perfetto amore* (1910)—the latter a sophisticated vehicle for a virtuoso actor and actress—his acutely developed social conscience is apparent in *Tragedy of a Soul* (1899), while his sympathy with feminism is shown in *Maternità* (1903). Some of his best work was done in naturalistic studies of Neapolitan lower life, as in *Don Pietro Caruso* (1895) and *Lost in the Shadows* (1901), but his most accomplished play, and the most significant historically, is *The Little Saint* (1909), remarkable for having anticipated the Theatre of *Silence.

Brace, Brace Jack, see FLAT.

Bracegirdle, Anne (*c*.1673–1748), one of the first English actresses, pupil and protégée of *Betterton. As a child of 6 or 7 she was already a member of his company, and in *Downes's *Roscius Anglicanus* (1708) she is mentioned as one of the leading young actresses at *Drury Lane in 1688. She went with Betterton to *Lincoln's Inn Fields Theatre in 1695, where she was much admired as Angelica in the opening production, *Love for Love* by *Congreve, whose mistress she was for some years. She was also considered outstanding as Millamant in his *The Way of the World* (1700). She left the stage at the height of her powers in 1707, wisely preferring not to be overshadowed by her younger rival Anne *Oldfield.

Brady, William Aloysius (1863–1950), American actor and theatre manager, who made his first appearance on the stage in San Francisco in 1882 and toured successfully for a long time as Svengali in Du Maurier's *Trilby*. From 1896 until its demolition in 1909 he managed the *Standard Theatre, and he also built and managed the Forty-Eighth Street Theatre and the *Playhouse. Many of his productions starred his second wife Grace *George. By his first wife Brady was the father of the actress **Alice Brady** (1892–1939), who first appeared on the stage in 1909 but spent her later years in films.

Brahm [Abrahamsohn], **Otto** (1856–1912), German director and literary critic, whose interest in contemporary drama, and particularly in the work of *Antoine at the *Théâtre Libre in Paris, led him to found in Berlin in 1889 the *Freie Bühne, where he successfully produced a number of new naturalistic dramas, opening with *Ibsen's *Ghosts*. In 1894 he took over the *Deutsches Theater where he continued to train his actors in a naturalistic style based on the teachings of *Stanislavsky which was well suited to the plays of Ibsen, *Hauptmann, and *Schnitzler, though less so to the classics. Under him the Deutsches Theater enjoyed 10 years of success and popular esteem. In 1904 he resigned his post to *Reinhardt and took over the Lessing Theater in Berlin, where he remained until his death. Although the naturalistic movement had by then lost its impact, Brahm had performed a great service to the German stage by clearing it of outmoded traditions and bringing it into the main current of European drama.

Braithwaite, Dame (Florence) **Lilian** (1873–1948), English actress. She made her first professional appearances in 1897 in South Africa with the Shakespeare company of her husband Gerald *Lawrence, playing several major roles. Her first appearance in London was as Celia in *As You Like It* to the Rosalind of Julia *Neilson, and she then joined Frank *Benson. She later appeared for several seasons with George *Alexander at the *St James's Theatre. Among the parts she created were Mrs Gregory in Vernon and Owen's *Mr Wu* (1913) and Florence Lancaster in *Coward's *The Vortex* (1924; NY, 1925). She gave some of her finest performances in her later years, notably as Elizabeth I in the English version of André Josset's *Élisabeth la femme sans homme* (1938) and Abby Brewster in Joseph Kesselring's *Arsenic and Old Lace* (1942).

Her daughter **Joyce Carey** (1892–   ), also an actress, is best remembered in a number of plays by Coward. Under the pseudonym Jay Mallory she wrote the popular play *Sweet Aloes* (1934) in which she also appeared.

Brandes, Johann Christian, see MONODRAMA.

Brasseur [Espinasse], **Pierre** (1905–72), French actor, who was on the stage very young and by 1929 was playing leading roles at a number of theatres in Paris. He then went into films, but in 1949 returned to the theatre and thereafter divided his time between stage and screen, appearing in the first productions of *Claudel's *Partage de midi* and *Sartre's *Le Diable et le bon Dieu* (both 1951). A romantic and passionate actor, he also made a great impression as Edmund *Kean in Sartre's recreation of the elder *Dumas's play (1953). Among his later successes were *Anouilh's *Ornifle* and *Montherlant's *Don Juan*, the French version of *Pinter's *The Homecoming*, *Shaw's *Don Juan* *in Hell* (from *Man and Superman*), and Shaw himself in Jerome Kilty's *Dear Liar*.

Brayton, Lily, see ASCHE.

Bread and Puppet Theatre, see PUPPET.

Brecht, Bertolt [Eugen Berthold Friedrich] (1898–1956), German poet and dramatist, who in 1918 was studying medicine in Munich when he wrote his first play *Baal*, not performed until 1923. His early writings are indebted rather to *Expressionism than to the Marxism which was to provide the political mainspring of his work. The first of his plays to reach the stage was *Drums in the Night* (1922). This sober and somewhat cynical study of a soldier returning from the war proved a great success. It was taken to Berlin, where in 1924 Brecht settled as assistant to *Reinhardt at the *Deutsches Theater. His next plays, *In the Jungle of Cities* (1923), *Edward II* (based on *Marlowe, 1924), and *Man is Man* (1926), were less successful, but marked the beginning of his attempt to develop his own form of *epic theatre, with its hotly debated 'distancing techniques' (*alienation). Throughout his career he undertook much theoretical writing, and his aesthetic position was given its most definitive form in *Kleines Organon für das Theater* (1949). In 1928 Brecht married his second wife Helene *Weigel, who appeared in many of his plays, and had his first great success in the theatre with *The Threepenny Opera* at the Berlin Schiffbauerdamm. This very free adaptation of *Gay's *The Beggar's Opera* had music by Kurt Weill, who also collaborated with Brecht on the operas *Rise and Fall of the City of Mahagonny* (1927) and *Happy End* (1929) and the opera-ballet *The Seven Deadly Sins* (1933). Weill's wife **Lotte Lenya** [Karoline Blamauer] (1900–81) appeared in most of these works. In the 1930s Brecht wrote a number of short didactic plays or *Lehrstücke*, the best of which were probably *The Exception and the Rule* (1938) and *St Joan of the Stockyards*, not staged until 1959 but considered by many the first work to show his full stature. When Hitler came to power in 1933 Brecht went into exile— first to Switzerland, then to Denmark and Finland, and finally in 1941 to the USA. During these years he wrote what are generally considered his best plays—*Mother Courage and Her Children* (1941), *The Life of Galileo*, *The Good Person of Setzuan* (both 1943), and *The Caucasian Chalk Circle* (1954)—which combine maturity of vision and depth of expression with a wider sympathy for the human predicament. During the same period he also wrote more overtly political plays such as *The Resistible Rise of Arturo Ui* (1958). In 1949 Brecht returned to East Germany and founded with Helene Weigel the *Berliner Ensemble. His last years were spent mainly on revivals of his own plays and adaptations of foreign ones, among them

Shakespeare's *Coriolanus* (as *Koriolan*) and *Far-quhar's *The Recruiting Officer* (as *Trumpets and Drums*). (All his adaptations were so different from the originals that they constituted virtually new works, particularly his Marxist version of *Gozzi's *Turandot*, begun in 1930, taken up again in 1954, and left unfinished at his death.)

While few would now dispute Brecht's greatness as a playwright, there remains strong disagreement between those who regard him as a great Marxist writer and those who see him as a great writer in spite of his Marxism; equally, his aesthetic theories are seen both as essential and as obstructive to his creative work. His presentation of 'epic' as a necessary alternative to 'dramatic' theatre remains, however, persuasive. Writers described by others as his disciples tend to disclaim his direct influence; but he has undeniably helped to liberate the English-speaking theatre from the constraints of the *well-made play. Most of Brecht's plays have now been seen in English, the first being *Señora Carrar's Rifles* in 1938.

Bredero, Gerbrand Adriaansz (1585–1618), Flemish dramatist, whose romantic plays, drawn mainly from tales of chivalry, are remarkable for the realism of the comic interludes, where characters from everyday life mingle with the more stereotyped knights and their ladies. Among them are *Griana* (1612) and *The Dumb Knight* (1618). His best work was done in the farces and comedies which laid the foundations of the Dutch comic tradition, among them *The Farce of the Cow* (1612), based on a German tale which tells how a rogue tricks an old peasant into selling him a cow and then returning him the money paid for it. In spite of its German origin it is purely Dutch in character, and Bredero, himself a man of the people, has caught admirably the spirit and dialogue of his times. His last and perhaps his best play was *The Spanish Brabanter* (1617), a series of comic scenes loosely held together by the two central characters, a master and his servant, which vividly creates on stage the widely differing worlds of the pretentious but penniless noble-man and the ordinary working people of Amsterdam. It became one of the most successful comedies of its day and was frequently revived.

Breeches Parts, see MALE IMPERSONATION.

Brenton, Howard (1942–), English dram-atist, a political dramatist like Howard *Barker who deals with large themes. He wrote his first play in 1965 while still at Cambridge, his first full-length play *Revenge* (1969) being seen at the *Royal Court, as was *Magnificence* (1973), which deals with urban terrorism. *Brassneck* (also 1973), written in collaboration with David *Hare, is about local government corruption, and *The Churchill Play* (1974), set in Britain in 1984 under a coalition government, shows political dissidents being brutally treated by the army. (A revised version was produced by the *RSC in 1988.) *Weapons of Happiness* (1976), the first new play to be produced in the *National Theatre's new building, uses a crisp factory as a setting for class warfare. It was followed by *Epsom Downs* (1977), set on Derby Day, and *Sore Throats* (1979), a violent study of marital friction. In 1980 *The Romans in Britain*, which draws a parallel between the Roman invasion of Britain and the British presence in Northern Ireland, was also produced at the National Theatre and aroused a good deal of controversy because of its violent scenes of bloodshed and homosexual rape (see CENSORSHIP). *Bloody Poetry* (1984; NY, 1987) is a study of *Byron and *Shelley's six-year relationship; *Pravda* (1985) another collaboration with Hare; and *Greenland* (1988) is set in a Utopian Britain 700 years hence. His translations of *Brecht's *Galileo* (1980) and *Büchner's *Danton's Death* (1982) were seen at the National Theatre.

Brice, Fanny, see REVUE.

Bridge, mechanical device (called an elevator in America) for raising heavy pieces of scenery or a tableau of actors from below up to stage level. The electrically controlled bridges of a large theatre may reach a high degree of engineering complexity.

Bridges-Adams, William (1889–1965), English director, who gained his experience in provincial repertory companies, including the *Birmingham Repertory Theatre, before in 1919 he was appointed director of the *Shakespeare Memorial Theatre in succession to Frank *Benson. There he directed 29 plays by Shakespeare and was instrumental in keeping the company together after the building was destroyed by fire in 1926, remaining with it until after it moved into the new theatre in 1932. He took the company several times on tour to Canada and the USA. After leaving Stratford in 1934 he worked with the *British Council, initiating the policy of sending theatrical companies abroad, mostly to Europe.

Bridie, James [Osborne Henry Mavor] (1888–1951), Scottish doctor and playwright. His initial effort, *The Sunlight Sonata* (1928), was staged under the pseudonym 'Mary Henderson' and made little impact; but its director was Tyrone *Guthrie, who not only directed the first play credited to 'James Bridie'—*The Anatomist* (1930)—but also took it in 1931 to London, where most of Bridie's later plays were seen, giving him a wider audience than most contemporary Scottish playwrights. Among them were *Tobias and the Angel* (also 1930) and *Jonah and the Whale* (1932), modern comedies based on biblical stories, *A Sleeping Clergyman* (1933), a disquisition on eugenics, *Storm in a Teacup* (1936), an adaptation of Bruno Frank's *Sturm*

im Wasserglass with a Scots setting, and *The King of Nowhere* (1938) seen at the *Old Vic with Laurence *Olivier in the chief part, Vivaldi. Alastair *Sim scored a great success in *Mr Bolfry* (1943), a religious play presenting the conflict between good and evil, also playing in and directing several more of his plays. In 1943 Bridie founded the *Citizens' Theatre in Glasgow, and in 1949 came his popular *Daphne Laureola*, with Edith *Evans in the title-role. For the *Edinburgh Festival he then wrote what many consider his best play, *The Queen's Comedy* (1950); his last two plays were produced posthumously by the Citizens' Theatre. Bridie remained in medical practice until 1938 and served as an army doctor in both world wars. He encouraged Scottish drama and was instrumental shortly before his death in forwarding the foundation of a drama department within the Royal Scottish Academy of Music in Glasgow, now the Royal Scottish Academy of Music and Drama.

Brieux, Eugène (1858–1932), French dramatist, whose plays are naturalistic dramas in the tradition of *Zola and *Becque. Each of his plays is a plea for the amelioration of a particular evil, and at his best he combined didacticism with a fierce pity for individuals. The most interesting plays are *Les Trois Filles de M. Dupont* (1898), on the dangers of a marriage of convenience; *La Robe rouge* (1900), which exposes the abuses of the judiciary; *Les Avariés* (1902), a study of venereal disease which as *Damaged Goods* created a sensation in London and New York; *Maternité* (1903), advocating legalized birth-control; and *La Femme seule* (1912), which deals with the financial dependence of married women.

Brighella, one of the *zanni* or servant roles of the *commedia dell'arte*. Originally a thief and a bully, with much in him of the Neapolitan street-corner boy, he later became a lackey. Through the establishment of the *Comédie-Italienne in Paris he influenced the development of the French valets of *Marivaux, and so the Figaro of *Beaumarchais.

Brighouse, Harold (1882–1958), English dramatist, usually referred to as one of the so-called 'Manchester School', as his early realistic comedies of North-country life were produced at the Gaiety Theatre there under Miss *Horniman. The only one of his works to survive on the stage is *Hobson's Choice* (1916), in which a cobbler's headstrong daughter insists on marrying her father's timid little employee. The play has been revived many times.

Brighton, Sussex. This famous seaside resort had its first permanent theatre, in North Street, in 1774; another theatre was erected in 1790 in Duke Street and closed in 1806, when a theatre was built in New Road which opened with

Charles *Kemble as Hamlet. It continued to function until 1866, when its lessee bought it outright, demolished it, and built the present Theatre Royal on the same site. Touring companies first visited the theatre in 1868, but it managed to retain a *stock company until 1873. The present theatre, seating 650, was once the most popular stopping-place for productions on their way to London, but its importance diminished with the reduction in touring and the increased use of the Theatre Royal, *Bath, for pre-London runs. It still presents a varied programme, though it cannot accommodate large-scale opera and ballet or big musicals.

Bristol, whose university was the first in Britain to have a *university department of drama, has a long theatre history. Elizabethan companies on tour played in the city, and Edward *Alleyn and Richard *Burbage appeared there. The first permanent theatre opened in 1729 at Jacob's Wells outside the city boundary. Mrs Hannah *Pritchard appeared there in the 1740s, as did Charles *Macklin. It closed during 1757, reopened a year later, and was used for the last time in 1765. In 1766 a larger theatre was built in King Street, with a new-style horseshoe-shaped auditorium. In 1778 it obtained a royal patent and became the Theatre Royal, being run in conjunction with the Theatre Royal at *Bath until 1817, a connection which brought prosperity to both houses. The *stock company in Bristol was a training-ground for young actors, among them Kate and Ellen *Terry; but as the centre of the city ceased to be a residential area the status of the theatre declined and it went through hard times, playing mainly *farce and *pantomime. When it closed after an air-raid in 1941 it seemed doomed to destruction; but in 1943 it started a new lease of life (see BRISTOL OLD VIC).

Bristol Hippodrome, built by Stoll in 1911–12 and intended to be the largest and best home of *variety outside London. It seated 2,000 people (now 1,975) and opened with a spectacular show, *Sands o' Dee*, which utilized the theatre's 40 ft. steel water tank. Later productions included opera, concerts, plays, circus, and, at the end of the First World War, a period of twice-nightly family *revue. It became a cinema in the 1930s but reverted to variety in 1938. In 1948 the stage area was destroyed by fire and the theatre was out of action for nearly a year. When it reopened it became a popular venue for musicals, *Guys and Dolls* and *The Music Man* having their British premières there. It is now a major touring house, with one of the largest stages in the country, 84 ft. wide and 60 ft. deep. It also presents ballet, opera, and plays.

Bristol Old Vic. In 1943 the old Theatre Royal in *Bristol, helped by the Council for the Encouragement of Music and the Arts (CEMA,

later the *Arts Council), became not only the oldest working theatre in England but also the first to be state subsidized. After reconstruction and redecoration it reopened with *Goldsmith's *She Stoops to Conquer*—first performed there in 1773—and three years later, through the joint efforts of CEMA and the London *Old Vic, a new resident company, the Bristol Old Vic, was launched under Hugh *Hunt, soon achieving a national and international reputation. One of the theatre's most popular productions was the musical play *Salad Days* (1954) by Dorothy Reynolds and Julian Slade, which moved to London for a long run. A production of *Love's Labour's Lost* staged in 1964 in honour of Shakespeare's Quatercentenary, toured all over the world. Other productions successful in London were *A Severed Head* (1963), based by J. B. *Priestley and Iris Murdoch on the latter's novel, Frank Marcus's *The Killing of Sister George* (1965), and Peter *Nichols's *Born in the Gardens* (1979). When the Arts Council relinquished the lease of the theatre in 1963 it was taken over by a Trust and in the early 1970s extensive redevelopments were carried out, including the building of a new stage, improved backstage facilities, and the acquisition of the adjacent Coopers' Hall, an 18th-century guildhall, to form a new entrance and foyer. The main theatre now seats 647. In 1989 the Bristol Old Vic was the first regional theatre to stage a co-production with the *National Theatre (of *Molière's *The Misanthrope*). A studio theatre, the New Vic, which opened in 1972, was built in the space occupied by the old entrance, but has been closed since 1989 for lack of local authority subsidy. From 1963 to 1980 the Bristol Old Vic also presented plays at the city-owned Little Theatre in the Colston Hall. An excellent Theatre School was founded at the same time as the company, and since its inception the Bristol Old Vic has worked in close conjunction with the Department of Drama at Bristol University.

Britannia Music-Hall, London, see ROTUNDA.

Britannia Theatre, London, in High Street, Hoxton, originally the Britannia Saloon at the rear of the Britannia Tavern, was opened by **Sam Lane** (1804–71) on Easter Monday 1841. It was used at first for free music-hall entertainments, the profits being made on the sale of refreshments; but the abolition of the monopoly of the *Patent Theatres in 1843 enabled Lane to turn it into a theatre and stage complete plays, mainly farces and strong drama. 'The Brit', as it was called, prospered; enlarged in 1850, it was rebuilt in 1858 to seat 3,923 people (reduced in 1866 to 2,972) in three tiers. Lane continued in management until his death in 1871 and was succeeded by his wife **Sara** (1823–99), an excellent actress, who managed the theatre until her death, no other theatre at that time having been so long under one management. It was a unique institution; authors wrote exclusively for it, actors joined the company as young men and remained until old age. It was supported mainly by the people of the surrounding district, its manageress playing in the annual *pantomime (which always ran until Easter) until she was in her 70s. After her death the control of the theatre passed to her nephews, but neither they nor later managers were successful, and in 1912 it became a cinema. It was destroyed by bombing in 1940.

British Actors' Equity Association, see EQUITY.

British Children's Theatre Association, see CHILDREN'S THEATRE.

British Council, organization founded in 1934 to promote a wider knowledge of the United Kingdom and the English language abroad. A drama department was founded in 1937 and, after amalgamating with the music department in 1961, again became a separate entity in 1977; it was renamed the drama and dance department in 1980, though dance had in fact always been within its jurisdiction. The department sponsors tours overseas by leading British companies; distributes copies of British plays with a view to their performance abroad; gives advice and information on all aspects of British theatre; and sponsors visits to Britain by theatre personalities from other countries for professional study and contacts with representatives of the British theatre.

British Drama League, British Theatre Association, see AMATEUR THEATRE.

British Theatre Museum Association, see THEATRE MUSEUM.

Britton, Hutin, see LANG.

Brizard [Britard], **Jean-Baptiste** (1721–91), French actor, who had been 10 years in the provinces when in 1756 he attracted the attention of Mlle *Clairon and Mlle *Dumesnil in Lyon, joining the *Comédie-Française in 1757. Owing to being prematurely white, he was able to play old men without a wig. Of a dignified presence, with a good voice and a natural style of acting, he was much admired as Henry of Navarre in Collé's *La Partie de chasse d'Henri IV* (1774). Brizard was also the first French actor to play King Lear, in the adaptation by *Ducis in 1782.

Broadhurst, George Howells (1866–1952), English-born American dramatist and theatre manager. His early plays were farcical comedies such as *What Happened to Jones* (1897) and *Why Smith Left Home* (1899), but *The Man of the Hour* (1906), about an idealistic mayor's struggle against political corruption, was in a more serious (if melodramatic) vein, in which he continued in *Bought and Paid For* (1911), *Today* (1913), and *The Law of the Land* (1914). In 1917

he opened a theatre under his own name (below), the first of his own plays to be seen there being *The Crimson Alibi* (1919). His later productions included an adaptation of Edgar Rice Burroughs's *Tarzan of the Apes* (1921).

Broadhurst Theatre, New York, on West 44th Street. George *Broadhurst opened this theatre, seating 1,185, in 1917 with the first production in New York of *Shaw's *Misalliance*. In 1926 *Broadway* by George *Abbott and Philip Dunning started a vogue for gangster plays, and in 1931 *Hamlet* was seen in a vast setting designed by Norman *Bel Geddes. The *Pulitzer Prize-winner *Men in White* (1933) by Sidney *Kingsley had a long run, as did *Housman's *Victoria Regina* (1935) with Helen *Hayes, Thomas Job's *Uncle Harry* (1942), and Jerome Lawrence and Robert E. Lee's *Auntie Mame*, based on Patrick Dennis's novel (1956). The musicals *Fiorello!* (1959), *Half a Sixpence* (1965), *Cabaret* (1966), and *Dancin'* (1978) also did well. Later productions included Peter *Shaffer's *Amadeus* (1980), with Ian *McKellen, Neil *Simon's female version of his *The Odd Couple* (1985), and his *Broadway Bound* (1986) and *Rumors* (1988).

Broadway, street running north–south of Manhattan Island, New York. It is used to denote the New York commercial theatre because many famous theatres stood on it, though the theatres are now mostly in side streets. The high cost and consequent unadventurousness of Broadway productions led to the establishment of smaller *Off-Broadway theatres away from the area.

Broadway Theatre, New York. 1. At 326 Broadway. Modelled on the *Haymarket Theatre in London, and resplendent in gold and white, this theatre opened in 1847 with Henry *Wallack in *Sheridan's *The School for Scandal*, his nephew Lester *Wallack making his first appearance in New York in the after-piece. When the old *Park Theatre was destroyed by fire in 1848 the Broadway took its place as a home for visiting stars. Edwin *Forrest was appearing there during the riot at the *Astor Place Opera House. This theatre also saw the first production of *Boker's epoch-making *Francesca da Rimini* (1855). Shortly afterwards excavations next door caused the theatre to collapse, and although it was quickly rebuilt it never regained its former eminence. It turned to *equestrian drama and finally closed in 1859.

2. At Broadway and West 41st Street. This opened as the Metropolitan Casino in 1880 and was rebuilt and renamed the Broadway, opening with Fanny *Davenport in *Sardou's *Tosca* in 1888. Several productions, including *Ben-Hur* (1899), adapted from Lew Wallace's novel, and an English *pantomime imported from London in 1901, had long runs; but in 1907 all the theatre's previous records were broken by the

immense popularity of Frances Hodgson Burnett's *Little Lord Fauntleroy*. Among the European visitors who played at this theatre were *Modjeska and *Salvini; Helen *Hayes also appeared there as a child. It became a cinema early in 1913 and was demolished in 1929.

3. A cinema at 1681 Broadway and 53rd Street which opened in 1924 is now used as a theatre, and since 1930 (with the exception of 1937–40) has been known as the Broadway. From 1930 to 1935 it housed a musical, *The New Yorkers, Earl Carroll's Vanities*, and *vaudeville. In 1943, after another period as a cinema, it was again used for musicals, with *Carmen Jones*; in 1959 *Gypsy*, starring Ethel *Merman, began a long run; *Cabaret* moved in from the *Imperial Theatre in 1968. Later successes included *Candide* (1974), based on *Voltaire, *The Wiz* (1977), transferred from the *Majestic, *Evita* (1979) by Tim Rice and Andrew Lloyd Webber, and *Les Misérables* (1987).

A building at 410 Broadway, originally called the Euterpean Hall, was named the Broadway Theatre for a brief period in 1837, as were Wallack's Lyceum in its last years, and *Daly's Theatre at 1221 Broadway in 1877–8.

Brochet, Henri see GHÉON.

Brockmann, Johann Franz Hieronymus (1745–1812), German actor, friend and pupil of *Schröder, to whose company he belonged in *Hamburg. He played Hamlet in the first production there of a German version of Shakespeare's play in 1776, with Schröder as the Ghost. He later went to the Burgtheater in *Vienna, where he was highly thought of and remained until his death.

Brook, Peter Stephen Paul (1925–), English director, already outstanding when in his late teens he directed *Marlowe's *Dr Faustus* and *Cocteau's *The Infernal Machine* on small stages in London. In 1945 his production of *King John* at the *Birmingham Repertory Theatre attracted a good deal of attention, and a year later he was at the *Shakespeare Memorial Theatre where he directed an enchanting *Love's Labour's Lost* costumed *à la Watteau*. In the same year he directed *Sartre's *Vicious Circle* with Alec *Guinness as Garcin, and in 1947 two further plays by the same author—*Men without Shadows* and *The Respectable Prostitute*—in a double bill. Later that year, at Stratford, he was responsible for a *Romeo and Juliet* which infuriated the critics, mainly on account of the clumsy cutting and speaking of the verse. He then went to direct opera at *Covent Garden, but returned to the theatre with excellent productions of *Anouilh's *Ring Round the Moon* (1950), with Paul *Scofield, and *Otway's *Venice Preserv'd* (1953), with Scofield and *Gielgud. Later productions included *Fry's *The Dark is Light Enough* (1954) with Edith *Evans,

Anouilh's *The Lark* (1955) with Dorothy *Tutin, and, at Stratford, *Titus Andronicus* (also 1955) with *Olivier. In the same year the Stratford company, with Brook's production of *Hamlet* starring Scofield, became the first English company to appear in Moscow since the Revolution. He also directed *Dürrenmatt's *The Visit* (NY, 1958; London, 1960) and *The Physicists* (London, 1963; NY, 1964). In 1962 Brook was appointed co-director, with Peter *Hall and Michel *Saint-Denis, of the *RSC. His first production at Stratford in his new capacity was *King Lear* starring Scofield, and at the *Aldwych Theatre he directed *Weiss's *Marat/Sade* (1964; NY, 1965) and an experimental documentary, *US* (1966), on the American intervention in Vietnam. After directing a modern-dress production of *Seneca's *Oedipus* for the *National Theatre in 1968, and at Stratford in 1970 (NY, 1971) an internationally acclaimed *A Midsummer Night's Dream* using stilts, trapezes, and other acrobatic tricks, Brook left England for Paris. Under the influence of *Grotowski, and with the encouragement of Jean-Louis *Barrault, he opened there his *International Centre of Theatre Research, to which the rest of his career has been almost entirely devoted.

Brooke, Gustavus Vaughan (1818–66), actor of Irish extraction, who appeared in Dublin at the age of 14 and then toured England and Scotland as the Dublin, or Hibernian, Roscius, playing adult roles including Young Norval in *Home's *Douglas*, Rolla in *Sheridan's *Pizarro*, and Virginius in Sheridan *Knowles's play of that name. He had a good presence and a fine voice, but his intemperance led to a chequered career during which he alternately triumphed and failed in such Shakespearian roles as Othello, Richard III, Hamlet, and Shylock, in London, the United States, and Australia, and he was finally imprisoned for debt in Warwick jail. On his release he set sail again for Australia and was drowned when his ship sank in the Bay of Biscay.

Brooks Atkinson Theatre, New York, on West 47th Street, between Broadway and 8th Avenue. Seating 1,088, it opened as the Mansfield Theatre in 1926, soon receiving a visit from the Moscow *Habimah Players in *Ansky's *The Dybbuk*. Marc *Connelly's *The Green Pastures* (1930) had an unexpected success, but after Ruth Gordon's nostalgic *Years Ago* (1946) the theatre was used for radio and television shows. It reopened in 1960 under its present name, given in honour of the critic Brooks *Atkinson. Neil *Simon's *Come Blow Your Horn* (1961) had a long run, and interesting later productions were Rolf *Hochhuth's *The Deputy* and Tennessee *Williams's *The Milk Train Doesn't Stop Here Anymore* (both 1964); Peter *Nichols's *A Day in the Death of Joe Egg* (1968) with Albert *Finney; Ronald *Millar's *Abelard and Heloise*

(1971) with Keith *Michell and Diana *Rigg; and Bernard Slade's *Same Time, Next Year* (1975), which ran for three years. Notable plays of the 1980s were Lanford *Wilson's *Talley's Folly* (1980) and Michael *Frayn's *Noises Off* (1983) and *Benefactors* (1985).

Brooks Theatre, Cleveland, see CLEVELAND PLAY HOUSE.

Brough, Fanny Whiteside (1854–1914), English actress daughter of the dramatist **Robert Brough** (1828–60) and niece of Lionel (below). She made her first appearance in Manchester in 1869 in a pantomime written by Lionel's elder brother William, and a year later was seen in London under Mrs John *Wood. Although she sometimes appeared in Shakespeare in her early years, she was at her best in modern comedy, and during her long career on the London stage appeared in a number of new plays, creating the parts of Mary Melrose in H. J. *Byron's *Our Boys* (1875), Mary O'Brien in Frances Hodgson Burnett's *The Real Little Lord Fauntleroy* (1888), and Lady Markby in Oscar *Wilde's *An Ideal Husband* (1895). In 1901 she appeared with Charles *Hawtrey in Anstey's *The Man from Blankley's*, and in 1909 made a great success as the Hon. Mrs Beamish in the sporting melodrama *The Whip*, by Cecil Raleigh and Henry Hamilton, at *Drury Lane. She made her last appearance, also at Drury Lane, at Christmas 1911, playing the Baroness Chicot in the pantomime *Hop o' My Thumb*.

Brough, Lionel (1836–1900), English actor, who made his first appearance in a play by his elder brother William. He was for a time in journalism, but returned to the stage in 1864, and in 1873 was principal low comedian at the *Gaiety under *Hollingshead. He was not so much a character actor as a clown in the best sense, his gift of improvisation and rich sense of humour making him an excellent player in *burlesque. Two of his finest parts were Tony Lumpkin in *Goldsmith's *She Stoops to Conquer* and Bob Acres in *Sheridan's *The Rivals*, but he was also good in Shakespeare's comic parts, among them Sir Toby Belch in *Twelfth Night* which he played with *Tree at *Her Majesty's Theatre. He toured extensively and was extremely popular in the United States and South Africa.

His daughter **Mary** (1863–1934) was an excellent actress who appeared in all the *Aldwych farces except the last.

Brougham, John (1810–80), American actor-manager and dramatist, born in Ireland, who made his first appearance in London in July 1830 in Egan's *Tom and Jerry*. After a long engagement with Mme *Vestris he became manager of the *Lyceum, until in 1842 he went to America, making his New York début at the *Park Theatre. In 1850 he opened his own theatre on Broadway as Brougham's Lyceum, hoping to rival the success of Mitchell's *Olympic. Brougham was a fine actor, an experienced manager, and a jovial popular personality, and his enterprise ought to have been successful; but the short burlesques and farces which had served Mitchell well for so long were now outmoded, and in spite of the hurried importation of such stars as *Florence and Charlotte *Cushman control of the theatre passed into the hands of the elder *Wallack. Brougham continued to act, spending several years in England before opening his second playhouse, the *Fifth Avenue Theatre 1, in 1869. This again was not a success. Brougham then retired from management and appeared with various *stock companies in New York, making his last appearance in 1879 at *Booth's Theatre. He was essentially a comedian and his best parts were stage Irishmen such as Sir Lucius O'Trigger in *Sheridan's *The Rivals* and Dennis Brulgruddery in the younger *Colman's *John Bull*, as well as Mr Micawber in *Dickens's *David Copperfield*, Captain Cuttle in his *Dombey and Son*, and Dazzle in *Boucicault's *London Assurance*.

Brown, Pamela (1917–75), English actress, who in a distinguished, though comparatively short, career won many admirers. She made her stage début in 1936 at the *Shakespeare Memorial Theatre as Juliet, later playing in London and in repertory at the *Oxford Playhouse. She made her first outstanding success in the title-role of Rose Franken's *Claudia* (1942). She played Ophelia in *Hamlet* (1944) and Goneril in *King Lear* (1946) for the *Old Vic, and in 1947 made her New York début as Gwendolen in *Wilde's *The Importance of Being Earnest*. On returning to London she played leading roles in Aldous Huxley's *The Gioconda Smile* (1948) and *Fry's *The Lady's not for Burning* (1949; NY, 1950), her haunting looks, far from conventionally beautiful, and husky voice giving a disturbing ambiguity to both characterizations. After appearing in Charles *Morgan's *The River Line* (1952) she joined *Gielgud's company at the *Lyric, Hammersmith, in 1953, being seen as Millamant in *Congreve's *The Way of the World* and Aquilina in *Otway's *Venice Preserv'd*. In the same year she appeared opposite Paul *Scofield in Wynyard Browne's *A Question of Fact*, returning to New York to play Lady Fidget in *Wycherley's *The Country Wife* (1957) and Lady Utterword in *Shaw's *Heartbreak House* (1959). She made her last stage appearance in Jack Ronder's *This Year, Next Year* (1960).

Browne, E(lliott) **Martin** (1900–80), English actor and director, closely connected with the revival of poetic, and particularly religious, drama in England. In 1935 he directed and played in T. S. *Eliot's *Murder in the Cathedral*, and thereafter directed all Eliot's plays. He was from 1939 to 1948 director of the Pilgrim

Players, who toured England in a repertory of religious plays, and in 1945 took over the *Mercury Theatre in London for the production of new plays by poets, among them *Fry's *A Phoenix Too Frequent* (1946). In 1951 he was responsible for the production in its native city, for the first time since 1572, of the York Cycle of *mystery plays which he revived several times. In 1945 he had succeeded Geoffrey Whitworth as Director of the British Drama League (later the British Theatre Association, see AMATEUR THEATRE), and he held this post until 1957, becoming in 1952 the first president of the International Amateur Theatre Association. He then went to the USA to inaugurate a programme of Religious Drama at the Union Theological Seminary in New York, where he remained as visiting professor until 1962. On his return to England he was appointed Honorary Drama Adviser to the rebuilt Coventry Cathedral; he was also President of Radius (the Religious Drama Society of Great Britain).

Browne, Maurice (1881–1955), actor, dramatist, and theatre manager. Born and educated in England, he made his name in the USA, where he is credited with having founded the Little Theatre movement (see AMATEUR THEATRE) by the establishment in 1912 of the Chicago Little Theatre, which he directed for several years. In 1920 he was directing on Broadway, and in 1927 he made his first appearance in London, as Adolf in *Strindberg's *The Creditors*. In 1929 he took over the management of the *Savoy Theatre and presented there with remarkable success R. C. *Sherriff's war play *Journey's End*, himself playing Lieutenant Raleigh. In the following year he produced *Othello* with Paul *Robeson, playing Iago himself. He was later responsible for the management of the *Globe and *Queen's theatres, presenting *Hamlet* (1930) with *Gielgud and seasons of *Moissi and the *Pitoëffs. Among later successful productions were *Fagan's *The Improper Duchess* (1931) with Yvonne *Arnaud, *Wings over Europe* (1932), of which he was part-author, Norman Ginsbury's *Viceroy Sarah* (1935) with Edith *Evans, and Esther McCracken's *Quiet Wedding* (1939). He retired at the end of 1939.

Browne, Robert, English actor, first mentioned as one of Worcester's Men in 1583. He became one of the best known of the *English Comedians travelling on the Continent at the end of the 16th century. In 1592 he toured Holland and Germany at the head of an English company which included in its repertory *jigs, biblical plays, early English comedies, and cut versions of plays by *Marlowe. In Aug. of that year Browne made the first of many visits to the great fair at *Frankfurt-Am-Main and in 1595 he and his companions took service with Maurice, Landgrave of Hesse-Kassel. The last

mention of him is in connection with the Easter fair at Frankfurt in 1620. He may be the Robert Browne who toured the English provinces with a puppet-show during the 1630s.

Browne, W. Graham, see TEMPEST.

Browning, Robert (1812–89), English poet, three of whose plays in verse were produced in the theatre: *Strafford* (1837), with *Macready, for whom it was written, in the leading role, at *Covent Garden; *A Blot in the 'Scutcheon* (1843) at *Drury Lane; and *Colombe's Birthday* (1853) at the *Haymarket. Helen *Faucit appeared in all of them. None of them was particularly successful, and they serve only to mark the great gap between poetry and the stage in the 19th century. Browning's other plays were written to be read and belong to literature rather than to the theatre, though the dramatic poem *Pippa Passes* has been performed occasionally.

Bruckner, Ferdinand [Theodor Tagger] (1891–1958), Austrian dramatist, whose first play *Krankheit der Jugend* (*Malady of Youth*, 1926) caused a sensation; it dealt with a group of medical students morbidly preoccupied with suicide, death being the one certain alternative to disillusion. Several of Bruckner's later plays portray modern problems under the guise of history, as in *Die Verbrecher* (*The Criminals*, 1928), set in Tudor England but dealing fundamentally with the causes of criminality and the workings of justice. He had his greatest success with a play on the same period, *Elisabeth von England* (1930), seen in London in 1931 in a translation by Ashley *Dukes. Like *Die Verbrecher* it made brilliant use of simultaneous action by juxtaposing on stage the Courts of Elizabeth I and Philip II of Spain. His *Timon: Tragödie vom überflüssigen Menschen* (*Timon: The Tragedy of the Superfluous Man*, 1933), ostensibly a reworking of Shakespeare's *Timon of Athens*, lays bare the weakness of the Weimar Republic in allowing the rise of Hitler, as does *Napoleon I* (1937). The earlier *Rassen* (*Races*, 1933) offers a vivid picture of the tragedies caused by racial persecution among German students. Bruckner left Europe in 1936 and lived in America until 1951, his most important plays during this period being *Heroische Komödie* (1938), which deals with Mme de Staël and Benjamin Constant, and *Simon Bolivar* (in two parts, 1943/5). After his return to Germany he attempted to revive classical tragedy in a modern form, using verse and a chorus in *Der Tod einer Puppe* (*The Death of a Doll*) and *Der Kampf mit dem Engel* (*The Fight with the Angel*) (both 1956). His last stage work was an adaptation of the old Indian play *Mṛcchakaṭikā* (known in English as *The Little Clay Cart*) under the title *Das irdene Wägelchen* (1957).

Brunelleschi, Filippo, see PARADISO.

Brunton, Anne, John, and Louisa, see MERRY.

Bruscambille [Jean Deslauriers] (fl. 1610–34), French actor at the Paris *fairs in the early 17th century who went to the Hôtel de *Bourgogne to play in farce. He won fame as a speaker of witty prologues and harangues to the rowdy audiences, which he composed himself. These are extant, and give an interesting picture of the tribulations of the actor before his profession became respectable, summed up in Bruscambille's oft-quoted epigram, 'une vie sans soucis et quelquefois sans six sous'.

Brustein, Robert, see HARVARD UNIVERSITY and YALE UNIVERSITY.

Bryan, Dora, see MILLAR, RONALD, MUSICAL COMEDY, and REVUE.

Bryant, Michael (1928–), English actor, first seen in London in *Ionesco's The New Tenant (1956). He came into prominence as the young tutor in *Shaffer's Five Finger Exercise (1958), in which he made his New York début (1959) and toured the USA. In London in 1961 he successfully took over from Alec *Guinness the part of T. E. Lawrence in *Rattigan's Ross, displaying the solitary and introspective qualities he had shown earlier. In 1964 he joined the *RSC, his roles including Selim Calymath in *Marlowe's The Jew of Malta and Teddy in *Pinter's The Homecoming (1965). He joined the *National Theatre in 1977 and played a wide variety of roles, outstanding among them *Ibsen's Brand (1978) and Gregers Werle in The Wild Duck (1979); Iago to Paul *Scofield's Othello (1980); *Chekhov's Uncle Vanya (1982); and Polonius in Hamlet (1989). He was also seen there in modern plays, notably as Lenin in Robert *Bolt's State of Revolution (1977) and Julius Caesar in Howard *Brenton's The Romans in Britain (1980).

Bryant's Opera House, New York, see KOSTER AND BIAL'S MUSIC HALL.

Bryden, Bill [William] **Campbell Rough** (1942–), Scottish director, who began his theatrical career as Assistant Director at the *Belgrade Theatre, Coventry, 1965–7, moving to the same post at the *Royal Court Theatre, 1967–9. From 1971 to 1974 he was Associate Director at the *Royal Lyceum Theatre in Edinburgh where he directed his own plays Willie Rough (1972) and Benny Lynch (1974). In 1975 he became an associate director of the *National Theatre, staging such plays as *Synge's The Playboy of the Western World (1975), *Osborne's Watch It Come Down (1976), and *O'Casey's The Plough and the Stars (1977). He used the flexibility of the *Cottesloe Theatre to stage excellent promenade productions (see THEATRE-IN-THE-ROUND) including The Passion (1978), drawn from the York *Mystery Plays,

and Lark Rise and Candleford (also 1978), based on a book by Flora Thompson. He also directed in 1980 a season of plays by Eugene *O'Neill and Arthur *Miller's The Crucible. In 1985 his promenade production of The Mysteries (The Nativity, The Passion, and The Doomsday) moved from the Cottesloe to the *Lyceum Theatre, London. He has directed three of David *Mamet's plays: American Buffalo (1978), Glengarry Glen Ross (1983), both at the National Theatre, and A Life in the Theatre (1989).

Buchanan, Jack [Walter John] (1890–1957), Scottish actor-manager, singer, dancer, and director, famous for his debonair manner, elegant dancing, and slightly croaky singing voice. In 1921 he starred in the Charlot revue A to Z, for which he also directed the sketches and staged the musical numbers. In the following year came the first of many ventures into management, Battling Butler, a musical farce in which he played the title-role. He made his début in New York, where he was to become well known, in 1924, co-starring in the Charlot Revue with Gertrude *Lawrence and Beatrice *Lillie. He was in the London production of Kern's musical Sunny (1926) and the British musical That's a Good Girl (1928), returning to New York in 1929 to star in another revue, Wake Up and Dream, this time with Jessie *Matthews. During the 1930s he appeared in London in such native musicals as Stand Up and Sing (1931), of which he was part-author, and This'll Make You Whistle (1936). In 1944, however, he played a straight role in *Lonsdale's The Last of Mrs Cheyney, and thereafter he was seen mainly in non-musical productions, including Lonsdale's Canaries Sometimes Sing (1947) and an adaptation of a play by Sacha *Guitry, Don't Listen Ladies! (NY, 1948; London, 1949). He directed many of the productions he acted in or presented, and at the time of his death controlled the *Garrick Theatre and the King's Theatre, Hammersmith.

Büchner, Georg (1813–37), German dramatist who died young of typhoid fever. Unknown during his lifetime, he has grown steadily in reputation since he was discovered by *Hauptmann in the 1890s and taken as a model by adherents of *Expressionism. His first play, Dantons Tod (Danton's Death), not acted until 1903, portrays the struggle for power between Robespierre and Danton during the early years of the French Revolution. It was produced in translation in New York by Orson *Welles in 1938 and in London in 1959 (also by the *National Theatre in 1971 and 1982). Büchner's only other complete play was a comedy, Leonce und Lena, first performed in 1895. Woyzeck, of which only an unfinished version remains, was first produced in 1913. Alban Berg made it the subject of his opera Wozzeck in 1925.

Buckingham, George Villiers, second Duke of, see DRYDEN and FLETCHER.

Buckley's Olympic, New York, see OLYMPIC THEATRE 3.

Buckstone, John Baldwin (1802–79), English actor and dramatist, who made his first appearance in the famous melodrama *The Dog of Montargis* in a barn in Peckham. In 1827 he was seen at the *Adelphi in his own play *Luke the Labourer*. He was for many years manager of the *Haymarket Theatre where most of his plays were given; he is said to haunt the building. Among his many melodramas and farces the best known are *Married Life* (1834), *Single Life* (1839), *The Green Bushes; or, A Hundred Years Ago* (1845), and *The Flowers of the Forest* (1847). He was a popular comedian of great breadth and humour; the mere sound of his voice heard off-stage, a mixture of chuckle and drawl, was enough to set an audience laughing.

Buen Retiro, Madrid, see CALDERÓN DE LA BARCA.

Buero Vallejo, Antonio (1916–), Spanish dramatist, whose *Historia de una escalera* (*Story of a Staircase*, 1949), set in a tenement slum, won the Lope de Vega Prize and marked the ascendancy of social realism in the Spanish post-Civil War theatre. It was followed by a study of blind people, *En la ardiente oscuridad* (*In the Burning Darkness*, 1950). After three historical plays, including *Las meniñas* (*The Ladies-in-Waiting*, 1959), he turned to symbolism in *El concierto de San Ovidio* (1962) and *El tragaluz* (*The Skylight*, 1967). *El sueño de la razón* (*The Dream of Reason*, 1970) is a hallucinatory interpretation of Goya. He also wrote such overtly political plays as *La doble historia del Dr Valmy* (*The Double Story of Dr Valmy*, 1968), first performed in England because it attacked the secret police, and *Jueces en la noche* (*Judges in the Night*, 1979). Among his later plays were *Diálogo secreto* (1984) and *Lázaro en el laberinto* (1986). His work is marked by a resigned yet hopeful humanism.

Buffalo Bill, see CODY.

Built Stuff, specially carpentered three-dimensional pieces of *scenery, from banks and *rostrums to columns, rocks, trees, porches, and complete scenes on *boat trucks.

Bulgakov, Mikhail Afanasyev (1891–1940), Soviet dramatist, whose first play was a dramatization of his own novel *The White Guard*, dealing with the Civil War in the Ukraine in 1918. As *The Days of the Turbins* it was produced by *Stanislavsky at the *Moscow Art Theatre in 1926. A second play on the same subject, *The Flight*, was not staged, while two later plays, a comedy entitled *The Red Island* (1928) and *The Cabal of Saintly Hypocrites* (1936),

which dealt with *Molière's difficulties with the censorship over *Tartuffe*, were seen only briefly. Bulgakov joined the staff of the Moscow Art Theatre, preparing for the company an excellent dramatization of *Gogol's *Dead Souls* (1930). His last play, based on *Cervantes' *Don Quixote*, was produced posthumously in 1941. *The White Guard*, directed by Michel *Saint-Denis, was seen in London in 1938 and revived in 1979 by the *RSC; as *The Days of the Turbins* it was produced in New York in 1977. An adaptation of *The Cabal of Saintly Hypocrites* as *Molière in Spite of Himself* was staged in New York in 1978. Adaptations of his novel *The Master and Margarita* were staged at the *Taganka Theatre, by *Lyubimov, in 1977 and in New York in 1978.

Bullins, Ed, see NEGROES IN THE AMERICAN THEATRE.

Bulwer-Lytton, Edward George Earle Lytton, Baron Lytton (1803–73), English novelist and dramatist, whose most successful play *The Lady of Lyons; or, Love and Pride* (1838) was first produced at *Covent Garden by *Macready, who played the hero Claude Melnotte with Helen *Faucit as Pauline. Its continued popularity led to a number of revivals, notably by Henry *Irving at the *Lyceum in 1879 with himself and Ellen *Terry. Though the plot is romantic and sentimental, it has a touching sincerity and wore well for many years. Bulwer-Lytton's next play *Richelieu; or, The Conspiracy* (1839) was also first produced by Macready, and frequently revived, Irving presenting it at the Lyceum no less than four times. *Money* (1840) struck a more serious and contemporary note. His half-dozen other plays are now forgotten.

Bunraku, see KABUKI and PUPPET.

Buontalenti, Bernardo, see SCENERY and WINGS.

Burbage [Burbadge, Burbege], James (c.1530–97), builder of the first permanent playhouse in London, the *Theatre. This opened in 1576 and was used by *Leicester's Men with whom Burbage, a carpenter by trade, had begun to act in about 1572. A man of violent temper, stubborn and unscrupulous and beset all his life by financial and other difficulties, he never wavered in his allegiance to the theatre, and though apparently not a good actor himself he was excellent at selecting and training his fellow players, among them his son Richard (below). In 1596 James took over part of an old monastery and adapted it as the second *Blackfriars Theatre, to use in winter in place of the unroofed Theatre, but he died before getting permission to open it.

The project was completed by his eldest son **Cuthbert** (c.1566–1636), who also helped to dismantle the Theatre and build the first *Globe Theatre.

Burbage, Richard (*c.*1567–1619), younger son of James (above) and the first outstanding English actor. Though short and fat, he created most of Shakespeare's great tragic roles— Hamlet, Othello, Lear, Richard III, the last being accounted his finest part. He appeared in a number of other contemporary plays, his roles including Hieronimo in the revised version of *Kyd's *The Spanish Tragedy* (1597) and Ferdinand in *Webster's *The Duchess of Malfi* (1614). He was associated with his brother Cuthbert in the management of the *Theatre and the building of the *Globe Theatre, scene of his greatest triumphs. His acting career began early, probably under the tuition of his father, the first mention of him being in 1590 when he was with the *Admiral's Men. He became the leading actor of the *Chamberlain's Men on their formation in 1594 and retained that position when in 1603 they were reorganized as the *King's Men. Burbage's name long remained synonymous with all that was best in English acting.

Bürgerliches Trauerspiel, domestic middle-class or bourgeois tragedy in prose inaugurated in Germany by *Lessing's *Miss Sara Sampson* (1755). Written under the influence of *Diderot and *Lillo, it refuted *Aristotle's assertion that tragedy belonged only to kings and princes and proved that it could equally well arise from a conflict between aristocratic depravity and middle-class morality. This Lessing showed again in *Emilia Galotti* (1772), and *Schiller even more forcibly in *Kabale und Liebe* (1784) which, while continuing the former criticism of absolutism, also suggests for the first time that the middle classes too may have their faults. One of the best plays of this type, *Hebbel's *Maria Magdalene* (1848), went so far as to dispense with aristocratic villainy; its middle-class heroine is seduced by an ambitious young clerk for the sake of her dowry, but her father's self-righteous obsession with respectability drives her to suicide. Though the term is no longer used, the theme became a staple of social drama.

Burgtheater, see VIENNA.

Burk, John Daly (*c.*1775–1808), Irish-born dramatist who went to America in 1796 and settled in Virginia around 1800. He is credited with seven plays: the first and best known, *Bunker-Hill; or, The Death of General Warren*, was first performed in Boston in 1797, opening in New York seven months later. The play, depicting a famous American victory and using bombastic verse rhetoric, was popular with patriotic audiences for 50 years and was regularly performed on holidays such as the Fourth of July. His *Female Patriotism; or, The Death of Joan of Arc* (1798) was a better play. Burk died in a

duel with a Frenchman after speaking insultingly of Napoleon.

Burke, Charles, see JEFFERSON, JOSEPH.

Burla (pl. *burle*) comic interlude or practical joke introduced, usually extempore, into a performance by the servant masks of the *commedia dell'arte*. Unlike the *lazzo, the *burla* involved some horseplay and could be developed at will into a small independent 'turn', the characters returning at its conclusion to the main theme of the plot. Although there is no adequate English translation of the word, the terms *burletta and *burlesque derive from it.

Burlesque, American, sex-and-comedy entertainment originally intended for men only. Known popularly as 'burleycue' or 'leg-show', it was devised by **Michael Leavitt** (1843–1935) in about 1868. Each performance was divided into three parts, the first combining chorus numbers and monologues with comedy sketches known as 'bits'; the second, known as the 'olio', was made up of *variety turns. The third part, which also had chorus numbers and 'bits', contained the show's only resemblance to the original burlesque in that it might feature a travesty on politics or a parody of a current theatrical success. The final number, known as 'the Extra Added Attraction', was usually a belly dance, or 'hootchy-kootchy'. One of the early stars of burlesque was **Lillian Russell** [Helen Louise Leonard] (1861–1922) who first appeared in 1881. In about 1920 strip-tease was introduced to counteract the competition of the films. This involved a complicated ritual of a woman undressing to music and exposing herself for a second completely nude except for a G-string. The most famous strip-tease artiste was **Gypsy Rose Lee** [Rose Louise Hovich] (1914–70). Heavy drinking contributed to the uninhibited atmosphere of burlesque houses, and with the enforcement of Prohibition in America burlesque lost its hold over the public and was finally banned from New York in 1942.

Burlesque, English, satirical play, usually based on some well-known contemporary drama or dramatic fashion that offered scope for parody. The prototype was the Duke of Buckingham's *The Rehearsal* (1671), which made fun of *Dryden and his *heroic drama, and the genre culminated in *The Critic* (1779), in which *Sheridan mocked the sentimental foibles of his day. In the meantime the traditions of burlesque had been upheld by *Gay with *The Beggar's Opera* (1728); by Henry Carey, who burlesqued both opera and drama; and by *Fielding in *The Tragedy of Tragedies; or, The Life and Death of Tom Thumb the Great* (1739).

In the 19th century a new type of burlesque flourished. It retained enough of its origins to choose as its target a popular play, but the element of criticism was lacking. One of its best writers was H. J. *Byron, with *Aladdin; or, The Wonderful Scamp* (1861), *The Corsican 'Bothers'; or, The Troublesome Twins* (1869), and *Robert MacMaire; or, The Roadside Inn Turned Inside Out* (1870). Possibly Byron's execrable puns, together with the reform of the stage initiated by T. W. *Robertson, finally killed the burlesque, though it survived until the early 1890s at the *Gaiety, with the famous 'quartet' headed by Nellie *Farren. Such *revues as the *Gate* (1938), the *Little* (1939), and the *Sweet and Low* series (1943–6) contained sketches in the burlesque tradition.

Burletta, originally a farce with interpolated music or a short comic opera. The name was extended in the mid-19th century to include plays put on at the minor theatres which, in order to circumvent the monopoly of the *Patent Theatres, contained at least five songs in each act, and sometimes instrumental interludes or solo dances. This made any play, including those of Shakespeare, legally a burletta, and so not subject to the licensing laws; it helps to explain many of the odd interpolations found in 19th-century productions of 'straight' plays.

Burnaby, Davy, see PIERROT TROUPES.

Burnand, F. C., see HOLBORN THEATRE.

Burton [Jenkins], **Richard Walter** (1925–84), Welsh actor of great power and presence, son of a miner, who took his stage name from that of his benefactor Philip Burton. He made his début in Emlyn *Williams's *The Druid's Rest* (1944) and later appeared in three plays by Christopher *Fry—*The Lady's not for Burning* (1949; NY, 1950), *The Boy with a Cart*, and *A Phoenix Too Frequent* (both 1950). In 1953–4 he was at the *Old Vic, his roles including *Hamlet*, the Bastard in *King John*, Sir Toby Belch in *Twelfth Night*, Caliban in *The Tempest*, and Othello and Iago alternately with John *Neville. In New York he was seen in *Anouilh's *Time Remembered* (1957), in the musical *Camelot* (1960), and in a *Hamlet* (1964) of 138 consecutive performances in *Gielgud's production, which broke Gielgud's own record. He appeared with the *Oxford University Dramatic Society in a production of *Marlowe's *Dr Faustus* (1966). In 1976 he starred in Peter *Shaffer's *Equus* in New York, where in 1983 he also played opposite his former wife Elizabeth Taylor in *Coward's *Private Lives*. His career latterly was mainly in films.

Burton, William Evans (1804–60), English-born actor-manager and dramatist, son of a printer, who inherited his father's business but gave it up to go on the stage. He first appeared in London in 1831 and in 1834 made his American début at the Arch Street Theatre,

*Philadelphia, first appearing in New York in 1837. After converting a circus in Philadelphia into a theatre and running it successfully as the National, he returned to New York and took over Palmo's Opera House. Renovated and redecorated, it opened in 1848 as Burton's Chambers Street Theatre and became one of the most important theatres of the day. Burton himself played there a number of richly comic parts, notably Captain Cuttle in *Brougham's adaptation of *Dickens's *Dombey and Son*, the theatre's first outstanding success. He presented several plays by Shakespeare, beginning with *The Winter's Tale* (1851). In the following year Burton began to suffer from the success of *Wallack's, which took some of his best actors from him, but before he finally deserted his theatre early in 1856—it closed in 1857—it had seen fine productions of *A Midsummer Night's Dream* and *The Tempest* (both 1854) with Burton as Bottom and Caliban. Later in 1856 Burton took over the *Metropolitan, but the competition of Wallack's was too keen and the theatre was too big to stand up to the financial crisis of 1857. It closed in 1858, Burton making his last appearance in New York at *Niblo's Garden in 1859. He was the author of several plays, of which none has survived.

Bury, John (1925–), stage designer, born in Wales, who began his career with *Theatre Workshop as an actor, his first work as a designer being for a production of *Ibsen's *An Enemy of the People* (1954). He later designed several of the company's most successful productions, including Brendan *Behan's *The Quare Fellow* (1956), Shelagh Delaney's *A Taste of Honey* (1958), and the musicals *Fings Ain't Wot They Used T'Be* (1959) and *Oh, What a Lovely War!* (1963; NY, 1964). In 1963 he was appointed associate designer with the *RSC, his work being seen in such productions as *Dürrenmatt's *The Physicists* (1963; NY, 1964), John *Barton's adaptation *The Wars of the Roses* (also 1963), *Pinter's *The Homecoming* (1965; NY, 1967) and *Old Times* (1971), and *Albee's *A Delicate Balance* (1969). He moved to the *National Theatre as Head of Design, 1973–85, and was responsible for the décor of *Hamlet* (1975), *Marlowe's *Tamburlaine the Great* (1976), Peter *Shaffer's *Amadeus* (1979) and *Yonadab* (1985), and *Coriolanus* (1984). His freelance work included the York Mystery Cycle (1963) and Ibsen's *Hedda Gabler* and *A Doll's House* in New York (1971).

Bury St Edmunds, Suffolk, small market town possessing the third oldest working theatre in England, the two older ones being the Theatre Royal, *Bristol, and the *Georgian Theatre, Richmond, Yorkshire. The first theatre in the town opened in the 1730s and was remodelled in the 1770s. By the end of the century it formed part of the Norwich *circuit, though it was licensed for only nine weeks in the year. In 1819 it was replaced by a new theatre in Westgate Street seating 780, which closed in 1843. Two years later it opened again, taking its present name of Theatre Royal though it does not appear to have had a royal patent. By the 1870s it was in constant use throughout the year but in 1925 it closed and was used as a barrel store. In 1961 a Trust was formed to restore the theatre as close as possible to its original form, and it reopened in 1965. It now seats 352, and no seat is more than 39 ft. from the stage in the horseshoe-shaped auditorium. It is a touring theatre presenting professional and amateur drama, dance, ballet, opera, pantomime, poetry, children's entertainment, and all kinds of music. It is the only working theatre in the country owned by the National Trust, which purchased it in 1975.

Bush Theatre, Shepherd's Bush, London, see FRINGE THEATRE.

Busker, itinerant open-air performer, usually in city streets. The term, derived from *buskin, included conjurors, jugglers, acrobats, *Punch and Judy operators, peep-show proprietors, and a wide variety of street musicians, from German bands to organgrinders and peripatetic one-man bands. Some buskers worked from a set pitch in a street or market-place, while others specialized in a much briefer turn for queues outside theatres, becoming an integral part of urban life throughout the Victorian and Edwardian periods. Many buskers led a hand-to-mouth existence, particularly during the winter months, and they were often virtually beggars. Many of them were Italian and the traditional buskers' language contained a curious mixture of Italian with 19th-century Cockney slang. After the 1930s the busker became a rarity, driven out by traffic congestion, increasing competition from portable radios, and the disappearance of theatre pits and galleries for which queues had to be formed. Recently a number of young amateur buskers have appeared in the streets, parks, and shopping precincts of big cities, and on Underground stations.

Buskin, long boot or *cothurnus worn by Greek actors in tragedy, the equivalent soft heelless shoe of the comic actor, the *soccus*, being known as the sock. In Elizabethan times the two words stood for tragedy and comedy respectively, but sock fell out of use and 'busking' became a general term for any kind of itinerant entertaining.

Butt, Sir Alfred, see REVUE and VICTORIA PALACE.

Butti, Enrico Annibale (1868–1912), Italian dramatist, a constructive pessimist in the Ibsenite sense, who, while recognizing the integrity of the progressives and sharing their views, none the less depicts the final triumph of conservative attitudes. Among his plays the most important are *Il vortice* (*The Vortex*, 1892), produced in the year of the first Italian version of *Ibsen's *Ghosts; L'Utopia* (1894), a consideration of mercy-killing; *La fine d'un ideale* (1898), in which he juxtaposes free love and the security of marriage; *Lucifero*, on agnosticism, and *La corsa al piacere* (*The Pursuit of Pleasure*), both 1900; and *La tempesta* (1901), a study of social reform through revolution. Butti's characters reach the conclusion that acquiescence in society's rules is better than struggling unavailingly against them.

Byam Shaw, Glen(cairn) **Alexander** (1904–86), English actor and director. He made his first appearance on the stage in 1923, and two years later was seen in London as Yasha in *Chekhov's *The Cherry Orchard*. He gained further experience in *Fagan's company at the *Oxford Playhouse. In 1937–8 he was a member of *Gielgud's repertory company at the *Queen's Theatre, and in 1939 played Horatio to Gielgud's Hamlet in Elsinore and at the London *Lyceum. After serving in the army during the Second World War he returned to the theatre as a director, having already had some experience before 1939. He directed *Rattigan's *The Winslow Boy* (1946; NY, 1947) and, for the first *Young Vic, *As You Like It* (1949) and *Henry V* (1951). In 1952 he was appointed co-director with Anthony *Quayle of the *Shakespeare Memorial Theatre company, becoming sole director, 1956–9, his productions including *Antony and Cleopatra* (1953) and *Hamlet* (1958), both with Michael *Redgrave, *Macbeth* (1955) with *Olivier, and *King Lear* (1959) with Charles *Laughton. He later directed several plays in London, among them Rattigan's *Ross* (1960; NY, 1961) and *Strindberg's *The Dance of Death* (1967, for the *National Theatre). From 1962 he was also involved in opera production. He was married to Angela *Baddeley.

Byron, George Gordon, Lord (1788–1824), English poet, author of several plays in verse which were staged with little success. Only one was produced during his lifetime, *Marino Faliero* (1821), seen at *Drury Lane. *Werner* (1830) was first produced by *Macready, also at Drury Lane; *Sardanapalus* (1834) and *The Two Foscari* (1838) likewise owed their first production to Macready. *Manfred*, a dramatic poem, was produced in 1834. *Cain*, written in 1821, does not appear to have been produced in England, but has been several times translated into German. Byron's letters are full of references to theatrical matters, but his plays read better than they act.

Byron, Henry James (1834–84), English actor and playwright, best known in his own day for a long series of *burlesques and *extravaganzas, usually staged at the smaller London theatres, which began in 1857 with *Richard of the Lion Heart* and ended in 1881 with *Pluto; or, Little Orpheus and his Lute*. Two of his more serious comedies were *War to the Knife* (1865) and *A Hundred Thousand Pounds* (1866). His most successful play and the only one now remembered was *Our Boys*. This was produced at the *Vaudeville Theatre in Jan. 1875 and ran till Aug. 1879, setting up a long-lasting record of 1,362 performances. He wrote about 150 works, based on themes from mythology, nursery tales, opera, legend, and topical events; his style, though ingenious, was overloaded with puns and in his more serious plays he tended to rely on stock types. He reflected the prevailing taste of the day, and that, and his own charming personality, doubtless accounted for much of his ephemeral success. As an actor he appeared mainly in his own plays, making his first appearance in London in 1869 in one of his early comedies and his last (exceptionally) in W. S. *Gilbert's *Engaged* in 1881.

C

Cabaret, in its modern sense, entertainment provided for an audience seated at tables at night-clubs, hotel-restaurants, and official dinners, and occupying only part of the evening. Strictly, however, it began in 1881 with the establishment of Le Chat Noir in Paris, where artists, poets, and musicians presented their work. Its many imitators toured widely and were influential on the Continent, particularly in Germany and Russia, where in 1908 *Balieff produced The Bat, later to become famous in Paris as Le Chauve-Souris. In Germany cabaret developed a political content after the First World War, especially in Berlin, reaching a peak in the 1930s and being banned by the Nazis in 1935. No post-war cabaret has achieved comparable political impact, though Berlin still has political cabaret and in other large cities *Fringe cabaret provides a platform for alternative comedians and musicians.

Café La Mama, see LA MAMA EXPERIMENTAL THEATRE CLUB.

Caffè Cino, New York, in Cornelia Street, theatre and coffee-house whose opening by Joe Cino in 1958 is often regarded as marking the beginning of the off-off-Broadway movement (see OFF-BROADWAY). Although the production of plays became a regular part of the café's arts programme only in 1961 and was curtailed after Cino's death in 1967, the venture had in that time become a vital part of the city's theatrical scene and had already presented, on a stage only 8 ft. square, a considerable body of new works by American dramatists including John *Guare's The Loveliest Afternoon of the Year and A Day for Surprises in 1966 and several of Lanford *Wilson's early plays.

Cahill, Marie, see VAUDEVILLE, AMERICAN.

Calderón de la Barca, Pedro (1600–81), Spanish dramatist, successor to Lope de *Vega. He wrote about 200 plays, the best of them dating from between 1625 and 1640. After being ordained priest in 1651 he wrote mainly *autos sacramentales*, which presented the abstract ideas of Catholic theology with intense lyricism and excellent stagecraft, and spectacle-plays for the Court theatre at Buen Retiro, with which he had been associated since its inception under Philip IV. The best known of his early religious plays are La cena del rey Baltasar (*Belshazzar's Feast, c.*1634), translated into English in 1969, and El gran teatro del mundo (*The Great World Theatre, c.*1641), on which Hugo von *Hofmannsthal based his Grosse Weltheater, first seen at the Salzburg festival in 1922 in a production by *Reinhardt. Of the secular plays the finest and best known is La vida es sueño (*c.*1638), a story of human regeneration which, as Life's a Dream, was produced in London in 1922. El alcalde de Zalamea (*The Mayor of Zalamea, c.*1640) is a study of prudent virtue and of the vices which spring from uncontrolled passions. As in all Calderón's plays, the ideas are transmitted not only through the surface movement but also through the images which parallel or subtly contradict it. The play was produced at the *National Theatre in 1981.

Calderón's plays are highly organized; he is always firmly in control of his medium in that the subplot becomes an aspect of the main plot and no character is superfluous, no action irrelevant, no detail extraneous. Among his other outstanding works are the *Faust-like El mágico prodigioso (*The Wonder-Working Magician*, 1637), based on the life of St Cyprian, part of which was translated by *Shelley; and the series based on the 'point of honour' so important in Spanish life: El médico de su honra (1635), translated as The Surgeon of His Honour in 1960, El pintor de su deshonra (*The Painter of His Dishonour*, 1637), and A secreto agravio, secreta venganza (1635), translated in 1961 as Secret Vengeance for Secret Insult.

Calderón had a considerable influence on European drama, many of his plays becoming known to English Restoration dramatists through French translations, and resulting in Tuke's Adventures of Five Hours (1663), based on Los empeños de seis horas (at the suggestion of Charles II), and Digby's Elvira; or, The Worst not Always True (*c.*1663), based on No siempre la peor es cierto. It is also probable that *Killigrew found some of The Parson's Wedding (1664) in La dama duende, a cloak-and-sword play written for production at Court in *c.*1629, and that *Wycherley took some of The Gentleman Dancing-Master (1672) from El maestro de danzar.

Caldwell, James H. (1793–1863), British-born American theatre manager, who developed and dominated the pioneering theatre of the Mississippi Valley. In 1816 he was engaged to play light comedy roles at the theatre in Charleston, and a year later was managing theatres in North Carolina and Virginia. In 1820 he was in New Orleans, where he opened the American Theatre on Camp Street in 1824. He established a number of touring companies and built theatres in such cities as Mobile, Nashville, and Cincinnati, finally extending his operations throughout the entire lower Mississippi Valley into parts of Ohio. The high point of his career was the construction of the St Charles Theatre in New Orleans, the first large and important theatre in the South. Caldwell's permanent company was a fine one, and every year attracted a number of visiting stars, but the financial panic of 1837 and his rivalry with Noah M. *Ludlow and Sol *Smith ruined him, and the destruction by fire of the St Charles Theatre in 1842 virtually put an end to his career.

Caldwell, Zoë [Ada] (1934–), Australian-born actress, who showed unusual maturity early in her career. She was an original member of the Union Theatre Repertory Company, Melbourne, making her professional début with them in 1953. She first appeared in England in 1958 at the *Shakespeare Memorial Theatre, returning in the following year to give outstanding performances as Helena in *All's Well That Ends Well* and Cordelia in *King Lear*. In 1961 she played Isabella in *Middleton and *Rowley's *The Changeling* at the *Royal Court Theatre, being seen in the same year as Rosaline in *Love's Labour's Lost* at the *Stratford (Ontario) Festival. After a visit to Australia in 1962, where she played *Shaw's St Joan, she was seen in the opening season at the *Guthrie Theater, Minneapolis, in 1963, and returned in 1965 when her roles included Millamant in *Congreve's *The Way of the World*. In the same year she made her New York début as the Prioress in John *Whiting's *The Devils*. Later roles in New York included a highly praised Jean Brodie in *The Prime of Miss Jean Brodie* (1968), based on Muriel Spark's novel, Alice in *Strindberg's *The Dance of Death* (1974), and Mary Tyrone in *O'Neill's *Long Day's Journey into Night* (1976). She returned to Stratford, Ontario, in 1967 to play Cleopatra (in *Antony and Cleopatra*) and other roles, and to London in 1970 to play Emma Hamilton in *Rattigan's *A Bequest to the Nation*. She also played Medea in *Euripides' play (NY, 1982) and Lillian *Hellman in William Luce's one-woman play *Lillian* (NY, 1986). One of the most eminent actresses to emerge from Australia, she brings a contemporary quality and original characterization to classic plays without any sacrifice in style.

Call Board, see STAGE DOOR.

Call-Boy, functionary in the English theatre, usually, though not necessarily, a young boy, whose duty it was to summon the players from their dressing-rooms in time for them to appear on stage on cue. He took the place of the Elizabethan *book-holder, who was also the *prompter. The word was already current at the end of the 18th century, since the Shakespearian Scholar Malone erroneously said that Shakespeare's first job in the theatre was that of 'call-boy'. It became obsolete in the mid-20th century, when the call-boy was replaced by a tannoy or loudspeaker placed in each dressing-room.

Call Doors, see PROSCENIUM DOORS.

Calmo, Andrea (1509–71), Italian playwright and amateur actor, a gondolier by profession, who seems to have had a considerable influence on the development of the *commedia dell'arte. A contemporary and rival of Angelo *Beolco, he specialized in playing old men, somewhat in the style of *Pantalone, though the name was not yet in use. His plays were edited by Rossi in 1888.

Cambridge, site of the second oldest university in England, has a long theatrical history. The first mention of *liturgical drama dates from 1350 in connection with the staging of *The Children of Israel*. By the 16th century the acting of plays in Latin was firmly established as part of the students' curriculum, as is evident from the production of Kirchmayer's *Pammachius* at Christ's College in 1545. Elizabeth I, on her only visit to Cambridge in 1564, was entertained by a performance of *Plautus' *Aulularia* in the chapel of King's College, which was also the setting for the first performance in England of a Latin translation of Guarini's famous *pastoral *Il pastor fido*. In 1613 Trinity men performed Bonarelli's *Filli di Sciro* in Latin. The visit of James I in 1615 was made memorable by the performance, among other plays, of Ruggle's *Ignoramus*.

All this activity was of course amateur and in Latin, but there appear to have been occasional visits by professional companies. The *Chamberlain's (or *King's) Men are believed to have played *Hamlet* in Cambridge in 1603, but non-academic acting was not encouraged, and as the Commonwealth approached even college productions grew fewer in number. The last recorded play before the closing of the theatres in 1642 was *Cowley's *The Guardian*. In the 18th century regular visits from the professional company on the Norwich *circuit were well received, though no acting was allowed within the university precincts. There was, however, a commodious wooden theatre at Stourbridge, and another at Barnwell which was eventually taken over by Terence *Gray to become the Festival Theatre. The founding in 1855 of the Amateur Dramatic Club (ADC) made the theatre again an integral part of undergraduate life. It is still active, as is the Footlights Club, founded in 1883 and well known for its annual revue.

In 1907 a group of undergraduates put on a production of *Marlowe's *Dr Faustus* with sufficient success to warrant the founding of the Marlowe Society a year later, with Rupert Brooke, who had played Mephistopheles, as its first president. An outstanding feature of its work has been its fine speaking of verse, and under the guidance of **George Rylands**

(1902–   ), who directed many of its productions as well as those of the ADC, it has recorded for the *British Council all the plays of Shakespeare. Women's parts were played by men until the production of *Antony and Cleopatra* in 1934, and a tradition of anonymity was maintained until the mid-1960s. Like the ADC, the Marlowe Society has given a number of fine actors to the professional stage, and a group of outstanding directors including Peter *Hall and John *Barton. In 1928 the Cambridge Mummers were founded with both men and women members.

The Theatre Royal was adapted in 1882 from the old St Andrew's Hall and demolished in 1896 to make way for the New Theatre, which became a cinema in the early 1930s. In 1936 the economist John Maynard Keynes built the Arts Theatre at his own expense. Under his direction and that of his wife, the Russian ballerina Lydia Lopokova, it did much to stimulate theatrical life in Cambridge. After Keynes's death in 1946 it went through a difficult period, but the skill and devotion of George Rylands, his successor, finally saved it. Run by a Trust, it houses not only the productions of the Footlights Club and the Marlowe Society but also visiting professional companies, including the touring Cambridge Theatre Company, financed by the *Arts Council and the City of Cambridge. An important theatrical event in Cambridge is the triennial production of a Greek play in the original which began in 1882 with *Sophocles' *Ajax*. After 1950 productions were directed by George Rylands. Since the Second World War many colleges have built their own small theatres, and the number of university productions has increased enormously. A Chair of Drama was established in 1974, though there is as yet no department of drama.

Cambridge Theatre, London, in Earlham Street, at Seven Dials. This theatre, seating 1,275 on three tiers, opened in 1930 with *Charlot's Masquerade*, starring Beatrice *Lillie. In 1934 the *Comédie-Française gave a season, and in 1943 there was a revival of *Shaw's *Heartbreak House*. Strauss's operetta *A Night in Venice* began a long run in 1944, while the successes of 1949 and 1950 were the revues *Sauce Tartare* and *Sauce Piquante*. Two long-running comedies were Verneuil's *Affairs of State* (1952) and William Douglas *Home's *The Reluctant Débutante* (1955), *Billy Liar* (1960) by Keith Waterhouse and Willis *Hall, with Albert *Finney, and the musical *Half a Sixpence* (1963) were popular, and in 1970 the *National Theatre company appeared. Brian *Rix took over the theatre for Michael Pertwee's *A Bit between the Teeth* (1974), and the musical *Chicago* transferred from the *Crucible Theatre, Sheffield, for a successful run (1979). After a brief attempt in 1984 to turn it into a home for illusion (as the Magic Castle), the theatre was closed until 1987, but scored a big success in 1989 with the rock-and-roll musical *Return to the Forbidden Planet*.

Cameri (Chamber) **Theatre,** company founded in Palestine in 1944 by a group of young actors who were in revolt against the conservatism of the two existing theatre groups, *Habimah and *Ohel, which still adhered to the earlier larger-than-life Russian style of acting, with exclusively realistic settings. The new company's first production was *Goldoni's *The Servant of Two Masters* (1945). Audiences were delighted by the light colourful style of acting and the impressionistic settings, and in 1948 the Cameri, which had done good work in presenting a number of contemporary European plays in Hebrew translations, presented Moshe Shamir's *He Walked through the Fields*, the first play to be produced in the independent state of Israel and the first in Hebrew to present native-born Israelis in a realistic manner. Cameri continued to present new Jewish plays and good plays from outside Israel. Two of its greatest successes were *Brecht's *The Good Woman of Setzuan*, presented at the *Théâtre des Nations in Paris in 1956, and Nathan Alterman's dramatization of Sammy Gronemann's *The King and the Cobbler*, seen in London during the *World Theatre Season of 1967. In 1961 the company moved into its own premises in the commercial and entertainment centre of Tel Aviv, and in 1970 it became the city's official municipal theatre.

Cameron, Beatrice, see MANSFIELD.

Caminelli, Antonio, see SACRA RAPPRESENTAZIONE.

Camp, see FEMALE IMPERSONATION.

Campbell, Bartley [Thomas] (1843–88), American dramatist, one of the first to make playwriting his only profession, giving up journalism when his first play, the melodrama *Through Fire*, was produced in 1871. He was associated for some years with a theatre in Chicago where several of his own plays were produced, among them *Fate* (1872) and *Risks* (1873). His best work, *My Partner* (1879), was a drama in which a man is wrongly accused of killing his rival in love. Among his other works were the emotional melodramas *The Galley Slave* (also 1879) and *The White Slave* (1882), the latter, though compared unfavourably to *Boucicault's *The Octoroon*, becoming his greatest success. Several of his other plays, however, which he directed and financed himself, lost the money which his popular plays had earned, and he died insane, of overwork and worry.

Campbell, Herbert, see LENO, DAN.

Campbell, Ken, see EVERYMAN THEATRE, Liverpool.

Campbell, Mrs Patrick [*née* Beatrice Stella Tanner] (1865–1940), English actress, who through her devastating wit and marked eccentricities became a legend in her lifetime. She made her first appearance in London in 1890 in Sheridan *Knowles's *The Hunchback*, and was already well known when she created the parts of Paula in *The Second Mrs Tanqueray* (1893) and Agnes in *The Notorious Mrs Ebbsmith* (1895), both by *Pinero. In 1895 she was also seen in the title-role of *Sardou's *Fédora*, and after playing Juliet to *Forbes-Robertson's Romeo in the same year created the title-role in *Magda* (1896; NY, 1902), the English version of *Sudermann's *Heimat* (1893). In 1898 she appeared as Mélisande in the English version of *Maeterlinck's *Pelléas et Mélisande*, repeating the role in French in 1904 to the Pelléas of Sarah *Bernhardt; in 1907 she was *Ibsen's Hedda Gabler. George Bernard *Shaw, who called her 'perilously bewitching', wrote for her the part of Eliza Doolittle in *Pygmalion* (1914) and exchanged letters with her over a long period. (Their correspondence was published after his death, and formed the basis of Jerome Kilty's dramatic dialogue *Dear Liar*, 1959.) Much of Mrs Campbell's later career was spent in revivals of her earlier successes. In spite of long and frequent absences from the stage, and a temperament which rendered her somewhat difficult to deal with, she was one of the outstanding stage personalities of her generation.

Camus, Albert (1913–60), French dramatist and novelist, who from 1936 to 1939 gained experience of the theatre with a left-wing group in Algiers, the Théâtre du Travail (later de l'Équipe), for which he directed and acted in a wide variety of plays. While editing a clandestine newspaper during the German occupation of France he wrote two important plays, *Le Malentendu* (1944) and *Caligula* (1945), the young Gérard *Philipe appearing with great success in the second. Both plays are concerned with Camus's own concept of absurdity as defined in his essay *Le Mythe de Sisyphe* (1942), a definition later applied to the Theatre of the *Absurd. The theme is developed further in *L'État de siège* (1947), based on his novel *La Peste* in collaboration with Jean-Louis *Barrault, in which plague is symbolic of the evil that thwarts mankind in its quest for freedom. Among Camus's later plays were *Les Justes* (1949) and adaptations of *Calderón, Dostoevsky, and William Faulkner which show him working towards a wider and freer theatrical technique. Three years before his tragically 'absurd' death in a car crash he was awarded the *Nobel Prize for Literature.

Canadian Actors' Equity Association, see EQUITY.

Canali, Isabella, see ANDREINI.

Cancan, see MOULIN ROUGE.

Candler Theatre, New York, see SAM H. HARRIS THEATRE.

Canterbury Music-Hall, London, in the Westminster Bridge Road, Lambeth. The first of the great *music-halls, it was erected by Charles *Morton in 1852 on the site of an old skittlealley adjacent to his Canterbury Tavern, and paid for out of the profits made on drink during free entertainment formerly offered there. The Canterbury Hall, as it was originally called, proved so popular that in 1854 Morton was able to replace it by the larger New Canterbury Music-Hall, which had a large platform stage and accommodation for 1,500. By 1867, when Morton left, the earlier programmes of light music and ballad singing had been dropped and comedy predominated. In 1876 the building was reconstructed as a three-tier theatre, its bar being for many years the favourite rendezvous of music-hall performers. The Canterbury was visited regularly by the Prince of Wales (later Edward VII), the Duke of Cambridge, and the Duke and Duchess of Teck. When the popularity of the 'halls' began to decline, drastic reductions in the price of seats brought audiences back for a while; but the heyday of the music-hall was over long before the Canterbury was destroyed by bombing in 1942.

Cantor, Eddie, see REVUE and VAUDEVILLE, AMERICAN.

Capa y Espada, Comedia de, see COMEDIA.

Čapek, Karel (1890–1938), Czech playwright, whose first play *The Brigand* (1920) was produced at the National Theatre in Prague. With his next play, *RUR* (1921)—the initials stand for Rossum's Universal Robots—Čapek became widely known outside Czechoslovakia and was responsible for adding the word 'robot' to the English language. *The Insect Play* (1922), like *RUR*, was much influenced by German *Expressionism and employed the techniques of *revue; both plays depict the horrors of a regimented technological world and the terrible end of the populace if they fail to rise against their oppressors. *The Insect Play*, written by Karel in collaboration with his brother **Josef** (1887–1945), an artist who died of typhus in Belsen, was immediately seen in London with the subtitle *And So Ad Infinitum* and in New York as *The World We Live In*. *The Makropulos Affair* (1923), on which Janáček based his 1926 opera, argues the case for longevity and decides against it. A sequel to *RUR*, *Adam the Creator* (1927), again written jointly with Josef, shows man endeavouring to rebuild the world destroyed by the robots; and two anti-Fascist plays,

The White Scourge (1937), seen in London as *Power and Glory* in 1938, and *The Mother* (1938; London, 1939), deal with the rise of dictatorship and the devastating effects of war. All Čapek's plays reflect the world he lived in and comment on its grosser follies. He also wrote an amusing short monograph, *How a Play is Produced*.

Capitano, Il, braggart soldier of the *commedia dell'arte*, vainglorious and cowardly, usually a Spaniard. One of the first to play him was *Andreini, as Capitano Spavento. In the 17th century Silvio *Fiorillo played the part as Capitano Matamoros, 'death to the Moors', anglicized as Captain Matamore Il Capitano may derive from the Miles Gloriosus of *Plautus, and perhaps contributed something to Shakespeare's Armado in *Love's Labour's Lost* and Pistol in *Henry IV, Part 2* and *The Merry Wives of Windsor*.

Capon, William (1757–1827), English architect and painter, who in 1791 was appointed scenic director of the new *Drury Lane Theatre by John Philip *Kemble. Of a plodding, pedestrian temperament, he was a painstaking antiquarian, which accorded well with Kemble's plans for scenic reform. He continued to work at Drury Lane until the theatre was burnt down in 1809, involving the loss of a large sum of money due to him for his scenery. He again worked for Kemble when he took over *Covent Garden Theatre, where many of Capon's elaborately detailed sets remained in use for many years.

Cardiff, capital of Wales, a country with a strong musical tradition but no long-standing theatrical one. The city's oldest theatre is the New Theatre, built in 1906 and now a *civic theatre seating 1,168 and housing touring productions. The Sherman Theatre, built in 1973 by University College with money provided by the Sherman Foundation, has two auditoriums. The larger, seating 472, is a proscenium theatre that plays host to a wide range of professional drama companies, Welsh-language theatre being well represented. It is used by ballet and opera companies and for concerts, films, lectures, recitals, and other activities. The theatre is also the home of the Sherman Theatre Company, a leading professional company in Wales. Attached is the smaller Arena Theatre, a flexible space with an octagonal stage and seating for a maximum of 200. It is used by smaller, more experimental professional companies and by amateurs. There is also the Chapter Arts Centre, a mixed media centre with a theatre which provides a home for experimental groups.

Carey, Joyce, see BRAITHWAITE.

Carl, Karl, see NESTROY.

Carlin, see BERTINAZZI.

Carney, Kate, see MUSIC-HALL.

Carnival Play, see FASTNACHTSSPIEL.

Carnovsky, Morris (1897–   ), highly respected American actor, mainly in supporting roles, who made his New York début in Sholom

*Asch's *God of Vengeance* (1922). He was with the *Theatre Guild, 1924–30, his roles including Alyosha in an adaptation of Dostoevsky's *The Brothers Karamazov* (1927), and *Chekhov's Uncle Vanya (1929). During the 1930s he was with the *Group Theatre and appeared in a number of modern plays, including Paul *Green's *The House of Connelly* (1931) and Clifford *Odets's *Awake and Sing* (1935), *Golden Boy* (1937; London, 1938), and *Rocket to the Moon* (1938). He was a member of the Actors' Laboratory Theatre in Hollywood, 1940–50, directing a number of productions there. After 1956 he appeared regularly at the *American Shakespeare Theatre, crowning a distinguished career by playing King Lear there in 1963, 1965, and 1975.

Carpenter's Scene, insertion into a *pantomime or spectacular *musical comedy, played in front of a *backcloth while elaborate scenery is set up behind out of sight of the audience. In Victorian times it was used also in serious plays, but was no longer needed when the practice of dropping the front curtain between scene changes became general.

Carpet Cut, long narrow stage-opening behind the front curtain, stretching nearly the whole width of the proscenium opening. It is closed by hinged flaps which trap the edge of a carpet or stagecloth, so that the actor cannot be tripped up by it.

Carretto, Galleotto del, see SACRA RAPPRESENTAZIONE.

Carriage-and-Frame (or **Chariot-and-Pole**), device for changing the *wings, devised by *Torelli in 1641 for the Teatro Novissimo in Venice; it was used extensively on the Continent, and can still be seen in operation at *Drottningholm. Each wing-piece was suspended just clear of the stage on a rectangular frame (or on a pole) which projected downward through a long slit in the floor, and was borne on a wheeled carriage running on rails in the *cellar. At each wing position this arrangement existed in duplicate, the two carriages being connected by ropes to a drum serving all the wing-sets in such a way that as one carriage of each pair moved off stage, its neighbour moved on. The withdrawn wing was then replaced by the wing needed for the next scene. In England a somewhat similar system was used to change the *book wings.

Carroll, Earl (1893–1948), American theatre manager and producer, originally a lyric writer, who settled in New York in 1912 and 10 years later built the first *Earl Carroll Theatre there. From 1923 until 1940 he produced nine editions of *Earl Carroll's Vanities*, modelled on the Follies of *Ziegfeld, all but the last two in his own theatres. They featured comedians alternating with chorus girls. He also presented two *Sketch*

Book revues and about 60 plays, of which the best known was Leon Gordon's *White Cargo* (1923). In 1931 he opened a second Earl Carroll Theatre on the same site. In 1938 he moved to Hollywood, where he opened a third theatre and was active in management until his death in an air accident.

Carroll, Paul Vincent (1900–68), Irish dramatist, who attacked the Catholic Church through his plays, and became as well known in England and the United States as in Ireland. He came into prominence with his second play, *Things that are Caesar's* (1932; London and NY, 1933), produced at the *Abbey Theatre, Dublin, as were *The Wise have not Spoken* (1933; London, 1946; NY, 1954) and *Shadow and Substance* (1934; London, 1943). The last, about the conflict between a canon and a liberal-minded schoolmaster, is perhaps his best play, and had a long run on Broadway (1937–8), as did *The White Steed* (NY, 1939; London, 1947). In 1942 a war play set in Glasgow, *The Strings, My Lord, are False*, was seen in Dublin at the *Olympia Theatre, where it was well received, and in New York. Among Carroll's later plays were *The Old Foolishness* (London, 1943), *The Devil Came from Dublin* (1952; London, 1953), and *The Wayward Saint* (NY, 1955). *We Have Ceased to Live* was published posthumously in 1972. Carroll was a schoolmaster in *Glasgow, 1921–37; he helped to found the Curtain Theatre there and later, with James *Bridie, the *Citizens' Theatre, for which he wrote *Green Cars Go East* (1940), set in the Glasgow slums. All Carroll's plays are marked by sympathy, humour, and subtle characterization; his studies of Irish priests are especially memorable.

Carroll, Sydney W. [George Frederick Carl Whiteman] (1877–1958), actor, dramatic critic, and theatre manager. Born in Melbourne, he settled in London and was drama critic of the *Sunday Times*, 1918–23, and was mainly responsible for the *Daily Telegraph*'s weekly theatre page, 1928–39. He also founded the *Open Air Theatre. Carroll was successively manager of a number of London theatres, among them the *Ambassadors, the *Criterion, His (now *Her) Majesty's, the New (now the *Albery), and the *Shaftesbury. He was part-author of several plays, published a volume of dramatic criticism, and wrote a book on acting.

Carter, Mrs Leslie [*née* Caroline Louise Dudley] (1862–1937), American actress, who in 1889, when her marriage broke up, asked *Belasco to launch her on an acting career. In spite of some initial opposition to her as a divorcée, she became one of his star players, being outstanding in Belasco's *The Heart of Maryland* (1895; London, 1898) and *Zaza* (1899; London, 1900). She then played the title-roles in *Du Barry* (1901) by Belasco and *Adrea* (1905) by Belasco

and John Luther Long. On her second marriage in 1906 she severed her connection with Belasco, touring under her own and other managements with some success until she retired in 1917. She returned to the stage in 1921 in *Maugham's *The Circle*.

Carton, Richard, see COMPTON, EDWARD.

Cartoucherie de Vincennes, Paris, see SOLEIL, THÉÂTRE DU.

Casarès, Maria (1922–), Spanish-born French actress, permanently resident in Paris since being exiled from her own country during the Spanish Civil War. After success in a French version of *Synge's *Deirdre of the Sorrows* (1942), she was associated with several existentialist plays, notably *Camus's *Le Malentendu* (1944), *L'État de siège* (1948), and *Les Justes* (1949), and *Sartre's *Le Diable et le bon Dieu* (1951), as well as playing Jeannette in *Anouilh's *Roméo et Jeannette* (1946). She was at the *Comédie-Française, 1952–4, making her début there in *Pirandello's *Six Characters in Search of an Author* and being outstanding as Phèdre in *Racine's play. She also played many leading roles in *Vilar's early productions at *Avignon, being particularly admired as Lady Macbeth (1954), and at the *Théâtre National Populaire, which she joined in 1955. Later she was notable as the mother in *Genet's *Les Paravents* (1966) at Barrault's Théâtre de France, a role she repeated in 1983. Other roles included *Brecht's Mother Courage (1968), Queen Victoria in *Bond's *Early Morning* (1970), and Shakespeare's Cleopatra (1975).

Case, Charlie, see VAUDEVILLE, AMERICAN.

Casino Theatre, New York, on the south-east corner of Broadway and 39th Street, for almost 50 years the leading musical comedy house of New York. Built in a massively Moorish style, it held 1,300, and had a roof garden for summer evening concerts. It opened in 1883, and among the musical plays produced there were *Florodora* (1900), *A Chinese Honeymoon* (1902), *Wildflower* (1923), and *The Vagabond King* (1925). Its last hit was *The Desert Song* (1926). It was pulled down in 1930.

The *Earl Carroll Theatre was renamed the Casino from 1932 to 1934.

Casson, Sir Lewis Thomas (1875–1969), English actor and director. He made his first professional appearance at the *Royal Court Theatre under the Vedrenne and *Granville-Barker management, playing mainly in Shakespeare and *Shaw. In 1908 he became a member of the repertory company at the Gaiety Theatre, Manchester, which he directed, 1911–14; there he met and married Sybil *Thorndike, with whom his career was later almost entirely associated. After the First World War he devoted much of his time to direction, though continuing to act occasionally, notably as Stogumber in Shaw's *Saint Joan* (1924), Buckingham in *Henry*

VIII at the *Old Vic (1926), professor Linden in *Priestley's *The Linden Tree* (1947), and Sir Horace Darke in Clemence *Dane's *Eighty in the Shade* (1959). He made his last appearance at the age of 90 in a revival of Kesselring's *Arsenic and Old Lace* (1966). He was a careful and painstaking director, one of his greatest assets as an actor being his clear and beautifully modulated voice.

Castro y Bellvís, Guíllen de (1569–1631), Spanish dramatist, chiefly remembered for his dramatization of the ballads celebrating the exploits of Spain's national hero the Cid. *Las mocedades del Cid*, in two parts, first published in 1618, formed the basis of *Corneille's *Le Cid* (1637). Castro wrote a number of other plays, including many comedies in the style of his friend and contemporary Lope de *Vega, and made three dramatizations of parts of *Cervantes' *Don Quixote* of which the best concentrates on the episode of Cardenio and Lucinda.

Catwalk, see FLIES.

Cave, Joe [Joseph], see MUSIC-HALL.

Cawarden, Sir Thomas, see MASTER OF THE REVELS.

Céleste, Mme Céline (1814–82), French dancer and *pantomime player, who appeared with much success on the Parisian stage as a child. In 1827 she went with a troupe of dancers to New York, and in 1830 she was seen in London. She scored a triumph in Haines's *The French Spy* (1837). Relying almost entirely on dumb show, she did not attempt a speaking part in English until 1838. She later created a number of parts in new plays, including Madame Defarge in a dramatization of *Dickens's *A Tale of Two Cities* (1860), and was briefly manager of the *Adelphi Theatre in association with Ben *Webster. She made her last appearance in New York in 1865, and in London in 1874, reappearing for one performance in a revival of *Buckstone's *The Green Bushes* as Miami, one of her best parts, at *Drury Lane in 1878. From her photographs she appears to have been extremely plain, but contemporary accounts leave no doubt of the beauty and expressiveness of her dancing and miming.

Celestina, La, the most important literary work of 15th-century Spain, published anonymously but now known to be the work of **Fernando de Rojas** (*c.*1465–1541), a Spanish novelist. Its eponymous character is an old bawd around whom are enacted amusing and licentious scenes of low life. Although written in dialogue form, it was probably intended to be read aloud rather than acted but a number of plays have been based on it, the earliest English version being John *Rastell's *Calisto and Meliboea* (*c.*1530). It was the source of Ashley *Dukes's *The Matchmaker's Arms* (1930) and of

Celestina produced by Joan *Littlewood in 1958. Dramatized versions in French and Italian have also proved successful, *Vitez directing it in *Avignon and Paris in 1989.

Cellar, space under the stage which housed the machinery necessary for *traps, scene-changing, and special effects. In British theatres the *grooves, which allowed tall framed pieces of scenery to be slid to the sides of the stage for a scene-change, meant that the cellar could be shallow. It consisted usually of a room below the stage floor, where most of the machines were housed, with a well in the central part into which the traps and *bridges descended, at the bottom of which were the *drum-and-shaft systems which worked them. In Continental theatres the cellar (or *dessous*) often descended four or five storeys.

CEMA (Council for the Encouragement of Music and the Arts), see ARTS COUNCIL.

Censorship, Dramatic. Specific statutory provision for theatrical censorship is now rare in democratic countries, one of the last to abolish the practice being Great Britain in 1968 (see LORD CHAMBERLAIN). In general, self-censorship is the rule, current levels of public tolerance being shrewdly, if unconsciously, assessed by writers, producers, and directors. Some governments retain the right to intervene in the interest of good order or public morality, as did the French government in the case of *Genet's *Les Paravents* (*The Screens*) in 1966, and most federal countries vest powers in regional or state authorities: the nude revue *Oh! Calcutta!* was banned in South Australia and Victoria, and plays by *Cocteau, Graham *Greene, and others have been banned in the Swiss canton of Valais. Throughout the world prosecutions may be laid *post hoc* under laws such as those relating to obscenity, blasphemy, and libel: examples include Tennessee *Williams's *The Rose Tattoo* in Eire in 1957; Mart Crowley's *The Boys in the Band* in Victoria, Australia, in 1969; the musical *Hair* in New Zealand in 1972. It has been rare for such prosecutions to succeed and, as the gulf widens between artistic attitudes to traditional morality and those of sections of the public, more arcane applications of common law to theatrical performances have been attempted. In 1981 a private prosecution brought against Michael *Bogdanov, who had directed Howard *Brenton's *The Romans in Britain*, was based on Section 13 of the Sexual Offences Act of 1956, the grounds for complaint being the stage enactment of an attempted homosexual rape; the case was withdrawn before any point of law could be established.

In the United States no separate legislation on the theatre has ever existed; information can be lodged with the police after the production of a play deemed immoral or obscene, and the actors arrested on stage and taken into custody to be tried under the federal or state laws dealing with literary works as a whole. In Communist countries and military autocracies pre-production censorship is usual, the severity of scrutiny varying with current orthodoxies. Political and social heresy is the censors' main concern, although attitudes to sexual matters are often more prudish than in libertarian societies. Here, too, an awareness of prevailing levels of acceptability will tend to ensure self-censorship.

Center Theatre Group, see MARK TAPER FORUM.

Centlivre, Mrs Susannah (1667–1723), English dramatist and actress, a masculine-looking woman who delighted in playing men's parts. She had already written several plays when she achieved her first success with *The Gamester* (1705), a sentimental drama based to some extent on *Regnard's *Le Joueur* (1696) but with the moral tone of Colley *Cibber and Richard *Steele. It was followed in the same year by a somewhat similar play, *The Bassett-Table*, which, like all her early works, was published under her second married name, Mrs Carroll. In 1706 she married as her third husband Joseph Centlivre, cook to Queen Anne, whose name she thereafter retained. Of her many comedies of intrigue, which in verve and ingenuity rivalled those of Aphra *Behn, the best were *The Busie Body* (1709), *The Wonder, a Woman Keeps a Secret* (1714), and *A Bold Stroke for a Wife* (1718). All three were frequently revived, the second, which owed much of its initial success to the acting of Anne *Oldfield as Violante, later providing an excellent vehicle for *Garrick in the part of Don Felix. The last, perhaps better classified as a comedy of manners, shows the influence of *Congreve.

Central School of Speech and Drama, see SCHOOLS OF DRAMA.

Central Theatre of the Soviet Army, see RED ARMY THEATRE.

Centre 42, see ROUND HOUSE and WESKER.

Centres Dramatiques, see DÉCENTRALISATION DRAMATIQUE.

Century Theatre, New York. 1. On Central Park West at 62nd Street, erected by Winthrop *Ames, in association with Lee *Shubert, to further his plan to establish a true *repertory theatre in New York. It opened in 1909 as the New Theatre, but Ames's venture was not a success and it closed, to reopen as the Century in 1911. It housed mainly musical shows, but in 1916 Shakespeare's Tercentenary was celebrated by a fine production of *The Tempest*, and in 1921 *Martin-Harvey appeared in *Hamlet*. In 1924 *Reinhardt's production of Karl Vollmöller's spectacle-play *The Miracle*, with striking

scenery by Norman *Bel Geddes, was a great success. Reinhardt returned in 1927 with his productions of *A Midsummer Night's Dream*, Hofmannsthal's version of *Everyman*, and *Büchner's *Dantons Tod*. The theatre closed in 1929 and was demolished a year later. On the roof was a small theatre known as the Cocoanut Grove or the Century Grove which staged intimate *revue and plays for children.

2. On 7th Avenue, between 58th and 59th Streets, large playhouse opened by the *Shuberts as the Jolson in 1921. In 1923 it housed an epoch-making visit by the company of the *Moscow Art Theatre under *Stanislavsky which had a lasting influence on the American stage. The following year saw a visit from *Gémier of the Paris *Odéon and the opening of Romberg's famous musical *The Student Prince*, which had a long run. In 1932 a company called the Shakespeare Theatre gave 15 of Shakespeare's plays at low prices to a mainly student audience. The theatre, which from 1934 to 1937 was known as the Venice, also housed for a time the *Federal Theatre Project. It was renamed the Century in 1944 and in 1946 saw a visit from the *Old Vic company headed by Laurence *Olivier and Ralph *Richardson. *Kiss Me, Kate*, Cole *Porter's musical, began a long run in 1948. The theatre was demolished in 1961.

Cervantes Saavedra, Miguel de (1547–1616), Spanish novelist and dramatist best remembered as the author of *Don Quixote de la Mancha*, a satirical romance published in two parts, in 1605 and 1615, which has been widely translated and dramatized in many languages. The first English version for the stage was performed at *Dorset Garden in 1694 with music by Purcell. In 1895 *Irving played Don Quixote at the *Lyceum in a version by W. G. Wills, and in 1969 *The Travails of Sancho Panza* by James *Saunders was the Christmas entertainment by the *National Theatre company. An American musical, *Man of La Mancha*, began a good run in New York in 1965 (London, 1968), and in France one of the most successful productions by *Baty was his own dramatization of an episode from the book, *Dulcinée* (1938). Cervantes himself said he wrote 30 plays, but apart from eight *comedias, of which the best, *Pedro de Urdemalas*, was translated in 1964 as *Pedro the Artful Dodger*, and eight *entremeses, all of which were translated in 1948 and again in 1964, the only full-length plays to have survived are *El cerco de Numancia* (*The Siege of Numancia*), a heroic tragedy which was revived by *Barrault in Paris in 1937, and *El trato de Argel* (*Business Affairs in Algiers*), based on the author's experiences there; these were probably produced between 1580 and 1590.

Český Krumlov, Czechoslovakia, castle about 100 miles from Prague which contains a small theatre built in 1766—the same year as that at

*Drottningholm. It has a small, horseshoe-shaped auditorium and a magnificent baroque proscenium. The theatre's most exciting asset is 10 sets of contemporary scenery, probably the oldest surviving in Europe, which are invaluable in showing how the sumptuous perspective scenes of such artists as the *Bibienas were translated into wood and canvas. One set is normally displayed on the stage, while others are set up in the adjacent riding-school. Performances are occasionally given, but the scenery is too fragile to be used often.

Chaikin, Joseph, see OPEN THEATER.

Chaillot, Palais de, Paris, Moresque building on the exhibition site opposite the Eiffel Tower, used by the first *Théâtre National Populaire from 1920 to 1934. It was demolished in 1934 to make way for the present building, part of the complex built for the International Exhibition in Paris in 1937. It houses two museums as well as two theatres built into the hillside. The larger auditorium, originally seating 3,000, was used by the United Nations General Assembly from 1946 to 1952 and for the next 20 years by the second Théâtre National Populaire. In 1972 this name was transferred to the theatre run by *Planchon, the restructured theatre at the Palais de Chaillot being reopened in 1973 as the Théâtre National de Chaillot. The new theatre failed to prosper until Antoine *Vitez took it over in 1981. He was succeeded in 1988 by **Jérôme Savary** (1942–), founder in 1966 of the experimental Grand Magic Circus, whose productions at the Chaillot included *Molière's Le Bourgeois gentilhomme, and who in 1989 staged the musical Metropolis in London. The large auditorium was named the Salle Jean-Vilar in 1989. The Salle Firmin-Gémier houses independent companies, and entertainments for children are given in the foyer.

Chairman, presiding genius of the old *music-hall. In full evening dress, he sat at a table at auditorium level with his back to the stage. A mirror enabled him to see what was going on there; a clock helped him to time the 'turns' and, if necessary, cut them short; and a gavel helped him to keep order. Usually an old 'pro', of ruddy countenance and commanding presence, with a stentorian voice, he needed a ready wit and a gift for impromptu repartee. There were many famous chairmen, among them Charles *Sloman. The chairman virtually disappeared by 1890 and had no place in the vast new Theatres of Variety.

Chamberlain's Men, company of actors with which Shakespeare was most closely associated, and for which he wrote most of his plays; it also performed plays by *Jonson and others. Founded in 1594, it took possession of the *Theatre, recently vacated by the *Admiral's Men, and remained there until the lease ran out

in 1599. It then migrated to the new *Globe Theatre on Bankside, and with Richard *Burbage as chief actor had a long and prosperous career apart from a little trouble over Shakespeare's Richard II which Elizabeth I, already angered by the Earl of Essex's unsuccessful rebellion, chose to regard as treasonable. However, the company is known to have played at Court not long before the Queen's death in 1603. With the accession of James I, the somewhat chaotic conditions of theatrical life in London were reorganized and the Chamberlain's Men, who were allowed to remain in sole occupation of their theatre, were appointed Grooms of the Chamber and renamed the *King's Men, coming under the direct patronage of the sovereign.

Chambers of Rhetoric, official organizations of rhetoricians which originated in northern France as early as the 12th century and spread in the 13th century into the Low Countries, where their members were known as Rederykers. They became a forceful element in civic cultural life and from the beginning of the 15th century were responsible for dramatic contests at which both serious and farcical plays were performed, some specially written for the occasion and afterwards published. Most of them corresponded to the *mystery, *miracle, and *morality plays, but the farces were mainly indigenous. The morality plays gave rise to a genre peculiar to the Low Countries, the spel van sinne, which featured a group of little imps, or sinnekens, who incited the characters to evil doing and so provided a satiric and comic element, somewhat akin to the Old *Vice of the English theatre. The development of the sinnekens eventually brought the rhetoricians in the southern half of the Low Countries into disrepute, but in The Netherlands they prospered until the beginning of the 17th century, after which they declined in popularity and disappeared, the last outstanding playwright with whom they were associated being *Vondel.

Chambers Street Theatre, New York, see BURTON, WILLIAM EVANS.

Champagne Charlie, see LEYBOURNE.

Champion, Harry, see MUSIC-HALL.

Champmeslé [Marie Desmares], **Mlle** (1643–98), French actress. She was in the company at the *Marais in 1669, and profited so much from her teaching that in six months she was playing leading roles. She went to the Hôtel de *Bourgogne in 1670, and became so popular that the rival troupes competed for her services. She became the leading lady in tragedy of the *Comédie-Française on its foundation in 1680, playing opposite *Baron and remaining with the company until her death. She created many famous tragic roles, including the title-roles in Bérénice (1670) and Phèdre (1677) by *Racine, whose mistress she was, and in the younger *Corneille's *Ariane* (1672). She favoured a chanting, sing-song declamation which she taught to the leading actresses of the next generation.

In 1665 she married as her second husband the actor **Charles Champmeslé** (1642–1701), author of many light comedies, among them *Crispin Chevalier* (1673), a one-act farce, and *Le Florentin* (1685). He acted in the same companies as his wife, but, though considered a good tragic actor in his youth, was soon overshadowed by her.

Champs-Élysées, Comédie des, see COMÉDIE DES CHAMPS-ÉLYSÉES.

Champs-Élysées, Théâtre des, see MARIGNY, THÉÂTRE.

Chancerel, Léon (1886–1965), French dramatist and director, a pupil of Jacques *Copeau. He was also with *Jouvet at the *Comédie des Champs-Élysées in 1926, and from 1929 to 1939 directed the Comédiens Routiers, a touring company which played to young audiences. In 1935 he founded in Paris a company of young actors to play to children, Le Théâtre de l'Oncle Sebastien. Its plays were improvised, in deliberate imitation of the *commedia dell'arte*. Chancerel, who succeeded Jouvet as President of the Société d'Histoire du Théâtre, was head of the Centre Dramatique in Paris (see DÉCENTRALISATION DRAMATIQUE). After his death a prize was established in his name to be awarded every two years to the writer of the best new play for children.

Chanfrau, Frank S. [Francis] (1824–84), American actor, best remembered for his playing of Mose the New York fireman, first seen in Benjamin A. *Baker's *A Glance at New York in 1848*. Later in the year Chanfrau continued Mose's adventures in *The Mysteries and Miseries of New York*, seen at the *Chatham Theatre 2 of which he was lessee for a time, renaming it the New National Theatre. He continued to appear as Mose for many years, in a series of plays written by Baker or himself. Among his other parts the best was the pioneer Kit Redding in Spencer and Tayleure's *Kit, the Arkansas Traveller* (1870), which he revived up to 1882. His son Henry revived a number of his father's parts, playing Kit until 1890.

Chanin's 46th Street Theatre, New York, see RICHARD RODGERS THEATRE.

Channing, Carol, see MUSICAL COMEDY.

Chapman, George (c.1560–1634), English poet and dramatist, author of the translation of Homer which inspired Keats's sonnet. He may be the original of Shakespeare's Holofernes in *Love's Labour's Lost* and Thersites in *Troilus and Cressida*. Little is known of his early life, but in 1596 he was on *Henslowe's payroll as accredited dramatist to the *Admiral's Men. Most of these early plays are lost, but enough remains of his later works, given mainly by the *boy companies, to ensure his acceptance as a major playwright. His best comedies are probably *May-Day* (1602) and *All Fools* (c.1604). Of the surviving tragedies the most important is *Bussy d'Ambois* (also 1604) with its sequel *The Revenge of Bussy d'Ambois* (1610). Chapman was one of the authors, with *Jonson and *Marston, of the comedy *Eastward Ho!* (1605) which gave offence to James I and resulted in a term of imprisonment for himself and Jonson; he narrowly escaped a second term of imprisonment for his *Charles, Duke of Byron* (1608) which upset the French Ambassador.

Chapman, William (1764–1839), one of the earliest and possibly the first of the American *showboat managers. Born in England, he made his first appearance on the London stage in 1803 and in 1827 went to New York, where he was first seen at the *Bowery Theatre. He then started on tour for the south-west, and in Pittsburgh persuaded a Captain Brown to build for him a 'floating theatre' on which he travelled up and down the Ohio and Mississippi rivers from 1831 until his death. His widow and sons continued to run the business until in about 1847 they sold out to Sol *Smith. In 1845 they had opened the first *Showboat in New York, known as Chapman's Temple of the Muses, and built on a flat-bottomed steam packet. When it opened it was anchored between Canal and Charlton Streets, but it moved from place to place, the audience locating it by the Drummond Light which was lit only on performance nights. It does not appear to have been used for more than about six months.

Chappuzeau, Samuel (1625–1701), French man of letters, author of *Le Théâtre françois* (1674), which consists of three parts: *De l'usage de la comédie*, a defence of the theatre; *Des auteurs qui soutiennent le théâtre*, a dictionary of dramatists; and *De la conduite des comédiens*, an apology for players, with lives of many of the best known. This third part contains a long chapter on *Molière, who had just died. Though somewhat inaccurate, the work as a whole is an important source-book of the 17th-century French theatre. An early farce by Chappuzeau, published in Lyons in 1656, seems to have had some influence on Molière's *Les Précieuses ridicules* (1658).

Chapter Arts Centre, see CARDIFF.

Charing Cross Music-Hall, see GATTI'S.

Charing Cross Theatre, London, see TOOLE.

Chariot-and-Pole, see CARRIAGE-AND-FRAME.

Charke, Charlotte, see CIBBER.

Charleston, SC, see DOCK STREET THEATRE.

Charlot, André, see REVUE.

Châssis à Développement, extension of the *falling flaps method of scene changing. In its simplest form, the set piece had half its subject painted on the normal surface and the other half on one face of a vertically hinged flap. By means of a trick-line, this flap was swung over, like the page of a book, to lie over the first half of the piece, revealing a new subject painted on the remaining portion and the reverse face of the hinged flap. In trick set-pieces a series of separate flaps were all spring-hinged together to lie compactly flat when folded; upon release all flew out to present not only a totally different subject but one which considerably exceeded the original in size.

Chatham Theatre, New York. **1.** Between Duane and Pearl Streets and originally known as the Pavilion, this opened in 1823. It was built by Barrière, a pastrycook, in a pleasure resort named Chatham Gardens. It seated 1,300, was the first theatre in New York to be lit by gas, and soon proved a serious rival to the old-established *Park Theatre 1. The original company included the first and second Joseph *Jefferson and also Henry *Wallack who, with James *Hackett, took over the management after Barrière died in 1826. Wallack was not successful, and relinquished the management in 1830, the theatre having been renamed the American Opera House in 1829. It was later run with a mixture of straight and *equestrian drama, finally closing in 1832 and becoming a Presbyterian chapel.

2. A New Chatham Theatre, on the south-east side of Chatham Street between Roosevelt and James Streets, opened in 1839. A large house, seating 2,200, it soon found itself in financial difficulties. In 1848 Frank *Chanfrau took the theatre over and it was henceforth known as the National. Edwin *Booth made his first appearance in New York here on 27 Sept. 1850. One of the great successes was Aiken's dramatization of Harriet Beecher Stowe's anti-slavery novel *Uncle Tom's Cabin* (1853). The building was severely damaged by fire in 1860, but continued in use until it was pulled down in 1862.

Chauve-Souris, see BALIEFF.

Cheek By Jowl, see FRINGE THEATRE.

Cheeseman, Peter, see NEW VICTORIA THEATRE.

Chekhov, Anton Pavlovich (1860–1904), Russian dramatist, the one best known outside Russia. He came of humble parentage, but graduated as a doctor from Moscow University in 1884, and always thought of himself as more of a doctor than a writer. Well known for his short stories even during his student days, he was early attracted by the theatre, and his first plays were one-act comedies—*The Bear* (1888), *The Proposal* (1889), *The Wedding* (1890)—for which he always retained an amused affection.

His first full-length plays, *Ivanov* (1887) and *The Wood-Demon* (1889), were unsuccessful, as was *The Seagull* (1896) when performed at the old-fashioned Alexandrinsky Theatre in St Petersburg (now the *Pushkin Theatre, Leningrad); it had nothing in common with the plays currently popular and was incomprehensible to actors trained in the old traditions. Chekhov might have given up the theatre entirely had not *Nemirovich-Danchenko persuaded him to let the newly founded *Moscow Art Theatre revive *The Seagull*. The production was a success, and was followed by *Uncle Vanya* (1899) (a revision of *The Wood-Demon*), *Three Sisters* (1901), and *The Cherry Orchard* (1904), in all of which Chekhov's wife Olga *Knipper played leading parts. Chekhov died shortly afterwards, at the height of his powers.

His plays portray the constant attrition of daily life and the waste, under the social conditions of Old Russia, of youthful energy and talent. Yet at the same time they contain a note of hope for the future. This hopefulness seems to accord with Chekhov's own view of his plays, but not with *Stanislavsky's, who wrote that he wept when he first read *The Cherry Orchard*. However one interprets the plays, they demand a subtlety of ensemble playing which was not available in Russia until the founding of the Moscow Art Theatre. The forces at work in Chekhov's dramas are those which produced the Russian Revolution, and the great speech by the student Trofimov at the end of Act II of *The Cherry Orchard*—the celebrated 'All Russia is our garden'—is not only a concise review of then recent Russian history, but an astonishingly accurate prediction of the Revolution itself.

As far as can be ascertained, the first of Chekhov's plays to be acted in English was *The Seagull* (*Glasgow Repertory Theatre, 1909, and London *Little Theatre, 1912). The *Stage Society was persuaded by *Shaw to produce *The Cherry Orchard* in 1911 and *Uncle Vanya* in 1914. These early productions failed, but helped to make Chekhov's work known to actors and critics. *Three Sisters* was produced in London in 1920, but the first production to be commercially successful was *The Cherry Orchard*, brought to London in 1925 by *Fagan from the *Oxford Playhouse. Even more successful was a series of productions by *Komisarjevsky at the *Barnes Theatre, which finally naturalized Chekhov in England. The early unfinished play *Platonov*, first produced in Russia in the 1920s, was seen in London in 1960, and as *Wild Honey*, in a translation by Michael *Frayn, at the *National Theatre in 1984 (NY, 1986). The Moscow Art Theatre played Chekhov in Russian on various visits to London, the first in 1928.

America's introduction to Chekhov was in Russian, Alla *Nazimova appearing in *The

Seagull in 1905. In an English translation, this play was produced by the *Washington Square Players in 1916, but American interest in Chekhov dates from the visits of the Moscow Art Theatre in 1923 and 1924, and subsequent American productions have been strongly influenced by Stanislavsky.

There have been a number of translations of Chekhov's plays into English. The standard acting versions for many years were those of Constance Garnett (1923), which have frequently been used as the basis of new adaptations. The Penguin Classics published versions of *The Cherry Orchard, Three Sisters,* and *Ivanov* by Elisaveta Fen. All Chekhov's plays are contained in the first three volumes of the Oxford Chekhov, in translations by Ronald Hingley based on the definitive Moscow edition of 1944–51.

Chéreau, Patrice (1944–), French director. After early work including Lenz's *Soldiers* (1967) and Shakespeare's *Richard II* in 1970, he became co-director with *Planchon of the Théâtre National Populaire at Villeurbanne, 1972–81. In 1982 he became Director of the Théâtre des Amandiers at Nanterre, which under his guidance, despite an unpromising location in an impoverished area 10 miles from Paris, developed into one of France's leading theatres. Notable among Chéreau's productions, which have great visual brilliance, are *Ibsen's *Peer Gynt* (1981), *Genet's *Les Paravents* (1983), and *Hamlet* (1989). He is also an opera director, famous especially for his productions of Wagner's *Ring* cycle at Bayreuth, 1976–80.

Cherkassov, Nikolai Konstantinovich (1903–66), Soviet actor, who began his career at the Leningrad Theatre of Young Spectators in 1926, where he played *Cervantes' Don Quixote. Between 1927 and 1932 he made his name in the music-halls of Moscow and Leningrad as an eccentric comedian, but in 1933 he joined the company at Leningrad's *Pushkin Theatre where he remained until his death. During the 1950s and early 1960s he appeared in a number of new Soviet plays, but was best known in classical roles. Internationally he became famous for his appearances as Alexander Nevsky and Ivan the Terrible in Eisenstein's films.

Cherry Lane Theatre, New York, in Commerce Street. This early 19th-century building was converted into a theatre in 1924. In 1927 the New Playwrights Theatre took it over and gave it their own name. The theatre reverted to its original name in 1928 and survived the Depression by a widespread use of reduced price ticket vouchers. In 1945 it was leased by an experimental company which opened with *O'Casey's *Juno and the Paycock.* Since then a number of companies have been seen there, including the *Living Theatre in the 1950s. In

the 1960s the first productions in New York of Edward *Albee's *The American Dream* and *The Death of Bessie Smith,* Samuel *Beckett's *Play,* and Harold *Pinter's *The Lover* were presented. Other notable productions included Edward *Bond's *Saved* (1970); the long-running *The Passion of Dracula* (1977), adapted from Bram Stoker's novel; and Sam *Shepard's *True West* (1982).

Chestnut Street Theatre, Philadelphia, America's first theatre in the English *proscenium style, built in 1793 for a company brought from England by Thomas *Wignell. It did not open until 1794 because of an epidemic of yellow fever. A copy of the Theatre Royal at *Bath, it was an elegant and impressive building holding about 2,000, with numerous dressing-rooms and two greenrooms. Wignell remained there until his death in 1803, and was succeeded by William *Wood and the elder William *Warren as joint managers. Under them the theatre enjoyed a monopoly of acting in Philadelphia until the opening of the *Walnut Street Theatre in 1811. Sometimes known, from its pre-eminent position in American theatrical life, as 'Old Drury', it was one of the first theatres in the United States to be lit by gas (in 1816). Wood's departure in 1826 and the success of the Walnut Street and Arch Street theatres finally forced the Chestnut Street Theatre into bankruptcy; after burning down in 1856 it was not rebuilt until 1863. The last performance given there was in 1913, and the building was finally demolished in 1917.

Chettle, Henry (c.1560–1607), prolific Elizabethan playwright, credited by *Henslowe in his diary with a long list of plays, mostly written in collaboration and now lost. Of those that survive, the only one Chettle appears to have produced unaided was *Hoffman,* known also as *Revenge for a Father* (c.1602), which follows the pattern of the *revenge tragedy set by *Kyd. Chettle was also a printer, in which capacity he published the pamphlets of *Nashe as well as *Greene's *A Groatsworth of Wit* (1592), which contains the famous attack on Shakespeare.

Chevalier, Albert (1861–1923), one of the outstanding stars of the British *music-hall. After a successful and varied career in comedy, *burlesque, and *melodrama, he was persuaded by his friend Charlie Coborn (see MUSIC-HALL), already a well-known name in the entertainment world, to go on the 'halls'. He made his first appearance at the *London Pavilion in 1891 singing his own composition 'The Coster's Serenade' and was immediately successful. Although he later presented other characters, including a Chelsea Pensioner, a yokel, and a curate, it was as a Cockney that he gained his immense popularity, with 'Knocked 'Em in the Old Kent Road', 'The Narsty Way 'E Sez It', 'It Gits Me Talked Abaht', ''Appy 'Ampstead', and above all 'My Old Dutch', for which he wrote the words himself, his brother supplying the music. Chevalier was also the author of a number of sentimental ballads and sketches and of one unsuccessful play, *The Land of Nod* (1898). He was one of the few music-hall stars who never appeared in *pantomime.

Chevalier, Maurice (1888–1972), French entertainer, singer, and actor, equally popular in Paris, London, and New York. He was with *Mistinguett at the *Folies-Bergère, 1909–13, and in 1919 made his first appearance in London in the revue *Hullo, America!* In 1929 he made his American début at the New Amsterdam Roof Garden for *Ziegfeld. His straw hat was his trade-mark, and its debonair flourish, combined with a slightly raffish appearance, a quizzical smile, and a seductive voice with a charming broken-English accent, helped to make him a popular idol, who could hold an audience's attention in a one-man show. Such songs as 'Mimi' and 'Louise' are associated with him. He also made many films.

Chiarelli, Luigi (1884–1947), Italian dramatist, the chief exponent of the Teatro *Grottesco, which took its name from Chiarelli's own description of his best-known play *The Mask and the Face* (1916, written in 1913) as 'grottesco in tre atti'. This study of man in modern society, where hypocrisy is rampant and all true feeling must be hidden under an expressionless mask, was seen in London and New York in 1924. In 1931 a play on a similar theme, *Fireworks* (1923), was produced in London as *Money, Money!* Chiarelli's other plays include *The Death of the Lovers* (1921), *K.41* (1929), a commercial melodrama, and *The Little Queen* (1931). *One Plus Two* (1935), a sophisticated light comedy, has great pace and unexpected depth, while *The Magic Circle* (1937) is a whimsical trifle. *To Be* was produced posthumously in 1953.

Chicago, Ill., was little more than a village until about 1830. Its first theatre, the Rialto, opened in 1838 under the management of John B. Rice and closed in 1840, the second Joseph *Jefferson being a member of the company. Rice's own theatre was built in 1847 and burned down three years later. In 1857 John H. McVicker built what was described as 'the most substantial, convenient, safe, and costly theatre building standing in the west', which was destroyed in the disastrous fire of 1871 and immediately rebuilt on the same site. Ten years later J. H. Haverly opened a theatre which he hoped would outdo McVicker's, though it seated only 2,000 against its rival's 2,600. In 1885 *Irving's company on its first American tour appeared at Haverly's, which was then rechristened the Columbia Theatre at a ceremony presided over by Ellen *Terry. The Auditorium, opened in 1889, was reputed to be the largest theatre in the United States at the

time, seating 4,500 and having outstandingly good acoustics. The vast Spectatorium planned by Steele *Mackaye for the Chicago Exposition of 1893 was abandoned owing to the financial panic of that year. From 1900 Chicago became almost totally dependent on New York for theatrical entertainment. The opening of the *Goodman Theatre in 1925 improved the city's theatrical reputation, though it tended to be regarded mainly as a staging post on the touring circuit until the 1970s, when a big growth in theatre audiences was stimulated by the formation of groups such as the St Nicholas Theatre Company and the Steppenwolf Theatre Company. The Chicago Theatre Company, founded in 1984, is a Black theatre group. A League of Chicago Theatres was founded in 1979 and a number of productions originating in the city achieved successful runs in New York, including the plays of David *Mamet.

Chicano Theatre, USA, name given to the Chicano (Mexican-American) theatre groups which cater for the Chicano community in a bilingual mixture of Spanish and English, with local colloquialisms added for regional flavour. Most of their productions are improvised, in the style of the *commedia dell'arte, and deal with such immediate problems as unemployment, housing, education, drug abuse, low wages, and racial discrimination. The movement began in 1965 with the founding in California of the Teatro Campesino (Fieldworkers' Theatre). A group of workers from the vineyards who were on strike toured in a programme of short sketches (*actos*) in an effort to recruit other workers to join the strike. Two years later they extended their activities to the whole USA, eventually breaking off their connection with the union in order to widen their scope and deal with issues that affected the whole Chicano community. In 1969 they took part in the Seventh World Theatre Festival in Nancy, France, and in 1970 they organized the first annual Chicano Theatre Festival. A year later the group moved to a permanent home in San Juan Bautista, Calif., where they devote their whole time to the movement, and in 1972 presented an improvised play based on Collective Creation entitled *La gran carpa de los rasquachis* (*The Great Tent of the Underdogs*). Although the first Chicano Theatre actors were field workers, later groups were formed mainly of Chicano students. One of the most important is the Teatro de los Barrios (*barrios* being Chicano communities or neighbourhoods), formed in 1969 by high school students in San Antonio, Tex. Other groups are the Teatro de la Gente (Theatre of the People), founded in 1970 in San José, and the Teatro de la Esperanza (Theatre of Hope), founded in 1971 by students of the University of California at Santa Barbara. In 1971 the various *teatros* of the Chicano Theatre movement formed an organization, the Teatro National de Aztlan, uniting all the groups from coast to coast. Known as Tenaz, it organizes yearly festivals and regional conferences and provides workshops. There are now approximately 100 *teatros* in the United States, most of them in the south-west.

Chichester Festival Theatre was opened in 1962, the money for its construction being raised mainly from private and commercial sources and from the *Arts Council. The hexagonal building, situated in Oaklands Park, has an auditorium seating 1,374 round a thrust stage, no one being more than 65 ft. from it. It has a semi-permanent balcony and a catwalk all round the interior wall. Four productions, often with major stars, are presented annually in a season running May–Oct. The first director was Laurence *Olivier, his company later forming the nucleus of the first *National Theatre Company. The most successful play of the first season was *Chekhov's *Uncle Vanya*, other productions being *Ford's *The Broken Heart* and *Fletcher's *The Chances*. The first new play to be given was *The Workhouse Donkey* (1963) by John *Arden, whose *Armstrong's Last Goodnight* (1965) was also first seen at Chichester, as were *Shaffer's *The Royal Hunt of the Sun* (1964) and *Black Comedy* (1965), and *Bolt's *Vivat! Vivat Regina!* (1970). In 1966 Olivier was succeeded by John *Clements, who was followed by Keith *Michell (1974–8) and others, including Michael *Rudman, 1989–91. Productions are now mainly revivals, including classics, more recent plays, and musicals. Many have subsequently been seen in London or on tour. A full programme of concerts, opera, ballet, and touring shows is maintained throughout the year.

The 250-seat Minerva Studio Theatre opened in 1989 with *Gorky's *Summerfolk*. It stages its own season concurrently with the main theatre and houses touring shows during the rest of the year.

Children of Blackfriars, of the Chapel, of Paul's, of the Revels, of Whitefriars, of Windsor, see BOY ACTORS.

Children's Theatre, performances given by adults for young people. The international organization for children's theatre is ASSITEJ (Association Internationale du Théâtre pour l'Enfance et la Jeunesse), based in Paris, with members in almost 40 countries. Scandinavian countries were early in the field, and Sweden is especially active. France made a good start with Léon *Chancerel's Théâtre de l'Oncle Sebastien, 1935–9, and now has a number of children's theatres such as the Théâtre des Jeunes Années in Lyons and the Théâtre du Gros Caillou in Caen. Children's theatre also flourishes in Japan and the Federal Republic of Germany, where the Gripstheater in West Berlin is outstanding, and where, as in most countries

where the theatre is heavily subsidized, young people are able to buy tickets for normal theatre-going very cheaply.

In Britain there have been many attempts to found a permanent children's theatre, but lacking sufficient funds they have usually ended with the retirement or death of the founder. Among them were Bertha Waddell's Scottish Children's Theatre, founded in 1927, which survived until 1968; Jean Sterling Mackinlay's afternoon seasons for six weeks over Christmas from 1914 to 1939; and the English School Theatre (partly educational), which was founded in 1936 but had to be disbanded during the Second World War. Still in existence is the Unicorn Theatre for Children, permanently established at the *Arts Theatre, London, since 1967, but originating as a touring group for both adults and children founded in 1947 by **Caryl Jenner** (1917–73), who ran the project until her death. Others are the Theatre Centre (formerly the London Children's Theatre), founded in 1954, which tours schools; the Little Angel Marionette Theatre in Islington, founded in 1961; and the Polka Children's Theatre in Wimbledon, founded in 1979. Both the *National Youth Theatre and the *Young Vic do excellent work, and many regional theatres run Theatre in Education schemes. The national association for children's theatre is NADECT (National Association for Drama in Education and Children's Theatre, formerly the British Children's Theatre Association), founded in 1959; the Young Theatre Association is concerned with amateur children's theatre.

Children's theatre in the USA receives—as in Britain—very little subsidy and has depended heavily on individuals such as the late Charlotte Chorpenning, who headed the Goodman Children's Theatre in Chicago for 21 years. Children's companies include the King-Coit school and theatre in New York; the Honolulu (Hawaii) Theatre for Youth founded in 1955; the Children's Theatre Company of Minneapolis (1961); and the Creative Arts Team (1974), an association of several companies affiliated with but separate from New York University. In 1976, the Empire State Institute for the Performing Arts (ERISA) was founded in Albany, New York, as the only state-authorized children's theatre in the USA. Many of the regional theatres, like their British counterparts, work in collaboration with the schools. Children's theatre thrives also in Canada.

In the former Communist countries children's companies—such as, in the Soviet Union, the Central Children's Theatre (founded in 1921) and the Moscow Theatre of the Young Spectator (1929)—are heavily subsidized, often having their own theatres and touring organizations, the actors being specially trained for their work. There are about 60 professional theatres for young people in the Soviet Union. Other former Communist countries active in children's theatre include Czechoslovakia and East Germany, where the Theater der Freundschaft in East Berlin is especially notable.

Chirgwin, George H., see MUSIC-HALL.

Chiswick Empire, London, see MUSIC-HALL.

'Chocolate-Coloured Coon' [G. H. Elliott], see MUSIC-HALL.

Chodorov, Jerome, see FIELDS, JOSEPH.

Choregus, in ancient Greece, the person responsible for equipping and paying a chorus for a tragic, comic, or dithyrambic contest. Hence the name *choregus*, or chorus-leader. The dramatic contests in the festivals were thus between *choregi* as well as between poets and, later, actors. It was important to the poet that his *choregus* should not be stingy, since the proper presentation of the play depended on him; *choregi* were therefore assigned to the poets by lot. In the Roman theatre the Latin form of the title (*choragus*) came to mean little more than property-man or stage manager.

Chorus, in Greek drama a group of actors who stand aside from the main action of the play and comment on it, as in the tragedies of *Aeschylus and *Sophocles; *Euripides made less use of the traditional chorus, concentrating some of its functions in one person who was much more closely concerned with the fate of the protagonists. In the comedies of *Aristophanes the chorus performs the same function as in the tragedies, but points the satire of each play by being dressed in some relevant disguise—birds, frogs, wasps—and by being more closely concerned with the intimate and bawdy aspects of the plot.

On the Elizabethan stage the Chorus was the speaker of an introductory prologue, a legacy from Euripides handed down via the Roman *closet dramas of *Seneca. He spoke either at the beginning of the play, as in *Henry VIII*, or before two or more acts, as in *Romeo and Juliet*, *Henry V*, and, in the person of Gower, in *Pericles Prince of Tyre*.

In the late 19th-century English theatre the chorus was the troupe supporting singers and dancers. They had always accompanied the principal actors in *burlesque, *extravaganza, and *pantomime, but it was not until about 1870 that the 'chorus girl' became a player in her own right. At first she was asked to do no more than wear lovely clothes and move gracefully in unison with her companions, often accompanied by handsome but fairly static 'chorus boys'. By the 1920s chorus girls were wearing fewer clothes and reached a high standard of precision dancing. The men were often dispensed with at this time, but after the Second World War, under the influence of the American musical, the chorus, both men and women, often took a much larger share in the plot.

Christiania Theatre, see KRISTIANIA THEATER.

Christie [Mrs Max Mallowan, *née* Miller], **Dame Agatha Mary Clarissa** (1890–1976), English novelist and playwright, appointed DBE in 1970, on her 80th birthday, publication day of her 80th book. Although chiefly remembered for her detective fiction, she has a place in theatrical history as the author of *The Mousetrap* (originally written for radio), which opened in London at the *Ambassadors Theatre on 25 Nov. 1952 and looks set to run for ever, transferring to the *St Martin's Theatre in 1974. It was seen in New York in 1960. Mrs Christie dramatized several of her novels, including *Ten Little Niggers* (1943), now known as *And Then There Were None*, produced in New York in 1944 as *Ten Little Indians*; *Appointment with Death* (1945); *Murder on the Nile* (1946); and *The Hollow* (1951). *Witness for the Prosecution* (1953) was still running when *Spider's Web* (1954) was presented, thus giving her, with *The Mousetrap*, three plays running concurrently in the West End. She later wrote *Verdict* and *The Unexpected Guest* (both 1958), and adapted *Go Back for Murder* (1960) from her novel *Five Little Pigs*. The most successful adaptations of her novels by other hands were *Love from a Stranger* (1936), based on *Philomel Cottage*, and *Murder at the Vicarage* (1950). *A Murder is Announced*, another adaptation, was produced posthumously in 1977.

Chronegk, Ludwig, see MEININGER COMPANY.

Churchill, Caryl, see ROYAL COURT THEATRE.

Churchill, Frank, see LEWES.

Cibber, Colley (1671–1757), English actor, theatre manager, and playwright. He went on the stage against the wishes of his family, and with few advantages of voice or person soon became an excellent comedian, being particularly admired in the fops of *Restoration comedy (such as Lord Foppington in *Vanbrugh's *The Relapse*, 1696) and of his own plays. In 1696 his first play *Love's Last Shift; or, The Fool in Fashion* was successfully produced at *Drury Lane. Striking a happy balance between the Restoration comedy of manners and the new vogue for sentiment and morality, this play is now considered the first sentimental comedy. In tragedy Cibber was less successful, though he achieved notoriety with his adaptation of Shakespeare's *Richard III* (1700), which remained the standard acting text until well into the 19th century. *She Would and She Would Not; or, The Kind Impostor* (1702) and *The Careless Husband* (1704) consolidated his reputation as a writer of comedies. A snob and a social climber. Cibber moved only on the fringe of the company he depicted; yet for over a century his comedies were taken as representative of English high society.

The greater part of Cibber's working life was spent at Drury Lane, first as actor, then from 1711 as joint manager with *Doggett and *Wilks and later Barton *Booth. Although he was in many ways a good manager, he was decidedly unpopular. Tactless, rude, conceited, and supremely self-confident, he was savagely ridiculed, notably by Pope in The Dunciad, by Dr *Johnson, and by *Fielding in several of his plays and his novel Joseph Andrews. Cibber was good at doctoring other authors' work, as can be judged by his adaptation of *Molière's Tartuffe as The Non-Juror (1717) and his completion of Vanbrugh's A Journey to London as The Provoked Husband (1728). In 1730 Cibber was appointed Poet Laureate, which caused no little dismay. He is best remembered now for his Apology for the Life of Mr Colley Cibber, Comedian (1740), which contains some admirable descriptions of Restoration acting.

His son **Theophilus** (1703–58) was a wild and undisciplined character who made his first appearance on the stage at 16. He seemed at first to have the makings of a good actor, particularly in such parts as Ancient Pistol in Henry IV, Part Two, but his imprudence and extravagance were his undoing. He was for a time manager of Drury Lane in succession to his father, but soon forfeited all respect by his insolence and complacency. Theophilus's sister **Charlotte** (1713–c.1760) was as eccentric and unpredictable as her brother. As Mrs **Charlotte Charke** she spent much of her time and energy in attempting unsuccessfully to run theatrical enterprises in opposition to her father, who disowned her.

Theophilus's second wife, whom he treated shamefully, was **Susanna Arne** (1714–66), sister of Dr Arne, the composer. She was an excellent actress, particularly in tragic roles such as the title-role in Aaron *Hill's Zara (1736). She was also good in comedy, making her last appearance as Lady Brute in Vanbrugh's The Provoked Wife.

Cincinnati Playhouse in the Park, Cincinnati, Ohio, is situated on a hilltop, and is the professional theatre for a three-state region of the Ohio River Valley. Of its two auditoriums, the Thompson Shelterhouse, seating 219, was converted from a gazebo in 1959, the Robert S. Marx Theatre, seating 627, being constructed in 1968. During the Sept.–Aug. subscription seasons, about six productions are presented in the larger theatre and four in the smaller. They have included 18 world premières and three American, as well as revivals of works by Shakespeare, *Molière, Samuel *Beckett, Tennessee *Williams, and other classical and contemporary playwrights. Playhouse productions tour to other American cities, and the subscription seasons are augmented by special productions from all over the United States, and

by community productions. The theatre also seeks new audiences through student seasons and special presentations, and is an important training resource.

Cinquevalli, Paul, see MUSIC-HALL.

Cinthio, Il, see GIRALDI.

Cintio, see COSTANTINI.

Circle-in-the-Square, New York. The first theatre of this name, in Sheridan Square, was designed specifically for *theatre-in-the-round productions by the Loft Players under José *Quintero. It opened in 1951 and its revival of Tennessee *Williams's Summer and Smoke (1952) was highly praised. The building was closed as a fire hazard in 1954, but reopened to become one of *Off-Broadway's most popular playhouses, staging in 1956 a revival of *O'Neill's The Iceman Cometh directed by Quintero. In 1960 the demolition of the building forced the company to move to the former New Stages Theatre in Bleecker Street, where they opened with *Wilder's Our Town. In 1972 the Circle-in-the-Square moved uptown to 1633 Broadway. The new theatre, the Joseph E. Levine, in the basement of the Uris Theatre (now the *Gershwin), was part of the first new theatre building to be erected on Broadway in 44 years. Seating 650, it has an adaptable stage and the seats are sharply tiered. It opened with O'Neill's Mourning Becomes Electra. Other interesting productions included Tennessee Williams's The Glass Menagerie (1975); *Ibsen's The Lady from the Sea (1976), with Vanessa *Redgrave, and John Gabriel Borkman (1980); and *Shaw's Heartbreak House (1983), with Rex *Harrison, and You Never Can Tell (1986), with Uta *Hagen. The old theatre continued as the Circle-in-the-Square Downtown.

Circle Repertory Company, off-off-Broadway company founded in a loft in 1969 by a group who had worked together at Café *La Mama, *Caffè Cino, and elsewhere. It makes an important contribution to the American repertory, its members numbering over 100, including writers and directors. Its initial season was remarkable for a double production of *Chekhov's Three Sisters in which the original and an experimental version alternated. The company is best known, however, for its productions of new American plays, including Mark Medoff's When You Comin' Back, Red Ryder? (1973), Feiffer's Knock Knock (1976), and Albert Innaurato's Gemini (1977), all of which transferred to the commercial theatre. Originally established at Broadway and 83rd Street, the company moved in 1974 to the former Sheridan Square Playhouse in 7th Avenue South, which became the Circle Repertory Company Theatre, a flexible performing space seating 100–150. Lanford *Wilson was one of the founders, and several of his plays were premièred by the company, which in the mid-1970s became professional and *Off-Broadway. Among other productions were Sam *Shepard's *Buried Child* (1979) and *Fool for Love* (1983).

Circuit, name given in the United Kingdom to a group of towns within a limited area which were served by the same travelling company. The system began early in the 18th century, when the proliferation of companies led to some unfortunate clashes, as when two groups of actors arrived together in Norwich and two companies gave performances of *Gay's *The Beggar's Opera* on the same night in Newcastle. The circuit was usually so arranged that the chief towns were visited several times a year, if possible during race or assize weeks, when large audiences could be expected. These towns usually had permanent theatre buildings, called into being by the circuit system. Smaller towns might be visited only once or twice a year, the actors playing in temporary accommodation. The companies had large repertories, the York circuit company performing as many as 100 plays and after-pieces in a year, while Mrs Sarah *Baker, lifelong manager of the Kent circuit, could reckon on a good 80. All the circuit companies became recognized training-grounds for young actors, and in their heyday it was rare for a player to begin his career anywhere else.

Circus, in Roman times a place of exhibition for chariot racing and athletic and gladiatorial contests. In its modern sense it dates from the mid-18th century. Mainly itinerant, it is performed in the central area of a tent (the Big Top) or in a specially adapted building. The programme is built up of separate turns featuring animal acts and acrobatics, controlled by a ringmaster and loosely connected by the antics of the *clown. Owing to the specialized nature of their work, circus performers tend to remain a class apart, with much intermarrying. The great names of the circus in the USA are Barnum and Ringling and in Britain Sanger, Mills, Smart, and Chipperfield; but the circus is universal and can be at home anywhere.

Citizens' Theatre, Glasgow, company founded in 1943 by James *Bridie, Paul Vincent *Carroll, and others. In 1945 they leased the Royal Princess's Theatre in the Gorbals which had been built in 1878 and renamed it the Citizens' Theatre (seating currently 657). Many of Bridie's plays were premièred at the theatre, two of them posthumously, *The Baikie Charivari* (1952) and *Meeting at Night* (1954). Tyrone *Guthrie's notable revival of *Lyndsay's *Ane Pleasant Satyre of the Thrie Estaitis* (1552) presented at the *Edinburgh International Festival of 1948 showed the high standard reached by the company in the first five years. Other important productions were William Douglas *Home's *Now Barabbas* (1947), John *Arden's

Armstrong's Last Goodnight (1964), and Peter *Nichols's *A Day in the Death of Joe Egg* (1967). There were also notable productions of plays by *Brecht including *Arturo Ui* (1967). Michael *Blakemore began his directing career here.

Under **Giles Havergal** (1938–), who has been Director with **Philip Prowse** (1937–) and **Robert David MacDonald** (1929–) since 1969, the theatre has presented a unique repertoire of plays from the British and European classical canon and some new plays and adaptations. Outstanding productions include Kraus's *The Last Days of Mankind* (1983), *Goethe's *Faust* (1985), an adaptation of Leo *Tolstoy's novel *Anna Karenina* (1987), and *Coward's *The Vortex* (1988). The company has presented eight new plays by Robert David MacDonald and toured extensively to the major European theatre festivals. Experimental works staged at the Close Studio Theatre from 1965 until its destruction by fire in 1973 included *Artaud's *The Cenci* and *Weiss's *Marat/Sade*. The Citizens' Theatre receives subsidy from the Scottish Arts Council, the City of Glasgow District Council, and Strathclyde Regional Council. The productions are cast from a pool of actors all of whom receive the same salary. The Citizens' Theatre for Youth Company founded in 1967 has now become the TAG Theatre Company, which does extensive middle-scale touring in Scotland and beyond and presents productions in primary, secondary, and special schools in Strathclyde Region.

City Center of Music and Drama, New York, on West 55th Street. Built as a Masonic temple in 1924, this building came into the possession of the city and opened as a theatre under its present name in 1943 with a revival of Rachel *Crothers's *Susan and God*. The building, seating 2,935 at prices kept as low as possible, also served as the home of the New York City Ballet and City Opera companies until they moved to the new Lincoln Center in the 1960s. The drama section continued to offer a wide variety of plays and musicals under such directors as José *Ferrer and Maurice *Evans until in 1972 the City Center Acting Company was formed by John *Houseman. Its productions included the work of European writers such as *Shaw, *Gorky, Brendan *Behan, and James *Saunders. The company severed its connection with the City Center in 1975 and since 1980 has been affiliated to the *John F. Kennedy Center in Washington, being now known as the Acting Company. The City Center still plays host to various dance companies and foreign troupes.

City Dionysia, see DIONYSIA.

City of London Theatre, in Norton Folgate, adjoining Bishopsgate, designed by Samuel *Beazley and seating 2,500, opened in 1837 with a dramatization of *Dickens's *Pickwick*

Papers before the book had completed its first serial publication. Its most successful period was 1848–68, with good value for money and the engagement of many leading actors. The Christy Minstrels were there in 1867, but a year later the house had reverted to crude *melodrama and romantic spectacle, and in 1871 it was destroyed by fire.

City Palace of Varieties, Leeds, one of Britain's two surviving full-time *music-halls. (The other is the *Players' Theatre.) It developed from modest beginnings in 1865 as a room attached to the White Swan public house. It now seats 713 and was restored to its earlier glory to house the television series 'The Good Old Days', which re-created Edwardian music-hall.

Ciulei, Liviu, see GUTHRIE THEATER.

Civic Repertory Theatre, New York, see LE GALLIENNE.

Civic Theatres, in Britain, theatres owned and subsidized by the local authority as a cultural amenity. Clause 132 of the Local Government Act of 1948 empowered local authorities to spend the proceeds of not more than a sixpenny rate on the provision of entertainment 'of any nature'; but it was 10 years before the opening of the *Belgrade Theatre paved the way, and true civic theatres are still a rarity. Two main types of theatre enjoy civic support. *Touring theatres, often providing facilities for large-scale opera and ballet, may accommodate both subsidized and commercial productions; they include the Alexandra and the Hippodrome in *Birmingham, the Empire in *Liverpool, the Grand Theatre and Opera House in Leeds, the *Opera House and *Palace Theatre, Manchester, and the Theatres Royal at *Bath, *Newcastle, Norwich, and Nottingham. *Repertory theatres usually receive local authority support but are often managed by a Trust to ensure artistic independence.

Clairon [Claire-Josèphe-Hippolyte Léris de la Tude], **Mlle** (1723–1803), French actress, who showed precocious ability and played minor parts with the *Comédie-Italienne before joining a provincial troupe. In 1743 she went to the Opéra, but her real talent was for acting and she was transferred to the *Comédie-Française. Her début as *Racine's Phèdre in 1743 was a triumph, and led to other important tragic roles. With the great actor *Lekain, encouraged by *Diderot, she was responsible for introducing some much-needed modifications into the costumes worn on the stage, mainly by adding some historical detail. She also abandoned in about 1753 her somewhat stiff and declamatory method of acting for a freer and more natural style. The tragic depth of her voice, though esteemed too low for high passion and not particularly felicitous in tender scenes, prevented

her from lapsing into triviality, and the new departure was a success. In 1765 Clairon, with other members of the Comédie-Française, was imprisoned for refusing to play with an actor who had brought disgrace upon the company; on her release she took refuge with *Voltaire, in many of whose plays she had appeared, thereafter acting only in his private theatre at Ferney or, on her return to Paris, at Court and in private theatricals. In 1773 she was invited to the Court of the Margrave of Anspach. She died in poverty, having lost her pension in the upheavals of the French Revolution, her benefactress having also died.

Clark, John Pepper (1935–), Nigerian playwright and poet. His first three plays make up a trilogy having much in common with Greek tragedy; *Song of a Goat* (Ibadan, 1961; London, 1965) is about a childless wife who takes her husband's brother as a lover; their child is the central figure of the second play, *The Masquerade* (1964; London, 1965), and in *The Raft* (also 1964; NY, 1978) the threads of the earlier plays are woven together in a discussion between four men adrift on the Niger. Clark's dramatic work arises from the myths and folklore of the Ijaw people; his *Ozidi* (1966) deals with the world of the imagination and the magical, being based on an Ijaw saga of the same name of which Clark prepared an English translation. *The Bikoroa Plays*, another trilogy (*The Boat, The Return Home, Full Circle*), were produced in Lagos in 1981. He now writes as J. P. Clark-Bekederemo.

Clarke, Austin (1896–1974), Irish poet, dramatist, and novelist, who kept interest in verse drama alive in Ireland during the 20 years following the death of *Yeats. His first play, *The Son of Learning*, was performed at the Cambridge Festival Theatre (see GRAY, TERENCE) in 1927 and the *Gate Theatre, Dublin, as *The Hunger Demon*, in 1930. Two short plays set in a convent, *The Flame* (1932) and *Sister Eucharia* (1939), were also staged at the Gate Theatre. *Black Fast*, a one-act farce based on a medieval debate, was produced by the *Abbey Theatre company in 1941. In 1944 Clarke established the Lyric Players to revive *poetic drama. Among his own plays produced by this group at the Abbey Theatre were *The Viscount of Blarney* (1944), *The Second Kiss* (1946), *As the Crow Flies* (1948), and *The Plot Succeeds* (1950), a 'poetic pantomime'. Although many of Clarke's plays are set in medieval or legendary Ireland and reflect the satirical tone of early Irish writers, his works comment thoughtfully and provocatively on contemporary issues, probing the conflict between faith and reason and between individual conscience and established teaching.

Clarke, John Sleeper (1833–99), American comedian and theatre manager, who made his début in 1851 in Boston, and later became joint lessee of the Arch Street Theatre, *Philadelphia. His comic roles included Dromio in *The Comedy of Errors*. He married a sister of Edwin *Booth, and with him managed the *Winter Garden. He was first seen in London in 1867, and in 1872 became manager of Charing Cross Theatre (see TOOLE), where he played Bob Acres in *Sheridan's *The Rivals*. He later took over, and appeared at, the *Haymarket and the *Strand.

Claudel, Paul-Louis-Charles-Marie (1868–1955), French poet, dramatist, and diplomat, whose early works were published anonymously for fear their ardent Catholicism should harm his diplomatic career. The first, *Tête d'or* and *La Ville* (pub. 1890 and 1893 respectively), were *poetic dramas intended for a form of total theatre far in advance of their time, and the first to be seen on stage were *L'Annonce faite à Marie* (1912) and *L'Échange* (1914), the former directed by *Lugné-Poë at the Théâtre de L'Œuvre, the latter by *Copeau at the *Vieux-Colombier. Lugné-Poë also directed *L'Ôtage* (1914). All three were produced in London, as *The Tidings Brought to Mary* and *The Exchange* by the Pioneer Players in 1915 and 1916, and as *The Hostage* in 1919, with Sybil *Thorndike. Many of Claudel's later plays were first seen outside France, among them *Le Pain dur* (1926) in Oldenburg, *Le Père humilié* (1928) in Dresden, both sequels to *L'Ôtage*, and *Christophe Colomb* (1930) in Berlin. *Le Repos du septième jour* (1928) had its première in Warsaw, *Protée* (1929) in Groningen, *La Ville* (1931) in Brussels, and the dramatic oratorio *Jeanne d'Arc au bûcher* (1938), with music by Honegger, in Basle. It was not until the Second World War that Jean-Louis *Barrault, working in close collaboration with Claudel, first brought his work to the notice of the general public in France with a production at the *Comédie-Française in 1943 of what is usually regarded as Claudel's masterpiece, *Le Soulier de satin; ou, Le Pire n'est pas toujours sûr*. The incidental music was again by Honegger. The impression it made was reinforced by further productions by Barrault at the *Marigny—*Partage de midi* (first seen in 1916) in 1948, with Edwige *Feuillère as Ysé, *Christophe Colomb* in 1953, and *Le Soulier de satin* again in 1965. All three productions were seen in London during visits from the *Renaud–Barrault company, in 1951, 1956, and 1965 respectively. The company was also seen in New York in 1957 in *Le Soulier de satin*, which, as *The Satin Slipper; or, The Worst is not the Surest*, was published in 1931 in an English translation; *Le Pain dur* and *Le Père humilié* were translated in 1946. *La Ville*, as *The City*, and *Tête d'or*, under its original title, have also been translated, and in 1972 *Partage de midi*, as *Break of Noon*, was produced

in England. *Jeanne d'Arc au bûcher* was first performed in London in 1954. Claudel's last play, *L'Histoire de Tobie et de Sara*, had its first production at the *Avignon Festival in 1947 under Jean *Vilar. Most of Claudel's plays were revised and rewritten several times and two or more versions exist of all his major works. All must be regarded as statements of his Christian faith, depicting the unending struggle between good and evil and the redemption of mankind through sacrifice.

Clements, Sir John Selby (1910–88), English actor-manager. He made his first appearance on the stage in 1930, and five years later founded the Intimate Theatre in Palmers Green, north London, where he presented over 200 plays in weekly repertory, most of which he acted in, directed, or both. It was the first theatre in London to reopen after the outbreak of war in 1939. Clements later starred in West End plays, including *Priestley's *They Came to a City* (1943), and for a time managed the *St James's Theatre, where he appeared with his second wife **Kay Hammond** [Dorothy Katharine Standing] (1909–80)—already well known for such roles as Diana Lake in *French without Tears* (1936) and Elvira in *Coward's *Blithe Spirit* (1941)— in *Dryden's *Marriage à la Mode* (1946). He also managed the *Phoenix Theatre, where *Farquhar's *The Beaux' Stratagem*, with Clements as Archer and his wife as Mrs Sullen, began a record run in 1949. In 1951 they appeared as John Tanner and Ann Whitefield in a revival of *Shaw's *Man and Superman* which Clements presented and directed. A year later they were seen in his own play *The Happy Marriage*, based on Jean-Bernard Luc's *Le Complexe de Philémon*, and he was in management at the *Saville Theatre, 1955–7, where he again presented some excellent revivals of classical comedies, including *Sheridan's *The Rivals* and *Congreve's *The Way of the World*, in both of which he starred with his wife. He later appeared in such productions as Benn W. Levy's *The Rape of the Belt* (1957) which he directed, playing Heracles to his wife's Hippolyte, and Ronald *Millar's *The Affair* (1961), *The Masters* (1963), and *The Case in Question* (1975), based on novels by C. P. Snow, also directing the last two. From 1966 to 1973 he was Director of the *Chichester Festival Theatre, where his productions included *Phillpotts's *The Farmer's Wife*, *Pinero's *The Magistrate*, and Shaw's *Heartbreak House*, in which he played Shotover; he also played Macbeth, Prospero in *The Tempest*, and Antony in *Antony and Cleopatra*. He returned to Chichester in 1979 to appear in Shaw's *The Devil's Disciple* and *Wilde's *The Importance of Being Earnest*.

Cleveland Play House, Cleveland, Ohio, the first resident professional theatre in the United States. Founded by an amateur group in 1915, it took over a disused church two years later and operated there on a small scale until the appointment in 1921 of its first director, who began to transform it into a professional company. In 1927 it moved into a new theatre incorporating two auditoriums, the Drury and the Brooks, seating 515 and 160 respectively. A third theatre, the Euclid-77th Street, seating 567, was added in 1949, when a nearby church was converted into a thrust-stage theatre which opened with *Romeo and Juliet*. At the Play House about a dozen productions are presented during a season which runs Oct.–May, with guest actors making occasional appearances. The repertory covers a wide range and over 70 world and American premières have been given, including the American premières of *Arbuzov's *Confession at Night* (1975) and *Gorky's *The Romantics* (1978). In 1980 the Play House erected the 612-seat Bolton Theatre on land adjacent to the main building; it opened in 1983 with a production of *The Tempest*. Later productions at the Play House included Tom *Stoppard's *On the Razzle* (1985) and John *Guare's *Bosoms and Neglect* (1986). The Play House also operates a Youth Theatre and a Play House Comes to School programme.

Clifton, Harry, see MUSIC-HALL.

Cline, Maggie, see VAUDEVILLE, AMERICAN.

Clive, Kitty [*née* Catherine Raftor] (1711–85), English actress, who at the age of 17 was playing minor parts at *Drury Lane under Colley *Cibber. She made her first success as Phillida in his *Love in a Riddle* (1729). Most of her career was spent at Drury Lane where she and *Garrick were constantly at loggerheads, mainly because he endeavoured to prevent her appearing in unsuitable parts. She was excellent in low comedy, *farce, and *burlesque, but quite unfitted for tragedy or even genteel comedy; though she made a success of Portia in *The Merchant of Venice*, mimicking famous lawyers of the day in the trial scene. Passionate and vulgar, she was always generous and quite without pride or ostentation. One of her greatest admirers was Horace Walpole, who on her retirement in 1769 presented her with a small house where her company and conversation were much relished by his friends, particularly Dr *Johnson. She was frequently painted, notably by Hogarth, and was the author of several short farces.

Cloak-and-Sword Play, see COMEDIA.

Close Studio Theatre, Glasgow, see CITIZENS' THEATRE.

Closet Drama, plays written to be read as opposed to those intended for production. Among the most important and influential were those of *Seneca. Many verse plays by established poets, particularly in the 19th century, though meant to be read were successful in production; while *Tennyson's have not been revived, de *Musset's have joined the permanent repertory. The term 'closet drama' also includes translations of plays, usually classics in their own countries, intended for reading and not for acting.

Cloth, any large unframed expanse of scenic material, made of widths of canvas or muslin seamed horizontally together and attached top and bottom to a sandwich *batten. (See under DROP for early English usage.) Among the specialized cloths are the *backcloth; the Cut-Cloth (Cut-Out Drop in America), with cut openings which, if elaborately fretted, need the reinforcement of a piece of netting glued on behind; the Gauze-Cloth (Scrim Drop in America), used for *cyclorama facing and for special effects such as a transparency (see TRANSFORMATION SCENE); the Sky-Cloth (Sky Drop), sometimes faced with gauze, used as a backdrop instead of a cyclorama; and the Stage-Cloth (Ground Cloth), with its variants such as the Sand-Cloth for desert and other outdoor scenes, an expanse of painted canvas laid on the stage as a floor covering, a function also performed by the *tragic carpet.

The Ceiling-Cloth differs from the others in being battened out before being suspended over the top of a *box-set; variations are the Roll Ceiling, which can be rolled up, and the Book (or Booked) Ceiling, consisting of two *flats hinged together, which takes up less space in the *flies.

Cloudings, permanent cloud borders used to mask the top of a scene. They could be drawn off sideways by hooked poles, and are mentioned as late as 1743. The detail of their arrangement is not clear, but they recall the form of border used by Inigo *Jones in his last masque, *Salmacida Spolia* (1640), which had also 'side-clouds'. These could presumably be drawn off to reveal another set behind, thus changing a stormy sky into a calm one, or vice versa.

Clown, in Elizabethan days a composite comic character, who might be a simpleton, a knave, or a Court Jester. Shakespeare provides examples of all three with Costard in *Love's Labour's Lost*, Autolycus in *The Winter's Tale*, and Touchstone in *As You Like It*. One of the formative elements of the clown was the Old *Vice of the *liturgical drama. The best-known players of clowns in Shakespeare's day were *Tarleton and *Kempe. The modern clown dates from the early 19th century, being a purely English character developed in the *harlequinade by *Grimaldi, the source of the clown's name of Joey and of his traditional costume. Once seen in theatres and *music-halls, clowns are today found mainly in the *circus, where their *slapstick is elaborated with *mime, skilful acrobatics, and juggling. (See also GROCK.)

Clunes, Alec [Alexander] **Sheriff de Moro** (1912–70), English actor, director, and theatre manager. Both his parents were on the stage, and after considerable experience as an amateur he joined Ben *Greet's company in 1934, and was then at the *Old Vic, where he played a wide variety of parts, and at the *Malvern Festival, later appearing in London in *Shaw's *In Good King Charles's Golden Days*. In 1942 he took over the *Arts Theatre in London where he remained until 1950, producing or directing numerous plays in many of which he appeared himself, playing Macbeth in the 100th production. Beginning with a revival of *Odets's *Awake and Sing*, the productions included revivals of English classics, such as *Farquhar's *The Constant Couple*, and new plays, among them Christopher *Fry's *The Lady's not for Burning* (1948) in which he created the part of Thomas Mendip. His chief roles after leaving the Arts were Claudius to the Hamlet of Paul *Scofield (1955), Caliban in *The Tempest* at the *Shakespeare Memorial Theatre (1957), Higgins in the musical *My Fair Lady* (1959), and the Bishop of Chichester in *Hochhuth's *Soldiers* (1968), his final appearance.

Clurman, Harold Edgar (1901–80), American director and critic, who made his first contact with the American theatre through the *Provincetown Players in 1924 and a year later joined the *Theatre Guild. In 1931 he co-founded the *Group Theatre, directing his first play on Broadway for them—*Odets's *Awake and Sing* (1935). He later directed Odets's other full-length plays for the Group, of which in 1936 he assumed sole management. After it disbanded in 1940 he directed a number of important new plays for other managements, including Carson McCullers's *The Member of the Wedding* (1950), William *Inge's *Bus Stop* (1955), *Anouilh's *Waltz of the Toreadors* (1957), *O'Neill's *A Touch of the Poet* (1958), and Arthur *Miller's *Incident at Vichy* (1964). He also directed abroad, including *Giraudoux's *Tiger at the Gates* in London (1955).

Coal Hole, see TERRY, EDWARD.

Coborn, Charlie, see MUSIC-HALL.

Coburg Theatre, London, see OLD VIC.

Cochran, Sir Charles Blake (1872–1951), English impresario, known in the theatre as 'Cocky'. After gaining experience in practically every branch of popular entertainment, he went into management on his own account with a production of *Ibsen's *John Gabriel Borkman* in New York in 1897. In London in 1911 he was responsible for *Reinhardt's production at Olympia of Vollmöller's *The Miracle*, which he revived at the *Lyceum in 1932, and in 1917 he presented in London *Brieux's controversial play *Damaged Goods*. In 1918 he began a series of remarkable productions at the *London Pavilion, including *Coward's *On with the Dance* (1925), the first show to feature Cochran's famous 'Young Ladies'. He brought many plays and players to London from abroad, arranging Sacha *Guitry's first visit (1920) and the first production in London of an *O'Neill play, *Anna Christie* (1923). He also presented Robert *Sherwood's *The Road to Rome* (1928), the Heywards' Negro tragedy *Porgy* (1929), *Moissi in *Hamlet*, and the *Pitoëffs in *Shaw's *Saint Joan* (both 1930). Closely associated with Coward, he was responsible for the production of his *Bitter Sweet* (1929), *Cavalcade* (1931), *Words and Music* (1932), and *Conversation Piece* (1934). He was manager of the Royal Albert Hall, 1926–38, and promoted rodeo shows and boxing matches. His longest-running production was A. P. Herbert and Vivian Ellis's musical *Bless the Bride* (1947).

Cockpit, London, public theatre in Drury Lane, built for cockfights in 1609 and converted in 1616 by Christopher *Beeston, on plans drawn up possibly by Inigo *Jones, into a roofed or 'private' theatre. A year later a band of London apprentices, merrymaking on Shrove Tuesday, sacked and set fire to it. It was immediately rebuilt and renamed the Phoenix, but the old name continued in use. The theatre is important as a link between the playhouses of Shakespeare's time and those of the Restoration. Although like all other London theatres it was closed in 1642, it must have been used for illicit performances, as it was raided and dismantled by Parliamentary soldiers in 1649. Two years later Beeston's son William refitted and reopened it, introducing some elementary machinery and effects which foreshadow the later use there by *Davenant of painted scenery. Two of Davenant's 'plays with music', or early operas, were seen at the Cockpit—*The Cruelty of the Spaniards in Peru* (1658) and *The History of Sir Francis Drake* (1659). When the theatres reopened on the return of Charles II in 1660, a troupe of young actors, including *Betterton, occupied the Cockpit, but it could not stand up to the competition of the *Patent Theatres, and closed for good in about 1665. It should not be confused with the Cockpit at Whitehall (1638–65), which was used only for private presentations of plays and masques before the Courts of Charles I and Charles II.

Cocteau, (Clement Eugène) **Jean Maurice** (1889–1963), French dramatist, poet, novelist, critic, artist, and film director, whose many-sided activities involved most of the outstanding artists and musicians of his day. As a dramatist he first came into prominence with a new version of *Sophocles' *Antigone* (1922), directed by *Dullin in a setting by Picasso. This was followed by *Orphée* (1924), directed by *Pitoëff.

In 1934 *Jouvet was responsible for the production of another excursion into Greek mythology with *La Machine infernale*, based on the Oedipus legend. Cocteau's other plays include *Les Mariés de la Tour Eiffel* (1921), an exercise in Surrealism first performed at the *Comédie-Française; *Les Parents terribles* (1938); *Les Monstres sacrés* (1940), on the nature of the theatrical experience; *La machine à écrire* (1941), a fairly straightforward thriller; two one-act monologues for an actress—*La Voix humaine* (1930) and *Le Bel Indifférent* (1941), written for *Piaf; an excursion into Arthurian legend, *Les Chevaliers de la Table Ronde* (1937); a tragic love-story in verse, *Renaud et Armide* (1943), produced at the Comédie-Française with Marie *Bell; and a romantic costume drama, *L'Aigle à deux têtes* (1946). Although several of Cocteau's plays have been translated into English and performed in London, the only one to achieve commercial success was *The Eagle Has Two Heads* (1946). It was also seen briefly in New York (1947) with Tallulah *Bankhead as the Queen, the part created by Edwige *Feuillère. On the whole Cocteau's plays have had less influence outside France than his films and ballets.

Cody, William Frederick (1846–1917), American showman, better known as Buffalo Bill. Born on a farm in Iowa, he was at the Colorado gold-mines as a boy, and then became a Pony Express rider and a Civil War scout. He appeared between 1872 and 1883 in more than a dozen plays written for him, the theatrical equivalent of Western dime novels. In 1883 he embarked on the Wild West show which made him famous all over the world. For 18 years the show starred the sharpshooter Annie Oakley (1860–1926), and the pair provided the basis for Irving *Berlin's musical *Annie Get Your Gun* (1946).

Coghill, Nevill, see OXFORD and OXFORD PLAYHOUSE.

Coghlan, Rose (1850–1932), American actress, born in England, who made her first appearance on the stage as a child, playing one of the witches in *Macbeth*. In 1869 she made her adult début under *Hollingshead at the *Gaiety Theatre in London, and two years later was seen in New York. Her success was such that she returned to the United States in 1877 and settled there, becoming an American citizen in 1902. She was for many years leading lady at *Wallack's Theatre, where she appeared with her brother **Charles Coghlan** (1842–99)—who also settled in America—playing Lady Teazle, one of her best parts, to his Charles Surface in *Sheridan's *The School for Scandal*. She also appeared in such modern productions as H. A. *Jones and Herman's *The Silver King*, Tom *Taylor's *Masks and Faces*, and *Sardou's *Diplomacy*, and was much admired as Rosalind in *As You Like It*. In 1893 she produced and starred

in the first American production of *Wilde's *A Woman of No Importance*, in which her brother played Lord Illingworth. (He had in the meantime played Antony and Macbeth opposite Lillie *Langtry in 1889.) Her career then began to decline, but though her style was outmoded her fine voice, stately presence, and technical ability caused her to be still in demand, and in 1916 she celebrated her stage jubilee. She was frequently seen in *vaudeville, and made her last major appearance in 1920 in *Belasco's production of Sacha *Guitry's *Deburau*.

Cohan, George M(ichael) (1878–1942), American actor, manager, and prolific song-writer, an outstanding figure in the American musical theatre who also wrote plays. He appeared as a child with his parents and sister in an act billed as 'The Four Cohans', and at 15 was writing songs and skits for performance in *vaudeville. In 1901 he made his first appearance in New York in *The Governor's Son*, for which he wrote book, music, and lyrics, and soon built up an enviable reputation with his musicals such as *Little Johnny Jones* (1904) and *George Washington, Jr.* (1906). In 1911 he opened his own playhouse, the *George M. Cohan Theatre, with his comedy *Get-Rich-Quick Wallingford*, appearing there later the same year in his musicals *The Little Millionaire* and *Forty-Five Minutes from Broadway*. Among his later successes were the comedy *Seven Keys to Baldpate* (1913) and *The Song and Dance Man* (1923), in which he played a second-rate *variety performer who thinks he is good. He was also a success in *O'Neill's *Ah, Wilderness!* (1933) and *Kaufman and Moss *Hart's *I'd Rather be Right* (1937). From 1904 to 1920 he ran a successful producing partnership with Sam H. Harris (see SAM H. HARRIS THEATRE). Among Cohan's best known songs are 'The Yankee Doodle Boy', 'Give My Regards to Broadway', and 'Over There'.

Cohan and Harris Theatre, New York, see SAM H. HARRIS THEATRE.

Coleridge, Samuel Taylor (1772–1834), English poet, critic, and philosopher, author of several plays in verse, one of which, *Remorse*, written in 1797 as *Osorio*, was produced at *Drury Lane in 1813 with moderate success. The others, which include several translations from the German, remained unacted, except for a Christmas entertainment which, with alterations by *Dibdin, was produced in 1818. Coleridge's chief importance in theatre history lies in his critical and editorial work on Shakespeare, though he was handicapped by ignorance of Elizabethan stage conditions, which the Shakespearian scholar Malone was only just bringing to light.

Coliseum, The, London, in St Martin's Lane, was originally a *music-hall, which opened under the management of Sir Oswald Stoll in

1904. It was the first London theatre to have a revolving stage, made up of three concentric circles. Seating 2,358 people in four tiers, it had a proscenium width of 49 ft. and a stage depth of 80 ft. Not at first successful, it closed for 18 months from June 1906, but from 1908 onwards its fortunes improved and many famous 'turns' were seen. Actresses appearing there included Ellen *Terry and Edith *Evans in 1917 in scenes from Shakespeare, Lillie *Langtry, and Sarah *Bernhardt. Diaghilev's Russian Ballet gave three seasons between 1918 and 1925. In 1931 variety gave way to *musical comedy, inaugurated by the long-running *White Horse Inn*. Subsequently the theatre was used for every kind of entertainment, including ice-shows, and a Christmas pantomime became an annual event for some years from 1936. In 1945 a spectacular *revue, *The Night and the Music*, began a long run, after which the theatre housed a series of American musicals—*Annie Get Your Gun* (1947), *Kiss Me, Kate!* (1951), *Guys and Dolls* (1953), *Can-Can* (1954), *The Pajama Game* (1955), *Damn Yankees* and *Bells are Ringing* (both 1957), and *The Most Happy Fella* (1960). The theatre then closed, reopening in 1961 as a cinema. In 1968 the building was modernized and taken over by the *Sadler's Wells Opera Company to become the permanent London home of the English National Opera.

Collective Creation, theatrical process whereby a group of persons working together develop a production from initial concept to finished performance through research, discussion, improvisation, writing, and rehearsal. This process, which became the typical method of the alternative theatre movement of the 1960s and early 1970s, appears to have much in common with the way the *commedia dell'arte* companies worked; and actor-playwrights such as Shakespeare and *Molière collaborated with other members of their companies in the productions of their plays.

At first some of the theatre groups of the 1960s were so idealistic that everyone did everything regardless of skill or interest. The process by which works were created was often considered more important than the finished product because it was in keeping with the anti-authoritarian and anti-specialization mood current among young people. However, as idealism gave way to greater pragmatism in the mid-1970s, the strict democratic principles of the method gave way to more practical ways of distributing responsibility.

Theatre collectives have varied greatly in their structures. From the beginning the *Living Theatre was directed by Julian Beck and Judith Malina, who made most of the important decisions, but other members were encouraged to participate in the creative process. Joseph Chaikin was the director of the *Open Theater throughout its 10 years, but the productions were created by the entire group under his guidance. The San Francisco Mime Troupe, a *political theatre troupe formed in 1959, no longer publicly identified actors, directors, or writers after its director was replaced by a collective organization; one or more members were assigned to write scripts and others to direct within guidelines determined by the group. The early productions of the Performance Group, founded in New York in 1967, were developed under its founder-director **Richard Schechner** (1934–), who guided improvisations which were the chief means used by the group to create material which was then shaped by Schechner. In England members of the People Show, formed in 1966, often worked independently on a forthcoming production and then integrated the separate contributions. In Welfare State International, formed in 1968, members of the company were responsible for sculptural elements within the general 'event' or 'celebration' devised by the two directors. In the Pip Simmons Theatre Group, also formed in 1968, Simmons served both as writer and director, but was open to the ideas of others in the company.

Collier, Constance (1878–1955), English actress, on the stage as a child, who in 1893 became one of the famous Gaiety Girls. She made a great success as Chiara the Gypsy in Esmond's *One Summer's Day* (1898), and later joined Fred *Terry to play Lady Castlemaine in *Kester's *Sweet Nell of Old Drury* (1900). Engaged by Beerbohm *Tree for *Her (then His) Majesty's Theatre, she remained with him from 1901 to 1908, appearing in all his major productions, and then made her first appearance in New York, dividing her time thereafter between London and the United States. She was much admired as Nancy in a dramatization of *Dickens's *Oliver Twist* (1905), as the Duchess of Towers in the dramatization of George Du Maurier's novel *Peter Ibbetson* (1915), and as the Duchesse de Surennes in *Maugham's *Our Betters* (1923); in 1925 she played Gertrude to John *Barrymore's Hamlet in London. The best of her later parts was Anastasia in the New York production of G. B. Stern's *The Matriarch* (1930). She was part-author with Ivor *Novello, who appeared in both plays, of *The Rat* (1924) and *Down Hill* (1926), which she directed herself.

Colline, Théâtre National de la, Paris, seating 800, opened in 1988 as the seventh French national theatre. Located, like the Théâtre National Populaire (see PLANCHON), in a working-class area, it is planned as a venue for new works. It opened with the first French production of *García Lorca's *The Public*, and has a small modular studio theatre.

Collins, Lottie, see MUSIC-HALL.

Collins [Vagg], Sam (1826–65), chimney-sweep who went on the 'halls', and was probably the first, and certainly the best remembered, of the Irish Comedians. He first appeared at *Evans's 'song-and-supper rooms', making his reputation with such ballads as 'The Rocky Road to Dublin', 'Paddy's Wedding', and 'The Limerick Races'. He starred at the *Canterbury and most of London's other major halls, and in 1858 opened the Marylebone Music-Hall. Later he took over the Lansdowne Tavern in Islington Green, which he opened as Collins's Music-Hall in 1863, his widow continuing to run it with great success. In 1897 it became a repertory theatre, and it was eventually demolished. On Collins's grave in Kensal Green cemetery are carved pictures of his billycock hat, his shillelagh, and the shamrock he always wore on stage.

Collins, (William) Wilkie, see DICKENS.

Colman, George the elder (1732–94), English dramatist and theatre manager, who was attracted to the theatre by his friendship with *Garrick, to whom his first play *Polly Honeycombe* (1760) was originally attributed. It was not acknowledged by its author until after the success of *The Clandestine Marriage* (1766), which represents Colman's best work. The play was frequently revived, the part of Lord Ogleby, which Garrick refused, considering it not suited to him and leaving it to Tom *King, being a favourite with many leading character actors. To show his displeasure with Garrick, Colman took over *Covent Garden, remaining there until 1774, and producing there a number of good plays, among them those of *Goldsmith. In 1776 he moved to the *Haymarket, where notable events of his excellent management were John *Gay's *Polly* (1777) and some of the early works of his own son George (below).

Colman, George the younger (1762–1836), English dramatist and theatre manager, son of the above. His early works were produced at the *Haymarket Theatre by his father, the most successful being the comic opera *Inkle and Yarico* (1787). In 1794 he succeeded him as manager, but was not a success, being reckless and extravagant and constantly involved in lawsuits; he was also hampered by a secret marriage he had contracted in 1784 with a young actress. But he was a good dramatist, excelling in the creation of comic characters, among them Dr Pangloss in *The Heir at Law* (1797) and Dennis Brulgruddery in *John Bull; or, The Englishman's Fire-side* (1803). The best of the others were probably *The Iron Chest* (1796), which later supplied *Kean and other tragedians with a fine part in Sir Edward Mortimer, and *Blue-Beard; or, Female Curiosity* (1798), based on a French original. These two were first produced at *Drury Lane. Colman was imprisoned for debt

in 1817, and in the same year wrote for the elder *Mathews *The Actor of All Work; or, First and Second Floor,* in which two rooms were shown on the stage simultaneously, an innovation. From 1824 until his death he was Examiner of Plays, in which capacity he showed an unexpected prudery.

Colonial Theatre, New York, see HAMPDEN.

Colum, Padraic (1881–1972), Irish dramatist and man of letters, three of whose early plays had a strong influence on the development of realistic Irish drama. *Broken Soil,* a moving and imaginative study of the vagrant artist in conflict with the practical demands of the life of a small farmer, was produced by the *Irish National Dramatic Society (1903) and at the *Abbey Theatre (1905). Revised as *The Fiddler's House* it was seen again in 1907 (Abbey, 1919). *The Land* (1905) is a firmly drawn study of farming life, with memorable characters, depicting the conflict between two generations and between the love of Irish soil and the longing for emigration. The setting of *Thomas Muskerry* (1910) is a small town, the drab life of whose inhabitants is meticulously delineated. Colum's relationship with the Abbey directors, particularly *Yeats, was always uneasy and in 1905 he left the theatre, emigrating to the United States in 1914. He achieved an enviable reputation as a poet and writer, but few of his later plays were successful.

Columbine, young girl of the English *harlequinade, daughter or ward of the old man Pantaloon and in love with *Harlequin, with whom she eventually elopes. She evolved from one of the maidservants in the *commedia dell'arte whose name, Colombina, was used by several actresses of the *Comédie-Italienne in Paris in the late 17th century and so passed into England through Weaver's 'Italian Night Scenes'.

Columbus Circle Theatre, New York, see MAJESTIC THEATRE I.

Comedia, in the Golden Age of Spanish drama a secular play, whether serious or comic, as distinct from the religious and devotional *auto sacramental.* There were two main categories—the *machine play (*comedia de tramoyas, de apariencia,* or, more expressively, *de ruido*) and the *comedy of wit (*comedia de ingenio*), which depended on intrigue rather than spectacle. The cloak-and-sword play (*comedia de capa y espada*) was a subdivision of this latter, the characters introducing the necessary complications into the plot by disguising themselves in the cloak and fighting the resultant duels with the sword. Spanish 19th-century dramatists were very much attracted to this type of play, and the name then came to mean any romantic costume play with a strong love interest and some sword-play.

In the foreword to his *Propalladia,* published

in 1517, *Torres Naharro distinguishes further between realistic comedy (*a noticia*) and the purely imaginative (*a fantasía*). The comedy of character (*a figurón*), in which *Moreto y Cabaña excelled, is concerned with universal rather than individual characteristics.

Comédie-Ballet, form of entertainment in vogue at the Court of Louis XIV, mingling speech, usually in rhymed couplets and originally of little importance, songs, and dancing, in which the King and his courtiers usually joined. Although the genre was not invented by *Molière, he gave it literary form, making the dialogue equal to or even more important than the music. His first *comédie-ballet* was *Les Fâcheux* (1661) but the finest example of it is *Le Bourgeois gentilhomme* (1671). Among Molière's collaborators in the preparation of these Court spectacles were Philippe *Quinault and, on one occasion, in *Psyché* (1671), the great *Corneille himself. The music was by Lully, who eventually obtained a monopoly of music in Paris which effectively put an end to the development of such 'plays with music', and so hastened the overwhelming popularity of opera and operetta.

Comédie-Canadienne, Théâtre de la, see GÉLINAS.

Comédie des Champs-Élysées, Paris. Built in 1912 in the Avenue Montaigne as an annexe to the concert hall of the same name, the reputation of this theatre was made in the 1920s with *Gémier's avant-garde productions of plays by *Strindberg and *Shaw. He was followed briefly by the *Pitoëffs, and then by *Jouvet, who began his collaboration with *Giraudoux at this theatre in 1928 with a production of *Siegfried*. During the 1950s the theatre was owned by *Anouilh, who staged *boulevard plays and revivals of his own works. The 1960s were similarly almost totally devoted to Anouilh, but later the work of a variety of authors was staged, notably a revival in 1983 of *Pirandello's *Each in His Own Way*.

Comédie-Française, La, foremost theatre of France, founded in 1680 by Louis XIV, who ordered the fusion of the company at the Hôtel de *Bourgogne with the already amalgamated troupes of *Molière (who had died in 1673) and the Théâtre du *Marais. Also known as Le Théâtre-Français and La Maison de Molière, it was apparently called the Comédie-Française to distinguish it from the Italian company who took over the Hôtel de Bourgogne as the *Comédie-Italienne. At first the new company continued to play in the theatre in the rue Guénégaud, with Mlle *Champmeslé, Armande *Béjart (Molière's widow), *Baron, and the elder *Poisson as its leading members. In 1689 they moved to a new theatre specially built for them in the tennis-court of the Étoile, St Germain-des-Prés, where they remained until

1770. After some years in the *Salle des Machines, at the Tuileries, the company moved in 1781 to a new theatre on the present site of the *Odéon. The Revolution caused a split, the more revolutionary actors headed by *Talma going to the *Palais-Royal renamed the Théâtre de la République, while the others under *Molé remained *in situ* in what was renamed the Théâtre de la Nation. The second group soon lost the favour of the public, who considered them 'aristos', and in 1799 the company of the Comédie-Française was reconstituted in the theatre occupied by Talma, though in the process it lost the monopoly it had enjoyed for so long.

The organization by Louis XIV of the Comédie-Française, formalized by Napoleon in 1812, resembled that of the medieval *Confrérie de la Passion. The company is a co-operative society in which each actor holds a share or, in the case of younger or less important actors, a half or quarter share. Admission depends on merit, and aspiring players are allowed to choose their own parts in tragedy and comedy for their début. If successful the newcomer is then on probation as a *pensionnaire*, drawing a fixed salary. After a time, which may vary from weeks to years, they may be admitted to the company as full members or *sociétaires*, taking the place of former members who have died or resigned. On retirement, which is not usually permitted under 20 years' service, the *sociétaires* are entitled to a pension for the rest of their life. The oldest actor in years of service is the nominal head of the company and is known as the *doyen* (though a director under contract now handles its day-to-day running). Some reorganization took place in 1945 and again in 1959, when the Odéon, till then the second theatre of France and attached to the Comédie-Française, was separated from it under Jean-Louis *Barrault, only to be returned in 1971. In the 1970s and 1980s the company's evolving repertoire included *Brecht, T. S. *Eliot, *Arrabal, and *Genet. Antoine *Vitez was its Director from 1988 until his death in 1990, staging *Beckett's *Fin de partie*.

Comédie-Italienne, La, Paris, name given in 1680 on the formation of the *Comédie-Française to the Italian *commedia dell'arte* actors who then took over the Hôtel de *Bourgogne. They had used the theatre ever since *Ganassa's first visit in 1570–1 and now settled in to play every day except Friday. In the company were Tiberio *Fiorillo, Domenico *Biancolelli and his two daughters, and Angelo *Costantini, known as Mezzetin. For some years before 1680 the company had been interpolating French songs and phrases, and even whole scenes, into their Italian texts, and Biancolelli persuaded Louis XIV to allow the Italians the free use of the French language. This soon led to the acting of some plays entirely in French, and contemporary dramatists such as *Regnard, took advantage of this new outlet. The acting, however, continued to be purely that of the commedia dell'arte, and even in French plays the actors figured under their own names and were allowed ample scope for improvisation. After several warnings they were expelled from France in 1697 because, it was said, they had offended Mme de *Maintenon by playing Lenoble's La Fausse Prude, which the audience took delight in applying to her. After Louis XIV's death in 1716 they returned under the leadership of the younger *Riccoboni, and again settled in the refurbished Hôtel de Bourgogne. They now used only French on stage and found that the younger dramatists were ready to write for them. They appeared in, among others, many of the finest plays of *Marivaux, no longer playing in the commedia dell'arte style but in a new and specialized manner in which foreign and native material were blended. In 1723, on the death of the Regent, they were given the title comédiens ordinaires du roi, with a yearly grant from public funds which is still paid to their successors. From this time they were Italian in name only, their productions ranging from true comedy to ballet-pantomime and the newly fashionable *vaudevilles. French actors finally ousted the Italians, of whom the last was Carlin *Bertinazzi. In 1752 the Comédie-Italienne ventured into opera buffa, with such success that they were able to absorb their rivals, though at the expense of their former repertory and individuality. In 1783 they left the Hôtel de Bourgogne and opened a fine new theatre called after them on the boulevard des Italiens. The success in 1789 of a rival theatre (known under the Revolution as the Théâtre Feydeau, while the Italians called themselves the Théâtre Favart) nearly ruined them, but in 1801 the two groups amalgamated as the Opéra-Comique, the name given to their present theatre, built in 1835 after the destruction of the old one by fire.

Comédie Larmoyante ('tearful comedy'), type of play popular in 18th-century France, with which the name of *La Chaussée is especially associated. While retaining the formal characteristics of stylized verse comedy, these plays presented virtually no traditional comic characters, the writing was sententious and sentimental, and the happy ending was reached only after many tears had been shed during five acts over the misfortunes of virtue unjustly persecuted. This type of drama made a strong appeal to contemporary sensibility.

Comedy, in its modern sense, differs from tragedy in having a happy ending, and from farce in containing some subtlety and character-drawing. The word, meaning 'revel-song' from the Greek comos and ode, was applied to the satiric plays of *Aristophanes and also to the *fabulae of *Terence and *Plautus. By medieval times it merely indicated any tale with a happy ending, particularly one written in a colloquial style and dealing with the love affairs of lesser folk. The Renaissance brought the term back to the theatre, but without its former satiric connotation. Comedy by its very nature resists translation, for it depends far more than tragedy on local and topical interest; innumerable comedies, successful in their own day, have soon been forgotten.

The Comedy of Humours, as practised by *Jonson and *Fletcher, was influenced by classical models. The Comedy of Intrigue, subordinating character to plot, originated in Spain and was practised in England by Mrs Aphra *Behn. With it may be classed Romantic Comedy, which also came from Spain and reached its highest point in France during the Romantic Revival led by Victor *Hugo and the elder *Dumas. It is marked by exaggeration and violence and by an overpowering use of local colour, costume, and scenery. The Comedy of Manners originated in France with *Molière's Les Précieuses ridicules (1658), and Molière himself said that 'the correction of social absurdities must at all times be the matter of true comedy'. Pushing it to its logical extreme, he produced the Comedy of Morals—the correction of abuse by the lash of ridicule—of which the greatest exemplar is Tartuffe.

In England, the Comedy of Manners is represented by the plays of *Congreve, *Farquhar, *Vanbrugh, and *Wycherley, later classed as Old Comedy but now generally known as *Restoration Comedy. The Comedy of Manners revived under *Sheridan with much wit and less indelicacy, but even Sheridan tended to be influenced by the prevalence of Sentimental Comedy, a type of pathetic play best represented by the work of *Steele, which reflected the sensibility of the rising 18th-century middle class. In France this led to the *comédie larmoyante which, after blurring the distinction between tragedy and comedy and ousting them from the stage, was in its turn eclipsed by the *drame bourgeois.

In the English-speaking theatre comedy during the 19th century was virtually replaced by *farce; modern playwrights such as Alan *Ayckbourn, Michael *Frayn, Neil *Simon, and Tom *Stoppard use verbal wit and robust humour to examine basically serious contemporary topics. The lack of true comedy is supplied by frequent revivals of plays by *Wilde and *Coward (perhaps the last writers of English comedy) and Restoration comedies.

The playing of comedy makes heavy demands on the actor, who must be able to suggest elegance, leisure, and a nimble wit. Its broader forms may demand great mobility of countenance and a command of dialect. All comedy acting needs an impeccable sense of timing. In the 19th century comedy was regarded as a specialized art and comedian and tragedian rarely trespassed on each other's territory; but the ability to succeed in both fields is now highly regarded.

For Greek comedy see OLD, MIDDLE, and NEW comedy.

Comedy Theatre, Leningrad, see AKIMOV.

Comedy Theatre, London, in Panton Street, off the Haymarket. This theatre, seating 820 on four tiers, opened in 1881 and housed a series of light operas until in 1887 Beerbohm *Tree took it over for his first venture into management. He was succeeded by Charles *Hawtrey and then by Comyns Carr, under whom Winifred *Emery and her husband Cyril *Maude appeared in Sydney Grundy's *Sowing the Wind* (1893) and *The New Woman* (1894) and *Pinero's *The Benefit of the Doubt* (1895). A lean period followed until in 1902 Lewis *Waller produced *Tarkington and Sutherland's *Monsieur Beaucaire* which ran for over 400 performances. Gerald *Du Maurier had a big success with Hornung's *Raffles* (1906), and Marie *Tempest was also successful in Somerset *Maugham's *Mrs Dot* (1908) and *Penelope* (1909). One of the longest runs was scored by John Hartley Manners's *Peg o' My Heart* (1914). During the First World War the Comedy was occupied by *revue, and its inter-war history was unremarkable. Though slightly damaged by bombing in the Second World War it continued in use. In 1956 it became a club to facilitate the staging of plays which had failed to obtain the *Lord Chamberlain's licence, among them Arthur *Miller's *A View from the Bridge* (1956), Robert *Anderson's *Tea and Sympathy* (1957), and Tennessee *Williams's *Cat on a Hot Tin Roof* (1958). In 1959 the club was disbanded and Peter *Shaffer's first play *Five Finger Exercise* was seen, followed by a dramatization of E. M. Forster's novel *A Passage to India* (1960). Peter *Nichols's *A Day in the Death of Joe Egg* (1967) was widely acclaimed, but farce kept the theatre open from 1968 to 1972. Later productions included Alan *Ayckbourn's *Time and Time Again* (1972) with Tom *Courtenay; the musical *The Rocky Horror Show* (1979); David *Storey's *Early Days* (1980) with Ralph *Richardson and Miller's *The Crucible* (1981), both transferred from the *Cottesloe Theatre; and *O'Neill's *A Touch of the Poet* (1988) with Vanessa *Redgrave.

Comedy Theatre, New York, on West 41st Street, between Broadway and Avenue of the Americas. Built by the *Shuberts as a small, intimate playhouse, it opened in 1909 with Zangwill's *The Melting Pot* and in 1912 saw the first production in New York of *Shaw's *Fanny's First Play*. The *Washington Square Players made some of their early appearances there, as did Ruth *Draper, who in 1928–9

created a record with five months' solo playing. In 1937–8, renamed the Mercury, the theatre housed a stimulating season by Orson *Welles which included *Julius Caesar* in modern dress, in which he played Brutus, *Dekker's *The Shoemaker's Holiday*, *Shaw's *Heartbreak House*, in which he played Captain Shotover, and *Büchner's *Danton's Death*, in which he played St Just. In 1939 the Artef Players appeared in *Gutzkow's *Uriel Acosta* and other Yiddish plays. The building was demolished in 1942.

Commedia dell'Arte, name usually given to the popular Italian improvised comedy first recorded in 1545, which flourished from the 16th to the early 18th centuries. Other names for it are *commedia a soggetto*, because it was acted in accordance with a scenario or pre-arranged synopsis; *all'improvviso*, because the actors made up their speeches as they went along; *dei *zanni*, from the comic servants who provided most of the humour; *dei maschere*, because most of the actors wore masks; and *all'italiana*, because it came from Italy. *Dell'arte*, the only phrase to survive in general use, means roughly 'of the profession', the actors being trained professionals. To distinguish it from this popular theatre, the written Italian drama of this time was known as the *commedia erudita*.

The chief companies of the *commedia dell'arte* in the 16th century were the *Gelosi, with the *Andreini family as their mainstay; the Desiosi, sometimes including Tristano *Martinelli; the *Confidenti; and the *Uniti, under Drusiano *Martinelli. The next generation carried on the tradition in the competing groups of the second Confidenti; the *Accesi; and the *Fedeli, under the younger Andreini known as Lelio.

Of the early patrons of the *commedia dell'arte* the Court of the Gonzaga at Mantua was the most important, followed in the latter part of the 16th century by the Courts of Modena and Parma. The companies soon took to the road, and in the 1570s *Ganassa was already leading a company in Spain, followed there a decade later by the Martinelli brothers. One of these, Drusiano, is the first Italian comedian known to have appeared in England (in 1577–8), and a troupe performed at the English Court in 1602. The proximity of London to Paris, where a permanent Italian company was already settled when *Molière arrived in 1658, meant that visits could easily be made. When the *Comédie-Italienne, as the troupe was called after 1680, was banished from Paris in 1697, some of the players may have come to England, but seem to have had no direct influence on the English theatre.

In a *commedia dell'arte* company each member had his or her own character or 'mask' and played nothing else, though the player of a youthful part might later graduate to an elderly one. The chief masks were adapted to suit successive generations of players, but the basic characteristics of each remained unaltered. The young lovers, whose desire for marriage and its constant thwarting by their elders supplied in general the plot of the play, did not wear masks. In league against them were the old men, fathers or guardians, of whom the most important was the Venetian *Pantalone (later the English Pantaloon), while the favourite mask for the second old man was the Bolognese lawyer, the *Dottore, usually known as Graziano. An independent role, though he could be a rival for the hand of the young girl, was the braggart *Capitano, a satire on alien soldiers and mercenaries currently occupying the country. Round these revolved the numerous servants who helped or hindered the lovers—the *zanni*. It was they who gave the *commedia dell'arte* its characteristic flavour and under various names have infiltrated the literature and theatre of the whole of Western Europe. They took over the functions of the slaves of classical comedy, discharging them with the physical skill of acrobats and the impudence of their immediate prototypes, the *facchini* or odd-job men who lounged about the piazza. Greed, shrewdness, and a love of mischief for its own sake were their outstanding characteristics, added to fertility of invention, an eye to the main chance, and a deep-rooted instinct for survival. Some of the lesser masks displayed a bovine stupidity which provided an amusing contrast to the quick wit of their companions, and all found infinite possibilities for surprise and twists of fortune in the *burla*, or practical joke, and the smaller piece of business known as a *lazzo* which formed the best part of their stock-in-trade. Not all the *zanni* survived, but among those that handed on their masks and are still remembered—often by other names—are *Arlecchino (Harlequin), *Pedrolino (Pierrot), and *Pulcinella (Punch); the only female servant to have survived is Colombina (*Columbine), originally an attendant on the leading lady. The French theatre adopted the masks of *Mezzetino (Mezzetin), *Pasquino (Pasquin), *Scapino (Scapin), and *Scaramuccia (Scaramouche), the last having originally something of the braggart soldier; and from a compound of other masks now forgotten, France created its own inimitable *Crispin.

The average company consisted of 12 to 15 members under an acknowledged leader, though other outstanding performers carried considerable weight, particularly in choice of scenario and methods of staging, the latter varying according to the status of the company and the nature of the place where they were to play. The smaller travelling troupes carried portable equipment which included basic props, costumes, and canvas scenery, and a platform stage for erection in such playing areas—open spaces and public squares—as opportunity

offered. When set up the stage would be at about head-height to standing spectators, the players performing before a painted canvas backcloth on which was usually depicted the traditional comic scene of a piazza or street with houses. For more prosperous companies, who could expect to rent a hall or theatre—as the Andreini did in Paris—a similar background could be provided by portable *wings. Some companies at the highest level might be accommodated in their patron's private theatre when not on the road, and have at their disposal the most sophisticated scenic equipment. Although the essence of the Italian style was improvisation, a skeleton plot, with indications of possible tricks and cues for music and dance, helped to ensure that the basic story developed in a way which would lead to the desired denouement. Successful extemporizing depended on the players' innate theatrical sense, their ability to supply and readily pick up cues, and a constant awareness of audience response. Although actors always played the same part, the details of it could vary enormously; preparation for it, even after years of experience, entailed study, including the accumulation of relevant material from any accessible sources, particularly the *zibaldoni* or commonplace books which contained speeches suitable for all occasions; the acrobatic and other spectacular comic scenes demanded constant testing and rehearsing. A number of collections of such skeleton plots survive, of which Flaminio Scala's *Teatro* (1611) is perhaps the most unusual. Other scenarios exist in manuscript in many Italian libraries, in Paris, and in Leningrad. In the larger collections farce, with some comedy, forms the largest part; much is drawn at second or third hand from classical and neo-classical plays; and in the 17th century the taste for Spanish drama (see *comedia) opened the way for a larger proportion of melodramatic and sentimental plots. *Pastorals with a strong infusion of buffoonery were also evidently popular.

By the 18th century the vitality of the *commedia dell'arte* was beginning to flag. France had absorbed much of it into her own drama, the German-speaking countries had drawn on it for their own clowns, among them *Hanswurst, and general decadence had set in. In an attempt to revive the splendour of the old days *Goldoni substituted a written text for improvisation. His contemporary and rival *Gozzi preferred to use the old masks and methods and with him the *commedia dell'arte* made a good end. Improvisation soon became a lost art and actors were no longer expected to be acrobats, dancers, and singers. Something of the tradition lingered on in *puppet-shows, in mask names, and in the English *harlequinade. *Commedia dell'arte* skills, even though no longer practised, remain a vital force in the modern theatre, which with

*Collective Creation is turning to a new form of improvisation.

Commedia Erudita, 'learned' Italian counterpart in the early 16th century to the 'popular' *commedia dell'arte*. Based on the Roman comedies of *Plautus and *Terence, it tended towards the creation of types rather than individuals. Typical of the genre are the comedies of *Della Porta, which are all reducible to a single plot—two young people crossed in love, whose difficulties are resolved by a servant; the plays usually end with a recognition scene, after it has been revealed that one of the characters has previously been carried off by pirates, sold into slavery, or captured by the Turks. The writers of learned comedies were convinced that they had breathed new life into classical comedy—as indeed they had, but mainly thanks to their observation of contemporary life and manners. Among the masterpieces of the *commedia erudita* are the comedies of *Aretino and *Ariosto, *La Calandria* (1506) by Cardinal Bibbiena, and *La Mandragola* (1520) by *Machiavelli.

Commonwealth Company, USA, see LUDLOW and SMITH, SOL.

Community Theatre, form of play-making which involves a mainly non-theatrical community in the production of a script usually based on a subject of local interest. It first appeared in England in the 1970s and is closely allied to *political theatre, often having a socialist commitment which may extend to feminism and gay liberation, which also have their specialist groups. Community plays are given in working men's clubs, public houses, village halls, or in streets and open spaces. The first community theatre in England was Inter-Action, founded in 1968, which ran Dogg's Troupe, presenting plays for children with audience participation on housing estates, and also a Fun Art Bus for other theatrical activities. Other community theatre groups included the Covent Garden in London, the Interplay in Leeds and West Yorkshire, the Second City in Birmingham and the West Midlands, and the Solent People's Theatre in Hampshire. A form of community theatre which has no political affiliations developed later, mainly in the Dorset–Devon area under Ann Jellicoe and her successor John Oram, employing a few professional actors with hundreds of local and usually non-theatrical players in a specially written large-scale drama based on local historical, or sometimes contemporary, subjects.

In the United States of America 'community theatre' is largely synonymous with *amateur theatre. (See also COLLECTIVE CREATION and HAPPENING.)

Compagnie des Quinze, see COPEAU and SAINT-DENIS.

Compagnons de Jeux, see GHÉON.

Compass, see QUAYLE.

Composite Setting, modern equivalent of the medieval *multiple setting where the action of a play takes place in several distinct areas, such as rooms, gardens, and streets, shown on stage at the same time. The lighting in each area intensifies as the action moves into it, dimming as it leaves. Usually only one area is lighted at any one time, but two or three areas may be lit and in use. In America the composite setting is known as the Simultaneous-Scene Setting.

Compton, Edward (1854–1918), English actor-manager, son of Charles Mackenzie who took his grandmother's maiden name when he went on the stage, and as **Henry Compton** became a well-known actor-manager. Edward was the founder of the Compton Comedy Company, which from 1881 to 1918 toured the English provinces in a repertory of plays by Shakespeare, *Sheridan, *Goldsmith, and other classic dramatists. An excellent actor, perhaps insufficiently appreciated in his own day, he appeared several times in London, playing opposite his wife Virginia, youngest daughter of H. L. *Bateman. They were the parents of the novelist Compton Mackenzie and of four other children, all on the stage, of which the best known was Fay (see below).

Edward's sister, the actress **Katherine Mackenzie Compton** (1853–1928), made her first appearance in London in 1877 as Julia in Sheridan's *The Rivals*. Adept at portraying society women of shrewd wit and few scruples, she married the dramatist **Richard Carton** [Critchett] (1856–1928) and acted almost exclusively in his plays from 1885 onwards. Carton's first plays were much influenced by *Dickens, but after the success of *Lord and Lady Algy* (1898) he continued to write comedies, bordering on farce, which poked discreet fun at the aristocracy.

Compton, Fay [Virginia Lillian Emmeline] (1894–1978), English actress, daughter of Edward *Compton. She had a long, varied, and distinguished career, making her début in the *Follies* of her first husband H. G. Pélissier (see REVUE) and being the first to play the title-role in *Barrie's *Mary Rose* (1920) and Phoebe Throssel in his *Quality Street* (1921), as well as Elizabeth in *Maugham's *The Circle* (1921) and Constance in his *The Constant Wife* (1927). She also appeared in Ashley *Dukes's *The Man with a Load of Mischief* (1925). Among her later successes were Fanny Grey in *Autumn Crocus* (1931) and Dorothy Hilton in *Call it a Day* (1935), both by Dodie Smith, Ruth in *Coward's *Blithe Spirit* (1941), and Martha Dacre in Esther McCracken's *No Medals* (1944).

At the first *Chichester Festival in 1962 she played Marya in *Chekhov's Uncle Vanya.

In the course of her long career she played Ophelia to the Hamlets of John *Barrymore (1925) and John *Gielgud (1939) and was seen in a number of other Shakespearian parts with the *Old Vic company, for which she also played Lady Bracknell in *Wilde's The Importance of Being Earnest (1959), and at the *Open Air Theatre. She appeared too in pantomime and variety.

Compton, Henry, see COMPTON, EDWARD.

Compton, Katherine Mackenzie, see COMPTON, EDWARD.

Concert Party, see PIERROT TROUPES and REVUE.

Condell, Henry (?–1627), Elizabethan actor, first mentioned in 1598 as playing in *Jonson's Every Man in His Humour. His only other known role is the Cardinal in *Webster's The Duchess of Malfi (1614), but he is believed to have appeared in Shakespeare's plays in such supporting roles as Horatio in Hamlet. Evidently a well-known and respected figure in Jacobean theatre society, and one of the *sharers in both the *Globe and *Blackfriars, he left the stage to devote himself to business matters in about 1616. With his fellow actor *Heminge he was jointly responsible for the collecting and printing in one volume—the so-called First Folio—of Shakespeare's plays, published in 1623, price 20 shillings. Although there is no record of how many copies were printed, a second edition was not called for until 1636.

Confidenti, company of the *commedia dell'arte, first mentioned in 1574, which toured Italy, France, and Spain, and by 1580 was under the control of Giovanni Pellesini. Some of the *Gelosi may have joined this company after the death of Isabella *Andreini in 1604, but its history remains obscure until it re-emerges in about 1610 with **Flaminio Scala** (fl. 1600–21) as its leader. It toured mainly in Italy and was disbanded in 1621, some of the actors joining the *Fedeli in Paris.

Confrérie de la Passion (Confraternity of the Passion), association of the burghers of Paris, formed in 1402 for the performance of religious plays. Their first permanent theatre was in the disused hall of the Guest-House of the Trinity outside the walls of Paris. In 1518 they were given a monopoly of acting in Paris which later proved a serious hindrance to the establishment of a permanent professional theatre. In 1548 the Confraternity built themselves a theatre, the Théâtre de l'Hôtel de *Bourgogne. No sooner was it ready for their occupation than they were forbidden to act religious plays, though still retaining their monopoly. Deprived of the major part of their repertory, they gave up acting, and from about 1570 onwards leased their theatre to travelling and foreign companies, always seeking to prevent any company from settling down. When a company finally became their permanent tenants, they still insisted on the payment of a levy for every performance. The first breach in their privileges was made when in 1595 the *fairs of St-Germain and St-Laurent were thrown open to provincial actors, but the monopoly lingered on until 1675, after the death of *Molière and only a few years before the foundation of the *Comédie-Française.

Congreve, William (1670–1729), English playwright, the greatest writer of *Restoration comedy. Born in England, he was educated in Ireland. In 1693 he had his first play, The Old Bachelor, produced at *Drury Lane with a fine cast headed by *Betterton and Mrs *Bracegirdle. It was well received and was quickly followed by The Double Dealer (1694), played by the same company. This was less successful, perhaps because of its convoluted plot. When in 1695 Betterton seceded from Drury Lane his first production was Congreve's Love for Love, the latter's most successful play and one which calls for a high degree of skill in the acting. Congreve's one tragedy The Mourning Bride (1697) failed to please the critics, though it was a success with the public and the players, the part of Almeria, first played by Bracegirdle, being long a favourite with tragedy queens. Congreve's last and perhaps best play was The Way of the World (1700); it was at first poorly received, and a combination of pique, laziness, and ill health kept the author thereafter out of the theatre. He died of injuries received when his coach overturned on the way to Bath, and was buried in Westminster Abbey. Congreve's three best-known comedies continue to hold the stage. *Gielgud played Valentine in Love for Love in 1943–4; Millamant in The Way of the World has been played by several notable actresses including Edith *Evans (1924), Geraldine *McEwan (1969), Maggie *Smith (1976, 1984), and Judi *Dench (1978).

Connelly, Marc(us) **Cook** (1890–1980), American dramatist, who was working in theatrical journalism in New York when he met George S. *Kaufman, with whom he collaborated in a number of successful plays including Merton of the Movies (1922), about a film-crazy innocent in Hollywood, and Beggar on Horseback (1924; London, 1925), a satire on contemporary American philistinism based on a German play. Connelly is chiefly remembered, however, for his own play, The Green Pastures (1930), awarded a *Pulitzer Prize. Based on Roark Bradford's Southern sketches Ol' Man Adam an' His Chillun, with a cast of Negro actors led by Richard B. *Harrison as De Lawd, this endeavoured to describe the image that black plantation workers in the Southern states had of God and Heaven. It proved immensely popular and ran for over five years, being several times revived. Connelly wrote several other plays but never again achieved such a success.

Conquest, George Augustus (1837–1901), English dramatist, acrobat, and pantomimist, the only son of the theatre manager **Benjamin Conquest** [née Oliver] (1805–72). George Augustus acted under his father's management at the *Garrick Theatre as a child and was then sent to Paris to study, his parents intending him for a musical career. He preferred, however, to return to the theatre, and alone or in collaboration wrote some 40 *pantomimes, celebrated for their brilliant flying ballets and acrobatic effects, and about 100 *melodramas, many of them adaptations from the French. These were mainly produced at the *Grecian Theatre, which he took over when his father died while in management there. A competent artist, Conquest designed, painted, and made scenery, properties, and masks for his productions, in many of which he appeared. He was a powerful character actor, and an excellent animal impersonator. Off-stage he stammered badly but this was never apparent in his acting. In 1879 he sold the Grecian and embarked on what was intended to be a lengthy tour of the United States, but a serious accident on stage soon after his first appearance forced him to return to England. In 1881 he took over the *Surrey Theatre and again made it famous for pantomimes and melodramas. He had three sons, all pantomimists.

Conservatoire, Paris, see SCHOOLS OF DRAMA.

Contact Theatre Company, see MANCHESTER.

Contat, Louise (1760–1813), French actress, who made her first appearance at the *Comédie-Française in 1776. In tragedy she was distinguished but cold, and it was soon evident that her bent was for high comedy. Her success as Suzanne in the first production of *Beaumarchais's Le Mariage de Figaro (1784), in which her younger sister, aged 13, appeared as Franchette, set the seal on her growing reputation. She was also particularly admired in the plays of *Marivaux. She retired at 50, having given up youthful parts in favour of elderly matrons owing to her increasing size.

Contemporary Theatre, A, see SEATTLE REPERTORY THEATRE.

Conti, Italia (1874–1946), English actress, who made her first appearance on the stage in 1891 at the *Lyceum and later toured extensively in England and Australia. In 1911 she was asked by Charles *Hawtrey to train the children, among them Noël *Coward, who were to appear with him in the fairy play Where the Rainbow Ends. This led her to found the Italia Conti School for the stage training of children, to which she devoted the rest of her life,

one of her outstanding pupils being Gertrude *Lawrence. She continued to act intermittently, being seen annually from 1929 to 1938 at the Holborn Empire as Mrs Carey in the Christmas production of Where the Rainbow Ends.

Conway [Rugg], William Augustus (1789–1828), Irish-born actor, known from his exceptional good looks and fine carriage as 'Handsome' Conway. He first appeared on the stage in Dublin in 1812 and then went to England. He was a good actor but morbidly sensitive, and is said to have thrown up an excellent part in London because of adverse criticism. In 1824 he went to New York and appeared at the *Park Theatre as Hamlet, Coriolanus, Romeo, and Othello with great success. He also played Edgar to the Lear of Thomas *Cooper, Faulconbridge to his King John, and Joseph Surface to his Charles Surface. An actor of the *Kemble school, Conway seemed on the threshold of a brilliant career; but he threw himself overboard on a voyage to Charleston and was drowned.

Cook, Peter, see REVUE.

Cooke, George Frederick (1756–1812), eccentric and unstable English actor, who spent more than 20 years as a strolling player in the provinces before in 1800 he appeared as Richard III at *Covent Garden. He was immediately successful and remained there for 10 years, playing a wide range of parts but constantly in trouble with the management, undisciplined, dissipated, usually in debt, and often in prison. He seemed to act better when drunk and was probably somewhat insane from constant inebriation. In 1810 he went to New York and appeared before an enthusiastic audience at the *Park Theatre, but soon proved as undependable as in London and lost his popularity. Careless in studying his parts, he picked them up quickly and played them intuitively. A powerful actor with fiery eyes and a lofty forehead, he was unequalled at expressing the darker passions.

Cooke, Thomas Potter (1786–1864), English actor, known as 'Tippy', who in 1804 forsook the Royal Navy and appeared at several minor theatres in London. He made his first outstanding success in 1820 as Ruthven in *Planché's The Vampire; or, The Bride of the Isles and was then seen as the Monster in Presumption; or, The Fate of Frankenstein (1823), based on Mary Shelley's novel; but it was as William in the 400 consecutive performances of *Jerrold's nautical melodrama Black-Ey'd Susan (1829) and as Harry Halyard in Edward Blanchard's Poll and My Partner Joe (1857) that he was best remembered.

Cooper, Dame Gladys (1888–1971), English actress, who excelled in drawing-room comedy but was by no means limited to this style. She made her first appearance in London in 1906 in The Belle of Mayfair, and was subsequently seen in both *musical comedy and straight plays with equal success. In 1916 she became joint manager of the *Playhouse (sole manager from 1928), presenting and often playing in a varied programme of old and new plays. She was particularly admired in revivals of *Pinero's *The Second Mrs Tanqueray* in 1922 and *Sudermann's *Magda* in 1923 and starred in *Maugham's *Home and Beauty* (1919), *The Letter* (1927), *The Sacred Flame* (1929), and *The Painted Veil* (1931). She occasionally appeared for other managements, playing opposite Gerald *Du Maurier at the *St James's Theatre in *Lonsdale's *The Last of Mrs Cheyney* (1925). After remaining at the Playhouse until 1933 she made her first appearance in New York in 1934. Her later stage career was divided between London and New York in roles ranging from Desdemona and Lady Macbeth to Fran Dodsworth in Sidney *Howard's *Dodsworth* (London, 1938). She was also seen in London in *Coward's *Relative Values* (1951) and on Broadway as Mrs St Maugham in Enid *Bagnold's *The Chalk Garden* (1955; London, 1956).

Cooper, Thomas Abthorpe (1776–1849), British-born actor, one of the first to become an American citizen. In 1795 he appeared at *Covent Garden as Hamlet and Macbeth, and a year later went with *Wignell to the *Chestnut Street Theatre in Philadelphia. In 1797 he was seen in New York as Jaffier in *Otway's *Venice Preserv'd*, always one of his best parts. He was also much admired in the title-role of *Kotzebue's *The Stranger* (1798). After a quarrel with Wignell he joined the *American Company under *Dunlap and in 1806 took a lease of the *Park Theatre in New York, but without much success. He then toured the eastern circuit—New York, Philadelphia, and Charleston, South Carolina—in the big tragic roles of Shakespeare. A handsome man, with a fine voice, much eloquence, and a dignified presence, he was a fine tragic actor, though he continued to act for too long and towards the end of his life his popularity declined.

Co-Optimists, see PIERROT TROUPES.

Copeau, Jacques (1879–1949), French actor and director, whose work had an immense influence on the theatre in Europe and America. Although interested in the innovations of *Antoine, he disliked the over-realistic theatre and in 1913 took over the Théâtre du *Vieux-Colombier in an effort to bring back 'true beauty and poetry' to the French stage. His repertory was based mainly on *Molière and Shakespeare. One of his earliest successes was in 1914 with *Twelfth Night*, translated as *Nuit des rois*. In 1921 he produced *The Winter's Tale* as *Conte d'hiver*, and later published French versions of Shakespeare's major tragedies. He trained his actors himself; in his original company were Charles *Dullin and Louis *Jouvet who later

founded their own theatres. During the First World War Copeau spent two years (1917–19) at the *Garrick Theatre in New York, and he returned to the Vieux-Colombier convinced that the training of young actors was an essential part of the reforms he envisaged. In 1924 he withdrew to his native Burgundy with a group of youngsters, later known as 'les Copiaus', with whom he worked on various facets of technical training. In 1930 several of his actors joined the Compagnie des Quinze, directed by his nephew and collaborator Michel *Saint-Denis. By 1936 the importance of Copeau's work had been recognized and he became one of the directors of the *Comédie-Française, retiring in 1941.

Coppin, George Selth (1819–1906), English-born actor-manager, usually referred to as 'the father of the Australian theatre'. The son of strolling players, he arrived in Australia in 1842 and at various times was in actor-management in Sydney, Tasmania, and Adelaide, finally settling in Melbourne. A low comic and clown in Shakespeare, he brought over 200 entertainers to Australia and built six theatres, the most famous, the Olympic in Melbourne (known as 'the Iron Pot'), having been prefabricated in Manchester. Actors brought to Australia by Coppin included Charles *Kean, Edwin *Booth, Mme *Céleste, and the one who was to have the greatest effect upon the Australian theatre, J. C. *Williamson.

Copyright in a Dramatic Work. 1. *Great Britain.* By the beginning of the 19th century, the right of the dramatist (as of other writers) to prevent for a limited period of time the making of copies of his work by unauthorized persons had been established by various Acts of Parliament. The Dramatic Copyright Act of 1833, commonly known as *Bulwer-Lytton's Act, finally gave to the author of 'any tragedy, comedy, play, opera, farce, or other dramatic entertainment', again for a limited period only, the sole right to perform it or authorize its performance. In 1842 the Literary Copyright Act consolidated the law relating to the protection of dramatic as well as literary and musical property and brought within the terms of a single statute the two rights so far recognized, 'copyright' or the right of 'multiplying' copies, and 'performing right' or the right of representation. Performances of dramatizations of non-dramatic works were not covered by the Act. It was generally believed, though with no clear legal support, that if a play was published before being performed the 'performing right' was lost. This led to the practice, much indulged in by *Shaw, for instance, of hiring actors to give a public reading of a manuscript play. These 'copyright performances' have often led to considerable confusion in the dating of first performances. Meanwhile the Berne Convention of 1886 had established the International Copyright Union, amended in 1896 and 1908, and it was deemed essential that Great Britain should keep abreast of international developments. This resulted in the Copyright Act of 1911, which for the first time merged the right of multiplying copies and the right of representation in the general term 'copyright'; and for the first time no formalities were necessary to obtain copyright. This Act remained in force until the Copyright Act of 1956, made necessary by the production of films, records, radio, and television, and by the provisions of the 1948 Brussels Convention and the Universal Copyright Convention of 1952.

The 1956 Act covered the reproduction in public of any original dramatic work written by a British subject or other 'qualified person', in any form, including adaptations into or from dramatic works. The copyright restrictions apply as much to a substantial part of a work as to the whole. Where there was no verbatim copying, both the plot and the dialogue and working out of the play must be regarded. Proof that a later author had no knowledge of an earlier author's work would be a good defence, however similar the two works might be. The normal period of copyright protection is 50 years from the death of the author, or the last surviving author.

2. *The United States.* American copyright derived initially from a British Act of 1710. The first comprehensive American Act was passed in 1790. During the 19th century various amendments were made, but by the early 20th century the mass of piecemeal copyright legislation and case law had reached the same confused state as in Great Britain, and a report on the position made by the first Recorder of Copyrights in 1903 prepared the way for the Copyright Act of 1909. This provided for two separate terms of copyright: a period of 28 years from publication, followed by a renewal period of a further 28 years. Proceedings for infringement of the copyright in a play cannot be instituted in the United States if the play has not been registered with the Library of Congress.

3. *International Copyright.* The Berne Convention of 1886, amended in 1896, 1908, 1914, 1928, 1948, 1967, and in Paris in 1971, established an International Copyright Union and in 1980 had more than 70 member states. As well as the UK and the Commonwealth and all European countries, it includes among its membership Japan, Mexico, and a number of states in Africa and South America. All adhere to the principle that a national of any member state of the Union enjoys in every other country of the Union the rights of a national of that country. No registration or other formality is required.

The most prominent absentees from the Union are the USA, the USSR, and China. The first two of these are, however, members of the Universal Copyright Convention of 1952, which has almost as many members as the Berne Union. It requires a minimum period of copyright protection of 25 years from the death of the author. This Convention is supplementary to and not in substitution for the Berne Union, and countries which are members of both Conventions are bound by the latter, which provides more comprehensive protection for a longer period.

Coquelin, Constant-Benoît (1841–1909), French actor, known as Coquelin *aîné* to distinguish him from his brother **Ernest-Alexandre-Honoré** (1848–1909), also an actor, known as Coquelin *cadet*. They were both at the *Comédie-Française, Coquelin *aîné* finally leaving the company in 1892 after a long absence on tour in Europe and America. Six years later he appeared at the *Porte-Saint-Martin as *Rostand's Cyrano de Bergerac, his son Jean playing Ragueneau the pastrycook. In 1900 he toured with Sarah *Bernhardt, with whom he had appeared at her own theatre in Rostand's *L'Aiglon* (1898), and towards the end of his life he was seen frequently in London where he was very popular. He died suddenly while rehearsing Rostand's *Chantecler*, in which he was to have played the lead, Coquelin *cadet* dying insane a fortnight later. A big man with a fine voice and presence, Coquelin *aîné* was outstanding in the great comic roles of *Molière and in romantic and flamboyant modern parts. He was the author of two books, *L'Art et le comédien* (1880) and *Les Comédiens par un comédien* (1882).

Corcoran, Katharine, see HERNE.

Corneille, Pierre (1606–84), French tragic dramatist, born in Rouen. His first plays were mainly comedies, produced in Paris 1629–33. His first tragedy *Médée* (1635) was well but not ecstatically received, and after a quarrel with *Richelieu he returned to Rouen. There he dipped into Spanish literature and on *Las mocedades del Cid* by *Castro y Bellvís he based *Le Cid*. Its production, probably early in 1637, is regarded as marking the beginning of the great age of French drama. First seen at the *Marais with *Montdory in the title-role, it was an immediate success; it was played in translation in London by Beeston's Boys (see BEESTON, WILLIAM) in the same year. A Spanish version was played in Madrid and Corneille's original script was soon translated into several other languages.

The success of *Le Cid* created a number of enemies for Corneille, and a fierce pamphlet war, the *querelle du Cid*, was waged. The public, however, remained faithful to the playwright and was ready to applaud his *Horace* (1640), *Cinna* (1641), and *Polyeucte* (1642). These were all given at the Marais with *Floridor in the title-roles. The year 1643 saw the production of another tragedy, *La Mort de Pompée*, and Corneille's finest comedy *Le Menteur*, based on *Ruiz de Alarcón's *La verdad sospechosa*, in which Floridor played Dorante with *Jodelet as Cliton. Its great success was not repeated with its sequel *La Suite du menteur* (1644). With *Rodogune*, *Théodore* (both 1645), and *Héraclius* (1646), Corneille was established as France's major dramatist, and in 1647 he was elected a member of the French Academy. *Nicomède* (1651), one of his best and most popular plays which *Molière chose for his reappearance in Paris in 1658, was preceded by the rather weak *Don Sanche d'Aragon* (1649) and by *Andromède* (1650), a *machine play written to show off *Torelli's new scenic devices at the *Petit-Bourbon. After the disastrous failure of *Pertharite* (1652) Corneille abandoned the theatre until 1659, when *Oedipe* was produced at the Hôtel de *Bourgogne. It was followed by another machine play *La Toison d'or* (1660), written for the marriage of Louis XIV. *Sophonisbe* (1663) and *Othon* (1664) had little success.

Corneille's diminishing appeal came about partly because of the increasing popularity of *Racine. *Agésilas* (1666) was hampered by the author's employment of a new verse form, while *Attila* (1667) was overshadowed by the success of Racine's *Andromaque* in the same year. *Attila* had been produced by Molière, one of Corneille's greatest admirers, who followed it by a production of Corneille's *Tite et Bérénice* (1670), given at the same time as Racine's play on the same subject at the rival Hôtel de Bourgogne. In Corneille's play the young *Baron had his first adult part as Domitien and Molière's wife Armande *Béjart was Bérénice. The play was moderately successful, and was followed by one of Corneille's most charming works, *Psyché* (1671), written in collaboration with Molière and *Quinault. *Pulchérie* (1672), at the Marais, and *Suréna* (1674), at the Hôtel de Bourgogne, were Corneille's last plays, and both, though successful for a short time, soon fell out of the repertory.

Corneille was brusque and shy with strangers, had none of Racine's easy graces—but was probably a far finer character—and was sometimes too pleased with himself and his work. Yet even Racine at his best could not easily prevail against the popularity of Corneille in his decline. His fame was a little eclipsed in the 18th century, but the 19th restored him to his true place and the best of his plays still hold the stage.

Corneille, Thomas (1625–1709), French dramatist, usually known as Corneille de l'Isle to distinguish him from his elder brother (above). He wrote over 40 plays, mostly successful at the time, though all are now forgotten. *Timocrate* (1656), his first tragedy, performed at the *Marais, was immensely popular. His best plays are considered to be *Ariane* (1672), in which Mlle *Champmeslé and later *Rachel were outstanding, and *Le Comte d'Essex* (1678), given at the Hôtel de *Bourgogne again with Champmeslé. Like his brother, the younger Corneille also wrote a number of comedies, some adapted from the Spanish and some on topical subjects.

Cornell, Katharine (1893–1974), outstanding American actress, daughter of a theatre manager in Buffalo, who made her first appearance with the *Washington Square Players in 1916. In 1919 in London she made a great success as Jo in a dramatization of Louisa M. Alcott's *Little Women*, and in New York she first came into prominence in Clemence *Dane's *A Bill of Divorcement* (1921) and *Will Shakespeare* (1923). In 1924 she made the first of several appearances as *Shaw's Candida, and in 1931 she appeared under her own management as Elizabeth Moulton-Barrett in Besier's *The Barretts of Wimpole Street*, her most famous role. It was directed by her husband Guthrie *McClintic who also co-produced it with her, a pattern to be repeated in all her later work. Her range was impressively wide, including Shakespeare's Juliet (1934) and Cleopatra (1947), Jennifer Dubedat in Shaw's *The Doctor's Dilemma* (1941), Masha in *Chekhov's *Three Sisters* (1942), and *Anouilh's modernized Antigone (1946). She was also seen in new plays as diverse as Sidney *Howard's drama *Alien Corn* (1933), *Behrman's comedy *No Time for Comedy* (1939), and *Fry's verse play *The Dark is Light Enough* (1955). Her final role was Mrs Patrick *Campbell in Jerome Kilty's *Dear Liar* (1960). She retired on the death of her husband in 1961, the recipient of many honorary degrees and other awards.

Cornish Rounds, remains of circular earthworks in Cornwall (particularly the western part) which are now known to have been used as open-air theatres for annual performances of medieval *mystery plays. These could accommodate on banks (usually seven) cut in the rising ground a large number of spectators grouped round a central area or 'playing-place' which must, from the stage directions, have contained *rostrums or stages of varying levels. Some of these theatres were still in use in the 17th century.

Coronet Theatre, New York, see EUGENE O'NEILL THEATRE.

Corpus Christi Plays, see MYSTERY PLAYS.

Cort Theatre, New York, on West 48th Street, between the Avenue of the Americas and 7th Avenue. This opened in 1912 with Laurette *Taylor in the immensely successful comedy *Peg o' My Heart* by her husband John Hartley Manners. In 1916 there was a revival of Hazelton and Benrimo's *The Yellow Jacket*, and in 1919 came *Drinkwater's *Abraham Lincoln*. In 1930 there was a notable revival of *Chekhov's *Uncle Vanya*, and in 1933 *The Green Bay Tree* by Mordaunt Shairp with Laurence *Olivier, in an elegant setting by Robert Edmond *Jones, had a good run. Also successful were the fine war play *A Bell for Adano* (1944), based on a novel by John Hersey; *Anouilh's *Antigone* (1946), with Katharine *Cornell; *The Diary of Anne Frank* (1955) adapted by Frances Goodrich and Albert Hackett; and Dore Schary's play about Franklin D. Roosevelt, *Sunrise at Campobello* (1958). In 1969 the theatre was leased for television productions but it reopened as a legitimate theatre in 1972, its productions including August *Wilson's *Ma Rainey's Black Bottom* (1984).

Cosmopolitan Theatre, New York, see MAJESTIC THEATRE I.

Costantini, Angelo (c.1655–1729), actor of the *commedia dell'arte*, who first appeared with the troupe of the Duke of Modena. In 1683 he joined the *Comédie-Italienne, ostensibly to share the role of Arlequin with *Biancolelli but playing it so seldom that he finally adopted and enlarged that of *Mezzetino. When the Italians were banished from Paris in 1697 Costantini went to Brunswick, where he had the misfortune to be the successful rival in love of the Elector of Saxony, which cost him 20 years in prison. His brother Giovan Battista was also an actor, who as Cintio played young lovers in the Paris troupe.

Costello, Tom, see MUSIC-HALL.

Costume. From earliest times costuming has been an essential part of the theatre. In Greek tragedy the actors wore *masks and long robes with sleeves, quite unlike the dress of the day. In the *Old Comedy of *Aristophanes the *chorus wore symbolic details like horses' heads and tails, or feathered wings, while the chief characters wore loose tunics, grotesquely padded, and a large red leather phallus. This was later abandoned, and in the *New Comedy of *Menander actors wore the ordinary clothes of the time. When the Romans took over Greek tragedy and comedy they adopted the existing costumes, with the addition of the *toga* and *stola* and also, for tragedy, the high boot (*cothurnus) and the exaggeratedly high peak (*onkos*) over the forehead. By the time of the Caesars, stage costumes, except for the popular *pantomimes, which were played almost in the nude, and the patchwork rags of the Atellan farce (see FABULA I: ATELLANA), had become very elaborate and colourful.

When the theatre, which had disappeared with the collapse of the civilized world, was reborn in *liturgical drama, the priests and choirboys wore their usual robes, with some simple additions, such as veils for the women characters, crowns for the Three Kings, or a rough cloak for a shepherd. But once the plays were moved from the church to the market-place and acted by laymen, costuming again became important, and sometimes the *mystery play of the later Middle Ages was expensively dressed, with golden robes for God and his angels, and fantastic leather garments for the Devil and his attendant imps. In the *morality play the allegorical figures representing the virtues and vices wore richly fantasticated versions of contemporary dress, except for the Devil, who kept the costume he had worn in the *miracle play. His attendant, the Old *Vice, was usually dressed as a fool or jester, with a long-eared cap decorated with bells, a cockscomb, and a parti-coloured close-fitting tunic and leggings, reminiscent of the costumes of the Roman farce-players. Much the same type of costume was worn by the masked actors of the *commedia dell'arte*, though except for the lovers, who wore contemporary dress, each character was recognizably from a different part of Italy: *Pulcinella from Naples, *Arlecchino from Bergamo, *Pantalone from Venice, while the *Capitano was a Spaniard, or sometimes a Greek. The gaudy rags of Arlecchino soon became stylized into the diamonds of brightly contrasted silk worn by *Harlequin today, the white belted and bloused suit of Pulcinella was stereotyped by Watteau into the familiar garb of *Pierrot, but otherwise the characters retained their individuality.

The main line of development was through the academic theatre of Renaissance Italy and the Court play. The Italian *intermezzi*, triumphs, and pageants, the French *ballets de cour*, the English *masques, gave immense scope for fantastic mythological costumes, often, for the men, based on the 'Roman' pattern, with a plumed head-dress, a breastplate moulded to the body, and some variation of the Roman kilt; this fashion was adopted on the public stage and continued to be used for tragic heroes for over 200 years. Women's costumes, as always, followed contemporary fashions. Many of the designs for costumes worn during the extravagant entertainments devised on the Continent, particularly in France to enhance the prestige of Louis XIV, still survive. Some of the earliest were by the great Italian stage designer *Torelli, but French designers soon took over, the most influential being Jean *Bérain, whose work, not only for the theatre but over the whole field of decorative art, synthesizes all the tendencies of the time. After the death of Louis XIV the elegant balance of *le style Bérain* declined into the fantasies of rococo, which achieved its finest development on stage in the designs of **Louis-René Boquet** (*fl.* 1760–82), who worked both for the Paris Opéra and for the Court. Like Bérain, he was content to work mainly in a contemporary style, suggesting character or period by some small decorative detail. His costumes, both male and female, are characterized by his use of wide paniers, forming a kind of ballet skirt covered with rococo detail. This panier-skirt reached as far as England, where *Quin wore it in Thomson's *Coriolanus* (1749). Among the reforms of *Garrick was the abolition of such garments in favour of contemporary styles, Macbeth, for instance, being acted in the scarlet of the King's livery. Garrick's leading actresses in tragedy also wore contemporary costume, including the high head-dresses of the 1770s with a crown or flowing veil, or, for Eastern potentates, a turban with a waving plume. Twenty years earlier *Voltaire, helped by the actor *Lekain, had begun a campaign for correct costuming, at least in classical plays. In his *Orphelin de la Chine* (1755) Lekain as the hero wore an embroidered robe in place of the usual panier-skirt, and the heroine, played by Mlle *Clairon, a simple sleeveless dress which was less incongruous than the exaggerated skirt of the time, held out by hip-pads. A great advance towards accuracy was made by the actor *Talma when in 1789 he appeared as Brutus, in Voltaire's play of that name, with bare legs and arms. The reaction of the public was unfavourable, but by the end of the century the Revolutionary passion for anything remotely connected with antiquity had made classical costumes acceptable on the stage.

During the early 19th century the popularity of Scott's novels, many of which were dramatized, led to a growth of interest in 'historical' costumes, which meant in practice the addition of Elizabethan, Stuart, and other details to contemporary dress. The resulting mixture is not without charm, but bears very little relation to historical accuracy. The first step towards 'antiquarian' detail in dress was made by *Planché with his costume designs for Charles *Kemble's production of *King John* at *Covent Garden in 1823. Charles *Kean strove for accuracy in his productions of Shakespeare at the *Princess's in the 1850s, though as Hermione in *The Winter's Tale* Mrs Kean wore a perfectly correct Grecian costume over a crinoline. Mrs *Bancroft, playing Peg Woffington in what she fondly imagined was 18th-century dress, now looks, in the photographs that survive, purely contemporary. Towards the end of the 19th century the triumph of *realism in the theatre meant on the one hand an almost pedantic accuracy in dress which was sometimes very untheatrical, and on the other an increase in the number of modern plays in which the actors wore the dress of the day, often their own clothes. Consequently the designer, excluded from the legitimate stage, let his fancy run riot in lighter musical productions and in opera.

From about this time costume, *scenery, and *lighting began to be seen as interlocking components of the overall design plan. Diaghilev's Russian Ballet, which startled Europe and America in 1911–16, introduced a brilliant non-naturalism in both settings and costumes which influenced all departments of theatre design. In contrast, *Appia and *Craig had been arguing for simplicity, grouping figures statuesquely against austere three-dimensional settings or neutral screens. Abstraction and symbolism swept through the avant-garde theatres of France, Germany, and, especially, Soviet Russia, where the Constructivists of the post-Revolutionary years favoured an impersonal approach with the actor treated almost as an icon. The more extreme examples of modernist costume design were self-defeating, the shape and movements of the human body conflicting with the elements of 'pure' design. Max *Reinhardt instituted a form of stylized realism that was strikingly successful in his large-scale productions, but the main line of development ran from *Piscator through the anti-illusionist work of *Brecht.

As in scenic design the boom in new materials since the 1960s, especially plastics and adhesives, has greatly increased the costume designer's range. For a contemporary piece the designer will often choose ready-made clothes; a play set in the past is rarely dressed exactly in the fashions of the time, but in versions modified to highlight character and mood. Plays by Shakespeare have been subjected to costuming in the style of almost every period and culture, from ancient Greek to futuristic and from Japan to the Wild West.

The production designer will often be responsible for both scenery and costumes, or each may have a separate designer. In either case, close collaboration with the lighting designer as well as with the director is a vital part of the design plan.

Cothurnus, from the Greek *kothornos*, a loose-fitting, thick-soled boot, coming high up the calf. It was worn by Greek women in everyday life, and in the Roman theatre was a distinctive feature of the tragic actor's costume, the soles becoming very much exaggerated, adding inches to the actor's stature and contributing to his padded, unnatural appearance.

Cottesloe Theatre, smallest of the three playhouses which make up the *National Theatre, opening in 1977. Named after the first Chairman of the South Bank Theatre Board, which is responsible for the structure, it is used mainly for testing new techniques and presenting new or little-known plays. It is a flexible theatre in which the position and shape of the stage can be changed, the seating capacity—maximum 400—varying according to the arrangement. Plays which have had their first productions in the Cottesloe include Julian Mitchell's *Half-Life* (1977), with John *Gielgud, Charles *Wood's

Has 'Washington' Legs? (1978), with Albert *Finney, and David *Storey's Early Days (1980), with Ralph *Richardson. David *Mamet's American Buffalo had its British première there (1978) and his Glengarry Glen Ross its world première (1983). In 1978 Lark Rise and Candleford, based on Flora Thompson's book, were presented as promenade productions (see THEATRE-IN-THE-ROUND). The same technique was adopted for Bill *Bryden's productions of parts of the York Cycle of *mystery plays, beginning with The Passion (1978, 1980). An *O'Neill season in 1980 consisted of Hughie, The Iceman Cometh, and The Long Voyage Home. Notable later productions were Sam *Shepard's True West (1981) and Fool for Love (1984); *Pinter's Other Places (1982); Arthur *Miller's The Crucible (1980), The American Clock (1986), and A View from the Bridge (1987); and Nicholas Wright's Mrs Klein (1988). Many Cottesloe productions have moved to the West End.

Council for the Encouragement of Music and the Arts, see ARTS COUNCIL.

Counterweight System, modern method of flying scenery by means of endless lines, pulleys, and counterweights, as opposed to the traditional system of lines from a fly-floor used in *hand working. In a counterweight house the lines, usually made of rope, are attached to *barrels from which the scenery is hung.

Country Playhouse, Westport, Conn., see LANGNER.

Courtenay, Tom [Thomas] Daniel (1937–), English actor whose haggard good looks were admirably suited to Konstantin in *Chekhov's The Seagull, in which part he made his professional début with the *Old Vic company in Edinburgh in 1960. After taking over from Albert *Finney as the fantasizing hero of Keith Waterhouse and Willis *Hall's Billy Liar (1961), he appeared with the *National Theatre company in Max *Frisch's Andorra (1964), and two years later began a long association with the 69 Theatre Company in Manchester (see ROYAL EXCHANGE THEATRE), his roles including Lord Fancourt Babberley in Brandon *Thomas's Charley's Aunt (1966), Hamlet (1968), Young Marlow in *Goldsmith's She Stoops to Conquer (1969), and *Ibsen's Peer Gynt (1970). He was seen in London in two new works by Alan *Ayckbourn—Time and Time Again (1972) and the trilogy The Norman Conquests (1974)—and made his New York début in Simon *Gray's Otherwise Engaged (1977). He returned to Manchester and London in Ronald Harwood's The Dresser (1980; NY, 1981) and *Molière's The Misanthrope (1981), and was also effective in the latter's The Hypochondriac (1987). Courtenay's voice lacks the richness needed for the great poetic roles, but he is an actor of great integrity

and versatility, being particularly good in comedy.

Court Fool, member of the Royal Household, also known as the King's Jester, not to be confused with the humbler *Fool of the *folk festivals. His origin has been variously traced to the Court of Haroun-al-Raschid, to the classical dwarf-buffoon, and to the inspired madman of Celtic and Teutonic legend. At some point in his career he adopted the parti-coloured costume of his rival, which led to confusion between the two types, but the Court Fool has nothing else in common with the folk tradition, whose Fool is nearer to the *clown. Shakespeare's Fools derive from the Court Fool, already a tradition in his time. In dramatic use they serve as vehicles for social satire.

Courtneidge, Dame (Esmeralda) **Cicely** (1893–1980), British comedienne of tremendous vitality, born in Sydney, who made her first appearance on the stage in Manchester at the age of 8 playing Peaseblossom in A Midsummer Night's Dream. After touring Australia with her father, the actor, manager, and producer Robert Courtneidge (1859–1939), she made her first appearance in London in 1907 and was seen in a variety of musical plays, including The Arcadians (1909) and The Mousmé (1911). She then went on to the *variety stage and into *pantomime and *revue, scoring a success in the revue By-the-Way (London and NY, 1925). It was directed by her husband **Jack Hulbert** (1892–1978) who also appeared in it. They were together in many subsequent productions including the revue The House That Jack Built (1929) and the musicals Under Your Hat (1938), Full Swing (1942), and Something in the Air (1943). She starred without her husband in musicals such as Her Excellency (1949) and *Novello's Gay's the Word (1951). She also appeared in Ronald *Millar's comedy The Bride and the Bachelor (1956) and its sequel The Bride Comes Back (1960), and played another straight comic part in Move Over, Mrs Markham (1971) by Ray Cooney and John Chapman—her last appearance in the West End.

Court Street Theatre, see MILWAUKEE REPERTORY THEATRE.

Court Theatre, London, see ROYAL COURT THEATRE.

Covent Garden, Theatre Royal, London, in Bow Street. There has been a theatre on the site of the present Royal Opera House, Covent Garden, since 1732, when John *Rich, holder of the patent granted to Sir William *Davenant (see PATENT THEATRES), built one on land owned by the Duke of Bedford that had once been part of a convent garden. It seated 1,897 in pit, amphitheatre, two galleries, and three tiers of side-boxes. The proscenium opening was 26 ft. and the stage depth 42 ft., with an apron stage giving an extra 13 ft. The new theatre opened in 1732 with a revival of *Congreve's *The Way of the World* acted by the company headed by James *Quin which Rich had brought with him from *Lincoln's Inn Fields. For several years plays alternated with the operas of Handel. Peg *Woffington, George Anne *Bellamy, Spranger *Barry, and for a short time *Garrick performed there. Rich's son-in-law, who inherited the patent in 1761, was interested only in opera and in 1767 sold it to George *Colman the elder and three partners, of whom one became sole manager in 1774 after a good deal of wrangling and some physical violence. In the previous year *Goldsmith's *She Stoops to Conquer* had had its first performance, and *Macklin had appeared as Macbeth. He was to be seen at Covent Garden again in his own play *The Man of the World* (1781) and his last appearance was made there in 1789. In 1775 *Sheridan's *The Rivals* had its first performance. In 1782 the auditorium was gutted and reconstructed, the whole building being again reconstructed within the original walls in 1792. George Frederick *Cooke was mainly associated with this theatre, making his London début there in 1800. Three years later John Philip *Kemble bought a sixth share of the patent, and appeared in company with his sister Sarah *Siddons. One of his first importations was the child prodigy Master *Betty. In 1808 the theatre was burnt down. Rebuilt with a pit and five galleries seating 2,800, the new theatre opened with *Macbeth*. Owing to the high cost of the rebuilding Kemble abolished the shilling gallery, which led to the famous OP (Old Prices) Riots. After disturbances every night for about two months, Kemble was forced to submit and restore the shilling gallery.

Between 1809 and 1821 most of the famous actors of the day, and many singers, appeared, as did pantomimists such as *Grimaldi. In 1812 Mrs Siddons made her farewell appearance, and in 1816 *Macready his first. Kemble retired in 1817 (in which year the theatre was first lit by gas) and his younger brother Charles *Kemble took over. In 1829 Fanny *Kemble embarked on a series of performances which filled the theatre for three years. Edmund *Kean made his last appearance on the stage there in 1833, and in 1835 a strong company included Charles *Kean, Macready, and Helen *Faucit, making her London début. Macready became manager in 1837, introducing limelight on to the stage long before it was in general use as a *lighting effect. His reign was marked by much internal dissension and he left in 1839, being succeeded by Mme *Vestris who with her husband Charles James *Mathews put on a number of beautifully staged productions; financially her greatest success was *Boucicault's first play *London Assurance* (1841). After Mme Vestris left in 1842 the theatre failed and was finally closed, to reopen

in 1847 as the Royal Italian Opera House, its seating having been increased following reconstruction to over 4,000. Smirke's building was burnt down in 1856 and the new theatre, with pit, stalls, three tiers of boxes, amphitheatre, and gallery, opened in 1858 with a performance of opera. Theatrical entertainments were in future limited to a handful of pantomimes, some revues, and in 1912 *Reinhardt's production of *Sophocles' *Oedipus the King*. Otherwise the house was entirely given over to opera, except for a brief existence during the Second World War as a Palais de Danse. In 1946 it became the joint home of London's chief opera and ballet companies, the latter coming from *Sadler's Wells.

Coventry, see BELGRADE THEATRE and HOCKTIDE PLAY.

Coventry Cycle, see MYSTERY PLAY.

Coward, Sir Noël Pierce (1899–1973), English actor, director, composer, librettist, and playwright. He was on the stage from childhood, making his first appearance in 1911. His early plays, some of which aroused controversy because of their characters' sexual sophistication, included *The Young Idea* (1923; NY, 1932), *The Vortex* (1924; NY, 1925), *Fallen Angels* (1925; NY, 1927), and *Hay Fever* (London and NY, 1925). He then wrote three *revues for C. B. *Cochran at the *London Pavilion. Further successes were the musical romance *Bitter Sweet* (London and NY, 1929), *Private Lives* (1930; NY, 1931), in which Coward starred memorably opposite Gertrude *Lawrence, the patriotic *Cavalcade* (1931), and a new revue *Words and Music* (1932). *Design for Living* (NY, 1933; London, 1939) aroused further controversy because of its light-hearted approach to a *ménage à trois* (played in New York by Coward and the *Lunts); it was followed by *To-Night at 8.30* (London and NY, 1936), consisting of six (later ten) one-act plays performed in groups of three. After the musical play *Operette* (1938) came the record-breaking *Blithe Spirit* (London and NY, 1941), which ran for 1,997 performances in London. In 1943 *Present Laughter* (NY, 1946) and *This Happy Breed* were presented in London in repertory, and Coward then returned to revue with *Sigh No More* (1945). *Peace in Our Time* (1947) was a serious play about England after a Nazi victory, but *Relative Values* (1951) marked a welcome return to sophisticated comedy and *Quadrille* (1952; NY, 1954) provided an excellent vehicle for the Lunts. John *Gielgud starred in the London production of *Nude with Violin* (1956; NY, 1957) and Vivien *Leigh in *South Sea Bubble* (also 1956) and the London production of *Look after Lulu* (London and NY, 1959), an adaptation of *Feydeau's *Occupe-toi d'Amélie*. After *Waiting in the Wings* (1960), set in a home for retired actresses, Coward produced his last original musical *Sail Away* (NY, 1961;

London, 1962) and then wrote the lyrics and music for *The Girl Who Came to Supper* (NY, 1963), a musical based on *Rattigan's *The Sleeping Prince*. His last work (in which he also made his last stage appearance) was *Suite in Three Keys* (1966), consisting of three plays of which two formed a double bill, the latter being seen in New York in 1974 as *Noël Coward in Two Keys*. Coward was an excellent actor who as well as taking the lead in many of his own plays appeared occasionally in plays by other authors, notably as Lewis Dodd in Margaret Kennedy's *The Constant Nymph* (1926) and King Magnus in *Shaw's *The Apple Cart* (1953).

Cowell, Joe Leathley [Joseph Hawkins-Witchett] (1792–1863), English actor, a fine low comedian. He was particularly noted for his playing of Crack in Knight's *The Turnpike Gate*, which he constantly revived both in England and in the United States, where he made his début in 1821 and became one of the most popular comedians of the day. He made his last appearance in New York in 1850 as Crack and then returned to England. He was connected by marriage with the *Siddons family and was the father-in-law of H. L. *Bateman. His second son Sam became a *music-hall star.

Cowell, Sam, see MUSIC-HALL.

Cowell, Sidney Frances, see BATEMAN.

Cowl, Jane (1884–1950), American actress and playwright, who made her first appearance on the stage under *Belasco in his *Sweet Kitty Bellairs* (1903). She subsequently played in a number of his productions, including his own *A Grand Army Man* (1907). Her finest part was probably Juliet, which she first played in 1923, but she was also seen as Cleopatra in 1924 and Viola in 1930. She was also much admired in revivals of the younger *Dumas's *The Lady of the Camellias*, as Camille (1931), and of *Shaw's *Captain Brassbound's Conversion* (1940) and *Candida* (1942). Among new plays in which she appeared were *Coward's *Easy Virtue* (1925; London, 1926), Robert *Sherwood's *The Road to Rome* (1927), Thornton *Wilder's *The Merchant of Yonkers* (1938), and *Van Druten's *Old Acquaintance* (1940). She appeared also in her own plays *Lilac Time* (1917), written in collaboration, *Information, Please* (1918), and *Smilin' Through* (1919). *Hervey House*, also written in collaboration, was produced in London in 1935, directed by Tyrone *Guthrie.

Cowley [*née* Parkhouse], **Hannah** (1743–1809), English playwright, whose works mark the transition from *Restoration comedy to the 18th-century comedy of manners. Her first play *The Runaway* (1776) was said to have been 'improved' by *Garrick, who produced it at *Drury Lane. The most successful was *The Belle's Stratagem* (1780), based on *Destouches's *La Fausse Agnès* (1759). First produced at

*Covent Garden with Mrs *Jordan as Letitia, it was many times revived, notably by Henry *Irving in 1881 with himself as Doricourt and Ellen *Terry as Letitia. It was one of the first English comedies to be seen in the New World, being in the repertory of the *Hallams and *Hodgkinson in New York in 1794. Mrs Cowley's later plays, which included *Which is the Man?* (1782), *A Bold Stroke for a Husband* (1783), and *The Town before You* (1794), were all given their first productions at Covent Garden.

Crabtree, Charlotte (1847–1924), American actress, known as Lotta. An attractive child, with black eyes and a mop of red hair, she toured the mining camps of Colorado from the age of 8, singing, dancing, and reciting, and becoming a well-known and much loved figure. In 1867 she made a great success in New York as Little Nell and the Marchioness in *Brougham's dramatization of *Dickens's *The Old Curiosity Shop*. Throughout her career she preserved a look of youthful innocence, even in her most daring dances and by-play. She was outstanding in *burlesque and *extravaganza and in slight plays specially written for her, such as Fred Marsden's *Musette*, in which she toured indefatigably. She retired unmarried in 1891, having amassed a large fortune which she left to charity.

Craig, (Edward Henry) Gordon (1872–1966), English scene designer and theorist. The son of Ellen *Terry and E. W. Godwin, he was for a time an actor. In 1903 he prepared some interesting sets for Fred *Terry's production of Calvert's *For Sword or Song* and for his mother's productions of *Much Ado about Nothing* and *Ibsen's *The Vikings*. He was also responsible for the designs for *Otway's *Venice Preserv'd* (*Das gerettete Venedig*) in Berlin (1905) and Ibsen's *Rosmersholm* for *Duse (1906). In 1908 he settled in Florence, where he founded and edited the *Mask*, a journal devoted to the art of the theatre, and also ran a school of acting. In 1911 he designed the costumes for *Yeats's *The Hour Glass* at the *Abbey Theatre, where his invention of screens as a background for lights to play on, which he used in his production of *Hamlet* for the *Moscow Art Theatre in 1912, was seen in public for the first time. He designed and produced Ibsen's *The Pretenders* in Copenhagen (1926), and his last designs for the theatre were for a New York production of *Macbeth* (1928). His theories on acting and stage settings are described under *scenery and amplified in his numerous publications, especially *On the Art of the Theatre* (1911). They had an immense influence on the theatre in Europe and the USA. Shortly before his death his vast theatrical library was bought by the French government.

His sister, **Edith Craig** (1869–1947), was an

actress and director. From 1911 to 1921 she directed the Pioneer Players in some 150 plays, including *Claudel's *L'Ôtage* as *The Hostage* (1919). She also inaugurated and ran until her death the annual Shakespeare matinée held on the anniversary of her mother's death in the barn at Small Hythe, where Ellen Terry spent her last years (now a museum).

Craig Theatre, New York, see ABBOTT.

Crane, William H., see ROBSON, STUART.

Craven, Hawes [Henry Hawes Craven Green] (1837–1910), English scene-painter, of a theatrical family, who served his apprenticeship under the scene-painter at the *Britannia Theatre, London. His first outstanding designs were for Wilkie Collins's *The Lighthouse* (1857) at the *Olympic Theatre, and he also worked at *Covent Garden, *Drury Lane, and at the Theatre Royal, Dublin. His finest work was for *Irving at the *Lyceum, where he was considered the equal of *Stanfield and *Beverley in craftsmanship, and their superior in his grasp of theatrical essentials. His scenery for *Hamlet*, *Romeo and Juliet*, W. G. Wills's *Faust*, and *Tennyson's *Becket* was particularly admired. As an innovator he ranks with de *Loutherbourg, and he was a pioneer in the skilful use of the newly introduced electric *lighting.

Crawford, Cheryl (1902–86), American director and producer, who in 1922 joined the *Theatre Guild, appearing in many of its later productions and being its casting manager from 1928 to 1930. She was also active in the foundation and running, with Harold *Clurman and Lee *Strasberg, of the *Group Theatre in 1930 (directing its first play, Paul *Green's *The House of Connelly*, 1931); of the American Repertory Theatre, in association with Eva *Le Gallienne and Margaret *Webster, in 1946; and of the Actors' Studio (see METHOD) in 1947. In 1950 she became one of the directors of the Anta play series produced annually by the *American National Theatre and Academy. She also presented or co-presented productions such as the musicals *Brigadoon* (1947) and *Paint Your Wagon* (1951), several plays by Tennessee *Williams, *Brecht's *Mother Courage and Her Children* (1963), and *Arbuzov's *Do You Turn Somersaults?* (1978).

Crazy Gang, The, association of seven British comedians, consisting of three double-act teams—**Bud Flanagan** [Robert Winthrop, originally Chaim Reuben Weintrop] (1896–1968) and **Chesney** [William] **Allen** (1894–1982); **Jimmie Nervo** [James Holloway] (1897–1975) and **Teddy Knox** [Albert Edward Cromwell-Knox] (1896–1974); **Charlie Naughton** (1887–1976) and **Jimmy Gold** [James McConigal] (1886–1967)—and **'Monsewer' Eddie Gray** (1897–1969), who appeared together in a number of 'crazy' shows at the *London Palladium

in the 1930s and the *Victoria Palace from 1947 to 1960. All the Gang had wide experience of variety and circus comedy techniques, and were able to make specific individual contributions to their shows, which included knockabout, satirical, sentimental, and farcical incidents, with a good deal of improvisation.

Crepidata, see FABULA 2: PALLIATA.

Crepuscolari, see MARTINI.

Crispin, character of French comedy who derives from the *commedia dell'arte* mask *Scaramuccia, gallicized as Scaramouche. As originally played in Paris by Raymond *Poisson he had in him also something of the braggart soldier (see CAPITANO), but successive generations of Poissons, playing the part until 1735, changed it to more of a quick-witted unscrupulous valet. The name was introduced into theatrical literature by *Scarron and was used by *Lesage. It was also the name of a character in *Regnard's *Le Légataire universel* (1708).

Criterion Theatre, London, in Piccadilly Circus. Originally an adjunct to Spiers and Pond's Criterion restaurant, this theatre was built entirely underground. A three-tier auditorium seating 675, it opened in 1874 and had its first outstanding success with *Pink Dominoes* (1877), adapted by James *Albery from a French farce. In the cast was Charles *Wyndham who took over the theatre in 1879. In 1883 the theatre was found to be unsafe and was reconstructed, electricity being installed at the same time. Many of Henry Arthur *Jones's plays were first produced at the Criterion, including *The Case of Rebellious Susan* (1894), *The Physician*, and *The Liar* (both 1897). Wyndham left in 1899 to open *Wyndham's Theatre, but remained lessee of the Criterion until his death. In 1902 the theatre was remodelled, reopening in 1903. One of the greatest successes of the First World War, Walter Ellis's *A Little Bit of Fluff*, began its run here in 1915. Between the wars came Ronald Mackenzie's *Musical Chairs* (1932) with John *Gielgud and *Rattigan's *French without Tears* (1936). During the Second World War the theatre became a BBC studio, reopening in 1945 with a revival of *Sheridan's *The Rivals* starring Edith *Evans. The post-war years brought a run of successes, including Warren Chetham-Strode's *The Guinea Pig* (1946), the revue *Intimacy at 8.30* (1954), Samuel *Beckett's *Waiting for Godot* (1955), and *Anouilh's *Waltz of the Toreadors* (1956). In 1963 *A Severed Head*, adapted by J. B. *Priestley and Iris Murdoch from the latter's novel, began a long run. Simon *Gray's *Butley* (1971) with Alan *Bates, *Ayckbourn's *Absurd Person Singular* (1973), Ray Cooney and John Chapman's *There Goes the Bride* (1974), and *Fo's *Can't Pay? Won't Pay!* (1981) were also popular. In 1989 the theatre closed temporarily because of building work on the surrounding site, its current production, Ray Cooney's long-running comedy *Run for Your Wife!*, moving to another theatre.

Croft, Michael, see NATIONAL YOUTH THEATRE OF GREAT BRITAIN.

Cronyn, Hume (1911–), Canadian actor, who trained for the stage in New York and made his début in Washington, DC, in 1931. A notable character actor, he made his name on Broadway in the 1930s with a series of outstanding performances including Elkus in Maxwell *Anderson's High Tor* (1937) and Andrei in *Chekhov's Three Sisters* (1939). In 1942 he married Jessica *Tandy and they frequently appeared together. In 1963 they starred in the first season of the *Guthrie Theater in Minneapolis, Cronyn playing Harpagon in *Molière's The Miser, Tchebutykin in *Three Sisters,* and Willy Loman in Arthur *Miller's *Death of a Salesman.* They were together again in the 1965 season, in which he played Richard III. He returned to Canada in 1969 to play the title-role in Peter Luke's *Hadrian VII* at the *Stratford (Ontario) Festival, where in 1976 he also played Shylock in *The Merchant of Venice* and Bottom in *A Midsummer Night's Dream.* In New York in 1972, and later on tour, he was the solo performer in *Beckett's *Krapp's Last Tape.* (For other appearances see TANDY.)

Crooked Mirror, Theatre of the [Krivoe Zerkalo], Leningrad, the best satirical theatre of pre-revolutionary Russia. It was founded in St Petersburg in 1908, its first outstanding success being *Vampuka; or, The Bride of Africa,* a hilarious skit on grand opera. It attracted a number of brilliant young talents, including *Evreinov, whose *Fourth Wall,* which ridiculed *verismo* opera, was produced there, as was *Andreyev's *The Pretty Sabines,* a biting social satire. One of the theatre's most successful productions was *Gogol's *The Government Inspector* shown as it might have been staged by a number of contemporary directors, including *Meyerhold and *Stanislavsky. The theatre closed in 1918, but managed to reopen in 1922 and in the following year the company made a brief visit to Moscow. After the death in 1928 of the critic A. P. Kugel, one of its original founders, it continued to function, but lost its special flavour and finally closed in 1931.

Crothers, Rachel (1878–1958), American dramatist, whose first short plays were produced while she was a student at a drama school. Her full-length plays, which she directed herself, included *The Three of Us* (1906); *A Man's World* (1910), an attack on moral double standards which was regarded as highly significant in its time; *A Little Journey* (1918), about a selfish woman's transformation by a railway accident; and *He and She* (1920, originally known as *The Herfords*). *Nice People* (1921) is a study of

post-war youth, while *Let Us be Gay* (1929), *As Husbands Go* (1931), which contrasts English and American marriages, and *When Ladies Meet* (1932), a deft study of feminine psychology, are well-observed comedies. Her last work, *Susan and God* (1937), starred Gertrude *Lawrence as a woman devoted to a new religious cult. Always in the vanguard of public opinion, she never allowed her feminist viewpoint to weaken the theatrical effectiveness of her writing.

Crouse, Russel, see LINDSAY.

Crow Street Theatre, see DUBLIN.

Crucible Theatre, Sheffield, opened in 1971, replacing the old Sheffield Playhouse, where a *repertory company was started in 1923. In 1967 a Trust was formed to organize the building of a new theatre, and the Crucible was financed mainly by the Arts Council, the City Council, and public donations. An important regional theatre, it seats 1,000 in one semi-circular tier round three sides of the thrust stage. It stages an enterprising mixture of classics, modern plays (including Botho *Strauss's *The Park,* 1988), and musicals such as *Chicago,* which later had a long run in London, *Lennon,* and *Carmen Jones.* It has also presented local documentaries, among them *The Stirrings in Sheffield on Saturday Night.* It has a studio theatre seating 250 and an award-winning Theatre in Education company.

The neighbouring Lyceum Theatre, a listed building seating 1,100, is the last extant theatre in the provinces designed by the Victorian architect Sprague. After 20 years of closure it reopened in 1990, newly refurbished as a No. 1 touring house, as part of the Crucible complex sharing the same administration.

Cruelty, Theatre of, see ARTAUD.

Cruz, Ramón de la, see SAINETE and ZARZUELA.

Cueva, Juan de la (1543–1610), Spanish dramatist whose most important plays were produced in Seville between 1578 and 1581. Writing in the period when classical influence was still paramount, Cueva owes much to *Seneca, but he was also the first writer of plays based on Spanish history. His dramas on Bernardo el Carpio and King Sancho and his tragedy *Los siete infantes de Lara* (*The Seven Princes of Lara*) paved the way for the great dramas of Lope de *Vega and his contemporaries. Cueva was also an innovator in the Spanish comedy of manners with his *Comedia del viejo enamorado* (*Comedy of the Infatuated Old Man*) and *El infamador* (*The Backbiter*).

Cumberland, Richard (1732–1811), prolific English dramatist. He wrote a number of poor tragedies, including an adaptation of *Timon of Athens* with a new fifth act (1771), and new versions of *Massinger's *The Bondman* and *The Duke of Milan* (both 1779), which have not survived. His best plays were sentimental domestic comedies such as *The Brothers* (1769) and *The West Indian* (1771), the latter proving his most successful work when produced by *Garrick at *Drury Lane. The only later play of interest was *The Jew* (1794), one of the first to plead the cause of Jewry on stage. It was frequently revived and translated into several languages. Cumberland, who was extremely sensitive to criticism, figures in *Sheridan's *The Critic* (1779) as Sir Fretful Plagiary.

Cummings, Constance (1910–), American-born actress, who made her New York début in 1928 in the chorus of the musical *Treasure Girl*. After her marriage to the English playwright **Benn W**(olfe) **Levy** (1900–73) she lived and worked mainly in England. She had roles in several of her husband's works, including the title-roles in *Young Madame Conti* (1936; NY, 1937), based on a play by Bruno Frank, and, in New York, *Madame Bovary* (1937), adapted from Flaubert's novel. She returned to London in 1938 to appear in James Hilton's *Goodbye, Mr Chips*, following it with Levy's production of his own play *The Jealous God* (1939). Later in the same year, for the *Old Vic, she showed great poise and emotional power as Juliet and as *Shaw's St Joan. She was seen in the London production of Robert *Sherwood's *The Petrified Forest* (1942) and in new plays by Levy such as *The Rape of the Belt* (1957). In 1964 she successfully took over from Uta *Hagen the role of Martha in *Albee's *Who's Afraid of Virginia Woolf?* in London, and in 1969 she played Gertrude in *Hamlet* in both London and New York. In 1971 she joined the company of the *National Theatre, where she gave a superb performance as the drug-addicted Mary Tyrone in *O'Neill's *Long Day's Journey into Night*, and was also much admired as Ranevskaya in *Chekhov's *The Cherry Orchard*. She went to the *Royal Court Theatre to appear with Michael *Hordern in Howard *Barker's *Stripwell* (1975), and was highly praised for her performance (mainly a monologue) as a woman recovering from a stroke in Arthur *Kopit's *Wings* (NY and NT, 1979).

Cup-and-Saucer Drama, see ROBERTSON, T. W.

Curtain, screen separating the stage from the auditorium. It was introduced into European theatres with the advent of enclosed theatre buildings in the 16th and 17th centuries. In the English theatre after 1660 it rose at the conclusion of the *prologue, which was spoken on the *forestage, and remained out of sight until the play was over. It was at first green and was occasionally, for special effects, dropped during a performance; it was not until the mid-18th century that it began to fall regularly to mark the end of an act and hide the stage during the interval. This function was shortly after transferred to the *act-drop. When *Irving revived *Boucicault's *The Corsican Brothers* in 1880 a crimson velvet curtain was introduced to hide changes of scene during an act.

The **front**, or **house, curtain,** so long a feature of all post-Restoration playhouses, is now often discarded, and no provision is made for it in many modern theatre buildings. It can be worked in a variety of ways, including the straightforward 'flying' of the **fly curtain,** which has largely replaced the formerly popular **french valance** (known in America as a **brail curtain**), raised vertically in a series of festoons. A variant of this was the **contour curtain.** There are also the centrally parting **draw** or **traverse curtain,** or **traveller,** and the bunching to outer top corners of the **tab** (short for **tableau**) **curtain.** The term tabs is now applied to any front curtain, and is sometimes misapplied to the *curtain set on the stage itself.

Two other curtains in the proscenium are the **advertisement curtain** and the **safety curtain,** or **iron.** The former appeared in the latter half of the 19th century in smaller theatres and music-halls, bearing in various panels painted notices of local shops and manufacturers. Advertisements thrown on the safety curtain by a slide projector later served the same purpose. The safety curtain itself was first installed at *Drury Lane in 1794 as a precaution against fire. It consisted of an iron or fireproof sheet which must by law be lowered once during every performance. This has now been made obsolete by modern water curtains and sprinklers.

Curtain-Music, Curtain-Tune, see ACT-TUNES.

Curtain-Raiser, one-act play, usually farcical, which in the 19th century served to whet the appetite of the audience before the main five-act drama of the evening, the last relic of the days when a full evening's entertainment included several plays to which latecomers were admitted on payment of a reduced fee. Like the *after-piece it disappeared in the 20th century.

Curtain Set (Drapery Setting in America), the simplest method of dressing the stage for a performance. It consists of one or more side curtains, a back curtain, *borders, and perhaps a draw *curtain, centrally divided and running off to the sides on a wire or railway. It is popular with amateurs and can be used with remarkable ingenuity. It has no scenic function but may be supplemented by small pieces of *built stuff placed before the back curtain and by the resourceful use of *lighting.

Curtain Theatre, Glasgow, see GLASGOW.

Curtain Theatre, London, in Curtain Close, Finsbury Fields, Shoreditch. This playhouse opened in 1577, a year after its close neighbour the *Theatre, which it resembled. It housed a number of distinguished companies including the *Chamberlain's Men. *Jonson's *Every Man in His Humour* (1598) and perhaps Shakespeare's *Romeo and Juliet* (c.1595) and *Henry V* (c.1599) were first seen there. Several attempts were made to suppress the Curtain, particularly after the building of the *Fortune, so as to limit the number of playhouses in London. From 1603 to about 1609 *Queen Anne's Men occupied it and its last recorded use was in 1622 by *Prince Charles's Men, though the building was standing in 1642 and perhaps as late as 1660.

Cusack, Cyril (1910–   ), Irish actor, director, and playwright. In 1932 he joined the company at the *Abbey Theatre, remaining there for 13 years and appearing in more than 65 plays including *Synge's *The Playboy of the Western World*, *Shaw's *Candida* (in which he played Marchbanks), and many of the major works of Sean *O'Casey, Lennox *Robinson, and Paul Vincent *Carroll. He made his first appearance in London in *O'Neill's *Ah, Wilderness!* (1936) and was subsequently seen there in a number of parts including Michel in *Cocteau's *Les Parents terribles* (1940) and Dubedat (playing opposite Vivien *Leigh) in a revival of Shaw's *The Doctor's Dilemma* (1942). In the same year he directed the Gaelic Players in Dublin in his own play, *Tareis an Aifreann* (*After the Mass*). In 1945 he left to form his own company with which he toured in Ireland and abroad in a repertory of Irish and European plays, giving the first performance of O'Casey's *The Bishop's Bonfire* in Dublin in 1955. He made several visits to the United States, and in 1960 was at the *Théâtre des Nations, where he received the International Critics' Award for his solo performance in *Beckett's *Krapp's Last Tape*. He was in the *RSC's production of *Dürrenmatt's *The Physicists* (1963), the *National Theatre company's production of *Frisch's *Andorra* (1964), and (in the 1968 *World Theatre Season) the Abbey Theatre's production of *Boucicault's *The Shaughraun*. In 1979 he played the title-role in *Chekhov's *Uncle Vanya* at the Abbey, and in 1980 he appeared with that company at the *Old Vic in Hugh *Leonard's *A Life*, in which he memorably portrayed an ageing, terminally ill civil servant.

Cushman, Charlotte Saunders (1816–76), great American tragedienne. She excelled in such parts as Lady Macbeth, which she played in 1836 for her début in New York, Mrs Haller in *Kotzebue's *The Stranger*, and above all Meg Merrilies in a dramatization of Scott's *Guy Mannering*. She also played Lady Gay Spanker in the first American production of *Boucicault's *London Assurance* (1841). In 1845 she was seen in London at the *Princess's in the above parts, as well as Rosalind in *As You Like It*, Beatrice in *Much Ado about Nothing*, and Portia in *The Merchant of Venice*. In 1839 she first played another famous role, Nancy in an adaptation of *Dickens's *Oliver Twist*. Being of a somewhat masculine cast of countenance, she also played a number of male roles, including Hamlet, Romeo to the Juliet of her younger sister Susan, Oberon in *A Midsummer Night's Dream*, and Claude Melnotte in *Bulwer-Lytton's *The Lady of Lyons*. During the last years of her life she gave many successful Shakespeare readings. In 1907 a Charlotte Cushman Club was established in Philadelphia which still flourishes.

Cyclorama, curved wall, or section of a dome, built at the back of the stage, with an unbroken surface upon which light can be thrown. The effects thus achieved can be striking, but since to present a perfect surface a cyclorama must be rigidly and heavily built, generally of hard cement, it has several drawbacks, interfering with access to the stage from the sides and with the suspension of scenery from above. Being immobile, it also becomes a potential obstacle when not in use, and limits the scenery to one style only. A moveable cyclorama has been invented, but never widely adopted. A partial, or shallow, cyclorama, or even a plain, uncurved, distempered back wall, has often been used instead and proved just as effective. In Germany, where the cyclorama was first used, it is called a *Rundhorizont*.

Cyrano de Bergerac, Savinien (1619–55), French dramatist, soldier, and celebrated swordsman, always ready to take offence if anyone dared to remark on his enormous nose. On leaving the army he settled in Paris and cultivated his undoubted gifts for literature and science. His friendship with *Molière and *Scarron turned his thoughts to the stage, but his tragedy, *La Mort d'Agrippine* (1653), was not a success, and a comedy, *Le Pédant joué*, written between 1645 and 1649, does not appear to have been performed. It must, however, have circulated in manuscript, since both Scarron and Molière show signs of its influence. The hero of *Rostand's great Romantic drama *Cyrano de Bergerac* (1898) is a mixture of fact and fiction, a swashbuckling Gascon with the soul of a poet, whereas the real Cyrano was a Parisian and probably much less given to altruism and introspection.

D

Dada, art movement of the 1920s which was later to have a vital influence on European drama. Irrational, nihilistic, and anarchic, it attacked the complacency of so-called 'society', and set out to arouse in its audiences the outrage later evoked by *Artaud's Theatre of Cruelty, exploiting also the sense of futility which gave rise to the Theatre of the *Absurd.

Dalberg, Wolfgang Heribert, see MANNHEIM.

Dallas Theater Center, Dallas, Tex., opened in 1959, professional regional theatre with an acting company of 16. Plays are performed at two locations: the Kalita Humphreys Theater, designed by Frank Lloyd Wright, located on Turtle Creek Boulevard, and the Arts District Theater in Dallas's downtown arts district. DTC produces six mainstage plays each season comprising international and American classics, contemporary and new plays. A series of challenging, intimately scaled plays are produced 'in the basement' of the Kalita Humphreys Theater. The Theater Center plays to 10,500 subscribers throughout the Sept.–May season. DTC also offers a Teen/Children's Theater for young audiences.

Daly, (Peter Christopher) **Arnold** (1875–1927), American actor, who came into prominence in 1903 with the first public performance in New York of *Candida*, which he directed as well as playing Marchbanks, at a special matinée. This was the first production in the United States of a play by *Shaw since *Mansfield's tour of *The Devil's Disciple* in 1897, and Daly followed its success with a production of *Mrs Warren's Profession* which led to an uproar and prosecution by the police. In spite of this he produced *You Never Can Tell, Candida, John Bull's Other Island, Arms and the Man,* and a double bill consisting of *The Man of Destiny* and a trifle written specially for him by Shaw, *How He Lied to Her Husband*. He also played this last sketch in *vaudeville. Opposition to the new 'theatre of ideas' finally defeated him and he was forced back into run-of-the-mill work, dying in a fire in his early fifties.

Daly, (John) **Augustin** (1839–99), American manager and dramatist, who while working as a dramatic critic wrote a number of plays, among them *Leah, the Forsaken* (1862), based on a German drama, about the doomed love of a Christian and a Jewess in 17th-century Germany; and a melodrama, *Under the Gaslight; or, Life and Love in These Times* (1867), containing a famous scene in which a man is tied to a railway track. In 1869 he took over the management of the *Fifth Avenue Theatre, which prospered until it was burnt down in 1873. He then opened another theatre, also on Fifth Avenue, and in 1879 opened *Daly's, with a fine company headed by John *Drew and Ada *Rehan. In 1884 it was seen in London with such success that after further visits between 1886 and 1890 Daly decided to open *Daly's Theatre there. Both in New York and in London Daly's first nights were important events, and he set a high standard in spite of his tendency to tamper with the text of established classics. He also presented new plays and wrote about 100 plays himself, mostly adaptations, his later works including *Horizon* (1871) and *Pique* (1875).

Daly's Theatre, London, in Cranbourn Street, off Leicester Square. This theatre, which had a seating capacity of about 600 in three tiers, was leased to Augustin *Daly by George *Edwardes, and opened in 1893 with *The Taming of the Shrew*, Ada *Rehan playing Katharina. This was followed by Sheridan *Knowles's *The Hunchback*, with Violet *Vanbrugh, and in 1894 by *Twelfth Night* and *As You Like It*. In the same year Daly presented Eleonora *Duse in the younger *Dumas's *La Dame aux camélias*. Later Edwardes transferred the musical comedy *A Gaiety Girl* there, Daly withdrawing in 1895. Daly's then became one of the most fashionable and successful theatres in London, its productions including *The Geisha* (1896), *San Toy* (1899), *A Country Girl* (1902), *The Merry Widow* (1907), and *The Dollar Princess* (1909). Edwardes's last production there was *The Marriage Market* (1913). In 1917 Oscar *Asche produced *The Maid of the Mountains* which ran for 1,352 performances; it was followed by *A Southern Maid* (1920). In 1927 Noël *Coward's *Sirocco* proved a failure, and a number of old musicals were revived with little success. The theatre closed in 1937.

Daly's Theatre, New York, on Broadway, below 30th Street. Originally known as Banvard's, and later Wood's, Museum, this opened in 1867. Although the repertory was mainly burlesque, variety, and melodrama, straight plays were occasionally performed after 1872. After several seasons as the Broadway and the New Broadway, the theatre was remodelled and redecorated and opened in 1879 as Daly's, to become one of the leading playhouses of New York. One of its most successful productions was *The Taming of the Shrew* with Ada *Rehan in

1887. *Daly continued to manage the theatre until his death in 1899, and it retained his name until it was demolished in 1920, having been a cinema since 1915.

Dame, female character in the English *pantomime, traditionally played by an actor, sometimes one recruited from the *music-halls such as Dan *Leno or George *Robey. Among the familiar Dame parts are Aladdin's mother Widow Twankey (a name taken by H. J. *Byron from a Chinese tea-exporting port), Idle Jack's mother in *Jack and the Beanstalk*, usually known as Dame Durden or Dame Trot, and Cinderella's ugly sisters. If the ugly sisters are played by women, their mother the Baroness is played by a man. Other Dame parts are the Cook in *Dick Whittington*, the Queen of Hearts, Mother Goose, and Mrs Robinson Crusoe.

Dancer, Ann, see BARRY, SPRANGER.

Danchenko, Vladimir Nemirovich-, see NEMIROVICH-DANCHENKO.

Dancourt, Florent Carton (1661–1725), French dramatist and actor who, like *Molière, was a man of the theatre. In 1680 he married **Marie Thérèse Lenoir** (1663–1725), giving up the study of law to join her on the stage. He had a good face and figure, and a natural liveliness of disposition which was an asset in comedy, though in tragedy he was judged cold and monotonous. After some years in the provinces he and his wife joined the *Comédie-Française and remained there until 1718. During this time he wrote more than 50 comedies based on small scandals or other topical events, his better plays showing some skill in etching the contemporary scene. He was particularly good in his delineation of the rising middle class. There is no bitterness in Dancourt's writing, but under the superficial wit and good humour much brutality and selfishness. His best play is *Le Chevalier à la mode* (1687), a portrait of contemporary life with much satire at the expense of parvenu financiers. Others are *La Maison de campagne* (1688) and the one-act *Les Vendanges de Suresnes* (1695), the most frequently revived of all his works.

Dane, Clemence [Winifred Ashton] (1888–1965), English dramatist and novelist, who took her pseudonym from the church of St Clement Danes in the Strand, London. She acted for a few years as Diana Cortis, but the success of her first play *A Bill of Divorcement* (1921), which deals sympathetically with the problem of divorce on the grounds of insanity, led her to abandon acting. The play gave Katharine *Cornell one of her first successes in New York. Among Clemence Dane's later plays were *Will Shakespeare* (also 1921), *Naboth's Vineyard*, *Granite* (both 1926), and *Wild Decembers* (1932), a play on the Brontës in which Diana *Wynyard played Charlotte, Beatrix *Lehmann Emily,

and Emlyn *Williams Branwell. Her last play, *Eighty in the Shade* (1958), was specially written for Sybil *Thorndike and Lewis *Casson, who played the leading parts. Several of her novels deal with the theatre. She was also an excellent sculptor, and her bust of Ivor *Novello stands in the foyer of the Theatre Royal, *Drury Lane.

D'Annunzio [Rapagnetta], **Gabriele** (1863–1938), Italian poet, novelist, and dramatist, who in 1919 captured the port of Fiume (now Rijeka in Yugoslavia) for Italy, and thereafter became a political rather than a literary figure. His plays, simple in structure but rich in poetry and sensuality, aroused much controversy when first produced, but none is the product of an emotionally mature mind and none, except perhaps *La figlia di Jorio* (1904), is truly dramatic. His stage directions reveal the extent and accuracy of his archaeological knowledge, but his people are puppets, driven by elemental passions, and his plays live mainly by their poetry. Among the best known are *La città morta* (1898), in which worship of the Nietzschean superman is already apparent; *La Gioconda* (1899), in which *Duse, who spent much time and money in furthering the production of D'Annunzio's plays, starred opposite *Salvini; and *Francesca da Rimini* (1902). None of these has remained in the repertory, but a religious play written in French, *Le Martyre de San-Sébastien* (1911), is sometimes revived in a cut version for the sake of Debussy's incidental music. D'Annunzio's medieval verse tragedies, such as *La fiaccola sotto il moggio* (*The Light under the Bushel*, 1905), best demonstrate his tempestuous yet curiously static style. His search for heroic transcendence, and frequent underlying sadism, reflect the mood which gave rise to Fascism.

Danske Skueplads, Den (The Danish Theatre), Copenhagen. This opened in Lille Grønnegade in 1722 and Ludvig *Holberg wrote 28 comedies for the theatre before it closed in 1727, thus establishing a truly Danish drama for the first time. The theatre was re-established under royal patronage in 1747, with Holberg acting informally as consultant. The following year the company moved to a new building in Kongens Nytorv. In 1770 the monarchy accepted financial responsibility for it, and two years later the building was officially designated the Kongelige Teater (see ROYAL THEATRE).

Danvers, Johnny, see LENO.

D'Arcy, Margaretta, see ARDEN.

Darlington, W(illiam) **A**(ubrey) (1890–1979), English dramatic critic and dramatist, who was drama critic of the *Daily Telegraph* from 1920–68. He continued to write theatrical articles for the paper until his death, and was also drama correspondent of the *New York Times* from 1939–60. His first play, and the one by which he is chiefly remembered, was *Alf's Button*

(1924), an extravaganza based on one of his own novels which had a great success and was several times revived and filmed. It was followed by *Carpet Slippers* (1930), *A Knight Passed By* (1931), a burlesque version of *Boucicault's *The Streets of London* (1932), and *Marcia Gets Her Own Back* (1938).

Darvas, Lili, see MOLNÁR.

Daubeny, Sir Peter Lauderdale (1921–75), English impresario and theatre manager. He started his theatrical career as an actor under William *Armstrong at the *Liverpool Playhouse, but soon moved to London, where he promoted a long series of productions, early successes including *Werfel's *Jacobowsky and the Colonel* (1945). He also arranged visits, often under great difficulties, by many outstanding foreign theatrical groups, notably the *Berliner Ensemble (1956) and the *Moscow Art Theatre (1958). Among his later productions were Michael *Redgrave's version of Henry *James's *The Aspern Papers* (1959), *Billetdoux's *Chin-Chin* (1960), and the *Living Theatre's staging of Jack Gelber's controversial *The Connection* (1961). Daubeny was also responsible for an annual *World Theatre Season from 1964 to 1973.

Davenant, Sir William (1606–68), English dramatist and theatre manager, born in Oxford. His father was a friend of Shakespeare, who often stayed at Davenant's tavern on his journeys to and from Stratford, and this later led some to believe that William was Shakespeare's son. He may have believed it himself; but it is now thought unlikely, though he was almost certainly his godson. Young Davenant went to London in 1622, and soon made his name as a playwright and as a writer of Court *masques in the style of Ben *Jonson, whom he later succeeded as Poet Laureate. His career was interrupted by the Civil War, during which he was knighted by Charles I. Towards the end of the Commonwealth he returned to the theatre, staging mainly 'dramatic concerts'. In 1656 he produced privately his own 'entertainment' *The Siege of Rhodes*, considered by some the first English opera, following it with *The Cruelty of the Spaniards in Peru* (1658) and *The History of Sir Francis Drake* (1659). When Charles II returned to London in 1660 Davenant obtained from him a *patent—still in force at *Covent Garden—for the opening of a playhouse in the former *Lincoln's Inn Fields Theatre, with a company headed by *Betterton and containing professional actresses for the first time in England. It opened in 1661, but soon proved too small, and Davenant was planning a more commodious building—*Dorset Garden—when he died, leaving his widow to complete and open it in 1671. Davenant was one of the first London managers to encourage the use of machinery, dancing, and music in the production of plays, and greatly influenced the development of the English theatre.

Davenport, Edward Loomis (1815–77), American actor, son of an innkeeper, who first appeared on the stage, billed as Mr E. Dee, playing a small part in *Massinger's *A New Way to Pay Old Debts* in 1837. After some years on tour he was seen in New York in 1843 in Mrs John *Drew's company, and in 1848, with Mrs *Mowatt, visited England, where he remained for several years, being much admired in such roles as Othello, Richard III, Sir Giles Overreach, Claude Melnotte in *Bulwer-Lytton's *The Lady of Lyons*, and the dual title-roles in *Boucicault's *The Corsican Brothers*. On his return to New York he played Hamlet, and then formed his own company. Most of his subsequent career was spent outside New York, and he made his last appearance in 1877 in W. S. *Gilbert's *Dan'l Druce, Blacksmith* at the National Theatre, Washington. Though an excellent actor, he was never a public favourite.

He married an English actress, **Fanny Elizabeth Vining** (1829–91), who regularly played opposite him. Of their nine children one, Blanche, became an opera singer; four others were on the stage, the best known being Fanny (see below). **May** (1856–1927) retired on her marriage; **Edgar** (1862–1918) and **Harry** (1866–1949) both played in their father's company as children. Edgar as an adult acted with such stars as Julia *Marlowe and Viola *Allen. Harry married the daughter of an actor and appeared with her in *vaudeville before going into films.

Davenport, Fanny Lily Gypsy (1850–98), American actress, daughter of Edward *Davenport, who was on the stage as a child in her father's company. She made her adult début in 1865 as Mrs Mildmay in Tom *Taylor's *Still Waters Run Deep*, and from 1869 to 1877 was Augustin *Daly's leading lady in New York. She then formed her own company, starring with it in the principal theatres of the USA. Her range of parts was wide, including Shakespeare's heroines and such contemporary women as Polly Eccles in *Robertson's *Caste* and Lady Gay Spanker in *Boucicault's *London Assurance*. At the end of her career, between 1883 and 1894, she appeared in four plays by *Sardou: *Fedora*, *Tosca*, *Cleopatra*, and *Gismonda*.

Davenport, Jean and **Thomas,** see LANDER; **Lizzie,** see MATHEWS, CHARLES J.

Davidson, Gordon, see MARK TAPER FORUM.

Davis, Fay, see LAWRENCE, GERALD.

Day, John (c.1574–c.1640), English dramatist, noted in *Henslowe's diary as writing plays for the *Admiral's Men, mainly in collaboration with *Chettle, with whom he wrote the first part of *The Blind Beggar of Bethnal Green* (1600).

His later works include the ill-fated *Isle of Gulls* (1606), played by the *boy company at *Blackfriars; because of its satire on the uneasy relations between England and Scotland under James I, it led to the imprisonment of some of those connected with its production. Day, however, continued to write, being author with William *Rowley of *The Travails of the Three English Brothers* (1607), on his own of *The Parliament of Bees* (1608), and with *Dekker of *The Bellman of Paris* (1623), apparently his last play.

Dean, Basil Herbert (1888–1978), English actor, dramatist, and theatre director, who made his first appearance on the stage in 1906, and was then for four years with Miss *Horniman's company in Manchester. In 1911 he became the first director of the repertory company at the *Liverpool Playhouse. From 1919 onwards he was active both as manager and director. Among the many important plays he directed were Clemence *Dane's *A Bill of Divorcement* (1921), James Elroy Flecker's *Hassan* (1923), John *Van Druten's *Young Woodley* (1928), and J. B. *Priestley's *Johnson over Jordan* (1939). He was a pioneer of stage *lighting, importing much new equipment from Germany and the USA and devising some of his own. He had already organized entertainments for the troops during the First World War, and during the Second he became director of the *Entertainments National Service Association. He returned to the London theatre in 1946 with a production of Priestley's *An Inspector Calls*, and among his later assignments were *The Diary of a Nobody* (1954), based on the Grossmiths' book, and Michael *Redgrave's adaptation of Henry *James's *The Aspern Papers* (1959). Dean also organized the first British Repertory Theatre Festival in 1948.

Dean, Julia (1830–68), American actress, granddaughter of Samuel *Drake. With her father and stepmother she appeared as a child under the management of *Ludlow and Sol *Smith, and in 1846 made her adult début in New York as Julia in Sheridan *Knowles's *The Hunchback*. A beautiful woman, with a gentle personality and great charm, she was at her best in roles of tenderness and pathos, such as Juliet, *Scribe's Adrienne Lecouvreur, and Mrs Haller in *Kotzebue's *The Stranger*. She made an unhappy marriage in 1855, and her acting declined. A tour of California in 1856 was a success, but she never regained her position in New York, where she returned after her divorce. She died in childbirth following her second marriage in 1867.

De Bar, Benedict (1812–77), British-born equestrian actor, manager, and theatre owner who, through his association with *Ludlow and Sol *Smith, did much to help establish New Orleans and St Louis as leading theatre cities in America. He made his American début in New Orleans in 1835 as Sir Benjamin Backbite in *Sheridan's *The School for Scandal*, and acted there for the next 20 years except for a few visits to New York and London. In 1856 he moved to St Louis where he remained for the rest of his life, becoming owner of the influential Bates Theatre, renaming it De Bar's. Six years later he also acquired the St Louis Varieties Theatre and rechristened it De Bar's Opera House. He had a fine comic gift and was highly praised as Shakespeare's Falstaff and *Chanfrau's Mose the Fireman.

De Bergerac, Cyrano, see CYRANO.

De Brie [Catherine Leclerc du Rozet], **Mlle** (*c.*1630–1706), French actress, who as Mlle de Rose joined *Molière's company in 1650, and was destined to create many of his finest women's parts, including Cathos in *Les Précieuses ridicules* (1658) and Agnès in *L'École des femmes* (1662). She retired in 1685, having been one of the original members of the *Comédie-Française.

Deburau, Jean-Gaspard [Jan Kašpar Dvořák] (1796–1846), famous French player of *pantomime, and creator of a new conception of *Pierrot based on the *commedia dell'arte* mask *Pedrolino, which has since remained a popular and constant figure in the public imagination. Born in Bohemia, member of an acrobatic family with whom he toured the Continent, he arrived in Paris in 1814 and in 1820 was engaged with his companions for the Théâtre des *Funambules on the *boulevard du Temple where he remained until his death. He was at first an inconspicuous member of the company, from whose Pierrot Deburau, eventually known as Baptiste, learned his art. He took over the role of Pierrot in 1826, and he developed, with great subtlety, his concept of the character as the white-clad, ever-hopeful, always disappointed lover, as child, prince, poet, and eternal seeker; all Paris flocked to see him. Deburau's appeal to the working-class public was based on a thorough knowledge of their lives and working conditions; with his lightning wit and occasional flashes of cruelty and malice he expressed their spirit of revolt. He was the inspiration behind all subsequent attempts to re-establish the art of *mime in the modern theatre.

De Camp, Maria Theresa, see KEMBLE, CHARLES.

Décentralisation Dramatique, term used to describe the renewal of theatrical life in the French provinces begun in 1945 as a result of private and government ventures and increasingly supported by local authorities. Plans for decentralization were implemented with the help of a group of directors, actors, producers, and authors. The whole pattern of dramatic innovation altered radically, with many new authors and theatrical groups appearing in the provinces and on the edges of the Paris urban area, all publicly sponsored and forming a national network within which productions and personnel circulated freely. The most important of the newly created organizations were the Centres Dramatiques, of which the first, covering eastern France, was founded in 1946, later becoming known as the Théâtre National de Strasbourg. The Centres Dramatiques must give at least 100 performances annually and send productions out on tour, or alternatively organize coach services to bring spectators in. By 1988 there were some 30 Centres Dramatiques, plus five for children and young people. They stage 70–80 new productions a year, of which about 30 are the work of contemporary French playwrights. For the Maisons de la Culture, established in a number of provincial towns and financed equally by the local council and the State, theatre is only part of the activities; on a smaller scale are the Centres d'Animation Culturelle. Also under the aegis of the Ministry of Culture are the Troupes Permanentes, provincial companies which have a permanent playhouse but are committed to constant touring in their area. The establishment of the Théâtre National Populaire at Villeurbanne in 1972 (see PLANCHON) formed part of the decentralization process.

De Courville, Albert, see REVUE.

De Filippo, Eduardo (1900–84), Italian actor and dramatist, illegitimate son of the Neapolitan actor Eduardo Scarpetta, in whose company he and his brother **Peppino** (1903–80) and sister **Titina** [Anastasia] (1898–1963) acted as children. In 1932 they opened their own playhouse in Naples—the Teatro Umoristico—and soon achieved a great reputation for their productions in the style of the *commedia dell'arte*. Many of their plays were the work of Eduardo. In 1945 Peppino left to found his own company, with which he was seen in London during the *World Theatre Season of 1964, while Titina remained with Eduardo until her death, appearing in many of his comedies, including *Filumena Marturano* (1946), in which a prostitute pretends to be dying so that her lover will marry her. Other outstanding works were *Napoli milionaria!* (1945), with which Eduardo visited the 1972 World Theatre Season, *La paura numero uno* (*Terror Number One*, 1950), and *Bene mio e core mio* (*My Goods and My Heart*, 1955). Always topical, his plays are usually set in Naples, their language reflecting the evolving Neapolitan dialect and the vocabulary of modern life. *Zeffirelli directed productions in English of *Saturday, Sunday, Monday* (1959), about a family quarrel spread over a weekend (*National Theatre, 1973; NY, 1974), and *Filumena* (1977; NY, 1980). *Le voci di dentro* (1948) was produced

as *Inner Voices* (1983) at the National Theatre, which also staged *Napoli Milionaria!* in 1991.

Déjazet, Pauline-Virginie (1798–1875), French actress, who was already well known for her *male impersonations in *vaudeville when in 1821 she appeared at the *Gymnase, where she remained for some time, playing male roles. In 1831 she went to the newly opened *Palais-Royal, where she stayed for 13 years and became one of the most popular actresses in Paris. After an argument over her salary she continued her triumphant career at the *Variétés and the *Gaîté, still playing masculine roles—soldiers, collegians, students— as well as great ladies and pretty peasant girls. Some idea of her versatility is given by the fact that her parts included Voltaire, Rousseau, Napoleon, and Henri IV. In 1859 she took over the *Folies-Nouvelles, which she renamed Théâtre Déjazet, making a great hit at the age of 62 in a male part in *Sardou's *Monsieur Garat*. She made her last appearance in Paris in 1870, and in the same year was seen in London at the *Opera Comique in a season of French plays.

Dekker, Thomas (c.1572–c.1632), English dramatist, who worked mainly in collaboration and is believed to have had a hand in more than 40 plays of which about 15 survive, several having been destroyed by *Warburton's cook. The most important of his own works is *The Shoemaker's Holiday* (1599), first played at the *Rose Theatre by the *Admiral's Men, which materially assisted the evolution of English comedy. Robust and full-blooded, it tells how Simon Eyre, a master shoemaker, became Lord Mayor of London, and shows a promise which was not fulfilled in Dekker's later works. He worked mainly in collaboration. With *Marston he wrote *Satiromastix* (1601), in which Ben *Jonson is satirized under the name of Horace; with *Middleton he wrote *The Honest Whore* (1604) and *The Roaring Girl* (1610); and with *Massinger a tragedy, *The Virgin Martyr* (1620). He is also believed to have been part-author of William *Rowley's *The Witch of Edmonton* (1621). In 1609 he published *The Gull's Handbook*, a satirical account of the fops and gallants of the day which gives some interesting information about the contemporary theatre.

Delacorte Theatre, New York, see PAPP.

Delaunay, Louis-Arsène (1826–1903), French actor, who after attracting attention by his excellent acting in a classical comedy at the *Odéon made his first appearance at the *Comédie-Française in 1848 and had a long and brilliant career, playing young lovers until he was 60 and never losing his hold on the affections of the public. He was the original Fortunio in Alfred de *Musset's *Le Chandelier* (1848), and was so much admired as Coelio in the same author's *Les Caprices de Marianne* (1851) that after his retirement in 1886 it went out of the repertory for many years. He appeared in revivals of *Marivaux and *Regnard and in many new plays.

Della Porta, Giambattista (1538–1613), Italian scientist, philosopher, and dramatist, exponent of the *commedia erudita*, 14 of whose plays have survived out of a possible 33. They are in prose, on subjects taken mainly from Boccaccio and *Plautus, though *Il due fratelli rivali* is based on the story by Bandello which also supplied Shakespeare with the plot of *Much Ado about Nothing*. Two of Della Porta's plays, *La fantesca* and *La Cintia*, were translated and performed at Trinity College, Cambridge, as *Leander* (1598) and *Labyrinthus* (1603). *La trappolaria* was used by Ruggle as the basis of his *Ignoramus* (1615).

De Loutherbourg, Philip James, see LOUTHERBOURG.

Delysia [Lapize], **Alice** (1889–1979), French actress and singer, who made her first appearance at the *Moulin Rouge in 1903 in the chorus of the musical *The Belle of New York*. She was later at the *Variétés and at the *Folies-Bergère with Yvonne *Printemps. In 1905 she was seen with Edna May at *Daly's Theatre, New York, in *The Catch of the Season*, and in 1914 C. B. *Cochran engaged her for London, where she made a great success in the revue *Odds and Ends*. She starred for many years under Cochran's management, notably in *On with the Dance* and *Still Dancing* (both 1925), and in his memoirs he constantly pays tribute to her loyalty and good nature. In 1930 she was in Pagnol's *Topaze*, and she made her last appearance in London in 1939 as Hortense in *The French for Love*, a light comedy by Marguerite Steen and Derek Patmore.

De Musset, Alfred, see MUSSET.

Dench, Dame Judi [Judith] **Olivia** (1934–), major English actress whose career has covered a wide variety of parts old and new. She made her first appearance in 1957 at the *Old Vic as Ophelia to the Hamlet of John *Neville, remaining with the company for four years before joining the *RSC to play Anya in *Chekhov's *The Cherry Orchard* followed by Isabella in *Measure for Measure*. She then appeared in her first modern play as Dorcas Bellboys in the 1962 revival of John *Whiting's *A Penny for a Song*. She made another incursion into modern drama in *Arbuzov's *The Promise* in 1967, afterwards scoring a success as Sally Bowles in her first musical, *Cabaret* (1968). In 1969 she returned to the RSC to play Viola in *Twelfth Night*, and in 1970 was seen both in the title-role of *Shaw's *Major Barbara* and as the bluestocking Grace Harkaway in *Boucicault's *London Assurance*. A year later she played the title-role in *Webster's *The Duchess of Malfi*.

Other roles with the RSC included Beatrice in *Much Ado about Nothing*, Lady Macbeth, Millamant in *Congreve's *The Way of the World*, Juno in *O'Casey's *Juno and the Paycock* and, in 1984, *Brecht's Mother Courage. Among her roles at the *National Theatre were a younger-than-usual Lady Bracknell in *Wilde's *The Importance of Being Earnest* in 1982, Cleopatra to Anthony *Hopkins's Antony in 1987, and Gertrude in *Hamlet* in 1989.

Denison, (John) **Michael Terence Wellesley** (1915–), English actor, who made his first appearance on the stage in Brandon *Thomas's *Charley's Aunt* in 1938 and his London début in *Troilus and Cressida* later the same year. In 1939 he married **Dulcie Gray** (1919–), who made her London début at the *Open Air Theatre in 1942. They appeared together for the first time in Peter Watling's *Rain on the Just* (1948), and then in several other plays including Jan de Hartog's *The Fourposter* (1950), Dulcie Gray's own play *Love Affair* (1955), which Denison directed, and revivals of *Lonsdale's *Let Them Eat Cake* (1959) and *Shaw's *Candida* (1960) and *Heartbreak House* (1961). After touring in a joint Shakespeare recital they co-starred again in revivals of *Wilde's *An Ideal Husband* (1965), Lonsdale's *On Approval* (1966), *Ibsen's *The Wild Duck* (1970), and *Sheridan's *The School for Scandal* (1983). Denison was also in other productions, including *The Black Mikado* (1975), as Pooh-Bah, and Shaw's *The Apple Cart* (1986) and *You Never Can Tell* (1987).

Depot, The, see DOWNSTAGE THEATRE.

Derwent, Clarence (1884–1959), English actor and producer. He made his first appearance on the stage in 1902, and was for five years with Frank *Benson's company. He subsequently spent two years at the Gaiety, Manchester, under Miss *Horniman's management, and then appeared in London in 1910 under *Tree. In 1915 he went to America, where he had a distinguished career on Broadway, making his début as Stephen Undershaft in *Shaw's *Major Barbara*. His only further appearance in London was in *The Late Christopher Bean* (1933) by Emlyn *Williams (NY, 1932). His last role was in *Giraudoux's *The Madwoman of Chaillot* (1948). He was responsible for many productions, and in 1945 instituted the Clarence Derwent awards given annually in London and New York for the best performances by players in supporting roles. He was the author of several plays, and in 1946 was elected President of American *Equity.

Deschamps, see PICARD.

Dessous, see CELLAR.

Destouches [Philippe Néricault] (1680–1754), French dramatist, important in the development of 18th-century *drame* from 17th-century comedy. He was a successor to *Molière, and had already written several moderately successful comedies when in 1716 he visited London, where he contracted a secret marriage with an Englishwoman which later supplied him with the material for one of his best works, *Le Philosophe marié* (1727), adapted by Mrs *Inchbald as *The Married Man* (1789). His stay in England may account for the mingling of sentiment and tragedy with comedy in his later plays, many of which were spoilt by sententiousness. His best play is *Le Glorieux* (1732), which portrays the struggle between the old nobility and the newly rich who are rising to power; it was translated as *The School for Arrogance* (1791). Destouches became a member of the French Academy in 1723.

Detail Scenery, small, changeable pieces of scenery used for a particular scene in or before a formalized Permanent Setting (known in America as a Unit, or Formal, Setting).

Deus ex Machina, literally 'the god from the machine', in classical drama the character, usually a god, who enters at the end of a play to resolve the complexities of a plot which would otherwise be insoluble. In Greek tragedy he was lowered from above by a crane (*mechane*). The use of this device was later criticized as showing the dramatist's lack of skill in resolving his plot naturally. By extension, the term has come to mean any arbitrary form of plot resolution.

Deuteragonist, see AGON.

Deutsches Theater, private play-producing society founded in Berlin in 1883 to stage good plays in repertory, as a protest against the deadening effect of long runs and outmoded theatrical tradition. A group of actors led by Josef *Kainz and Agnes *Sorma presented classical historical plays in the style of the *Meininger company. In 1894 the enterprise was given a new direction by its affiliation to the *Freie Bühne. It enjoyed another period of fame under Max *Reinhardt, who went to it in 1905 from the Neues Theater with a band of young actors trained in his own methods. After the First World War the theatre was in the doldrums, but it revived somewhat under **Heinz Hilpert** (1890–1967), who from 1934 to 1944 managed the theatre with complete integrity in the face of Nazi intransigence. It was closed in 1944, reopened the following year, and in 1946 became the National Theatre of East Berlin. It housed the *Berliner Ensemble from its foundation until 1954. The theatre knew further periods of success under *Besson, 1961–9, and from 1978 under the actor Alexander Lang, whose artificial-grotesque productions attracted international attention.

Devine, George Alexander Cassady (1910–65), English actor and theatre director, who made his professional début in 1932 and later joined *Gielgud's company at the *Queen's Theatre. He was on the staff of the London Theatre Studio founded by Michel *Saint-Denis, 1936–9. After six years in the army he joined Saint-Denis at the *Old Vic school, being also a director of the *Young Vic company. In 1954 he gave an outstanding performance as Tesman in *Ibsen's *Hedda Gabler* to the Hedda of Peggy *Ashcroft, and in 1955 at the *Shakespeare Memorial Theatre directed an interesting *King Lear* with décor by the Japanese artist Isamu Noguchi. A year later he was appointed Artistic Director of the newly formed *English Stage Company. In addition to directing a number of plays for it he gave some fine performances, particularly as Mr Shu Fu in *Brecht's *The Good Woman of Setzuan* (1956), Mr Pinchwife in *Wycherley's *The Country Wife*, and the Old Man in *Ionesco's *The Chairs* (both 1957). In 1966 a George Devine Award was instituted in his memory to provide financial encouragement to young workers in the theatre.

Devrient, Ludwig (1784–1832), German actor, first of a famous 19th-century theatrical family. In 1814 he was engaged by *Iffland for *Berlin, where he became the leading German actor of the Romantic period, evoking comparisons with Edmund *Kean by his wild-eyed, unbridled, passionate acting. His rendering of King Lear's madness was celebrated, as was his incandescently evil Franz Moor in *Schiller's *Die Räuber*. He was also an excellent comedian, Falstaff in *The Merry Wives of Windsor* being considered his finest part; he was also much admired as Shylock in *The Merchant of Venice* and as Harpagon in *Molière's *L'Avare*.

His three nephews were all distinguished actors. **Karl** (1797–1872) became a leading actor at the Court Theatre in Dresden, much admired as Hamlet and, later in life, as *Goethe's Faust, Shylock, and King Lear. His younger son **Max** (1857–1929) was for many years at the Vienna Burgtheater. A handsome man of commanding presence, he excelled in big tragic roles, but could also play comedy, his Petruchio in *The Taming of the Shrew* being much admired. Ludwig's second nephew **Eduard** (1801–77) was an excellent actor but, overshadowed at Dresden by his younger brother **Emil** (1803–72), took to directing, and also published some good translations of plays by Shakespeare. Emil excelled in tragic parts, to which his fine speaking voice and heroic gestures were well suited. His Hamlet was well received in London.

Dewhurst, Colleen (1924–91), Canadian-born actress, who made her New York début in 1946 while still a drama student, and in 1952 had a small part in a production of *O'Neill's *Desire under the Elms*, in which in 1963 she was to play the leading role of Abbie Putnam. A passionate and powerful actress, she starred in other plays by O'Neill—*A Moon for the Misbegotten* (in 1958 at the Spoleto Festival in Italy and in 1973 with Jason *Robards); the first New York production of *More Stately Mansions* (1967); and revivals of *Mourning Becomes Electra* (1972), *Long Day's Journey into Night*, and *Ah, Wilderness!* (both 1988). She was seen to advantage in several Shakespearian roles for Joseph *Papp: Tamora in *Titus Andronicus*, Katharina in *The Taming of the Shrew* (both 1956), Lady Macbeth (1957), Cleopatra (1959 and 1963), and Gertrude in *Hamlet* (1972). She was also in plays by *Albee— *The Ballad of the Sad Café* (1963), *All Over* (1971), and as Martha in *Who's Afraid of Virginia Woolf?* (1976). Other outstanding performances were in Athol *Fugard's *Hello and Goodbye* (1969), *Brecht's *The Good Woman of Setzuan* (1970), as Lillian *Hellman in Eric *Bentley's *Are You Now or Have You Ever Been . . . ?* (1978), and in *Betti's *The Queen and the Rebels* (1982).

Dexter, John (1925–90), English director, originally an actor. After joining the *English Stage Company in 1957 he came into prominence with his productions of Arnold *Wesker's trilogy *Chicken Soup with Barley* (1958), *Roots* (1959), and *I'm Talking about Jerusalem* (1960). Also in 1960 he directed the West End production of Lillian *Hellman's *Toys in the Attic*, and after another Wesker play *Chips with Everything* (1962; NY, 1963) he was responsible for the musical *Half a Sixpence* (1963). He was an associate director of the *National Theatre, 1963–6, for which he mounted revivals of *Shaw's *Saint Joan* with Joan *Plowright, Harold *Brighouse's *Hobson's Choice*, *Othello* with Laurence *Olivier, and Peter *Shaffer's *The Royal Hunt of the Sun* (1964; NY, 1965). After *Rattigan's double bill *In Praise of Love* (1973) he staged Shaffer's *Equus* (also 1973) for the National Theatre (NY, 1974) and a revival of Shaw's *Pygmalion* (1974). He was Director of Production at the Metropolitan Opera House in New York, 1974–81, but he also directed plays, including *Phaedra Britannica* (1975) based on *Racine and *Brecht's *Life of Galileo* (1980), both for the National Theatre. His work in the 1980s ranged from Christopher *Hampton's *The Portage to San Cristobal of A.H.* (1982) and the musical *Gigi* (1985) in London to David Henry Hwang's *M. Butterfly* (NY, 1988; London, 1989).

Dibdin, Charles (1745–1814), English actor, dramatist, and composer of many ballads, including 'The Lass that Loved a Sailor' and 'Tom Bowling', the latter in memory of his brother, a naval officer. He also wrote a number of *ballad operas, one of which, *The Waterman* (1774), was successful enough to pass into the repertory of the *toy theatre. He was a good actor, making his first success as Mungo in *Bickerstaffe's *The Padlock* (1768), for which he wrote the music. Handsome, but quarrelsome,

he was constantly at odds with the London managers, particularly *Garrick. Two of his sons by the actress Harriet Pitt were on the stage. The elder, **Charles Pitt** (1768–1833), was a successful writer of plays and pantomimes. The younger, **Thomas Pitt** (1771–1841), also an actor and dramatist, took the name Dibdin in 1800 in order to annoy his father, whom he accused of neglect. His most successful work was the pantomime *Harlequin and Mother Goose* (1806) in which *Grimaldi first played *Clown.

Dickens, Charles John Huffam (1812–70), the great English novelist, was all his life intimately connected with the stage and had an immense influence on it through the numerous dramatizations of his books. Although there is no proof that he was ever an actor, both *Nicholas Nickleby* and *Great Expectations* show an intimate knowledge of an actor's life between 1837 and 1844, and those who saw Dickens in his many amateur appearances thought he would have made a fine eccentric comedian. His famous readings from his own works were in a way solo dramatic performances, as has been clearly shown by Emlyn *Williams's reconstruction of them. In his later years he had in his London home a small private theatre, perfectly equipped, where with his friends and family he gave private performances before a distinguished audience. Two of Wilkie Collins's plays were produced there before they were seen at the *Olympic, and Dickens also collaborated with Collins in the writing of *No Thoroughfare* (1867), in which *Fechter and Ben *Webster appeared.

It would be impossible to catalogue here all the plays based on Dickens's novels, many of which were seen before the books had finished appearing in fortnightly parts. The most persistent adapters were W. T. *Moncrieff and Edward Stirling, but many versions were performed anonymously. Dickens entrusted Albert *Smith with the dramatization of *The Cricket on the Hearth* at Christmas 1845, when 12 different versions were being given at London theatres, all to be superseded later by Dion *Boucicault's excellent *Dot* (1862), and Dickens himself wrote the script for *The Old Curiosity Shop*, not seen until 1884 at the *Opera Comique. His own version of *Great Expectations*, in which he hoped *Toole would appear, was never used, and the first to be staged was that prepared by W. S. *Gilbert (1871). Many dramatizations were also prepared for the American stage, but owing to the absence of *Copyright laws Dickens received nothing for them.

Dickens's characters are so vivid, his plots so dramatic, that it is not surprising they did well on the stage. Many famous actors had their favourite Dickens characters, *Irving appearing as Jingle (from *The Pickwick Papers*), *Tree as Fagin (from *Oliver Twist*), and, most successful of all, *Martin-Harvey as Sidney Carton in *The*

Only Way (based on *A Tale of Two Cities*). Betsey Prig and Sairey Gamp (from *Martin Chuzzlewit*) were for a long time acted by men. Three of Dickens's novels, *Oliver Twist, The Pickwick Papers*, and *A Tale of Two Cities*, were turned into musicals, as *Oliver!* (1960) by Lionel *Bart, *Pickwick* (1963), and *Two Cities* (1969). A version of *The Pickwick Papers* was an outstanding success on the Soviet stage, being seen at the *Moscow Art Theatre in 1934. The *RSC's production of a two-part, eight-hour adaptation of *Nicholas Nickleby* was a phenomenal success in London (1980) and New York (1981).

Diderot, Denis (1713–84), French man of letters. Compared with his Encyclopaedia (1751–77) and his numerous other works Diderot's dramatic writings are only a minor feature of his literary life; but they were an important expression of the Enlightenment, and had a considerable influence on *Lessing and through him on the European drama of the 19th century. Diderot was an exponent of *drame bourgeois*, as can be gauged by the titles of some of his plays—*Le Fils naturel, Le Père de famille*. Some of them were published long before they were acted, others were seen only in private theatres. Yet such was Diderot's reputation that during his lifetime they were translated into German, English, Dutch, and Italian. The best of Diderot's work for the theatre is to be found in his *Observations sur Garrick, Essai sur la poésie dramatique, Entretien avec Dorval*, and especially *Paradoxe sur le comédien*, published posthumously in 1830, a dialogue on the art of acting dealing notably with the relative merits of spontaneous sensibility and intellectual control. One of Diderot's stories in dialogue, *Le Neveu de Rameau*, written 1761–74, was staged in Paris in the 1960s with Pierre *Fresnay in the title-role.

Digges, Dudley (1879–1947), Irish-American actor, trained for the stage by Frank J. *Fay, who appeared in some early plays by *Yeats, being particularly good as the Wise Man in *The Hour-Glass* (1903). In 1904 he went with the *Abbey Theatre company to New York, and remained there until his death, making his first appearance on Broadway in 1904 in *Shaw's *John Bull's Other Island*. For some years he acted as stage-manager for George *Arliss and he then joined the *Theatre Guild company, remaining with it until 1930 and being seen in most of their productions, notably *Rice's *The Adding Machine* (1923) and *O'Neill's *Marco Millions* (1928), and also directing, among other plays, Shaw's *Candida* (1925), *Pygmalion* (1926), and *The Doctor's Dilemma* (1927). His last part and one of his finest was Harry Hope in O'Neill's *The Iceman Cometh* (1946).

Dimmer, see LIGHTING.

Dingelstedt, Franz von, see MUNICH.

Dionysia, ancient Greek festival in honour of the god *Dionysus, closely associated with the development of comedy and tragedy, which, under the influence of *Thespis, evolved from the *dithyramb or ceremonial hymn. The most important of the Greek festivals was the City or Great Dionysia, held in Athens in Mar.–Apr., when the city was normally full of visitors. It was reorganized on a grand scale by the tyrant Pisistratus in the 6th century BC, and lasted for five or six days, opening with a splendid Dionysiac procession to the Parthenon. On the next three days a tragic trilogy (see TETRALOGY), with its attendant *satyr-drama, was presented in the morning, while from the early 5th century onwards comedies, first seen only at the smaller and more intimate winter festival, the Lenaea, were given in the afternoon. The festivities concluded with a competitive contest for 10 dithyrambs. Plays given at the City Dionysia were usually revived later at the Rural Dionysia.

Dionysus, Greek nature-god (Bacchus being the roughly equivalent Latin name), associated mainly but not exclusively with wine. As Dionysus was a vegetation-spirit, who died and was reborn each year, he was associated not only with fertility rites but with mystery-religions, of which an important part was their teaching about death, purgation, and rebirth. The death, sufferings, and rebirth of Dionysus were commonly presented in a quasi-dramatic ritual (the same sequence of events is also found in the English *mumming play), and drama in Athens, whether tragic, satyric, or comic, was always strictly associated with the *Dionysia and similar festivals. In *Old Comedy the phallus (as a symbol of fertility) and the Dionysiac *komos* or revel-song are constant features; another element, the *agon*, is by some scholars derived from the 'agony' of Dionysus. In *satyr-drama, too, direct Dionysiac influence is obvious, the satyrs, or 'horse-men', imaginary creatures of the wild, coming to be regarded as attendants on Dionysus. With *tragedy the connection is less clear and may be more indirect, and it might be argued that tragedy actually became itself by getting rid of the Dionysiac elements.

Director, person responsible for the general interpretation of a play, and for the conduct of rehearsals; known on the Continent as the *régisseur*. In England until 1956 he was called the *producer, but it was then officially decided to adopt the usage of the American stage and the cinema. Until the beginning of the 20th century rehearsals were conducted by the author, the chief actor, the *stage-manager, or the *prompter, and they were brief and perfunctory. Many actors appeared on stage with no preliminary rehearsal, trusting to the older members of the company to carry them. The first to insist on long detailed rehearsals was the actress-manageress Mme *Vestris. Dion *Boucicault, who worked with Vestris, took her methods to the United States, where he directed his own plays and influenced David *Belasco, America's first outstanding director in the modern style. Meticulous rehearsals, especially of crowd scenes, had already found favour in Germany with the *Meiningen company, whose influence was apparent on Otto *Brahm in Germany, on *Antoine in France, and on *Stanislavsky in Russia. Among their disciples were *Reinhardt in Germany, and in Russia *Meyerhold. Meanwhile in England *Granville-Barker had taken over the management of the *Royal Court Theatre and trained actors in the new way. He was much admired by the French actor-manager *Copeau, the first of a new generation which also produced *Jouvet, *Baty, *Dullin, and, from Russia via Switzerland, *Pitoëff. In Germany the work of the director was extended by *Piscator and *Jessner; in Russia Meyerhold gave way to *Taïrov, *Okhlopkov, and *Vakhtangov; and in England French influence was brought to bear when Copeau's nephew Michel *Saint-Denis settled in London. America developed a body of truly American directors with Eva *Le Gallienne followed by *Strasberg, *Kazan, and *Clurman. England was represented by Basil *Dean and by the brilliant Shakespearian director Tyrone *Guthrie; outstanding later directors were Peter *Brook and Peter *Hall. On the Continent the main currents of play production were influenced by *Brecht, *Vilar, and later *Planchon and *Stein. The styles of some directors may be immediately recognized. There is no lack of young and original directors in the theatre today; and there are courses for directors as well as actors.

Discovery Space, see INNER STAGE.

Disguising, term used in the 15th and 16th centuries to describe a form of Court entertainment in England which had something in common with the *mumming play and, through the use of masks, with the later Court *masque. The word was already falling out of use by 1544 and being replaced by 'mask'. The Disguising was essentially an amateur production, involving the royal household and attendant nobility, and was often given in honour of a distinguished guest, ending with an exchange of gifts and a dance in which everyone joined. Music and dancing were originally more important than the text.

Dithyramb, hymn originally in honour of the Greek god *Dionysus, and then of other gods. It was performed by a *chorus of 50 and would normally relate some incident in the life of the deity to whom it was addressed. The leader of

the chorus later became the solo *Protagonist, and the ensuing dialogue between him and the rest of the chorus may have helped in the development of early drama.

Dmitrevsky, Ivan Afanasyevich (1733–1821), Russian actor, who appeared with the amateur company founded by Fedor *Volkov, and went with him to play before the Court in 1752. He was later a member of the company organized by *Sumarokov. After the death of Volkov in 1763 Dmitrevsky was appointed Inspector of Theatres, and took a leading part in the running of the state playhouses. Between 1765 and 1768 he twice went abroad to complete his theatrical education, spending most of the time in Paris with the leading French actors of the day. On his return he occupied the highest position in the St Petersburg theatre, as both actor and administrator, appearing with equal success in tragic and comic parts. His best performances were considered to be the title-roles of *Molière's Le Misanthrope and Sumarokov's Dmitri the Impostor, and Starodum in The Minor by *Fonvizin.

Döbbelin, Karl Theophil, see BERLIN.

Dockstader, Lew, see VAUDEVILLE, AMERICAN.

Dock Street Theatre, Charleston, SC. The first theatre of this name opened with *Farquhar's The Recruiting Officer in 1736 but was not a success, and no performances are recorded there after 1738. In 1937 a second Dock Street Theatre, seating 463, was built there, and again opened with The Recruiting Officer, which was produced by the Footlight Players, Charleston's oldest amateur theatre group, founded in 1931. Since 1958 the Players have been sole lessee of the theatre, which is owned by the City of Charleston. During the interval between the two Dock Street theatres Charleston had several other theatres; but the town is not now important theatrically.

Documentary Theatre, theatre based, usually for propaganda purposes, on fact, as documented in material such as records, films, newspapers, official reports, and transcripts of trials. It is also known (especially in the USA) as the Theatre of Fact. Examples were seen in post-Revolutionary Russia (see AGIT-PROP) and in the work of *Piscator in Germany in the 1920s, which included film projections. In the 1930s the government-sponsored *Living Newspaper in the USA was responsible for six documentary productions on contemporary themes. In the 1960s Piscator produced a series of documentary dramas which were also seen in London and New York. Other examples were *Theatre Workshop's production of Oh, What a Lovely War! (1963), about the First World War, and the *RSC's production of US (1966), about the US Vietnam involvement. (See also POLITICAL THEATRE.)

Dodsley, Robert (1703–64), English playwright and publisher. He was working in London when his literary gifts attracted the attention of Pope and Defoe. Helped by them, and by the success of his first play *The Toy Shop* (1735), he established himself as a bookseller and publisher in Pall Mall, where he issued works by such authors as Pope and Dr *Johnson, and also published a *Select Collection of Old Plays*, later revised and edited by *Hazlitt. The best known of his own plays was *The King and the Miller of Mansfield* (1737), which, with its sequel *Sir John Cockle at Court* (1738), was first seen at *Drury Lane and frequently revived. It provided the basis for Collé's *Partie de chasse d'Henri IV*, and became part of the repertory of the 19th-century *toy theatre. Dodsley also wrote the libretto of a *ballad opera, *The Blind Beggar of Bethnal Green* (1741).

Dog Drama, type of spectacular entertainment which probably had its origin in the performing dog troupes of the circuses. It became extremely popular in Europe in the 19th century, the immediate cause of the craze for dogs on stage in London being a little *after-piece at *Drury Lane in which a real dog, Carlos, rescued a child from a tank of real water—an echo of the equally fashionable *aquatic drama. The most famous dog drama was *Pixérécourt's *Le Chien de Montargis; ou, La Fôret de Bondy*, first seen in Paris at the Théâtre de la *Gaîté in 1814, and in an English adaptation as *The Dog of Montargis; or, The Forest of Bondy*, by William Barrymore, at *Covent Garden in the same year.

Doggett, Thomas (c.1670–1721), English actor, a fine low comedian for whom *Congreve wrote the parts of Fondlewife in *The Old Bachelor* (1693) and Ben in *Love for Love* (1695). He was himself the author of a comedy, *The Country-Wake* (1690), which Colley *Cibber later used as the basis of his *Hob; or, The Country Wake* (1711). In 1711 Doggett became joint manager of *Drury Lane with *Wilks and Cibber, but retired in disgust when Barton *Booth, whose politics he disapproved of, was given a share in the patent. In 1714, in honour of the accession of George I, Doggett instituted the Doggett Coat and Badge for Thames watermen, a trophy which is still rowed for annually and which plays an important part in the plot of *The Waterman* (1774), a *ballad opera by Charles *Dibdin.

Dogg's Troupe, see COMMUNITY THEATRE.

Dolphin Theatre Company, see NATIONAL YOUTH THEATRE OF GREAT BRITAIN.

Domestic Drama, see DRAME BOURGEOIS.

Dominique, see BIANCOLELLI.

Don Juan, character derived from an old Spanish legend, who first found vital expression in *Tirso de Molina's *El burlador de Sevilla y convidado de piedra* (*The Trickster of Seville and the Stone Guest, c.*1630) and has since become a constantly recurring figure in European literature. There is no evidence for his historical existence. Tirso's play combines two plots derived from separate sources, the first being concerned with the character and activities of the hero, the second with his mocking invitation to dinner to the marble statue, who accepts it and brings retribution upon Don Juan by supernatural means for his many crimes.

Among the many works on the same theme are Mozart's opera *Don Giovanni* (1787), *Byron's poem *Don Juan* (1819–24), *Molière's *Le Festin de pierre* (1665), an Italian version by *Goldoni, and a Russian one by *Pushkin, several works in Spanish, the best being Zorrilla y Moral's *Don Juan Tenorio* (1844), *Rostand's *La Dernière Nuit de Don Juan* (c.1910), Horváth's *Don Juan Comes Back from the War* (1937), *Frisch's *Don Juan, oder Die Liebe zur Geometrie* (1953), and Ronald Duncan's *Don Juan* and *The Death of Satan* (both 1956). Don Juan also appears in the third act of *Shaw's *Man and Superman* (1905), and in Tennessee *Williams's *Camino Real* (1953).

Donmar Warehouse, originally a rehearsal studio built by Donald Albery (see ALBERY, BRONSON) for Margot Fonteyn, whom he managed; hence its present name. As the Warehouse, seating 200, it became the *RSC's studio theatre in London, opening in 1977 to show plays which had proved successful at the *Other Place as well as the works of young British playwrights, among them Howard *Barker, Howard *Brenton, and Willy Russell (see EVERYMAN THEATRE, Liverpool), whose *Educating Rita* (1980) was its most popular presentation. The RSC left the Warehouse on the transfer of its London base to the Barbican in 1982, and the theatre became the Donmar Warehouse, one of the new management's early successes being Jonathan *Miller's studio version of *Hamlet*. In 1984 it began to stage seasons by the best British *Fringe touring companies such as Cheek By Jowl. It also staged the musical Show People Seasons which featured such entertainments as *Kern Goes to Hollywood* (1985) and *Blues in the Night* (1987). The theatre presented a 'Pick of the Fringe' Festival in Sept.–Oct., with nine of the best shows from the Fringe of the *Edinburgh Festival. It closed in 1990, but will reopen in 1992 with short runs of its own productions.

Doone, Rupert, see GROUP THEATRE, London.

Doors of Entrance, see PROSCENIUM DOORS.

Dorset Garden Theatre, London. This theatre, known also as the Duke's, or Duke of York's, House, fronted the Thames, to the south of Salisbury Court. It was planned in c.1669 by *Davenant for his company the Duke's Men, but he died before it was completed and his widow, with the actors Henry *Harris and *Betterton, took control. The building, by tradition attributed to Wren, was 57 ft. wide by 140 ft. long; it was uncomfortable and inconvenient, its long narrow auditorium giving a tunnel-like view of the stage. It opened in 1671 with *Dryden's Sir Martin Mar-All, a tried favourite. The first new play was Crowne's King Charles VIII of France; opera, for which Dorset Garden later became famous, began with Davenant's adaptation of Macbeth (1673). An operatic version of The Tempest by *Shadwell was given a spectacular production, as was his own play Psyche (both in 1674). Betterton's performances in The Libertine (1675) and The Virtuoso (1676), both by Shadwell, enhanced his already considerable reputation, while Mrs *Barry made her reputation in *Otway's The Orphan (1680) and Venice Preserv'd (1682).

Under Betterton's capable management the theatre flourished and from 1672 until 1674, while *Drury Lane was out of action, it was London's only first-class playhouse. By 1682, however, both houses were in financial difficulties, and their companies amalgamated, choosing Drury Lane as their headquarters. Dorset Garden was deserted, except for occasional performances of opera and, towards the end, of acrobatic and wild animal shows. It was demolished in 1709.

Dorval [née Delaunay], **Marie-Thomase-Amélie** (1798–1849), French actress, child of strolling players, who was on the stage from her earliest years, and as an adult actress first came into prominence as Amélie in the melodrama Trente Ans; ou, La Vie d'un joueur (1827), playing opposite *Frédérick. In 1835 she made her first appearance at the *Comédie-Française, giving a remarkable performance as Kitty Bell in Chatterton by Alfred de Vigny, whose mistress she was at the time. Later in the same year she appeared with success in *Hugo's Angelo and might have remained with the company, but she found the restrictions irksome and the jealousy of Mlle *Mars insupportable. She therefore returned to the popular theatres, where she again played with Frédérick. In 1842 she was seen at the *Odéon in *Racine's Phèdre, with marked success, but her health failed after a gruelling provincial tour and she returned to Paris to die in poverty.

Dotrice, Roy (1923–), British actor, who became interested in the theatre while a prisoner of war in Germany in the Second World War. He made his first appearance in 1945 in a revue performed by ex-prisoners-of-war, after which he acted in various repertory companies. In 1958 he was at the *Shakespeare Memorial Theatre, and there and at the *Aldwych Theatre in London played a wide variety of parts, including Father Ambrose in *Whiting's The

Devils (1961), Simon Chachava in *Brecht's The Caucasian Chalk Circle (1962), Caliban in The Tempest (1963) and Shallow in Henry IV, Part Two (1964), and the title-role in Brecht's Puntila (1965). He is best known for his solo performance as John Aubrey in Brief Lives, an admirable re-creation of the malicious 17th-century diarist and his age (*Hampstead and NY, 1967; London, 1969). After a notable Peer Gynt in *Ibsen's play at *Chichester (1970) he acted mainly in America. In New York in 1980 (London, 1981) he gave another one-man performance, as Abraham Lincoln, and he was also seen in New York in Hugh *Leonard's A Life (1980). He played Henry V and Falstaff in Henry IV at the *American Shakespeare Theatre (1981), and in 1982 played Winston Churchill, again as a solo.

Dottore, Il, second elderly character of the *commedia dell'arte, forming a pair with *Pantalone. He was usually depicted as a Bolognese lawyer named Graziano, given to long expositions and pedantic utterances which in the coarser playing of the character lapsed into spropositi (exaggerations) and nonsensical tongue-twisters, the original use of local dialect giving way to the habit of 'saying everything the wrong way round'. He was distinct from the stage pedant, who belonged to the *commedia erudita, but they had many characteristics in common and were later often confounded by foreigners. Unlike Pantalone, whom he resembled in his greed, gullibility, and amorous pretensions, he did not pass into the English *harlequinade, but he was adopted on to the French stage through the plays of *Molière.

Double Masque, see MASQUE.

Douglass, David (?–1786), American actor-manager who in 1758 met and married the widow of the elder *Hallam in Jamaica. Amalgamating his actors with hers, he took them back to New York, named them the *American Company, and built first a temporary theatre on Cruger's Wharf, another in Beekman Street, and finally a permanent one in *John Street. He was also responsible for the erection of the first permanent theatre in the United States, the *Southwark in *Philadelphia, and for theatres in a number of towns which he visited. Under Douglass's management the American Company staged Thomas *Godfrey's The Prince of Parthia (1767), the first American play to have a professional production.

Downes, John (fl. 1661–1710), author of Roscius Anglicanus, a volume of scattered theatrical notes which is one of the rare sources of information on the early Restoration theatre. Downes wanted to be an actor, but his first appearance in *Davenant's The Siege of Rhodes (1661) was such a failure that he gave up, and worked back-stage as *prompter and *book-keeper.

Roscius Anglicanus, first published in 1708, was reprinted in 1886, in 1930, and in 1987 appeared in a new edition.

Down Stage, see STAGE DIRECTIONS.

Downstage Theatre, Wellington, New Zealand, company founded in 1964 by actors and writers. New Zealand's longest running professional theatre, it began life as a small café-theatre and developed into a highly expert ensemble staging many plays by local authors. In 1973 the company moved into its new flexible Hannah Playhouse, and in the following year it put on the first full season of plays entirely by New Zealand playwrights. In 1981 Downstage opened a second performance space, The Depot, now a flourishing theatre in its own right, encouraging emergent writers and contemporary Maori and Polynesian theatre and dance. Downstage has a reputation for radical interpretations of classic plays (*Brecht being particularly popular) and commissioning new work from New Zealand playwrights.

Drag, see FEMALE IMPERSONATION.

Drake, Alfred, see PORTER, COLE.

Drake, Samuel [Samuel Drake Bryant] (1768–1854), American actor-manager, born in England, who left for America in 1810 with his large and talented theatrical family. They performed in Boston for several years, moving to Albany in 1813 to join the company of John *Bernard. They left for Kentucky in 1815, and after a long and dangerous journey performed in Louisville, Lexington, and Frankfort. Drake visited Missouri in 1820, and then went to Cincinnati. His was probably not the first company to appear in the West, but it was the most professional and accomplished and he established several theatres in Kentucky and elsewhere. His granddaughter Julia *Dean became an outstanding American actress.

Drama, term applied loosely to the whole body of work written for the theatre or to a group of plays related by their style, content, or period, as Restoration drama, realistic drama. It is also applied specifically to any situation in which there is conflict and, for theatrical purposes, resolution of that conflict. This implies the co-operation of at least two actors, or, as in early Greek drama, a *Protagonist and *chorus, and rules out narrative and monologue. The dramatic instinct is inherent in man, and the most rudimentary dialogue with song and dance may be classed as drama. In a narrower sense the word is applied to plays of high emotional content, which at their best may give us literary masterpieces and at their worst degenerate into *melodrama. The term dramatist is not necessarily restricted to a writer of such dramas, but serves, like playwright, to designate anyone writing for the theatre.

Drama and Comedy, Theatre of, Moscow, see TAGANKA THEATRE.

Drama Schools, see SCHOOLS OF DRAMA.

Dramaten, Stockholm, see ROYAL DRAMATIC THEATRE.

Dramatic Censorship, see CENSORSHIP, DRAMATIC.

Dramatic Copyright, see COPYRIGHT IN A DRAMATIC WORK.

Dramatis Personae, characters in a play, usually listed in the introductory pages of a printed text, or on programmes. In Latin, *persona* (from the Etruscan) replaced the Greek *prosopon*, or *mask, which concealed the identity of the actor and allowed him to 'become' the personage he was portraying.

Drame Bourgeois, name given by *Diderot to a type of play of which his own works are significant examples. He envisaged it as a blending of the outmoded forms of comedy and tragedy which would deal seriously with the domestic problems of the middle-class audiences now frequenting the theatre. It differed from the earlier *tragédie bourgeoise* as exemplified by some of the plays of *Voltaire in that it usually ended happily, or at least peacefully, with a reconciliation after repentance. Rooted in the *comédie larmoyante* of *La Chaussée, it was strongly influenced from England by the novels of Richardson and the domestic dramas of *Lillo and *Moore. Among its other exponents was *Mercier.

Draper, Ruth (1884–1956), American actress who achieved world-wide fame with dramatic monologues which she wrote herself. She first employed her gift for mimicry in short sketches performed at private parties and charity performances, and it was not until 1920 that she made her first professional appearance at the Aeolian Hall in London. She quickly established herself as an international figure and for the rest of her life toured continuously, elaborating and adding to her repertory but never changing the basic formula—a bare stage, a minimum of props, and herself as one person responding to invisible companions (as in 'Opening a Bazaar' or 'Showing the Garden') or as several people in succession (as in 'Three Generations', 'Mr Clifford and Three Women', or 'An English House Party'). The texts of her monologues served as *aides-mémoire* only, since she varied them at every performance. Her career was a long series of triumphs on the Continent, in New York, where she often remained in the same theatre for four or five months, and in England.

Drapery Setting, see CURTAIN SET.

Dresden, capital of the Electorate (later the Kingdom) of Saxony until 1918. The earliest dramatic performances there took place at

Court, where the prevailing taste was for opera. It was not until the 1770s that subsidized German companies settled in the town, playing in the Kleines Hoftheater built in 1755. Ludwig *Tieck, who was attached to the theatre from 1835 to 1841, established a literary repertoire with plays by *Kleist, *Calderón, and above all Shakespeare. The Königliches Hoftheater, built by Semper in 1841, was burnt down in 1871 and replaced by another Hoftheater in 1878. Carl Zeiss, coming there in 1901, renovated not only the repertory but also the building, installing Linnebach's elaborate lifts and revolving stage. Dresden became one of the major theatrical centres of the Weimar Republic, Oskar Kokoschka directing there the first German production of his *Mörder, Hoffnung des Frauen (Murderer, Hope of Women)* in 1917. A year later Hasenclever's *Expressionist play *Der Söhn*, previously banned in Germany, was also seen. The Staatstheater, constructed in 1984, replaced the Hoftheater, destroyed by bombing in 1945. After lengthy reconstruction work a replica of the famous Semper building was completed in 1985. Its excellent facilities, together with the already established reputation of the Staatstheater, make Dresden today an important cultural centre.

Dressler, Marie, see VAUDEVILLE, AMERICAN.

Drew, Georgiana (1856–93), American actress, daughter of Mrs John *Drew. At the age of 16 she appeared with her mother's company in Philadelphia, and was later with Augustin *Daly in New York. An actress of great ability whose career was hampered by ill health, she married in 1876 Maurice *Barrymore, with whom she appeared in *Diplomacy*, a version of *Sardou's *Dora*. She also played with *Modjeska, Lawrence *Barrett, and Edwin *Booth, and appeared under the management of Charles *Frohman. Her three children, Ethel, John, and Lionel *Barrymore, were all on the stage.

Drew, John (1853–1927), American actor, son of Mrs John *Drew, under whom he first appeared on the stage in Philadelphia. In 1875 he was engaged by Augustin *Daly to play opposite Fanny *Davenport and later Ada *Rehan. During the 1880s he was several times seen in London, being much admired as Petruchio in *The Taming of the Shrew* and in other parts. From 1892 to 1915 he was in a series of modern comedies under the management of Charles *Frohman, often playing opposite Maude *Adams and making frequent tours of the United States. A handsome, distinguished-looking man, he gave a fine performance in 1916 as Major Arthur Pendennis in a dramatization of Thackeray's novel. He was last seen on tour in *Pinero's *Trelawny of the 'Wells'*. He was the third president of the *Players' Club, and in 1903 presented the library of the theatre historian and bibliographer Robert W. Lowe, which he

had acquired, to *Harvard, thus laying the foundation of the fine theatre collection there.

Drew, Mrs John [*née* Louisa Lane] (1820–97), American actress and theatre manager, daughter of English actors who could trace their theatrical ancestry back to Elizabethan days. She went on the stage in London as a small child and in 1827 was taken by her widowed mother to New York, where she appeared as many characters in one play. She was also seen with the elder *Booth, with the first Joseph *Jefferson, and with Edwin *Forrest. By the age of 16 her roles included Lady Macbeth and the Widow Melnotte in *Bulwer-Lytton's *The Lady of Lyons*, and she was appearing all over the United States. In 1850 she married as her third husband **John Drew** (1827–62), an Irish actor who was admired in eccentric parts such as Sir Lucius O'Trigger in *Sheridan's *The Rivals* and Handy Andy in the stage version of Samuel Lover's novel; she was thereafter known as Mrs John Drew. From 1861 to 1892 she managed the stock company at the Arch Street Theatre, *Philadelphia, and from 1880 to 1892 was constantly seen on tour as Mrs Malaprop in *The Rivals*, one of her best parts, with Joseph *Jefferson as Bob Acres. A woman of strong, almost masculine, personality, she ruled her theatre and family with firmness and energy, and contributed greatly to the establishment of the American theatre during the 19th century. Two of her children, John and Georgiana (above), were on the stage. *The Royal Family* (1927) by George S. *Kaufman and Edna Ferber was drawn from the personalities of the Drew family and the succeeding *Barrymore generation.

Drinkwater, John (1882–1937), English poet and dramatist, who was for some years actor and general manager at the *Birmingham Repertory Theatre, where most of his plays were first produced. The most successful of his early verse plays was the anti-war $X = O$; *A Night of the Trojan War* (1917), after which he wrote in prose. His best work was *Abraham Lincoln* (1919); transferred to London, it ran for over a year and in 1920 was seen in America. Other chronicle plays—*Mary Stuart* (1922), *Oliver Cromwell*, and *Robert E. Lee* (both 1923)—were less successful, and the comedy *Bird in Hand* (1927; London, 1929) was the most popular of his later plays.

Droll, short comic sketch, usually a scene taken from a longer play. It originated in London during the *Puritan Interregnum (1642–60), when the actors, forbidden to act, nevertheless managed to provide a modicum of entertainment. For their illicit purposes they invented the 'droll'—short for Droll Humours or Drolleries—rounding it off with a *jig. Some of the most famous drolls are taken from Shake-speare—'Bottom the Weaver' from *A Mid-summer Night's Dream* and 'The Gravemakers' from *Hamlet*. Others were from biblical sources. Droll was also the name applied to early puppet-shows, and to collections of humorous or satiric verse, as in *Westminster Drolleries* (1672). It was sometimes used to designate actors, particularly players of humorous parts, and men of quick wit and good company: Pepys uses it in this sense of *Killigrew. In the late 19th century it was applied to such comediennes as **Louie Freear** (1871–1939), who was equally successful in minstrel shows, music-halls, musical comedy, and Shakespeare.

Drop, early English name for *cloth, still retained in America. In England the word is first found in about 1690 to indicate an unframed canvas *backcloth which offered a plain surface for painting, free from the central join which marked the earlier 'pair of flats' (see FLAT). It was rolled up on a bottom roller, which furled up the drop as it rose. Swords, cloaks, and trains frequently got caught up in the ascending roller and an alternative method known as 'tumbling' was then used. A *batten was fixed across the back of the drop a third of the way up, and it was taken away in coils, with a loose roller, or 'tumbler', inside to weight it and keep the coil straight. As it was not possible to have a practicable door or window in an unframed drop, it was later provided with battens at the back to which doors and windows could be fixed, thus making it, in effect, a single flat which could be flown. The lack of height above the stage made this impracticable in the early theatres.

Drottningholm Theatre and Museum, Swe-den, part of a royal palace on an island near Stockholm, built in 1766. With the palace it became state property in 1777 and its most brilliant period was from then until 1792, during the reign of Gustaf III (1772–92). During the 19th century it fell into disuse, but its employ-ment as a lumber-room saved it from demolition and in 1921 it was restored, the only alteration being the substitution of electric light for the former wax candles. The theatre seats 450 on a single, stepped tier. The stage is about 57 ft. deep and 27 ft. wide at the footlights. The 18th-century stage machinery on the *carriage-and-frame system is still in working order, and there are more than 30 sets of usable scenery of the same period. The theatre is mainly used for early opera and ballet, but the *Royal Dramatic Theatre has also staged productions there. The museum exhibits, which include a rich deposit of 17th-century French stage designs, are housed in the former royal apartments.

Drum–and–Shaft System, early method of moving theatre scenery by means of a rope fixed to a cylindrical shaft. A lever was inserted through the shaft and the whole twisted like a corkscrew, drawing the piece of scenery to the required position. An improvement on this method was achieved by the use of a circular drum, or *barrel, built round the shaft and considerably larger in diameter. The shaft was then rotated by pulling on a line wound round the drum, and in this way the piece was moved more easily and steadily. The system was used to move *borders, *bridges, *cloudings, *drops, and *traps. By using drums of different diameters on the same shaft several pieces of scenery could be moved at the same time, and by attaching them singly to different drums they could be made to move at different speeds. The Glory, a scenic effect frequently employed in Renaissance and Baroque spectacles in which a number of clustered cloudings gradually expanded into a great aureole, was controlled by the drum-and-shaft method. It is now obsolete, all pieces of flown scenery today being worked independently.

Druriolanus, Augustus, see HARRIS, AUGUSTUS HENRY GLOSSOP.

Drury Lane, Theatre Royal, London's most famous theatre, the present building being the fourth on the site. The first was erected by *Killigrew under a charter granted by Charles II in 1662, making it one of London's two *Patent Theatres. Seating about 700, it occupied a site 112 ft. by 58 ft., and being hemmed in by other buildings could be reached only by a narrow passage from Brydges Street. It opened in 1663 (as the Theatre Royal, Brydges Street) with *Beaumont and *Fletcher's *The Humorous Lieutenant*. From June 1665 to Nov. 1666 it remained closed because of plague and the Great Fire, and it was itself burnt down in 1672. The second theatre on the site, known as the Theatre Royal in Drury Lane, is believed to have been designed by Sir Christopher Wren, though there is no documentary proof. It had a pit, amphitheatre, and two galleries, holding 2,000 people; the stage depth from the front of the apron was no less than 130 ft. It opened in 1674 but lost its younger actors to the company at the more popular *Dorset Garden Theatre and was forced to close. By 1682 it was evident that London could support only one theatre, so a combined company under *Betterton settled at Drury Lane. In 1690 the patent was bought by Christopher *Rich; his clashes with some of the actors led to their departure to *Lincoln's Inn Fields, again headed by Betterton. Rich lost his charter in 1709, and the theatre closed. A triumvirate consisting of Colley *Cibber, Rob-ert *Wilks, and Thomas *Doggett then took over, and with Anne *Oldfield as their leading lady inaugurated a period of prosperity which lasted until 1733. Serious difficulties then arose between the theatre's management and the leaseholders, which were resolved only when complete control was gained by **Charles Fleet-wood** (?–*c*.1745). He ran into difficulties himself when he abolished the Footmen's *Gallery, an action which led to rioting in 1737, and being an inveterate gambler he soon plunged the theatre into debt. The one noteworthy event of this period was *Macklin's performance of Shylock in *The Merchant of Venice* in 1741, in effect inaugurating the new school of inter-pretative acting which *Garrick was later to popularize. Garrick himself made his first appearance at Drury Lane in 1742, as Chaumont in *Otway's *The Orphan*. Five years later, with James Lacy, he took over the theatre and rescued it from bankruptcy. His first production was again *The Merchant of Venice* with Macklin; and with a good company he brought prosperity to the theatre for the next 30 years. He was responsible for major alterations to the building by Robert Adam in 1775, the year in which the 20-year-old Mrs *Siddons made an unsuccessful début as Lady Anne to his Richard III. Garrick himself retired a year later and was succeeded by *Sheridan. In 1780 the theatre was damaged during the Gordon Riots, and a company of Guards was then posted there nightly, a custom not abolished until 1896. In 1788 Sheridan, preoccupied with politics, handed over the effective management of the theatre to John Philip *Kemble while still retaining his share of the patent. The theatre was rebuilt in 1791 on a site 155 ft. by 300 ft., its capacity being increased to 3,611. This third Drury Lane opened in 1794 with a concert. Plays began with Kemble and Mrs Siddons in *Macbeth*, followed by an epilogue during which an iron safety curtain was lowered to prove that the theatre was now protected against fire. This device, however, did not prevent it burning down again in 1809, after a stormy period during which Sheridan's mismanagement caused Kemble to leave for *Covent Garden, taking Mrs Siddons with him, and their place had been taken by *melodramas and spectacles which brought real elephants and performing dogs on to the stage. After the fire there were no funds for rebuilding, but the situation was saved by the brewer Samuel Whitbread, a shareholder with Sheridan in the patent, who raised £400,000.

This fourth Drury Lane, which still stands, is a four-tier house with a capacity of 2,283, a proscenium opening of 42 ft., and a stage depth of 80 ft. It opened in 1812 with *Hamlet*. Edmund *Kean appeared there from 1814 until 1820; but even his success could not keep pace with rising costs. Whitbread committed suicide in 1815, and *Elliston, who took over in 1819 and under whose management the portico was added in 1820 and the interior reconstructed in 1822, finally went bankrupt. He was succeeded in 1826 by the American impresario Stephen *Price, under whom Charles *Kean made his

first, and *Grimaldi his last, appearance. From then onwards the history of Drury Lane was one of unmitigated artistic and financial disaster, apart from the brief reign of *Macready from 1841 to 1843. The theatre closed in 1878 but in 1879 Sir Augustus *Harris reopened it with a revival of *Henry V* and embarked on a series of spectacular shows, realistic melodramas, and an annual *pantomime of great splendour. He was succeeded in 1896 by Arthur Collins, who remained until 1923. He continued Harris's policy, staging explosions, earthquakes, ava-lanches, chariot races, shipwrecks, and, in *The Whip* (1909), a horse race with real horses; but he also had to his credit *Irving's last London season, Ellen *Terry's Jubilee (both 1905), and *Forbes-Robertson's farewell appearance (1913). In 1921–2 the interior was reconstructed. From 1924 to 1931 there was a run of great *musicals including *Rose-Marie* (1925), *The Desert Song* (1927), *Show Boat* (1928), *New Moon* (1929), and *The Land of Smiles* (1931). Ivor *Novello next occupied the theatre with *Glamorous Night* (1935), *Careless Rapture* (1936), *Crest of the Wave* (1937), and *The Dancing Years* (1939). Drury Lane then became the headquarters of the *Entertainments National Service Association (ENSA), under Basil *Dean, reopening with Noël *Coward's *Pacific 1860* (1946), followed by a series of popular musicals by Rodgers and *Hammerstein—*Oklahoma!* (1947), *Carousel* (1950), *South Pacific* (1951), and *The King and I* (1953). *My Fair Lady* (1958) ran for nearly five years, and the musicals *Camelot* (1964), *Hello, Dolly!* (1965), *A Chorus Line* (1976), *42nd Street* (1984), and *Miss Saigon* (1989) were all successful.

Drury Theatre, Cleveland, see CLEVELAND PLAY HOUSE.

Druten, John Van, see VAN DRUTEN.

Dryden, John (1631–1700), English critic, poet, satirist, and one of the outstanding dramatists of the Restoration period, though his best work was done in other fields. The first of some 30 plays, written alone or in collaboration, was *The Wild Gallant* (1663), seen at *Drury Lane, as was *The Rival Ladies* (1664), based on a Spanish original. His first *heroic drama, a genre of which he was the chief exponent, was *The Indian Queen* (also 1664). Its sequel *The Indian Emperor* (1665) was followed by one of his most successful plays, *Secret Love; or, The Maiden Queen* (1667), based partly on Mlle de Scudéry's famous novel *Le Grand Cyrus*, the parts of Florimel and Celadon having sometimes been considered the prototypes of *Congreve's Millamant and Mirabell in *The Way of the World*. Also in 1667 came the first production of *Sir Martin Mar-All; or, The Feign'd Innocence*, followed a year later by another comedy, *An Evening's Love; or, The Mock Astrologer*, which combined elements from *Corneille, *Molière, and *Quinault. Another heroic drama, *Tyrannic Love; or, The Royal Martyr* (1669), was followed by Dryden's finest work in this style, *Almanzor and Almahide*, usually known by its subtitle *The Conquest of Granada*. This vast and complicated play, of which the first part was performed in 1670 and the second in 1671, contains all the elements, good and bad, of heroic drama—rant, bombast, poetry, vigour, battle, murder, and sudden death. It was satirized unmercifully by **George Villiers, second Duke of Buckingham** (1628–87), in *The Rehearsal* (1671), and Dryden returned to comedy with *Marriage à la Mode* (also 1671) and *The Assignation; or, Love in a Nunnery* (1672). He then produced two more heroic tragedies in *Amboyna; or, The Cruelties of the Dutch to the English Merchants* (1673) and *Aureng-Zebe* (1675). From the restraints of rhymed couplets he returned to blank verse for *All For Love; or, The World Well Lost* (1677), his masterpiece, a retelling of the story of Antony and Cleopatra which takes only its plot from Shakespeare. It is well constructed, contains some fine poetry, and observes more strictly than any other English tragedy the *unities of time, place, and action. It was frequently revived in the 18th century. Dryden's later plays were less important; they included rewritings of *Troilus and Cressida* (1679) and *Plautus'* *Amphitryon* (1690), and, his last play, *Love Triumphant; or, Nature Will Prevail* (1693), a tragi-comedy. Dryden also wrote a large number of prologues and epilogues, then very much in fashion, both for his own and for other people's plays, which are not only mines of information about theatrical matters but also notable contributions to English poetry. In his prefaces and critical writings he contributed largely to contemporary literary and theatrical controversies.

Dubé, Marcel (1930–), the leading French-Canadian playwright of his day, born in Montreal and educated there and in Paris. At first he wrote for his own troupe and then for several seasons was resident playwright with the Théâtre du *Nouveau Monde. His early plays, which established him as a popular writer strongly influenced by contemporary American social realism, included *Le Bal triste* (1950), *De l'autre côté du mur* and *Zone* (both 1952), *Un simple soldat* and *Le Temps des lilas* (both 1958), the last set in a seedy lodging-house. He was also responsible for the French adaptations of several modern American plays, among them Arthur *Miller's *Death of a Salesman*. He then began to write in a more elegant and poetic style, producing *drames bourgeois* such as *Florence* (1960) and *Les Beaux Dimanches* (1965). One of his few plays to be translated into English (as *The White Geese*) was *Au retour des oies blanches* (1966), a near-classical tragedy which many consider his best work. A smooth and cynical phase—typified by *Virginie* (1968) and *Le Coup de l'étrier* (*One for the Road*, 1969)—was succeeded by lighter, more satirical work including *Dites-le avec des fleurs* (1976), written in collaboration, and *L'Amérique à sec* (1986), whose hero is a whisky smuggler. His output of nearly 40 plays contains many which originated on radio or television.

Dublin. The first theatre in Dublin was built in 1637. It closed in 1642, under the Commonwealth, and was replaced at the Restoration by the Smock Alley (or Orange Street) Theatre which was used mainly for plays imported from London with guest stars, and which was at its best from about 1730 to 1760. It then declined and was finally converted into a corn store. Almost its only rival was the Crow Street Theatre, which opened in 1758 under Spranger *Barry and flourished until Harris, an actor from *Covent Garden, opened the Theatre Royal in 1819. The Crow Street Theatre then closed, being pulled down in the early 1830s. The *Olympia Theatre opened, as a music-hall, in 1855. The Theatre Royal perished by fire in 1880, and a much larger theatre of the same name replaced it, eventually becoming a cinema. It housed touring companies from England in spectacular drama, as did the Queen's Theatre. The smaller and more intimate Gaiety—opened in 1871 and still housing a variety of entertainment from opera, ballet, and musicals to plays and pantomime—was used by the *Irish Literary Theatre from 1900 until the opening of the *Abbey Theatre. The Abbey quickly turned to realism, and it was not until 1944, with the founding by Austin *Clarke of the Lyric Players, that pure *poetic drama returned to Dublin. The *Gate Theatre was founded in 1928, and a number of small theatres, seating 30–50, opened in Dublin during the 1950s. They were known locally as 'basement theatres', and a number of them were indeed housed in the basements of Georgian houses. Most notable both for repertoire and survival were the Pike Theatre Club, 1953–9, where Brendan *Behan's work was first seen, and the Lantern Theatre, 1957–72, which specialized in poetic drama. Other groups included the Players Theatre and Globe Theatre Productions.

Dublin Roscius, see BROOKE.

Dublin Theatre Festival, annual event, usually in Oct., when the Dublin theatres combine in a festival of new plays, revivals, musicals, mime shows, and *Fringe events. The first, in 1957, involved opera, concerts, ballet, and international folk-dancing, as well as theatrical events which included the *Théâtre National Populaire from Paris and a visit from an English company in *Wilde's *The Importance of Being Earnest*. At local level there were revivals of *Johnston's *The Old Lady Says 'No!'* at the *Gate Theatre; *O'Casey's *Juno and the Paycock* and *Synge's

The Playboy of the Western World at the *Abbey Theatre; seven of *Yeats's plays by the Globe Theatre company; and, at the Pike Theatre, a notorious production of Tennessee *Williams's *The Rose Tattoo*. No festival was held in 1958, following a row over O'Casey's *The Drums of Father Ned* which resulted in O'Casey banning professional productions of his plays in Ireland, and from 1959 the festival narrowed in scope, concentrating mainly on Irish material and performers, though visits from such foreign companies as the *Living Theatre continued to form an important part of the programme. Since 1960 the policy of encouraging new Irish writing has led to the discovery of many new playwrights, among them Brian *Friel and Hugh *Leonard.

Duchess Theatre, London, in Catherine Street, off the Aldwych. This small theatre, seating 491 in two tiers, with a proscenium opening of 25 ft., opened in 1929 and housed some early productions by the People's National Theatre company under Nancy *Price. In 1934 J. B. *Priestley, whose *Laburnum Grove* had had a successful run at the theatre the previous year, took it over and produced his plays *Eden End* (1934) and *Cornelius* (1935). Emlyn *Williams then appeared in his own thriller *Night Must Fall* (also 1935), which ran for a year and was followed by the first West End staging of T. S. *Eliot's *Murder in the Cathedral*. Another long-running Priestley play was *Time and the Conways* (1937). In 1938 Emlyn Williams returned in his play *The Corn is Green*, which was still running when the theatre closed on the outbreak of war in Sept. 1939. It reopened shortly afterwards, but closed again until 1942, when *Coward's *Blithe Spirit* began a long run. Later productions included Priestley's *The Linden Tree* (1947), *Rattigan's *The Deep Blue Sea* (1952), William Douglas *Home's *The Manor of Northstead* (1954), and Ronald *Millar's *The Bride and the Bachelor* (1956). Two plays by Agatha *Christie, *The Unexpected Guest* (1958) and *Go Back for Murder* (1960), were seen there, as was *The Reluctant Peer* by William Douglas Home (1964). The nude revue *Oh! Calcutta!* ran from 1974 to 1980 and Richard Harris's *The Business of Murder* (1981) began its long run here. From 1987 to 1990 the building was occupied by the *Players' Theatre while its new theatre was built, the Duchess then reopening with a transfer of Ray Cooney's comedy *Run for Your Wife*.

Ducis, Jean-François (1733–1816), French dramatist, the first adapter for the French stage of the plays of Shakespeare, which he probably read in one of the deplorable translations since he almost certainly knew no English. Encouraged by the current French Anglomania, he altered *Hamlet* (in 1769) followed by *Romeo and Juliet*, *King Lear*, and *Macbeth*, so that little remained of them but their titles. He knew the tastes of his time and realized audiences would accept Shakespeare only with modifications. His *Othello* (1792) owed much of its success to the fine acting of *Talma, and after seeing his last effort, *King John*, a contemporary critic deplored the waste of his talents on such rubbish. None of Ducis's own plays, however, has survived.

Ducrow, Andrew, see ASTLEY'S AMPHITHEATRE.

Duff, Mrs [*née* Mary Anne Dyke] (1794–1857), American actress, born in London, who first appeared on the stage in Dublin, and in 1810 went with her second husband, an Irish actor named **John Duff** (1787–1831), to America. She made a great reputation in Boston and Philadelphia, and in 1823 was seen in New York as Hermione in *The Distrest Mother* by Ambrose *Philips. A tall, dark, graceful woman, she was at her best in tragic or pathetic parts, among them Belvidera in *Otway's *Venice Preserv'd*. Widowed a second time, and left with seven small children, she inadvisedly married the Australian-born American actor **Charles Young** (?–1874), but the marriage was never consummated and was almost immediately annulled. In 1835 she married again, a lawyer named Seaver from New Orleans, where she appeared in the following year as Jane Shore in *Rowe's tragedy of that name, and as Portia in *The Merchant of Venice*. She is said to have appeared in Toronto at the Royal Lyceum in the summer of 1850, billed as 'the American Siddons', but this seems unlikely.

Duke of York's Theatre, London, in St Martin's Lane. As the Trafalgar Square Theatre, seating 900 in three tiers, it opened in 1892, and in 1893 saw the first performances in England of *Ibsen's *The Master Builder*, with Elizabeth *Robins. In 1895 the theatre took its present name, and in 1896 it had its first success with the musical comedy *The Gay Parisienne*, in which the quaint *droll Louie Freear, who had been on the stage since early childhood, first made a name for herself. In 1897 Charles *Frohman began his tenancy, introducing many well-known American actors to London, among them Maxine *Elliott in 1899. 'Dot' *Boucicault was appointed resident manager in 1901, and under him several plays by *Barrie had their first production, including *The Admirable Crichton* (1902), *Peter Pan* (1904), and *What Every Woman Knows* (1908). In 1910 Frohman tried unsuccessfully to introduce a *repertory system to London with an outstanding programme of modern plays which included *Galsworthy's *Justice*, *Shaw's *Misalliance*, and *Granville-Barker's *The Madras House*. Somerset *Maugham's *The Land of Promise* (1914) was the last new play put on by Frohman. In 1923 came Charlot's successful *revue *London Calling*, written mostly by Noël *Coward, whose *Easy Virtue* was produced there three years later.

Matheson *Lang appeared in *Such Men are Dangerous* (1928) and *Jew Süss* (1929), both adapted by Ashley *Dukes. From 1933 to 1936 Nancy *Price's People's National Theatre used the theatre occasionally. The theatre closed in 1940 because of enemy action, reopening in 1943 with Paul Vincent *Carroll's *Shadow and Substance*. Later productions included *The Happy Marriage* (1952) with John *Clements (who adapted it from a French play), Frank Marcus's *The Killing of Sister George* (1965), Alan *Ayckbourn's *Relatively Speaking* (1967), and Arthur *Miller's *The Price* (1969). In 1978 the theatre was bought by Capital Radio and major structural changes were made, including the replacement of the gallery by a recording and broadcasting studio. The theatre reopened in 1980 with Andrew Davies's *Rose*, with Glenda *Jackson, who later starred in a revival of *O'Neill's *Strange Interlude* (1984). Richard Harris's *Stepping Out* (also 1984) ran for three years.

Dukes, Ashley (1885–1959), English dramatist, theatre manager, and dramatic critic. He worked for *Vanity Fair*, the *Star*, and the *Illustrated Sporting and Dramatic News*, and was also for many years English editor of the American *Theatre Arts Monthly*. He had a wide knowledge of modern Continental drama and adapted a number of foreign plays for the London stage, among them Neumann's *Der Patriot* as *Such Men Are Dangerous* (1928) and Feuchtwanger's *Jew Süss* (1929), both for Matheson *Lang. His very free version of an episode from Fernando de Rojas's *La *Celestina* as *The Matchmaker's Arms* (1930) provided an excellent part for Sybil *Thorndike, and Vivien *Leigh made her first outstanding appearance in his *The Mask of Virtue* (1935), based on *Sternheim's *Die Marquise von Arcis*. In 1933 he opened the *Mercury Theatre where he did excellent work with productions of new and foreign plays, particularly verse plays. His translation of *Machiavelli's *Mandragola* was seen there in 1939. Among his own plays the most successful was *The Man with a Load of Mischief* (1924), in which Fay *Compton made a great success as the Lady.

Duke's House, Duke of York's House, Duke's Men, see DORSET GARDEN THEATRE.

Duke's Theatre, London, see HOLBORN THEATRE.

Dullin, Charles (1885–1949), French actor and producer, a pupil of *Gémier, who, after some appearances in melodrama, joined *Copeau when he first opened the *Vieux-Colombier. In 1919 Dullin formed his own company and took it on a long provincial tour. Back in Paris, confronted by many difficulties and always short of money, he finally succeeded in establishing the company in the Théâtre de l'Atelier, a suburban playhouse built in 1822 and used mainly for melodramas and vaudevilles. It had

become a cinema in 1914, but after Dullin took it over in 1922 it soon gained a reputation as one of the outstanding experimental theatres of Paris. The plays produced there included the classics of France, the comedies of *Aristophanes, translations of famous foreign plays, among them *Calderón's *La vida es sueño* (1922), Shakespeare, Ben *Jonson, *Pirandello for the first time in France, and such new French plays as *Cocteau's *Antigone* (also 1922) and the works of *Achard and *Romains. Himself an excellent actor, Dullin ran a school of acting connected with his theatre, and in 1936 was invited to become one of the directors at the *Comédie-Française. During the occupation of France he toured the unoccupied zone with *Molière's *L'Avare*, and in 1943 he was responsible for the first production of *Sartre's *Les Mouches*.

André *Barsacq succeeded Dullin at the Atelier after the war, moving in with his own company and, in spite of growing financial difficulties, upholding the theatre's reputation. After Barsacq's death in 1973 it was managed by his son André-Alexis, who in 1974 handed over to Pierre Franck.

Dumas [Davy de la Pailleterie], Alexandre, père (1802–70), French novelist and playwright, of Creole parentage, Dumas being the name of his West Indian grandmother. He is now chiefly remembered for his novels, but his plays were important in the French Romantic movement. His historical drama *Henri III et sa cour* (1829) was the first triumph of the Romantic theatre, delighting an audience starved for years of colour and movement. A play on Napoleon was followed by *Antony* (1831), seen at the Théâtre de la *Porte-Saint-Martin, where several of Dumas's more melodramatic pieces were first produced, including the famous *La Tour de Nesle* (1832), which for terror and rapidity of action, not to mention the number of corpses, surpassed anything seen on the French stage since the days of Alexandre *Hardy. Later *Kean; ou, Désordre et génie* (1836), with *Frédérick in the title-role, had a tremendous success (as did an adaptation of it by *Sartre in 1953). Dumas then proceeded, alone or in collaboration, to dramatize his own historical novels, with varying success. Some were produced at the Théâtre Historique, which he built and financed and which nearly ruined him when it failed in 1850. (In spite of his enormous earnings, Dumas was extravagant and an easy prey to harpies.) He was eventually rescued from his creditors by his daughter, who, with the help of her brother Alexandre (below), took over the management of her father's affairs in 1868 and remained with him until his death.

Dumas, Alexandre, fils (1824–95), French dramatist, natural son of the above, who approached the theatre by way of a dramatization of his own novel *La Dame aux* *camélias* (1848). First acted in 1852, this became one of the outstanding theatrical successes of the second half of the 19th century, and is still occasionally revived. As *Camille*, it was equally popular in England, America, and Italy; but in spite of its success it was destined to remain the younger Dumas's only Romantic play. He turned to social drama and, though himself an agnostic, sought to enforce Christian virtues and conventional morality by using the stage as a pulpit, as in *Les Idées de Madame Aubray* (1867). Dumas had little liking for the bohemian society in which his childhood had been passed and which he summarized in the title of his play *Le Demi-Monde* (1855). The bitterness of his own illegitimacy found expression in *Le Fils naturel* (1858) and *Un père prodigue* (1859), while the question of sexual morality was ventilated in such plays as *La Femme de Claude* (1873), *L'Étrangère* (1876), and his last play *Francillon* (1887). Only occasionally, as in *La Question d'argent* (1857), did he deal with social issues of more general interest, and most of his plays have disappeared with the conditions which gave rise to them. In his own day a popular and powerful social dramatist, he is now remembered only by his least typical work, *La Dame aux camélias*, on which Verdi based his opera *La Traviata* (1853).

Du Maurier, Sir Gerald Hubert Edward (1871–1934), English actor-manager. He was with *Tree at the *Haymarket Theatre, playing a small part in *Trilby* (1895), a dramatization of the novel by his father George. After remaining with Tree for some years, he came into prominence at the *Duke of York's Theatre, under Charles *Frohman, in *Barrie's *The Admirable Crichton* (1902). He was the first to play Mr Darling and Captain Hook in Barrie's *Peter Pan* (1904), and made a great success at the *Comedy Theatre in E. W. Hornung's *Raffles* (1906), following it with *Arsène Lupin*, from the French of F. de Croisset, in 1909, in which year he directed the stirring patriotic play *An Englishman's Home*, produced anonymously, by his elder brother **Guy** (1865–1916) who was killed in the First World War. In 1910 Du Maurier again scored an immense success with Paul Armstrong's *Alias Jimmy Valentine*, and then took over the joint management of *Wyndham's Theatre, making it a home of light comedy for many years. His range of parts was limited; but within those limits he was seldom excelled, and he could on occasion step beyond them, as in his portrayal of Mr Dearth in Barrie's *Dear Brutus* (1917). A more typical part was the title-role in Sapper's *Bulldog Drummond* (1921). Shortly afterwards he left Wyndham's and went to the *St James's Theatre, where he made his last outstanding appearances in *Lonsdale's *The Last of Mrs*

Cheyney (1925) and Roland Pertwee's *Interference* (1927).

His daughter **Daphne** (1907–89), best known as a novelist, was also the author of three plays, *Rebecca* (1940), based on one of her own novels, *The Years Between* (1945), and *September Tide* (1949).

Dumb Ballet, see TRICKWORK.

Dumesnil, Marie-Françoise (1713–1803), French actress, who joined the *Comédie-Française in 1737 and soon showed herself an excellent actress, particularly in passionate roles. She was considered by *Voltaire superior even to Adrienne *Lecouvreur in high pathos, and he attributed much of the success of his *Mérope* (1743) to her acting. Unlike her contemporary and rival Mlle *Clairon, she took no interest in the reform of theatrical costume and aimed at magnificence rather than correctness, being always robed in rich stuffs made in contemporary styles and loaded with jewels. She retired in 1776 but remained in full possession of her faculties until the end of her long life, and was able to pass on to younger players traditions temporarily lost during the French Revolution.

Dunlap, William (1766–1839), American dramatist and theatre manager, the dominating force of the American stage from 1790 to 1810. He went to England in 1784 to study art, but neglected his work for the theatre. On his return to the USA in 1787 he wrote a comedy for the *American Company which was accepted but never acted. A second comedy, *The Father; or, American Shandyism*, was produced at the *John Street Theatre in 1789. In 1796 he became one of the American Company's managers in partnership with the younger *Hallam and *Hodgkinson, strengthening the company by the inclusion of the first Joseph *Jefferson. In 1797 Hallam withdrew from active management, and Hodgkinson and Dunlap opened the first *Park Theatre in 1798 with *As You Like It*. One of the first plays staged there was *André*, a tragedy which Dunlap based on an incident in the War of Independence—the first native tragedy on American material. Hodgkinson played André, and the part of his friend Bland was taken by Thomas Abthorpe *Cooper, who was to succeed Dunlap as lessee and manager of the theatre.

In 1798 Hodgkinson left the Park and Dunlap continued on his own, producing adaptations of French and German plays, mainly those of *Kotzebue of which the most successful were *The Stranger, Lovers' Vows, The Wild Goose Chase, The Virgin of the Sun* and its sequel *Pizarro*, as well as some of his own plays, among them *Leicester* and *The Italian Father*. Many of them were performed also in Boston, where Dunlap had leased the Haymarket Theatre, and in Philadelphia under Warren and Wood at the

*Chestnut Street Theatre. Meanwhile Dunlap struggled on at the Park, hampered by temperamental actors and recurrent epidemics of yellow fever, until in 1805 he went bankrupt. A year later he became assistant stage manager at the Park under Cooper. In 1812 he accompanied George Frederick *Cooke on his American tour, and then retired from the theatre to devote himself to literature and painting.

Dunlop, Frank (1927–), English director, who in 1961 became Director of the *Nottingham Playhouse, where he remained until 1964, seeing the new theatre building through its first season. In 1966 he directed the Pop Theatre in *The Winter's Tale* and *Euripides' Trojan Women* for the *Edinburgh Festival, and in the following year he joined the *National Theatre company in London, where his productions included *Brecht's *Edward II*, *Maugham's *Home and Beauty* (both 1968), and *Webster's *The White Devil* (1969). In 1970 he founded the *Young Vic, initially as part of the National Theatre, and directed its first production *Scapino* (NY, 1974), which he helped to adapt from *Molière's *Les Fourberies de Scapin*. Among his other productions for the company were *The Comedy of Errors* (1971), *Genet's *The Maids* (1972), and the musical *Joseph and the Amazing Technicolor Dreamcoat* (also 1972; NY, 1976). When the Young Vic became independent in 1974 he continued as Director. He also worked extensively as a freelance, staging the *RSC's revival of William *Gillette's *Sherlock Holmes* (1974), which he also directed in New York, other productions there including Alan *Bennett's *Habeas Corpus* (1975). In 1980 he returned to the Young Vic to direct *King Lear*. He was Director of the *Edinburgh Festival, 1983–91.

Dunlop Street Theatre, see GLASGOW.

Dunnock, Mildred [Dorothy] (1900–91), American actress, formerly a schoolteacher, who after studying with Lee *Strasberg and Elia *Kazan made her New York début in 1932. She was seen in a variety of roles including Queen Margaret in *Richard III* (1943) and Lavinia Hubbard in Lillian *Hellman's *Another Part of the Forest* (1946) before scoring a big success as Linda Loman in Arthur *Miller's *Death of a Salesman* (1949). This was followed by major roles such as Gina in *Ibsen's *The Wild Duck* (1951) and Big Mama in Tennessee *Williams's *Cat on a Hot Tin Roof* (1955). She appeared at the Festival of Two Worlds in Spoleto, Italy, in another Tennessee Williams play, *The Milk Train Doesn't Stop Here Anymore* (1962), and as Hecuba in *Euripides' *Trojan Women* (1963), playing both roles later in New York. She continued to make regular stage appearances in her seventies, including leading roles in Marguerite *Duras's *A Place without Doors* (1970) and *Days in the Trees* (1976).

Dunsany, Edward John Moreton Drax Plunkett, Lord (1878–1957), Irish man of letters connected with the early years of the *Abbey Theatre, where his first plays were produced—*The Glittering Gate* (1909) and *King Argimenes* (1911). These were also seen in London, as were *The Gods of the Mountain* (1911) and *The Golden Doom* (1912). His plays were mainly fantasies with their own self-made mythology. Most of the later ones were seen only in Ireland or in amateur productions, though *If* had a long run in London in 1921. He also wrote one-act farces (*Cheezo*), satire (*The Lost Silk Hat*), tragedy (*Alexander*), and comedy (*Mr Faithful*).

Dunville [Wallen], **Thomas Edward** (1868–1924), eccentric English comedian who as a young man went to London with a small provincial troupe of acrobatic entertainers. He made his first solo appearance in 1889, and was seen at *Gatti's-Under-the-Arches and the *Middlesex Music-Hall with immediate success, soon becoming well known for his short comic songs consisting of telegraphic phrases delivered in an explosive manner in the style of 'Little Boy, Pair of Skates, Broken Ice, Heaven's Gates'. For many years he was top of the bill wherever he appeared, wearing a long black coat which accentuated his height, with a white Puritan collar, a small bowler hat, baggy trousers, and a Dutch-doll wig over a red-nosed face. When he tried to vary his appearance audiences forced him to return to his usual style. In the early 1920s, when music-hall began to lose its appeal, he became very depressed and, after overhearing himself referred to as a fallen star, drowned himself in the Thames.

Duodrama, see MONODRAMA.

Du Parc, Mlle, see RACINE.

Durante, Jimmy, see VAUDEVILLE, AMERICAN.

Duras, Marguerite (1914–), French dramatist, film-maker, and novelist, whose plays belong basically to the Theatre of the *Absurd but temper the vision of life as fundamentally ridiculous by means of a surface realism. Her first play *Le Square* (1956) deals with the impossibility of direct communication between human beings. In *Les Viaducs de la Seine-et-Oise* (1963) and *L'Amante anglaise* (1968) she dramatizes the same true murder story. As *The Viaduct*, the first of these was seen in England in 1968. The second version, as *A Place without Doors*, was seen in New York in 1970, and as *The Lovers of Viorne* in London in 1971, starring Mildred *Dunnock and Peggy *Ashcroft respectively. The original version was followed in 1965 by *Les Eaux et les fôrets*, a bitter comedy on the theme of ingratitude, and by a study of divorce, *La Musica*, which with *Le Square* was seen in London in 1966. In that year Duras's first full-length play, *Des journées entières dans* *les arbres* (also 1965), an account of a curious, abortive reconciliation between an estranged mother and son, originally produced by *Barrault at the *Odéon, was staged by the *RSC as *Days in the Trees* with Peggy Ashcroft as the Mother, played in New York in 1976 by Mildred Dunnock. In 1973 the RSC also put on *Suzanna Andler*, which portrays the anguish of a woman, married for 17 years, as she debates whether to continue an affair. The play's presentation was due largely to the advocacy of Eileen *Atkins. Duras's later plays include *L'Éden cinéma* (1977), set in a French Indo-China plantation in the 1930s; *Vera Baxter* (1982; London, 1989), reminiscent of *Suzanna Andler*; and *Savannah Bay* (1983; London, 1988), an elegant, lyrical encounter between a young woman and Madeleine, an ageing actress. The last role was written for Madeleine *Renaud, who appeared in several of Duras's plays.

Dürrenmatt, Friedrich (1921–90), Swiss dramatist, whose work, though influenced by *Brecht, *Wedekind, and *Expressionism, has an unmistakable flavour of its own, with its mixture of the grotesque and the macabre and its clever use of modern dramatic techniques. His first success came with a mock-heroic comedy *Romulus der Grosse* (1949), set on the Ides of March 476, when the Emperor calmly receives the news of the barbarians' final victory. *Die Ehe des Herrn Mississippi* (1952), seen briefly in London in 1959 as *The Marriage of Mr Mississippi*, and *Ein Engel kommt nach Babylon* (1953) made little impact outside the German-speaking countries, but he won international fame with *Der Besuch der alten Dame* (1956), which as *The Visit* was seen widely in America in 1958 and England in 1960 in a production by Peter *Brook with the *Lunts. The story concerns a fabulously wealthy woman who bribes the citizens of her impoverished native town to kill her former seducer. After *Frank V*, an unsuccessful musical play, Dürrenmatt achieved another world-wide success with *Die Physiker* (1962), staged in London (by the *RSC) in 1963 and a year later in New York as *The Physicists*, both productions again directed by Peter Brook. The scene is a lunatic asylum in which a nuclear physicist has taken refuge after burning his papers to prevent his researches being used for destructive purposes. Dürrenmatt's later plays have been less successful. *Porträt eines Planeten* (1971) presents a synopsis of man's life on earth from its beginnings to its destruction by a cosmic catastrophe. Dürrenmatt also adapted several plays, among them *Strindberg's *The Dance of Death* under the title *Play Strindberg* (1969; NY, 1971; London, 1973) and Shakespeare's *King John* and *Titus Andronicus*. All his plays are fundamentally pessimistic, reflecting the anxieties of their time.

Du Ryer, Pierre (*c.*1600–58), French dramatist, whose earliest plays were *tragi-comedies in the style of *Hardy, three being presented in the year 1628–9. They were spectacular, calling for elaborate staging in the old-fashioned *multiple setting, and ignored the *unities of time and place. Several comedies followed, containing good parts for the comedian *Gros-Guillaume, of which the best was *Les Vendanges de Suresne* (1633). The most important of several tragedies, which included some on biblical subjects, was *Scévole* (1644), produced at the Hôtel de *Bourgogne like most of Du Ryer's plays; it was in the repertory of *Molière's *Illustre-Théâtre, was produced by him several times in the provinces before he went to Paris, and remained in the repertory of the *Comédie-Française for over 100 years. Du Ryer did more than anyone except *Mairet and *Corneille to establish French classical tragedy.

Duse, Eleonora (1858–1924), great Italian actress, born into a theatrical family. She was on the stage at the age of 4, and at 14 appeared as Juliet. As an adult actress her first notable success came in *Zola's *Thérèse Raquin* (1879). Equally successful was her performance as Santuzza in *Verga's *Cavalleria rusticana* (1884), a challenging and rewarding role which reinforced her growing interest in modern drama. In 1885 she made her first foreign tour, in Latin America, and on her return to Italy founded her own company, the Città di Roma. During a tour of Russia in 1891 *Chekhov saw her as Shakespeare's Cleopatra; it has been suggested that he had her in mind while he was writing the part of Madame Arkadina in *The Seagull*. She made the first of four visits to the United States in 1893 and first visited England in 1894, where she played Mirandolina in *Goldoni's *La locandiera* in a command performance at Windsor before Queen Victoria. A year later she and Sarah *Bernhardt were both in London playing in *Sudermann's *Heimat*, which gave the London critics the opportunity of comparing their styles, *Shaw vigorously championing Duse while Clement *Scott much preferred Bernhardt. (Their rivalry was renewed in Paris in 1897.) From her late thirties Duse devoted much of her time and money to promoting the plays of *D'Annunzio, who first came into prominence as a dramatist when in 1898 she appeared in his *La Gioconda* and *La città morte*. She retired in 1909 but financial difficulties forced her back to the stage, and she reappeared in Turin in 1921 as Ellida in *Ibsen's *The Lady from the Sea*. Two years later she started on her last international tour, dying in Pittsburgh in 1924; her body was returned to Italy and buried at Asolo. A slender, graceful woman, with dark eyes and expressive, mobile features, melancholy in repose, she was noted for the beauty of her gestures. She had a statuesque way of playing, a slowness and subtlety; her moments of immobility revealed her greatness. Sustained application went into the preparation of her roles; above all, her technical expertise and electrifying presence enabled her to transcend the handicap of a somewhat frail physique. She was noted for her refusal to wear make-up on stage, having apparently the ability to blush or turn pale at will. She was much admired in such big emotional parts as *Sardou's Fédora, Tosca, and Théodora, and Camille in the younger *Dumas's *La Dame aux camélias*, and was outstanding in Ibsen—particularly as Hedda Gabler, and as Rebecca West in *Rosmersholm* in the production Gordon *Craig designed for her in Florence in 1906.

E

Eagle Theatre, New York, see STANDARD THEATRE.

Earl Carroll Theatre, New York, on the south-east corner of 7th Avenue and 50th Street. Built and managed by Earl *Carroll, who staged there from 1923 several editions of his *revue *Earl Carroll's Vanities*, it opened in 1922. A number of straight plays were also seen there, including Leon Gordon's *White Cargo* (1923). A second Earl Carroll Theatre opened on the same site in 1931. It seated 3,000 and boasted one of the most technically innovatory stages in the country. The economic depression, however, forced Carroll to close it, and it reopened in 1932 as the Casino with the musical *Show Boat*. In 1934 it became a theatre restaurant.

Ecclesiastical Drama, see LITURGICAL DRAMA.

Echegaray [Eizaguirre], **José** (1832–1916), Spanish dramatist, awarded the *Nobel Prize for Literature in 1905. His plays retain the verse-form and much of the imagery of the Romantics, but deal mainly with social problems, and though enthusiastically received caused fierce controversy. They had a great influence not only in Spain but on the European theatre generally. The best-known are *O locura o santidad* (*Madman or Saint*, 1877), *El loco Dios* (*The Divine Madman*, 1900), *El hijo de Don Juan* (*The Son of Don Juan*, 1892), a study of inherited disease which owes something to *Ibsen's *Ghosts*, and, most important, *El gran Galeote* (1881), produced in England as *Calumny* and in the United States as *The World and His Wife*, in which a woman wrongfully accused of being a poet's mistress finally becomes so.

Eddy, Edward, see BOWERY THEATRE I.

Edinburgh. In spite of the opposition of the Church, efforts appear to have been made in the 17th and 18th centuries to establish a theatre in Edinburgh. The scanty records indicate a succession of performances there and in nearby towns between 1663 and 1689, including the first known Scottish performance of *Macbeth* in 1672, while an advertisement of 1715 reveals the existence of an established company. The best documented of early managements is that of Tony *Aston from 1725 to 1728. In 1736 the poet Allan Ramsay opened a theatre in Carrubber's Close, only to have it closed by the Licensing Act of 1737. From 1741 there were theatrical seasons every year, the law being evaded by the device of charging not for the play but for an introductory concert. In 1747 a concert hall in the Canongate opened as a theatre which in 1756 saw the first performance of John *Home's *Douglas*. A patent was obtained for it in 1767, and it then became Edinburgh's first Theatre Royal. Two years later a new Theatre Royal was built at the east end of Princes Street. Many distinguished players appeared there, including, in 1784 and 1785, Mrs *Siddons, whose younger brother Stephen Kemble managed the theatre from 1791 to 1800. Her son Henry Siddons took over in 1809, but so mismanaged his financial affairs that when he died in 1815 his widow and her brother William Murray, joint managers with him, were left heavily in debt. They nevertheless managed to keep the theatre open, and up to 1851, when Murray retired, it had a period of almost unbroken success. It was pulled down in 1859, after a further period of success from 1857, when Henry *Irving was a member of the resident *stock company. A new Theatre Royal, named the Queen's in honour of Queen Victoria, opened in 1859. It stood on the site of an earlier theatre which had been known by many names before, as the Adelphi, it was destroyed by fire in 1853. It was burnt down again in 1865, 1875, and 1884. The *Royal Lyceum, opened in 1883, still survives, having taken over in 1965 the assets of the Gateway which opened in 1946. The internationally famous *Traverse Theatre dates from 1963, and the King's Theatre is used by touring productions. The *Edinburgh Festival of Music and Drama brings an annual influx of world-famous companies and performers to the city.

Edinburgh Civic Theatre Company, see ROYAL LYCEUM THEATRE, Edinburgh.

Edinburgh Festival of Music and Drama, international event presented annually since 1947 for three weeks, opening usually in mid-August. Although the main emphasis is on music, some distinguished theatrical productions have been seen, including those of a number of foreign companies, since it has always been the policy of the Festival's directors to invite visitors from overseas. The first, in 1947, was *Jouvet's company. In 1948 and 1957 the *Renaud–*Barrault company appeared and in 1955 Edwige *Feuillère was seen in the younger *Dumas's *La Dame aux camélias*. Other companies to visit Edinburgh include the *Comédie-Française, the *Théâtre National Populaire, the *Berliner Ensemble, the *Piccolo Teatro della Città di Milano, and the *Ninagawa

Company. Marcel *Marceau has also appeared several times, and T. S. *Eliot's last three plays were commissioned for performance there. In 1948, 1949, and 1959 Tyrone *Guthrie directed the old Scottish morality play by Sir David *Lyndsay, *Ane Pleasant Satyre of the Thrie Estaitis*, last seen in 1552. It was revived again for the 1973 Festival in a new version directed by Bill *Bryden. Other performances have included the *Old Vic company in *Romeo and Juliet*, *Jonson's *Bartholomew Fair*, and *Schiller's *Mary Stuart*, and the *Stratford (Ontario) Festival Theatre company in *Sophocles' *Oedipus the King* and *Henry V*. In 1950 the Glasgow *Citizens' Theatre was seen at the Festival, returning in 1968 with Michael *Blakemore's production of *Brecht's *Arturo Ui*. Scottish companies performing at the Festival also include the Edinburgh Civic Theatre Company from the *Royal Lyceum Theatre and the companies from Perth and Dundee. Other companies to visit the Festival include *Birmingham Repertory Theatre, the *Bristol Old Vic, the *Abbey Theatre, and the *Nottingham Playhouse. The *English Stage Company has made several visits, presenting the British premières of *O'Casey's *Cock-a-Doodle-Dandy* (1959) and *Ionesco's *Exit the King* (1963), and in 1962 the *RSC made its first visit with the British première of *Fry's *Curtmantle*; they returned in 1974 with *Marlowe's *Dr Faustus*. A memorable event of later years was the adaptation of *Richard III* staged in 1979 by the Rustaveli company from Georgia, and the following year the *National Theatre company made its first visit to the Festival.

The term *'Fringe' was first coined to describe performances not included as part of the official Festival. From these unofficial beginnings the Fringe has grown to encompass diverse events from every area of the performing arts. The essence of the Edinburgh Festival Fringe, and every Fringe festival around the world that has sprung from it, is that no artistic control is exerted by the organizers. The Fringe is now the largest arts festival in the world, over 500 groups from many different countries presenting 1,000 productions and giving 9,000 performances. Participation by university groups has always been strong, providing a unique platform for emerging talent. Trevor *Nunn, Maggie *Smith, Derek *Jacobi, Tom *Stoppard, and comedians such as Rowan Atkinson and John Cleese all performed on the Fringe in their university days. New work is another strong element, over 300 plays being given their British premières on the Fringe each year. Experimental work, *Collective Creation, alternative comedy, and small-scale productions are other important components. The *Traverse Theatre, which expands its year-round programme of events during the Fringe, acts as a cultural catalyst to the proceedings.

Edwardes, George (1852–1915), Irish-born theatre manager, who introduced *musical comedy to London. He first worked as a theatre manager in Ireland in 1875, and in 1881 became manager of the newly opened *Savoy Theatre in London. In 1885 he went into partnership with John *Hollingshead at the *Gaiety Theatre, a year later becoming sole manager and producing a fine series of musical comedies with the famous chorus of 'Gaiety Girls', whom he chose and trained himself. In 1893 he built *Daly's Theatre in London for Augustin *Daly of which he later took control, making it as successful as the Gaiety. He also managed the *Prince of Wales and the *Apollo. Always known as 'The Guv'nor', Edwardes was tall, good-looking, and burly, with a very soft voice. He had an extraordinary flair for knowing what the public wanted, and spared no cost in providing it.

Edwards, Hilton (1903–82), actor and director born in London, who began his career at the *Old Vic in 1922, where he appeared in all but two of Shakespeare's plays. In 1927 he went with Anew *McMaster to Ireland, where he remained and with Micheál *MacLiammóir founded in 1928 the Dublin *Gate Theatre. There and at the Gaiety Theatre he directed over 300 plays, in many of which he appeared himself, including new works by Irish authors, among them Brian *Friel, Denis *Johnston, and W. B. *Yeats. In 1937 he directed *Hamlet* at Elsinore, playing Claudius. He took the Gate Theatre company on many tours in Europe, North Africa, and the United States and in 1947 was seen again in London with his company, notably in MacLiammóir's *Ill Met by Moonlight*. In 1970, while the Gate was closed for repairs, he directed a production of *Chekhov's *The Seagull* at the *Abbey Theatre. His association with the Gate continued until his death.

Effects Man, see NOISES OFF.

Efremov, Oleg, see MOSCOW ART THEATRE and SOVREMENNIK THEATRE.

Eisenhower Theater, see JOHN F. KENNEDY CENTER FOR THE PERFORMING ARTS.

Ek, Anders (1916–79), Swedish actor, who worked in the city theatre of Gothenburg, 1946–50, the *Royal Dramatic Theatre, 1952–60, and subsequently with the city theatre of Stockholm. A powerful actor, he was seen in such parts as Othello, Macbeth, Jean in *Strindberg's *Miss Julie*, and the title-roles in Eugene *O'Neill's *The Emperor Jones* and *Camus' *Caligula*. In 1965 he returned to the RDT, and one of his last roles was Tessier in *Becque's *Les Corbeaux*, directed by Alf *Sjöberg. He worked often with Ingmar *Bergman in both theatre and films.

Ekhof, Konrad (1720–78), German actor. Being short and ungraceful, with no pretensions to good looks, he was not at first highly regarded, but he had a fine speaking voice and studied and perfected his art, discarding the earlier declamatory tone of Carolina *Neuber in favour of a more natural style hitherto unknown in Germany. He was at the height of his powers when in 1767 he joined *Ackermann, who had just brought together an excellent company at *Hamburg, destined to become Germany's first National Theatre. Unfortunately, after five years Ekhof could no longer endure the jealousy and arrogance of the young *Schröder and left, spending several miserable years travelling before going to *Weimar, where later he was to meet and act with *Goethe, to whom he imparted some of the theatrical reminiscences found in *Wilhelm Meister*. After a disastrous fire at Weimar in 1775, Ekhof became director and chief actor of the company at Gotha, where one of his last official acts was to engage *Iffland as a member of the company. He made his last appearance as the Ghost in *Hamlet* in an adaptation by Schröder, now famous and at the head of his profession. Among Ekhof's outstanding parts, in the mingled tragi-comic and pathetic roles of the new drama, were Old Barnwell in *Lillo's *The London Merchant*, the title-role in *Diderot's *Le Père de famille*, and Odoardo in *Lessing's *Emilia Galotti*. He lived to see his fellow actors in Germany, mainly through his own exertions, raised from the misery of strolling players to the dignity of an assured and well-regarded profession.

Ekkyklema, piece of ancient Greek stage machinery which has occasioned a great deal of controversy among modern scholars. Meaning literally something 'rolled out', it was obviously used to bring forward some object, character, or grouping important in the play's context. This was once thought to have been a movable platform which was either pushed on stage or revolved on a turntable to show an interior scene. It is now thought to have been nothing more than a couch on wheels, or a grouping arranged within a pair of double doors which opened to reveal it, as the doors open in *Aeschylus' *Libation Bearers* to reveal the bodies of Clytemnestra and Aegisthus.

Elen, Gus, see MUSIC-HALL.

Elevator, see BRIDGE.

Eliot, T(homas) S(tearns) (1888–1965), poet and dramatist, American by birth but English by adoption, who initiated a revival of *poetic drama in England with a play on the murder of Thomas à Becket, *Murder in the Cathedral* (1935). First acted in the Chapter House of Canterbury Cathedral, it was subsequently revived several times with great success in commercial theatres in Britain and the United States. A later play, *The Family Reunion* (1939), based on the *Oresteia* of *Aeschylus, was considered less successful, mainly because Eliot failed to integrate the ritualism of the *chorus with the realism of the setting, but it has had occasional revivals. Of his last three plays, commissioned for performance at the *Edinburgh Festival, *The Cocktail Party* (1949) and *The Confidential Clerk* (1953) were based on *Euripides' *Alcestis* and *Ion* respectively; *The Elder Statesman* (1958) on *Sophocles' *Oedipus at Colonus*. In them Eliot moved closer to a mannered realism, disguising his serious purpose under the form of modern drawing-room comedy, and discarding the closely wrought poetic style of the earlier plays for a plain undecorated verse. All his plays were directed by E. Martin *Browne.

Elizabethan Playhouse. Before James *Burbage built the *Theatre in London in 1576, English actors had played on temporary stages set up in the open air, in noblemen's houses, or in *innyards, whose characteristic features of a large open unroofed space surrounded by galleries giving access to bed-chambers can be found not only in the Theatre but in the later *Curtain, the *Fortune, the famous *Globe, the *Hope, the *Rose, and the *Swan. These were all wooden structures, roughly circular or octagonal, with a large platform-stage, backing on to one wall and jutting out into the central space. Round this space, still called a 'yard', ran two or three galleries with thatched roofs. The first probably continued behind the stage and so formed an upper room, reminiscent of the musicians' gallery in a nobleman's Great Hall. This could be used for instrumentalists, battlements, or first-storey windows. Underneath it may have been the *'inner stage'. Other features were the tiring-house, or actors' dressing-room, behind the stage wall. This permanent architectural façade was pierced by doors of entrance, over which projected a roof—the Heavens, supported by pillars. Above the balcony at the back of the stage was a hut to house the machinery used for raising or lowering actors or properties on to the stage, and above this a tower from which a trumpeter announced the start of a performance and a flag flew during it. Under the stage, hidden by boarding or drapery, a cellar held the machinery for projecting ghosts and devils through the trap doors, which also served for Ophelia's grave in *Hamlet* and the witches' cauldron in *Macbeth*. The stage itself was partly railed, and privileged spectators could pay to sit on it (see AUDIENCE ON THE STAGE). The galleries were furnished with wooden stools or benches, and round three sides of the stage stood the groundlings. In all parts of the house the audience, which entered through one main door, amused itself by cracking and eating nuts and munching apples, often throwing the cores at the actors. The so-called *private theatres (see BLACKFRIARS) were roofed. They approximated more to the *tennis-court theatres of Paris, and it was from them, rather than from the public theatres, that the Restoration playhouses developed after the *Puritan Interregnum of 1642–60.

Elizabethan Stage Society, see POEL.

Elizabethan Theatre Trust, Australia, body founded in 1955 to mark the visit of Queen Elizabeth II to Australia the previous year, and to establish the principle of government subsidies for the performing arts throughout the country. It was hoped that among its activities would be the training of young actors, the encouragement of promising young playwrights, and eventually the founding of an Australian National Theatre. Its initial preoccupation was however to provide touring companies covering the whole country, and with this in mind a Board of Trustees under Hugh *Hunt took over the former Majestic Theatre in Melbourne (built in 1917 but later used as a cinema) as a base for the Trust's work. Renovated and redecorated, it opened as the Elizabethan Theatre in 1955 with a cast of guest artists from England in *Rattigan's *The Sleeping Prince*. The Trust's own company, headed by Judith *Anderson, made its début in Canberra the following Sept. in *Euripides' *Medea*, and then embarked on a long tour. Among the new Australian plays presented by the Trust were Ray *Lawler's *The Summer of the Seventeenth Doll*, Alan Seymour's play about Anzac Day *The One Day of the Year*, and Douglas Stewart's verse drama *Ned Kelly*. The Trust also had its own ballet and opera companies which later became autonomous; the theatre company was then disbanded and the subsidy work of the Trust largely taken over by what is now the Australia Council.

Elliot, Michael, see ROYAL EXCHANGE THEATRE.

Elliott, G. H., see MUSIC-HALL.

Elliott, Maxine [Jessie Dermot] (1868–1940), American actress, who in 1895 joined the company of Augustin *Daly, appearing with him in New York and in London, where she was first seen as Sylvia in *Two Gentlemen of Verona*. After touring Australia with her husband Nat *Goodwin, from whom she was later divorced, she returned to America and appeared with him in a number of productions, including *Sheridan's *The Rivals* in 1897, *The Merchant of Venice* in 1900, and two new plays by Clyde *Fitch, *Nathan Hale* and *The Cowboy and the Lady* (both 1899). In 1908 she opened her own theatre in New York (see MAXINE ELLIOTT'S THEATRE), but had little success there, and in 1911 she retired to England, where she had

many friends. She made several reappearances, the last in 1920.

Her sister, (May) **Gertrude** (1874–1950), also an actress, married in 1900 Johnston *Forbes-Robertson, being his leading lady until his retirement in 1913, after which she toured under her own management, returning to New York in 1936 to play Gertrude to Leslie Howard's Hamlet.

Elliston, Robert William (1774–1831), English actor, who made his London début in 1796 at the *Haymarket Theatre. Charles *Dibdin wrote a number of entertainments for him, and he was frequently seen at *Drury Lane, being one of the most popular actors of the day, second only to *Garrick in tragedy. He was also much esteemed in comedy, playing Doricourt in Mrs *Cowley's *The Belle's Stratagem*, Charles Surface in *Sheridan's *The School for Scandal*, Rover in *O'Keeffe's *Wild Oats*, and Ranger in Hoadly's *The Suspicious Husband* with equal success. In later life one of his finest parts was Falstaff in *Henry IV, Part One* in which he had previously played Hotspur. He managed a number of provincial theatres, the *Surrey, and the *Olympic, and in 1819 achieved his ambition of managing Drury Lane where he opened with Edmund *Kean in *King Lear* and in 1821 put on *Byron's *Marino Faliero* in the face of much opposition. But his resources could not meet his expenditure and his outside speculations, and in 1826 he went bankrupt. He then returned to the Surrey Theatre, where he made a substantial profit from *Jerrold's *Black-Ey'd Susan* (1829) starring T. P. *Cooke. He made his last appearance about a fortnight before his death.

Eltinge, Julian [William Dalton] (1883–1941), American actor, who developed the modern style of *female impersonation. Abandoning the grotesque representations of comic servants and elderly women, he presented himself as a convincingly glamorous young lady, paying particular attention to his walk, hand movements, make-up, and clothes. He made his first appearance in a female part in the musical comedy *Mr Wix of Wickham* (1904), and from 1911 to 1914 toured in the dual roles of Mrs Monte and Hal Blake in *The Fascinating Widow*, a musical comedy specially written for him. He continued to appear in *vaudeville and in films until his retirement in 1930, returning briefly in 1940 to work in night-clubs.

The Eltinge Theatre, on West 42nd Street, between 7th and 8th Avenues, was named after him, opening in 1912 with a melodrama. In 1930 it became a home of American *burlesque, notorious for the daring of its strip-tease acts and its dubious jokes. It was closed in 1942, and a year later became a cinema.

Emery, John (1777–1822), English actor, son of the actor **Mackle Emery** (1740–1825). Even as a boy he had a remarkable talent for playing old men and at 15 he was already considered an outstanding actor. In 1798 he was engaged by *Covent Garden to take the place of the comedian John *Quick and apart from a short engagement at the *Haymarket remained there until his sudden death. His Caliban in *The Tempest* was highly praised, and he was good as Sir Toby Belch in *Twelfth Night*, the First Grave-digger in *Hamlet*, and Dogberry in *Much Ado about Nothing*. He was also an artist, and between 1801 and 1817 exhibited frequently at the Royal Academy.

His son **Sam**(uel) (1817–81) was also an actor, much admired in several dramatizations of *Dickens and as Sir Peter Teazle in *Sheridan's *The School for Scandal*.

Emery, (Isabel) **Winifred Maud** (1862–1924), English actress, granddaughter of John *Emery. She began her long career in 1870 at Liverpool and four years later was in *pantomime in London. In 1879 she made her adult début at the *Imperial Theatre. She was later with Wilson *Barrett, John *Hare, and the *Kendals, and then joined the *Lyceum Theatre company under Henry *Irving whom she accompanied on his American tours in 1884 and 1887, playing leading roles in Shakespeare and other productions. She returned to London in 1888 and married Cyril *Maude, becoming his leading lady when he went into management at the *Haymarket Theatre in 1896. A beautiful woman of great charm, she was one of the most versatile and popular actresses of her day; among her finest parts were Lady Windermere in *Wilde's *Lady Windermere's Fan* (1892) and Lady Babbie in *Barrie's *The Little Minister* (1897), both of which she created.

Emney, Fred (1900–80), leading English comedian in *musical comedy and *revue, son of the *music-hall star **Fred Emney** (1865–1917), whose best known sketch was 'A Sister to Assist 'Er' and who died of injuries sustained during the *pantomime *Cinderella* at *Drury Lane after slipping on the soapsuds during a knockabout comedy scene. The younger Fred made his first appearance in London in 1915, and from 1920 to 1931 played in *vaudeville in the United States. On his return to England he played a few 'halls', but then turned to light opera and musical plays. He gave several excellent performances in straight comedy, particularly as Ormonroyd, the photographer, in *Priestley's *When We Are Married* (1970). He was an excellent raconteur and in his solo performance of 'songs at the piano' his heavy-weight personality, imperturbable face and manners, his eyeglass, and his eternal cigar made up an unforgettable image of good-tempered lethargy.

Empire, The, London, in Leicester Square, famous *music-hall, originally a theatre which from 1884 to 1886 housed *burlesques and *extravaganzas. In 1887 it opened as the Empire Theatre of Varieties under the joint management of George *Edwardes and Augustus *Harris, whose spectacular ballets rivalled those of the *Alhambra. One of the Empire's most popular features was its promenade, which was attacked as a 'haunt of vice' by Mrs Ormiston Chant in her 1894 'Purity Campaign'. Screens were erected to separate the promenade and its prostitutes from the auditorium but the audience rioted and tore them down. Shortly before the First World War *revue made its appearance, and in 1918 the first of a series of successful *musical comedies, *The Lilac Domino*, began a long run, followed by *Irene* (1920), *The Rebel Maid* (1921), and *Lady, Be Good!* (1926). In 1927 the Empire closed and was demolished, being replaced by a cinema.

Empire Theatre, Liverpool, see LIVERPOOL.

Empire Theatre, New York, on the south-east corner of Broadway and 40th Street. Built for Charles *Frohman, this handsome edifice with its red and gold rococo interior opened in 1893 with *Belasco's *The Girl I Left behind Me*. Let in the summer months to visiting touring companies, it housed during the winter not only Frohman's *stock company but also companies led by John *Drew and Maude *Adams, who in 1905 was seen as *Barrie's Peter Pan. One of the most popular plays seen at the Empire was *Kester's *When Knighthood was in Flower* (1901) with Julia *Marlowe. Other famous actresses who visited this theatre were Ellen *Terry in *Shaw's *Captain Brassbound's Conversion* (1907); Ethel *Barrymore in *Pinero's *Mid-Channel* (1910) and Zoë *Akins's *Déclassée* (1919); and Jane *Cowl in *Coward's *Easy Virtue* (1925). In 1926 the Empire was temporarily closed by the authorities after the production of *The Captive*, a translation of Édouard Bourdet's controversial play *La Prisonnière*. It soon reopened and continued its successful career, Katharine *Cornell appearing in Besier's *The Barretts of Wimpole Street* (1931) and Shaw's *Candida* (1937). Two important productions were Elmer *Rice's *We, the People* (1933), with settings by Aline *Bernstein, and *Ibsen's *Ghosts* (1936) with *Nazimova. *Hamlet* (also 1936), with John *Gielgud, ran for 132 performances, breaking the records set up by Edwin *Booth and John *Barrymore. In 1939 came *Lindsay and Crouse's *Life with Father*, which occupied the theatre for six years. After its transfer in 1946 the Empire staged *Rattigan's *O Mistress Mine* with the *Lunts. The theatre was demolished in 1953.

Encina, Juan del (1469–c.1539), Spanish playwright, who with his later contemporaries *Torres Naharro and Gil *Vicente founded and secularized Spanish Renaissance drama. He was director of entertainments in the household of the Duke of Alba for several years from about 1492, and for his patron wrote eight eclogues, or pastoral dialogues, in the Italian style, though the content was purely Spanish. Three of these are on religious themes—Christmas, the Passion, and the Resurrection; the others are secular, with simple but well-constructed plots and amusing dialogue, often in peasant dialect. Into all of them Encina introduced *villancicos*, or rustic songs, and his work in general owed much to his musical talent. The plays were probably first acted by talented amateurs of high rank, or their servants, either at Court or in the private houses of the nobility; but with the rise of professional actors they eventually found their way into the repertories of the strolling players.

English Aristophanes, see FOOTE.

English Comedians, troupes of English actors who toured the Continent during the late 16th and early 17th centuries. Although a group of English musicians is known to have visited the Danish Court in 1579–80, the first actors to appear abroad seem to have been those who accompanied the Earl of Leicester to Holland and Denmark in 1584–6. Among them was the famous clown Will *Kempe who with his fellow actors was invited to *Dresden. Thereafter references to visits from English actors can be found in the archives of many German towns. The first actor to become widely known was Robert *Browne, who toured from 1592 to 1619. One of Browne's original actors, **Thomas Sackville,** broke away to form a company of his own which was for a time at *Wolfenbüttel in the household of Heinrich Julius, Duke of Brunswick. Sackville was probably the first to create an indigenous comic character, Jan Bouschet (Jean Posset), designed specifically for an audience whose knowledge of English was very slight. He combined the often bawdy antics of the *Clown with the wit of the *Fool and relied mainly for his effects on pidgin German laced with English and Dutch phrases. He was the forerunner of such German-speaking clowns as Hans Stockfisch, created by John Spencer who with headquarters in Berlin travelled as far afield as The Hague and Dresden, and Pickelhering, the prototype of such purely indigenous characters as *Hanswurst and *Thaddädl.

It seems likely that the English Comedians first appeared in *jigs and short comic sketches whose humour was broad enough to appeal even to a foreign audience; but a collection of texts from their repertory printed in 1620 shows that by that time they were relying more on pirated editions of full-length plays, the dialogue pruned to the minimum and the plots reduced to a series of dramatic incidents. The titles in this collection are all of English plays, but a further collection printed in 1630 shows a preponderance of German titles, and it is probable that by then the plays were given mainly in German. Even in serious plays a good deal of fooling by the clowns was added for the benefit of the groundlings, with additional music and dancing for the more sophisticated. From the literary point of view the influence of the English Comedians on German drama was negligible, but they helped to give German audiences the habit of theatre-going, and the brevity of their dialogue may have helped to counteract the native tendency to excessive discussion. Once the companies began to engage German actors the use of English was discontinued and indigenous drama, in the shape of the *Haupt- und Staatsaktion, took over. In spite of the troubles of the Thirty Years War, some English actors continued to tour the Continent, particularly during the *Puritan Interregnum, the last authentic record of them dating from 1659 and the last notable player in the tradition being George *Jolly; but such was the prestige of the 'Englische Komödianten' that the name was used in Germany for publicity purposes as late as the middle of the 18th century.

English Opera House, London, see LYCEUM THEATRE 1, PALACE THEATRE, and ROYAL THEATRE 2.

English Shakespeare Company, see BOGDANOV.

English Stage Company, organization formed in 1956, mainly to present plays by young and experimental dramatists but also to put before the public the best contemporary plays from abroad. After an unsuccessful attempt to reopen the bomb-damaged *Kingsway Theatre, the company, under the artistic direction of George *Devine, took over the *Royal Court Theatre in Sloane Square, where it still functions. For a time it played an indispensable role, promoting the work of such playwrights as *Osborne, *Wesker, and *Arden and making the Royal Court the most exciting theatre in London. Its function is now shared by the *National Theatre, the *RSC, *Fringe theatres, and some of the regional theatres, but it still plays a useful part in the London theatrical scene. For individual productions see ROYAL COURT THEATRE.

ENSA, see ENTERTAINMENTS NATIONAL SERVICE ASSOCIATION.

Enters, Angna (1907–89), American solo entertainer whose programme consisted of a series of wordless mime and dance sketches. Her range was wide, and with a gesture she could evoke an Impressionist picture of a wood on a hot summer's day or the sad ennui of provincial life in France in 1910, an ageing prostitute or a

Byzantine ikon. She made her first appearance in New York in 1924 and in London in 1927, and made extensive tours of America and other parts of the world. An accomplished painter, she also wrote a number of books, in one of which, *Artist's Life* (1958), she described her professional life in the theatre.

Entertainments National Service Association (ENSA), organization formed in 1938–9 to provide entertainment for the British and allied armed forces and war workers during the Second World War. It was directed by Basil *Dean and had its headquarters at *Drury Lane Theatre. Working closely in collaboration with the Navy, Army, and Air Force Institute (NAAFI), which was responsible for the financial side, it provided all types of entertainment, from full-length plays and symphony orchestras to concert parties and solo instrumentalists, not only in the camps, factories, and hostels of Great Britain but on all war fronts, from the Mediterranean to India and from Africa to The Faeroes.

Enthoven, Gabrielle, see SOCIETY FOR THEATRE RESEARCH and THEATRE MUSEUM.

Entremés, Spanish term, deriving originally from the French *entremets*, applied to a diversion, dramatic or otherwise, which took place between the courses of a banquet. The name was first applied in Spain to the short dramatic interludes which enlivened the Corpus Christi processions in Catalonia, and was later applied in Castilian to a short comic interlude, often ending in a dance, which was performed, sometimes incongruously, between the acts of a play given in the public theatres. The *entremés* reached the height of its popularity in the 16th and 17th centuries, and most of the well-known dramatists of the Golden Age wrote them, including Lope de *Vega, *Cervantes, and *Calderón. In modern times the genre was revived by the brothers *Álvarez Quintero.

Epic Theatre, phrase taken from *Aristotle, where it implies a series of incidents presented without regard to theatrical conventions, and used in the 1920s by such pioneers as *Brecht and *Piscator of episodic productions designed to appeal more to the audience's reason than to its emotions, thus excluding sympathy and identification. (See also ALIENATION.) It employs a multi-level narrative technique and places the main emphasis on the social and political background of the play. Typical of epic theatre were the several productions by Piscator of a dramatized version of *Tolstoy's novel *War and Peace*.

Epidauros, one of the finest surviving examples of a Greek theatre building, dating from the 4th century BC. In its original form it had a full circular *orchestra, an elaborate stone stage building or skene, and a raised stage with ramps leading to ground level. The orchestra and the seating, though not the buildings, have been restored, and the theatre is in regular use for annual summer festivals of ancient drama performed by the Greek National Theatre. The auditorium holds about 14,000 spectators and is notable for its superb acoustics.

Epilogue, see PROLOGUE.

Epiphany Play, see LITURGICAL DRAMA.

Equestrian Drama, form of entertainment, popular in London in the first half of the 19th century, which evolved out of the feats of horsemanship shown at *Astley's Amphitheatre. The first was *The Blood-Red Knight; or, The Fatal Bridge* (1810) by J. H. Amherst, who was also responsible for such later spectacular shows as *The Battle of Waterloo* (1824) and *Buonaparte's Invasion of Russia; or, The Conflagration of Moscow* (1825), in both of which horses and their riders figured largely. One of the most successful plays adapted for Astley's was *Richard III*, with the leading role allotted to Richard's horse White Surrey; but the most famous of all equestrian dramas was *Mazeppa*, based by H. M. Milner on *Byron's poem and first seen in London in 1823. The part of Mazeppa was afterwards associated entirely with Adah Isaacs *Menken, who first appeared in it in the United States in 1863. Although the fashion for equestrian drama soon died out, real horses were used on stage at *Drury Lane as late as 1909.

Equity (American, British, and Canadian Actors' Equity Associations), trades unions formed to oversee and regulate the pay and conditions of those working in the theatre. American Actors' Equity, which deals only with performers in the legitimate theatre, was founded in 1913. It called its first strike in 1919, and was successful in gaining official recognition for itself and better contracts for its members. A second strike in 1960 resulted in further improvements in members' working conditions; in 1961 a policy of racial non-discrimination was set on foot. The union later suffered some reverses, but continues to press for the best possible conditions for its members.

British Actors' Equity was founded in 1929, the earlier Actors' Association, founded in 1891 with *Irving as its President, and the Actors' Union, founded in 1905, both being short-lived. It now looks after the interests of all those working in theatre, in film, on radio, and in television. *Variety artists originally had their own Federation, but this amalgamated with Equity in 1968.

Canadian Actors' Equity was originally a branch of the American body, with an office in Toronto which opened in 1935. It became independent in 1976. A West Coast office opened in Vancouver in 1979.

Erckmann-Chatrian, pseudonym of two French authors who wrote novels and plays in collaboration, **Emile Erckmann** (1822–99) and **Louis-Gratien-Charles-Alexandre Chatrian** (1826–90). They are best remembered for their melodrama *Le Juif polonais* (*The Polish Jew*, 1869) which, in an English adaptation as *The Bells*, provided *Irving with the fine part of Mathias. He played the part many times, and after his death it entered the repertory of his elder son, H. B. *Irving, and was also revived by *Martin-Harvey.

Erlanger's, New York, see ST JAMES THEATRE.

Ervine, (John) **St John Greer** (1883–1971), Irish dramatist and critic, who eventually settled in England, where he became dramatic critic for a number of papers, including the *Morning Post* and the *Observer*. Most of his early plays, with their realistic Northern Irish characters, were first seen at the *Abbey Theatre, among them *Mixed Marriage* (1911) and *John Ferguson* (1915). *Jane Clegg* (1913), in which Sybil *Thorndike gave an outstanding performance as a woman witnessing her weak husband's deterioration, was given its first production by Miss *Horniman's repertory company in Manchester. Later two successful comedies were seen in London—*The First Mrs Fraser* (1929), with Marie *Tempest, and *Anthony and Anna* (1935), which ran for over two years, having first been seen in Liverpool in 1926. In more serious vein were *Robert's Wife* (1937), with Edith *Evans in the title-role, which discusses Christianity and birth control, and *Private Enterprise* (1947), the last of Ervine's plays to be seen in London.

Espert I Romero, Nuria (1936–   ), Spanish actress, the foremost of her generation. A professional from the age of 12, she played Juliet at 16 and the title-role in *Euripides' *Medea* at 19. In 1959 she formed with her husband the Nuria Espert Company, which became the most prestigious in Spain, its productions, in many of which she appeared, ranging from *O'Neill, *Brecht, and *Sartre to the classical Spanish dramatists, Euripides, and Shakespeare. (She played Hamlet and both Ariel and Prospero in *The Tempest.*) Her performances as Claire in *Genet's *Les Bonnes* (1969) and in the title-role of *García Lorca's *Yerma* (1971) were seen in London in the *World Theatre Seasons of 1971 and 1972 respectively, her Yerma being especially memorable. The company has performed throughout the world, and accompanied her when in 1980–1 she was Director of the Spanish National Theatre. In 1986 in London she made her directorial début with García Lorca's *The House of Bernarda Alba*, and she has since directed several opera productions in Britain.

Esslair, Ferdinand (1772–1840), German actor, who made his début at Innsbruck and toured for many years until in 1820 he was appointed leading actor and manager of the Court theatre in *Munich. His passionate, inspirational acting, allied to exceptional graces of person and voice, made him one of the most popular actors in Germany. He was seen at his best in *Schiller's plays, particularly as Wilhelm Tell and Wallenstein.

Estébanez, Joaquín, see TAMAYO Y BAUS.

Estelle R. Newman Theatre, New York, see PUBLIC THEATRE.

Ethel Barrymore Theatre, New York, on West 47th Street, between Broadway and 8th Avenue. Built by Lee *Shubert, with a seating capacity of 1,099, it opened in 1928 with Ethel *Barrymore, in whose honour it was named, in *Martínez Sierra's *The Kingdom of God*. Its subsequent history has been one of almost unvaried success. Among its outstanding productions were *Coward's *Design for Living* (1933) and *Point Valaine* (1935); *Night Must Fall* (1936) by Emlyn *Williams, who also appeared in it; and Clare Boothe's scathing comedy *The Women* (also 1936). In 1947 Tennessee *Williams's *A Streetcar Named Desire*, with Marlon Brando and Jessica *Tandy, won a *Pulitzer Prize, as did Ketti Frings's adaptation of Wolfe's novel *Look Homeward, Angel* ten years later. Other productions included Frederick Knott's *Wait until Dark* (1966); Peter *Shaffer's double bill *Black Comedy* and *White Lies* (1967); and Christopher *Hampton's *The Philanthropist* (1971) with Alec *McCowen. A musical about Black life, *Ain't Supposed to Die a Natural Death* (1971), scored a great success. Later came the double bill *Noël Coward in Two Keys* (1974) with Jessica Tandy and Hume *Cronyn, David *Mamet's *American Buffalo* (1977), *Rabe's *Hurlyburly* (1984), and *Lillian* (1986), based on Lillian *Hellman's writings.

Etherege, Sir George (1634–91), English dramatist, the first to attempt the social comedy of manners imported from Paris and later perfected by *Congreve and *Sheridan. His first play, *The Comical Revenge; or, Love in a Tub* (1664), was a serious verse drama with a comic prose subplot. Etherege explored this latter style further in his two later comedies, *She Would if She Could* (1668) and *The Man of Mode* (1676). The latter is a picture of a society living exclusively for amusement, with a tenuous plot of entangled love-affairs offering an opportunity for brilliant dialogue and character-drawing. It contains the 'prince of fops', Sir Fopling Flutter, and the heartless, witty Dorimant, often considered a portrait of Lord Rochester, just as the poet Bellair is supposed to be Etherege himself.

Ethiopian Opera, turn in the *minstrel show, consisting of burlesques on Shakespeare and opera, with Negro melodies inserted. It was modelled on the early burlesques of T. D. *Rice (Jim Crow), of which the best were *Bone Squash Diavolo* (on 'Fra Diavolo'), *Jumbo Jim*, *Jim Crow in London*, and a burlesque of *Othello*.

Euclid–77th Street Theatre, see CLEVELAND PLAY HOUSE.

Eugene O'Neill Theatre, New York, on West 49th Street, between Broadway and 8th Avenue. Seating 1,075, it opened as the Forrest in 1925. Its history was uneventful until in 1934 *Tobacco Road*, based on a novel by Erskine Caldwell, was transferred there from the *John Golden Theatre where it had opened in 1933; it ran until 1941. In 1945 the theatre closed, and after extensive alterations reopened as the Coronet, its first success being a revival of Elmer *Rice's *Dream Girl*, which was followed by Arthur *Miller's *All My Sons* (1947). After a revival of Eugene *O'Neill's *The Great God Brown* (1959) the theatre took its present name, reopening with *Inge's *A Loss of Roses*. In 1969 it housed *The Last of the Red Hot Lovers* by Neil *Simon, who became owner of the theatre until 1982, several more of his plays being later seen there: *The Prisoner of Second Avenue* (1971), *California Suite* (1976), *Chapter Two* (1979), and *I Ought to be in Pictures* (1980). In 1985 the musical *Big River*, based on Mark Twain's *Huckleberry Finn*, was staged.

Eugene O'Neill Theatre Center, see O'NEILL, EUGENE.

Euripides (484–406 BC), ancient Greek tragic dramatist, born probably on the Athenian island of Salamis. Little is known of his life, but tradition represents him as a recluse living in a cave on Salamis; this may have arisen from his reluctance to take an active part in civic life. Unlike his contemporaries *Aeschylus and *Sophocles he seems to have given all his time to writing plays, and he stands out as an individualist in an age which still venerated the ideal of duty towards one's fellow citizens. He was also labelled a misogynist, in spite of the obvious sympathy with women's rights and problems displayed, for example, in his *Medea*.

Euripides is said to have written 92 plays, of which there survive 16 tragedies, one *satyr-drama (the *Cyclops*, a burlesque version of Odysseus's adventures in Sicily), and a large number of fragments, testimony to his later popularity. The extant tragedies, with dates where known or conjectured, are: *Medea* (431), *Hippolytus* (428), the *Children of Heracles* (?also 428), *Hecuba* (?426), the *Trojan Women* (415), *Iphigenia in Tauris* (?414), *Helen* (412), the *Phoenician Women* (411), *Orestes* (408), and, of unknown date, the *Madness of Heracles*, the *Suppliant Women* (with a plot different from that of Aeschylus' tragedy of the same name), *Ion*, and *Electra*. The *Bacchae* and *Iphigenia in Aulis* were both produced posthumously. One other play, *Alcestis* (438), though usually classed among the tragedies, contains pronounced satyric elements. The plays fall into two clearly marked categories—tragedies in the modern sense, and plays which may be variously called tragi-comedies, black comedies, romantic dramas, melodramas, or even high comedies. The tragedies have often been much misunderstood and criticized because of their apparently episodic plot structure. Sometimes (as in *Hecuba* and the *Madness of Heracles*) the play is based on two stories which seem to have little in common except the leading characters. The first half of the play illuminates a set of incidents from one point of view, the second from a different point of view; and the audience is asked to judge between the two.

In his lifetime Euripides aroused great interest and great opposition with his realism, his interest in abnormal psychology, his portraits of women in love, his new and emotional music, his unorthodoxy, and his argumentativeness. He was critical, sceptical, interested less in the community than in the individual, dealing less with broad questions of morality and religion than with personal emotions and passions—love, hate, revenge—and with specific social questions such as the suffering of the individual in war. He tried to bring tragedy down to earth by using more colloquial language, popular forms of music, and characters who, though still drawn from myth and legend, had recognizable counterparts in 5th-century Athens.

Euripides was described by *Aristotle as 'the most tragic of the poets', and it is in such tragedies as *Medea*, *Hippolytus*, the *Bacchae*, the *Trojan Women*, and *Hecuba* that his best work is probably to be found. *Electra* and *Orestes*, in which some critics have found a shrewish heroine and a pusillanimous hero, are powerful studies in morbidity and insanity; *Alcestis*, *Ion*, and *Iphigenia in Tauris* are excellent tragicomedy or romantic drama; *Helen* is delightful high comedy; and the *Phoenician Women* is perhaps best described as a pageant-play. All except the last show Euripides as a master of non-tragic dramatic writing.

Euripides continued to use the three actors and *chorus as finally established by Sophocles, with one important innovation. The opening song of the chorus as used by Aeschylus, or the scene in dialogue with which all the extant plays of Sophocles begin, were replaced by a formal *prologos* spoken sometimes by a character in the play, sometimes by an external god, which summarizes the story up to the point at which the action begins. Together with the progressive detachment of the chorus from the main action, inevitable once the play was concerned with private rather than public issues, this eventually, via *Seneca, gave rise to the Elizabethan idea of an extraneous person called the Chorus speaking a *prologue.

Euston Music-Hall, London, see REGENT THEATRE.

Evans, Dame Edith Mary (1888–1976), distinguished English actress, who made her London début as Cressida in *Poel's production of *Troilus and Cressida* in 1912. After further experience in London and on tour she was at the *Birmingham Repertory Theatre, where in 1923 she created the parts of the Serpent and the She-Ancient in *Shaw's *Back to Methuselah*. A fine performance as Millamant in *Congreve's *The Way of the World* at the *Lyric, Hammersmith, in 1924 consolidated her reputation as one of the outstanding actresses of her generation. In 1925–6 she was at the *Old Vic, playing Rosalind in *As You Like It*, Katharina in *The Taming of the Shrew*, and the Nurse in *Romeo and Juliet*, all parts she was to repeat successfully in later years. She was seen in her only *Ibsen role, Rebecca West in *Rosmersholm*, in 1926, and a year later she returned to *Restoration comedy as Mrs Sullen in *Farquhar's *The Beaux' Stratagem*. In 1929 (NY, 1931) she scored a great success in Reginald Berkeley's *The Lady with the Lamp*, playing Florence Nightingale, also creating at the *Malvern Festival the role of Orinthia in Shaw's *The Apple Cart* and playing Lady Utterword in a revival of his *Heartbreak House*. She returned to the Old Vic in 1932, playing Emilia in *Othello* and Viola in *Twelfth Night*, and was then in a series of new plays, including *Evensong* (1932; NY, 1933) by Edward *Knoblock and Beverley Nichols; Emlyn *Williams's *The Late Christopher Bean* (1933); Norman Ginsbury's *Viceroy Sarah* (1934); and Rodney Ackland's *The Old Ladies* (1935), based on a novel by Hugh Walpole. In 1936 she was seen as Arkadina in *Komisarjevsky's production of *Chekhov's *The Seagull*, and with the Old Vic company played Lady Fidget in *Wycherley's *The Country Wife*, which was followed in 1937 by Katharina in *The Taming of the Shrew* to the Petruchio of Leslie *Banks and a long run in St John *Ervine's *Robert's Wife*. In 1939 she first played her most famous role, Lady Bracknell in *Wilde's *The Importance of Being Earnest*, in which her beautifully modulated voice was used to excellent comic effect. During the Second World War she was in *Van Druten's *Old Acquaintance* (1941) and again in Shaw's *Heartbreak House* in 1943, this time as Hesione Hushabye. She proved an excellent Mrs Malaprop in *Sheridan's *The Rivals* (1945) and a superb Katerina Ivanovna in Dostoevsky's *Crime and Punishment* (1946), though Shakespeare's Cleopatra (also 1946) was outside her natural range. Her 'old peeled wall' of a Lady Wishfort in *The Way of the World* (1948), again with the

Old Vic, was as outstanding as her earlier Millamant, and she was much admired, in the same year, as Ranevskaya in Chekhov's *The Cherry Orchard*. In her sixties new plays provided her with some of her finest roles—Lady Pitts in *Bridie's *Daphne Laureola* (1949; NY, 1950); Helen Lancaster in N. C. *Hunter's *Waters of the Moon* (1951); Countess Rosmarin in *Fry's *The Dark is Light Enough* (1954); and Mrs St Maugham in Enid *Bagnold's *The Chalk Garden* (1956). In 1958 she was at the Old Vic to play Queen Katharine in *Henry VIII*, and in 1959 at the *Shakespeare Memorial Theatre as the Countess of Rousillon in *All's Well that Ends Well*. She returned to modern drama with *Bolt's *Gentle Jack* (1963), Enid Bagnold's *The Chinese Prime Minister* (1965), and *Anouilh's *Dear Antoine* (*Chichester, 1971), and made her last appearance on the stage at the *Haymarket Theatre in 1974 in a programme entitled *Edith Evans . . . and Friends*.

Evans, Maurice Herbert (1901–89), English actor, who made his first success as Raleigh in *Sherriff's *Journey's End* (1928). He joined the *Old Vic company in 1934 and was seen in a number of Shakespeare roles, including Hamlet in its entirety. Faced with the growing popularity of his contemporaries *Gielgud, *Richardson, and *Olivier in London, he decided to go to America, and in 1935 made his first appearance on Broadway as Romeo to the Juliet of Katharine *Cornell. He followed it during the next six years with a series of Shakespearian parts—Richard II, Hamlet, Falstaff, Malvolio, and Macbeth—directed by Margaret *Webster, whom he was instrumental in introducing to the American stage. Becoming an American citizen in 1941, he entertained the troops during the war with his so-called *GI Hamlet*, a shortened version which was subsequently published. He also played a number of leading roles in *Shaw: John Tanner in *Man and Superman*, Dick Dudgeon in *The Devil's Disciple*, King Magnus in *The Apple Cart*, and Captain Shotover in *Heartbreak House*. Among his other parts were Crocker-Harris in *Rattigan's *The Browning Version* (1949), Hjalmar Ekdal in *Ibsen's *The Wild Duck* (1951), and Tony Wendice in Frederick Knott's *Dial 'M' for Murder* (1952). He retired from the stage in 1967, but continued to appear in films and on television.

Evans, Will (1875–1931), star of British *music-hall and *pantomime, son of an acrobat and clown with whom he made his first appearance in 1881 in *Robinson Crusoe* at *Drury Lane. He then toured with his father's troupe, returning to London in 1890 to appear on the 'halls' with his wife and, after her death, as a solo eccentric comedian. He specialized in 'scenes of domestic chaos', of which the most popular were 'Building a Chicken House', 'Whitewashing the Ceiling', and 'Papering a House'.

For many years he imported the same style of knockabout humour into kitchen scenes in Drury Lane pantomimes. He was part-author of *Tons of Money* (1922), which became the first of a long series of *Aldwych farces. Evans also collaborated with R. Guy Reeve in the farce *The Other Mr Gibbs* (1924) and was the author of many of his own music-hall songs and sketches.

Evans's, in King Street, Covent Garden, probably the best known of the convivial 'song-and-supper rooms' which preceded the *music-hall, and were themselves the successors of the old singing-clubs where entertainment of a mixed kind but mostly comic songs and monologues was provided free of charge for those who came to eat and drink, in the style of the modern *cabaret. Evans's was first situated in the basement of a large tavern, whose lessee before W. C. Evans, a comedian from *Covent Garden, took over in about 1820 was a Mr Joy; hence its first name 'Evans's, late Joy's'. Under the influence of the new music-halls, the original simple 'sing-song' gave way to highly organized entertainment, though retaining the old separate tables and the presiding *Chairman. In the 1840s, Evans's heyday, the programme included such entertainers as Charles *Sloman, and varied from somewhat risqué songs to the boys of the Savoy Chapel in glees and extracts from opera. By the 1860s Evans's was feeling the competition of the music-halls, and in 1880 it lost its licence and closed. Its premises were taken over in 1934 by the *Players' Club.

Everyman Theatre, Hampstead, London, small theatre which had a brief blaze of glory from 1920 to 1926 under **Norman Mac-Dermott** (1890–1977). Originally a drill hall, it opened in 1920 with the first British production of *Benavente's *The Bonds of Interest*. The success of *Shaw's *You Never Can Tell* two weeks later led to many more Shaw productions, including the first public performance in London of *The Shewing-Up of Blanco Posnet* (1921). Other foreign plays first seen at this theatre included *Bjørnson's *Beyond Human Power* (1923), *Chiarelli's *The Mask and the Face* (1924), and *Pirandello's *Henry IV* (1925). Chiarelli's play transferred to the West End, as did Noël *Coward's *The Vortex* (1924). Eugene *O'Neill's work was seen for the first time in London with *In the Zone* (1921) and other one-acters. Among other productions were *Ibsen's *A Doll's House*, *John Gabriel Borkman* (both 1921), *Hedda Gabler* (1922), *The Wild Duck*, and *Ghosts* (both 1925). The venture ended in 1926 with a dramatization of Chesterton's *The Man Who was Thursday*. The scenery and costumes for most of the plays were designed by MacDermott himself. After his departure the Everyman was taken over by Raymond *Massey, followed by Malcolm Morley, under

whom the first English production of *Ostrovsky's *The Storm* was given in 1929. The building has been a repertory cinema since 1933.

Everyman Theatre, Liverpool, originally a chapel, opened in 1964. It was intended by its founders, who included Terry *Hands, for the production of serious modern plays which it was felt were being ignored by the *Liverpool Playhouse, and among the dramatists presented during the first six years were *Beckett, *Bond, Peter *Nichols, and Harold *Pinter. The theatre became known for its treatment of local themes and its contemporary and innovative interpretations of the classics. Plans were made for rehousing it, but in the mid-1970s it was rebuilt on its existing site. Owned by an independent Trust, it now seats 430 and has an open-end stage with a small thrust capacity. **Ken Campbell** (1941–) was Artistic Director from 1980 to 1983. He was well known as director of the Science Fiction Theatre of Liverpool, which, in makeshift premises, presented works which combined imagination, fantasy, and outlandish comedy. Its first production, *Illuminatus* (1976), a group of five plays of which Campbell was joint author and director, was seen in London in 1977 as the opening production at the *Cottesloe Theatre. Though the authorities eventually forbade the use of its premises for theatrical purposes, the Science Fiction Theatre of Liverpool retained its title, presenting Neil Oram's *The Warp* (1979) in London. It later became Campbell's first offering at the Everyman, being shown in 10 weekly instalments as a promenade production (see THEATRE-IN-THE-ROUND). The Everyman attracts a predominantly young audience and has its own thriving Youth Theatre, while the Hope Street Project provides theatre training for all ages. The 1989–90 season celebrated the theatre's 25th anniversary with a programme of Greek, Jacobean, and contemporary Russian plays.

The theatre has supported several local authors, including **Willy Russell** (1947–), whose musical about the Beatles, *John, Paul, George, Ringo . . . and Bert* (London and NY, 1974), originated there. Russell's other successes include another musical, *Blood Brothers* (1983); the two-character *Educating Rita* (1980; NY, 1987), about a drunken Open University lecturer and his working-class pupil; and the one-woman *Shirley Valentine* (1988; NY, 1989), a study of a discontented Liverpool housewife, which also originated at the Everyman.

Evreinov, Nikolai Nikolaivich (1879–1953), Russian dramatist, an exponent of *Symbolism, whose anti-naturalistic theatre is exemplified in *The Theatre of the Soul* (1912), a *monodrama in which various aspects of the same person appear as separate entities. His *Revisor* (also 1912) illustrated amusingly what such directors as *Stanislavsky and *Reinhardt might do to

*Gogol's masterpiece *The Government Inspector*, and in *The Fourth Wall* (1915) Evreinov portrayed an ultra-realistic director let loose on *Goethe's *Faust*. *The Chief Thing* (1921), which deals with the power of illusion, has some affinity with the work of *Pirandello. Evreinov's theories were swept aside by the *Socialist Realism of the Soviet era. Two of his short plays were translated and published in *Five Russian Plays* (1916) as *A Merry Death* and *The Beautiful Despot*.

Exeter, Devon. The first theatre was erected in about 1735 by a company from *Bath. Closed by the Theatres Act of 1737, it became a Methodist chapel but reverted later to its original use as part of the Bath *circuit. It was managed for a time in the 1760s by Thomas Jefferson, great-grandfather of Joseph *Jefferson. In 1787 a new theatre was built, based on the plans of *Sadler's Wells, the London stars who appeared there including Stephen *Kemble, Mrs *Siddons, Master *Betty, and Edmund *Kean. It was destroyed by fire in 1820, the only part to survive being the colonnade which was incorporated into a new building. This opened in 1821 with *The Merchant of Venice* and was visited by Mme *Vestris, Charles and Fanny *Kemble, *Macready, and *Phelps. *Hearts of Oak* (1880) by Henry Arthur *Jones was the first play by a local man. The theatre, which although it had no royal patent was known locally as the Theatre Royal from about 1828, burned down in 1885. A new Theatre Royal, opened in 1886, was also destroyed in 1887 in one of the worst conflagrations in the history of the English theatre, 186 lives being lost. The new theatre, opened in 1889, survived the financial vicissitudes of the 1920s and 1930s and escaped damage by bombing during the Second World War, but although after the war it became well known for its long-running Christmas pantomimes, it closed and was demolished in 1962. Exeter was then without a theatre until the opening in 1967 of the *Northcott Theatre.

Existentialism, see SARTRE.

Expressionism, movement that began in Germany in about 1910, and is best typified by the plays of Georg *Kaiser and Ernst *Toller. The term was first used in 1901 by Auguste Hervé to describe some of his paintings conceived in reaction against Impressionism in art. It was later used of art, music, and literature as well as plays that displayed reality as seen by the artist looking out from within, instead of, as with Impressionism, reality as it affects the artist inwardly. The Expressionist theatre was a theatre of protest, mainly against the contemporary social order and the domination of the family. Most of its dramatists were poets who used the theatre to further their ideas, and it was partly their use of poetic language that led to the collapse of the movement. It was too personal,

as was the concentration on the central figure, the author-hero whose reactions are 'expressed' in the play. Among the forerunners of Expressionism were *Strindberg and *Wedekind; the first drama of the Expressionist movement is usually considered to be *Der Bettler*, by Reinhard *Sorge. One of the few dramatists outside Germany to be influenced by Expressionist drama was Eugene *O'Neill, particularly in *The Emperor Jones* (1920) and *The Hairy Ape* (1922).

Extravaganza, spectacular and brilliantly costumed dramatic entertainment which flourished in England in the mid-19th century. Distinguished from *burlesque by its lack of a satiric target, it was usually based on a well-known story from mythology or folk-lore and characterized by its witty, punning use of rhymed couplets and clever songs. It was first offered as an alternative to *pantomime at Easter and Whitsun in 1824 and, with a *harlequinade attached and sometimes billed as a pantomime, was gradually introduced into the theatre on Boxing Night (26 Dec.). Its leading exponents were J. R. *Planché, H. J. *Byron, and E. L. Blanchard.

Eyre, Richard Charles Hastings (1943–), English director, who became known for his work at the *Royal Lyceum Theatre, Edinburgh, first as Associate Director, 1967–70, then as Director of Productions, 1970–2. As Artistic Director of *Nottingham Playhouse, 1973–8, he promoted the work of new playwrights. In 1981 he became an associate director of the *National Theatre, his productions including an enormously successful revival of the musical *Guys and Dolls*, *Gay's *The Beggar's Opera* (both 1982), and *Gogol's *The Government Inspector* (1985). He directed several plays for the *Royal Court, among them David *Mamet's *Edmond* (1985) and *The Shawl* (1986); in 1986 he also directed *High Society*, a musical version of Philip *Barry's *The Philadelphia Story* for which he wrote the book. He twice tackled *Hamlet*, in 1980 at the Royal Court and in 1989 at the National Theatre, of which he became Artistic Director in 1988. In 1990 his highly praised productions of *Richard III*, with Ian *McKellen, and David *Hare's *Racing Demon* were staged there.

Eysoldt, Gertrud (1870–1955), German actress, who made her first appearance in *Henry IV* (1890) with the *Meininger company. After touring in Germany and Russia she appeared in *Berlin in 1899, and later played under *Reinhardt. An extremely clever and subtle player, she was at her best in modern realistic parts, particularly in the works of *Wedekind, in which she played opposite the author, and was also good in *Ibsen, in such parts as *Wilde's Salome, and in the plays of *Maeterlinck. She was in the first German-language production of *Schnitzler's *Reigen* (1921) at the Kleines Schauspielhaus in Berlin, and at the subsequent court hearing defended the play against the charge of obscenity.

Eytinge, Rose (1835–1911), American actress, who made her début as Melanie in *Boucicault's *The Old Guard* (1852). A beautiful Jewess, she was greatly admired in both high comedy (as Lady Gay Spanker in Boucicault's *London Assur-* *ance* and as Beatrice in *Much Ado About Nothing*) and tragedy (as Ophelia in *Hamlet*, Desdemona in *Othello*, and especially Cleopatra in *Antony and Cleopatra*). She appeared with Edwin *Booth in *Bulwer-Lytton's *Richelieu* (1864) and other plays. Perhaps her most popular role was that of Nancy in *Dickens's *Oliver Twist* (1867) with E. A. *Davenport as Bill Sikes and the younger James *Wallack as Fagin. Her later career suffered from her increased proneness to temperamental outbursts both on and off stage.

F

59 Theatre Company, see ROYAL EXCHANGE THEATRE.

Fabbri, Diego (1911–80), Italian dramatist influenced both by *Pirandello and by his older contemporary Ugo *Betti. His plays, unlike those of the dialect dramatists of his day, were written in Italian; they are markedly religious and discuss contemporary social and moral problems in the light of the Catholic faith. His best known work, *Processo a Gesù* (*The Trial of Jesus*, 1955; NY, as *Between Two Thieves*, 1960), was first produced for the *Piccolo Teatro della Città di Milano. *La Bugiarda* (*The Liar*, 1953) was seen in London during the *World Theatre Season of 1965. His other plays include *Inquisizione* (*The Inquisition*, 1950), *Il sedutore* (*The Seducer*, 1951), *Processo di famiglia* (*The Family on Trial*, 1953), and *Veglia d'armi* (*The Armed Vigil*, 1956). He also made excellent dramatizations of Dostoevsky's *The Devils* and *The Brothers Karamazov*, published in 1961 in a volume entitled *I demoni*.

Fabula, generic Latin name for a play, under which many different types of drama were grouped. Among them were:

1. The **Atellana,** short impromptu entertainment popular in early Roman times. As the surviving titles show, these sketches were intended to amuse the crowd on market days and holidays—*The Farmer, The Vine-Gatherers, The Woodpile, The She-Goat*—and were acted by a few stock characters—Bucco (the fat-faced Fool), Dossennus (the Hunchback), Maccus (the Idiot), Manducus (the Glutton), Pappus (the Old Man)—who all figure in the 1st-century BC works of two authors, Pomponius and Novius. They were responsible for a short period of popularity for the written, as opposed to the extempore, Atellan farces, which gave way eventually to the equally vulgar but more sophisticated *mime.

2. The **Palliata,** play translated into Latin from the Greek; from *pallium*, a Greek cloak. An alternative name for such plays was *crepidata*, from *crepida*, a Greek shoe always worn with the *pallium*. Writers of such plays include *Terence and *Plautus.

3. The **Praetexta,** original play in Latin on a theme taken from Roman legend or contemporary history; from *toga praetextata*, the purple-bordered toga worn by magistrates.

4. The **Togata,** Latin comedy based on incidents of contemporary daily life in Roman towns; from *toga*, the long garment worn by a Roman citizen. Its alternative name was *tabernaria*, from *taberna*, a poor man's house. The cast of a *togata* was smaller, and the plot simpler, less restrained by social conventions, and more overtly sexual, than in the imported *palliata*.

Fact, Theatre of, see DOCUMENTARY THEATRE.

Fagan, J(ames) **B**(ernard) (1873–1933), playwright and director, born in Ulster, who began his stage career in Frank *Benson's company in 1895. He then played two seasons under Beerbohm *Tree at *Her Majesty's Theatre, and later became manager of the *Royal Court Theatre, where he gave some notable productions of Shakespeare. In 1923 he took over the old Big Game Museum in Oxford and made it into a small theatre, the precursor of the *Oxford Playhouse, installing there a notable repertory company. In 1929 he became director of the Festival Theatre, *Cambridge, and he was also responsible for many productions by the Irish Players. His excellent dramatization of R. L. Stevenson's *Treasure Island* was first seen in London in 1922, and then annually as a Christmas play up to 1931. Of his own plays, the most successful were *And So to Bed* (1926), based on *Pepys's Diary, and *The Improper Duchess* (1931), in both of which Yvonne *Arnaud gave fine performances.

Fair, William B., see MUSIC-HALL.

Fairs. The big continental trade fairs throughout Europe were from their earliest days associated with theatrical enterprises. Itinerant actors with their portable theatres were glad to find a ready-made audience, and in France particularly they flocked to the fairs of Saint-Germain, in the spring, and Saint-Laurent, in late August. Some companies were able in due course to replace their *booths by permanent theatres which stayed open before and after the official time-span of the actual fair. Many such theatres had a balcony built out above the main entrance, on which some of the actors would appear before the show itself, whetting the appetite of passers-by with a display of acrobatics, mime, and singing. This became known as the *parade, and provided a tenuous link with the early days of the *commedia dell'arte. It was from such backgrounds that many of the best known farce-players of the official Parisian theatres made their way. But the bulk of the *forains*, as they were called, remained faithful to the fairgrounds until in the latter part of the 18th century they migrated to the small permanent theatres on the *boulevard du Temple, where they remained until the building of the great boulevards, in the 1860s, swept them away.

In London the fairs of St Bartholomew (immortalized by Ben *Jonson), Smithfield, and Southwark, as well as the smaller Greenwich and May fairs, were always connected with theatrical entertainments, particularly *puppet-shows, and when the London theatres closed during the summer, many of the actors were glad to migrate to the show-booths. Elkanah *Settle is said to have ended his days as a green dragon in a Southwark booth, and he also wrote *drolls and sketches for Bartholomew Fair; but, owing perhaps to the inclemency of the weather, there is no record of permanent theatres being built on English fairgrounds.

Of the German fairs, those of Leipzig and Frankfurt-am-Main are the best known in connection with theatrical matters; it was at Frankfurt that the *English Comedians appeared most frequently, while Leipzig saw the reforms of *Gottsched first put into practice by Carolina *Neuber. Fairs also played a large part in the development of the early Russian theatre, which however developed somewhat differently from the rest of Europe. In Italy the itinerant *commedia dell'arte* troupes, and in Spain the travelling companies of such actor-managers as Lope de *Rueda, were in evidence wherever a captive audience could be found.

Falckenberg, Otto, see MUNICH.

Falling Flaps, method of making quick changes of scenery and trick effects which dates from the early 19th century. It consisted of a series of double flat scenes, framed and moving on hinges. One side of these was painted to represent, for example, an indoor scene and the other an outdoor. The stage was thus set with a complete picture in sections, each piece being kept in place by catches. When these were simultaneously released, the flaps fell by their own weight, presenting an immediate change of scene. (See also CHÂSSIS A DÉVELOPPEMENT.)

False, or **Inner, Pros(cenium),** temporary structure set behind the *proscenium arch to diminish the size of the opening. In America it is known as a Portal Opening, and in France as *le manteau d'Harlequin*. It is particularly useful on tour, when scenery has to be accommodated on stages of varying sizes, and consists usually of a pair of *wings, opened to a right angle or more and set on either side of the stage, with a deep *border connecting the two sides across the top. The whole is cut and painted to simulate hanging drapery. In some modern proscenium theatres the wings are often replaced by a fixed pair of narrow *flats covered in black velvet (tormentors) linked by a plain border (a teaser). This arrangement can be used with a *cyclorama to reduce the amount of scenery needed to dress the stage. A false pros, consisting of a pair of

flats and a decoratively profiled border (Show Portal in America), is often used in *revue and *musical comedy.

Fan Effect, see TRANSFORMATION SCENE.

Farce, form of popular comedy in which laughter is raised by horseplay and bodily assault in contrived and highly improbable situations. It must, however, retain its hold on humanity, even if only in depicting the grosser faults of mankind, otherwise it degenerates into *burlesque. It deals with the inherent stupidity of man at odds with his environment, and originated in the great submerged stream of folk-drama, of which few written records remain. It stands at the beginning of classical drama (see FABULA I: ATELLANA) as well as of modern European drama, and was especially popular in France in the later Middle Ages. Among the many medieval farces that were long current, the greater part no doubt transmitted orally, the best known is *Maître Pierre Pathelin* (c.1470), the portrait of a rascally lawyer. Traditional farce survived in France until well into the 17th century, particularly in the provinces; it died out in Paris in the 1640s with the disappearance of the gifted trio of farce-players *Turlupin, *Gros-Guillaume, and *Gaultier-Garguille. In his early career as an actor *Molière played in farce, and its tradition exercised a great influence on his career as a dramatist. There were elements of farce in early English biblical plays, and farcical *interludes were later written by scholars for production in schools and other places, but as in Italy and Germany the influence of the French farce was paramount in England, culminating in the works of John *Heywood.

In the 18th and 19th centuries short one-act farces were popular on the English and American stages, usually as part of a bill which also included a five-act *tragedy. They were ephemeral productions, though some of them achieved a great success, mainly through the acting of some particular comedian. In modern usage the word farce is applied to a full-length play dealing with some absurd situation, generally based on extra-marital adventures—hence 'bedroom farce'. An early exponent of modern farce in England was *Pinero, several of whose early plays in this genre have been successfully revived. A full-length farce which still holds the stage is *Charley's Aunt* (1892) by Brandon *Thomas. In the 1920s and early 1930s there was a series of successful farces at the *Aldwych Theatre, mostly written by Ben *Travers, and in the 1950s and 1960s a similar series was produced by Brian *Rix at the *Whitehall. Thanks to its robust character, farce survives translation better than *comedy, as is shown by the recent success in England of the farces of *Feydeau, some of the best of their kind.

Farjeon, Herbert, see REVUE.

Farquhar, George (1678–1707), Irish-born dramatist who is usually classed among the writers of *Restoration Comedy, though he stands a little apart from them, both chronologically and in showing more variety of plot and depth of feeling. He was for a short time an actor at the Smock Alley Theatre in *Dublin, but went to London, where his first play *Love and a Bottle* (1698) was produced at *Drury Lane. With his second, *The Constant Couple; or, A Trip to the Jubilee* (1699), he established himself as one of the foremost dramatists of the day. The play held the stage throughout the 18th century, the hero Sir Harry Wildair, first played by *Wilks, being a favourite breeches part (see MALE IMPERSONATION) with Peg *Woffington and Dorothy *Jordan. A sequel, *Sir Harry Wildair* (1701), proved less successful, as did a revision of *Beaumont and *Fletcher's *The Wild Goose Chase* (1652) as *The Inconstant; or, The Way to Win Him* (1702), and two new farces, *The Twin Rivals* (1703) and *The Stage Coach* (1704). Having contracted a marriage with a penniless lady whom he believed to be an heiress, Farquhar left London for a time, but returned to the theatre with a fine comedy *The Recruiting Officer* (1706), based on his own experiences in that capacity in Shropshire. An immediate success, it was followed by his finest play *The Beaux' Stratagem* (1707). A realistic comedy with a wholesome, open-air humour, it was Farquhar's last play, written in six weeks while he was lying ill and penniless in mean lodgings. Both it and *The Recruiting Officer* are still revived. The latter, on which *Brecht based his *Pauken und Trompeten* (1955), was seen at the *Old Vic in 1963 with *Olivier as Brazen; the former at the Old Vic in 1970 with Maggie *Smith as Mrs Sullen, which was also one of Edith *Evans's best parts.

Farr, Florence [Mrs Edward Emery] (1860–1917), English actress and director, chiefly connected with the introduction of plays by *Ibsen, *Shaw, and *Yeats to the London stage. She played Rebecca West in the first London production of Ibsen's *Rosmersholm* (1891) and appeared as Blanche in Shaw's *Widowers' Houses* (1892), the first of his plays to be staged. Financed by Miss *Horniman, she then produced at the Avenue Theatre (later the *Playhouse) his *Arms and the Man* (1894), creating the part of Louka. It was preceded by Yeats's one-act play *The Land of Heart's Desire*, again the first of his plays to be seen in London. She was later associated with Yeats in Dublin where she played Aleel the Minstrel in *The Countess Cathleen* (1899); back in London she arranged the music and trained the chorus for *Granville-Barker's productions of *Euripides' *The Trojan Women* (1905) and *Hippolytus* (1906), in which she played the Leader of the Chorus and the Nurse respectively.

Farrant, Richard, see BLACKFRIARS THEATRE.

Farren, Elizabeth (1759–1829), English actress, the child of strolling players and on the stage from her earliest years. As an adult she made her first appearance in London in 1777 at the *Haymarket Theatre as Kate Hardcastle in *Goldsmith's *She Stoops to Conquer*, and then joined the company at *Drury Lane. She was not immediately successful, owing to the outstanding popularity of Mrs *Abington, but she became recognized as an excellent player of fine ladies, for which her natural elegance, tall, slim figure, and beautifully modulated voice rendered her particularly suitable. She made her last appearance on the stage in 1797 as Lady Teazle in *Sheridan's *The School for Scandal* and then married the recently widowed Earl of Derby, who had been in love with her for many years.

Farren, Nellie [Ellen] (1848–1904), English actress, descendant of a long line of actors, her great-grandfather **William** (1725–95) having appeared as Careless and Leicester in the first productions of *Sheridan's *The School for Scandal* (1777) and *The Critic* (1779) respectively, while his son **William** (1786–1861) was unequalled in his day as Sir Peter Teazle, also in *The School for Scandal*, and in many Shakespearian roles. Nellie's father **Henry** (1826–60) was on the stage in London before going to America, where he was manager of a theatre in St Louis at the time of his early death. She herself was a favourite at the old *Gaiety Theatre from 1868 until her retirement in 1891. Being small and slight, she specialized in the playing of boys' parts—Smike in *Dickens's *Nicholas Nickleby*, Sam Willoughby in Tom *Taylor's *Ticket-of-Leave Man*, and the cheeky Cockney lads in H. J. *Byron's burlesques and extravaganzas.

Fastnachtsspiel, German Carnival or Shrovetide play of the 15th century, in one act, performed mainly in Nuremberg by students and artisans. It shows, in its somewhat crude couplets, a mingling of religious and popular elements interesting in the light of later developments in German drama. It developed, independently of religious drama, from a sequence of narrated episodes in which each character endeavoured to outdo the others in absurdity or obscenity. The earliest known writer of *Fastnachtsspiele* was **Hans Rosenplüt** (c.1400–c.1470). The subjects of the Carnival plays are those which would appeal to mainly urban audiences—the weaknesses and venial sins of lawyers and their clients, doctors and their patients, clerics and their female parishioners. In most of these farces the *Narr or fool is the central character, sometimes with a dull-witted companion to serve as a butt for his practical jokes. At first the plays were presented in the simplest possible way, rather in the style of the English *mumming play, but later, when the town guilds took them over, a raised stage and hangings, with a few properties, became general. Many of the Carnival plays were written by the *Mastersingers.

Fate Drama (*Schicksalstragödie*), type of early 19th-century German play inaugurated by *Werner's *Der 24 Februar* (1809), in which a malignant fate dogs the footsteps of the chief character, driving him, by a chain of fortuitous circumstances, to commit a horrible crime, often the unwitting murder of a son by his own father. The genre was further exploited by Adolf Müllner in *Der neunundzwanzigste Februar* (*The Twenty-Ninth of February*, 1812) and *Die Schuld* (*Guilt*, 1813) and by the young *Grillparzer in his first play *Die Ahnfrau* (1817).

Faucit, Helen(a) **Saville** (1817–98), English actress, trained by Percival Farren, great-uncle of Nellie *Farren. A woman of great charm and beauty, she did her best work in revivals of Shakespeare and in new verse plays, many of them specially written for her. She first appeared in the provinces in 1833, and in 1836 made her début in London as Julia in Sheridan *Knowles's *The Hunchback*. A year later she was seen in *Browning's *poetic drama *Strafford* and then had a great success in one of her best parts, Pauline in *Bulwer-Lytton's *The Lady of Lyons* (1838), playing opposite *Macready. She was also in Browning's *A Blot in the 'Scutcheon* (1843) and *Colombe's Birthday* (1853). Her finest Shakespearian roles were Juliet, Portia in *The Merchant of Venice*, and Desdemona in *Othello*. In later life she acted almost entirely for charity and became the friend and guest of Queen Victoria.

Faust, medieval legendary character in league with the devil who became linked with the name of Johann Faust (*c.*1480–1540), a wandering conjurer and entertainer. The story of his adventures was published in a Frankfurt chapbook in 1587, and in an English translation provided the material for *Marlowe's *The Tragical History of Doctor Faustus* (*c.*1589–92). The legend returned to Germany via the *English Comedians, who stressed the spectacular and farcical elements, and lived on in a puppet-play until once more taken up seriously by *Lessing in 1759. Only fragments of his work remain, but it is evident that he envisaged Faust as a scholar whose inquiring mind finds itself in conflict with the limits imposed by God on human knowledge. During the period of *Sturm und Drang* the Faust legend made a strong appeal, being used by Friedrich Müller and particularly by *Goethe, whose play on the subject, in two parts, engaged his attention from 1774 to his death in 1832. In it Faust, though tempted by Mephistopheles and guilty of the death of Gretchen, defies the devil and escapes him, his soul, as in Lessing's version, being eventually borne up to heaven.

Fauteuil, see STALL.

Favart, Théâtre, see COMÉDIE-ITALIENNE.

Faversham, William (1868–1940), American actor-manager, born in London, where he made a brief appearance on the stage before going to New York in 1887. There he played small parts at the *Lyceum with Daniel *Frohman's company, and was for two years with Mrs *Fiske. In 1893 he was engaged by Charles *Frohman for the *Empire, remaining there for eight years and playing a wide variety of parts including Romeo to the Juliet of Maude *Adams. He appeared in a number of modern plays, and in 1909 made his first independent venture with the production at the *Lyric Theatre, New York, of Stephen *Phillips's *Herod*, in which he played the title-role. His production of *Julius Caesar* (1912), with himself as Mark Antony, set the seal on his reputation both as actor and director. With a fine cast, the play ran for some time in New York and then went on tour, being followed by *Othello* and *Romeo and Juliet*. After his appearance in *Shaw's *Getting Married* (1916), which he directed and produced, his career was unremarkable, though he continued to act until the early 1930s.

Fay, Frank J. (1870–1931) and **William George** (1872–1947), Irish actors, who in 1892 formed the Ormonde Dramatic Society, playing in Dublin and the surrounding country in a repertory of sketches, short plays, and farces. Among their fellow actors were Dudley *Digges and Sara *Allgood, who were associated with them in 1898 in the *Irish National Dramatic Society and in 1904 went with them to the *Abbey Theatre. Frank, who was interested in verse-speaking, was responsible for the company's speech training, while W. G. acted as stage-manager. Both brothers appeared in most of the plays produced at the Abbey, W. G. playing Christy Mahon in *Synge's *The Playboy of the Western World* (1907) and Frank his rival Shawn Keogh. In 1908, after a disagreement with *Yeats and Lady *Gregory over artistic control, the Fays left the Abbey and went to America, where they directed a repertory of Irish plays for Charles *Frohman, W. G. making his first appearance on Broadway in Yeats's *The Pot of Broth*. Moving to London in 1914, W. G. was seen in several new plays and was successively a director at the Nottingham and *Birmingham Repertory theatres. Among his later parts were the Tramp in Synge's *In the Shadow of the Glen* (1928); Mr Cassidy in *Bridie's *Storm in a Teacup* (1936), which he also directed; and the title-role in Paul Vincent *Carroll's *Father Malachy's Miracle* (1945). Frank returned to Dublin in 1918 and became a teacher of elocution, going back briefly to the Abbey in 1925 to play in a revival of Yeats's *The Hour-Glass*.

Feast of Fools, generic name for the New Year revels in European cathedrals and collegiate churches, when the minor clergy usurped the functions of their superiors and burlesqued the services of the Church. The practice may have arisen spontaneously, as an outlet for high spirits, or may be an echo of the Roman Saturnalia. It appears to have originated in France in about the 12th century, and from the beginning evidently included some form of crude drama. The proceedings opened with a procession headed by an elected 'king'—in schools a *boy bishop—riding on a donkey, a detail which was taken over by the *liturgical drama for the scenes of Balaam's Ass, the Flight into Egypt, and perhaps Christ's Entry into Jerusalem. The Feast of Fools lingered on in France until the 16th century, by which time the festivities had moved out of the church, and was eventually absorbed into the merrymaking of the *sociétés joyeuses*; but in England, where it is known to have taken place at St Paul's in London as well as at Lincoln, Beverley, and Salisbury, it died out some time in the 14th century. Part of its functions, though without the burlesque church services, survived during the Christmas season at Court and in the colleges of the universities, the 'king' being replaced by an Abbot or Lord of Misrule; at the Inns of Court in London the custom of appointing a Christmas Lord of Misrule lingered on intermittently until the 1660s.

Fechter, Charles Albert (1824–79), actor born in London but educated in France, who played in French and English, both in America and in Europe, with equal success. He made his début at the *Comédie-Française in 1840, but soon left to make his reputation elsewhere, and became an outstanding *jeune premier*, being the first actor to play Armand Duval in the younger *Dumas's *La Dame aux camélias* (1852). At that time he was co-director of the *Odéon, but chafing at the restrictions imposed on him in favour of the Comédie-Française he returned to London, where he had appeared some years earlier, in an English translation of *Hugo's *Ruy Blas*. Though his accent was never very good, he played with so much fire and fluency that the audience was carried away, and he became very popular. He followed this success with a revolutionary production of *Hamlet* in 1861, in which the subtlety and depth of his performance impressed even those who clung to the traditional style. His Othello was not so successful, and in a revival he played Iago. In 1863 he took over the management of the London *Lyceum, where he appeared in a series of melodramas, and in 1870 he made his début in New York. After touring for some time, he opened the old

Globe Theatre in Boston under his own name, making many improvements and innovations, and appeared there in new plays as well as in some of his former successes. His quarrelsome and imperious nature made him many enemies and after a last visit to England he retired in 1876 to a farm near Philadelphia where he died.

Fedeli, company of actors of the Italian *commedia dell'arte* formed in about 1598 by Giovann Battista *Andreini. It first came into prominence in 1603, and later took in actors from the *Gelosi and *Accesi. For some time the company's *Arlecchino was the famous Tristano *Martinelli who paid prolonged visits to Paris, his acting, and that of Andreini, being much admired by Louis XIII. The Fedeli also toured extensively in Italy, and seems to have continued in existence until shortly after 1644. The company was officially under the protection of the Duke of Mantua, succeeding the Gelosi.

Federal Street Theatre, see BOSTON.

Federal Theatre Project, the first theatre scheme in the USA to be officially sponsored and financed by the Government. It was intended to give socially useful employment to members of the theatrical profession who were out of work. Inaugurated in 1935, it was under the control of the Works Progress Administration, and run from Washington by **Hallie Flanagan** (1890–1969), a theatre historian who was at the time Professor of Drama and Director of Experimental Theatre at Vassar. At its peak the Project had over 10,000 employees, ran theatre groups in 40 different states, and played to audiences, particularly of young people, which totalled millions. It covered a wide range of projects, from Greek to modern drama, musical comedy, children's plays, dance drama, puppet shows, and special programmes for Black, Catholic, and Jewish audiences. From 1936 to 1939 it ran a number of shows on Broadway, and one of its most successful innovations was the *Living Newspaper. The socialist content of many of the productions led to criticism, and in the summer of 1939, in spite of a public outcry, the Federal Project was disbanded by the United States Government.

Félix, Élisa, see RACHEL.

Female Impersonation. In the Greek and Elizabethan theatres female parts were invariably played by boys and men, as they were for centuries in China and Japan, where the actress is a comparatively late importation. Some actors, including Ned *Kynaston in London and *Mei Lanfang in China, became renowned for their skilled portrayal of women. In England the playing of the *dame in *pantomime by a man is more a form of *burlesque or caricature than an expression of the art of impersonation, though some famous Dames, notably Dan *Leno, raised their characterization of elderly women to great heights of artistry. Today the term 'female impersonator' is reserved for those actors, often solo performers, whose appearance in women's dress, or 'in drag', is so perfect as almost to convince the spectators that they are in fact women. Among the first modern exponents of 'drag' were the Americans **Tony Hart** [Anthony Cannon] (1855–91) and Julian *Eltinge. In England **Arthur Lucan** (1887–1954) toured for many years as Old Mother Riley, and an acknowledged master of the art is **Danny La Rue** [Daniel Patrick Carroll] (1928–). Outstanding also are the mock-prim singing duo Hinge and Bracket (Dr Evadne Hinge and Dame Hilda Bracket, really George Logan and Patrick Fyffe) and the Australian **Barry Humphries** (1934–), whose *alter ego*, a 'housewife superstar', is known as Dame Edna Everage.

Fenn, Ezekiel (1620–?), boy actor with *Queen Henrietta's Men under Christopher *Beeston at the *Cockpit, where in 1635 he played the chief female part in Nabbes's *Hannibal and Scipio*. He must already have had a good deal of experience with a *boy company, as at about the same time he played an even more exacting role, Winifred in *Dekker and *Rowley's *The Witch of Edmonton*. When the Queen's Men disbanded in 1637 Fenn stayed on with Beeston, probably as one of the older members of his new company Beeston's Boys, playing his first male role in 1639. Fenn's history after the closing of the theatres in 1642 is unknown.

Fennell, James (1766–1816), English actor who gained his experience in the provinces and in 1792 went to America, where he was very popular. He was a member of the *American Company and also of *Wignell's company in Philadelphia. His best part was Othello, though he appeared in most of the tragic roles of the current repertory, including Jaffier in *Otway's *Venice Preserv'd*, which he played at the *Park Theatre, New York, in 1799. An excellent actor, over 6 ft. tall, with a mobile, expressive face, he retired in 1810.

Fenton, Lavinia, see GAY.

Ferber, Edna, see KAUFMAN and MUSICAL COMEDY.

Fernández de Moratín, see MORATÍN.

Ferrari, Paolo (1822–89), Italian dramatist, who was practising law when the first of his numerous plays was produced. He came into prominence in 1853 when his play on *Goldoni—*Goldoni e le sue sedici commedie*—won a prize offered by a Florentine dramatic academy. Of his later works, which deal with social problems but are intended to reinforce rather than criticize current attitudes towards them, the most successful were *Il duello* (1868), which argues in favour of duelling; *Il suicido* (1875), which held the stage for many years; and *Due dame* (*The Two Ladies*, 1877), which supports the prevailing prejudice against marriages between gentlemen and women of the lower classes. Ferrari's best-known work is *La satira e Parini* (1856), on the life of the poet Giuseppe Parini, in which the Marchese Colombi is as familiar to Italian audiences as Mrs Malaprop in *Sheridan's *The Rivals* is to English.

Ferreira, António (1528–69), Portuguese dramatist, author of the finest play of the Portuguese Renaissance and also the first to combine classical models with native material, the *Tragédia de Dona Inêz de Castro* (pub. 1587), which tells of the murder in 1355 of the mistress of Pedro, eldest son of Alfonso IV, a subject which has inspired many other dramatists, including Henri de *Montherlant in *La Reine morte* (1942). Ferreira also wrote two moral but witty comedies, *O Cioso* and *Bristo* (pub. 1622), called after their chief characters.

Ferrer, José [José Vincente Ferrer de Otero y Cintron] (1912–), Puerto Rican-born American actor, producer, and director, who originally studied architecture. He made his stage début in 1934 in showboat melodramas touring Long Island Sound, and a year later was seen in New York in *Lindsay and Runyon's *A Slight Case of Murder*. After appearing in Maxwell *Anderson's *Key Largo* (1939) he played Lord Fancourt Babberley in a revival of Brandon *Thomas's *Charley's Aunt* in 1940, Iago to Paul *Robeson's record-breaking Othello in 1943, and gave an outstanding performance in the title-role of *Rostand's *Cyrano de Bergerac* in 1946. In 1948 he was appointed Director of the *City Center drama company, playing such roles as *Jonson's Volpone, Jeremy and Face in his *The Alchemist*, and Richard III. The last part he created in New York was the Prince in the musical *The Girl Who Came to Supper* (1963), based on *Rattigan's *The Sleeping Prince*. He produced plays at the City Center and elsewhere, notably *Stalag 17* (1951), by Donald Bevan and Edmund Trzcinski, and Joseph Kramm's *The Shrike* (1952), directing them both and playing the leading role of a mental patient in the latter. He directed films as well as plays and was an important film actor.

Festival Theatre, Cambridge, see GRAY, TERENCE.

Feuillère, Edwige [Caroline Cunati] (1907–), French actress, who, as Cora Lynn, first appeared in light comedies at the *Palais-Royal. She made her début at the *Comédie-Française in 1931 but left to go into films, returning to the Paris stage in 1934. Her quality as an actress became apparent in *Becque's *La Parisienne* (1937) and her playing of Marguerite Gautier in the younger *Dumas's *La Dame aux camélias* (1940). She created the parts of Lia in *Giraudoux's *Sodome et Gomorrhe* (1943), Paola in his

Pour Lucrèce (1953), and the Queen in **Cocteau's L'Aigle à deux têtes* (1946). In 1948 she joined the **Barrault–*Renaud* company to play Ysé, one of her finest parts, in **Claudel's Partage de midi*. In 1957 she played the Queen in a French version of **Betti's La regina e gli insorti*, and she was in Paris productions of **Kopit's Oh Dad, Poor Dad* (1963), Giraudoux's *La Folle de Chaillot* (1965), Tennessee **Williams's Sweet Bird of Youth* (1971), and **Dürrenmatt's Der Besuch der alten Dame* in 1976. She scored an outstanding success in 1977 in **Arbuzov's Old World*, known in French as *Le Bateau pour Lipaïa*, and in 1984 was in **Anouilh's Léocadia*. She was seen in London in 1951 and 1968 in *Partage de midi*, in 1955 in *La Dame aux camélias*, and in 1957 with her own company at the **Palace Theatre*, in a repertory that included *La Parisienne* and **Racine's Phèdre*.

Feydeau, Georges-Léon-Jules-Marie (1862–1921), French dramatist, son of the novelist Ernest-Aimé Feydeau, and author of more than 60 farcical comedies. He had his first success with *Tailleur pour dames* (1887) and continued writing until his death, his last play, *Cent million qui tombent*, being produced posthumously. Among his more notable plays were *Le Système Ribadier* (1892), *Le Dindon* (1896), *La Dame de Chez Maxim* (1899), and some one-act farces, including *Feu la mère de Madame* (*My Late Mother-in-Law*, 1908); *On purge bébé* (*Time for Baby's Medicine*, 1910); *Mais n'te promène pas toute nue!* (*For God's Sake, Get Dressed!*, 1912); and *Hortense a dit: J'm'en fous!* (*Hortense Couldn't Care Less*, 1916). Feydeau was an outstanding writer of classic farce, whose reputation has been consolidated by productions of his plays at the **Comédie-Française* and elsewhere. Several of his masterpieces have been seen in London in translation, among them *L'Hôtel du Libre Échange* (1899) as *Hotel Paradiso* (1956), starring Alec **Guinness*; *Occupe-toi d'Amélie* (1908) as *Look after Lulu* (1959) by Noël **Coward*, with Anthony **Quayle* and Vivien **Leigh*; *Une puce à l'oreille* (1907) as *A Flea in Her Ear* (1965), *Un fil à la patte* (1908) as *Cat among the Pigeons* (1969), and *The Lady from Maxim's* (1977), all adapted by John **Mortimer*.

Feydeau, Théâtre, see COMÉDIE-ITALIENNE.

Ffrangcon-Davies, Dame Gwen (1891–   ), English actress of Welsh extraction, who trained as a singer and appeared with great success as Etain in Rutland Boughton's 'music drama' *The Immortal Hour* (1920). She was however drawn to straight acting, and already had considerable experience when she appeared as Phoebe Throssel in a revival of **Barrie's Quality Street* in 1921. She then joined the **Birmingham Repertory Theatre company*, where she played many roles, including Eve and the Newly Born in **Shaw's Back to Methuselah* (1923). She was much admired as Juliet to **Gielgud's Romeo*

in London in 1924, and was with him again as Anne of Bohemia in Gordon Daviot's *Richard of Bordeaux* (1932), as Olga in **Chekhov's Three Sisters* in 1937, as Gwendolen in his memorable production of **Wilde's The Importance of Being Earnest* in 1939, and as Lady Macbeth to his Macbeth in 1942. Other outstanding roles were Elizabeth Moulton-Barrett in Rudolf Besier's *The Barretts of Wimpole Street* (1930) and Mrs Manningham in Patrick Hamilton's *Gaslight* (1939). In 1950, after several years in South Africa, she was with the **Shakespeare Memorial Theatre company*, and three years later with the **Old Vic company*. Among the best of her later roles were Ranevskaya in Chekhov's *The Cherry Orchard* in 1954, Mary Tyrone in **O'Neill's Long Day's Journey into Night* in 1958, and Amanda in Tennessee **Williams's The Glass Menagerie* in 1965. Her New York début was in 1963 in **Sheridan's The School for Scandal*. She made her last appearance on the stage in 1970, in Chekhov's *Uncle Vanya*, giving a beautifully controlled and moving performance as Madame Voynitsky at the **Royal Court Theatre*.

Fiabe, see GOZZI.

Field, Nathan (1587–1620), English actor and playwright, who was taken from St Paul's School to become a **boy actor* with the Children of the Chapel at the second **Blackfriars Theatre*, where he appeared in *Cynthia's Revels* (1600) and *The Poetaster* (1601), both by **Jonson* in whose *Epicoene; or, The Silent Woman* (1609) he was later to appear. He was an excellent actor, particularly admired in such parts as the title-role in **Chapman's Bussy d'Ambois* (c.1604), and in 1615 joined the **King's Men*, possibly in succession to Shakespeare. He wrote two plays, *A Woman is a Weathercock* (1609) and *Amends for Ladies* (1611), and collaborated in several works with **Massinger* and **Fletcher*.

Field, Sid, see MUSIC-HALL and PRINCE OF WALES THEATRE.

Field Day Theatre Company, Irish company founded in 1980 by Brian **Friel* and the actor Stephen Rea, the only major touring company in Ireland. Based in Derry, it is funded by both north and south and operates throughout Ireland, aiming to promote Irish writing through theatre, poetry, and literary publications. The first of its annual productions, Friel's *Translations*, was followed by other plays by Irish authors, including several more by Friel, and adaptations of **Sophocles* and **Molière*. The company also publishes pamphlets and an anthology of Irish literature.

Fielding, Henry (1707–54), English novelist and dramatist. His first three plays were comedies aimed at contemporary and literary follies, as was *Tom Thumb the Great; or, The Tragedy* of *Tragedies* (1730), which satirized the conventions of the *heroic drama. With his *ballad opera *The Welsh Opera; or, The Grey Mare the Better Horse* (1731) he openly attacked both political parties, and also brought the Royal Family, thinly disguised, on to the stage. The repercussions were sufficiently alarming for Fielding to drop politics for a time and write for *Drury Lane five plays dealing light-heartedly with the contemporary social scene. In 1736 he again attacked the government in *Pasquin*, followed in 1737 by *The Historical Register for the Year 1736*. He also put on plays by other authors which attacked the administration, and it was probably the cumulative effect that finally decided the government to curtail the liberty of the theatres. Adopting as a pretext the scurrility of a play entitled *The Golden Rump* (see GOODMAN'S FIELDS THEATRE) the authorities rushed through the Licensing Act (see LORD CHAMBERLAIN). The closing of the unlicensed *Haymarket Theatre, where all Fielding's plays had been produced, hit him hard, and with a wife and growing family to support he deserted the stage for the more lucrative career of novel-writing, including *Tom Jones*, published in 1749.

Fields, Dame Gracie [Grace Stansfield] (1898–1979), English comedienne and singer, for some years the most popular entertainer in Britain, known as 'our Gracie'. She became famous in the revue *Mr Tower of London*, which opened at the *Alhambra in 1923. A big music-hall attraction, she reached the height of her fame in the 1930s, being loved for her down-to-earth Lancashire personality and her comic songs such as 'The Biggest Aspidistra in the World'. Her singing voice was excellent. During the Second World War she gave numerous concerts for the troops and raised vast sums in America for the war effort. Her signature tune was 'Sally' and during the war she became closely associated with 'Wish Me Luck as You Wave Me Goodbye'. She settled in Capri after her third marriage in 1951.

Fields, Joseph (1895–1966), American dramatist, son of Lew Fields (see WEBER AND FIELDS), who collaborated with **Jerome Chodorov** (1911–) in several plays, including *My Sister Eileen* (1940; London, 1943) and *Anniversary Waltz* (1954; London, 1955). He also collaborated with him on the book for *Wonderful Town* (1953; London, 1955), the musical version of *My Sister Eileen*; with Anita Loos on that for *Gentlemen Prefer Blondes* (1949; London, 1962); and with Oscar *Hammerstein on that for *Flower Drum Song* (1958).   His brother **Herbert** (1897–1958) and sister **Dorothy** (1905–74) were responsible for the lyrics and book of such musicals as Cole *Porter's *Mexican Hayride* (1944) and Irving *Berlin's *Annie Get Your Gun* (1946; London, 1947). Herbert was also sole author of the book for other shows, among them Rodgers and *Hart's *Dearest Enemy* (1925) and *A Connecticut Yankee* (1927); he co-authored Porter's *Du Barry was a Lady* (1939) and *Panama Hattie* (1940). Dorothy wrote the lyrics for shows such as *A Tree Grows In Brooklyn* (1951) and *Sweet Charity* (1966).

Fields, Lew, see WEBER AND FIELDS.

Fields [Dukenfield], **W**(illiam) **C**(laude) (1879–1946), American actor and *vaudeville star, born in Philadelphia, where he made his first appearance at an open-air theatre as a 'tramp juggler' in 1897. A year later he was in New York, where he juggled and appeared in a *slapstick comedy routine, soon becoming a 'topliner'. In 1900 he was in London and thereafter toured Europe. He gradually began to introduce comic monologues into his juggling act and finally established himself as a comedian, appearing in 1905 in a loosely connected series of vaudeville sketches entitled *The Ham Tree*. He abandoned vaudeville permanently in 1915 to appear as an eccentric comedian in the *Ziegfeld Follies* and *George White's Scandals*. Stout, with small, cold eyes and a bulbous nose, a rasping voice, and a grandiose vocabulary and manner, he usually played petty and inept crooks, or anarchic characters devoid of any conventionally decent feelings. In 1923, after appearing as a strolling carnival swindler in the musical *Poppy*, he abandoned the theatre for films, in which he had a brilliant career.

Fifth Avenue Theatre, New York. **1.** On 24th Street near Broadway. Originally the Fifth Avenue Opera House, specializing in Negro *minstrel shows, it opened as a theatre in 1867, but closed abruptly after a fight in the auditorium in which a man was killed. It reopened in 1869 with John *Brougham in one of his own plays, and on 16 Apr. became *Daly's Fifth Avenue Theatre, with a good company that included Mrs *Gilbert and Fanny *Davenport. Its first outstanding success was Daly's own version of *Frou-Frou* by Meilhac and Halévy. Being anxious to encourage American dramatists Daly staged Bronson *Howard's *Saratoga* (1870), which had a long run. In 1873 the theatre was burnt down and not rebuilt. The site was later used for the Fifth Avenue Hall which became the *Madison Square Theatre.

2. Daly opened his second Fifth Avenue Theatre, at Broadway and 28th Street, in 1874 with an elaborate and costly production of *Love's Labour's Lost*, not previously seen in New York, which failed. Success came with the production in 1875 of Daly's own play *The Big Bonanza*, in which John *Drew made his New York début. The following season saw the New York production of the popular London success *Our Boys* by H. J. *Byron, in which Georgiana *Drew made her first appearance in New York. The first night of *Sheridan's *The School for

Scandal on 5 Dec., with Charles *Coghlan, was marred by the disastrous Brooklyn Theatre fire (see FIRES IN THEATRES). By 1878 Daly was finding the financial loss on the theatre too great and left to take over Banvard's Museum, which he opened as *Daly's Theatre. The Fifth Avenue came under new management and was ultimately leased to various travelling companies; after many changes of name it was pulled down in 1908.

Fifty-Eighth Street Theatre, New York, see JOHN GOLDEN THEATRE I.

Fifty-First Street Theatre, New York, see MARK HELLINGER THEATRE.

Fifty-Fourth Street Theatre, New York, see ABBOTT, GEORGE.

Filippo, Eduardo De, see DE FILIPPO.

Finlay, Frank (1926–   ), English actor, who came into prominence as Harry Kahn, the ageing Jewish paterfamilias in Arnold *Wesker's *Chicken Soup with Barley* (1958), which he played at the *Belgrade Theatre and later at the *Royal Court, where he also appeared as Attercliffe in John *Arden's *Serjeant Musgrave's Dance* (1959), Libby Dobson in Wesker's *I'm Talking about Jerusalem* (1960), and Corporal Hill in his *Chips with Everything* (1962). In 1963 he scored a triumph as the thieving Alderman Butterthwaite in another Arden play, *The Workhouse Donkey*, at *Chichester, and he then joined the *National Theatre company, where his roles included Willie Mossop in *Brighouse's *Hobson's Choice* and Iago to *Olivier's Othello (1964), Giles Corey in Arthur *Miller's *The Crucible* (1965), and Joxer Daly in *O'Casey's *Juno and the Paycock* (1966). After playing in David *Mercer's *After Haggerty* (1970) for the *RSC he returned to the National Theatre in *De Filippo's *Saturday, Sunday, Monday* (1973) with Joan *Plowright. He was later in *Osborne's *Watch It Come Down* and Howard *Brenton's *Weapons of Happiness* (both 1976, for the National Theatre), and De Filippo's long-running *Filumena* (1978; NY, 1980), succeeding Colin *Blakely. In 1981 he succeeded Paul *Scofield as Salieri in *Shaffer's *Amadeus* at the National Theatre, continuing the role in the West End. He was also seen in *Chekhov's *The Cherry Orchard* (1983) and the musical *Mutiny!* (1985). Essentially a fine character actor, he is especially strong in moments of watchful stillness.

Finn, Henry James (1785–1840), Canadian-born actor and playwright, who studied law at Princeton but deserted it for the stage, joining the company at the *Park Theatre, New York, as a prop boy. After performing in London, New York, and Charleston he played Hamlet and other tragic roles in New York in 1820. Only moderately successful, he turned to comedy in which he was inimitable, notably in such

parts as Aguecheek in *Twelfth Night* and Sir Peter Teazle in *Sheridan's *The School for Scandal*. In 1825 he appeared in his own play, *Montgomery; or, The Falls of Montmorency*, as Sergeant Welcome Sobersides, an amusing Yankee character who was later incorporated into *The Indian Wife* (1830), which he may also have written, and played by James H. *Hackett. Finn died in the destruction by fire of the SS *Lexington* in Long Island Sound.

Finney, Albert (1936–   ), English actor, who gained his early experience at the *Birmingham Repertory Theatre from 1956 to 1958. He then went to the *Shakespeare Memorial Theatre to play Edgar to *Laughton's Lear, and also took over the part of Coriolanus from Laurence *Olivier. He returned to London in a new musical *The Lily-White Boys* and in Willis *Hall and Keith Waterhouse's *Billy Liar* (both 1960). His reputation was further enhanced by his performance in *Osborne's *Luther* (1961; NY, 1963), his thickset physique and a hint of surly stubbornness perfectly fitting the part. In 1965 he joined the *National Theatre company, creating at *Chichester the title-role in *Arden's *Armstrong's Last Goodnight*, seen later in the year at the *Old Vic. During the same season he also appeared in *Shaffer's *Black Comedy* and *Strindberg's *Miss Julie* in a double bill, and in *Feydeau's *A Flea in Her Ear*. In 1967 (NY, 1968), he co-presented Peter *Nichols's *A Day in the Death of Joe Egg*, also playing the leading role in New York. In 1972 he gave a powerful performance as the embittered husband in E. A. Whitehead's *Alpha Beta* at the *Royal Court, of which he was Associate Artistic Director, 1972–5. He starred in the West End in another play by Nichols, *Chez nous* (1974), and in 1975 rejoined the National Theatre company to play Hamlet and the title-role in *Marlowe's *Tamburlaine the Great*. Among later roles at the National were Horner in *Wycherley's *The Country Wife* (1977) and Macbeth (1978). In 1984 he directed and appeared in Arden's *Serjeant Musgrave's Dance* at the Old Vic, and in 1989 he scored a success in a dual role in Ronald Harwood's *Another Time*. Finney is often credited with having introduced into British acting a new spirit of working-class earthiness, but he has several times proved himself an actor in the true heroic tradition. He is also a major film actor.

Fiorillo, Silvio (?–*c*.1633), actor of the *commedia dell'arte, the original *Capitano Matamoros and probably the first *Pulcinella. He had a company of his own in Naples, his birthplace, in the last years of the 16th century, and later appeared with other companies, notably the *Accesi on their visits to France in 1621 and 1632. He was the author of several plays and of scenarios in the *commedia dell'arte* tradition. His son **Battista** (*fl.* 1614–51) was married to the actress **Beatrice Vitelli** (*fl.* 1638–54), and is known to have played *Scaramuccia. Silvio may also have been the father of Tiberio, below.

Fiorillo, Tiberio (1608–94), *commedia dell'arte* actor who may have been the son of Silvio, above, but whose name is also found as Fiorilli and Fiurelli. He was already well known in Paris in the 1640s, and in 1658 he was with the Italian company which shared the *Petit-Bourbon with *Molière, an arrangement repeated at the *Palais-Royal in 1661 with a new company. Although he was not the first to play *Scaramuccia, who as Scaramouche became a stock character in French comedy, he was certainly its most skilled interpreter. Fiorillo visited London at least twice (in 1673 and 1675) and was popular with Court and public alike. A musician, dancer, singer, acrobat, and pantomimist, Fiorillo retained his agility well into his 80s, when he was still so supple that he could tap a fellow-actor's cheek with his foot.

Fires in Theatres. Fire was once a major hazard in the theatre. Both the London Patent Theatres, *Drury Lane and *Covent Garden, were burnt down twice, the first in 1672 and 1809, the second in 1808 and 1856. The first recorded theatre fire in America was that of the Federal Street Theatre in Boston in 1798, and between then and 1876, when the Brooklyn Theatre went up in flames with the loss of about 300 lives, including two members of the cast, over 75 serious fire disasters were reported. In Richmond, Va., 70 people were killed in 1811 when the candles of a stage chandelier set fire to the scenery. The introduction of gas lighting, coupled with the vogue for very large theatres, caused heavy death-rolls in 19th-century disasters: in the Lehman Theatre in St Petersburg in 1836 there were 800 casualties; in Quebec in 1846, 50; in Karlsruhe a year later, 631; in Leghorn in 1857, 100. The greatest disaster of all time, however, was probably the fire in a Chinese theatre in 1845 which killed 1,670.

Even after the introduction of more stringent fire regulations in Britain in 1878, the Theatre Royal, *Exeter, was burnt down in 1885, rebuilt, and again destroyed by fire in 1887, with 186 killed. In America, where strong safety measures were taken after the Brooklyn fire, the supposedly fireproof Iroquois Theatre in Chicago was the scene in 1903 of the worst such catastrophe in the history of the American theatre. A fire in an overcrowded auditorium, though quickly controlled, led to a panic which resulted in the loss of over 600 lives. With the introduction of electric lighting, fireproofing of stage materials, and a comprehensive code of fire regulations which laid the onus for prevention of fire on theatre managers, conditions improved, and fires in theatres are now rare.

First Floor Theatre, see WESTMINSTER THEATRE 2.

First Folio, see CONDELL, HEMINGE, and SHAKESPEARE.

First Workers' Theatre, see TRADES UNIONS THEATRE.

Fiske, Minnie Maddern [Marie Augusta Davey] (1865–1932), American actress, who appeared with her parents at the age of 3 under her mother's maiden name of Maddern, and at 5 went to New York, where she played juvenile parts, including the Duke of York in *Richard III* and Prince Arthur in *King John*. At 13 she graduated to adult parts, being seen as the Widow Melnotte in *Bulwer-Lytton's *The Lady of Lyons*. She retired from the stage in 1890 on her marriage to Harrison Grey Fiske, writing several plays before returning to star in her husband's play *Hester Crewe* (1893). In 1894 she was seen as Nora in *Ibsen's *A Doll's House*. Her playing of Tess in a dramatization of Hardy's novel in 1897 established her reputation, and she was also much admired as Becky Sharp in Langdon *Mitchell's dramatization of Thackeray's *Vanity Fair* (1899). Since her husband's opposition to the *Theatrical Syndicate prevented her from appearing in their theatres, she rented the Manhattan (formerly the *Standard Theatre) in 1901, and remained there for five years with an excellent company in a series of fine plays, including Ibsen's *Hedda Gabler* (1903). She was later seen in Ibsen's *Rosmersholm* (1907), as Rebecca West, and *Pillars of Society* (1910), and in the title-role of *Sheldon's *Salvation Nell* (1908). After an unsuccessful spell in films she returned to the New York stage in a series of light comedies, and then gave excellent performances as Mrs Malaprop in *Sheridan's *The Rivals* (1925) and Mrs Alving in Ibsen's *Ghosts* (1927); she made one of her last appearances as Beatrice in *Much Ado about Nothing*. Although not beautiful, she could totally involve audiences with her subtle playing.

Fitch, (William) **Clyde** (1865–1909), American dramatist, author of nearly 60 plays, of which the first, *Beau Brummell* (1890), was commissioned by Richard *Mansfield. The best of Fitch's early works were *Nathan Hale* (1898) and *Barbara Frietchie* (1899), based on American history and mingling personal and political problems in marked contrast to his subsequent light comedies. In 1901 *The Climbers* and *Lovers' Lane*, social comedies of life in New York, and *Captain Jinks of the Horse Marines*, in which Ethel *Barrymore first appeared as a star, were running simultaneously in New York, while in London *Tree was producing *The Last of the Dandies*. Many of Fitch's plays were written to order for certain stars, among them *The Moth and the Flame* (1898) and *The Cowboy and the*

Lady (1899), both melodramas, *The Stubbornness of Geraldine*, dealing with the American abroad, *The Girl with the Green Eyes*, about jealousy (both 1902), and *The Truth* (1907), sometimes considered his best play. One of America's best-loved playwrights, Fitch was excellent at pin-pointing certain aspects of contemporary and domestic life; but his social observation was marred by his deference to the prevailing fashion for melodrama.

Fitzball [Ball], **Edward** (1792–1873), English dramatist, author of a vast number of immensely popular melodramas, now forgotten. Several, notably *The Red Rover; or, The Mutiny of the Dolphin* (1829), based on a novel by Fenimore Cooper, and *Paul Clifford* (1835), from a novel by *Bulwer-Lytton, were successful enough to be included in the repertory of the *toy theatre. He also helped the development of *nautical drama with the dramatization of another of Cooper's novels as *The Pilot; or, A Tale of the Sea* (1825) and such works as *Nelson; or, The Life of a Sailor* (1827), also subtitled *Britannia Rules the Waves*. His *Jonathan Bradford; or, The Murder at the Wayside Inn* (1833), based on a sensational murder case, made a fortune for the manager of the *Surrey Theatre, where it was first produced.

Flanagan, Bud, see CRAZY GANG.

Flanagan, Hallie, see FEDERAL THEATRE PROJECT.

Flanders, Michael, see REVUE.

Flat, timber frame generally covered with canvas, though modern stage designers use a variety of surfaces. The standard full-sized flat is 18 ft. high; in small theatres it may be less and in large ones may reach 24 ft. Its width may vary from 1 ft. to 6 ft., exceptionally to 8 ft. For widths above this two or more flats are hinged together, known as book (or booked) flats (in America two- or three-folds). Flats may be plain or contain openings; they may be straight-edged or bear profiling boards, fixed to the side and sawn to the required shape. They may be used as *wings, and then carry on their on-stage edge an extension of profiling known as a flipper. This is hinged so that it faces the audience even if the flat is set at an angle, and can be folded flat for packing. Flats are used to form the three walls of a *box-set. When those at the back are battened together so that the whole wall can be flown it is called a French flat. Variants of the flat are the *groundrow, the *set piece, and the backing flat, set outside a door or other opening to block the view beyond. A free-standing flat is supported by a stage brace—a wooden rod hooked to the back of the flat and screwed into the floor, or held down by a weight made to fit over the foot of the brace. Sometimes the flat is supported by a French brace (brace jack in America), a right-angled triangle framework

of wood hinged to the back of the piece and opened out and weighted as needed. (See also FALLING FLAPS and SCENERY.)

Fleck, Johann Friedrich Ferdinand (1757–1801), German actor, who joined the company at the *Berlin National Theatre shortly before *Iffland arrived in 1796. He first made his mark as Gloucester to *Schröder's Lear, his fine presence, resonant voice, and fiery artistic temperament also making him an ideal interpreter of the stormy heroes of the *Sturm und Drang* period. He was especially admired in *Schiller's plays, as Don Carlos, Karl Moor in *Die Räuber*, and Wallenstein—his last and greatest role. Although his capricious character was a trial to his fellow actors and an irritation to his audience (who never knew whether they would see the 'big' Fleck or the 'little' Fleck), his Romanticism was a promisingly modern counterbalance to Iffland's cautious and old-fashioned repertory.

Fleetwood, Charles, see DRURY LANE.

Fleschelles, see GAULTIER–GARGUILLE.

Fletcher, John (1579–1625), English poet and dramatist, who spent most of his working life actively writing for the stage, either alone or in collaboration. His name is so closely connected with that of Sir Francis *Beaumont that at one time more than 50 plays were attributed to them, of which some six to 16 may be their joint work. Among the best of these are *Philaster* (c.1609), *The Maid's Tragedy* and *A King and No King* (both c.1611), and *The Scornful Lady* (c.1613). After that Fletcher seems to have written either alone or in collaboration with other authors such as *Massinger, *Middleton, and *Rowley. Among the plays attributed solely to him the first was *The Faithful Shepherdess* (1608); others were *The Wild Goose Chase* (1621), on which *Farquhar based his comedy *The Inconstant* (1702); and *The Chances* (1623), rewritten by the second Duke of Buckingham and acted in 1666 with some success. Fletcher is believed by many critics to have had a hand in the writing of Shakespeare's *Henry VIII* (1613) (also known to its contemporaries as *All is True*) and to be the joint author with Shakespeare of *The Two Noble Kinsmen* (c.1613).

Flexible Staging, type of staging which goes back, in essentials, to the Greek and Elizabethan theatres in an effort to break away from the conventions of the *proscenium or picture-frame stage. The initial aim was to re-create a sense of intimacy and immediacy between actors and audience by abolishing the proscenium arch. This led to the evolution of the all-purpose theatre which provides a variety of seating plans and adaptable levels and shapes for the stage. Such theatres can be used for *theatre-in-the-round, for productions with the audience on three sides of the acting area (the *open, or thrust, *stage), for staging at one end of a hall, usually with a large *forestage, or for proscenium productions. Pioneering English examples include the university theatre at Bristol, the *Questors Theatre at Ealing, London, the experimental theatre belonging to the London Academy of Music and Drama, and the *Cottesloe Theatre.

Flies, space above the stage, hidden from the audience, where scenery can be lifted clear from the stage, or 'flown', by the manipulation of ropes. In the traditional system of *hand working, the stage hands in charge of the flies were known as fly-men and worked on fly-floors, or galleries running along each side of the stage. Between the fly-floors there were formerly catwalks, narrow communicating bridges 2 to 3 ft. wide over each set of *wings, slung on iron stirrups from the *grid to which the ropes, or lines, were attached. Large Continental theatres and opera houses sometimes had as many as three pairs of fly-floors, but English theatres rarely had more than one pair. Scenery is now generally flown by the *counterweight system.

Flipper, see FLAT.

Florence, William Jermyn (*or* James) [Bernard Conlin] (1831–91), American actor, who made his New York début in 1850 at *Niblo's Garden in dialect impersonations. Two years later he married the actress **Malvina Pray** (1831–1906), touring North America with her in a repertory of Irish plays. Among Florence's best parts were Captain Cuttle in *Dickens's *Dombey and Son*, Bob Brierly in Tom *Taylor's *The Ticket-of-Leave Man*, which he was the first to play in America, and Bardwell Slote in Woolf's *The Mighty Dollar* (1876), which remained in his repertory for many years. He was good in burlesque and as a comedian ranked with Joseph *Jefferson, playing Sir Lucius O'Trigger to his Bob Acres in *Sheridan's *The Rivals* and Zekiel Homespun to his Dr Pangloss in the younger *Colman's *The Heir-at-Law*. Florence was responsible for the first production in New York of *Robertson's *Caste* (1867), which he had memorized in London, staging it (no copy being available) only four months after its first appearance, with the realistic scenery and atmosphere of the *Bancroft production. This caused a good deal of controversy, as Lester *Wallack had bought the American rights of the play, but in the absence of *copyright laws he had no redress. Probably fearing to challenge comparison with Florence's excellent production, Wallack did not himself stage the play until 1875.

Floridor [Josias de Soulas, Sieur de Primefosse] (*c.*1608–71), French actor, leader of a troupe of strolling players whom he took to London in 1635, appearing before the Court and at the *Cockpit in Drury Lane. Three years later he toured the French provinces and then joined the company of the Théâtre du *Marais. Some of his best performances were given in the plays of Pierre *Corneille, and it may have been Floridor's move to the Hôtel de *Bourgogne in 1647 which induced Corneille to give his later plays to that theatre rather than the Marais. Floridor soon became the leader of the company, his quiet, authoritative acting being in marked contrast to the bombastic style of his colleague *Montfleury, and he was the only actor spared by *Molière in his mockery of the rival troupe in *L'Impromptu de Versailles* (1663).

Florindo, see ANDREINI, GIOVANNI.

Fly-Floor, Fly-Men, see FLIES.

Flying Effects have been achieved in the theatre since the earliest times. The Greeks had the *deus ex machina; in *liturgical drama, and later in the Renaissance theatre, flying effects varied from the simple rise and fall of figures on a movable platform concealed behind *cloudings, as in the *Paradiso, to the elaborate undulating flight of a character across the stage. The mechanism needed to produce such effects was well known in England by the Restoration period, and by the end of the 18th century there were diagrams of the procedure to be followed for complicated flights. By the mid-19th century the machines in use included a complicated arrangement of pulleys and counterweights for controlling the circulatory gyrations of a pair of flying figures. In the *pantomimes of the early 20th century a group known as Kirby's Flying Ballet gave many brilliant and graceful performances; the lines to which players are attached are still known as Kirby Lines. In such effects, where no clouding or chariot is used for the character being flown, a line is attached to a hook at the player's back which forms part of harness worn under the costume; a safety device prevents it from leaving the hook during flight. One of the few 20th-century plays in which flying is essential is J. M. *Barrie's *Peter Pan* (1904).

Fo, Dario (1926–), Italian dramatist, actor-manager, and mime, who began his career by collaborating in satirical revues, among them *A Finger in the Eye* (1953) and *Certified in Possession of Their Faculties* (1954). With his wife Franca Rame he founded in 1957 a company which staged mainly his own pieces, including *Archangels Don't Play the Pin-Tables* (1959) and *He had Two Pistols with Black-and-White Eyes* (1960). After the abolition of censorship in 1962 Fo's hitherto good-natured comedies adopted more explicitly left-wing themes. He attacked capitalism and government corruption in such plays as *Seventh: Thou Shalt Steal a Little Less* (1964) and *It's Always the Devil's Fault* (1965). In 1967 he established his own group, La Nuova Scena, to seek working-class audiences. It was succeeded by the co-operative group La Comune, formed in 1970, for which he wrote the highly successful socialist farces *Accidental Death of an Anarchist* (1970; London, 1979; NY, 1984), about the fall from a window of a prisoner in custody; and *We Won't Pay! We Won't Pay!* (1974; NY, 1980), seen in London as *Can't Pay? Won't Pay!* (1981). In *Trumpets and Raspberries* (1981; London, 1984) a car worker is confused with his boss after plastic surgery. Fo also wrote a number of short one-woman plays with Franca Rame, groups of which were performed in London (1981, 1983, 1989) and New York (1983). His remarkable one-man show *Mistero buffo* (1969) has been widely performed.

Fogerty, Elsie, see SCHOOLS OF DRAMA.

Folger, Henry, see WASHINGTON.

Folies-Bergère, the first major Parisian music-hall, which opened in 1869 as a *café-spectacle*, with a mixed bill of light opera and pantomime. Its main attraction, however, apart from its delightful winter garden, was the immense promenade (copied at the *Empire, London) where young men about town could meet their friends in comfort. It catered largely for French and foreign visitors, and one of its main features was a bevy of beautiful young women who paraded either stark naked or clad only in inessentials, with enormous feathered head-dresses and long spangled trains. Its other turns consisted of acrobats, singers, and sketches, the last either extremely vulgar or incredibly beautiful, the scenic resources of the theatre being vast, though its stage is only 11 by 6½ metres. Its name has become synonymous with French 'naughtiness', now somewhat out of date.

Folies-Dramatiques, Théâtre des, Paris, on the *boulevard du Temple. It opened in 1831 and had a successful career, catering mainly for a local audience with short runs of patriotic and melodramatic plays. In 1834 *Frédérick had one of his first successes there in his own play *Robert Macaire*. When the boulevard du Temple was demolished in 1862 a new Folies-Dramatiques was built in the rue de Bondy, which became a home of light musical shows.

Folk Festivals, seasonal celebrations connected with the activities of the agricultural year, particularly seed-time and harvest. They appear to date from the first organized communities and vary considerably in form. Some consist merely of a processional dance, as in the morris dance. In England the simpler forms of folk festival were connected with Plough Monday (the first Monday after Twelfth Night); May-Day (1 May), with its sports, Maypole, May Queen, and *Robin Hood play; Midsummer Day (24 June), with its fires derived from the Celtic festival of Beltane; and Harvest Home,

which celebrated the final gathering into barns of the year's grain harvest. Such festivals as these, found all over Europe, survived the rise and fall of Greece and Rome and the coming of Christianity, although the Church, considering them an undesirable pagan survival, tried either to suppress them or to graft them on to its own festivals. Wandering minstrels or other nomadic entertainers may have influenced them, but in the main they relied on local talent. The most elaborate form of folk festival is the *mumming play.

Folk Play, rough-and-ready dramatic entertainments given at village festivals by the villagers themselves. They were derived from the dramatic tendencies inherent in primitive *folk festivals, and were given on May-Day, at Harvest Home, or at Christmas, when to the central theme of a symbolic death and resurrection, which comes from remotest antiquity, were added the names and feats of local worthies. Later, though not before 1596, these were replaced by the Seven Champions of Christendom or other heroes, probably under the influence of the village schoolmaster (cf. Holofernes in *Love's Labour's Lost*). As patron saint of England, St George may have figured among them from the earliest times. With some dramatic action went a good deal of song and dance. Practically no written records of the folk play survive, and it contributed very little to the main current of modern drama, but its early influence should not on that account be entirely disregarded.

Follies, see REVUE and ZIEGFELD.

Folz, Hans, see MASTERSINGERS.

Fontanne, Lynn, see LUNT.

Fonvizin, Denis Ivanovich (1744–92), Russian dramatist, whose first attempt at playwriting was a comedy, *The Minor*, which pointed the contrast between the uneducated noblemen of the provinces and the cultured nobility of the city. He put it aside unfinished in favour of *The Brigadier-General*, which he read before the Court in 1766 with great success. It satirized the newly rich illiterate capitalists and also attacked the current craze for praising everything from Western Europe. It was not until 1781 that Fonvizin again took up *The Minor*, and virtually rewrote it, sharpening his satire on the landowners and their politics; this version is still in the repertory of the Soviet theatre. As the landowning party was then in the ascendant, Fonvizin's daring was his downfall and he was forced into premature retirement. Although he wrote in the comic tradition of *Molière and the French 18th century, he imparted to his work a native element of Russian folk comedy which was to be further exploited by his successors. His later plays included *A Friend of*

Honest People, or Starodum (1788) and *The Choice of a Tutor* (1790).

Fool, licensed buffoon of the medieval *Feast of Fools, later an important member of the *sociétés joyeuses* of medieval France, not to be confused with the *Court Fool. The traditional costume of the Fool, who was associated with such *folk festivals as the morris dance and the *mumming play (especially the Wooing Ceremony), was a hood with horns or ass's ears, and sometimes bells, covering the head and shoulders; a parti-coloured jacket and trousers, usually tight fitting; and sometimes a tail. He carried a *marotte* or bauble, either a replica of a fool's head on a stick or a bladder filled with dried peas, and he sometimes had a wooden sword or 'dagger of lath', like the 'Old *Vice' in the *morality plays, from whom he may have borrowed it. It is surmised that this costume was a survival of the old custom of animal disguising, when worshippers in primitive religious ceremonies wore the head and skin of a sacrificial animal, while the ass's ears were taken from the donkey who figured in the Feast of Fools.

Fools, Feast of, see FEAST OF FOOLS.

Foote, Samuel (1720–77), British actor and dramatist, well born and well educated, but so extravagant that in 1744 lack of money drove him to adopt the stage as a profession. In 1747 he took over the *Haymarket Theatre where, with great ingenuity, he evaded the Licensing Act of 1737 by inviting his friends to a dish of tea or chocolate, their invitation card giving them admittance to an entertainment in which Foote mimicked his fellow actors and other public figures. In 1749, having inherited a second fortune, he went to Paris, spent it, and returned to London to take up a life of hard work as actor-manager and playwright. He had already written a few farces but his first successful play was *The Englishman in Paris* (1753), with its sequel *The Englishman Returned from Paris* (1756). He then took over the Haymarket again and staged there his best play, *The Minor* (1760), a satire on the Methodists in which he himself played Shift, a character intended to ridicule Tate *Wilkinson. In 1766 while at the still unlicensed Haymarket he lost a leg through some ducal horseplay and the Duke of York, who was present at the accident, procured him a royal patent for the Haymarket in compensation. He was then able to present summer seasons of 'legitimate' plays, the first in 1767. He finally disposed of his patent to the elder *Colman, dying shortly afterwards on his way to France. Foote had a bitter wit, and his plays were mainly devised to caricature people he disliked, particularly *Garrick and Walpole, but he had such wonderful powers of mimicry and repartee that even his victims found themselves laughing. Although he had a keen eye for character, and wrote brilliant sketches of contemporary manners which caused him to be nicknamed 'the English Aristophanes', his plays were successful only through their topicality and have not survived. Short, fat, flabby, with an ugly but intelligent face, he was at once feared and admired by his contemporaries.

Footlight Players, see DOCK STREET THEATRE.

Footlights, see LIGHTING.

Footlights Club, see CAMBRIDGE.

Footlights Trap, long rectangular opening at the front of the stage before the curtain, with a post below at either end. Between the posts a framework slid up and down, and upon this the lamps of the footlights were arranged. They were lowered (according to an account of 1810) not only to enable a stage hand in the cellar to trim them, but so that the stage might, upon occasion, be darkened.

Footmen's Gallery, see GALLERY.

Forbes-Robertson, Jean (1905–62), English actress, second daughter of Johnston *Forbes-Robertson. She made her first appearance in 1921 in South Africa, touring with her mother under the name of Anne McEwen. In 1926 in London she was seen as Sonia in *Komisarjevsky's production of *Chekhov's *Uncle Vanya* under her own name which she thenceforth retained. In 1928 she scored a success as Helen Pettigrew in *Berkeley Square*, based on Henry *James's novel *The Sense of the Past*. From 1927 until 1934 she played the title-role in the Christmas production of *Barrie's *Peter Pan*, for which her slight, boyish figure made her eminently suitable, as it did also for Puck in *A Midsummer Night's Dream* (1937) and Jim Hawkins in Stevenson's *Treasure Island* (1945). She was for some time with the *Old Vic company, where she played a number of Shakespeare's heroines, and was much admired in the plays of *Ibsen, particularly as Hedda Gabler and as Rebecca West in *Rosmersholm* (both 1936). Among her other outstanding parts were Lady Teazle in *Sheridan's *The School for Scandal* and Marguerite in the younger *Dumas's *The Lady of the Camellias*. She also appeared in a number of modern plays, among them Rodney Ackland's *Strange Orchestra* (1932) and *Priestley's *Time and the Conways* (1937). After the war she was seen less often, making her last appearance as Branwen Elder in Priestley's *The Long Mirror* (1952).

Forbes-Robertson, Sir Johnston (1853–1937), English actor-manager. He studied art but gave it up for the stage, learning his perfect elocution from Samuel *Phelps. He made his first outstanding success at the *Haymarket Theatre in W. S. *Gilbert's *Dan'l Druce, Blacksmith* (1876). Two years later he joined the *Bancrofts, and was then engaged by Wilson *Barrett to play opposite *Modjeska. In 1882 he went to the *Lyceum under *Irving, later touring with Mary *Anderson with whom he made his first appearance in New York as Orlando to her Rosalind in *As You Like It* (1885). He returned to the Lyceum to give outstanding performances as the Duke of Buckingham in Irving's production of *Henry VIII* (1892), Romeo to the Juliet of Mrs Patrick *Campbell (1895), and, for the first time, Hamlet (1897), his voice, fine ascetic features, and graceful figure being ideally suited to the part. An outstanding later role, in the style of the time, was the Stranger in Jerome K. *Jerome's *The Passing of the Third-Floor Back* (1908), with which he inaugurated his final season at the *St James's. He made his last appearance at *Drury Lane as Hamlet in 1913. He married in 1900 Gertrude Elliott (see ELLIOTT, MAXINE) who acted with him until 1913. Their daughter Jean *Forbes-Robertson was also on the stage, as were Forbes-Robertson's three brothers.

Ford, John (1586–1639), English dramatist, many of whose plays are lost, some destroyed by *Warburton's cook. Ford was part-author with *Dekker and William *Rowley of *The Witch of Edmonton* (1621), and with Dekker alone of *The Sun's Darling* (1624). His own extant works include four romantic dramas in melancholy vein, written in a highly personal style of blank verse: *Love's Sacrifice* (1627), *The Lover's Melancholy* (1628), *The Broken Heart* (1629), and, his best-known work, *'Tis Pity She's a Whore* (c.1630). The last was revived in London in 1661 but not again until 1923, since when there have been several revivals, *Visconti directing it in Paris in 1961 under the title *Dommage qu'elle soit putain*.

Ford, John Thomson, see FORD'S THEATRE.

Forde, Florrie, see MUSIC-HALL.

Ford's Theatre, Washington, DC, originally a Baptist church, was taken over by **John T. Ford** (1829–94) and opened in 1862 as Ford's Atheneum. Destroyed by fire, it was rebuilt and reopened in 1863 as Ford's Theatre, becoming an immediate success, some of the finest American players of the time soon appearing there. On 14 Apr. 1865, however, President Abraham Lincoln was assassinated there during a performance by Laura *Keene's company of Tom *Taylor's *Our American Cousin*. The assassin was John Wilkes *Booth, brother of Edwin *Booth. After the assassination Ford and his brother were imprisoned for 39 days, but later acquitted of complicity in the crime. The theatre was officially closed and prevented from reopening by public outcry. The government therefore bought the property and used it for storage and office space. Part of the theatre collapsed in 1893 and the building remained derelict for some time, but in 1932 it became a Lincoln museum,

which it still remains, and in 1968 reopened as a theatre. The remodelled building outwardly resembles the original structure but with its capacity reduced to 741 from almost 2,000. Ford's Theatre now presents new or touring productions each season, many of which have gone on to Broadway success, including the Black musicals *Don't Bother Me, I Can't Cope* (1971 and 1974) and *Your Arms Too Short to Box With God* (1976). The management is committed to the development and preservation of musical theatre as a uniquely American art, although it has not forsaken non-musical works.

Forestage, small area of the stage in front of the *proscenium arch and the front *curtain, actors approaching it either from the main stage or through the *proscenium doors. Known as the apron stage, itself a modified version of the Elizabethan platform stage, it was the main acting area in Restoration times. Its loss proved a great handicap in staging revivals of Elizabethan and Restoration plays. In some new theatre buildings, such as the *Barbican Theatre, forestages were built out in front of the proscenium, or, as at the *Old Vic and the *Shakespeare Memorial Theatre, added to existing stages. In the *Lyttelton Theatre the proscenium arch can be adjusted to give a narrow forestage.

Formalism, theatrical method popular in Russia soon after the October Revolution, and practised by *Meyerhold, by *Akimov in his earlier period, and to a lesser degree by *Taïrov. It entailed the subjection of the actor so that he became nothing more than a puppet in the hands of the director, and insistence on exterior symbolism at the expense of inner truth. Though it originally helped to clear the stage of the old falsities and conventions of pre-Revolutionary days, when pushed too far it resulted in a complete lack of harmony between actors and audience which eventually caused it to be abandoned. It was succeeded by *Socialist Realism.

Formal Setting, see DETAIL SCENERY.

Formby, George, see MUSIC-HALL.

Forrest, Edwin (1806–72), American tragedian, who at 14 played Young Norval in *Home's *Douglas* at the *Walnut Street Theatre in Philadelphia, his birthplace. His early years were overshadowed by poverty and thwarted ambition, but in the end he became the acknowledged head of his profession in the USA for nearly 30 years. Even then the defects of his character made him as many enemies as friends, though no one denied the power of his acting. In his early years he was much criticized for 'ranting'; this he later cured to some extent, and was then outstanding as Lear, Hamlet, Macbeth, Othello, and Mark Antony in *Julius Caesar*, and in two parts especially written for him: Spartacus in *Bird's *The Gladiator* (1831) and the title-role in Stone's *Metamora* (1829). Among his other parts were Jaffier in *Otway's *Venice Preserv'd*, Rolla in *Sheridan's *Pizarro*, and the title-role in *Knowles's *Virginius*. He appeared in London in 1836 with some success, but in 1845 met with a hostile reception which he attributed to the machinations of *Macready. Their quarrel led eventually to a riot at the *Astor Place Opera House in New York in 1849. This caused Forrest to be ostracized by more sober members of the community, but he was the idol of the masses, who looked on him as their champion against the tyranny of the British. From 1865 his career declined and he died lonely and unhappy; he gave his last performance at the Globe Theatre, Boston, in 1872 in *Bulwer-Lytton's *Richelieu*.

Forrest Theatre, New York, see EUGENE O'NEILL THEATRE.

Fort, Paul, see SCENERY.

Fortune Theatre, London. 1. In Golden Lane, Cripplegate. This was built in 1600 for Edward *Alleyn and Philip *Henslowe to house the *Admiral's Men, of whom Alleyn was the leading actor. It was erected by Peter Street, who also built the first *Globe, and the contract for it is still in existence; it gives a good deal of valuable information, though a number of details are dismissed with the words 'like unto the Globe'. It is clear however that it was an 80-ft.-square timber building, almost certainly constructed within the walls of a former inn or dwelling house, and it had two galleries reputed to hold 1,000 people. It took its name from a statue of the Goddess of Fortune over the entrance. It was a popular playhouse, and drew a fashionable audience of nobles and distinguished foreign visitors. In 1621 it burnt down, and with the building perished also the wardrobe and all the playbooks. Two years later, rebuilt in brick and with a repertory of 14 new plays, it reopened, still tenanted by the same company. Though not as successful as before, the Fortune continued in use until all the theatres were closed in 1642. Even then it was used occasionally for illicit performances, until in 1649 it was raided by Commonwealth soldiers, who dismantled the interior. The building was finally demolished in 1661.

2. The old name was revived when in 1924 the present Fortune Theatre, in Russell Street, Covent Garden, opened under Laurence Cowen with his own play *The Sinners*. It seats 424 in three tiers, and has a proscenium width of 25 ft. It had few successes in its early years, except for *Lonsdale's *On Approval* (1927), which ran for 469 performances, and a season of plays by *O'Casey presented by J. B. *Fagan. During the Second World War it was occupied by ENSA. It reopened in 1946 but during the next 10 years housed more amateur productions than professional, though *Joyce Grenfell Requests the Pleasure* (1954) proved popular. In 1957 Michael Flanders and Donald Swann presented their musical entertainment *At the Drop of a Hat*, which had a long run, as did the revue *Beyond the Fringe* (1961). Other successes were Francis Durbridge's thriller *Suddenly at Home* (1971), the revival of the musical *Mr Cinders* (1983), John Godber's *Up 'n' Under* (1985), and *Re: Joyce* (1988), Maureen Lipman's re-creation of Joyce *Grenfell.

Fortuny, Mariano, see LIGHTING.

Forty-Fourth Street Theatre, New York, on West 44th Street. It opened in 1912 as the New Weber and Fields Music-Hall, seating 1,463, and was taken over a year later by the *Shuberts, who staged there a successful musical revival, *The Geisha*. In 1915 there was a season of classical repertory, but later the theatre reverted to light opera and musicals. The Marx Brothers had a long run in *Animal Crackers* (1928), and the outstanding production of 1930 was an adaptation of *Aristophanes' *Lysistrata*. High-lights of the 1930s were J. B. *Priestley's *The Good Companions* (1931), a four-week season by Walter *Hampden in 1935, and the short-lived but memorable *Johnny Johnson* (1936) by Paul *Green with incidental music by Kurt Weill. One of the last productions was Moss *Hart's musical documentary on aviation *Winged Victory* (1943). The theatre was demolished in 1945.

Forty-Ninth Street Theatre, New York, on West 49th Street. This small house opened in 1921 and in the following year presented *Balieff's La Chauve-Souris. A series of successful new plays followed, though *Wedekind's *Lulu* and *Coward's *Fallen Angels* both failed in 1927. A fine revival of *Ibsen's *The Wild Duck* (1928) with Blanche *Yurka was followed by her Hedda in his *Hedda Gabler* (1929) and by *Mei Lanfang (1930). Among later productions were *Drinkwater's *Bird in Hand* (1930), a revival of *Strindberg's *The Father* (1931) with Robert *Loraine, a season of Yiddish plays with Maurice *Schwartz, and three productions by the *Federal Theatre Project. The last play in the theatre was *The Wild Duck* in modern dress in 1938, after which it became a cinema, later demolished.

Forty-Sixth Street Theatre, New York, see RICHARD RODGERS THEATRE.

Forum Theatre, Manchester, see LIBRARY THEATRE.

Forum Theatre, New York, see VIVIAN BEAUMONT THEATRE.

Fox, G. L. [George Washington Lafayette] (1825–77), American actor and pantomimist, was on the stage as a child, and from 1850 to 1858 was at Purdy's National Theatre (see CHATHAM THEATRE 2) where in 1853 he persuaded the management to put on Aiken's adaptation of Harriet Beecher Stowe's *Uncle Tom's Cabin*, which was a great success. In 1858 he became joint manager of the *Bowery, where from 1862 to 1867 he staged a long series of successful *pantomimes. He then moved to the *Olympic, giving in 1867 a fine performance as Bottom in *A Midsummer Night's Dream*, and a year later producing his pantomime *Humpty-Dumpty* and a travesty of *Hamlet* which Edwin *Booth is said to have enjoyed. His *Wee Willie Winkie* (1870) also had a long run. Considered the 'peer of pantomimists', he continued to appear in successive editions of *Humpty-Dumpty* until his death.

Much of the pantomime 'business' in his shows was devised by his brother **Charles Kemble Fox** (1835–75), who was also a child actor. He was in *Uncle Tom's Cabin*, as was his mother, and played Pantaloon in his brother's pantomimes.

Foy, Eddie [Edwin Fitzgerald] (1856–1928), American actor and *vaudeville player. He was a singer and entertainer from childhood, and in 1878 toured the Western boom towns with a *minstrel show. He was later seen in comedy and *melodrama, and from 1888 to 1894 played the leading part in a long series of *extravaganzas in Chicago. He was acting in the Iroquois Theatre there in 1903 when fire broke out, and tried unsuccessfully to calm the audience. Foy, an eccentric comedian with many mannerisms and a distinctive clown make-up, played in *musical comedy until 1913 and then went into vaudeville, accompanied by his seven children, with whom he made his last appearance in 1927.

Fragson, Harry, see MUSIC-HALL.

France, Théâtre de, Paris, see ODÉON.

Frankfurt-am-Main, German city whose Easter and Autumn *fairs made it attractive to early touring companies, the *English Comedians playing there many times between 1592 and 1658. It had, however, an undistinguished theatrical history until it became a stronghold of *Expressionism with such productions at the Schauspiel as *Sternheim's *1913* (1919) and Bronnen's *Vatermord* (*Parricide*, 1922). In 1919 it saw the first performance of Hasenclever's *Antigone*, and with first productions at the Neues Theater of plays by *Kaiser—*Die Bürger von Calais* (1917), *Gas I* and *II* (1918/1920), and *Hölle Weg Erd* (*Hell, Path, Earth*, 1919)—Frankfurt became a serious rival to *Berlin. Between 1951 and 1968 productions of a number of *Brecht's plays when they were mainly being ignored by other West German theatres restored some of Frankfurt's earlier importance, and in the 1970s there were experiments in *Collective Creation under *Palitzsch and Rainer Werner Fassbinder.

Franz, Ellen, see MEININGER COMPANY.

Fraser, Claude Lovat (1890–1921), English artist and stage-designer, whose designs for Nigel *Playfair's productions of *As You Like It* and, particularly, *Gay's *The Beggar's Opera* at the *Lyric Theatre, Hammersmith, in 1920 inaugurated a new era in stage design. He took his inspiration from the 18th and early 19th centuries, and his work embodied a gay, brightly coloured romanticism. His influence, considering the brevity of his career, was phenomenal.

Frayn, Michael (1933–), English playwright and novelist, who was already well known for his novels when his collection of four short plays *The Two of Us* (1970) was produced in London. His full-length plays, mainly comedies with serious undertones, began with *The Sandboy* (1971). He achieved his first big success with *Alphabetical Order* (1975), set in the cuttings library of a provincial newspaper. *Donkeys' Years* (1976) deals with the reunion after 20 years of six former university students and *Clouds* (also 1976) is set in Cuba, its two main characters being rival male and female press reporters. Frayn achieved both popularity and critical praise with *Make and Break* (1980), about a work-obsessed salesman at a foreign trade fair; *Noises Off* (1982; NY, 1983), which hilariously depicts an incompetent theatrical troupe; and *Benefactors* (1984; NY, 1985), in which an architect's plans for an inner-city site affect his relationships with his wife and friends.

His translations of *Chekhov's *The Cherry Orchard* and *Tolstoy's *The Fruits of Enlightenment* were produced at the *National Theatre in 1978 and 1979 respectively; *Wild Honey*, his translation of Chekhov's *Platonov*, was produced there in 1984 (NY, 1986). He also translated Chekhov's *The Seagull* (1986), *Three Sisters* (1987), and *Uncle Vanya* (1988), and a group of his one-act plays and short stories presented as *The Sneeze* (also 1988). Other translations included *Anouilh's *Number One* (1984).

Frazee Theatre, New York, see WALLACK'S THEATRE 2.

Frédérick [Antoine-Louis-Prosper Lemaître] (1800–76), French actor, who embodied all the glory and excesses of the Romantic drama, many of whose heroes he created. He never appeared at the *Comédie-Française, but spent most of his career in the theatres on the *boulevard du Temple, having made his first appearance there as the lion in a pantomime, *Pyrame et Thisbé*, at the age of 15. He then went to the *Funambules, where he was a contemporary of the great mime *Deburau, and also attended classes at the Conservatoire. In 1820 he became a member of the company at the *Odéon but, finding the audiences apathetic and opportunities for sustained acting too few, returned thankfully to the popular stage. In 1823 he made his first appearance as Robert Macaire in *L'Auberge des Adrets*, a part ever after associated with him. The play had been written as a serious melodrama, but Frédérick carried it to success by burlesquing it. He was equally successful in a sequel, *Robert Macaire* (1834), much of which he wrote himself, and which had political repercussions; it was said to have contributed to the downfall of Louis-Philippe. At the *Porte-Saint-Martin in 1827, Frédérick had made a tremendous impression in *Trente Ans; ou, La Vie d'un joueur*, a play on the evils of gambling which he took to London on his first visit in 1828. Among his later successes were Othello, in *Ducis's translation, and several leading roles in the plays of the elder *Dumas, notably the title-role in *Kean; ou, Désordre et génie* (1836), which was specially written for him, as was Balzac's *Vautrin* (1840). In 1838 he gave an electrifying performance in *Hugo's *Ruy Blas*. His last years were unhappy, for the taste of the public changed and he could find only trivial or unsuitable plays to appear in; but he left his mark on the theatre, and when *Bateman first saw *Irving he could bestow no higher praise than to say that his acting was equal to that of Frédérick.

Fred Miller Theatre, see MILWAUKEE REPERTORY THEATRE.

Fredro, Aleksander (1793–1876), Polish poet and playwright, one of the few who saw his plays produced in Poland. The son of an aristocratic family, he served in Napoleon's Grand Army, and then settled down on his family estates. His comedies, written in strict classic form and impeccable verse, deal with the life of the country gentleman as he knew it or as he had heard of it from his elders. Among the best are *Mr Moneybags* (1821), the story of a social climber which somewhat resembles *Molière's *Le Bourgeois gentilhomme; Husband and Wife* (1822; London, 1957); *Ladies and Hussars* (1825), an excellent farce about the invasion of officers' quarters by a group of determined women, seen in translation in New York in 1925; and *The Life Annuity* (1835), again in the style of Molière, about a usurer's troubles with a young rake, presented in London during the *World Theatre Season of 1964. After a silence of some years, due to unjustified attacks on him by the young Romantics of the time, Fredro resumed writing, but none of his last dozen or so plays approached the excellence of his early comedies and they were not staged in his lifetime, though several, including *The Candle Has Gone Out*, were seen later.

Freear, Louie, see DROLL and DUKE OF YORK'S THEATRE.

Freedley, George Reynolds (1904–67), American theatre historian, founder and director of the Theatre Collection of the New York Public Library (see MUSEUMS AND COLLECTIONS). A graduate of George Pierce *Baker's '47 Workshop' at Yale, he was for a short time an actor and stage manager on Broadway, where he first appeared in 1928, being associated a year later with the *Theatre Guild. In 1931 he joined the staff of the New York Public Library to administer the theatre section based on the recently presented David *Belasco Collection, becoming its curator in 1938. He was also one of the founders of the Theatre Library Association in 1939 and was closely associated with the running of the *American National Theatre and Academy. From 1938 until his death he was drama critic and drama feature writer of the New York *Morning Telegraph*, and lectured and wrote extensively on theatre history.

Freie Bühne [Free Stage], private theatre club founded in *Berlin in 1889 by ten writers and critics under Otto *Brahm for the production of plays by the new writers of the school of *Naturalism, on the lines of *Antoine's *Théâtre Libre in Paris. It had no home of its own, and played only matinées in different theatres. Its first production was *Ibsen's *Ghosts*, and in 1890 it produced Holz's *Die Familie Selicke*, a sordid picture of lower middle-class life played in a realistic manner against an equally realistic background. The greatest achievement of the Freie Bühne was the discovery of Gerhard *Hauptmann, whose first play *Vor Sonnenaufgang* was produced in 1889. Several more of his plays, up to *Die Weber* (1893), also had their first performances there. Other authors staged included *Bjørnson, *Tolstoy, *Zola, *Strindberg, and *Anzengrüber. Along with the promotion of contemporary drama, the Freie Bühne also developed a new, realistic style of ensemble playing. When in 1894 Brahm was appointed director of the *Deutsches Theater, the Freie Bühne was attached to it as an experimental theatre. It continued to present new plays up to the turn of the century, but had by then fulfilled its mission as a platform for naturalistic drama.

Freie Volksbühne, Berlin, see VOLKSBÜHNE.

French Brace, French Flat, see FLAT.

Fresnay [Laudenbach], **Pierre** (1897–1975), French actor, trained at the Conservatoire, who made his stage début at the *Comédie-Française in 1915 in *Marivaux's *Le Jeu de l'amour et du hasard*, playing Britannicus in *Racine's play in the same year and making his first appearance in London with the company in 1922. He remained with it until 1926, and thereafter acted at a number of theatres. In 1934 he was seen in London with the Compagnie des Quinze (see SAINT-DENIS) and took over Noël *Coward's role in the first production of *Conversation Piece*

(NY, 1934), acting for the first time opposite his wife Yvonne *Printemps, with whom he starred in London again in Ben *Travers's *O Mistress Mine* (1936). With her he took over in 1937 the management of the Théâtre de la Michodière, where they appeared together in a long succession of light comedies. In 1935 he played the title-role in English in *Obey's *Noé* in New York. He was also a major film actor.

Freytag, Gustav (1816–95), German writer who, although best known as the author of sociological and historical novels, began his literary career as a dramatist, and became the German exponent of the *well-made play. His best work was a comedy, *Die Journalisten* (1852), a good-humoured portrayal of party politics in a small town during an election; his attempts at serious problem-plays and a historical tragedy in verse are less attractive. In 1863 he published his *Technik des Dramas*, with its famous pyramid, or diagrammatic plot, of a 'well-made' play.

Fridolin, see GÉLINAS.

Friedrichsstadt–Palast, see GROSSES SCHAU-SPIELHAUS.

Friel, Brian (1929–), Ulster dramatist, whose favourite theme is the interaction between various kinds of institutional failure—of government, Church, class, and family—and that of the individual personality. His play *This Doubtful Paradise* was produced by the Ulster *Group Theatre, Belfast, in 1959, but his first outstanding success was *Philadelphia, Here I Come!* (Dublin, 1964; NY, 1966; London, 1967), in which a despairing young Irishman contemplating emigration to America seeks some sign of regret from an unresponsive father. It was followed by *The Loves of Cass McGuire* (NY, 1966; Dublin, 1967); *Lovers* (Dublin, 1967; NY, 1968; London, 1969), consisting of two plays *Winners* and *Losers*; and *Crystal and Fox* (Dublin, 1968; NY, 1973). In *The Freedom of the City* (London and Dublin, 1973; NY, with Kate *Reid, 1974) three innocent people are shot dead during a Derry demonstration. *Faith Healer* (NY, 1979; Dublin, 1980; London, 1981) was followed by two of his finest plays: *Aristocrats* (Dublin, 1979; London, 1988), about the decline of a family of Catholic gentry; and *Translations*, premièred in Derry in 1980 by *Field Day Theatre Company of which he was co-founder. Produced in New York and London in 1981, it is set in a 'hedge school' in Donegal in 1833, telling with humour and compassion of the resentment caused by the arrival of a British Army unit to make the first maps, the rationalization of Celtic place-names into English symbolizing the cultural rape of Ireland. Friel's translation of *Chekhov's *Three Sisters* was staged in Derry in 1981, and his adaptation of *Turgenev's novel *Fathers and Sons* at the *National Theatre in 1987. He also wrote *The Communication Cord* (1983) and

Making History (1988), a companion piece to *Translations* which re-examines the life of an Irish national hero. *Dancing at Lughnasa* (NT, 1990) movingly shows five spinster sisters coping with poverty and frustration in Ireland in the 1930s.

Fringe Theatre in Britain offers a platform for productions that because of their political content, experimental nature, or unfamiliarity are unlikely to succeed in more conventional surroundings. The term dates from the late 1960s and derives from the activities on the 'fringe' of the *Edinburgh Festival. There are around 70 Fringe theatres in London, mostly away from the centre where rents are lower. They are usually small, their seating ranging from 40 to 200, and few were built as theatres: mostly they are found in converted warehouses or factories, in basements, or in rooms in public houses. Fringe theatres are less formal, less expensive, and usually less comfortable than ordinary theatres, and their audiences tend to be younger and more anti-Establishment. Being short of money and often housed in temporary accommodation, they tend to be transitory. The better established ones include the King's Head at Islington, the Soho Poly, the Orange Tree at Richmond (where many of James *Saunders's plays were presented), the Bush at Shepherds Bush, West London, and the Gate, Notting Hill, which specializes in rarely performed classics. Outside London Fringe theatre is housed in the studio theatres attached to many *civic theatres, the multi-purpose arts centres opened in the 1970s and 1980s, and sometimes in more makeshift quarters.

Numerous fringe groups tour these locations, among the most important being **Shared Experience** (founded 1975), best known for its adaptations of novels, such as *Zola's *Germinal*; **Cheek By Jowl** (1981), which presents lively productions of classical plays with minimal scenery, including British premières such as *Corneille's *The Cid* and *Ostrovsky's *It's All in the Family*; **Joint Stock** (1974) (see AUKIN, GASKILL, and HARE, DAVID), **Paines Plough** (1975), and the feminist **Monstrous Regiment** (1975), all committed to new work. Joint Stock works by *Collective Creation, and other fringe groups use it at times. (See also COMMUNITY THEATRE.)

Frisch, Max Rudolf (1911–91), Swiss playwright, a disciple of *Brecht, whose influence was apparent in *Nun singen sie wieder* (*Now They Sing Again*, 1945) and *Die chinesische Mauer* (*The Great Wall of China*, 1946), both episodically constructed, the first dealing with the sorrows and squalor of war, the second with the atom bomb. In complete contrast were the more romantic plays such as *Santa Cruz* (1946), a blend of dream and reality, and *Don Juan; oder, Die Liebe zur Geometrie* (1953), in which women are attracted to *Don Juan because he has no time for them, his only interest being in geometry. His short play *Biedermann und die Brandstifter* (1958) ran at the *Royal Court as *The Fire Raisers* (1961) and was seen in New York as *The Firebugs* (1963). It offers the paradox of a man worried about arson who nevertheless incites it. Frisch's next play *Andorra* (1961) is the story of a non-Jewish illegitimate child brought up as a Jew who is murdered by an anti-Semitic invader. It was produced in New York in 1963 and twice performed in London in 1964, in German by the Schillertheater company during the *World Theatre Season and by the *National Theatre company. *Biografie* (1967), seen in New York as *Biography: A Game* (1979), is a play on one of Frisch's recurrent themes, the problem of identity. The world première of his *Triptychon* was directed by *Axer in 1980.

Frohman, Charles (1860–1915), American manager, who like his brothers Daniel and Gustave (below) inherited a passion for the theatre. From selling souvenirs and programmes he graduated to various managerial posts. He first visited England in 1880 as business manager of Haverly's Minstrels, for whom he achieved royal patronage. His first great success came in 1888–9 with Bronson *Howard's *Shenandoah*, which laid the foundations of his fortune since he had shrewdly bought the American rights. In 1893 he opened the *Empire Theatre with a fine *stock company which he ran for many years. He soon had a controlling interest in many other theatres, in New York and elsewhere, and helped to organise the *Theatrical Syndicate. In 1896 he had his first London success with *Klein's *A Night Out* at the *Vaudeville and then joined forces with the *Gattis, with George *Edwardes, and with *Barrie, taking Edwardes's musical comedies and plays by Barrie and other notable playwrights to America. He took a long lease of the *Duke of York's Theatre in London and mounted many notable productions there, including for the first time Barrie's *Peter Pan* (1904); at one time he controlled five London theatres. Frohman died when the *Lusitania* was torpedoed in 1915: a memorial to him stands in the churchyard at Great Marlow, Bucks. A small, odd-looking man, charming and kindly, who never used a written contract, he left a name for fair dealing and a record of remarkable theatrical achievements.

Frohman, Daniel (1851–1940), American manager, who started work as a journalist but in 1880 became business manager of the *Madison Square Theatre. He went into independent management in 1885 with his brother **Gustave** (1855–1930) at the *Lyceum Theatre; here he brought together an excellent *stock company and was responsible for a number of

outstanding productions, including plays by *Pinero and H. A. *Jones. E. H. *Sothern scored his earliest successes under his management. When the old Lyceum closed in 1902 Frohman became owner of the new theatre bearing the same name and he was also manager of *Daly's New York theatre, 1899–1903. On the death of his youngest brother Charles (above) he took over the administration of his affairs in America.

Front of House, all those parts of a theatre used by the audience, as distinct from the performers, who are *back stage. They include the auditorium, passages, lobbies, foyers, bars, cloakrooms, refreshment-rooms, and the box-office or pay-box for the booking of seats, the whole being under the control of a front-of-house manager who also acts as host to important guests and is at all times concerned with the comfort and well-being of the audience.

Front-of-House (FOH) *lighting is any stage-lighting equipment placed on the auditorium side of the *proscenium arch. Its purpose was originally to illuminate the *forestage or far down-stage area by means of carbon arcs and, later, spotlights mounted on the front of the balcony or gallery. Modern theatre buildings have in addition provision for spotlights suspended from the ceiling of the auditorium, and many have permanent lighting positions in the walls as well as a lighting bridge in the ceiling outside the proscenium line.

Fry [Harris], **Christopher** (1907–   ), English dramatist, who took his stage name from his grandmother's maiden name. Originally a schoolmaster, he gained his early experience of the theatre in Oxford, and had already written a couple of pageants, a religious play about St Cuthbert entitled *The Boy with a Cart* (1937), and, for Tewkesbury, *The Tower* (1939), when he first attracted attention with *A Phoenix Too Frequent* (1946), a one-act *jeu d'esprit*. In 1948 *The Firstborn* (on the early life of Moses) was seen at the *Edinburgh Festival, *Thor, with Angels* at Canterbury, and *The Lady's not for Burning* at the *Arts Theatre in London, with Alec *Clunes as Thomas Mendip. This last (the 'spring' play in a planned tetralogy of the seasons), transferred to the *Globe with *Gielgud as Thomas, consolidated Fry's reputation, and seemed to herald a renaissance of *poetic drama on the London stage. It was followed in 1950 by *Venus Observed*, the 'autumn' play, which starred Laurence *Olivier, and a translation of *Anouilh's *L'Invitation au château* as *Ring round the Moon*, in which Paul *Scofield played the dual role of the hero and the villain. Fry then reverted to his earlier biblical vein with *A Sleep of Prisoners* (1951), planned for performance in a church and first seen at St Thomas's, Regent Street. In spite of the superb acting of Edith *Evans as the Countess Rosmarin, the 'winter' play, *The Dark is Light Enough* (1954), did not have the hoped-for success, and for some time Fry busied himself with translations, among them Anouilh's *L'Alouette* (on Joan of Arc) as *The Lark* (1955); *Giraudoux's *La Guerre de Troie n'aura pas lieu* as *Tiger at the Gates* (London and NY, 1955), with Michael *Redgrave as Hector; and Giraudoux's *Pour Lucrèce* as *Duel of Angels* (1958; NY, 1960).

It was by now obvious that the trend of the English theatre was away from poetry, and Fry's next play *Curtmantle* (on Henry II and Becket) was first produced in The Netherlands in 1961. It was seen in London a year later, but had only a qualified success, as did the 'summer' play *A Yard of Sun* (1970). Two further translations, of *Ibsen's *Peer Gynt* and *Rostand's *Cyrano de Bergerac*, were performed at the *Chichester Festival Theatre in 1970 and 1975 respectively: both employed Fry's characteristic form of free but regularly stressed verse with richly imaginative word-play.

Fugard, Athol (1932–), South African playwright, director, and actor, born of Anglo-Irish and Afrikaner parents, who studied philosophy and social anthropology at the University of Cape Town. His first play, *No Good Friday* (1959), was closely followed by *Ngogo* (translatable as 'a 35-cent woman'), the story of a mine-worker's whore. *The Blood Knot* (1961; London, 1963; NY, 1964) tackled the bitter topic of the temptation to pass for white in a racially segregated society. In 1963 Fugard and his wife were invited to help a group of non-white actors in the African township of New Brighton, Port Elizabeth, to form an amateur drama group, the Serpent Players. His first production there, *Machiavelli's *The Mandrake*, was followed by productions of plays by *Brecht, *Büchner, *Sophocles, and others, and work with the group concurrently enriched his own writing. *Hello and Goodbye* (1965; NY, 1969; London, 1973) and *Boesman and Lena* (1969; NY, 1970; London, 1971) form with *The Blood Knot* a trilogy about the lives of poor people in Port Elizabeth; *People are Living There* (1969; NY, 1971; London, 1972) is set in a cheap boarding house in Johannesburg. Working with the Serpent Players, Fugard explored *improvisation and *Collective Creation to put on plays presenting the Black South African point of view. This resulted in *The Coat* (1971), *Sizwe Bansi is Dead* (1972), dealing with the effects of the pass laws, and *The Island* (1973), a dialogue between two prisoners on Robben Island. The last two, played by the two actors credited as co-authors, John Kani and Winston Ntshona, were seen at the *Royal Court Theatre and in New York in 1974. The *Edinburgh Festival of 1975 commissioned *Dimetos*, a poetic allegory with an unlocalized setting; a production at the

*Nottingham Playhouse in 1976, with Paul *Scofield, transferred to London. *A Lesson from Aloes* (NY and *National Theatre, 1980) was a triumphant return to the South African environment. *Master Harold . . . and the Boys* (NY, 1982; NT, 1983) portrays a conflict between White and Black South African teenage friends; *The Road to Mecca* (NT, 1985) is about an elderly Afrikaner sculptress; but *A Place with the Pigs* (NT, 1988), in which a Red Army deserter hides in a pigsty for 41 years, marks a sharp change of setting and theme.

Fuller, Charles, see NEGROES IN THE AMERICAN THEATRE.

Fuller, Isaac (1606–72), English scene painter, who studied in Paris under François Perrier, probably at the new Academy there. He worked for the Restoration theatre and in 1669 painted a scene of Paradise for *Dryden's *Tyrannic Love*, later suing *Drury Lane for payment. He was awarded £335. 10s. 0d.—a large sum in those days—but his scene may also have been used for other plays.

Fuller, Rosalinde (1901–82), English actress who made her first success in America, where in 1922 she played Ophelia to the Hamlet of John *Barrymore. She first appeared in London in 1927, among her outstanding parts being the Betrothed in Raynal's *The Unknown Warrior* and Irina in *Chekhov's *Three Sisters*. She was also seen in a number of *Shaw plays, and in 1938–40 she was with Donald *Wolfit's Shakespeare company, playing many of Shakespeare's heroines. In 1950 she first appeared in a solo programme entitled *Masks and Faces*, consisting of her own adaptations of a number of short sketches from such authors as *Dickens, Maupassant, Henry *James, and Katherine Mansfield, performed in costume but with a minimum of scenery. She developed and specialized in this form of entertainment, touring all over the world, sometimes under the auspices of the *British Council, and adding new programmes to her repertoire, among them *Subject to Love* (1965) and *The Snail under the Leaf* (1975), the latter based on the life and work of Katherine Mansfield.

Full Scenery, stage set where all the parts of the stage picture belong to the current scene only and must be changed for another scene, as opposed to *detail scenery, where some elements only are changed against a Permanent Setting. Because of its cost, both in materials and labour, full scenery is now generally confined to single-set plays.

Fulton Theatre, New York, see HELEN HAYES THEATRE I.

Funambules, Théâtre des, Paris, playhouse on the *boulevard du Temple which derived its name from the Latin for rope-dancers. It began as a *booth for acrobats and pantomime, but in 1816 a permanent theatre was built on the site which under the existing laws had to have a *barker outside, while the actors, even in more serious roles, had to indulge in somersaults and handsprings. In the year it opened the famous actor *Frédérick made his stage début there, in *harlequinades, but its great days began in the 1830s with the appearance of *Deburau as *Pierrot, and lasted until his death. He was always surrounded by a good company, and the pantomimes in which he starred were given with wonderful scenery, *transformation scenes, and *trickwork, the stage being excellently equipped. The theatre was finally demolished by Haussmann in 1862 in his building of the great boulevards.

Furtenbach, Josef, see LIGHTING.

Futurism, artistic movement which emerged in Italy during the first decade of the 20th century. Embracing painting, sculpture, poetry, and the theatre, its main concern was to introduce contemporary ingredients into art, and notably to reflect the dynamic impact of technology. During the 1920s its theories became identified with Fascism, though in many of its experiments with language, and in its breaking-up of action, it prefigured the Theatre of the *Absurd. A number of futurist plays were performed at the Teatro degli Indipendenti.

Fyffe, Will, see MUSIC-HALL.

G

Gabrielli, Francesco, see SCAPINO.

Gaff, 19th-century term for an improvised theatre, in the poorer quarters of London and other large towns, on whose stage an inadequate company dealt robustly with a repertory of *melodrama. The entrance fee was a penny or twopence. The lowest type of gaff was known as a blood-tub. In Scotland it was called a 'geggie'.

Gaiety Theatre, London, at the east end of the Strand. It opened as the Strand Musick Hall in 1864, but its mixed bill of serious music and music-hall turns proved unsuccessful and it closed in 1866. A new building was then erected covering a larger area and seating 1,126 in three tiers. It opened as the Gaiety in 1868. The company included Nellie *Farren and Madge Robertson (the future Mrs *Kendal). The manager, John *Hollingshead, instituted a number of reforms, establishing regular morning matinées and doing away with the old *benefit system by paying his actors better salaries. One of the theatre's early successes was H. J. *Byron's *Uncle Dick's Darling* (1869) in which Henry *Irving played Reginald Chenevix, and in 1871 came the first joint work by *Gilbert and Sullivan—*Thespis; or, The Gods Grown Old*. The great feature of the Gaiety was its *burlesques, many of them written by Burnand and Byron, featuring the famous 'quartet'—Nellie Farren, E. W. Royce, Edward *Terry, and Kate Vaughan—first seen in Byron's *Little Don Cesar de Bazan* (1876). Fred Leslie joined the company in 1885 and his partnership with Nellie Farren lasted until 1891. Just before Hollingshead retired in 1885 he was joined by George *Edwardes, burlesques still occupying the theatre until in 1892 Edwardes transferred from the *Prince of Wales Theatre a new type of show called *In Town*, now considered the first English *musical comedy. It was followed by a series of similar shows—*The Shop Girl* (1894), *The Circus Girl* (1896), *A Runaway Girl* (1898)—all featuring in the chorus the 'Gaiety Girls', chosen for their good looks as well as their singing and dancing ability.

The Gaiety was a popular and successful theatre, but it was destined to fall victim to the Strand widening scheme. The theatre closed on 4 July 1903, its place being taken by a second Gaiety nearby, built by Edwardes on an island site at the west end of the Aldwych. A four-tier house holding 1,267, it opened on 26 Oct. 1903 with a musical comedy *The Orchid* starring Gertie *Millar. Among its later successes were *The Girls of Gottenberg* (1907) and *Our Miss Gibbs* (1909). The long run of musical comedies was broken when in 1921 *Maeterlinck's *The Betrothal* was staged, but it began again later in the year with *The Little Girl in Red*. A later success was *Love Lies* (1928) and in 1936 and 1937 Fred *Emney and Leslie *Henson had good runs in *Swing Along* and *Going Greek*. The last production was *Running Riot* (1938) which closed on 25 Feb. 1939. The building then stood empty until in 1945 Lupino *Lane bought it, hoping to reopen it; but in 1950 he was forced to sell it and in 1957 it was demolished.

Gaiety Theatre, Manchester, see HORNIMAN.

Gaîté, Théâtre de la, Paris. This playhouse became famous in the mid-19th century, when *Frédérick performed there regularly. Until 1862 it was situated on the *boulevard du Temple, having been opened by *Nicolet in 1764 for the performance of puppet-shows. Later it specialized in mime-plays and acrobatics. Named Théâtre de la Gaîté in 1795 after the upheavals of the French Revolution, and rebuilt in 1805, it was assigned to *melodrama by Napoleon's decree on theatres, and staged productions of many works by *Pixérécourt, its manager from 1823 to 1834. When the boulevard du Temple disappeared in the 1860s, the Gaîté was rebuilt nearby, adjacent to the boulevard Sebastopol, where it still stands and is known as the Théâtre de la Musique.

Galanty Show, see SHADOW-SHOW.

Galimafré, see BOBÈCHE.

Gallery, in the 19th-century theatre building the highest and cheapest seats in the house, usually unbookable. The seating generally consisted of wooden benches, in some cases without backs. The occupants of the gallery were in about 1752 nicknamed 'the Gods'; they often formed the most perceptive and always the most vociferous part of the audience. (Later the term was applied to the gallery itself rather than its occupants.) In the Restoration period the usual charge for the upper gallery was one shilling, but at *Dorset Garden and later at *Drury Lane footmen waiting for their masters were admitted free at the end of the fourth act. From 1697 Christopher *Rich, hoping to curry favour with the rougher element in the audience, allowed footmen to occupy the Drury Lane gallery without payment from the opening of the performance, which led to its being known as the Footmen's Gallery. The custom led to so much noise and disorder that it was finally abolished in 1737. The gallery survived until after the Second World War, but virtually all theatre seats are now bookable.

Galli-Bibiena, Galli-Bibbiena, Galli da Bibbiena, see BIBIENA.

Gallienne, Eva Le, see LE GALLIENNE.

Galsworthy, John (1867–1933), English novelist and dramatist, awarded the *Nobel Prize for Literature in 1932, mainly for his novels. His plays deal with questions of social justice in the fashion of his time. *The Silver Box* (1906; NY, 1907) highlights the inequality before the law of a rich thief and a poor one, and *Strife* (London and NY, 1909) explores the effect of an industrial strike. Both were well received and several times revived, as was his best-known play *Justice* (1910; NY, 1916), which includes a scene showing the effect of solitary confinement on its chief character, and was at the time credited with having led to a reform of prison practice. Of his later plays, in which the reformer seemed to have triumphed over the dramatist, the most successful were *The Skin Game* (London and NY, 1920), *Loyalties* (London and NY, 1922), and *Old English* (London and NY, 1924); his last play *The Roof* (1929; NY, 1931) failed. *Strife* was revived at the *National Theatre in London in 1978.

Galvin, Sidney, see LENO.

Galway Theatre, see MACLIAMMÓIR.

Gambon, Michael John (1940–), Irish-born British actor who trained for seven years as an engineer before turning to the theatre, making his début in Dublin in 1962, and playing small parts at the *National Theatre, 1963–7. At the *Birmingham Repertory Theatre, 1967–9, he began to play major roles (Macbeth, Othello, Coriolanus), and from there moved to the *RSC, 1970–1. He was seen in several West End productions, including *Ayckbourn's *The Norman Conquests* (1974) and *Just between Ourselves* (1977), but it was not until 1980, as *Brecht's Galileo at the National Theatre, that he achieved stardom, appearing there in the same year in *Pinter's *Betrayal*. In 1982 he played King Lear, and Antony in *Antony and Cleopatra*, for the RSC. By now widely acclaimed, he returned to the National to give excellent comic performances in Ayckbourn's *A Chorus of Disapproval* (1985) and *A Small Family Business* (1987), and in a revival of the old farce *Tons of Money* (1986). He was also impressive as the tragically jealous father in Arthur *Miller's *A View from the Bridge* (NT and *Aldwych, 1987) and as *Chekhov's Uncle Vanya (1988) in the West End, where in 1990 he starred in another Ayckbourn comedy, *Man of the Moment*.

Ganassa, Zan (?–c.1583), actor-manager of the *commedia dell'arte, whose real name was probably Alberto Naseli. He is first heard of in about 1568, and his company was one of the earliest to leave Italy and travel abroad. He was in Paris in 1571, and a year later took part in the festivities held in honour of the marriage of the King of Navarre (later Henri IV) to Marguerite de Valois. A curious painting now in the Bayeux Museum is believed to represent a performance by Ganassa's company assisted by the French King and members of the royal household. It was, however, in Spain that Ganassa made his most successful tours, being found there frequently during the 1570s, and there seems to be no doubt that he exerted a considerable influence on the nascent Spanish professional theatre.

García Lorca, Federico (1898–1936), Spanish poet and playwright, executed by firing squad at the opening of the Spanish Civil War. As a child he made and played with puppets in his own miniature theatre, and later produced puppet-plays in his native Granada. His first full-length play *El maleficio de la mariposa* (*The Butterfly's Curse*) was produced by *Martínez Sierra in 1920 at the Teatro Eslava. It was followed by a historical drama, *Mariana Pineda* (1927), and several light comedies, among them *La zapatera prodigiosa* (*The Shoemaker's Amazing Wife*, 1930). His fame rests on his tragic folk trilogy *Bodas de sangre* (1933), about a wife's elopement on her wedding night and the husband's revenge; *Yerma* (1934), in which a childless wife murders her husband; and *La casa de Bernarda Alba* (produced posthumously in 1945 in Buenos Aires), a powerful indictment of a society in which natural impulses are frustrated by Catholic morality. These have been presented all over the world, the first to be seen in an English translation being *Bodas de sangre*, as *Bitter Oleander* in New York (1935) and as *The Marriage of Blood* in London (1939); it has since been widely revived as *Blood Wedding*. *Yerma* was produced in London in 1957 and *The House of Bernarda Alba* in New York in 1951; it was not produced professionally in London until 1973. García Lorca's influence on the Spanish theatre has been most important, both through his own plays and productions and through his association with La *Barraca.

Garden Theatre, New York, in Madison Avenue. It opened in 1890 and in the following year saw the reappearance in New York of Sarah *Bernhardt in *Sardou's *La Tosca*. Other productions included *Gilbert's *The Mountebanks* (1893), a dramatization of Ouida's *Under Two Flags* (1901), and *Hauptmann's *The Weavers* (1915). In 1910 a season of Shakespeare's plays was given by Ben *Greet and his company, and in 1919 the theatre became the Jewish (later

Yiddish) Art Theatre. It was demolished in 1925.

Garnier, Robert (*c.*1545–90), French Renaissance dramatist, by profession a lawyer, who with his seven tragedies on classical models prepared the way for the great tragic writers of the next generation. The best is usually considered *Les Juives* (*c.*1580), based on the Old Testament story of Nebuchadnezzar's cruelty to Zedekiah after the fall of Jerusalem. All his plays give evidence of wide reading, keen perceptions, and great lyric gifts. Garnier's choruses, a traditional feature which he retained from classical tragedy, are particularly fine; and at its best his writing is forceful, imaginative, and fluent. He was also the author of the first French tragi-comedy, *Bradamante* (1582), on a theme from *Ariosto's *Orlando furioso*, which was still being acted in the 17th century.

Garrick, David (1717–79), one of the greatest English actors, who effected a radical change in the style of acting by replacing the formal declaration favoured by James *Quin with an easy, natural manner of speech. Of Huguenot descent, he showed an early inclination for the stage, and at the age of 11 appeared as Sergeant Kite in a schoolboy production of *Farquhar's *The Recruiting Officer*. Later he was sent to study under Dr *Johnson at Lichfield, accompanied him to London, and there indulged in amateur theatricals at the expense of his business career in the wine trade, which he soon abandoned. In 1741 he was playing small parts at *Goodman's Fields Theatre, where on 19 Oct. he made his formal début as Richard III with such success that he was soon drawing crowds. He was then engaged for *Drury Lane, embarking in 1742 on a triumphant career which continued until his retirement in 1776. A small man, with a clear though not resonant voice, he appeared unsuited to tragedy, but was nevertheless unsurpassed in the tragic heroes of contemporary works as well as in such great parts as Hamlet, Macbeth, and Lear. He was not at his best as Romeo, a part he soon resigned to Spranger *Barry, nor as Othello, partly because blacking his face deprived him of the marvellously expressive play of his mobile features. He was much admired in comedy, one of his earliest and most acclaimed performances being Abel Drugger in *Jonson's *The Alchemist*; he was also good as Benedick in *Much Ado about Nothing*. Garrick's fiery temper, vanity, and snobbishness, as well as his sudden rise to fame, brought him many enemies, among them Samuel *Foote who lampooned him mercilessly. He also had to contend with the petulance of unacted authors and disappointed small-part actors.

Garrick became joint manager of Drury Lane in 1747, and sole manager in 1774. His management was marred by two serious riots, the first in 1755, occasioned by the appearance of French dancers as war between France and England was about to break out, and the second in 1762, when the concession of 'half-price after the third act' was abolished but had to be restored. This led Garrick to retire for a time, and from 1763 to 1765 he travelled on the Continent with his wife, a Viennese dancer he married in 1749 after her appearances at the *Haymarket as Mlle Violette (also found as Violetta and Violetti). They were well received everywhere, particularly in France. Garrick's 'Shakespeare Jubilee' at *Stratford-upon-Avon in 1769 was remarkable for the number of odes, songs, speeches, and other effusions by Garrick and for the complete absence of anything by Shakespeare.

In spite of the cares of management and constant appearances on stage, Garrick was a prolific dramatist, vivacious and competent. The best of his works were the farces *Miss in her Teens; or, The Medley of Lovers* (1747), in which he played Fribble, and *Bon Ton*, usually known from its subtitle as *High Life above Stairs* (1775). Much of his energy was expended on rewriting old plays, the most successful being *The Country Girl* (1766), a bowdlerization of *Wycherley's *The Country Wife* (1675) which held the stage for many years. He was also a great writer of prologues and epilogues, for both his own and other men's plays; he was responsible for a production of *Hamlet* with the Grave-diggers omitted, of *King Lear* without the Fool, and of a *Romeo and Juliet* which allowed the lovers a scene together in the tomb before dying; and he concocted a *Katharine and Petruchio* and a *Florizel and Perdita* (both 1756) from *The Taming of the Shrew* and *The Winter's Tale* respectively. He had, however, to contend with the taste of the time and the prevailing ignorance and lack of appreciation of Shakespeare's genius.

Garrick made his last appearance on the stage in 1776 as Don Felix in Mrs *Centlivre's *The Wonder, a Woman Keeps a Secret*, which he had first played 20 years earlier and in which he was unequalled for spirit and vivacity. His death, felt as a personal loss by many admirers, drew from Dr Johnson the memorable epitaph: 'I am disappointed by that stroke of death which has eclipsed the gaiety of nations and impoverished the public stock of harmless pleasure.' His younger brother George, for many years his right-hand man at Drury Lane, died a few days later because, said the wits of the time, 'Davy wanted him'.

Garrick Club, London, gentleman's club, with membership restricted to 700, which has strong theatrical associations. Its first committee meeting was held in 1831, the club, with the Duke of Sussex as its patron, opening in the same year, though its premises in Probatt's Family Hotel, King Street, Covent Garden, were not ready for the use of members until 1832. The present club-house, opened in 1864, stands on part of old Rose Street. It has a fine collection of theatrical portraits, of which an annotated catalogue was prepared in 1908, and other stage memorabilia.

Garrick Theatre, London. **1.** In Leman Street, Whitechapel. This theatre, which took its name from its proximity to the old *Goodman's Fields Theatre where *Garrick made his début, opened in 1831. Burnt down in 1846, it was rebuilt and reopened in 1854. It held a very low position even among East End theatres, and was practically a *gaff. After the theatre had again been reconstructed, to hold 462 people on two tiers, it staged Bazin's *opéra bouffe Le Voyage en Chine* (1879) as *A Cruise to China* in which Beerbohm *Tree made his first professional appearance. The building fell into disuse after 1881.

2. In Charing Cross Road, near Trafalgar Square. This theatre, with a three-tier auditorium seating 800, opened in 1889 under John *Hare with *Pinero's *The Profligate*. Grundy's comedy *A Pair of Spectacles* had a long run in 1890, and five years later *The Notorious Mrs Ebbsmith*, also by Pinero, with Mrs Patrick *Campbell in the title-role, caused a sensation. Hare left after this production, and standards declined until in 1900 Arthur *Bourchier took over, and with his wife Violet *Vanbrugh inaugurated a long and brilliant period of productions ranging from Shakespeare to farce, among them *Knoblock's *Kismet*, starring Oscar *Asche, and *The Merry Wives of Windsor*. After Bourchier left in 1915 the theatre had no regular policy or management, and was used mainly for *revue. Closed in 1939, it opened again in 1941 and scored successes with Thomas Job's *Uncle Harry* (1944), Garson Kanin's *Born Yesterday* (1947), and the revues *La Plume de ma tante* (1955) and *Living for Pleasure* (1958). A transfer from *Theatre Workshop of the musical *Fings Ain't Wot They Used T'Be* in 1960 was followed by a long run of Charles Dyer's *Rattle of a Simple Man* (1962). From 1967 Brian *Rix presented and appeared in a series of farces. Ira Levin's thriller *Death Trap* ran from 1978 until 1981. Anthony Marriott and Alistair Foot's *No Sex Please—We're British* transferred there in 1982 for a long run, being followed by *Coward's *Easy Virtue* (1988) and *The Vortex* (1989).

Garrick Theatre, New York, in West 35th Street. This opened as Harrigan's Theatre in 1890. In 1895 Richard *Mansfield took it over and renamed it, opening with *Shaw's *Arms and the Man*, the longest run under his management being that of William *Gillette in his own play *Secret Service*. Gillette also appeared as Sherlock Holmes, a part always associated with him. Among later productions were *Captain Jinks of the Horse Marines* (1901) with Ethel *Barrymore,

The Stubbornness of Geraldine (1902), and *Her Own Way* (1903), all by Clyde *Fitch, while 1905 saw a successful run of Shaw's *You Never Can Tell*. From 1917 to 1919 the theatre was occupied by a French company under Jacques *Copeau, who named it after his theatre in Paris, the *Vieux-Colombier, and inaugurated his tenancy with a performance of *Molière's *Les Fourberies de Scapin*. In 1919 the *Theatre Guild opened at the Garrick, which had reverted to its old name; the Guild's productions were presented there until it opened its own playhouse, now the *Virginia Theatre, in 1925. The *Provincetown Players made their last appearance at the Garrick in 1929; it was demolished in 1932.

Gascoigne, George (c.1542–77), scholar of Cambridge who helped to prepare the entertainments given before Elizabeth I at Kenilworth and Woodstock in 1575. He had earlier prepared translations of Lodovico Dolce's *Giocasta* (based on *Euripides' *Phoenician Women*) as *Jocasta*, and of *Ariosto's *I suppositi* (based on *Plautus's *Captivi*) as *The Supposes*, both performed at Gray's Inn in 1566. *The Supposes*, which provided Shakespeare with the sub-plot of Bianca and her suitors in *The Taming of the Shrew*, was revived at Trinity College, Oxford, in 1582.

Gascon, Jean (1921–88), French-Canadian actor and director, born in Montreal. Abandoning a career in medicine he began acting with amateur companies, and from 1946 to 1950 studied in Paris with Ludmilla *Pitoëff and at the *Vieux-Colombier. In 1951 he co-founded the Théâtre du *Nouveau Monde (TNM), serving as its Artistic Director until 1966. Between 1960 and 1963 he was also the founding Director-General of the National Theatre School of Canada. At TNM he starred in *Molière's *L'Avare* and *Don Juan* and *Strindberg's *The Dance of Death*, and was known for his Rabelaisian productions of plays by such European authors as *Brecht, *Chekhov, *Beaumarchais, *Feydeau, *Musset, and John *Webster. As Artistic Director of the *Stratford (Ontario) Festival, 1967–74, he repeated many of these productions in English, as well as introducing such lesser-known plays of Shakespeare as *Cymbeline* and *Pericles*. He also instituted the company's Third Stage for more experimental and youthful work. In 1975 he directed a revival of John Coulter's *Riel*, about the Canadian folk-hero, for the National Arts Centre in Ottawa, and became its Director of Theatre, 1977–84. In the early 1980s he acted also as a consultant for Edmonton's Citadel Theatre complex. He died while directing rehearsals for Stratford's production of the musical *My Fair Lady*.

Gaskill, William (1930–), English director, who spent several years as an actor and stage manager in repertory before going to the

*Royal Court Theatre, where he directed N. F. Simpson's double bill *A Resounding Tinkle* and *The Hole* (1958) and his *One-Way Pendulum* (1959). After *Richard III, Cymbeline*, and *Brecht's *The Caucasian Chalk Circle* for the *RSC in 1961–2 he became an associate director of the *National Theatre company, his productions including *Farquhar's *The Recruiting Officer* (1963), Brecht's *Mother Courage and Her Children*, and *Arden's *Armstrong's Last Goodnight* (both 1965). Returning to the Royal Court as Artistic Director, 1965–72, he staged several controversial new plays, among them Edward *Bond's *Saved* (1965), *Early Morning* (1968), *Lear* (1971), and *The Sea* (1973). In 1974 he helped to found Joint Stock with the aim of involving dramatists in collaborative preparatory work with their actors. His intermittent association with the National Theatre continued with Farquhar's *The Beaux' Stratagem* (1970), *Middleton and *Rowley's *A Fair Quarrel* (1979), and *Pirandello's *Man, Beast and Virtue* (1989). He also staged *Congreve's *The Way of the World* (*Chichester and London, 1984) and *Marivaux's *Infidelities* (1987), in his own translation.

Gassman, Vittorio (1922–), Italian actor-manager, who made his first appearance on the stage in 1943 and soon proved himself an excellent actor in both comedy and tragedy, having a good presence and a finely modulated voice. In the late 1940s he played roles as diverse as Antony in the elder *Dumas's play of that name, Orlando in *As You Like It*, Kowalski in Tennessee *Williams's *A Streetcar Named Desire*, and the title-role in *Alfieri's *Oreste*, the last being accounted one of his best parts. Among his own productions, in most of which he starred himself, were *Ibsen's *Peer Gynt* at the Teatro Nazionale (1950–1) and *Hamlet* at the Teatro dell'Arti (1952–3). In 1954–5, with his own company the Teatro Popolare Italiano, he played two of his most famous roles, *Sophocles' Oedipus and *Kean in *Sartre's version of the elder Dumas's play; in 1956 he alternated the parts of Othello and Iago. The company, modelled somewhat on *Vilar's *Théâtre National Populaire, was one of the first in Italy to perform in unconventional locations: in 1960 he appeared with it in a circus tent in Rome in Manzoni's *Adelchi*, and it was seen in London in 1963 in a programme entitled *The Heroes*, incorporating excerpts from his extensive repertory. In later years Gassman excelled in comedy, especially in films.

Gate Theatre, Dublin, founded in 1928 by Micheál *MacLiammóir and Hilton *Edwards to fill the gap left by the *Abbey Theatre's concentration on Irish drama in the naturalistic mode. The company played for two seasons at the Abbey's Peacock Theatre, presenting seven plays during their first season including *Ibsen's

Peer Gynt, *O'Neill's *The Hairy Ape*, and *Wilde's *Salomé*, the last then banned in England. In 1930 the company moved to the Rotunda Buildings, seating 400, its first production there being *Goethe's *Faust*. Over the next few years the productions were equally ambitious, including *Shaw's *Back to Methuselah* in its entirety and *Aeschylus' *Oresteia*. The latter was translated by the sixth **Earl of Longford** [Edward Arthur Henry Pakenham] (1902–61), the Irish dramatist and director, who with his wife became patron of the Gate in 1931. The best of Lord Longford's own plays was probably *Yahoo* (1933), which, with Hilton Edwards as Jonathan Swift, had a great success in both Dublin and London, where the Longford company played seasons in 1935, 1936, and 1937. In 1936 a split developed between the founders of the theatre and the Longfords, as a result of which two separate companies played at the Gate for six months each until Lord Longford's death.

In 1940 Edwards and MacLiammóir, unable to make their usual overseas tours because of the Second World War, presented their plays at the larger Gaiety Theatre and after the war the Gaiety continued to be used. The Gate, under the general management of Lady Longford, was let out to various companies throughout the 1960s, and for a time was even closed. In 1969 a government subsidy was awarded for the renovation of the Gate Theatre under Edwards, MacLiammóir, and Lady Longford, and in 1971 it officially reopened with a production of *Anouilh's *Ornifle* as *It's Later than You Think*. The resident company mounted productions for six or eight months and made the theatre available to other companies for the remainder of the year. Particularly popular in the early 1970s were the vintage productions of *Goldsmith, Shakespeare, Shaw, and Wilde in which the company's style was probably seen at its best; Irish authors were far from dominant in the Gate's programmes, which were drawn from the full range of classic and contemporary drama. The theatre's golden jubilee in 1978 was marked by the publication of a history of the enterprise, *Enter Certain Players* by Peter Luke. After MacLiammóir's death in the same year the theatre went into decline, but under new management from 1984 it found new prosperity. It marked its diamond jubilee with a second production of Wilde's *Salomé*.

Gate Theatre, London, small experimental theatre club whose first home was in Floral Street, Covent Garden, on the top floor of a warehouse. Accommodating 96, it opened in 1925 with Susan *Glaspell's *Bernice* and had its first success a year later with *Kaiser's *From Morn to Midnight* translated by Ashley *Dukes. The company mounted 32 productions in two years before moving to new premises in Villiers

Street, off the Strand, where it occupied part of the site formerly known as *Gatti's-Under-the-Arches. The second Gate Theatre opened in 1927. The venture began well, with such productions as *Toller's *Hoppla* and *O'Neill's banned play *Desire under the Elms*, with Eric *Portman and Flora *Robson; but by 1933 it was obviously failing. In 1934 Norman *Marshall took over, presenting an enterprising programme of plays and the annual Gate *revue. The last play to be produced was Beckwith's *Boys in Brown* (1940), a picture of Borstal life. In 1941 the theatre was extensively damaged by bombing and it was not reopened.

Gate Theatre, Notting Hill, London, see FRINGE THEATRE.

Gatherer, functionary in the Elizabethan playhouse whose task it was to collect the spectators' entrance fees from that part of the theatre, usually the upper gallery and the boxes, allotted to the *housekeepers, or owners of the actual theatre building, as distinct from the *sharers in the assets of the company.

Gatti's, London music-hall. In 1856 the brothers Carlo and Giovanni Gatti, with Giuseppe Marconi, had a restaurant in Hungerford Market. When this was demolished to make way for Charing Cross Station, they used the compensation money to open another restaurant in the Westminster Bridge Road, which in 1865 was licensed as a music-hall. Meanwhile the arches under the new station were being let, and in 1866 the Gattis acquired two of them, which they opened a year later as another music-hall. To avoid confusion it was known as Gatti's-Under- (or In-) the-Arches, the other being nicknamed Gatti's-Over-the-Water or In-the-Road. This was rebuilt in 1883 to hold 1,183 people in two tiers and reopened as Gatti's Palace of Varieties; it was there that Harry *Lauder made his first London appearance in 1910. It closed in 1924 and was demolished in 1950. Gatti's-Under-the-Arches was renamed the Hungerford Music-Hall in 1883, and later became the Charing Cross Music-Hall. It closed in 1903 and in 1910 became a cinema. From 1927 to 1941 part of the site was occupied by the *Gate Theatre and in 1946 another part of the premises was taken over by the *Players' Theatre.

The Gattis also ran a restaurant, the Adelaide Gallery, a former theatre for which they were unable to get a performing licence, and for some years owned the *Vaudeville Theatre.

Gaultier-Garguille [Hugues Guéru] (c.1573–1633), French actor, chief farce-player, with *Gros-Guillaume and *Turlupin, of the company at the Hôtel de *Bourgogne to which he probably graduated from the Paris *fairs. He was a tall, thin man with a dry humour much appreciated by Parisians. Under the name of Fleschelles he also played serious parts, particularly kings in tragedy, but he is best remembered as a low comedian. He figures with other members of the company in Gougenot's *La Comédie des comédiens*, produced at the Hôtel de Bourgogne in 1631.

Gaussin [Gaussem], **Jeanne-Cathérine** (1711–67), French actress, daughter of *Baron's valet, who made her first appearance at the *Comédie-Française in 1731 as Junie in *Racine's *Britannicus*. She was at her best in roles demanding tenderness and grief rather than the sterner passions; with her dark, languorous eyes and rich voice (which, as a contemporary critic said, 'had tears in it'), she never lost her youthful look, playing young girls until her retirement in 1763. She appeared in several of *Voltaire's plays, notably *Zaïre* (1732), and was considered unrivalled in the sentimental comedies of *La Chaussée. In 1759 she made an unhappy marriage with a dancer at the Opéra, and retired from the stage four years later.

Gay, John (1685–1732), English poet and satirist, author of *The Beggar's Opera*, first given at *Lincoln's Inn Fields in 1728 under John *Rich, thus, it was said, 'making Gay rich and Rich gay'. A light-hearted mixture of political satire and burlesque of Italian opera, then a fashionable craze, this *ballad opera, one of the best of its kind, had music selected from well-known airs. It was constantly revived and the music rearranged, notably by Benjamin Britten for *Sadler's Wells in 1948. In 1928 a German adaptation was made by *Brecht as *Die Dreigroschenoper*, with music by Kurt Weill. *Polly*, a sequel to *The Beggar's Opera*, was not produced for many years owing to political censorship, but was finally given at the *Haymarket in 1777 with alterations by the elder George *Colman. Gay was also the author of several comedies.

The part of Polly in *The Beggar's Opera* was created by **Lavinia Fenton** (1708–60). She briefly became one of the best-known actresses in London, but retired in the same year.

Geary Theatre, San Francisco, see AMERICAN CONSERVATORY THEATRE.

Geddes, Norman Bel, see BEL GEDDES.

Geggie, see GAFF.

Gelber, Jack, see LIVING THEATRE.

Gélinas, Gratien (1909–   ), first outstanding actor of the French-Canadian theatre, known as Fridolin from the leading character in the 'Fridolinons', revues which he wrote for production at the Monument National, Montreal, from 1938 to 1946. In 1948 his first play *Tit-Coq* (*Lil' Rooster*) began a run of more than 450 performances in French and English which broke all Canadian records; it was filmed in

1952. In 1958 Gélinas founded the Théâtre de la Comédie Canadienne, housing it in a former burlesque theatre bought and redecorated with a donation from a brewery. The opening production was *Anouilh's *L'Alouette* (*The Lark*) in both French and English. Another of his own plays, the immensely successful *Bousille et les justes* (1959), was played in both languages in Montreal and on tour, as was *Hier, les enfants dansaient* (1966), about a family divided by separatism. The company also staged plays by other Canadian authors as well as a number imported from abroad. In 1956 Gélinas appeared with other French-Canadian actors at the *Stratford (Ontario) Festival, playing Charles VI in *Henry V* and Dr Caius in *The Merry Wives of Windsor*. He left the live theatre in 1970 to join the Canadian Film Development Corporation.

Gelosi, one of the earliest and best known of the *commedia dell'arte* companies, which after an initial visit to France in 1571 was summoned to play before the French king, Henri III, at Blois in 1577, and from there went to Paris, thus inaugurating the visits of the Italian players which resulted in the establishment of the *Comédie-Italienne. In the company, which was attached to the household of the Duke of Mantua, were Francesco *Andreini and his wife Isabella. After constant travelling the Gelosi returned to Paris in 1603, but on their way back to Italy in 1604 Isabella died at Lyons and her husband disbanded the troupe.

Gémier [Tonnerre], **Firmin** (1869–1933), French actor, director, and manager, who played a major part in the theatrical revival of the 1920s. A pupil of *Antoine, he made his first success as Ubu in *Lugné-Poë's production of *Jarry's *Ubu Roi* (1896). He was an inspired, if somewhat slapdash, actor, with great powers of rhetoric and ample gestures, and he had a remarkable ability for directing crowd scenes in which he also appeared. His interest in popular theatre was first shown in his production of Romain *Rolland's *Le 14 juillet* (1902), and in 1911 he founded the Théâtre Ambulant—a scheme which entailed moving around France by rail the entire equipment of a 1,700-seat theatre. During the First World War he helped to cement the Franco-British alliance by producing Shakespeare in Paris, and his sense of public responsibility, heightened by an important meeting with Max *Reinhardt, led him to attempt with the help of Gaston *Baty a series of gigantic dramatic events at the Cirque d'Hiver. In 1920 he became manager of the first *Théâtre National Populaire; at the same time he was managing the *Comédie des Champs-Élysées, and from 1921 to 1928 the *Odéon. During this last period he successfully directed spectacular performances of foreign classics by such authors as *Gorky, *Kleist, and *Tolstoy, as well as Shakespeare, and presented for the first time an eclectic assortment of new playwrights. Towards the end of his life he left the Odéon and returned to his original idea of a travelling theatre. Much of his pioneer work later bore fruit in the emergence of a second Théâtre National Populaire under Jean *Vilar and in the work of Jean-Louis *Barrault.

General Utility, see STOCK COMPANY.

Género Chico, or *teatro por horas*, generic term applied in Spain to a form of light dramatic entertainment dating from about 1868, which reached the height of its popularity towards the end of the 19th century. Deriving from the *sainete, it consisted at first of one-act scenes of everyday life, usually set in Madrid and heightened to the point of caricature. As the fashion for musical accompaniment became more widespread, *género chico* became synonymous with the one-act *zarzuela. Though still to be seen in Madrid and the provinces, it has been largely superseded by the *astracanadas*, sketches with wildly improbable plots and full of untranslatable puns and plays on words.

Genet, Jean (1910–86), French dramatist and poet, whose view of theatre as an act of revolt against society was conditioned by an early life spent largely in correctional institutions and prisons. His plays are now generally recognized as important not only for their extreme beauty of language but also for their creation of a disciplined, realistic form akin to that of the Theatre of Cruelty advocated by *Artaud. His first play *Les Bonnes*, directed by *Jouvet in 1947, introduced his conception of a play as ceremony and masquerade. Under the mask the characters act out their dreams and secret desires, thus demonstrating the nullity of what is usually termed 'reality'. The ceremony, or ritual, imposed on the masquerade is designed, on the analogy of the Catholic mass, to unite the spectators in a metaphysical experience beyond normal conceptions of good and evil. This conception was developed and strengthened in *Haute surveillance* (1949); *Le Balcon* (1956), a sexual-political ceremony in which a brothel becomes the focus for a revolution; and *Les Nègres* (1959; NY and London, as *The Blacks*, 1961). *Les Paravents*, first performed in Berlin in 1961, deals obliquely yet perceptively with the Algerian struggle for independence; it had a stormy reception at the *Odéon in 1966. As *The Screens* it was given an abbreviated workshop production by Peter *Brook in 1964. *Les Bonnes*, as *The Maids*, was seen in New York in 1955 and in London in 1956; *Haute surveillance*, as *Deathwatch*, appeared in New York in 1958 and in London in 1961; *Le Balcon* was produced as *The Balcony* in London in 1957, but not seen in France until 1960, when it was also seen in New York.

Gentleman, Francis (1728–84), Irish critic, dramatist, and actor, the best of whose criticisms were published anonymously in 1770 as *The Dramatic Censor* in two volumes. Gentleman's remarks on contemporary actors at the end of each article are more valuable than his somewhat verbose criticisms of the plays. From 1752 to 1755 he acted as 'Mr Cooke' in Bath, where two of his tragedies, *Osman* and *Zaphira*, were produced. He was seen as Othello in Edinburgh, where his most successful play, *The Tobacconist* (*c.*1760), an adaptation of *Jonson's *The Alchemist*, was first staged, being seen at the *Haymarket in 1771. In 1761 he appeared in his own play *The Modish Wife* at Chester, playing opposite *Macklin. Probably his best work, this was seen at the Haymarket in 1773.

George, Grace (1879–1961), American actress, wife of the manager William A. *Brady. She played the lead in many of her husband's productions, and was particularly admired as Lady Teazle in *Sheridan's *The School for Scandal* (1909), Barbara Undershaft in the first American production of *Shaw's *Major Barbara* (1915), and the title-role in St John *Ervine's *The First Mrs Fraser* (1929). Her only appearance in London was in 1907 as Cyprienne in *Sardou's *Divorçons*, which many considered her finest part. She was last seen with Katharine *Cornell in *Maugham's *The Constant Wife* (1951).

George Abbott Theatre, New York, see ABBOTT.

George M. Cohan Theatre, New York, at 1482 Broadway. This was opened in 1911 by *Cohan, who staged several of his own works there. Among later successes were the Jewish comedy *Potash and Perlmutter* (1913) by Charles *Klein and Montague Glass, the farces *It Pays to Advertise* (1914) by Walter Hackett, and *Come Out of the Kitchen* (1916) by A. E. Thomas, and several musical comedies, including *Two Little Girls in Blue* (1921). Clemence *Dane's *A Bill of Divorcement* (also 1921) established Katharine *Cornell as a star and was followed in the same year by Ed Wynn in his own musical comedy, *The Perfect Fool*, from which he took his nickname. The last years of the theatre were uneventful, the final production being a musical, *The Dubarry* (1932), with Grace Moore. A year later the building became a cinema, and in 1938 it was demolished.

Georgian Theatre Royal, Richmond, Yorkshire, one of the three oldest working theatres in England, the others being in *Bristol and *Bury St Edmunds. Small, with a rectangular auditorium originally seating 400, it is unique in still having its original proscenium. It opened in 1788, and in 1811 assumed the title of Theatre Royal apparently without justification. It closed in 1848, the lower part becoming a wine vault, the upper an auction room, and the sunken pit

being boarded over. In 1960 a Trust was formed to restore and redecorate it. It reopened in 1963 and now seats 229, the stage being 23 ft. wide including wings and 21 ft. deep, with a proscenium opening of 15.5 ft. The theatre is open on some days each week from mid-March to early December and there is also a Christmas show. It presents plays and concerts and is occasionally used by amateurs. A theatre museum was opened in 1979.

Germanova, Maria Nikolaevna (1884–1940), Russian actress, who made her début at the *Moscow Art Theatre in 1904 as Calpurnia in *Julius Caesar*. Her intensity and delicate appearance struck *Nemirovich-Danchenko as ideal for the more *Expressionist dramas with which he was trying to counterbalance *Stanislavsky's naturalism, and among her many successful roles were Agnes in *Ibsen's *Brand* (1906), the Fairy in *Maeterlinck's *The Blue Bird* (1908), Grushenka in Dostoevsky's *The Brothers Karamazov* (1910), and the title-role in *Andreyev's *Katerina Ivanovna* (1912), which was written for her. From 1922 to 1929 she was on tour with the Prague Group of the Moscow Art Theatre, visiting Paris in 1926 and London in 1928. She retired in 1930.

Gershwin, Ira [Israel Gershvin] (1896–1989), American lyricist, whose collaboration with his brother **George** [Jacob] (1898–1937) as composer produced a series of memorable musicals. Their first published collaboration was the song 'Waiting for the Sun to Come Out' in 1920, Ira using the pseudonym Arthur Francis to avoid trading on his brother's reputation, already established with songs such as Al Jolson's hit 'Swanee'. George continued to write with other lyricists until 1924, when the score for *Lady, Be Good!* (London, 1926) triumphantly renewed their partnership, Ira's colloquial and innovative lyrics combining with George's jazz-inspired score in such enduring songs as 'Fascinating Rhythm'. Their later work included *Oh, Kay!* (1926; London, 1927); *Funny Face* (1927; London, 1928), with ''S Wonderful' and 'My One and Only'; *Strike up the Band* and *Girl Crazy*, with 'I Got Rhythm' (both 1930). *Of Thee I Sing* (1931) was the first musical to win the *Pulitzer Prize, and in 1935 came their finest (and last) work, the Negro folk opera *Porgy and Bess* (London, 1952). After his brother's death Ira wrote the lyrics for only three more shows, the most notable being *Lady in the Dark* (1941) to the music of Kurt Weill. The brothers also wrote for the cinema, and George composed music for the concert hall.

Gershwin Theatre, New York, on 51st Street, west of Broadway. This theatre, seating 1,900, originally the Uris, was designed under new building laws which allow playhouses to be built within another type of structure. (The *Circle-in-the-Square theatre is in the basement.) It opened in 1972 with *Via Galactica*, a musical set in 2972, which ran for only a week, and for some years had a luckless career, though the *Gershwins' *Porgy and Bess* (1976) was performed there with virtually the complete score for the first time. *Sondheim's *Sweeney Todd* (1979) had a *succès d'estime*, and in 1981 the theatre scored a big hit with a production by *Papp of *Gilbert and Sullivan's *The Pirates of Penzance*. After changing its name in 1983 it housed a season by the *RSC (1984) and the musicals *Singin' in the Rain* (1985) and *Starlight Express* (1987).

Ghelderode, Michel de (1898–1962), Belgian dramatist, writing in French, who created in his plays a world which recalls the Flemish fairgrounds painted by Breughel and Bosch and the theatre of *Artaud. In grotesque decaying settings the characters act out a burlesque of the human condition, engulfed by the obscene deformities of the flesh which end in death. Several of his most important works, beginning with *Barabbas* (1928), one of a number with biblical subjects, were originally produced in Flemish by the Théâtre Populaire Flamand. Ghelderode was little known outside Belgium until after the Second World War, when he was 'discovered' by the avant-garde in Paris, which led to revivals of *Hop! Signor* (1935) in 1947, *Escorial* (1927) in 1948, and *Mademoiselle Jaïre* (1934) and *Fastes d'enfer* (1929) in 1949. The last caused such a scandal that it was eventually withdrawn. His numerous other plays include *Christophe Colomb* (1927) and *La Mort du Docteur Faust* (1928), part of a group inspired by *burlesque and *music-hall, and *Les Femmes au tombeau* (*The Women at the Tomb*, 1928), another biblical play. His work, which includes early plays for puppets and shows the influence of *Maeterlinck, is still almost unknown in Britain.

Ghéon [Vanglon], **Henri** (1875–1943), French religious dramatist. His early plays, among them *Le Pauvre sous l'escalier* (1913), were produced by *Copeau at the *Vieux-Colombier; but from 1920 he was engaged in writing and directing almost 100 plays on religious and biblical themes for schools, colleges, and parish churches. The best known of these were *L'Histoire du jeune Bernard de Menthon* (1925), translated by Barry *Jackson as *The Marvellous History of St Bernard* and produced at the *Birmingham Repertory Theatre and the *Malvern Festival in 1925, and *Le Noël sur la place* (1935), translated as *Christmas in the Market Place*, which had many amateur productions in Britain. It was originally presented by Les Compagnons de Jeux, a semi-amateur company organized in 1932. Like the best of Ghéon's work, it combines poetry and theatricality with a simplicity appropriate to his intended audiences.

Ghost Effects, see LIGHTING; PEPPER'S GHOST; SLOTE; and TRAPS.

Ghost Glide, see TRAPS.

Giacometti, Paolo (1816–82), Italian dramatist who wrote mainly on historical and social subjects. Among his early plays the most successful was *Elisabetta regina d'Inghilterra* (1853); of his social plays the best known is *La morte civile* (*The Outlaw*, 1861), which helped the Italian theatre to move away from the rhetorical tradition towards a more naturalistic manner. Like many of Giacometti's plays, which were designed to appeal to the new bourgeois audiences, it was essentially an actor's vehicle, being concerned with an escaped convict who returns to his wife and daughter after 15 years in prison; he finds them comfortably settled in a new life and rather than bring shame on them he commits suicide. A fusion of realistic and romantic elements, this remained a stock piece in the Italian theatre for some 50 years, and provided a major part for *Salvini, who toured in it extensively.

Giacosa, Giuseppe (1847–1906), Italian playwright, after *Verga the most important dramatist of the Italian school of *verismo*. Although an accomplished craftsman and an alert observer of contemporary life, his range was narrow, his psychological insight limited, and his ideas essentially derivative; but, seeking to assert the value of middle-class morality, he unconsciously painted an accurate and pathetic portrait of the decadence of the Italian bourgeoisie. A kindly and humane man, Giacosa was most at home in such charming and successful comedies as *Una partita a scacchi* (*A Game of Chess*, 1871), *La zampa del gatto* (*The Cat's Paw*, 1883), and *L'onorevole Ercole Malladri* (1884). With the historical play *Signora di Challant* (1891) he provided an excellent vehicle for both *Bernhardt and *Duse.

Gibbs, Wolcott (1902–58), American author who in 1939 replaced Robert Benchley on the *New Yorker* and soon became recognized as a valuable addition to the ranks of American dramatic critics. A shrewd and alert playgoer, he often wrote more brilliantly of bad plays than of good ones; but his irony was a tonic which Broadway badly needed, and though he wrote with humorous and sardonic detachment he never descended to 'wise-crack' reviewing. He was an excellent writer of parodies, and his play *A Season in the Sun* (1952) contains a dozen biting burlesques of contemporary novelists and playwrights.

Gide, André (1869–1951), French novelist and dramatist, whose early plays *Le Roi Candaule* (1899) and *Saül* (1902) made little stir, but whose *Œdipe* (1932) successfully reinterpreted *Sophocles' drama as a quest for humane values in a clerically dominated society. Although he reworked several other classical themes, his translations of *Antony and Cleopatra* (1920) and particularly *Hamlet* (1946), with which the *Renaud–*Barrault company opened their first season at the Théâtre *Marigny in 1946 and two years later visited the *Edinburgh Festival, were his best works for the theatre. It was also for Barrault that Gide prepared *Le Procès* (1947), a dramatization of Kafka's novel which Barrault revived in 1962 at the *Odéon.

Gielgud, Sir (Arthur) **John** (1904–), English actor and director, the foremost Shakespearian actor of his day, with a magnificent speaking voice. The grandnephew of Ellen *Terry, he made his first appearance on the stage at the *Old Vic in 1921 as the Herald in *Henry V*. In 1924 he became a member of *Fagan's repertory company (see OXFORD PLAYHOUSE), and in 1925 in London he took over the part of Nicky Lancaster in *Coward's *The Vortex* from the author, also succeeding Coward as Lewis Dodd in Margaret Kennedy's *The Constant Nymph* (1926). After Oswald in *Ibsen's *Ghosts* (1928) he joined the company at the Old Vic in 1929 to play among other parts Hamlet, his greatest role, which he was eventually to play over 500 times. In 1930 he played another part important in his career—John Worthing in *Wilde's *The Importance of Being Earnest*—and he then returned to the Old Vic. On the reopening of *Sadler's Wells Theatre in 1931 he played Malvolio in *Twelfth Night*, and later in that year he appeared in *Priestley's *The Good Companions*, the following year achieving his first popular success with Gordon Daviot's *Richard of Bordeaux*, which he also directed. In 1935 he alternated the parts of Romeo and Mercutio with Laurence *Olivier in a production which he also directed, achieving the longest run on record for this play. He followed it, after a brilliant performance as Trigorin in *Chekhov's *The Seagull* in 1936, with *Hamlet* in New York in the same year, whose 132 consecutive performances broke John *Barrymore's record. In 1937 he took over the *Queen's Theatre, where he headed a distinguished *repertory company in which his own roles were Richard II, Joseph Surface in *Sheridan's *The School for Scandal*, Vershinin in Chekhov's *Three Sisters*, and Shylock in *The Merchant of Venice*, also directing the first and last plays. Before and after playing Hamlet at the *Lyceum Theatre in 1939 he again appeared in *The Importance of Being Earnest*, which he also directed. In a repertory season at the *Haymarket (1944–5) he played Oberon in *A Midsummer Night's Dream*, Felix in *Webster's *The Duchess of Malfi*, Hamlet, and Valentine in *Congreve's *Love for Love*. He gave an outstanding performance as Raskolnikoff in Dostoevsky's *Crime and Punishment* (1946; NY, 1947), and then scored another personal triumph as Thomas Mendip in *Fry's *The Lady's not for

Burning (1949; NY, 1950). His first appearance at the *Shakespeare Memorial Theatre in 1950 was memorable for his Lear, Benedick (in *Much Ado about Nothing*), and Angelo (in *Measure for Measure*). They were followed by a memorable Leontes in *The Winter's Tale* (1951) in the West End and an excellent repertory season at the *Lyric, Hammersmith (1952–3), his roles there including Jaffier in *Otway's *Venice Preserv'd* and Mirabell in Congreve's *The Way of the World*; the latter he also directed.

Having played most of the great classic roles, he then turned his attention to modern plays, appearing in N. C. *Hunter's *A Day by the Sea* (1953) and Coward's *Nude with Violin* (1956), both of which he directed, and Graham *Greene's *The Potting Shed* (1958). In 1957 there was a superb Prospero in *The Tempest* at Stratford, and in 1958 an equally impressive Wolsey in *Henry VIII* at the Old Vic. In the same year he first presented his Shakespeare recital *Ages of Man* in North America, bringing it to London in 1959 and subsequently taking it on extensive world tours. He directed two productions of Chekhov—*The Cherry Orchard* (1961), in which he played Gaev, and *Ivanov* (1965; NY, 1966), in which he played the title-role—and appeared in new plays, among them *Albee's *Tiny Alice* (NY, 1964), Alan *Bennett's *Forty Years On* (1968), Peter *Shaffer's *The Battle of Shrivings* (1970), and, at the *Royal Court, David *Storey's *Home* (London and NY, 1970) and Charles *Wood's *Veterans* (1972). In 1974 he inaugurated Peter *Hall's directorship of the *National Theatre by appearing again as Prospero in *The Tempest*, but returned to modern drama with Edward *Bond's *Bingo* (also 1974), again at the Royal Court. The seedy Spooner in *Pinter's *No Man's Land* (1975; NY, 1976), also for the National Theatre, was one of his best modern roles. He was seen at the National in 1977 in the title-role of *Julius Caesar* and in Julian Mitchell's *Half-Life*. After the latter's West End run (1978) he was absent from the stage for 10 years, returning in Hugh Whitemore's *The Best of Friends* (1988).

Giffard, Henry, see GOODMAN'S FIELDS THEATRE.

Gilbert, Mrs George H. [*née* Ann Hartley] (1821–1904), American actress, who was in her youth a ballet dancer in London, where in 1846 she married George Gilbert, an actor and dancer. They emigrated to Wisconsin, but failed to make a success of farming and in 1850 returned to the stage, touring the larger cities. Gilbert was injured in a fall through a *trap, and though he continued to work in the theatre was unable to appear on the stage, dying in 1866. His wife found her niche playing comic elderly women, in which capacity she was with Mrs John *Wood's company at the *Olympic, New York, in 1864. Her first important part was the Marquise de St Maur in *Florence's pirated production of *Robertson's *Caste* (1867). The period of her greatest fame, however, was from 1869 to 1899, when she was in *Daly's company with Ada *Rehan and John *Drew. When Daly died she was engaged by *Frohman, with whom she remained until her death. Her angular body and homely features were assets in her skilled playing of eccentric spinsters and elderly dowagers.

Gilbert [Gibbs], **John** (1810–89), American actor, who made his début in Boston in 1828. He then toured the Mississippi river towns until 1834 and returned to the Tremont Theatre, Boston, remaining there until it closed, playing character roles. An excellent comic actor, he was particularly good as Sir Anthony Absolute and Sir Peter Teazle in *Sheridan's *The Rivals* and *The School for Scandal*. In 1847 he had a successful season in London and then returned to play at the *Park Theatre, New York, until it was destroyed by fire. He was with Lester *Wallack's company from 1861 until it was disbanded in 1888, dying the following year while touring with *Jefferson in *The Rivals*.

Gilbert, Sir William Schwenck (1836–1911), English dramatist, whose name is always associated with that of Sir Arthur Sullivan, for whose music he wrote the libretti of the *Savoy operas. Their collaboration began with *Thespis; or, The Gods Grown Old* (1871), and lasted for over 20 years, ending with *The Grand Duke* (1896), though *The Gondoliers* (1889) was the last of those which still hold the stage. The others are: *The Sorcerer* (1877), *HMS Pinafore* (1878), *The Pirates of Penzance* (1880), *Patience* (1881), *Iolanthe* (1882), *Princess Ida* (1884), *The Mikado* (1885), *Ruddigore* (1887), and *The Yeoman of the Guard* (1888). *Utopia Limited* (1893) was never revived. Gilbert wrote libretti for other composers but without success, just as Sullivan wrote music for other librettists but failed to recapture the brilliance of the Savoy operas. Yet their partnership was not a happy one, Gilbert being a man of irascible temperament and a martinet at rehearsals. He was already known as a dramatist before joining forces with Sullivan, having been encouraged by T. W. *Robertson to write for the stage. His early works were mainly *burlesques and *extravaganzas, the first being *Dulcamara; or, The Little Duck and the Great Quack* (1866). Of his more serious plays the most successful were *The Palace of Truth* (1870), *Pygmalion and Galatea* (1871), *Dan'l Druce, Blacksmith* (1876), whose title-role was long a favourite with character actors, and *Engaged* (1877). He continued writing almost up to his death, but of all his works only the librettos for Sullivan have survived.

Gilder, (Janet) **Rosamond de Kay** (1891–1986), American drama critic, daughter of the distinguished poet and editor Richard Watson Gilder, whose letters she edited in 1916. She

was actively concerned with the foundation and work of the *Federal Theatre Project, the *American National Theatre and Academy (ANTA), and the Institute for Advanced Studies in the Theatre Arts, and for many years headed the US delegations to the Congresses of the *International Theatre Institute, of which she was one of the founders. She was also its President, 1963–7, and in 1968 was nominated first President of the independent International Theatre Institute of the United States, Inc. From 1924 to 1945 she was on the staff of the distinguished journal *Theatre Arts Monthly*, of which she was editor, 1945–8, when it changed hands, and wrote copiously for it on all aspects of American and international theatre. She was the author of several works on the theatre.

Gill, Peter (1939–   ), Welsh director. He was an actor from 1957 to 1965, the first play he directed being D. H. *Lawrence's *A Collier's Friday Night* at the *Royal Court in 1965, one of a trilogy of Lawrence's plays he directed there. He worked mostly at the Royal Court until in 1977 he was appointed the first director of *Riverside Studios. His productions there included the highly praised opening production of *Chekhov's *The Cherry Orchard* (1978), *Middleton and *Rowley's *The Changeling* (also 1978), and *Measure for Measure* (1979). In 1980 he joined the *National Theatre as Assistant Director, his productions including *Turgenev's *A Month in the Country* (1981), *Shaw's *Major Barbara* (1982), *Shepard's *Fool for Love* (1984), and *O'Casey's *Juno and the Paycock* (1989). He was Director of the National Theatre Studio, 1984–90, and remains an associate director of the National Theatre. He has also written and directed his own plays, including the partly autobiographical *The Sleepers' Den* (1965) and *Over Gardens Out* (1969), *Small Change* (1976), and *Mean Tears* (1987), about a fraught homosexual relationship.

Gillette, William (1855–1937), American actor and dramatist, author of a number of adaptations and dramatizations of novels in most of which he appeared himself, among them Conan Doyle's *Sherlock Holmes* (1899), with which his name is always associated. He played Holmes with outstanding success in both England and America, and frequently revived it up to his retirement in 1932. Of his original plays the best were melodramatic spy stories of the Civil War, *Held by the Enemy* (1886) and *Secret Service* (1896). He also wrote a comedy, *Too Much Jonson* (1894). He appeared in *Barrie's *The Admirable Crichton* and *Dear Brutus*; but his best work was done in his own plays, the only one of which to have been revived is *Sherlock Holmes*, which had a notably successful production by the *RSC in London and New York in 1974.

Gingold, Hermione, see REVUE.

Giraldi, Giovanni Battista (1504–73), Italian humanist, known as 'il Cinthio'. He was the author of a number of 'horror' tragedies written under the influence of *Seneca, of which the first, *Orbecche* (1541), was performed before the Duke of Ferrara at the house of Alfonso Della Viola. Two of Giraldi's later plays, *L'altile* (1543) and *Epizia* (1547), were based on stories which he later included in his *Hecatommithi* (1565), a collection of *novelle* in the style of Boccaccio's *Decameron*. It was from this collection that Shakespeare, who had already used the story of Epizia (translated as *Promos and Cassandra,* 1578) in his *Measure for Measure*, took the plot of *Othello*. For him, as for other English dramatists of the time, Giraldi opened up chivalric sources and provided romantic situations suitable for tragedy and tragi-comedy; but in Italy his work served to strengthen the stranglehold of the neo-classical form.

Giraudoux, (Hippolyte) **Jean** (1882–1944), French novelist and dramatist, whose fruitful collaboration with the actor-director *Jouvet produced some of the finest plays of the period. Giraudoux was already a well-known novelist when his first play *Siegfried* (1928) was produced at the *Comédie des Champs-Élysées. After *Amphitryon 38* (1929), which, in an English adaptation by S. N. *Behrman, had a great success in England and America with the *Lunts as Jupiter and Alcmena, and *Judith* (1931), a biblical tragedy, came the enchanting *Intermezzo* (1933), a mixture of fantasy and realism seen in London in 1956 in a production by the *Renaud–*Barrault company. In *La Guerre de Troie n'aura pas lieu* (1935), Giraudoux traced the causes of the Trojan War to a lie and a misunderstanding. As *Tiger at the Gates* (with Michael *Redgrave playing Hector) this made a belated appearance in London and New York in 1955 in a translation by Christopher *Fry. In 1939, *Ondine*, a retelling of the legend of the water-nymph, began a successful run which was cut short in 1940; it was not revived until 1949. It was seen in French in London in 1953 and in English in an *RSC production in 1961. In 1942 *L'Apollon de Bellac* was produced in Rio de Janeiro by Jouvet, who directed its first production in France in 1947. In 1943 came the première in Paris of *Sodome et Gomorrhe*, in which Lia, the leading role, was played by Edwige *Feuillère. *La Folle de Chaillot*, a fantasy produced posthumously in 1945, re-established Giraudoux's magic. The play was an instant success in Paris and, as *The Madwoman of Chaillot*, was seen in New York (1948) and London (1951). Giraudoux's last play was *Pour Lucrèce* (1953), directed by Barrault at the *Marigny and staged in London as *Duel of Angels* (1958) with Vivien *Leigh as Paola, played originally by Edwige Feuillère. Revivals of some of Giraudoux's earlier plays have shown that their distinctive blend of verbal and theatrical fluency continues to be attractive.

Gitana, Gertie, see MUSIC-HALL.

GITIS, Moscow, see SCHOOLS OF DRAMA.

Glasgow. Like *Edinburgh, with which it shared theatrical managements until the early years of the 19th century, Glasgow had a long struggle to achieve a permanent playhouse. The first, erected in 1753, was soon dismantled in the face of religious criticism; a fire was started in the second on the eve of the opening performance in 1764. Damage was slight, and the theatre survived until burnt down in 1780. The famous Dunlop Street Theatre was opened in 1782, but was eclipsed when the Queen Street Theatre Royal was built in 1805, where Edmund and Charles *Kean first played together. The Dunlop Street Theatre became a warehouse. Part of the building was, however, used for miscellaneous entertainments and in 1824 it was all brought into use again, as the Caledonian Theatre. It became the Theatre Royal in 1829, after the Queen Street theatre was burnt down. In 1849 a false alarm of fire caused a disastrous panic in which at least 65 lost their lives. Burnt down in 1863, the theatre was rebuilt, but demolished in 1869. Other theatres in Glasgow were the Adelphi, opened in 1842 and destroyed by fire in 1848, the City, opened and burnt down in 1845, the Prince's (1849), and the Royalty (1879). In 1867 a music-hall was built which in 1869 became the Theatre Royal, and was rebuilt on the same site after its destruction by fire in 1879 and again in 1895. The *Glasgow Repertory Theatre, 1909–14, was succeeded by the *Scottish National Players, whose work was taken over in the 1930s by the amateur Curtain Theatre company. The last became the main vehicle for the production of new plays in the vernacular, among them those of Paul Vincent *Carroll, its co-founder, then living in Glasgow. It was disbanded in 1940. The *Unity Theatre, founded in 1941, was active in the 1940s. Present Glasgow theatres include the King's Theatre, the Theatre Royal, now a touring theatre and the home of Scottish Opera, the *Citizens' Theatre, and the *Tron Theatre. In 1988 a former tram depot was converted into a large performance space, the Tramway, which opened with Peter *Brook's production of the *Mahabharata*.

Glasgow Repertory Theatre opened in 1909 with a production of *Shaw's *You Never Can Tell* in the Royalty Theatre. For five seasons the excellent company produced works by many British and Continental dramatists, including *The Seagull* (1909), claimed as the first performance of *Chekhov in English. In its initial efforts to foster a purely Scottish drama it produced a few plays in the vernacular, but its Scottish commitment was never very strong. The First World War forced it to close, the remaining funds being transferred to the St Andrew Society and later used to help launch the *Scottish National Players.

Glaspell, Susan (1882–1948), American novelist and dramatist, active in the formation of the *Provincetown Players. Several of her one-act plays, some written in collaboration with her husband, were produced by the company, among them *Suppressed Desires* (1915) and *Tickless Time* (1918) which respectively mocked psychoanalysis and excessive idealism. Her full-length plays include *Bernice* (1919), *The Inheritors* (1921), which compares two generations of Americans, and *The Verge* (also 1921), an experimental study of female stress. Her finest work for the theatre was *Alison's House* (1930; London, 1932), based partly on the life of the American poet Emily Dickinson. Produced by Eva *Le Gallienne at the Civic Repertory Theatre in New York, this was awarded the *Pulitzer Prize for drama.

Globe Theatre, London. **1.** On the south side of Maiden Lane, Bankside, in Southwark, the theatre most intimately associated with Shakespeare, who was one of the *Housekeepers, or owners of a share in it. It was built in 1599 by Cuthbert *Burbage of materials taken from the *Theatre, built by his father, and was the largest and best known of the Elizabethan playhouses. It housed the company known as the *Chamberlain's Men, led by Richard *Burbage. Being largely open to the elements, it was used only during the summer months, the company transferring after 1613 to the *Blackfriars Theatre for the winter. All that is known of its interior is that it had a thatched roof over the upper gallery and that some of its dimensions were the same as those of the *Fortune. In 1613, during or shortly after a performance of *Henry VIII*, it caught fire and was burned down. Rebuilt in substantially the same style, but with a tiled roof in place of the thatch in which the fire was believed to have originated, it reopened in 1614 and remained in constant use until all the London theatres were closed in 1642. When the Burbages' lease ran out in 1644 the building was demolished. A reconstruction close to the original site brought about by the initiative of Sam *Wanamaker is scheduled to open in 1992. (See also TOKYO GLOBE.)

2. A four-tiered auditorium holding about 1,800 which stood in Newcastle Street at the east end of the Strand near the *Opera Comique, the two theatres being known as the Rickety Twins. It opened in 1868 and had a success in 1876 with *Jo*, a play based on *Dickens's *Bleak House*. In 1884 *Hawtrey's *The Private Secretary*, with *Penley in the lead, achieved immense popularity, becoming one of the classic stage farces of all time. In 1893 Penley returned to

the Globe, which had in the meantime housed *Mansfield in 1889 and Frank *Benson's company in 1890, with another classic farce, Brandon *Thomas's *Charley's Aunt*, which ran for four years. In 1898 John *Hare became manager, his most interesting production being *Pinero's *The Gay Lord Quex* (1899), with Irene *Vanbrugh. The last production, in 1902, was a revival of *Kester's *Sweet Nell of Old Drury*, with Fred *Terry and Julia *Neilson, after which the theatre, long considered a fire hazard, was demolished as part of the Strand widening scheme.

3. In Shaftesbury Avenue, built for Seymour *Hicks, whose name it first bore, having a three-tier auditorium, a proscenium width of 30 ft., and stage depth of 36 ft. It opened under the management of Charles *Frohman in 1906 and, renamed in 1909, remained Frohman's London headquarters until his death in 1915. Later successes included *Lonsdale's *Aren't We All?*, *Maugham's *Our Betters* (both 1923), and *Coward's *Fallen Angels* (1925). The appearance in 1930 of *Moissi in *Hamlet* was followed by the *Pitoëffs in *Shaw's *Saint Joan*, *Fagan's *The Improper Duchess* (1931) with Yvonne *Arnaud, and St John *Ervine's *Robert's Wife* (1937). *Rattigan's *While the Sun Shines* (1943) was a wartime hit and later came *Fry's *The Lady's Not for Burning* (1949) with John *Gielgud; *Anouilh's *Ring Round the Moon* (1950) in which *Scofield scored his first West End success; Graham *Greene's *The Complaisant Lover* (1959); and *Bolt's *A Man for All Seasons* (1960), with Scofield as Sir Thomas More. In 1966 Terence Frisby's *There's a Girl in my Soup* began a long run. The theatre also staged the work of Peter *Nichols (*Chez nous*, 1974, and *Born in the Gardens*, 1980); Michael *Frayn (*Donkeys' Years*, 1976); and Alan *Ayckbourn (*The Norman Conquests*, 1974, *Ten Times Table*, 1978, and *Joking Apart*, 1979). Glenda *Jackson and Joan *Plowright starred there in *García Lorca's *The House of Bernarda Alba*, and Maggie *Smith in Peter *Shaffer's *Lettice and Lovage* (both 1987).

Globe Theatre, New York, see LUNT–FONTANNE THEATRE.

Glory, see DRUM-AND-SHAFT SYSTEM.

Gobo, see LIGHTING.

Godfrey, Charles, see MUSIC-HALL.

Godfrey, Thomas (1736–63), American dramatist, who became the first playwright of the United States when in 1759 he wrote a tragedy entitled *The Prince of Parthia* which he sent to *Douglass, manager of the *American Company. It was received too late for production during the current season in Philadelphia, and Godfrey died before it was produced by Douglass for one night only at the *Southwark Theatre in 1767. It was then not acted again until its revival at the University of

Pennsylvania in 1915. It was published in 1765, and shows the influence of the plays which were in the repertory of the elder *Hallam in about 1754.

Gods, see GALLERY.

Godwin, Edward William, see TERRY, DAME ELLEN.

Goethe, Johann Wolfgang von (1749–1832), German man of letters, statesman, scientist, and philosopher, who devoted much of his time and genius to the theatre. He had already written a couple of unimportant comedies when he was introduced to the works of Shakespeare. Immediately captivated by them, he expressed his enthusiasm in the first German play to be written in the Shakespearian style, *Götz von Berlichingen mit der eisernen Hand.* Produced in Berlin in 1773, this somewhat idealized portrait of a robber baron became the spearhead of the *Sturm und Drang* movement, a model for young dramatists, particularly *Schiller, and the prototype of a wave of *Ritterdrama* which swept across Germany. Later plays included *Stella* (1776); the domestic tragedy *Clavigo* (1779), set in Spain and based on a true incident in the life of *Beaumarchais, seen in London in the *World Theatre Season of 1964; and *Egmont,* not completed until 1787. Dealing with the revolt of The Netherlands against Spain in 1567, the play was first performed, with extensive alterations by Schiller, in 1791. The original version was not seen until 1810.

In 1775 Goethe accepted an invitation from the reigning duke to settle in *Weimar, where he remained for the rest of his life. One of his duties was to organize ducal entertainments, and among the plays he directed was the first (prose) version of his *Iphigenie auf Tauris* (1779). During two years' sojourn in Italy, 1786–8, he recast it in verse, superimposing on the situation of *Euripides' Iphigenia the moral vision of the 18th century. The play, one of the masterpieces of European drama, is an expression of Goethe's belief that the salvation of mankind can come only through humanity and renunciation. In its new form it was produced in 1802. The visit to Italy was also responsible for *Torquato Tasso* (1807), which portrays a poetic temperament in conflict with the world of action. These two classical plays are cast in iambic verse of exquisite mellowness and have an intimate, personal appeal lacking in Goethe's earlier works. After his visit to Italy the first signs of the Olympian detachment which was such a feature of his old age began to appear. In 1791 he was appointed director of a professional company established in the Court theatre, which became known throughout Europe.

The crowning achievement of Goethe's career, as both playwright and poet, and in some ways as philosopher, was his *Faust.* Begun in the early 1770s and inspired by a puppet-play seen in his youth, this was to occupy him throughout his whole life. The original draft, which he abandoned, was discovered and published as *Ur-Faust* in 1887. A second version, which had not been acted, was included by Goethe in an edition of his works published in 1790. It was Schiller who in about 1799 persuaded Goethe to return to the play and finish it. Part I, which contains the seduction and desertion of the innocent Gretchen and her execution for infanticide, appeared in 1808; Part II, in which occurs the scene of the raising of Helen of Troy used by *Marlowe in his *Dr Faustus,* was completed in the last year of Goethe's life and appeared posthumously. Taken together, the two parts are a distillation of an old man's wisdom, accumulated over a long and active career and embracing all aspects of human life, and insisting throughout on unremitting activity and endeavour as the true aim of man's life on earth. Owing to its complexity the play has always proved difficult to stage and many directors fall back on the simpler *Ur-Faust.* Part I was not seen in Germany until 1829; Part II in 1854. The first production of both parts was in 1876 at Weimar, and there have been notable later productions by *Reinhardt and *Gründgens.

Although most of Goethe's works, including his plays, have been translated into English, very little of his dramatic work has been seen on stage; but in 1963 the *Bristol Old Vic put on a version of *Götz von Berlichingen* by John *Arden as *Ironhand,* and *Iphigenia in Tauris* was seen in Manchester and London in 1975.

Gogol, Nikolai Vasilievich (1809–52), Russian writer and dramatist, the first great realist of the Russian theatre. He was already known for his short stories when in 1832, after making the acquaintance of *Shchepkin, he began work on a play but abandoned it when he realized that as a satire on bureaucracy it would not pass the censor; the manuscript was later destroyed. Two other plays, both satires, were begun but not finished until 1842. Gogol's dramatic masterpiece *Revizor* was produced at the Court theatre in the presence of the Tsar in 1836. Its theme was official corruption in a small town, where an impecunious impostor is mistaken for a government official and fêted accordingly. It came at an opportune moment when the authorities were reorganizing municipal affairs, but the satire bit too deep and the play was viciously attacked. Gogol, already broken in health, left Russia, not to return until 1848. The play had an immense influence in Russia and also became well known in translation, being first seen in London in 1920 as *The Government Inspector* and in New York in 1923 as *The Inspector-General.* In 1966 an outstanding revival by the *RSC was directed by Peter *Hall, with Paul *Scofield as Khlestakov. An English translation of an amusing but less important comedy *Marriage* (1842) was seen in London as *The Marriage Broker* in 1965. Gogol's other dramatic works include the farce *Gamblers* (1842) and a sketch entitled *On Leaving the Theatre after a Performance of a New Comedy*, which analyses audience reaction in St Petersburg in the same year. In 1928 Gogol's great novel *Dead Souls* was dramatized by *Bulgakov and presented by *Stanislavsky at the *Moscow Art Theatre, and this version was seen in London in 1964 when the company appeared during the *World Theatre Season.

Gold, Jimmy, see CRAZY GANG.

Golden, John, see JOHN GOLDEN THEATRE.

Golden Theatre, New York, see JOHN GOLDEN THEATRE and ROYALE THEATRE.

Goldfaden [Goldenfodim], **Abraham** (1840–1908), the first important Yiddish dramatist and the first to put women on the stage in Yiddish plays. His early folk songs and dramatic sketches were performed by the Brody Singers, who in 1876 also put on a two-act musical entertainment in Yiddish which Goldfaden had prepared for them. Its success encouraged him to form a company for the production of his own plays, of which he wrote about 400. Among the best known are *The Recruits* (1877), *The Witch* (1879), *The Two Kune Lemels*, a version of *Romeo and Juliet*, *Shulamit* (both 1880), *Dr Almosado* (1882), and *Bar Kochba* (1883). Goldfaden gave his audiences what they wanted—a mixture of song and dance, with plots and music borrowed from all over Europe, racy dialogue and broad characterization, much action and little analysis. At the time of his death his last play *Son of My People* (1908) was running at the Yiddish People's Theatre in New York, where he had settled in 1903, opening a school of drama. Goldfaden was also the author of the first Hebrew play seen in New York, *David at War* (1904).

Goldoni, Carlo (1709–93), Italian dramatist, destined for the law, but drawn to the theatre by his desire to reform the now moribund *commedia dell'arte* by replacing the old-style improvisation on an outline plot with fully written comedies of character. In this he was to be bitterly opposed not only by some of the actors but by his contemporary fellow-dramatist Carlo *Gozzi, who had his own plans for reform. Goldoni began slowly, writing out one part fully in his *Momolò Cortesan* (1738). It was followed by what was to become one of his most popular works, *Il servitore di due padrone* (*The Servant of Two Masters*, 1743). For its first production he left some of the scenes to be improvised by Truffaldino, the 'servant' of the title, but wrote them out in full when the play was published in 1755. It was a great success, and was followed by *La vedova scaltra* (*The Wily Widow*, 1748), *La buona moglie* (*The Good Wife*) and *Il cavaliere e la dama* (*The Knight and the Lady*, both 1749), and *Il teatro comico* (1750), his manifesto for the reform of the stage; all of these four plays were staged at the Teatro Sant'Angelo in Venice. Goldoni was now at the height of his success, as was evinced by the popularity of *Il bugiardo* (*The Liar*) and *La bottega del caffè* (*The Coffee House*, both 1750), and particularly of what is usually regarded as his masterpiece, *La locandiera* (*The Landlady*, 1751), a picture of feminine coquetry which delighted audiences all over Europe and later America, more especially when Mirandolina, the landlady of the title, was played by Eleonora *Duse. With his next play Goldoni left the Teatro Sant'Angelo for the Teatro San Luca, which was too big for his intimate comedies, and where he had to contend with the hostility not only of Gozzi but of a mediocre playwright, Pietro Chiari. Between them they drove Goldoni to Paris, where from 1761 he wrote plays in both Italian and French for the *Comédie-Italienne, which were not as successful as he had hoped. Caught up in the turmoil of the French Revolution, he died in poverty. By employing the masks of the old *commedia dell'arte* in a realistic manner, he paved the way for the later *drame bourgeois* of *Diderot, and much of his work survives not only on the Italian stage but in the international repertory.

Goldsmith, Oliver (1730–74), Irish poet, novelist, and dramatist, whose two plays *The Good-Natured Man* (1768) and *She Stoops to Conquer* (1773) stand, like the works of *Sheridan his contemporary, far in advance of the drama of his time. The first, produced by the elder *Colman at *Covent Garden, had a cool reception, but the second, also at Covent Garden, was an immediate success. Satirizing London snobbery, it has little in common with the genteel comedy of the day and has been constantly revived. Goldsmith wrote nothing more for the theatre, but in 1878 his novel *The Vicar of Wakefield* was made into a charming play, *Olivia* by W. G. Wills, with Ellen *Terry as the young heroine.

Goliard, name given to the wandering scholars and clerks of the early Middle Ages who, impatient of discipline, joined with the itinerant entertainers of the time and were often confused with them, as in an order of 1281 that 'no clerks shall be *jongleurs*, goliards or buffoons'. They imparted a flavour of classical learning to the often crude performances of their less erudite fellows, and even when, as happened in the 14th century, the word was used for 'minstrel' without any clerical association, the goliard is still shown rhyming in Latin, as in Langland's late 14th-century poem *Piers Plowman*.

Gomersal, Edward Alexander, see ASTLEY'S
AMPHITHEATRE.

Gonzaga, Pietro, see SCENERY.

Goodman's Fields Theatre, London. There
were two theatres of this name, both in Ayliffe
Street, Whitechapel. The first opened in 1727
in a converted shop with a performance of
*Farquhar's *The Recruiting Officer*; in 1730
*Fielding's second play *The Temple Beau* was
produced. **Henry Giffard** (c.1695–1772) was
manager for a time, his last production being
seen in 1732. The theatre provided exhibitions
of acrobatics and rope-dancing, and a brief
return to drama, until it closed in 1751. It then
became a warehouse and was burnt down in
1802.

The second theatre was built by Giffard not
far from the first, and opened in 1732 with
Henry IV, Part One. The passing of the Licensing
Act of 1737, for which Giffard was mainly
responsible (since he had passed on to Walpole,
the Prime Minister, the script of a scurrilous
play, *The Golden Rump*, which gave occasion
for legislation), cost him his licence, and the
theatre had to close. By various subterfuges
Giffard was able to reopen it in 1740, and the
following year he put on *The Winter's Tale*,
'not seen in London for over 100 years', with
himself as Leontes. It was at this theatre that in
the same year *Garrick made his first professional appearance in London, as Richard III,
and a year later this second theatre finally closed.

Goodman Theatre, Chicago, Ill., opened in
1925. After the *Cleveland Play House this is
the oldest regional theatre in the United States,
donated to the Art Institute of Chicago by the
parents of the playwright Kenneth Sawyer
Goodman in memory of their son. Seating 683,
it was intended to combine a school of acting
with a resident professional acting company;
but the Depression forced the company to close,
though the school continued to flourish. In
the 1950s a subscription series of plays was
introduced during which professional guest artists performed with students. In the 1960s it
once again housed a fully professional company,
the student productions being moved to the
adjacent Goodman Theatre Studio, seating 135,
until in 1977 the theatre was incorporated as
the Chicago Theatre Group Inc. and the school
ceased to be connected with it. The Mainstage
theatre now offers six productions a year, classic
revivals and works by new authors, while the
studio theatre houses smaller productions. David
*Mamet was for a time playwright in residence.

Goodwin, Nat [Nathaniel] **Carl** (1857–1919),
American actor, who made his début in Boston,
his birthplace, in 1874. His career as a *vaudeville comedian began at Tony *Pastor's Opera
House, New York, in 1875, and the following
year he scored an immense success in *Off the*

Stage at the New York *Lyceum with imitations
of the popular actors of the day. Although best
known as an actor of light comedy, he was also
successful in such serious plays as Augustus
*Thomas's *In Mizzoura* (1893) and Clyde
*Fitch's *Nathan Hale* (1899), which was written
for him and in which he played opposite his
third wife, Maxine *Elliott. In 1901 he played
Shylock to her Portia and in 1903 appeared as
Bottom, but Shakespeare proved beyond his
range and he returned to modern comedy. The
trade mark of his mature years was a drily
humorous manner which made him popular as
a comedian and as a vaudeville raconteur.

Gordin, Jacob (1853–1909), Jewish dramatist,
born in the Ukraine, who in 1891 emigrated to
New York and was drawn into the orbit of the
newly founded Yiddish theatre there. He wrote
in all about 80 plays, of which *The Jewish King
Lear* (1892), *Mirele Efros* (1898), which portrays
a feminine counterpart of Shakespeare's Lear,
and *God, Man and Devil* (1900), based to some
extent on *Goethe's *Faust*, are the best known.
Like most of his contemporaries, Gordin took
much of his material from non-Jewish plays,
giving them Jewish characters against a Jewish
background, setting his face against improvisation and insisting on adherence to the
written text. In its simplicity, seriousness, and
characterization, his work marks a great advance
on what had gone before. He was one of the
directors of *Goldfaden's drama school in New
York, and towards the end of his life fought
against the 'star' system, which he foresaw
would be a danger to the theatre.

Gordone, Charles, see NEGROES IN THE AMER-
ICAN THEATRE.

Gordon-Lennox, Cosmo, see TEMPEST.

Gorky, Maxim [Alexei Maximovich Pyesh-
kov] (1868–1936), Russian dramatist, the only
one to belong equally to the Tsarist and Soviet
epochs. He had a hard and unhappy childhood
and a youth overshadowed by brutality: his
pseudonym, 'bitter', was well chosen. Painfully,
during a series of menial jobs, he educated
himself, and in 1892 his first short story was
published. It was followed by a succession of
works in which he became the outspoken
champion of the underdog. They brought him
fame, and some money, but also imprisonment
and eventually exile to Italy; he did not return
to Russia for good until 1931. It was *Chekhov
who in 1902 persuaded the *Moscow Art
Theatre to stage his first play, *Scenes in the House
of Bersemenov* (known also as *The Smug Citizens*).
This was followed by what is generally considered his best play, *The Lower Depths*, which
depicted with horrifying realism the lives of
some of the inhabitants of Moscow's underworld, huddled together in the damp cellar of
a doss-house. In a translation by Laurence

*Irving it was seen in London in 1903 and has been revived several times, notably by the *RSC in 1972. In New York it was first produced in 1930 as *At the Bottom* and revived in 1964 as *The Lower Depths*. It was followed by many lesser plays, all dealing with the class struggle, four of which—*Summerfolk* (1904), *Children of the Sun* (1905), *Enemies* (written in 1906 but not staged in Russia until 1933), and *The Zykovs* (1913)—were staged by the RSC in the 1970s; *Enemies* and *The Zykovs* were seen in New York in 1972 and 1975 respectively. In 1931 Gorky began a trilogy on the decay of the Russian bourgeoisie; of this only two parts were staged—*Yegor Bulychov and Others* (1932) and *Dostigayev and Others* (1933). The last part, *Somov and Others*, was left unfinished. Some of Gorky's novels were successfully dramatized by other hands, notably *Foma Gordeyev* (1899) and *The Mother* (1907). Though he later suffered some disillusionment, Gorky welcomed the Revolution and by his work helped to bring about the establishment of the new régime. In acknowledgement of this his birthplace Nizhny-Novgorod was renamed Gorky, as was the theatre in Leningrad where his plays were first seen outside Moscow (see below).

Gorky Theatre, Grand, Leningrad, also known as the Bolshoi Dramatic Theatre, founded in 1919 and named after *Gorky, who took a great interest in it. Its first production was *Schiller's *Don Carlos*, which was followed by a number of other foreign classics and some new plays. A period of *Expressionism featured plays by *Toller, *Kaiser, *Shaw, and *O'Neill. From 1925 onwards Soviet plays were introduced and in 1933 Gorky's *Enemies* was first seen at this theatre, being produced in Moscow two years later. Among the theatre's scene designers was *Akimov. After the Second World War the theatre, which had been severely damaged, was repaired, and the company embarked on an ambitious programme, including the revival of a fine production of *King Lear* which had been cut short by the outbreak of war. *Tovstonogov became Artistic Director in 1956, making the company one of the country's best. It appeared in London during the *World Theatre Season of 1966 in a musical entitled *Grandma, Uncle Iliko, Hilarion, and I*, and in Dostoevsky's *The Idiot*, dramatized and directed by Tovstonogov, with *Smoktunovsky as Prince Myshkin. Russian classics presented at this theatre included *Griboyedov's *Woe from Wit* (or *Wit Works Woe*) (1962), *Chekhov's *Three Sisters* (1965), and *Gogol's *The Government Inspector* (1972). Productions of foreign plays included *Brecht's *The Resistible Rise of Arturo Ui* (1963), O'Neill's *A Moon for the Misbegotten* and *Miller's *The Price* (both 1968), and a composite version of the two parts of *Henry IV* (1969). *The Story of a Horse* (1975),

which Tovstonogov co-adapted from a short story by Leo *Tolstoy, was staged in New York as *Strider: The Story of a Horse* in 1979. He presented a musical version of *Sukhovo-Kobylin's *Tarelkin's Death* in 1984.

Gôt, Edmond-François-Jules (1822–1901). French actor, who passed the whole of his long and honourable career at the *Comédie-Française, of which he was the 29th *doyen*. He made his début in comedy roles in 1844, playing Mascarille in *Molière's *Les Précieuses ridicules* and, in 1848, the Abbé in *Musset's *Il ne faut jurer de rien*, becoming a member of the company in 1850. He was one of the finest and most dependable actors of his day, creating 200 parts as well as playing in most of the classic repertory.

Gottsched, Johann Christoph (1700–66), German literary critic who tried to reform the German stage on the lines of the French classical theatre. He was helped by the actress Carolina *Neuber, for whom he drew up a model repertory, later published in 1740–5, of adaptations of French plays, with a few German works of which the best were those of J. E. *Schlegel. Some of them were acted by Carolina Neuber's company in place of the traditional farces featuring *Hanswurst, whom she banished from the stage. She also produced Gottsched's own plays, of which *Der sterbende Cato* (1732), based on *Addison's *Cato* (1713), was the most successful.

Gow, Ronald, see HILLER.

Goward, Mary Ann, see KEELEY, MARY ANN.

Gozzi, Carlo (1720–1806), Italian dramatist, who tried to reform the moribund *commedia dell'arte* in the mid-18th century by using its characters and methods, but not its subject-matter, for a new type of play which he called *fiabe*—a mixture of fantasy and fooling in a set text which nevertheless allowed plenty of room for improvisation. In opposition to the realistic and bourgeois comedies of his contemporary and rival *Goldoni, Gozzi used stories of magicians, fabulous animals, and fairy-tale characters, and wrote in 'pure Tuscan' as a counterblast to Goldoni's use of local Italian dialects. The best were *L'amore delle tre melarance* (*The Love of Three Oranges*) and *Il corvo* (*The Raven*) (both 1761), *Il re cervo* (*King Stag*, 1762), *Turandot* (based on a Chinese fairy-tale and Gozzi's best work), and *L'augellino belverde* (*The Beautiful Green Bird*) (both 1765). Perhaps because his vein of fantasy became a more general preoccupation (he anticipated *Pirandello in his use of myth and the working of the subconscious), there were some notable 20th-century productions, particularly of *Turandot* directed by *Vakhtangov in Moscow (1922) and *King Stag* directed by George *Devine for the *Young Vic in London (1946). A good deal of incidental music was written for the plays, and two were used as opera libretti: *The Love of Three Oranges* by Prokofiev (1921) and *Turandot* by Puccini (1926).

Grabbe, Christian Dietrich (1801–36), German poet and dramatist who with *Büchner was the most notable playwright associated with the 'Young Germany' movement in the early 19th century. His grotesque satirical play *Scherz, Satire, Ironie und tiefere Bedeutung (Joke, Satire, Irony, and Deeper Significance)*, written in 1822 but not produced until 1892, is now seen as a forerunner of the Theatre of the *Absurd. The first of his plays to be produced was the ambitious *Don Juan und Faust* (1829), in which he strove to emulate both Mozart and *Goethe; it has some striking scenes, as has his *Napoleon* (1835) which, though consisting of little more than a series of sketches loosely strung together in the style of the *Sturm und Drang* movement, was the first German drama in which the mob was the main character. Grabbe left several unfinished plays and his *Hermannschlacht*, written in 1836, was not produced for 100 years; he led an unhappy, harassed life and died young. He was later made the hero of a play, *Der Einsame (The Lonely One*, 1925) by Hanns Johst.

Gracioso, comic servant or peasant in Spanish plays of the Golden Age. Though he may ultimately derive from Latin comedy, his immediate ancestor appears to be the *bobo* or rustic clown of Lope de *Rueda's interludes, or *pasos*. He first appears in the plays of Gil *Vicente and his contemporaries, particularly *Torres Naharro. In works by Lope de *Vega, who did not as he claims introduce the *gracioso* to the Spanish stage, he is nevertheless an important element, parodying the actions of his master in lively and popular language. By *Calderón he is used to present yet another facet of the moral or doctrinal lesson implicit in the play, whether *comedia* or *auto sacramental*. The *gracioso* reached his apotheosis in the plays of Agustín de *Moreto, where he and his female counterpart set on foot and maintain the complicated intrigues.

Gramatica, Emma (1875–1965), Italian actress, the child of strolling players, and on the stage from her earliest years. Though originally considered less promising than her elder sister Irma (below) she later gave notable interpretations of the leading characters in the realistic plays of *Ibsen, *Pirandello, and particularly *Shaw, appearing in the Italian versions of *Mrs Warren's Profession, Pygmalion, Saint Joan,* and *Caesar and Cleopatra*. She was also seen as Marchbanks in *Candida*. An actress of great versatility, she was equally admired as Sirenetta in *D'Annunzio's *La Gioconda*; in the title role of *Rostand's *La Samaritaine*; and in Rosso di San Secondo's *Tre vestiti che ballano (Three Dresses which Dance)*. In the 1930s she toured Europe, and she went to

America in 1945. She was an excellent linguist and played in German and Spanish.

Gramatica, Irma (1873–1962), Italian actress, elder sister of the above, and like her on the stage from childhood. As a young girl she became well known for her skilful playing of the heroines of contemporary French plays in translation, among them Meilhac and Halévy's *Froufrou* and *Sardou's *Odette*. She was also much admired in such new Italian plays as *Giacosa's *Come le foglie (Like the Leaves*, 1900) and *Verga's *Dal tuo al mio (From Thine to Mine*, 1903). She was an excellent Katharina in *The Taming of the Shrew*, and made her last appearance in 1938 as Lady Macbeth to the Macbeth of *Ruggeri. An actress of great charm and vivacity, she could make even trivial parts appealing.

Grand Gorky Theatre, Leningrad, see GORKY THEATRE.

Grand Guignol, see GUIGNOL.

Granovsky, Alexander, see MOSCOW STATE JEWISH THEATRE.

Granville-Barker, Harley (1877–1946), English theatre scholar, actor, director, and playwright, one of the outstanding figures of the progressive theatre at the beginning of the 20th century. In 1891 he joined the stock company at Margate, and later toured with well-known actor-managers and was seen in London, notably in *Poel's production of *Richard II* (1899). In 1900 *Shaw chose him to play Marchbanks in the first London production, by the *Stage Society, of *Candida*, and he appeared also in their productions of *Captain Brassbound's Conversion* (also 1900), *Mrs Warren's Profession* (1902), and *Man and Superman* (1905), in which his first wife Lillah *McCarthy played Ann Whitefield. In 1904, with **J. E. Vedrenne** (1867–1930), he assumed the management of the *Royal Court Theatre, where he embarked on an extensive programme of new plays, including his own *The Voysey Inheritance* (1905), in which the hero finds that the firm he has inherited achieved its wealth dishonestly; the social criticism, as in his other plays, is effectively made. An earlier play, *The Marrying of Ann Leete* (1901), had been successfully produced by the Stage Society, but their later production of *Waste* in 1907 fell foul of the *Lord Chamberlain because it contained mention of an abortion, and the play was not licensed until 1936. His only other play of any importance was *The Madras House* (1910; NY, 1921). His experiences at the Royal Court, which had been artistically rather than financially rewarding, made him a fervent advocate of a subsidized theatre, for which he campaigned ceaselessly. His own approach to Shakespeare, whom he naturally considered the foundation stone of any British National Theatre, was conditioned by his early association with Poel, and his productions at the *Savoy Theatre in 1912–14 of *The Winter's Tale, Twelfth Night,* and *A Midsummer Night's Dream* were later considered epoch-making in their simplicity and poetic beauty. Barker was at the height of his career in England when a visit to America directed his energies into fresh channels. Divorced from Lillah McCarthy, he married as his second wife the American Helen Huntingdon. At her instigation he gave up all contact with the theatre backstage, hyphenated his name, and settled down to translate, with her help, the plays of *Martínez Sierra and the *Álvarez Quintero brothers, and to write the *Prefaces to Shakespeare* (1927–47) on which his posthumous fame chiefly rests. *The Marrying of Ann Leete* was revived by the *RSC in 1975, and *The Madras House* and *The Voysey Inheritance* were produced by the *National Theatre in 1977 and 1989 respectively.

Grassi, Paolo, see PICCOLO TEATRO DELLA CITTÀ DI MILANO.

Grasso, Giovanni (1873–1930), Italian actor, son and grandson of Sicilian puppet-masters, who was encouraged to go on the stage by the actor-manager *Rossi, under whom he trained. He was seen all over Europe with his own company, which specialized in Sicilian and dialect plays. He was considered a fine actor of the *verismo* school, being at his best in the plays of *Pirandello. He was several times seen in London, where in 1910 he played Othello.

Gray, Dulcie, see DENISON.

Gray, 'Monsewer' Eddie, see CRAZY GANG.

Gray, Simon James Holliday (1936–   ), English dramatist, whose first stage play *Wise Child* (1967; NY, 1972), starring Alec *Guinness (in London) in a travesty role, was followed by *Dutch Uncle* (1968) and *Spoiled* (1971). He had his first outstanding success with *Butley* (also 1971; NY, 1972), about a university lecturer facing the breakdown of both his marriage and his homosexual relationship. It starred Alan *Bates, as did *Otherwise Engaged* (1975), in which he played a publisher who avoids emotional entanglements; it was seen in New York with Tom *Courtenay in 1977. *Molly* (1977) and *The Rear Column* (London and NY, 1978) had only short runs. *Close of Play* (*National Theatre, 1979; NY, 1981) centres on a family reunion at which a distinguished academic, played by Michael *Redgrave, remains silent while relationships disintegrate. Gray resumed his successful partnership with Bates in the comedy-thriller *Stage Struck* (1979), and again in *Melon* (1987), in which the hero, another publisher, experiences a mid-life mental crisis. *Quartermaine's Terms* (1981; NY, 1983), considered by some to be his best play, is a study of anguish and non-communication in a

Cambridge language school. *The Common Pursuit* (1984) follows the post-graduation fortunes of a group of university friends; a revised version was staged in New York in 1986 (London, 1988). The popular *Hidden Laughter* (1990) covers the 10-year tenure of a week-end cottage by a writer and her family. His adaptation of Dostoevsky's *The Idiot* was produced by the National Theatre company in 1970.

Gray, Terence (1895–   ), co-founder in 1926 of the Festival Theatre, Cambridge (formerly the Barnwell), which during his brief management had an enormous influence, particularly on the Continent. Like *Craig, whose theories on lighting and stage-craft were the basis of his experiments, Gray fertilized the theatre more by his ideas than by his achievements. He abolished the proscenium arch and footlights, and built out a forestage connected with the auditorium by a staircase, broken by platforms on different levels which offered exceptional opportunities for significant groupings. With Maurice *Evans as his leading man, and with interesting experiments in lighting, Gray produced the *Oresteia* of *Aeschylus, following it with a number of English and foreign classical and modern plays, including some by Shakespeare. In 1929 a company headed by Flora *Robson occupied the theatre, and in 1932 Norman *Marshall ran a season there. Gray returned intermittently but finally abandoned his project in 1933, after the first performance in English of Aeschylus' *Suppliant Women*. The theatre was then bought by a commercial management and restored in a conventional style. It was later owned by the Trustees of the *Cambridge Arts Theatre and used as a workshop and costume store.

Graziano, see DOTTORE, IL.

Grease-Paint, see MAKE-UP.

Great Dionysia, see DIONYSIA.

Grecian Theatre, London, in Shepherdess Walk, City Road, Shoreditch. This opened in c.1830 as a two-tier concert-hall for the production of light opera. In spite of excellent actors and singers the venture was a failure. After 1851, when Benjamin *Conquest became manager and obtained a full theatre licence, a wide range of plays was presented, including Shakespeare's. The theatre was reconstructed in 1858 with four tiers to hold 3,400. Drama, ballet, and *pantomime drew audiences from all parts of London. George *Conquest was responsible for many years for the Grecian Christmas pantomimes. He succeeded to the management of the theatre on the death of his father in 1872, and four years later the theatre was again rebuilt with a three-tier auditorium to hold 1,850. It was sold in 1879, but lost a great deal of money and was eventually bought by the Salvation Army in 1881.

Greek Art Theatre, see KOUN.

Greek Drama. The classical Greek drama which reached its maturity in the 5th and 4th centuries BC was in fact Athenian drama: for although every Greek city and many a large city elsewhere came to have its own theatre, and although some dramatic forms, such as *mime, originated and flourished elsewhere, Athens established and maintained a complete pre-eminence among the Greek states, and all the Greek drama that we possess was written by Athenians for Athens. Both *tragedy and *comedy were part of the religious celebrations in honour of *Dionysus and addressed to a whole community, which came to the theatre as a community, not as individuals; a community which was its own political master and its own government. In both forms of drama the *chorus, the communal element, was originally very prominent.

Preparations for drama were in the hands of the state. Plays were given only at the city festivals, the *Dionysia and the Lenaea, and early in the official year, which in Athens began at midsummer, officials in charge of the festivities would choose from among the many applicants three poets whose works would be performed in the festival. It is surmised that the lesser known were asked for a complete script, established writers for a synopsis only. The chosen three were then assigned a *choregus, who became responsible for all the expenses in connection with the production except for the chorus and the three statutory actors paid by the state. The poet not only wrote the play but also composed the music and arranged the dances. In earlier times he trained the chorus and acted the chief part himself, but later these functions were handed over to specialists, and in the 5th century individual actors seem to have been associated closely with particular poets. When later the importance of the actor increased it was thought fairer, at least in the case of the *Protagonist, to assign them, like the *choregus*, by lot.

Very little is known about individual Greek actors, but with the decline of tragedy in the 4th century they became more prominent. Actors known by name include, in the 4th century, Polus, said to have taught Demosthenes elocution, Theodorus, who had a reputation for adapting plays to suit his own personality, and Aristodemos, known to have been sent as an envoy from Athens to the Court of Macedonia. Actors, who were usually men of good repute and members of guilds such as the Artists of Dionysus, were often employed on secret missions, since their semi-religious function gave them a degree of diplomatic immunity. From the 5th century have survived the names of Nicostratos, famous for his delivery of messenger speeches, and Callippides, often the butt of comic writers because of his high-flown style.

Playgoing in Greece was a civic duty, and still retained traces of its religious origins, as did the theatres. These were all open-air, cut into the side of a hill (see THEATRE BUILDINGS), and had to be big enough to contain a vast number of spectators during a day-long session—*Epidauros could seat 14,000—while seats had to be provided for distinguished visitors and officials, the seat of honour in the centre of the front row being reserved for the priest of Dionysus. Originally all other seats were free; later a charge of 2 obols (about 4p or 8 cents) was made. Those citizens too poor to pay were given a grant. The acting area, with its stage-wall and vast circular *orchestra, probably duplicated the original playing-place in front of a temple wall, and the orchestra still retained its *thymele* or altar to Dionysus. When the importance of the chorus diminished and a raised stage or *logeion* was placed against the back wall for the actors, that too had its altar, which could also be used if necessary as a tomb or other holy shrine.

There was no scenery in the early Greek theatres, colour and splendour being supplied by the rich robes of the actors, who all wore *masks, and the multifarious costumes of the chorus, particularly in comedy. Later, easily changed backcloths helped to diversify the permanent set, and *periaktoi, or movable screens, indicated a change of scene. There were also mechanical devices, such as the *mechane, or crane, and the *ekkyklema*, or wheeled platform. (See also OLD, MIDDLE, and NEW COMEDY.)

Green, Paul Eliot (1894–1981), American dramatist, born on a farm, who gained a knowledge of Negro folk-life from working in the fields, and used it in his plays. Early in his career he wrote almost 40 one-act plays, mainly produced by the Carolina Playmakers, on the problems of Negroes and poor whites in the American South. His first full-length play *In Abraham's Bosom* (1926), which deals with the frustrated attempts of an ambitious but illiterate Negro, son of a white man, to start a school for Negro children, culminating in his murder at the hands of an infuriated mob, was awarded a *Pulitzer Prize for its imagination, sympathy, and power. Other full-length plays include *The Field God* (1927), on religious repression, *Tread the Green Grass* (1929), *The House of Connelly* (1931), the first independent production by the *Group Theatre, *Johnny Johnson* (1936), with music by Kurt Weill, and an adaptation of Richard Wright's novel *Native Son* for a production by Orson *Welles in 1941. In 1937 Green wrote *The Lost Colony* (produced at Roanoke Island, NC), the first of 15 'symphonic dramas' celebrating American history, all designed to be performed out-of-doors in specially built amphitheatres and using professionals with local amateurs. They include *The Founders* (Williamsburg, Va., 1957), *Trumpet in the Land* (New Philadelphia, Ohio, 1970), and *Louisiana Cavalier* (Natchitoches, La., 1976). Many of these productions have become annual events in their localities.

Green-Coat Men, footmen in green livery who, in the early Restoration theatre, placed or removed essential pieces of furniture in full view of the audience.

Greene, (Henry) **Graham** (1904–91), distinguished English novelist and dramatist, whose novel *Brighton Rock* was successfully dramatized in 1943, followed by *The Heart of the Matter* in 1950 and *The Power and the Glory* in 1956. Greene's first play written directly for the theatre was *The Living Room* (1953; NY, 1954), in which Dorothy *Tutin as a young girl in love with an older married man gave a fine performance. It was followed by *The Potting Shed* (NY, 1957; London, with John *Gielgud, 1958), about an atheist's pact with God; *The Complaisant Lover* (London, with Ralph *Richardson, 1959; NY, with Michael *Redgrave, 1961), a comedy of manners in which a husband and lover knowingly share a wife's favours; and *Carving a Statue* (1964; NY, 1968), in which Richardson again played the lead in London. In 1975 *The Return of A. J. Raffles* was presented by the *RSC. Greene's work bears the strong impress of his Roman Catholicism, though his expression of it is not always acceptable to the authorities.

Greene, Robert (c.1560–92), English dramatist, who led a wild and dissolute life, and shortly before his early death published his famous recantation *A Groatsworth of Wit Bought with a Million of Repentance,* which contains the first known reference to the emergence of Shakespeare as a playwright—'an upstart Crow, beautified with our feathers . . . in his own conceit the only Shake-scene in a country'. Greene was a prolific writer, and as the friend of *Marlowe, *Nashe, and *Peele—the so-called *University Wits—may have helped in the writing and rewriting of some of their works. How far he collaborated with Shakespeare—he is believed, for instance, to have had a hand in the trilogy of *Henry VI*—is still hotly debated. Certainly Shakespeare took the plot of *The Winter's Tale* (1611) from one of Greene's prose romances, *Pandosto; or, The Triumph of Time* (1588). Of Greene's own eight plays, the most successful seem to have been *James IV of Scotland* and *The Honourable History of Friar Bacon and Friar Bungay* (both c.1591). The latter, a study of white magic, was probably intended as a counterblast to Marlowe's black magic in *Dr Faustus* (c.1589).

Green Room, after the Restoration in 1660, a room behind the stage in which actors and actresses gathered before and after the performance to chat or to entertain their friends. It has almost disappeared from the modern English theatre, but still exists in a modified form at *Drury Lane. The first reference to it occurs in *Shadwell's *A True Widow* (1678) and it is mentioned in Colley *Cibber's *Love Makes a Man* (1700). It seems probable that it got its name simply because it was hung or painted in green. It was also known as the Scene Room, a term later applied to a room where scenery was stored, and it has been suggested that 'green' is a corruption of 'scene'. In the larger early English theatres there was sometimes more than one green room; they were then strictly graded according to the salary of the player, who could be fined for presuming to use a green room above his rank.

Green Room Club, London, in Adam Street, social club formed in 1866 when the professional members of the old Junior *Garrick Club were outvoted. It remains a notable meeting place for the actors and actresses who comprise 75 per cent of its membership, and maintains reciprocal arrangements with the *Lambs in New York.

Greenwich Theatre, London. 1. The first permanent theatre in Greenwich was opened in 1709, probably in Church Street. Colley *Cibber's *Love Makes a Man* is known to have been produced there the following year, but there is no further mention of it after 1712.

2. A second Greenwich Theatre, at 75 London Road, opened in 1864. It held 721 in a three-tier auditorium, and later came under the control of William Morton, who gave it his own name, as did his successor Arthur Carlton in 1902. It was converted into a cinema in 1910 and demolished in 1937.

3. The present Greenwich Theatre, in Crooms Hill, was originally a concert-hall attached to the Rose and Crown public house. The theatre was rebuilt in 1871 and became Crowder's Music-Hall. After several changes of name and management it became the Greenwich Hippodrome in 1911, and was successively a variety theatre, a cinema, a repertory theatre, and an antiques store. By 1962 it was derelict and was acquired by the council for demolition, but it was saved by a group of enthusiasts from the *Bristol Old Vic. The present theatre, opened in 1969, was erected within the walls of the old building, with a new façade. It holds 423 in a single-tier auditorium, and has an open stage. Many notable new works have been staged, including John *Mortimer's *A Voyage round My Father* (1970), Peter *Nichols's *Forget-Me-Not Lane* (1971), and Alan *Ayckbourn's trilogy *The Norman Conquests* (1974), all successfully transferred to the West End.

Jonathan *Miller directed several plays there, and the theatre housed the British premières of *O'Neill's *More Stately Mansions* (1974) and Hugh *Leonard's *Da* (1978). Among later West End transfers were *Coward's *Private Lives* (1980), *Present Laughter* (1981), and *Design for Living* (1982), Julian Mitchell's *Another Country* (also 1982), and *Chekhov's *Three Sisters* (1987).

Greenwich Village Theatre, New York, see PROVINCETOWN PLAYERS.

Greet, Sir Ben [Philip Barling] (1857–1936), English actor-manager. He first appeared on the stage in 1879, as Philip Ben, and three years later was in London, where he appeared with a number of outstanding players including Lawrence *Barrett and Mary *Anderson. In 1886 he gave the first of his many open-air productions of Shakespeare, and formed a company with which he toured incessantly in the United Kingdom and the USA, rivalling Frank *Benson as a trainer of young actors. He spent several years in America, but returned in 1914 and helped to found the *Old Vic, where between 1915 and 1918 he produced 24 of Shakespeare's plays, including *Hamlet* in its entirety, as well as a number of other English classics. Many London schoolchildren owed their introduction to the theatre to his visits in the 1920s and 1930s to local authority and other centres with a repertory in which Shakespeare predominated. In his later years he concentrated mainly on productions for schools and open-air performances. He celebrated his stage jubilee in 1929, and continued working until his death.

Gregory, (Isabella) **Augusta, Lady** (1852–1932), Irish landowner, who entered the theatre in middle age with an unsuspected gift for comedy-writing which, but for her contact with *Yeats and the Irish dramatic movement, might never have been realized. She proved an indefatigable worker in the movement which led to the founding of the *Abbey Theatre in 1904. In 1909 she won a notable victory for the Abbey by frustrating attempts to suppress the production of *Shaw's *The Shewing-Up of Blanco Posnet*, and in 1911 took the company on a stormy and triumphant visit to America. The best known of her numerous masterly short plays are comedies of peasant life: *The Pot of Broth* (with Yeats, 1902), *Spreading the News* (1904), *Hyacinth Halvey* (1906), and *The Workhouse Ward* (1908); but she is also known for two fine patriotic plays, *Cathleen ni Houlihan* (with Yeats, 1902), and *The Rising of the Moon* (1907), and for one brief peasant tragedy, *The Gaol Gate* (1906). Later she wrote fantasies of mingled humour, pathos, and poetic imagination—*The Travelling Man* (1910), *The Dragon* (1919), *Aristotle's Bellows* (1921), and others. She also contributed to the Abbey repertory many translations, of which *The Kiltartan Molière*, a version of several of *Mol-

ière's plays transplanted to the west of Ireland, is the best known.

Gregory, Johann Gottfried (1631–75), German pastor, who in 1658 was appointed to the Lutheran Church in Moscow where 10 years later he founded a school. In 1672 he was responsible for the first organized dramatic entertainment given before the Russian Court on the orders of Alexei, father of Peter the Great. This was *The Play of Artaxerxes*, taken from the repertory of the *English Comedians and performed in German by German students in Moscow. It lasted 10 hours, and was enlivened by songs, music, dancing, and comic interludes. Gregory later produced *The Comedy of Young Tobias* (1673) and, with Russian student-actors whom he had trained himself, *The Comedy of Holofernes* (1674). He is portrayed in *A Comedian of the Seventeenth Century* by *Ostrovsky. His theatre, a hastily erected wooden building in the summer palace at Preobrazhen, survived until Alexei's death in 1676, when it was pulled down.

Grenfell [*née* Phipps], **Joyce Irene** (1910–79), English *diseuse*, who made her first appearance on the professional stage in *The Little Revue* (1939) in a selection of the monologues with which she had for some time entertained her friends. She was immediately successful, proving herself an excellent mimic and an accurate though kindly satirist of contemporary manners, particularly of schoolmistresses and middle-class wives and daughters. During the Second World War she toured the service hospitals, but returned to London in 1945 to appear in Noël *Coward's *revue *Sigh No More*, which was followed by *Tuppence Coloured* (1947) and *Penny Plain* (1951). In 1954 (NY, 1955) she was seen in *Joyce Grenfell Requests the Pleasure*, and thereafter toured world-wide in her one-woman entertainments of sketches and songs, many of which she wrote herself.

Grévin, Jacques (1538–70), French poet, a humanist of the school of Ronsard. His tragedy *La Mort de César* (1561)—important as the first original French tragedy to be performed and published—shows the influence of *Seneca, and portrays idealized Roman virtues in the style later adopted by the neo-classical *Corneille and his contemporaries. It is, however, for his comedies *La Tresorière* (1559) *and Les Esbahis* (1561) that Grévin is chiefly remembered.

Griboyedov, Alexander Sergeivich (1795–1829), Russian diplomat and dramatist, assassinated in Tehran while acting as Russian Minister to Persia. His comedies were translated from the French or written in collaboration with his friends, the exception being his classic play *Gore ot Uma*, whose title has been variously rendered as *Woe from Wit*, *Wit Works Woe*, *Too Clever by Half*, *The Misfortunes of Reason*,

The Trouble with Reason, and *The Disadvantages of Being Clever*. It deals with the struggles of a young man, arriving in Moscow full of liberal and progressive ideas, against the stupidity and trickery of a corrupt society. Classic in form yet realistic and satiric in content, this first dramatic protest against tsarist society is one of the great plays of the Russian theatre. Griboyedov worked on it for many years, and although it was banned from the stage during his lifetime it circulated freely in manuscript and was printed in a cut version four years after his death. It was first performed at the Bolshoi Theatre in St Petersburg in 1831, published in full in 1861, and revived by the *Moscow Art Theatre in 1906 with *Kachalov and *Moskvin as Chatsky and Famusov, which they played again in a production in 1938. The part of Chatsky is to young Russian actors what Hamlet is to English, and Famusov, the heroine's conservative father, has long been a favourite role with older actors.

Grid, or Gridiron, open framework from 5 ft. to 10 ft. below the stage roof from which scenery or lights can be hung. It can be made of wood or metal, and should be built at a height three times that of the proscenium opening.

Grieg, Nordahl (1902–43), Norwegian dramatist, poet, and novelist, whose untimely death in a bombing raid over Berlin during the Second World War was a great loss to the European theatre. After some years as a student (which included a year at Oxford), he began by publishing volumes of prose and poetry. His first play, *Barabbas* (1927), deals with the civil war in China. *Vår ære og vår makt* (*Our Honour and Our Power*, 1935) is an anti-war play of overwhelming force and originality; *Nederlaget* (*The Defeat*, 1937), set in the Paris Commune of 1871, explores the same theme more deeply and is believed to have inspired *Brecht's *Die Tage der Kommune* (1956).

Grieve, William (1800–44), English scene designer, who as a young man worked at *Covent Garden, where his father **John Henderson Grieve** (1770–1845) was responsible for the scenery of spectacle plays and pantomimes under John Philip *Kemble. William later went to *Drury Lane, where he did his best work, being considered after the retirement of *Stanfield the finest scenic artist of the day. His moonlit scenes were particularly admired, and he was the first theatre artist to be called before the curtain by the applause of the audience. His elder brother **Thomas** (1799–1882), who in 1839 was at Covent Garden under Mme *Vestris, also worked at Drury Lane, being assisted by his son **Thomas Walford Grieve** (1841–82), whose painting was remarkable for the brilliance of its style and the artistic beauty of its composition.

Griffiths, Trevor, see NOTTINGHAM PLAYHOUSE.

Grillparzer, Franz (1791–1872), Austrian dramatist, the major figure of the Romantic period, whose first play *Die Ahnfrau* (*The Ancestress*, 1817) was followed by two plays on classical themes, *Sappho* (1818) and *Das goldene Vliess* (1820). A historical play in the style of *Schiller, *König Ottokars Glück und Ende*, was banned by the censor for two years because of the resemblance between the career of its hero and that of Napoleon. Finally produced in 1825, it became one of the acknowledged masterpieces of the German-speaking theatre. Of Grillparzer's other plays, which include *Ein treuer Diener seines Herrn* (*His Master's Faithful Servant*, 1826) and *Des Meeres und der Liebe Wellen* (1829), on the story of Hero and Leander, the most important is his adaptation of *La vida es sueño* by *Calderón as *Der Traum ein Leben* (*Life is a Dream*, 1834). Grillparzer's only comedy, *Weh' dem, der lügt* (*Thou Shalt Not Lie*, 1838), failed on its first production at the Burgtheater, with the result that Grillparzer turned from the theatre and wrote only for his own amusement. *Ein Bruderzwist in Habsburg, Die Jüdin von Toledo*, and *Libussa* were published posthumously, being produced in 1872, 1888, and 1874 respectively.

Grimaldi, Joseph (1778–1837), English actor, the creator of the English *clown, the only character in the *harlequinade not derived from the Italian *commedia dell'arte. In his honour all later clowns were nicknamed Joey. The illegitimate son of Giuseppe Grimaldi, ballet-master at *Drury Lane, he made his first appearance on the stage at *Sadler's Wells at the age of 2½. As a boy he played in *pantomime, and by the time he became a regular member of the Sadler's Wells company in 1792 he was already well trained in acrobatic and pantomimic skills. In 1796 he was at Drury Lane where he played a number of parts besides Clown, being seen as Aminadab in Mrs *Centlivre's *A Bold Stroke for a Wife* and as Robinson Crusoe and Blue Beard in pantomime. In 1805, on the recommendation of Charles *Dibdin, he was engaged for *Covent Garden, where, with occasional returns to Sadler's Wells, he remained until his retirement. At these theatres his inventive comic genius—making a man out of vegetables, a coach out of four cheeses, a cradle, and a fender—was given full scope. Making Clown a rustic booby on the model of Pantaloon's servant, Grimaldi gave him his traditional costume—baggy breeches, a livery coat with scarlet patches, fantastic wigs, usually a turned-up pigtail, and a white face with scarlet triangles and exaggerated eyebrows. His acrobatics were characterized by dynamic energy which finally wore him out, and in 1823 his place at Covent Garden was taken by his dissolute and sottish son, **Joseph S.** (1802–32). He made his last appearance at Sadler's Wells in 1828. He then retired to Pentonville and was buried in the churchyard of St James's Chapel on Pentonville Hill, now a public garden, where his tombstone can still be seen.

Gringore, Pierre, see SOTIE.

Gripsholm Theatre, Sweden, small and extremely beautiful playhouse in one of the round towers of Gustavus Vasa's fortress, built in 1535. The theatre opened in 1782 and its scenery and machinery still remain intact, one set of scenes reproducing exactly the pillared décor of the semi-circular auditorium. The interior has been carefully restored, and the theatre is used for occasional performances in the summer months.

Grock [Karl Adrien Wettach] (1880–1959), the supreme *clown of his generation, for many years a much-loved star of the British *music-hall. Born in Switzerland, he toured as a boy with a circus, changing his name when he appeared in a double act (with Brick). He later appeared in Berlin, and in 1911 was engaged by *Cochran for the *Palace Theatre in London. His clowning was so expressive that it needed no words. He was constantly at the mercy of inanimate objects: any chair he sat on would collapse and entangle him; his enormous double-bass case would be found to contain a tiny violin; when he began to play the piano he discovered he was wearing thick gloves, and was delighted at the improvement in his performance when he removed them. An accomplished performer on at least 20 instruments, he made his hour-long act a series of musical excerpts, which at first he failed to grapple with but finally, to his own simple delight, managed to master. He always wore the same make-up and costume, a pale, egg-shaped face with a large red mouth, a massive tail-coat with collar and tie, tight trousers, clumsy boots, a bowler hat, and a bald wig. He played in London almost continuously up to 1924, mainly at the *Coliseum, after which he returned to the Continent, giving his farewell performance at Hamburg in 1954.

Grooves, characteristic of British, as opposed to Continental, stage machinery, by means of which *wings and *flats were slid on and off stage in full view of the audience, the top and bottom of each flat running in a groove between two strips of timber built into the stage structure. The origin of the groove can be found in the Court *masque, as seen in the designs of Inigo *Jones. Although stage directions in plays from 1660 onwards refer indirectly to the use of grooves, the first direct mention of them dates from 1743, at *Covent Garden. They were also used in early theatres in America, where in 1897 they were referred to as 'old-fashioned'. During the time they remained in use, several innovations were made to enable the scenery to be

changed more quickly, the most efficient being the *drum-and-shaft system. Even so grooves had many disadvantages. As they always had to run parallel to the front of the stage, because of the difficulty of placing them obliquely on a raked floor, masking was poor, and spectators in side boxes could see deeply into the wings. Because of the rake the wing flats became progressively shorter upstage, so each could be used only in one position and interchangeability was impossible. Sometimes the scenes stuck in the grooves, or moved raggedly. These factors combined to bring about the abolition of the grooves system in favour of the Continental *carriage-and-frame, which was first installed at Covent Garden in 1857. The last London theatre to use grooves was the *Lyceum, where they were removed in 1880 by *Irving. Grooves remained, however, for some years in smaller theatres, and a pivoted variant is found in the 1880s which enabled wing and groove to be twisted to any angle. Eventually even this modification gave way to forks, in which the tops of the wings were held as by an inverted garden-fork. The modern system of supporting flats by braces superseded all these earlier methods.

Gros-Guillaume [Robert Guérin] (*fl.* 1598–1634), French actor, who under the name of La Fleur played serious parts in tragedy, but is best remembered as a farce-player, with *Turlupin and *Gaultier-Garguille, in the permanent company at the Hôtel de *Bourgogne. He probably played at the Paris *fairs originally, as he was already known as an actor by 1630. A fat man, with black eyes and a very mobile face (which he covered with flour for comic parts, thus giving rise to the tradition that he was a baker before going on the stage), he figures as himself with his companions in Gougenot's *La Comédie des comédiens* (1631). His wife and daughter were both actresses, the latter marrying an actor, **François Juvenon** (*c.*1623–74), who took his father-in-law's stage name and as La Fleur played kings in tragedy in succession to *Montfleury, and Gascons or the ranting *Capitano in comedy.

Großes Schauspielhaus, Berlin, converted from the Zirkus Schumann for Max *Reinhardt by the German architect **Hans Poelzig** (1869–1936). Intended to achieve the maximum unity of stage and auditorium, it had seating for over 3,000 on three sides of a U-shaped forestage made up of three movable segments. This could be shut off by a movable wall from a main stage 30 metres deep and 20 metres wide, flanked by banks of steps, and having an 18-metre revolve. Behind this was a huge domed *cyclorama illuminated from the flies by banks of spotlights in five colours, and pierced with stars which were echoed in the light-tipped stalactites decorating the dome above the auditorium.

This early attempt to get away from the proscenium arch was a seminal design for later arena theatres. The Großes Schauspielhaus, which opened in 1919 with Reinhardt's production of *Aeschylus' *Oresteia*, was later modified and renamed the Friedrichsstadt-Palast.

Grotowski, Jerzy (1933–   ), Polish director, who studied in Cracow, graduated from the Moscow State Institute of Theatre Arts (GITIS), and visited China to study the theatre there. On returning to Poland he ran an experimental Laboratory Theatre in Wrocław, which on its tours abroad in the 1960s made a deep impression. The company's seminal productions were *Wyspiański's *Akropolis*, *Calderón's *The Constant Prince* translated and adapted by *Słowacki, *Apocalypsis cum figuris*, a compilation of biblical texts and liturgical chants interspersed with quotations from Dostoevsky, T. S. *Eliot, and others, together with Słowacki's own *Kordian* and *Marlowe's *Tragical History of Dr Faustus*. Trained in the theories of *Stanislavsky, Grotowski saw the actor as paramount, using all his physical and mental powers to achieve a close fusion of meaning and movement. He envisaged a theatre stripped of all such inessentials as scenery, costume, lighting, and music, as described in *Towards a Poor Theatre*, a compilation of his own and his associates' ideals published in English in 1968 with a foreword by Peter *Brook, a director strongly influenced by Grotowski. In 1976, after 10 years' touring, Grotowski disbanded his company and retired to work with a group of students behind closed doors. In 1988 he reappeared at a seminar at the University of California to propound his new theories, the effects of which remain to be seen.

Grottesco, Teatro, Italian dramatic movement which emerged during the First World War, and took its name from *Chiarelli's *La maschera e il volto* (*The Mask and the Face*), described as a *grottesco in tre atti*, which, though written in 1913, was not staged until 1916. As the title implies, the central concern of the movement, which arose in opposition to the heroics of *D'Annunzio, was with the contradiction between social behaviour and personal reality, between the hypocrisy of the bourgeois conformist hiding behind his mask and the primitive passions which sometimes tore it from his face. This idea was developed by such writers as *Bontempelli and **Pier Maria Rosso di San Secondo** (1887–1956), the most lyrical of the writers of the school, in whose best-known play, *Marionette, che passione!* (1917), the characters are puppets. The movement was short-lived, but proved an important element in the development of *Pirandello.

Groundrow, originally a strip of gas lights laid flat along the stage to illuminate the foot of a back scene. As a term in stage *lighting it was later applied to the rows of electric bulbs fixed on a *batten and sunk into the stage floor, or masked. The name is now given to all low cut-out strips of scenery, made of canvas stretched on wood, like a *flat laid on its side and cut along its upper edge to represent, for instance, a hedge with a stile in it or a bank topped by low bushes. Series of groundrows, set one behind the other with their top edges cut to represent waves, and so arranged as to allow the passage of a stage boat between them, are known as Set Waters, or Sea or Water Rows.

Group Theatre, Belfast, intimate theatre sited in the Minor Hall of the Ulster Hall, Bedford Street. It was opened in 1932 as the Little Theatre and offered weekly repertory for five years; as the Playhouse it continued until 1939. A year later the hall was leased as the Group Theatre by three amateur groups, the Carrickfergus Players, the Northern Ireland Irish Players, and the Ulster Literary Theatre, initially to share theatre facilities while presenting independent production programmes. The last group, founded in 1904 in emulation of the *Abbey Theatre, Dublin, had achieved a considerable reputation internationally and had mounted first productions of nearly 100 plays. Eventually the three became fully amalgamated and by 1950 the company was a professional one, though working on a minimal budget. James *Bridie, stationed in Ulster during the Second World War, was closely associated with the Group Theatre, other playwrights whose work was successful there including St John *Ervine, George *Shiels, and Joseph Tomelty. Important plays from England and America were also produced by a company that included at various times Tomelty, Colin *Blakely, and J. G. Devlin. The Group Theatre was seen in London in 1953, and at the *Edinburgh Festival in 1958 with a bitterly sectarian play *The Bonefire* by Gerald McLernon. The production of another contentious play about the 'troubles', Sam Thompson's *Over the Bridge* (also 1958), was hastily withdrawn amid controversy. The arrival of television in Northern Ireland affected box-office takings, and the company came near to collapse. It was rescued by two popular local comedians, James Young and Jack Hudson, who as joint managing directors from 1960 to 1972 staged productions of enormously successful local comedies, many specially commissioned from Sam Cree. Audiences then dwindled again owing to civil disturbances, and the theatre closed, but the Belfast City Council carried out refurbishment and in 1978 reopened it as a venue mainly for amateur companies. The theatre hosts over 30 productions annually, encouraging new Ulster writers as well as staging old favourites. (See also BELFAST CIVIC ARTS THEATRE and LYRIC PLAYERS, Belfast.)

Group Theatre, London, private play-producing society founded in 1933 to present modern non-commercial plays and revivals of experimental work. It had its headquarters at the *Westminster Theatre, where most of its productions were staged. These included *Timon of Athens* in modern dress and T. S. *Eliot's *Sweeney Agonistes* (both 1935); the poetic plays of W. H. *Auden and Christopher Isherwood; and Stephen Spender's *Trial of a Judge* (1938). Most of the plays were directed by **Rupert Doone** (1904–66) on a bare stage with little scenery and few props but with the occasional use of masks. Disbanded during the Second World War, the society was re-formed and gave as its first production a translation of *Sartre's *Les Mouches* as *The Flies*. It continued to function spasmodically for a couple of years, but finally disappeared in about 1953.

Group Theatre, New York, production company formed in 1931 by Harold *Clurman, Cheryl *Crawford, and Lee *Strasberg which broke away from the *Theatre Guild because of the latter's alleged lack of political commitment. It had high ideals and a democratic mode of operation, and was intended to present works of serious social content free from the pressures of commercial theatre. Its first production was Paul *Green's *The House of Connelly*, and among its early ventures were Maxwell *Anderson's *Night over Taos* (1932) and Sidney *Kingsley's *Men in White* (1933). The most important dramatist discovered by the Group Theatre was, however, Clifford *Odets, a member of the company. Clurman assumed responsibility when the Group was reorganized in 1936 and Strasberg and Crawford ceased to direct. Other major productions included Paul Green's *Johnny Johnson* (1936) and William *Saroyan's *My Heart's in the Highlands* (1939). A permanent repertory company was built up, dedicated to the principles of group acting as formulated by *Stanislavsky, which produced a number of outstanding actors. In 1941, however, beset by financial difficulties and worn down by conflicts of personality, the Group Theatre ceased production.

Gründgens, Gustav (1899–1963), German actor and director, who made his first appearances in *Hamburg and from 1928 until the end of the Second World War was at various theatres in *Berlin, where he directed and acted in a wide range of classical plays. Tall, blond, and extremely good-looking, he made a striking Hamlet at Elsinore in 1938. He was also much admired in plays by *Shaw. After a brief period in a Soviet internment camp he was released and went to Düsseldorf, where he remained until 1955, making it one of the outstanding theatre centres of the Federal Republic. He spent his last years in Hamburg, where he added many modern works to the repertory, including *Brecht's *Die heilige Johanna der Schlachthöfe* (1959) and John *Osborne's *The Entertainer* (1957). He revived both parts of *Goethe's *Faust* in 1957/8, playing Mephistopheles, the role in which he had first attracted attention in 1922. This was received with acclaim and he was negotiating for it to be seen in South America when he died, apparently by his own hand, in Manila.

Gryphius, Andreas (1616–64), German baroque dramatist, author of a number of tragedies written in lofty poetic prose with scenes of horror and bloodshed. His heroes included Charles I of England, who had just been beheaded (1649), viewed with strong royalist sympathies. Gryphius was sufficiently in advance of his time to write what deserves to rank as the first domestic drama, *Cardenio und Celinde* (1647), which centres on the passions of ordinary men and women and the final miraculous transformation of the repentant sinners. He was also the author of a number of comedies, the most successful being *Horribilicribrifax* (1663), which gives a lively picture of contemporary follies—bombastic conceit, pedantry, self-seeking—as seen by a naturally austere mind with a sense of humour. Although a staunch Protestant, Gryphius owed much to the *Jesuit drama, as well as to the contemporary secular plays in France; a comedy of intrigue with farcical scenes in Low German also shows that he was acquainted with the work of the *English Comedians. He was the first German dramatist to handle his material with conscious artistry; but there was no permanent theatre in which his plays could be given, and only sporadic performances, mostly by schoolboys, are recorded.

Guare, John (1938–), American dramatist, who first attracted attention when *The Loveliest Afternoon of the Year* and *A Day for Surprises* were produced at the *Caffé Cino in 1966. His professional début was in 1968, when *Muzeeka* was performed in a double bill *Off-Broadway (London, 1969). After his double bill of *Cop-Out* and *Home Fires* he became widely known with *The House of Blue Leaves* (1970; London, 1988), his first full-length play. Centring on the effect of the Pope's visit to New York on a middle-aged zookeeper's family and friends, it showed his gift for savage farce. He also helped to adapt, and wrote the exuberant lyrics for, a successful musical version of *The Two Gentlemen of Verona* produced by Joseph *Papp (NY and London, 1973). Later works, both produced by Papp, were *Rich and Famous* (1976), in which a playwright fantasizes on his first opening night, and *The Landscape of the Body* (1977), about a young woman encountering madness and murder in the big city. In *Bosoms and Neglect* (1979) (a comedy in spite of its theme) a couple have a long discussion about themselves after finding that the man's mother is dying of cancer. The heroine of *Lydie Breeze* (1982) and *Gardenia* (also 1982; London, 1983) is a nurse who persuades three male patients to join her in a Utopian community in late 19th-century America. His *Six Degrees of Separation* (1990) is based on an actual event in which a young Black man hoaxed some East Side New Yorkers.

Guignol, French puppet which originated in Lyons, probably in the last years of the 18th century, and may have been invented by a puppet-master, grafting native humour on to *Polichinelle. In Paris the name attached itself to cabarets which, like the Théâtre du Grand Guignol, specialized in short plays of violence, murder, rape, ghostly apparitions, and suicide. In a modified form these made their appearance in London in 1908 and have been seen sporadically ever since, notably in the seasons of Grand Guignol at the *Little Theatre, 1920–2, with Sybil *Thorndike and Lewis *Casson. British Grand Guignol never reached the intensity of the French, however, and its true home is in the small theatres of Montmartre.

In the French theatre a 'guignol' is also a *quick-change room.

Guilbert, Yvette [Emma Laure Esther] (1865–1944), French singer, known as 'la diseuse fin-de-siècle', and an outstanding figure of the *music-hall and *cabaret. She made her début in 1887, and soon became an established favourite. She was thin to the point of emaciation, and her hennaed hair, white mask-like face with a vivid gash of a mouth, and wide range of facial expressions and gestures were immortalized by Toulouse-Lautrec in many paintings and sketches, as were the long black gloves, originating in her early poverty. Her first songs, on themes from Parisian low life, full of crude language, were delivered with an air of utter innocence—one of her admirers described her as the exponent of 'depraved virginity'—but later she added to her repertoire some thousands of *chansons*, ranging from the 14th to the 19th century, with which she illustrated the many lecture-recitals she gave on tour between 1901 and 1914. She was in the United States, 1915–18, reappearing in London in a series of recitals in the 1920s, when she settled again in France. She was also an accomplished actress, her outstanding role being Mrs Peachum in the French version of *Brecht's *The Threepenny Opera* in 1937.

Guildhall School of Music and Drama, see SCHOOLS OF DRAMA.

Guild Theatre, New York, see VIRGINIA THEATRE.

Guinness, Sir Alec (1914–), English actor, first seen on the stage in 1934, whose earliest appearances, notably as Osric to John *Gielgud's *Hamlet* and as Yakov in *Chekhov's *The

Seagull, presaged a brilliant career. He first came into prominence as Sir Andrew Aguecheek in *Twelfth Night* at the *Old Vic in 1937, in which year he joined Gielgud's repertory company at the *Queen's Theatre, his parts there including an unusually tender and poetic Lorenzo in *The Merchant of Venice*. Returning to the Old Vic, he was seen in 1938 as Bob Acres in *Sheridan's *The Rivals* and then as Hamlet in an uncut modern-dress production by Tyrone *Guthrie. During the Second World War he served in the Royal Navy, being temporarily released in 1942 to appear in New York in *Rattigan's *Flare Path*. Returning to London in 1946, he played Mitya in his own adaptation of Dostoevsky's *The Brothers Karamazov* and Garcin in *Sartre's *Vicious Circle (Huis-Clos)*, and then rejoined the Old Vic company, where, in addition to a number of Shakespearian parts including Richard II and the Fool in *King Lear*, he was seen as Khlestakov in *Gogol's *The Government Inspector* and Abel Drugger in *Jonson's *The Alchemist*. At the *Edinburgh Festival in 1949 (NY, 1950) he appeared in T. S. *Eliot's *The Cocktail Party*, and in 1951 he played Hamlet in London in his own Elizabethan-style production. After his Richard III in the inaugural production at the *Stratford (Ontario) Festival in 1953 he turned to modern plays: as the imprisoned cardinal in Bridget Boland's *The Prisoner* (1954), in *Feydeau's farce *Hotel Paradiso* (1956), as T. E. Lawrence in Rattigan's *Ross* (1960), and in *Ionesco's *Exit the King* (1963). His Dylan Thomas in Sidney Michaels's *Dylan* (NY, 1964) was followed by Arthur *Miller's *Incident at Vichy* (London, 1966). He returned briefly to Shakespeare with *Macbeth* at the *Royal Court (also 1966), and then continued to display his protean versatility: in Simon *Gray's *Wise Child* (1967), in a travesty role; in John *Mortimer's *A Voyage round My Father* (1971), as the Father; in Alan *Bennett's hilarious farce *Habeas Corpus* (1973); and in Julian Mitchell's adaptation of Ivy Compton-Burnett's novel *A Family and a Fortune* (1975). In 1976 he was Dean Swift in *Yahoo*, of which he was co-deviser, and in 1977 a British defector living in Soviet Russia in Alan Bennett's *The Old Country*. After Shylock in *The Merchant of Venice* at *Chichester in 1984 he returned to London in 1988 as a Soviet diplomat in Lee Blessing's two-character play *A Walk in the Woods*. He is also a major screen actor.

Guitry, Sacha [Alexandre-Pierre-George] (1885–1957), French actor and dramatist, son of the actor-dramatist **Lucien-Germain Guitry** (1860–1925), for several years manager of the *Renaissance, where Sacha made his first appearance in Paris in 1902. Sacha began writing for the stage at 17 and produced 130 plays and some 30 film scripts. From his first success *Le Veilleur de nuit* (1911) until his death he enjoyed uninterrupted popularity, except for a short period when he was suspected of pro-German sympathies during the Second World War. His work was typical of French light *boulevard plays during their last great period. In his lightweight social comedies—*La Jalousie* (1915) or *Mémoires d'un tricheur* (1935)—he showed himself heir to *Sardou and *Feydeau, while his quasi-historical comedies, dealing with anecdotal episodes on the fringe of great events—*Histoires de France* (1929) or *N'écoutez pas, Mesdames* (1942), seen in London in 1948 as *Don't Listen Ladies!*—and his plays on the private lives of great men—Pasteur, Mozart, *Deburau—continued a tradition established by *Augier and *Dumas *fils*. He was five times married to actresses, his third wife Yvonne *Printemps appearing with him in revivals of several of his plays, notably *Nono* (London, 1920) and *Mozart* (London and NY, 1926).

Guizards, Guizers, see MUMMING PLAY.

Gulbenkian Studio, see NEWCASTLE PLAYHOUSE.

Gundersen, Laura (1832–98), the first great actress in the Norwegian theatre. She made her début at the *Kristiania Theater in 1850, and remained there with a short break until her death. A tragedienne on the grand scale, she was excellent in Shakespeare and *Bjørnson (especially as Mary Stuart). She was also effective in *Ibsen, ranging from early works such as *Lady Inger* and *The Vikings at Helgeland* to *The Lady from the Sea* (as Ellida) and *John Gabriel Borkman* (as Mrs Borkman).

Guthrie, Sir (William) **Tyrone** (1900–71), English actor and director, who through his mother Norah Power was the great-grandson of the Irish actor Tyrone *Power. He made his first appearance on the stage under J. B. *Fagan in Oxford in 1924, and directed plays at the Festival Theatre, Cambridge, 1929–30, his first London production being *Bridie's *The Anatomist* (1931). Much of his finest work was in Shakespeare for the *Old Vic. In 1933 he directed an interesting *Measure for Measure* with Charles *Laughton as Angelo, repeating it in 1937 with Emlyn *Williams. In 1937 he also directed Laurence *Olivier in *Hamlet*, a production later seen at Elsinore; at Christmas 1937 and 1938 he was responsible for delightful productions of *A Midsummer Night's Dream* with Robert *Helpmann as Oberon and Mendelssohn's music. Among his other productions were *Hamlet* (1938) in modern dress, starring Alec *Guinness, and *Ibsen's *Peer Gynt* (1944) with Ralph *Richardson. He was Administrator of the Old Vic and *Sadler's Wells theatres, 1939–45. For the *Edinburgh Festivals of 1948 and 1949 he directed *Lyndsay's *Ane Pleasant Satyre of the Thrie Estaitis* and Allan Ramsay's *The Gentle Shepherd*. From 1953 to 1956 he ran

the *Stratford (Ontario) Festival theatre, largely his own creation. In 1963 he became Artistic Director of the Minneapolis Theater, later the *Guthrie Theater, contributing his own productions of *Chekhov's *Three Sisters* and another modern-dress *Hamlet* to the first season. His productions there also included *Henry V* and *Jonson's *Volpone* (1964), and *Richard III* and Chekhov's *The Cherry Orchard* (1965). Though no longer Director after 1966 he returned every year to direct until 1969. A creative artist who was not afraid to experiment, Guthrie was at his best in the handling of crowd scenes. He worked in many European countries, including Germany and Finland, and in Israel. In 1967 he returned to the Old Vic to direct the *National Theatre company in *Molière's *Tartuffe*.

Guthrie Theater, Minneapolis, originally the Minneapolis Theater, was planned in 1958 by Tyrone *Guthrie as a fully professional classical repertory theatre free from commercial pressure. It opened under his direction in 1963, seating 1,441, of whom none was more than 52 ft. from the centre of the seven-sided thrust stage designed by Guthrie in conjunction with Tanya *Moiseiwitsch, who was closely associated with the theatre for several years. It originally presented a summer repertory season only. Many outstanding actors appeared there, including Hume *Cronyn, Jessica *Tandy, and Zoë *Caldwell, all seen in the opening season, when the plays produced were *Hamlet*, *Chekhov's *Three Sisters*, Arthur *Miller's *Death of a Salesman*, and *Molière's *The Miser*. In 1971 the theatre was renamed in honour of Guthrie, who had remained Artistic Director until 1966. The theatre was run from 1971 to 1977 by the English director **Michael Langham** (1919–   ), who had succeeded Guthrie at the *Stratford (Ontario) Festival. Under his direction the season expanded gradually to 42 weeks, and the touring programme also grew. The Romanian **Liviu Ciulei** (1923–   ), former Artistic Director of the Bulandra Theatre in Bucharest, ran the theatre from 1981 to 1985. He redesigned the stage to give greater flexibility, and opened with an acclaimed production of *The Tempest*, earlier staged at the Bulandra. Other admired productions were *Ibsen's *Peer Gynt* and *Beaumarchais's *The Marriage of Figaro*. Ciulei was succeeded in 1986 by Garland Wright, who staged works by Molière, Shakespeare, and in 1990 the musical *Candide*.

Gutzkow, Karl Ferdinand (1811–78), German writer, a prominent member of the 'Young Germany' movement, mainly remembered as the author of the great Jewish play *Uriel Acosta*, a moving and terrible picture of the struggle for intellectual freedom, written in 1847, which has become a recognized classic of world drama and is in the repertory of almost all Jewish theatres. It was first seen in England in 1905 and has been played, both in the original and in translation, in most European countries and in America.

Gwynn, Nell [Eleanor] (1650–87), English actress, who was an orange-girl, probably under Mrs *Meggs at *Drury Lane, when she attracted the attention of one of the actors. With his help she made her first appearance on the stage in *Dryden's *The Indian Emperor* (1665). She was not a good actress, and owed her success in comedy to her charm and vivacity. Her best part seems to have been Florimel in Dryden's *Secret Love* (1667), in which she was much admired in male attire. She was first noticed by Charles II when speaking the witty epilogue to Dryden's *Tyrannic Love; or, The Royal Martyr* (1669). She then became his mistress and left the stage, her last part being Almahide in Dryden's *The Conquest of Granada* (1670). Tradition has it that the founding of Chelsea Hospital was due to her influence. She became the subject of a number of plays, one of the best known being Paul *Kester's *Sweet Nell of Old Drury* (1900), in which the title-role was played for many years by Julia *Neilson.

Gymnase, Théâtre du, Paris. Built in 1820 on what is now the boulevard Bonne-Nouvelle, this theatre was originally intended as a training-ground for young actors. This proved impractical, and in 1830 the theatre began to present the plays of Eugène *Scribe with such success that he was offered a life annuity and bonuses for the sole and permanent rights in his work. Scribe was succeeded by *Sardou, *Augier, and others as resident playwrights, and in 1852 the younger *Dumas's *La Dame aux camélias* had its first night there. From then until the end of the century the theatre ranked among the theatres of Paris only after the *Comédie-Française and the *Odéon. Henri *Becque's first plays were staged there, *Bernhardt and Lucien *Guitry began their careers there. Between the two world wars it continued to offer popular comedy and melodrama. In more recent times, under the management of Marie *Bell from 1959, it tended more towards social drama and the established classics of comedy. After Marie Bell's death in 1985 Jacques Bertin took over the management.

H

Habimah (Stage) **Theatre,** company founded in Moscow in 1917 to perform plays in Hebrew. Its first performance took place in 1918 and attracted the attention of *Stanislavsky, who sent his pupil Vakhtangov (see VAKHTANGOV THEATRE) to direct the young company in *Ansky's *The Dybbuk*, which opened in 1922 and became world famous. In 1926 Habimah left Moscow for a world tour, and achieved a sensational success in Europe. In the USA they were less successful and a split developed, a few of the actors remaining in America while the rest returned to Europe and set up a temporary base in Berlin. They were never to return to Russia, but having always intended to settle in Palestine (as it then was), they made their first visit in 1928, being seen in *Aleichem's *The Treasure* and *Calderón's *The Hair of Absolom*. After a further tour of Europe, during which Habimah presented its first Shakespeare production, *Twelfth Night* staged by Michael Chekhov in Berlin, the company finally settled in Tel Aviv in 1932 and in 1945 moved into its own theatre there, where it remained until 1970, being declared a National Theatre in 1958. From 1969 the company, already partly subsidized, was fully supported by the Israeli Government, whose appointees replaced the previous collective administration. In 1970 a performance of *Dekker's *The Shoemaker's Holiday* in Hebrew inaugurated a new, comfortable, and well-equipped National Theatre building which stages Israeli and foreign plays.

Hackett, James Henry (1800–71), American actor, who made his first appearance in 1826 and became famous for his portrayal of Yankee characters, particularly Nimrod Wildfire in Paulding's *The Lion of the West* (1831). He was the first American actor to appear in London as a star, playing Falstaff, one of his best parts, in 1833, and several of his Yankee characterizations. He was manager of the *Astor Place Opera House on the occasion of the *Macready riot. A scholarly and hard-working actor, he influenced the development of the American theatre by his encouragement of native playwrights. He was the father of the romantic actor J. K. *Hackett.

Hackett, James Keteltas (1869–1926), American actor, son of J. H. *Hackett. He was a romantic actor who played leading Shakespeare and *Sheridan roles under *Daly in 1892 and in 1895 joined Daniel *Frohman's company at the *Lyceum in New York, where he appeared in Hope's *The Prisoner of Zenda* and similar romantic plays. In 1905 he briefly opened his own theatre in New York, and later, with the proceeds of a legacy, put on an *Othello* with sets by Joseph *Urban which marked an important step forward in the history of American stagecraft and scenic design.

Hagen, Uta Thyra (1919–), German-born American actress, who first appeared on the stage in 1935 in *Coward's *Hay Fever*, and in 1937 played Ophelia to the Hamlet of Eva *Le Gallienne in Dennis, Mass. She made her New York début with the *Lunts in 1938 as Nina in *Chekhov's *The Seagull*, and starred with her first husband José *Ferrer in Maxwell *Anderson's *Key Largo* (1939). In 1943 she was Desdemona to Ferrer's Iago and Paul *Robeson's Othello, and she played in German in *Goethe's *Faust* (1947) and *Ibsen's *The Master Builder* (1948). Later in 1948 she took over from Jessica *Tandy the part of Blanche Du Bois in Tennessee *Williams's *A Streetcar Named Desire*, which she played for two years. In 1950 she gave a highly praised performance in the title-role of *Odets's *The Country Girl*, other notable roles including Joan in *Shaw's *Saint Joan* (1951), Tatiana in *Sherwood's *Tovarich* (1952), Natalia in *Turgenev's *A Month in the Country* (1956), and the dual leading role in the first New York production of *Brecht's *The Good Woman of Setzuan* (also 1956). In 1962 she was first seen in her most famous role, Martha in *Albee's *Who's Afraid of Virginia Woolf?* (London, 1964). In 1968 she appeared as Ranevskaya in Chekhov's *The Cherry Orchard*. Later she taught acting and her own performances became rare, though she starred in Shaw's *Mrs Warren's Profession* in 1985 and his *You Never Can Tell* in 1986.

Haines, Joseph, see HARLEQUIN.

Halevy, Moshe, see OHEL THEATRE.

Hall, Sir Peter Reginald Frederick (1930–), English theatre manager and director. He directed his first professional production in 1953 at the Theatre Royal, Windsor. A year later he was in London, at the *Arts Theatre, where among other new plays he directed *Beckett's *Waiting for Godot* (1955) and *Anouilh's *The Waltz of the Toreadors* (1956). He was responsible for an excellent *Love's Labour's Lost* (also 1956) at the *Shakespeare Memorial Theatre, where he later directed *Cymbeline* (1957), *Twelfth Night* (1958), *A Midsummer Night's Dream* with Charles *Laughton, and *Coriolanus* with *Olivier (both

1959). In 1960 he became Managing Director of the newly formed *RSC, where his productions included Anouilh's *Becket* (1961), *The Wars of the Roses* (based on *Henry VI* and *Richard III*, 1963), and *Pinter's *The Homecoming* (1965; NY, 1967). He resigned in 1968, though he remained Co-Director and continued to direct plays for the company, notably *Albee's *A Delicate Balance* (1969) and Pinter's *Old Times* (London and NY, 1971). From 1973 to 1988 he was Director of the *National Theatre, himself directing *Hamlet* (1975) and *Marlowe's *Tamburlaine the Great* (1976), the opening productions at the *Lyttelton and *Olivier theatres respectively. He also directed such outstanding productions as Pinter's *No Man's Land* (1975; NY, 1977); *Shaffer's *Amadeus* (1979; NY, 1980); a masked, all-male version of *Aeschylus' *Oresteia* (1981); and *Antony and Cleopatra* (1987). His work in the commercial theatre included Tennessee *Williams's *Cat on a Hot Tin Roof* (1958) and *Orpheus Descending* (1988), and *The Merchant of Venice* (1989); the last two, the first productions of his own company, were seen in New York in 1989. He is also a major director of opera.

Hall, Willis (1929–   ), English playwright, most of whose later plays have been written in collaboration with **Keith Waterhouse** (1929–   ). The first of Hall's own plays to be staged, *The Royal Astrologers* (1958), was written for children, his first play for adults being the highly successful *The Long and the Short and the Tall* (1959; NY, 1962), a study of the diverse members of a patrol lost in 1942 in the Malayan jungle. After a double bill, *Last Day in Dreamland* and *A Glimpse of the Sea* (also 1959), both plays set in a seaside resort at the end of the season, Hall began his collaboration with Waterhouse, their first joint work—one of their most successful—being *Billy Liar* (1960; NY, 1963), based on Waterhouse's novel about a fantasizing North Country boy. It was followed by *Chin-Chin* (also 1960), based on a play by *Billetdoux, and *England, Our England* (1962), a revue. More typical of their collaboration, however, were the comedy *All Things Bright and Beautiful* (1962) and the farce *Say Who You Are* (1965), seen in America in 1966 as *Help Stamp Out Marriage*. They then wrote the book for the musical *The Card* (1973), based on Arnold *Bennett's novel, and had a big success at the *National Theatre with *Saturday, Sunday, Monday* (also 1973), based on a play by Eduardo *De Filippo, which in 1974 was transferred to the West End and seen in New York. In 1977 he again collaborated with Waterhouse in *Filumena* (NY, 1980), based on another play by De Filippo. His later stage work has been for children. Waterhouse also wrote the highly successful *Jeffrey Bernard is Unwell* (1989).

Hallam, Lewis, the elder (1714–56), English actor closely connected with the establishment of the professional theatre in the United States. Son of the actor Adam Hallam, he had already had considerable experience on the stage in London when in 1752 he took his wife and children, with a company of 10 actors, to *Williamsburg, Va., where they appeared in *The Merchant of Venice* and *Jonson's *The Alchemist*. A year later Hallam refurbished and reopened the Nassau Street Theatre, where his company, in the face of some opposition, appeared in a wide variety of plays. Later, after a visit to Philadelphia and Charleston, he went to Jamaica, where he died, his widow marrying David *Douglass. Hallam's youngest daughter Isabella (1746–1826), left behind in England, became famous on the English stage as **Mrs Mattocks**.

Hallam, Lewis, the younger (c.1740–1808), son of the above, who went with his father to Williamsburg in 1752 and in 1757 became leading man of the combined companies of his mother and stepfather, going with them to New York, where in 1758, as the *American Company they played in a temporary theatre on Cruger's Wharf. Lewis was an excellent actor and appeared in *Godfrey's *The Prince of Parthia* (1767), the first American play to be given a professional production. After the death of *Douglass in 1786 he took over the American Company, first with *Henry and later with *Hodgkinson and *Dunlap until 1797, when he retired from management, though he continued to act until his death.

Hall Keeper, see STAGE DOOR.

Ham, term of derision applied to the old-fashioned rant and fustian which is supposed to have characterized 19th-century acting, particularly in *melodrama. The derivation of the word is uncertain. It was current in America from the 1880s, and seems to have found its way to England after the First World War. In essence, 'ham' acting is devoid of inner truth or feeling, covering its deficiencies with a veneer of over-worked technical tricks, bombast, showy but meaningless gestures, and bad diction.

Hamburg, city important in German theatrical history, since it was there that the first National Theatre was established with *Lessing as its accredited dramatic critic and *Ekhof as its leading actor under *Ackermann. It opened in 1767 and closed two years later. *Schröder, leader of Ackermann's company after 1771, made Hamburg a vital theatrical centre where from 1785 *Iffland's plays were staged. Schröder's successor took his company in 1827 to a new theatre, now the Staatsoper. A second playhouse, the Thaliatheater, opened in 1843 for the production of popular comedy. In

1905 *Jessner took over the theatre where he remained until 1915, introducing plays by *Wedekind, *Büchner, and *Ibsen. The company moved to a new building in 1912, which was destroyed in 1945. Rebuilt in 1960, it reopened with *Shaw's Saint Joan, and now competes successfully with the Deutsches Schauspielhaus (opened in 1900), the largest playhouse in West Germany, which was especially successful from 1955 to 1963 under *Gründgens. Michael *Bogdanov was its Artistic Director, 1989–91, launching 22 productions in the theatre's three spaces in his first season. In addition to the state theatres there are numerous private ones.

Hammerstein, Oscar II (1895–1960), American librettist and lyricist, son and nephew of theatrical managers and grandson of the impresario **Oscar Hammerstein** (1847–1919), who built theatres in London and New York. Oscar II was one of the most prolific writers of his generation, though most of his earlier work was done in collaboration. After several now forgotten pieces he worked on three 'American Viennese' operettas, Rose-Marie (1924; London, 1925), The Desert Song (1926; London, 1927), and The New Moon (1928; London, 1929). More important, however, was his association with the composer Jerome Kern, beginning with Sunny (1925; London, 1926) and followed by the masterly Show Boat (1927; London, 1928), for which Hammerstein wrote all the book and lyrics. His partnership with Kern continued with Sweet Adeline (1929), Music in the Air (1932; London, 1933), and Very Warm for May (1939). In 1943, with Oklahoma! (London, 1947), he began his even more famous collaboration with Richard Rodgers, with whom, apart from an updated version of Bizet's Carmen as Carmen Jones (also 1943; London, 1991), all his remaining work was done. Their most popular shows were Carousel (1945; London, 1950), South Pacific (1949; London, 1951), The King and I (1951; London, 1953), Flower Drum Song (1958; London, 1960), and The Sound of Music (1959; London, 1961).

Hammond, Kay, see CLEMENTS.

Hampden, Walter [Walter Hampden Dougherty] (1879–1955), American actor, born in New York, who first appeared on the stage in England, where he was a member of Frank *Benson's company and later played leading parts in London. In 1907 he returned to the United States, appearing with *Nazimova in a series of plays by *Ibsen and other modern dramatists. Among his most successful parts was the title-role in *Rostand's Cyrano de Bergerac, in which he first appeared in 1923, reviving it several times. He was also seen in a wide range of Shakespearian parts, which included Caliban, Hamlet, Macbeth, Oberon, Othello, Romeo, and Shylock. In 1925 he took over the Colonial

Theatre at 1887 Broadway, which he renamed Hampden's, and appeared there in an interesting repertory which included Henry V, *Benavente's The Bonds of Interest, and *Bulwer-Lytton's Richelieu. He remained in his own theatre until 1930, and then toured, mainly in revivals of his previous successes. In 1939 he played the Stage Manager in Thornton *Wilder's Our Town, and in 1947 was seen in Henry VIII, the first production of the American Repertory Theatre, playing Cardinal Wolsey to the Queen Katharine of Eva *Le Gallienne. His last Broadway appearance was in Arthur *Miller's The Crucible in 1953.

Hampstead Theatre, London, opened at the Moreland Hall, Hampstead, in 1959, the first London productions of *Pinter's The Room and The Dumb Waiter being seen there in 1960. In 1962 the theatre moved to its own prefabricated premises at Swiss Cottage; among many notable productions on its open-end stage were Laurie Lee's Cider with Rosie (1963) and Roy *Dotrice's solo performance as John Aubrey in Brief Lives (1967). In 1970 the building was moved to a new site near by, where the theatre, seating 157, enhanced its reputation, its numerous West End transfers including Michael *Frayn's Alphabetical Order (1975), James *Saunders's Bodies (1978), and Pinter's The Hothouse (1980). It also staged the British premières of such foreign plays as Tennessee *Williams's Small Craft Warnings and Peter *Handke's The Ride across Lake Constance (both 1973), both of which transferred to the West End, *Dürrenmatt's Play Strindberg (1973), Sam *Shepard's Buried Child (1980), and Arthur *Miller's double bill Danger: Memory! (1988). Two of Brian *Friel's plays were seen there, Translations (1981) moving on to the *Lyttelton, and Aristocrats (1988) having its British première.

Hampton, Christopher James (1946–), English dramatist, whose first play When Did You Last See My Mother?, written while he was still an undergraduate, was seen in the West End in 1966 and in New York a year later. Hampton became resident dramatist at the *Royal Court, 1968–70. His next play, Total Eclipse (1968), dealing with the relationship between Rimbaud and Verlaine, was followed by The Philanthropist (1970; NY, 1971), in which Alec *McCowen played an amiable but dispirited don, whose good intentions are constantly defeated by stronger personalities. Savages (1973; NY, 1977), which starred Paul *Scofield in London, is an ambitious non-naturalistic play about the kidnapping of a British diplomat by South American guerrillas. It was followed by the more conventional Treats (1976; NY, 1977). The Portage to San Cristobal of A. H. (1982), adapted from George Steiner's novel, and again starring McCowen, posits the discovery of Hitler in his nineties in a South American jungle; and Tales from Hollywood

(*National Theatre, 1983) the exile in Los Angeles of a group of European literary figures escaping from Hitler. His greatest success was *Les Liaisons dangereuses*, based on Laclos's epistolatory novel (*RSC, 1985; *Ambassadors, 1986; NY, 1987), which ran in London until 1990.

Hampton is also a notable translator of plays, including *Chekhov's *Uncle Vanya* and *Ibsen's *Hedda Gabler* (both 1970); Ibsen's *A Doll's House* (1971), *Molière's *Don Juan* (1972), and Ibsen's *The Wild Duck* (1979), the last two both seen at the National Theatre; and Molière's *Tartuffe* (RSC, 1983).

Hampton translated three plays by the Hungarian playwright and novelist **Ödön von Horváth** (1901–38), the first two being staged at the National Theatre. *Tales from the Vienna Woods* (1931; London, 1977), his best play, depicts a wartime love affair which evokes Viennese viciousness and degeneracy. *Don Juan Comes Back from the War* (1937; London, 1978) reinterprets the *Don Juan story; while *Faith, Hope, and Charity* (London, 1989), about an unemployed young girl who commits suicide, was banned by the Nazis in 1933. Horváth's plays, written in short scenes in a vein of ironic humour, often in Viennese dialect, were virtually unperformed for several decades. He was the main character in Hampton's play *Tales from Hollywood*.

Hancock, Sheila (1933–), highly versatile English actress. She had early experience in weekly *repertory, on tour, and with *Theatre Workshop, in whose musical *Make Me an Offer* (1959) she appeared in the West End. The revues *One to Another* (also 1959) and *One over the Eight* (1961) were followed by her first outstanding success, the prostitute in Charles Dyer's *Rattle of a Simple Man* (1962). She made her New York début in Joe *Orton's *Entertaining Mr Sloane* (1965) and on her return was seen in Charles *Wood's *Fill the Stage with Happy Hours* (1967) and, for the *RSC, two plays by *Albee, *A Delicate Balance* (1969) and *All Over* (1972). After *Ayckbourn's *Absurd Person Singular* (1973) she returned to revue with *Déjà Revue* (1974) and starred in the London productions of two major American musicals, *Annie* (1978) and *Sondheim's *Sweeney Todd* (1980). In 1981 she played Tamora in *Titus Andronicus* and Paulina in *The Winter's Tale* for the RSC, and in 1985 Ranevskaya in *Chekhov's *The Cherry Orchard* at the *National Theatre. She scored a great success in 1989 as a teachers' training college principal in Andrew Davies's *Prin*.

Handke, Peter (1942–), Austrian dramatist and novelist, who first attracted attention with his provocatively anti-theatrical *Publikumsbeschimpfung* (*Offending the Audience*, 1966). Other undramatic, plotless, characterless one-act pieces followed, until later works established Handke's serious concern with the problem of individual expression in a world overstocked with readymade concepts. In *Kaspar* (1968; London, 1973) the chief character, an innocent *tabula rasa*, is imprinted with conformist language and behaviour; in *Der Ritt über den Bodensee* (1971; seen in NY, 1972, and London, 1974, as *The Ride across Lake Constance*) the characters try to communicate through stereotyped speeches and gestures. In the more conventional *Die Unvernünftigen sterben aus* (*The Foolish Ones Die Out*, 1974) the banality of his business life stifles the sensitivity of the capitalist protagonist. A dramatic monologue reflecting on his mother's suicide was seen in New York in 1977 as *A Sorrow beyond Dreams*. *The Long Way Round* (1981; *National Theatre, 1989), about an inheritance dispute between three children after their parents' death, is a 'dramatic poem' in which the characters declaim but do not interact.

Hands, Terry [Terence] **David** (1941–), English director, one of the founders in 1964 of the *Everyman Theatre in Liverpool. In 1966 he joined the *RSC, initially as artistic director of Theatregoround, its travelling company, becoming an associate director of the main company in 1967 and its joint artistic director, with Trevor *Nunn, in 1978. Outstanding among his productions at the *Royal Shakespeare Theatre were *The Merry Wives of Windsor* (1968 and 1975), *The Merchant of Venice* (1971), the Stratford Centenary productions of *Henry IV* and *Henry V* (1975), and *Henry VI* (1977), the last three all starring Alan *Howard. After *Richard II* and *Richard III* (1980), again with Howard, he had directed all Shakespeare's history plays in five years. His productions at the *Aldwych Theatre included Triana's *The Criminals* (1967), the first Cuban play to be seen in London; *Genet's *The Balcony* (1971), *Arbuzov's *Old World* (1976), with Peggy *Ashcroft and Anthony *Quayle, and *Gorky's *The Children of the Sun* (1979). He was also consultant director of the *Comédie-Française, 1975–7, and directed several plays there. His productions of *Much Ado about Nothing* (1982) and *Rostand's *Cyrano de Bergerac* (1983) were staged in New York in 1984. In 1987 he was appointed sole Artistic Director and Chief Executive of the RSC, but he relinquished the post in 1991.

Hand Working, traditional practice of raising scenery by hand-lines, as against the modern *counterweight system. The piece of scenery to be flown is hung by a set of lines, generally three, which pass over three pulley blocks in the *grid and are made fast to a cleat on the fly rail. A complete *box-set can be battened together, with property furniture attached to its walls, and the whole unit flown entire on a number of sets of lines.

A hand-worked theatre was formerly known as a Rope House, and sometimes as a Hemp House.

Hanlon-Lees, a troupe of acrobatic actors who became internationally famous for their outstanding *trickwork in Dumb Ballets, of which the best known was probably *Le Voyage en Suisse,* seen in Paris at the Théâtre des *Variétés in 1879, and in London, at the *Gaiety Theatre, a year later. It included a bus smash, a chaotic scene on board ship during a storm, an exploding Pullman car, a banquet which turned into a juggling party, and one of the cleverest drunk scenes ever presented on the stage. The six Hanlon brothers began their career some time after 1860 in partnership with a famous acrobat, 'Professor' John Lees, and in 1883 (by which time Thomas had died) were in New York in a show called *Fantasma,* presented at the Fifth Avenue Theatre. This was revived in 1889, without Frederick, who died in 1886, and Alfred, who had retired, dying in 1892. Of the three survivors, William died in 1923, George, the eldest in 1926, and Edward in 1931.

Hannah Playhouse, see DOWNSTAGE THEATRE.

Hannen, Nicholas, see SEYLER.

Hansberry, Lorraine, see NEGROES IN THE AMERICAN THEATRE.

Hanswurst, comic character from German folklore, who borrowed some of his knavish tricks from the servant-roles of the *commedia dell'arte,* and became the equivalent of the Pickelhering of the *English Comedians. It was against his intrusion into serious plays that *Gottsched and Carolina *Neuber waged war in the 1730s. In Vienna the character was immortalized by *Stranitsky, instantly recognizable by his loose red jacket, blue smock, yellow pantaloons, a conical hat, and a wooden sword stuck into a wide leather belt. When Stranitsky retired in 1725 the part was taken over by **Gottfried Prehauser** (1699–1769) who made it less boisterous and clownish. On Prehauser's death in 1769 Hanswurst was replaced by Bernadon, and disappeared from the stage.

Happening, form of theatrical or artistic expression which is presumed to take the actors, as well as the audience, by surprise. It can also take place, usually for propaganda purposes, in the street or at a meeting. Though some groups, notably the *Living Theatre, were influenced by it, the phenomenon was a transient one.

Hardwicke, Sir Cedric Webster (1893–1964), English actor. He had had some pre-war acting experience when in 1922 he returned from war service and joined the *Birmingham Repertory Theatre, where his most successful parts were Churdles Ash in *Phillpotts's *The Farmer's Wife* and Caesar in *Shaw's *Caesar and Cleopatra,* in which he was subsequently seen in London. At the *Malvern Festival he created the parts of Magnus and the Burglar in Shaw's *The Apple Cart* (1929) and *Too True to be Good* (1932) and Edward Moulton-Barrett in Besier's *The Barretts of Wimpole Street* (1930), which he also played during a long run in London and on tour. He first went to the United States in 1936 and returned there in 1937 to play the title-role in Barré Lyndon's *The Amazing Dr Clitterhouse.* After a season at the *Old Vic in 1948 during which he played Sir Toby Belch in *Twelfth Night,* the title-role in *Marlowe's *Doctor Faustus,* and Gaev in *Chekhov's *The Cherry Orchard,* he settled permanently in New York, where in 1959 he made a great success as Koichi Asano in Spigelgass's *A Majority of One.*

Hardy, Alexandre (c.1575–c.1631), the first professional French playwright, attached to the company under Valleran-Lecomte which settled at the Hôtel de *Bourgogne, where his plays, of which about 40 survive from a possible 600–700, were given in the old-fashioned simultaneous settings designed by *Mahelot. They represent every type of contemporary drama but are usually classified as tragi-comedies, though many of them verge on melodrama. With a spark of genius Hardy might have changed the course of French dramatic literature, but his facility and easy success told against him and he lived to see the triumphs of *Corneille, who first made contact with the theatre through his plays, and of the *unities, which he had disregarded.

Hare, David (1947–), English playwright, who in 1968 co-founded the Portable Theatre, a travelling *Fringe group. He became literary manager (1969–70) and resident dramatist (1970–1) at the *Royal Court Theatre, where his play *Slag* (London and NY, 1971), about three mistresses in a girls' school, was presented. *Brassneck,* written in collaboration with Howard *Brenton, was first produced at the *Nottingham Playhouse in 1973 while Hare was resident dramatist. It covers three generations of an unscrupulous Midlands family, highlighting its commercial and political corruption. *Knuckle* (1974; NY, 1975) uses a conventional thriller format to make another attack on capitalist corruption. In 1974 Hare became co-founder of another Fringe company, Joint Stock, which in the following year presented *Fanshen,* his adaptation of a book about the Chinese revolution. *Teeth 'n' Smiles* (Royal Court, 1975; NY, 1979) starred Helen *Mirren in London as the drunken lead singer of a disintegrating rock group. His later plays were staged by the *National Theatre, beginning with *Plenty* (1978; NY, 1982), in which an intelligent and honest woman seeks in the post-war world for an outlet for her wartime idealism. Later came *A Map of the World*

(1983; NY, 1985), a debate between a left-wing journalist and a right-wing Indian novelist against the background of a UNESCO conference; and *Pravda* (1985), again with Brenton, a satirical portrait of a South African-born Press tycoon. After a double bill, *The Bay at Nice* and *Wrecked Eggs* (1986; NY, 1987), he produced his most lauded works, *The Secret Rapture* (1988; NY, 1989), a study of two contrasting sisters, a graphic designer and a Tory politician, in the aftermath of their father's death; and *Racing Demon* (1990), set in a South London parish, an examination of conflicting contemporary attitudes in the Church of England. In *Murmuring Judges* (1991) he scrutinizes the law. He has also directed several of his own and other people's plays.

Hare [Fairs], **Sir John** (1844–1921), English actor-manager. In 1865 he was engaged by the *Bancrofts to appear with them at the Prince of Wales Theatre in the plays of T. W. *Robertson. He was manager of the *Royal Court from 1875 to 1879 and then went into partnership with *Kendal at the *St James's, where one of his outstanding successes was *Pinero's *The Money Spinner* (1881). From 1889 to 1895 he was at the *Garrick, where he first played Benjamin Goldfinch in Grundy's *A Pair of Spectacles* (1890), a part with which his name is always associated. He made his first appearance in New York in 1895 in Pinero's *The Notorious Mrs Ebbsmith* and then revived several of his former successes, touring as Old Eccles in Robertson's *Caste*, in which he had played Sam Gerridge on its first production in 1867, and finally retiring in 1911.

Hare, J. Robertson (1891–1979), English comedian and a consummate player of farce, who made his first appearance on the stage in 1911 in Carton's *The Bear Leaders*, and spent many years touring in the provinces before becoming associated with Ralph *Lynn at the *Aldwych Theatre in a series of farces by Ben *Travers. Hare always played the 'little' man, swept along by the succession of outrageous mishaps which constituted the typical Aldwych farce. He was seen in Vernon Sylvaine's *Aren't Men Beasts!* (1936), but a year later was back with Travers in *Banana Ridge*, followed by *Spotted Dick* (1940), *She Follows Me About* (1943), *Outrageous Fortune* (1947), and *Wild Horses* (1952), being reunited with Ralph Lynn in the last two. He also appeared in Sylvaine's *One Wild Oat* (1948), and in Ronald *Millar's *The Bride and the Bachelor* (1956) and *The Bride Comes Back* (1960). In 1963 he played Erronius in the American musical *A Funny Thing Happened on the Way to the Forum*, and in 1968 appeared in John Chapman's *Oh, Clarence!*, based on stories by P. G. Wodehouse.

Harlequin, young lover of *Columbine in the English *harlequinade. His name, though not his status, derives from the *Arlecchino of the *commedia dell'arte*, where he was one of the quick-witted, unscrupulous serving-men; and so he remained in Italy. But in France *Marivaux turned him into a pretty simpleton, while in the English harlequinade he was first a romantic magician and later a languishing, lackadaisical lover, foppishly dressed in a close-fitting suit of bright silk diamonds (derived from the patches on his original rags), sometimes with lace frill and ruffles. He retained from his origins the small black cat-faced mask, and a lath or bat of thin wood which acted as a magic wand (see TRANSFORMATION SCENE). One of the first English Harlequins was **Joseph Haines** (?–1701), in *Ravenscroft's adaptation of *Molière's *Les Fourberies de Scapin* (1677).

Harlequinade, important element in the development of the English *pantomime. It resulted from the fusion of the dumbshow of the actors at the Paris fairgrounds with the convention that the trickster *Harlequin could turn himself into someone else. When the 'Italian Night Scenes' in which he figured were brought to London by **John Weaver** [Wever] (1673–1760), this convention was not understood, and it was the unhappy lover, or later the hero of the fairy-tale opening, who turned into Harlequin and so gave his name to the entire performance. Once the change had been effected the rest of the evening was devoted to the various stratagems by which the young lovers Harlequin and *Columbine managed to escape from the old man Pantaloon. Not until the early 19th century, when *Grimaldi made the purely English character *Clown into the chief personage of the Harlequinade, did Harlequin lose his premier position, and the love scenes between him and Columbine dwindled into short displays of dancing and acrobatics. As these in their turn began to pall, the fairy-tale opening increased in length, and the harlequinade dwindled into a short epilogue. For a time there was a pretence of changing the fairy-tale hero (the *principal boy, played always by a woman) into Harlequin by means of traps, but at last this too was abandoned, and the harlequinade, placed after the Grande Finale of the pantomime, lost all meaning. It finally disappeared altogether during the Second World War.

Harrigan, Ned [Edward] (1845–1911), American actor, manager, and dramatist, whose partnership with the female impersonator **Tony Hart** [Anthony Cannon] (1855–91) first brought him into prominence in New York in 1872. As 'Harrigan and Hart' they produced many successful shows, including *The Mulligan Guards*, a series in which Harrigan played Dan Mulligan and Hart his wife Cordelia. They parted company after a fire in 1884 had destroyed the theatre in which they were appearing, but Harrigan continued to act, appearing in revivals of some of his own plays, among them *Old Lavender* (first produced in 1877) and *The Major* (first produced in 1881). In 1890 he opened a theatre under his own name (see GARRICK THEATRE 3), where he remained for five years. He composed a number of songs and over 80 *vaudeville sketches, his characters being recognizable types of old New York, chiefly Irish- and German-Americans and Negroes.

Harrigan's Park Theatre, New York, see HERALD SQUARE THEATRE; **Harrigan's Theatre,** New York, see GARRICK THEATRE.

Harris, Augustus Glossop (1825–73), English actor and theatre manager, son of Joseph Glossop, first manager of the Royal Coburg Theatre (see OLD VIC), and an opera singer named Mme Féron. Why Harris assumed the name by which he is known is uncertain. He was first seen on the stage in America at the age of 8 and later went to London, where he appeared at the *Princess's Theatre, taking over the management when Charles *Kean retired in 1859. It was under him that *Fechter made his first appearance in London. But it is as the manager responsible for opera and ballet at *Covent Garden for 27 years that he is chiefly remembered. He also directed opera in Madrid, Paris, Berlin, and St Petersburg (Leningrad), proving himself an excellent stage manager, with a good eye for grouping and colour. He was the father of the theatre manager Sir Augustus *Harris.

Harris, Sir Augustus Henry Glossop (1852–96), son of Augustus *Harris, manager for many years of *Covent Garden. The son took over *Drury Lane in 1879 and remained there until his death, his devotion to his theatre earning him the nickname of 'Augustus Druriolanus'. He made a speciality of spectacular melodramas and elaborate Christmas shows, and although some critics held him responsible for the vulgarization of the *pantomime by the introduction of music-hall turns, particularly knockabout comedians, he had a feeling for the old *harlequinade, providing for it lavish scenery and machinery and engaging excellent clowns and acrobats.

Harris, Henry (c.1634–1704), English actor of the Restoration period, accounted by some contemporary critics superior even to *Betterton. He joined *Davenant's company at the *Lincoln's Inn Fields Theatre in 1661, where he played the title-roles in *Orrery's *Henry V* (1664) and *Mustapha, Son of Solyman the Magnificent* (1665), and also several Shakespearian parts, including Romeo, Sir Andrew Aguecheek in *Twelfth Night*, and Wolsey in *Henry VIII*. A pastel drawing of him in this last part hangs in Magdalen College, Oxford. In 1668, on the death of Davenant, Harris became joint manager of the theatre with Betterton, and he was also a shareholder in the new *Dorset Garden Theatre, built in 1670–1. He made his last appearance on the stage in 1681, playing the Cardinal in Crowne's *Henry VI, The First Part*.

Harris, Julie [Julia] Ann (1925–), American actress, who made her first appearance in New York in 1945, and a year later, as a member of the *Old Vic company in New York, was seen in *Henry IV, Part Two* and *Sophocles' *Oedipus*. Expert at portraying vulnerability, she was outstandingly successful as the adolescent Frankie Addams in Carson McCullers's *The Member of the Wedding* (1950) and the bohemian Sally Bowles in *Van Druten's *I am a Camera* (1951). She was later in two plays by *Anouilh, *Mademoiselle Colombe* (1954) and (as Joan of Arc) *The Lark* (1955). She was also seen in classical roles, playing Margery Pinchwife in *Wycherley's *The Country Wife* (1957), Juliet in *Romeo and Juliet* at the *Stratford (Ontario) Festival (1960), and Ophelia in *Hamlet* for Joseph *Papp (1964). Later roles were in Jay Allen's *Forty Carats* (1968), James Prideaux's *The Last of Mrs Lincoln* (1972), and *Rattigan's *In Praise of Love* (1974), and she gave a highly praised solo performance as Emily Dickinson in William Luce's *The Belle of Amherst* (1976; London, 1977). She is also well known in films and television.

Harris, Robert (1900–); English actor, who made his first appearance in 1923 in *Barrie's *The Will*, his fine voice and excellent presence later proving valuable assets in such parts as Oberon in *A Midsummer Night's Dream* (1924) and the title-role in *Ghéon's *The Marvellous History of St Bernard* (1925), in which year he also made his New York début in *Coward's *Easy Virtue*. In 1931 he joined the *Old Vic company, playing Hamlet and Mark Antony in *Julius Caesar*. He appeared as Marchbanks in *Shaw's *Candida* in New York (1937), and in the same year as Orin Mannon in *O'Neill's *Mourning Becomes Electra* in London (both 1937); and in *Priestley's *Dangerous Corner* (1938) and *Music at Night* (1939). In 1946 and 1947 he was at the *Shakespeare Memorial Theatre, where he played Prospero in *The Tempest*, Richard II, and the title-role in *Marlowe's *Dr Faustus*. After two important *Ibsen roles, Gregers Werle in *The Wild Duck* (1948) and Rosmer in *Rosmersholm* (1950), and *Hochwälder's *The Strong are Lonely* (1955), he returned to the Shakespeare Memorial Theatre and then to the Old Vic. In 1963–4 he toured the USA as More in *Bolt's *A Man for All Seasons*, and in 1966 he returned to the West End in the title-role of *Kipphardt's *In the Matter of J. Robert Oppenheimer*.

Harris, Rosemary (1930–), English-born actress, equally at home in America, being seen in New York in *Hart and Mittelholzer's *The Climate of Eden* (1952) before appearing for the first time in London in George Axelrod's *The Seven Year Itch* (1953). After playing major roles with the *Bristol Old Vic and the *Old Vic, she went with the Old Vic company to New York in 1956 to play Cressida in Tyrone *Guthrie's production of *Troilus and Cressida* in modern dress. Remaining in the United States, she became an important member of the Association of Producing Artists (see PHOENIX THEATRE, New York), and from 1960 to 1962 toured in such roles as Ann Whitefield in *Shaw's *Man and Superman* and Arkadina in *Chekhov's *The Seagull*. She returned to England in 1962 to appear in the first season at *Chichester, being seen as Constantia in *Fletcher's *The Chances* and Penthea in *Ford's *The Broken Heart*. She returned there in 1963, and later in the year played Ophelia in the *National Theatre's inaugural production of *Hamlet*. After appearing in New York in a revival of *Kaufman and Hart's *You Can't Take it with You* (1965) for the APA and as Eleanor of Aquitaine in James Goldman's *The Lion in Winter* (1966), she scored a big success in London as three different characters in Neil *Simon's *Plaza Suite* (1969). In America she continued to show her versatility in *Pinter's *Old Times* (1971) and in revivals of Shaw's *Major Barbara* (1972) (at the *American Shakespeare Theatre), Tennessee *Williams's *A Streetcar Named Desire* (1973), and Kaufman and Edna Ferber's *The Royal Family* (1975). In London in 1981 she gave a highly praised performance in a revival of *Miller's *All My Sons*, which she followed there with Shaw's *Heartbreak House* (1983; NY, 1983) and *The Best of Friends* (1988) by Hugh Whitemore, in whose *Pack of Lies* she starred in New York in 1985.

Harris, Sam H., see SAM H. HARRIS THEATRE.

Harrison, Sir Rex Carey (1908–90), English actor, who made his first appearance at the *Liverpool Playhouse in 1924 and was first seen in London in 1930. His good looks and elegant presence led almost inevitably to his being cast in a succession of featherweight parts in comedy, though his performances in *Rattigan's *French without Tears* (1936) and *Coward's *Design for Living* (1939) hinted at reserves of irony and anger. During the Second World War he was in the RAF, and he then went into films. He returned to the stage to play Henry VIII in Maxwell *Anderson's *Anne of the Thousand Days* (NY, 1948) and the Uninvited Guest in T. S. *Eliot's *The Cocktail Party* (London, 1950). He then starred in New York in *Van Druten's *Bell, Book and Candle* (1951; London, 1954), *Fry's *Venus Observed* (1952), and *Ustinov's *The Love of Four Colonels* (1953). In 1956 he was seen in his most famous role, Henry Higgins in *My Fair Lady*, a musical based on *Shaw's *Pygmalion*, which he played for two years on Broadway before repeating the part in London; he showed exactly the right tetchy authority and charming intolerance for Shaw's language-loving professor. Later he played the General in *Anouilh's *The Fighting Cock* (NY, 1959) and the title-roles in *Chekhov's *Platonov* (London, 1960) and *Pirandello's *Henry IV* (NY, 1973; London, 1974). He also appeared in Shaw's *Caesar and Cleopatra* (NY, 1977) and *Heartbreak House* (London and NY, 1983); *Lonsdale's *Aren't We All?* (London, 1984; NY, 1986); *Barrie's *The Admirable Crichton* (London, 1988); and *Maugham's *The Circle* (NY, 1989). He was also a major film star.

Harrison, Richard Berry (1864–1935), American Negro actor, the son of slaves who escaped to Canada. He returned as an adult to Detroit, was befriended by L. E. Behymer, and after some training in elocution toured the Behymer and Chautauqua circuits with a repertory of Shakespearian and other recitations. He was working as a teacher of drama and elocution when he was persuaded to play 'De Lawd' in Marc *Connelly's *The Green Pastures* (1930), in which he made an immediate success, appearing in the part nearly 2,000 times. Of medium build, with a soft but resonant voice, he was for the greater part of his life a lecturer, teacher, and arranger of festivals for Negro schools and churches. A modest man, he was surprised by his success as 'De Lawd'; but he enjoyed his contact with the professional theatre, his one great regret being that he never appeared in Shakespeare, whose works he knew so well.

Hart, Christine, see SCHRÖDER, F. L.

Hart, Lorenz (1895–1943), American lyricist, renowned for the musicals he wrote with Richard Rodgers as composer. He began his career translating plays for the *Shuberts. His first collaboration with Rodgers was 'Any Old Place with You' interpolated into the musical *A Lonely Romeo* (1919). They were also asked to write the songs for *Poor Little Ritz Girl* (1920) but most of their numbers were replaced by others. They collaborated on the revue *The Garrick Gaieties* (1925) but their first musical was *Dearest Enemy* (also 1925). It was followed by many other hits including *A Connecticut Yankee* (1927; London, 1929, as *A Yankee at the Court of King Arthur*), *On Your Toes* (1936; London, 1937), *Babes in Arms* (1937), *The Boys from Syracuse*, based on *The Comedy of Errors* (1938; London, 1963), and *Pal Joey* (1940; London, 1954). The revue *One Dam Thing after Another* (1927) and the musical *Ever Green* (1930) were seen in London but not New York. The cleverness and wit of Hart at his best—as in 'Manhattan' and 'The Lady is a Tramp'—have rarely been equalled. He was succeeded as Rodgers' lyricist by Oscar *Hammerstein.

Hart, Moss (1904–61), American dramatist and director, who began his career as office boy to Augustus Pitou, a theatre impresario to whom he sold his first play *The Beloved Bandit*. His second, *Once in a Lifetime* (1930; London, 1933), was bought by Sam Harris, and successfully produced after extensive rewriting by George S. *Kaufman, which led to a long collaboration between the two. Hart also wrote plays on his own, of which the most interesting were *Winged Victory* (1943), about the Air Force, *Christopher Blake* (1946), and *Light up the Sky* (1948). His last play was *The Climate of Eden* (1952) based on Edgar Mittelholzer's novel *Shadows Move among Them* and set on a missionary colony in British Guiana. His work for the musical stage included the sketches for the Irving *Berlin revue *As Thousands Cheer* (1933), the libretto for the operetta *The Great Waltz* (1934), and the book for Kurt Weill's *Lady in the Dark* (1941). He directed a number of his own works, and, among those by others, the musicals *My Fair Lady* (1956; London, 1958) and *Camelot* (1960).

Hart, Tony, see HARRIGAN.

Harvard University, Cambridge, Mass., the oldest institution of higher learning in the USA. The first plays known to have been acted there were *Addison's *Cato*, Whitehead's *The Roman Father*, and *Otway's *The Orphan*, performed surreptitiously by students. Acting was later encouraged, and the Hasty Pudding Club was formed, its productions now being mainly musicals. The Harvard Dramatic Club was started in 1908 and for many years produced only plays written by students or graduates of Harvard or Radcliffe (the women's college in Cambridge). It later concentrated on foreign plays. In 1905 Professor G. P. *Baker inaugurated his influential '47 Workshop' to stimulate the writing and production of new American plays. The Loeb Drama Center, seating 556, was built in 1960, the first fully automated flexible theatre in the United States, offering a proscenium, thrust, or arena stage. There is also an experimental theatre seating 120. The Center is used by the Harvard-Radcliffe Dramatic Club and by the American Repertory Theatre, a professional company under the directorship of the distinguished critic **Robert Brustein** (1927–) which took up residence in 1980. Its opening seasons included highly praised productions of *A Midsummer Night's Dream* and *Brecht and Weill's musical *Happy End*. The Harvard Theatre Collection, begun in 1901, is one of the finest performing arts research libraries in the world.

Harvey, Sir John Martin-, see MARTIN-HARVEY.

Harwood, H. M., see AMBASSADORS THEATRE.

Harwood, John Edmund (1771–1809), American actor. He was for some years at the *Chestnut Street Theatre, Philadelphia, under *Wignell, and was with the company when it appeared in New York in 1797, being much admired in low-comedy parts. He was engaged by *Dunlap for the *Park Theatre in 1803, and remained there until his death, making his first great success as Dennis Brulgruddery in the younger *Colman's *John Bull; or, An Englishman's Fireside*. He was celebrated for his portrayal of Falstaff in *Henry IV*, which he first played at the Park Theatre in 1806 with *Cooper as Hotspur. He later changed his style of acting, appearing as polished gentlemen, for which his fine presence and handsome countenance made him eminently suitable.

Harwood, Ronald, see ROYAL EXCHANGE THEATRE.

Hasenhut, Anton, see THADDÄDL.

Hauptmann, Gerhart (1862–1946), German dramatist, the chief playwright of *naturalism, who in 1912 was awarded the *Nobel Prize for Literature. A Silesian by birth, he often used his native dialect to heighten the authenticity of his plays, of which the first, *Vor Sonnenaufgang* (*Before Dawn*, 1889), was produced in Berlin by the *Freie Bühne. Its depiction of a farmer's family which suddenly becomes wealthy, but sinks into degradation through alcoholism and sexual promiscuity, delighted some sections of the audience but outraged others by its uncompromising presentation of human and social misery. It was followed by two middle-class psychological dramas, *Das Friedensfest* (*The Coming of Peace*, 1890) and *Einsame Menschen* (*Lonely People*, 1891), which examines the problem of marital incompatibility. Hauptmann's next play *Die Weber* (*The Weavers*, 1892), based on the revolt of the Silesian weavers in 1844, was unusual in having as its hero a group of men instead of a single individual. Its naturalism displays dramatic qualities seldom achieved in this form: it is probably Hauptmann's best, as well as his best-known, play. It was followed by a satirical comedy *Der Biberpelz* (*The Beaver Coat*, 1893), in which, and in its sequel *Der rote Hahn* (*The Red Rooster*, 1901), Hauptmann created some fine character parts, notably Mutter Wolffen, a scurrilous but appealing Berlin washerwoman who cunningly outwits Prussian bureaucracy. *Florian Geyer* (1896) is a naturalistic excursion into historical drama based on the Peasants' Revolt in the time of Luther. Of other naturalistic plays written at this time, *Rose Bernd* (1903) shows a simple girl driven by circumstances and instinct to infanticide; *Führmann Henschel* (1898) and *Die Ratten* (1911) paint even darker pictures of man's enslavement to environment and circumstance, both ending with the suicide of the main character. But Hauptmann's plays were not always naturalistic. *Hanneles Himmelfahrt* (*The Assumption of Hannele*, 1893), with its interpolated dream sequence, marks the beginning of a transition to a more poetic and symbolist approach, while *Die versunkene Glocke* (*The Sunken Bell*, 1896) explores the Romantic theme of the creative artist's relationship with reality. Hauptmann's last dramatic work, written in the shadow of the Second World War, was a blank verse tetralogy on the doom of the Atrides. The theme of senseless bloodshed exacted by inscrutable powers reflects Hauptmann's pain at witnessing Europe disembowelling itself for the second time in his lifespan.

Several of Hauptmann's plays have been performed in English translation: *The Sunken Bell* (NY, 1900; London, 1907); *Hannele* (NY, 1910; London, 1924); *The Weavers* (NY, 1915; London, 1980). In 1933 Miles *Malleson's adaptation of *Vor Sonnenuntergang* (1932) as *Before Sunset* was seen in London with Peggy *Ashcroft.

Haupt- und Staatsaktionen, plays acted in German by the *English Comedians and later by professional strolling players between 1685 and 1720, so named by *Gottsched to distinguish them from the comic *after-piece. They were usually concerned with the strong passions and elaborate intrigues of kings, emperors, and great warriors. The plots were taken from all available European sources and rewritten to provide an unsophisticated audience with a feast of rhetoric, often crude and bombastic, and spectacular effects featuring executions, ghostly apparitions, weddings, and coronations, the whole interrupted at frequent intervals by comic interludes featuring *Hanswurst.

Havel, Václav (1936–), Czech dramatist, influenced by Kafka, who worked as a stagehand and lighting technician before turning to writing. His most famous play *The Memorandum* (1965; NY, 1968; London, 1977), a satire on bureaucracy and Big Brother, shares a common theme with *Garden Party* (1963) and *The Increased Difficulty of Concentration* (1968; London, 1978). Other works included three short plays, *Audience* (*Interview* in New York), *A Private View*, and *Protest* (1975; NY, 1983; London, 1989); they are known as *The Vanek Plays* after an almost silent playwright character who appears in them. In *Largo Desolato* (1984; NY, 1986; London, 1987), a distinguished academic author is reduced to impotent despair by political pressure; the translation used in London was by *Stoppard. *Temptation* (1985; *RSC, 1987) was another attack on political repression, in the form of a Faustian allegory. Havel was imprisoned several times because of the political content of his plays and other writings. Even when nominally free he was under constant surveillance and his plays were banned in Czechoslovakia from 1968. His long record of protest against the Communist regime led to his being elected President after its overthrow in 1989.

Havergal, Giles, see CITIZENS' THEATRE.

Hawtrey, Sir Charles Henry (1858–1923), English actor-manager. A far better actor than his popularity in what became known as a 'Hawtrey' part ever allowed him to demonstrate, he was a fine light comedian and had no equal in the delineation of the man-about-town of his day. He made his first appearance on the stage in 1881, under the name of Bankes, and soon went into management on his own account. In 1883 he adapted a German comedy by Von Moser as *The Private Secretary*. When tried out at the Prince of Wales (later the *Scala) Theatre in 1884 it was not a success, but transferred to the *Globe Theatre, with W. S. *Penley in the title-role, first played by *Tree, it ran for two years and was frequently revived. Hawtrey himself played Daniel Cattermole in this production. Among other plays which he directed and appeared in the most successful were R. C. Carton's *Lord and Lady Algy* (1898), Ganthony's *A Message from Mars* (1899), Anstey's *The Man from Blankley's* (1901), George A. Birmingham's *General John Regan* (1913), and Walter Hackett's *Ambrose Applejohn's Adventure* (1921). Hawtrey was for many years responsible for the production at Christmas of the children's play *Where The Rainbow Ends* by Clifford Mills and John Ramsey, first seen at the *Savoy Theatre in 1911.

Hayes [Brown], **Helen** (1900–), American actress, a tiny woman who became a theatrical *grande dame*. She made her first appearance on the stage at the age of 5, played the lead in Frances Hodgson Burnett's *Little Lord Fauntleroy* at 7, the dual lead in Mark Twain's *The Prince and the Pauper* at 8, and graduated easily to adult roles from about 1920 onwards by way of Eleanor Porter's *Pollyanna* and Margaret in *Barrie's *Dear Brutus*. She scored a great success in *Kaufman and *Connelly's *To the Ladies* (1922) and appeared for the *Theatre Guild as Cleopatra in *Shaw's *Caesar and Cleopatra* (1925). In 1933 she was again outstanding as Mary Queen of Scots in Maxwell *Anderson's *Mary of Scotland*. Her greatest triumph, however, was the title-role in Laurence *Housman's *Victoria Regina* (1935) in New York and later throughout the USA. She was much admired in 1940 as Viola in *Twelfth Night* and made her first appearance in London in 1948 as the mother in Tennessee *Williams's *The Glass Menagerie*. She gave a moving performance as Nora Melody in *O'Neill's *A Touch of the Poet* (1958), and then played mainly in revivals.

In 1962 she and Maurice *Evans presented *Shakespeare Revisited*, a programme of scenes from the plays, on a nation-wide tour. She retired from the stage in 1970.

Haymarket Theatre, Leicester. This theatre, opened in 1973, was intended to take the place of the Phoenix, opened 10 years earlier on a site expected to be available for only five years. The theatres worked together, however, until 1979, when the Phoenix became a mixed programme arts centre, known from 1988 as Phoenix Arts. The Haymarket is larger, seating 710, and in its first season presented two highly successful musicals, *Cabaret* and *Joseph and the Amazing Technicolor Dreamcoat*, together with a revival of *Farquhar's *The Recruiting Officer*, John Hopkins's *Economic Necessity*, *Brecht's *The Caucasian Chalk Circle*, and a pantomime, *Aladdin*. Later seasons showed a similarly wide range, including productions of the musicals *Oliver!*, *My Fair Lady*, *Oklahoma!*, *Me and My Girl*, and *High Society*, all of which later had long London runs, as well as two vastly different trilogies—*Sophocles' *Theban Plays* and *Ayckbourn's *The Norman Conquests*. The studio theatre has presented mainly modern plays. The Haymarket company tours for five or six weeks annually, during which the theatre houses visiting companies. The theatre enjoys international acclaim for its bold and innovative policy of presenting world drama on a broad base. In 1989 it launched John *Dexter's production of David Henry Hwang's *M. Butterfly*, later to have a big success in London, and Yuri *Lyubimov's *Hamlet*, destined for a major world tour.

Haymarket Theatre, London. In 1720 John Potter, a carpenter, built a 'Little Theatre in the Hay' on the site of the old King's Head tavern, the first recorded performance being given in 1720 by a visiting French company. In 1729 the theatre, which was not licensed and could not stage 'legitimate' drama, had an unexpected success with a wild burlesque entitled *Hurlothrumbo; or, The Supernatural*, which ran for 30 nights. In the 1730s the satires of Henry *Fielding, which attacked the Government and the Royal Family, brought notoriety to the theatre and led indirectly to the passing of the Licensing Act of 1737 which caused it to be closed down (see CENSORSHIP). It stood empty until in 1747 Samuel *Foote took it over, evading the law by various ingenious methods. A feature of his entertainments was his mimicry of well-known persons, which soon became all the rage. He had set his heart on obtaining a patent for his theatre, which by accident he did in 1766, and although it was valid for the summer months only the Haymarket became a Theatre Royal, a title it still retains. Foote sold out in 1776 to the elder *Colman, who made many improvements and launched the theatre

on a period of prosperity, all the great actors of the day appearing there in the summer when *Drury Lane and *Covent Garden were closed. In 1794 Colman was succeeded by his son, who was always in financial difficulties. In 1817 he was imprisoned for debt; his brother-in-law and partner carried on alone, and in 1820 built the present Haymarket a little to the south of the old. The new building, designed by Nash, whose pedimented portico of six Corinthian columns extending over the pavement has survived reconstructions, opened in 1821 with *Sheridan's *The Rivals* and in 1825 had a great success with *Liston, always a prime favourite in comedy, in Poole's farce *Paul Pry*. In 1837 Benjamin *Webster became manager, and under him the theatre was substantially altered, gas lighting being installed (the Haymarket was the last theatre in London to use candles) and the forestage and proscenium doors abolished. The theatre prospered, *Phelps making his début there in 1837, and most of the great players of the day being seen there later. One of the good new plays, in which Webster himself appeared, was *Masks and Faces* (1852) by Tom *Taylor and Charles *Reade. A year later Webster was succeeded by *Buckstone, an excellent comedian whose ghost is said to haunt the theatre. As Drury Lane at this time was little better than a showbooth and Covent Garden was given over to opera, the Haymarket became the leading playhouse of London. Under Buckstone's management Edwin *Booth made his first appearance in London in 1861, in which year E. A. *Sothern also came from the USA to appear as Lord Dundreary in Taylor's *Our American Cousin*. Sothern made a further success in 1864 in *David Garrick*, by the then unknown T. W. *Robertson. Buckstone retired in 1879 and the *Bancrofts took possession. They remodelled the interior of the theatre, doing away with the pit and taking the stage back behind the proscenium arch, making it the first picture-frame stage in London, and opened in 1880 with a revival of *Bulwer-Lytton's *Money*. They ran the theatre most successfully, adding immensely to its prestige, until they retired in 1885. In 1887 the theatre passed into the hands of *Tree, under whom Oscar *Wilde's *A Woman of No Importance* (1893) and *An Ideal Husband* (1895) were first produced. His greatest success, however, was George *Du Maurier's *Trilby* (also 1895). A year later Tree moved to *Her Majesty's and Cyril *Maude took over with Frederick Harrison, opening in 1896 with *Under the Red Robe* from Stanley Weyman's novel. After a long period of success Maude withdrew in 1905, having overseen the reconstruction of the interior, but Harrison carried on until his death in 1926. Among the outstanding successes were *Barrie's *Mary Rose* (1920) with Fay *Compton, who returned in 1925 in Ashley *Dukes's *The Man with a Load of Mischief*;

Yellow Sands (1926) by Eden and Adelaide *Phillpotts; St John *Ervine's *The First Mrs Fraser* (1929) with Marie *Tempest; and *Ten-Minute Alibi* (1933) by Anthony Armstrong. The theatre escaped damage from enemy action during the Second World War and in 1944–5 housed a fine company in repertory under John *Gielgud. In 1948 Helen *Hayes made her first appearance in London in Tennessee *Williams's *The Glass Menagerie*. Notable later productions included *The Heiress* (1949), a dramatization of Henry *James's novel *Washington Square*; N. C. *Hunter's *Waters of the Moon* (1951) and *A Day by the Sea* (1953); Enid *Bagnold's *The Chalk Garden* (1956); Robert *Bolt's *Flowering Cherry* (1957); and *Rattigan's *Ross* (1960) with Alec *Guinness as T. E. Lawrence. All-star revivals held the stage until in 1971 Guinness returned in John *Mortimer's *A Voyage round My Father*. A year later Royce Ryton's *Crown Matrimonial*, on the abdication of Edward VIII, had a long run. The theatre housed transfers of such *Chichester revivals as *Maugham's *The Circle* (1976), *Waters of the Moon* (1978), *Shaw's *The Millionairess* (also 1978), and *Congreve's *The Way of the World* (1984). In the 1980s there were more star revivals plus new plays such as Simon *Gray's *Melon* (1987). In 1988 the theatre staged Tennessee Williams's *Orpheus Descending*, the first production of the Peter *Hall Company, with Vanessa *Redgrave, and in 1989 *Mamet's *A Life in the Theatre*.

Hazlitt, William (1778–1830), English essayist and critic, and the first of the great dramatic critics, contemporary of Leigh *Hunt, *Coleridge, and *Lamb. From 1813 to 1818 he reviewed plays for the *Examiner*, the *Morning Chronicle*, the *Champion*, and *The Times*, and it was his good fortune to be living in an age of great acting, displayed chiefly in revivals of Shakespeare, since there were few good new plays. In his zeal for bygone dramatists Hazlitt was once led to say he loved the written drama more than the acted, but he nevertheless took a vivid delight in acting and much enjoyed the society of actors.

Heavy Father, Woman, see STOCK COMPANY.

Hebbel, Friedrich (1813–63), German dramatist, whose work bears the imprint of the bitter struggles of his early years. His intuitive insight into feminine psychology resulted in a series of subtle portraits, as in *Judith* (1840), his first play, *Maria Magdalena* (1844), a powerful middle-class tragedy which anticipates the later naturalism of *Ibsen, and *Herodes und Mariamne* (1850), a fierce tragedy of jealousy. He was also the author of *Gyges und sein Ring* (1856) and the trilogy of *Die Nibelungen* (1861), his last work. He was fortunate in finding an excellent interpreter of his heroines in his wife.

Hebrew Theatre. Although plays in Hebrew formed part of the repertory of *Jewish drama from the 16th century onwards, there was no permanent Hebrew theatre until the 20th. The popularity of Yiddish drama made it clear that the revival of spoken Hebrew aimed at by the Zionists could be encouraged by the provision of Hebrew plays, and in 1907 two groups were founded for this purpose, in Warsaw and Jaffa. This second group, consisting of amateurs who called themselves 'Lovers of Dramatic Art', hoped to encourage the use of the Hebrew language among the small Jewish community in their city; they flourished until the outbreak of the First World War in 1914, and the consequent expulsion of the Jews from Jaffa by the Turkish authorities. The most important Hebrew theatre, the *Habimah, was founded in Moscow in 1917, and was ultimately, with theatres such as *Ohel and *Cameri, to contribute to the flourishing theatre of the new state of Israel. This theatre is attended regularly by over half the population, partly because all theatres are called upon to travel; the patronage of the kibbutzim also contributes to the theatre's popularity.

In the 1960s **Nissim Aloni** (1926–) emerged as the country's leading playwright, directing his own play, *Bigdey Ha'melech* (*The Emperor's Clothes*), in 1961. His work, influenced by the Theatre of the *Absurd, is complex and enigmatic, but some of his later plays, such as *Doda* [*Aunt*] *Liza* (1969), dealing with three generations of an Israeli family, seem to be moving away from fantasy to a locally based reality. Aloni is also a notable director.

Hedgerow Theatre, Moylan, Pa., founded in 1923, presented a wide range of classical and modern works in *repertory with frequent changes. With a semi-professional company working as a co-operative, it performed all the year round in a converted grist mill dating from 1840 seating 136. It staged the world premières of several American plays, toured throughout the United States, and ran a theatre school. The repertory grew to over 200 plays, and the theatre was at times the only one in the USA to operate the system; it was, however, abandoned in 1956. The building was burnt down in 1985, but is being reconstructed.

Heiberg, Gunnar Edvard Rode (1857–1929), Norwegian dramatist, whose work shows the influence of *Ibsen's later style. He was for some years (1884–8) Director of Det Norske Teatret in Bergen. His first play, *Tante Ulrikke* (1884), was a deftly satirical contribution to contemporary 'problematic' literature. The two plays by which he is best known—*Balkonen* (*The Balcony*, 1894) and *Kjærlighedens Tragedie* (*The Tragedy of Love*, 1904)—explore the nature of sexual passion. His ability to write light witty dialogue is perhaps best in evidence in *Kunstnere*

(*The Artists*, 1893) and *Gerts have* (*Gert's Garden*, 1894). His later plays did not greatly add to his stature, except for the last, *Paradesengen* (1913)—*The Catafalque*—a bitter satire on the unseemly behaviour of a great man's heirs quarrelling over their heritage.

Heiberg, Johan Ludvig (1791–1860), Danish poet, dramatist, and critic. After leaving college he spent six years abroad before settling in Copenhagen, where he became one of the leaders of the city's intellectual and cultural life and was for a time director of the *Royal Theatre. His dramatic output began as early as 1814 with *Marionetteater*, and included over the following years popular successes such as *Julespøg og Nytaarsløjer* (*Christmas and New Year Fun and Games*, 1816) and the vaudeville *Kong Salomon og Jørgen hattemager* (*King Solomon and Jørgen the Hatter*, 1825). His reputation as a dramatist reached its peak in 1828 with *Elverhøj* (*Elfin Hill*), a romantic blend of realism and fantasy, which retains its popularity today. Of his later works, *Fata Morgana* (1838), *Syvsoverdag* (*Day of the Seven Sleepers*, 1840), and *En sjæl efter døden* (*A Soul after Death*, 1841)—the last an elegantly satirical verse drama and the pinnacle of his achievement—are the most noteworthy. He married in 1831 the actress **Johanne Pätges** (1812–90) who until her retirement in 1864 was supreme among Danish actresses. She was the author of several light but popular vaudevilles.

Heijermans, Herman (1864–1924), Dutch dramatist, and the first since *Vondel to become well known outside his own country. Under the influence of the new *naturalism, he set out to depict the hypocrisy of contemporary bourgeois morality and the miseries of racial minorities and the working class. His first successful plays *Ahasverus* (1893) and *Ghetto* (1898), seen in London and New York in 1899, both dealt with the sufferings of the Jewish people in Russia and Amsterdam—Heijermans was himself a Jew. Among his later one-act plays was *In der Jong Jan* (1903), written for the Dutch actor Henri de Vries, who took it to New York in 1905 as *A Case of Arson*, playing all seven witnesses of the crime himself. Heijermans's masterpiece was undoubtedly the three-act drama of the sea, *Op Hoop van Zegen* (1900), which dealt with the exploitation of poor Dutch fishermen by greedy shipowners. As *The Good Hope* it was seen in London in 1903, with Ellen *Terry, and in 1927 was produced for the Civic Repertory Company by Eva *Le Gallienne.

Heinrich Julius, Duke of Brunswick, see WOLFENBÜTTEL.

Helburn, Theresa, see LANGNER.

Helen Hayes Theatre, New York. 1. On West 46th Street, between Broadway and 8th Avenue. Originally a theatre restaurant which

failed, it was remodelled and opened as the Fulton Theatre, seating 1,160, in 1911. Its first success was Hazelton and Benrimo's *The Yellow Jacket*; equally successful was *Brieux's *Damaged Goods* (1913). *Abie's Irish Rose* (1922) by Anne Nichols began its long run there before moving, and among later productions were *Lonsdale's *The High Road* (1928), the Stokes's *Oscar Wilde* (1938) with Robert *Morley, and Kesselring's *Arsenic and Old Lace* (1941). A play on the Negro problem, *Deep are the Roots* (1945) by Arnaud d'Usseau and James Gow, was also seen. In 1955 the Fulton was renamed in recognition of the stage jubilee of Helen *Hayes, who appeared there in 1958 as Nora in *O'Neill's *A Touch of the Poet*. In 1961 Jean Kerr's *Mary, Mary* started a three-year run, to be followed by the same author's *Poor Richard* (1964). An adaptation of Muriel Spark's novel *The Prime of Miss Jean Brodie* (1968) starred Zoë *Caldwell, and Alec *McCowen played the title-role in Peter Luke's *Hadrian VII* (1969). *The Me Nobody Knows* (1970) was an unusual musical based on the writings of New York tenement children. Later productions included Royce Ryton's *Crown Matrimonial* (1973), a revival of George S. *Kaufman and Edna Ferber's *The Royal Family* (1975), and *Strider: The Story of a Horse* (1979), a musical based on a short story by *Tolstoy. The theatre was demolished in 1982, together with the nearby *Morosco Theatre.

2. On West 44th Street, between Broadway and Eighth Avenue. Built by Winthrop *Ames as a try-out theatre, it opened as the Little Theatre in 1912. Productions in the first year included *Schnitzler's *Anatol* with John *Barrymore, *Prunella* by Laurence *Housman and Harley *Granville-Barker, *Shaw's *The Philanderer*, and a revival of Clyde *Fitch's *Truth*. The *Shuberts later took control, enlarging the seating capacity from 300 to 600, but the theatre was never very successful and after several changes of name and use the *New York Times* bought it in 1941 and used it for lectures, recitals, and television. In 1963 it reverted to live theatre, and in 1964, when Frank D. Gilroy's *The Subject was Roses* was transferred there, the name was changed to the Winthrop Ames Theatre; but a year later it reverted to its original name and was used for television. It became a theatre again in 1974 and Albert Innaurato's *Gemini* ran from 1977 until the early 1980s. The name was changed in 1983 during the run of Harvey Fierstein's *Torch Song Trilogy*.

Hellinger, Mark, see MARK HELLINGER THEATRE.

Hellman, Lillian (1905–84), American dramatist, whose first play *The Children's Hour* (1934; London, 1936) aroused extraordinary interest with its story of a neurotic schoolgirl who falsely accuses her teachers of lesbianism. In *The Little Foxes* (1939; London, 1942), a study of a predatory family of industrial entrepreneurs, and *Watch on the Rhine* (1941; London, 1942), which implies that America will soon have to join the fight against Fascism, Lillian Hellman fulfilled the promise of her début. In *Another Part of the Forest* (1946) she returns to the antecedent history of her 'little foxes'. *The Autumn Garden* (1951) is a powerful group study in frustration and *Toys in the Attic* (New York and London, 1960) a searing study of failure and possessiveness. She also adapted a number of novels for the stage, and William Luce's one-woman play *Lillian* (1986) is based on her autobiographical work *Scoundrel Time* (1976).

Helpmann, Sir Robert Murray (1909–86), Australian-born ballet dancer, choreographer, actor, and director who successfully made the transition from dancing to acting, bringing to both the same meticulous attention to detail and artistic integrity. He first appeared, as dancer and actor, in Sydney, and in 1933 joined the Vic-Wells Ballet, being principal dancer of the Sadler's Wells Ballet from its inception until 1950. He first came into prominence as an actor in London when he played Oberon in *A Midsummer Night's Dream* at the *Old Vic at Christmas 1937, repeating the part the following Christmas. In 1944 he appeared with the Old Vic company at the New Theatre (now the *Albery) as Hamlet, and in 1947, in partnership with Michael Benthall, took over the *Duchess Theatre, where he was seen as Flamineo in *Webster's *The White Devil* and Prince in *Andreyev's *He Who Gets Slapped*. He then went to the *Shakespeare Memorial Theatre, where in 1948 he again played Hamlet, as well as Shylock in *The Merchant of Venice* and King John. He returned to the London stage in 1951 (NY, 1951) to play Apollodorus in *Shaw's *Caesar and Cleopatra* and Octavius Caesar in *Antony and Cleopatra*, with Laurence *Olivier and Vivien *Leigh. He was seen as the Doctor in Shaw's *The Millionairess* (London and NY, 1952) on its first West End appearance, playing opposite Katharine *Hepburn. With the Old Vic company, which he directed in a number of plays including *Eliot's *Murder in the Cathedral* (1953), he appeared in a wide variety of Shakespearian parts, his versatility encompassing in 1955 Petruchio in *The Taming of the Shrew* and Angelo in *Measure for Measure*, and in 1956–7 Launce in *The Two Gentlemen of Verona* and Pinch in *The Comedy of Errors* as well as the title-role in *Richard III*. Among his non-Shakespearian parts were Georges de Valera in *Sartre's *Nekrassov* (1957) and Sebastien in *Coward's *Nude with Violin* which he took over from *Gielgud in the same year. He also directed *Giraudoux's *Duel of Angels* (1960) and the musical *Camelot* (1964). From 1965 to 1976 he was Director of the Australian Ballet, and in 1970 he was appointed Artistic Director of the Adelaide Festival, returning briefly to London in 1971 to direct a new production, several times revived, of *Barrie's *Peter Pan*.

Heminge [Heminges, Hemmings], John (1556–1630), English actor, a member of the *Chamberlain's Men and probably the first to play Falstaff. When the company was reorganized in 1603 as the *King's Men he appears to have acted as their business manager, and probably for this reason was instrumental with his fellow actor Henry *Condell in arranging for the printing of Shakespeare's plays in the one-volume edition—the First Folio, containing 36 plays—published in 1623. He and his joint editor have been criticized on several counts, including their arbitrary division of each play into five acts, on the classic model favoured by Ben *Jonson, and other sins of omission and commission which have been at the root of much literary controversy; but without their efforts the plays might well have been lost, as the author himself made no move to preserve them and no manuscript indisputably by Shakespeare has been found.

Hemp House, see HAND WORKING.

Henry, John (1738–94), Irish-born American actor who was for many years one of the leading men of the *American Company, which he joined in 1767 at the *John Street Theatre, New York, under *Douglass. On the return of the American Company from the West Indies, where they took refuge during the War of Independence, Henry assumed the management, jointly with the younger *Hallam. He accepted and played in *The Father; or, American Shandyism* (1789), the first of *Dunlap's plays to be produced, and was responsible in 1792 for the importation from England of John *Hodgkinson. He was twice married, his first wife, an actress, being lost at sea before the company came back from the West Indies. After living for some time with her sister Henry married a younger member of the family.

Henry Miller's Theatre, New York, on West 43rd Street, east of Broadway. This opened in 1918 with *The Fountain of Youth*, a translation of *Flor de la vida*, by the brothers *Álvarez Quintero. After achieving a modest success, the theatre embarked on the 'sophisticated' drama with which its name became connected, beginning with Noël *Coward's *The Vortex* (1925). In 1929 *Sherriff's *Journey's End* had a long run, and in 1932 *The Late Christopher Bean* by Sidney *Howard had its first production. The *Theatre Guild produced *O'Neill's *Days without End* in 1934; a revival of *Wycherley's *The Country Wife*, with designs by Oliver *Messel, was seen in 1936. Later successes were Thornton *Wilder's *Our Town* (1938), Norman Krasna's *Dear Ruth* (1944), T. S. *Eliot's *The Cocktail Party* (1950), and Saul Levitt's *The Andersonville Trial* (1961). The *RSC's recital *The Hollow

Crown was staged in 1963. In 1969 the building became a cinema.

Hensler, Karl, see RITTERDRAMA.

Henslowe, Philip (?–1616), English impresario, who unlike most theatre people of his time was never an actor but derives his importance in the history of the Elizabethan stage from being the owner of the *Fortune, *Hope, and *Rose playhouses. His stepdaughter Joan Woodward married Edward *Alleyn, who on his father-in-law's death inherited his property and papers, the latter now being housed in Dulwich College. Among them is Henslowe's 'diary', a basic document for the study of Elizabethan theatre organization. Since some of the actors in the companies which used his theatres were contracted to Henslowe personally, and not, as was usually the case in Elizabethan companies, to their fellow actors, and as he paid the dramatists for their work, it follows that he had a large say in the choice of play and method of presentation. That his relations with his actors were not always cordial is proved by a document drawn up in 1615, in which he is accused of embezzling their money and unlawfully retaining their property. There is no note of how the controversy ended, but evidently Henslowe kept actors and dramatists in his debt in order to retain his hold over them. This arrangement was not as good, nor did it make for such stability, as that in force among other companies like the *Chamberlain's Men, where the actors, led by their chief player Richard *Burbage, were joint owners of their own theatre, responsible only to each other.

Henson, Leslie Lincoln (1891–1957), English comedian, best remembered for his appearances in *musical comedy, who had been on the stage for some time before appearing in the West End in *Nicely, Thanks* (1912) at the *Strand Theatre. In 1914 he went to New York, playing Albert in *To-Night's the Night*, and returned to London to play the same part a year later, scoring an instantaneous success. After serving in the First World War he returned to the stage in a series of musicals beginning with *Kissing Time* (1919). He made a great success in *Funny Face* (1928), an apt description of his mobile indiarubber features, and was also seen in *It's a Boy* (1930) and *It's a Girl* (1931). From 1935 to 1939 he appeared in and directed a series of musicals at the *Gaiety containing parts specially written for him. After the Second World War, which he mostly spent entertaining the troops, he was seen in *Bob's Your Uncle* (1948), and as Pepys in a musical version of *Fagan's *And So to Bed* (1951); he made his last appearance in London as Mr Pooter in the Grossmiths' *Diary of a Nobody* (1955), after which he played Old Eccles in a musical version of *Robertson's *Caste* at Windsor.

Hepburn, Katharine Houghton (1909–), American star best known for her work in films but also an excellent stage actress. She made her first appearance in 1928 in Baltimore and her New York début the same year. From 1932 she worked mainly in Hollywood until 1939, when she made a triumphant return to Broadway in Philip *Barry's *The Philadelphia Story*. She again worked in films until 1950, when she played Rosalind in *As You Like It* in New York. Two years later she made her London début as the Lady in *Shaw's *The Millionairess* (NY, 1952). In 1955 she was seen in a series of Shakespearian parts in Australia with the *Old Vic company; and in 1957 she played Portia in *The Merchant of Venice* and Beatrice in *Much Ado about Nothing* at Stratford, Conn. (see *AMERICAN SHAKESPEARE THEATRE), returning in 1960 with her Viola in *Twelfth Night* and Cleopatra in *Antony and Cleopatra*. In New York in 1969 she starred in *Coco*, a musical based on the life of the French fashion designer Chanel. Later starring roles were in Enid *Bagnold's *A Matter of Gravity* (1976) and Ernest Thompson's *West Side Waltz* (1981).

Herald Square Theatre, New York, at the north-west corner of Broadway and 35th Street. This opened as the Colosseum in 1874. Renamed the Criterion in 1882, Harrigan's Park in 1885, and the Park in 1889, it had an undistinguished career until, rebuilt as Herald Square Theatre, it opened in 1894 with Richard *Mansfield in *Shaw's *Arms and the Man*. It was for some years an important link in the opposition to the powerful *Theatrical Syndicate, but was converted to *vaudeville by the *Shuberts, who had taken control in 1900, and was demolished in 1915.

Herbert, Sir Henry, see MASTER OF THE REVELS.

Herbert, Jocelyn (1917–), English stage designer. She trained under Michel *Saint-Denis in 1936–7 but did not begin practising until she joined the *English Stage Company in 1956. She played an important part in promoting the work of new playwrights with her settings for such productions as *Wesker's *Roots* and *Chips with Everything*, *Osborne's *Luther*, *A Patriot for Me*, and *Inadmissible Evidence*, and *Storey's *Home* and *The Changing Room*. She also worked for the commercial theatre, the *RSC, and the *National Theatre. The spare asceticism of her designs was seen to advantage in plays by *Beckett, *Brecht (including *Mother Courage* and *Galileo* for the National), and *Aeschylus' *Oresteia* (also for the National). Among her Shakespearian designs were *Richard III* for both the RSC and the *Royal Court, and *Hamlet* at the *Round House. She designed Storey's *The March on Russia* for the National Theatre in 1989.

Her Majesty's Theatre, London, in the Haymarket. The first theatre on this site was called the Queen's, after Queen Anne. Designed by Sir John *Vanbrugh, it opened in 1705 under the management of William *Congreve, one of the first plays to be given there being Vanbrugh's *The Confederacy.* The house proved unsuitable for drama and became London's first opera-house. On the death of Queen Anne in 1714 the theatre changed its name to the King's, in honour of George I. After a fire in 1789 a new theatre devoted entirely to opera and ballet opened in 1791. On the accession of Queen Victoria in 1837 it was renamed Her Majesty's, a name it retained until the accession of Edward VII in 1901, when it became His Majesty's, reverting to Her Majesty's on the accession of Elizabeth II in 1952. It had a consistently successful career until it was again burnt down in 1867. Although it was rebuilt between 1868 and 1869, it did not reopen until 1877, and never regained its former popularity. It finally closed in 1891 and was demolished, only the Royal Opera Arcade being left standing.

Some years later Beerbohm *Tree acquired part of the site, and with the profits from his production of Du Maurier's *Trilby* (1895) at the *Haymarket built a new theatre there, retaining the old name. Designed by C. J. Phipps, it held 1,283 in four tiers. It opened in 1897 with Gilbert Parker's *The Seats of the Mighty,* which was followed by a series of excellent productions including a number of Shakespeare's plays and new works. In his rooms in the dome of this theatre Tree instituted in 1904 a drama school which eventually moved to other premises to become the Royal Academy of Dramatic Art. After Tree's departure in 1915 the theatre achieved a new kind of success with the opening in 1916 of *Chu-Chin-Chow,* a musical fantasy by Oscar *Asche which ran for 2,238 performances. Later productions were *Cairo* (1921), also by Asche, Flecker's *Hassan* (1923) with music by Delius, *Coward's *Bitter Sweet* (1929), *The Good Companions* (1931), a dramatization of J. B. *Priestley's bestselling novel of that name, and a musical, *The Dubarry* (1932). In 1935 George *Robey played Falstaff in *Henry IV, Part One,* and in 1939 the Greek actress Katina *Paxinou made her first appearance in London in *Sophocles' *Electra* with the Greek National Theatre company, subsequently playing Gertrude to the Hamlet of her husband Alexis *Minotis. During the Second World War the theatre housed a series of revivals, but in 1947 Robert *Morley had a considerable success with *Edward, My Son,* which he co-authored. The musical *Brigadoon* (1949) inaugurated a successful series of American productions which included John Patrick's *The Teahouse of the August Moon* (1954) and the musicals *Paint Your Wagon* (1953) and *West Side Story* (1958). The musical *Fiddler on the Roof* (1967), in which *Topol scored a great success,

ran for five years, and later musicals included *Company* by *Sondheim, *Applause* (both 1972), the British musical version of *The Good Companions* (1974), and the African musical *Ipi Tombi* (1975). Peter *Shaffer's *Amadeus* (1981) with Frank *Finlay, transferred from the *National Theatre, had a long run, and Andrew Lloyd Webber's enormously successful musical *The Phantom of the Opera* (1986) looked set to run for the foreseeable future.

Hermann, David, see VILNA TROUPE.

Herne [Ahearn], James A. (1839–1901), American actor and dramatist, who made his first appearance on the stage in 1850. As manager of Maguire's New Theatre in San Francisco he appeared in a number of adaptations of *Dickens's novels, which appear to have had a great influence on his own writing. He also worked with *Belasco at the neighbouring Baldwin Theatre, collaborating with him on the production of his own *Hearts of Oak* (1879). This was pure *melodrama; Herne's later work showed more realism and sobriety, mainly under the influence of his wife **Katharine Corcoran** (1857–1943), a fine actress who played most of his heroines, including the leading role in *Margaret Fleming* (1890), a sombre drama of infidelity. In 1892 Herne put on *Shore Acres,* which after a slow start became one of the most popular plays of the day, mainly through the character of Uncle Nathaniel Berry. It was followed by *The Reverend Griffith Davenport* (1899), of which no complete copy has survived, and *Sag Harbour* (1900), in which Herne was acting at the time of his death.

Heroic Drama, style of playwriting imported into England during the Restoration in imitation of French classical tragedy. Written in rhymed couplets, and observing strictly the *unities of time, place, and action, it deals mainly with the Spanish theme of 'love and honour' on the lines of *Corneille's *Le Cid.* Its chief exponent was *Dryden, whose *Conquest of Granada* (1670/1) best displays both its faults and virtues. Its vogue was short-lived.

Heron, Matilda Agnes (1830–77), American actress, who made her début in Philadelphia in 1851. She later achieved recognition as Marguerite Gautier in *Dumas *fils*'s *The Lady of the Camellias,* which she had seen while on a visit to Paris, subsequently making a fairly accurate version of it in which she toured all over the United States. She was not, however, the first to play the part in America, Jean *Lander having forestalled her in 1853. Miss Heron's version was particularly successful in New York and she made a fortune out of it, most of which she spent or gave away. She later played Medea in *Euripides' tragedy and Nancy in a dramatization of *Dickens's *Oliver Twist,* and was seen in several of her own plays.

Among the actresses she trained for the stage was her daughter by her second marriage, Hélène Stoepel, known as **Bijou Heron** (1862–1937), who married Henry *Miller.

Heywood, John (c.1497–1580), English dramatist, author of a number of interludes which mark the transition from medieval plays to the comedy of Elizabethan times. He married Elizabeth Rastell, niece of Sir Thomas More, a lover of plays and playing to whom he may have been indebted for advice and encouragement. Heywood's best known work is *The Playe called the foure P.P.; a newe and a very mery enterlude of a palmer, a pardoner, a potycary, a pedler* (c.1520), each of whom tries to outdo the others in lying. The palmer wins when he says that in all his travels he never yet knew one woman out of patience. The text was published some 20 years later by Heywood's brother-in-law William Rastell, as were *The Play of the Wether* and *The Play of Love* (both 1533). *The Dialogue of Wit and Folly*, which probably dates from the same year, remained in manuscript until 1846, when it was issued by the Percy Reprint Society. Two further interludes are sometimes attributed to Heywood, *The Pardoner and the Frere* and *Johan Johan*, both published anonymously by Rastell in 1533, while he may also be the author of *Thersites*, sometimes attributed to *Udall.

Heywood, Thomas (c.1570–1641), English actor and dramatist, who may have been related to John *Heywood. He was with the *Admiral's Men, and later with *Queen Anne's Men, until they disbanded in 1619, after which he does not seem to have acted again. As a dramatist he is first mentioned in 1596, when *Henslowe recorded in his diary an advance payment made to him for an unnamed play. In 1599 he was apparently part-author with *Chettle and others of a chronicle play, *Edward IV*. The following year he produced on his own a romantic drama *Four Prentices of London*, later satirized by *Beaumont in *The Knight of the Burning Pestle* (1607). Mainly for Henslowe, he wrote or had a hand in over 200 plays, many of which are lost, as he only troubled about publishing them when forced to it by piracy. His masterpiece is undoubtedly the domestic tragedy *A Woman Killed with Kindness* (1603), which has been many times revived. Other good extant plays include *The Wise Woman of Hogsdon* (1604); *If You Know Not Me, You Know Nobody* (1604/5), a rambling account of Elizabeth I's early reign; *The Rape of Lucrece* (1607), obviously designed to profit from the popularity of the poem of the same name by Shakespeare, from whom Heywood made frequent borrowings; and *The Fair Maid of the West* (1610). He also wrote, from 1631 to 1639, a series of civic *pageants for the Lord Mayor's Show. The last play

attributed to him was *Love's Masterpiece*, a comedy now lost.

Hicks, Sir (Edward) **Seymour** (1871–1949), English dramatist and actor-manager. He began his stage career in 1887, and was for a long time with the *Kendals, both in England and America. He also produced the first *revue seen in London, *Under the Clock* (1893). In the course of his long career he topped the bill in the *music-halls and appeared with equal success in *musical comedy and straight plays, being particularly admired as Valentine Brown in *Barrie's *Quality Street* (1902). Among his numerous plays the most successful were *Bluebell in Fairyland* (1901), frequently revived at Christmas up to 1937; *The Gay Gordons* (1907), in which he played Angus Graeme; *Sleeping Partners* (1917), in which his own performance was a *tour de force* of silent acting; and *The Man in Dress Clothes* (1922), which, like many of his plays, was taken from the French. He was manager of several London theatres, including the *Aldwych, which he built and opened in 1905; the *Globe (first named the Hicks), which he built for Charles *Frohman and opened in 1907 with his own play *The Beauty of Bath*; and *Daly's, where he inaugurated his management by playing Charles Popinot in *Vintage Wine* (1934). He was the first actor to take a party of entertainers to France in the First World War, and also the first to go to France during the Second World War. He married the actress Ellaline *Terriss, who appeared with him in many of his productions.

Hicks Theatre, see GLOBE THEATRE 3.

Hill, Aaron (1685–1750), English playwright satirized by Pope in *The Dunciad* (1728). He is now chiefly remembered for his association with Handel, for whose first London opera, *Rinaldo* (1711), he wrote the libretto, and for the three translations, of *Zaïre* (1732), *Alzire* (1736), and *Mérope* (1743), by which he introduced the plays of *Voltaire to London audiences. Hill's own plays include several forgotten tragedies in the current French style and a farce, *The Walking Statue; or, The Devil in the Wine Cellar* (1710), which was extremely successful and frequently revived up to 1845. His letters and journals contain a good deal of entertaining information about the theatre of his time.

Hill, Jenny, see MUSIC-HALL.

Hiller, Dame Wendy (1912–), English actress. She began her career at the *Manchester Repertory Theatre in 1930, and scored an instant success five years later in the London production of *Love on the Dole* (NY, 1936), adapted from Walter Greenwood's novel by the author and **Ronald Gow** (1897–), later her husband and the author of several other plays, including *Ma's Bit o' Brass* (1938). After some years in films she returned to the West End in 1944 in a revival of *Martínez Sierra's *The Cradle Song*, and a year later was seen as Princess Charlotte to Robert *Morley's Prince Regent in Norman Ginsbury's *The First Gentleman*. At the *Bristol Old Vic in 1946 she played Tess in Gow's adaptation of Hardy's novel, repeating the part in London, where she was also seen in Gow's adaptation of H. G. Wells's *Ann Veronica* (1949) and in N. C. *Hunter's *Waters of the Moon* (1951). This ran for two years, after which she joined the *Old Vic company for a wide variety of Shakespearian parts. In New York she portrayed the downtrodden heroine of *The Heiress* (1947), based on Henry *James's novel *Washington Square*, and Josie Hogan in *O'Neill's *A Moon for the Misbegotten* (1957). Among her later parts were Carrie Berniers in Lillian *Hellman's *Toys in the Attic* (1960); Miss Tina in Michael *Redgrave's adaptation of Henry James's *The Aspern Papers* (NY, 1962); and Susan Shepherd in an adaptation of James's *The Wings of the Dove* (1963). In 1967 she was in a revival of *Maugham's *The Sacred Flame*, and three years later in Peter *Shaffer's *The Battle of Shrivings*. The finest part of her later career was a remarkably faithful portrait of Queen Mary in Royce Ryton's *Crown Matrimonial* (1972). For the *National Theatre in 1975 she played Gunhild Borkman opposite Ralph *Richardson in *Ibsen's *John Gabriel Borkman*. She also reappeared in *Waters of the Moon* (*Chichester, 1977; London, 1978) and *The Aspern Papers* (1984), in different roles from before.

Hilpert, Heinz, SEE DEUTSCHES THEATER.

Hippodrome, London, in Cranbourn Street, Leicester Square. This opened in 1900 as a circus with a large water-tank which was used for aquatic spectacles. It later became a *music-hall and in 1909 was reconstructed internally, the circus arena being covered by stalls. Ballet and variety were seen there, and from 1912 to 1925 Albert de Courville staged a number of successful *revues. These were followed by an equally successful series of musical comedies, among them *Sunny* (1926), *Hit the Deck* (1927), *Mr Cinders* (1929), and *Please, Teacher* (1935). Later productions were the revue *The Fleet's Lit Up* (1938), Ivor *Novello's musical *Perchance to Dream* (1945), and Herman Wouk's play *The Caine Mutiny Court-Martial* (1956). In 1958, after complete reconstruction of the interior, the building opened as a combined restaurant and cabaret, the Talk of the Town, which closed in 1982.

Hippodrome Theatre, New York, on the Avenue of the Americas, between 43rd and 44th Streets. This theatre, the largest in America, seating 6,600, opened in 1905 with a lavish spectacle entitled *A Yankee Circus on Mars*, and a year later was taken over by the *Shuberts. Every kind of entertainment was given, including grand opera. In 1923, as B. F. Keith's

Hippodrome, it became a *vaudeville house, and in 1928, as the RKO Hippodrome, a cinema. Closed in 1932, it reopened in 1933 as the New York Hippodrome, and in 1935 was taken over by Billy Rose, whose spectacular musical *Jumbo* marked the end of the Hippodrome as a theatre. It was finally demolished in 1939.

Hirsch, John Stephen [Janos Stefan] (1930–89), Canadian director, born in Hungary. He emigrated to Canada in 1947 and after graduating from the University of Manitoba became interested in children's theatre, writing several plays for his own Muddiwater Puppets Theatre. He co-founded Rainbow Stage, an outdoor summer *musical comedy theatre, and Theatre 77, a semi-professional group. In 1958 he also co-founded the Manitoba Theatre Centre, which with its resident *stock company set the pattern for a chain of regional theatres across Canada. He was Artistic Director, 1958–66, his productions including *Brecht's *Mother Courage* in 1964 and *Frisch's *Andorra* in 1965, and on one of his many return visits he directed his own adaptation of *Ansky's *The Dybbuk* in 1973. With Jean *Gascon he was co-director of the *Stratford (Ontario) Festival theatre, 1967–9, where he staged *Henry VI*, *Richard III*, *A Midsummer Night's Dream*, and *Hamlet*, as well as *Chekhov's *The Cherry Orchard* and a spoof version of *The Three Musketeers*. He directed the first production of James *Reaney's *Colours in the Dark* (1967), and in 1976 a memorable production of Chekhov's *Three Sisters*. His first productions outside Canada were of *García Lorca's *Yerma* and *Brecht's *Galileo* at the *Vivian Beaumont Theatre in 1967. He headed the Canadian Broadcasting Corporation's TV Drama Department, 1974–8, and was Consultant Artistic Director at the *Seattle Repertory Theatre, 1979–81. He was Artistic Director of the Stratford (Ontario) Festival, 1981–5, where his productions included *The Tempest* and *Molière's *Tartuffe*.

Hirschbein, Peretz (1880–1949), Jewish actor and dramatist, and founder of the first Yiddish Art Theatre in Odessa. Although his first play *Miriam* (pub. 1905) was written in Hebrew, he founded his company in 1908 for the production of plays in Yiddish, in which language he wrote his later works, translating them into Hebrew himself. Among the most important are *The Blacksmith's Daughters* (1915) and *Green Fields* (1919), both idylls of Jewish country life, the latter being considered one of the finest works of Yiddish dramatic literature.

His Majesty's Theatre, London, see HER MAJESTY'S THEATRE.

Hobart, see THEATRE ROYAL, Hobart.

Hobby Horse, form of *animal impersonation common to many *folk festivals throughout Europe, and particularly popular in England, the most famous being the Hooden Horse of Kent, the Old Hoss of Padstow, in Cornwall, and the Minehead Soulers' Horse. It is probably a survival of the primitive worshipper dressed in the skin of a sacrificial animal, and usually appears as a wicker framework in the shape of a horse covered with a long green cloth which conceals the legs of the man animating it. It can, however, be merely a carved horse's head attached to a long stick, in which form it has always been a popular toy for children. The Hobby Horse usually accompanies the morris dance, and is often found in the *mumming play. In the Mascarade of La Soule a horse is the chief character, and dies and is resurrected in token of the return of spring after the death of winter.

Hobson, Sir Harold (1904–   ), English dramatic critic, who began his career in 1931 as dramatic critic of the *Christian Science Monitor*. He always championed the cause of the avant-garde and experimental theatre and was also passionately devoted to the modern French theatre. He was on the staff of the *Sunday Times*, 1944–76, becoming chief dramatic critic in 1947. His assessments often differed sharply from those of his fellow critics, notably in his favourable reaction to *Pinter's first full-length play *The Birthday Party* (1958).

Hochhuth, Rolf (1931–   ), German dramatist, resident in Switzerland. His first play, *Der Stellvertreter* (1963), first seen in Berlin in a production by *Piscator, indicted Pope Pius XII for criminal non-intervention in the Nazis' extermination of the Jews. Though far too long, over-ambitious, and muddled in its aims, it remains an impressive dramatic treatment of the Nazi era, and can be seen as the beginning of a post-war renaissance in German drama. It was staged in London by the *RSC as *The Representative* (1963) and in New York as *The Deputy* (1964). Hochhuth's second play, *Soldaten: Nekrolog auf Genf* (1967), also too long and overlaid with documentation, accuses Winston Churchill of causing the death of his Polish ally General Sikorski and of criminal inhumanity in the bombing of Dresden. The play became a *cause célèbre* in Britain long before it was produced in London as *Soldiers* in 1968, following the abolition of stage censorship. It had previously been seen in English in Toronto and New York, also in 1968. Hochhuth's next plays, such as *Guerrillas* (1970), a wholly fictional account of a modern revolution in America, *Die Hebamme* (*The Midwife*, 1972), a comedy based on the real life of a social worker, *Lysistrate und NATO* (1974), and *Juristen* (*Lawyers*, 1983) had no influence abroad. *Judith*, in which a Vietnam veteran plans the assassination of an American President, received its world première at the *Citizens' Theatre in 1984.

Hochwälder, Fritz (1911–   ), Austrian dramatist, born in Vienna, but resident since 1938 in Switzerland, where he had his first success in the theatre with *Das heilige Experiment* (1945), which as *The Strong are Lonely* was seen in New York in 1953 and in London, with *Wolfit in the chief part, in 1955. It deals with the destruction of the Jesuit Community in Paraguay in the 18th century, and displays considerable theatrical skill in dramatizing a moral issue. After *Der Flüchtling* (*The Fugitive*, 1945), set in modern times, Höchwalder again used a historical setting for *Der öffentliche Ankläger* (1947; London, also 1947, as *The Public Prosecutor*), in which the eponymous character, briefed to conduct a case against an anonymous 'enemy of the people', finds that he himself is the accused. *Donadieu* (1953) depicts a man giving shelter to his wife's murderer; *Die Herberge* (*The Inn*, 1956) shows the investigation of a theft leading to the discovery of a far greater crime; in *Der Unschuldige* (*The Innocent Man*, 1958), a comedy with serious undertones, a man realizes that though innocent of the murder of which he is accused he might very well have committed it; *Der Himbeerpflücker* (*The Raspberry Picker*, 1965) satirizes the latent Nazism of a small Austrian town; and in *Der Befehl* (*The Command*, 1968) Hochwälder returns to the situation of *The Public Prosecutor*. His *Die Prinzessin von Chimay* was performed in 1981. Hochwälder is concerned with a conflict of ideas, emphasized not by the use of modern stage techniques but by a strict adherence to the classical *unities.

Hocktide Play, survival of one of the early English *folk festivals, given in Coventry on Hock Tuesday (the third Tuesday after Easter Sunday), and revived as a pleasant antiquity in the Kenilworth Revels prepared for the visit of Queen Elizabeth in July 1575. It began with a Captain Cox leading in a band of English knights, each on a *hobby horse, to fight against the Danes, and ended with the leading away of the Danish prisoners by the English women. It was intended to represent the massacre of the Danes by Ethelred in 1002, but this is probably a late literary assimilation of an earlier folk-festival custom, traceable in other places (Worcester, Shrewsbury, Hungerford), by which the women 'hocked' or caught the men and exacted a forfeit from them on one day, the men's turn coming the following day. The practice was forbidden at Worcester in 1450. This ceremony may have been a symbolic re-enactment of the capture of a victim for human sacrifice.

Hodgkinson [Meadowcroft], **John** (1767–1805), English actor, who had had some experience in the provinces when in 1792 he accepted an offer from John *Henry to join the *American Company at the *John Street Theatre in New York, and spent the rest of his life in the United States. He soon became extremely popular, ousting Henry from management as well as from public favour, and becoming joint manager with the younger *Hallam and *Dunlap of the *Park Theatre when it first opened. A handsome man, with a fine stage presence, he excelled both in tragedy and comedy, among his best parts being the title-role in Dunlap's *André* and Rolla in his adaptation of *Kotzebue's *Pizarro*.

Hofmannsthal, Hugo von (1874–1929), Austrian poet and dramatist, who as a young man was in the forefront of the reaction against *naturalism. His early verse-plays, of which *Der Tor und der Tod* (*Death and the Fool*) was produced at Munich in 1898, reveal his delight in beauty and in poetic and mystical intuitions, but in his later plays his exquisite poetry was diluted in an effort to enhance the dramatic content of his dialogue and relate his characters to a social framework. He turned to subjects which were obviously theatrical, and his first success in the theatre came with *Elektra* (1903), produced in Berlin by *Reinhardt, who directed most of his plays. It was followed by *Ödipus und die Sphinx* (1905), *König Ödipus* (1907), and *Alkestis* (1909). Also to this period belongs the first of his adaptations, which in his hands became almost new plays—*Das gerettete Venedig* (1905), based on *Otway's *Venice Preserv'd*, produced with sets by Gordon *Craig. In 1911 Hofmannsthal produced what is perhaps his best known play, *Jedermann*, based on the old Dutch *morality play *Elckerlyc* (*Everyman*) and incorporating some elements from Hans *Sachs. After its first production in Berlin it was transferred to the Salzburg Festival, founded in 1917 by Reinhardt and Hofmannsthal, and it has since been seen at all the festivals there, in an open-air setting in front of the Cathedral. For the festival in 1922 Hofmannsthal adapted *Calderón's *El gran teatro del mundo* as *Das Salzburger grosse Welttheater*. In two plays of the same period, set in and near Vienna, *Der Schwierige* (*The Difficult Man*, 1921) and *Der Unbestechliche* (*The Incorruptible Man*, 1923), Hofmannsthal reverted to the vein of comedy which had been apparent in *Der Abenteurer und die Sängerin* (*The Adventurer and the Singer*, 1898) and *Cristinas Heimreise* (*Cristina's Journey Home*, 1909). Both are pleas for marital fidelity as the natural and desirable sequel to the equally natural philanderings of immaturity, but *Der Schwierige* in particular stands out as a masterpiece, combining high comedy of subtle human relationships with irony and social satire. Hofmannsthal's last play was again based on Calderón, an adaptation, as *Der Turm* (1925), of *La vida es sueño*, which takes up once more the conflict between material power and spiritual integrity.

Hofmannsthal's plays are little known in the English-speaking world, where he is re-membered mainly as the librettist of several of Richard Strauss's operas, including *Elektra* (1909), based on his play.

Hoist, see SLOTE.

Holberg, Ludvig (1684–1754), historian, philosopher, satirist, and playwright, who, though born in Norway, spent most of his working life in Denmark, being for many years Professor of Metaphysics at the University of Copenhagen. His connection with the theatre was brief, but far-reaching in effect. When in 1721 he was appointed director of the *Danske Skueplads, he brought to his task a knowledge of and love for the drama unusual in Denmark at that time, his favourite playwrights being *Plautus and *Molière. There were at this time no Danish plays, only French and German. To remedy this he wrote plays in Danish with a mixture of Danish and Norwegian scenes and characters not previously seen, creating in the vernacular a tradition of comedy which was upheld by his successors. His first two productions, in 1722, were a translation of Molière's *L'Avare* and his own *Den politiske Kandestøber* (*The Political Tinker*). He wrote in all 32 comedies, six of them after 1747, when the Danske Skueplads reopened after 20 years of inactivity. The best of his plays belong to the earlier period. Although many of Holberg's plays became known throughout Europe in translation, they have made little impact in England or the USA. Apparently the only one to be professionally produced was *Jeppe paa Bjerget, eller den forvandlede Bonde* (1722), which, as *Jeppe of the Mountains*, was seen at the *Pitlochry Festival Theatre in 1966.

Holborn Empire, London, in High Holborn. As Weston's Music-Hall, this opened in 1857, and proved the first serious rival to the *Canterbury. It had a consistently successful career, being renamed the Royal Music-Hall in 1868. Rebuilt in 1887, it opened as the Royal Holborn and saw the débuts of many famous music-hall stars, among them Bessie Bellwood and J. H. Stead; its *Chairman for many years was W. B. Fair, singer of 'Tommy, Make Room for Your Uncle'. Round about 1900 its popularity began to decline, and it closed in 1905. Completely rebuilt, it reopened in 1906 as the Holborn Empire, and was occasionally used for plays, Sybil *Thorndike appearing there early in 1920 in the title-roles of *Shaw's *Candida* and *Euripides' *Medea*, and as Hecuba in the latter's *Trojan Women*. From 1922 to 1939 the children's play *Where the Rainbow Ends* was given an annual Christmas production by Italia *Conti. In 1932 Nellie *Wallace appeared in *The Queen of Hearts*, a *revue by Tom Arnold, and at the end of 1939 another revue, *Haw-Haw!*, was staged. In 1941 the building was severely damaged by enemy action; it was finally demolished in 1960.

Holborn Theatre, London, in High Holborn. Built by Sefton Parry, this opened successfully in 1866 with *Boucicault's *The Flying Scud*. Subsequent productions and several managements failed until in 1875 the theatre reopened as the Mirror, in the same year having a great success with *All for Her*, the first dramatization of *Dickens's *A Tale of Two Cities*. In 1876 the theatre again changed hands, opening under F. C. Burnand as the Duke's, with his own burlesque of *Jerrold's *Black-Ey'd Susan*, transferred from the *Opera Comique. After a further undistinguished period a modern drama by Paul Merritt, *The New Babylon* (1879), proved acceptable, and had just returned to the theatre after a provincial tour when the building was destroyed by fire in 1880.

Holcroft, Thomas (1744–1809), English dramatist, usually credited with the introduction of *melodrama to the English stage with his *A Tale of Mystery* (1802), an adaptation of *Pixérécourt's *Coelina; ou, L'Enfant de mystère* (1800). He was already well known as the author of *The Road to Ruin* (1792), with its excellent roles of Goldfinch and Old Dornton, favourite parts with many elderly character actors in both London and in the USA. Holcroft, who was entirely self-taught, was a very good French scholar and had a phenomenal memory, a combination which enabled him, while in Paris, to learn by heart *Beaumarchais's *Le Mariage de Figaro* from hearing it on stage a number of times, and to produce it in London in 1784 as *The Follies of a Day*. Among his other comedies the most successful was *Love's Frailties* (1794), based on a German original, Gemminger's *Der deutsche Hausvater*. He was also a friend of Charles *Lamb and editor of *The Theatrical Recorder*, 1805–6.

Holland, George (1791–1870), English actor, born in London, who after seven years on the stage there went to New York and founded a family of American actors, his sons **Edmund Milton** (1848–1913) and **Joseph Jefferson** (1860–1926) both being well-known light comedians. George Holland made his first appearance in New York at the *Bowery Theatre in 1827, and soon became a popular comedian. He toured extensively and was well known in the South, being for some time in management with *Ludlow and Sol *Smith, and he was for six years at Mitchell's *Olympic in *burlesque. From 1855 to 1867 he played character parts in Lester *Wallack's company, being outstanding as Tony Lumpkin in *Goldsmith's *She Stoops to Conquer*, which he was still playing at the age of 75.

Hollingshead, John (1827–1904), English theatre manager, whose name is chiefly associated with the *Gaiety Theatre. He opened it in 1868 and remained there for 18 years, being succeeded by George *Edwardes, who was to make it the home of musical comedy. Under Hollingshead it had been used mainly for *burlesque. Hollingshead is credited with the introduction of matinées and with being the first manager to use electric light outside the theatre (in 1886). In 1880 he staged a translation by William *Archer of *Samfundets Støtter* as *Quicksands; or, The Pillars of Society*, the first play by *Ibsen to be seen in London.

Holt, Maud, see TREE, SIR HERBERT.

Home, the Revd John (1722–1808), Scottish minister, author of the tragedy *Douglas* (1756), produced in Edinburgh with Dudley *Digges as the hero, Young Norval. It caused a great deal of controversy, as many members of the Church of Scotland were horrified that one of their number should write for the theatre; but it was rapturously received by the audience. In 1757 it was accepted by John *Rich for *Covent Garden, where Spranger *Barry ('six feet high and in a suit of white puckered satin,' says Doran) played Young Norval to the Lady Randolph of Peg *Woffington. The play was constantly revived, Lady Randolph being a favourite part with Sarah *Siddons, while many young actors in England and America, including Master *Betty and John Howard *Payne, delighted in Young Norval. The speech beginning 'My name is Norval' became a regular recitation piece, and the play eventually found its way into the repertory of the *toy theatre. It was revived at the *Edinburgh Festival in 1950, with Sybil *Thorndike as Lady Randolph. Home wrote other tragedies, none of them successful.

Home, William Douglas (1912–   ), English dramatist, who was for a time on the stage, making his first appearance in London in Dodie Smith's *Bonnet over the Windmill* (1937). After service in the Second World War he devoted his energies to playwriting and acted only occasionally in his own plays, which are mainly light comedies with an upper-class background. He first achieved success with a political comedy *The Chiltern Hundreds* (1947), which starred A. E. *Matthews, following it with a serious play about prison life, *Now Barabbas . . .* (also 1947). Among his later plays were *The Bad Samaritan* (1953), *The Manor of Northstead* (1954), a sequel to *The Chiltern Hundreds* which again starred Matthews, and the highly popular *The Reluctant Débutante* (1955; NY, 1956). *The Reluctant Peer* (1964), with Sybil *Thorndike, *The Secretary Bird* (1968), and *The Jockey Club Stakes* (1970; NY, 1973), a comedy about racing, all had long runs. During the next few years Home's plays attracted major stars—Ralph *Richardson and Peggy *Ashcroft in *Lloyd George Knew My Father* (1972), Michael *Denison and Dulcie Gray in *At the End of the Day* (1973); Celia *Johnson in *The Dame of Sark* (1974); and Richardson with Celia Johnson in *The Kingfisher* (1977; NY, 1978, with Rex *Harrison). His many subsequent plays were unsuccessful.

Hope Theatre, London, on Bankside. This brick and wood building, of the same shape and size as the *Swan Theatre, was built by Philip *Henslowe for the *Lady Elizabeth's Men. It replaced the old Bear Garden, and had a removable stage so that it could still be used for bull- and bear-baiting, and for cockfights. The contract for its construction dated 29 Aug. 1613 still survives and *Jonson's *Bartholomew Fair* was shown there before being presented at Court. After Henslowe's death in 1616 a new agreement was made between *Alleyn and the company, now known as *Prince Charles's Men; but, having for a time attracted the audience which had formerly frequented the *Globe, burnt down in 1613, the Hope lost it again when the rebuilt Globe opened in 1614, and its fortunes continued to decline. It was used by a few minor companies until 1617, after which it reverted to bear-baiting and to its old name of Bear Garden. It was finally demolished in or after 1682.

Hopkins, Anthony (1937–), Welsh actor. After his London début at the *Royal Court in 1964 he joined the *National Theatre company, at first in supporting roles but eventually playing Coriolanus (1971) and Macbeth (1972). After his Petruchio in *The Taming of the Shrew* (1972) at *Chichester he worked mainly in films and television for some years, living in America and appearing on stage there in such productions as *Shaffer's *Equus* (1974) and *Pinter's *Old Times* (NY, 1984). He returned to the London stage in 1985 in *Schnitzler's *The Lonely Road* (*Old Vic) and as the South African press baron in *Hare and *Brenton's *Pravda*, which were followed by his King Lear (1986) and Antony (in *Antony and Cleopatra*, 1987). The last three roles, all at the National Theatre, placed him in the front rank of contemporary actors. In 1989 he starred in the West End in David Henry Hwang's *M. Butterfly*. He is also well known in films.

Hopkins, Priscilla, see KEMBLE, JOHN.

Hordern, Sir Michael Murray (1911–), English actor, an amateur with the St Pancras People's Theatre for several seasons before making his professional début in 1937. After serving in the Navy during the Second World War he returned to the stage in 1946 as Torvald Helmer in *Ibsen's *A Doll's House* and also played Bottom in Purcell's masque *The Fairy Queen* at Covent Garden. He was at *Stratford-upon-Avon at Christmas 1948 and 1949 to play Mr Toad in *Toad of Toad Hall*, A. A. Milne's adaptation of Kenneth Grahame's *The Wind in the Willows*, and was then seen in the title-role

of *Chekhov's *Ivanov* (1950) and in John *Whiting's *Saint's Day* (1951). In 1952 he joined the company at Stratford, establishing himself as a first-rate classical actor and going from there to the *Old Vic, where in 1954 he gave an excellent performance as Malvolio in *Twelfth Night*. After appearing in the West End as Sir Ralph Bloomfield Bonington in *Shaw's *The Doctor's Dilemma* (1955), he greatly enhanced his reputation as the senile barrister, cavorting with mincing pomposity, in John *Mortimer's *The Dock Brief* (1958), and then returned to the Old Vic to play Macbeth, and Pastor Manders in Ibsen's *Ghosts*. He followed *Dürrenmatt's *The Physicists* for the *RSC (1963) with Alan *Ayckbourn's *Relatively Speaking* (1967) and Tom *Stoppard's *Enter a Free Man* (1968). He then returned to the RSC in *Albee's *A Delicate Balance* (1969), and in the same year was a memorable King Lear at *Nottingham Playhouse under Jonathan *Miller's direction. After his lecherous, agnostic clergyman in David *Mercer's *Flint* (1970) he gave for the *National Theatre one of his finest performances as George, the word-spinning, God-obsessed metaphysician in Stoppard's *Jumpers* (1972), and was also excellent in an adaptation of Evelyn Waugh's *The Ordeal of Gilbert Pinfold* (Manchester, 1977; London, 1979) and as Prospero in *The Tempest* at Stratford (1978). He later appeared in *Sheridan's *The Rivals* (National Theatre, 1983) and Shaw's *You Never Can Tell* (1987). He is a remarkable actor who has few peers in the portrayal of quixotic eccentrics.

Horniman, Annie Elizabeth Fredericka (1860–1937), theatre patron and manager, one of the seminal influences in the Irish and English theatres at the beginning of the 20th century. The daughter of a wealthy Victorian tea-merchant, she had no connections with the theatre, but from her travels abroad came to realize the importance of subsidized *repertory theatres. She therefore made funds available in 1894 for a repertory season at the Avenue Theatre (later the *Playhouse) which included *Arms and the Man*, the first play by *Shaw to be seen in the commercial theatre, and *The Land of Heart's Desire*, the first play by *Yeats to be seen in London. Her introduction to Yeats through this season led her to take an interest in the new *Irish Literary Theatre, which in turn led her to build the *Abbey Theatre in Dublin in 1904. In 1908 she bought and refurbished the Gaiety Theatre, Manchester, where from 1908 to 1917 she maintained an excellent repertory company, putting on more than 200 plays, at least half of which were new works, among them the plays of the so-called 'Manchester School' (see MANCHESTER) and an early play by St John *Ervine, *Jane Clegg* (1913). Most of the productions were directed by Lewis *Casson, who married a member of the company, Sybil *Thorndike. The venture was not a success financially and the company was disbanded in 1917, the building becoming a cinema in 1921. Miss Horniman lived long enough, however, to see her pioneer work bear fruit with the spread of the *repertory theatre movement.

Horses, see EQUESTRIAN DRAMA.

Horváth, Ödön von, see HAMPTON.

Hôtel d'Argent, de Bourgogne, see ARGENT, BOURGOGNE.

Houghton, (William) **Stanley** (1881–1913), English playwright, one of the best of the so-called 'Manchester School' (see MANCHESTER) and much influenced by *Ibsen. His plays, dealing with the revolt against parental authority and the struggle between the generations, were first seen at the Gaiety Theatre in Manchester under Miss *Horniman. They include *The Dear Departed* (1908) and *The Younger Generation* (1910), both of which later proved popular with amateur and repertory companies. His best known play *Hindle Wakes* (1912) was, however, first seen in London, played by actors from the Gaiety, Manchester, and directed by Lewis *Casson, their resident producer. It was an immediate success, though its plot, in which Fanny Hawthorn, a working girl, refuses to marry the cowardly, vacillating rich man's son who has seduced her, portrayed a reversal which took contemporary playgoers by surprise and caused a good deal of controversy. Houghton subsequently had time to write only a couple of one-act plays before his untimely death.

Housekeepers, name given in the Elizabethan theatre to the men, whether actors or not, who owned part (or in some cases the whole) of the actual building in which the company was working. They were responsible for the maintenance of the building and for payment of the ground rent, but, unlike the *sharers, were not part-owners of the wardrobe and playbooks.

Houseman, John [Jacques Haussmann] (1902–88), American producer, director, and actor, born in Romania. He became known in 1935 when he produced Archibald *MacLeish's *Panic* in New York, and in 1936 was the most important figure in the Negro theatre established by the *Federal Theatre Project. In 1937–8 he collaborated with Orson *Welles in running the Mercury Theatre (see COMEDY THEATRE, New York). He later became a film producer, but still directed plays on Broadway and elsewhere, becoming Artistic Director of the American Shakespeare Festival, 1956–9 (see AMERICAN SHAKESPEARE THEATRE), and of the Theatre Group of the University of California at Los Angeles, 1959–64 (see MARK TAPER FORUM). He also headed the Drama Division of the Juilliard School of the Performing Arts, 1968–76, and in 1972 founded the City Center Acting Company at

the *City Center of Music and Drama. Well known also as an opera producer, in the 1960s he began a new career as a film and television actor.

House of Ostrovsky, see MALY THEATRE, Moscow.

Housman, Laurence (1865–1959), English author and playwright, brother of the poet A. E. Housman. His first plays, *Bethlehem* (1902), directed by Gordon *Craig, and *Prunella* (1904), a *commedia dell'arte* fantasy with music, were full-length, but most of his work consisted of one-act plays. The best known of them, dealing with Queen Victoria and her Court, were banned by the censor. A selection of them, as *Victoria Regina* (1935), were seen privately at the *Gate Theatre, and Helen *Hayes scored a big success in them in New York in the same year. It was not until Edward VIII intervened that they were licensed for public performance in Britain, and produced at the *Lyric Theatre in 1937. A further series of one-act plays entitled *The Little Plays of St Francis*, though often performed by amateurs, does not appear to have been produced professionally.

Howard, Alan Mackenzie (1937–   ), English actor, nephew of the film and stage star Leslie Howard, whose good looks he inherited. He made his first appearance in the initial production at the *Belgrade Theatre, Coventry, the musical *Half in Earnest* (1958), going with the company to London to appear in *Wesker's *Roots* (1959) at the *Royal Court, where in 1960 he was in the complete Wesker trilogy. He first attracted attention in Julian Mitchell's adaptation of Ivy Compton-Burnett's novel *A Heritage and Its History* (1965). In 1966 he joined the *RSC, remaining with it for many years and playing a wide variety of parts, among them Lussurioso in *Tourneur's *The Revenger's Tragedy* (1966), Benedick in *Much Ado about Nothing* (1968), Hamlet (1970), and doubling Theseus and Oberon in Peter *Brook's production of *A Midsummer Night's Dream* (also 1970; NY, 1971), achieving a triumph of physical dexterity and meticulous verse-speaking. Over the next few years a series of Shakespearian parts—Prince Hal, Henry V, Richard II, Richard III, Coriolanus, and Antony (to the Cleopatra of Glenda *Jackson)—brought him great critical praise, and he was also much admired as Jack Rover in the revival of *O'Keeffe's *Wild Oats* (1976). One of his most acclaimed performances was in the title-role of all three parts of *Henry VI* (1977). He appeared in Gorky's *Enemies* (1971) and *The Children of the Sun* (1979) and was seen in the British première of *Ostrovsky's *The Forest* and the world première of C. P. *Taylor's *Good* (both 1981), starring in the last in the West End and New York in 1982.

Howard, Bronson (1842–1908), American playwright, one of the first to make use of native material with any skill, and also to earn his living solely by playwriting. He worked as a journalist in Detroit until success came to him with *Saratoga* (1870), a farcical comedy produced by Augustin *Daly. As *Brighton* (1874), it was adapted for London, Charles *Wyndham playing the hero. Howard wrote several other comedies, including *The Banker's Daughter* (1878), which had previously been seen in 1871 as *Lilian's Last Love*, and, as *The Old Love and the New*, was successfully produced in London in 1879. His most important play, however, was *Young Mrs Winthrop* (1882), the first to be produced in London (in 1884) without alteration or adaptation. The most successful of his later plays were *The Henrietta* (1887), a satire on financial life, and *Shenandoah* (1888), a drama of the Civil War.

Howard, Eugene and **Willie**, see VAUDEVILLE, AMERICAN.

Howard, Sidney Coe (1891–1939), American dramatist. His first play to be produced, a romantic verse drama entitled *Swords* (1922), was a failure; but success came with the *Pulitzer Prize-winner *They Knew What They Wanted* (1924), a drama about a middle-aged Italian and his mail-order bride set in grape-growers' country in Howard's native state of California. It was followed by *Lucky Sam McCarver* (1925), the portrait of a night-club proprietor, *Ned McCobb's Daughter* and *The Silver Cord* (both 1926), the first a sympathetic tale of a New England woman at odds with rum-runners, the second a study of maternal possessiveness. The position of the artist in an unsympathetic community was the theme of *Alien Corn* (1933), a somewhat melodramatic piece starring Katharine *Cornell; *Yellow Jack* (1934) was a factually accurate account of the fight against yellow fever; *The Ghost of Yankee Doodle* (1937), though only moderately successful in production, was perhaps the most satisfactory of Howard's plays, showing how in all classes of society economic considerations overcome the normal aversion to war. Howard had just finished the first draft of *Madam, Will You Walk?* when he was killed in an accident. It was produced in 1953 but, lacking the author's revisions, was not a success.

Howard was a prolific translator and adaptor, being responsible for the American versions of Vildrac's *Le Paquebot Tenacity* in 1922 and René Fauchois's *Prenez garde à la peinture* in 1932, among others. As *The Late Christopher Bean*, the latter had a great success, and was further adapted for British audiences by Emlyn *Williams. Howard also dramatized Sinclair Lewis's novel *Dodsworth* in 1934 and was the joint adaptor of a 14th-century Chinese play as *Lute Song*, eventually staged as a musical, with Mary *Martin, in 1946.

Howard Athenaeum, see BOSTON.

Hoyt's Theatre, New York, see MADISON SQUARE THEATRE.

Hroswitha [Hrotsvitha, Roswitha], Benedictine abbess of Gandersheim in Saxony, who in the 10th century, finding herself drawn by the excellence of his style to read the pagan plays of *Terence and fearing their influence on a Christian world, set out to provide a suitable alternative. This she did in six original prose plays modelled on those of Terence, but dealing with subjects drawn from Christian history and morality—*Paphnutius, Dulcitius, Gallicanus, Callimachus, Abraham*, and *Sapientia*. They were intended for reading rather than production, but the use of miracles and abstract characters links them with the later *mystery and *morality plays. The Latin is poor, but the dialogue is vivacious and there are some elements of farce. The plays were published in 1923 in an English translation, and *Paphnutius*, which deals with the conversion of Thaïs, was produced in translation in London in 1914 by Edith Craig (see CRAIG, GORDON).

Hubris, literally, 'insolence'; in Greek tragedy the type of pride or presumption in a mortal which offends the gods and causes them to punish the hubristic hero by encompassing his downfall—a situation exemplified in the English proverb 'Pride goes before a fall'.

Hudson Theatre, New York, on West 44th Street, between Broadway and the Avenue of the Americas. This handsome and elegant playhouse opened in 1903 with Ethel *Barrymore in H. H. Davies's *Cousin Kate*. Among its early successes were the first productions in New York of *Shaw's *Man and Superman* (1905), Henry Arthur *Jones's *The Hypocrites* (1906), and Somerset *Maugham's *Lady Frederick* (1908). Later productions included the sharply contrasted *Pollyanna* (1916), from the children's book by Eleanor Porter, and Maugham's *Our Betters* (1917). In 1919 Booth *Tarkington's comedy *Clarence*, with Helen *Hayes and Alfred *Lunt, had a long run, and in 1927 Sean *O'Casey's *The Plough and the Stars* had its American première. From 1934 to 1937 the building was used for broadcasting, but returned to live theatre with Cedric *Hardwicke in Barré Lyndon's *The Amazing Dr Clitterhouse*, and a year later Ethel Barrymore returned in Mazo de la Roche's *Whiteoaks*. In 1940 there was a revival of *Congreve's *Love for Love* and in 1945 came *The State of the Union*, a *Pulitzer Prize-winner by Howard *Lindsay and Russel Crouse. The theatre was again used for broadcasting from 1949, but reopened in 1959 and Lillian *Hellman's *Toys in the Attic* was produced there in 1960. In 1963 the Actors' Studio presented a revival of *O'Neill's *Strange Interlude*, and five years later the Hudson became a cinema. Renovated and adapted as a 1,000-seat cabaret in 1978, and renamed the Savoy, it reopened in 1981 with a rock concert.

Hugo, Victor-Marie (1802–85), one of France's greatest poets, and the leader of the French Romantic movement. He was also a dramatist, whose plays mark the entry of *melo-drama into the serious theatre. All alike suffer from overwhelming rhetoric, too much erudi-tion, and not enough emotion, yet by their vigour and their disregard of outworn con-ventions they operated a revolution in French theatre history. The best is *Ruy Blas* (1838), which has two excellent acts, the second and the fourth, and a superb ending. Of Hugo's other plays, *Cromwell* (published in 1827) was not intended for the stage and would take six hours to act. It was a battle-cry, and its preface became the manifesto of the new Romantic movement. *Marion Delorme* was banned on political grounds and not acted until 1831, a year after *Hernani*, whose first night led to a riot at the *Comédie-Française. *Lucrèce Borgia*, *Marie Tudor* (both 1833), and *Angelo, tyran de Padoue* (1835) are prose melodramas. *Le Roi s'amuse* (1832), banned after one performance, was used for the libretto of Verdi's opera *Rigoletto* (1851). The vogue for Romantic theatre was bound to be short-lived. The failure of Hugo's last play, *Les Burgraves*, in 1843 showed that the tide had turned in favour of prose and common sense, and he withdrew from the stage, but not before he had brought back to it dramatic verse of a quality unknown since *Racine, and of a totally different inspiration—lyrical, elegiac, colourful, and moving, con-veying tragic and comic effects with equal skill.

Hulbert, Jack, see COURTNEIDGE.

Humphries, Barry, see FEMALE IMPERSONA-TION.

Huneker, James Gibbons (1860–1921), Amer-ican dramatic critic, born in Philadelphia, of Irish-Hungarian extraction. In 1890 he became music and drama critic of the *Morning Advertiser* and the *New York Recorder* and in 1902 joined the staff of the *Sun*, leaving in 1912 for the *New York Times*. He was perhaps more of an interpreter than a critic, battling in print with William *Winter over *Ibsen and *Shaw, and his gusto and worldly knowledge shocked the puritans of the day. He edited a two-volume edition of Shaw's criticisms from the *Saturday Review*, and through his contacts with Europe—as a young man, he had lived in Paris for some years—was able to introduce and explain foreign dramatic literature to his compatriots.

Hungerford Music-Hall, see GATTI'S.

Hunt, Hugh Sydney (1911–), English theatre director and playwright. He already had some varied experience in play-production when in 1935 he became a director of the

*Abbey Theatre, where his first play *The Invincibles*, written in collaboration, was pro-duced in 1938. After serving throughout the Second World War he became the first director of the *Bristol Old Vic (1945), leaving it in 1948 to go to the *Old Vic in London. In 1955 he was appointed director of the *Elizabethan Theatre Trust in Australia, where he remained for five years. He was the first Professor of Drama at Manchester University, 1961–73, and also Artistic Director of the Abbey Theatre, 1969–71. His *In the Train* (1958), again written in collaboration, was produced in London. (See also MANCHESTER.)

Hunt, (James Henry) **Leigh** (1784–1859), Eng-lish poet and essayist, and one of the pioneers of modern dramatic criticism. He probably had a keener appreciation of acting than any of his contemporaries, and his criticisms—the best of which are contained in *Dramatic Essays* (1894) edited by William *Archer—recreate the art of the great players who brought distinction to the theatre of his day. He was the first regular critic of any importance to report upon all the principal theatrical events of his time, both in the *News* from 1805 to 1807 and in his own paper the *Examiner*, which he edited from 1808 to 1821, continuing to supervise it even while in prison in 1813 for having published in it some criticisms of the Prince Regent. In 1840 his only play *A Legend of Florence* was produced at *Covent Garden, and in the same year he published an edition of the dramatic works of *Sheridan and the Restoration dramatists with biographical notes, which inspired Macaulay to publish in the *Edinburgh Review* his famous essay on 'The Comic Dramatists of the Restoration'.

Hunter, N(orman) **C**(harles) (1908–71), English dramatist. He wrote a number of light comedies, of which *All Rights Reserved* (1935), *Ladies and Gentlemen* (1937), and *A Party for Christmas* (1938) reached the West End, before scoring an outstanding success with two plays which showed the influence of *Chekhov in their atmosphere and characterization—*Waters of the Moon* (1951), starring Sybil *Thorndike, Edith *Evans, and Wendy *Hiller, and *A Day for the Sea* (1953), in which Sybil Thorndike again appeared, partnered by Irene *Worth, John *Gielgud, and Ralph *Richardson. Michael *Redgrave and his daughter Vanessa were seen in *A Touch of the Sun* (1958). Of his later plays, the only one presented in London was *The Tulip Tree* (1962). *Waters of the Moon* was revived at *Chichester in 1977 (*Haymarket, London, 1978).

Hurry, Leslie (1909–78), English stage designer, most of whose work was done for ballet and opera. He was, however, responsible for the décor of many Shakespeare plays given at the *Old Vic and at the *Shakespeare

Memorial Theatre, and also for productions of *Marlowe's *Tamburlaine the Great* (1951), Graham *Greene's *The Living Room*, *Otway's *Venice Preserv'd* (both 1953), Tennessee *Williams's *Cat on a Hot Tin Roof* (1958), and *Webster's *The Duchess of Malfi* (1960). His work was characterized by a sombre magnificence which imparted a brooding air of tragedy to his settings, shot through with sudden gleams of red and gold, and he was at his best in designing for plays which like *Venice Preserv'd* called for the conjuring-up of mystery and a sense of space, together with poetic imagery.

Hutt, William (1920–   ), Canadian actor-director, who was a student at Hart House Theatre in the University of Toronto, his birthplace, before playing in summer stock in Ontario and with the Canadian Repertory Theatre in Ottawa. In 1953, in its first season, he joined the *Stratford (Ontario) Festival company, with which he has been almost continuously associated as an actor, director, and Associate Director (13 seasons). He toured extensively, initially with the Canadian Players in Canada and the USA and then with the Stratford company, being first seen in New York in Tyrone *Guthrie's production of *Marlowe's *Tamburlaine* in 1956. For Stratford he appeared at the *Edinburgh Festival (1956),

*Chichester (1964), and led the company throughout Europe (1973) and Australia (1974). He played with the *Bristol Old Vic in 1959 in *O'Neill's *Long Day's Journey into Night* and in London in *Waiting in the Wings* (1960) by Noël *Coward, in whose *Sail Away* he toured North America a year later. In 1964 he created the part of the lawyer in the Broadway production of *Albee's *Tiny Alice* and in 1968 he was seen in New York in *Shaw's *Saint Joan*, which he directed at Stratford in 1975. He returned to London in 1969 as Caesar in Shaw's *Caesar and Cleopatra*. Among other plays he directed for Stratford are *Beckett's *Waiting for Godot* in 1968, *Much Ado about Nothing* in 1971, *As You Like It* in 1972, and *Turgenev's *A Month in the Country* in 1973. From 1976 to 1979 he was Artistic Director of Theatre London (Ontario). His prodigious range is exemplified in three King Lears (1961, 1972, 1988), the title-roles in *Chekhov's *Uncle Vanya* and in *Titus Andronicus* in 1978 and *Timon of Athens* in 1983, Wolsey in *Henry VIII* in 1986, and even Lady Bracknell in *Wilde's *The Importance of Being Earnest* in 1975. His work in the 1980s included leading roles in Peter *Shaffer's *Equus*, Ronald Harwood's *The Dresser*, Bernard Pomerance's *The Elephant Man*, and Robert *Bolt's *A Man for All Seasons*. In 1989 he made his first appearance with the *Shaw Festival.

I

Ibsen, Henrik Johan (1828–1906), Norwegian dramatist and poet, the most important theatrical figure of his generation, whose plays completely changed the main current of European dramatic literature. At one time regarded purely as a depicter of small-town provincial life, he is now recognized as a universal genius who infused even his prose works with a profoundly passionate and tragic poetic spirit. His early years were unhappy, his first plays, *Catilina* (1850) and *Fru Inger til Østraat* (*Lady Inger of Østraat*, 1854), unsuccessful. *Gildet paa Solhaug* (*The Feast at Solhaug*, 1855) was the first to achieve recognition; *Hærmændene paa Helgeland* (*The Warriors*—or *Vikings*—*at Helgeland*, 1857), set in the heroic age of the sagas, shows what remarkable progress Ibsen had made during the time he had spent working at the theatre in Bergen, as assistant to Ole Bull. All these early plays were historical. After this Ibsen used the past as a setting for his plays only twice more—in *Kongsemnerne* (*The Pretenders*, 1864) and *Kejser og Galilæer* (*Emperor and Galilean*) completed in 1873.

In 1862 the theatre in Bergen went bankrupt and Ibsen moved to Christiania (Oslo), where his first play on contemporary life, a satire in verse entitled *Kjærlighedens komedie* (*Love's Comedy*), was produced with some success. A year later he received a travelling fellowship which enabled him to visit Italy and Germany, and in 1864 he settled in Rome, where he wrote his great poetic drama *Brand*. This established his reputation throughout Europe and earned him a state pension. It was followed by his last play in verse, *Peer Gynt*, written in 1867 and produced in a revised stage version, with incidental music by Grieg, in 1876.

The four plays that followed are realistic portrayals of ageless and universal parochialism set in the small-town life of Ibsen's own day: *Samfundets støtter* (*Pillars of Society*, 1877) is a study of public life based on a lie; *Et dukkehjem* (*A Doll's House*, 1879) of the insidious destruction of domestic life by another lie; *Gengangere* (*Ghosts*, 1881) of the lingering poison in a marriage based on a lie; *En folkefiende* (*An Enemy of the People*, 1882) of a man of truth in conflict with the falsity of society. All have the structural economy and simplicity of a skilled writer at the height of his powers, and all, in thought and technique, exercised an immense influence on the contemporary theatre. Ibsen's last plays, in which symbolism plays an increasingly large part and the interest shifts gradually from the

individual in society to the individual isolated and alone, include *Vildanden* (*The Wild Duck*, 1884), *Rosmersholm* (1886), *Fruen fra havet* (*The Lady from the Sea*, 1888), *Hedda Gabler* (1890), a subtle study of feminine psychology, *Bygmester Solness* (*The Master Builder*, 1892), which is concerned with the dual nature of the man and the artist, *Lille Eyolf* (*Little Eyolf*, 1894), a study of marital relations, *John Gabriel Borkman* (1896), a study of unfulfilled genius in relation to society, and *Naar vi døde vaagner* (*When We Dead Awaken*, 1899), Ibsen's last pronouncement on the artist's relation to life and truth.

The first play by Ibsen to be seen in London, in 1880, was *Samfundets støtter*, translated by William *Archer (who quickly established himself as the main translator of Ibsen) as *Quicksands*; under the better-known title, *Pillars of Society*, it was staged in 1889. *Et dukkehjem*, as *Breaking a Butterfly*, with *Tree as Krogstad (renamed Dunkley), was seen in 1884; as *Nora*, in 1885; and as *A Doll's House*, with Janet *Achurch as Nora, in 1889. Florence *Farr and Frank *Benson appeared in *Rosmersholm* in 1891. In the same year came the private production of *Gengangere*, as *Ghosts*, which aroused a storm of abuse, and the first performance of *The Lady from the Sea*. In 1893 came *The Master Builder* and *An Enemy of the People*, again with Tree. *The Wild Duck* in 1894, *Little Eyolf* in 1896, *John Gabriel Borkman* in 1897, and *The League of Youth* in 1900 were followed by a *Stage Society production of *When We Dead Awaken* in 1903. In the same year came Gordon *Craig's production of *The Vikings at Helgeland* with his mother, Ellen *Terry, as Hiordis. *Lady Inger* came in 1906, *Peer Gynt* and *Olaf Liljekrans* in 1911, *The Burial Mound* (as *The Hero's Mound*) and *Brand* in 1912, *The Pretenders* in 1913, *St John's Night* in 1921. In 1936 Donald *Wolfit appeared in *Catiline*. A production outside London deserving mention is *Love's Comedy*, produced as early as 1909 at the Gaiety Theatre in *Manchester. Most of Ibsen's plays have been revived, often in new translations.

In 1882 the world première of *Gengangere* was given in Chicago, in Norwegian, a year before its first production in Europe, and in the same year America saw its first translated Ibsen play, *Et Dukkehjem* as *The Child Wife*; it was presented in 1883 in Louisville, Ky., as *A Doll's House* with *Modjeska as Nora. *Ghosts* followed in 1894 and *John Gabriel Borkman* in 1897. The critical response was generally hostile; but leading ladies of the calibre of Mrs *Fiske persisted in playing Ibsen's heroines. *Hedda Gabler* was first produced in America in 1898, *The Master Builder* in 1900, *Rosmersholm* and *The Pillars of Society* in 1904, and *When We Dead Awake(n)* in 1905. By the year of Ibsen's death, opinion had changed in his favour; Richard *Mansfield presented *Peer Gynt* in that year, and when *Nazimova mounted a season of Ibsen plays in 1906–7 the important literary critics James *Huneker and William Dean Howells came to his defence. The first American productions of *Brand*, *Little Eyolf*, and *The Lady from the Sea* came in 1910, and of *The Wild Duck* in 1918. Eleonora *Duse's national tour with *Ghosts* and *The Lady from the Sea* in 1923 paved the way for the Actors' Theater and Eva *Le Gallienne's Civic Repertory Company, both in New York, to champion Ibsen's plays alongside other European works. Walter *Hampden gave *An Enemy of the People* its American première in 1927 and by 1929 there were six Ibsen productions on Broadway, with Eva Le Gallienne and Blanche *Yurka appearing simultaneously as Hedda Gabler. *The Vikings* had its first American production in 1930. A shift in the taste of American audiences towards new and purely American plays led to 20 years of comparative neglect; but with Arthur *Miller's adaptation of *An Enemy of the People* in 1950 and Lee *Strasberg's production of *Peer Gynt* in 1951 this came to an end, and Ibsen's work has since ranked highly.

Iden, Rosalind, see WOLFIT.

Iffland, August Wilhelm (1759–1814), German actor and playwright, who virtually controlled the National Theatre at *Mannheim from its foundation in 1778 until 1796, playing Franz Moor in the first production of *Schiller's *Die Räuber* (1781) and appearing in many of his own plays. Though now forgotten, these were very successful in their own day and several were translated into English, among them his best-known work, *Die Jäger* (1785), as *The Foresters* in 1799. Iffland catered for a popular audience, for whom he turned domestic tragedies into sentimental family dramas with happy endings. As an actor he had a fine technique but no depth, and he was at his best in elderly witty roles in dignified comedy. In 1796 he left Mannheim for Berlin, where he remained until his death, training a number of young actors, including Ludwig *Devrient, not in his own virtuoso style but in the serious, sober methods of *Schröder.

Illustre-Théâtre, the company with which *Molière, drawn into it by his friendship with Madeleine *Béjart, made his first appearance on the professional stage. The contract drawn up between the first members, among whom were three of the Béjart family, is dated 30 June 1643, and was modelled upon that of the *Confrérie de la Passion. In essentials, it has remained the basic constitution of the *Comédie-Française ever since. The company played for a time in the provinces, possibly at Rouen, opening in 1644 in Paris, but without much success. By 1645 the Illustre-Théâtre had come to an ignominious end, and vanished without leaving a trace in contemporary records. Its repertory included plays by *Corneille, *Du Ryer, and *Tristan L'Hermite, and some specially written by a member of the company, all of which had the word *illustre* in the title.

Ilyinsky, Igor Vladimirovich (1901–87), Soviet actor, who made his first appearance in Petrograd (later Leningrad) in 1918, in which year he also played in the first 'Soviet' play, *Mayakovsky's *Mystery-Bouffe*. Two years later he joined *Meyerhold's company, appearing in all his major productions until 1935 including Crommelynck's *The Magnificent Cuckold* in 1922, which made his name, and Mayakovsky's *The Bed Bug* (1929). In 1938 he went to the *Maly Theatre, where he remained, as actor and director, for the rest of his life, establishing himself as a fine interpreter of leading roles in *Ostrovsky's plays, such as *The Forest* in 1974, which he directed. He was also seen in *Korneichuk's *In the Steppes of the Ukraine* (1942) and *Vishnevsky's *The Unforgettable 1919* (1951). He played Lenin in *John Reed* (1967), an adaptation of Reed's *Ten Days that Shook the World*, and was responsible in 1958 for a production based on Thackeray's *Vanity Fair* and in 1962 for Sofronov's *Honesty*. He was also well known in films.

Imperial Theatre, London, in Tothill Street, originally the Aquarium Theatre, part of an amusement and exhibition palace known as the Royal Aquarium Winter Garden, which stood on the site of the present Central Hall, Westminster. It opened in 1876, and *Phelps made his last appearance there in 1878. In 1879 the name was changed to the Imperial. It closed in 1899 and Lillie *Langtry, who had already appeared there in 1882 in Tom *Taylor's *An Unequal Match*, took over, opening in 1901 with Berton's *A Royal Necklace*. In spite of good reviews and sumptuous costumes and scenery the play was not a success, and in 1903 Mrs Langtry withdrew. Ellen *Terry then presented *Ibsen's *The Vikings at Helgeland*, and *Much Ado about Nothing*, with herself as Hiordis and Beatrice, both productions being designed and directed by her son Gordon *Craig. Lewis *Waller then presented a series of romantic plays of which the most successful, apart from the perennial *Monsieur Beaucaire* by Booth *Tarkington, was Conan Doyle's *Brigadier Gerard* (1906). The last play seen at this theatre was Dix and Sutherland's *Boy O'Carrol* (also 1906), with *Martin-Harvey. The theatre was then dismantled and taken to Canning Town, where it was re-erected as the Imperial Palace. It later became a cinema, and was destroyed by fire in 1931.

Imperial Theatre, New York, on West 45th Street, between Broadway and 8th Avenue. Seating 1,452, it opened in 1923 and was intended by the *Shuberts for musical shows, among its early successes being *Rose-Marie* (1924) and *Oh, Kay!* (1926). In 1936 Leslie Howard appeared there in *Hamlet*, competing unsuccessfully with the record-breaking production starring *Gielgud at the *Empire. It then reverted to musicals, Ethel *Merman starring in *Annie Get Your Gun* (1946) and *Call Me Madam* (1950), which were followed by *The Most Happy Fella* (1956), *Oliver!* (1963), *Fiddler on the Roof* (1964), and *Zorba* (1968). Two rare straight plays, John *Osborne's *A Patriot for Me* (1969) and Neil *Simon's *Chapter Two* (1977), preceded the musical *They're Playing Our Song* (1979) for which Simon wrote the book. Later musicals were *Dreamgirls* (1981) and *The Mystery of Edwin Drood* (1985), based on *Dickens.

Impersonation, see ANIMAL, FEMALE, and MALE IMPERSONATION.

Improvisation, impromptu performance by an actor or group of actors, which may be an element in actor-training, a phase in the creation of a particular role, or part of a staged production. For much of its history the theatre relied heavily on the actor's ability to improvise on a given theme, as in the comic scenes of the *mystery play, in early farces and comedies deriving from folk tradition which gave ample scope for the actor's initiative, and above all in the productions of the *commedia dell'arte. The practice continued in *melodrama and *pantomime and in the *music-hall. *Stanislavsky recognized the value of improvisation during lengthy rehearsal periods to help an actor explore a character's background and motivation. His ideas were widely adopted, especially by *Method actors, and exercises in improvisation aimed at the release of personal inhibitions and the development of physical and vocal skills became a part of all actors' training. Meanwhile, in a reaction against the set text, the influence of *Dada and Surrealism was encouraging spontaneous activity and the rejection of the achieved 'work of art'. This philosophy resulted in the emergence of *Collective Creation from which developed *happenings and similar environmental events.

Inchbald [*née* Simpson], **Elizabeth** (1753–1821), English actress and one of the first English women dramatists. She was for a time on the provincial stage, where she acted Cordelia to the King Lear of her husband **Joseph Inchbald** (?–1779), an inoffensive little man who painted and acted indifferently, and survived his marriage to the beautiful and spirited Elizabeth for only seven years. She was acting in Tate *Wilkinson's company when her husband died, and later appeared at *Covent Garden, retiring in 1785 to devote herself to playwriting after the moderate success of her first plays, which included *I'll Tell You What* (1785), *The Widow's Vow* (1786), and, from the French, *The Midnight Hour; or, War of Wits* (1787). She was a capable writer of sentimental comedy, and her plays, though not revived, were successful in their own day. Among the best were *Wives as They Were and Maids as They Are* (1797), and her last comedy, *To Marry, or Not to Marry* (1805). She also made English versions of *Destouches's *Le Philosophe marié* as *The Married Man* (1789) and of *Kotzebue's *Das Kind der Liebe* as *Lovers' Vows* (1798).

Incidental Music, music written expressly for a dramatic performance, which seldom survives the play for which it was intended. It can be seen first in Elizabethan drama and also in classical Spanish drama. Shakespeare's plays demanded a good deal of music, not only for interpolated songs, for sonnets (a word probably derived from *sonata*), and for tuckets (from *toccata*), but also for interludes and dances. There must have been music at the opening of *Twelfth Night* and in the last acts of *A Midsummer Night's Dream* and *Much Ado about Nothing*. In Spain, the plays of *Calderón, Lope de *Vega, and many other playwrights continually call for music of various sorts; yet they have too much action and spoken dialogue to be classed as operas. In English Restoration plays Purcell's music for such dramatists as *Beaumont and *Fletcher, *Congreve, and *Dryden marks an important historical advance in the use of music in the theatre. During the 18th century the popularity of *ballad opera and other types of light operatic entertainment in which the music, though often trivial, was none the less essential prepared the way for developments in the 19th by enlarging and improving the orchestral resources of the theatre. Overtures and musical interludes between the acts became customary, and it was not long before playwrights inserted not only songs and dances and processions to music, but scenes of excitement or pathos in which the spoken words were accompanied by an orchestral undercurrent. The *melodrama developed from this.

From the beginning of the 19th century, major composers were supplying scores which have also survived independently, including Beethoven's overture for the 1810 production of *Goethe's *Egmont*, Schubert's music for Helmine von Chézy's ephemeral *Rosamunde* (1823), Schumann's score (1848–9) for *Byron's *Manfred*, and Mendelssohn's full incidental music for *A Midsummer Night's Dream*. By the second half of the century incidental music was established as a separate category of some importance, with such works as Bizet's music for Daudet's

L'*Arlésienne* (1872) and Grieg's for *Ibsen's *Peer Gynt* (1876). Everywhere *poetic drama, in particular, was provided with some form of music, often by famous composers. In pre-Revolutionary Russia, Balakirev wrote music for *King Lear* (1861) and Tchaikovsky for *Hamlet* (1891); and the Soviet Union has maintained the traditions of using large-scale orchestral forces in the serious theatre. The cinema has claimed the best of modern American incidental music, although early productions of plays by Eugene *O'Neill had interesting scores.

In England incidental music was rarely taken seriously until the 20th century, though music was commissioned from Sir Arthur Sullivan for productions of Shakespeare between 1862 and 1888, as well as for *Tennyson's *The Foresters* in 1892 and Comyns Carr's *King Arthur* in 1895. A revival of *The Tempest* in London in 1921 in which some of Sullivan's music was used, together with some new and very striking music by Arthur Bliss, showed how good Sullivan could be at theatre music of this kind. Other music of quality was produced mainly for academic performances—Stanford's for *Aeschylus' *Eumenides* (1885) and *Sophocles' *Oedipus Tyrannus* (1887) in *Cambridge; Parry's for productions of *Aristophanes in *Oxford between 1883 and 1914; and Vaughan Williams's for Aristophanes' *Wasps* (1909), also in Cambridge. Stanford also provided good incidental music for two London plays, Tennyson's *Queen Mary* (1876) and *Becket* (1893), in which Henry *Irving appeared.

The incidental music by Norman O'Neill for plays at the *Haymarket was slight but combined a special aptitude for the requirements of the stage with graceful, sometimes fanciful invention, particularly in the scores for *Maeterlinck's *The Blue Bird* (1909), *Barrie's *Mary Rose* (1920), and Ashley *Dukes's *The Man with a Load of Mischief* (1925). Elgar wrote music for the London production of *Yeats's *Diarmuid and Grania* (1902) and for *The Starlight Express* (1915). Other outstanding composers of incidental music at this time were Armstrong Gibbs—for Maeterlinck's *The Betrothal* (1921); Eugene Goossens—for *Maugham's *East of Suez* (1922), Margaret Kennedy's *The Constant Nymph* (1926), and Dodie Smith's *Autumn Crocus* (1931); Frederic Austin—for the *Čapeks' *The Insect Play* (1923) and a revival of *Congreve's *The Way of the World* in 1924; and, most important of all, Delius—for Flecker's *Hassan* (1923). Benjamin Britten wrote the music for the London productions of *Priestley's *Johnson over Jordan* (1939), Ronald Duncan's *This Way to the Tomb*, and *Webster's *The Duchess of Malfi* (both 1945); and before turning his attention to film music during the war years, William Walton provided a score for John *Gielgud's production of *Macbeth* in 1942.

In recent years companies who regularly use incidental music have tended to employ a small instrumental ensemble or some form of recorded score. Until it closed in 1963 the *Old Vic had a flexible chamber group and commissioned music from a number of distinguished contemporary composers. These include John Gardner (*Marlowe's *Tamburlaine*, 1951); Malcolm Arnold (*The Tempest*, 1954); Peter Maxwell Davies (*Richard II*, 1959); Thea Musgrave (*A Midsummer Night's Dream*, 1960); Michael Tippett (*The Tempest*, 1962); Elizabeth Lutyens (*Julius Caesar*, 1962; also Aeschylus' *Oresteia* for the *Oxford Playhouse company at the Old Vic, 1961). The company's resident composer George Hall wrote incidental music for Ibsen's *Peer Gynt* in 1962: the National Theatre in Oslo had also commissioned a score from Harald Saeverud in 1947. When Peter *Hall became director of the *RSC in 1960, Raymond Leppard was appointed music adviser and there too outstanding composers were employed: Lennox Berkeley (*The Winter's Tale*, 1960), Humphrey Searle (*Troilus and Cressida*, 1960), and Roberto Gerhard (*Macbeth*, 1962). As resident composer Stephen Oliver provided an effective score for the RSC's adaptation of *Dickens's *Nicholas Nickleby* (1980) as well as for a number of Shakespeare productions.

Taped electronic music has proved increasingly useful. *Brook's chilling accompaniment to his production of *Titus Andronicus* (Stratford, 1957) and Leppard's enchanted music for *The Tempest* (Stratford, 1963) are notable examples.

Independent Theatre Club, see KINGSWAY THEATRE.

Independent Theatre Society, see ROYALTY THEATRE 2.

Inge, William (1913–73), American dramatist, whose first success was *Come Back, Little Sheba* (1950), with Shirley Booth as the garrulous, pathetic, lonely wife of an alcoholic. *Picnic* (1953), which won the *Pulitzer Prize, depicts the effect of an unemployed wanderer with powerful sexual magnetism on the women he meets in a Kansas town. It was followed by *Bus Stop* (1955), Inge's most cheerful work, again set in Kansas and bringing together a group of characters in a café used by bus passengers; and *The Dark at the Top of the Stairs* (1957), originally produced in Dallas in 1947 as *Farther Off from Heaven*, and dealing with the sexual and other problems of a travelling salesman and his family in a small town in Oklahoma in the early 1920s. After these plays—all hits in New York, though little known in Britain except as films—Inge's career went into decline. *A Loss of Roses* (1959) was a failure, as were the plays which followed. Inge, though he had a vivid sense of the theatre, was somewhat too elementary in his approach to psychological problems.

Ingegneri, Angelo, see LIGHTING.

Inner Pros(cenium), see FALSE PROSCENIUM.

Inner Stage, presumed feature of the Elizabethan theatre which has aroused a great deal of controversy. It was formerly thought to be either a large curtained recess behind the back wall or a structure projecting on to the stage from the back wall as in the *Terence-stage. Some form of concealment must have existed, as stage directions in a number of Elizabethan plays demand 'a discovery' by the drawing of a curtain (for example, Ferdinand and Miranda playing chess in *The Tempest*); but it is now thought that the 'inner stage' was nothing more than part of a narrow corridor behind the stage-wall which could be made visible through an opening usually closed by a door or curtain. As it would not have been possible for many people in the audience to see a scene played inside this recess, it seems probable that actors so 'discovered' came forward on to the main stage to join in the action of the play. There may also have been small structures—caves, tents, monuments, even simple rooms—either standing free on the stage, as they did in the *masque, or abutting on to the back wall. Some such arrangement would have been necessary for the monument scene in *Antony and Cleopatra*. Stage directions prove the existence of a similar 'inner stage' in the Spanish *auto sacramental*, where a curtain might be drawn back to reveal a crib, or a statue of the Virgin, or the empty tomb of the Resurrection. Because of the uncertainty over its size and position, the term 'inner stage' is now being abandoned in favour of the expression 'discovery space', which allows for a wide variety of interpretations.

Innyards Used as Theatres, London. These may have been converted, or merely equipped with a trestle stage at one end of the yard. The best-known were the **Bel Savage** on Ludgate Hill in the City of London, where plays were performed from 1579 to 1588 and occasionally thereafter; the **Bell** in Gracious (now Gracechurch) Street in the City of London, which was used for plays in 1576 and 1583; the **Boar's Head** in Aldgate, where a 'lewd' play called *A Sack Full of News* was suppressed in 1557, the players being kept under arrest for 24 hours; another **Boar's Head** somewhere in Middlesex, which was in use between 1602 and 1608; the **Bull** in Bishopsgate Street in the City of London, which was used for plays before 1575 and until some time after 1594; the **Cross Keys**, also in Gracious Street, where plays were performed before 1579 and up to about 1596; the **Red Lion** in Stepney, where a play called *Samson* was performed in 1567; and the **Saracen's Head** in Islington, where, according to Foxe's *Book of Martyrs* (1563), a play was being performed when the dissenter John Rough was arrested.

Inset, small set, usually a corner of a room or an attic, lowered from the *flies, or set inside (i.e., in front of) a full set that does not have to be struck. An Inset Scene is a similar set piece put behind an opening in a *flat to mask the view beyond.

Interlude, in early English drama a short dramatic sketch, from the Latin *interludium*. It appears to have some affinity with the Italian *tramesso*, signifying something extra inserted into a banquet and so an entertainment given during a banquet. By extension it came to indicate short pieces played for light relief between the acts of a longer and more serious play. For these Renaissance Italy adopted the term *intermedio* or *intermezzo*, the former term giving rise to the French *entremets* or *intermède*, meaning a short comedy or farce. In Spain the *entremés*, while having a somewhat similar origin, became a distinct dramatic genre. The first English dramatist to make the interlude an independent dramatic form was John *Heywood. The Players of the King's Interludes (Lusores Regis) were first recorded in 1493 and disappeared under Elizabeth, the last survivor dying in 1580.

Intermezzo (*Intermedio*), interpolation of a light, often comic, character performed between the acts of serious dramas or operas in Italy in the late 15th and early 16th centuries. They usually dealt with mythological or classical subjects, and could be given as independent entertainments for guests at royal or noble festivals, on the lines of the English *disguising and dumb-show, the French *momerie* and *entremets*, or the Spanish *entremés*.

International Amateur Theatre Association, see AMATEUR THEATRE.

International Centre of Theatre Research, organization founded as a result of *Barrault's invitation to Peter *Brook to conduct in Paris in 1968 a workshop for actors, writers, and directors from diverse backgrounds and cultures. This experience inspired Brook to open two years later, in a former tapestry factory, a theatrical centre for questioning, experimentation, and discovery, his colleagues being chosen from over 150 theatre workers from all over the world. One of the Centre's first projects was a retelling of the myth of Prometheus in a new language called Orghast invented by the poet Ted Hughes. It was staged on a mountain top in front of the tomb of King Artaxerxes, overlooking the ruins of Persepolis. In 1972 Brook took his company to Africa, where they toured through five countries, playing on a carpet in temples, houses, squares, forests, and on dirt roads. Much of their work was improvisational, as in *The Conference of the Birds*, based on a work by the 12th-century Persian poet Farid Uddin Attar, revived in the USA in 1973.

In 1974 the Centre moved to the derelict Théâtre des Bouffes du Nord in Paris, deliberately left in its dilapidated state. The first production there, a French version of *Timon of Athens*, was followed in 1975 by *The Ik*, the true story of an African tribe (London, 1976). Later productions were Alfred *Jarry's *Ubu-Roi* (1977); *Chekhov's *The Cherry Orchard* (1981; NY, 1988); *La Tragédie de Carmen* (1981; NY, 1983; Glasgow, 1989), a pared-down version of Bizet's opera; and *Mahabharata* (*Avignon and Paris, 1985; NY, 1987; Glasgow, 1988), a nine-hour adaptation of a pre-Christian Indian epic.

International Federation for Theatre Research, founded in July 1955 at a meeting in London attended by delegates from 22 countries at the invitation of the English *Society for Theatre Research. The Federation, which now has 34 member countries, is devoted to the promotion of international liaison between organizations and individuals devoted to theatre research. Its constitution was drafted at a meeting in Paris in 1956 and accepted at a world conference in Venice in 1957. In association with Oxford University Press it publishes three times a year a journal, *Theatre Research International*, with résumés in French of the articles, which are in English. Membership is open to institutions engaged in theatre research and there is also a category of individual members. Meetings of its committee are held annually in various European and North American cities and are usually accompanied by a symposium at which experts are invited to speak on a subject chosen by the host country. Every four years a world conference open to the general public is held, the proceedings of which are published in volume form. A close relationship is maintained with its branch organization the International Association of Libraries and Museums of the Performing Arts. The Federation has established an International Institute for Theatre Research in the Casa Goldoni in Venice, where international summer courses are held, and a Universities Commission especially charged with research within universities and the promotion of the teaching of Theatre Studies throughout the world. The Federation hosts the periodic Edward Gordon Craig Memorial Lecture.

International Shakespeare Globe Centre, see WANAMAKER.

International Theatre, New York, see MAJESTIC THEATRE I.

International Theatre Institute (ITI). Founded as a branch of Unesco in Prague in 1948, the ITI exists to promote international co-operation and exchange of ideas among all workers in the theatre. It works through national centres, with headquarters in Paris, and publishes a quarterly illustrated journal in French and English, *International Theatre Information*. A world congress is held every two years, together with conferences and colloquia on such questions as the training of the actor and theatre architecture. The British Centre, one of the first to be opened, helps those who come from abroad to study the theatre in Britain and also serves as a central information bureau, particularly about the theatre abroad. The other centres, including the American Centre, now an independent body, cover much the same ground. A number of international organizations connected with the theatre are affiliated to the ITI.

Intima Teatern, Stockholm, see STRINDBERG.

Intimate Theatre, Palmers Green, London, see CLEMENTS.

Ionesco [Ionescu], **Eugène** (1912–), French dramatist of Romanian origin, exponent and virtual founder of the Theatre of the *Absurd. Rejecting both the realistic and the psychological theatre, Ionesco's plays stress the impotence of language as a means of communication, the oppression of physical objects, and the incapacity of man to control his own destiny. Many are in one act: they include *La Cantatrice chauve* (1950), *La Leçon* (1951), and *Les Chaises* (1952), seen in London as *The Bald Prima Donna* (1956; NY, as *The Bald Soprano*, 1958), *The Lesson* (1955; NY, 1958), and *The Chairs* (1957); and *Le Nouveau Locataire* (1955, seen in London as *The New Tenant* in 1956 and in New York in 1960, but not staged in Paris until 1967). Ionesco's first full-length play, *Amédée; ou, Comment s'en débarrasser* (1954), seen in New York in 1955 and in London in French in 1963, was only partially successful. He endeavoured to rectify this by introducing into his next full-length play *Tueur sans gages* (1959), seen in New York in 1960 and in London in 1968 as *The Killer*, a well-intentioned if ultimately impotent 'little man', Élie Berenger, who reappears in *Rhinocéros* (1960), *Le Roi se meurt* (1962), and *Le Piéton de l'air* (1963). The first two were produced at the *Royal Court Theatre in London with Laurence *Olivier in *Rhinoceros* in 1960 and Alec *Guinness in *Exit the King* in 1963 (NY, 1961 and 1968, the first starring Zero *Mostel). In 1966 a new play, *La Soif et la faim*, was produced at the *Comédie-Française. Ionesco's later work, which received less attention in the English-speaking world, included *Jeux de massacre* (1971) and *Macbett* (1972; London, 1973), a reworking of Shakespeare's play in his own image, *The Mire* (1972), an experimental montage of speech, image, and sound which broke new ground without departing from Ionesco's characteristic *angst*-ridden themes, and *Ce formidable bordel* (*This Dreadful Mess*, 1973). *Voyages chez les morts* (*Journeys among the Dead*, 1981; London, 1987) is an autobiographical play about a writer's dreams and memories.

Ipsen, Bodil, see REUMERT.

Ireland, William Henry (1775–1835), brilliant but eccentric Englishman, who at the age of 19 forged a number of legal and personal papers purporting to relate to Shakespeare which for a time deceived even the experts. He persuaded *Sheridan to put on at *Drury Lane, in 1796, a forged play, *Vortigern and Rowena*. Although John Philip *Kemble and Mrs *Jordan played the title-roles it failed, thus preventing Ireland from bringing forward further forgeries, of which *Henry II* was already written and *William the Conqueror* nearly completed. All Ireland's forgeries were published by his father, a dealer in prints and rare books who never ceased to believe in them, saying his son was too stupid to have composed them. They were exposed and Ireland was forced to confess. He spent the rest of his life in ignominious retirement, doing nothing but hack work.

Irish Comedians, see COLLINS.

Irish Literary Theatre, founded by Lady *Gregory and W. B. *Yeats, the first manifestation in drama of the Irish literary revival. The first performances were of Yeats's *The Countess Cathleen* and Martyn's *The Heather Field* at the Antient Concert Rooms, Dublin, in 1899. In 1901 a cast headed by Frank *Benson came from England to appear in *Diarmuid and Grania*, by Yeats and *Moore, and the first Gaelic play, Douglas Hyde's *Casadh an tSugáin*, was performed by Gaelic-speaking amateurs. The Irish Literary Theatre was then taken over by the *Irish National Dramatic Society.

Irish National Dramatic Society, company founded by Frank and W. G. *Fay, which in 1902 produced AE's *Deirdre*, *Yeats and Lady *Gregory's *Cathleen ni Houlihan* and *The Pot of Broth*, and four other plays, including one in Gaelic. These were given in small halls in Dublin, but in 1903 and 1904 the company appeared in a larger hall, where, as the National Theatre Society, they were seen in Yeats's *The Hour-Glass*, *The King's Threshold*, and *The Shadowy Waters*, as well as in Lady Gregory's *Twenty-Five*, *Synge's *In the Shadow of the Glen* and *Riders to the Sea*, and Padraic *Colum's *Broken Soil*. The company was invited to London in 1903, appearing there in two performances of five short plays each. It is believed that these productions finally decided Miss *Horniman to finance the *Abbey Theatre. In 1904 an Ulster branch of the society was formed with the support of the parent company, whose members took part in an initial production of *Cathleen ni Houlihan*. This group merged with the *Ulster Group Theatre in 1939.

Irish Theatre, New York, see PROVINCETOWN PLAYERS.

Iron, see SAFETY CURTAIN.

Irving, Sir Henry [John Henry Brodribb] (1838–1905), English actor-manager, knighted in 1895, the first actor to be so honoured. Of Cornish extraction but born in Somerset, he went to London at the age of 10 and while at school had elocution lessons to help overcome a stutter. In 1856 he played Romeo in an amateur production at the Soho Theatre and, encouraged by his success, accepted an invitation to join the *stock company at the new Royal Lyceum Theatre in Sunderland, where he made his first professional appearance on 29 Sept. In 1857 he was in Edinburgh and in the autumn of 1859 returned to London, where he played four small parts including Osric in *Hamlet*. Feeling that he still had a lot to learn, he returned to the provinces and was not seen in London again until in 1866 he appeared successfully as Doricourt in Mrs *Cowley's *The Belle's Stratagem* and Rawdon Scudamore in *Boucicault's *Hunted Down*. In 1867 he played for the first time with Ellen *Terry in *Garrick's *Katharine and Petruchio*, a one-act version of *The Taming of the Shrew*. He was then seen in H. J. *Byron's *Uncle Dick's Darling* (1869) and had a personal success as Digby Grant in *Albery's *Two Roses* (1870), and in 1871 appeared for the first time at the *Lyceum Theatre, under the management of the American impresario H. L. *Bateman. This theatre had long been considered unlucky, and Irving's Jingle, in Albery's adaptation of *Dickens's *The Pickwick Papers*, did nothing to restore its fortunes. Bateman, almost in despair, agreed to let Irving appear in *The Bells*, a study in terror adapted by Leopold Lewis from *Erckmann-Chatrian's *Le Juif polonais*. The theatre was almost empty on the first night; but by next morning Irving was famous and he was to dominate the London stage during the last 30 years of Queen Victoria's reign. *The Bells* was followed by Wills's *Charles I* (1872) and *Eugene Aram* (1873). In the same year he was seen in *Bulwer-Lytton's *Richelieu*. In all these productions he successfully pitted his own conception of acting against that of the current school of *Macready. In 1874 he appeared as Hamlet, presenting him as a gentle prince who fails to act not from weakness of will but from excess of tenderness, and it was with the revival of this play that he inaugurated his own management at the Lyceum on 30 Dec. 1878.

Irving was a good manager and employed only the best players, painters, and musicians of his day, but he was never free from critical attack. At the height of his renown there were still people who found his mannerisms unsympathetic, even faintly ludicrous, yet were drawn to see him because his acting was overwhelming in its intensity. Once under his spell, they found his peculiar pronunciation, crabbed elocution, halting gait, and the queer intonations of his never very powerful or melodious voice part of the true expression of a strange, exciting, and dominating personality. Occasionally he chose to depict tenderness, as when he played Dr Primrose in Wills's dramatization of *Goldsmith's *The Vicar of Wakefield* as *Olivia* in 1885, but in his finest and most powerful parts, Charles I, Richelieu, Wolsey in *Henry VIII*, which he first played in 1892, and *Tennyson's *Becket* in 1893, he was able to give free rein to his individual genius. Although his tenancy of the Lyceum is mainly remembered for his productions of Shakespeare, from *The Merchant of Venice* in 1879 to *Cymbeline* in 1896, he included in his repertory revivals from his earlier days and also a number of new plays, usually written for him and now forgotten. Some of these, together with revivals of *The Bells* and *Charles I*, were seen during his tours of America and Canada, which he visited eight times from 1883 to 1903, returning in 1904 to make his last appearance in New York in Boucicault's *Louis XI*. His last years were unhappy. His health was failing and he was beset by financial difficulties, exacerbated by a disastrous fire in 1898 which destroyed the costumes and settings for 44 plays. He gave up the Lyceum a year later after appearing there in *Peter the Great* by his son Laurence *Irving. Under another management he returned in 1901 as Coriolanus, and made his final appearance there in 1902 as Shylock. His last appearance in London, on 10 June 1905, was as Tennyson's Becket. He then set out on a farewell tour which ended with his death in Bradford on 13 Oct. He was buried in Westminster Abbey. The two sons of his unhappy marriage in 1869, which ended in separation after only two years, were both on the stage.

Irving, Henry Brodribb (1870–1919), English actor-manager, elder son of Henry *Irving. He made his first appearance in 1891 with John *Hare at the *Garrick Theatre as Lord Beaufoy in *Robertson's *School*. He was for a time with Ben *Greet, and later with George *Alexander, and in 1902 created the title-role of *Barrie's *The Admirable Crichton*. In 1906 he formed his own company with which he toured England and America, reviving many of his father's famous parts, and he later managed both the *Queen's and the *Savoy theatres. He married in 1896 the actress **Dorothea Baird** (1875–1933), who a year previously had created the title-role in Du Maurier's *Trilby*.

Irving, Laurence Henry Forster (1897–1988), English stage designer and theatre historian, son of H. B. *Irving, who in 1951 published the definitive biography of his grandfather, *Henry Irving: The Actor and his World*, continuing the story of the Irving family in *The Successors* (1967) and *The Precarious Crust* (1971). As artist and scenic designer he was responsible for the décor of many London productions, including *Priestley's *The Good Companions* (1931) and *I Have Been Here Before* (1937) and T. S. *Eliot's *Murder in the Cathedral* (1935). He designed Michael *Redgrave's *Hamlet* at the *Old Vic in 1950 and later worked for John *Clements (including *Ibsen's *The Wild Duck*, 1955) and Donald *Wolfit. He was also the first chairman of the British Theatre Museum Association.

Irving, Laurence Sidney Brodribb (1871–1914), English actor and playwright, younger son of Henry *Irving, who made his first appearance in 1891 in Frank *Benson's company. In 1898 he was at the *Lyceum with his father, for whom he wrote *Peter the Great* (1898), an epic poem rather than a play, and also adapted *Sardou's *Robespierre* (1899) and *Dante* (1903). Of his later plays the most successful was *The Unwritten Law* (1910). In 1913 he made a great success as Earle Skule in *Ibsen's *The Pretenders* at the *Haymarket Theatre, and a year later left with his wife to tour Canada and the USA. They were both drowned when the *Empress of Ireland* sank after a collision in the St Lawrence.

Irving, Washington (1783–1859), the first American author to gain recognition abroad. Chiefly remembered as a historian and as the writer of romantic sketches and tales, he also served as a diplomat. In 1802–3 he published, in the New York *Morning Chronicle*, a series entitled 'The Letters of Jonathan Oldstyle', several of which give a vivid picture of the contemporary New York stage. He was co-author of *Salmagundi* (1807–8), which includes a number of satiric letters on the state of the theatre. He later collaborated with John Howard *Payne in several plays. Their *Charles II* (1824) and *Richelieu* (1826), both adapted from French originals, were particularly successful. His short story 'Rip Van Winkle' (first published in 1819) was adapted for the stage by James *Hackett in 1825. Other adaptations followed, the most successful being that of Joseph *Jefferson III and Dion *Boucicault in 1865, in which Jefferson scored his greatest success.

Irving Place Theatre, New York, see SCHWARTZ.

Isherwood, Christopher, see AUDEN and VAN DRUTEN.

Israeli Theatre, see HEBREW THEATRE.

Italian Night Scenes, see HARLEQUINADE.

ITI, see INTERNATIONAL THEATRE INSTITUTE.

Ivanov, Vsevolod Vyacheslavovich (1895–1963), one of the first Soviet dramatists. He began writing while in the army and in 1920, with the help of *Gorky, settled in Leningrad.

His first play, *Armoured Train 14-69*, was an adaptation of one of his own novels, dealing with the capture of a trainload of ammunition during the Civil War. Produced by the *Moscow Art Theatre in 1927 with *Kachalov in the lead, it was the first successful Soviet play and now has a permanent place in the Russian repertory. A later play about events in the Far East, *The Doves See the Cruisers Departing* (1938), also had some success. His plays on more remote history included *Twelve Youths from a Snuffbox* (1936), on the assassination of Tsar Paul I, and *Lomonosov* (1953), on the 'father' of Russian science.

J

Jack-in-the-Green, see ROBIN HOOD.

Jackson, Anne, see MATHEWS, CHARLES.

Jackson, Sir Barry Vincent (1879–1961), English director and wealthy amateur of the theatre. He was trained as an architect, but in 1907 founded an amateur company, the Pilgrim Players, which became professional when in 1913 he built and opened for it the *Birmingham Repertory Theatre in his birthplace. Classics and new plays, tragedy and farce, pantomime and ballet, opera, and even marionettes were seen on its stage, and Jackson maintained it with his own money for 22 years as a creative force in the English theatre, often in the face of local hostility and indifference. Among the many plays he directed were several of his own, including *The Christmas Party* (1913), a children's play which was many times revived. He helped to establish the reputation of George Bernard *Shaw with his production of *Back to Methuselah* (1923), and also presented his own versions of *Ghéon's *The Marvellous History of St Bernard* (1925), *Beaumarchais's *The Marriage of Figaro*, and *Andreyev's *He Who Gets Slapped* (both 1926). Considering the theatre as a workshop for artistic experiment rather than a museum for the preservation of tradition, Jackson produced *Cymbeline* in 1923, *Hamlet* in 1925, and *Macbeth* in 1928 in modern dress and in 1929 founded the *Malvern Festival, mainly as a shop window for Shaw's plays. In 1935 he transferred the Birmingham Repertory Theatre, whose company had proved an excellent training ground for many young players, to a Board of Trustees, but remained associated with it, among his later productions being versions of Wyss's *The Swiss Family Robinson* (1938), *Dickens's *The Cricket on the Hearth* (1941), and *Fielding's *Jonathan Wild* (1942). In 1945 he was appointed Director of the *Shakespeare Memorial Theatre at Stratford-upon-Avon, where he remained until 1948, helping the theatre over the difficult years which followed the Second World War and instituting several salutary reforms. Among the highlights of his management were *Love's Labour's Lost* directed by Peter *Brook; *Hamlet* in Victorian dress, with Paul *Scofield and Robert *Helpmann alternating in the title-role; *The Winter's Tale* directed by Anthony *Quayle, who was to succeed Jackson as Director of the theatre; and *Othello* with Godfrey *Tearle in the name part.

Jackson, Glenda (1936–), English actress of strong personality and high intelligence, who made her first appearance in 1957 and was seen in London later the same year. In 1964 she joined the *RSC, appearing in their Theatre of Cruelty season (see ARTAUD) in London, and first came into prominence as a strikingly erotic Charlotte Corday in the company's production of *Weiss's *Marat/Sade* (NY, 1965), directed by Peter *Brook, which sprang directly from that season. Her Masha in *Chekhov's *Three Sisters* at the *Royal Court Theatre in 1967 was remarkable for its tough, uninhibited passion. She was then only in films for some years, but returned to the stage as the waspish wife in John *Mortimer's *Collaborators* (1973). In 1975 she toured Britain, the USA, and Australia for the RSC in the title-role of *Ibsen's *Hedda Gabler*, and in 1976 was seen at the *Old Vic in *Webster's *The White Devil*, following it with the title-role in Hugh Whitemore's *Stevie* (1977), about the poet Stevie Smith. She was in Stratford in 1978 to play Cleopatra in *Antony and Cleopatra*, and scored a big success as the discontented, married, middle-aged schoolteacher in Andrew Davies's *Rose* (1980; NY, 1981). She used her high reputation to increase the audience for such commercially risky productions as Botho *Strauss's *Great and Small* (1983), *O'Neill's *Strange Interlude* (1984; NY, 1985), *Racine's *Phedra* (1984), and *García Lorca's *The House of Bernarda Alba* (1986). In New York in 1988 she was Lady Macbeth to Christopher *Plummer's Macbeth, and in 1990 she played *Brecht's Mother Courage in Glasgow and London. She is equally famous as a film star.

Jackson, Joe, see VAUDEVILLE, AMERICAN.

Jacobi, Derek George (1938–), highly regarded English actor, who had some experience with the *National Youth Theatre and at *Cambridge before making his professional début at the *Birmingham Repertory Theatre, 1960–3. After a season at *Chichester he joined the *National Theatre company, 1963–71, making his London début as Laertes in *Hamlet* and appearing in such roles as Cassio in *Othello* in 1964 and Tusenbach in *Chekhov's *Three Sisters* in 1967. In 1972 he joined the *Prospect Theatre Company, playing the title-roles in Chekhov's *Ivanov* in 1972 and in *Pericles* in 1973, and repeating for the same company in 1975 his Chichester performance as Rakitin in *Turgenev's *A Month in the Country*. Again for Prospect, in 1977 and 1979 he gave an acclaimed performance as Hamlet, later seen at Elsinore and on a world tour, and in 1978 he was seen as Thomas Mendip in a revival of *Fry's *The Lady's not for Burning*. For the *RSC he played Benedick in *Much Ado about Nothing*, the title-role in *Ibsen's *Peer Gynt*, and Prospero in *The Tempest* (all 1982), and the title-role in *Rostand's *Cyrano de Bergerac* in 1983; in 1984 he took his Benedick and Cyrano to New York. After a highly acclaimed performance in Hugh Whitemore's *Breaking the Code* (1986; NY, 1987) he returned to Shakespeare in *Richard II* (1988) and *Richard III* (1989), ultimately playing them in repertory.

James, David, see VAUDEVILLE THEATRE.

James, Henry (1843–1916), American novelist, who spent most of his life in England, and in 1915 became a British subject. He had a great love for the theatre, but after the hostile reception which greeted his *Guy Domville* (1895) when it was produced by George *Alexander at the *St James's Theatre, he withdrew, and nothing further of his was seen on the stage until *Forbes-Robertson played Captain Yule in *The High Bid* (1909) at *Her (then His) Majesty's Theatre; it was revived in 1967 at the *Mermaid Theatre, where *The Other House*, written in 1893, had its first performance in 1969. *The Outcry*, written in 1909, was performed posthumously by the *Stage Society in 1917 and was revived at the *Arts Theatre in 1968. The theatrical success James had longed for came after his death through dramatizations of his novels by other hands—*Berkeley Square* (1928), based on the unfinished *The Sense of the Past*; *The Heiress* (1947), based on *Washington Square*; *The Innocents* (1950), based on *The Turn of the Screw*; *The Aspern Papers* (1959), adapted by Michael *Redgrave, who also played 'H.J.' in the London production, Maurice *Evans playing the role in New York in 1962; *The Wings of the Dove* (1963); *A Boston Story*, based on *Watch and Ward*, and *The Spoils*, based on *The Spoils of Poynton* (both 1968).

Janauschek, Francesca Romana Maddalena (1830–1904), Czech actress, born in Prague, where she made her début in 1846 in light comedy, leaving two years later to become leading lady at the theatre in Frankfurt, where she appeared mainly in classic revivals. After 10 years, during which time she also toured extensively in Germany and Austria and even as far afield as Russia, she went to the *Dresden Court Theatre, and within a short time had built up an enviable reputation as a tragic actress in such parts as *Euripides' Medea and Iphigenia, *Schiller's Joan of Arc and Mary Queen of Scots, and Shakespeare's Lady Macbeth. In 1867 she went for the first time to the United States, where, in the fashion of the time, she played in German opposite Edwin *Booth and other leading American actors. She studied English intensively, however, and from 1873 played

mainly in that language, usually in parts she had appeared in already. She was one of the last of the great international actresses in the grand tragic style.

Janis, Elsie, see REVUE and VAUDEVILLE.

Japanese Theatre, see KABUKI, NŌ, and PUPPET.

Jarry, Alfred (1873–1907), French poet and dramatist, whose *Ubu-roi* is now considered the founder-play of the modern avant-garde theatre. Written when Jarry was 15 years old and first performed in 1888 as a marionette play, it was given its first live stage production by *Gémier at the Théâtre de l'Œuvre (see LUGNÉ-POË) in 1896. A savagely funny, anarchic revolt against society and the conventions of the naturalistic theatre, it scandalized audiences in 1896 and still has considerable impact, as *Vilar's successful revival in 1958 at the *Théâtre des Nations amply demonstrated. Père Ubu, vicious, cowardly, coarse, pompously cruel, and unashamedly amoral, is a farcical prototype of the later anti-hero of the nuclear age. Many of the marionette elements in the play, expressly demanded by Jarry in his stage instructions to Gémier, have become common currency in the work of playwrights such as *Genet and *Ionesco and directors such as *Brecht and *Planchon: the use of masks, skeleton sets, crude pantomime, and stylized speech to establish character, gross farce and slapstick elements, placards indicating scene changes, cardboard horses slung round actors' necks, and similar unrealistic props. Jarry returned to Mère and Père Ubu in other plays—*Ubu cocu* (completed in 1898, but not published or performed until 1944), *Ubu enchaîné* (1899), and the marionette play, *Ubu sur la butte* (1901)—but without recapturing the same creative spark. In 1966 an English translation of *Ubu-roi* was produced at the *Royal Court Theatre.

Jeans, Isabel (1891–1985), English actress, first seen on the stage in 1909. She had had a good deal of varied experience before beginning an important association with the *Phoenix Society, appearing for them in such roles as Cloe in *Fletcher's *The Faithful Shepherdess* (1923) and Margery Pinchwife in *Wycherley's *The Country Wife* (1924). She also played in such modern works as Ivor *Novello and Constance *Collier's *The Rat* (1924) and Robert E. *Sherwood's *The Road to Rome* (1928). In 1931 she starred in *Counsel's Opinion* by her second husband Gilbert Wakefield. She was in revivals of *Lonsdale's *On Approval* (1933), Somerset *Maugham's *Home and Beauty* (1942), and *Shaw's *Heartbreak House* (1943), and was much admired as Arkadina in *Chekhov's *The Seagull* (1949). In 1945 she had a great success as Mrs Erlynne in *Wilde's *Lady Windermere's Fan*, which was followed by roles in his *A Woman of No Importance* (1953) and as Lady

Bracknell in his *The Importance of Being Earnest* (1968). She was last seen in two plays by Anouilh, a revival of *Ring round the Moon* (1968) and *Dear Antoine* (1971). Her exuberant vitality and gift for high comedy kept her in constant demand for both classic and modern plays.

Jefferson, Joseph (1774–1832), British-born American actor, whose father **Thomas Jefferson** (1732–1807) was an actor at *Drury Lane under *Garrick, and for some years manager of a theatre in Plymouth. Four of his five children were on the stage, Joseph, the second son, trained by his father, going in 1795 to America. He joined the company at the *John Street Theatre in New York and later went to the *Park Theatre, where he was popular with the company and the public, though somewhat held back by the pre-eminence of *Hodgkinson. In 1803 he moved to the *Chestnut Street Theatre in Philadelphia, where he remained until about 1830. He married an actress whose sister was married to William *Warren, thus uniting two families of great importance in American stage history. His seven children all went on the stage, including a second Thomas, John, and four daughters; the best remembered is the second **Joseph** (1804–42), who inherited his father's happy temperament but not his theatrical talent, being a better scene-painter than actor. By his marriage to the actress **Cornelia Thomás** (1796–1849), mother by a former marriage of the actor **Charles Burke** (1822–54), he became the father of the third Joseph (below).

Jefferson, Joseph (1829–1905), American actor, third of the name (see above). He made his first appearance on stage in Washington, at the age of 4, being tumbled out of a sack by T. D. *Rice whose song and dance he then mimicked. With his family he toured extensively, losing his father at 13 and living the hard life of pioneer players. By 1849 he had achieved some eminence, and in 1856 made his first visit to Europe. On his return he joined the company of Laura *Keene, making a great success as Dr Pangloss in the younger *Colman's *The Heir-at-Law* and as Asa Trenchard in Tom *Taylor's *Our American Cousin*. He was later at the *Winter Garden Theatre, where in 1859 he played Caleb Plummer in *Dot*, *Boucicault's version of *Dickens's *The Cricket on the Hearth*, and Salem Scudder in the same author's *The Octoroon; or, Life in Louisiana*. After the death of his first wife in 1861 he went on a four-year tour of Australia. On a subsequent visit to London in 1865 he played the part with which he is always associated—Rip Van Winkle. There had been several dramatizations of Washington *Irving's story before this, but it was in Boucicault's version that Jefferson made his greatest success, altering the text so much as he continued to appear in it that in the end it was virtually

his own creation, and lived only as long as he did. Not until 1880 did he appear in something else, reviving *Sheridan's *The Rivals* and making Bob Acres a little more witty and a little less boorish than his predecessors had done. With Mrs John *Drew as Mrs Malaprop and a good supporting company this production toured successfully for many years. Jefferson, whose charming, humorous personality made him typical of all that was best in the America of his time, made his last appearance in 1904 as Caleb Plummer and then retired, having been 71 years on the stage. In 1893 he had succeeded Edwin *Booth as President of the *Players' Club, and so become the recognized head of his profession. As his second wife he married the granddaughter of the first William *Warren, and so strengthened the ties between the two families.

His eldest son by his first marriage, **Charles Burke Jefferson** (1851–1908), was for many years his manager and also an actor, as was **Thomas** (1857–1931), the third of the name. His only daughter was the mother of the composer Harry Farjeon and the novelists and playwrights Eleanor, Joseph, and Herbert Farjeon (see REVUE). By his second marriage Jefferson had four sons, of whom two went on the stage.

Jefford, Barbara Mary (1930–), English actress, who made her début in Brighton in 1949, and in 1950 played Isabella in *Measure for Measure* to *Gielgud's Angelo at Stratford-upon-Avon. After further seasons at Stratford in 1951 and 1954, she appeared in the West End as Andromache in *Giraudoux's *Tiger at the Gates* (1955; NY, 1955). She was with the *Old Vic company, 1956–62, establishing herself as an outstanding classical actress in roles including Beatrice in *Much Ado about Nothing*, Portia in *The Merchant of Venice* (both 1956), Tamora in *Titus Andronicus* (1957), and Viola in *Twelfth Night* (1958). In 1960 she toured with the company as Lady Macbeth, Gwendolen Fairfax in *Wilde's *The Importance of Being Earnest*, and *Shaw's St Joan, a role in which she was particularly memorable, communicating both peasant earthiness and a radiant sense of the ineffable. She was seen as Lavinia Mannon in *O'Neill's *Mourning Becomes Electra* (1961), and the Stepdaughter in *Pirandello's *Six Characters in Search of an Author* (1963). In David *Mercer's *Ride a Cock Horse* (1965) she made one of her rare excursions into modern drama. In 1976 she joined the *National Theatre company, appearing as Gertrude in the initial production of *Hamlet* and in *Marlowe's *Tamburlaine*; in 1977–8 she was with the *Prospect company at the Old Vic. She took over Joan *Plowright's part in *De Filippo's *Filumena* (1979) in the West End, where in 1983 she was in a revival of *Rattigan's *The Winslow Boy*. In 1989 she returned to Stratford as Volumnia in *Coriolanus*.

Jenner, Caryl, see CHILDREN'S THEATRE.

Jerome, Jerome Klapka (1859–1927), English humorist, now chiefly remembered for his novel *Three Men in a Boat, not to Mention the Dog* (1899). He wrote also a number of plays, of which the only one still in the repertory is *The Passing of the Third-Floor Back* (1908), in which *Forbes-Robertson scored a signal triumph as the mysterious and Christlike stranger whose sojourn in a Bloomsbury lodging-house changes the lives of all its inhabitants.

Jerrold, Douglas William (1803–57), English actor, playwright, and journalist, who from its foundation in 1841 until his death was associated with the humorous journal *Punch*. As a playwright he had a good deal of contemporary success, though none of his plays has survived. Among them were the melodrama *Fifteen Years of a Drunkard's Life* (1828), produced at the Coburg (later the *Old Vic) Theatre, and the *nautical drama *Black-Ey'd Susan; or, All in the Downs* (1829), which provided an excellent vehicle for T. P. *Cooke. Some of his later plays, notably *The Rent Day* (1832), were first produced at *Drury Lane. In 1836 he took over the *Strand Theatre, where he put on a number of his own plays, including *A Gallantee Showman; or, Mr Peppercorn at Home* (1837). His last play, a comedy entitled *St Cupid; or, Dorothy's Fortune* (1853), was first seen at Windsor Castle and later the same year at the *Princess's Theatre. Jerrold's son **William Blanchard Jerrold** (1826–84) was the author of a farce, *Cool as a Cucumber* (1851), which provided the younger *Mathews with one of his best parts.

Jessner, Leopold (1878–1945), German theatre director, who abandoned the use of representational scenery in favour of a bare stage on different levels connected by stairways (*Jessnertreppe* or *Spieltreppe*). During his years as Director of the Staatliche Schauspiele in Berlin (1919–30), where he had Fritz *Kortner as his leading actor, he was considered one of the most advanced exponents of *Expressionism and his work greatly influenced the contemporary theatre. Among his outstanding productions were *Schiller's *Wilhelm Tell* in 1919, Shakespeare's *Richard III* and *Wedekind's *Der Marquis von Keith* in 1920, and *Hauptmann's *Die Weber* in 1928. He left Germany in 1933, and after a short stay in London went to the USA, where he died.

Jester, see COURT FOOL.

Jesuit Drama, term used to describe a wide variety of plays, mainly in Latin, written to be acted by pupils in Jesuit colleges. Originally, as in other Renaissance forms of *school drama, these were simply scholastic exercises, but over the years, particularly in *Vienna under the influence of opera and ballet, they became full-scale productions involving elaborate scenery, machinery, costumes, music, and dancing, as well as an almost professional technique in acting and diction.

The earliest mention of a play produced in a Jesuit college dates from 1551, when an unspecified tragedy was performed at the Collegio Mamertino in Messina. In 1555 the first Jesuit play was seen in Vienna, *Euripus sive de inanitate verum* by Levinus Brechtanus (Lewin Brecht), a Franciscan from Antwerp. This was followed by productions at Cordoba in 1556, at Ingoldstadt in 1558, and in *Munich in 1560. There were already 33 Jesuit colleges in Europe when Ignatius de Loyola, the founder of the Order, died in 1556: by 1587 there were 150 and by the early 17th century about 300. For over two centuries at least one play a year was performed in each college. The total of plays specially written for these performances was enormous. Only the best were published, but recent researches have brought to light a great many manuscripts. The early plays were based mainly on classical or biblical subjects—Theseus, Hercules, David, Saul, Absalom—but later, stories of saints and martyrs—Theodoric, Hermenegildus—were also used, as well as personifications of abstract characters—Fides (Faith), Pax (Peace), Ecclesia (the Church). The popularity of plays based on the stories of women—Judith, Esther, St Catharine, St Elizabeth of Hungary—led to the early abolition of the rule against the portrayal on stage by the boy pupils of female personages. The use of Latin was less easily disregarded, bound up as it was with its use in class and in daily conversation between masters and pupils. The vernacular seems first to have been used, in conjunction with Latin, in Spain, but the *Christus Judex* (1569) of **Stefano Tuccio** (1540–97) was translated into Italian in 1584 and into German in 1603. During the 17th century many plays appeared in French or in Italian, and by the beginning of the 18th century most Jesuit plays were written in the language of the country in which they were to be produced. Parallel with the increased use of the vernacular went the introduction of operatic arias, interludes, and ballets. Of all the splendid productions given in Vienna the most memorable appears to have been the *Pietas Victrix* (1659) of **Nikolaus Avancini** (1611–86) which had 46 speaking characters as well as crowds of senators, soldiers, sailors, citizens, naiads, Tritons, and angels. The technical development reached by Jesuit stagecraft can be studied in the illustrations to the published text of the play, which was acted on a large stage equipped with seven *transformation scenes. Lighting effects became increasingly elaborate; though the plays often began in daylight, which came through large windows on each side of the stage, they usually ended by torchlight, while in the course of the action sun, moon, and stars, comets, fireworks, and conflagrations were regularly required. All this, added to the splendour of the costumes and the large choruses and orchestras—often employing as many as 40 singers and 32 instrumentalists—made the Jesuit drama a serious rival to the public theatres. In Paris in the 17th century the three theatres in the Lycée Louis-le-Grand, where Louis XIV and his Court often watched the productions, were better equipped than the *Comédie-Française and almost on a par with the Paris Opéra. Jesuit drama continued to flourish in such conditions all over Catholic Europe until the Order was suppressed in 1773. It left its mark on the developing theatres wherever it was played, notably through the works of such authors as Avancini in Austria, *Bidermann in Germany, and Tuccio in Italy, and through its influence on pupils who were to become playwrights, among them *Calderón, *Corneille, *Goldoni, *Lesage, *Molière, and *Voltaire.

Jewish Art Theatre, New York, see GARDEN THEATRE and SCHWARTZ.

Jewish Drama. Unlike other dramas, that of the Jewish communities had originally no territorial limits. Its boundaries were linguistic, comprising Hebrew, the religious and historical language which never ceased to be written and has now been reborn as a living tongue in Israel; Yiddish, the vernacular of the vast Jewish communities which lay between the Baltic and the Black Seas, spread by emigrants all over the world; and Ladino (Judaeo-Spanish), the speech of the Sephardic Jews of the Middle East, who never achieved a permanent stage, their plays being for reading only.

Drama was not indigenous to the Jew. Deuteronomy 22: 5 expressly forbade the wearing of women's clothes by men, and the connection between early drama and the idol-worship of alien religions was a strong argument against the theatre as late as the 19th century. Yet the classical theatre exercised a strong attraction, and Jewish actors were found in Imperial Rome, while Ezekiel of Alexandria wrote a Greek tragedy on the Exodus. From the Jewish itinerant musicians and professional jesters, and from the questions and responses in the synagogue services, the Jewish theatre slowly evolved, the *Purim plays being an important influence. Yet the establishment of a truly *Hebrew theatre did not come until well into the 20th century, with the founding of the state of Israel, and the Yiddish theatre was destined to go ahead first, mainly in an attempt to check the growing vulgarity of the Purim plays.

In 1876 Abraham *Goldfaden founded the first permanent Yiddish theatre, and others followed. This activity was however brought to a sudden stop by the anti-semitic measures following the assassination of Tsar Alexander II in 1881. All plays in Yiddish were forbidden,

and the Yiddish theatre existed precariously in Russia until the Revolution of 1917. Most of the actors and dramatists left the country for Britain and America, and New York, where a Yiddish theatre had been founded in 1883, became the new centre of Yiddish drama. But the stock themes of old Jewish life in Europe, which had been acceptable to the first bewildered and largely illiterate immigrant audiences, together with the broad farce and sentimental melodrama which had provided the only alternatives, became increasingly outmoded as Americanization proceeded; even Goldfaden, when he first visited New York in 1887, found himself out of touch with the new audiences. It was left to Jacob *Gordin to revitalize the American Yiddish theatre. He was followed by other writers such as **Halper Leivick** [Leivick Halpern] (1888–1962), considered by many critics the best Yiddish writer of his time. A number of companies were founded under the influence of *Schwartz, among them Artef, the workmen's studio theatre, which adopted the methods of the *Moscow State Jewish Theatre and staged works by Soviet-Jewish authors. The widespread adoption of English in Yiddish-speaking homes, and the slackening in immigration, were potent factors in the continued decline of the Yiddish theatre, which was not arrested by Schwartz's efforts to enlarge the repertory by playing European classics in Yiddish and Yiddish plays in English.

Nevertheless the period between the two world wars was a flourishing time for the Yiddish theatre. The Argentine, home of a large Jewish community, had two permanent Yiddish theatres; London, which had its first Yiddish theatre in Whitechapel in 1888, had two in the 1930s, both in the East End and both playing Yiddish classics. Paris, where Goldfaden founded a company in 1890, also had a company which played the more popular operettas from the New York Yiddish stage. Vienna had several Yiddish theatres, and in the 1920s New York alone had 12 and there were others scattered throughout the country. The Nazi holocaust of Jews on the European continent, and the progress of assimilation in Western countries not so affected, led inevitably after 1946 to the decline of the Yiddish theatre, and though isolated pockets may have lingered on it finally disappeared from the international scene. Yiddish actors migrated to the national stage of the country in which they found themselves, or went into films; most theatres were closed, one of the last being the Polish State Jewish Theatre.

Jewish Theatre Studio, see MOSCOW STATE JEWISH THEATRE.

Jig, Elizabethan *after-piece, given in the smaller public theatres only. (Shakespeare says of Polonius: 'He's for a jig, or a tale of bawdry.')

It consisted of rhymed dialogue, usually on the frailty of women, sung and danced to existing tunes by three or four characters, of whom the *Clown was always one. The best known exponents of the jig were *Kempe and *Tarleton. It disappeared from the legitimate theatre at the Restoration, but remained in the repertory of strolling players and from the late 16th century onwards became popular in Germany, being taken there by the *English Comedians who toured the Continent. Very few texts survive, and those mostly in German translations. The jig is thought to have been a formative element in the development of the German *Singspiel* (see BALLAD OPERA).

Jim Crow, see RICE, THOMAS.

Jodelet [Julian Bedeau] (?1590–1660), French comedian. He joined the company of *Montdory shortly before the opening of the Théâtre du *Marais in 1634, and with several of his companions was transferred by order of Louis XIII to the Hôtel de *Bourgogne. He returned to the Marais in 1641, where he played Cliton in *Corneille's *Le Menteur* (1643). He also appeared in a series of farces written for him, mostly with his name in the title, by, among others, *Scarron (*Jodelet; ou, Le Maître-valet*, 1643) and Thomas *Corneille (*Jodelet prince*, 1655). He was extremely popular with Parisian audiences—he had only to show his flour-whitened face to raise a laugh—and he frequently added jokes of his own to the author's lines. When *Molière established himself in Paris in 1658 he induced Jodelet to join his company, thus assuring the co-operation of the one comedian he had reason to fear, and wrote for him the part of the Vicomte de Jodelet in *Les Précieuses ridicules* (1659). He probably intended to give him the title-role in *Sganarelle; ou, Le Cocu imaginaire* (1660), but Jodelet died just before the production opened and Molière played the part himself.

Jodelle, Étienne (1532–73), French Renaissance poet and dramatist, a member (with Ronsard and others) of the famous 'Pléiade'. His *Cléopâtre captive* (1552) was the first French tragedy to be modelled on *Seneca. Together with a comedy, *Eugène*, also an attempt to acclimatize a classical dramatic form, it was given before Henri II and his Court with Jodelle, not yet 21, as Cleopatra. The Pléiade, delighted at the success of one of their members, organized a festival in Jodelle's honour, presenting him with a goat garlanded with ivy; the Church took umbrage at this pagan revival and Jodelle bore the brunt of its displeasure. Of his later plays only *Didon* (1558) survives, but he deserves credit as the precursor of a great dramatic tradition.

John F. Kennedy Center for the Performing Arts, Washington, DC, national cultural centre opened in 1971. It contains three theatres, the largest being the Eisenhower, seating 1,142, which opened with a revival of *Ibsen's *A Doll's House* starring Claire *Bloom. It houses touring productions and also originates its own, and many famous actors have appeared there. Its premières have included those of the musical *Annie* (1977) and Arthur *Kopit's *Wings* (1978), and the American première of Tom *Stoppard's *Travesties*. Among notable revivals were Lillian *Hellman's *The Little Foxes* in 1981, with Elizabeth Taylor, and *Kaufman and *Hart's *You Can't Take it with You* in 1983, both of which went on to Broadway. The Terrace Theater, seating 512, and the Theater Lab, seating 120, stage small-scale and experimental works, the latter offering many free presentations of the Center's educational programme. There are also a concert hall, an opera-house (which also takes musicals, including the American première of *Les Misérables*), and a Performing Arts Library.

John Golden Theatre, New York. 1. On West 58th Street, between Broadway and 7th Avenue. Named after the actor, song-writer, and theatre director **John Golden** (1874–1955), for whom it was built, it opened in 1926 and was almost immediately taken over by the *Theatre Guild, who staged there two plays by Sidney *Howard—*Ned McCobb's Daughter* and *The Silver Cord*. In 1928 the Theatre Guild returned with a successful run of *O'Neill's *Strange Interlude*, but after a short period as the Fifty-Eighth Street Theatre in 1935–6 it became a cinema, making a brief return to live entertainment as the Concert Theatre in 1942, with intimate *revue. It is now used for television shows.

2. On West 45th Street, between Broadway and 8th Avenue. A small theatre, seating 799, it opened as the Masque in 1927 with a translation of Rosso di San Secondo's *Marionette, che passione!* as *Puppets of Passion*. Kirkland's long-running *Tobacco Road* opened in 1933, and four years later John Golden took over, naming the theatre after himself. His first successful productions were Paul Vincent *Carroll's *Shadow and Substance* (1938) and Patrick Hamilton's *Angel Street* (1941). Later the theatre housed a series of revues which included *At the Drop of a Hat* (1959), *Beyond the Fringe* (1962), and the White African revue *Wait a Minim* (1966), all imported from London. More recent productions included a double bill by Robert *Anderson, *Solitaire/Double Solitaire* (1971), D. L. Coburn's *The Gin Game* with Jessica *Tandy and Hume *Cronyn (1977), David *Mamet's *Glengarry Glen Ross* (1984), and Athol *Fugard's *Blood Knot* (1985).

Johnson, Dame Celia (1908–82), English actress. She made her début in *Shaw's *Major Barbara* in Huddersfield in 1928 and was first seen in London in 1929. She had a long run in

Merton Hodge's *The Wind and the Rain* (1933) and was much admired for her playing of Elizabeth Bennet in an adaptation of Jane Austen's *Pride and Prejudice* (1936) and in Daphne Du Maurier's *Rebecca* (1940). In 1947 she played the title-role in Shaw's *Saint Joan* for the *Old Vic company, and in 1951 gave a fine performance as Olga in *Chekhov's *Three Sisters*. She made a successful appearance in modern comedy in William Douglas *Home's *The Reluctant Débutante* (1955) and gave a sensitive portrayal of Isobel Cherry in Robert *Bolt's *Flowering Cherry* (1957). Over the next few years she was in a succession of new plays including Hugh and Margaret Williams's *The Grass is Greener* (1958) and *Billetdoux's *Chin-Chin* (1960). She then broadened her range by joining the *National Theatre company to play Mrs Solness in *Ibsen's *The Master Builder* (1964) opposite *Olivier, and took over the part of Judith Bliss in its revival of *Coward's *Hay Fever* (1965), which, like Alan *Ayckbourn's *Relatively Speaking* (1967), gave her another opportunity to display her virtuosity in light comedy. In contrast, she was highly effective at *Chichester in 1966 as Ranevskaya in Chekhov's *The Cherry Orchard*, and in Nottingham in 1970 (London, 1971) as Gertrude to Alan *Bates's *Hamlet*. She was later in three more plays by William Douglas Home, *Lloyd George Knew My Father* (1972), *The Dame of Sark* (1974), and *The Kingfisher* (1977).

Johnson, Dr Samuel (1709–84), the great English lexicographer, was the author of a five-act tragedy, *Irene*, which his friend and fellow townsman David *Garrick produced at *Drury Lane in 1749, with little success. After its failure Johnson never again essayed the stage, though he made more money from the proceeds of the third, sixth, and ninth nights of his play than by anything he had previously done. His edition of the plays of Shakespeare is valuable for the light it throws on the editor rather than on the author, since Johnson had little knowledge of Elizabethan drama or stage conditions, and was not temperamentally a research worker. He should not be confused with Samuel Johnson of Cheshire, author of *Hurlothrumbo* (1729) and other burlesques.

Johnston, (William) **Denis** (1901–84), Irish dramatist. His first play, *The Old Lady Says 'No!'* (1929), is a satirical review of certain dominant elements in Irish life. Refused by Lady *Gregory for the Abbey (hence its title), it was produced by the Gate Theatre during its second season. *The Moon in the Yellow River* (1931) earned wide popularity for the richness of its characterization and the precision with which the author diagnoses the mood of mid-1920s Ireland. *A Bride for the Unicorn* (1933) shows the influence of *Expressionism. *Blind Man's Buff*, Johnston's adaptation of *Toller's *Die blinde*

Göttin, had a successful run at the Abbey Theatre in 1936. Later works were *The Golden Cuckoo* (1939), a comedy of individual rebellion; *The Dreaming Dust* (1940), on the life of Dean Swift; *Strange Occurrence on Ireland's Eye* (1956), a successful court-room drama; and *The Scythe and the Sunset* (1958).

John Street Theatre, New York, on John Street, west of Broadway. The first permanent playhouse in New York and the third to be built by David *Douglass, this opened in 1767 with *Farquhar's *The Beaux' Stratagem*; it was described by *Dunlap, whose first play *The Father; or, American Shandyism* was produced there in 1789, as 'principally of wood, an unsightly object, painted red'. There is also a reference to it in Royall *Tyler's *The Contrast*, a landmark in American theatre history produced at John Street two years earlier, where Jonathan, the country bumpkin, describes his first visit to a theatre. Until the War of Independence the *American Company gave regular seasons there, mounting the first productions in New York of such plays as *The Merchant of Venice*, *Macbeth*, *King John*, *Jonson's *Every Man in His Humour*, and *Dryden's *All for Love*, as well as a large repertory of contemporary plays and *after-pieces. During the war the playhouse was rechristened Theatre Royal and used for productions by the officers of the English garrison, among them Major John André—later the subject of a play by Dunlap—whose scene-painting was much admired. Just before the British evacuated New York a professional company came from Baltimore and stayed for a time at the John Street Theatre, but without much success.

Two years after the British evacuation, in 1785, the American Company, now under the control of the younger *Hallam and John *Henry, took possession of the theatre again, the company being reinforced shortly afterwards by Thomas *Wignell. During the next few years regular seasons were given, with productions for the first time in New York of *The School for Scandal* and *The Critic* by *Sheridan, *Much Ado about Nothing*, *As You Like It*, and *Garrick's version of *Hamlet*. George Washington visited John Street three times in 1789, the year of his inauguration.

Soon after this Wignell and Mrs Morris left John Street to found their own *Chestnut Street Theatre in Philadelphia and John *Hodgkinson, newly arrived from England, joined the John Street company, soon becoming so popular, and so grasping, that he ousted both Hallam and John Henry from management and from the affections of the public. Henry and his wife withdrew from the company in 1794, Hallam in 1797, leaving Hodgkinson in command with Dunlap. He had been added to the management the previous year, which had also seen the first appearance in New York, with the John Street Company, of the first Joseph *Jefferson. The John Street Theatre was used for the last time in 1798 and was later sold by Hallam for £115, the company moving into their new *Park Theatre.

Joint Stock, see FRINGE THEATRE.

Jolly, George (*fl.* 1640–73), English actor, the last of the notable *English Comedians in the German theatre. He may have been at the *Fortune Theatre in London in 1640, and is first found in Germany in 1648. He was particularly active in *Frankfurt, where Prince Charles (later Charles II) probably saw him act. He appears to have anticipated *Davenant's use of music and scenery on the public stage, and already had women in his company in 1654. He returned to England at the Restoration, and got permission to open the *Cockpit, where the French theatre historian Chappuzeau saw him in 1665; but *Killigrew and Davenant managed to deprive him of his patent and he had to content himself with overseeing the *Nursery, a training school for young actors.

Jolson, Al, see MUSICAL COMEDY and REVUE.

Jolson Theatre, New York, see CENTURY THEATRE 2.

Jones, Henry Arthur (1851–1929), English dramatist, whose first London production, after several of his one-act plays had been performed in the provinces, was *A Clerical Error* (1879) staged by Wilson *Barrett, who later made an immense success in *The Silver King* (1882), a melodrama written by Jones in collaboration with Henry Herman which was many times revived. Although regarded by his contemporaries as one of the new school of dramatists whose plays belonged to the so-called 'theatre of ideas', it was the melodramatic element rather than the social criticism which drew the public to such plays as *Saints and Sinners* (1884), *The Dancing Girl* (1891), in which Julia *Neilson made a spectacular success, *The Case of Rebellious Susan* (1894), and particularly *Michael and His Lost Angel* (1896), which had to be withdrawn after 10 performances, mainly on account of the scene in a church in which a priest, standing before the altar, makes a public confession of adultery after having some years before exacted a similar penance from a young woman in his congregation. *The Liars* (1897) and *Mrs Dane's Defence* (1900) also contain a strong melodramatic strain, though the third act of the latter is still considered a masterpiece. With *Pinero and *Shaw, Jones was an important playwright at the time when *Ibsen's influence was beginning to make itself felt in the English theatre, but there is no reason to doubt his own assertion that he was in no way consciously indebted to Ibsen. Shaw praised Jones at the expense of Pinero. Posterity has

reversed this judgement, and Pinero's plays still hold the stage while those of Jones are forgotten. His skill in naturalistic dialogue and in the creation of dramatic tension remains impressive, but his social and moral criticism lacked a firm philosophical basis and the redeeming gift of humour.

Jones, Inigo (1573–1652), English architect and artist, the first to be associated with scenic decoration on the English stage. Having studied in Italy and worked in Denmark, he was in 1604–5 attached to the household of Prince Henry, eldest son of James I. In addition to his work as an architect, he was given complete control of the staging of *masques presented at Court, the first being Ben *Jonson's *The Mask of Blackness* (1605). In the same year he supervised the staging of the plays seen at Christ Church Hall, Oxford, where he first used revolving screens in the Italian manner. Later he was to use as many as five changes of scenery, with *backcloths, shutters, or *flats painted and arranged in perspective. The flats ran in *grooves and were supplemented by a turntable (*machina versatilis*) which presented to the audience different facets of a solid structure. Jones also introduced into England the picture-stage framed in a *proscenium arch. It is now thought that a collection of plans in the British Library by Jones's pupil and successor John *Webb may include those on which Christopher *Beeston rebuilt the *Cockpit in 1616, but the main bulk of Jones's work is preserved in the library of the Duke of Devonshire at Chatsworth.

Jones, James Earl (1931–), American actor, the finest Negro player of his generation, with an imposing appearance and a deep voice. He was closely associated with Joseph *Papp's New York Shakespeare Festival, for which he played Oberon in *A Midsummer Night's Dream* (1961), Caliban in *The Tempest* (1962), Othello (1964), Macbeth (1966), Claudius in *Hamlet* (1972), and King Lear (1973). He was also associated with the plays of *Fugard—*The Blood Knot* (1964), *Boesman and Lena* (as Boesman, 1970), *A Lesson from Aloes* (1980), *Master Harold . . . and the Boys* (1982)—and appeared in works as diverse as *Genet's *The Blacks* (1961), James *Saunders's *Next Time I'll Sing to You* (1963), and *Büchner's *Danton's Death* (1965). His portrayal of the boxer in Howard Sackler's *The Great White Hope* (1968) was widely acclaimed, and other roles included Lopahin in an all-Black production of *Chekhov's *The Cherry Orchard*, Hickey in a revival of *O'Neill's *The Iceman Cometh* (both 1973), and Lennie in a revival of *Steinbeck's *Of Mice and Men* (1974). He was seen in the two-character play *Paul Robeson* by Phillip Hayes Dean in New York and London in 1978. In 1981, at the *American Shakespeare Theatre, he again played Othello, repeating the role in New York in 1982. In 1987 he starred in August *Wilson's *Fences*.

Jones, Leroi, see NEGROES IN THE AMERICAN THEATRE.

Jones, Margo (1913–55), American director and producer. After studying at the Southwestern School of the Theatre in Dallas she worked with the Ojai Community Players and at the Pasadena Playhouse. In 1939 she was associated with community and university drama in Houston, Texas, and in 1943 she staged an early play by Tennessee *Williams, *You Touched Me*, at the *Cleveland Play House. In 1945 she founded an experimental theatre in Dallas, where she encouraged the work of new playwrights and gave experimental productions of older plays. Her work was first seen on Broadway in the same year, when she co-directed Tennessee Williams's *The Glass Menagerie*. Later productions included Williams's *Summer and Smoke*, Maxwell *Anderson's *Joan of Lorraine*, and Owen Crump's *Southern Exposure*. After her death the theatre she founded continued to function under different directors until the end of 1959, the last production being *Othello*.

Jones, Robert Edmond (1887–1954), American writer, lecturer, director, and above all scene designer, whose first designs, for Ashley *Dukes's *The Man Who Married a Dumb Wife* in 1915, began a revolution in American scene design. His ability to integrate his designs with all the aspects of the play made his work memorable, particularly when he also directed the actors, as he did in *The Great God Brown* (1926) by *O'Neill, with all of whose early plays he was connected through his association with the *Provincetown Players. He was responsible for the décor of a number of Shakespearian productions, his *Othello* in 1937 being much admired, and of such modern works as Marc *Connelly's *The Green Pastures* (1930), Maxwell *Anderson's *Night over Taos* (1932), and Sidney *Howard's adaptation of a Chinese play, *Lute Song* (1946).

Jongleurs, medieval itinerant entertainers, who flourished throughout Europe, often along the traditional pilgrim routes. The term embraced most kinds of performers—acrobats, jugglers, bear-leaders, ballad-singers—much of their patter and songs being provided by the *trouvères*, who usually performed only their own works. Although some *jongleurs* worked in family or friendly groups, most were independent; both men and women followed the profession.

Jonson, Ben(jamin) (1572–1637), English poet and playwright, contemporary and friend of Shakespeare, and like him despised by the *University Wits for not having been educated at Oxford or Cambridge. In 1598 his first play *Every Man in His Humour* was produced,

Shakespeare playing Kno'well. It was followed by *Every Man out of His Humour* (1599) and *Cynthia's Revels* (1600), performed by the Children of the Chapel at Blackfriars. They also appeared in *The Poetaster* (1601), in which Jonson vented his spleen on several of his contemporaries. In 1605 he collaborated with *Chapman and *Marston in *Eastward Ho!*, whose satirical references to James I's Scottish policy caused the authors to be sent to prison, where Jonson had already spent some time for his part in a lost play *The Isle of Dogs* (1597). He also found himself in trouble with the authorities over his tragedy *Sejanus* (1603), which was considered 'seditious' though it dealt with Roman rather than English history. A more peaceful period resulted in some of his best plays, among them four comedies of deception and gullibility: *Volpone, or, The Fox* (1606), *Epicoene; or, The Silent Woman* (1609), *The Alchemist* (1610), whose Abel Drugger provided many actors, including *Garrick and *Guinness, with a splendidly comic part, and *Bartholomew Fair* (1614), whose slight plot links a number of scenes portraying a typical London crowd on holiday. All these have been many times revived. Jonson's later plays, after an absence from the stage of 10 years, were less successful and have not been seen again. Jonson, an excellent poet, wrote between 1605 and 1612 the texts of eight *masques given at Court with scenery and effects by Inigo *Jones.

Jordan, Dorothy [Dorothea] (1761–1816), English actress, who excelled as high-spirited hoydens and in breeches parts (see MALE IMPERSONATION). She was the illegitimate daughter of an actress by a gentleman named Francis Bland; her brother, also an actor, called himself **George Bland** (?–1807), but she made her first appearance in Dublin in 1777 as Miss Francis. In 1780 she was engaged by Daly for the Smock Alley Theatre, but two years later, after being seduced by him, she fled secretly to England with her mother and sister. There she was befriended by Tate *Wilkinson. On his advice, since she was clearly pregnant, she changed her name to Mrs Jordan, though she was never married. In 1785, probably on the recommendation of William ('Gentleman') *Smith, who saw and admired her at York during race week, she was engaged by *Sheridan for *Drury Lane, and in spite of the preferences of audiences of the time for tragedy in the style of Mrs *Siddons (who was also in the company, and for a time thought poorly of her acting), she chose to make her début as Peggy in *Garrick's *The Country Girl*. She was immediately successful and, wisely abandoning tragedy altogether, continued to delight audiences in such parts as Priscilla Tomboy in *The Romp*, in which character she was painted by Romney, as well as Miss Hoyden in Sheridan's *A Trip to Scarborough*, Sir Harry

Wildair in *Farquhar's *The Constant Couple*, and Miss Prue in *Congreve's *Love for Love*. Early in her career at Drury Lane Mrs Jordan became entangled with a young man by whom she had four children, leaving him in 1791 to become the mistress of the Duke of Clarence, later William IV, by whom she had 10 children. She continued to act intermittently; she was in the company which performed *Ireland's Shakespeare forgery *Vortigern and Rowena* (1796) at Drury Lane, and in 1800 she appeared as Lady Teazle in Sheridan's *The School for Scandal*. She parted from the Duke, to whom she had been a faithful and affectionate companion, in 1811 when he had perforce to make a dynastic marriage. Her last appearances in London were in 1814, when she was seen in a revival of *As You Like It* and in a new play by James Kenney, *Debtor and Creditor*. After a final appearance at Margate she retired to Paris, where she died in poverty. Her grave was swept away during rebuilding in the early 1930s.

Jornada, name given in Spain to each division of a play, corresponding to the English 'act'. It probably comes from the Italian *giornata*, and in its present form was first used by *Torres Naharro.

Josefstädter Theater, see VIENNA.

Joseph, Stephen, see NEW VICTORIA THEATRE and STEPHEN JOSEPH THEATRE IN THE ROUND.

Joseph E. Levine Theatre, see CIRCLE-IN-THE-SQUARE.

Jouvet, Louis (1887–1951), French actor and director, one of the most important figures of the French theatre in the years before the Second World War, who in 1913 joined *Copeau at the *Vieux-Colombier as actor and stage manager. His best work there was done in Shakespeare, in such roles as Aguecheek in *Twelfth Night* and Autolycus in *The Winter's Tale*. Jouvet went with Copeau to America in 1917–19, but in 1922 left him to take over the

*Comédie des Champs-Élysées. Here he was responsible for the success in 1923 of *Romains's *Knock; ou Le Triomphe de la médecine*, which he directed, playing also the chief part himself. In 1928 he began his long association with *Giraudoux, in *Siegfried*. In 1934 he moved to the larger Théâtre de l'Athénée, to which he added his own name. He joined the staff of the Conservatoire in 1935, and a year later was appointed one of the advisers to the *Comédie-Française, becoming a director in 1940, only to be forcibly retired under the German occupation. In 1945 he returned to the Athénée, where he staged *Genet's first play *Les Bonnes* in 1947. The great passion of his life was for *Molière, his finest part being Géronte in *Les Fourberies de Scapin*.

After his death several managers tried to continue his work, but without success. Since 1982 the Athénée has housed visiting companies in its two auditoriums.

Juvarra, Filippo (1676–1736), Italian architect, who designed scenery for several private theatres in Rome, his most important work being done for Cardinal Ottoboni, a great lover of plays and opera for whom he designed and built a small theatre (probably for rod *puppets). Two plans for this are extant; it is not known which was finally used, as the theatre was demolished when Ottoboni died and no trace of it remains. There are still in existence, however, a number of scene-designs which Juvarra made for productions there between 1708 and 1714, including those in an album now at the Victoria and Albert Museum in London which probably dates from 1711.

Juvenile Drama, see TOY THEATRE.

Juvenile Lead, Tragedian, see STOCK COMPANY.

Juvenon, François, see GROS-GUILLAUME.

K

Kabuki, the popular theatre of Japan, as opposed to the aristocratic *nō play. As its name implies—*ka* = singing, *bu* = dancing, *ki* = acting—it combines the three main theatrical arts, allied to an astonishing virtuosity, particularly in the playing of female parts, by highly trained actors. In its present form *kabuki* dates from about the middle of the 17th century, though some of its elements go back a thousand years. Since 1945 the many small *kabuki* troupes which used to tour the countryside have been disbanded and performances are given only in the larger cities. The plays are performed on a wide, shallow platform which since *c.*1760 has incorporated a revolving stage, later adopted by the Western theatre. Another characteristic of the *kabuki* theatre, taken from the *nō* play, is the *hana-michi*, or 'flower way', running along the left-hand wall of the auditorium to the stage at the level of the spectators' heads. Along this the actors make their entrances and exits, or withdraw for an aside. They wear rich brocaded costumes for historical parts, plain dress for scenes from daily life and, unlike the *nō* actors, are not masked. Music is provided by a small group of instrumentalists placed inconspicuously behind a lattice on the right of the stage. There are also two stage hands, the *kurogo* and the *kōken*, one hooded, the other not, who are by tradition invisible. They date from the time when each of the chief actors had a 'shadow' who crouched beside him holding a light on the end of a bamboo pole to illuminate the play of his features. The *kabuki* plays, which have no particular literary value, being frameworks for the display of technical accomplishments, take their subjects from many sources—history, myth, daily life, even from the *nō* plays, or from the puppet-theatre or *bunraku*. For a performance based on a puppet-play a singer, or *jōruri*, seated on the stage, recites the story which the actors are miming; for a *nō* play, the *samisen* players are joined by a group of *nō* musicians. In the old days performances by *kabuki* actors could extend over several days, and they still last from midday to midnight, the audience, in family groups, coming and going from the boxes in the auditorium, eating and talking during the less interesting parts of the performance and giving all their attention to the great set speeches and dance-dramas which, together with comic episodes and historical set-pieces, make up the programme.

Kachalov [Shverubovich], **Vasili Ivanovich** (1875–1948), Russian actor, whose career covered the transition from Tsarist to Soviet Russia. After three years' apprenticeship in the provinces he joined the company at the *Moscow Art Theatre, and played many leading roles, including the name parts in *Julius Caesar* and *Hamlet* and *Ibsen's *Brand*. He was also seen as Ivan Karamazov in a dramatization of Dostoevsky's *The Brothers Karamazov* and as the Reader in Leo *Tolstoy's *Resurrection*, considered one of his best parts. He created the role of Vershinin, the hero in *Ivanov's *Armoured Train 14-69* (1927), and in 1938 appeared as Chatsky in *Griboyedev's *Woe from Wit*, the part he had played when it was first seen at the Moscow Art Theatre in 1906.

Kahn, Florence, see BEERBOHM.

Kainz, Josef (1858–1910), German actor, famed for the richness of his voice and the purity of his diction. He trained with the *Meininger company and made his first appearance on the stage in 1874, in *Vienna, where in 1899 he returned to end his days as a leading member of the Imperial Theatre. He was for some time in *Munich, where he was the friend and favourite actor of King Ludwig II of Bavaria, and in 1883 played opposite Agnes *Sorma in the newly founded *Deutsches Theater in Berlin. He toured extensively in America in many of his best parts, which included Romeo, Hamlet, and the heroes of *Grillparzer. He was also good as *Molière's Tartuffe, Oswald in *Ibsen's *Ghosts*, and *Rostand's Cyrano de Bergerac.

Kaiser, Georg (1878–1945), German dramatist, a major exponent of *Expressionism. His early plays were satirical comedies directed against Romanticism, but the first of his works to attract attention was *Die Bürger von Calais* (written in 1913 but not performed until 1917). This is generally considered his best play, though not so well known as *Von morgens bis mitternachts*, seen later the same year. The latter, a sombre history of a bank clerk whose bid for freedom from the futility of modern civilization leads to suicide, was translated by Ashley *Dukes as *From Morn to Midnight*, and was produced by the *Stage Society in 1920. It was given its first public production in London in 1926, having been seen in New York in 1922. It was followed in Germany by the powerful trilogy *Die Koralle* (1917) and *Gas, I and II* (1918 and 1920), a symbolic picture of modern industrialism crashing to destruction and taking with it the civilization it has ruined. Except for the melodramatic *Der Brand im Opernhaus* (1919), Kaiser's later works—he wrote about 70 plays—made less impact, but he collaborated with the musician Kurt Weill in several operas; in 1938 he left Germany for Switzerland, where he died.

Kalita Humphreys Theater, see DALLAS THEATER CENTER.

Kameri Theatre, Israel, see CAMERI THEATRE.

Kamerny Theatre, Moscow, chamber, or intimate, theatre, founded in 1914 by Alexander *Taïrov as an experimental theatre for those to whom the naturalistic methods of the *Moscow Art Theatre no longer appealed. Here he sought to work out his theory of 'synthetic theatre', which, unlike the 'conditioned theatre' of *Meyerhold, made the actor the centre of attention, combining in his person acrobat, singer, dancer, pantomimist, comedian, and tragedian. Taïrov's first successful production was *Vishnevsky's *An Optimistic Tragedy* (1934), in which his wife, Alisa Koonen, played the heroine. The theatre then became important for its productions of non-Russian plays, providing a link with Western drama at a time when it was badly needed. After Taïrov's death in 1950 the theatre was reorganized and lost its identity, many of the company joining the newly opened *Pushkin Theatre.

Karatygin, Vasily Andreyevich (1802–53), Russian tragic actor, who made his professional début in 1820. He was noted for the care with which he studied his roles, returning where possible to the original sources, and labouring for historical accuracy in costume and décor, though he was opposed to the realistic style of acting and the innovations of *Shchepkin and was criticized for a certain frigidity and mechanical quality. In contrast to *Mochalov he developed a subtle and calculated technique which enabled him to play the most varied roles, though his preference was always for classical tragedy, and he was especially admired in the patriotic drama of the day. He created the part of Chatsky in *Griboyedov's *Woe from Wit* (1831).

Kasperle, originally a German puppet, somewhat akin to the English *Punch. Imported into the live theatre, he took on many of the characteristics of *Hanswurst, and in the hands of **Johann Laroche** (1745–1806) developed into an important element in Viennese popular comedy.

Katayev, Valentin Petrovich (1897–1986), Soviet dramatist. His most successful play was *Squaring the Circle* (1928), a comedy about two ill-assorted couples compelled to live in one room because of the housing shortage who finally change partners. First produced at the *Moscow Art Theatre, it was seen in New York in 1935 (London, 1938). Katayev wrote a number of other comedies, including *The Primrose Path* (1934), which the *Federal Theatre Project produced in New York in 1939, and an amusing trifle called *The Blue Scarf* (1943). Among his more serious plays are *Lone White Sail* (1937), adapted from his novel, which, in a revised version, had a successful run in 1951; *I, Son of the Working People* (1938); a play for children, *Son of the Regiment* (1946); and *All Power to the Soviets* (1954), the last two being based on his own novels. *Violet* (1974) deals with the moral problems of those who lived through Stalin's rule.

Kaufman, George S. (1889–1961), American dramatist and director, whose first plays were written in collaboration with Marc *Connelly. After venturing on his own with *The Butter and Egg Man* and *The Cocoanuts* (both 1925), Kaufman collaborated with **Edna Ferber** (1887–1968) in *The Royal Family* (1927) based on the lives of the *Drews and the *Barrymores; this was seen in England in 1934 as *Theatre Royal*. They later wrote other plays together, such as *Dinner at Eight* (1932; London, 1933) and *Stage Door* (1936). Collaboration with Moss *Hart produced a number of light-hearted comedies, among them *Once in a Lifetime* (1930; London, 1933), a satire on Hollywood; *Merrily We Roll Along* (1934); *You Can't Take it with You* (1936; London, 1937), the most successful, which won the *Pulitzer Prize; and *The Man Who Came to Dinner* (1939; London, 1941), in which the leading character is reputedly based on Alexander *Woollcott. Kaufman also wrote (usually in collaboration) many works for the musical stage, including the book for the *Gershwins' *Of Thee I Sing* (1931), Rodgers and *Hart's *I'd Rather be Right*, with Moss Hart (1937), and Cole *Porter's *Silk Stockings* (1955). Known as 'the Great Collaborator', Kaufman was an expert technician and an excellent director, with a keen sense of satire and a thorough knowledge of the theatre. He directed not only many of his own works but also such varied productions as *Steinbeck's *Of Mice and Men* (1937), Loesser's musical *Guys and Dolls* (1950), and *Ustinov's *Romanoff and Juliet* (1957).

Kazan [Kazanjoglou], **Elia** (1909–), American director and actor, born in Turkey, who made his first appearances on stage with the *Group Theatre, being seen in Clifford *Odets's *Waiting for Lefty* (1935) and *Golden Boy* (1937). He came into prominence as a director with his staging of Thornton *Wilder's *The Skin of Our Teeth* (1942). Renowned for his ability to establish rapport with actors, he helped in 1947 to found the Actors' Studio, a workshop where professional actors could experiment and study their art according to the *Method, of which he was an outstanding exponent. He was particularly associated with the two leading American playwrights of the time: Arthur *Miller (*All My Sons*, 1947, *Death of a Salesman*, 1949, *After the Fall*, 1964); and Tennessee *Williams (*A Streetcar Named Desire*, 1947, *Camino Real*, 1953, *Cat on a Hot Tin Roof*, 1955, and *Sweet Bird of Youth*, 1959). He also directed Archibald *MacLeish's *J.B.* (1958). Leaving the Actors' Studio in 1962, he was appointed co-director of the Lincoln Center repertory company (see VIVIAN BEAUMONT THEATRE). He resigned in 1965 and has not since worked in the theatre. He was also an outstanding film director.

Kean, Charles John (1811–68), English actor-manager, son of Edmund *Kean, who sent him to Eton with the idea of detaching him from the stage. But when, at the time of his father's break with *Drury Lane, he was offered an engagement there, he accepted it and made his first appearance in 1827 as Young Norval in *Home's *Douglas*. Realizing that he lacked experience he then went into the provinces, and in 1828 was seen on stage with his father in Glasgow. They were not to play together again until Edmund Kean's last appearance. Charles had none of his father's genius, but was a serious, hardworking man, somewhat priggish, but with plenty of application and common sense. With his wife **Ellen Tree** (1806–80), a good actress who played opposite him in many important productions, he ran an excellent company, which from 1850 to 1859 appeared at the *Princess's Theatre in a series of carefully chosen and well-rehearsed plays, set and costumed lavishly, as the fashion of the time demanded, but with some attempt at historical accuracy. Queen Victoria was a frequent visitor to the Princess's, as was the Duke of Saxe-Meiningen, husband of one of the Queen's nieces, and Kean's work probably inspired many of the reforms attributed to the *Meininger company. It was under Kean that Ellen *Terry, as a child of 9, made her first appearance on stage.

Kean, Edmund (1789–1833), English tragedian, whose undoubted genius was offset by wild and undisciplined behaviour, an ungovernable temper, and habitual drunkenness. Very little is known about his early life. He was apparently the illegitimate son of a small-part actor and drunkard, related to the Savile family, and Ann Carey, granddaughter of Henry Carey, composer of the ballad 'Sally in Our Alley', also related to the Saviles. Left fatherless at 3, with a mother who had apparently abandoned him, Edmund was cared for by an elderly member of the *Drury Lane company. She planned to make an actor of him (knowing no other life) and had him taught singing, dancing, fencing, and elocution. By 8 years old he was already something of a prodigy and had appeared in several small parts at Drury Lane.

Anxious to exploit his talents for her own benefit, his mother then reclaimed him, and with her he became a strolling player, acting Hamlet before Nelson and Lady Hamilton in Carmarthen, and giving a command performance in Windsor Castle before George III. In 1804, tired of earning money for his mother to spend, he set out on his own, going from one provincial company to another. During the next nine years he endured not only all the privations of a strolling player's life but added to his burdens a wife, a mediocre actress, and two sons, of whom the elder died young.

On 26 Jan. 1814 Kean finally achieved his ambition and appeared at Drury Lane in a major role, playing Shylock in *The Merchant of Venice* not in the traditional red wig and beard of the low comedian, which even *Macklin had not dared to discard, but as a swarthy embittered fiend with a butcher's knife in his grasp and blood-lust in his eyes. The audience acclaimed him, and he continued for a time to delight them in a series of villainous parts. The technical novelty of his acting is revealed by a contemporary comment that 'by-play' was one of his greatest excellencies; he relied less on his naturally harsh voice than on facial expression. In spirit the change was even greater. Deficient in dignity, grace, and tenderness, he needed a touch of the malign, of murderous frenzy, to inspire him. Comedy he seldom essayed; he was seen as Abel Drugger—*Garrick's favourite comic part, in *Jonson's *The Alchemist*—only three times. Mild villainy made no appeal to him, and he rejected the part of Joseph Surface, in *Sheridan's *The School for Scandal*, with scorn. Even his magnificent Othello was too often overplayed, too constantly on the rack. But as Macbeth he was heartrending, and he gave his finest performances as the arch-villains Richard III and Iago. Two of his other great masterpieces were Sir Giles Overreach, with his ruthless frenzy of miserliness, in *Massinger's *A New Way to Pay Old Debts*, and the barbarous fiend Barabas, in *Marlowe's *The Jew of Malta*. After his enthusiastic reception at Drury Lane, Kean seemed about to embark on a career of unequalled splendour; but the wildness in his blood and the intemperate habits of the lost years overpowered him. He was seldom sober; his frequent absences from the stage and his poor showing when drunk, added to the many scandals which attached themselves to his name—particularly his affair with the wife of one of the members of the Drury Lane general committee—alienated the audiences who had been attracted by his superb acting; he could win them back time and again with a fine performance, only to lose them through his bad behaviour. In the United States, where he made his first appearance in 1820, the same thing happened. Audiences who had awaited him on tiptoe with expectation, and applauded his first appearances with fervour, turned against him when he fell back into his usual state of arrogance and unreliability. He returned to England, but his powers were waning, his wife had left him, and his health was deteriorating. He continued to act intermittently, and in 1831 took on the management of the King's Theatre at Richmond, near London. He even went on tour occasionally, and on his better days returned to the stage of Drury Lane, where on 25 Mar. 1833, while playing Othello to the Iago of his son Charles *Kean, he finally collapsed, dying a few weeks later.

Keane, Doris, see SHELDON.

Keeley [*née* Goward], **Mary Ann** (1806–99), English actress, who was trained as a singer, but turned to acting and had already made a name for herself when in 1829 she married **Robert Keeley** (1793–1869), a good low comedian who made his name as Jemmy Green in *Moncrieff's *Tom and Jerry; or, Life in London.* He was also outstanding as Dogberry (in *Much Ado about Nothing*) and as *Dickens's Sarah Gamp. His wife was undoubtedly the better player of the two. She was at her best in pathetic, appealing parts such as Nydia in *Buckstone's *The Last Days of Pompeii* (1834), or Smike in Stirling's *The Fortunes of Smike; or, A Sequel to Nicholas Nickleby* (1840), adapted from Dickens. Her greatest success was in the title-role of the highwayman in Buckstone's version of *Jack Sheppard* (1839).

Keene, Laura (?1826–73), American actress and theatre manager, born in England. The date of her birth has been variously given as 1820, 1826, 1830, and 1836, and her real name may have been Moss, Foss, or Lee. She is said to have been trained for the stage by an aunt, Mrs Yates, and to have made her first appearance as Juliet at the *Richmond Theatre, Surrey, in 1851. Shortly afterwards she was seen at the *Olympic Theatre in London, and in 1852 went to New York on her way to tour Australia. She returned to New York in 1855 and spent the rest of her life in the USA, where she was the first woman to become a theatre manager. In 1856 she opened her own theatre in New York with *As You Like It*, in which she played Rosalind. She presented a repertory of good foreign and American plays with an excellent *stock company until after the outbreak of the Civil War, which forced her to lower her standards. She gave up her own theatre in 1862, but continued to act on tour and in other New York theatres. Among her outstanding actors were the third Joseph *Jefferson and E. A. *Sothern, who were jointly responsible for the success of Tom *Taylor's *Our American Cousin* (1858). It was her company that was playing it at *Ford's Theatre in Washington on the night Abraham Lincoln was assassinated there.

Keith, Benjamin, see VAUDEVILLE.

Kelly, George Edward (1890–1974), American dramatist, whose first full-length play was *The Torchbearers* (1922), a satire on pretentious amateur theatricals. He expanded a *vaudeville skit, *Poor Aubrey*, into a hilarious satire entitled *The Show-Off* (1924), and was awarded the *Pulitzer Prize for *Craig's Wife* (1925), a study of feminine possessiveness in a loveless marriage. None of his later plays was so successful and in 1931 Kelly withdrew from the theatre. His return in 1936 with a comedy, *Reflected Glory*, was a failure, and it was not until 1945 that his work was seen on stage again with *The Deep Mrs Sykes*, a satire on feminine intuition which was followed in 1946 by his last play, *The Fatal Weakness*, about feminine romanticism.

Kelly, Hugh (1739–77), English playwright, whose sentimental comedy *False Delicacy* (1768) was produced at *Drury Lane by *Garrick in the hope of emulating the success of *Goldsmith's *The Good-Natur'd Man* at *Covent Garden. Though now forgotten, for a time it eclipsed its rival: it was produced in the provinces, revived in London, and translated into French and German. Of Kelly's other plays the best was *The School for Wives* (1773), which most nearly approached the true spirit of the comedy of manners.

Kemble, Charles (1775–1854), English actor-manager, youngest brother of John Philip *Kemble. As a child he acted in the provincial company of his father Roger Kemble and at the age of 17 he was seen as Orlando in *As You Like It*, later one of his best parts. He made his first appearance in London in 1794, playing Malcolm to the Macbeth and Lady Macbeth of his brother and sister (Mrs *Siddons). Not at first a good actor, being somewhat awkward, with a weak voice, he in time overcame his difficulties and was considered an accomplished player in such parts as Benedick, Mercutio, Charles Surface, and Mirabell. He wisely left tragedy to his eldest brother, whom he assisted in the management of *Covent Garden, taking over when John retired in 1817. He was saved from bankruptcy by the acting of his daughter Fanny (below). Troubled by increasing deafness, Charles retired in 1832 to become Examiner of Plays and to give Shakespeare readings. He also visited America, where his courtesy and affable manner caused him to be considered a typical 'English gentleman'.

He married in 1806 the actress **Maria Theresa De Camp** (1774–1838), who retired in 1819, having been much admired as Edmund in Kenney's *The Blind Boy* (1807) and as Lady Elizabeth Freelove in her own one-act comedy *The Day after the Wedding; or, A Wife's First Lesson* (1808). She was also good as Mrs Sullen in *Farquhar's *The Beaux' Stratagem* and as

Madge Wildfire in Daniel Terry's musical version of Scott's *The Heart of Midlothian* (1818).

Kemble, Fanny [Frances] **Anne** (1809–93), English actress, daughter of Charles *Kemble. With no particular desire for a theatrical career, she was persuaded to appear at *Covent Garden in 1829 by her father, who hoped she would help to avert the bankruptcy which threatened his management. She was first seen as Juliet in *Romeo and Juliet*, and was immediately successful. For three years she brought prosperity to the theatre and everyone connected with it, being seen as Portia in *The Merchant of Venice*, Beatrice in *Much Ado about Nothing*, Lady Teazle in *Sheridan's *The School for Scandal*, and a number of tragic parts formerly associated with her aunt Mrs *Siddons, among them Isabella in *Southerne's *The Fatal Marriage*, Euphrasia in *Murphy's *The Grecian Daughter*, Calista in *Rowe's *The Fair Penitent*, and Belvidera in *Otway's *Venice Preserv'd*. She also created the part of Julia in Sheridan *Knowles's *The Hunchback*. Unlike the rest of her family she seems to have been good in both tragedy and comedy. In 1832 she went with her father to America, and was received everywhere with acclamation. In 1834 she left the stage on her marriage to an American, whom she divorced in 1848. After living in retirement in Lennox, Mass., for 20 years, she returned to London, where she died.

Kemble, John Philip (1757–1823), English actor, eldest son of **Roger Kemble** (1722–1802) and his wife Sarah Ward, strolling players. He appeared on the stage as a child with his parents, but was then sent to Douai to study for the priesthood. He returned to the theatre, however, and after acting in the provinces for several years made his London début in 1783 at *Drury Lane as Hamlet, in which character he was painted by Lawrence. He gave an unusual rendering of the part which at first puzzled but finally captivated the audience by its gentleness and grace. During his long career he steadily improved, playing all the great tragic parts of the current repertory—Wolsey in *Henry VIII*, Brutus in *Julius Caesar*, the title-role in *The Stranger*, an adaptation of *Kotzebue's *Menschenhaß und Reue*, Rolla in *Sheridan's *Pizarro*, *Addison's *Cato*, and above all Shakespeare's *Coriolanus*, the part in which he took leave of the stage in 1817. He managed both Drury Lane and *Covent Garden, where his raising of the prices of admission after the disastrous fire of 1808 caused the famous OP (Old Prices) riots. Financially his managements were not a success and as an actor he was limited in his choice of parts, being unfit, by reason of a harsh voice and stiff gestures, for any form of comedy. Nor was he good in romantic parts, in spite of his handsome presence. Even in tragedy, in which he excelled, he eschewed sudden bursts of pathos

or passion, achieving his effects by a studied and sustained intensity of feeling. More respected than loved, he was nevertheless much admired, and exercised a salutary influence on the theatre of his time.

He married in 1787 the actress **Priscilla Hopkins** (1755–1845), who, in her early years, had been a member of *Garrick's company. John Philip's brothers Charles (above) and Stephen (below) were also on the stage, as were his brother Henry and his four sisters: the eldest became famous as Mrs *Siddons; Frances and Ann had unremarkable careers; Elizabeth emigrated with her husband to America, and became well known there as **Mrs Whitlock** (1761–1836). Descendants of Roger and Sarah Kemble are still to be found in the theatrical and literary worlds of England and the United States.

Kemble, Stephen (1758–1822), English actor, son of Roger and Sarah Kemble, strolling players. He was born in a theatre in Herefordshire, his mother having just completed her part in the evening's entertainment. As a child he acted with his parents. He then left the theatre, but returned when his elder sister Mrs *Siddons became famous, hoping to profit by her popularity. He was a useful actor, but always overshadowed by his elder brother John Philip (above). In later life he became very fat, which enabled him to play Falstaff without padding, so that it was said that *Drury Lane, under the management of John Philip, had the great Kemble, and *Covent Garden, where Stephen was appearing, the big Kemble. He led a roving life, managing provincial theatres and a company in Ireland, and eventually returned to London to manage Drury Lane, with little success.

Kemp, Robert (1908–67), Scottish dramatist, until 1947 a producer with the BBC. His first play, *Seven Bottles for the Maestro*, was staged by the Dundee Repertory company in 1945 and his last, *Scotch on the Rocks*, at the Theatre Royal in Windsor in 1967. He wrote fluently in both Scots and English, and his work ranged from *pageants—*The Saxon Saint* (1949) and *The King of Scots* (1951), both written for performance in Dunfermline Abbey—through history—*John Knox* (1960), written to mark the fourth centenary of the Reformation—to a number of light pieces. The most effective of these was possibly *The Penny Wedding*, seen at the Edinburgh Gateway Theatre in 1957, which satirizes those Scottish nationalists who insist on the revival of the Scots language. Kemp was also responsible for the adaptation of *Lyndsay's *The Thrie Estaitis* performed under *Guthrie's direction at the 1948 and subsequent *Edinburgh Festivals. His best play is *The Other Dear Charmer* (1951), in which he makes use of Robert Burns's affair with Mrs MacLehose to explore the problem of identity in Scottish life and art.

Kempe, William (?–c.1603), one of the best known of Elizabethan clowns and a great player of *jigs. A member of the company which toured Holland and Denmark in 1584–6, he probably joined *Queen Elizabeth's Men in succession to *Tarleton. He is known to have been in the 1592 production of the anonymous *A Knack to Know a Knave*, of which he may have been part-author. On his return from a provincial tour with them in 1594 he joined the *Chamberlain's Men on their formation and was the original Dogberry in *Much Ado about Nothing* and Peter in *Romeo and Juliet*. He also appeared in Ben *Jonson's *Every Man in His Humour* (1598). He was famous for his improvisations, and it may have been Shakespeare's dislike of extempore gagging (as shown in his slighting reference in *Hamlet* to clowns that speak 'more than is set down for them') that caused him to leave the company in 1600. In that year, for a wager, he danced his famous nine-day morris dance from London to Norwich, of which he published an account in his *Kemps Morris to Norwiche*. He then went back to the Continent, but returned to London in 1602, when he borrowed some money from *Henslowe (duly recorded in the latter's diary), and is last heard of as one of Worcester's Men at the *Rose Theatre.

Kempson, Rachel, see REDGRAVE, MICHAEL.

Kendal, Dame Madge [Margaret Sholto Robertson] (1848–1935), English actress, twenty-second child of an actor-manager and the younger sister of T. W. *Robertson. She first appeared on the stage at the age of 5, and in 1865 made her adult début as Ophelia in *Hamlet*. In 1874 she married **William Hunter Kendal** [Grimston] (1843–1917), an English actor-manager who had made his first appearance on the stage in 1861 and had been playing at the *Haymarket for eight years, taking leading parts in old and new plays. After their marriage their careers were inseparable. Together they went to the *Royal Court Theatre under *Hare, and after appearing with the *Bancrofts at the Prince of Wales (later the *Scala) Theatre in a revival of *Boucicault's *London Assurance* in 1877 and in *Diplomacy* (1878), based by Clement *Scott and B. C. Stephenson on *Sardou's *Dora*, the Kendals went to the *St James's in partnership with Hare, remaining with him until 1888 and playing leading roles in many productions. Under other managements they played together until they retired in 1908. Kendal was somewhat overshadowed by his wife's brilliance but was none the less a good reliable actor, better in comedy than in tragedy, and an excellent business manager. Mrs Kendal was a fine *comédienne*, though she could on occasion play in a more gentle mood, as in *The Elder Miss Blossom* (1898) by Ernest Hendrie and Metcalfe Wood, which she frequently revived and took on tour.

The Kendals formed an ideal partnership, both on and off stage, and did much to raise the status of the acting profession.

Kennedy Center, see JOHN F. KENNEDY CENTER FOR THE PERFORMING ARTS.

Kenny, Sean (1932–73), stage designer and director, born in Ireland, where he trained as an architect. He designed his first stage set in 1957 for a revival of *O'Casey's *The Shadow of a Gunman*. His collaboration with Joan *Littlewood on Brendan *Behan's *The Hostage* (1958) and with Lindsay *Anderson on Cookson's *The Lily-White Boys* (1960) launched him on a career which in the next 14 years resulted in over 35 major West End productions. As the first resident Art Director at the *Mermaid Theatre, he helped to arrange the stage and auditorium and designed its first four shows, including the successful musical *Lock up Your Daughters* (1959). Other musicals included Lionel *Bart's *Oliver!* (1960; NY, 1963), for which he designed ingenious moving sets, and Anthony Newley and Leslie Bricusse's *Stop the World—I Want to Get Off* (1961; NY, 1962) and *The Roar of the Greasepaint, the Smell of the Crowd* (NY, 1965). For *Chichester he designed the sets for *Chekhov's *Uncle Vanya* (1962) and *Ibsen's *Peer Gynt* (1970). In 1963 he also replanned the interior of the *Old Vic to suit the needs of the new *National Theatre company and designed the décor for their opening production of *Hamlet*. He returned to the Mermaid in 1968 to adapt, design, and direct Swift's *Gulliver's Travels*. In 1970 he directed Siobhán *McKenna's one-woman show *Here are Ladies*, and when he died he was directing a revival of O'Casey's *Juno and the Paycock* for the Mermaid. Kenny's ingenious set designs and revaluation of the relationship between playwright, designer, and audience did much to revitalize the art of scene design in Britain; many of these ideas were expressed in his design of the *New London Theatre, which opened in the year of his death.

Kerr, Walter Francis (1913–), American drama critic, who from 1945 to 1949 was Associate Professor of Drama at the Catholic University. He was dramatic critic of the Catholic weekly, *The Commonwealth*, for two years, and then went to the *Herald-Tribune*. After the latter ceased publication in 1966 he moved to the *New York Times*, and in 1978 he won the *Pulitzer Prize for his writings on the theatre. In 1949 he collaborated with his wife **Jean** [née Collins] (1923–) in the revue *Touch and Go* (1949; London, 1950), and with her he wrote the book of the musical *Goldilocks* (1958). Jean Kerr herself was the author of a number of bitingly witty plays, of which the best known are *Mary, Mary* (1961; London, 1963) and *Finishing Touches* (1973).

Kester, Paul (1870–1933), American dramatist, whose most successful play was *Sweet Nell of Old Drury* (1900), first seen in London with Fred *Terry and his wife Julia *Neilson, for whom it was written. It remained in their repertory for over 30 years. In the USA in 1901 the part of Nell *Gwynn was played by Ada *Rehan. Other plays by Kester in the same vein were based on novels by Charles Major, *When Knighthood Was in Flower* (1901; London, 1906), written for Julia *Marlowe, and *Dorothy o' the Hall* (1903; London, 1906), the latter also being a favourite with Fred Terry and his wife.

Khmelev, Nikolai Pavlovich (1901–45), Soviet actor and director, who joined the *Moscow Art Theatre in 1919, where his first role was Fire in *Maeterlinck's *The Blue Bird*. He subsequently played many important parts, including Alexei Turbin in *Bulgakov's *Days of the Turbins*, Karenin in an adaptation of *Tolstoy's *Anna Karenina*, Peklevanov in V. *Ivanov's *Armoured Train 14-69*, Zabelin in *Pogodin's *The Kremlin Chimes*, and Tusenbach in *Chekhov's *Three Sisters*. In 1937 he became director of the *Yermolova Theatre, remaining there until 1943, when he replaced *Nemirovich-Danchenko at the Moscow Art Theatre.

Killigrew, Thomas (1612–83), English dramatist and theatre manager. Before the closing of the theatres in 1642 he had already written several tragi-comedies, including *The Prisoners* (1635), *Claracilla*, *The Princess; or, Love at First Sight* (both 1636), and *The Parson's Wedding* (1641), based on a play by *Calderón. This last, when revived in 1664 with a cast consisting of women only, made even *Pepys blush. It is not, however, as a dramatist that Killigrew is important in the history of the English theatre, but as the founder of the present *Drury Lane, which he opened as the Theatre Royal in 1662 under a Charter granted by Charles II. With Sir William *Davenant, holder of a similar Charter for the Duke's House, later transferred to *Covent Garden, Killigrew thus held the monopoly of serious acting in Restoration London, his company including *Mohun, *Hart, and, for a short while, Nell *Gwynn. He also established a training school for young actors at the Barbican, and in 1673 was appointed *Master of the Revels. He was, according to Pepys, a 'merry droll' and a great favourite at Court, but he was not as good a business manager as his rival Davenant and was often in financial difficulties. His brother **Sir William** (1606–95) and his son **Thomas** (1657–1719) both wrote plays, while another son, **Charles** (1665–1725), took over the management of the Theatre Royal in 1671, assisted by his half-brother Henry.

King [Pratt], **Dennis** (1897–1971), British-born actor and singer who spent most of his life in America. He made his first appearance in 1916 in a small part at the *Birmingham Repertory Theatre. His first London appearance was in 1919, and in 1921 he made his New York début with Ethel and John *Barrymore in *Clair de Lune*, based on Victor *Hugo's novel *L'Homme qui rit*. He played several further roles in New York before achieving stardom there in two famous musicals by Friml: as Jim Kenyon in *Rose-Marie* (1924) and François Villon in *The Vagabond King* (1925). In 1928 he was seen as D'Artagnan in another Friml musical *The Three Musketeers*, based on the novel by *Dumas *père* (*Drury Lane, 1930). He played the title role in a revival of *Peter Ibbetson* (1931) based on the novel by George Du Maurier, and Gaylord Ravenal in a revival of Kern and *Hammerstein's *Show Boat* (1932). Thereafter he appeared mainly in straight plays, ranging from *Ibsen, *Chekhov, and *Shaw (General Burgoyne in *The Devil's Disciple*, 1950) to modern parts such as Richard II in Gordon Daviot's *Richard of Bordeaux* (1934). He was also seen in N. C. *Hunter's *A Day by the Sea* (1955), *Ustinov's *Photo Finish* (1963), and *Osborne's *A Patriot for Me* (1969). He made rare returns to the musical stage in Rodgers and *Hart's *I Married an Angel* (1938) and Kern's *Music in the Air* (1951).

King, Hetty, see MUSIC-HALL.

King, Tom (1730–1804), English actor, who made his first appearance at *Drury Lane in 1748 under *Garrick. He was not suited to tragedy, and deciding to confine himself entirely to high comedy he went to Dublin, worked for a time under Thomas Sheridan, and returned to Drury Lane a finished comedian. His Malvolio in *Twelfth Night* and Touchstone in *As You Like It* were both admirable, but Lord Ogleby in *The Clandestine Marriage* (1766) by the elder *Colman and Garrick made his reputation. He was the first to play Sir Peter Teazle in *The School for Scandal* (1777) and Puff in *The Critic* (1779), both by R. B. *Sheridan, and he took his leave of the stage as Sir Peter in 1802. He was manager of *Sadler's Wells from 1772 until his death.

King of Misrule, see FEAST OF FOOLS.

King's Concert Rooms, London, see SCALA THEATRE.

King's Head Theatre, Islington, London, see FRINGE THEATRE.

King's Jester, see LENO.

Kingsley, Sidney (1906–), American dramatist, whose first play *Men in White* (1933; London, 1934), set in a hospital, was awarded a *Pulitzer Prize. The economic depression of the 1930s moved him to write *Dead End* (1935), a bleak but provocative study of crime-breeding slum conditions, while his *Ten Million Ghosts* (1936) excoriated the international munitions cartels that had profited from the First World War. *The World We Make* (1939), based on a novel, gave a moving account of a neurotic rich girl's discovery of comradeship and hope among the poor. *The Patriots* (1943) was a chronicle of the formative years of American democracy, and *Detective Story* (1949) a powerful indictment of excessive righteousness covering one day in a New York police station. In 1951 Kingsley dramatized Arthur Koestler's anti-Communist novel *Darkness at Noon*, and in 1954 he broke new ground with a farce, *Lunatics and Lovers*, set in a hotel suite. His last play was *Night Life* (1962). He directed *Dead End* and all his own later plays.

King's Men, company of actors known as the *Chamberlain's Men until they received their new name on the accession of James I in 1603. They remained in their own theatre, the *Globe, which they rebuilt after a fire in 1613, and continued to appear in the plays of Shakespeare, a member of the company, as they were written, their chief actor Richard *Burbage usually playing the leading parts. On his death in 1619 his place was taken by Joseph *Taylor, who with John *Lowin also replaced *Heminge and *Condell as business managers. Among the plays produced by the King's Men before the death of Shakespeare in 1616 were *Philaster* (*c.*1610) by *Beaumont and *Fletcher and *Webster's *The Duchess of Malfi* (1614). Other notable events in the company's history were the taking over of the second *Blackfriars in 1608 for use as a winter indoor theatre, and the publication in 1623 of the First Folio of Shakespeare's plays. The following year saw the production of *Middleton's *A Game at Chess*, a popular success which gave offence to the Spanish Ambassador. The players were admonished and fined and the play shelved. On the death of James I in 1625 the company came under the patronage of his son Charles I, who with his queen continued the interest shown in them by his father. Among their later dramatists were *Massinger and *Shirley. From the time of their foundation they were the leading London company, their only serious rival being the company under *Alleyn at the *Fortune. In spite of growing Puritan opposition they continued to act at the Globe until in 1642 all the theatres were closed and the company disbanded.

King's Theatre, London, see HER MAJESTY'S.

Kingsway Theatre, London, in Great Queen Street, Holborn, opened in 1882 as the Novelty Theatre, and closed the same month. Renamed the Folies-Dramatiques, it reopened in March 1883 with a little more success, and in 1888 was called the Jodrell after its manageress. It had

reverted to its original name when in 1889 the first production in English of *Ibsen's *A Doll's House*, translated by William *Archer, was given with Janet *Achurch as Nora. It became the New Queen's Theatre in 1890, and the Eden Palace of Varieties in 1894. It then stood empty until it was taken over by W. S. *Penley, who reopened it in 1900 as the Great Queen Street Theatre. An important event was the first London performance of *Synge's *The Playboy of the Western World* by the *Abbey Theatre company in 1907, and in 1908 Lena *Ashwell took over the management. Outstanding events of the next few years were Arnold *Bennett's *The Great Adventure* (1913) with Henry *Ainley, under the direction of *Granville-Barker, and a visit in 1925 by the *Birmingham Repertory Theatre company, under Barry *Jackson, during which his modern-dress production of *Hamlet* was seen. The theatre then continued on its erratic course until 1932, when it became the home of the Independent Theatre Club for the production of plays banned by the *Lord Chamberlain. The venture was not a success, and the theatre reverted to normal use. In 1940 Donald *Wolfit brought his touring company to London for the first time in a season of Shakespeare, and the last play to be produced at this theatre was a revival of Anthony Kimmins's farce *While Parents Sleep*. This lasted only 10 days as the building was bombed. It was demolished in 1956.

Kipphardt, Heinar (1922–82), German dramatist, whose first plays were produced in East Berlin; in 1959 he moved to the Federal Republic, where he became the leading exponent of *documentary theatre. The first of his works to attract attention, *Der Hund des Generals* (1962), was based on one of his own short stories, and is a bitter satire on war. He achieved international fame with *In der Sache J. Robert Oppenheimer* (1964), which deals with the trial of the well-known American nuclear physicist for suspected treason. As *In the Matter of J. Robert Oppenheimer* it was seen in London in 1966 (NY, 1969). A further documentary, *Joel Brand* (1965), recalls the offer of the Nazis during the Second World War to spare the lives of a million Jews in exchange for 10,000 lorries. In all these plays, depicting the anguish of minority groups caught between contending world powers, Kipphardt showed remarkable skill in turning documented fact into effective theatre. A later comedy, *Die Nacht, in der Chef geschlachtet wurde* (*The Night the Boss was Slaughtered*, 1967), portrayed the technologically advanced society of the West as a nightmare.

Kirby's Flying Ballet, see FLYING EFFECTS.

Kitchen-Sink Drama, expression coined in the British theatre to describe *Osborne's *Look Back in Anger* (1956), Shelagh Delaney's *A Taste of Honey* (1958), and other plays. The characters were less affluent than those of the conventional drawing-room drama; they had no servants and were seen doing mundane things such as ironing and washing up. With the decline of the genre the term is now seldom used.

Klein, Charles (1867–1915), British-born American dramatist, who went to the United States at the age of 15, and because of his short stature was able for some years afterwards to continue playing juvenile roles. He first came into prominence as co-author of *Heartsease* (1897). Klein then had an immense success with two sentimental plays written for and produced by *Belasco, *The Auctioneer* (1901) and *The Music Master* (1904), in which David *Warfield became famous in the title-roles. In 1913 he helped to dramatize some Jewish short stories which as *Potash and Perlmutter* were successful both in New York and in London a year later. He also wrote several plays on social issues, such as *The Lion and the Mouse* (1905) and *The Third Degree* (1909), about police malpractice. Klein was a play reader for Charles *Frohman and was drowned with him in the sinking of the *Lusitania*.

Kleist, (Bernd Wilhelm) **Heinrich Von** (1777–1811), German dramatist, born into a Prussian military family, who was himself in the army and saw active service. In 1799 he resigned his commission in order to devote himself to literature and philosophy; but he failed to achieve his ambitions and committed suicide. His first play, *Die Familie Schroffenstein* (1804), was a *fate drama involving the destruction of two feuding families; *Die Hermannsschlacht*, a patriotic drama written in 1808 but not performed until 1871, dealt with the defeat of the Romans by Arminius, and was intended to serve as a warning to Napoleon; *Penthesilea*, written at about the same time and produced in a shortened version in 1876, was based on the story of the Amazon Queen; *Käthchen von Heilbronn* (1810), a *Ritterdrama, was revived by *Reinhardt for the opening of the *Deutsches Theater in 1905; and *Prinz Friedrich von Homburg* (1821), in which the hero disobeys a military command but is brought to accept the necessity of discipline and his own death sentence (whereupon he is reprieved), was seen in Berlin with Josef *Kainz in the chief part. It was not a success at the time, but is now considered an important part of the German repertory. In 1951, in a French translation, with Gérard *Philipe as Friedrich and produced by Jean *Vilar, it was one of the outstanding successes of the *Avignon Festival. As *The Prince of Homburg* the play was produced in New York in 1976 (*National Theatre, London, 1982). Kleist was also the author of two comedies, an adaptation of *Molière's *Amphitryon*, published in 1807, and *Der zerbrochene Krug*, in which a village magistrate with a Falstaffian virtuosity

in lying tries a case in which he is himself the culprit. It was produced at *Weimar by *Goethe in 1808, but its merits were not apparent until its revival at *Hamburg in 1820. It is now regarded as one of the best short comedies in the German language, and has been seen in English as *The Broken Jug* (or *Pitcher*) and in Scots as *The Chippit Chantie*.

Knepp, Mary (?–1677), one of the first English actresses. Trained by *Killigrew, she was in his company at the first Theatre Royal, a friend and contemporary of Nell *Gwynn. She was also a friend of *Pepys, in whose diary she often appears, usually as the source of back-stage gossip. According to him she was a merry, lively creature, at her best in comedy and an excellent dancer. She was also much in demand by authors to speak the witty prologues and epilogues which the fashion of the time demanded.

Knickerbocker Theatre, New York, on Broadway, at the north-east corner of 38th Street. As Abbey's Theatre, named after its first manager, it opened in 1893 with *Irving and Ellen *Terry in *Tennyson's *Becket*, and later saw the New York débuts of such overseas stars as *Mounet-Sully, *Réjane, and John *Hare. In 1896 it was renamed the Knickerbocker, but continued its policy of housing famous visitors, among them Wilson *Barrett in his own melodrama *The Sign of the Cross* and Beerbohm *Tree in Gilbert Parker's *The Seats of the Mighty*. Among the American stars who appeared there were Maude *Adams in *Rostand's *L'Aiglon* in 1900 and *Barrie's *Quality Street* in 1901, and Otis *Skinner, who in 1911 scored an instantaneous success in *Knoblock's *Kismet*. The last production at this famous theatre was Philip Dunning's *Sweet Land of Liberty*, which closed after eight performances. The theatre was demolished in 1930.

The *Bowery Theatre was called the Knickerbocker for a short time in 1844.

Knipper-Chekhova, Olga Leonardovna (1870–1959), Russian actress, who joined the company of the *Moscow Art Theatre on its foundation in 1898 and became famous for her interpretations of the heroines of *Chekhov, whom she married in 1901. When the theatre revived *The Seagull*, which had been unsuccessful elsewhere, she was seen as Arkadina, and she created the roles of Elena Andreyevna in *Uncle Vanya* (1899), Masha in *Three Sisters* (1901), and Ranevskaya in *The Cherry Orchard* (1904). She played this last part again in 1943, on the occasion of the 300th performance of the play. After Chekhov's death in 1904 she remained with the Moscow Art Theatre, and though mainly considered at her best in poetic and delicately psychological roles, she was also much admired in such comedy parts as Shlestova in *Griboyedov's *Woe from Wit*, which she played in a revival in 1925.

Knoblock [Knoblauch], **Edward** (1874–1945), playwright born and bred in the United States, who spent much of his life in England and on the Continent and eventually became a British citizen. He was for a short time an actor, and had a thorough knowledge of the stage, which he applied with much skill to the dramatization of other writers' novels, sometimes in collaboration with them, and was considered an admirable and reliable 'play carpenter' rather than an original dramatist. Of his own fairly numerous plays the most successful were *Kismet* (1911), an oriental fantasy produced in London by Oscar *Asche and in New York (also 1911) by Otis *Skinner, and *Marie-Odile* (1915), a tale of the Franco-Prussian war which in New York was directed by David *Belasco. He collaborated with Arnold *Bennett on *Milestones* (1912), with Seymour *Hicks on *England Expects* (1914), and with Bennett again on *London Life* (1924) and *Mr Prohack* (1927). The novels of which he made stage versions include *Priestley's *The Good Companions*, Vicki Baum's *Grand Hotel* (both 1931), Beverley Nichols's *Evensong*, and A. J. Cronin's *Hatter's Castle* (both 1932).

Knowles, James Sheridan (1784–1862), Irish-born playwright, a cousin of R. B. *Sheridan and a close friend of *Coleridge, *Hazlitt, and Charles *Lamb. In 1808 he was a member of the Crow Street Theatre in Dublin, where he proved himself a passable actor and also wrote a melodrama, *Leo; or, The Gypsy* (1810), for Edmund *Kean, who was a member of the company. He later offered Kean *Virginius; or, The Liberation of Rome* (1820), but it was refused. *Macready accepted it and played it with great success at *Covent Garden. Knowles's most popular play was *The Hunchback* (1832), whose heroine Julia, first played by Fanny *Kemble, was a favourite part with many young actresses. He himself appeared in the play after many years off the stage, playing Master Walter, the part in which he made his first appearance in New York in 1834. Knowles wrote his tragedies on classical subjects in the light of 19th-century domesticity, and was more concerned with the emotions of his characters than with their actions. None of his works has survived on the stage.

Knowles, Richard, see MUSIC-HALL.

Knox, Teddy, see CRAZY GANG.

Koch, Heinrich Gottfried (1703–75), German actor, who in 1728 joined Carolina *Neuber's company. He was a good scene-painter, a good translator and adapter of plays, and a competent actor in the new style sponsored by *Gottsched and Neuber, particularly popular in classical comedy and acceptable in tragedy as long as the declamatory style remained in fashion. After the break-up of the Neuber company he started on his own. He treated his actors well and was one of the few managers of his time, apart from *Ackermann, to be generally esteemed. He travelled continuously, but made Leipzig his headquarters.

Koltai, Ralph (1924–), German-born stage designer, mainly a freelance, who in 1962 was engaged to design the sets for the *RSC's production of *Brecht's *The Caucasian Chalk Circle*. He was Associate Designer to the company, 1963–6, working on its production of *Hochhuth's *The Representative* (1963), with its gas-chamber setting, and revivals such as *Marlowe's *The Jew of Malta* (1964). Returning to the same post in 1976, he designed *Wild Oats* (1977) by *O'Keeffe, *Romeo and Juliet*, *Hamlet* (both 1980), and a memorable *Much Ado about Nothing* (1982) with Derek *Jacobi. He also designed several *National Theatre productions, among them the all-male *As You Like It* (1967), *Ibsen's *Brand* (1978) and *The Wild Duck* (1979), and Brecht's *Baal* (also 1979). His work elsewhere included Hochhuth's *Soldiers* (London and NY, 1968) and the musicals *Billy* (1974) and *Bugsy Malone* (1983). Although based in England, he worked all over the world, in opera and dance as well as the theatre. From 1965 to 1973 he was head of the School of Theatre Design at the Central School of Art and Design in London. After demonstrating his boldness and use of modern materials in the machine-dominated set for the musical *Metropolis* (1989), he moved his working base to France.

Komisarjevskaya, Vera Fedorovna (1864–1910), Russian actress and theatre manager, sister of Theodore *Komisarjevsky. She made her début in 1891 as Betsy in Leo *Tolstoy's *The Fruits of Enlightenment* and in 1896 went to the Alexandrinsky (see PUSHKIN THEATRE) where she played Nina in the ill-fated first production of *Chekhov's *The Seagull*. She left in 1904 to found her own theatre, where in the midst of the social upheavals of 1905 her productions included plays by *Gorky, Chekhov, and *Ibsen. A year later she came under the influence of *Symbolism, inviting *Meyerhold to produce in her theatre, but soon breaking with him. On tour again she caught smallpox and died. She never appeared in England, but in 1908 played a season at *Daly's in New York. She was at her best in such parts as Gretchen in *Goethe's *Faust*, Rosy in *Sudermann's *The Battle of the Butterflies*, and Ibsen's Nora, in *A Doll's House*, and Hedda Gabler.

Komisarjevsky, Theodore [Fedor] (1882–1954), Russian director and designer, brother of Vera *Komisarjevskaya. Born in Venice, he was brought up in Russia and gained his initial experience in the pre-Revolutionary theatre, where from 1907 he produced a number of plays and operas, first at his sister's theatre and from 1910 in Moscow, in his own studio and elsewhere, including *Goethe's *Faust* in 1912. In 1919 he emigrated to England, where he worked initially as a theatre designer, and was first noticed during the Russian season at the *Barnes Theatre in 1926. Although his best work was done in productions of and designs for Russian plays, particularly those of *Chekhov, he was also responsible for a number of controversial productions at the *Shakespeare Memorial Theatre, notably *Macbeth* (1933) with aluminium scenery, *The Merry Wives of Windsor* (1935) in the style of a Viennese operetta, a widely acclaimed *King Lear* (1936), *The Comedy of Errors* (1938), and *The Taming of the Shrew* (1939). In London during these years he also produced a wide variety of plays, including Chekhov's *The Seagull* with Edith *Evans in 1936; the last was *Barrie's *The Boy David* (also 1936). He subsequently went to the United States, where he remained until his death. He was the second husband of Peggy *Ashcroft.

Kongelige Teater, Copenhagen, see ROYAL THEATRE.

Koonen, Alisa, see TAÏROV.

Kopit, Arthur (1938–   ), American dramatist, whose first plays were produced at *Harvard while he was a student there. They included *Sing to Me through Open Windows* (1959) and *Oh Dad, Poor Dad, Mamma's Hung You in the Closet and I'm Feelin' So Sad* (1960), which achieved an international reputation. A brilliant black farce parodying the Oedipus complex, it was seen in London in 1961 and 1965, in New York in 1962, where it had a long run, and in Paris in 1963 with Edwige *Feuillere as the possessive mother, Madame Rosepettle. Later works include *The Day the Whores Came out to Play Tennis*, satirizing social-climbing middle-class Americans, in which a group of whores demolish an exclusive country club with their tennis balls. It was presented in New York in 1965 with a revised version of *Sing to Me through Open Windows*. *Indians* (1968), set in the Wild West with Buffalo Bill (W. F. *Cody) as its central character, attacks America's treatment of its Indian population. Its world première was given by the *RSC, and a revised version produced in New York in 1969. In *Wings* (NY and London, 1979) Constance *Cummings gave an outstanding performance as a stroke victim. Kopit wrote the book for the musical *Nine* (1982), based on Fellini's film *8½*, his adaptation of *Ibsen's *Ghosts* being staged in the same year. In the black comedy *End of the World* (1984) a millionaire commissions a play about nuclear warfare.

Korneichuk, Alexander Evdokimovich (1905–72), Ukrainian dramatist, whose first successful play, *The Wreck of the Squadron* (1934),

dealt with the sinking of their fleet by Red sailors to prevent its capture by White Russians. It was followed by *Platon Krechet* (1935), the story of a young Soviet surgeon, and by *Truth* (1937), which shows a Ukrainian peasant led by his search for truth to Petrograd and Lenin at the moment of the October Revolution. *Bogdan Hmelnitsky* (1939) depicts a Ukrainian hero who in 1648 led an insurrection against the Poles. *The Front* (1943) was a popular war play, while a satirical comedy, *Mr Perkins's Mission to the Land of the Bolsheviks*, in which an American millionaire visits Russia to discover for himself the truth about the Soviet regime, was produced in 1944 by the Moscow Theatre of *Satire. Later plays included *Come to Zvonkovo* (1946), *Makar Dobrava* (1948), an inimitable portrait of an old Donetz miner, *Why the Stars Smiled* (1958), and *On the Dnieper* (1961), both of which centre on smaller-scale domestic problems.

Kortner, Fritz (1892–1970), Austrian actor, who after some experience in the provinces achieved an immediate success in *Berlin in 1919 as the obsessed young hero of *Toller's *Die Wandlung* (*Transfiguration*). In the same year he consolidated his position with a satanic, sadistic portrayal of Gessler in *Schiller's *Wilhelm Tell*, and in the 1920s he established himself as the ideal interpreter of the new *Expressionism, beginning at this time a long and fruitful partnership with the director *Jessner. After playing the title-roles in *Richard III* and *Wedekind's *Der Marquis von Keith*, in which he used a rapid, high-pressure delivery exploiting the strident power of his voice, he began in 1921, in *Othello*, to make use of quieter and more controlled speech, with greater variety of gesture. This new realism was apparent in his Macbeth, Coriolanus, and Shylock, as well as in *Ibsen's *John Gabriel Borkman* in 1923 and *Sophocles' *Oedipus the King* in 1929. He left Germany in 1933 and played small parts in a number of American films, returning home in 1949. Working mainly in Berlin and *Munich, he then became one of the most important directors in post-war Germany with a series of meticulously detailed productions. Notable among them were plays by Shakespeare and *Molière, *Strindberg's *The Father* in 1950, *Beckett's *En attendant Godot* in 1953, and such German classics as Schiller's *Kabale und Liebe* in 1965, *Goethe's *Clavigo* in 1969, and *Lessing's *Emilia Galotti* in the year of his death.

Koster and Bial's Music Hall, New York, at the corner of West 23rd Street and the Avenue of the Americas. Originally Bryant's Opera House for *minstrel shows and *variety, it was enlarged by Koster and Bial for musical entertainments in general and reopened in 1879. In 1881 it began importing outstanding *vaudeville stars from abroad and became a famous house of light entertainment. It closed in 1893 when Koster and Bial moved to their New Music Hall on the north side of 34th Street, between Broadway and 7th Avenue, which had been opened in 1892 by Oscar Hammerstein as the Manhattan Opera House. They intended to present there a better-class and more expensive brand of variety, but their fortunes declined, and in 1901 the theatre closed, Macy's department store being built on the site. Meanwhile their old music hall had reopened as the Bon Ton, but it too was unsuccessful and soon closed, being eventually demolished in 1924.

Kotzebue, August Friedrich Ferdinand Von (1761–1819), German dramatist, who in his day was more popular than his contemporary *Schiller with the new audiences of the Revolutionary period, not only in Germany but all over Europe. He wrote over 200 melodramas, the most successful being *Menschenhaß und Reue* (1789), in which an erring wife obtains forgiveness from her husband by a life of atonement. As *The Stranger* it was produced at *Drury Lane in 1798, the heroine, Mrs Haller, providing an excellent part for Mrs *Siddons, playing opposite her brother, John Philip *Kemble. It was frequently revived up to the end of the 19th century. Equally successful in the following year, with the same leading players, was *Pizarro*, an adaptation by R. B. *Sheridan of *Die Spanier in Peru*. In America Kotzebue's plays, in adaptations by *Dunlap, led to a vogue for melodrama which tended to eclipse more serious works and pandered to a craving for sensationalism. The best of Kotzebue's comedies is *Die deutschen Kleinstadten* (1803), an entertaining skit on provincialism.

Koun, Karolos (1908–87), Greek director, who while teaching English in Athens during the 1930s mounted a remarkable series of productions of the classics performed by young people. In 1942, during the German occupation, he founded the Art Theatre (Theatro Technis) in Athens, where he produced the plays of ancient and modern Greece as well as such works as *Ibsen's *The Wild Duck* (1942), *García Lorca's *Blood Wedding* (1948), and Tennessee *Williams's *The Glass Menagerie* (1947) and *A Streetcar Named Desire* (1949). The theatre was closed because of his Communist sympathies between 1949 and 1954. His work for other companies included Arthur *Miller's *Death of a Salesman* (1950) and for the Greek National Theatre *Pirandello's *Henry IV* and *Chekhov's *Three Sisters*. In 1962 he took his company to the *Théâtre Des Nations in Paris, where their performance of *Aristophanes' *Birds* was much admired. It was seen during the *World Theatre Season of 1964, and again in 1965 and 1967 together with *Aeschylus' *Persians* and Aristophanes' *Frogs* respectively; in 1969 the company presented Aristophanes' *Lysistrata* there. In

1967 Koun also directed *Romeo and Juliet* for the *RSC. Koun's school of drama, attached to the Art Theatre, supplied the Greek stage with a number of leading actors. The company moved to a new building in 1981.

Krasnya Presnya Theatre, Moscow, see REALISTIC THEATRE.

Krejča, Otomar (1921–), Czech director, who was at the Vinohrady Theatre, 1946–51. He then went to the National Theatre, becoming its Artistic Director in 1956, the sensitive style of his first productions setting the tone for his later work. He staged some remarkable revivals of the classics, including *Hamlet* and *Chekhov's *The Seagull*, and set up an experimental workshop to help young writers. In 1965 he founded the Theatre Behind the Gate, and the company was seen in London in the *World Theatre Season of 1969 in Chekhov's *Three Sisters*, *The Single-Ended Rope* based on several farces by *Nestroy, *Schnitzler's *Der grüne Kakadu*, and a new play by Josef Topol, *An Hour of Love*. The theatre was officially disbanded in 1972, Krejča having been dismissed the previous year. His career has also included productions of *Romeo and Juliet*, *Measure for Measure*, Chekhov's *Ivanov* and *Platonov*, and *Beckett's *Waiting for Godot*. He has latterly worked a good deal abroad.

Kristiania Theater, Norway. The theatre in Christiania (now Oslo) traces its origins back to 1827, when the first permanent playhouse was established in the Teatergata by a Swede, who was quickly succeeded by a Dane. In 1835 the theatre burnt down, reopening in 1837 in the Bankplassen. Until well after the mid-19th century Danish influence and control over the theatre was supreme, despite demonstrations from patriotic Norwegians.

After 1850, led by *Ibsen and Bjørnstjerne *Bjørnson, a genuinely Norwegian drama of considerable power began to emerge, encouraging the emergence of Norwegian players and directors. In 1852, Det Norske Teatret began operating in Møllergata, Christiania, in opposition to the older theatre, but in 1863 the two were combined. Ibsen briefly held an appointment there, succeeded from 1865 to 1867 by Bjørnson, and between 1884 and 1893 Bjørn Bjørnson was an energetic director there. In 1899 virtually the whole company moved to the new *Nationaltheatret, with Bjørnson as its director.

Kungliga Dramatiska Teatern, see ROYAL DRAMATIC THEATRE.

Kuppelhorizont, see LIGHTING.

Kwanami, see NŌ PLAY.

Kyd, Thomas (1558–94), Elizabethan dramatist, intimate friend of *Marlowe, with whom he was implicated in accusations of atheism and subsequently imprisoned. His reputation as a playwright rests almost entirely upon hearsay, as all the plays attributed to him, with the exception of a translation of Robert *Garnier's *Cornelia* (1594), were published anonymously; but he is definitely known to be the author of *The Spanish Tragedy; or, Hieronimo is Mad Again* (c.1589), one of the most popular productions of its day and the forerunner of many similar *revenge tragedies. It was constantly revived and revised, in one instance by Ben *Jonson, and survived into Restoration times, being seen on stage by *Pepys in 1668. Among the other plays almost certainly ascribed to Kyd is *Soliman and Perseda* (c.1589–91), a tragedy at one time erroneously attributed to *Peele. A domestic tragedy, *Arden of Feversham* (pub. 1592), once thought to be by Shakespeare, has now tentatively been assigned almost entirely to Kyd, though he is no longer credited with a forepiece to *The Spanish Tragedy* entitled *The First Part of Jeronimo*, or with the authorship of early plays entitled *Hamlet* and *The Taming of a Shrew* on which Shakespeare was said to have based his own versions.

Kynaston, Ned [Edward] (c.1640–1706), one of the last boy-players of women's roles. After seeing him as the Duke's sister in a revival of *Fletcher's *The Loyal Subject*, *Pepys said: 'He made the loveliest lady that ever I saw in my life', and it was the delight of fashionable ladies to take him, in his petticoats, driving in the Park after the play. Colley *Cibber recounts that on one occasion Charles II had to wait for the curtain to rise at the theatre because Kynaston, who was playing the tragedy queen, was being shaved. In later life he fulfilled the promise of his youth and made many fine appearances in dignified heroic roles. He was particularly admired as the ageing king in *Henry IV*.

L

Labiche, Eugène (1815–88), French dramatist, who between 1838 and 1877 wrote more than 150 light comedies, alone or in collaboration. The best known is *Un chapeau de paille d'Italie* (1851), of which W. S. *Gilbert, who translated a number of Labiche's comedies, made two versions—as *The Wedding Guest* in 1873, and as *Haste to the Wedding* in 1892. There were several successful revivals in London between the two world wars, under the title *An Italian Straw Hat*, and in 1936 an American version, as *Horse Eats Hat*, scored a great success in New York. Labiche's success with his contemporaries may have been due to a revolt against the serious problem-plays of younger dramatists such as *Dumas *fils*. With his broad humour and predictable but well-presented situations, he gave new life to the *vaudeville, and raised French farce to a height which it has since attained only with *Feydeau. Labiche's other plays include *Le Voyage de M. Perrichon* (1860) and *La Cagnotte* (*The Kitty*, 1864).

Lacey, Catherine (1904–79), English actress, who made her first appearance in Brighton with Mrs Patrick *Campbell in 1925 and her London début later the same year. She first came into prominence as Leonora Yale in *The Green Bay Tree* (1933) by Mordaunt Shairp, and in 1935 was at the *Shakespeare Memorial Theatre, where she was seen as Cleopatra in *Antony and Cleopatra* and Katharina in *The Taming of the Shrew*. A sensitive actress with great reserves of emotional strength, she made a deep impression as Amy O'Connell in *Granville-Barker's *Waste* (1936), and was also much admired as the heroine of J.-J. *Bernard's *The Unquiet Spirit* (1937). She played Agatha in the first production of T. S. *Eliot's *The Family Reunion* (1939) and again in 1946. In 1951, with the *Old Vic company, she played Clytemnestra in *Sophocles' *Electra*. She was seen as Hecuba in *Giraudoux's *Tiger at the Gates* (London and NY, 1955), and returned to the Old Vic as Elizabeth I in *Schiller's *Mary Stuart* (1958). After *Bolt's *The Tiger and the Horse* (1960) at the *Queen's she appeared as Clytemnestra again in *Aeschylus' *Oresteia* (1961). In the Old Vic's last season her roles included Aase in *Ibsen's *Peer Gynt* (1962), and she was last seen in 1970 in Robert *Anderson's *I Never Sang for My Father*.

La Chaussée, Pierre Claude Nivelle de (1692–1754), French dramatist, and the chief exponent of 18th-century *comédie larmoyante*. Though he was already a well-known figure in literary society La Chaussée was 40 before he wrote his first play, *La Fausse Antipathie* (1733). This and *Le Préjugé à la mode* (1735) were both well received, and may be said, with their mingling of pathos and comedy, to typify 'domestic drama'. La Chaussée was a wealthy man with somewhat frank and licentious tastes, and it is clear from his prologue to *La Fausse Antipathie* that he wrote as he did less out of conviction than because he saw that this was what his audience wanted. He produced a number of later plays in the same vein, including *Mélanide* (1741), *Paméla* (1743), adapted from Richardson's novel, and *La Gouvernante* (1747), a foretaste of *East Lynne*. In his own day La Chaussée was immensely popular; he was elected to the French Academy in 1736, and his plays were translated into Dutch, English, and Italian.

Lackaye, Wilton (1862–1932), American actor, who was intended for the Church, but adopted the stage after a chance visit to *Madison Square Theatre on his way to Rome. Lawrence *Barrett gave him a part in his revival of *Boker's *Francesca da Rimini* at the *Star Theatre, New York, in 1883, and he later appeared many times with Fanny *Davenport. In 1887 he made a success in a dramatization of Rider Haggard's novel *She*. He was thereafter constantly in demand, appearing in a number of Shakespeare plays and in many new productions. He also played Jean Valjean in his own dramatization of *Hugo's *Les Misérables* but his greatest part was undoubtedly Svengali in Du Maurier's *Trilby* (1895), which he played for two years and then in many revivals. He founded the Catholic Actors' Guild and helped to organize the American Actors' *Equity Association.

Lacy, John (?–1681), English actor, originally a dancing-master, who took to the stage when the theatres reopened in 1660 and became a great favourite with Charles II. A painting by Michael Wright which still hangs in Hampton Court shows him in three different parts—as Teague (in which he was much admired by *Pepys) in Howard's *The Committee* (1662), as Mr Scruple in John Wilson's *The Cheats* (1663), and as Monsieur Galliard in a 1662 revival of Cavendish's *The Variety* (1639). He was the first to play Bayes in Buckingham's *The Rehearsal* (1671), and was judged to have hit *Dryden off to the life. His Falstaff was also much admired. He wrote several plays which have not survived on stage.

Lady Elizabeth's Men, company under the patronage of James I's daughter. Formed in 1611, they were seen at Court a year later and in 1613 were merged with the Revels company, Nathan *Field being their leading actor. They were at the *Hope under the management of *Henslowe, who built it for them, in 1614, and on his death two years later several of the company joined *Prince Charles's Men under *Alleyn; but Field was taken on by the *King's Men, probably in the place of Shakespeare, who died in the same year. The rest of the Lady Elizabeth's Men went into the provinces, but in 1622 a new company under the same name took possession of the *Cockpit under Christopher *Beeston. They were successful for a time, and appeared in a number of plays by distinguished new dramatists—*The Changeling* by *Middleton and *Rowley, *Massinger's *The Bondman*, and Thomas *Heywood's *The Captives; or, The Lost Recovered* among them. The great plague of 1625 caused them to break up and on the accession of Charles I their place was taken by a new company, *Queen Henrietta's Men.

La Fleur, see GROS-GUILLAUME.

Lagerkvist, Pär (1891–1974), Swedish poet, novelist, and dramatist, who was awarded the *Nobel Prize for Literature in 1951. Although best known outside Sweden for his novels and short stories, he was probably the most remarkable playwright of the modern Swedish theatre as well as an influential theorist and critic. He repudiated *naturalism, and followed the theories of *Strindberg, being an adherent of *Expressionism. The first of his plays to be performed, *Himlens hemlighet* (*The Secret of Heaven*), was staged in 1921. This play and those that followed up to 1928 were clouded with deep despair, but in later plays Lagerkvist shows a compassionate vision expressed in an increasingly realistic style, though with symbolic overtones. His most successful play was *De vises sten* (*The Philosopher's Stone*, 1948), first seen at the *Royal Dramatic Theatre under Alf *Sjöberg. Set in the Italian Renaissance, it deals with problems of faith, knowledge, and social responsibility. *Barabbas* (1953) is based on one of his own novels.

La Grange [Charles Varlet] (1639–92), French actor who joined the company of *Molière in 1659, playing young lovers. A methodical man, he kept a register of all the plays presented by Molière at the *Palais-Royal and the receipts from each, interspersed with notes on the internal affairs of the company which have proved invaluable to later students of the period. He was active in forwarding the company's affairs after Molière's death, editing and writing a preface to the first collected edition of Molière's plays, published in 1682. His wife **Marie** (1639–1737), known as Marotte from the part she played in *Les Précieuses ridicules*,

was the daughter of the pastrycook Cyprien Ragueneau, immortalized by *Rostand in his play *Cyrano de Bergerac* (1897). One of the original members of the *Comédie-Française, she retired on the death of her husband.

Lahr, Bert [Irving Lahrheim] (1895–1967), American *vaudeville and *burlesque player, who became a leading actor in both musical shows and straight plays. Basically a clown, he first played the vaudeville circuits at the age of 15, and within a few years had developed his own solo act. After serving in the Navy during the First World War he teamed up in a double act with Mercides Delpino, later his wife, and made his Broadway début in the revue *Delmar's Revels* (1927), his first outstanding success being the prizefighter Gink Schiner in the musical *Hold Everything* (1928). During the 1930s he was in a number of musical shows, including the revue *Life Begins at 8.40* (1934) and the Cole *Porter musical *Du Barry was a Lady* (1939). In 1939 he also became internationally famous for his performance as the Cowardly Lion in the film *The Wizard of Oz*. His subsequent stage career included Skid, a *passé* burlesque comic, in a revival of *Burlesque* in 1946, Gogo (Estragon) in *Beckett's *Waiting for Godot* (1956), and Boniface in *Hotel Paradiso* (1957), based on a farce by *Feydeau. He gave fine performances as Bottom in *A Midsummer Night's Dream* in the American Shakespeare Festival production of 1960, and in *Foxy*, a musical version of *Jonson's *Volpone*, in 1964.

La Mama Experimental Theatre Club, New York, avant-garde theatre group formed in 1961 and originally known as the Café La Mama, from the basement room under an Italian restaurant where new and uncommercial plays by young playwrights were presented for a week's run. After two moves the group settled in its own premises and took its present name, supporting a permanent company formed in 1964. The seminal rock musical *Hair* was staged in 1967. Further experiments led to the merging of classical forms of music with rock genres, employing also such skills as tap-dancing and acrobatics. In 1973 financial stringencies forced the company to stop working full-time and its members formed a pool from which performers could be elected for specific projects. La Mama then became an international movement, with off-shoots in Britain, Europe, Australia, Canada, and South America, all founded by people who at one time worked with the group in New York. La Mama also produced such experimental groups as Mabou Mines, which (taking its name from a disused coal mine in Nova Scotia) began work in 1970, preparing by *Collective Creation productions which the participants described as Animations or Collaborations, the former, such as *Red Horse*

Animation (1971), being non-literary presentations using images of animals as metaphors for human conditions. La Mama also housed the productions of Chaikin's *Open Theater.

Lamb, Charles (1775–1834), English critic and essayist, whose anthology of dramatists contemporary with Shakespeare, published in 1808, revealed to his contemporaries the little-known beauties of such playwrights as *Marlowe, *Webster, and *Ford. In the *Essays of Elia* (1823–33) he recalls many happy moments spent in the theatre and in the company of actors. His love for Shakespeare in particular led him to write in collaboration with his sister Mary *Tales from Shakespeare* (1807), which, though now considered out of date, served as a pleasant and easy introduction to many of the plays for several generations of schoolchildren. Of his own four plays, a farce, *Mr H*—(1806), was produced, unsuccessfully, at *Drury Lane. The others were published but never played.

Lambs, The, London supper club founded in the 1860s by a group which included Squire *Bancroft, John *Hare, and Henry *Irving. Limited to 24 members under a chairman known as the Shepherd, it had no premises of its own but met regularly for many years at the Gaiety Restaurant and subsequently the Albemarle Hotel. It survived until the late 1890s. Meanwhile an American actor, Henry Montague, who had been one of the founder-members, had returned to New York, where in 1875 he started a similar club, with himself as the first Shepherd. On the dissolution of the London club the Shepherd's crook, bell, and badge were presented to the American Shepherd. In 1904, after two moves, the club settled in premises at 128 West 44th Street, which had to be enlarged several times to accommodate a growing membership. These were disposed of in 1945, and the club moved to new quarters in the Women's Republican Club on 51st Street. The Lambs in its present form is roughly the equivalent of the London Savage Club, as the *Players is of the *Garrick Club.

Lamda, see SCHOOLS OF DRAMA.

Lanchester, Elsa, see LAUGHTON.

Lander [*née* Davenport], **Jean Margaret** (1829–1903), American actress, daughter of the English actor-manager **Thomas Davenport** (1792–1851), on whom *Dickens is believed to have based Vincent Crummles in *Nicholas Nickleby*—in which case Jean, who at the age of 8 was playing such unsuitable parts as Richard III and Shylock, would be the original Infant Phenomenon. She went with her parents to America in 1838 and was exploited as a child prodigy for several years, making her adult début in 1844. She was then seen in roles such as *Scribe and Legouvé's Adrienne Lecouvreur

and the younger *Dumas's Lady of the Camellias, in several plays by Sheridan *Knowles, as *Bulwer-Lytton's The Lady of Lyons, and as *Reade's Peg Woffington. She left the stage on her marriage in 1860, but on her husband General Lander's death in the Civil War returned and continued to act until her retirement in 1877. A small, well-formed woman with a sweet face, clear voice, and graceful figure, she was an actress of great talent and intellectual judgement, but lacked fire.

Lane, Louisa, see DREW, MRS JOHN.

Lane, Lupino [Henry George Lupino] (1892–1959), English actor, nephew of Barry and Stanley *Lupino, who took his stage name from his great-aunt Sara (see BRITANNIA THEATRE). As Nipper Lane he made his first appearance on the stage at the age of 4 and as an adult actor he toured extensively in *variety, *musical comedy, and *pantomime. He made a great success as Bill Snibson in *Me and My Girl*, a musical comedy in which he created the dance known as 'The Lambeth Walk'. In the cast with him was his brother Wallace Lupino, also an actor, dancer, and pantomimist, and his son Lauri, who later went into films. Surviving silent films of Lupino Lane's stage act are a good record of the tumbling and acrobatics of the *harlequinade.

Lane, Sam and **Sara,** see BRITANNIA THEATRE.

Lang, (Alexander) **Matheson** (1879–1948), actor-manager and dramatist, born in Canada but brought up in Scotland. Though he came from a clerical family—his cousin Cosmo Lang became Archbishop of Canterbury—and was himself destined for the Church, his determination to go on the stage was strengthened by visits to the productions of *Benson and *Irving when they toured Scotland. He made his first appearance in Wolverhampton in 1897, and later joined Benson, appearing with him in London in 1900 and going on tour with him to the West Indies. In 1904 he was back in London, where he played under the management of Vedrenne and *Granville-Barker at the *Royal Court Theatre in *Ibsen and *Shaw. He scored his first outstanding success in Hall Caine's *The Christian* (1907) at the *Lyceum, where he also gave good performances as Romeo and Hamlet. He then took his own company to South Africa, Australia, and India, playing Shakespeare and modern romantic drama with much success. On his return to London he appeared in *Mr Wu* (1913), an improbable Anglo-Chinese melodrama by Harry Vernon and Harold Owen, which was seen all over the world and frequently revived. In 1914 Lang, with his wife **Hutin** [Nellie] **Britton** (1876–1965), who had been with him in Benson's company and subsequently toured as his leading lady, inaugurated the Shakespeare productions at the *Old Vic under Lilian *Baylis with *The Taming of the Shrew*, *Hamlet*, and *The Merchant of Venice*. Four years later he was in his own adaptation of a French romantic comedy as *The Purple Mask*, and in 1920 he was seen in E. Temple Thurston's *The Wandering Jew*, which ran for a year and was several times revived. *The Chinese Bungalow* (1925) by Marion Osmond and James Corbet was followed by *Dukes's adaptations of two German works— *Such Men Are Dangerous* (1928), based on Neumann's *Der Patriot*, and *Jew Süss* (1929), based on a novel by Feuchtwanger. In his later years Lang appeared mainly on tour in his most successful parts, and in 1941 he moved to South Africa.

Langham, Michael, see GUTHRIE THEATER, Minneapolis, and STRATFORD (ONTARIO) FESTIVAL.

Langner, Lawrence (1890–1962), successful patent agent in New York who was also a potent force in the American theatre. Born in Wales, he was educated in London and emigrated to the USA in 1911, becoming an American citizen. In 1914 he helped to organize the *Washington Square Players, who produced his first play; he was also instrumental in helping the group to re-form as the *Theatre Guild, of which he became a director, and for which, with the playwright **Theresa Helburn** (1887–1959), he supervised the production of over 200 plays. He was the founder and first president of the *American Shakespeare Theatre at Stratford, Conn., and also founded and directed from 1931 to 1933 the New York Repertory Company, building for it the Country Playhouse, Westport, Conn.

Langtry [*née* Le Breton], **Lillie** [Emilie] **Charlotte** (1853–1929), English actress, daughter of the Dean of Jersey, known from her surpassing beauty as 'the Jersey Lily'. She married at the age of 22 a wealthy Irishman, being an intimate friend of the Prince of Wales (later Edward VII). She was one of the first society women to go on the stage, making her début in 1881 under the *Bancrofts at the *Haymarket Theatre as Kate Hardcastle in *Goldsmith's *She Stoops to Conquer*. She caused a great sensation, more on account of her looks and social position than her acting, which for many years was not taken seriously by the critics. She eventually organized her own company and proved herself a good manageress, playing at the *Imperial and other London theatres and touring the provinces and the United States. Although never a great actress, she became in time a good one, particularly in such parts as Rosalind in *As You Like It* and Lady Teazle in *Sheridan's *The School for Scandal*.

Larivey, Pierre De (c.1540–1612), French dramatist, of Italian extraction—his name is a pun on the family name Giunti, 'the newly arrived'. In 1577 he saw the Italian *commedia dell'arte* company the *Gelosi play at Blois, and inspired by them he wrote nine comedies, six of which were published in 1579, and three in 1611. Though based on Italian models, they are in no sense translations but adaptations which often contain much new material. They were played extensively in the provinces, and also in Paris. The best known, *Les Esprits* (*The Ghosts*), is taken from a comedy by Lorenzino de' Medici, itself based on material from *Plautus and *Terence; in its turn it provided material for both *Molière—*L'École des maris* (1661) and *L'Avare* (1668)—and *Regnard—*Le Légataire universel* (1708). Larivey was the most substantial writer of comedy in France before *Corneille.

Laroche, Johann, see KASPERLE.

La Rue, Danny, see FEMALE IMPERSONATION.

Lashline, see BOX-SET and FLAT.

Late Joys, see PLAYERS' THEATRE.

Laterna Magika, see LIGHTING and SVOBODA.

Laube, Heinrich, see VIENNA.

Lauder, Sir Harry [Hugh MacLennan] (1870–1950), one of the most famous stars of the British *music-hall and the first to be knighted, in 1919. Although he was to become the best-known and loved of the Scottish comedians on the 'halls', he made his début as a singer of Irish songs, first at the Argyle in Birkenhead and then in London in 1900 at *Gatti's Palace of Varieties (Gatti's-Over-the-Water). When in response to a demand for encores he ran out of material, he turned to Scottish songs and found them immediately successful. Thereafter he remained faithful to them, appearing invariably in a kilt and glengarry and carrying a crooked stick. Among his most successful numbers were 'I Love a Lassie', 'Roamin' in the Gloamin'', 'A Wee Deoch-an'-Doris', 'It's Nice to Get up in the Morning', and 'Stop yer Ticklin', Jock!' He was well known in all the 'halls' of London and the provinces, and made numerous tours of the United States, South Africa, and Australia. He also appeared in *revue and in at least one straight play, Graham Moffat's *A Scrape o' the Pen* (1909).

Laughton, Charles (1899–1962), stage and film actor, an American citizen since 1940, but born in England, where he made his first appearance on the stage in 1926 in *Gogol's *The Government Inspector* at the *Barnes Theatre. He then played the title-role in *Mr Prohack* (1927) by Arnold *Bennett and Edward *Knoblock, and in 1931 starred in *Payment Deferred*, based on a novel by C. S. Forester, in London and New York. With his round moon face and bulky frame he was destined for character roles, at which he excelled. His season at the *Old Vic in 1933, when he played Lopakhin in *Chekhov's *The Cherry Orchard*, Henry VIII, Macbeth, Prospero in *The Tempest*, and Angelo in *Measure for Measure*, greatly enhanced his reputation. After his Captain Hook in *Barrie's *Peter Pan* at the *London Palladium (1936) he was seen in 1937 at the *Comédie-Française in *Molière's *Le Médecin malgré lui*, being the first English actor to appear there. He was then seen only intermittently on stage, being engrossed in his outstanding film career, but he returned to the theatre in Los Angeles in 1947 in *Brecht's *Galileo*, having adapted the English text in collaboration with the author. In New York in 1951 he directed and played the Devil in the *Don Juan in Hell* section of *Shaw's *Man and Superman*. He directed his own adaptation of Stephen Vincent Benét's poem *John Brown's Body* (1952), Wouk's *The Caine Mutiny Court-Martial* (1954), and Shaw's *Major Barbara* (1956), in which he also appeared. He was seen once more in London in Jane Arden's *The Party* (1958), which he also directed, and in 1959 played King Lear, and Bottom in *A Midsummer Night's Dream* at the *Shakespeare Memorial Theatre.

His wife **Elsa Lanchester** (1902–86) appeared with him in *Payment Deferred*, *The Tempest*, *Peter Pan* (in the title-role), and *The Party*, and gave the solo performance *Elsa Lanchester—Herself* (1961) directed by her husband.

Laura Keene's Theatre, New York, see KEENE and OLYMPIC THEATRE 2.

Lautenschläger, Karl, see SCENERY.

Lawler, Ray (1921–), Australian dramatist and actor, who had already written several light comedies when in 1949 his *Cradle of Thunder* was awarded first prize in a national competition. In 1955 the Union Theatre Repertory Company (see MELBOURNE THEATRE COMPANY) put on his *Summer of the Seventeenth Doll*. Produced in London (1957) and New York (1958), this story of two cane-cutters whose informal domestic arrangements erupt after 17 years is by far his best-known play. He later completed a 'Doll' trilogy with *Kid Stakes* (1975) and *Other Times* (1977), about the characters' earlier lives. Other plays included *Piccadilly Bushman* (1959), *The Man Who Shot the Albatross* (1971), about Captain Bligh, and *Godsend* (1982).

Lawrence, D(avid) **H**(erbert) (1885–1930), English novelist, poet, and playwright, whose novels, considered shocking and highly controversial when first published, made him famous during his lifetime. Of his eight plays, *The Widowing of Mrs Holroyd* and *David*, a biblical drama, were given Sunday-night productions by the *Stage Society in 1926 and 1927 respectively. The others remained unproduced until in 1965 *A Collier's Friday Night* (written in about 1906) was seen at the *Royal Court Theatre in a 'production without décor', and in 1967 *The Daughter-in-Law* (written during, and based on, the coal strike of 1912) was given a public

showing, also at the Royal Court. It was named as one of the best new plays of the year, and led to a D. H. Lawrence season at the Royal Court in 1968, when the three plays (excluding *David*), done in repertory to emphasize their common setting of family life in a mining background, revealed Lawrence as a starkly realistic playwright of great subtlety and vigour.

Lawrence, Gerald (1873–1957), English actor, who served his apprenticeship with *Benson and later joined Henry *Irving at the *Lyceum. He was playing Henry II the night Irving made his last appearance as Becket (in *Tennyson's play) in 1905. He then went to America, and on his return in 1909 directed a number of Shakespeare's plays at the *Royal Court Theatre; in 1912 he gave an outstanding performance as Brassbound in a revival of *Shaw's *Captain Brassbound's Conversion*. After serving in the Royal Navy during the First World War, he returned to the stage in 1919, playing de Guiche to the Cyrano of Robert *Loraine in *Rostand's *Cyrano de Bergerac*. He also directed and played in Louis N. *Parker's *Mr Garrick* (1922), and in 1923 toured the provinces in a revival of Booth *Tarkington's *Monsieur Beaucaire*. He made his last appearance on the stage in 1938. His first wife was Lilian *Braithwaite. His second wife, who appeared with him in many of his later productions, was **Fay Davis** (1872–1945), an American actress who came to London in 1895 and made a great success as Flavia in Anthony Hope's *The Prisoner of Zenda*.

Lawrence [Klasen], **Gertrude** (1898–1952), English actress, who trained under Italia *Conti and made her first appearance in *pantomime at the age of 12. Continuing her career uninterruptedly, mainly as a dancer and later as a leading lady in *revue, she appeared in several editions of *Charlot's Revue* in London and New York in 1924–5, and was seen as Kay in the musical *Oh, Kay!* (NY, 1926; London, 1927). She made her first outstanding success as Amanda Prynne in Noël *Coward's *Private Lives* (1930; NY, 1931) with Coward himself as her leading man, appearing with him again in his *To-Night at 8.30* (London and NY, 1936). After playing in New York in the latter she remained there for some time, starring in Rachel *Crothers's *Susan and God* (1937), Moss *Hart and Kurt Weill's musical *Lady in the Dark* (1941 and 1943), and a revival of *Shaw's *Pygmalion* in 1945. She returned to London in Daphne Du Maurier's *September Tide* (1948), but was back in New York in 1951 starring in Rodgers and *Hammerstein's *The King and I* when she became ill, dying shortly afterwards.

Lawrence, Slingsby, see LEWES.

Lazzo (pl. *lazzi*), slight piece of byplay indulged in by the comic servants of the *commedia dell'arte*, consisting of a small, unexpected ornamentation by word or gesture superimposed on the main plot. There is no satisfactory English translation of the word, which can be rendered as 'antic', 'trick', 'gambol', 'pun', 'patter', according to the context. The longer comic episode involving a practical joke or some horseplay was known as the *burla.

Leavitt, Michael, see BURLESQUE, AMERICAN.

Leblanc, Georgette, see MAETERLINCK.

Lecouvreur, Adrienne (1692–1730), French actress, who made her first appearance at the *Comédie-Française in 1717 in the title-role of Crébillon's *Électre*. Her immediate popularity aroused much jealousy among her fellow actresses, but she continued to triumph with the public. She was better in tragedy than comedy and, disliking the declamatory style which had come down from Mlle *Champmeslé, she succeeded, in the teeth of their opposition, in introducing a much simpler and more natural form of delivery. Her reign was a brief one, and she died suddenly after only 13 years. As an actress she was refused Christian burial, and was interred secretly by night in a marshy corner of the rue de Bourgogne. The English actress Anne *Oldfield, who died in the same year, was buried in Westminster Abbey. *Voltaire, in some of whose plays Lecouvreur had appeared, contrasted bitterly the respect shown to the English actress with the harsh treatment of her French counterpart.

Lederer Theatre Complex, Providence, RI, see TRINITY REPERTORY COMPANY.

Lee, Canada [Leonard Lionel Cornelius Canegata] (1907–52), American Negro actor, who made his début in 1928 and received stage training with the Negro Unit of the *Federal Theatre Project, appearing in its all-Negro production of *Macbeth* in 1936. In 1939 he played with Ethel Waters in Dorothy Heyward's *Mamba's Daughters*, and he made a major impact as the chauffeur in Paul *Green's outstanding tragedy of Negro life, *Native Son* (1941), based on the novel by Richard Wright. Among other productions in which he appeared were the musical *South Pacific* (1943), Philip Yordan's *Anna Lucasta* (1944), and Dorothy Heyward's *Set My People Free* (1948). In 1945 he won high praise as Caliban in *The Tempest*, directed by Margaret *Webster, and in 1946 played Bosola, in a white make-up, in John *Webster's *The Duchess of Malfi*.

Lee, Gypsy Rose, see BURLESQUE, AMERICAN.

Lee, Nathaniel (c.1653–92), English dramatist, author of a number of tragedies on subjects taken from ancient history, of which the best and most successful was *The Rival Queens; or, The Death of Alexander the Great* (1677), dealing with the jealousy of Alexander's wives Roxana and Statira. It owed much of its initial success to the acting of *Betterton and Mrs *Barry,

and held the stage for over 100 years, being in the repertory of John Philip *Kemble, Edmund *Kean, and Mrs *Siddons. Lee, who collaborated with *Dryden in *The Duke of Guise* (1682), was one of the most popular dramatists of his day and his plays were frequently revived and reprinted. Their ranting verse and plots, which left the stage encumbered with corpses or lunatics, betrayed, however, a streak of insanity which led to his being confined in Bedlam where he died.

Leeds, see CITY PALACE OF VARIETIES and WEST YORKSHIRE PLAYHOUSE.

Le Gallienne, Eva (1899–1991), American actress and director, daughter of the poet Richard Le Gallienne, who as well as playing many of the classical roles made great efforts to increase the audience for serious drama. Born in London, she appeared there in several small parts before going to New York in 1915, where her great success as Julie in *Molnár's *Liliom* (1921) was followed by five years of starring roles on Broadway. In 1926 she founded the Civic Repertory Theatre to present important foreign plays at low admission prices, directing and appearing in many of its productions, playing such roles as Masha in *Chekhov's *Three Sisters* and the White Queen in her own adaptation of Lewis Carroll's *Alice in Wonderland*. Before the enterprise collapsed in the depression of 1933, Le Gallienne had mounted 37 productions, including plays by Shakespeare, *Rostand, *Tolstoy, *Molière, *Goldoni, *Giraudoux, and especially *Ibsen: she also toured the USA in 1934–5 in her own versions of his *A Doll's House*, *Hedda Gabler*, and *The Master Builder*. With Cheryl *Crawford and Margaret *Webster she founded the American Repertory Theatre, active 1946–8, playing Queen Katharine in *Henry VIII*, the initial production, and Ella Rentheim in Ibsen's *John Gabriel Borkman*. From 1961 to 1966 she directed and acted with the National Repertory Theatre, touring productions of classic plays coast to coast and receiving great critical acclaim for her performances in Maxwell *Anderson's *Elizabeth the Queen* and *Euripides' *Trojan Women*. She appeared in New York in 1968 in *Ionesco's *Exit the King*, and two years later was at the *American Shakespeare Theatre in *All's Well that Ends Well*. She starred on Broadway in a revival of *Kaufman and Edna Ferber's *The Royal Family* in 1976, and played the White Queen again in 1983.

Legislation, Theatre, see COPYRIGHT IN DRAMATIC WORKS, DRAMATIC CENSORSHIP, FIRES IN THEATRES, LICENSING OF THEATRES, and LORD CHAMBERLAIN.

Legitimate Drama—sometimes abbreviated to 'legit'—term which arose in the 18th century during the struggle of the Patent Theatres

*Covent Garden and *Drury Lane against the illegitimate playhouses springing up all over London. It covered in general those five-act plays (including Shakespeare's) which depended entirely on acting, with little or no singing, dancing, or spectacle. In the 19th century the term was widely used by actors of the old school as a defence against the encroachments of *farce, *musical comedy, and *revue.

Leg-Show, slang term for a spectacular musical entertainment, largely designed to display chorus-girls in a series of scanty costumes and energetic dances.

Lehmann, Beatrix (1903–79), English actress of great power and intensity. She made her first appearance in 1924, and in 1929 attracted attention as Ella Downey in *O'Neill's *All God's Chillun Got Wings*. Other notable roles were as Susie Monican in *O'Casey's *The Silver Tassie* (also 1929), Emily Brontë in Clemence *Dane's *Wild Decembers* (1933), and Stella Kirby in *Priestley's *Eden End* (1934). In 1936 she consolidated her position as one of London's major actresses with her Winifred in the *Old Vic production of William *Rowley's *The Witch of Edmonton*, which was followed by a number of important productions including O'Neill's *Mourning Becomes Electra* (1937) and *Desire under the Elms* (1940), *Ibsen's *Ghosts* (1943), and Thomas Job's *Uncle Harry* (1944). In 1947 she was at the *Shakespeare Memorial Theatre, where she gave one of her finest performances as Isabella in *Measure for Measure*, also playing Viola in *Twelfth Night*. Returning to contemporary plays, she was seen in Ted Willis's *No Trees in the Street* (1948), *Ustinov's *No Sign of the Dove* (1953), and *Anouilh's *The Waltz of the Toreadors* (1956). She was in Harold *Pinter's first play *The Birthday Party*, played Lady Macbeth at the Old Vic (both 1958), and gave a remarkable performance as the 100-year-old Miss Bordereau in Michael *Redgrave's *The Aspern Papers* (1959), based on a story by Henry *James; but she was thereafter seen less frequently and mostly in supporting roles. Her last appearance on the stage was in T. S. *Eliot's *The Family Reunion* at the *Royal Exchange Theatre in 1979.

Leicester, see HAYMARKET THEATRE.

Leicester's Men, earliest organized company of Elizabethan actors, founded in 1559 and noted as playing at Court a year later. In the company in 1572 was James *Burbage, who built for it the *Theatre, the first theatre in London. Leicester's Men remained in high favour with the Queen from about 1570 until the formation in 1583 of *Queen Elizabeth's Men, into which several of them were drafted. They formed part of the Earl of Leicester's household and continued to act under his patronage until his death in 1588 when they were amalgamated

with the provincial company belonging to the household of Lord Strange, later the 6th Earl of Derby.

Leigh, Vivien [Vivian Mary Hartley] (1913–67), English actress, who first came into prominence as Henriette in *The Mask of Virtue* (1935), an adaptation by Ashley *Dukes of Carl *Sternheim's *Die Marquise von Arcis*. In 1937 she played Ophelia to the Hamlet of Laurence *Olivier at Elsinore and Titania in *A Midsummer Night's Dream* to Robert *Helpmann's Oberon at the *Old Vic. In 1940 she made her New York début as Juliet to the Romeo of Olivier, whom she married the same year as her second husband. Back in London, she gave excellent performances as Jennifer Dubedat in *Shaw's *The Doctor's Dilemma* in 1942, and as Sabina in *Wilder's *The Skin of Our Teeth* (1945). In 1949 for the Old Vic she played Antigone in *Anouilh's play, which, with her Blanche Du Bois in Tennessee *Williams's *A Streetcar Named Desire* in the same year, established her as a powerful stage actress. In 1951 she appeared with Olivier in Shaw's *Caesar and Cleopatra* and Shakespeare's *Antony and Cleopatra* in London and New York. She again appeared with Olivier in *Rattigan's *The Sleeping Prince* (1953), and two years later was with him at the *Shakespeare Memorial Theatre, where she played Lady Macbeth, Viola in *Twelfth Night*, and Lavinia in *Titus Andronicus*, the last being seen in London in 1957. She had further successes in *Coward's *South Sea Bubble* (1956) and *Look after Lulu* (1959) and *Giraudoux's *Duel of Angels* (1958; NY, 1960). In 1963 she was seen in New York in a musical version of *Sherwood's *Tovarich*; her last appearance, also in New York, was in 1966 in *Chekhov's *Ivanov*. She also had an outstanding career in films.

Leighton, Margaret (1922–76), stylish and elegant English actress, who first appeared at the *Birmingham Repertory Theatre in 1938 and later joined the *Old Vic company in London. She emerged as a leading actress with her Celia Coplestone in T. S. *Eliot's *The Cocktail Party* (1950), her Masha in *Chekhov's *Three Sisters* in 1951, and a season at the *Shakespeare Memorial Theatre (1952), where she played Lady Macbeth, Ariel in *The Tempest*, and Rosalind in *As You Like It*. In a revival of *Shaw's *The Apple Cart* in 1953 she played Orinthia to the King Magnus of Noël *Coward, and in the same year was seen as Lucasta Angel in Eliot's *The Confidential Clerk*. She was much admired in *Rattigan's *Separate Tables* (1954; NY, 1956), and in the same author's *Variations on a Theme* (1958). After her Beatrice in *Much Ado about Nothing* to the Benedick of John *Gielgud in New York in 1959, she returned to London in John *Mortimer's *The Wrong Side of the Park* (1960) and in a revival of Ibsen's *The Lady from the Sea* (1961). Back in New

York she gave a fine performance as Hannah Jelkes in Tennessee *Williams's *The Night of the Iguana* (also 1961). She remained in the United States for some years, appearing in *Billetdoux's *Chin-Chin* (1962), Enid *Bagnold's *The Chinese Prime Minister* (1964), and a revival in 1967 of Lillian *Hellman's *The Little Foxes*. Returning to England, she was seen at *Chichester in 1969 as Cleopatra in *Antony and Cleopatra* and again in 1971 as Mrs Malaprop in *Sheridan's *The Rivals* and Elena in *Sherwood's *Reunion in Vienna*, repeating the latter part in London a year later. She made her last appearance in Julian Mitchell's dramatization of Ivy Compton-Burnett's novel *A Family and a Fortune* (1975).

Leivick, Halper, see JEWISH DRAMA.

Lekain [Caïn], Henri-Louis (1729–78), French actor, who in 1748 was seen by *Voltaire while playing the leading role in an amateur production. Much impressed by the excellence of his acting, Voltaire befriended him, and eventually built him a theatre in which he himself played with his two nieces. It was immediately successful and Lekain was accepted into the company of the *Comédie-Française, playing Titus in a revival of Voltaire's *Brutus*. After a slow start he became immensely popular, and was frequently compared to *Garrick (his grandfather was English). He instigated many reforms in the theatre, notably the suppression of seats on the stage and the introduction of some trace of historical costume, in which he was nobly supported by Mlle *Clairon. Together they did away with hip-pads and panniers, and in 1755 introduced some touches of *chinoiserie* into the costumes for Voltaire's *Orphelin de la Chine*. Lekain also insisted on more mobility on stage, doing away with the old tradition of delivering long speeches down stage centre front. After giving a magnificent performance as Vendôme in Voltaire's *Adélaïde du Guesclin*, he went out into the chill night air, caught cold, and died just as his benefactor and greatest admirer was returning to Paris after 30 years of exile. The news of his funeral was the first thing Voltaire heard on his arrival.

Lelio, see ANDREINI.

Lemaître, A.-L.-P., see FRÉDÉRICK.

Leno, Dan [George Galvin] (1860–1904), one of the best loved and most famous stars of the English *music-hall and *pantomime. Born in London of itinerant entertainers, he made his first appearance on the stage at the age of 4, and for many years played in the provinces in *slapstick sketches with 'the Leno family'—the stage name of his stepfather, whose real name was Wilde. He later did a double dancing act with his elder brother Jack and then with his young uncle **Johnny Danvers** (1860–1939), later a Mohawk Minstrel and a player in pantomime. Working mainly in the north of England, he specialized in clog-dancing and in 1880, at Leeds, won the World Championship. Returning to London in 1885 with his young wife, a 'comedy vocalist' whom he married in 1883 and by whom he had six children, he found that his dancing was of no account, whereas his comic songs and patter were rapturously received. He therefore abandoned the clogs to concentrate on comic songs, and was soon playing at several 'halls' every night. He first appeared in pantomime in 1886, as Jack's mother in *Jack and the Beanstalk*. Two years later he was at *Drury Lane as the Baroness in *Babes in the Wood*. He was then seen in every Drury Lane pantomime up to 1903, when he made his last appearance as Queen Spritely in *Humpty-Dumpty*, partnered by **Harry Randall** (1860–1932), who was to prove a worthy successor to him at Drury Lane. Leno had previously been partnered by another music-hall comedian, **Herbert Campbell** [Story] (1844–1904), whose vast bulk proved the perfect foil to Leno's diminutive quicksilver figure. The pantomimes usually ran until the spring, and during the rest of the year Leno, in common with other pantomime stars, returned to the 'halls'. He created a wide range of music-hall characters, two of the most popular being the Shop-Walker and the Beefeater, though some audiences preferred the Railway Guard. With his husky voice and worried little face, he always remained a man of the people, carrying the stamp of the poverty and privations from which he had emerged. He usually made his entrance in a rush, stopping suddenly and darting suspicious looks at the audience while starting a song, and then leaving it to indulge in long, rambling monologues, with muttered asides and sudden bursts of step-dancing. He worked alone, without much in the. way of props, and built up his characters with brilliant use of mime and rapid gestures. In 1901 he was commanded by Edward VII to Sandringham, which earned him the nickname of 'the King's Jester'. Towards the end of his life he broke down from overwork and died insane. Three of his children went on the 'halls', **Sidney Paul Galvin** (1891–1962), who looked very like him, using some of his material and billing himself as Dan Leno junior.

Lenoble, Eustache (1643–1711), French lawyer who was imprisoned for forgery, and later became a literary hack. Among his miscellaneous works were some unsuccessful plays, including three produced at the Hôtel de *Bourgogne. One of these, *La Fausse Prude*, was taken as a reflection on Mme de *Maintenon and was used by the authorities as an excuse for the banishment of the Italian troupe from Paris in 1697.

Lenoir, Marie, see DANCOURT.

Lenormand, Henri-René (1882–1951), French dramatist, who in reaction against the realism of *Antoine based his own plays on Freud, all his work being marked by a vision of man as the victim of a fate determined in infancy and by the force of his own unconscious. His first play, *Le Temps est un songe* (1919), was produced by *Pitoëff in Geneva and then Paris, and was seen in New York in 1924 (London, 1950) as *Time is a Dream*. Pitoëff also produced *Les Ratés* (1920; NY, as *The Failures*, 1924) and *Le Mangeur des rêves* (1922; London, as *The Eater of Dreams*, 1929). The last was seen at London's *Gate Theatre, which in 1927 had given a Sunday-night performance of *Simoun*, usually considered Lenormand's best play, and first staged by *Baty at the Théâtre Montparnasse in 1920.

Lensky, Alexander, see MALY THEATRE, Moscow.

Lenya, Lotte, see BRECHT and MUSICAL COMEDY.

Leonard, Hugh [John Keyes Byrne] (1926–), Irish dramatist and critic, whose first play *The Italian Road* (1944) was produced by an amateur company in Dublin. His second, *The Big Birthday* (1956), was seen professionally at the *Abbey Theatre, as was *A Leap in the Dark* (1957). For the *Dublin Theatre Festival (of which he was director in 1978) he wrote *The Passion of Peter Ginty* (1961), based on *Ibsen's *Peer Gynt*, and two plays based on the works of James Joyce: *Stephen D.* (1962; London, 1963; NY, 1967) and *Dublin One* (1963). Later plays included *The Poker Session* (also 1963; London, 1964; NY, 1967), in which a young man is released from a mental hospital and tries to discover why he was sent there; *The Au Pair Man* (1968; London, 1969; NY, 1973), a two-character allegory about Britain and Ireland; *The Patrick Pearse Motel* (Dublin and London, 1971); and *Da* (1973; London and NY, 1978), an autobiographical play in which a middle-aged man remembers his father on the day of the latter's death. In *Da*'s companion piece, *A Life* (1979; London, with Cyril *Cusack, and NY, 1980), one of the characters with six months to live assesses his life. *The Mask of Moriarty* (1985) is based on Conan Doyle's characters. Leonard's technical facility is sometimes given free rein at the expense of characterization; he is by his own admission obsessed with the theme of betrayal, frequently presented through clever satires on the Irish *nouveaux riches*.

Leonidov, Leonid Mironovich (1873–1941), Russian actor and director, who joined the Korsh Theatre in Moscow in 1901 and in 1903 became a member of the company at the *Moscow Art Theatre. His most brilliant performance as a tragic actor was as Dmitri Karamazov in a dramatization of Dostoevsky's

The Brothers Karamazov. Among his other outstanding roles were *Ibsen's Peer Gynt, Lopakhin in *Chekhov's *The Cherry Orchard* and Solyony in his *Three Sisters*, Professor Borodin in *Afinogenov's *Fear*, and Plushkin in the dramatization of *Gogol's *Dead Souls*. At the time of his death he was engaged with *Nemirovich-Danchenko on the production of *Pogodin's *The Kremlin Chimes*.

Leonov, Leonid Maximovich (1899–), Soviet dramatist, whose *Untilovsk* (1928), one of the first Soviet plays to be produced by the *Moscow Art Theatre, was not entirely successful, though its successor *Skutarevski* (1934), based on his own novel and produced at the *Maly Theatre, was warmly received; it deals with the problems of an elderly scientist torn between his work and his family, and between the old and new régimes. Among Leonov's later plays was the Chekhovian *The Orchards of the Polovtsi* (1938), which as *The Apple Orchards* was produced at the *Bristol Old Vic in 1948. *The Wolf* (1939), about the impact of the Soviet régime on personal problems, was well liked; but his reputation was established by *Invasion* (1942), which recounts with great force and pathos the story of a Soviet village under Nazi rule. In 1957 *Gardener in the Shade* was produced at the *Mayakovsky Theatre by *Okhlopkov, and a revised version of an earlier play, *Golden Chariot*, was seen at the Moscow Art Theatre. *The Snowstorm*, first written in 1939 but officially banned, was revised and finally published in 1963.

Leopoldstädter Theater, see MARINELLI and VIENNA.

Léotard, Jules, see MUSIC-HALL.

Lermontov, Mikhail Yurevich (1814–41), Russian lyric poet and the author of three romantic plays, most of whose work fell foul of the censor. His first play *The Spaniards*, written in 1830, deals ostensibly with the Spanish Inquisition but is in reality aimed at the despotism of the Tsar; it was banned, not to be performed in Russia until after the Revolution in 1917. His next, written two years later, was given the German title *Menschen und Leidenschafter* to indicate its kinship with the *Sturm und Drang* movement and particularly with the works of *Schiller. Based on a family conflict which recalls Lermontov's own unhappy home life, it is again an indictment of contemporary society. In a later rewritten version Lermontov gave it a wider application, with less stress on the family and more on the struggles of the rebellious hero confronting a hostile world. His most important play and the only one by which he is now remembered is *Masquerade*, written in 1835. Showing clearly the influence of Shakespeare—the theme is very similar to that of

Othello—and **Byron, it depicts the tragedy of a man who murders the wife he loves because he suspects her of infidelity. In deference to the censor the play was later given a happy ending, but it was not produced until 1852 and then only in a mutilated text. The full version was first seen in 1864, and then not again until **Meyerhold produced it in 1917 at the Alexandrinsky, where it was the last play to be performed before the October Revolution. It now forms part of the permanent repertory of the Soviet theatre. In a translation by Robert David MacDonald it was seen at the **Citizens' Theatre in Glasgow in 1976.

Lerner, Alan Jay, see MUSICAL COMEDY.

Lesage, Alain René (1668–1747), French novelist and dramatist, orphaned when young and left penniless. Little is known of his early years; by 1694 he was married and established in Paris. His brilliant literary career began with translations from the Spanish and it was under the influence of Spanish literature in general that he wrote the two novels which constitute his main claim to fame: *Le Diable boiteux* (1707) and *Gil Blas* (1715). Though pre-eminently a novelist, Lesage is by no means negligible as a dramatist. His first success in the theatre was a one-act comedy, *Crispin rival de son maître* (1707). Two years later a one-act play, *Les Étrennes*, which the actors had refused, was remodelled as *Turcaret* (1710), one of the outstanding comedies in the history of French drama. It is a satire on the vulgar parvenu, the tax-farmer who has enriched himself by exploiting the poor, and it reflects that bitterness against taxation which came to a head under Louis XVI. It met with considerable opposition, and those whom it satirized tried to bribe the actors to suppress it; but it was eventually put on, and was not only successful at the time but has remained in the repertory. It was produced, for instance, in 1960 by Jean **Vilar for the **Théâtre National Populaire, with a musical accompaniment by Duke Ellington. A quarrel between Lesage and the actors of the **Comédie-Française led him to break off his association with the official theatre, and for many years he wrote farces for the theatres of the Paris **fairs, alone or in collaboration.

Lescarbot, Marc (*c*.1570–1642), French lawyer, author of the first play to be performed in Canada, which he directed himself. This was a marine masque, *Le Théâtre de Neptune en la Nouvelle France*, seen in 1606 at Port Royal, Acadia (today Annapolis Royal, Nova Scotia). Lescarbot had been persuaded by a friend to accompany him to New France. He served there as an instructor in religious matters, and was left in charge of the settlement when his friend and a companion went on an exploratory trip to the south. He prepared the masque to welcome them home. Subsequently he returned to France to resume his legal career. In 1609 he published a collection of poems and writings, *Les Muses de la Nouvelle France*, that contained the script of *Le Théâtre de Neptune*, which in 1611 was translated into English and published in London.

Lessing, Gotthold Ephraim (1729–81), German playwright and dramatic critic, the opponent of the reforms of **Gottsched. In 1767 he became official critic to **Ackermann's short-lived National Theatre, for which he published the *Hamburgische Dramaturgie*, advocating a theatre based on Shakespeare rather than on the fashionable French classic style. His own works were more akin to those of **Dryden and the novelist Richardson, whose influence shows clearly in his first major play, *Miss Sara Sampson* (1755), a tragedy. His early plays had been light comedies in verse, but *Minna von Barnhelm* (1767) is an admirable prose comedy, and *Emilia Galotti* (1772) a second tragedy. Towards the end of his life Lessing engaged in a prolonged struggle against narrow-minded orthodoxy, which finds expression in his final work, *Nathan der Weise*, a noble plea for religious tolerance, written in blank verse. First produced two years after his death, its merits were not recognized until **Goethe revived it at **Weimar in 1801. Translated into many languages, it was given its first professional production in English in 1967 at the **Mermaid Theatre.

Le Tourneur, Pierre, see SHAKESPEARE IN TRANSLATION.

Levy, Benn W., see CUMMINGS.

Lewes, George Henry (1817–78), philosopher, dramatist, and dramatic critic. He was the grandson of the actor (Charles) **Lee Lewes** (1740–1803), who created the role of Young Marlow in **Goldsmith's *She Stoops to Conquer* (1773), and was himself an amateur actor, appearing with a group formed by Charles **Dickens in such parts as Sir Hugh Evans in *The Merry Wives of Windsor* and Old Kno'well in **Jonson's *Every Man in His Humour*. In 1849 he appeared in Manchester as Shylock in *The Merchant of Venice* with some success, anticipating **Irving's conception of the part as 'a noble nature driven to outlawry by man'. He also played the chief part in his own play *The Noble Heart* when it was produced in Manchester in 1849 before its appearance in London, where he does not appear to have acted professionally. Under the pseudonyms of Slingsby Lawrence and Frank Churchill he wrote several other plays, of which *The Game of Speculation* (1851) was based on Balzac's *Mercadet*. *A Chain of Events* (1852) and *A Strange History in Nine Chapters* (1853) were written for and in collaboration with the younger Charles **Mathews and produced at the **Lyceum, as were all Lewes's other plays except *Buckstone's Adventure* with a *Polish Princess* (1855), seen at the *Haymarket, and *Stay at Home* (1856), produced at the *Olympic. Lewes, who in 1854 began a lifelong liaison with the novelist George Eliot, was the founder and editor of the *Leader*, for which as 'Vivian' he wrote dramatic criticism from its foundation in 1850 to 1854, sometimes reviewing his own plays. He particularly disliked Charles *Kean, whom he harried whenever possible, but was one of the first to recognize the genius of Henry Irving.

Lew Fields Theatre, New York, see WALLACK'S THEATRE 2.

Lewis, Mabel Terry-, see TERRY-LEWIS.

Lewis, Matthew Gregory (1775–1818), English novelist and dramatist, usually known as 'Monk' Lewis from the title of his most famous novel, *Ambrosio; or, The Monk* (1795). This provided material for a number of sensational plays, which, together with *The Castle Spectre* (1797) and *Timour the Tartar* (1811), found their way into the repertory of the 19th-century *toy theatre. Lewis's work, most of which is now forgotten, was deliberately concocted to appeal to the prevailing taste for *melodrama and spectacle. Somewhat crude but offering great scope for effective acting and lavish scenery enhanced by incidental music, it shows very clearly the influence of *Kotzebue.

Lewisohn, Alice and **Irene,** see NEIGHBORHOOD PLAYHOUSE.

Leybourne, George [Joe Saunders] (1842–84), famous *music-hall star. He gained his initial experience in the 'free-and-easys' of provincial and East End taverns, and after appearing in one or two minor London halls was engaged for the *Canterbury, probably about 1865. Tall and handsome, always immaculately dressed as a 'man-about-town', with monocle, whiskers, and a fur-collared coat, he represented the second generation of music-hall entertainers, whose 'heavy swell' was a complete contrast to the working-class characters of their predecessors. Unfortunately Leybourne's nickname of 'Champagne Charlie' was well deserved, and his last years were a constant struggle with disillusionment and ill health. He made his last appearance at the Queen's in Poplar shortly before his death.

Liberty Theatre, New York, on the south side of 42nd Street, between 7th and 8th Avenues. This theatre was for a long time famous for its spectacular farces. It opened in 1904, and among its early productions was the great horse-racing drama *Wildfire* (1908), written by George Broadhurst and George Hobart for Lillian Russell. The season of 1909–10 included two important productions, *Tarkington's *Springtime* and the famous musical comedy *The Arcadians*, seen in London the previous year. A further successful importation

from England was *Milestones* (1912), by Arnold *Bennett and Ernest *Knoblock. The great Negro musical *Blackbirds of 1928* had part of its long run at this theatre, and the last production was given there in 1933, after which the building became a cinema.

Library Theatre, Manchester. In 1934 the basement of the Public Library was adapted by the City Council to serve as a lecture hall seating 308, and from 1946 it was used for a varied programme of entertainments including drama. In 1952 the Libraries Committee formed its own resident repertory company which opened with *Wilde's *The Importance of Being Earnest*. In 1971 a second theatre, the Forum, was opened at Wythenshawe eight miles away as part of a leisure complex. It seats 483 and is run in tandem with the Library Theatre under one director: but after an initial period during which productions at the two theatres were interchangeable they now house separate productions. The Library concentrates on modern plays and the Forum on more family-orientated shows and musicals, notably the European première of *Sondheim's *Pacific Overtures*. Costumes, props, and sets for both theatres are made at Wythenshawe. This is the only theatrical enterprise in Britain to be operated directly by a local authority.

Libretto, see BOOK.

Licensing of Theatres. Stage plays in Britain may not be publicly performed except on premises licensed as suitable for the purpose. Under the Theatres Act of 1843 the monopoly of the *Patent Theatres was abolished, and the powers of the *Lord Chamberlain to license theatres (as distinct from his responsibility for the *censorship of plays) were limited to London, Brighton, Windsor, and other places of royal residence. Detailed requirements for licensing were evolved, mainly relating to the safety and comfort of the public. These regulations were confirmed by the Theatres Act of 1968, which transferred the licensing authority for the central London area to the Greater London Council. The two former Patent Theatres, *Covent Garden and *Drury Lane, although nominally exempt from the need to be licensed by the GLC, were made subject to the same requirements as other London theatres. Licensing authorities are forbidden under the Act to stipulate conditions concerning the nature of the plays performed or the manner of their performance. After the abolition of the GLC in 1986 authority devolved to individual boroughs.

LIFT, see LONDON INTERNATIONAL FESTIVAL OF THEATRE.

Lighting. The open-air performances in Greece, Rome, medieval Europe, and Elizabethan England—also in Spain's Golden Age— took place by daylight, with torches used for the final scenes. It was only in the enclosed private playhouses, which originated in Renaissance Italy and gradually spread across Europe, that artificial lighting was needed, and even there large windows could throw light on to the stage for afternoon performances. However, the value of artificial light in the theatre was soon apparent, particularly as an additional element in the stage picture, and as early as 1545 *Serlio advised placing candles or lamps behind coloured glass in windows, or behind bottles filled with coloured liquid to give a jewel-like effect. He also advocated the use of barbers' basins behind the lamps to act as reflectors. The auditorium was at first brightly lit by lamps or candelabra. In the 1560s the Jewish dramatist **Leone di Somi** (1527–92), in his *Dialogues on Stage Affairs*, insisted on reducing the amount of light in the auditorium in order to intensify the effect of the stage lighting. He also used mirror reflectors behind lamps fixed to the backs of the side wings, and extinguished many of the stage lamps at the first tragic moment in a play, which apparently had a profound effect upon the spectators. A later writer on the theatre, **Angelo Ingegneri** (c.1550–c.1613), wanted to darken the auditorium entirely, but this was not achieved for a very long time, possibly because a fashionable audience wanted to be seen as well as to see. The theatres of the 17th and 18th centuries continued to use candelabra hung over the stage and auditorium, with concealed lamps behind the wings and below the backcloth, and also footlights which, as can be inferred by a reference to them in the *Architectura civilis* of the German architect **Josef Furtenbach** (1591–1667), were in use by 1628, a practice further confirmed by the Italian architect **Nicola Sabbattini** (1574–1654) in his book *Pratica di fabricar scene e machine ne'teatri* (1638). Footlights are first shown in use in an English theatre in the frontispiece to Kirkman's *The Wits* (1672). The unconcealed chandeliers over the stage threw a painful glare into the eyes of the spectators in the gallery, as *Pepys recorded as early as 1669, and *Garrick at *Drury Lane in 1765, in imitation of the French theatre, was the first in England to remove them entirely, relying on extra lamps concealed behind the wings, in addition to the footlights, to give a sufficiently strong light on stage. The same improvement was made at *Covent Garden, but with rather smelly oil-lamps instead of wax candles. Gas was first used to light the stage at the *Lyceum in Aug. 1817. The *Olympic in 1815 was possibly the first theatre to use gas inside the playhouse, but the first London theatre to be entirely given over to gas-lighting was Drury Lane in Sept. 1817. Covent Garden followed two days later. The new lighting was considered a great improvement, and in the next 10 years practically all the more important theatres in

London and the provinces were converted to it, the *Haymarket being one of the last (1843). Some playgoers regretted the loss of wavering candlelight, just as some later audiences preferred the soft glow of gas to the hard brilliance of electricity. Candles, oil, and gas were all equally lethal (see FIRES IN THEATRES), but at least gas could be more easily controlled at the source, thus allowing the creation of beautiful effects of sunrise, twilight, and moonshine. Yet there were complaints that in general gas-lighting was too bright, the steady glare being fatal to the stage illusion. This was partly offset by sinking the footlights below the stage level, as planned by *Fechter in 1863, and by the reforms of *Irving at the Lyceum. Among other innovations, he arranged for all the lights to be regulated from the prompt corner and for the first time darkened the auditorium throughout the performance, as Ingegneri had suggested nearly 300 years earlier. Associated with the use of gas-lighting is limelight, which lived on into the age of electricity. The lime, or calcium flare, first used in 1816, gave a brilliant white light much used for 'realistic' beams of sun, moon, or lamp light through doors or windows. But its main use was to spotlight the chief actor and follow him about on stage, whence the expression 'in the limelight'. Electric arc lights were in use as early as 1846 at the Paris Opéra, and were first used throughout a theatre at the Paris Hippodrome in 1878. Carbon arcs gave brilliant illumination but were noisy and apt to flicker, and limelighting continued in use in many theatres until the invention of the incandescent bulb. The first American theatre to be lit by electricity was the California in San Francisco in 1879, and the first English theatre was the *Savoy in 1881. Although the Paris Opéra was not completely lit by incandescent bulbs until 1886, electricity was installed in most theatres in Europe and America by that time.

Electrification led to the development of modern stage lighting. German theatres pioneered the use of the *Kuppelhorizont* invented by **Mariano Fortuny** (1871–1949) in 1902. This 'sky-dome' of coloured silk reflected the beams of high-powered arcs down on to the stage, giving a bright, diffused light; it was expensive in terms of current and was replaced by the *Rundhorizont*, a half-dome of silk or plaster which developed into the *cyclorama. Adolphe *Appia and Gordon *Craig did their most innovative work in Germany. They rejected the conventions of painted scenery in favour of a three-dimensional setting: the flat illumination of the late 19th-century stage had consequently to be replaced by directional lighting, casting shadows in the manner of real light to convey the mass and moulding of both actor and set. Appia insisted on the importance of lighting as an integral part of a unified artistic conception, as did Craig, who used changing lights against severely simple settings to build up an effective non-naturalistic stage picture. Special effects in the field of lighting can call on a multiplicity of techniques. Apart from imaginative colour changes, one of the simplest uses the Gobo, a cut-out mask slotted into a lantern housing to throw the shadow of leaves, architectural units, or other off-stage features on to the scene. Shadow projection, by the direct-beam projector, is a refinement of the technique introduced in 1916 by **Adolph Linnebach** (1876–1963) at the Court Theatre, Dresden. The Linnebach and direct-beam projectors create magical effects by means of shadowgraphs. Optical lens projectors, or stereopticons, can create up to a complete setting using large format glass slides specially painted or photographed for the purpose. Moving effects such as clouds, snow, or flames are achieved by effects projectors (sciopticons in America) carrying slides in a motor-driven turntable or drum. Such effects can be particularly successful when projected on to a gauze *cloth, or unbroken stretch of fine net, in front of the action. Lighting units and dimmers have become increasingly refined and easy to operate. The spotlight, precise and economical, is the most important element in lighting, floodlighting being seldom used. Remote control was introduced in the 1930s. In the 1940s some theatres used light consoles, resembling organs, and in the 1960s lighting systems became computerized. Film projection has a place in stage production and has been used to great acclaim in the Laterna Magika of Prague, devised by Josef *Svoboda, where slide and film projection are wittily combined with a few live performers to convey the illusion of a stage packed with scenery and action. Ghostly effects can be created with ultra-violet light, under which only objects coloured by fluorescent paint or dye can be seen. The pulsing flashes of strobe (stroboscopic) light can 'freeze' action into a series of jerks; the technique became popular due to the disco vogue and can be powerful when used with discretion. Stage lighting has achieved great subtlety in suggesting mood and scene, creating both natural and non-naturalistic effects. The lighting designer plays a vital role in the modern theatre.

Lilian Baylis Theatre, see SADLER'S WELLS.

Lillie, Beatrice Gladys [Lady Robert Peel] (1898–1989), Canadian comedienne, born and educated in Toronto, but equally well known and admired in Britain and the USA. She made her first appearance in England in 1914, at the Chatham Music Hall, and in the same year was at the *London Pavilion. She then played in a series of *revues, and in 1924 made her début in New York in a production by André Charlot, alternating between London and New York in his revues until 1926. In 1928 she was in New York in *Coward's revue This Year of Grace,

which had a long run. In 1932 she played the straight part of the Nurse in the première in New York of *Shaw's Too True to be Good, but she was later seen in several more revues: At Home Abroad (1935), The Show is On (1936), and Coward's Set to Music (1939), all in New York, and All Clear (also 1939) in London. During the Second World War she entertained the troops tirelessly with a programme of revue sketches and one-act plays, including Coward's To-Night at 8.30, returning to New York to star in the revues The Seven Lively Arts (1944) and Inside USA (1948). From 1952 to 1956 she toured the world in a solo entertainment, An Evening with Beatrice Lillie. In London in 1958 she gave a superb performance in Jerome Lawrence and Robert E. Lee's adaptation of Patrick Dennis's novel Auntie Mame, a part she had already played in New York. Her last appearance, in New York in 1964, was as Madame Arcati in High Spirits, a musical version of Coward's Blithe Spirit. Past mistress of the elegant double entendre and tongue-in-cheek riposte, heightened by the manipulation of an enormously long cigarette-holder, she was described in her heyday as the funniest woman on earth.

Lillo, George (1693–1739), English dramatist, best remembered for his play The London Merchant; or, The History of George Barnwell (1731). Based on an old ballad, it tells how a good young man's passion for a bad woman leads him to murder his old uncle for money, the murderer and his accomplice being subsequently hanged. It was immensely successful and was frequently revived, notably by Mrs *Siddons. It was known well enough to be the butt of several burlesques, and was also the play performed by the Crummles family in When Crummles Played (1927), based by Nigel *Playfair on *Dickens's novel Nicholas Nickleby. It had a great vogue on the Continent where it influenced the development of domestic tragedy, particularly in Germany. Lillo wrote several other plays, of which the most important was Guilt Its Own Punishment; or, Fatal Curiosity (1736), also based on an old ballad about a murder done in Cornwall. First produced at the *Haymarket Theatre by Henry *Fielding, it was chosen by Mrs Siddons for her *benefit in 1797, her brothers Charles and John Philip *Kemble appearing with her. This play also had a great influence on the German *fate drama, inspiring *Werner's Der 24 Februar (1810). Lillo also wrote a version of the anonymous Arden of Feversham, produced in 1759.

Limelight, see LIGHTING.

Lincoln Center for the Performing Arts, see VIVIAN BEAUMONT THEATRE.

Lincoln Cycle, see MYSTERY PLAY.

Lincoln's Inn Fields Theatre, London, in Portugal Street. This was originally Lisle's Tennis-Court, built in 1656. It was leased in

1660 by Sir William *Davenant, who enlarged it for use as a theatre, making it the first playhouse in England to have a proscenium arch behind the apron stage. It opened in 1661 with Davenant's own play *The Siege of Rhodes, Part I*, the second part appearing shortly afterwards. Thomas *Betterton, who was to be closely associated with this theatre, made his first appearance there playing Hamlet. There was also a revival of *Romeo and Juliet*, with the original ending one day and the alternative happy ending by James Howard the next. Among the outstanding plays first seen at this theatre were Tuke's *The Adventures of Five Hours* (1663), based on *Calderón, and *Dryden's *Sir Martin Mar-all* (1667), in which the comedian *Nokes scored a great success. Davenant died in 1668, and his widow, with the assistance of Betterton, kept the theatre going until *Dorset Garden, the new playhouse begun by her husband, was completed. The company gave its last performance at Lincoln's Inn Fields Theatre in 1671. Two months later *Killigrew took over the empty building, following a fire which had destroyed the first *Drury Lane Theatre, and remained there until 1674, after which the building reverted to its former use as a tennis-court, until in 1695 Betterton, who had seceded from the United Company formed by the amalgamation of the actors at Drury Lane and Dorset Garden, took a company which included Elizabeth *Barry and Anne *Bracegirdle to the old theatre. He financed the restoration of the building by public subscription, and reopened it with *Congreve's new play *Love for Love*. Though somewhat handicapped by the smallness of the stage and the limited accommodation in the two-tier auditorium, the company remained in occupation for 10 years, moving in 1705 to the new Queen's Theatre in the Haymarket, built by *Vanbrugh. The old building then ceased to be used as a theatre until in 1714 Christopher *Rich took it over and put in hand extensive renovations, dying before they were completed. It was left to his son John *Rich to finish the alterations, which gave the theatre a handsome auditorium seating more than 1,400 spectators and lighted by six overhead chandeliers, and a stage, with mirrors each side, larger than that at Drury Lane. All the scenery was new. The opening production was a revival of *Farquhar's *The Recruiting Officer*, and in 1728 came the first night of Rich's most important new production, *Gay's *The Beggar's Opera*. It was also at this theatre that Rich first appeared as *Harlequin. In 1732, for reasons that are not yet fully understood, Rich undertook the building of a new theatre in *Covent Garden, and moved there in the autumn of that year. Lincoln's Inn Fields was then used mainly for music and opera, except in the season of 1736–7 and again in 1742–3 when Giffard was there after the

closure of *Goodman's Fields Theatre. The final performance took place in 1744, and the old theatre then became, among other things, a barracks, an auction room, and finally the Salopian China Warehouse. It was pulled down in 1848.

Lincoln's Men, small company of Elizabethan actors who were in the service of the first Earl of Lincoln and of his son Lord Clinton, whose name they sometimes took. They appeared at Court before Queen Elizabeth I several times between 1572 and 1575 and were active in the provinces up to 1577, as was a later company of the same name from 1599 to 1610.

Lindberg, August (1846–1916), Swedish actor, whose travelling company did much to bring *Ibsen to provincial audiences all over Scandinavia. He was the first actor to play Oswald in *Ghosts* in 1883 and among his other Ibsen roles were Brand, Peer Gynt, John Gabriel Borkman, and Solness in *The Master Builder*. He also appeared as Hamlet and King Lear. From 1906 he worked with the *Royal Dramatic Theatre in Stockholm.

Lindsay, Howard (1889–1968), American actor, dramatist, and director, who already had many successes to his credit when he collaborated with **Russel Crouse** (1893–1966) in a dramatization of Clarence Day's book *Life with Father*. Produced in 1939, with Lindsay as Father, this ran for seven years. They continued their collaboration in the *Pulitzer Prize-winner *State of the Union* (1945) and a dramatization of Day's *Life with Mother* (1948), which also had a long run, Lindsay again playing Father.

Lindstrom, Erik (1906–74), Swedish actor, who after training at the *Royal Dramatic Theatre's drama school joined the company there and remained with it during his whole career, making his début in 1927 in *Shaw's *You Never Can Tell*. For a time he specialized in comedy, one of his earliest successes being Sir Peter Teazle in *Sheridan's *The School for Scandal*. He was also much admired in a revival of *Holberg's *Den stundesløse* (*The Fussy Man*) in 1968. But even in 1929 he had shown that he could play serious drama, appearing as Stanhope in *Sherriff's *Journey's End* and later in such parts as *Ibsen's Peer Gynt, *Goethe's Faust, *Pirandello's Enrico IV, and Shakespeare's Macbeth and Iago (with Mogens *Wieth as Othello). In 1948 he appeared at Elsinore as Hamlet. One of his outstanding successes in later years was the title-role in *Strindberg's *Gustav III*.

Linnebach, Adolph, see LIGHTING.

Lisle's Tennis-Court, see LINCOLN'S INN FIELDS.

Liston, John (1776–1846), English comedian, who in spite of a nervous and melancholic turn of mind in private life had only to appear on the stage to set the audience laughing. After extensive experience in the provinces he was seen at the *Haymarket in 1805, playing Sheepface in *The Village Lawyer*, an adaptation of the old French farce of *Maître Pierre Pathelin*. For the next 30 years he was one of the leading players of London, where he made a fortune for several managers and authors and was the first comic actor to command a salary greater than that of a tragedian. (He occasionally aspired to play tragedy himself, but without success.) His Paul Pry in Poole's comedy of that name, first produced at the Haymarket in 1826, was so successful that it was imitated, dress and all, by such later players of the part as *Toole in 1866.

Liston, Victor, see MUSIC-HALL.

Little Angel Theatre, Islington, London, see CHILDREN'S THEATRE and PUPPET.

Little Drury Lane Theatre, London, see OLYMPIC.

Little Theatre, London, in John Adam Street, Adelphi. This small theatre, which held about 300, opened in 1910 with Laurence *Housman's adaptation of *Aristophanes' *Lysistrata*, under the management of Gertrude Kingston, who played the title-role. In 1911 Noël *Coward, aged 11, made his first appearance on the stage as Prince Mussel in a fairy play called *The Goldfish* and in the same year *Shaw's *Fanny's First Play*, directed by the author, was performed anonymously and then ran for a year, as did G. K. Chesterton's *Magic* (1913). The theatre was severely damaged by bombing in 1917, but was repaired and reopened in 1920 with *Knoblock's *Mumsee*. Later in the year there was a season of Grand *Guignol with Sybil *Thorndike, followed by *The Nine O'Clock Revue* (1922) and *The Little Revue Starts at Nine* (1923). Two 'horror' plays were produced, both based on novels, Bram Stoker's *Dracula* (1927) and Mary Shelley's *Frankenstein* (1930). In 1932 Nancy *Price made the theatre the headquarters of her People's National Theatre, the most successful productions being *Lady Precious Stream* (1934) by S. I. Hsiung, and *Mazo de la Roche's Whiteoaks* (1936). Herbert Farjeon's *Nine Sharp* (1938) and *Little Revue* (1939) were both successful, the latter being adapted to war conditions by playing from Sept. 1939 onwards in the afternoon. The last production, in Apr. 1940, was a revival of *Wycherley's *The Country Wife*. The theatre was severely damaged by enemy action in 1941 and demolished in 1949.

Little Theatre, New York, see HELEN HAYES THEATRE 2.

Little Theatre Guild of Great Britain, see AMATEUR THEATRE.

Little Theatre in the Hay, London, see HAYMARKET THEATRE.

Little Theatres, see AMATEUR THEATRE.

Little Tich [Harry Relph] (1868–1928), English *music-hall comedian, who got his name from his supposed likeness as a baby to the claimant in the famous Tichborne Case. He made his first appearance as a child of 12 in blackface at one of London's last pleasure grounds, the Rosherville near Gravesend, and was first seen on the 'halls' in 1884 at the *Middlesex. After playing in *vaudeville in the USA for some years he returned to England and was seen in a number of *pantomimes and *burlesques. In 1902 he was at the *Tivoli and there, and in a number of other 'halls' throughout the country, made a great success with his impersonations, which ranged from grocers, blacksmiths, and sailors on leave to fairy queens and Spanish dancers, and invariably ended, at least until his last years, with a fantastic dance in which he balanced on the tips of his preposterous boots, which were over half as long as he was high. He toured extensively, and was as popular in Paris as in London. His act was essentially pantomimic, and his whole personality was reminiscent of the *Court Fool or dwarf of medieval aristocratic households. Though unusually small (just over 4 ft.) he was not deformed except for an extra finger on each hand.

Littlewood, (Maudie) **Joan** (1914–), English director, born in London of working-class parents, who studied for the stage at the Royal Academy of Dramatic Art. Although she did well there, her training, combined with a naturally aggressive and experimental nature, left her impatient with what she regarded as the inanities of the normal West End theatrical routine, and she moved to Manchester. While working there in radio she founded with her husband an amateur group, Theatre Union, which soon made a name for itself with unconventional productions, in halls and in the open air, of experimental plays. The group dispersed on the outbreak of war in 1939, but came together again in 1945 as *Theatre Workshop, with Joan Littlewood as Artistic Director, and in 1953 took over the lease of the *Theatre Royal at Stratford, London. There, working on a system entirely her own (though it seems to have some affinities with those of *Stanislavsky and *Brecht), she became responsible for a series of successful productions. The constant drain on the company's resources, and a feeling that commercial success was inimical to her ideals, led her to leave Theatre Workshop in 1961 to work elsewhere, though she returned in 1963 to undertake the production of *Oh, What a Lovely War!* From 1965 to 1967 she was at the Centre Culturel, Hammamet, Tunisia, after which she again returned to Theatre Workshop. Since 1975 she has worked in France.

Liturgical Drama, plays based on the life, death, and resurrection of Jesus Christ. Considering the drama inherent in its subject, it is not surprising that the mimetic instinct in man, driven underground by the Early Fathers, should have broken out again in the celebration of the Mass, particularly at Easter and Christmas. The service itself provided the bare bones of a plot, and the introduction of antiphonal singing, which may have owed something to the memory of the Greek *chorus, paved the way for the use of dialogue. As church services became more elaborate, the vocal additions took on dialogue forms. The best known and most important of these, from a dramatic point of view, was the *Quem quaeritis?* (Whom Seek Ye?) sung on Easter morning. From a short scene in front of the empty tomb sung by four male voices, this soon developed into a small drama of three or four scenes covering the main events of the Resurrection. Later detached from the Mass and performed separately, it was followed by the *Te Deum*, and so merged into Matins.

An important step forward in dramatic evolution was taken with the introduction of extraneous lyrics to be sung by the three Marys as they approached the tomb, and an even greater one when comic characters, with no scriptural basis, were introduced in the persons of the merchants who sold spices to the women for the embalming of Christ's body. They probably appeared first in Germany, and were a counter-influence on the late liturgical play from the secular vernacular drama which had developed alongside it and partly under its influence.

A further Easter play, enacted at Vespers and perhaps modelled on the *Quem quaeritis?*, was the *Peregrinus*, which showed the Risen Christ with the disciples at Emmaus, sometimes accompanied by the three Marys and Doubting Thomas. This led by degrees to a drama which extended from the preparations for the Last Supper to the burial of Christ. This was given in a rudimentary form as long as it remained within the church, mostly in dumb show with passages from the Vulgate, but once performed outside, it coalesced and took on more substance to form the all-important and widespread *Passion play.

Meanwhile the services of Christmas had given rise to a play on the Nativity, centring on the crib, with Mary, Joseph, the ox and ass, shepherds, and angels. This never attained in liturgical drama the status of the Easter play, and, together with a short scene dealing with Rachel and the Massacre of the Innocents, was soon absorbed into an Epiphany play, in which the interest centred on the Wise Men and their gifts and on the wickedness of Herod who was to be so important in later secular religious plays.

Another Christmas play, and the most important for the future development of the drama, was based on a narrative sermon attributed to St Augustine and known as the Prophet play. It listed all the prophecies from the Old Testament concerning the coming of the Messiah, and included also those of Virgil and the Erythraean Sibyl. At some time in the 11th century it became a metrical dramatic dialogue, introducing Balaam and his Ass and the Three Children in the Fiery Furnace. The ass was probably an importation from the *Feast of Fools, and its use may mark a determined effort by the Church to canalize the irrepressible licence of Christmas merrymaking by incorporating into its own more orderly proceedings a slight element of buffoonery.

As long as the play remained within the church it was part of the liturgy, and the actors were priests, choirboys, and perhaps, later on, nuns. The dialogue, entirely in Latin, was chanted, not spoken, and the musical interludes were sung by the choir alone, with no participation by the congregation. By the end of the 13th century the evolution of liturgical drama was complete, and in its final phase it was not necessarily connected with a church service. From it came the vernacular *mystery play, and eventually the Church, having given back to Europe a regular and coherent form of drama, withdrew and prepared to do battle with the art form which it had engendered, and whose development henceforward was to become part of the theatrical history of each separate European country.

Liturgy, in ancient Greece, a public service required of wealthy citizens. One such duty was the staging of a play by a *choregus. In Europe the liturgy, or form of worship, of the early Christian Church gave rise to the performance of plays in Latin, or *liturgical drama.

Liverpool, which in the 1740s had theatrical entertainments given by visiting Irish players who appeared in a converted room known as the Old Ropery Theatre, had its first permanent theatre building, in Drury Lane, in 1750. Used by actors from London during the summer months, it had originally no boxes. These were added in 1759, and in 1767 a green room and dressing-rooms were also provided. It became a Theatre Royal in 1771, the name being transferred to a new theatre erected a year later. The interior was rebuilt in 1803 to provide a horseshoe-shaped auditorium which was later adapted as a circus, and eventually the building became a storage depot. In its heyday Liverpool, which was independent of any *circuit, had a theatre season which lasted practically all the year round. The Star Theatre, which opened in 1866, became the home of lurid melodrama until 1911, when it was taken over as a *repertory theatre, known since 1916 as the *Liverpool Playhouse. In 1866 also the present Empire Theatre, taken over by the Merseyside County Council in 1979 and extensively renovated, opened as the New Prince of Wales, changing its name to the Alexandra a year later. Seating 2,312, it is the largest two-tier theatre in the country, providing entertainment ranging from touring drama, opera, and ballet to variety and pop music. The enterprising *Everyman Theatre has achieved a high reputation, and appeals particularly to young people.

Liverpool Playhouse, Britain's oldest surviving *repertory theatre, opened partly in emulation of the venture so ably run by Miss *Horniman. When an experimental season run by Basil *Dean at another theatre showed that Liverpool was willing to accept good plays under repertory conditions, the old Star Theatre, a home of lurid melodrama built in 1866, was taken over and completely reconstructed. It opened in 1911 as the Liverpool Repertory Theatre and was closed only for short summer vacations, being given its present name in 1916. It set out to cater for every type of playgoer, though under William *Armstrong, its Artistic Director 1922–44, its programmes were perhaps somewhat less adventurous than those of Barry *Jackson at the *Birmingham Repertory Theatre. In recent years, however, classical and established works have been offset by a number of controversial modern plays by such writers as David *Hare, Harold *Pinter, Tom *Stoppard, and Alan *Bennett. A musical is presented annually at Christmas. The Playhouse, which seats 758, was extensively renovated in 1968, but the auditorium still retains the spaciousness of earlier times, with wide circles and balconies curving round the stage. Many well-known players started their careers there, among them Michael *Redgrave, Rex *Harrison, Diana *Wynyard, and more recently Ian *McKellen. There is also a studio theatre, the Playhouse Upstairs, seating 100, which presents mainly new work.

Living Newspaper, form of stage production which employed documentary sources to present subjects of current social importance, usually in a sequence of short scenes with individualized dialogue alongside more abstract, often didactic, presentation. The term is primarily associated with a unit of the *Federal Theatre Project, whose members were theatre and newspaper workers. Together they produced six Living Newspapers, of which the first, *Ethiopia*, a montage of newspaper reports about the current war in Abyssinia, was cancelled before its public showing under strong pressure from the US State Department. Others included *Triple-A Plowed Under* (1936), which dealt with farming conditions during a widespread drought and profiteering from food distribution during the depression; *Power* (1937), which advocated government ownership of the electrical power industry to save it from the menace of financial manipulation; and *One-Third of a Nation* (1938), which advocated federal and state development of low-cost housing. Following the example of the Unit, other Federal Theatre companies outside New York wrote and presented their own Living Newspapers, often with local or regional relevance. The outspokenness of the Living Newspaper productions was among the complaints made against the Federal Theatre Project as a whole, and one of the main causes of its being disbanded by the Government in 1939. The techniques influenced other companies and productions, and during the Second World War the British army used them to keep the troops informed about the wider issues of the conflict and conditions at home. Post-war *documentary theatre also owes a considerable debt to the form.

Livings, Henry (1929–), English dramatist and actor, who appeared in *Theatre Workshop's production of Brendan *Behan's *The Quare Fellow* (1956). His first plays, *Stop It, Whoever You Are*, about a lavatory attendant in a factory, and *Big Soft Nellie*, about a 'mother's boy', were both seen in London in 1961. *Nil Carborundum* (1962), set in the kitchen of an RAF station, and *Eh?* (1964; NY, 1966) were presented by the *RSC. His other plays include *Kelly's Eye* (1963), *This Jockey Drives Late Nights* (1972), an adaptation of Leo *Tolstoy's play *The Power of Darkness*, and *Jug* (1975), an adaptation of *Kleist. A revised version of *Jug* was produced at the *Theatre Royal, Stratford East (1986), which in 1987 staged *This is My Dream*, Livings's play about Josephine *Baker. He uses comedy, farce, and fantasy to depict the trials of people who appear to be underdogs but are not necessarily so, and though he was born and lives in Lancashire and his plays have a Lancashire 'tone', they are hardly ever specifically set there.

Living Stage, Washington, see ARENA STAGE.

Living Theatre, experimental *Off-Broadway group formed in New York in 1951 by **Julian Beck** (1925–85) and his wife **Judith Malina** (1926–), who had been one of *Piscator's students. Their company first appeared in public at the *Cherry Lane Theatre, one of the first productions of their original and iconoclastic career being Gertrude Stein's *Dr Faustus Lights the Lights*. From 1954 until 1956 they performed in a loft at Broadway and 100th Street, but it was closed by the Building Department because its 65 seats were considered too many for safety. In 1959 they opened a theatre seating 162 at 14th Street and Sixth Avenue, and used it until 1963, when they were evicted for non-payment of taxes. During this period they produced *The Connection* (1959; London, 1961) by **Jack Gelber** (1932–), a play about drug addiction involving some improvisation in performance. The final production before the company left for Europe was Kenneth Brown's *The Brig* (1963; London, 1964), based on his experiences in a US Marine Corps prison.

During the 1960s the Living Theatre was the most influential of the experimental groups which emerged from America. In 1961 it won three first prizes at the *Théâtre des Nations in Paris, and from 1964 to 1968 it toured Europe with four pieces made by *Collective Creation in which they played themselves and interacted with the spectators. *Mysteries and Smaller Pieces* (1964) was a series of exercises and improvisations; *Frankenstein* (1965) dealt with the creation of an artificial man through acculturation; *Antigone* (1967), based on *Brecht, justified civil disobedience as a protest against the war in Vietnam; and *Paradise Now* (1968) was intended to inaugurate a non-violent anarchist revolution by freeing the individual. In 1968 and 1969 the group toured the USA with these four pieces, which they presented in London in 1969. In 1970 Beck, Judith Malina, and several other members of the group went to Brazil with the intention of working with oppressed minorities, but were arrested and imprisoned on drug charges and deported to the USA. They continued developing new performance techniques, mainly in the service of political manifestations. The company is still active in New York under the leadership of Judith Malina and her present husband.

Lloyd, Marie [Matilda Alice Victoria Wood] (1870–1922), idol of British *music-hall audiences for many years. She first appeared in 1885, billed as Bella Delmere, a name she soon discarded for the one under which she became famous. She made her first great success at the *Middlesex, singing Nellie Power's song, 'The Boy I Love is Up in the Gallery', and she was then engaged for a year at the *Oxford. For three years, from 1891 to 1893, she appeared in *pantomime at *Drury Lane, but the 'halls' were her true home and she eventually returned to them for good. In her work she was wittily improper, but never coarse or vulgar, and her reputation for *double entendre* lay less in her material than in her delivery of it, with appropriate actions and an enormous wink. Her cheery vitality and hearty frankness won over all but the most captious critics. In her late forties she lost her looks, and went into semi-retirement, emerging in 1920 to take part in a revival of old-style music-hall. Among her best-loved songs were 'Oh, Mr Porter!', 'My Old Man Said Follow the Van', 'A Little of What You Fancy Does You Good', and 'I'm One of the Ruins that Cromwell Knocked Abaht a Bit', which she sang on her last appearance a few days before her death.

Loa, prologue or compliment to the audience which preceded the early Spanish theatrical performance. It ranged from a short introductory monologue to a miniature drama having some bearing on the play which was to follow. Águstin de *Rojas, in his *El viaje entretenido* (*The Entertaining Journey*, 1603), says that a strolling company generally had a variety

of *loas* which could be fitted to any play in any town.

Loeb Drama Centre, see HARVARD UNIVERSITY.

Loesser, Frank, see MUSICAL COMEDY.

Logan, Joshua Lockwood (1908–88), American director and producer, who trained as an actor under *Stanislavsky but turned to direction and had a big success with Paul Osborn's *On Borrowed Time* (1938). He became famous as a director of musicals such as *I Married an Angel* (1938), *Annie Get Your Gun* (1946), *South Pacific* (1949; London, 1951), and *Fanny* (1954), being co-librettist and co-producer of the last two. He also wrote *The Wisteria Trees* (1950), an adaptation of *Chekhov's *The Cherry Orchard* to an American setting which he directed and co-produced. His biggest success in the non-musical theatre was *Mister Roberts* (1948; London, 1950), which he directed and jointly adapted with Thomas Heggen from the latter's novel. Other plays he directed included Norman Krasna's *John Loves Mary* (1947), *Inge's *Picnic* (1953), and Paddy Chayevsky's *Middle of the Night* (1956), all of which he produced or co-produced. His film work included versions of his stage successes.

London Academy of Music and Dramatic Art, see SCHOOLS OF DRAMA.

London Casino, see PRINCE EDWARD THEATRE.

London Coliseum, see COLISEUM.

London Hippodrome, see HIPPODROME.

London International Festival of Theatre (LIFT), biennial three-week summer festival started in 1981, presenting a programme of theatre, dance, workshops, and open-air events. Companies from many countries perform in a variety of forms and styles, from the classical to the avant-garde, and in their own language. Many theatres are used, most of them small and away from the centre. Some of the events are free, and some involve audience mobility. A festival club provides late night cabaret and musical entertainment.

London Palladium, in Argyll Street. The first building on this site was a circus which ran successfully from 1871 to 1887. Its fortunes then declined, and it was reconstructed as a *music-hall in three tiers, with a proscenium width of 47 ft., a stage depth of 40 ft., and a revolving stage with three concentric rings. Its seating capacity of 2,325 is the largest of any live theatre in the West End. (The *Coliseum, which has more seats, now ranks as an opera-house.) The theatre opened as the Palladium on 26 Dec. 1910 with a *variety bill which included Nellie *Wallace and also *Martin-Harvey in a one-act play. Variety was soon replaced by *revue, and shows such as *Rockets* (1922) and *The Whirl of the World* (1923) had considerable success. The theatre was used as a cinema for three months in 1928, but then came under the control of George Black, who brought there the *Crazy Gang in a series of shows which included *Life Begins at Oxford Circus* (1935), *O-Kay for Sound* (1936), *These Foolish Things* (1938), and *The Little Dog Laughed* (1939). From 1930 until 1938 there was also a Christmas revival of *Barrie's *Peter Pan*. The theatre was officially renamed the London Palladium in 1934. During the Second World War more revues followed, including Irving *Berlin's *This is the Army*, but after Black's departure in 1946 there was a policy of twice-nightly variety with a number of acts. Later, following a general trend, the Palladium changed to once-nightly performances by a top international star with only one or two supporting acts. In 1968 there was a short break in continuity when Sammy Davis Junior appeared in a musical version of Clifford *Odets's *Golden Boy*. After 1970 musicals and star appearances both featured in the schedules. In 1974 the Christmas pantomime was replaced by a musical based on the life of Hans Andersen, with Tommy Steele in the title-role. There was a highly successful revival of *The King and I* in 1979, and in 1981 *Barnum* began a long run. *Singin' in the Rain* (1983 and 1989), again with Tommy Steele, based on the film, had the longest run in the Palladium's history, but the theatre had two of its rare failures with the spectacular musicals *La Cage aux folles* (1986) and *Ziegfeld* (1988). The theatre has been used more than any other for the annual Royal Variety Show since its inception in 1930.

London Pavilion, famous *music-hall which began as a 'song-and-supper' room attached to the Black Horse Inn in Tichborne Street at the top of the Haymarket, the stable-yard being roofed in for the purpose. After extensive renovation it opened in 1861 as a music-hall seating 2,000, and many famous music-hall stars appeared there. The hall was demolished in 1885 and a new Pavilion arose on the site. It retained the old-style separate tables and chairs, with a presiding *chairman. They were abolished a year later, when normal theatre seating was installed, but music-hall 'turns', sometimes as many as 20 in one evening, still reigned supreme, until in 1918 *Cochran took over and the Pavilion became a theatre, famous for its *revues. Among the best were *Dover Street to Dixie* (1923), with Florence Mills, who reappeared with the all-Black *Blackbirds* (1926); *Coward's *On with the Dance* (1925), *This Year of Grace* (1928, with Jessie *Matthews), and *Cochran's 1931 Revue*; *One Dam Thing after Another* (1927) and *Wake Up and Dream* (1929), both also starring Jessie Matthews. After Cochran left in 1931 the theatre turned to non-stop *variety, the last performance being in 1934,

when the building became a cinema. It was closed for demolition in 1981.

London Theatre Studio, see SAINT-DENIS.

Longacre Theatre, New York, on West 48th Street, between Broadway and 8th Avenue. With a seating capacity of 1,115, it opened in 1913 and was used mainly for *musical comedy, though *Kaufman's farce *The Butter and Egg Man* (1925) had a long run. Later the theatre housed short runs of several notable plays, among them G. B. Stern's *The Matriarch* (1930) and *Obey's *Noé* (1935) with Pierre *Fresnay. Several plays by *Odets were produced under the auspices of the *Group Theatre, but in 1944 the building was taken over for broadcasting. It returned to use as a theatre in 1953, Lillian *Hellman's adaptation of *Anouilh's *The Lark* being staged there in 1955 with Julie *Harris as Joan of Arc. Emlyn *Williams was seen as Dylan Thomas in *A Boy Growing Up* (1957), and Zero *Mostel starred in *Ionesco's *Rhinoceros* (1961). Robert *Anderson's *I Never Sang for My Father* (1968) was followed in the 1970s by Julie Harris's solo performance as Emily Dickinson in *The Belle of Amherst* and *Gielgud and *Richardson in *Pinter's *No Man's Land* (both 1976); a musical tribute to Fats Waller, *Ain't Misbehavin'* (1978); and John *Guare's *Bosoms and Neglect* (1979). Mark Medoff's *Children of a Lesser God* was a big success in 1980, and two plays by Peter *Nichols were seen: *Passion* (1983; *Passion Play* in London) and a revival of *A Day in the Death of Joe Egg* as *Joe Egg* (1985).

Longford, Edward Arthur Henry Pakenham, 6th Earl of, see GATE THEATRE, Dublin.

Long Wharf Theatre, New Haven, Conn., is situated in a meat and produce terminal. A thrust-stage playhouse seating 484, it opened in 1965 and now has one of the most highly regarded resident companies in the United States. During seasons running from Oct. to June it presents classic and modern plays, including world and American premières. Some of its productions have moved practically unaltered to Broadway, among them David *Storey's *The Changing Room* (1973); Athol *Fugard's *Sizwe Bansi is Dead* (1975); Michael Cristofer's *The Shadow Box* and D. L. Coburn's *The Gin Game* (both 1977); and David *Mamet's *American Buffalo* (1980). In 1982 it staged the world première of Arthur *Miller's double bill *Two by A.M.*; in 1985 it presented Simon *Gray's *The Common Pursuit*, and in 1987 *Stoppard's *Dalliance*. It has an intimate Stage Two.

Lonsdale [Leonard], **Frederick** (1881–1954), English dramatist, who was responsible for the librettos of several *musical comedies, among them *The Maid of the Mountains* (1917), *Monsieur Beaucaire* (1919), and *Madame Pompadour* (1923). He is best remembered, however, as the author of a number of comedies of contemporary manners somewhat in the style of Somerset *Maugham, though with less subtlety. The best known are *Aren't We All?* (London and NY, 1923), *Spring Cleaning* (NY, 1923; London, 1925), *The Last of Mrs Cheyney* (London and NY, 1925), *On Approval* (NY, 1926; London, 1927), *The High Road* (1927; NY, 1928), and *Canaries Sometimes Sing* (1929; NY, 1930). All gave scope for good, brittle, sophisticated acting, and stars of the calibre of Gladys *Cooper and Yvonne *Arnaud appeared successfully in them. Their amusing situations, easy and effective dialogue, and rich, worldly, and well-bred characters made them immensely popular in their day. Lonsdale's later plays, which include *Another Love Story* (NY, 1943; London, 1944), *The Way Things Go* (1950), and *Let Them Eat Cake* (produced posthumously in 1959), were not so popular.

Lope de Rueda, see RUEDA, LOPE DE.

Lope de Vega, see VEGA CARPIO, LOPE DE.

Loraine, Robert (1876–1935), English actor-manager, who in 1889 made a great success as D'Artagnan in a dramatization of the elder *Dumas's *The Three Musketeers.* In 1911, after a visit to America, he took over the *Criterion Theatre in London, opening with a revival of *Man and Superman* by *Shaw, whose Don Juan he had played in the first production of *Don Juan in Hell* (1907). During the First World War he made a great reputation as an aviator and was awarded the MC and the DSO for gallantry in action. Essentially a romantic actor, he returned to the stage in 1919 in the title-role of *Rostand's *Cyrano de Bergerac,* which had a long run. Among his later parts were Deburau in Sacha *Guitry's play of that name in 1921, the dual role of Rassendyl and Prince Rudolf in a revival of Hope's *The Prisoner of Zenda* in 1923, the Nobleman in the New York production of Ashley *Dukes's *The Man with a Load of Mischief* in 1925, and Adolf in *Strindberg's *The Father* in 1927. He also played a number of Shakespearian roles including Petruchio in *The Taming of the Shrew* and Mercutio in *Romeo and Juliet* (both in 1926).

Lorca, Federico García, see GARCÍA LORCA.

Lord Chamberlain, officer of the British Royal Household under whom the *Master of the Revels was first appointed in 1494 to supervise Court entertainments. After the Restoration in 1660 the Lord Chamberlain himself began to intervene directly in the regulation of theatres and in *censorship, mainly in relation to political and religious issues, his powers being legally formulated by the Licensing Act of 1737. These powers, particularly in regard to the *licensing of theatres, were modified by the Theatres Act of 1843. New plays, and new matter added to existing plays, had to be submitted to the Lord Chamberlain's office for scrutiny by the Examiner of Plays. At the time of the abolition of his powers the Lord Chamberlain had three English readers and one Welsh, who reported on the work submitted, with particular reference to indecency, impropriety, profanity, seditious matter, and the representation of living persons. Plays written before the passing of the 1843 Act could be suppressed under Section 14 of the Act. Play-producing societies and theatre clubs came within the Lord Chamberlain's powers, although it was customary to treat bona fide societies with lenience. His power to withhold or withdraw a licence was absolute and he was under no legal obligation to disclose the reasons; but in practice the Lord Chamberlain's office was normally ready to indicate changes in the text that would enable a licence to be issued. There was no appeal against his decision. The Theatres Act of 1968 took all theatrical matters out of the hands of the Lord Chamberlain, vesting theatre licensing in local authorities and abolishing theatrical censorship, since when stage plays have been subject to common law such as those Acts pertaining to obscenity, blasphemy, libel, and breach of the peace.

Lord of Misrule, of Unreason, see FEAST OF FOOLS.

Lortel, Lucille, see LUCILLE LORTEL THEATRE.

Los Angeles, see LOS ANGELES THEATRE CENTER and MARK TAPER FORUM.

Los Angeles Theatre Center, Los Angeles, opened as Los Angeles Actors' Theatre in 1975. Its productions, beginning with *O'Neill's *The Hairy Ape,* were noted for their controversial content and presentation, over 200 world, American, and West Coast premières being produced in 10 years. The Theatre Center, a new building opened in 1985, has at its core the Security National Bank building dating from 1916, whose marble walls enclose the Center's Grand Lobby. Occupying eight floors, the complex contains four theatres: the Tom Bradley Theatre, seating 503, an open stage; Theatre Two (296), a proscenium theatre; Theatre Three (323), a thrust stage; and Theatre Four (99), a flexible black box. The Center mounts 14 productions a year, of which over half are world premières, and stages an annual festival of new plays. It also presents new interpretations of classics and numerous co-productions, and is an important training centre.

Lotta, see CRABTREE.

Loutherbourg, Philip James de (1740–1812), German painter, who had for some time made a special study of stage illusion and mechanics before visiting London in 1771. There he met *Garrick, who two years later appointed him scenic director at *Drury Lane, a position he retained under *Sheridan. He was particularly successful in producing the illusion of fire, volcanoes, sun, moonlight, and cloud-effects, and invented strikingly effective devices for thunder, guns, wind, the lapping of waves, and the patter of hail and rain. He was the first designer to bring a breath of naturalism into the artificial scenic conventions of the day, and paved the way for the realistic detail and local colour favoured by Charles *Kemble. A visit to the Peak district in 1779 resulted in an *act-drop depicting a romantic landscape—possibly the earliest example of a scenic curtain in Western Europe—which remained in use until Drury Lane was destroyed by fire in 1809. For Sheridan's *The Critic* (1779), in which Mr Puff refers to him by name, he executed a striking design of Tilbury Fort, and he was also responsible for some excellent new transparencies used in a revival of *The Winter's Tale* in the same year. He was probably the first designer in England to break up the scene by the use of perspective, and also the first to use *built stuff, though somewhat sparingly. Much of his best work was lavished on unremarkable plays, to which he gave a momentary popularity. Shortly after preparing the scenery for the first dramatization of Defoe's *Robinson Crusoe* (1781) by Sheridan—the first act alone had eight changes—Loutherbourg left the theatre, mainly on account of a dispute over his salary, and devoted much of his time to a remarkable scenic exhibition, the 'Ediophusikon', whose influence lingered on until 1820, when *Elliston attempted to reproduce in *King Lear* the powerful effects of storm and tempest which Loutherbourg had created for his exhibition.

Low Comedian, see STOCK COMPANY.

Lowin, John (1576–1653), English actor, and an important link between the Elizabethan and Restoration stages, since he was said to have passed on to *Davenant the instructions he had received for the playing of Henry VIII from Shakespeare himself, Davenant passing them on to his chief actor under Charles II, Thomas *Betterton. It is possible that Lowin, who is always referred to as a big, bluff man, may have been the first to play the part in 1613; he is also believed to have been much admired as Falstaff in revivals of *Henry IV*, and as the eponymous hero of Ben *Jonson's *Volpone* (1606), though he may not have been its creator. He is first mentioned as an actor in 1602, and a year later joined the *King's Men on their foundation, his name being found in the cast-lists of Jonson's *Sejanus* (1603) and *Marston's *The Malcontent* (1604), probably as a player of small parts. He created the part of Melantius in *Beaumont and *Fletcher's *The Maid's Tragedy* (1611), and probably that of Bosola in *Webster's *The Duchess of Malfi* (1614); he also appeared in several of *Massinger's plays. He remained with the King's Men until the closing of the theatres in 1642, and was one of the actors caught and

punished under the Commonwealth for playing surreptitiously at the *Cockpit in 1649. In his old age he kept the Three Pigeons at Brentford.

Lucan, Arthur, see FEMALE IMPERSONATION.

Lucille Lortel Theatre, New York, on Christopher Street in Greenwich Village. This *Off-Broadway playhouse, run by the actress **Lucille Lortel** (1902–) as the Theatre de Lys, opened in 1952, and from 1954 to 1961 housed *The Threepenny Opera*, Marc Blitzstein's English version of *Brecht's *Die Dreigroschenoper*; it was followed by *Brecht on Brecht* (1962), which also had a long run. Later productions included John *Arden's *Serjeant Musgrave's Dance* (1966) and *The Deer Park* (1967), adapted by Norman Mailer from his own novel. Notable productions of the 1970s were a revue, *Berlin to Broadway with Kurt Weill* (1972), David *Mamet's *A Life in the Theatre* (1977), and Sam *Shepard's *Buried Child* (1978). Robert Harling's *Steel Magnolias* began a long run in 1987.

Lucille Lortel, after whom the theatre took its name in 1981, has been called 'the Queen of Off-Broadway'. She has produced hundreds of plays, beginning in 1947 at the White Barn Theatre, Conn., a summer theatre in the grounds of her home. She has supported the work of such playwrights as *O'Casey, Athol *Fugard, Samuel *Beckett, *Genet, *Ionesco, and many African playwrights, and has won almost every theatre award.

Ludlow, Noah Miller (1795–1886), American actor-manager of the pioneering days, who in 1815 was engaged by Samuel *Drake to go to Kentucky and later founded his own company, with which in 1817 he gave English plays in New Orleans. He travelled extensively, with his own or other companies, often being the first actor to penetrate to some of the more remote regions in the South and West. In 1828 he was induced to take over the old *Chatham Theatre in New York with *Cooper, but failed to make it pay, and moved to St Louis. From 1835 to 1853 he was in partnership with Sol *Smith in the American Theatrical Commonwealth Company, and ran several theatres simultaneously in St Louis, New Orleans, Mobile, and other cities, often engaging outstanding stars. He was himself an excellent actor, particularly in comedy.

LuEsther Hall, New York, see PUBLIC THEATRE.

Lugné-Poë, Aurélien-François (1869–1940), French actor, director, and theatre manager, who in 1893 took over the Théâtre d'Art from Paul Fort, renaming it the Théâtre de l'Œuvre. He remained there until 1899 and returned to it from 1912 to 1929. In his early years he continued Fort's reaction against the realism of the *Théâtre Libre by staging poetic and *Symbolist plays. His first production was *Pelléas et Mélisande* by *Maeterlinck, several of whose later plays were also first seen at this theatre, and among the other foreign authors he presented were *Ibsen, *Bjørnson, *Hauptmann, *D'Annunzio, and *Echegaray. In 1895 he staged *Wilde's Salome, and he brought *Claudel before the public with L'Annonce faite à Marie in 1912 and L'Ôtage in 1914. In 1896 he had been responsible for the riotous first performances of *Jarry's Ubu-roi. Between 1919 and 1929 he introduced *Shaw and *Strindberg to Paris, and encouraged such new French playwrights as *Achard. After he left the Théâtre de l'Œuvre he continued to direct elsewhere, being associated in 1932 with the production of *Anouilh's first play L'Hermine, and in 1935 with L'Inconnue d'Arras by *Salacrou. Himself an excellent actor, Lugné-Poë appeared in many of his own early productions, and in 1908 was seen in London in Jules Renard's Poil de Carotte.

Lun, see RICH; **Lun junior,** see WOODWARD.

Lunacharsky, Anatoli Vasilevich (1875–1933), first Commissar for Education in Soviet Russia, an able and cultured man to whom the USSR owes the preservation and rejuvenation of those Imperial theatres—notably the *Moscow Art Theatre—which survived the Revolution. He was also responsible for the organization of the new Soviet theatres which sprang up in vast numbers, many of which are named after him. He realized that the new audiences would inevitably demand new plays and methods of production, but that the old plays, both Russian and European, were part of the cultural heritage of the new world and must not be discarded. With this end in view he endeavoured, while supporting the move towards *Socialist Realism, to counter some of the post-Revolutionary experimental excesses by a return to the past, skilfully adapted to meet contemporary requirements. He was himself the author of a number of plays mainly on revolutionary themes, one of which, Oliver Cromwell (1921), pays tribute to a fellow-revolutionary and draws an interesting parallel between the establishment of the Commonwealth in England in the 17th century and the Russian Revolution of 1917.

Lunacharsky State Institute of Theatre Art, see SCHOOLS OF DRAMA.

Lunt, Alfred (1892–1977), American actor, who made his début in Boston in 1912 and had his first big success in 1919, playing the title-role in Booth *Tarkington's Clarence. In 1922 he married **Lynn** [Lillie Louise] **Fontanne** (1887–1983), who after studying with Ellen *Terry and making her début in pantomime in 1905 had pursued a successful stage career in London and New York. The Lunts first appeared together in *Kester's Sweet Nell of Old Drury (1923). They then joined the *Theatre Guild

company, and over the next five years were in a succession of distinguished plays, including *Shaw's Arms and the Man and Pygmalion, *Molnár's The Guardsman, and *Werfel's Goat Song. Later successes were Robert *Sherwood's Reunion in Vienna (1931; London, 1934); *Coward's Design for Living (1933); *Giraudoux's Amphitryon 38, adapted by S. N. *Behrman (1937; London, 1938); and *Rattigan's Love in Idleness (1944; NY, as O Mistress Mine, 1946). They were at their best in intimate modern comedy, playing together with a subtlety and sophistication which gave life and strength even to the flimsiest dramatic material. In 1958 they starred at the Globe Theatre, New York, renamed the *Lunt–Fontanne in their honour, in *Dürrenmatt's The Visit, with which they had previously toured in England under the title of Time and Again. The Visit was the first production at the new *Royalty Theatre (3) in London in 1960, their last appearance. In 1964 the Lunts were awarded the US Medal of Freedom by President Johnson.

Lunt–Fontanne Theatre, New York, on West 46th Street. This attractive playhouse was opened as the Globe in 1910, mainly for musical shows, though at the end of the year the company of Sarah *Bernhardt appeared in a repertory of French plays. The Ziegfeld Follies were there in 1921 and George White's Scandals in 1922 and 1923. After a successful musical, The Cat and the Fiddle (1931), it became a cinema; but in 1958, completely remodelled and redecorated, with a seating capacity of 1,714, it reopened as the Lunt–Fontanne with Alfred *Lunt and Lynn Fontanne in *Dürrenmatt's The Visit. From 9 Apr. to 8 Aug. 1964 Richard *Burton appeared in Hamlet, setting up a new record for the run of the play in New York. In the 1970s the theatre staged revivals of the musicals My Fair Lady (1976), Hello, Dolly! (1978), and Peter Pan (1979). In 1981 it had another success with Sophisticated Ladies, a compilation of Duke Ellington numbers. It housed revivals of *Coward's Private Lives in 1983, with Richard Burton and Elizabeth Taylor, and of *O'Neill's The Iceman Cometh in 1985 with Jason *Robards.

Lupino, Barry (1882–1962), English comedian, one of a numerous family (including Lupino *Lane) descended from an Italian puppet-master who emigrated to England early in the 17th century. He made his first appearance as a baby in Cinderella, and as an adult actor was for some years the leading comedian and pantomimist at the *Britannia Theatre, being connected by marriage with the lessees Sam and Sara Lane. He toured extensively, was seen in *musical comedy, and played the *Dame in innumerable *pantomimes, of which he himself wrote more than 50. Two of his finest parts were Dame Durden in Jack and the Beanstalk and the Widow

Twankey in *Aladdin*, which he was still playing in his seventies. He made his last appearance as Dame Sarah in *Dick Whittington* in 1954.

Lupino, Stanley (1893–1942), English comedian, younger brother of Barry (above). He was on the stage as a child, as a member of an acrobatic troupe, and appeared for many years in *pantomime at *Drury Lane. He was also seen in *musical comedy and *revue and was part author of *So This is Love* (1928), in which he played Potty Griggs, *Love Lies* (1929), in which he played Jerry Walker, and *Room for Two* (1932), which he directed. He wrote a number of plays on his own, notably *The Love Race* (1930), *Hold My Hand* (1931), *Over She Goes* (1936), and *Crazy Days* (1937), in all of which he also appeared.

Lyceum Theatre, Edinburgh, see ROYAL LYCEUM THEATRE.

Lyceum Theatre, London, in Wellington Street, just off the Strand, a theatre indissolubly linked with *Irving. There was a place of entertainment on the site as early as 1765, but it was not until 1809, when the *Drury Lane company moved there after the destruction by fire of their own theatre, that it was licensed for plays. When the new Drury Lane opened in 1812 the Lyceum was rebuilt to designs by *Beazley and licensed for the summer only, being used for mixed entertainments of opera and plays. In 1830 it was burnt down, and in 1834 a new building, whose frontage still stands, opened as the Royal Lyceum and English Opera House. This too was designed by Beazley and was the building in which Irving made his name. It had a chequered career until the passing of the Licensing Act of 1843 enabled it to stage legitimate drama. Robert *Keeley ran it, 1844–7, after which Mme *Vestris and the younger *Mathews took over and a series of brilliant productions followed, mostly of extravaganzas by *Planché. After their bankruptcy the theatre was occupied by the company from *Covent Garden, which had been burnt down in 1856. In 1871 *Bateman took over, engaging as his leading man the young Henry Irving. Seven years later Irving took control, and with Ellen *Terry inaugurated a series of fine productions which made the Lyceum the most notable theatre in London. Their last appearance there was in 1902 in *The Merchant of Venice*, and after their departure the fortunes of the theatre declined. It was partly demolished and rebuilt as a music-hall which opened in 1904, but under the *Melvilles from 1910 to 1938 it became famous as the home of melodrama and pantomime. In 1939 the theatre was scheduled for demolition and six farewell performances of *Hamlet*, with John *Gielgud, took place. The outbreak of the Second World War led to the abandonment of the scheme and the theatre stood empty until 1945, when it became a dance hall, which it remained until 1985. In that year it reverted temporarily to theatrical use for Bill *Bryden's production of *The Mysteries*. It is now closed, and its future is under discussion.

Lyceum Theatre, New York, on 45th Street east of Broadway, seating 995, one of the city's most glamorous playhouses. Built by Daniel *Frohman, it opened in 1903, after an earlier Lyceum which he had run on 4th Avenue had been demolished. The first new play produced there was *Barrie's *The Admirable Crichton*, and later Charles *Wyndham brought his London company for an eight-week season. From 1916 there were a number of productions by *Belasco, Frohman's first stage manager at the Lyceum, and in 1922 David *Warfield gave a fine performance as Shylock in *The Merchant of Venice*. *Berkeley Square* (1929) by John Balderston and J. C. Squire repeated its London success, and two comedy hits were *Kaufman and *Hart's *George Washington Slept Here* (1940) and Garson Kanin's *Born Yesterday* (1946). Later productions included *Odets's *The Country Girl* (1950) and three further plays from London— *Osborne's *Look Back in Anger* (1957), Shelagh Delaney's *A Taste of Honey* (1960), and *Pinter's *The Caretaker* (1961). During the late 1960s the theatre housed the company from the *Phoenix Theatre. In 1976 it staged a musical, *Your Arms Too Short to Box with God*, based on the Gospel according to St Matthew. Arthur *Kopit's *Wings* was seen in 1979, with Constance *Cummings, and in 1980 Paul Osborn's *Morning's at Seven* had a successful revival. It was declared a landmark in 1975.

Lyceum Theatre, Sheffield, see CRUCIBLE THEATRE.

Lyly, John (*c.*1554–1606), English novelist and playwright, whose involved and allusive style in his *Euphues: The Anatomy of Wit* (1579) and *Euphues and his England* (1580) gave rise to the expression 'euphuism'. His plays were written mainly for a courtly audience, which delighted in his grace and artificiality, and in his many sly allusions to current scandals. They were mostly acted by the Children of Paul's, whose vice-regent Lyly became in *c.*1590. Among them were *Campaspe* and *Sapho and Phao* (both *c.*1584), two comedies, *Midas* and *Mother Bombie* (both *c.*1586), the latter a 'comedy of errors' in the style of *Terence, and his most important play, *Endimion, the Man in the Moon*, first published in 1591, and possibly acted a couple of years earlier. Lyly, whose new use of language undoubtedly influenced Shakespeare, may have had a hand in other plays of his time, but nothing can be ascribed to him with any certainty, except for one or two *pastorals on mythological subjects, of ephemeral interest.

Lyndsay, Sir David (1490–*c.*1554), Scottish poet, and the author of the only surviving example of a Scottish medieval *morality play *Ane Pleasant Satyre of the Thrie Estaitis*. First performed at Cupar in 1552 in the presence of James V, this was a serious attack on the established Church and the authority of the Pope, interspersed with comic episodes highlighting clerical follies and abuses of the time. The complete play, which was added to and revised in 1554, is very long, but an abbreviated and modernized version was successfully produced at the *Edinburgh Festival in 1948 under the direction of Tyrone *Guthrie and has since been revived several times.

Lynn, Ralph (1882–1962), English actor, who specialized in 'silly ass' parts, with monocle, protruding teeth, a winning smile, and sweet though asinine reasonableness in the most trying circumstances. He made his first appearance on the stage in 1900, and after many years in the provinces and a visit to New York in 1913 achieved his first outstanding success in London in 1922, playing Aubrey Henry Maitland Allington in Will *Evans's farce *Tons of Money*. First seen at the *Shaftesbury Theatre, this was transferred to the *Aldwych, where Lynn was to remain for nearly 10 years, starring with Robertson *Hare and Tom *Walls in the so-called 'Aldwych farces' by Ben *Travers. After leaving the Aldwych, Lynn appeared in the long-running farcical comedy *Is Your Honeymoon Really Necessary?* (1944) by E. Vivian Tidmarsh, followed by two more farces by Travers, in which Lynn again teamed up with Robertson Hare—*Outrageous Fortune* (1947) and *Wild Horses* (1952). He made his last appearance in London in 1958, but continued to tour in some of his old successes until shortly before his death.

Lyric Hall, Lyric Opera House, London, see LYRIC THEATRE, Hammersmith.

Lyric Players, Dublin, see CLARKE, AUSTIN.

Lyric Players Theatre, Belfast, founded in 1951 for the performance of *poetic drama. Its first productions, mainly of short plays by *Yeats and other Irish writers, were given in a private house, but in 1952 a small theatre was built and nearly 200 plays, covering the whole range of poetic drama from the Greeks to modern times, were performed on a stage 8 by 10 ft. In 1956 the theatre was enlarged to accommodate a drama school, reopening with *Chekhov's *The Seagull*, and in 1968 a new theatre was built for the company, which is now run as a non-profit-making association by the Lyric Players Trust. The theatre is subsidized by the Arts Council of Northern Ireland and produces approximately 10 plays each year, covering a wide spectrum of international drama. Visits have been paid to the Republic of Ireland, Great Britain, France, Germany, and the USA, and actors, directors, and designers from the Republic and Britain are encouraged to work with the company. The Lyric Theatre is Northern Ireland's only professional repertory theatre.

Lyric Theatre, Hammersmith, West London, in King Street, opened as the Lyric Hall in 1888. It was reconstructed and reopened as the Lyric Opera House in 1890, rebuilt in 1895, its seating capacity being increased from 550 to 800, and further improved in 1899. The home of a resident *stock company, it drew a large local audience to its *melodramas and annual *pantomime, but its fortunes declined and it housed mainly touring companies until in 1918 it was taken over by Nigel Playfair. He redecorated it, renamed it the Lyric Theatre, and made it prosperous and fashionable, drawing large audiences from the West End. The opening production was A. A. Milne's *Make-Believe*; in 1919 *Drinkwater's chronicle play *Abraham Lincoln*, brought from the *Birmingham Repertory Theatre for two weeks, began a run of 466 performances. A notable revival of *Gay's *The Beggar's Opera*, with décor by Lovat *Fraser, opened in 1920 and ran for 1,463 performances. Edith *Evans was a fine Millamant in *Congreve's *The Way of the World* (1924) and returned as Mrs Sullen in *Farquhar's *The Beaux' Stratagem* (1927). In between Ellen *Terry made her last appearance on the stage, in De la Mare's *Crossings*, and there was a notable revival of *Chekhov's *The Cherry Orchard* with *Gielgud as Trofimov (both 1925). A number of new light operas were staged, including A. P. Herbert's *Riverside Nights*, and in 1930 *Wilde's *The Importance of Being Earnest* was seen in a black-and-white décor. Playfair left in 1933, after which the theatre was used only intermittently, but in 1946 it achieved success with a production by Peter *Brook of Dostoevsky's *The Brothers Karamazov* starring Alec *Guinness. A number of new plays and revivals were then staged, several of which were transferred to the West End, among them *Cocteau's *The Eagle has Two Heads* (also 1946), *Vanbrugh's *The Relapse* (1947), Arthur *Miller's *All My Sons*, and *Sartre's *Crime passionel* (both 1948). Gielgud ran a successful repertory season, 1952–3, directing Paul *Scofield in *Richard II* and appearing himself. Later productions were *Anouilh's *The Lark* (1955) with Dorothy *Tutin as Joan of Arc; the revue *Share My Lettuce* (1957), in which Maggie *Smith made her London début; *Pinter's first full-length play *The Birthday Party* (1958); and in 1959 *Büchner's *Danton's Death* and *Ibsen's *Brand*.

A period of stagnation then set in and the theatre finally closed in 1966. It was demolished in 1972, but plaster castings of the original Victorian interior were taken. A new theatre

20 yards away, seating 540, with the original interior, opened in 1979 with *Shaw's *You Never Can Tell*. Its many West End transfers included Michael *Frayn's *Make and Break* (1980) and *Noises Off* (1982) and *García Lorca's *The House of Bernarda Alba* (1986). The small studio theatre seating 130 houses small touring companies and stages rarely performed work.

Lyric Theatre, London, in Shaftesbury Avenue, a four-tier house holding 1,306 (later reduced to 948). The theatre opened in 1888, and was devoted mainly to musicals, except when in 1893 *Duse made her London début in the younger *Dumas's *La Dame aux camélias*. It had its first outstanding success in 1896, when Wilson *Barrett played Marcus Superbus in his own play *The Sign of the Cross*. Seasons of French plays were given in 1897 by *Réjane and in 1898 by Sarah *Bernhardt, and in 1902 *Forbes-Robertson appeared in *Hamlet* and *Othello*. Successful musicals were *Florodora* (1899), *The Duchess of Dantzig* (1903), *The Chocolate Soldier* (1910), and *Lilac Time* (1922). In 1919 *Romeo and Juliet* was staged with Ellen *Terry as Juliet's nurse. A succession of good modern plays included *Priestley's *Dangerous Corner* (1932), Robert *Sherwood's *Reunion in Vienna* (1934), *Housman's *Victoria Regina* (1936), *Giraudoux's *Amphitryon 38* with the *Lunts, and Charles *Morgan's *The Flashing Stream* (both 1938). During the Second World War the Lunts were seen again in *Love in Idleness* (1944) by Terence *Rattigan, whose *The Winslow Boy* (1946) provided the theatre's first post-war success. Equally successful was Roussin's *The Little Hut* (1950) with Robert *Morley, which ran until T. S. *Eliot's *The Confidential Clerk* took over in 1953. Two musicals, *Grab Me a Gondola* (1956) and *Irma la Douce* (1958), did well, as did *Robert and Elizabeth* (1964), a musical version of Besier's *The Barretts of Wimpole Street*. Alan *Ayckbourn's *How the Other Half Loves* (1970) with Robert Morley, Alan *Bennett's farce *Habeas Corpus* (1973) with Alec *Guinness, and the musical *John, Paul, George, Ringo . . . and Bert* (1974) all prospered. An attempt in 1975 to found a repertory company was abandoned after only two productions, *Chekhov's *The Seagull* and Ben *Travers's *The Bed before Yesterday*, though the latter had a long run on its own. Later successes were William Douglas *Home's *The Kingfisher*, *De Filippo's *Filumena* (both 1977), Hugh Whitemore's *Pack of Lies* (1983), and Lanford *Wilson's *Burn This* (1990).

Lyric Theatre, New York, on West 42nd Street. This opened in 1903 with *Old Heidelberg*, starring Richard *Mansfield, who for some years returned regularly with his company, the theatre being used also for opera and musical plays. One of its earliest successes was *The Taming of the Shrew*, with Ada *Rehan and Otis *Skinner, and in its second season *Réjane and *Novelli appeared with their companies. Plays by *Sudermann and *Hauptmann were seen and the musical comedy *The Chocolate Soldier* (1909) had a long run. In 1911 came the first production in the United States of *Ibsen's *The Lady from the Sea*. Florenz *Ziegfeld filled the theatre for many years with a series of musical comedies, and the last production in 1933, before the building became a cinema, was a Negro drama with music, Hall Jonson's *Run, Little Chillun*.

New York's *Criterion Theatre was known as the Lyric for its first four years.

Lyttelton Theatre, the first of the three theatres inside the *National Theatre building to be opened, the others being the *Cottesloe and the *Olivier. Named after Lord Chandos (Oliver Lyttelton), the first Chairman of the National Theatre Board, it is a traditional picture-frame theatre seating 890 in two tiers, and has an adjustable proscenium with an opening which can range from 34 to 45 ft., a stage height of from 16 up to 29 ft., and a depth of 51 ft. Although its opening production in 1976 was *Hamlet* with Albert *Finney, it is used mainly for new plays and modern classics, the former including *Ayckbourn's *Bedroom Farce* (1977), later seen in the West End, and *Way Upstream* (1982); David *Hare's *Plenty* (1978), *A Map of the World* (1983), and *The Secret Rapture* (1988); Alan *Bennett's *Single Spies* (1988); and David *Mamet's *Speed-the-Plow* (1989). Outstanding revivals have included Ben *Travers's *Plunder* (1978); Arthur *Miller's *Death of a Salesman* (1979); Terence *Rattigan's *The Browning Version* (1980) with Alec *McCowen and Geraldine *McEwan; and Tennessee *Williams's *Cat on a Hot Tin Roof* (1988). There have also been notable productions of older classics such as *Vanbrugh's *The Provok'd Wife* (1980) with Dorothy *Tutin and Geraldine McEwan; *Wilde's *The Importance of Being Earnest* (1982) with Judi *Dench; and *Chekhov's *Wild Honey* (1984) with Ian *McKellen. In 1990 it staged *Sondheim's *Sunday in the Park with George*.

Lytton, Edward George Earle Lytton Bulwer-Lytton, Baron, see BULWER-LYTTON.

Lyubimov, Yuri Petrovich (1917–), outstanding Soviet director, who joined the company at the *Vakhtangov Theatre after the Second World War. A production of *Brecht's *The Good Woman of Setzuan* in 1962, which he staged with a group of his acting students, was so successful that he was made director of the *Taganka Theatre in 1964, transforming it into a thriving centre for avant-garde productions. Although he was nominally the company's Artistic Director, it functioned in a tradition of *Collective Creation that owed much to

*Meyerhold, Vakhtangov, and Brecht. In 1984 he was deprived of his citizenship and went into exile, living in Israel and directing plays and operas in several Western countries, including adaptations of Dostoevsky's novels *Crime and Punishment* and *The Possessed* in both England and America. (He had originally directed the former at the Taganka in 1983.) His citizenship restored in 1989, he directed at the Taganka *A Feast in the Year of the Plague*, based on *Pushkin's writings, which he had staged at Stockholm in 1986. Later in 1989 his production of *Hamlet* at the *Haymarket, Leicester, preceding a world tour, used the image of a vast curtain swinging across the stage which had dominated his earlier *Hamlet* in 1974.

M

Mabou Mines, see LA MAMA EXPERIMENTAL THEATRE CLUB.

McCallum, John, see WITHERS.

McCarthy, Lillah (1875–1960), English actress, who made her first appearance on the stage in 1895 and later in the year joined Ben *Greet's company to play leading roles in Shakespeare. In 1896 she was with Wilson *Barrett, playing Mercia in his play *The Sign of the Cross*, in which part she made her first appearance in the USA. She remained with him until 1904, when she joined *Tree at *Her (then His) Majesty's Theatre. It was then that she first met *Shaw, who became her ardent admirer and a great friend; and for whom she played Nora in *John Bull's Other Island*, Ann Whitefield in the first public performance of *Man and Superman* (both 1905), Gloria in a revival of *You Never Can Tell*, and Jennifer Dubedat in *The Doctor's Dilemma* (both 1906). Also in 1906 she married *Granville-Barker, who had directed all these plays at the *Royal Court Theatre. In 1911 she took over the *Little Theatre, where she presented *Ibsen's *The Master Builder*, appearing in it as Hilda Wangel, and played Margaret Knox in the first production of Shaw's *Fanny's First Play*. In 1912 she was seen as Jocasta in *Sophocles' *Oedipus Rex*, with *Martin-Harvey, and as Iphigenia in *Euripides' *Iphigenia in Tauris*. She created the part of Lavinia in Shaw's *Androcles and the Lion* in 1913. After divorcing Granville-Barker in 1918, she took over the management of the *Kingsway Theatre. Her second marriage took her to Oxford and she virtually left the theatre.

McClintic, Guthrie (1893–1961), American actor, producer, and director, who was for a time a member of Jessie *Bonstelle's stock company in Buffalo and later became assistant stage director to Winthrop *Ames at the *Little Theatre, New York. He then went into management on his own account, and from 1921 until his death he was active in the New York theatre. In 1921 he also married the actress Katharine *Cornell, directing her in Michael Arlen's *The Green Hat* (1925), which first brought him into prominence, and Rudolf Besier's *The Barretts of Wimpole Street* (1931), which marked the beginning of a permanent partnership with her as director and co-producer. In 1936 he directed *Hamlet* on Broadway with John *Gielgud.

McCormick, F. J. [Peter Judge] (1891–1947), Irish actor, considered the most versatile and intelligent player of character parts ever to appear at the *Abbey Theatre in Dublin, where he was first seen in 1918, and remained (except for a brief period in 1921–2) until his death. Among the many roles he created were Seumas Shields in *The Shadow of a Gunman* (1923), Joxer Daly in *Juno and the Paycock* (1924), and Jack Clitheroe in *The Plough and the Stars* (1926), all by Sean *O'Casey; Oedipus in *Yeats's versions of *Sophocles' *Oedipus the King* (1928) and *Oedipus at Colonus* (1934), and the leading role in *The King of the Great Clock Tower* (1935), also by Yeats; and the title-role in George *Shiels's *Professor Tim* (1925).

McCowen, Alec [Alexander] **Duncan** (1925–   ), English actor, who gained his early experience in repertory and was first seen in London in 1950. He was particularly admired as Claverton-Ferry in T. S. *Eliot's *The Elder Statesman* (1958) and was with the *Old Vic company, 1959–61, where he played such varied parts as Richard II, Touchstone in *As You Like It*, and Malvolio in *Twelfth Night*. For the *RSC he was seen as the Fool in *King Lear* (1962) and Father Riccardo Fontana in *Hochhuth's *The Representative* (1963). In 1968 he made an international reputation in the title-role of Peter Luke's *Hadrian VII* (NY, 1969), following it with an outstanding Hamlet at the *Birmingham Repertory Theatre (1969). A year later he was back in London in Christopher *Hampton's *The Philanthropist* (NY, 1971), and in 1973 he played Alceste in *Molière's *The Misanthrope* (NY, 1977) and Martin Dysart in Peter *Shaffer's *Equus* for the *National Theatre. In 1974 he was an excellent Higgins to Diana *Rigg's Eliza in *Shaw's *Pygmalion*; in 1977 he played Antony in *Antony and Cleopatra* for *Prospect and in 1978, in London and New York, gave a remarkable solo performance in which he recited the whole of St Mark's Gospel from memory. He returned to the National Theatre in revivals of *Rattigan's double bill *The Browning Version* and *Harlequinade* in 1980 and *Beckett's *Waiting for Godot* in 1987. Other notable performances were as Hitler in Christopher Hampton's *The Portage to San Cristobal of A.H.* (1982), which included a 25-minute monologue, and another solo performance, as Kipling (1984).

Macdermott, the Great, see MUSIC-HALL.

MacDermott, Norman, see EVERYMAN THEATRE, Hampstead.

MacDonald, Robert David, see CITIZENS' THEATRE.

McEwan [McKeown], **Geraldine** (1932–), English actress, who first appeared with the repertory company at Windsor, where she was born, and from 1951 to 1956 was seen in and around London in a series of light comedies. In 1957 she gave an interesting performance at the *Royal Court as the unhappy and tiresome child Frankie Addams in Carson McCullers's *The Member of the Wedding*, and in 1958 she joined the company at the *Shakespeare Memorial Theatre, going with it to Russia and returning to Stratford in 1961 to play Beatrice in *Much Ado about Nothing* and Ophelia in *Hamlet*. In complete contrast was her role as the suburban prostitute-wife in Giles Cooper's *Everything in the Garden* (1962). In the same year she again proved her versatility by taking over the part of Lady Teazle in *Sheridan's *The School for Scandal*, in which she made her New York début a year later, starring there also in Peter *Shaffer's double bill *The Private Ear* and *The Public Eye*. She then joined the *National Theatre company at the *Old Vic, where she gave an excellent and spirited performance as Angelica in *Congreve's *Love for Love* (1965). She remained with the company for some years, playing in *Feydeau's *A Flea in Her Ear* (1966) with the precision and neatness essential to French farce, in *Strindberg's *Dance of Death* (1967), in *Maugham's *Home and Beauty* (1968), and in Congreve's *The Way of the World* (1969), in which she was an excellent Millamant. In *Webster's *The White Devil* (also 1969), she radiated flamboyant sexuality. After appearing with the National Theatre company again in 1971 in *Giraudoux's *Amphitryon 38*, she was seen in the West End in Jerome Kilty's *Dear Love* (1973), in which she played Elizabeth Barrett Browning, Peter *Nichols's *Chez nous* (1974), and a revival of *Coward's *Look after Lulu* (*Chichester and London, 1978). She returned to the National Theatre in *Rattigan's double bill *The Browning Version* and *Harlequinade*, *Vanbrugh's *The Provok'd Wife* (both in 1980), and in 1983 in Sheridan's *The Rivals*, as Mrs Malaprop. In 1988 she took over from Maggie *Smith in Shaffer's *Lettice and Lovage*. Her highly mannered, clearly articulated mode of speech is ideal for high comedy.

MacGowran, Jack (1918–73), Irish actor, who made his first appearance at the Dublin *Gate Theatre in 1944 in Hilton *Edwards's revival of John *Drinkwater's *Abraham Lincoln*. During the next six years he appeared in numerous productions both at the Gate and at the *Abbey Theatre, and also directed plays at the Abbey's experimental Peacock Theatre. In 1954 he was responsible, with Cyril *Cusack, for a production of *Synge's *The Playboy of the Western World* seen in Paris at the *Théâtre des Nations, and in the same year he made his London début as Young Covey in a revival of *O'Casey's *The Plough and the Stars*. He was later seen in a number of plays by Samuel *Beckett, including *Endgame* (1958), in which he played Clov, and a revival of *Waiting for Godot* (1964), in which he played Lucky (he was Vladimir in a later revival), both at the *Royal Court Theatre. His association with Beckett, on stage, radio, and television, resulted in three solo programmes derived from the playwright's works—*End of Day* (*Dublin Theatre Festival, 1962), *Beginning to End* (Lantern Theatre, Dublin, 1965), and a final and much revised version of *Jack MacGowran in the Works of Samuel Beckett*, produced by Joseph *Papp in New York in 1970. He also gave intelligent and sensitive performances in the title-role of *Ionesco's *Amédée* (1957) and as Harry Hope in *O'Neill's *The Iceman Cometh* (1958). In 1960 he spent a season with the *RSC.

Machiavelli, Niccolò di Bernardo dei (1469–1527), Florentine statesman and political philosopher, whose most famous work is *Il principe* (*The Prince*, 1513). Exiled from the service of the Medicis on suspicion of conspiracy, he gave some of his time and genius to writing comedies of which the best is *La mandragola* (*The Mandrake*), written between 1513 and 1520 and considered one of the masterpieces of the *commedia erudita*. Its sharp, precise prose contains pungent criticism of Florentine society: it portrays the gradual betrayal of its lovely heroine by her credulous husband, her ardent but unscrupulous lover, and her scheming mother, aided by the evil machinations of the corrupt priest Fra Timoteo. In a translation by Ashley *Dukes this was successfully given in London in 1940, and in 1965, in a modernized Italian version, it was revived by Peppino *De Filippo, who played Fra Timoteo. Machiavelli's *Clizia* (1525), derived from the *Casina* of *Plautus, was more strictly neo-classical in form. Its theme of amorous dotage may have been an ironic self-criticism of the author's own feelings for the actress Barbara Salutati.

Machine Play, type of 17th-century French spectacle which made excessive use of the mechanical contrivances and scene-changes developed in connection with the evolution of opera, particularly by *Torelli. The subjects were usually taken from classical mythology, the first French example being *Corneille's *Andromède* (1650); the genre reached its peak with *Molière's *Amphitryon* (1668) and *Psyché* (1671). Most of the machine plays, though not necessarily the best, were produced at the Théâtre du *Marais, which had a large and excellently equipped stage suitable for their production.

Machinery, Stage, see STAGE MACHINERY.

Mackaye, (James Morrison) **Steele** (1842–94), American actor, playwright, theatre designer, pioneer, and inventor. After studying in London and Paris he played Hamlet in London in 1873, repeating the role later in the year in Manchester with Marion *Terry, who was making her first stage appearance, as Ophelia. Later he appeared in New York with a group of students whom he had trained. In order to carry out his ideas he remodelled the old Fifth Avenue Theatre, opening it in 1879 as the *Madison Square Theatre. Although the venture failed, it was at this theatre that Mackaye's own play *Hazel Kirke* (1880) was produced for a long run. It was frequently revived and was seen in London in 1886. After leaving the Madison Square Theatre Mackaye built his own theatre, the *Lyceum, and established there the first school of acting in New York, later known as the American Academy of Dramatic Art. Mackaye, who was thin, dark, nervous, and dynamic, was everything by turns—actor, dramatist, teacher, lecturer. Shortly before his death he planned a vast 'Spectatorium' for the Chicago World Fair, which was never built; but later theatre architects were indebted to its plans for the introduction of many new ideas and methods. He influenced the theatre more by what he thought, dreamed of, and fought for than by what he actually achieved, and none of his plays has survived.

McKellen, Sir Ian Murray (1939–   ), English actor, who made his first appearance at the *Belgrade Theatre as Roper in Robert *Bolt's *A Man for All Seasons* (1961). After major roles in the provinces, such as Henry V, he made his début in London in 1964 in James *Saunders's *A Scent of Flowers* (1964), which was followed by a season with the *National Theatre. He then gave a remarkable performance as the young hero of Donald Howarth's *A Lily in Little India* (1966), wrestling with his nascent sexuality. After *Arbuzov's *The Promise* (London and NY, 1967) he was excellent in Peter *Shaffer's double bill *White Liars* and *Black Comedy* (1968). He set the seal on his growing reputation by playing Richard II (1968) and *Marlowe's *Edward II* (1969) with the *Prospect Theatre Company, showing Richard gradually coming to terms with his own vulnerability and Edward as a wayward and impetuous youth hungering for sensual contacts. He played both parts on tour and in London, as he did Hamlet with the same company in 1971, and in 1972 he helped to found the Actors' Company, in which the actors chose their own plays and shared equal pay, billing, and leading roles. Joining the *RSC in 1974, he laid claim to be *Olivier's successor with his performances as Marlowe's Dr Faustus, as Romeo and Macbeth (both 1976), and in

*Ibsen's *Pillars of the Community* (1977). In 1979 he was seen at the *Royal Court and in the West End in Martin Sherman's play *Bent*, about two homosexuals in a Nazi prison camp, and in 1980 he scored a big success in New York as Salieri in Shaffer's *Amadeus*. Later roles at the National Theatre included Coriolanus (1984) and Platonov in *Chekhov's *Wild Honey* (also 1984; LA and NY, 1986–7). After playing in *Ayckbourn's *Henceforward . . .* (1988) he was highly praised as Iago in *Nunn's intimate *Othello* (1989) for the RSC and as Richard III at the National (1990). He was also seen intermittently, in the USA, London, and elsewhere, in his one-man entertainment 'Acting Shakespeare'.

McKenna, Siobhán (1923–86), Irish actress, who made her first appearance in 1940 with the An Taibhdearc Theatre in Galway, where for two years she acted leading roles in Gaelic, sometimes in English plays translated by herself. From 1944 to 1946 she acted in both Gaelic and English at the *Abbey Theatre, and in 1947 appeared for the first time in London, as Nora Fintry in *Carroll's *The White Steed*. By 1951, when she was seen at the *Edinburgh Festival in Lennox *Robinson's *The White-Headed Boy* and as Pegeen Mike in *Synge's *The Playboy of the Western World*, she had established herself as one of the leading interpreters of Irish parts. Avril in *O'Casey's *Purple Dust* followed in 1953, but her outstanding role at this time was *Shaw's St Joan in London in 1954. She made her New York début in 1955 as Miss Madrigal in Enid *Bagnold's *The Chalk Garden*, returning as St Joan in 1956, and at the *Stratford (Ontario) Festival in 1957 played Viola in *Twelfth Night*. At the *Dublin Theatre Festival in 1960, and later on a European tour and in London, she again played Pegeen Mike, and a year later she was seen as Joan Dark in *Brecht's *St Joan of the Stockyards* (London, 1964). In the 1967 Festival she starred as Cass in Brian *Friel's *The Loves of Cass McGuire*, and the following year at the Abbey she was an excellent Ranevskaya in *Chekhov's *The Cherry Orchard*. In 1970 in London, and again at the *Gate Theatre, Dublin, in 1975, she presented her one-woman show *Here are Ladies*, in which she added to her repertoire parts by *Yeats and *Beckett and characters from the novels of James Joyce. In 1973 in London she gave a moving portrayal of Juno in O'Casey's *Juno and the Paycock*, which she also played in 1966 and 1980. Later roles included Josie in *O'Neill's *A Moon for the Misbegotten* at the Gate, Bessie Burgess in O'Casey's *The Plough and the Stars* at the Abbey and in New York (both in 1976), and Agrippina in *Racine's *Britannicus* in London in 1981. Perhaps the outstanding characteristic of this strongly individualistic actress was her ability, reminiscent of Sara *Allgood, to capture the common humanity beneath the poetic fabric of her Irish heroines. She also directed many plays.

Mackinlay, Jean Sterling, see CHILDREN'S THEATRE and WILLIAMS, HARCOURT.

Macklin [M'Laughlin], Charles (c.1700–97), Irish actor, who in 1716, after a wild and restless boyhood, joined a company of strolling players. Four years later he was in Bath, and in 1725 was engaged by Christopher *Rich for *Lincoln's Inn Fields Theatre. There his natural manner of speaking, which preceded *Garrick's reforms in stage delivery, was unacceptable in the high-toned tragedies of the day, and he returned to the provinces and minor theatres, playing *Harlequin and *Clown at fairs and at *Sadler's Wells. In 1730 he again appeared at Lincoln's Inn Fields, and two years later was engaged for *Drury Lane, where he played secondary comic parts before persuading the management in 1741 to revive The Merchant of Venice, with himself as Shylock. He became famous overnight, rescuing the character from the clutches of the low comedian, and making him a dignified and tragic figure, drawing from Alexander Pope the memorable couplet: 'This is the Jew / That Shakespeare drew'.

With advancing years Macklin became extremely quarrelsome and jealous (he had already killed another actor in a fight over a wig), and moved erratically from one theatre to another, causing trouble backstage and engaging in constant litigation. Apart from Shylock and his Iago to the Othello of Garrick and Spranger *Barry, his most memorable part was Macbeth, which he first played at *Covent Garden in 1773 in something approximating to the dress of a Highland chieftain in place of the red military coat favoured by Garrick. He was the author of several plays, of which two survived well into the 19th century—Love à la Mode (1759), in which he himself played the leading role, Sir Archy McSarcasm, and The Man of the World (1781), in which, in spite of his great age, he again played the lead, Sir Pertinax McSycophant. He made his last appearance on the stage in 1789, when he essayed Shylock but was unable to finish it.

MacLeish, Archibald (1892–1982), American poet and dramatist, who continued the attempts made by Maxwell *Anderson to write plays in poetry adapted to the rhythms of everyday American speech. He had a distinguished career as a poet and lecturer on poetry, but is best remembered in the theatre for J.B., a retelling of the story of Job in modern terms which was seen on Broadway (1958), in London (1961), and in many European countries. Though not wholly successful in portraying living people in terms of a cosmic myth, it nevertheless has moments of authentic tragedy; it was awarded the *Pulitzer Prize for drama. An earlier play, Panic, was staged on Broadway in 1935.

MacLiammóir, Micheál (1899–1978), Irish actor, designer, and dramatist, who as Alfred Willmore appeared on the London stage as a child. He then studied art, travelled widely in Europe, and in 1927 joined Anew *McMaster's Shakespeare company. In 1928 he founded, with Hilton *Edwards, the Galway Theatre (which he directed until 1931) for the production of Gaelic drama, and later the same year the Dublin *Gate Theatre, which opened at the *Abbey's experimental Peacock Theatre with *Ibsen's Peer Gynt. In addition to acting in and designing over 300 productions for the Gate, including Denis *Johnston's first play, The Old Lady Says 'No!' (1929), he acted in London, New York (where at the invitation of Orson *Welles he played Hamlet in 1934), and on tour with the Gate company in Egypt and the Balkans. In 1960 came his first one-man show, on Oscar *Wilde, The Importance of Being Oscar, and in 1963 he presented a second one-man entertainment I Must be Talking to My Friends; a third, Talking about Yeats, was first seen in 1965. He also wrote a number of plays, including Diarmuid agus Grainne (1929), seen in Irish and English; Ill Met by Moonlight (1946), seen in London in 1947; and Prelude in Kazbek Street, which received its première at the *Dublin Theatre Festival in 1973.

McMaster, Anew (1894–1962), Irish actor, who first appeared on the stage in 1911 in Fred *Terry's company. In 1920 he played opposite Peggy O'Neill in Paddy the Next Best Thing, and a year later went to Australia, where among other parts he played Iago to the Othello of Oscar *Asche. In 1925 he founded a company to present Shakespeare on tour, acting, managing, and directing the plays himself. A tall, handsome man, with an imperious face crowned, in his later years, by a fine head of white hair, he 'took the stage' with great dignity, his Shylock, Richard III, Coriolanus, Lear, and Othello being outstanding. He appeared at the *Shakespeare Memorial Theatre in 1933, playing Hamlet and Coriolanus, and then took his company on a tour of the Near East. After the Second World War he toured indefatigably throughout Ireland, and at the time of his death was preparing to play Othello at the *Dublin Theatre Festival to the Iago of Micheál *MacLiammóir.

MacNamara, Brinsley, see ABBEY THEATRE.

Macowan, Michael (1906–80), English actor and director, son of the actor-dramatist **Norman Macowan** (1877–1961), whose plays included The Blue Lagoon (1920), The Infinite Shoeblack (1929), and Glorious Morning (1938). He made his first appearance in 1925 in Charles Macdona's repertory season of *Shaw, but in 1931 abandoned acting for direction. His reputation was established in 1936–9, when he directed many plays at the *Westminster

Theatre, including *Hamlet* (1937), *Troilus and Cressida* (1938), and plays by *Ibsen, *Chekhov, *Shaw (*Heartbreak House*, 1937, *You Never Can Tell*, 1938, *Candida*, 1939), and *O'Neill (notably *Mourning Becomes Electra*, 1937). He was particularly associated with *Priestley's plays, among them *Dangerous Corner* (1938), *Music at Night* (1939), both at the Westminster, and *The Linden Tree* (1947). He directed some of *Fry's early plays, as well as Ibsen's *Rosmersholm* in 1950 and Charles *Morgan's *The River Line* (1952) and *The Burning Glass* (1954). While head of the London Academy of Music and Dramatic Art (LAMDA), 1954–66, he continued to work in the commercial theatre, directing such plays as Chekhov's *The Seagull* in 1956 and Graham *Greene's *The Potting Shed* (1958).

Macqueen-Pope, Walter James (1888–1960), English actor, manager, and theatre historian, who traced his ancestry back to the 18th-century actress Jane Pope of *Drury Lane, and even further back, to Thomas Pope of the *Globe, friend and associate of Shakespeare, and Morgan Pope of the Bear Garden. Though these claims were never fully substantiated, they gave him much pleasure and were seldom challenged. Certainly he merited a distinguished theatrical pedigree, for his devotion to the London theatre (he ignored all others) was whole-hearted and he spent his life in its service. He was at some time in his career business or front-of-house manager of almost every theatre in London, and for 21 years was in charge of publicity at Drury Lane. A stimulating companion, 'Popie' excelled in communicating the glamour and excitement of Edwardian theatre-going, and his inaccuracies were forgiven for the sake of the brilliant light he threw on personalities and events in the theatre of his lifetime. After his death a plaque to his memory was unveiled by Sir Donald *Wolfit in St Paul's Church, Covent Garden.

Macready, William Charles (1793–1873), English actor, one of the finest tragedians of his day. The son of a provincial actor-manager, he was forced by his father's financial difficulties to cut short his schooldays at Rugby and find work in a touring company. In 1810 he was playing Romeo in Birmingham and in 1811 Hamlet in Newcastle. In 1816 he joined the company at *Covent Garden and by 1819 was firmly established both there and at *Drury Lane as the only rival of the great Edmund *Kean. He was much admired in new plays but it was his Hamlet, Lear, Macbeth, and, later, Othello that were universally acclaimed. In 1826 he visited America, making his first appearance at the *Park Theatre in New York, and in 1828 he played Macbeth in Paris, returning later the same year to play Hamlet and Othello. In 1838 he appeared at Covent Garden with Helen

*Faucit in *Bulwer-Lytton's *The Lady of Lyons*. Their fine acting did much to ensure the success of the play, which was frequently revived. Another new play which owed its appearance to Macready's encouragement and initiative was Lord *Byron's *The Two Foscari* (1838), and he also scored a personal triumph in Bulwer-Lytton's *Richelieu* (1839).

In the late 1830s Macready, who had an ungovernable temper, became the implacable rival of the American actor Edwin *Forrest, their mutual animosity leading eventually to the *Astor Place riot in New York in 1849, when several people were killed. Macready never acted in America again. He made his last appearance on the stage at Drury Lane in 1851 as Macbeth. Although he never wavered in his dislike of the profession into which he had been forced, he was scrupulous in performing his theatrical duties both as actor and as manager at both *Patent Theatres, insisting on full rehearsals, particularly for supers and crowd scenes.

Maddermarket Theatre, Norwich, replica of an Elizabethan theatre interior built inside a former chapel in 1921 by **Nugent Monck** (1877–1958) to house the Norwich Players, an amateur company which he founded in 1911. It was altered and enlarged in 1953 and again in 1966 and now holds about 300. Monck, whose career had begun as stage-manager for William *Poel, directed most of the Maddermarket productions himself until 1952. Since its inception the company, still amateur, has staged all 37 of the plays associated with Shakespeare, as well as the full range of British and world drama from ancient Greek to the present day. A new production is put on every month for nine performances, the actors by tradition appearing anonymously.

Maddern, Minnie, see FISKE.

Maddox, Michael, see MEDDOKS.

Madison Square Theatre, New York, on the south side of 24th Street, west of 5th Avenue. This was built on the site of *Daly's first *Fifth Avenue Theatre and was adapted by Steele *Mackaye for use as a repertory theatre on Continental lines, with a company made up of students whom he had trained himself. It opened in 1879, but soon failed and was taken over by Daniel *Frohman who reopened it in 1880 with Mackaye's *Hazel Kirke*, a domestic drama which ran for nearly two years. Viola *Allen made her first appearance in New York here in 1882, in *Hugo's *The Hunchback of Notre Dame*, and in the same year Bronson *Howard's *Young Mrs Winthrop*, one of the first good American plays, was seen. Albert M. *Palmer took over the management in 1885, and in 1889–90 Richard *Mansfield appeared in a series of revivals and new plays, including the first straight-forward translation of *A Doll's House* which, with Beatrice Cameron as Nora, first brought *Ibsen to the attention of the American public. In 1891 the theatre was taken over by Charles Hoyt, and from 1905 it was known by his name. The building was demolished in 1908.

Maeterlinck, Maurice (1862–1949), Belgian poet and dramatist, who in 1911 was awarded the *Nobel Prize for Literature. A *Symbolist, and a forerunner of the Theatre of *Silence, he wrote a number of plays which proved immensely popular in his lifetime, among them *Pelléas et Mélisande* (1893), in which Mrs Patrick *Campbell played in French opposite Sarah *Bernhardt in London. It originally had incidental music by Fauré, but was later used as a basis for an opera by Debussy. Well received in London also was *Le Bourgmestre de Stilmonde* (1918), which provided an excellent part for *Martin-Harvey, who frequently revived it. In America in 1922 Eva *Le Gallienne appeared as Aglavaine in *Aglavaine and Seylsette*, the part originally written for **Georgette Leblanc** (1867–1941), who created most of Maeterlinck's heroines between 1896 and 1910. The universal favourite among Maeterlinck's plays was *L'Oiseau bleu* (1909), first directed in Moscow by *Stanislavsky, and only then in Paris and (as *The Blue Bird*) in London; a sequel, *Les Fiançailles* (1911), seen in London as *The Betrothal* in 1921, was only a modified success. Among other works were *Les Aveugles* (1891), *Intérieur* (1895), and *Ariane et Barbe-Bleu* (1901), used as the libretto of an opera by Dukas.

Mahelot, scenic designer, scene painter, and director at the theatre of the Hôtel de *Bourgogne in Paris, whose *Mémoire*, now in the Bibliothèque Nationale, contains notes on and sketches for the settings of plays performed at the theatre in 1633 and 1634, at a time when the old *multiple setting derived from the *liturgical drama was giving way to the single set of French classical comedy and tragedy. It also gives valuable information on about 71 published plays, and the titles and dates of many unpublished which might otherwise have remained unknown. From the *Mémoire*, which was continued by Laurent, it has been possible to reconstruct some of the settings made for specific plays, among them *Pyramus et Thisbé* by Théophile.

Maintenon, Madame de [*née* Françoise d'Aubigné] (1635–1719), second wife of Louis XIV, whom she married secretly in about 1684. Her first husband was the dramatist *Scarron, on whose household and dramatic works she had a salutary effect. Her ascendancy over the Court was held to be largely responsible for the austerity which reigned there during the 1680s and 1690s, when, in contrast to the encouragement earlier given to *Molière and others,

the public theatre came in for severe disapproval, the King even banishing the actors from the *Comédie-Italienne in 1697 for having put on *La Fausse Prude* by Étienne *Lenoble, which offended her. Private theatricals, however, were another matter, and Madame de Maintenon not only persuaded *Racine, after 12 years' abstention from the theatre, to write the poetic and religious dramas *Esther* (1689) and *Athalie* (1691) for performance at her school in Saint-Cyr for impoverished young ladies, but herself composed for them a number of one-act sketches illustrating well-known proverbs, a type of entertainment very popular in society at the time and later made famous by Alfred de *Musset. These, preserved in manuscript, were not published until 1820.

Mairet, Jean (1604–86), French dramatist, whose reputation suffered in later years from his opposition to *Corneille in the literary war over *Le Cid* (1637). He was, however, the foremost dramatist of his time and the first to formulate and use the theory of the *unities, newly developed in Italy. His first play, *Chryséide et Arimand* (1625), was a tragi-comedy which figured in the repertory of both the Hôtel de *Bourgogne and the *Marais. In the preface to a *pastoral, *Silvanire* (1630), published in 1631 he first put forward his theory of the unity of time and place, and in his most important play, *Sophonisbe* (1634), a tragedy on a Roman theme, he laid the foundations of French classical tragedy. The play already possessed the simplicity, refinement, and concentration needed for such a work, and it had an immediate influence. Mairet remained attached as dramatist to *Montdory's troupe until 1640, then retired from the theatre and entered the diplomatic service.

Maison de Molière, see COMÉDIE-FRANÇAISE.

Maisons de la Culture, see DÉCENTRALISATION DRAMATIQUE.

Majestic Theatre, New York. 1. On Columbus Circle. As the Cosmopolitan, it opened in 1903 with a musical version of *The Wizard of Oz*, followed by another successful musical *Babes in Toyland*. In 1911 it was renamed the Park, reopening with *The Quaker Girl*. Three years later Mrs Patrick *Campbell was seen there in *Shaw's *Pygmalion*, and in 1917 Constance *Collier and Herbert *Tree appeared in a notable revival of *The Merry Wives of Windsor*. A season of light opera followed, which included a long run of *Gilbert and Sullivan's *Ruddigore*, but from 1923 to 1944 the building was used as a cinema. Renamed the International in 1944, it opened again as a theatre and in 1945, as the Columbus Circle Theatre, housed Maurice *Evans in his *GI Hamlet*. It was again known as the International, 1946–9, after which it became a television studio. It was pulled down in 1954, an exhibition hall, the New York Coliseum, being built on the site.

2. On West 44th Street, between Broadway and 8th Avenue. With a seating capacity of 1,655, it opened in 1927 with an ephemeral production which soon gave way to *musical comedy, and in 1928 John *Gielgud made his first appearance in New York as the youthful Grand Duke Alexander in Neumann's *The Patriot*, which had only eight performances. After several more failures the theatre reverted to musical comedy, though in 1935 it housed Michael Chekhov and his Moscow Art Players in a series of Russian plays, and it was also used for several thrillers. In 1945 the *Theatre Guild presented *Carousel*, a musical version of *Molnár's *Liliom*, which had a long run; other successful musicals included *South Pacific* (1949), *The Music Man* (1957), and *Camelot* (1960). In 1963 Gielgud returned successfully as Joseph Surface in his own production of *Sheridan's *The School for Scandal*, and in 1967 *Weiss's *Marat/Sade* had its New York première, the musical *Fiddler on the Roof* moving in from the *Imperial later the same year. Further musicals included the all-Black *The Wiz* (1975), based on *The Wizard of Oz*, and the highly successful *42nd Street*, which transferred from the *Winter Garden in 1981.

Make-Up. Make-up was used in the theatre from the earliest times by all those not wearing masks, their main purpose being to make their faces appear older or younger, more or less ferocious, more or less godlike. The two extremes were the gilding of God's face in *liturgical drama and the elaborate painting of the actors' faces in the *kabuki* plays of Japan. The use of everyday cosmetics to enhance personal beauty was probably brought into the European theatre by actresses. Even in the open-air theatre of the Elizabethans there is evidence of some form of character make-up— black for Negroes, umber for sunburnt peasants, red noses for drunkards, chalk-white for ghosts. In the Restoration theatre the grotesque make-up of some characters was probably nothing more than a desire to mock the extremes of fashion. The general standard of make-up in the 18th century in London theatres appears to have been somewhat low, and one of the excellences attributed to *Garrick was his skill in making up his face to suit the age and character of his part, particularly when he played old men. Before the introduction of modern grease-paints, all make-up was basically a powder, compounded with a greasy substance or with some liquid medium, which was often harmful to the skin and sometimes extremely dangerous, particularly if white lead was used in its composition. By the beginning of the 19th century gas-lighting had recently led to changes in the art of making-up, and although the paints

used were still powder-based, some form of grease was used as a foundation. Grease or oil was used to remove the paint after the performance. Not everyone, however, approved of the use of greasy substances, though the use of cold cream after the performance seems to have been generally adopted by about 1866. A revolution in make-up was achieved in the second half of the 19th century by the introduction of grease-paint, invented in about 1865 by Ludwig Leichner, a Wagnerian opera-singer. He opened his first factory in 1873, and his round sticks, numbered and labelled from 1, light flesh colour, to 8, a reddish brown for Indians (later increased to 20, and by 1938 to 54), were soon to be found in practically every actor's dressing-room. The first sticks were imported into London between 1877 and 1881, and a London branch of L. Leichner Ltd. opened in 1928. The introduction of electric light again caused fundamental changes in theatrical make-up, which has also more recently been influenced by the techniques of film and television. In the modern theatre, where the make-up that embellishes and the make-up that disguises are of equal importance, the art of making-up has reached a high standard, and covers more than the simple painting and lining of the face. To age an actor or actress from 20 to 60 in one evening is now a commonplace, though in many cases such a transformation is achieved not by the player but by a make-up expert. This is perhaps truer of films and television than of the theatre, where the individual actor usually still attends to his own make-up, inventing and discarding his own methods within the limits of the materials available.

Male Impersonation. Since in the early theatre parts were mainly played by men and boys, the question of male impersonation did not arise until actresses had become firmly established. In England this dated from 1660, and it was not long before young actresses, inspired perhaps by their success in the temporary assumption of male attire when playing such parts as Rosalind in *As You Like It*, Viola in *Twelfth Night*, or Silvia in *Farquhar's *The Recruiting Officer*, took over some of the male leads in contemporary comedy, which became known as 'breeches parts'. The most famous of these was Sir Harry Wildair in Farquhar's *The Constant Couple*, which Peg *Woffington played with immense success in 1740. Others successful in breeches parts were Nell *Gwynn, Mrs *Bracegirdle, and particularly Mrs *Mountfort, the last being considered outstanding as Lothario in *Rowe's *The Fair Penitent* and Macheath in *Gay's *The Beggar's Opera*. This fashion for playing *en travesti*, as it was later called, formed an essential part of Regency spectacle and Victorian *extravaganza, the great exponent of the latter being

Mme *Vestris, and it was one of the formative elements in the development of the *principal boy in pantomime. Apart from these appearances in light male roles a number of intrepid actresses essayed such tragic roles as Hamlet, Romeo, and Richard III, among them Charlotte *Cushman and Sarah *Bernhardt, who in 1900 also appeared in a 'serious' breeches part, the young Duc de Reichstadt in *Rostand's L'Aiglon. In a more realistic age such courageous feats are seldom attempted.

In the 19th century and afterwards the term 'male impersonator' was used mainly of women on the *music-halls who sang comic songs in a variety of male costumes, from the man-about-town to the Cockney urchin, with particular emphasis on the more glamorous uniforms of the armed forces. Outstanding among them were Vesta *Tilley, Ella Shields, and Hetty King. The woman dressed as a man is still an effective feature of satirical European cabaret, though the other aspects of male impersonation do not seem to have flourished outside Britain, the USA, and France.

Malina, Judith, see LIVING THEATRE.

Malleson, (William) **Miles** (1888–1969), English actor, who made his first appearance in Liverpool in 1911, and two years later was in London. In 1918 he was seen as Aguecheek in Twelfth Night, and soon established a reputation as an eccentric comedian with his playing of Sir Benjamin Backbite in *Sheridan's The School for Scandal and Launcelot Gobbo in The Merchant of Venice in 1919; Peter Quince in A Midsummer Night's Dream in 1920; and Trinculo in The Tempest in 1921. In 1925 he was seen as Filch in a long run of *Gay's The Beggar's Opera, and he played Scrub in *Farquhar's The Beaux' Stratagem in 1927 and Wittol in *Congreve's The Old Bachelor in 1931. One of his best parts was Foresight in Congreve's Love for Love; he was also much admired as Sir Fretful Plagiary in Sheridan's The Critic at the *Old Vic in 1945, and, in more serious vein, as Old Ekdal in *Ibsen's The Wild Duck in 1948. His best work was done in his own very free adaptations of *Molière, which began in 1950 with The Miser (L'Avare), and continued with Tartuffe, Sganarelle, The School for Wives (L'École de femmes), The Slave of Truth (Le Misanthrope), The Imaginary Invalid (Le Malade imaginaire), and The Prodigious Snob (Le Bourgeois gentilhomme). Although they aroused some controversy, they had the merit of bringing Molière to the English stage in versions which appealed to the average playgoer, and proved invaluable to repertory and provincial theatres. He made one of his rare appearances in a modern play when he was seen as Mr Butterfly in *Ionesco's Rhinoceros (1960).

Malthouse, The, see MELBOURNE THEATRE COMPANY.

Malvern Festival, founded in 1929 by Barry *Jackson, who provided much of the money to initiate and support it, as well as actors from his *Birmingham Repertory Theatre company, distinguished players being imported for special parts. Jackson's long association with Bernard *Shaw led him to devote the first year's programme entirely to his plays, with the first English production of The Apple Cart (with Cedric *Hardwicke and Edith *Evans) and revivals of Back to Methuselah, Caesar and Cleopatra, and Heartbreak House. Shaw became the patron-in-chief of the festival, and more than 20 of his plays were presented there, Geneva and In Good King Charles's Golden Days having their first productions in 1938 and 1939 respectively. Too True to be Good (1932) and The Simpleton of the Unexpected Isles (1935) were also given their first English productions at Malvern. Apart from Shaw, the productions of the festival ranged over 400 years of English drama, from Hickscorner (c.1513) to contemporary plays by *Bridie, *Drinkwater, *Priestley, and others, the first non-Shavian play to be seen being Besier's The Barretts of Wimpole Street (1930). After the 1937 season Jackson withdrew and the festival was run by the manager of the Malvern Theatre. After the Second World War the festival was revived only once, in 1949, when Shaw's Buoyant Billions had its first English production. In 1964 the theatre, seating 799, was taken over by a Trust, and since 1977 the festival has been revived as a Shaw and Elgar Festival each spring, the rest of the year being taken up by visits from touring companies and productions by a resident amateur company. The theatre is now part of the Winter Gardens complex operated by the local authority, but the festival is organized by an independent voluntary body. Shaw's Pygmalion was staged in 1989 to mark the 60th anniversary.

Maly Theatre, Leningrad, seating 460, is situated in a turn-of-the-century block of flats. Founded in 1944, when most other theatre companies had been evacuated during the siege of Leningrad, it presented a wide range of plays. The company now comprises around 50 actors with a repertory of 15 plays and giving 500 performances a year in Leningrad and the surrounding towns and villages. In 1988 it took its production of Alexander Galin's Stars in the Morning Sky to Glasgow and London, its first visit to the West, which was followed by visits to New York and Paris. The company returned to Glasgow in 1990 (London, 1991) with Brothers and Sisters, a two-part play based on a trilogy of novels by Fyodor Abramov.

Maly Theatre, Moscow. This theatre (maly, small, as opposed to bolshoi, big) opened in 1824 with a company which had been in existence since 1806. It is the oldest theatre in the city and the only one to keep the traditional drop-curtain. With its unbroken history it has played an important part in the development of Russian drama, particularly in the 1840s, when *Shchepkin was appearing in such plays as *Gogol's *The Government Inspector* and *Griboyedov's *Woe from Wit*, and *Mochalov in translations of Shakespeare's tragedies made directly from the original and not, as hitherto, from the French. In 1854 the Maly first produced a play by *Ostrovsky, and so began a brilliant partnership which lasted until 1885. The actor who first played many of the leading roles was Prov *Sadovsky, whose descendants continued the connection with the theatre. Other great names connected with the Maly are Maria *Yermolova and **Alexander Lensky** (1847–1908), an actor who joined the company in 1876 and eventually became its Director, being responsible for introducing the plays of *Ibsen to Russian audiences. Having survived the Russian Revolution, the theatre took its place in the theatrical life of Soviet Russia with the production of *Trenev's *Lyubov Yarovaya* in 1926. It has since included many new Soviet plays in its repertoire, but has not neglected the classics, one of its outstanding productions being *Othello* with the veteran actor **Alexander Ostushev** (1874–1953) in the title-role. *Sudakov was Artistic Director, 1937–59, and was followed by Yevgenyi Simonov (see SIMONOV, REUBEN). For many years the theatre was known as the House of Ostrovsky, and the 150th anniversary of its opening was celebrated in 1974 with a revival of Ostrovsky's *The Storm*. Productions of the 1980s included *King Lear* and *Pushkin's *Little Tragedies*. In 1990 the theatre reopened after reconstruction on the same site.

Mamet, David (1947–   ), American dramatist, born in *Chicago, where many of his plays were first staged. He was a co-founder of the St Nicholas Theatre Company, for which he was also playwright in residence and, 1973–6, Artistic Director; among his early plays were the one-act *Duck Variations* and *Sexual Perversity in Chicago* (1974; NY, 1975). In 1978 he became Associate Artistic Director and playwright in residence at the *Goodman. He made his Broadway début in 1977 with his comedy *American Buffalo*, about the abortive plans of two minor crooks to steal a coin collection (*National Theatre, 1978). In *A Life in the Theatre* (NY, 1977; London, 1989), two actors of different generations have various backstage encounters during a repertory season, parts of plays in which they are currently appearing being seen. *The Water Engine* (1977; London, 1989) portrays the hostility of business interests to the invention of a water-driven engine; and *Edmond* (NY, 1982; London, 1985) the self-sought debasement of a middle-class man

in after-dark New York. *Glengarry Glen Ross* (NT, 1983; NY, 1984) casts a beady eye on real estate wheeler-dealers, and *Speed-the-Plow* (NY, 1988; NT, 1989) on Hollywood producers. He also continued to write one-acters such as *Prairie du Chien* and *The Shawl*, shown in a double bill (NY, 1985; London, 1986). Mamet is a major playwright whose success is based on his colloquially American yet lyrical use of language.

Mamoulian, Rouben (1897–1987), American director, born in Georgia, USSR, who worked at the *Vakhtangov in Moscow before going to New York, where between 1923 and 1927 he directed a number of operettas and musicals. He then directed DuBose and Dorothy Heyward's Negro play *Porgy* (1927; London, 1929) for the *Theatre Guild. His later productions, also for the Guild, included *O'Neill's *Marco Millions*, *Wings over Europe* by Maurice *Browne and Robert Nichols (both 1928), and *Turgenev's *A Month in the Country* (1930), in his own adaptation. He returned to the musical with *Porgy and Bess* (1935), the *Gershwins' folk opera based on *Porgy*, Rodgers and *Hammerstein's *Oklahoma!* (1943; London, 1947) and *Carousel* (1945; London, 1950), all for the Theatre Guild, and Kurt Weill and Maxwell *Anderson's *Lost in the Stars* (1949). He was also a distinguished film director.

Manager, see PRODUCER.

Manchester. The first permanent theatre in the city was erected in 1758 and from 1760 to 1775 was used regularly by a company on its way from Leeds to Worcester. It closed when a new theatre was built at the junction of York Street and Spring Gardens, being finally demolished in 1869. The new building, with a royal patent, opened as the Theatre Royal in 1775. Among the well-known players who appeared there were John Philip *Kemble, who at 19 was seen as Othello to the Desdemona of his elder sister Mrs *Siddons, and Mrs Elizabeth *Inchbald. The building was burned down in 1789, but rebuilt and reopened in 1790. It proved too small for the new mass audiences and in 1807 was replaced by a much larger theatre, which soon ruined its first manager, the father of William *Macready. It too was destroyed by fire, in 1844, and the last Theatre Royal to be built in Manchester, in Peter Street, opened in 1845, became a cinema in 1929, was then used for bingo, and is now a discothèque. In 1891 the *Palace Theatre was erected for the use of touring companies, and in 1908 the Gaiety Theatre, a former music-hall, was opened as the first repertory theatre in Britain by Miss *Horniman. The dramatists connected with this theatre, including Harold *Brighouse and Stanley *Houghton, were known as the Manchester School of Drama, the chief characteristic of which was its *realism; their works formed part of the 'new drama' which came into prominence shortly before the First World War. The theatre closed in 1917. The *Opera House was built in 1912, and the *Library Theatre has had its own company since 1952. A university theatre, directly connected with the Manchester *University Department of Drama, opened under Hugh *Hunt in 1966, and for some years housed the 69 Theatre Company, now the *Royal Exchange Theatre Company. Since 1973 the university theatre has housed the Contact Theatre Company, a professional company founded by Hugh Hunt, which also visits schools and youth clubs and runs educational drama workshops. The theatre is used for part of the year by amateur groups from the university. It has an adaptable auditorium seating between 250 and 350.

Manet, see STAGE DIRECTIONS.

Manhattan Opera House, see KOSTER AND BIAL'S MUSIC HALL.

Manners, John Hartley, see TAYLOR, LAURETTE.

Mannheim, city important in the history of the German stage. It already had a long tradition of opera when the first Mannheim National Theatre opened in 1778 under **Wolfgang Heribert, Baron von Dalberg** (1750–1806), who made it one of the foremost playhouses in the country, particularly when after *Ekhof's death later the same year he took over his troupe with *Iffland at its head. Dalberg's greatest service to the German theatre was undoubtedly his support of the young *Schiller, whose *Die Räuber* had its first production at Mannheim in 1782, followed by *Fiesko* and *Kabale und Liebe* in 1784, in which year one of the best plays of the *Sturm und Drang* movement, Leisewitz's *Julius von Tarent*, was also produced. In 1796, partly owing to the rigours of war, the company was disbanded. In 1884 the National Theatre reopened and there was an upsurge of theatrical activity, culminating in 1916 with the first production of Hasenclever's *Der Sohn*, inaugurating a period of *Expressionism. The National Theatre was completely destroyed in 1943, but a new building opened in 1957 with a production of *Die Räuber*, directed by *Piscator, which marked the 175th anniversary of its first production, since when the repertory has included a wide selection of new European plays as well as German classics.

Mansfield, Richard (1854–1907), American actor, born in Berlin. Educated in England and on the Continent, he made his first appearance on the stage in London, and then toured in light opera. In 1882 he made his first appearance in New York, again in light opera; but it was as Baron Chevrial in Feuillet's *A Parisian Romance* (1883) that he first attracted attention. Some of his outstanding parts were the dual title-roles in a dramatization of Stevenson's *Dr Jekyll and Mr Hyde* (1887) and the title-roles in

Clyde *Fitch's *Beau Brummell* (1890), specially written for him, *Rostand's *Cyrano de Bergerac* (1898), and Booth *Tarkington's *Monsieur Beaucaire* (1901). He was first seen as Richard III, later accounted one of his finest parts, during a visit to London in 1889, and among his other Shakespearian roles were Shylock in *The Merchant of Venice*, Brutus in *Julius Caesar*, and Henry V, making a spectacular appearance after Agincourt on a white horse. He was instrumental in introducing *Shaw to America, playing Bluntschli in *Arms and the Man* in 1894 and Dick Dudgeon in *The Devil's Disciple* in 1897, Raina in the first and Judith Anderson in the second being played by his wife **Beatrice Cameron** (1868–1940), whom he married in 1892. She had already played Nora in *Ibsen's *A Doll's House* in 1889, and it may have been under her influence that Mansfield, who in general had little sympathy with the 'new drama' though he much admired Ibsen's poetic prose, put on in his last season, 1906–7, the first production in English of Ibsen's *Peer Gynt*, with himself in the title-role.

Mansfield Theatre, New York, see BROOKS ATKINSON THEATRE.

Manteau d'Harlequin, see FALSE (PROS)CENIUM.

Mantle, Robert Burns (1873–1948), American dramatic critic, who in 1898 became dramatic editor of the *Denver Times*. Later he worked as Sunday editor of the *Chicago Tribune*, and from 1922 until his retirement in 1943 he was dramatic critic of the *New York Daily News*. He edited till his death a series of play anthologies which began in 1920 with *The Best Plays of 1919–1920*; each volume contains extracts from 10 of the season's productions, together with an annotated index of every play produced in New York and at leading regional theatres during the year. It was supplemented by three more volumes covering the years 1894 to 1919. This useful chronicle of the modern American theatre continues with the subtitle *The Burns Mantle Theater Yearbook*. Mantle was also the author of *American Playwrights of To-Day* (1935) and co-edited *A Treasury of the Theatre* (1938).

Marais, Théâtre du, Paris, early French theatre, which is thought to have opened on 31 Dec. 1634 in a converted tennis-court in the rue Vieille-du-Temple, with a company under *Montdory which had been appearing in *Corneille's early comedies, and was responsible early in 1637 for the production of his great tragedy *Le Cid*. Among other notable productions at this time was *Tristan l'Hermite's *La Mariane*, in which Montdory gave a powerful rendering of Herod. After he left, the theatre went through some bad times. Its best actors joined the rival company at the Hôtel de *Bourgogne, to which Corneille also gave his new plays, and those that were left were forced to revert to the playing of popular farces. They were however lucky in the return of *Jodelet, for whom some excellent new farces were written, and later they specialized in spectacular performances with a good deal of imported Italian scenery and machinery. The theatre never regained the place it had held previously in public esteem, particularly after a bad fire in 1644 burnt down the old building. The contract for its replacement on the same site indicates that its overall dimensions were about 114 ft. by 36 ft. The newly housed company survived until 1673, when it was amalgamated with the company of *Molière, who had just died, and by the combined companies' fusion in 1680 with that of the Hôtel de Bourgogne became part of the *Comédie-Française, the Marais stage being finally abandoned.

Marble, Danforth (1810–49), American actor, originally a silversmith. After some experience as an amateur, he joined a professional company in 1831, and proved himself admirable in Yankee dialect parts. In 1837 he made a great success in the title-role of *Sam Patch*, an anonymous play which he may have written or arranged himself. It proved to be so popular, especially along the Mississippi and at the *Bowery Theatre in New York, that it was followed by two other 'Sam Patch' plays. In 1844 Marble visited London and the English provinces, being received with enthusiastic applause and going also to Glasgow and Dublin, in such typical pieces from his repertory as the younger *Colman's *Jonathan in England* and Joseph Jones's *The People's Lawyer*, which gave full scope to his inimitable assumption of Yankee characteristics. He married an actress, one of the four daughters of William *Warren the elder, and died young at the height of his popularity.

Marceau, Marcel (1923–   ), French actor, the finest modern exponent of *mime or, as he prefers to call it, mimodrama. He joined the company of Jean-Louis *Barrault in 1945, but a year later abandoned conventional acting to study mime, basing his work on the character of the 19th-century French Pierrot and evolving his own Bip, a white-faced clown with sailor trousers and striped jacket. In this part, which he first played at the tiny Théâtre de Poche in Paris in 1946, he toured all over the world, accompanied by supporting players he had trained himself. He also evolved short pieces of concerted mime, including one based on *Gogol's *The Overcoat*, and longer symbolic dramas such as his own *The Mask-Maker*, and with the aid of a screen contrived to appear almost simultaneously as two sharply contrasted characters—David and Goliath, or the Hunter and the Hunted. His work gave immense impetus to the study of mime by young actors, who also benefited from the many demonstrations of his technique which he gave to students. His École de mimodrame de Paris opened at the Théâtre de la *Porte-Saint-Martin in 1978.

Mardzhanov [Mardzhanishvili], **Konstantin Alexandrovich** (1872–1933), Soviet director, the virtual founder of the modern theatre in his native Georgia. After some experience in the provinces, he joined the *Moscow Art Theatre in 1910, remaining there for three years and directing *Ibsen's *Peer Gynt* in 1912. In 1913 he organized the so-called Free Theatre, having *Taïrov and Alisa Koonen among his actors. His methods of staging had a noticeable influence on Taïrov's subsequent work at the *Kamerny Theatre. After the Revolution Mardzhanov was in Kiev, but in 1922 returned to Tbilisi to reorganize the Georgian theatre along Soviet lines. Mardzhanov encouraged the writing and production of new Georgian plays, many of which he directed himself; but he did not neglect the older classics, one of his most successful productions being *The Merry Wives of Windsor*. In 1928 he went to Kutaisi, where he founded the Second State Georgian Theatre which was transferred to Tbilisi in 1930 and renamed the Mardzhanov after his death. The plays he directed there included works by *Pogodin, *Afinogenov, *Toller, and Shakespeare.

Marigny, Théâtre, Paris, theatre built in 1850 in the gardens of the Champs-Élysées. The son of *Deburau took it over in 1858 and opened it under his own name, appearing there in many of his father's old parts, but it was not a success and soon became the Théâtre des Champs-Élysées (not to be confused with the later *Comédie des Champs-Élysées). Success continued to elude it until as the Folies-Marigny it became a home of *vaudeville after the demolition in 1862 of the *boulevard du Temple. It was pulled down in 1881, and a circular building intended for a panorama was erected on the site. This became a music-hall in 1896 and took its present name in 1901. In 1925 it was completely redecorated under a new owner and became the most elegant playhouse in Paris, with spacious approaches, a luxurious auditorium, comfortable seating, and an impressive cupola. The *Comédie-Française used it for matinées and special performances, and moved there temporarily during the rebuilding of its own theatre in 1937 and again in 1974. From 1946 to 1956 it housed the company of Jean-Louis *Barrault and his wife Madeleine *Renaud. After renovation in 1964 it reopened as a home for plays, musical comedies, and operettas. Its productions have included *Shaffer's *Amadeus*, *Sartre's *Kean*, and *Rostand's *Cyrano de Bergerac*.

Marinelli, Karl (1744–1803), Austrian actor, dramatist, and impresario, who in his play *Der Ungar in Wien* (*The Hungarian in Vienna*, 1773) first introduced the figure of the light-hearted romantic Magyar, later a stereotype of Viennese folk-comedy and operetta which survived long enough to be ridiculed by *Shaw in *Arms and the Man* (1894). When Joseph II, as part of his rationalist reforms, banished the old impromptu burlesques from the Burgtheater, Marinelli provided a new home for them at the Leopoldstädter Theater, which opened in 1781, and there the great comedian Johann Laroche established the comic character of *Kasperle. Marinelli was also the first to recognize the talent of Anton Hasenhut, later famous as *Thaddädl; and with his own plays and adaptations he helped to establish the popularity of the traditional Viennese folk theatre.

Marionette, see PUPPET.

Marivaux, Pierre Carlet de Chamblain de (1688–1763), French dramatist, the dialogue of whose plays is characterized by a peculiarly paradoxical and sensitive style later known as *marivaudage*, first contemptuously, then in admiration of its superb subtlety. His first success came with a one-act comedy, *Arlequin poli par l'amour* (1720), seen at the *Comédie-Italienne. He continued to write for that theatre with such plays as *La Surprise de l'amour* (1722); *La Double Inconstance* (1723); *Le Jeu de l'amour et du hasard* (1730); *Les Fausses Confidences* (1737); and *L'Epreuve* (1740). He also wrote occasionally for the *Comédie-Française, where *La Seconde Surprise de l'amour* (1727) and *Le Legs* (1736) were well received.

Marivaux's delicate, psychological theatre, in which the major emphasis is on the female roles, was not on the whole popular with his contemporaries, who preferred the cruder emotions of *La Chaussée's *comédie larmoyante*. Lost sight of during the period preceding the French Revolution, he came back into favour again with the Restoration, when his plays had a considerable influence on the work of Alfred de *Musset. It was not, however, until the 20th century that his work was fully appreciated. His plays were then frequently revived at the Comédie-Française, and he had an unexpected success at the *Marigny Theatre in the 1940s–1950s, when Madeleine *Renaud proved herself the perfect interpreter of Marivaux's heroines. Although French companies have performed Marivaux's plays both in London and in New York with some success, the subtlety of his dialogue makes it almost impossible to translate him adequately into English.

Market Theatre Company, Johannesburg, non-racial company founded in 1973 by a group of actors. It converted the old Indian fruit market building into an arts complex which opened in 1976 and now contains three theatres. Without government sanction or funding, it stages the work of South African writers, directors, and actors both in Johannesburg and

all over the world. Notable productions have included several plays by *Fugard; *Woza Albert* by Percy Mtwa and Mbongeni Ngema, who wrote and appeared in it, in which Christ visits modern South Africa; and *Poppie Nongena* by Sandra Kotze and Elsa Joubert, about the life of a Black woman in South Africa. International and classical theatre is also presented.

Mark Hellinger Theatre, New York, in West 51st Street, between Broadway and 8th Avenue, seating 1,581. Originally a cinema which opened in 1930, it became the Fifty-First Street Theatre on two occasions, first in 1936 with *Sweet River*, a new version of *Uncle Tom's Cabin* by George *Abbott which was taken off after five performances, and again in 1940 for a short run of *Romeo and Juliet* with Laurence *Olivier and Vivien *Leigh. In 1949 it became a theatre again and was named after **Mark Hellinger** (1903–47), a newspaper columnist, dramatist for stage and screen, and film producer, reputedly the first newspaperman to write a column solely concerned with Broadway. From 1956 to 1962 the theatre was occupied by the musical *My Fair Lady*. Later musicals seen there included *On a Clear Day You Can See Forever* (1965); *Coco* (1969), starring Katharine *Hepburn as Chanel; and *Jesus Christ Superstar* (1971). In 1979 *Sugar Babies*, a tribute to the great days of *burlesque, began a long run.

Mark Taper Forum/Center Theatre Group, Los Angeles, one of the most important and enterprising theatre organizations in Los Angeles, a city which with several notable commercial theatres, including the James A. Doolittle Theatre (formerly the Huntington Hartford) and the Shubert Theatre, ranks as one of the most exciting theatrical areas in the United States outside New York. Center Theatre Group began with the founding of the Theatre Group in 1959 under the auspices of the University of California at Los Angeles and Artistic Director John *Houseman. Its first full productions, at UCLA in 1960, were T. S. *Eliot's *Murder in the Cathedral* and *Chekhov's *Three Sisters*, both directed by Houseman. In 1966 the Theatre Group moved downtown to become the Center Theatre Group at the Music Center of Los Angeles. The non-profit organization now had a home in two theatres, the Ahmanson (2,071 seats) and the Mark Taper Forum. The Taper, which has a thrust stage and a semicircular auditorium seating 760, has no permanent company, but assembles excellent casts and designers from all over the country. Under **Gordon Davidson** (1933–), its Artistic Director/Producer since 1965, the Taper has been responsible for the world premières of several important plays, including Daniel Berrigan's *The Trial of the Catonville Nine* (1970; NY, 1971), Michael Cristofer's *The Shadow Box* (1975; NY, 1977), and Mark Medoff's *Children of a Lesser God*

(1980; NY, 1980; London, 1981), all of which were directed by Davidson himself. Other Taper programmes include Taper, Too, an intimate 90-seat theatre for new and experimental works; ITP (Improvisational Theatre Project), the Taper's theatre for young people; and Sundays at the Itchey Foot, the Taper's literary cabaret, a series of Sunday afternoon entertainments at the Itchey Foot Ristorante. New Theatre For Now, Taper Lab, and the Taper Lab New Work Festival are designed to help create and explore new work in progress.

Marlowe, Christopher (1564–93), playwright of the English Renaissance and an important figure in the development of the Elizabethan stage, which he helped to liberate from the influence of medieval drama and the Tudor *interlude. His first play, *Tamburlaine the Great, Part I*, was produced in about 1587 by the *Admiral's Men with Edward *Alleyn in the title-role. *Part II* was seen in the following year. Written in flamboyant blank verse of great poetic beauty, these two plays had an undoubted influence on Shakespeare, who must have seen them soon after his arrival in London. *Tamburlaine* was followed by *The Tragical History of Dr Faustus*, based on a German medieval legend (see FAUST). Though written and probably produced in 1589, this was not printed until 1604, and then only in a fragmentary condition, with comic scenes, featuring the Devil, interpolated by a later hand. As originally planned, it had much in common with the earlier *morality play, and even in its mangled condition was extremely popular. Marlowe's other plays were *The Massacre at Paris*, of uncertain date, which survives only in a much-mutilated text; *The Jew of Malta* (*c.*1590), whose chief character, Barabas, may have contributed something to Shakespeare's Shylock in *The Merchant of Venice*; and *Edward II* (*c.*1592), which has affinities with Shakespeare's *Richard II*. Marlowe, who was often in danger of arrest for his outspoken and atheistical opinions, died young in a tavern brawl, probably assassinated because of his secret-service activities. The theory that he survived the brawl and was spirited away, to return later and write most of Shakespeare's plays, is now discredited.

Marlowe, Julia [Sarah Frances Frost] (1866–1950), American actress, who was on the stage as a child and made her adult début in New York in 1887, as Parthenia in G. W. Lovell's *Ingomar*. She was immediately successful and began a long career as a leading actress, being at her best as Shakespeare's heroines. She was also good in standard comedy, playing Lydia Languish in *Sheridan's *The Rivals* with *Jefferson and Mrs John *Drew, Julia in Sheridan *Knowles's *The Hunchback*, and Pauline in *Bulwer-Lytton's *The Lady of Lyons*. She married as her second husband E. H. *Sothern,

playing Juliet to his Romeo in 1904. Three years later she made her first appearance in London, being well received in Shakespearian and other parts. She retired for a time in 1915, but returned to play mainly in Shakespeare until her final retirement in 1924.

Marlowe Society, see CAMBRIDGE.

Marquis Theatre, New York, on Broadway at 46th Street. Seating 1,600 in two tiers, it opened in 1986, the first new musical theatre on Broadway in 13 years, and only the third built since the 1930s. It is housed in the Marriott Marquis Hotel, whose erection had caused controversy because it entailed the demolition of the *Helen Hayes, *Morosco, and Bijou theatres. The theatre's auditorium features undulating designs, avoiding sharp angles, and no seat is more than 80 ft. from the stage. The theatre opened with concerts by Shirley Bassey, and the first musical was *Me and My Girl* imported from London.

Mars, Mlle [Anne-Françoise-Hippolyte Boutet] (1779–1847), French actress, younger daughter of the actor-dramatist Monvel by a provincial actress. She appeared on the stage as a child, playing in Paris and at Versailles under Mlle *Montansier, and in 1795 made her first appearance at the *Comédie-Française. When the company was reconstituted after the upheavals of the Revolution, she was again a member of it, and embarked on a long and successful career. She was at her best in the comedies of *Molière, but was also much liked in such dramas as *Henri III et sa cour* (1829) by *Dumas *père* and *Hugo's *Hernani* (1830). She retired in 1841, making her last appearances on 31 Mar. as Elmire in Molière's *Tartuffe* and Silvia in *Marivaux's *Le Jeu de l'amour et du hasard*. A beautiful woman with a lovely voice, she continued to play young parts until she was over 60.

Marshall, Norman (1901–80), English director and theatre manager. After gaining some experience on tour, he was appointed in 1926 one of the directors under Terence *Gray of the *Cambridge Festival Theatre, where he was responsible for some interesting productions. In 1932, under his own management, he directed there the first production in England of *O'Neill's *Marco Millions*. Two years later he took over the direction of the *Gate Theatre in London, where he presented an annual Gate *revue, as well as a varied programme of new and uncommercial plays, several of which—among them Elsie Schauffler's *Parnell* (1936), *Housman's *Victoria Regina* (1937), and *Steinbeck's *Of Mice and Men* (1939)—were later seen in the West End. After serving in the army during the Second World War he returned to London to direct Robert *Sherwood's *The Petrified Forest* (1942), and then formed his own company with which he toured extensively, at the same time continuing to direct in London such new plays as Beckwith's *A Soldier for Christmas* (1944), Ginsbury's *The First Gentleman* (1945), with Robert *Morley as the Prince Regent, and *Ustinov's *The Indifferent Shepherd* (1948). Under the auspices of the British Council he then toured Europe and India (where he was born) in abridged versions of Shakespeare's plays, and in 1952 was invited to direct Ben *Jonson's *Volpone* at the *Cameri Theatre in Tel Aviv. He continued for some years to direct plays in London, and was active in the planning of the *National Theatre, as well as lecturing and writing on modern drama.

Marshall Theatre, see RICHMOND, Va.

Marston, John (*c*.1575–1634), English dramatist, satirized by Ben *Jonson as Crispinus in *The Poetaster* (1601). His first plays were Italianate tragedies, *Antonio and Mellida* and *Antonio's Revenge*, acted by the Children of Paul's (see BOY ACTORS) in 1599. They may also have put on a comedy, *What You Will* (1601), which was followed by the best of Marston's works, *The Malcontent* and *The Dutch Courtesan* (both 1604), and another tragedy, *Sophonisba* (1606). A *droll based on *The Dutch Courtesan* was published in Kirkman's *The Wits* (1662), as *The Cheater Cheated*, and was later adapted by Aphra *Behn as *The Revenge; or, A Match in Newgate* (1680). In 1605 Marston was implicated in the trouble over *Eastward Ho!*, of which he seems to have been the chief author, with Jonson and *Chapman, and only escaped imprisonment by a hasty trip abroad. He returned to the theatre with a tragedy, *The Insatiate Countess* (1610), of which he was probably not sole author, and, having taken orders in 1609, became vicar of Christchurch, Hampshire, 1616–31.

Martin, Mary (1913–90), American singer and actress, who first attracted attention in the Broadway musical *Leave it to Me* (1938), in which she sang 'My Heart Belongs to Daddy', a song with which she was ever after associated. After several years in Hollywood she returned to Broadway in another musical, *One Touch of Venus* (1943), and made her first appearance in London at *Drury Lane as Elena Salvador in *Coward's *Pacific 1860* (1946). After touring the United States in *Annie Get Your Gun* (1947–8), she created the role of Ensign Nellie Forbush in *South Pacific* (NY, 1949; London, 1951), and was then seen in her first non-singing role, in Krasna's *Kind Sir* (1953). She subsequently starred in a musical version of *Barrie's *Peter Pan* (1954), and in a revival of Thornton *Wilder's *The Skin of Our Teeth* (1955). In 1959 she created another famous role, Maria Rainer in the musical *The Sound of Music*, and in 1965 at Drury Lane she played Dolly Levi in another musical, *Hello, Dolly!* In the following year she starred on Broadway in *I Do! I Do!*, the musical version of Jan de Hartog's two-character play

The Fourposter, which ran for over a year. After a year's tour, 1968–9, she virtually retired from the stage, though she was seen briefly in New York in 1978 in *Arbuzov's *Do You Turn Somersaults?*

Martin Beck Theatre, New York, in West 45th Street, between 8th and 9th Avenues. With a seating capacity of 1,280, it opened in 1924 with *Madame Pompadour*, and has since housed many other successful musicals. It was also used by the *Theatre Guild for Robert Nichols and Maurice *Browne's *Wings over Europe* (1928); for the first American production of *Shaw's *The Apple Cart* (1930); and by the *Group Theatre for its initial productions, *Green's *The House of Connelly* and *Sherwood's *Reunion in Vienna* (both 1931). Katharine *Cornell was seen in 1934–5 in a repertory which included *Romeo and Juliet*, Besier's *The Barretts of Wimpole Street*, and *Buckstone's *The Flowers of the Forest*. Other productions included Maxwell *Anderson's *Winterset* (1935) and *High Tor* (1937), Lillian *Hellman's *Watch on the Rhine* (1941), *O'Neill's *The Iceman Cometh* (1946), and John Patrick's *The Teahouse of the August Moon* (1953). Edward *Albee's adaptation of *The Ballad of the Sad Café* (1963), from a novella by Carson McCullers, and his own plays *A Delicate Balance* (1966) and *All Over* (1971) were produced here. Later came Alan *Bennett's *Habeas Corpus* (1975) with Donald *Sinden, a revival of Lillian Hellman's *The Little Foxes* (1981) starring Elizabeth Taylor and Maureen *Stapleton, and the musical *The Rink* (1984).

Martinelli, Drusiano (?–1606/8) and **Tristano** (*c*.1557–1630), Italian actors. Drusiano may have been a member of the first *commedia dell'arte troupe which visited England, in 1577–8. He was a good actor, but eclipsed by the excellence of his wife Angelica Alberigi, who with her husband directed a company known as the *Uniti. Tristano, who is believed to have been the first actor to play *Arlecchino, is first found as a member of the *Confidenti, but appears to have been of a restless and quarrelsome disposition, and is found with many different groups. He was very popular in Paris, where he appeared on several occasions.

Martínez Sierra, Gregorio (1881–1948), Spanish dramatist, whose works are more notable for delicacy and quiet humour than for action or excitement, and have lately been neglected. He was much influenced by *Benavente, and by *Maeterlinck, whose plays he translated into Spanish. He was himself fortunate to find sympathetic translators in Harley and Helen *Granville-Barker, whose version of Sierra's best known play *Canción da cuna* (1910), as *The Cradle Song*, was seen in New York in 1921, and in London in 1926 in a double bill with *The Lover* (*El enamorado*, 1913). *El reino de Dios* (1916), which as *The Kingdom of God* was seen in London in 1927, is best remembered as the play with which Ethel *Barrymore, in a fine and moving performance, opened in 1928 the theatre in New York named after her. More important than his plays was Sierra's work as a director in charge of the Teatro Eslava from 1917 to 1925, when he introduced to Madrid audiences the new techniques imported into the peninsula by the Catalan Adriá Gual. He also staged for the first time many contemporary plays in translation, and new or little known Spanish works, including the first of *García Lorca's plays La maleficio de la mariposa (The Butterfly's Curse, 1920).

Martin-Harvey, Sir John (1863–1944), English actor-manager. He made his first appearance on the stage in 1881 and a year later joined *Irving's company at the *Lyceum, where he remained for 14 years. He left the Lyceum in 1896 to appear with other managements, notably in *Maeterlinck's Pelléas and Mélisande (1898) with Mrs Patrick *Campbell, but returned there as manager in 1899. His first production was the overwhelmingly successful The Only Way, a dramatization of *Dickens's A Tale of Two Cities, in which he played the hero, Sydney Carton. He became closely identified with the part, and found himself constantly forced to revive it to satisfy the demands of the public, thus limiting the time and energy available for other and perhaps more worthwhile roles. In 1904 he played Hamlet for the first time, and he was later seen as Richard III, Henry V, and Petruchio in The Taming of the Shrew. In 1912 he gave a magnificent performance in *Reinhardt's production of *Sophocles' Oedipus Rex at *Covent Garden. In later years he revived many of his old parts, but was also seen in new plays, among them Maeterlinck's The Burgomaster of Stilemonde (1918), and in the title-role of the old morality play Everyman in 1923. During the 1920s he toured North America, particularly Canada, with great success, and on his return to London added to his repertory two plays by *Shaw, being seen in The Shewing-Up of Blanco Posnet in 1926 and as Richard Dudgeon in The Devil's Disciple in 1930. His last appearances, up to 1939 when he retired, were mainly in revivals of his best known parts. A handsome man, with clear-cut features and a distinguished presence, he was regarded by many as the lineal descendant of Irving, and his death broke the last link with the Victorian stage. He married in 1889 the actress Angelita Helena Margarita de Silva Ferro (1869–1949), who as **Nina de Silva** was his leading lady for many years.

Martini, Fausto Maria (1886–1931), Italian dramatist, an exponent of the Crepuscolari or 'twilight' school of drama, whose plays reflect an increasingly hopeless acceptance of bourgeois values. In Il giglio nero (The Black Lily, 1913) he portrays the disturbing effect of city attitudes on a quiet provincial couple, and in Il fiore sotto gli occhi (The Bloom behind the Eyes, 1921) he preaches the value of monotony and drabness. Between these two plays came a very successful venture into the commercial theatre with Ridi, pagliaccio! (1919), a commonplace tale of frustration leading to suicide which as Laugh, Clown, Laugh! was staged in New York in 1923; but with L'altra Nanetta (The Other Nanette, 1923) Martini reveals the influence of *Pirandello, and in his next two plays, La facciata (The Face, 1924) and La sera del 30 (The Night of the 30th, 1926), he approaches the Theatre of *Silence of J.-J. *Bernard.

Martinson Hall, New York, see PUBLIC THEATRE.

Mask, see MASQUE.

Masking Units, in a theatre, single elements of *scenery specially designed to conceal the backstage area and fly-gallery from the audience. The same effect can be achieved by the use of black or dark-coloured draperies. An actor is said to be 'masking' another if, inadvertently or otherwise (and if done on purpose, except on the instructions of the director, it is a serious fault), he gets between him and the audience, so that he cannot be seen properly.

Masks, Theatrical. The wearing of masks in the theatre derives from the use of animal skins and heads in primitive religious rituals. In the Greek theatre, masks served, in an all-male company, to distinguish between the male and female characters and to show the age and chief characteristic of each—hate, anger, fear, cunning, stupidity. In tragedy the mask gave dignity and a certain remoteness to demi-gods and heroes, and also enabled one actor to play several parts by changing his mask. In comedy the mask helped to unify the *chorus (which, as can be seen in the plays of *Aristophanes, wore identical masks of such creatures as frogs, birds, and horses) and served as an additional source of humour, particularly with the comic masks of slaves. The Roman theatre took over the use of masks from the Greeks, adopting for tragedy the later exaggerated form with a high peak (onkos) over the forehead. Many fine copies of classical masks, in marble, still survive and can be seen in museums. They are also shown in several wall-paintings and bas-reliefs. The golden masks worn by God and the archangels in some versions of the medieval *mystery play may have been a survival of the Greek tragic mask or an independent discovery of the new European theatre, but the devils' masks, though often comic in intention, seem nevertheless by their horrific animal forms to be linked to early primitive religious usage. They were usually made of painted leather, and many excellent specimens can be seen in European museums, particularly in Germany. The comic actors of the later *commedia dell'arte* companies always wore masks, usually a small black 'cat-mask' which left the lower part of the face bare. Otherwise masks, which continued to be an essential factor in the Japanese *nō* play and other Far Eastern theatres, were discarded in Europe, and they are seldom seen on stage, though they were sometimes used for special effects by such writers as *Yeats and *O'Neill, and later by John *Arden in *The Happy Haven* (1960); by Peter *Shaffer in *The Royal Hunt of the Sun* (1964) and *Equus* (1973); and in the *National Theatre production of *Aeschylus' *Oresteia* (1981). Apart from such isolated examples, the main use of masks at present is in the training of drama students, on whom they seem to have a liberating effect, particularly in improvisation. The making of masks has also provided a useful subject for handicraft in schools and elsewhere. The old English name for the black mask used by the early Tudor actor or 'guizard' in the Court *masque was 'visor'.

The Latin word *persona*, meaning mask, was used by *Terence in the sense of 'character', whence our expression *dramatis personae*, 'the characters in the play'. In the *commedia dell'arte* the word was used of the mask *and* of the person wearing it. Masks were originally made of carved wood or painted linen, later of painted cork, leather, or canvas, and later still of papier mâché or lightweight plastics.

Mask Theatre, see BELFAST CIVIC ARTS THEATRE.

Mason, Marshall, see CIRCLE REPERTORY COMPANY.

Masque (originally Mask, the French spelling being first used by Ben *Jonson), spectacular entertainment which combined music and poetry with scenery and elaborate costumes. It derived originally from a primitive folk ritual featuring the arrival of guests, usually in disguise, bearing gifts to a king or nobleman, who with his household then joined the visitors in a ceremonial dance. The presentation of the gifts soon became an excuse for flattering speeches, while the wearing of outlandish or beautiful costumes and *masks, or visors, led to miming and dancing as a prelude to the final dance. The early, relatively simple, form of the masque was known as a *disguising, and is part of the folk tradition that includes the *mumming play. In Renaissance Italy, mainly under the influence of Lorenzo de' Medici, it became a vehicle for song, dance, scenery, and machinery, one of its non-dramatic offshoots being the elaborate Trionfo, or Triumph. At the French Court it gave rise to the simple *ballet de cour* and the more spectacular *mascarade* (from which is derived 'masquerade'), and eventually the *comédie-ballet. In the 16th century it came back under its new name to Tudor England, where maskers played before the king in elaborate

dresses, with all the appurtenances of scenery, machinery, and rich allegorical speech. In Elizabethan times the formula proved useful for the entertainment of the Queen, either in her own palace or during her 'progresses' throughout the land. Shakespeare makes fun of a simple country masque in *Love's Labour's Lost*, and uses the form seriously for typical early 'disguisings' in *Timon of Athens* and *The Tempest*. This latter already shows some of the elaboration reached by the Court masques prepared for James I and Charles I by Ben Jonson (appointed Court Poet in 1603) and the scenic designer Inigo *Jones. Their first joint work was the Twelfth Night masque of 1605, their best probably *Oberon the Fairy Prince* in 1611. One of Jonson's innovations was the anti-masque, known also as the antemasque, because it preceded the main entertainment, or the antic masque, because it employed earlier elements of antic or grotesque dancing. First introduced in 1609, the anti-masque provided a violent contrast to the main theme, as Hell before Heaven, War before Peace, Storm before Calm. The simplicity of the early masque, in which the performers appeared in one guise only, later gave way to the double masque, in which they were seen in two different groups of characters—fishermen and market-women, for instance, or sailors and milkmaids. In time the literary content of the masque diminished, and the spectacular aspect, particularly the dancing, in which Charles I and Henrietta Maria became performers after the fashion of Louis XIV, became more important. This led Jonson, after constant altercations with Inigo Jones, to withdraw, his last masque being performed in 1634. The Civil War put an end to the masque, which was never revived; but it had provided the means of introducing into England the new Italian scenery, and the Restoration theatre was to take over many of its spectacular effects. The decorative frame set up for the masque in a ballroom became the *proscenium arch, behind which Inigo Jones's movable shutters or *wings, trebled or quadrupled, ran in *grooves to open or close in front of a painted backcloth, or, less often, what Jones called a 'sceane of releeve', consisting of cutout pieces on various planes. As this had to be prepared in advance and shown to the audience by drawing back the shutters, it was termed a Set Scene, whence the modern use of the word *set for the scenic components of a play.

Milton's *Comus* (1634), though entitled 'a masque', is in reality a *pastoral, and was probably called a masque to distinguish it from the plays given in the public theatre.

Masque Theatre, New York, see JOHN GOLDEN THEATRE 2.

Massey, Raymond Hart (1896–1984), American actor and director, born in Toronto. Through his brother he became associated with the University of Toronto's Hart House Theatre after the First World War, in which he served with the Canadian forces, and in 1922 he made his début there as Rosmer in *Ibsen's *Rosmersholm*. He then went to London and made his professional début as Jack in *O'Neill's *In the Zone* at the *Everyman Theatre, returning there in 1926 as joint manager. He made his New York début as Hamlet in 1931 and for several years was seen in both New York and London, his roles including Ethan Frome in the dramatization of Edith Wharton's novel of that name (NY, 1936) and Harry Van in *Sherwood's *Idiot's Delight* (London, 1938). His finest performance was as Abraham Lincoln in Sherwood's *Abe Lincoln in Illinois* (NY, 1938), a part which made good use of his dark brooding looks, rangy physique, and distinctive voice, and which he played until 1940. His later stage work, confined almost entirely to the USA, included notable appearances in New York in revivals of *Shaw's *The Doctor's Dilemma* (1941), *Candida* (1942), and *Pygmalion* (1945), and *Strindberg's *The Father* (1949). In the opening season at Stratford, Conn., in 1955 (see AMERICAN SHAKESPEARE THEATRE) he played Brutus in *Julius Caesar* and Prospero in *The Tempest*, and among his later roles were Mr Zuss in Archibald *Macleish's *J.B.* (NY, 1958) and Tom Garrison in Robert *Anderson's *I Never Sang for My Father* (London, 1970). He also directed many plays, and is the author of the play *The Hanging Judge* (1952), based on Bruce Hamilton's novel.

His children **Daniel** (1933–   ) and **Anna** (1937–   ) were born in England and are both well known on the stage. Daniel played Charles Surface in *Sheridan's *The School for Scandal* (1962), Captain Absolute in his *The Rivals* (1966), and John Worthing in *Wilde's *The Importance of Being Earnest* (1968). He was later seen in several roles at the *National Theatre, among them the title-role in Horváth's *Don Juan Comes Back from the War* (1976), Robert in *Pinter's *Betrayal* (1978), John Tanner in the full-length version of Shaw's *Man and Superman*, and the title-role in *Molière's *The Hypochondriac* (both 1981). He also starred in the musicals *She Loves Me* (NY, 1963) and *Sondheim's *Follies* (London, 1987). Anna Massey achieved a big success in her first stage appearance, in William Douglas *Home's *The Reluctant Débutante* (London, 1955; NY, 1956). Notable later roles were Annie Sullivan in William Gibson's *The Miracle Worker* (1961), Lady Teazle in Sheridan's *The School for Scandal* (1962), Jennifer Dubedat in Shaw's *The Doctor's Dilemma* (1963), and Laura Wingfield in Tennessee *Williams's *The Glass Menagerie* (1965). She appeared for the National Theatre in several productions including Shaw's *Heartbreak House* (1975) and Simon *Gray's *Close of Play* (1979).

Massinger, Philip (1583–1640), English playwright, who is believed to have written, either alone or in collaboration, about 40 plays, of which perhaps half survive. Eight are known to have been destroyed by *Warburton's cook, who used the pages of their unique copies to line her pie-dishes. Early in his career, Massinger is believed to have worked with *Fletcher on *The Two Noble Kinsmen* and Shakespeare's *Henry VIII* (both 1613), but the first of his own surviving works was a tragedy, *The Duke of Milan* (1620). This was followed by his best-known play, a comedy entitled *A New Way to Pay Old Debts* (1623). Its chief character, Sir Giles Overreach, provided a fine part for a succession of leading actors, including, after its revival in the late 17th century, Edmund *Kean in 1816, and Donald *Wolfit in 1950. Among Massinger's later plays the best are *The Roman Actor* (1626), also revived by Kean, and two comedies, *The City Madam* (1632) and *The Guardian* (1633).

Master of the Revels, official of the Royal household, first appointed in 1494 to serve under the *Lord Chamberlain in connection with entertainments at the Court of Henry VII. At first a temporary position, it became in 1545 a permanent appointment held by Sir Thomas Cawarden, who supervised the revels for the coronation of Elizabeth I in 1558. The Revels Office then became concerned mainly with censorship. As unofficial Master, Thomas Blagrove served the Revels Office faithfully for 57 years, but eventually the post went to a courtier who did little but license plays for performance and collect the fees charged for all public productions. The best-known Master was **Sir Henry Herbert** (1596–1673), appointed in 1622, whose records are an important source of information on the theatre of his time. On the passing of the Licensing Act of 1737 the censorship of plays (abolished in 1968) became the direct responsibility of the Lord Chamberlain, and the old office of Master of the Revels became extinct.

Mastersingers, members of the musical and literary guilds which flourished in the larger towns of southern Germany, particularly in Nuremberg, in the 15th and 16th centuries. The activities of such a guild are portrayed in Wagner's opera *Die Meistersinger von Nürnberg* (1868), one of the main characters being the musician and dramatist Hans *Sachs. Other Mastersingers who are known to have written plays—mainly short farces—are **Hans Rosenplüt** (*fl.* 15th century) and **Hans Folz** (c.1450–1515). The Mastersingers were the lineal descendants of the medieval Minnesingers.

Mathews, Charles (1776–1835), English actor and entertainer, who from childhood showed amazing powers of mimicry, joined to a most retentive memory and an intense desire to go on the stage. He made his first appearance in Dublin in 1794 and after some years in the English provinces, mostly under Tate *Wilkinson at York, he appeared in London, at the *Haymarket Theatre, in 1803, soon gaining an enviable reputation as an eccentric comedian. He later appeared at both *Covent Garden and *Drury Lane, and was much admired as Falstaff in *Henry IV*, Sir Archy MacSarcasm in *Macklin's *Love à la Mode*, and Sir Peter Teazle in *Sheridan's *The School for Scandal*. In 1808 he conceived the idea of the one-man entertainment in which he appeared in London and the provinces, as well as in America, for more than 20 years. The first of these was *The Mail Coach Adventure; or, Rambles in Yorkshire*, in which Mathews's second wife, the actress **Anne Jackson** (1782–1869), played a small part. From 1812 Mathews appeared alone in a series of entertainments which from 1817, when Mathews was in Brighton, became known as *Mr Mathews at Home*. These 'At Homes', given in London in the winter and throughout the provinces during the summer, became immensely successful. Originally a programme of comic songs linked together with depictions of eccentric characters based partly on observation, partly on intuition, they developed into short plays to which many writers of comedy contributed; though the overall product was always attributable to Mathews himself, except in such cases as the younger *Colman's *The Actor of All Work; or, First and Second Floor* (1817), in which a manager is shown interviewing a number of applicants for a place in his company. This gave Mathews the opportunity of displaying his powers of mimicry in a bewildering series of totally dissimilar characters. He continued to appear on stage from time to time, in spite of being lame from a carriage accident which occurred in 1814, among his later parts being Goldfinch in *Holcroft's *The Road to Ruin* and Dr Pangloss in the younger Colman's *The Heir at Law*. He returned to America for the last time in 1834, but was already in poor health and died at Liverpool on the return journey. By his second wife he was the father of the actor Charles James *Mathews.

Mathews, Charles James (1803–78), English actor-manager, the only son of Charles *Mathews. Trained as an architect, he had had little connection with the theatre until on the death of his father in 1835 he briefly succeeded him as joint manager with Frederick *Yates of the *Adelphi Theatre. Later the same year he made his first professional appearance at the *Olympic. Like his father he was an excellent mimic, and one of his most popular sketches was *Patter v. Clatter* (1838), in which he played five parts. Also in 1838 he married Mme *Vestris, and on their return from a visit to New York they took over *Covent Garden,

where they staged some fine productions including *Boucicault's *London Assurance* (1841), in which Mathews played Dazzle, always one of his best parts. The venture was not a success financially, and hoping to recover their losses they moved to the *Lyceum Theatre, which proved even more disastrous. Mme Vestris died in the midst of their bankruptcy, and Mathews continued to act, but for other managements. He also visited America again, returning with his second wife, the actress **Lizzie Davenport** (?–1899). With her help he eventually extricated himself from his financial embarrassments, and embarked on a more successful, though less adventurous, career which lasted until his death. Tragedy and pathos were outside his range, but he was inimitable not only as Dazzle but in such parts as Puff in *Sheridan's *The Critic* and Plumper in W. B. *Jerrold's *Cool as a Cucumber* (1851). He also appeared with his second wife in an entertainment, reminiscent of his father's, called *Mr and Mrs Mathews at Home*. He had not the solid gifts of the older Mathews, but much charm and delicacy tempered his high spirits and made him, within certain limits, one of the best light comedians of his day.

Mathurins, Théâtre des, Paris, in the rue des Mathurins. This was built in 1906 for Sacha *Guitry, and became a leading house for *boulevard plays until 1934, when it was taken over by Georges and Ludmilla *Pitoëff, who had already had a season there in 1927–8. They remained there until Georges's death in 1939, one of the first plays they introduced to Paris being *Pirandello's *Questa sera si recita a soggetto* as *Ce soir, on improvise*; Pitoëff himself played the part of the director Hinkfuss. He and his wife appeared frequently in the productions, making the Mathurins one of the outstanding theatres of Paris. The last part Georges played there was Doctor Stockman in *Ibsen's *An Enemy of the People*. Maria *Casarès made her début at the Mathurins in 1942, and frequently acted there. From 1951 to 1981 the theatre was directed by the widow of the actor Harry Baur. Since 1984 it has been run by Gérard Caillaud, whose productions have included plays by Brian *Friel and Sławomir *Mrożek, and *Claudel's *Partage de midi*. A Théâtre des Petits Mathurins was opened in 1988.

Matthews, A(lbert) **E**(dward) (?1870–1960), English actor, son of a Christy Minstrel (see MINSTREL SHOW) and grand-nephew of *Grimaldi's pupil, the *Clown Tom Matthews. He began his career as a call-boy at the *Princess's Theatre in 1886, and continued to act almost up to the time of his death. He never appeared in Shakespeare or in classical plays, which he rightly considered outside his range, but in his own line he was inimitable, with a sure technique which enabled him to seem at his most careless when he was most in control. He excelled in farce, and in his early years toured England, South Africa, and Australia in such plays as *Pinero's *Dandy Dick* and *The Magistrate*, *Hawtrey's *The Private Secretary*, and Brandon *Thomas's *Charley's Aunt*. He returned to England in 1896, was in the first production of R. C. Carton's *Lord and Lady Algy* (1898), and then created a wide range of parts, including Cosmo Grey in *Barrie's *Alice Sit-By-The-Fire* and Eustace Jackson in St John Hankin's *The Return of the Prodigal* (both 1905). He made his first appearance in New York in 1910, returning many times, and was as popular there as in London. Among his last and most successful parts were the Earl of Lister in *The Chiltern Hundreds* (1947) and *The Manor of Northstead* (1954), both by William Douglas *Home. In private life Matthews, who was known affectionately as 'Matty', was an eccentric of dry humour, who refused to take his success seriously and posed as the bluff country gentleman. In youth he added 10 years to his age for fear of seeming too young for the parts he wanted, and in old age cut off 10 years in case he was thought too old to go on acting, but it now seems certain that he was 90 at the time of his death.

Matthews, Jessie [Margaret] (1907–81), English singer and dancer. She made her début at the age of 12 in the children's play *Bluebell in Fairyland* by Seymour *Hicks, and appeared in her teens in revue in London and New York. She then starred for *Cochran in the revues *One Dam Thing after Another* (1927), *This Year of Grace* (1928), and *Wake Up and Dream* (1929), in the last of which she was also seen in New York, returning to London to appear in her first musical comedies, *Ever Green* (1930), by Rodgers and *Hart, and *Hold My Hand* (1931). During this period she introduced such famous songs as *Coward's 'A Room with a View', Cole *Porter's 'Let's Do It', and Rodgers and Hart's 'My Heart Stood Still' and 'Dancing on the Ceiling'. By the early 1930s she had become Britain's highest paid theatrical star, and she then became an international film star. She returned to the West End in 1940 in the revue *Come Out to Play*, and in 1942 played Sally in Jerome Kern's musical *Wild Rose*. After the Second World War she continued to appear in revue but began to be seen more often in straight parts, mostly outside London, in such plays as *Rattigan's *The Browning Version* and *Harlequinade* (1949), *Shaffer's *Five Finger Exercise* (1960), and Frank Marcus's *The Killing of Sister George* (1971).

Mattocks, Mrs, see HALLAM, LEWIS, the elder.

Maude, Cyril Francis (1862–1951), English actor-manager, intended for the army or the Church, who preferred the theatre and while travelling in America in 1884 made his first appearance on the stage at Denver, Colo., in

East Lynne. After appearing in New York he returned to England, where he scored his first outstanding success as Cayley Drummle in *Pinero's *The Second Mrs Tanqueray* (1893) at the *St James's Theatre with George *Alexander. In 1896, in partnership with Frederick Harrison, he took over the *Haymarket, where he put on a number of excellent productions with a distinguished company led by his wife Winifred *Emery. On leaving the Haymarket in 1905 Maude took over the Avenue Theatre (later the *Playhouse), where he was seen in Clyde *Fitch's *Toddles* (1907) and Austin Strong's version of *Rip Van Winkle* (1911). While on tour in America in 1913 he made a great success as Andrew Bullivant in Hodges's *Grumpy*, which he repeated in London a year later. He gave up the Playhouse in 1915 and again went on tour, reappearing in London in Sydney Blow and Douglas Hoare's *Lord Richard in the Pantry*. He retired in 1927, returning briefly from time to time and making his final appearance in 1933. On his 80th birthday he emerged from retirement in Devon to play Sir Peter Teazle in *Sheridan's *The School for Scandal*, always one of his best parts, for charity. He was for many years President of the Royal Academy of Dramatic Art.

Maugham, W(illiam) **Somerset** (1874–1965), English novelist and dramatist. He trained as a doctor and was well known as a novelist before his first play *A Man of Honour* was produced in 1904. The height of his achievement as a playwright was reached in 1908, when he had four plays running in London—*Lady Frederick* (NY, 1908), *Jack Straw* (NY, 1908), *Mrs Dot* (NY, 1910), and *The Explorer* (NY, 1912). For the next 25 years he was prolific and, at least with the comedies of manners which formed the bulk of his output, fashionable. The comedies grew more pointed in their social and sexual observation with *Our Betters* (NY, 1917; London, 1923), about social climbing; *The Circle* (London and NY, 1921), usually considered his best play, in which a woman who deserted her son finds that his wife is about to do the same 30 years later; *The Constant Wife* (NY, 1926; London, 1927); and *The Breadwinner* (1930; NY, 1931). He also wrote the melodrama *The Letter* (London and NY, 1927), based on his own short story, and serious plays such as *The Sacred Flame* (NY, 1928; London, 1929), about mercy killing, and *For Services Rendered* (1932; NY, 1933), about post-war disillusionment. After the comparative failure of *Sheppey* (1933; NY, 1944), another serious play in which a barber who wins a fortune tries to put Christ's precepts on wealth-sharing into action, Maugham stopped writing for the theatre. Several of his non-dramatic works were adapted for the stage by other writers, notably his short stories *Rain*

(1922) and *Jane* (1947), the latter by S. N. *Behrman.

Maurstad, Toralv (1926–), Norwegian actor, who from his début in 1949 until 1967 worked mainly at the Oslo Nye Teater and the *Nationaltheatret, in plays which ranged from Shakespeare, *Holberg, and *Ibsen to those of more recent dramatists. In 1967 he became manager of the Oslo Nye Teater, where he directed a number of productions and where his performances included the Master of Ceremonies in the musical *Cabaret* in 1968 and the young James Tyrone in *O'Neill's *A Moon for the Misbegotten*. He was manager of the Nationaltheatret, 1978–87, his productions there including Peter *Shaffer's *Amadeus* in 1980.

His father **Alfred Maurstad** (1896–1967), also an actor, made his début in 1920, and from 1931 was one of the leading players with the Nationaltheatret in Oslo. He is remembered mainly for his roles in Holberg, *Bjørnson, and Ibsen. Alfred's wife **Tordis Maurstad** (1901–), an actress, worked mainly with the *Norske Teatret. She was a player of passion, well suited to such strongly defined classical roles as the Antigone of *Sophocles and the Medea of *Euripides, as well as many parts in Shakespeare, Ibsen, *Strindberg, *Chekhov, and O'Neill.

Maxine Elliott's Theatre, New York, on West 39th Street, between Broadway and 6th Avenue. This was built for Maxine *Elliott, and opened in 1908. The first outstanding success, in the following April, was Jerome K. *Jerome's *The Passing of the Third-Floor Back*. *Synge's *The Playboy of the Western World*, played by the *Abbey Theatre company on their first American tour, caused a riot in 1911. In 1922 a dramatization of *Maugham's short story *Rain* began a run of 648 performances, and another Maugham success was *The Constant Wife* (1926), with Ethel *Barrymore. Later productions were *Twelfth Night* (1930), with Jane *Cowl as Viola, and *Pirandello's *As You Desire Me* (1931), with Judith *Anderson. Lillian *Hellman's *The Children's Hour* was staged in 1934, and two years later the theatre was taken over for a season by the *Federal Theatre Project with *Horse Eats Hat* (a translation of *Labiche's *Un chapeau de paille d'Italie*) and *Marlowe's *Doctor Faustus*, both directed by Orson *Welles. *Separate Rooms* (1940) by Carole and Dinehart began its long run at this theatre, but from 1941 it was used only for broadcasting, and it was demolished late in 1959.

Mayakovsky, Vladimir Vladimirovich (1894–1930), Soviet poet, dramatist, painter, actor, director, and film scenarist, after whom many public institutions have been named. As a youth he was several times arrested for revolutionary activity, and in 1917 used his undoubted talents in support of the new régime.

His first full-length play in verse, *Mystery-Bouffe*, was staged by *Meyerhold in Petrograd on the first anniversary of the October Revolution, and is usually regarded as the first Soviet play. His next important play, a satire entitled *The Bed-Bug*, was commissioned by Meyerhold in 1929; it was staged by him at his own theatre with music by Shostakovich, but was not an unqualified success. It was not seen professionally in London until 1962. Mayakovsky's next play, *The Bath House*—'Drama in Six Acts with a Circus and Fireworks'—again produced by Meyerhold, in 1930, was even less well received. Worn out by the constant strain of meeting deadlines—his literary output in his last years was prodigious—and by an unhappy love affair, he committed suicide. Several of his plays were later revived, notably by *Taïrov.

Mayakovsky Theatre, Moscow, founded in 1922 as the Theatre of the Revolution for the production of propaganda plays on Soviet themes. Its first outstanding director was *Popov, who raised its standards considerably, introducing many new and worthwhile plays such as *Pogodin's *Poem about an Axe* (1931), as well as excellent productions of European classics. His policy was continued from 1943 by *Okhlopkov, under whom, in 1954, the theatre was renamed in honour of *Mayakovsky. Okhlopkov remained in charge until his death in 1967, his outstanding productions including a controversial *Hamlet* (1954) and *Brecht's *Mother Courage* (1960). He was succeeded by Andrei Goncharov, and later productions included Tennessee *Williams's *A Streetcar Named Desire* (1970), *Ostrovsky's *It's All in the Family* (1975), and works by *Chekhov, *Bulgakov, and Arthur *Miller. Goncharov's production of Boris Vassiliev's *Tomorrow was War* (1986) was seen in London a year later at the *Lyttelton Theatre.

Mayne, Rutherford [Samuel Waddell] (1878–1967), Irish dramatist, chiefly associated with the Ulster *Group Theatre for which he began to write soon after its foundation in 1904. Most of his plays have been seen only in Ulster, although *The Drone* (1908) was produced in the USA in 1913; *The Troth* (1908) had its first production in London; and *Red Turf* (1911), *Peter* (1930), and *Bridgehead* (1934) were first seen at the *Abbey Theatre in Dublin. Like most Ulster dramatists Mayne concentrated on local issues, but some of his plays achieve a broad view of the problems of rural life and have been translated into Dutch, Norwegian, and Swedish.

Mazarine Floor, see MEZZANINE FLOOR.

Meadow Players, see OXFORD PLAYHOUSE.

Mechane, in the ancient Greek theatre, piece of stage machinery in the form of a large crane which enabled characters, and objects such as chariots, to appear to fly through the air. It was also used to lower a character, usually a god, from the top of the stage building so that he could resolve the complexities of the plot—the *deus ex machina.

Meddoks, Mikhail Egrovich [Michael Maddox] (1747–1825), British-born Russian impresario, who in 1767 arrived in St Petersburg with an exhibition of mechanical dolls. Ten years later he was in Moscow, where he was engaged by Prince Ourusov to manage the theatre, built under a licence from the Tsar which gave him the sole right to provide dramatic entertainments in the capital for a period of 10 years. After the destruction of his theatre by fire in 1780, Ourusov ceded his monopoly to Maddox, who built a theatre on Petrovsky Square (where the Bolshoi Theatre now stands), and opened with a company drawn mainly from the private theatres of the nobility. It flourished until its destruction by fire in 1795; it was not reopened. One unexpected result of Maddox's work was the adoption of the name Vauxhall (voksal), which he gave to the amusement park he opened in Moscow, as the Russian word for a railway station, the first one in the city being built on part of Maddox's park. His own name, first transmuted to Meddoks, later became Medok.

Medley, Matt, see ASTON.

Medwall, Henry (fl. 1490–1514), early English dramatist whose work was practically unknown until in 1919 the manuscript of his *Fulgens and Lucrece* came to light in a London saleroom. An *interlude, performed in two parts as an entertainment at a banquet, it was probably acted in 1497, and as an example of secular drama is much earlier than anything hitherto known. With its story of the wooing of Lucretia and comic subplot of the wooing of her maid, it foreshadows the mingling of romantic and comic elements which was to be a feature of later Elizabethan drama.

Meggs, Mrs Mary (?–1691), known as Orange Moll, a well-known figure in the early days of the Restoration theatre. A widow, living in the parish of St Paul's, Covent Garden, she was granted a licence in 1663 to sell oranges and other eatables in the Theatre Royal, *Drury Lane. For this she paid £100 down and 6s. 8d. for every day the theatre was open, which seems to show that the business was a lucrative one. The fire which destroyed the theatre in 1672 was believed to have started under the stairs where she kept her wares. In his *Diary* *Pepys refers several times to Orange Moll and her orange-girls, of whom Nell *Gwynn was one, usually in connection with items of theatrical scandal. Towards the end of her life she was frequently involved in trouble with the management, and when in 1682 the companies of Drury Lane and *Dorset Garden were amalgamated a new orange-woman was appointed. This led to constant disputes, and the matter was still unsettled when Mrs Meggs died.

Mei Lanfang (1894–1961), Chinese actor, the only one to become known outside his own country. The son and grandson of actors, he began his stage training at the age of 9 and made his first public appearance at 14. Between 1919 and 1935 he visited Japan, the United States, Europe, and Russia, where he enjoyed the friendship of *Stanislavsky and *Nemirovich-Danchenko. Renowned for his exquisitely delicate playing of dan, or female characters, he was the first to combine the dramatic techniques of the five roles in *Peking Opera into which they are divided, and his meteoric rise to fame after the First World War gave the female role the predominant place formerly held by the laosheng, or elderly male role. During the Chinese civil war Mei Lanfang lived in retirement, signifying his resolution not to appear on stage by allowing his beard and moustaches to grow; but in 1949 he returned to Peking and was appointed President of the Research Institute of Chinese Drama. He then returned to the theatre and in 1958 celebrated his stage jubilee, retiring a year later.

Meininger Company, troupe of actors resident at the Court of George II, Duke of Saxe-Meiningen, led by his morganatic wife the actress **Ellen Franz** (1839–1923). The Duke, who directed the plays himself and also designed the costumes and scenery, was ably assisted by the actor **Ludwig Chronegk** (1837–91), who joined him in 1866 and became responsible for the general direction and discipline of the company. The innovations for which the Meininger company later became famous had a great influence throughout Europe, firmly establishing the creative and interpretive role of the director and pioneering ensemble acting. Historically accurate scenery and costumes were used, and the setting of the chief actors within the scene took the place of formal groupings, while the handling of crowd scenes was revitalized by making the crowd a personage of the drama—every member an actor in his own right, yet the whole responding to the needs of the moment in a unified way. By the use of steps and rostrums the action was kept moving on different levels, and the inadequacy of the conventional painted set was overcome by moving from two-dimensional to three-dimensional scenery, making use of the *box-set. Though the Duke never sought to abolish stage waits and so give Shakespeare's plays in one continuous, rapid flow, his other reforms were exemplary, and even included the requirement that star actors should from time to time play minor roles. In 1874 the Meininger company appeared for the first time outside its own town when it visited *Berlin and in 1881 it was seen in London, appearing at *Drury Lane in *Julius Caesar, Twelfth Night*, and *The Winter's Tale*, all in German, as well as in a number of German and other classics. In the following years up to 1890, when Chronegk's health broke down and the tours were discontinued, the company visited 38 cities in Europe; *Stanislavsky saw the Meiningers in Moscow on their second visit there in 1890, and *Antoine saw them in Brussels; thus the two men who were to become the greatest exponents of stage *realism both came under the Meiningen influence, which through them spread far into the 20th century.

Melbourne Theatre Company, Australia's largest and most active professional drama company. It was founded in 1953 as the Union Theatre Repertory Company for the University of Melbourne, at whose Union Theatre it first appeared. The company was responsible in 1955 for the first production of Ray *Lawler's *Summer of the Seventeenth Doll*. It began performing full-time in the Russell Street Theatre in 1966 when the Union Theatre was no longer available. Two years later it was adopted its present name. In 1973 the company took over the St Martin's Theatre, moving in 1977 to the Athenaeum Theatre, where in 1979 it opened the Athenaeum 2 for the presentation of 'alternative' experimental plays. In 1984 the company moved to the Victorian Arts Centre and made its home in the Playhouse, still retaining the Russell Street Theatre as a venue. Almost half its repertoire is Australian.

Melbourne also has the Playbox company, which has staged over 100 Australian plays since its foundation in 1976. Its original home was burnt down in 1984, and in 1990 it occupied a refurbished three-storey building, The Malthouse, containing the three-tiered Merlyn, seating 500, and the Beckett, seating 200.

Mellon, Harriot (1777–1837), English actress, who was with a strolling company at Stafford in 1795 when *Sheridan saw her and engaged her for *Drury Lane, where she remained until her retirement in 1815 on her marriage to the banker Thomas Coutts. She was at her best as the light impertinent chambermaids of comedy. After her first husband's death she married the Duke of St Albans, leaving the vast Coutts fortune to the daughter of Sir Francis Burdett, later the Baroness Burdett-Coutts, friend and patroness of Sir Henry *Irving.

Melmoth, Mrs Charlotte (1749–1823), American actress. Born in England (her maiden name is unknown), she ran away from school with an actor named **Courtney Melmoth** [Samuel Jackson Pratt] (1749–1814), and appeared with him in the provinces. They soon separated, but she continued to act under her married name, and was seen both at *Covent Garden and *Drury Lane before leaving for New York in 1793, where she made her début with the *American Company as Euphrasia in Arthur *Murphy's *The Grecian Daughter*. The excellence of her acting, particularly as Lady Macbeth, made her a universal favourite, and caused many more tragedies to be added to the current repertoire. She was one of the leading actresses at the *Park Theatre, New York, when it first opened in 1798, and after *Dunlap's bankruptcy was at the *Chestnut Street Theatre in Philadelphia. She retired in 1812, and became a teacher of elocution in New York until her death.

Melodrama, type of play popular all over Europe in the 19th century. The term derives from the use of *incidental music in spoken dramas, which became customary in German theatres during the 18th century, and from the French *mélodrame*, a dumb show accompanied by music; its application to Gothic tales of horror and mystery, vice, and virtue triumphant stems from the early works of *Goethe (*Götz von Berlichingen*, 1773) and *Schiller (*Die Räuber*, 1782), and its most important authors on the Continent were *Kotzebue and *Pixérécourt. It was first introduced into England through translations of their plays, particularly those made by Thomas *Holcroft, whose *A Tale of Mystery* (1802), based on Pixérécourt's *Coelina; ou, L'Enfant de mystère* (1800), was the first work in England to be labelled a melodrama. Gradually the music became less important, and the setting of the plays less Gothic. *The Brigand* (1829) by *Planché was one of the last of the old-fashioned musical melodramas; the setting of *Jerrold's *Fifteen Years of a Drunkard's Life* (1828) heralded an era of domestic melodrama, which ran concurrently with a vogue for plays based on real-life or legendary crimes—the anonymous *Maria Marten; or, The Murder in the Red Barn*, which became a classic of melodrama in the 1830s; *Fitzball's *Jonathan Bradford; or, The Murder at the Roadside Inn* (1823); and Dibdin Pitt's *Sweeney Todd; or, The Fiend of Fleet Street* (1847).

The growth of a middle-class audience produced a new type of melodrama, notably at the *Adelphi Theatre under *Buckstone. While the rougher elements on the Surrey side enjoyed the horrors of real life borrowed from *Les Bohémiens de Paris* (1843), with its glimpses of the Paris or London underworld in slums and sewers, prosperous merchant families enjoyed the equally spectacular but less violent domestic tragedies of the elder *Dumas, among them *Pauline* (1840), seen by Queen Victoria at the *Princess's Theatre in 1851, and *The Corsican Brothers* (1852), the latter adapted by *Boucicault. Among his other adaptations was one of *Les Pauvres de Paris*, which was first seen in Liverpool in 1864 as *The Poor* (or *The Streets*) *of Liverpool*. A new phenomenon at this time was the sudden success of the numerous dramatizations of popular novels by women writers—Harriet Beecher Stowe's *Uncle Tom's Cabin* (1852), Mrs Henry Wood's *East Lynne* (1861), and Miss Braddon's *Lady Audley's Secret* (1862). Few of the prolific dramatists of the time bothered to concoct their own plots, though the exercise of the *copyright laws in the 1860s began to inhibit their wholesale piracy. None the less, all the melodramas staged by *Irving at the *Lyceum, from Leopold Lewis's *The Bells* in 1871 to Boucicault's *The Corsican Brothers* in 1880, originated on the Continent. Other actor-managers had their greatest successes with dramatizations of novels—*Tree with Du Maurier's *Trilby* (1895), *Alexander with Anthony Hope's *The Prisoner of Zenda* (1896), *Martin-Harvey with *Dickens's *A Tale of Two Cities*, retitled *The Only Way* (1899), and Fred *Terry with Baroness Orczy's *The Scarlet Pimpernel* (1903). Some exceptions over the years were *The Silver King* (1882) by Henry Arthur *Jones and Herman; *The Sign of the Cross* (1895) by Wilson *Barrett; and the nautical melodramas popularized by William *Terriss at the Adelphi. The turn of the century saw spectacular melodramas staged at *Drury Lane, with shipwrecks, railway accidents, earthquakes, and horse-racing, and the joint productions of the *Melville brothers with *The Worst Woman in London* (1899) and *The Bad Girl of the Family* (1909). Melodrama had come a long way from its original simplicity, which equated poverty with virtue and wealth with villainy. The day of true melodrama was over, and occasional revivals of such classic examples as *Maria Marten*, *East Lynne*, and *The Streets of London* have been played as comic caricatures, but melodramatic elements continue to flourish in the theatre as they have done since the time of *Euripides.

Melodramma, Italian play with music, each element being equally important. It evolved during the 18th century from the earlier *pastoral, the chief writers connected with it being Apostolo Zeno and Metastasio, whose librettos have since been used by innumerable composers of opera. The term is sometimes used for the *monodrama popular in Germany during the late 18th century.

Melpomene, see MUSES.

Melucha Theatre, see MOSCOW STATE JEWISH THEATRE.

Melville, Alan, see REVUE.

Melville, Walter (1875–1937) and **Frederick** (1879–1938), two brothers who for 25 years were joint proprietors of the *Lyceum Theatre in London, where they produced annually a spectacular *pantomime, usually written by Fred. Elaborately produced, these filled the theatre to capacity for several months. The Melvilles, who jointly built the Prince's (later

the *Shaftesbury) Theatre, were also successful writers of highly coloured *melodramas.

Menander (*c*.342–293 BC), classical Greek dramatist, the leading writer of Athenian *New Comedy, as distinct from the *Old Comedy of *Aristophanes. In spite of his evident popularity, very little of his work survived, except in fragments and passages quoted by later writers, such as St Paul (Corinthians 15: 33): 'Evil communications corrupt good manners'. But in 1905, 1955, and 1965 considerable portions of a number of plays were discovered. Of these, *The Rape of the Locks* and *The Arbitration* were edited and translated by Gilbert *Murray in 1941 and 1945. It is in these plays that the stock characters—the irascible old man, the greedy slave, the cruel guardian, the garrulous pedant—first appear in guises which, in the opinion of some critics, link them to the 'masks' of the *commedia dell'arte. But it is a debatable point. All we know for certain is that Menander was responsible for the evolution of a new type of comedy which did away with the *chorus and was based on minute observation of contemporary man and manners. It was this aspect of Menander's work which appealed most strongly to the Hellenistic age. Menander was praised most highly by ancient critics for his fidelity to nature. If nowadays this 'fidelity' is questioned on the grounds that his plots are contrived and his characters stereotypes, it should be remembered that he stands at the head of a long succession of comic writers, beginning with the Romans, whose works have been brilliantly adapted to their own times, through *Molière and *Goldoni to *Pinero, *Coward, *Ayckbourn, and Neil *Simon. Menander is, in a very real sense, the father of modern comedy.

Menken, Adah Isaacs [Dolores Adios Fuertes] (1835–68), American actress, whose theatrical reputation rests entirely on her playing of Mazeppa, in a dramatization of *Byron's poem of that name, 'in a state of virtual nudity while bound to the back of a wild horse'. This *equestrian drama was first seen at the Coburg (later the *Old Vic) in London in 1823, and continued to be popular for the next 50 years, though it was not until 1859 that a woman played the part of the hero. Menken, who in 1856 had married John Isaacs Menken, and kept his name through all her subsequent matrimonial and other adventures, first played Mazeppa in 1861 in Albany, and after appearing in the part in New York was seen at Astley's Amphitheatre in London in 1864. She appears to have exercised a fatal fascination over 'literary gentlemen', including *Dickens, the elder *Dumas, and Swinburne. She died in Paris, where she had made her first appearance in 1866.

Mercer, David (1928–80), English playwright, whose plays reflect an interest in both politics—he was a disillusioned Marxist—and schizophrenia and madness. He was already well known for his television plays when in 1965 the *RSC staged his one-act political play *The Governor's Lady*. In the same year Peter *O'Toole appeared in *Ride a Cock Horse* as the working-class anti-hero alienated from his background by his success as a writer; it was produced in New York in 1979. Two further plays, *Belcher's Luck* (1966) and *After Haggerty* (1970), the latter about a dramatic critic with a sense of professional impotence, were produced by the RSC. Michael *Hordern played the title-role of *Flint* (also 1970), a lecherous and agnostic clergyman who elopes with a pregnant Irish girl. The eponymous hero of *Cousin Vladimir* (1978), a fugitive from Soviet Russia confronted by the moral degeneracy of modern England, perhaps mirrored Mercer's own perplexities; and *No Limits to Love* (1980) concerns the sexual and other relationships between four intellectuals, a woman and three men. Both were presented by the RSC.

Merchant, Vivien, see PINTER.

Mercier, Louis–Sébastien (1740–1814), French dramatist, an exponent of the *drame bourgeois* popularized by *Diderot. His plays, in which he gives elaborate directions for scenery, were more popular outside France, particularly in Germany, on account of their unimpeachable morality and declamatory style. A characteristic example of his work is *La Brouette du vinaigrier* (1784), the story of a marriage between a wealthy young girl and the son of a working-class man. He had earlier adapted *Lillo's *The London Merchant* (1731) as *Jenneval* (1768), allowing the hero to escape punishment by a last-minute conversion. In the same spirit of optimism he gave his translation of *Romeo and Juliet* a happy ending and reduced *King Lear* to the tale of a bourgeois household quarrelling over the misdeeds of the servants. Having taken refuge in Switzerland because of his political views, Mercier returned to Paris on the outbreak of the Revolution; as a deputy he voted against the execution of Louis XVI and in favour of a life sentence. He was later imprisoned, but was saved from the guillotine by the fall of Robespierre in 1794.

Mercury Theatre, Auckland, New Zealand. Built as a cinema in 1910, this was bought in 1968 by the Auckland Theatre Trust to serve the theatrical needs of the city; the main auditorium seats 672, the smaller studio theatre 184. The two theatres have enabled the Mercury to present a wide repertory from opera to musicals, drama including *García Lorca's *Yerma*, Arthur *Miller's *A View from the Bridge*, and *Chekhov's *The Cherry Orchard*, and contemporary plays such as David *Hare's *The Secret Rapture*. The theatre also stages a number of new local plays. It has a full-time acting company of 16.

Mercury Theatre, London, at Notting Hill Gate, small but well-equipped playhouse holding about 150, opened in 1933 by Ashley *Dukes for the production of new and uncommercial plays and to serve as a centre for his wife's Ballet Rambert. The first production was an adaptation of *Molière's *Amphitryon*. The most important event of the theatre's early years was the first London production, in 1935, of T. S. *Eliot's *Murder in the Cathedral*, transferred from the Chapter House at Canterbury Cathedral. Two years later *Auden and Isherwood's poetic play *The Ascent of F.6* was first seen in London, and in 1943 *O'Neill's *Days without End*. In 1945–6 the Mercury became the home of *poetic drama under the direction of E. Martin *Browne. In 1947 *Saroyan's *The Beautiful People* and O'Neill's *SS Glencairn* both had their London premières there; but from 1956 the theatre was used mainly by the Ballet Rambert School until sold in 1987 for private development.

Mercury Theatre, New York, see COMEDY THEATRE.

Merlyn Theatre, see MELBOURNE THEATRE COMPANY.

Mermaid Theatre, London, originally a small private theatre in the garden of Bernard *Miles's house in St John's Wood. Designed on Elizabethan lines, it opened in 1951 with a performance of Purcell's opera *Dido and Aeneas*, followed a week later by *The Tempest*. In 1953, to celebrate the coronation of Elizabeth II, the Mermaid was re-erected in the City of London for performances of *As You Like It*, *Macbeth*, and *Jonson, *Marston, and *Chapman's *Eastward Ho!* The success of his venture encouraged Miles to build a permanent professional theatre, which opened in 1959 with his own musical adaptation of *Fielding's *Rape upon Rape* as *Lock up Your Daughters*, a huge success which, like several later productions, transferred to the West End. The new Mermaid, financed by public subscription and erected within the walls of an old bombed warehouse at Puddle Dock, near Blackfriars Bridge, seated nearly 500 in one steeply raked tier, and had an open Elizabethan-style stage. Miles himself directed many of the productions and gave some notable performances there. Among the plays seen were several revivals of *Shaw (1961–2), a series of plays by *O'Casey (also 1962), and Peter Luke's *Hadrian VII* (1968). Ian *McKellen gave much-acclaimed performances in the title-roles of *Richard II* and *Marlowe's *Edward II* (1969), and two interesting revivals were Joyce's *Exiles* (1970) and *Sherriff's *Journey's End* (1972). A musical biography of Noël *Coward entitled

Cowardy Custard opened in 1972 and ran for a year, being followed by similar treatments of Cole *Porter (*Cole*, 1974) and Stephen *Sondheim (*Side by Side by Sondheim*, 1976). In 1978, at the end of the run of *Stoppard's *Every Good Boy Deserves Favour*, the theatre closed for reconstruction. It reopened in 1981 seating 610 and with a much larger stage. The opening productions, however, were failures, and the consequent financial problems caused the curtailment of future plans, in spite of the successful transfer of Mark Medoff's *Children of a Lesser God* to the *Albery Theatre. Miles's involvement ended in 1982 and the theatre has since had a varied history. The *RSC's production of Stephen Poliakoff's *Breaking the Silence* moved there in 1985 and the *National Theatre's of David *Mamet's *Glengarry Glen Ross* in 1986. The RSC occupied the theatre as an extra London base in 1987, but withdrew after a year. *Brecht's *Mother Courage* was staged in 1990, with Glenda *Jackson.

Merman [Zimmermann], **Ethel** (1909–84), American actress and singer, who was seen in cabaret in 1928 and in *vaudeville a year later. Her first role on Broadway was Kate Fothergill in George *Gershwin's musical *Girl Crazy* (1930). Her performance as Reno Sweeney in the Cole *Porter musical *Anything Goes* (1934) established her as a star, and she went on to play the leading roles in four more Porter musicals—*Red, Hot and Blue!* (1936), *Du Barry was a Lady* (1939), *Panama Hattie* (1940), and *Something for the Boys* (1943). Her biggest success was as Annie Oakley in Irving *Berlin's *Annie Get Your Gun* (1946), which ran for three years, and she had another long run as Sally Adams, American ambassador to a mythical European country, in Berlin's *Call Me Madam* (1950). *Gypsy* (1959), in which she scored a big hit as Gypsy Rose Lee's mother, was her last new musical, though she starred in Broadway revivals of *Annie Get Your Gun* (1966) and *Hello, Dolly!* (1970). She was seen in London in 1964 in cabaret. An extrovert personality, with a loud, brassy voice, she brought tremendous attack to every role she played.

Merry (*née* Brunton), **Anne** (1769–1808), English actress, eldest child of a provincial theatre manager named **John Brunton** (1741–1822). She first appeared in Bath in her father's company in 1784 with such success that a year later she was engaged for *Covent Garden, where she remained until she left the stage in 1792 to marry **Robert Merry** (1755–98). Soon after he lost all his money, and she accepted an offer from Thomas *Wignell to go to the *Chestnut Street Theatre in Philadelphia. She made her first appearance there in 1796 as Juliet in *Romeo and Juliet*, and was soon acclaimed as the finest actress of the time, both in Philadelphia and in New York, where she first appeared in

1797. Many contemporary accounts testify to her beauty and fine acting, and she was particularly admired in Mrs *Siddons's great part of Belvidera in *Otway's *Venice Preserv'd*. Widowed in 1798, she married Wignell in 1803, and he died shortly afterwards. Three years later she married William *Warren. She made several successful appearances in New York under William *Dunlap, who had been in the audience when she made her début at Covent Garden, and died in childbirth at the height of her success. Her brother **John Brunton** (1775–1849) was also an actor, as was her sister **Louisa** (1779–1860), who retired in 1807 to marry the Earl of Craven.

Merson, Billy, see MUSIC-HALL.

Messel, Oliver Hilary Sambourne (1905–78), English artist and stage designer, who first attracted attention with his masks and costumes for *Cochran's revues in the late 1920s. In 1932 he was given the task of converting the stage and auditorium of the *Lyceum Theatre into a cathedral for a revival of Vollmöller's *The Miracle* directed by *Reinhardt, and three years later he designed the settings for Ivor *Novello's musical *Glamorous Night*. One of his most memorable settings was that for *Anouilh's *Ring round the Moon* (1950), translated by Christopher *Fry, and he provided the sets for two of Fry's own plays, *The Lady's not for Burning* (1949) and *The Dark is Light Enough* (1954). He also designed the setting for the production of the latter play in New York, where his work was later seen in Fay and Michael Kanin's *Rashomon* (1959), based on Ryunosuke Akutagawa's Japanese stories, and Anouilh's *Traveller without Luggage* (1964). He was active also in opera, ballet, and film, and was responsible for the interior decoration of Sekers's private theatre in Whitehaven, and of the Billy Rose Theatre (now the *Nederlander Theatre) in New York.

Method, introspective approach to acting based on the system evolved by *Stanislavsky for the actors at the *Moscow Art Theatre. It first came into prominence in the USA during the 1930s, when it was adopted by the *Group Theatre in its reaction against what were considered the externalizing, stereotyped techniques current on the contemporary New York and London stages. Its notoriety rested mainly on its adoption by the Actors' Studio, founded in 1947 by Elia *Kazan and others, later including Lee *Strasberg. The Method's aim was to create a character from within by imagination and intuition. The result could be a more lifelike portrayal, with improvised dialogue, hesitations, mumblings, scratchings, and other naturalistic features; but the system could lead to self-absorption, to the exclusion of the audience and even of other actors. It achieved its greatest success in modern American plays, particularly those of Tennessee *Williams. The Method is now seen as only one of many valid methods of approaching a part. It is less suited to the classics and probably more appropriate to the cinema, where its supreme exponent was Marlon Brando.

Metropolitan Casino, New York, see BROADWAY THEATRE 2.

Metropolitan Music-Hall, London, in the Edgware Road. In 1862 a small concert hall attached to the White Lion Inn was replaced by a new building, which as Turnham's, named after its proprietor, opened as a *music-hall and was renamed the Metropolitan two years later. It gradually achieved popularity, and under a number of different managers had a prosperous career until 1897 when it was rebuilt. The new three-tier house, holding 1,855, opened with a star-studded bill and prospered for a time. When the popularity of *variety began to decline, seasons of opera and visits by touring companies were tried, but without success, and the hall closed except for occasional shows and wrestling matches on Saturdays. It eventually became a television studio, which closed down in 1962, the building being demolished a year later.

Metropolitan Theatre, New York, on Broadway. Built on the site of Tripler Hall, which was burnt down on 7 Jan. 1854, this theatre opened unsuccessfully on 20 Sept., and soon became a circus. It was occasionally used for plays, and in Sept. 1855 *Rachel, with a French company, appeared there in *Racine's Phèdre and *Scribe and Legouvé's Adrienne Lecouvreur. In the following December Laura *Keene took over and in spite of prejudice against a woman manager was doing well when she lost her lease on a technicality. She was succeeded by *Burton, and in 1859 by *Boucicault, who reopened the theatre as the Winter Garden with Dot, his own dramatization of *Dickens's The Cricket on the Hearth. He also presented a stage version of Nicholas Nickleby and several of his own plays during his first season. An early version of Joseph *Jefferson's Rip Van Winkle was seen at the Winter Garden, and important American actors who appeared there included John Sleeper *Clarke, Charlotte *Cushman, *Sothern, and the *Florences. Edwin *Booth took over the theatre in 1864, appearing there on 25 Nov. as Brutus in a performance of Julius Caesar with his brothers Junius Brutus and John Wilkes as Cassius and Antony respectively, the only time they are believed to have appeared together. Booth's record run of 100 performances of Hamlet took place here in 1864–5. On 23 Mar. 1867, just as Booth was about to appear as Romeo, the theatre burnt down and was not rebuilt.

Meyerhold, Vsevolod Emilievich (1874–1940/3), Russian actor and director, who after training under *Nemirovich-Danchenko joined the *Moscow Art Theatre on its foundation in 1898, making his first appearance as Treplev in *Chekhov's The Seagull. In 1902 he left to found his own company, with which he toured the provinces until 1905, in which year he was invited by *Stanislavsky to take charge of a newly opened studio theatre. It was soon apparent that his conception of the actor as a puppet to be controlled from outside by the director (a concept he formulated under the influence of Gordon *Craig) was totally opposed to the *naturalism of the Moscow Art Theatre style, and the studio closed. From 1906 to 1907 Meyerhold worked with Vera *Komisarjevskaya, but again came into conflict with the actors when he tried to put his theories into practice, and was forced to leave. For some years he directed productions at the Imperial theatres in St Petersburg. After the Revolution he was the first theatre director to offer his services to the new government. He was also the first to stage a Soviet play, Mystery-Bouffe (1918) by *Mayakovsky, also directing his The Bed-Bug (1929) and The Bath House (1930). In 1936, with the establishment of *Socialist Realism, Meyerhold, who was already out of favour because of his *Formalism, was dismissed, and in 1938 he was imprisoned. He was rehabilitated in 1955 and is now recognized as one of the most important directors of his time. The date and circumstances of his death are still unknown.

Mezzanine (or **Mazarine**) **Floor,** in large theatres the first level of the *cellar, which housed most of the *traps and other machinery, and also served as a retiring-room for the members of the orchestra.

Mezzetino, Mezzetin, servant of the *commedia dell'arte, with many of the characteristics of *Brighella or *Scapino, though more polished. Towards the end of the 17th century (c.1682) the role was altered and elaborated by Angelo *Costantini, who adopted red and white as the distinguishing colours of his costume as opposed to the green and white stripes of Scapino. The character passed into the French theatre by way of the *Comédie-Italienne; one of the finest French Mezzetins was the actor Préville of the *Comédie-Française, who was painted by Van Loo in the part.

Michell, Keith (1928–), Australian-born actor and singer, formerly an art teacher, who made his stage début in Adelaide in 1947, and then trained at the *Old Vic Theatre School. He appeared as Charles II in the musical version of *Fagan's And So to Bed (1951), and then toured Australia with the company from the *Royal Shakespeare Theatre, returning to Stratford for the seasons of 1954 and 1955. At the Old Vic in 1956 he was seen as Benedick in Much Ado about Nothing and Antony in Antony and Cleopatra. He then went to the West End to star in another musical, *Irma la Douce* (1958), in which he made his début in New York in 1960, returning there after playing Don John in *Fletcher's *The Chances* at the first *Chichester Festival in 1962 to appear in *Anouilh's *The Rehearsal* in 1963. Two other musical roles were Robert Browning in the long-running *Robert and Elizabeth* (1964), based on Besier's *The Barretts of Wimpole Street*, and Don Quixote in *Man of La Mancha* (1968), based on *Cervantes. He scored a great success in Ronald *Millar's *Abelard and Heloise* (1970; NY, 1971), and again played Browning in Kilty's *Dear Love* (1973). He was Artistic Director of the Chichester Festival, 1974–7, his roles there including the title-role of *Rostand's *Cyrano de Bergerac*, Iago in *Othello*, and King Magnus in *Shaw's *The Apple Cart*, the last also in London. After leaving Chichester he starred in the musicals *On the Twentieth Century* (1980) in London and *La Cage aux folles* (1984–5) in San Francisco, New York, Sydney, and Melbourne.

Mickiewicz, Adam (1798–1855), the outstanding poet of Poland, and the founder of the Romantic movement in Polish literature. Born and educated in Vilna, he was exiled in 1829, and lived thereafter mainly in Paris. His finest dramatic work, *Forefathers' Eve*, was composed in a fragmentary manner, not being intended for the stage. Parts II and IV were published in 1823; Part III in 1832. Part I, which he left unfinished, was published after his death. The four parts were first co-ordinated and staged by *Wyspiański in Cracow in 1901, and the play was many times revived before being officially banned after the Second World War, during the ascendancy of *Socialist Realism. In 1955 it was revived at the Polski Teatr in Warsaw, and an outstanding arena production at the Stary Theatre in Cracow was brought to London during the *World Theatre Season of 1975 and staged in Southwark Cathedral.

Middle Comedy, term applied to the last two plays of *Aristophanes, the *Ecclesiazusae* (*Women in Parliament*) and *Plutus*, and those of his immediate successors in the early and middle 4th century BC, from which much of the spirit of revelry present in *Old Comedy has disappeared. The *chorus shrinks, the importance of the plot grows, and dramatic illusion is taken more seriously. The earlier Middle Comedy plays are still concerned, though to a lesser degree, with politics, but they are less personal and fantastic, and even the obscenity becomes less obvious. Judging from the remaining fragments, later Middle Comedy in general was social rather than overtly political, with a background of private life. By the middle of the 4th century it had passed into *New Comedy.

Middlesex Music-Hall, London, in Drury Lane, originally the Mogul Saloon, known as the 'Old Mo', a nickname it subsequently retained. At the end of 1847 it began to feature the new-style *music-hall turns, becoming the Middlesex Music-Hall in 1851. Many famous music-hall stars made their débuts there, including, it is said, Dan *Leno. Reconstructed in 1872, it was rebuilt in 1891 and again in 1911, when, as the New Middlesex Theatre of Varieties, it reopened with a capacity of 3,000, a far cry from its original 500. Music-hall turns continued to predominate in the programme, though interspersed with a series of sensational French *revues, until the theatre closed in 1919, to reopen as the *Winter Garden Theatre. The site is now occupied by the *New London Theatre.

Middleton, Thomas (1580–1627), English dramatist, who worked mainly for *Henslowe, writing often in collaboration with other playwrights of the time. Many of the plays in which he had a hand are now lost, but he is known to have worked with *Dekker on *The Honest Whore* (1604) and *The Roaring Girl* (1610); with William *Rowley on *A Fair Quarrel* (c.1615), *The Changeling* (1622), one of the plays by which they are chiefly remembered, and *The Spanish Gipsy* (1623), based on two plays by *Cervantes. Among the more important of Middleton's own plays are *A Trick to Catch the Old One* (1604), to which *Massinger may be indebted for the plot of his *A New Way to Pay Old Debts* (c.1623); *A Mad World, My Masters* (1607); *A Chaste Maid in Cheapside* (1611); *Women Beware Women* (1621), showing that mingling of fine poetry, melodrama, and insight into feminine psychology which characterize the best of Middleton's work; and finally *A Game at Chess* (1624). This last was a political satire dealing with the fruitless attempts being made at the time to unite the royal houses of England and Spain, and may have earned Middleton a short spell in prison on the complaint of the Spanish Ambassador. Middleton is credited by some critics with *The Revenger's Tragedy* (1606), formerly attributed to *Tourneur, but the question so far remains undecided. A prolific and versatile writer, Middleton was also responsible for the text of a number of *masques and *pageants, now lost.

Mielziner, Jo (1901–76), American scene designer, who was for a short time an actor. His first stage designs were done for the *Lunts in *Molnár's *The Guardsman* (1924), after which he was responsible for the décor of a wide variety of plays, among them *Romeo and Juliet* in 1934 for Katharine *Cornell and *Hamlet* in 1936 for John *Gielgud; *Winterset* (1935), *The Wingless Victory* (1936), and *High Tor* (1937) by Maxwell *Anderson; *The Glass Menagerie* (1945), *A Streetcar Named Desire* (1947), *Summer* and Smoke (1948), and Cat on a Hot Tin Roof (1955), by Tennessee *Williams; Death of a Salesman (1949), by Arthur *Miller; and a number of musicals, including The Boys from Syracuse (1938), Finian's Rainbow (1947), Guys and Dolls (1950), and The King and I (1951). He was also responsible for the design of the *Washington Square Theatre and for the décor of its productions of Miller's After the Fall and *Behrman's But for Whom Charlie (both 1964); with the Finnish architect Eero Saarinen he designed the *Vivian Beaumont Theatre at the Lincoln Center.

Mikhoels [Vovsky], **Salomon Mikhailovich** (1890–1948), Jewish actor, who in 1919 joined the studio which later became the *Moscow State Jewish Theatre. Mikhoels, who first came into prominence with a fine performance in 1921 in Sholom *Aleichem's Agents, became the company's leading actor and in 1927 took over the running of it. He remained in charge until his death; one of his finest performances was as King Lear in a production by *Radlov in 1935.

Miles, Sir Bernard (1907–91), English actor and director. He made his first appearance on the stage in 1930, and spent several years in repertory before joining the company at the *Players' Theatre in Late Joys. He appeared several times with the *Old Vic company, both in London and on tour, and in 1951 founded the first *Mermaid Theatre, where he was seen as Caliban in The Tempest. A year later he played Macbeth, and in 1953 was seen at the second Mermaid in As You Like It, which he also directed, and as Slitgut in *Jonson's Eastward Ho! After opening the third Mermaid he devoted all his time to it, directing many of the plays seen there and playing, among other parts, Long John Silver in his own dramatization of Stevenson's Treasure Island, first seen in 1959 and revived many times. He also played the title-roles in *Brecht's Galileo (1960), *Ibsen's John Gabriel Borkman (1961), Brecht's Schweyk in the Second World War (1963), and *Sophocles' Oedipus the King and Oedipus at Colonus (1965). He appeared as Falstaff in both parts of Henry IV in 1970, as Iago in Othello in 1971, and staged several one-man shows. The replanning and rebuilding of his theatre occupied much of the 1970s, but his involvement with the theatre ended in 1982, soon after its reopening.

Miles's Musick House, London, see SADLER'S WELLS THEATRE.

Millar, Gertie (1879–1952), one of the most famous of George *Edwardes's 'Gaiety Girls', an excellent dancer, with a beautiful face and figure and a small but sweet voice. Born in Bradford, the daughter of a mill-worker, she first appeared on the stage at the age of 13, playing the Girl Babe in The Babes in the Wood

in Manchester. In 1899 she appeared as Dandini in Cinderella, and two years later she made her first appearance at the *Gaiety, singing 'Keep off the Grass' in The Toreador. She remained there till 1908, when she made her first appearance in New York in The Girls of Gottenburg, marrying the composer Lionel Monckton in the same year, and then returned to the Gaiety until 1910, when she was seen at the *Adelphi as Prudence in The Quaker Girl; two years later she was at *Daly's in Gipsy Love, and in 1914 she appeared at the *Coliseum in *variety. She made her last appearance on the stage at the *Prince of Wales in Flora (1918), and then retired, marrying the 2nd Earl of Dudley after Monckton's death in 1924.

Millar, Sir Ronald Graeme (1919–   ), English actor and playwright, who made his first appearance on the stage in 1940, and was seen in a number of productions before devoting himself entirely to playwriting. His first play was a thriller, Murder from Memory (1942), and The Other Side (1946) was based on Storm Jameson's novel of the same name. His own play Frieda was also produced in 1946, and was followed by Champagne for Delilah (1949) and Waiting for Gillian (1954), based on Nigel Balchin's novel A Way through the Wood. His first outstanding successes were The Bride and the Bachelor (1956) and its sequel The Bride Comes Back (1960), farces in which Cicely *Courtneidge and Robertson *Hare appeared. He then dramatized three novels by C. P. Snow, The Affair (1961; NY, 1962), The New Men (1962), and The Masters (1963), featuring academic life in Cambridge, and wrote a musical on the Brownings, Robert and Elizabeth (1964), following it with Number Ten (1967), set in Downing Street with Alastair *Sim as the Prime Minister, and They Don't Grow on Trees (1968), a comedy in which Dora Bryan appeared in nine different roles. His Abelard and Heloise (1970; NY, 1971), with Keith *Michell and Diana *Rigg in the title-roles, had a long run, and he then returned to Snow's novels, adapting In Their Wisdom as The Case in Question (1975). A Coat of Varnish (1982) was another Snow adaptation.

Miller, Arthur (1915–   ), American dramatist, who while a student at Michigan University won three drama prizes, one of which he shared with Tennessee *Williams. His first success was All My Sons (1947; London, 1948), in which a wartime profiteer realizes he has caused the death of his own son. Death of a Salesman (NY and London, 1949), about a salesman's refusal to face the failure of his career and family relationships, received the *Pulitzer Prize for Drama and contained some of his best work. In 1950 Miller prepared a new translation of An Enemy of the People by *Ibsen, a dramatist by whom he is believed to have been much influenced. The Crucible (1953; London, 1956), a powerful play dealing with the witch-trials of the 17th century, was widely interpreted as an attack on the political witch-hunting then prevalent in the USA. Two one-act plays, *A View from the Bridge* and *A Memory of Two Mondays*, were seen in 1955. Rewritten in three acts, *A View from the Bridge* was produced in London in 1956 as a club production because it concerns a false attribution of homosexuality, the revised version being produced in New York in 1965. It was widely believed, though Miller denied it, that his marriage to the film star Marilyn Monroe provided the background to his next play *After the Fall* (1964; *National Theatre, 1990). Later in 1964 came *Incident at Vichy* (London, 1966), about a group of men undergoing examination by the Nazis to see if they are Jews. Neither of these plays pleased the critics, and in 1968 Miller returned to the scene of his earlier successes—American family life—with *The Price* (London, 1969). His later works—*The Creation of the World and Other Business* (1972), about Adam and his family, *The Archbishop's Ceiling* (1977; *RSC, 1986), not given a New York production, and *The American Clock* (1980; NT, 1986), set in the Depression—found little favour. He made a partial return to his earlier quality, however, with *The Ride down Mt Morgan* (London, 1991), a 'serious comedy' about bigamy.

Miller, Henry (1860–1926), British-born American actor and manager, who made his début in Toronto in 1878, and then went to the United States, where after touring with many leading actresses of the time and with *Boucicault he became leading man of the *Empire Theatre *stock company in New York. In 1899 he appeared as Sydney Carton in *The Only Way*, an adaptation of *Dickens's *A Tale of Two Cities* which had a long run, and in 1906 he went into management, producing and playing in William Vaughn *Moody's *The Great Divide*, in which he was also seen in London the same year. In 1916 he opened *Henry Miller's Theatre in New York. He died during rehearsals for Dodd's *A Stranger in the House*. He married the daughter of the actress Matilda *Heron, and their son **Gilbert Heron Miller** (1884–1969) was also a theatre manager in London and New York, fostering the exchange of new plays across the Atlantic. From 1918 until its demolition in 1958 he was lessee of the *St James's Theatre in London.

Miller, Jonathan Wolfe (1934–   ), English director, who qualified as a doctor at *Cambridge, where he also acted in several Footlights revues. After the unexpected success at the *Edinburgh Festival of the 1960 revue *Beyond the Fringe*, of which he was co-author, he appeared in it in London and New York. The first play he directed was John *Osborne's one-act *Under Plain Cover* (1962) at the *Royal Court Theatre. Among his major productions were *King Lear* in 1969 at Nottingham, with Michael *Hordern, and *The Merchant of Venice* (1970), with *Olivier as Shylock, at the *National Theatre, where he later directed *Büchner's *Danton's Death* in 1971 and *Sheridan's *The School for Scandal* in 1972. In 1972 he also directed *The Taming of the Shrew* for *Chichester, following it with *Chekhov's *The Seagull* in 1973. Appointed associate director of the National Theatre, 1973–5, he directed *Beaumarchais's *The Marriage of Figaro*, *Measure for Measure*, and Peter *Nichols's *The Freeway* (all 1974). He directed several productions at *Greenwich, including *Hamlet* and *Ibsen's *Ghosts* in 1974, and in 1976 was responsible for a memorable West End production of Chekhov's *Three Sisters*. He 'retired' from the theatre in 1979, but returned to direct Jack Lemmon in *O'Neill's *Long Day's Journey into Night* (NY and London, 1986). From 1988 to 1990 he was Artistic Director of the *Old Vic, where his productions included George *Chapman's *Bussy d'Ambois* in 1988 and *Corneille's *The Liar* in 1989. He is also a well-known director of opera.

Miller, Marilyn, see MUSICAL COMEDY and REVUE.

Mills, Florence, see LONDON PAVILION.

Milton, Ernest (1890–1974), American-born actor who spent most of his life in England. He made his first appearance in the USA in 1912, and was first seen in London in Montague Glass and Charles *Klein's *Potash and Perlmutter* (1914) and *Potash and Perlmutter in Society* (1916). After some varied experience, during which he played Oswald in *Ibsen's *Ghosts* and Marchbanks in *Shaw's *Candida*, he joined the company at the *Old Vic in 1918, and then and on subsequent visits gave interesting and highly idiosyncratic performances of many of Shakespeare's leading characters, especially Hamlet, Shylock, Macbeth, and Richard II. He was at his best in fantastic or sinister roles, and was admirable as *Pirandello's *Henry IV* (1925), a part he played again in 1929 when the play was renamed *The Mock Emperor*. He was also excellent as Channon in *Ansky's *The Dybbuk* (1927); as Rupert Cadell in Patrick Hamilton's *Rope* (1929; NY, as *Rope's End*, 1929); as a gaunt and somewhat frightening Pierrot in Laurence *Housman's *Prunella* (1930); and as Lorenzino de' Medici in *Night's Candles* (1933), an adaptation of Alfred de *Musset's *Lorenzaccio*. In 1933 he was also seen in New York in *The Dark Tower* by George S. *Kaufman and Alexander *Woollcott, and after returning to London appeared in the title-roles of his own play *Paganini* and of *Timon of Athens* (both 1935). After playing King John for the Old Vic (1941) he was less often seen, though he gave interesting performances at the *Lyric Theatre, Hammersmith, as Lorenzo

Querini in *Hochwälder's *The Strong are Lonely* (1955) and as Pope Paul in *Montherlant's *Malatesta* (1957). He joined the *RSC in 1962 to play the Cardinal in *Middleton's *Women Beware Women*.

Milwaukee Repertory Theater, Milwaukee, Wis., founded in 1954 as the Fred Miller Theater, was originally a community theatre using local actors supplemented by stars. Gradually a fine company evolved which officially became the Milwaukee Repertory Theater in 1964 and in 1969 moved into a new civic centre home, the Wehr Theater. In 1974 a second, experimental theatre, the Court Street, was established, and in 1983 a research and development department known as the Lab. In 1987 the MRT moved into an all-inclusive downtown facility consisting of the 720-seat mainstage theatre, the 216-seat Stiemke Theater, and the 100-seat Stackner Cabaret. The MRT is a non-profit-making company supporting new playwrights, presenting neglected classics, and commissioning new translations of classic and contemporary foreign theatre. It undertakes state, regional, and international tours. In 1976 it began to stage occasional large-scale productions at the 1,393-seat century-old Pabst Theater.

Mime, in Ancient Rome a spoken form of popular, farcical drama which, unlike the *fabula atellana* and the silent acting of the *pantomimus*, was played without masks. Soon after the introduction of Greek drama to the Roman stage in the mid-3rd century BC the Floralia, or festival of Flora, was founded, and this became a favourite occasion for the performance of mimes, where licence even went so far as to sanction the appearance of mime-actresses naked on stage. The influence of mime on the development of Latin comedy must have been considerable; much of the jesting and buffoonery which *Plautus introduced into his adaptations of Greek comedy would have been quite appropriate in the mime.

The mime differed from more conventional forms of drama by its preoccupation with character-drawing rather than plot. The distinctive costume of the mime-player was a hood or *ricinium*—whence the name *fabula riciniata* for a mime—which could be drawn over the head or thrown back, a patchwork jacket, tights, and the phallus; the head was shaven and the feet bare. The companies were small; Ovid speaks of a cast of three to take the roles of the foolish old husband, the erring wife, and the dandified lover. The plots were simple, the endings often abrupt.

In the 1st century BC the mime achieved a precarious status as a literary form when fixed scripts for mime-players were written. The fragments of their works that survive show that they had much in common with the *fabulae atellanae* which the new mime-plays seem to have absorbed or replaced. But the popular mimes which all but drove other forms of spoken drama from the stage under the Roman Empire were largely impromptu, with dialogue in prose which the chief actor—the *archimime*—was free to cut or expand at will. The sordid themes and startling indecency of the language, judged by some later fragments which have survived, seem to be characteristic of the mime in general. Not only was adultery a stock theme, but the Emperor Heliogabalus appears to have ordered its realistic performance on stage, and if the plot included an execution it was possible, by substituting a condemned criminal for the actor, to give the spectators the thrill of seeing the execution actually take place. It is hardly surprising that however popular the actors were, socially they were beneath contempt. They countered the opposition of the Church by mocking Christian sacraments; but gradually the Church got the upper hand and in the 5th century succeeded in excommunicating all performers in mime. Yet the mime lived on. Its simple requirements could be supplied in any public place or private house, and in such settings it continued to entertain audiences who were now officially Christian. Though forced to drop its habit of burlesquing the sacraments, it still scandalized the Fathers of the Church by its indecency and the immorality of its performers.

How far the mime survived the fall of Rome and the onslaught of the barbarians is doubtful. So simple a type of performance might arise independently at different ages and in different countries; yet precisely because of its primitive character it is hard to be sure that the classical mime ever became wholly extinct in Europe. The Middle Ages had their mime-players, who may have taken over from the *mimi*, those last representatives of classical drama, something of their traditions, and handed them on to their descendants in the modern world.

Mime, Modern, of which the outstanding exponent is Marcel *Marceau, has nothing in common with the Roman mime (above), and approximates far more closely to the art of the Roman *pantomimus*, being entirely dependent on gesture and movement, usually accompanied by music, but wordless. The word *pantomime has now taken on such an entirely different meaning from its original one that modern exponents of dumb-show call their work 'mime' to differentiate it from the popular English Christmas show.

Minack Theatre, Porthcurno, Cornwall. This open-air theatre, hewn out of the granite cliffs near Penzance, was founded by Rowena Cade. Work on it began in 1932, and from 1933 to 1939 a play was produced there every two years by local amateur companies. During the Second

World War the theatre was covered with barbed wire, but productions began again in 1949 with a performance of *Euripides' *Trojan Women* by pupils from two Penzance schools. In 1959 a Minack Theatre Society was formed to assist and publicize the venture, which has since flourished with weekly productions from June to September by companies from all over Britain, plays ranging from Shakespeare and *Restoration comedy to the works of such writers as *Anouilh, Christopher *Fry, and Alan *Ayckbourn. Many improvements have been made to the theatre, which seats 600 and is now administered by a Trust, receiving no grant. An Exhibition Centre, opened in 1988, tells the story of the theatre's creation.

Minerva Studio Theatre, see CHICHESTER FEST-IVAL THEATRE.

Minneapolis Theater, see GUTHRIE THEATER.

Minotis, Alexis (1900–90), Greek actor and director, who made his first appearance on the stage in 1925, and in 1930 made his New York début as Orestes in a Greek production of *Euripides' *Electra*. Returning to Greece, he played a wide range of ancient and modern parts for the National Theatre, with which he was associated from 1930 to 1967, succeeding Fotos *Politis as director and directing many of the plays as well as acting in them. He was first seen in London in 1939, when he played Hamlet. Among his other Shakespearian roles were Shylock in *The Merchant of Venice*, King Lear, and Richard III. From 1955 he was responsible for the production of many classical tragedies in the open-air theatre at *Epidauros. Modern plays with which he was connected included *Fry's *The Dark is Light Enough* (1957) and *Dürrenmatt's *The Visit* (1961) and *The Physicists* (1963). He appeared again in London during the 1966 *World Theatre Season, playing Oedipus to the Jocasta of his wife Katina *Paxinou in *Sophocles' *Oedipus Rex*, Tal-thybius to her Hecuba in Euripides' *Hecuba*, and Oedipus again in Sophocles' *Oedipus at Colonus*. From 1967 to 1974 he ran his own company, presenting mainly modern European plays, returning to the National Theatre as General Director from 1974 to 1981. In 1982 he gave a memorable performance as the Cardinal of Spain in *Montherlant's play, and he made his last appearance in 1988 in *Beckett's *Endgame*, which he also directed.

Minstrel, Ménestrel, term dating from the 12th century, derived from the Latin *minister* (servant, official), and originally used to dis-tinguish those performers who were in the regular employment of a particular lord from their itinerant fellows the *jongleurs*, but later descriptive of both. The origin of these pro-fessional entertainers, who flourished from the 11th to the 15th centuries, must be sought in the fusion of the Teutonic *scôp*, or bard, with the floating débris of the Roman theatre, particularly the *mimus* (see MIME). The process went on obscurely from the 6th to the 11th centuries, helped by the *goliards, wandering scholars who brought to the mixture a measure of classical erudition. Dressed in bright clothes, with flat-heeled shoes, clean-shaven face, and short hair—legacies of Rome—and with their instruments on their backs, they tramped, alone or in company, all over Europe, often harassed by the hostility of the Church and the restrictions of petty officialdom. In spite of this, they enlivened the festivities of religious fraternities, and performed wherever they could gather an audience—in noblemen's halls, in market-places, along pilgrim routes—and kept alive many traditions handed down from Greece and Rome. They may even have had a share in the development of *liturgical drama.

Minstrel Show, entertainment which ori-ginated in the Negro patter songs of T. D. *Rice (known as Jim Crow), and from his burlesques of Shakespeare and opera, to which Negro songs were added. From 1840 to 1880 the Minstrel Show was the most popular form of amusement in the United States, whence it spread to England. Unlike the *music-hall, which was intended for adults only, it was essentially a family entertainment, given in a hall and not in a theatre. The performers were at first White men with their faces artificially blacked, whence the name Burnt-cork Minstrels, but later they were true Negroes. Sitting in a semi-circle with their primitive instruments, banjos, tambourines, one-stringed fiddles, bones, etc., they sang plaintive coon songs and sen-timental ballads interspersed with soft-shoe dances and outbursts of back-chat between the two 'end-men', Interlocutor and Bones. Their humour was simple and repetitive, and after a great burst of popularity the Minstrels gradually faded away, some, like Chirgwin and Stratton, to the music-halls, some to stroll along the beach at seaside resorts in the summer in traditional minstrel costume—tight striped trousers and waistcoat and tall white hat or straw boater—singing and playing their banjos. Among the most famous troupes were the Christy Minstrels, the Burgess and Moore, and the Mohawks.

Miracle Play, in English medieval times a synonym for *mystery play; in France the term was used for plays in which the central incident was a miraculous intervention in human affairs by the Virgin Mary or one of the saints, St Nicholas being a favourite subject. The chief surviving examples of these anonymous *saint plays, as they were called in England, are found in a 14th-century collection of forty *Miracles de Notre-Dame par personnages*.

Mirren, Helen (1946–   ), English actress, one of the most glamorous and talented of her generation. Her early appearances with the *National Youth Theatre led to her playing an extremely youthful Cleopatra, in *Antony and Cleopatra*, at the *Old Vic in 1965. Two years later she joined the *RSC, remaining with it for five seasons, her roles including Cressida in *Troilus and Cressida*, Ophelia in *Hamlet*, Tatyana in *Gorky's *Enemies*, and the title-role in *Strindberg's *Miss Julie*. After working with the *International Centre of Theatre Research she played Lady Macbeth for the RSC in 1974, and a year later was seen at the *Royal Court in David *Hare's *Teeth 'n' Smiles*, in which she gave an outstanding performance as an alcoholic singer with a rock group. In complete contrast were her excellent Nina in *Chekhov's *The Seagull* and Ella in Ben *Travers's farce *The Bed before Yesterday* (both 1975). She returned to the RSC in 1977 in the three parts of *Henry VI*, and was seen as Isabella in *Measure for Measure* at the *Riverside Studios (1979) and in the title-role of *Webster's *The Duchess of Malfi* with the *Royal Exchange Theatre company (1980; *Round House, 1981). In 1981 she also appeared in Brian *Friel's *Faith Healer* at the Royal Court Theatre, and in 1983 for the RSC she again played Cleopatra. She was absent for several years in films before returning to London in Arthur *Miller's double bill *Two-Way Mirror* (1989).

Mistinguett [Jeanne-Marie Bourgeois] (1875–1956), Flemish-born French music-hall artist, with beautiful (and highly insured) legs, mischievous good looks, and a smart line in repartee. 'Miss', as she was nicknamed, was a symbol of Paris to millions of tourists, and even to Parisians themselves, for over half a century. In her early years she was an eccentric comedienne, specializing in character sketches of low-life Parisian women, but, reversing the usual pattern, she later became almost exclusively a dancer and singer, appearing first at the *Moulin Rouge, of which she was for some time part-proprietor, and, with Maurice *Chevalier as her partner, at the *Folies-Bergère, where she was seen in some sensational dances, descending with superb panache vast staircases, wearing enormous hats and trailing yards of feathered train. She became world famous in spite of hardly ever appearing outside Paris, her one appearance in London being at the London Casino (later the *Prince Edward Theatre) in 1947 at the age of 72.

Mitchell, Langdon Elwyn (1862–1933), American poet and playwright, who in 1899 dramatized Thackeray's *Vanity Fair* for Mrs *Fiske under the title *Becky Sharp*. It was a great success, and was frequently revived. His finest play, however, was *The New York Idea* (1906), a satire on divorce, also produced by Mrs Fiske

with a remarkable cast. It was translated into German (directed by Max *Reinhardt), and into Dutch, Swedish, and Hungarian. In the same year Mitchell translated *The Kreutzer Sonata* from the Yiddish of Jacob *Gordin, and in 1916 he dramatized Thackeray's *Pendennis* for John *Drew. He left a number of unfinished plays in manuscript. In 1928 he became the first Professor of Playwriting at the University of Pennsylvania.

Mitchell, Maggie [Margaret] **Julia** (1832–1918), American actress, who first appeared on the stage at 13, playing children's parts at the *Bowery Theatre and being particularly admired as *Dickens's Oliver Twist. After some years on tour she became leading lady of *Burton's company in 1857 and in 1861 played the part with which she is mainly associated—Fanchon in an adaptation of George Sand's novel *La Petite Fadette*. Although she was good in other parts, notably as Charlotte Brontë's Jane Eyre and as Pauline in *Bulwer-Lytton's *The Lady of Lyons*, she was constantly forced by an admiring public to return to Fanchon, which she played for over 20 years, remaining to the end a small, winsome, sprite-like figure, overflowing with vitality.

Mitchell, William, and **Mitchell's Olympic**, New York, see OLYMPIC THEATRE 1.

Mitzi E. Newhouse Theatre, New York, see VIVIAN BEAUMONT THEATRE.

Mnouchkine, Ariane, see SOLEIL, THÉÂTRE DU.

Mochalov, Pavel Stepanovich (1800–45), Russian tragic actor, the leading exponent in Moscow of the 'intuitive' school of acting. He made his début in 1817, and was soon playing leading roles, his finest parts being Hamlet, King Lear, and the heroes of *Schiller's *Die Räuber* and *Kabale und Liebe*. With great gifts of temperament and passion, Mochalov was antipathetic to any rational method, relying entirely on inspiration and so being extremely uneven in his acting and in its effect on his audiences. With the advance of realism his romantic appeal became out of date, and when heavy drinking impaired his powers of concentration and weakened the flow of inspiration, he had no technique to fall back on. His influence persisted for a time, but was eventually to succumb to the reforms of *Shchepkin, which he had resisted to the end.

Modena, Gustavo (1803–61), Italian actor, accounted one of the finest of his day. The son of **Giacomo Modena** (1766–1841), who in his youth was well known for his excellent acting in *Goldoni's comedies, he first came into prominence in 1824, when he played David to his father's Saul in *Alfieri's tragedy of that name. He soon established himself as a leading actor and in 1829 formed his own company with which he toured Italy. During the 1830s he became involved in revolutionary activities and left Italy, ending up in London, where he supported himself by teaching Italian and giving Dante recitals. He returned to Italy in 1839 and in 1843 formed a company of young actors in Milan which included the 14-year-old *Salvini. He was now at the summit of his profession, and in 1847 he produced *Othello*, the first play by Shakespeare to be seen in Italy. It was not a great success, but it undoubtedly paved the way for the later triumphs in the part of Salvini and Ernesto *Rossi. He made his last appearance in 1860, having done much to forward the reform of the Italian stage and give it a social purpose. Though not handsome, he had a fine figure and a nobly rugged and expressive face.

Modjeska [Modrzejewska, *née* Opid], **Helena** (1840–1909), Polish actress, who in 1860 joined a company in Cracow, adopting the name by which she is best known and retaining it throughout her career. In 1868 she moved to Warsaw, where for eight years she played a wide range of parts, from classical tragedy to contemporary productions in the style of *Sardou. In 1876 she emigrated to the USA, making her first appearance there a year later in San Francisco, and scoring an immense success, in spite of her poor command of the English language. She had a great love for Shakespeare, having 14 parts in his plays in her repertory of 260 roles, some of which she had already played in Warsaw. She first appeared in London from 1880 to 1882, and again in 1890, and in 1881 achieved her ambition of playing Shakespeare in English in London, when after intensive study of the language she appeared at the *Royal Court Theatre as Juliet to the Romeo of *Forbes-Robertson. One of her finest parts was Lady Macbeth, particularly in the sleep-walking scene. She toured extensively in America and Europe, and before she gave her farewell performance at the Metropolitan Opera House in New York in 1905 she had achieved an outstanding reputation as one of the great international actresses of her time, at her best in tragic or strongly emotional parts.

Mogul Saloon, London, see MIDDLESEX MUSIC-HALL.

Mohun, Michael (c.1620–84), English actor, who as a boy acted under William *Beeston, and was playing adult parts when the theatres closed in 1642. He then joined the Royalist army, and returned to the stage in 1660. With Charles *Hart, he was leading man of *Killigrew's company in 1662 when it took possession of the first *Drury Lane theatre. His Iago in *Othello* was much admired, and he created many roles in Restoration tragedy, including Abdelmelech in *Dryden's *The Conquest of Granada* (1670) and the title-role in *Lee's *Mithridates, King of Pontus* (1678).

Moiseiwitsch, Tanya (1914–   ), English stage designer, daughter of the pianist Benno Moiseiwitsch and his first wife, the violinist Daisy Kennedy. She laid the foundations of her career at the *Abbey Theatre, Dublin, where between 1935 and 1939 she was responsible for the settings of more than 50 productions. After working as resident designer for the *Oxford Playhouse, 1941–4, she joined the *Old Vic, and in 1949 moved to the *Shakespeare Memorial Theatre, where her first work was a permanent set with no front curtain for *Henry VIII*. She also provided the settings for a cycle of history plays performed there in 1951 for the Festival of Britain. Two years later she was in Canada, where she designed the first, tentlike theatre for the *Stratford (Ontario) Festival and also the settings for its first productions, *Richard III* and *All's Well that Ends Well*. During her association with the theatre she designed over 20 of its productions, as well as its new theatre in 1957. She was closely associated also with the *Guthrie Theater in Minneapolis from its opening season in 1963—to which she contributed the settings for *Hamlet*, *Molière's *The Miser*, and *Chekhov's *Three Sisters*—until 1967. Her work for the *National Theatre in London included settings for revivals of *Jonson's *Volpone* (1968), Molière's *The Misanthrope* (1973; NY, 1975), and *Congreve's *The Double Dealer* (1978). She was also responsible for the décor of a number of West End productions, among them the musical *Bless the Bride* (1947), R. C. *Sherriff's *Home at Seven* (1950), *Rattigan's *The Deep Blue Sea* (1952), and John *Mortimer's *The Wrong Side of the Park* (1960).

Moissi, Alexander (1880–1935), German actor of Italian extraction, born in Albania where a drama school has been named after him. He played his first speaking part in German at Prague in 1902, and remained there until 1905, when he went to *Berlin. There his fine presence, and above all his rich, musical, speaking voice, soon brought him into prominence. At the *Deutsches Theater under *Reinhardt he played a number of Shakespearian parts. He was also seen as both Faust and Mephistopheles in *Goethe's *Faust*; as Oswald in *Ibsen's *Ghosts*; as Louis Dubedat in *Shaw's *The Doctor's Dilemma*, and as Marchbanks in his *Candida*. His playing of *Sophocles' Oedipus and *Aeschylus' Orestes in productions by Reinhardt in Vienna was accounted outstanding. In 1930 he visited London, appearing in a German version of *Hamlet* at the *Globe Theatre while *Gielgud was playing it in English at the *Queen's Theatre next door.

Molander, Olof, see ROYAL DRAMATIC THEATRE.

Molé [Molet], **François-René** (1734–1802), French actor, who first appeared at the *Comédie-Française in 1754, and after six years in the provinces became a member of the company, playing young heroes and lovers. He soon became immensely popular. He excelled in comedy, and was usually content to leave the tragic roles to others, but ironically he was destined to be the first French Hamlet, in 1769, in an adaptation made by *Ducis. This was nothing like Shakespeare's play, but audiences liked it, and Molé was considered good in the part. In later life he became fat, but lost nothing of his agility of mind and body. An adherent of the Revolution, he escaped prison, and went to play revolutionary drama at the theatre of Mlle *Montansier. When the Comédie-Française reopened in 1799, he returned with his wife, who was also a member of the company, and remained there until his death.

Molière [Jean-Baptiste Poquelin] (1622–73), French dramatist and actor, author of some of the best comedies in the history of the European theatre. Son of a prosperous upholsterer of Paris, attached to the service of the King, he was intended to succeed his father, but in 1643 he renounced his succession and changed his name. (No reason for the adoption of 'Molière' has ever been found.) He joined a family of actors named *Béjart—perhaps because he had fallen in love with the eldest daughter, Madeleine—in their efforts to establish a theatre group, *Illustre-Théâtre, in Paris in a deserted tennis-court. The enterprise failed, but the intrepid actors, probably urged on by Madeleine, decided to take to the provinces. From 1645 to 1658 the company, of which Molière soon became the acknowledged leader, toured in a repertory of old French comedies and partly extemporized farces in the style of the *commedia dell'arte. Few of these have survived, and the company's records consist mainly of lists of the towns they visited.

Encouraged by his success in the provinces, Molière decided in 1658 that the time had come to return to Paris, and on 24 Oct. he appeared, under the patronage of Monsieur, the King's brother, at the Louvre, before the young Louis XIV and his Court, in *Corneille's tragedy Nicomède. This was a mistake; Molière was no tragedian, all the excellence of his acting lying in comedy. In any case, most of the audience would probably have seen, and preferred, the appearance as Nicomède at the Hôtel de *Bourgogne of the actor *Montfleury (who was in the audience); Molière's quieter and more natural delivery would not have appealed to them. The future of the new company hung in the balance. Then Molière asked permission to put on a short farce of his own, Le Docteur amoureux. It was an immediate success, and Molière's troupe was given permission to share the theatre at the *Petit-Bourbon with an Italian company already established there. The relationship between the rival groups was most cordial, and Molière always acknowledged how much he

had learned from the Italians. The production later in 1658 of the one-act Les Précieuses ridicules set the seal on his growing reputation.

By the time he had been in Paris for four or five years, his pre-eminence was virtually unchallenged; and with the success of his comedies, tragedy almost fell out of the repertory. Until his death from overwork Molière enjoyed the appreciation of his audiences and the constant support of Louis XIV, who himself, with members of his Court, danced in several of the ballets (with music by Lully) introduced into the Court entertainments. These began with Les Fâcheux (1661) and included Le Bourgeois gentilhomme (1670) and, in the same year, Les Amants magnifiques, with which Louis XIV, dancing as Neptune and Apollo, finally ended his theatrical career. The last Court plays were Psyché and La Comtesse d'Escarbagnas (both 1671).

These Court entertainments, shorn of much of their splendid scenery and costumes, were also seen on the stage of the *Palais-Royal, which Molière had occupied since 1661. Among the plays written for this stage were L'École des femmes (1662) and its sequel, La Critique de L'École des femmes (1663). In this the part of the heroine was played by the youngest child of the Béjart family, Armande, whom Molière had married the previous year. This led Molière's enemies, particularly Montfleury, to accuse him before the King of having married his own daughter. The King replied to this by commissioning L'Impromptu de Versailles (also 1663), which ridicules Montfleury and his company, and by standing godfather to the couple's first child, who was born and died in 1664. Louis also supported the playwright through the troubles caused by his Don Juan; ou, Le Festin de pierre (1665), which many considered blasphemous, and his Tartuffe (1667), a study of hypocrisy which infuriated the bigots of Paris. But none of this endangered the success of such plays as Le Misanthrope and Le Médecin malgré lui (both 1666), L'Avare (1668), Les Fourberies de Scapin (1671), Les Femmes savantes, and Le Malade imaginaire (both 1672). He was appearing in the leading role in the last on the night he died. Being only an actor, he was buried hastily in unconsecrated ground at dead of night and left in an unmarked grave.

Molière's great achievement was that by his own efforts he raised French comedy to the heights attained in tragedy. His great roles, like those of Shakespeare in England, serve to demonstrate the excellence of successive generations of players, appearing in what is known in his honour as La Maison de Molière. He is not easy for English audiences to understand, either in French or in translation. But he had a great influence on the playwrights of the English Restoration, and later returned to captivate new audiences wherever a worthy translator could be found, as witness the success of such free renderings as Lady *Gregory's *The Kiltartan Molière*, set in the west of Ireland, and the free but spirited renderings of Miles *Malleson.

Molina, Tirso de, see TIRSO DE MOLINA.

Molnár [Neumann], **Ferenč** (1878–1952), Hungarian playwright, and one of the best known outside his own country. He first attracted attention with some light-hearted farces in which he exploited the humours of Hungarian city life. These were followed by a variation on the theme of *Faust, *The Devil* (1907), which a year later was seen in translation at two theatres in New York simultaneously. His best-known play is *Liliom* (1909), a failure when first produced in Budapest but a success on its revival some 10 years later. It was seen in New York in 1921 (London, 1926), and was used as the basis of the successful musical *Carousel* (1945). Among Molnár's later plays, several of which starred his third wife, the Hungarian actress **Lili Darvas** (1902–74), were *The Guardsman* (1910), which in 1924 provided an excellent vehicle for the *Lunts in a production by the *Theatre Guild; *The Red Mill* (1923), seen in New York in 1928 as *Mima*; and the Cinderella-story of a Budapest servant-girl's romance, *The Glass Slipper* (1924). *The Play in the Castle* (also 1924) showed his mastery of sophisticated comedy, as did *Olimpia* (1927). (The former was seen in New York in 1926 as *The Play's the Thing* in an adaptation by P. G. Wodehouse and at the *National Theatre in 1984 as *Rough Crossing* in a translation by *Stoppard.) In *The Good Fairy* (1931) he reverted to his former whimsical vein with the adventures of a romantic usherette in a Budapest theatre. In 1940 Molnár, who had emigrated some years before to the USA, where he died, became an American citizen. His success abroad for a time militated against his acceptance in Hungary, where his work was largely ignored, but after his death a reaction set in; a revival of *Olimpia* at the Madách Theatre in 1965, with Lili Darvas, set the seal on his growing reputation, and many of his plays have since been revived in Budapest.

Momus, Greek god of ridicule, and so by extension of *clowns, who appears in the 'secular masque' provided by *Dryden for a revival of *Fletcher's *The Pilgrim* in 1700. The name was later used frequently to denote a clown, as in *Grimaldi's reference to himself as 'the once Merry Momus', and became attached to one of the characters in the *harlequinade.

Monck, Nugent, see MADDERMARKET THEATRE.

Moncrieff, William George Thomas (1794–1857), English dramatist and theatre manager, author of about 200 plays, mainly burlesques and melodramas written for the minor theatres. The most successful was *Tom and Jerry; or, Life in London* (1821), the best of many adaptations of the book by Pierce Egan. In 1823 Moncrieff's *The Cataract of the Ganges* was produced at *Drury Lane with the added attraction of real water. He adapted *Dickens's *Pickwick Papers* for the stage as *Sam Weller; or, The Pickwickians* (1837) almost before the last instalment had appeared, which aroused the anger of the author, who satirized him as 'the literary gentleman' in *Nicholas Nickleby* (1838).

Monodrama, short solo piece for one actor or actress supported by silent figures or by a chorus. It is sometimes referred to as a *melodrama on account of its musical accompaniment. It was popularized in Germany between 1775 and 1780 by the actor **Johann Christian Brandes** (1735–99) who developed the lyrical element introduced into German drama by the poet Klopstock and intensified by the influence of *Rousseau's *Pygmalion*. The Duologue, a similar compilation, had two speaking characters. Both types of entertainment, which were useful in filling out the triple bill then in vogue, frequently consisted of scenes extracted and adapted from longer dramas, somewhat in the style of the earlier English *droll.

Monstrous Regiment, see FRINGE THEATRE.

Montansier [Marguerite Brunet], **Mlle** (1730–1820), French actress-manageress who, having fallen in love with a young actor, **Honoré Bourdon de Neuville** (1736–1812), decided to go on the stage. With the help of powerful friends she took over the management of the theatre at Rouen, with such success that she was soon managing several others, Neuville acting as her business manager. While in charge of the theatre at *Versailles, she was presented to Marie-Antoinette, who was so charmed with her gaiety and wit that she invited her to act at Court. In 1777 she built a new theatre at Versailles (demolished in 1886) which was used to try out aspirants to the *Comédie-Française. On the outbreak of the Revolution she went to Paris and presided over a salon, where Napoleon, then an officer in the artillery, first met *Talma and formed a lasting friendship with him. Accused of Royalist sympathies she was arrested, but was saved from the guillotine by the fall of Robespierre. She immediately married Neuville and returned to the theatre she had recently opened at the *Palais-Royal, giving it her own name and remaining there until 1806, after which it became a café and was frequently visited by its former owner.

Montdory [Guillaume Desgilberts] (1594–1651), French actor, the friend and interpreter of *Corneille, whose Rodrigue in *Le Cid* (1636) he was the first to play. A good business man, he built up a fine company at the *Marais in the face of much hostility, training and directing his actors well. Among them were the parents of Michel *Baron, first leading man of the *Comédie-Française. Montdory, who was an excellent actor in the old declamatory style, seldom appeared in comedy and never in farce. He was at his best in tragedy, one of his finest roles being Herod in *Tristan's *La Mariane* (1636).

Montfleury [Zacharie Jacob] (*c.*1600–67), French actor, who was at the Hôtel de *Bourgogne, where he was second only to Belleville (see TURLUPIN) and later *Floridor. An enormously fat man, with a loud voice and pompous delivery, he was considered by his contemporaries a fine tragic actor and was much sought after by aspiring authors. He was satirized by *Molière in *L'Impromptu de Versailles* (1663). There was no love lost between them, for it was Montfleury who accused Molière before Louis XIV of having married his own daughter. He was also much disliked by *Cyrano, who is said to have ordered him off the stage on one occasion, an incident made use of by *Rostand in his play *Cyrano de Bergerac* (1898). His son **Antoine** (1639–85) was an actor and dramatist, and his daughters **Françoise** (*c.*1640–1708), known as Mlle d'Ennebault, and **Louise** (1649–1709) were also on the stage.

Montherlant, Henri de (1896–1972), French writer, known chiefly for his novels until in 1942 the success of his first play, *La Reine morte* (based on the story of Inés de Castro), at the *Comédie-Française turned his thoughts seriously to the theatre. Three more plays were produced at the Comédie-Française—*Port Royal* (1954), *Brocéliande* (1956), and *Le Cardinal d'Espagne* (1960). His other plays include *Le Maître de Santiago* (1948), about the struggle between love and religion, *Celles qu'on prend dans ses bras* (1950), about sexual obsession, *Malatesta* (also 1950), and *Don Juan* (1958). The most interesting, *La Ville dont le prince est un enfant*, was written in 1951 but not performed until 1967. It deals with a platonic friendship between two boys which is destroyed by a priest not, as he thinks, out of kindness but out of jealousy. This portrayal of spiritual agony, of the conflict between love and religion, is typical of Montherlant's plays, which contain little external action and are written in a sonorous prose that makes few concessions to realism. Although they have a religious context, Montherlant did not consider himself a 'Catholic' writer, preferring to describe himself as a 'psychological' dramatist. His plays have had little success in English, though in 1957 Donald *Wolfit appeared in *The Master of Santiago* and *Malatesta*.

Montparnasse, Théâtre, see BATY.

Moody, William Vaughn (1869–1910), American dramatist, whose work is important in the development of indigenous drama in the United States. Poet, scholar, and educationist, Moody had been for many years a teacher when he decided that his real vocation lay in the theatre. The first of his plays to be staged, by Henry *Miller, was *The Great Divide* (1906), originally known as *The Sabine Woman*. It was successfully produced in London in 1909, the year in which another of Moody's plays, *The Faith Healer*, was seen in New York. Both these are written in a dignified, poetic style, and mark the arrival on the American scene of the serious social dramatist, still somewhat melodramatic but moving away from French farce and adaptations of sentimental novelettes. None of his long poetic plays was produced in his lifetime.

Mooney, Ria (1903–73), Irish actress, teacher, and director for many years with the *Abbey Theatre, Dublin, where she first appeared in *Shiels's *The Retrievers* (1924), achieving prominence as Rosie Redmond in *O'Casey's *The Plough and the Stars* (1926). In the following year she played Mary Boyle with Arthur *Sinclair's Irish Players in their New York production of O'Casey's *Juno and the Paycock*; but by 1932 she was once again with the Abbey company, appearing with them on their American tour in the winter of 1932–3. From 1933 to 1935 she played at the *Gate Theatre, Dublin, in such roles as Catherine Earnshaw in her own adaptation of Emily Brontë's *Wuthering Heights*, and Bride in Denis *Johnston's *Storm Song* (1934). She returned to the Abbey in 1935, and two years later was in charge of its experimental Peacock Theatre where she directed *Yeats's *The Dreaming of the Bones* and other plays. In 1948 she was appointed the first woman director of the Abbey Theatre, a position she held until her retirement in 1963.

Moore, Edward (1712–57), English dramatist, author of the fashionably sentimental plays *The Foundling* (1747) and *The Gamester* (1753). The first bridges the gap between the works of Colley *Cibber and Richard *Steele and those of *Cumberland and Mrs *Inchbald. The second, partly written by *Garrick who staged it at *Drury Lane, is a domestic tragedy in the style of *Lillo. Translated into French, it had a marked influence on the development of the *tragédie bourgeoise*.

Moore, Maggie, see WILLIAMSON, JAMES.

Moore, Mary, see ALBERY, JAMES and WYNDHAM.

Moral Interlude, short play, intended mainly for educational purposes, which in the 16th century was derived from the longer *morality play. It had a good deal of robust humour, some of the earlier Vices, notably Pride and Gluttony, and such abstract personifications as Free-Will and Self-Love, being transformed into purely comic characters. More space was also given to the antics of the Devil and the old *Vice, or buffoon. Notable English examples of the moral interlude are the anonymous

Hickscorner (*c*.1513) and R. Wever's *Lusty Juventus* (*c*.1550).

Morality Play, medieval form of drama which, with the Creed and Paternoster plays, aimed to teach through entertainment, as the Dominicans and Franciscans were beginning to do at the beginning of the 13th century. Sermons already contained the elements of drama, and it was not difficult for actors to inject into them the animation needed to convert abstract ideas into vivid verbal images, readily understood by the largely illiterate congregations of the time. Even supernatural and miraculous events could be made comprehensible, with specific Vices and Virtues, led by a recognizable Devil and a Good Angel, battling realistically for man's soul. The morality play, unlike the plays derived from the Liturgy, was not tied to any church service, but could be played at any time, given a sufficient number of actors and spectators. It provided excellent material for the use of the professional actors emerging from the many amateur groups by the middle of the 15th century; there must have been many more scripts of such plays than the handful which have survived. The only one to be seen on stage today is the Dutch *Elckerlyc* (*c*.1495), known in English as *Everyman*. This adaptation, prepared by William *Poel, was produced in London in 1903, and has frequently been revived. As *Jedermann*, in a German version by Hugo von *Hofmannsthal, it has played an important part in the Salzburg Festival since its inception.

Moratín, Leandro Fernández de (1760–1828), Spanish poet and dramatist, son of **Nicolás Fernández de Moratín** (1737–80), who disapproved of Lope de *Vega and his followers and of the *auto sacramental*, and based his three tragedies on the strict laws of French neo-classicism. His son was also a neo-classicist, but unbent enough to produce an amusing satire on the extravagant drama popular in Spain at the time, entitled *La comedia nueva o el café*, aimed at the popular playwright Comella. Leandro's most successful play was *El sí de las niñas* (*When a Girl says Yes*, 1806), in which he defends a woman's right to choose for husband the man she loves. Leandro, who was exiled from Spain for supporting Joseph Bonaparte, lived in France for some years. He was a great admirer of *Molière, and made excellent though somewhat free translations of two of his plays. He was also the first to translate *Hamlet* directly from the English text into Spanish.

Moreto y Cabaña, Agustín (1618–69), Spanish playwright, author of a number of good religious and historical plays, but mainly remembered for his *comedias*, of which the best are considered to be *El desdén con el desdén* (*Scorn for Scorn*, pub. 1654) and *El lindo Don Diego* (pub. 1662). The first, based on several comedies by Lope de *Vega, who with *Tirso de Molina

was the main influence on Moreto's work, shows a young noblewoman who has disdained her suitors being in turn disdained by the man she loves—a situation later used by *Molière in *La Princesse d'Élide* (1664). The second is an amusing trifle about a foppish young man who spends most of the day getting dressed, but whose arrival in Madrid nevertheless threatens the happiness of two young lovers. It is an excellent example of the *comedia de figurón*, in which Moreto excelled. An inventive and engaging *gracioso* figures in many of his plays.

Morgan, Charles Langbridge (1894–1958), English novelist and essayist, who succeeded A. B. *Walkley as dramatic critic of *The Times* in 1926, a position he held until 1939. He was also the author of three plays—*The Flashing Stream* (1938), *The River Line* (1952), and *The Burning Glass* (1954)—written with the fastidious care and nervous vitality that distinguished all his work.

Morley, Robert (1908–   ), English actor and playwright, son-in-law of Gladys *Cooper. He made his first appearance on the stage in 1928, and was in *Fagan's repertory company at Oxford. Later he joined the Cambridge Festival Theatre company under Norman *Marshall, under whose direction he made his first outstanding success in the title-role of Leslie and Sewell Stokes's *Oscar Wilde* (1936) at the London *Gate Theatre (NY, 1938). In 1937 he played Professor Higgins in *Shaw's *Pygmalion* at the *Old Vic, and among his later successes were Sheridan Whiteside in *Kaufman and *Hart's *The Man Who Came to Dinner* (1941), the Prince Regent in Norman Ginsbury's *The First Gentleman* (1945), and Arnold Holt in *Edward, My Son* (1947; NY, 1948), of which he was co-author. He was then seen in two plays by André Roussin, *The Little Hut* (1950), which ran for three years, and *Hippo Dancing* (1954), which he adapted himself. He made his début in a musical play as Panisse in *Fanny* (1956), and played the Japanese Mr Asano in Spigelgass's *A Majority of One* (1960). He was also effective in *Ustinov's *Halfway up the Tree* (1967), *Ayckbourn's *How the Other Half Loves* (1970), in which he also played in North America and Australia, *A Ghost on Tiptoe* (1974), which he again co-authored, and a revival of Ben *Travers's *Banana Ridge* (1976). Of his own plays, the best known is *Goodness, How Sad!* (1938). He is limited by his burly form and rich, booming voice to mainly arrogant, domineering roles, but he is an excellent actor and widely and deservedly popular.

Morosco Theatre, New York, on West 45th Street, between Broadway and 8th Avenue. Built by the *Shuberts, with a seating capacity of 1,009, it opened in 1917 with *Canary Cottage* by Elmer Harris and Oliver Morosco, the latter a well-known West Coast play producer after whom the new theatre was named. In 1920 the success of *The Bat* by Mary Roberts Rinehart and Avery Hopwood started a fashion for mystery thrillers. George *Kelly's *Craig's Wife* (1925) won a *Pulitzer Prize, and the *Theatre Guild's production of *Call it a Day* by Dodie Smith ran for six months in 1935. Later successes included two plays by John *Van Druten, *Old Acquaintance* (1940) and *The Voice of the Turtle* (1943), and *Coward's *Blithe Spirit* (1941). Arthur *Miller's *Death of a Salesman* (1949) and Tennessee *Williams's *Cat on a Hot Tin Roof* (1955) were also Pulitzer Prize-winners, and Gore Vidal's comedy of American political life *The Best Man* (1960) did extremely well. In 1963 Peter *Shaffer's double bill *The Private Ear* and *The Public Eye* was seen. Arthur Miller's *The Price* (1968) had a long run, and David *Storey's *Home* (1970), with *Gielgud and *Richardson, repeated its London success. Other plays from London were Simon *Gray's *Butley* (1972) with Alan *Bates, David Storey's *The Changing Room* (1973), *Rattigan's *In Praise of Love* (1974), and *Ayckbourn's *The Norman Conquests* (1975). In 1977 Cristofer's *The Shadow Box* was yet another Pulitzer Prize-winner. The theatre was demolished in 1982, together with the nearby *Helen Hayes Theatre.

Morris [Morrison], **Clara** (1846–1925), American actress who is believed to have been on the stage as a child, appearing in pantomime at the Royal Lyceum in Toronto, her birthplace. At 16 she joined the stock company in Cleveland, Ohio, and a few years later toured North America, before joining *Daly's company in New York. She played a wide range of parts, and in spite of a strong accent and an extravagant, unrestrained manner was popular both in New York and on tour. Though not accounted a good actress, she had an extraordinary power of moving an audience, and could always be relied on to fill the theatre, particularly in such parts as Marguerite Gautier in the younger *Dumas's *The Lady of the Camellias*. Among her more serious roles were Lady Macbeth, Julia in *Knowles's *The Hunchback*, and a number of modern heroines. Ill health forced her into semi-retirement, but she was seen in Canada again in 1889, and in 1892 appeared for the first time in Montreal in *Sardou's *Odette*. She made a return to the theatre in 1904, playing Sister Geneviève in an all-star revival of Oxenford's *The Two Orphans*, and subsequently appeared in vaudeville until her eyesight failed in 1909.

Mortimer, John Clifford (1923–), English dramatist, by profession a barrister, whose aim, in his own words, is to chart 'the tottering course of British middle-class attitudes in decline'. His first plays, the one-act *The Dock Brief* and *What Shall We Tell Caroline?* (1958), were produced in a double bill (NY, 1961); they were followed by another one-acter, *Lunch Hour* (1961). His

first full-length plays were *The Wrong Side of the Park* (1960), in which Margaret *Leighton played a woman whose illusions about her deceased first husband prevent her from making a success of her second marriage, and *Two Stars for Comfort* (1962), about the involvement with a young woman of the middle-aged owner of a riverside inn. Mortimer then prepared an English version of *Feydeau's farce *Une puce à l'oreille* as *A Flea in Her Ear* (1965) for the *National Theatre, and after a somewhat non-naturalistic play, *The Judge* (1967), and a second translation from Feydeau—*Un fil à la patte* as *Cat among the Pigeons* (1969)—returned to the one-act form with *Come as You Are* (1970), four playlets set in different parts of London. They were followed by an autobiographical play, *A Voyage round My Father*, produced at the *Haymarket Theatre in 1971, where the role of the Father was played first by Alec *Guinness and then by Michael *Redgrave. In the same year an adaptation of *Zuckmayer's *The Captain of Köpenick* was seen at the National Theatre. *I, Claudius* (1972) was based on two novels by Robert Graves, and later plays were *Collaborators* (1973), in which a husband and wife collaborate in writing a film about their marriage, *Heaven and Hell* (1976), another double bill, and *The Bells of Hell* (1977), which depicts the Devil arriving at the home of a trendy South London vicar. Two more translations from Feydeau, *The Lady from Maxim's* (1977) and *A Little Hotel on the Side* (1984), were also seen at the National Theatre.

Morton, Charles (1819–1904), early and extremely able *music-hall manager, who from his association with the early pioneering days of music-hall was known as 'the father of the halls'. He opened the first of the major London music-halls, the *Canterbury, in 1852, and in 1861 inaugurated a new era with the opening of the *Oxford. He was later called in to revive the flagging fortunes of several other well-known halls, among them the Philharmonic (later the Grand Theatre) in Islington in 1870, the *Palace in Cambridge Circus in 1892, and the *Tivoli in the Strand in 1893. He was successful mainly because he was a hard worker, set a high standard, was good at choosing his performers, and though appreciating honest vulgarity was opposed to innuendo and salacity.

Morton, James J., see VAUDEVILLE, AMERICAN.

Morton, Thomas (c.1764–1838), English dramatist, whose first play, a *melodrama entitled *Columbus; or, A World Discovered* (1792), was produced at *Covent Garden. He wrote several sentimental comedies—*The Way to Get Married* (1796), *A Cure for the Heart-Ache* (1797), and *Secrets Worth Knowing* (1798)—before writing the play by which he is best remembered, *Speed the Plough* (1800). It became famous for the character of Mrs Grundy, who never appears but is frequently referred to as the embodiment of British respectability. Among Morton's later plays the most important were *The School of Reform; or, How to Rule a Husband* (1805), an amusing comedy, *The Slave* (1816), and *Henri Quatre; or, Paris in the Olden Time* (1820), in both of which *Macready appeared. One of Morton's most spectacular melodramas was *Peter the Great; or, The Battle of Pultawa*, seen at *Drury Lane in 1829. His son **John Maddison Morton** (1811–91) was a prolific writer of one-act farces. His *Box and Cox* (1847), first produced in 1843 as *The Double-Bedded Room*, provided the libretto for Sullivan's first operetta, *Cox and Box* (1867).

Moscow Art Theatre, famous theatre, now dedicated to Maxim *Gorky, founded by *Stanislavsky and *Nemirovich-Danchenko in 1898. It soon gave a successful production of *Chekhov's *The Seagull*, which had recently failed at another theatre. It was followed by *Uncle Vanya* (1899), *Three Sisters* (1901), and *The Cherry Orchard* (1904). The repertory of pre-Revolutionary plays included a number of European classics, but the only play by Shakespeare to be performed was *Julius Caesar*. During the upheavals of the Revolution the theatre was saved from extinction by the efforts of *Lunacharsky, and after a long tour of Europe and America in 1922 and 1923 the company returned to Moscow. Some tentative productions of new plays followed, including *Bulgakov's *The Days of the Turbins* (1926), and the theatre finally found its feet in the new world in 1927 with the production of *Ivanov's *Armoured Train 14-69*. The theatre was now firmly established, many actors, among them *Kachalov, *Moskvin, and Chekhov's widow Olga *Knipper, spending the rest of their lives with the company, and being succeeded by actors trained in the theatre's own dramatic school. From this school developed also several important individual groups, among them those at the *Realistic and *Vakhtangov theatres. The Moscow Art Theatre visited London in 1958, 1964, 1970, and 1989, its productions of plays by Chekhov being of particular interest. In 1971 the artistic directorship of the theatre was assumed by Oleg Efremov, and two years later the company moved into new premises on the Tverskoy Boulevard, with a large auditorium seating nearly 1,400. The company also retained its old premises, which, with the dramatic school, gave it three bases from which to work. Because of its status as a showcase theatre, the Moscow Art Theatre is particularly prone to bureaucratic control, leading to ossification and overmanning. Efremov's attempts at reform partly succeeded when in 1985 the government divided the theatre into two companies sharing the same name but with different artistic policies.

Moscow State Jewish Theatre, known in Yiddish as Melucha, and in Russian as Goset. This theatre was founded in Leningrad in 1919 by **Alexander Granovsky** [Abraham Ozark] (1890–1937) as the Jewish Theatre Studio for the production of plays in Yiddish; it later moved to Moscow, opening at the small Chagall Hall in 1921 in a programme of short comedy sketches by Sholom *Aleichem, and finally achieved its own theatre, seating 766. Granovsky was succeeded in 1927, during a European tour, by his leading actor Salomon *Mikhoels, who continued much of Granovsky's work with the addition of new plays by such authors as **David Bergelson** (1884–1952) and Peretz Markish. The leading stage designer at this theatre was for some time Isaac Rabinovich, who later worked for the *Vakhtangov and *Maly theatres, and one of its artistic directors was *Radlov. After Mikhoels's death in 1948 the company disbanded.

Moskvin, Ivan Mikhailovich (1874–1946), Russian actor, who joined the *Moscow Art Theatre on its foundation in 1898, playing the lead in its initial production of A. K. *Tolstoy's *Tsar Feodor Ivanovich* and Epihodov and Lvov in *Chekhov's *The Cherry Orchard* and *Ivanov*. Among his later roles were Luka in *Gorky's *The Lower Depths*, Nozdrev in *Gogol's *Dead Souls*, and the merchant Pribitkov in *Ostrovsky's *The Last Sacrifice*. In 1942 he played a leading part in *Pogodin's play about Lenin, *The Kremlin Chimes*.

Mossoviet Theatre, Moscow, founded in 1923 to encourage young playwrights to tackle Soviet and European political themes. The theatre's earliest and best dramatist was *Bill-Belotserkovsky, whose *Storm* (also known as *Hurricane*) was produced there in 1926. The company was led by Lyubimov-Lanskoy until 1940, when it came under the direction of *Zavadsky, who in 1951 directed a revised version of *Storm* which proved a landmark in the theatre's history, as were his productions of *The Merry Wives of Windsor* in 1957 and of *Lermontov's *Masquerade* in 1965. An excellent production of *King Lear* in 1958 further enhanced the reputation of the theatre, which continued also to produce new Soviet plays, among them Virta's *In Summer the Sky is High* (1961), as well as plays by such European dramatists as *Shaw, *García Lorca, and Heinrich Böll. In 1965 the company was seen at a theatre festival in Paris in plays by Lermontov and *Rozov, and in an adaptation of Dostoevsky's *Uncle's Dream*. The theatre celebrated its 60th anniversary in 1983 with a new production of *Gorky's *Yegor Bulychov and Others*, having earlier staged many of his other plays.

Mostel, Zero [Samuel Joel] (1915–77), American actor, a huge, moon-faced man of enormous talent, who gained his theatrical experience in

Greenwich Village night clubs before appearing in *vaudeville on Broadway in 1942. In 1946 he was seen as Hamilton Peachum in *Beggar's Holiday*, an adaptation of *Gay's *The Beggar's Opera* with music by Duke Ellington, and in 1952 he played Argan in *The Imaginary Invalid*, his own adaptation of *Molière's *Le Malade imaginaire*. Among his later parts were Shu Fu in *Brecht's *The Good Woman of Setzuan* in 1956, Leopold Bloom in a dramatization of part of Joyce's *Ulysses* as *Ulysses in Nighttown* in 1958, John in *Ionesco's *Rhinoceros* in 1961, and Prologus in the highly successful musical *A Funny Thing Happened on the Way to the Forum* in 1962. In 1964 he scored a great success in the musical *Fiddler on the Roof*. He died while on a pre-Broadway tour.

Motion, name given in the 16th and 17th centuries to the puppet-plays of the itinerant showmen. The earliest dealt with biblical subjects, and Shakespeare refers in *The Winter's Tale* to 'a motion of the Prodigal Son'. Later the range of subjects was extended, and episodes were used from medieval romance, mythology, and contemporary history.

Moulin Rouge, Paris, well-known dance-hall which opened in 1889, including in its amenities a large garden used in the summer for dancing and entertainments. A constant feature of its programmes has been a cabaret show, in which the cancan made its first appearance, the dancers in 1893 being Grille d'Égout, la Goulue, la Môme Fromage, and Nini-patte-en-l'air. *Mistinguett was for some years part-proprietor of the Moulin Rouge and frequently appeared there, as did most of the stars of variety and music-hall.

Mounet-Sully [Jean Sully Mounet] (1841–1916), famous French actor who made his début at the *Comédie-Française in 1872 as Oreste in *Racine's *Andromaque*. His fine physique, beautifully modulated voice, and sombre, penetrating gaze, added to fiery, impetuous acting and great originality, soon brought him into prominence. His career was one of unclouded success and he appeared in all the great tragic roles of the French classical repertory. He was also outstanding in the plays of Victor *Hugo when they were finally given at the Comédie-Française.

Mountfort, William (1664–92), English actor, who specialized in 'fine gentlemen' parts, being an excellent Sparkish in *Wycherley's *The Country Wife* and creating the title-role in John *Crowne's *Sir Courtly Nice; or, It Cannot Be* (1685). He was also the author of several plays, including a *harlequinade 'in the Italian Manner', *The Life and Death of Doctor Faustus . . . With the Humours of Harlequin and Scaramouche* (1685), and a comedy, *Greenwich Park* (1691), which was extremely successful and frequently

revived. He was brutally murdered at the instigation of a certain Captain Hill, who suspected him of being a successful rival in the affections of Mrs *Bracegirdle, with whom Hill was passionately but unsuccessfully in love.

Mountview Theatre School, see SCHOOLS OF DRAMA.

Mowatt [*née* Ogden], **Anna Cora** (1819–70), American author and actress, mainly remembered for her social comedy *Fashion* (1845), considered to be one of the best of the early satires on American life. It was first produced at the *Park Theatre, so successfully that Mrs Mowatt was encouraged to go on the stage herself, making her début as Pauline in *Bulwer-Lytton's *The Lady of Lyons* in 1845. She then formed her own company, with E. L. *Davenport as her leading man, with which she visited London in 1848–51. In 1854 she retired to live quietly in London, where she died.

Mrożek, Sławomir (1930–), Polish playwright, who with *Witkiewicz and Gombrowicz represents in Polish drama the Theatre of the *Absurd, though with more political content than is usual with exponents of this style. In 1964 two one-act plays, *What a Lovely Dream!* and *Let's Have Fun!* (both 1962), were seen in London during the first *World Theatre Season. Mrożek's longer plays, which include *The Turkey Cock* (1961) and *The Enchanted Night* (1963), culminated in the world-famous satire on totalitarianism, *Tango*, produced first in Belgrade in 1965, and later the same year in Warsaw by *Axer. The first modern Polish play to make a decided impact on the theatres of Western Europe and America, it was seen in London in 1966 in an *RSC production directed by Trevor *Nunn, and in New York in 1969. In 1968 Mrożek went into voluntary exile, but most of his recent plays have been seen in Poland, *The Hunchback* (1976) having been given its first production simultaneously in Cracow and Łódź. Later plays include *Émigrés* (1974; NY, 1982), about the response of exiles to freedom, and *A Summer's Day* (London, 1985), in which two men contemplate suicide, one because of failure, the other because success has left him no further goals.

Müller, Heiner (1929–), playwright of the German Democratic Republic, whose first plays, *Der Lohndrücker* (*The Scab*, 1950) and *Die Korrektur* (*The Correction*, 1958), were propaganda works in the style of *Brecht designed to instil socialist attitudes in his worker-audiences. *Cement* (1972) is an adaptation of Gladkov's novel on the Russian Revolution. He also uses classical subjects, taken mainly from *Sophocles, such as *Philoktet* (1968), to depict with less overt socialism the conflicting claims the community makes on the individual. *Medeamaterial* (1982;

London, 1989) uses *Euripides' *Medea* to reflect present disasters. He also translated and adapted *Shwartz's *The Dragon* and Shakespeare's *Macbeth* (1972) and *Hamlet* (*Hamletmachine*, 1979; NY, 1984; London, 1987). *Quartet* (1982; London, 1983; NY, 1985) places the characters of Laclos's novel *Les Liaisons dangereuses* in a bunker after World War III as well as in the pre-French Revolution salon of their original setting. *The Task* (London, 1989), set in Jamaica in 1794, again uses the past as a metaphor for the present. His works, with their terse, dense, finely chiselled language, are highly esteemed throughout Germany.

Multiple Setting, term applied to the stage décor of the medieval play (known in France as *décor simultané* and in Germany as *Standort-* or *Simultanbühne*), inherited from the *liturgical drama with its 'mansions' or 'houses' disposed about the church. When biblical dramas were first performed out of doors, the 'mansions' were disposed on three sides of an unlocalized *platea* or acting space, but by the 16th century, at any rate in France, they were set in a straight line or on a very slight curve. In England the different scenes of a *mystery play were on perambulating *pageants, and the multiple setting was not needed. It continued in France, and possibly in Germany, for a long time, and was still in use at the Hôtel de *Bourgogne in Paris in the early 17th century. It is even possible that *Corneille's early plays, produced at the *Marais, were staged in a multiple setting, which was finally ousted by the development of the single set used for classical tragedy, as in the plays of *Racine. The Elizabethan public stages such as the *Globe Theatre did not employ multiple settings, though something of the kind may have been used in the early days of the private roofed playhouse (see BLACK-FRIARS) and was certainly a feature of the elaborate Court *masque. The modern equivalent of the multiple setting is the *composite setting.

Mumming Play, English folk-play which first emerges in the 18th century, and certainly in the 1930s could still be seen in remote villages in England, particularly in the Cotswolds, at Christmas. It is essentially a dramatization of the death of winter and the springtime revival of the earth. It was traditionally played by men only, dressed in rags or shredded paper, and the texts were transmitted orally. The players met in secret to rehearse and for the performance blacked their faces with soot; there may be here a connection with the morris dancers. The size of the company varied considerably, but essentially consisted of Father Christmas, the Turkish Knight, sometimes known as the Bold Smasher, and St (sometimes King) George. There is plenty of scope for the introduction of secondary characters. The play opens with Father Christmas with his broom clearing a space in a room or hall for the entry of his fellow players. He is followed by St George, who boasts of his mighty deeds and challenges anyone to take up arms against him. The Turkish Knight advances and the two men fight. The Turkish Knight is killed, and Father Christmas, declaring that the dead man is his son, turns into the Doctor, and eventually, by magic spells and a good deal of mumbo-jumbo, brings the Knight to life again. The character of the Doctor, accompanied by his faithful servant Jack Finney or Johnny Jack, modelled on the quack doctor of many itinerant groups, seems also to indicate some influence from the *commedia dell'arte*. Most of the *slapstick comedy of the mumming play is concentrated in this scene, as the Doctor abuses his servant, boasts of the wonderful cures he has been responsible for, and finally returns the Turkish Knight to life. (In some versions it is St George who is slain and returned to life, thus offering occasion for another battle.) The play then ends with a procession introducing several new characters, who may or may not indulge in byplay of their own—Beelzebub, the Bessie (a man dressed as a woman), even historical characters such as Charlemagne or Oliver Cromwell. The ceremony ends with a collection, taken up by Father Christmas, or by Little Devil Dout, who with his broom sweeps everyone out of the room—his name being a corruption of 'Do Out!'

There are vestiges of two other mumming plays, which in general resemble the one detailed above. The Sword Play, which includes a complicated sword dance, and The Wooing Ceremony, in which the Lady (played by a man) sees her accepted wooer killed by rival suitors and then brought back to life. The actors in the mumming play were sometimes known as Guizards or Guisers, from their disguise.

Munday, Anthony (1560–1633), English hack-writer, pamphleteer, ballad-maker, translator, and playwright. His first extant play, *Fedele and Fortunio* (1584), translated from the Italian, may have been used by Shakespeare in writing *Much Ado about Nothing* (1598), while parts of his play on a contest between two wizards, *John a Kent and John a Cumber* (c.1590), probably suggested the Bottom scenes in *A Midsummer Night's Dream* (1595). Munday is known to have collaborated in several plays, now lost, about *Robin Hood and Sir John Oldcastle, and worked mainly for *Henslowe. He certainly had a hand in the fragmentary *Sir Thomas More*, since most of the extant manuscript is in his handwriting.

Munden, Joseph Shepherd (1758–1832), English comedian, who after several years' hard work in the provinces joined the company at *Covent Garden in 1790. His first outstanding success was as Old Dornton in the initial production of *Holcroft's *The Road to Ruin* (1792). He was the favourite actor of Charles *Lamb, who left an admirable description of him, and seems to have been particularly admired in drunken scenes, for which he invented new 'business' at every performance. The broadness of his acting would scarcely be conceivable today and even in his own time he was censured for caricature, but he made the fortune of many a poor play. He remained at Covent Garden for over 20 years, and then went to *Drury Lane, retiring in 1824.

Munich, capital of Bavaria, had a rich cultural life from the 16th century onwards, but although *Jesuit drama, first seen in 1560, flourished from 1606 to 1614 with plays written and directed by Jakob *Bidermann, who in 1609 staged a splendid revival of his *Cenodoxus,* the first half of the 17th century was mainly given over to opera. It was not until 1769 that a German company first became resident in the city, and began gradually to oust the French players who had been so influential there. A National Theatre opened in 1811, where during the 19th century **Franz von Dingelstedt** (1814–81), together with *Possart at the Court Theatre, tried to build up a repertoire of German classics in the teeth of considerable opposition. During Possart's years as Artistic Director of the Court Theatre Josef *Kainz appeared there—from 1881 to 1883—often in private performances commanded by the eccentric King Ludwig II. The National Theatre was destroyed by bombs in 1943, reopening in 1963 after rebuilding. The Künstlertheater, founded in 1908, was leased by *Reinhardt from 1911 to 1913. Under the direction of **Otto Falckenberg** (1873–1947) the Kammerspiele, founded in 1911 as the Lustspielhaus and already known as the home of *Expressionism, became the first theatre to stage a play by *Brecht, *Trommeln in der Nacht* (1922), and a year later his *Im Dickicht der Städte* was also first produced in Munich. In 1923 Brecht made his début as a director by producing at the Kammerspiele his free adaptation of *Marlowe's *Edward II.* Among the other theatres in Munich one of the most interesting, and certainly the most beautiful, is Cuvilliés' Residenztheater, built 1751–3. Though it was bombed in 1943–4, its superb rococo interior was later salvaged and re-erected in the palace, where it reopened in 1958 and is now used for festival productions of intimate drama and opera. Ingmar *Bergman joined the company in 1977, remaining for several years.

Munk [Petersen], **Kaj Harald Leininger** (1898–1944), Danish priest and playwright, whose first substantial play, *En idealist* (*Herod the King,* 1928), was disastrously received on its first production and only fully established itself as a major work at its revival in 1938. One of his earliest successes was *Cant* (1931), a verse play about Henry VIII and Anne Boleyn. It was followed by *Ordet* (*The Word,* 1932), which takes as its theme the nature of miracles. After a brief flirtation with Fascism in the early 1930s, Munk became increasingly and openly antagonistic to the movement, as was shown in his next plays, *Sejren* (*The Victory,* 1936), which condemned Italian aggression in Abyssinia, and *Han sidder ved smeltediglen* (*He Sits by the Melting Pot,* 1938), an attack on Nazi anti-Semitism. During the Second World War Munk continued fearlessly to write with *Niels Ebbensen, Før Cannae,* and *Ewalds død.* He was shot by the Nazis to silence his outspoken hostility to their occupation of his country.

Murdoch, James Edward (1811–93), American actor, considered by many of his contemporaries the finest light comedian of his day. He was especially noted for his clear diction. He began his career at the Arch Street Theatre in *Philadelphia, where he was born, in 1829 and appeared at the *Chestnut Street Theatre with Fanny *Kemble in 1833. Intermittent ill health kept him off the stage on several occasions, but he always returned, to be warmly received in such parts as Charles Surface in *Sheridan's *The School for Scandal,* Mirabell in *Congreve's *The Way of the World,* and Benedick in *Much Ado about Nothing.* He was also admired in such serious roles as Hamlet and Claude Melnotte in *Bulwer-Lytton's *The Lady of Lyons.* He finally retired in 1889, having achieved a national reputation in spite of acting mostly in Philadelphia.

Murphy, Arthur (1727–1805), English actor and dramatist, who gave up acting, for which he had no particular talent, to devote all his time to writing. His tragedy *The Grecian Daughter* (1772) provided Mrs *Siddons with one of her favourite parts in the heroine Euphrasia, but his chief strength lay in comedy. He had little originality, but was adept at choosing and combining the best elements from the works of others, particularly *Molière, and several of his works, including *The Way to Keep Him* (1760) and *The School for Guardians* (1767), contributed to the revival of the comedy of manners which culminated in *Sheridan's *The School for Scandal* (1777).

Murray, Sir (George) **Gilbert Aimé** (1866–1957), English classical scholar, poet, humanist, and philosopher, whose verse translations of the plays of *Euripides were staged in London at the beginning of the 20th century. Though they have now been superseded, they were in their day superior theatrically to anything heard previously, and revealed the beauties of classical tragedy to those who had no Greek. The first, under the direction of Harley *Granville-Barker at the *Royal Court Theatre, was *Hippolytus* in 1904. It was followed by the *Trojan Women* in

1905 and *Electra* in 1906. *Medea* was seen at the *Savoy Theatre in 1907, the *Bacchae*, in which Lillah *McCarthy played Dionysus, again at the Court, and *Iphigenia in Tauris*, with McCarthy in the title-role, at the *Kingsway. The *Alcestis* and the *Rhesus* have had only amateur productions. Murray also translated *Sophocles' *Oedipus Rex*, *Antigone*, and *Oedipus at Colonus*, and *Aeschylus' the *Oresteia*, the *Suppliant Women*, *Prometheus Bound*, the *Persians*, and the *Seven against Thebes*, somewhat less successfully; but his versions of *Aristophanes' the *Birds*, the *Frogs*, and the *Knights* were eminently actable, as were his reconstructions of *Menander's *Perikeiromenê* as *The Rape of the Locks* in 1941 and of the *Epitrepontes* as *The Arbitration* in 1945.

Murray, Thomas Cornelius (1873–1959), Irish dramatist, whose subjects were drawn from the peasant and farming life of his native county of Cork, and were distinguished for their sympathetic perception of the deeply religious and sometimes mystical quality of the peasant mind. He made his name with *Birthright*, a tragedy of rivalry and family jealousy produced at the *Abbey Theatre in 1910. Among the plays which followed it were *Aftermath* (c.1922), a full-length tragedy on the theme of an arranged marriage; and *Autumn Fire* (1924), another tragedy of mismating, probably the author's best play. *The Pipe in the Fields* (1927) is a fine, brief study of the sudden flowering of the mind of an artist. *The Blind Wolf* (1928), the first of Murray's plays to be set outside Ireland, is again a peasant tragedy. His last play was *Illumination* (1939).

Muses, The. Three of the nine Muses in Greek mythology, daughters of Mnemosyne the goddess of memory, were particularly connected with the theatre—Melpomene, the Muse of tragedy; Terpsichore, the Muse of dancing; and Thalia, the Muse of comedy.

Museums and Collections, Theatrical. These fall roughly into three categories: those of general national and international interest; those attached to or dealing with a particular theatre; and those devoted to a particular player or playwright, or a particular subject.

Among the first are the Bakhrushin Museum in Moscow, founded in 1894; the Clara Ziegler Foundation in Munich (1910); the Bucardo Museum in Rome (1931); the Oslo Theatre Museum (1939); and the Toneel Museum in Amsterdam (1960). In London there is the *Theatre Museum (1974), and the Bodleian Library at Oxford has the Douce Collection. There is a good deal of theatrical material scattered throughout the various departments of the British Museum and, for the French theatre, in the Bibliothèque Nationale in Paris. The Bibliothèque de l'Arsenal in Paris also has valuable private collections, and in Vienna the National Library has a separate theatre section.

In the USA, the Theatre Collection of the Museum of the City of New York came into existence in 1923. It contains a collection relating to live theatrical entertainments and personalities in New York City dating back to the 18th century, and 2,500 theatrical costumes dating back to the mid-19th century. The dance, music, theatre, and recorded sound collections of the New York Public Library were relocated to Lincoln Center in 1965 to form the Library-Museum of the Performing Arts for circulating material and the Performing Arts Research Center for research materials. The theatre collection (see FREEDLEY) was established as a research unit of the Public Library in 1931 and covers stage, film, radio-television, vaudeville, circus, magic, puppetry, and miscellaneous forms of performing arts. Over 80 per cent of the collection is of non-book materials. In 1970 a new unit called Theatre on Film and Tape (TOFT) was set up within the collection to videotape live theatrical performances throughout the USA for archival purposes. The collection was named the Billy Rose Theatre Collection in 1979, in recognition of a gift from the Billy Rose Foundation. The Hampden-Booth Theatre Library (1957) is available to research workers at the *Players' Club. The San Francisco Performing Arts Library and Museum grew out of a private collection started in 1947. It covers performing arts from circus to grand opera, with primary emphasis on the west coast of the USA.

Of collections dealing with a particular theatre probably the most important is that at *Drottningholm. A smaller museum in the same genre is attached to the former Court theatre in Copenhagen. There is also one in Meiningen devoted to the work of the *Meininger company. In Russia most of the well-known theatres have their own museums, including the *Moscow Art and the *Vakhtangov theatres. The *Shubert Archive, housed in the *Lyceum Theatre, New York, documents the activities of this important producing organization. Work on sorting and cataloguing material began in 1975, and in 1986 the Archive opened for public use.

Pride of place among the museums devoted to an individual must go to the Folger Shakespeare Library in *Washington. London has the Shakespeare Globe Museum opened by Sam *Wanamaker, and there are libraries devoted to Shakespeare in Stratford-upon-Avon and Birmingham. Material relating to individuals can be found in such collections as the Victor *Hugo Museum in Paris, the Casa *Goldoni in Venice, and the former homes of *Stanislavsky and *Chekhov in Russia, *Lessing in Germany, *Pixérécourt in France, and Edwin *Booth in New York. In England there is a collection devoted to Ellen *Terry in her former home at Smallhythe in Kent. In Stockholm there is the Hamilton Collection of English plays. Leningrad has a museum devoted to the circus, and in Moscow and Munich there are specialized puppet museums.

Among the collections of theatre material in private hands, the most important is the collection of over 400 drawings by Inigo *Jones at Chatsworth, Derbyshire, the home of the Duke of Devonshire. The Mander and Mitchenson Collection in Beckenham, Kent, contains illustrations relating to the British theatre in the 19th and 20th centuries. Bristol University Department of Drama Theatre Collection has been built on the nucleus of the Richard Southern Collection, and there is material on *Henslowe at Dulwich College (see ALLEYN). American universities have vast theatrical deposits which are continually being augmented, particularly at *Harvard and *Yale.

Most art galleries contain theatrical paintings and portraits not catalogued separately. The *Garrick Club in London and the Players' Club in New York own fine collections. There are also a number in the Drama Department of Harvard University of which a catalogue in four volumes was published in 1930–2. In 1961 the *Society for Theatre Research issued a catalogue, *Theatrical Portraits in London Public Collections*.

Music, Incidental, see INCIDENTAL MUSIC.

Musical Comedy, or **Musical,** entertainment in which a story is told by a combination of spoken dialogue and musical numbers. Originally the plot was very slight, but with the importation of more serious themes the word 'comedy' was dropped, and the genre is now known simply as the 'musical'. In the earlier examples the music was often irrelevant, but later it tended to become more integrated with the plot. Dancing too came to play a more important role.

Initially the better examples of the genre originated in England, with productions by George *Edwardes at the *Gaiety Theatre and elsewhere. Osmond Carr's *In Town* (1892; NY, 1897) is often considered the first English musical comedy. It was followed by Sidney Jones's *A Gaiety Girl* (1893; NY, 1894), Ivan Caryll's *The Shop Girl* (1894; NY, 1895), and Lionel Monckton's *A Country Girl* (London and NY, 1902) and *The Quaker Girl* (1910; NY, 1911). As musical comedy became established, productions were increasingly exchanged across the Atlantic; Gustave Kerker's *The Belle of New York* (1897; London, 1898), though American in origin, achieved its major success in London, where it became the first American musical to run for over a year. On the other hand, Leslie Stuart's *Florodora* (1899; NY, 1900), which began its successful career in London, ran even longer in New York. Although the prestige of the American musical stage was materially

enhanced by such entertainments as Victor Herbert's *Naughty Marietta* (1910) and Rudolf Friml's *The Firefly* (1912), which, like the 'Savoy operas' of *Gilbert and Sullivan in London, deserve perhaps to be classed as light operas rather than musical comedies, musical shows from Europe were still dominant in the early years of the 20th century. Monckton and Talbot's *The Arcadians* and Monckton and Caryll's *Our Miss Gibbs*, both seen in London in 1909, demonstrated the continuing strength of the English musical comedy. Viennese musicals, however, reigned supreme, Franz Lehár's *The Merry Widow* (Vienna, 1905) being seen in London and New York in 1907, and Oscar Straus's *The Chocolate Soldier* (Vienna, 1908)— based without permission on *Shaw's *Arms and the Man*—appearing on the New York stage in 1909 (London, 1910).

Soon afterwards a number of talented American musicians, writing in a distinctively native idiom, made their appearance, among them Jerome Kern, Irving *Berlin, and Cole *Porter, the last two being their own lyricists. Kern's first outstanding success was *Very Good, Eddie!* (1915), which marked a trend towards everyday characters and realistic situations. One of the collaborators on the book was **Guy Bolton** (1884–1979), born in England of American parents, who was later to collaborate in many famous musicals, including *Kissing Time* (1919), *She's My Baby* (1927), *The Fleet's Lit Up* (1938), and *Follow the Girls* (1945). Emmerich Kalman's *Miss Springtime* (1916) had lyrics by one of Bolton's most eminent collaborators, **P. G.** [Sir Pelham Grenville] **Wodehouse** (1881–1975), a famous English writer of humorous novels, who also worked with Kern and Bolton on *Oh, Boy!* (1917), known in England as *Oh, Joy!* (1919). The pre-war type of musical comedy still found an audience with the success of *Maytime* (1917) by the Hungarian-born Sigmund Romberg who had settled in America. He also wrote most of the music for *Sinbad* (1918), in which the great star of musical comedy and *revue, **Al Jolson**, well known for his singing, in blackface, of coon songs, interpolated two of his best-known melodies, 'My Mammy' and 'Swanee'. In 1919 Harry Tierney's *Irene*, with its hit song 'Alice Blue Gown', was the musical success of the season, being seen in London in 1920, in which year Broadway saw *Always You*, the first musical to have book and lyrics by Oscar *Hammerstein II. The same year also saw the production of *Sally* (London, 1921) by Kern and Bolton, which during its three-year run on Broadway starred **Marilyn Miller** [Mary Ellen Reynolds] (1898–1936), who became one of Broadway's leading musical comedy stars.

A comparative dearth of musicals during the early 1920s was followed by a period of enormous activity during which the American musical achieved a lasting supremacy, even though the first outstanding success of the time, Friml's *Rose-Marie* (1924; London, 1925), seemed to mark a return to the Viennese tradition, as did Romberg's *The Student Prince* (1924; London, 1926), *The Desert Song* (1926; London, 1927), and *The New Moon* (1928; London, 1929). There were, however, distinct traces of American influence in the music of *Rose-Marie*, and the *Gershwins' *Lady, Be Good!* (1924; London, 1926) employed jazz rhythms. In Sept. 1925 four outstanding musical shows opened on Broadway within a week. The first was Vincent Youmans's *No! No! Nanette!*, already seen in London in March, with its hit songs 'I Want to be Happy' and 'Tea for Two'; then came *Dearest Enemy*, Lorenz *Hart's first collaboration with Richard Rodgers on a full musical score; Friml's *The Vagabond King* followed, based on Justin McCarthy's play *If I were King* (London, 1927); and finally Jerome Kern's *Sunny*, which again starred **Marilyn Miller** (London, 1926). In the theatrical season of 1927–8 the number of musicals opening in New York reached a record total of over 50, among them Kern's *Show Boat* (London, 1928), based on a novel by **Edna Ferber**, which with its strong plot, believable characters, and integrated songs enlarged the boundaries of the musical.

The best musicals of the 1930s came mainly from familiar sources. But in 1941 *Lady in the Dark* had music by Kurt Weill, who with his Austrian-born wife, the singer and actress **Lotte Lenya** [Karoline Blamauer], had emigrated to America from Germany; both are known best for their association with Bertolt *Brecht. *On the Town* (1944; London, 1963) was the first venture into the theatre of the composer and conductor Leonard Bernstein; and *Brigadoon* (1947; London, 1949) was the work of a new team, **Alan Jay Lerner** (1918–86) as author of the book and lyrics, and Frederick Loewe as composer. Jule Styne's *Gentlemen Prefer Blondes* (1949), based on Anita Loos's novel, made a star of **Carol Channing** (1921–), her part of Lorelei being played in London in 1962 by **Dora Bryan**. *Guys and Dolls* (1950; London, 1953) had music and lyrics by **Frank Loesser** (1910–69) and a book based on the stories and characters of Damon Runyon. After their *Paint Your Wagon* (1951; London, 1953), Lerner and Loewe wrote *My Fair Lady* (1956; London, 1958), based on Shaw's *Pygmalion*, in which Rex *Harrison and **Julie Andrews** [Julia Elizabeth Wells] (1935–) starred both in New York and in London. Bernstein composed *Wonderful Town* (1953; London, 1955), based on the play *My Sister Eileen* by Joseph *Fields and Jerome Chodorov; *Candide* (1956; London, 1959), based on *Voltaire's satire; and *West Side Story* (1957; London, 1958), a retelling of the story of *Romeo and Juliet* in terms of gang warfare in contemporary New York. The lyrics for the

last marked the début on the musical comedy scene of Stephen *Sondheim, who also wrote the lyrics of Styne's *Gypsy* (1959; London, 1973). Loesser contributed the book and music of *The Most Happy Fella* (1956; London, 1960), based on Sidney *Howard's play *They Knew What They Wanted*; and Meredith Willson's *The Music Man* (1957; London, 1961), with its popular hit '76 Trombones', was a big success.

The years between the wars had produced little of note in London, apart from the musical plays of Noël *Coward and Ivor *Novello, who continued the old tradition of European operetta. The English musical had however been showing signs of revival with a gentle parody of the 1920s musical in Sandy Wilson's *The Boy Friend* (1953; NY, 1954, with **Julie Andrews**) and with Julian Slade's *Salad Days* (1954). Neither composer, however, was to achieve such a success again, and it was not until 1960 that Lionel *Bart raised the English musical from the doldrums with *Oliver!* (NY, 1963), an adaptation of *Dickens's *Oliver Twist*. David Heneker's *Half a Sixpence* (1963; NY, 1965), based on H. G. Wells's novel *Kipps*, provided a lively vehicle, in both London and New York, for **Tommy Steele** [Hicks] (1936–), a former pop star who became an all-round entertainer, even appearing at the *Old Vic in *Goldsmith's *She Stoops to Conquer* in 1960.

In New York, Loewe collaborated again with Lerner in the Arthurian *Camelot* (1960; London, 1964), based on T. H. White's *The Once and Future King*, starring Richard *Burton and **Julie Andrews** in New York. Loesser's *How to Succeed in Business without Really Trying* (1961; London, 1963) was his last work to reach New York. Styne's *Funny Girl* (1964; London, 1966) was based on the career of revue star Fanny Brice, and made a star of Barbra Streisand. Jerry Herman's *Hello, Dolly!* (1964; London, 1965), based on *Wilder's play *The Matchmaker*, owed much of its success to its title song, elaborate staging, and the appeal of **Carol Channing**. In London the part was first played by Mary *Martin and then by **Dora Bryan**. The greatest success of the 1960s, however, was Jerry Bock's *Fiddler on the Roof* (1964; London, 1967). Based on Sholom *Aleichem's *Tevye the Milkman*, it ran in New York for almost eight years, at that time the longest run ever achieved by a Broadway musical. Perhaps the most satisfying musical of the decade artistically was John Kander's *Cabaret* (1966; London, 1968), based on *Van Druten's play *I am a Camera*. Some of its songs showed the influence of Kurt Weill, whose widow **Lotte Lenya** was in the American production. *Cabaret* was directed by **Hal** [Harold] **Prince** (1928–), a producer who became a well-known director, especially of musicals, including most of Sondheim's. The most unorthodox success of the decade was probably Galt MacDermot's *Hair* (NY and

London, 1968), which brought rock music and total nudity to the stage.

By the late 1960s the American musical had lost much of its former appeal. Though the cinema had long been a serious competitor for talent, creators of successful stage musicals could normally assume that a film version, providing extra earnings and a much larger audience, would follow. Hollywood's financial troubles, however, reduced the number of film musicals. In any case the music of stage musicals, which had once provided many of the popular songs of the day, no longer held any appeal for film audiences addicted to the fashionable rock and beat music; while the latter was not much liked by theatre audiences. Theatre music almost disappeared from the hit parade, and musical shows therefore received less publicity. New talent was in any case scarce, and as production costs rose profitable shows became increasingly rare. The desperate need to fill empty theatres on Broadway led to bizarre choices of subject, and musicals were made from films, plays, novels, and even comic strips.

Fortunately there was still Stephen Sondheim, the only major American composer still writing regularly for the musical stage. Probably the best non-Sondheim American musical of the 1970s was Marvin Hamlisch's *A Chorus Line* (1975; London, 1976), which in 1989 became the longest running Broadway musical of all time (but see RUN). The outstanding new talent of the decade was that of the English composer Andrew Lloyd Webber, who wrote *Jesus Christ Superstar* (NY, 1971; London, 1972) and *Evita* (1978; NY, 1979).

A big hit of the 1980s was an adaptation of the 1933 film musical *42nd Street* (NY, 1980; London, 1984). Another long-runner was Claude-Michel Schönberg and Alain Boublil's *Les Misérables* (London, 1985; NY, 1987), based on *Hugo's novel. Andrew Lloyd Webber wrote *Cats* (1981; NY, 1982), based on T. S. *Eliot's *Old Possum's Book of Practical Cats*; *Song and Dance* (1982; NY, 1985); *Starlight Express* (1984; NY, 1987); *The Phantom of the Opera* (1986; NY, 1988), directed by **Hal Prince**; and *Aspects of Love* (1989; NY, 1990). Sondheim also was still active, but otherwise the musical by the 1990s was at a low ebb.

Music Box, New York, on West 45th Street between Broadway and 8th Avenue. Built by Irving *Berlin and Sam H. Harris (see SAM H. HARRIS THEATRE) and seating 1,010, this theatre opened in 1921 with the *Music Box Revue* which ran into four editions. After a further series of revues came *Kaufman and *Hart's *Once in a Lifetime* (1930) and *Of Thee I Sing* (1931), a collaboration between Kaufman and Morris Ryskind. Other successes were Kaufman and Ferber's *Dinner at Eight* (1932), *Steinbeck's *Of Mice and Men* (1937), *The Man Who Came to Dinner* (1939), again by Kaufman and Hart, and *I Remember Mama* (1944) by John *Van Druten. After Kurt Weill's musical *Lost in the Stars* (1949) the theatre reverted to straight plays including *Inge's *Picnic* (1953), *Rattigan's *Separate Tables* (1956), and *Rashomon* (1959), based on a play by Akutagawa. Four productions from London were Anthony Shaffer's *Sleuth* (1970), *Ayckbourn's *Absurd Person Singular* (1974), Trevor Griffiths's *Comedians* (1976), and the musical *Side by Side by Sondheim* (1977). Ira Levin's *Deathtrap* (1978), with John *Wood, continued into the 1980s, and was followed by such plays as *Kopit's *End of the World* (1984) and *Hampton's *Les Liaisons dangereuses* (1987).

Music-Hall, type of *variety entertainment which flourished in England during the second half of the 19th century. It evolved from the 'sing-songs' held in taverns, and arose from the desire of tavern landlords to emulate the success of the 'song-and-supper rooms' of which *Evans's was typical. It was at first housed in a hall or annexe adjoining some popular public-house, where the customers sat at small tables to eat and drink while enjoying the comic turns, ballad-singers, acrobats, and jugglers, all controlled by a *chairman.

As music-hall became even more popular, special theatres or 'palaces of *variety' were built to accommodate it. The first was the *Canterbury, built by Charles *Morton, known as 'the father of the halls'. Later outstanding ones in London were the *Alhambra and the *Empire, famous also for their ballets, the *Holborn Empire, the *London Pavilion, the *Bedford, the *Metropolitan, the *Middlesex or 'Old Mo', the *Oxford, *Gatti's, and the *Tivoli. At one time every town and suburb had its Empire or Hippodrome. The last to be built was the Chiswick Empire in 1912, its three-tier auditorium, typical of many all over London, seating 2,000.

The stars who made the 'halls' and were made by them were as many and diverse as the buildings themselves. At first singers such as Sam *Collins and **Sam Cowell** (1820–64) came in from the 'song-and-supper rooms', which also gave the music-hall its first comedian in 'blackface', **Joe** [Joseph] **Cave** (1823–1912), later manager of the Marylebone Music-Hall, as well as **Harry Clifton** (1832–72), who wrote the words and music of 'Pretty Polly Perkins of Paddington Green' and sang 'motto' songs such as 'Paddle Your Own Canoe', and **Victor Liston** (1838–1913), who popularized Fred Albert's 'Shabby Genteel' ('Too proud to beg, too honest to steal'), which was much admired by Edward VII when Prince of Wales. It was not long before the music-hall was making its own stars and its own stereotypes. One of the latter, which replaced for a time the earlier stereotype of the tough, good-hearted working-man, was the 'man-about-town', the 'West End toff'. This was the invention of George *Leybourne and the **Great Vance** (**Alfred Vance** [Alfred Peck Stevens], 1840–88), whose versatility covered also Cockney and 'motto' songs.

One of the interesting features of the early music-hall was its employment of women singers, and later comediennes, in its programmes. Women had been seen rarely, if ever, at the Coal Hole and Cyder Cellars and suchlike male resorts, probably not even at Evans's, which reluctantly, in the late 1860s, had allowed women with a male escort to sit in the auditorium behind a grille. But from the early days of the Canterbury, and later at the Oxford, Morton engaged **Emily Soldene** (1840–1912) to sing excerpts from light opera. She was the forerunner of such almost forgotten stars as **Bessie Bellwood** [Elizabeth Ann Katherine Mahony] (1847–96), who was turned down at the Holborn for being 'too quiet', but later learned to belt out such Cockney songs as 'Wot Cheer, 'Ria!', and **Jenny Hill** [Elizabeth Pasta] (1851–96), a diminutive, sharp-featured little creature, billed as 'the Vital Spark', whose early privations, combined with the frenzy with which she sang and danced once she became famous, contributed to her early death. The work of such performers as these came to perfection in Marie *Lloyd, to many the finest of all music-hall's comic women singers.

Although, understandably, it is the great solo performers of the music-hall who are mainly remembered, the many 'turns' on a normal evening's bill—sometimes as many as 20 to 25 at a time—were all different, comprising alone or in groups every type of acrobat, juggler, conjurer, ventriloquist, speciality dancer, slapstick or knockabout comedian, and singers who ran the gamut from vulgar, often suggestive, comic songs to serious ballads. The old 'song-and-supper rooms' had featured little but singers, with an occasional acrobat, conjurer, or ventriloquist, and it is to the music-hall of the second half of the 19th century onwards that the greatest of these last three groups belong, among them **Jules Léotard** (1830–70), **Charles Blondin** [Jean François Gravelet] (1824–97), **Paul Cinquevalli** [Kestner] (1859–1918), and **Fred Russell** [Thomas Frederick Parnell] (1862–1957). Léotard, first seen in London in 1861 at the Alhambra, where his grace and agility captured all hearts, has been immortalized in the song 'The Daring Young Man on the Flying Trapeze', and his name is now given to the sober one-piece practice costume, adopted by acrobats and ballet-dancers, which he always wore. The only person to approach him in daring and popularity was Blondin, who also appeared at the Alhambra, in 1862. Cinquevalli, a German by birth, also began as an acrobat, but after an accident became probably the

greatest juggler of all time, his skill with everything from a billiard ball to a cannon ball being universally recognized. His contemporary Fred Russell, who made his first appearance with his doll 'Coster Joe' in 1896, soon became 'top of the bill' wherever he appeared.

The 1870s and 1880s saw the rise to popularity of many stars now forgotten, or remembered for one or two songs only, such as **William B. Fair** (1851–1909), with 'Tommy, Make Room for Your Uncle', which he sang at as many as six 'halls' in one night, and **Charles Coborn** [Colin Whitton McCallum] (1852–1945), with 'Two Lovely Black Eyes' and 'The Man Who Broke the Bank at Monte Carlo'. Among those who came to the music-halls from the *minstrel show, retaining their 'blackface' make-up, were **George H. Chirgwin** (1854–1922), originally known as 'the White-Eyed Musical Moke', but from 1877 billed as 'the White-Eyed Kaffir' because of the white lozenge-shaped patch round his right eye, who accompanied himself on the one-string fiddle while he sang sentimental 'coon' songs, of which the favourites were 'The Blind Boy' and 'My Fiddle is My Sweetheart', sung in a high-pitched, piping voice; and slightly later **Eugene Stratton** [Eugene Augustus Ruhlman] (1861–1918), who also sang 'coon' songs, the best remembered being 'Lily of Laguna', to which he whistled a refrain while dancing a 'soft-shoe shuffle' in a spotlight on a darkened stage. **Harry Champion** (1866–1942), on the other hand, soon discarded his 'blackface', and after making a hit with such songs as 'Ginger, Ye're Barmy' and 'I'm 'Enery the Eighth I am, I am', is mainly remembered for his songs in praise of food—'Boiled Beef and Carrots', 'Hot Tripe and Onions'—which he sang at terrific speed and with great zest and vitality up to the day of his death.

At a time when Britain was at the height of her prosperity, patriotic songs, particularly when sung by a handsome man in army or naval uniform, formed part of the staple fare of the halls. Among the best-known singers of such songs was **The Great Macdermott** [Gilbert Hastings Farrell] (1845–1901), whose 'We Don't Want to Fight, but By Jingo! if we do' added the word 'Jingoism' to the English language. **Charles Godfrey** [Paul Lacey] (1851–1900), who came from the theatre to sing 'Masher King' in 1880, dressed in silk knee-breeches and buckled shoes, but also put on uniform for the descriptive *scena* songs in which he excelled, among them 'Poor Old Benjamin', about a veteran of the Crimean War. He reversed the process adopted by **Arthur Roberts** (1852–1933), who left the music-halls in 1880 to go on the stage, but returned to them in 1903 to become a 'Veteran of Variety'. Another handsome singer of naval songs, **Tom Costello** (1863–1945), was famous in his own day for 'Comrades', but is now better remembered as the henpecked husband in 'At Trinity Church I Met Me Doom'. Undoubtedly one of the outstanding figures of the 'halls' in the last 15 years of the 19th century was Dan *Leno, who made his first adult appearance in a London music-hall in 1885, the same year as Marie Lloyd. For many people these two epitomize the spirit of the 'halls' at their best, and also highlight that connection between music-hall and pantomime which was to be a marked feature of the next 30 years or more.

The 1890s saw the arrival of many stars who were to survive into the 20th century, among them **Mark Sheridan** (?–1917), who in top hat, frock coat, and bell-bottomed trousers strapped below the knee sang Cockney songs with rousing choruses, of which the best-remembered are 'Oh, I Do Like to be beside the Seaside' and 'Here We Are, Here We Are, Here We Are Again', to which the troops marched during the First World War. He was first seen in London in 1891, as were **Richard Knowles** (1858–1919), born in Canada, who billed himself as 'the very peculiar American comedian', and wore a shabby black frock-coat, opera hat, and white duck trousers as he strode about the stage singing at the top of his voice songs such as 'Girly-Girly' and 'Brighton'; and the great comedian Albert *Chevalier, who brought back the homespun Cockney comedian. In the same year **Gus** [Ernest Augustus] **Elen** (1862–1940), known till then as a 'blackface' comedian, left off his Negro make-up and reinforced Chevalier's picture of the rollicking costermonger with 'Never Introduce Yer Donah to a Pal'. Some of his characterizations were truly Dickensian, among them ''E Dunno Where 'E Are', 'If It Wasn't For the 'Ouses in Between', and 'Wait Till the Work Comes Round'. Among the feminine equivalents of Chevalier was the much-loved **Kate Carney** (1868–1950), who, dressed in a coster suit of 'pearlies' and a vast hat with towering feathers, sang 'Liza Johnson' and 'Three Pots a Penny'.

As Cockney songs returned to the 'halls', the elegant silhouette of the 'man-about-town', which had been in partial eclipse since the days of Leybourne and Vance, also returned, but this time in the form of *male impersonation, of which **Nellie Power** (1855–87) and **Bessie Bonehill** (?–1902) were early examples, and Vesta *Tilley the supreme exponent. Other examples were **Ella Shields** (1879–1952), who took Vesta Tilley's West End Burlington Bertie and made him a pathetic East-Ender striving after gentility, and **Hetty King** (1883–1972), an exuberant artiste with impeccable timing, who continued to perform up to the time of her death. In her early years she suffered from the hostility of Vesta Tilley's husband, who was reluctant to let her perform in the music-halls which he controlled, but she nevertheless achieved considerable eminence, one of her

best-loved songs being 'All the Nice Girls Love a Sailor', and, in the popular image of the young man of the day, 'Follow the Tramlines' and 'I'm Afraid to Come Home in the Dark'.

The last decade of the 19th century was not without its share of eccentric humorists, among them Thomas Edward *Dunville, the ineffable *Little Tich, and **Lottie Collins** (1866–1910), whose 'Ta-Ra-Ra-Boom-De-Ay', which had failed when first produced in New York, became the rage of London. She first performed it in *Dick Whittington*, the Islington pantomime for 1891, and it was then introduced into the burlesque *Cinder-Ellen Up Too Late* at the *Gaiety.

Out-topping them all were a pair as famous as Dan Leno and Marie Lloyd—George *Robey and Nellie *Wallace. They continued to carry the spirit of the true music-hall into the 20th century, as did such stars as **Bransby Williams** [Bransby William Pharez] (1870–1961) with his musical monologues, of which the best known were 'The Green Eye of the Little Yellow God' and 'The Whitest Man I Know'. **Harry Tate** [Ronald Macdonald Hutchison] (1872–1940) took his stage name from the firm of Henry Tate & Sons, Sugar Refiners, who once employed him, and from his first appearance in 1895 built up a series of sketches on golfing, motoring, fishing, and so on. **Wilkie Bard** [Billie Smith] (1870–1944), with his high, domed forehead, reminiscent of Shakespeare (hence the 'Bard'), fringed with sparse hairs and with two black spots over the eyebrows, helped to popularize tongue-twisters like 'She Sells Sea-Shells on the Sea-Shore', and played Pantaloon in the *harlequinade in pantomime, reviving much of its old spirit in company with Will *Evans. **Florrie Forde** [Florence Flanagan] (1876–1940) came from Australia and was best remembered as a massive *principal boy and as the singer of 'Down at the Old Bull and Bush', 'Has Anybody Here Seen Kelly?', and 'Hold Your Hand Out, Naughty Boy'. **Charles R. Whittle** (c.1870–1947) favoured songs with 'Girl' in the title, but is best remembered for 'Let's All Go Down the Strand'.

The new century opened auspiciously with the first appearance in London of Harry *Lauder, master of the daft eccentricity, the pawky humour, and the occasional streak of pathos which later marked the work of **Will Fyffe** (1885–1947), creator of a whole picture-gallery of Scottish worthies, and singer of 'I Belong ta Glasgae', who made his first appearance in London in 1921. It was in a series of sketches written for but refused by Lauder that Fyffe first went on the 'halls' after many years in melodrama and revue, and established himself as a prime favourite. Among Lauder's more immediate contemporaries were **Albert Whelan** [Waxman] (1875–1961), who came from Australia to appear at the Empire in 1901, and was the first to use a signature tune ('Lustige Brüder'), which he whistled as he strode on stage in immaculate evening dress, nonchalantly placing on the top of the piano his stick, top-hat, overcoat, white gloves, scarf, and wristwatch, the whole process being reversed at the end of his act. G(eorge) H(enry) **Elliott** (1884–1962) was another recruit from the Minstrel Show who was first seen in London in 1902 and, billed as 'the Chocolate-Coloured Coon', became the successor of Eugene Stratton, equally admired for his soft-shoe dancing and with a similar repertory of 'coon' songs to which he added 'I Used to Sigh for the Silvery Moon'. **Billy Merson** [William Henry Thompson] (1881–1947), a circus clown and acrobat, put on an auburn wig with a bald patch at the back and two coy curls across the top of his high forehead to sing about Alphonso Spagoni, 'The Spaniard that Blighted My Life', and 'The Good Ship Yacki-Hicki-Doola'. **Gertie Gitana** [Gertrude Mary Ross, *née* Astbury] (1889–1957) first appeared on stage at the age of 4 with a troupe dressed as gipsies from whom she took her stage name, and by the time she was 16 was topping the bill all over the country with songs such as 'Nellie Dean'. Fred Emney was a stalwart of slapstick comedy, whose son, also Fred *Emney, delighted the 'halls' with his songs at the piano and his talent as a raconteur. **Harry Fragson** [Potts] (1869–1913), son of an English mother and a Belgian father (who shot him in a fit of insanity), played in Paris from 1887 onwards with a Cockney accent and in 1905 came to London to play with a French accent; when he appeared as Dandini in *Cinderella* the character was rechristened Dandigny as a tribute to his Anglo-French reputation.

They were all singers and comedians; but when it came to music and comedy combined, none of them was as great as *Grock, the supreme clown of his generation. The music-hall, attacked by new forms of entertainment, was already dying when he left London in 1924, and only a few names are attached precariously to its last flickers, since what had been a full-time profession now had to be split up, not only between *revue and *musical comedy, but between the cinema, the radio, and above all television. Among them are Maurice *Chevalier, Gracie *Fields, and her fellow-Lancastrian **George Formby** (1905–61), who sang and pattered to the ukulele. **Sid Field** (1904–50) was a *droll who was essentially a comic actor, in sketches—the best involving the 'spiv', Slasher Green—which relied more on characterization and situation than on slapstick and verbal gags. He had already built up an immense reputation before he first appeared in London in 1943 in the revue *Strike a New Note*, and he made his last appearance on stage as Elwood P. Dowd in Mary Chase's *Harvey* (1949). He was, like his contemporaries, a man of many parts; the

dwindling music-halls could no longer contain them, and they sought other worlds to conquer.

In its heyday music-hall presented the type of entertainment most loved by the ordinary people. It was gay, raffish, and carefree, vulgar but not suggestive, dealing amusingly with the raw material of their own lives, their emotions, their troubles, their rough humour. Sophistication and subtlety were its undoing. Yet, paradoxically, the main force ranged against it, television, has revived something approaching the earlier form of music-hall, which seems to be springing up again as informal entertainment in the place where it all began, the local public-house. The only surviving full-time music-halls are the *Players' Theatre in London and the *City Palace of Varieties in Leeds, which was restored to its earlier glory in order to house BBC Television's music-hall series 'The Good Old Days'.

Musset, (Louis-Charles) **Alfred de** (1810–57), French poet and playwright of the Romantic era, and the one man who might have been able to fuse the new Romantic drama with the best of the classical tradition in the French theatre. Unfortunately the failure of his first play *La Nuit vénitienne* (1830) turned him against the stage, and for many years he wrote only plays to be read. It was not until 1847 that the *Comédie-Française put on *Un caprice*, an immediate success which was followed in 1848 by *Il faut qu'une porte soit ouverte ou fermée* (*A Door Should be Either Open or Shut*) and *Il ne faut jurer de rien* (*One Can Never be Sure of Anything*). Although this encouraged Musset to write further plays for performance, and to revise some of his earlier ones for the same purpose, his real importance as a dramatist was realized only after his death. His plays show a delicacy and restraint quite unlike the work of his contemporaries Victor *Hugo and *Dumas *père*, suggesting rather the influence of *Marivaux. They are either bitter-sweet comedies mingled with fantasy, such as *Les Caprices de Marianne* (perf. 1851) and *On ne badine pas avec l'amour* (*There's no Trifling with Love*) (perf. 1861), or scenes of social life usually illustrating some well-known saying, such as *On ne saurait penser à tout* (*One Cannot Think of Everything*, 1849). Musset's most important dramatic work, however, is the historical drama *Lorenzaccio*, written in 1834 after his tragic liaison with George Sand, about the assassination of Alexander de' Medici by his young cousin Lorenzo. In a drastically cut and rearranged version it was first staged in 1896 with *Bernhardt in the leading role, a part often compared to Hamlet. There have been some important 20th-century productions of this play, notably in 1952 at the *Avignon Festival, when Gérard *Philipe both directed the play and appeared as Lorenzo, thus breaking the tradition which assigned the part to a woman. In 1933 an English version was produced in London as *Night's Candles*. Generally speaking, however, Musset's plays remain little known in England, probably because, like Marivaux, he does not translate easily.

Mystery Play, medieval religious play which derives from *liturgical drama, but differs in being wholly or partly in the vernacular and not chanted but spoken. Also it was performed out of doors—in front of the church, in the market square, or on perambulating *pageants. The earlier English name for it was *miracle play, now seldom used, and a better name would be Bible-histories, since each play was really a cycle of plays based on the Bible, from the Creation to the Second Coming. Substantial texts of English 'cycles' of such plays have survived from Chester, Coventry, Lincoln, Wakefield, and York. Simultaneously with the English mystery play there arose in Europe, in the vernacular, the French *mystère*, the German *Mysterienspiel*, the Italian *sacra rappresentazione*, and the Spanish *auto sacramental*, to name only the most important. Traces of similar plays are found in Russia, in the states of Central Europe, and also in Denmark.

N

Naharro, Bartolomé de Torres, see TORRES NAHARRO.

Nares, Owen (1888–1943), English actor, for many years one of the most popular 'matinée idols' of the London stage, his charm and good looks somewhat obscuring his gifts as an actor. He made his first appearance in 1908, and first came into prominence with an excellent performance as Lord Monkhurst in Arnold *Bennett and Edward *Knoblock's *Milestones* (1912). He was also much admired in the title-role of Du Maurier's *Peter Ibbetson* and as Thomas Armstrong in the London production of Edward *Sheldon's *Romance* (both 1915). In 1923 he was at the *St James's Theatre (where he had first appeared with *Alexander in 1909), giving an interesting performance as Mark Sabre in a dramatization of A. S. M. Hutchinson's novel *If Winter Comes* . . ., and then toured South Africa with *Romance* and *Sardou's *Diplomacy*. In later years he gave signs of greater depth of feeling and reserves of strength, particularly in Dodie Smith's *Call it a Day* (1935), St John *Ervine's *Robert's Wife* (1937), in which he played opposite Edith *Evans, and Daphne Du Maurier's *Rebecca* (1940), but he died suddenly before he had had time to develop his new-found powers.

Narr, German equivalent of the early English *Fool, who appears in the 16th-century *Fastnachtsspiel* or Carnival Play. Though not originally a comic character (the fools in Sebastian Brant's poem *Das Narrenschiff* (*The Ship of Fools*, 1494) are those who live foolishly), he was made to appear so on stage. He could also be a comic peasant, like the *clown in Shakespeare's plays, as opposed to the more sophisticated Court Jester, and his assumed stupidity could often cover slyness. He bequeathed most of his characteristics to Jan Bouschet (originally John Posset), the creation of one of the *English Comedians, Sackville, who allowed him, though a secondary character in the play, to hold the stage and improvise at will. Similar types were created by other leading actors—Hans Stockfisch, Pickelhering, and eventually the native *Hanswurst and *Kasperle.

Nashe, Thomas (1567–1601), English pamphleteer and playwright, friend of *Greene and *Lyly, who collaborated with *Marlowe in *Dido, Queen of Carthage* (1587/8) and was also author, or part-author with *Jonson who appeared in the play, of the ill-fated 'seditious' comedy *The Isle of Dogs* (1597), now lost, which

caused Jonson, and perhaps Nashe also, to be put in prison. Nashe's only extant dramatic work is *Summer's Last Will and Testament* (1592/3), designed for performance in the house of a nobleman, probably Archbishop Whitgift at Croydon.

Nathan, George Jean (1882–1958), American dramatic critic, who in 1905 joined the staff of the *New York Herald* and began his fight for the 'drama of ideas', denouncing the works of such contemporary American playwrights as David *Belasco and Augustus *Thomas, whose popular plays occupied the playhouses to the exclusion of almost everything else. Nathan introduced to America the modern dramatists of Europe—*Hauptmann, *Ibsen, *Shaw, *Strindberg, and a host of others whose early work he published in *The Smart Set*, a magazine which he edited with H. L. Mencken. His most important discovery was Eugene *O'Neill, some of whose earliest plays also appeared in *The Smart Set*. Later Nathan championed Sean *O'Casey, being largely responsible for the New York production of *Within the Gates* (1934), and William *Saroyan, whose first work for the theatre, *My Heart's in the Highlands* (1939), he praised enthusiastically when there was still much doubt as to its value.

Nation, Théâtre de la, see COMÉDIE-FRANÇAISE.

National Association for Drama in Education and Children's Theatre, see CHILDREN'S THEATRE.

Nationale Scene (National Theatre), Bergen, was established in 1850 under the name Norske Theater (Norwegian Theatre). The aim was to present a truly Norwegian theatre in opposition to the Danish influence which had hitherto dominated the theatre in Norway. *Ibsen was the Norske Theater's first resident dramatic author, 1851–7, and *Bjørnson was its Director, 1857–9; both gained valuable experience of dramatic practice there. The theatre was forced to close for financial reasons in 1863, reopening under its present name in 1876. Gunnar *Heiberg was Director, 1884–8. The theatre enjoyed a particularly lively period from 1934 to 1939, when Hans Jacob *Nilsen was its Director, premièring some of Nordahl *Grieg's pioneering plays. For most of the first 50 years of the 20th century, the Nationale Scene was the breeding ground for dramatic talent in the country. In 1967 a studio was added, and in 1982 a third middle-sized auditorium, so that the regular repertoire involves an average of three performances a day. Under the present Artistic Director the theatre has become a home for new Norwegian writing, as well as a centre of international collaboration and co-production.

National Operatic and Dramatic Association, see AMATEUR THEATRE.

National Playwrights' Conference, see O'NEILL, EUGENE.

National Repertory Theatre, see LE GALLIENNE.

National Theatre, London (officially renamed the Royal National Theatre of Great Britain on its Silver Jubilee in 1988). The establishment of a permanent state-subsidized theatre in London, on the lines of the *Comédie-Française, was first suggested by David *Garrick in the 18th century, and in the 19th century both *Irving and *Bulwer-Lytton were enthusiastic supporters of the idea. It was not until 1908, however, that a committee was set up to investigate the possibility of opening such a theatre in 1916 to celebrate the tercentenary of Shakespeare's death. A large sum of money had been subscribed and a foundation stone laid on a site in Gower Street when the outbreak of the First World War in 1914 brought the project to a standstill. In 1938 another site in South Kensington was acquired and a second foundation stone laid by G. B. *Shaw. The outbreak of the Second World War a year later caused further delay, and it was not until 1951 that the idea was taken up again. A more ambitious plan was launched, and the site moved to the South Bank of the Thames, where a third foundation stone was laid. In 1961 a decision was taken to found a National Theatre company under Laurence *Olivier, to be housed in the *Old Vic Theatre pending the erection of a new building on a site downstream from the third foundation stone, below Waterloo Bridge. Performances began at the Old Vic in 1963.

The new building, designed by Denys Lasdun, is a vast complex housing three theatres with extensive backstage accommodation, rehearsal rooms, and workshops, dressing rooms for 150 actors, and a large foyer for exhibitions and informal entertainment. Work began on the site in 1969 and on 16 Mar. 1976 the first theatre, the *Lyttelton, gave its opening performance, followed on 25 Oct. by the *Olivier and on 4 Nov. 1977 by the *Cottesloe. The South Bank Theatre Board is responsible for the structure. In 1973, after Olivier had resigned owing to ill health, Peter *Hall took over, remaining until 1988. He was succeeded by Richard *Eyre as Artistic Director, working jointly with David *Aukin as Executive Director until the latter's departure in 1990. A full UK touring programme was initiated in 1989 with a joint production of *Molière's The

Misanthrope with the *Bristol Old Vic. The National Theatre Studio opened in 1984 under Peter *Gill. An experimental workshop which encourages new writing, it is based in the Old Vic Annexe.

National Theatre, New York, see NEDERLANDER THEATRE.

National Theatre, Washington, see WASHINGTON.

National Theatre Conference, USA, honorary professional association limited to 120 members drawn from academic, community, and non-profit-making professional theatres throughout the United States. Its purpose is the exchange of experience and ideas in the field of non-commercial theatre, especially at the Conference's annual meeting.

National Theatre of the Deaf, see O'NEILL, EUGENE.

National Theatres. The oldest national theatre is the *Comédie-Française, founded in 1680 by Louis XIV; there are six other French national theatres: the *Odéon (1781), the *Théâtre National Populaire (1920), and the post-Second World War Théâtre National de *Chaillot, Théâtre National de la *Colline, Théâtre National de Strasbourg, and Théâtre National de Marseille. Other long-established national theatres are the *Royal Theatre (Denmark, 1772); the Burgtheater (Austria, 1776), established by decree of Josef II; the *Royal Dramatic Theatre (Sweden, 1788); the Dutch National Theatre (1870); the Finnish National Theatre (1872); and the *Nationaltheatret (Norway, 1899). The *Abbey (1904) has became Ireland's unofficial national theatre. The Greek National Theatre, strong on Greek classical drama, dates from 1930; in 1961 it was supplemented by the National Theatre of Northern Greece. The Belgian National Theatre was established in 1945. The British *National Theatre was founded only in 1961, and the *RSC, also founded in 1961, virtually ranks as a second national theatre. *Habimah is the national theatre of Israel. In Eastern Europe national theatres have survived in Hungary (founded 1840), Romania (1854), Yugoslavia (Serbian National Theatre, Belgrade, 1869), Bulgaria (1907), and Czechoslovakia (Bratislava, 1920).

National Theatre Society, see ABBEY THEATRE, Dublin.

Nationaltheatret, Oslo, playhouse which in 1899 took over from the old *Kristiania Theater in newly built premises, and has since held pride of place as the premier theatre of Norway. Its first director was Bjørn *Bjørnson, and during the first quarter of the 20th century it contributed a brilliant page to the history of the Norwegian theatre. A second generation of exceptional talent came to the fore in the 1930s, with Tore *Segelcke, Alfred *Maurstad and others. During the Second World War the theatre suffered many setbacks. Its leading members were sent to concentration camps, and its official productions, approved by the Occupation authorities, were boycotted by the general public. After the war its problems were mainly economic; eventually the State and the municipality of Oslo came to the rescue and subsidies from public funds now play a major role. In 1963 a smaller theatre, the Amfiscenen, was opened inside the Nationaltheatret's own premises. With the appointment of a new director in 1967 the Nationaltheatret took on a new lease of life, technically, administratively, and architecturally. He was succeeded by Toralv *Maurstad, 1978–87. The main theatre was badly damaged by fire in 1980 but reopened in 1985. The Teatret på Torshov in eastern Oslo also forms part of the Nationaltheatret.

National Youth Theatre of Great Britain, organization founded in 1956 by Michael Croft, a master at Alleyn's School, Dulwich, who had formerly been an actor, for the production of Shakespeare's plays. Drawn originally from Alleyn's and Dulwich College, the players eventually came from schools all over Britain. The first production was *Henry V*, which was followed by *Troilus and Cressida* (at the *Edinburgh Festival), *Hamlet* (in London and on tour), and *Antony and Cleopatra* (at the *Old Vic). The first contemporary play to be presented was David Halliwell's *Little Malcolm and His Struggle against the Eunuchs* at the *Royal Court in 1965. Many well-known actors, among them Derek *Jacobi and Helen *Mirren, gained their first experience with the NYT, and the company formed a particularly productive relationship with Peter *Terson, who wrote several plays for it. In 1971 the company took over the Shaw Theatre, originally erected as a conference centre and adapted by the local authority as a theatre to hold 510. The NYT established the Dolphin Theatre Company, a professional company aiming to provide high quality theatre related to the needs of schools and the interests of young people, including plays by new writers. The NYT company itself, which used the Shaw Theatre for its summer productions and made regular tours abroad, remained amateur. In 1981 its grant from the *Arts Council was withdrawn. It was eventually replaced by commercial sponsorship, though the NYT no longer has its own theatre and the professional company had to be disbanded. The NYT continues to stage plays in London and undertakes much provincial and international touring, including the performance of T. S. *Eliot's *Murder in the Cathedral* at the *Moscow Art Theatre in 1989.

Nations, Théâtre des, see THÉÂTRE DES NATIONS.

Nativity Play, see LITURGICAL DRAMA.

Naturalism, movement in the theatre of the late 19th century which carried a step further the revolt against the artificiality of contemporary forms of playwriting and acting initiated by the selective *realism of *Ibsen. *Thérèse Raquin* (1873), dramatized by *Zola from his own novel, was the first consciously conceived naturalistic drama, *Strindberg's *Miss Julie* (1888) its first masterpiece. But it was *Antoine who established it in the theatre. The influence of his *Théâtre-Libre led to the foundation of the *Freie Bühne in Germany and Grein's Independent Theatre in London, and the movement finally attained world recognition in the work of *Stanislavsky, particularly with his production of *Gorky's *The Lower Depths* (1902). In Spain naturalism is represented by *Benavente's *La malquerida* (1913) and in the United States by the early works of *O'Neill and the dramatized novels of *Steinbeck.

Naughton, Charlie, see CRAZY GANG.

Nautical Drama, type of romantic *melodrama popular in England in the late 18th and early 19th centuries, which had as its hero a 'Jolly Jack Tar', a lineal descendant of the sailor characters in the novels of Smollett, who was himself the author of one of the earliest plays of this kind, *The Reprisal; or, The Tars of Old England* (1757). The Jack Tar was further popularized by the elder Charles *Dibdin and by the naval victories of Nelson, before being given its final form in the noble-hearted William, hero of *Jerrold's *Black-Ey'd Susan; or, All in the Downs* (1829), in which T. P. *Cooke made his name. The character continued to flourish in the minor 'transpontine' theatres in London, particularly at the *Surrey, until well into the 1880s, in spite of being burlesqued in *Dickens's *Nicholas Nickleby* (1838) and *Gilbert and Sullivan's *HMS Pinafore* (1878).

Nazimova, Alla (1879–1945), Russian actress, who studied with *Nemirovich-Danchenko, acted for a season with the *Moscow Art Theatre, and in 1904 was the leading lady of a theatre company in St Petersburg. She toured Europe and America with a Russian company and, having learnt English in less than six months, made her first appearance in an English-speaking part—*Ibsen's Hedda Gabler—in 1906 at the Princess Theatre in New York under the management of the *Shuberts, who built and named for her the Nazimova Theatre, later the Thirty-ninth Street Theatre, which she opened in 1910 with Ibsen's *Little Eyolf*. A superb actress, vibrant, passionate, yet subtle, she astonished Broadway audiences with the variety of her characterizations; but by 1918 her popularity began to wane and she spent the next 10 years in films. She then returned to the stage, appearing with Eva *Le Gallienne's Civic Repertory Theatre and for the *Theatre Guild in Ibsen, *Chekhov, *Turgenev, and *O'Neill, in whose *Mourning Becomes Electra* (1931) she created the part of Christine Mannon. In 1935 she directed and starred in her own version of Ibsen's *Ghosts*, on Broadway and then on a national tour.

Nederlander Theatre, New York, on West 41st Street between 7th and 8th Avenues. As the National Theatre, seating 1,162, this opened in 1921 and housed a number of successful productions, among them Clemence *Dane's *Will Shakespeare* (1923), *Coward's *To-Night at 8.30* (1936), Lillian *Hellman's *The Little Foxes* (1939), and Emlyn *Williams's *The Corn is Green* (1940). After extensive structural alterations, the theatre reopened in 1941 with Maurice *Evans in *Macbeth*, and continued its successful career until 1958, when it was taken over by the producer, composer, and lyricist **Billy Rose** [William Samuel Rosenberg] (1899–1966), who gave it his own name and opened it in 1959 with a revival of *Shaw's *Heartbreak House*, again with Evans, followed by *Albee's *Who's Afraid of Virginia Woolf?* (1962). Other productions were *Pinter's *Old Times* (1971) and *Stoppard's *Jumpers* (1974). In 1979 the theatre was bought by the Nederlander Organization and renamed the Trafalgar, as a showcase for successful plays from London; but after Brian Clark's *Whose Life is it Anyway?* and Pinter's *Betrayal* (1980), the venture was deemed unsuccessful, and the theatre was renamed after its owners, who are second only to the *Shuberts in the number of theatres they control in the USA. Later productions included Peter *Ustinov's *Beethoven's Tenth* (1984) and a revival of *O'Neill's *Strange Interlude* in 1985, with Glenda *Jackson.

Negro Ensemble Company, see NEGROES IN THE AMERICAN THEATRE.

Negroes in the American Theatre. Black characters were rare on the early American colonial stage, since Negroes lacked social standing and in any case the repertory came from England; the main examples were Othello and the African prince in *Southerne's *Oroonoko* (1695). Isaac *Bickerstaffe and Charles *Dibdin's comic opera *The Padlock* (1769) featured a drunken slave, Mungo, the precursor of many comic Negroes in American drama. Stage versions of Defoe's *Robinson Crusoe* turned Man Friday into a similar clown. Such Black characters as did appear—even at first those in the *minstrel show—were normally played by Whites. The first outstanding Negro actor was Ira *Aldridge, but he lived and worked mainly in Europe. The most effective anti-slavery play was the successful dramatization of Harriet Beecher Stowe's *Uncle Tom's Cabin* (1852). Dion

*Boucicault's *The Octoroon* (1859) sympathetically portrayed its heroine, though without criticizing slavery; and Ned *Harrigan's *vaudeville sketches contained some realistic depictions of Black people. In the 1900s musicals such as *In Dahomey* (1903), written and performed by Blacks and staged before White audiences, were popular; Bert *Williams starred in several.

*O'Neill's *The Emperor Jones* (1920) broke new ground in creating a major Black character for the stage; it was followed by his study of intermarriage *All God's Chillun Got Wings* (1924). Paul *Green, however, made a more extensive use of Negro folk-life, and Marc *Connelly's *The Green Pastures* (1930) had an enormous success. The *Federal Theatre Project (1935–9) provided opportunities for Black writers and performers. The 1920s brought a resurgence of Black musical shows, such as *Blackbirds* (1928). Jerome Kern's famous musical *Show Boat* (1927) had several Black characters; and the *Gershwins' folk opera *Porgy and Bess* (1935)—based on the play *Porgy* (1927) by DuBose and Dorothy Heyward—had an all-Black cast, as did the musical fantasy *Cabin in the Sky* (1940), starring Ethel *Waters, and *Carmen Jones* (1943) based on Bizet's *Carmen*.

By the 1940s Blacks were appearing in Broadway productions as ordinary cast members rather than speciality turns, as in Sidney *Kingsley's *Detective Story* (1949), which featured a Negro policeman without reference to his colour. Plays dealt with Black problems, James Gow and Arnaud d'Usseau's *Deep are the Roots* (1945) being the most successful; even musicals, notably *South Pacific* (1949), attacked racial prejudice. Black actors, including Canada *Lee and especially Paul *Robeson, achieved stardom. The **American Negro Theatre** supported Black theatre from 1940 until the mid-1950s, its most popular production being Philip Yordan's *Anna Lucasta* (1944; London, 1947). Blacks also began to appear in roles written for Whites. The 1964 musical of Clifford *Odets's *Golden Boy* (1937) starred the Black singer Sammy Davis Jnr.; famous musicals such as *Guys and Dolls* (1950) and *Hello, Dolly!* (1964) were restaged with Black casts; and the outstanding Black actor James Earl *Jones played not only Othello but Macbeth and King Lear. The 1950s saw the emergence of Black playwrights, among them **Lorraine Hansberry** (1930–65), whose *A Raisin in the Sun* (1959) was a big Broadway success. The civil rights movement of the 1960s increased the pressure for Black representation. The **Negro Ensemble Company,** formed in 1967, provided an *Off-Broadway non-commercial outlet for Black talent, originating substantial works such as *No Place to be Somebody* (1969) by **Charles Gordone** (1925–) and *A Soldier's Play* (1981) by **Charles Fuller** (1939–), both of which won the *Pulitzer Prize.

Off-Broadway also offered a favourable environment for plays of protest by other Black writers: (Everett) **Leroi Jones** (1934–) originally wrote plays such as *Dutchman* (1964) for mixed audiences, but after changing his name in 1965 to **Amiri Baraka** became more militantly pro-Black (*A Black Mass*, 1966) and later wrote as a Maoist revolutionary (*The Motion of History*, 1977); **Ed Bullins** (1935–) located most of his plays in Black slums (*Clara's Ole Man*, 1965), and in his best-known work *The Taking of Miss Janie* (1975) the hostility of Black towards White was expressed in physical assault. In the 1980s the plays of August *Wilson marked the emergence of a major Black playwright of the mainstream.

Neighborhood Playhouse, New York, on Grand Street, on the Lower East Side. This was built and endowed by Alice and Irene Lewisohn, who designed, choreographed, and directed most of the productions seen there. It opened in 1915 with a dance-drama entitled *Jephthah's Daughter*, and before it closed in 1927 with the fifth annual edition of the revue *Grand Street Follies* it had staged many productions, including plays by *Dunsany, *Chekhov, Sholem *Asch, *Yeats, and *Shaw, and the first dramatic rendering of *Browning's *Pippa Passes* (1917). Among later productions were a number of new dance-dramas and ballets and such varied plays as *Galsworthy's *The Mob*, *Granville-Barker's *The Madras House*, *O'Neill's *The First Man*, *Sheridan's *The Critic*, the Hindu drama *The Little Clay Cart*, and *Ansky's *The Dybbuk*. After the theatre closed, a school of acting under the same name opened on East 54th Street.

Neil Simon Theatre, New York, on West 52nd Street between Broadway and 8th Avenue, a handsome building with an Adam-style interior seating 1,344. It opened in 1927 as the Alvin with the *Gershwins' *Funny Face*, taking its name from the first syllables of the names of Alex A. Aarons and Vinton Freedley, who built it and retained control until 1932. It was used mainly for musical shows with such stars as the Astaires, Ginger Rogers, and Ethel *Merman, but also for occasional straight plays, among them *Mary of Scotland* by Maxwell *Anderson (1933). The Gershwins' folk opera *Porgy and Bess* was first seen there (1935), and *Kaufman and *Hart's *I'd Rather be Right* (1937) inaugurated a series of successful productions, including Rodgers and *Hart's *The Boys from Syracuse* (1938) and the *Lunts in *Sherwood's *There Shall be no Night* (1940). In 1945 Margaret *Webster produced *The Tempest* with the Negro actor Canada *Lee as Caliban. Later successes were Ingrid Bergman in Maxwell Anderson's *Joan of Lorraine* (1946); Sidney *Kingsley's dramatization of Koestler's novel *Darkness at Noon* (1951); *Sondheim's *A Funny Thing Happened on the Way to the Forum* (1962)

and *Company* (1970); and *The Great White Hope* (1968) by Howard Sackler. The musical *Annie* began a long run in 1977, and in 1983, while Neil *Simon's *Brighton Beach Memoirs* was playing there, the theatre was renamed. His *Biloxi Blues* followed in 1985, and in 1989 Vanessa *Redgrave starred in Tennessee *Williams's *Orpheus Descending*.

Neilson, (Lilian) **Adelaide** [Elizabeth Ann Brown] (1846–80), English actress, daughter of a strolling player, who had an unhappy childhood. In 1865 she went on the stage, making her first appearance as Julia in Sheridan *Knowles's *The Hunchback*, always one of her favourite parts. After several years in London and the provinces, where she was much admired in Shakespearian parts and in dramatizations of Scott's novels, she made her first visit to the USA in 1872, touring the country with a fine repertory. She became exceedingly popular with American audiences, and had just returned from a second extended and highly successful tour of the States when she died suddenly. A beautiful woman, with dark eyes and a most expressive countenance allied to a fine speaking voice, she was considered at her best as Juliet in *Romeo and Juliet*, though her Viola in *Twelfth Night* was also much admired.

Neilson, Julia Emilie (1868–1957), English actress, wife of Fred *Terry, whom she married while they were both appearing in Henry Arthur *Jones's *The Dancing Girl* (1891), in which Julia scored a sensational success. From 1900 until Fred's retirement in 1929 she played opposite him. Originally intended for a musical career, she was studying at the Royal Academy of Music when she was persuaded by W. S. *Gilbert to go on the stage, her first appearances being in revivals of several of his plays. One of her outstanding performances before becoming her husband's leading lady was as Hester Worsley in Oscar *Wilde's *A Woman of No Importance* (1893), and she was also much admired as Princess Flavia in George *Alexander's production of Anthony Hope's *The Prisoner of Zenda* (1896). After Fred's death in 1932 she was seen with Seymour *Hicks in his *Vintage Wine* (1934), and in 1938 she celebrated her stage jubilee. After some years in retirement she made a final appearance in 1944.

Neilson-Terry, Phyllis, see TERRY, FRED.

Nemirovich-Danchenko, Vladimir Ivanovich (1859–1943), Russian dramatist and director, who worked under both the Imperialist and Soviet régimes. In his early years he wrote about a dozen light comedies, successfully produced at the *Maly Theatre, and he was in charge of the Drama Course of the Moscow Philharmonic Society, where Olga *Knipper, *Meyerhold, and *Moskvin were among his pupils, when in 1897 a meeting with *Stanislavsky resulted in the founding of the *Moscow Art Theatre a year later. He was responsible for the literary quality of the theatre's repertory, as Stanislavsky was for the acting, and it was he who persuaded *Chekhov to allow *The Seagull* to be revived after its disastrous first performance at the Alexandrinsky (later *Pushkin) Theatre. He himself directed a number of the theatre's outstanding successes, both classic and modern, the last being *Pogodin's play about Lenin, *The Kremlin Chimes* (1942).

Nero, Roman Emperor from AD 54 to 68. He was passionately fond of the theatre and appeared frequently on stage, not only as a dancer in *pantomime but as a solo tragic actor in such parts as the Mad Hercules, the Blind Oedipus, the Matricide Orestes, even Canace in Travail. On such occasions he wore a mask, but the features were always modelled on his own or on those of his current mistress. From his famous theatrical tour of Greece in AD 66–7 he returned with 1,808 triumphal crowns. Even his worst crimes do not seem to have shocked conservative opinion in Rome as much as these antics—a fact which illustrates the low status of professional entertainers under the Empire.

Nervo, Jimmy, see CRAZY GANG.

Nesbitt, Cathleen (1888–1982), English actress, who made her first appearance, in London, in 1910 in a revival of *Pinero's farce *The Cabinet Minister*. A year later she joined the Irish Players, with whom she made her New York début in *Synge's *The Well of the Saints*. In London she played Deirdre in the same author's *Deirdre of the Sorrows*, Phoebe Throssel in *Barrie's *Quality Street* (both 1913), and such classic roles as *Webster's Duchess of Malfi (1919) and Belvidera in *Otway's *Venice Preserv'd* (1920). Being extremely versatile as well as intelligent she moved easily from comedy to drama, from verse to prose. She was in such famous plays as *Lonsdale's *Spring Cleaning* (1925), Margaret Kennedy's *The Constant Nymph* (1926), and Clemence *Dane's *A Bill of Divorcement* (1929). She was equally good as Katharina in *The Taming of the Shrew* (1935) and as Thérèse Raquin in *Zola's *Thou Shalt Not . . .* (1938), and in 1940 played Goneril to *Gielgud's King Lear for the *Old Vic. After the New York production of T. S. *Eliot's *The Cocktail Party* (1950) she was seen more in the United States than in England, appearing on Broadway in 1956 as Mrs Higgins in the musical *My Fair Lady* and as the Grand Duchess in *Rattigan's *The Sleeping Prince*. Her last London role was the Dowager Lady Headleigh in Robin Maugham's *The Claimant* (1964), but in New York she played Madame Voynitsky in *Chekhov's *Uncle Vanya* in 1973 and in her nineties she again played Mrs Higgins in an American revival of *My Fair Lady*.

Nestroy, Johann Nepomuk (1801–62), Austrian actor and dramatist, who in his satiric comedies reflects the rising tide of liberalism and social discontent which was to result in the Revolution of 1848. The foundations of his success were laid in *Vienna, his birthplace, where he first appeared in 1829. His gift for improvisation and his immense facility (he wrote at least 83 plays) soon made him a popular figure. His plots came from wherever he could find them, but he altered them so much that they were hardly recognizable. He excelled in parody, his main target being Wagner, and showed considerable courage in attacking social and political targets. He was the last outstanding exponent of Viennese popular theatre, which after him declined into operetta.

In 1826 he embarked on a long and fruitful partnership with **Karl Carl** [Karl Andreas von Bernbrunn] (1789–1854), who had adapted his comic *persona* *Staberl and made it extremely popular in Germany. Returning to Vienna he managed the Theater an der Wien and the new theatre which in 1847 replaced the old Leopoldstädter Theater. He knew exactly what his audiences wanted and gave it to them, particularly the local farces, or *Posse*. He died a millionaire. In 1842 he scored a big success with Nestroy's adaptation of John Oxenford's farce *A Day Well Spent* as *Einen Jux will er sich machen*, later used by Thornton *Wilder as the basis for his play *The Merchant of Yonkers* (1938), and by Tom *Stoppard for his *On the Razzle* (*National Theatre, 1981).

Nethersole, Olga Isabel (1870–1951), English actress and theatre manager, who made her first appearance in London in 1887 and two years later was at the *Garrick Theatre, where her portrayal of the betrayed country girl Janet Preece in *Pinero's *The Profligate* (1889) quickly brought her recognition as an actress of unusual emotional power. In 1893 she scored a further triumph with her Countess Zicka in a revival of *Sardou's *Diplomacy*, and then took over the management of the *Royal Court Theatre, where she directed and played in A. W. Gattie's *The Transgressor* (1894), in which she made her New York début later the same year. For the next 20 years she divided her time between London and the USA, being equally popular in both. Her intense and realistic characterizations of fallen women in such plays as the younger *Dumas's *The Lady of the Camellias* (*Camille*), *Sudermann's *Magda*, and Pinero's *The Second Mrs Tanqueray* and *The Notorious Mrs Ebbsmith* shocked some older playgoers, but to the younger generation she became a symbol of the revolt against prudery, particularly when in 1900 she was arrested by the New York police for alleged indecency while appearing as Fanny Legrand in Clyde *Fitch's *Sapho*. Defended by William *Winter, she was eventually acquitted.

She managed several London theatres, including *Her Majesty's in 1898, the *Adelphi in 1902, and the *Shaftesbury in 1904. One of her last outstanding parts was the title-role in *Maeterlinck's *Mary Magdalene* (1910). She retired in 1914.

Neuber (*née* Weissenborn), (Frederika) **Carolina** (1697–1760), one of the earliest and best-known of German actress-managers. After an unhappy childhood she eloped with a young clerk, **Johann Neuber** (1697–1759), and with him joined a theatrical company. Ten years later they formed a company of their own. 'Die Neuberin', as she was called, was at this time at the height of her powers and had already attracted the notice of *Gottsched, who enlisted her help in his projected reform of the German stage, persuading her in 1727 to stage French classical plays instead of the old improvised comedies, farces, and harlequinades. High-spirited and intolerant of restraint, however, she soon found herself in conflict with his rigid principles and they parted in 1739, after which the fortunes of her company declined. Even the collaboration of the young *Lessing was of no avail, and after struggling along until the outbreak of the Seven Years War, which reduced them to poverty, husband and wife died within a year of each other. Carolina Neuber's association with Gottsched is generally regarded as the starting-point of the modern German theatre. She did a great deal for her profession, ruling her company with a firm hand, and insisting on regularity and order. Her style of acting, in its day, was a vast improvement on the old clowning and farcical horse-play and prepared the way for the subtle and more natural methods of *Ekhof and *Schröder.

Neuhaus Arena Stage, see ALLEY THEATRE.

Neuville, Honoré de, see MONTANSIER, MLLE.

Neville, John (1925–), English actor, director, and manager, who gained his early experience in repertory and was with the *Bristol Old Vic, 1950–3. He then spent some years with the *Old Vic in London in a wide variety of parts; in 1956 he alternated Othello and Iago with Richard *Burton and went with the company to the United States in 1956 and 1958. After leaving it in 1959 he made several appearances in London before going in 1961 to the *Nottingham Playhouse where he remained until 1968, though in the first season at *Chichester he was seen in *Fletcher's *The Chances* and *Ford's *The Broken Heart*, and in 1963 he was at the *Mermaid in the title-role of Naughton's *Alfie*. He became Joint Artistic Director at Nottingham in 1963, taking sole charge in 1965, and under him the theatre gained an excellent reputation, his roles there including Coriolanus in the opening production

at the new Playhouse in 1963. He also directed a number of plays, including an adaptation of *Calderón's *The Mayor of Zalamea* in 1964. In 1968 he returned to London, directing *Livings's *Honour and Offer* (1969) and appearing as King Magnus in *Shaw's *The Apple Cart* in 1970. He then worked mainly in Canada, running the Citadel Theatre in Edmonton, 1973–8, the Neptune Theatre in Halifax, 1978–83, and the *Stratford (Ontario) Festival, 1985–9. He was seen in a revival of *Beckett's *Happy Days* at the *National Theatre in London in 1977, as Pastor Manders in *Ibsen's *Ghosts* (NY, 1982), and, again at the National, as Sir Peter Teazle in *Sheridan's *The School for Scandal* in 1990.

New Amsterdam Theatre, New York, on West 42nd Street, between 7th and 8th Avenues. This opened in 1903 with Nat *Goodwin in *A Midsummer Night's Dream*, while later in the year came the *Drury Lane pantomime *Mother Goose*. Among visiting stars were Mrs Patrick *Campbell in *Sardou's *The Sorceress* (1904) and the *Forbes-Robertsons in *Shaw's *Caesar and Cleopatra* (1906). *Brewster's Millions* (also 1906) by McCutcheon, Smith, and Ongley and the operetta *The Merry Widow* (1907) were both successful. Beerbohm *Tree came in *Henry VIII* (1916). The theatre, however, was mainly occupied by musical comedy and revue, being best known for the *Ziegfeld Follies*, seen annually from 1913 to 1924. In 1933 Eva *Le Gallienne brought an adaptation of Carroll's *Alice in Wonderland* and *Chekhov's *The Cherry Orchard* from the Civic Repertory Theatre for a successful run. The last production was *Othello* in 1937, with sets by Robert Edmond *Jones, after which the building became a cinema.

Newcastle Playhouse, Newcastle-upon-Tyne. The first theatre of this name was originally called after Flora *Robson, who appeared there in Emlyn *Williams's *The Corn is Green* in 1962. The Tyneside Theatre Company presented a varied programme of plays old and new until the theatre was demolished in a road-widening scheme. It was replaced in 1970 by the present Playhouse, owned by the University of Newcastle and first known as the University Theatre. Seating 480 in a single steeply rising tier, it has a proscenium arch which deliberately fails to conceal the backstage workings. The Tyneside Theatre Company continued to occupy it until 1977, when financial troubles led to its being used for touring companies. A year later a new company, the Tynewear, supported by the local authority and the *Arts Council, took possession, and the theatre's name was changed. The company presented a number of interesting productions, including *Brecht's *The Resistible Rise of Arturo Ui*, *The Merchant of Venice*, the musical *Cabaret*, and *And a Nightingale Sang . . .* by C. P. *Taylor. In 1987 the company moved to the Tyne Theatre and Opera House (see NEWCASTLE-UPON-TYNE), changing its name to the Tyne Theatre Company. After a few months' closure the Playhouse reopened under private management and soon became an important venue for national touring companies, including the *RSC on its annual visit to Newcastle, and regional companies. Music is also prominent, varying from local bands to full-scale opera productions. Behind the theatre is the Gulbenkian Studio, with movable seating for 120 to 200, which is run by the university.

Newcastle-under-Lyme, see NEW VICTORIA THEATRE.

Newcastle-upon-Tyne. There is little evidence of theatrical activity in this area much before the late 17th century, when actors are known to have played in rooms near the Quayside in the then centre of the town. They were compelled to act secretly, however, on account of the hostility of the Puritan element among the citizens. The first permanent theatre was built in Mosely Street in 1788, and having a royal patent was known immediately as the Theatre Royal. It was demolished in the 1830s to make way for Grey Street, in which a new Theatre Royal was built and still stands. It opened in 1837 with *The Merchant of Venice*. Reconstructed and enlarged in 1895, it was extensively damaged by fire in 1899. Rebuilt and enlarged by the addition of properties from the adjoining Shakespeare Street, the theatre reopened yet again in 1901. It housed touring companies until 1973, when it was bought by the Newcastle City Council. It is now managed by a Trust, and continues to be one of the country's major touring centres. Seating 1,400, it has since 1977 provided a regional base for the *RSC, which presents seasons there every year. The theatre underwent major refurbishment in 1987.

Other theatres include the *Newcastle Playhouse and the Tyne Theatre and Opera House, seating 1,200, built in 1867. A cinema from 1916 to 1974, it was renovated, and finally bought, by local amateurs. It reopened in 1986 as a home for amateur productions and those of the Tyne Theatre Company which moved there from the Playhouse in 1987, opening with *Ayckbourn's *Woman in Mind*. The arrangement, however, was not a success, and the company was forced to leave in 1989. It changed its name to the Northern Stage Company and has no resident performing venue. The Tyne Theatre is now mostly used by touring companies, though it continues to stage the productions of its own amateur company.

New Chelsea Theatre, London, see ROYAL COURT THEATRE I.

New Comedy, term used to describe the last period of ancient Greek comic drama, which in the later 4th and 3rd centuries BC developed from the transitional *Middle Comedy. It later became the model and the quarry for Roman comedy, and so influenced later European comic dramatists, particularly *Molière. Its finest exponent was *Menander, who was immensely popular in antiquity but little more than a name in modern European literature until some of his work was discovered among papyri from Egypt. New comedy was pure comedy of manners. It used stock characters and conventional turns of plot; but these are treated, by Menander in particular, with a delicacy of feeling and observation which make a drama of great charm. The *chorus survives from the earlier type of comedy, but has nothing to do with the plot.

Newington Butts Theatre, London. There are many references to performances at Newington in *Henslowe's diary, but it is still uncertain whether they were given in an innyard or in an open-air enclosure. A theatre may have been built in about 1576, but the first reference to plays being given there is found in a letter from the Privy Council desiring that the Surrey justices should prohibit performances at Newington for fear of spreading the plague. A similar letter was sent to the Lord Mayor of London in 1586. In 1594 Henslowe recorded 10 days of performances given by the *Chamberlain's Men in association with the *Admiral's Men, among them *Marlowe's *The Jew of Malta* and Shakespeare's *Hamlet*. Thereafter the theatre fell into disuse. Its accepted site is the present Elephant and Castle.

New London Theatre, in Drury Lane, designed by Sean *Kenny and erected on the site of the *Winter Garden Theatre, forming part of a complex which includes a multi-storey car park and a restaurant. The auditorium, reached by lifts from the car park, and by an escalator from the foyer, is on two levels, one-third of the lower level comprising a 60 ft. revolve which holds the stage, the orchestra pit, and the first rows of the stalls. The stage can thus be used for either proscenium or open-stage productions, 206 of the total of 911 seats moving with the stage to complete the transformation. It proved difficult to find plays which would suit the new set-up, however, and after the opening production of *Ustinov's *The Unknown Soldier and His Wife* (1973) and a rock-and-roll musical, *Grease* (also 1973), it was used as a television studio from 1977 until 1980. In 1981 it returned to theatrical use with the enormously successful musical *Cats*.

New Theatre, London, see ALBERY THEATRE.

New Theatre, New York, see CENTURY THEATRE I.

New Victoria Palace, London, see OLD VIC.

New Victoria Theatre, Newcastle-under-Lyme, Staffordshire. In 1962 Stephen Joseph, whose visits to Newcastle-under-Lyme with his touring *theatre-in-the-round company had aroused great interest, converted a cinema in Stoke-on-Trent into Britain's first permanent theatre-in-the-round, the Victoria Theatre, seating 347 (389 from 1971). He was helped by **Peter Cheeseman** (1932–) who became Theatre Director. In 1986 the company moved into the 605-seat New Victoria Theatre, Europe's first purpose-built theatre-in-the-round, a project planned since 1959. The new theatre continued the company's policy of presenting a broad programme of plays with a resident company, in a form of *repertory. In the old Victoria Theatre over a third of the productions were original works, outstanding among which were the famous local musical documentaries, a form pioneered by this theatre and embodying Cheeseman's commitment to community involvement. Beginning with *The Jolly Potters* (1964), about 19th-century potworkers, they have included *The Staffordshire Rebels* (1965), on the local experience of the Civil War; *The Knotty* (1966), about the life and death of the local railway—the most popular show ever presented at the theatre; and *Fight for Shelton Bar!* (1974), part of a campaign to save the local steelworks. The playwright most associated with the Victoria Theatre is Peter *Terson, the theatre's first resident playwright. Alan *Ayckbourn was a founder member of the company and two of his plays, *Christmas v. Mastermind* (1962) and *Mr Whatnot* (1963), were given their first productions there. The New Victoria Theatre, set in a wooded garden, was built and is sustained by grants from the *Arts Council and the local authorities, plus the proceeds of a public appeal. It is administered by an independent Trust including civic representatives. The programme continues to give a high priority to new plays, together with major theatre classics and at least one children's play in the normal annual programme of 10 major productions.

New York Repertory Company, see LANGNER.

New York Shakespeare Festival, see PAPP.

New Zealand Players, company formed in 1953 which gave its first production at the Wellington Opera House. It then went on tour two or three times a year and did excellent work, but remained a travelling troupe with no permanent home. It presented a representative selection of plays from overseas in all parts of the country, including *Anouilh's *Ring round the Moon*, *Ustinov's *The Love of Four Colonels*, *Fry's *The Lady's not for Burning*, with Keith *Michell and Barbara *Jefford specially brought out from England, *Shaw's *Saint Joan*, and Willis *Hall's *The Long and the Short and the Tall*. The company also appeared in occasional plays by New Zealanders and gave a workshop production of Bruce Mason's *The Pohutukawa Tree*, the first important modern New Zealand play. In spite of its artistic success, however, the constant touring proved too costly, particularly as audiences in general remained small, and the company finally disbanded in 1960, ironically enough while playing to full houses in Napier in *Miller's *A View from the Bridge*.

Niblo's Garden, New York, on the north-east corner of Broadway and Prince Street, summer resort opened by William Niblo on the site of the Columbia Garden. Here, in 1827, he opened the small Sans Souci Theatre used, in the summer only, for concerts and also for plays, particularly after the destruction by fire in 1828 of the *Bowery (to whose displaced company the theatre was leased). Rebuilt in 1829 and renamed Niblo's Garden, the theatre prospered until it was burnt down in 1846. In 1849 a new theatre was built as part of (and entered through) the Metropolitan Hotel. It was improved and enlarged in 1853, and two years later saw the last appearance in New York of *Rachel. In 1866 came the first performance of *The Black Crook*, a fantastic mixture of drama and spectacle, which ran for 475 performances and was followed by a similar spectacle, *The White Fawn*, which was less successful. Later productions included *Boucicault's melodrama *After Dark; or, London by Night* (1868) and a version of the younger *Dumas's *La Dame aux camélias* entitled *Heartsease* with Charlotte *Crabtree. In 1872 it was again burnt down. Rebuilt, it opened on 30 Nov., but its great days were over, and it was too far downtown for a front-rank theatre. It served for some time as a home for visiting companies but was finally demolished in 1895.

Nichols, Mike [Michael Igor Peschowsky] (1931–), American actor, director, and producer, born in Berlin, who for some years played a double act in night clubs with Elaine May based on satirical material most of which they wrote themselves, transporting it to the *John Golden Theatre, New York, in 1960, as *An Evening with Mike Nichols and Elaine May*. Three years later Nichols directed Neil *Simon's *Barefoot in the Park*, which was followed by three more of Simon's plays—*The Odd Couple* (1965), *Plaza Suite* (1968), and *The Prisoner of Second Avenue* (1971)—and other new plays such as Ann Jellicoe's *The Knack*, Murray Schisgal's *Luv* (both 1964), David *Rabe's *Streamers*, and Trevor Griffiths's *Comedians* (both 1976). He also directed revivals of Lillian *Hellman's *The Little Foxes* (1967) and *Chekhov's *Uncle Vanya* (1973), co-adapting the latter. In 1977 he produced the musical *Annie* and D. L. Coburn's *The Gin Game*, directing the latter both in New York and, in 1979, in London. One of the most successful directors in America, he later worked on *Stoppard's *The Real Thing* and Rabe's *Hurlyburly* (both

1984), and *Beckett's *Waiting for Godot* in 1988. He is also a major film director.

Nichols, Peter Richard (1927–), English playwright, who had written a number of television plays before he had an unexpected success in the theatre with *A Day in the Death of Joe Egg* (1967; NY, 1968), transferred from the *Citizens' Theatre, Glasgow. This portrayal of the stresses imposed on parents by the care of a spastic child was followed by *The National Health* (1969; NY, 1974), set in a men's ward for incurables and including fantasy scenes in the style of a television serial; *Forget-Me-Not Lane* (1971), in which the action switches between past and present as the middle-aged narrator looks back on his adolescence in the 1940s; and the more conventional *Chez nous* (1974; NY, 1977), whose London cast included Albert *Finney and Geraldine *McEwan. *The Freeway* (also 1974), like *The National Health*, was produced by the *National Theatre, and *Privates on Parade* (1977), a musical about an army concert party, by the *RSC. In *Born in the Gardens* (1980) an eccentric lady, newly widowed, and her middle-aged son still living at home are contrasted with her two other, more ambitious, children, returned for their father's funeral. One of Nichols's most highly praised plays, *Passion Play* (RSC, 1981; NY, as *Passion*, 1983), a complex study of adultery, shows great technical virtuosity in bringing on stage the *alter egos* of the two main characters. After writing the book and lyrics of *Poppy* (RSC, 1982), a *pantomime-style musical about the Opium War between Britain and China, he announced his retirement from the theatre. *A Piece of My Mind* (1987) was a possibly autobiographical play about a playwright with writer's block.

Nickinson, John (1808–64), actor-manager, born in London, who played an important part in the development of the theatre in Canada, particularly in Toronto. Joining the army at 15, he was sent to Canada, where both in Quebec and in Montreal he acted with the Garrison Amateurs, appearing at Montreal's Theatre Royal in 1833. Three years later he left the army and became a professional actor, specializing in dialect comedy and playing such parts as Havresack in *Boucicault's *The Old Guard*, Pickwick in a dramatization of *Dickens's novel, and Sir Peter Teazle in *Sheridan's *The School for Scandal*. He made his first appearance in New York in 1837 and for some years acted there in the winter, spending the summer on tour in Canada. He was for six seasons at the *Olympic Theatre in New York, and when it closed in 1850 he formed his own company, with which he toured the Great Lakes, becoming manager of the Royal Lyceum Theatre in Toronto in 1853. He remained there until 1858, when he

was succeeded by his son-in-law, and returned to America where he died.

Nicolet, Jean-Baptiste (c.1728–96), French acrobat and entertainer, son of a puppet-master, who played at the *fairs of St-Germain and St-Laurent. In 1760 he had a *booth on the *boulevard du Temple, which he soon transformed into a small permanent theatre, with a good troupe of acrobats and animal turns, notably monkeys. Nicolet himself played young lovers and *Harlequins when he replaced his puppets by living actors, and his theatre flourished in spite of the opposition of the *Comédie-Française. In 1772 he was summoned to Court by Louis XV, who allowed him to call his theatre the Spectacle des Grands Danseurs du Roi, a title which it retained until 1795, when it became the Théâtre de la *Gaîté. The freedom of the theatres under the Revolution allowed Nicolet to play the repertory of the Comédie-Française, which he did, choosing for preference the lighter pieces, until in 1795 he retired.

Nicoll, Allardyce (1894–1976), Scottish theatre historian, successively Professor of English Language and Literature in London and Birmingham Universities. As head of the Department of Drama at *Yale University, he began a vast and comprehensive file of photographs of theatrical material from all over Europe which is constantly being enriched, forming a deposit of valuable material for the theatre research worker. He was the author of a number of useful and well-illustrated books on specialized aspects of the theatre, including a series of nine volumes on the English theatre from the Restoration, each period having an invaluable hand-list of plays arranged under dramatists. Nicoll was for many years Director of the Shakespeare Institute. He was also an excellent lecturer, and President of the *Society for Theatre Research from 1958 until his death.

Nicostratos, SEE GREEK DRAMA.

Nigger Minstrels, SEE MINSTREL SHOW.

Nilsen, Hans Jacob (1897–1957), Norwegian director, actor, and theatre manager. He began his career in the 1920s as an actor, joining the company at the *Nationaltheatret in 1928. After 1933, however, he worked mainly as a director, first at the *Nationale Scene in Bergen, 1934–9, then at the *Norske Theatret in Oslo, 1946–50, and finally at the Folketeatret, 1952–5. Among his productions the most noteworthy were those of *Holberg's plays, and a startlingly 'anti-romantic' production of *Ibsen's *Peer Gynt* in 1948, in which he played the title-role. He also promoted more recent works such as the *Čapeks' *The Insect Play* (1939). His acting roles included Hamlet.

Ninagawa, Yukio (1936–), internationally acclaimed Japanese director. He was active in student theatre in the 1960s, founding the avant-garde troupe Modern Man's Theatre in 1969. In 1974 he moved into mainstream theatre with *Romeo and Juliet*, the first commercial production of Shakespeare in Japan, whose nubile Juliet and use of pop music evoked strong reactions. It was followed by *King Lear* (1975) and *Sophocles' Oedipus Rex* (1976), with a cast of 160, none of whom left the stage. The Ninagawa Company were seen in his productions of *Macbeth* (1980), *Euripides' Medea* (1985), and *The Tempest* (1987) at the *Edinburgh Festival in 1985, 1986, and 1988 respectively. The first two were also staged at the *National Theatre in 1987, and *Macbeth* in New York and other North American cities in 1989. In 1988 he directed his first American play, Tennessee *Williams's *A Streetcar Named Desire*. Over two-thirds of his work, however, is Japanese, including a version of Chikamatsu's *kabuki* puppet plays (1979), staged at the National Theatre as *Suicide for Love* in 1989, and Shimizu's *Tango, at the End of Winter* (1984), which he directed in Edinburgh and London in English, with British actors, in 1991. Ninagawa's productions are memorable for their magnificent costumes, spectacular sets, large casts, and use of colour, especially red.

Noah, Mordecai Manuel (1785–1851), American playwright, whose first play, *The Wandering Boys* (1812), was a translation of *Pixérécourt's *Le Pèlerin blanc* (1801). First produced in Charleston, this was later taken at *Covent Garden with alterations, and in its amended form returned to New York, where it remained popular for many years. Noah's later plays were produced at the *Park Theatre 1; it was after the third night of *The Siege of Tripoli* in 1820 that the theatre was destroyed by fire. In a later play, *The Grecian Captive* (1822), the hero and heroine made their entrances on an elephant and a camel respectively, a spectacular product, no doubt, of the fertile brain of the manager, Stephen *Price. Noah's plays are simply written, with a good deal of action and sustained interest, and with the aid of lavish scenery, transparencies, and illuminations they held the stage for many years.

Nobel Prize for Literature, has been awarded in recognition of their dramatic writings to the following: *Beckett, Samuel, 1969; *Benavente, Jacinto, 1922; *Bjørnson, Bjørnstjerne, 1903; *Echegaray, José, 1904; *Hauptmann, Gerhart, 1912; *Maeterlinck, Maurice, 1911; *O'Neill, Eugene, 1936; *Pirandello, Luigi, 1934; *Sartre, Jean-Paul (who declined it), 1964; *Shaw, George Bernard, 1925; *Soyinka, Wole, 1986.

Noble, Adrian Keith (1950–), English director. After beginning his career as an associate director with the *Bristol Old Vic, 1976–9, he

was guest director, 1980–1, at the *Royal Exchange Theatre, Manchester, his highly praised production of *Webster's *The Duchess of Malfi* with Helen *Mirren being seen also at the *Round House. In 1980 he joined the *RSC as resident director, becoming an associate director in 1982 and soon receiving acclaim for his strikingly imaginative productions of Shakespeare, among them *King Lear* (1982) with Michael *Gambon and Anthony *Sher, *Henry V* (1984) with Kenneth Branagh, *Macbeth* (1986), and *The Plantagenets* (1988), adapted from *Henry VI* and *Richard III*. His other work ranged from *Massinger's *A New Way to Pay Old Debts* (1983) and the new play *The Art of Success* (1986) by Nick Dear to the musical *Kiss Me, Kate* (1987) and *Ibsen's *The Master Builder* (1989). In 1991, as part of the new governing triumvirate, he became Artistic Director.

Noises Off, see SOUND EFFECTS.

Nokes, James (?–1696), English actor, and a member of *Davenant's company. He was a fine comedian, of whom Colley *Cibber has left a masterly pen-portrait, and usually played foolish old husbands and clumsy fops, as well as a few ridiculous old ladies. His best part, among the many that he created in Restoration drama, seems to have been the title-role in *Dryden's comedy *Sir Martin Mar-All* (1667).

Nō Play, lyrical drama of Japan, established by **Kwanami** (1333–84) and his son **Zeami** (1363–1443), though the courtly language and formal style of the works attributed to them suggest an earlier derivation, as far back even as the end of the 12th century. The *kyōgen*, or comic interludes, which form part of the *nō* plays, are in the vernacular of the mid-16th century, and the genre reached its point of perfection a hundred years later, since when it has changed very little. The *nō* play draws its material mainly from the Buddhist scriptures and the mythology of China and Japan, and its form from the ritual dances of the temples with accretions from folk-dances of the countryside. It is essentially a drama of soliloquy and reminiscence, and unlike Western drama has no development through conflict. Having been for a long time reserved for the amusement of the ruling class, its audiences tend still to be elderly and educated, with a sprinkling of foreign visitors. *Nō* plays are acted on a raised resonant stage of polished wood, with a temple roof over it supported on four pillars, and the audience on two sides. The actors are supported by the eight members of the chorus, four musicians—a flute and three drums—and two stage hands, who by tradition are invisible. The performers enter along the *hashigakari*, a passage about 40 ft. long which runs at an angle backward from the right-hand edge of the stage. There is no scenery, but the costumes are sumptuous, particularly those of the First Actor (*shite*), who is introduced by the

Second Actor (*waki*), and as god or hero, wearing a mask, performs the ritual dances which are the heart of the play. (See also KABUKI.)

Norske Teatret, Oslo, opened in 1913 in Bøndernes Hus. Its prime objective was to provide a stage for the performance of plays in Nynorsk, the officially standardized Norwegian language, to offset the Dano-Norwegian influence of *Ibsen and *Bjørnson. It also aimed to stimulate Nynorsk dramatists into providing a continuous supply of new plays in the official language. Although the theatre lives on as a lively component of the Oslo theatrical scene, and has mounted many significant productions, among them an 'anti-romantic' version of Ibsen's *Peer Gynt* directed in 1948 by Hans Jacob *Nilsen, the Nynorsk contribution to Norwegian drama remains meagre. The theatre's new building, with three stages, which opened in 1985 is one of the most modern theatre buildings in Europe. It is now able to stage large-scale musical productions such as *Jesus Christ Superstar*, *Cats*, and *Les Misérables*.

Norske Theater, Bergen, see NATIONALE SCENE.

Northcott Theatre, Exeter, Devon, named after its chief benefactor, a Devon business man. It was erected in 1967 on a site donated, together with a large sum of money, by the University of Exeter, within whose precincts it stands. It has a proscenium stage and the auditorium, fan-shaped and steeply raked, seats 433. In-house productions run for approximately four weeks and are a mixture of classic, contemporary, and commissioned plays and musicals. Since 1986 the theatre's rehearsal hall has been developed as the Northcott Studio Theatre and offers seasons of plays when the main theatre is used by amateurs and visiting companies.

Northern Stage Company, see NEWCASTLE-UPON-TYNE.

Norton, Thomas (1532–84), member of the Inner Temple, who in collaboration with a fellow-student, **Thomas Sackville** (1536–1608), later the first Earl of Dorset and Lord Treasurer under Elizabeth I and James I, wrote *Gorboduc; or, Ferrex and Porrex*, the first surviving example of a regular five-act tragedy in the style of *Seneca in English dramatic literature. Apparently Norton wrote the first three acts and Sackville the last two. The play was performed before Elizabeth I on New Year's Day 1562 during an entertainment in the Inner Temple Hall; its theme resembles that of *King Lear*, with Gorboduc, king of Britain, dividing his kingdom between his two sons, who quarrel over it and are both killed. Norton does not appear to have written anything else, but Sackville contributed to the second edition of *A Mirror for Magistrates* (1563) the Induction

and 'The Complaint of Buckingham', the only contributions having any literary merit.

Norwich, see MADDERMARKET THEATRE.

Norworth, Jack, see VAUDEVILLE, AMERICAN.

Nottingham Playhouse. This theatre, run by a Trust since 1948, was originally a converted cinema seating 467. The new Playhouse, one of the first modern regional theatres, opened in 1963, funded by private contributions and Nottingham City Council. Its cylindrical auditorium gives a feeling of intimacy and seats 500 in the stalls and 250 in the circle. It has a large and well-equipped stage on which elaborate productions can be mounted. Frank *Dunlop (1961–4), John *Neville (1965–8), and Richard *Eyre (1973–8) were among its Artistic Directors. Outstanding presentations included *Brecht's *The Resistible Rise of Arturo Ui*, *Shaw's *Widowers' Houses*, *Hamlet* with Alan *Bates, and *King Lear* with Michael *Hordern, all of which were seen in London. Notable among numerous premières were Peter Barnes's *The Ruling Class* (1968) and his *Lulu* (1970), adapted from *Wedekind, Christopher *Fry's *A Yard of Sun* (also 1970), Howard *Brenton and David *Hare's *Brassneck* (1973), and Brenton's *The Churchill Play* (1974). Trevor Griffiths's *Comedians* (1975; London, 1976), set in an evening class for training comics, also had its première there. In the 1980s it staged a highly acclaimed production of Tolkien's *The Hobbit*. A Theatre in Education scheme is run by the Roundabout Company.

Nouveau Monde, Théâtre du, French-Canadian company founded in Montreal in 1951 by a group of actors, under Jean *Gascon, who had been associated with Father Émile Legault's Compagnons de Saint-Laurent. They opened with *L'Avare* by *Molière, the author of most of their early successes. A trio of his farces, played in the style of the *commedia dell'arte*, was an outstanding success at the *Théâtre des Nations in Paris in 1955, and brought the new company an international reputation. In the following year Gascon and several of his companions joined the company at the *Stratford (Ontario) Festival theatre to play the French roles in *Henry V*. Although for over 20 years the group had no permanent home, it evolved into a civic repertory theatre with high standards, presenting an international repertoire varied with a new Canadian play each season; one of its most successful authors was Marcel *Dubé. The company suffered a severe setback when in 1963 fire destroyed their offices, records, and rehearsal space, and again in 1966 when the Orpheum cinema, which they had rented in 1957, was scheduled for demolition. They moved in 1967 to the Port-Royal in the new Place des Arts complex. Ill at ease, however, in these lavish surroundings, they found a permanent home in 1973 at the Théâtre de la Comédie-Canadienne, left vacant by the departure of Gratien *Gélinas and his company. An immediate innovation there was the introduction of Lunch-Time Theatre (*Théâtre-midi*), and among the new plays produced was the controversial *Les Fées ont soif* (*The Fairies are Thirsty*, 1978) by Denise Boucher, an attack on male chauvinism and the restraints imposed on women by the Catholic Church which provoked accusations of blasphemy and resulted in subsidies being withheld by the civic authorities. The company toured abroad in 1965 and 1971 and from 1965 to 1975 also supported Les Jeunes Comédiens, a touring company of young actors which began as a graduation class from the National Theatre School of Canada. Crippling deficits and protracted labour disputes forced the company to cancel their 1984–5 season and sell their theatre; but their fortunes rose again with strong and imaginative productions of *Othello* (1986), *A Midsummer Night's Dream* (1988), and *Brecht's Galileo (1989).

Novelli, Ermete (1851–1919), Italian actor, who made his first appearance at the age of 18 and though not at first well received became one of the outstanding actors of the Italian stage. A large man, weighing some 18 stone, he was nevertheless light on his feet and quick in action, with fine, expressive features and a mobile countenance. At first he played exclusively in comedy, but it was finally in tragedy that he made his reputation, and he toured extensively in such Shakespearian parts as Othello, Lear, Shylock in *The Merchant of Venice*, Macbeth, and Hamlet. In the last-named he is said to have played the death-scene with brutal realism—indeed, from contemporary accounts his acting appears to have been exceedingly forceful and melodramatic—and in the scene with the Ghost he seemed to impart to the audience the certainty of a visitation from another world. Another part in which he excelled was the title-role in a translation of Aicard's *Le Père Lebonnard*, a play first seen at *Antoine's Théâtre-Libre. In 1900 Novelli attempted to found a permanent theatre in Rome, La Casa di *Goldoni, but the enterprise failed through lack of public support and had to be abandoned.

Novello [Davies], (David) **Ivor** (1893–1951), actor-manager, dramatist, and composer, the son of musical parents, his mother being a choral conductor. During the First World War he was responsible for part of the score of several successful *musical comedies. He also wrote the popular song 'Keep the Home Fires Burning'. He made his first appearance on the stage in Sacha *Guitry's *Deburau* (1921), and three years later was in his own play *The Rat*, written in collaboration with Constance *Collier. He subsequently wrote more than 20 comedies and musical plays, composing also the

scores for the latter and starring in most of the productions himself. He was the author, composer, and leading man of four successive musicals at *Drury Lane: *Glamorous Night* (1935), *Careless Rapture* (1936), *Crest of the Wave* (1937), and—after he had appeared in a spectacular revival of *Henry V* in 1938 at the same theatre—*The Dancing Years* (1939). The last was revived at the *Adelphi Theatre in 1942 and was one of the big successes of the Second World War. Novello also wrote, composed, and played in another great success, *Perchance to Dream* (1945). He was appearing in his own musical *King's Rhapsody* (1949) at the time of his death. During its run he had written and composed *Gay's the Word* (1951), which had a long run starring Cicely *Courtneidge. His straight plays, far less well known than his musicals and now largely forgotten, included *The Truth Game* (1928), *I Lived with You* (1932), *Full House* (1935), and *We Proudly Present* (1947).

Novelty Theatre, London, see KINGSWAY THEATRE.

Nunn, Trevor Robert (1940–   ), English director, who acted in and directed several plays for the Marlowe Society while at *Cambridge. In 1962 he went to the *Belgrade Theatre in Coventry as a trainee director, remaining there until in 1965 he joined the *RSC, where his first production was *The Thwarting of Baron Bolligrew*, a children's play by Robert *Bolt. A year later he directed the first revival for 300 years of *Tourneur's *The Revenger's Tragedy*, and in 1967 he was responsible for excellent revivals of *Vanbrugh's *The Relapse* and *The Taming of the Shrew*. In 1968 he was appointed the company's Artistic Director, directing in the same year productions of *Much Ado about Nothing* and *King Lear*. Among his other notable productions were *The Romans* (1972), which included *Coriolanus*, *Julius Caesar*, *Antony and Cleopatra*, and *Titus Andronicus*; his own adaptation of *Ibsen's *Hedda Gabler* (1975); a musical version of *The Comedy of Errors*, for which he wrote the lyrics, and a studio *Macbeth*, with Ian *McKellen and Judi *Dench (both 1976). Between 1978 and 1987 he was Chief Executive and Joint Artistic Director with Terry *Hands, his productions including a revival of *Kaufman and *Hart's *Once in a Lifetime* in 1979; *The Life and Adventures of Nicholas Nickleby*, based on *Dickens's novel, in 1980 (NY, 1981); and a highly praised Edwardian *All's Well that Ends Well* (also 1981). In the same year he directed Andrew Lloyd Webber's *Cats*, demonstrating his ability to handle large-scale musicals in his first freelance production since becoming the RSC's Artistic Director. He later directed the musicals *Starlight Express* (1984; NY, 1987), *Aspects of Love* (1989; NY, 1990), both by Lloyd Webber, *Chess* (1986; NY, 1987), and *The

Baker's Wife (1989). He was also joint adaptor and director of the RSC's own highly successful musical *Les Misérables* (1985; NY, 1987), and in 1989 the RSC staged his small-scale production of *Othello*. His first wife was Janet *Suzman.

Nurseries, name given during the Restoration period to training schools for young actors, the best known being that set up by *Killigrew in Hatton Garden in about 1662. This moved in 1668 to the Vere Street Theatre, where it flourished until 1671, when *Davenant's widow opened a new Nursery in the Barbican. This was still in use in 1682, as it is referred to in *Dryden's poem *MacFlecknoe*, published in that year. A third Nursery is known to have opened briefly on Bun Hill, in Finsbury Fields, some time during 1671, and Nursery companies occasionally performed in borrowed playhouses. Little is known of their work, although they must have provided at least a minimal training for young players.

O

Oakley, Annie, see CODY.

Oberammergau, see PASSION PLAY.

Obey, André (1892–1975), French dramatist, whose plays *Noé*, *Le Viol de Lucrèce*, and *La Bataille de la Marne* were all produced in 1931–2 by *Saint-Denis for the Compagnie des Quinze. The first, produced in London and New York in 1935 with *Gielgud and Pierre *Fresnay respectively as Noah, was remarkable for the liveliness of its beasts; the second, which made use of a modified Greek chorus, was one of the sources of the libretto of Britten's opera *The Rape of Lucrece* (1946). Continuing to employ a distanced technique, often involving the use of a narrator and heightened language, Obey later wrote *Le Trompeur de Séville* (1937), on the theme of *Don Juan, and *L'Homme de cendre* (1950); also, on biblical themes, *Lazare* (1952) and *Plus de miracles pour Noël* (1957), seen in London in 1965 as *Frost at Midnight*.

O'Casey, Sean [John Casey] (1880–1964), Irish dramatist, whose best plays, set in the slums of Dublin, show how intimately he knew the people of whom he wrote and the events of 1915–22 from which he drew his material. His treatment of his themes is closely related to that of the Irish realists before him—grim, clear-cut, and satirical; but in his work the comedy of satire points directly to tragic implications. His first play, produced at the *Abbey Theatre in Dublin, was *The Shadow of a Gunman* (1923; London, 1927; NY, 1932), a melodramatic story of the war in 1920 and its effects on the lives of a group of people in a Dublin tenement house. This anticipates, in its subject and setting and in some of its objective commentary, the finer plays that followed; the men who talk, live, and die for an idea are contrasted with the women who live and die for actualities. O'Casey's next play, *Juno and the Paycock* (1924; London, 1925; NY, 1926), a moving, realistic tragedy, set in 1922 with much the same background, was popular with both English and Irish audiences, though for different reasons, as was *The Plough and the Stars* (Dublin and London, 1926; NY, 1927), a play on the Easter Rising of 1916 which caused a riot in the Abbey Theatre when it was first produced there. The consequent refusal of the Abbey to produce his next play, *The Silver Tassie* (London and NY, 1929), which was not seen in Dublin until 1935, led O'Casey to leave Ireland and settle in England. His next play, *Within the Gates* (London and NY, 1934), is set in London, and gives further evidence of the influence on his work of *Expressionism, already apparent in *The Silver Tassie*. Although remarkable in many ways, it is doubtful if the extended use of stylization and symbolism helped O'Casey to master his material, nor was he particularly successful with his Cockney characters. His next play, *The Star Turns Red* (1940), was first seen at the *Unity Theatre in London, but in 1943 O'Casey had his first Irish premières for 17 years with *Purple Dust* (NY, 1955; London, 1962) and *Red Roses for Me* (London, 1946; NY, 1955). In these, and again in *Oak Leaves and Lavender*, seen in London in 1947, he again uses *Symbolist and Expressionistic devices to reinforce ideals expressed by their Marxist heroes. *Cock-a-Doodle Dandy*, seen in Newcastle-upon-Tyne in 1949, in New York in 1958, and at the *Royal Court Theatre in London in 1959, stimulated a new interest in O'Casey's later plays. In 1964 the Abbey Players were seen in London in the *World Theatre Season at the *Aldwych in *Juno and the Paycock* and *The Plough and the Stars*. In his last plays, *The Bishop's Bonfire* (1955; London, 1961), *The Drums of Father Ned* (1958), *Behind the Green Curtains* (1961), and *Figuro in the Night* (also 1961; NY, 1963), O'Casey contrasts the repressive forces of the clergy and the moneyed classes in modern Ireland with the yearnings of Irish youth for artistic, sexual, and political freedom.

Octagon Theatre, Bolton, the first fully flexible professional theatre to be erected in Britain, its octagonal shape being chosen as best suited to the needs of a theatre which can be adapted to open-end, thrust, *theatre-in-the-round, and promenade staging, seating 362, 334, 420, and 212 (plus promenaders) respectively. The money required for this was raised mainly by public subscription and the site was donated by Bolton Corporation. The Octagon opened in 1967, only 18 months after the campaign to build it began, a remarkable achievement for a town with no great theatrical tradition, and which had been without a theatre since 1961. The company presents a wide selection of classic and modern plays, recently establishing a reputation for world premières of both plays and musicals. The theatre was able to develop further in 1987 with the opening of the Octopus Studio. It has an attractive Youth Theatre (11–18) and Young Company (18–25). Many small-scale touring companies bring their productions to the theatre, which has also introduced a programme of cabaret evenings, 'Live Fridays'.

Odéon, Théâtre National de l', Paris, second theatre of France, ranking next to the *Comédie-Française, which in 1781 occupied the first building to be erected on the present site of the Odéon. Following the vicissitudes of the early Revolutionary period, the theatre acquired its present name in 1795. Rebuilt in 1816, and again after its destruction by fire two years later, it was managed by Picard from 1816 to 1821; but it was only in 1829 that light comedies and operettas gave way to a classical and contemporary repertoire. André *Antoine was its Director, 1906–16, *Gémier 1921–8. During the reorganization of the Parisian theatres under André Malraux in 1959 the Odéon was removed from the control of the Comédie-Française and renamed the Théâtre de France, under the direction of Jean-Louis *Barrault. He presented there a number of important plays, including *Genet's Les Paravents (1966), which led to rioting in the theatre. Barrault also created a small studio for experimental work, and provided a home for the *Théâtre des Nations. After the demonstrations of May 1968, for which the Odéon provided a focal point, Barrault was dismissed, and the theatre reverted to the control of the Comédie-Française, and its old name, in 1971. Within this new association, the Petit Odéon was created to promote the work of new playwrights. In 1983 the Odéon was again separated from the Comédie-Française, housing the Théâtre de l'Europe which staged foreign-language productions, the Odéon also continuing to stage its own productions. After a further period of control by the Comédie-Française, 1986–9, the Odéon regained its independence. **Giorgio Strehler** was Artistic Director of the Théâtre de l'Europe from 1983 to 1990, in which year it became a full-time occupant of the Odéon. His most notable productions there include The Tempest, *Strindberg's The Storm, and *Corneille's L'Illusion.

Odets, Clifford (1906–63), American dramatist. Born in Philadelphia but growing up in New York, he became an actor after graduation from high school. In 1935 he attracted attention with a long one-act play about a taxi-drivers' strike, Waiting for Lefty, first produced by the *Group Theatre of which he was a member (London, 1936). For a Broadway production he added another multi-scened one-acter, Till the Day I Die (London, 1940). Almost simultaneously, the Group Theatre produced an earlier full-length play, Awake and Sing (London, 1938), which Odets rewrote for the occasion. Portraying the conflicts in a Jewish family in the Bronx caused by poverty and a domineering mother, it is notable for its realism, contrapuntal technique, and mingling of humour with explosive passion. Although his next play, Paradise Lost (also 1935), about the false ideals instilled in the

success-orientated young, met with a tepid reception, Odets retrieved his reputation with Golden Boy (1937; London, 1938), the story of a sensitive Italian youth's deterioration after economic pressures turn him from music to professional boxing. Rocket to the Moon (1938; London, 1948) failed as a social parable in spite of excellent characterization and considerable pathos. Night Music (1940), an extravaganza depicting the struggles of disorientated youth, also failed, and Clash by Night (1941), in which Odets attempts to create a political allegory out of the personal humiliations of an unemployed labourer, emerges as a heavy-handed domestic triangle. He worked in films for some years, which gave rise to an abrasive indictment of Hollywood, The Big Knife (1949; London, 1953). The Country Girl (1950; London, as Winter Journey, 1952) deals sympathetically with the relationship between a drunken actor and his loyal wife. Odets's last play, The Flowering Peach (1954), is a retelling of the biblical story of Noah as a nostalgic Bronx Jewish fable. Odets's reputation rests on his plays of social protest of the 1930s; an unproduced play from this period, The Silent Partner (1937), received a posthumous *Off-Off-Broadway production in 1972.

Oenslager, Donald Mitchell (1902–75), American scene designer, who worked under George Pierce *Baker at *Harvard. In 1923 he went to Europe to study scenic production there, and on his return to America worked with the *Provincetown Players and the Greenwich Village Theatre. With Robert Edmond *Jones, Jo *Mielziner, and Lee *Simonson, he may be said to have contributed to the creation of a new age of stagecraft in the United States, and he left a permanent mark on the contemporary theatre there. He designed sets for a number of major drama productions (as well as operas and ballets), among them *Steinbeck's Of Mice and Men (1937); *Shaw's The Doctor's Dilemma in 1941 for Katharine *Cornell, Pygmalion in 1945 for Cedric *Hardwicke, and Major Barbara in 1956 for Charles *Laughton; also for *Ibsen's Peer Gynt in 1951, Coriolanus in 1954, and Spigelgass's A Majority of One (1959). He was active up to the time of his death, and was for many years Professor of Scenic Design in the *Yale Department of Drama.

Œuvre, Théâtre de l', Paris, see LUGNÉ-POË.

Off-Broadway, term used collectively of theatres and plays outside the orbit of the New York mainstream theatre located on or near *Broadway. Off-Broadway arose in the 1950s because of the high cost of Broadway productions, the lower overheads and sometimes non-profit basis away from the centre enabling risks to be taken. Off-Broadway productions may transfer to Broadway, though as costs escalate there is a tendency for even the more popular ones to remain in their original locations, especially if a transfer would require the acquisition of star names. Off-Broadway theatres are generally smaller and less well equipped than Broadway ones, but even these became increasingly subject to commercial pressures. Since the early 1960s a further, largely non-professional, movement known as Off-Off-Broadway has arisen, which presents experimental drama in lofts and other unconventional locations.

Ohel (Tent) **Theatre,** theatrical company of the Jewish Labour Federation, founded in Palestine in 1925 by **Moshe Halevy** (1895–1974), a former member of *Habimah. The actors were at first encouraged to divide their time between the stage and their daily work in field and factory, so as not to sever their ties with the working class, but this proved impractical and Halevy soon found that it was necessary for them to devote their whole time to acting if they were to reach professional standards. The company's first public performance, given in Tel Aviv in 1926, consisted of dramatizations of stories by Isaac Leib *Peretz. These, and biblical plays, of which Stefan Zweig's *Jeremiah* (1929) was one, proved immensely popular with the kibbutz members who made up most of Ohel's audiences as the troupe travelled throughout the country, and were also well received in Europe on tour. By 1958, however, the theatre had lost sight of its original purpose and, its standards having declined, the Labour Federation withdrew its support. From then on Ohel led a precarious existence and in spite of occasional successes, such as revivals of Hašek's *The Good Soldier Schweik* and Ephraim Kishon's *The Marriage Contract* (1961), the company was eventually forced to disband in 1969.

O'Keeffe, John (1747–1833), Irish-born dramatist, who was for about 12 years an actor in Dublin where his first plays, including *The Shamrock; or, St Patrick's Day* (1777), were produced. In his twenties his sight began to fail, which stopped his acting but did not prevent him from writing a vast number of plays, mainly farces and light operas, the latter containing many well-known songs. The first of his plays to be seen in London was the farce *Tony Lumpkin in Town* (1778), produced at the *Haymarket, as were his next three 'operatic farces'. The most successful of his later plays was *The Poor Soldier* (1783), which had a great vogue in the United States, as did the comedy *Wild Oats; or, The Strolling Gentleman* (1791), first seen at *Covent Garden. One of O'Keeffe's farces, *The Little Hunch-Back; or, A Frolic in Bagdad* (1789), was included in the repertory of the *toy theatre.

Okhlopkov, Nikolai Pavlovich (1900–67), Soviet actor and director, whose first production, in 1921, was a May-Day spectacle in the central square of Irkutsk (his birthplace), of which he was author, director, and chief actor. In it can be seen the beginnings of that original style of production which, after four years with *Meyerhold, 1923–7, he was able to develop more fully on his appointment in 1930 as Artistic Director of the *Realistic Theatre. There he produced a number of new plays, among them an adaptation of *Gorky's *The Mother* in 1933 and *Pogodin's *Aristocrats* (1934), for each of which he set up new types of staging, on movable platforms, drawing the audience, whose location also varied, into the development of the action. His work, influenced by his experience as a film actor and director, was necessarily experimental and its limited appeal led to an unsuccessful attempt to merge his company with *Taïrov's in 1938. Okhlopkov then worked for a while at the *Vakhtangov Theatre, where he directed a notable production of *Rostand's *Cyrano de Bergerac*, and in 1943 became director of the Theatre of the Revolution (later the *Mayakovsky Theatre), where his revivals of *Ostrovsky's *Fear* in 1953 and of Shakespeare's *Hamlet* in 1954 were much admired. He also continued his policy of producing new Soviet plays, among them A. P. Shteyn's *Hotel Astoria* (1956), *Arbuzov's *Endless Distance* (1958), and Pogodin's *The Little Student* (1959).

Old Comedy, term used to distinguish the early comedy of ancient Greece—specifically of Athens in its prime—from its subsequent development into *Middle Comedy and *New Comedy. (See COMEDY for comedy in the modern sense.) It had a much stricter and more complex form than Attic *tragedy, though it too was a blend of the choral and the histrionic. The name 'comedy' is derived from *komos* and *ode*, meaning a 'revel song'. One form of revel was associated with fertility rites: it was a mixture of singing, dancing, scurrilous jesting against bystanders, and ribaldry; *Aristotle derives comedy from this, and certainly comedy contained all these elements, including the use of the phallus, the symbol of fertility. Another form of *komos*, well represented on vase-paintings, was the Masquerade, in which revellers disguised themselves as animals or birds. Since the comic *chorus was often of this type (as in the *Wasps*, *Birds*, and *Frogs* of *Aristophanes) the influence of this kind of revel on comedy seems clear enough. Old Comedy (represented now only by Aristophanes) is the most local form of drama that has ever reached literary rank. It was a sort of national lampoon, in which anything prominent in the life of the city, whether persons or ideas, was unsparingly ridiculed—a unique mixture of fantasy, criticism, wit, burlesque, obscenity, parody, invective, and exquisite lyricism. The atmosphere of the whole is well suggested by the story that during the performance of

Aristophanes' *Clouds* Socrates rose from his seat to give the audience an opportunity of comparing the mask of the stage Socrates with the man himself.

In the late 18th and 19th centuries the term Old Comedy was used in Britain, and more particularly in North America, to denote the repertory of comedies from Shakespeare to *Sheridan.

Old Drury, see CHESTNUT STREET THEATRE, Philadelphia; DRURY LANE THEATRE, London; and PARK THEATRE I, New York.

Oldfield, Anne [Nance] (1683–1730), English actress, who succeeded Mrs *Bracegirdle as one of the leading players at *Drury Lane. She first came into prominence when Colley *Cibber, struck by her playing of Leonora in a revival of Crowne's *Sir Courtly Nice*, cast her as Lady Betty Modish in his own play *The Careless Husband* (1704). From then on her career was one of unbroken triumph. Lovely in face and figure, she had an exceptionally beautiful voice and clear diction, and *Voltaire said of her that she was the only English actress whose speech he could follow without difficulty. She was particularly admired as Silvia in *The Recruiting Officer* (1706) and as Mrs Sullen in *The Beaux' Stratagem* (1707), both by *Farquhar, who first encouraged her to go on the stage, but she was also considered outstanding in tragedy, playing with majesty and power such parts as Andromache in *Philips's *The Distrest Mother* (1712) and the title-role in *Rowe's *Jane Shore* (1714). She much preferred comedy, however, and in the year of her retirement gave a highly acclaimed performance as Lady Townly in *The Provoked Husband* (1728), based by Cibber on *A Journey to London*, an unfinished play by *Vanbrugh. She made her last appearance in *Fielding's first play, *Love in Several Masques* (also 1728), and after her death two years later was buried in Westminster Abbey, near *Congreve, but was not allowed a monument over her grave, because she had had two illegitimate sons.

Old Globe Theatre, San Diego, see SHAKE-SPEARE FESTIVALS.

Old Man, Woman, see STOCK COMPANY.

Old Mo, see MIDDLESEX MUSIC-HALL.

Old Tote Theatre, Sydney, company founded in 1963, its name being derived from its first premises, the refurbished tote box of a race-track. Its opening production was *Chekhov's *The Cherry Orchard*, and it presented more classical plays than any comparable Australian company, featuring top local actors and directors and sometimes an overseas guest, as when Tyrone *Guthrie directed *Sophocles' *Oedipus the King* in 1970. In its first 10 years the Old Tote presented 18 Australian plays, 12 of them being first performances, and it was the first

drama company to perform in the *Sydney Opera House, appearing in *Richard II*, *Brecht's *The Threepenny Opera*, and David *Williamson's *What if You Died Tomorrow?* (The last was seen in London in 1974.) The company depended on substantial subsidies from the Australia Council and the Government of New South Wales; these were withdrawn in 1978 and the company was replaced by the Sydney Theatre Company.

Old Vic Theatre, London, in the Cut, off the Waterloo Road. At first called the Royal Coburg, after the husband of Princess Charlotte, it opened in 1818 with a melodramatic spectacle by William Barrymore, *Trial by Battle; or, Heaven Defend the Right*, based on a recent notorious murder trial. The journey across the river was too hazardous for a fashionable audience, but a series of melodramas of the most sensational kind soon attracted a large local audience, particularly as the plays were well staged with well-known actors. In 1833 the theatre was redecorated, reopening with a revival of *Jerrold's *Black-Ey'd Susan*, being renamed the Royal Victoria Theatre in honour of Princess (later Queen) Victoria. It soon became affectionately known as the Old Vic, but its standards noticeably declined and it finally sank to the level of a penny *gaff. The audience was very rough, and in 1858 16 people lost their lives in a panic due to a false alarm of fire. In 1871, after a short period as a *music-hall, the theatre closed. It was sold by auction and renamed the New Victoria Palace. This closed early in 1880, and the building was then bought by Emma Cons, a social worker and the first woman member of the London County Council, with the intention of turning it into a temperance music-hall. Renamed the Royal Victoria Hall and Coffee Tavern, it opened in 1880 under the management of William *Poel, who remained until 1883. Intended as a place of inexpensive family entertainment, it prospered in spite of initial misgivings, much helped by the support of a Bristol MP, Samuel Morley, whose efforts to promote the arts in South London included the founding of Morley College which for a time occupied part of the building. In 1900 the first full length opera, Balfe's *The Bohemian Girl*, was given, and scenes from Shakespeare were introduced into the concert programmes. In 1912 Emma Cons's niece Lilian *Baylis, who had been assisting her aunt since 1898, took over and two years later presented the first full Shakespeare season. It was successful enough to warrant further productions of his plays, in spite of the outbreak of war later that year, and in the next nine years all the plays in the First Folio were performed, the completion of the project, with a staging of *Troilus and Cressida* in 1923, coinciding with the tercentenary of its first publication. Ben *Greet was in charge of productions from 1915, during which time Sybil *Thorndike played most of Shakespeare's heroines and some of his heroes. In 1927 the LCC insisted on the theatre's closure for urgent and essential alterations. The company moved to the *Lyric Theatre, Hammersmith, and returned in 1928 with a performance of *Romeo and Juliet*. In 1931 Lilian Baylis opened the renovated *Sadler's Wells Theatre, and mixed seasons of opera, ballet, and drama alternated between the two theatres. This proved impracticable, and two years later it was decided to use Sadler's Wells for opera and ballet and the Old Vic for drama. A succession of fine actors, among them John *Gielgud, Laurence *Olivier, and Ralph *Richardson, made the theatre famous.

The Old Vic closed on the outbreak of war in 1939; it reopened a year later, but was badly damaged by bombing in 1941 and closed again, the company moving to the New Theatre (now the *Albery). Repaired and redecorated, it reopened in 1950 with *Twelfth Night*. Between 1953 and 1958 a 'five-year' plan resulted in a second presentation of the 37 plays in the First Folio, opening with Richard *Burton in *Hamlet* and ending with Gielgud, Edith *Evans, and Harry *Andrews as Wolsey, Katharine, and the King in *Henry VIII*. The Old Vic had from early times presented some plays by dramatists other than Shakespeare, and two interesting works produced in 1962 were *Ibsen's *Peer Gynt* and *Guthrie's modernized version of *Jonson's *The Alchemist*. Later that year it was decided that the theatre should house the *National Theatre's company under Laurence Olivier pending the erection of their own theatre, and in June 1963, after a final performance of *Measure for Measure*, the theatre closed. It reopened on 22 Oct. (after extensive alterations) with the National Theatre company in *Hamlet*, starring Peter *O'Toole, the other actors being mainly from the *Chichester company, which was also under the direction of Olivier. G. B. *Shaw's *Saint Joan* and *Chekhov's *Uncle Vanya*, both transferred from Chichester, followed, the latter proving one of the company's most successful achievements in ensemble acting. The first notable revival was *Farquhar's *The Recruiting Officer*, and the first foreign play Max *Frisch's *Andorra*. Shakespeare's Quatercentenary was celebrated in 1964 by Olivier in *Othello*, possibly the most controversial *tour de force* of his career. The company's first world première was Peter *Shaffer's *The Royal Hunt of the Sun*. In the seasons that followed the company's greatest successes were with comedy, particularly with *Feydeau's *A Flea in Her Ear* (1965). In 1967 Tom *Stoppard's *Rosencrantz and Guildenstern*

are Dead introduced a new playwright of considerable talent, while the classics were prominent with an interesting all-male *As You Like It* and *Molière's *Tartuffe*, with Gielgud. The abortive attempt by the company's literary manager Kenneth *Tynan to introduce *Hochhuth's controversial play *Soldiers* into the repertory, and the extension in 1970 of the company's activities to West End theatres while the Old Vic was let to visiting companies, attracted some adverse criticism in spite of the excellence of some of the plays. The return to the Old Vic was generally welcomed, *Scofield giving an excellent comic performance in *Zuckmayer's *The Captain of Köpenick* in 1971, and old and new plays being equally successful, with a revival of *O'Neill's *Long Day's Journey into Night*, also in 1971, Stoppard's *Jumpers* in 1972, and Molière's *The Misanthrope* and Shaffer's *Equus* in 1973. The last play to be staged under Olivier's management was Trevor Griffiths's *The Party*, after which Peter *Hall took over, and in 1974 staged a varied repertory in anticipation of the move to the National Theatre, ending with *Pinter's *No Man's Land* (1975). The last production before the move took place on 28 Feb. 1976 was a special charity performance of *Tribute to the Lady*, a celebration of the life and work of Lilian Baylis. A year later the theatre, which was being used by the Old Vic Youth Theatre (which retained its headquarters there until 1982), became the London home of the *Prospect Theatre Company, which changed its name to the Old Vic Company in 1979. In 1981 the company was forced to disband and the theatre was bought by a Canadian entrepreneur and handsomely refurbished, reopening in 1983 with limited runs of its own productions, such as *Racine's *Phèdre* in 1984 with Glenda *Jackson. It staged the *RSC's production of the musical *Kiss Me, Kate* in 1987, and in 1988 Jonathan *Miller was appointed Artistic Director, presenting seasons of European and British classics until 1990. The musical *Carmen Jones*, based on Bizet's *Carmen*, had its London première there in 1991.

Olio, see BURLESQUE, AMERICAN.

Olivier, Laurence Kerr (1907–89), English actor, director, and manager, knighted in 1947, created a life peer in 1970 (as Baron Olivier of Brighton, the first actor to be made a peer), and CH in 1981. He made his first appearance on the stage in 1922, playing Katharina in a schoolboy production of *The Taming of the Shrew*, and his professional début in London in 1924, in a small part in Alice Law's *Byron*. From 1926 to 1928 he was with the *Birmingham Repertory company. He was the first Captain Stanhope in *Sherriff's *Journey's End* when it was tried out in 1928, and in 1929 he made his New York début in Frank Vosper's *Murder *on the Second Floor*. He had already attracted attention in *Coward's *Private Lives* (1930; NY, 1931) and Edna Ferber and George S. *Kaufman's *Theatre Royal* (1934) when he was seen in his first major Shakespearian roles, alternating Romeo and Mercutio with *Gielgud at the New Theatre (now the *Albery) in 1935. During a subsequent season at the *Old Vic, where he played Hamlet in its entirety (being later seen in the part at Elsinore), Henry V, Macbeth, and Sir Toby Belch in *Twelfth Night*, he emerged as an actor of the front rank, returning to the Vic in 1938 as Iago to the Othello of Ralph *Richardson and as Coriolanus. After appearing in New York in S. N. *Behrman's *No Time for Comedy* (1939) and as Romeo in 1940, he spent four years with the Fleet Air Arm, rejoining the Old Vic company in 1944 as a co-director and remaining with it until 1949. He gave some remarkable performances, notably as a superb Richard III (1944) and King Lear (1946) and in a double bill (1945; NY, 1946) in which he appeared in the title-role of *Sophocles' *Oedipus the King* and as Mr Puff in *Sheridan's *The Critic*. In 1950 he was in *Fry's *Venus Observed*, written specially for him, and in 1951, in London and New York, he starred with his second wife Vivien *Leigh in *Antony and Cleopatra* and *Shaw's *Caesar and Cleopatra*, playing Antony and Caesar. He was also with her in *Rattigan's *The Sleeping Prince* (1953), and spent a season at the *Shakespeare Memorial Theatre which produced memorable portrayals of the title-roles in *Titus Andronicus* and *Macbeth* (1955), both opposite his wife. He scored his greatest success in a modern role as the second-rate music-hall comedian Archie Rice in John *Osborne's *The Entertainer* (1957; NY, 1958) opposite Joan *Plowright, later to become his third wife. After Coriolanus at Stratford (1959) he played Berenger in *Ionesco's *Rhinoceros* (1960) in London, and in New York played first the title-role (1960) and then Henry II (1961) in *Anouilh's *Becket*. Later the same year he was appointed the first Director of the *Chichester Festival Theatre. His starring role in David Turner's *Semi-Detached* (1962) was his last in the commercial theatre; a year later he became the first Director of the state-subsidized *National Theatre company. He directed the opening production of *Hamlet* at the Old Vic, and during the first season appeared as Brazen in *Farquhar's *The Recruiting Officer* and Astrov in *Chekhov's *Uncle Vanya* (which he also directed, in a production transferred from Chichester). In 1964 he was a controversial but memorable Othello and a superb Solness in *Ibsen's *The Master Builder*. A year later he handed Chichester over to John *Clements and concentrated all his attention on the National Theatre, where he was active as both actor and director. Among his important roles were Tattle in *Congreve's *Love for Love* in 1965, Edgar in

*Strindberg's *The Dance of Death* in 1967, Shylock in a Victorian production of *The Merchant of Venice* in 1970, and James Tyrone in *O'Neill's *Long Day's Journey into Night* in 1971. In 1973 he was succeeded by Peter *Hall. Blessed with a strong interpretative intelligence, he was commonly regarded as the supreme actor of his generation. He was also an outstanding film actor.

Olivier Theatre, largest of the three theatres housed in the *National Theatre complex in London, and the second to be opened. Named after Laurence *Olivier, the first Director of the National Theatre, it seats 1,165 in two stepped tiers and has an open stage 56 ft. wide and 72 ft. deep at the centre with a revolve 40 ft. in diameter. It is mainly used for plays which require larger casts or more spectacular effects than can be accommodated in the other theatres. The Olivier opened in 1976 with *Marlowe's *Tamburlaine the Great*, Albert *Finney playing the title-role. Other notable productions included *Wycherley's *The Country Wife* (1977) with Finney, *O'Casey's *The Plough and the Stars* (also 1977) with Cyril *Cusack, *Congreve's *The Double Dealer* (1978) with Dorothy *Tutin, and *Ibsen's *The Wild Duck* (1979). Among modern plays were *Bond's *The Woman* (1978), *Shaffer's *Amadeus* (1979) with Paul *Scofield, and Howard *Brenton's controversial *The Romans in Britain* (1980). Highlights of the next decade were *Aeschylus' *Oresteia* (1981), a striking production with masks, the musical *Guys and Dolls* (1982), *Sheridan's *The Rivals* (1983) with Michael *Hordern and Geraldine *McEwan, *Hare and Brenton's *Pravda* (1985), and *Antony and Cleopatra* (1987), both with Anthony *Hopkins.

Olympia Music-Hall, London, see STANDARD THEATRE.

Olympia Theatre, Dublin, one of the oldest surviving theatres in Dublin, situated off Dame Street. A music-hall opened on the site in 1855, which became Dan Lowrey's Star of Erin Music Hall in 1879. All the famous English performers of the time were seen here, and it was enormously successful. In 1897 it was completely rebuilt, providing substantially the building that survives today. It was known as the Empire Palace of Varieties until 1922, when it was renamed the Olympia. For many years legitimate drama was part of the repertoire, especially after 1940, when the Second World War prevented the engagement of visiting performers and Irish plays were performed regularly. From 1951 to 1964 the theatre had its own company, but also saw the return of English touring companies. In 1964 it was sold, but was leased by a group of theatre managers dedicated to preserving it, and up to 1974 successfully blended traditional variety with legitimate theatre, often in association with the *Dublin Theatre Festival.

In 1974 a serious structural collapse closed the theatre, but after generous financial support from both public and private sources it reopened in 1977 with Willy Russell's musical about the Beatles, *John, Paul, George, Ringo . . . and Bert*. Brian *Friel's *Translations* had a notable production there in 1980. The theatre continues to present a mixture of musicals, variety, and straight drama.

Olympic Theatre, London, in Wych Street, Strand. This was erected by Philip Astley while *Astley's Amphitheatre was being rebuilt in 1803, and was constructed mostly of timber from a French warship, with a little brickwork and a tin roof, in the shape of a tent. It opened as the Olympic Pavilion, housing circus acts and performances of horsemanship in an arena. It was not a success, and in 1813 was bought by R. W. *Elliston, who changed its name to the Little Drury Lane Theatre. This was objected to by *Drury Lane as infringing its patent, and Elliston's licence was withdrawn. However, by the end of the year he had obtained a new licence for *burletta only, and he reopened the theatre as the Olympic. It did well for the next five years with a mixed programme of pantomime, ballet, farce, and melodrama, and was then reconstructed and reopened with an excellent company which attracted a fashionable audience and made so much profit that Elliston was able to purchase the patent of Drury Lane on the proceeds. He leased the Olympic to a series of lessees, the majority of whom went bankrupt as did Elliston himself. The theatre was put up for sale and bought by the owner of the *Adelphi, who used it for melodramas until at the end of 1830 Mme *Vestris leased it from him and opened in 1831 with a programme which included *Planché's *Olympic Revels*. Her policy of low prices and beautifully staged light entertainment made the Olympic a success. In 1835 Charles J. *Mathews made his first appearance there and three years later he and Mme Vestris were married, leaving the Olympic in 1839 to go to *Covent Garden. The theatre then led a precarious life until it was burned down in 1849. Rebuilt to seat 1,750, it reopened, but had to close hurriedly when the manager was arrested on charges of defalcation and forgery. William *Farren then took over, with Frederick *Robson as his star and later co-manager. It was under him that Tom *Taylor's *The Ticket-of-Leave Man* had a successful run in 1863. Robson died prematurely the following year, and his successor introduced a series of new plays with Henry Neville and Kate *Terry. Neville himself became manager for six years, during which Wilkie Collins's *The Moonstone* (1877) was first produced. Charles *Wyndham appeared in 1880 in Bronson *Howard's *Brighton* with great success, and in 1883 Geneviève *Ward was seen in her own productions. A

succession of managements came and went until the theatre closed in 1889, and after reconstruction opened under Wilson *Barrett two years later. It was never again successful, and closed for the last time in 1897, being demolished in 1904.

Olympic Theatre, New York. **1.** At 444 (later 442) Broadway, between Howard and Grand Streets. This handsome theatre opened in 1837 with a mixed bill, but was not a success, and passed through many hands before in 1839 **William Mitchell** (1798–1856) opened it as a home of light entertainment. As Mitchell's Olympic it flourished for over 10 years with burletta, burlesque, and extravaganza, and survived even the depression of 1842–3 which proved fatal to many other enterprises. It was the first theatre in New York to play a weekly matinée. Among its outstanding successes were *Hamlet Travestie* by John Poole, and burlesques of *Richard III*, of *Boucicault's *London Assurance*, and of the opera *The Bohemian Girl* as *The Bohea-Man's Girl*. The season of 1847–8 saw *Planché's *The Pride of the Market*, and Frank *Chanfrau as Mose, his famous fireman character, in Benjamin *Baker's *A Glance at New York in 1848*. A year later the theatre was redecorated; but Mitchell was failing in health and took the easy way of importing foreign stars. This proved disastrous, and in 1850 the Olympic closed abruptly. After a short spell under William *Burton it housed plays in German and finally closed in 1851. It was burnt down in 1854.

2. At 622–4 Broadway, above Houston Street. This opened in 1856 with Laura *Keene as Rosalind in *As You Like It*; as Laura Keene's Varieties it was the first American theatre to have a woman manager. It flourished until 1863 when Laura Keene left, and reopened on 8 Oct. of that year as the Olympic under Mrs John *Wood, who ran it until 1866. Among later productions were *A Midsummer Night's Dream* and *Fox's famous pantomime *Humpty-Dumpty*, which opened in 1868 and ran for 483 performances. In 1872 the theatre became a home of *variety. It finally closed in 1880 and was demolished, shops being built on the site.

3. The New Olympic Theatre, at 585 Broadway, was opened in 1856 by Chanfrau who intended to revive the mixed bills of Mitchell's Olympic. He was unsuccessful, and after a few weeks the theatre was taken over by a *minstrel show, Buckley's Serenaders, and became Buckley's Olympic. A later manager was Tony *Pastor, from 1875 to 1881.

Ombres Chinoises, see SHADOW-SHOW.

O'Neill, Eliza (1791–1872), Irish actress, who made her first appearance on the stage in her birthplace, Drogheda, where her father was manager of the local theatre. Going to Belfast and Dublin, she soon made a name for herself and in 1814 was engaged for *Covent Garden. Her first appearance as Juliet in *Romeo and Juliet* was overwhelmingly successful, as was her Lady Teazle in *Sheridan's *The School for Scandal*, and for five years she had a career of unbroken triumph, being considered a worthy successor to Mrs *Siddons, with less nobility, perhaps, but more sweetness and charm. In 1819 she made her last appearance on the stage as Mrs Haller in *Kotzebue's *The Stranger*, and then retired on her marriage.

O'Neill, Eugene Gladstone (1888–1953), American playwright, born in New York City, son of the actor James *O'Neill. His education was fragmentary, including a year at Princeton, after which he signed on as a seaman on several voyages. He was working as a reporter on a newspaper in New London, Conn., when his health broke down, and during six months spent in a sanatorium he began writing his first play, *The Web*. In 1914–15 he studied under George Pierce *Baker at *Harvard, and in 1916 became connected with the *Provincetown Players who, with the Greenwich Village Theatre, first presented many of his early plays. His first full-length play *Beyond the Horizon* (1920), produced on Broadway, was a starkly effective study of character set in rural New England. Awarded a *Pulitzer Prize, it established him as a playwright of genuine talent and considerable skill, and was followed almost immediately by productions of the one-act *Exorcism*; *Diff'rent*, a grim bit of dramatic irony in two acts; and *The Emperor Jones* (seen in London in 1925), which uses a powerful Negro as its central figure to represent the violent urges of human nature: Paul *Robeson gave an electrifying performance in London in the title-role. *Anna Christie* (1921; London, 1923) tells the story of a prostitute who is, presumably, 'purified' by the love of a man. In quick succession other O'Neill plays were brought to the stage—*Gold* (1921), *The Straw* (also 1921), drawing on his experiences in a sanatorium and an expression of the indomitable human spirit, and *The First Man* (1922)—each a failure with the public, yet each revealing new aspects of the author's preoccupations. *The Hairy Ape* (1922) stemmed, according to the author, from *The Emperor Jones* rather than from the work of the European *Expressionists, which it in many ways resembles. In 1924 three new plays were produced—*Welded*, *All God's Chillun Got Wings*, and *Desire under the Elms*. The first, a compact and rather bloodless study in marriage, was a quick failure; the second (seen in London in 1926), dealing with racial intermarriage, verged on the sentimental, but was threatened with demonstrations by racialist factions; the last (seen in London in 1931) showed a new maturity, using a powerful tale of sexual passion,

incest, and infanticide to comment on contemporary American society. *The Fountain* (1925) was short-lived. *The Great God Brown* (1926) remains one of the most tortuous and complicated of O'Neill's plays. Making use of elaborate masks, it studies the conflict between man's material and spiritual needs. *Marco Millions* (1928; London, 1938) was more serene, pleasantly ironic, and full of comedy and romantic colour, even though it too is a bitter satire on the aggressive business man who has lost touch with beauty and the eternal verities. *Strange Interlude* (1928; London, 1931), a play in nine acts, is a work of extraordinary power which, with the copious use of asides and soliloquies, seeks to expose the motives of human character. In this play and in others that followed, O'Neill seemed to be clarifying his ideas on the soul of man and the destiny of the human race to the detriment of his art as a playwright. *Lazarus Laughed* (1928), first produced by the *Theatre Guild, tells the story of the resurrection of Lazarus and his ultimate triumph over death. In 1929 came *Dynamo*, planned as the first part of an uncompleted trilogy on man's efforts to find a lasting faith. It was unsuccessful, but is remembered for its exciting set design by Lee *Simonson. Another trilogy, *Mourning Becomes Electra* (1931; London, 1937), is in many respects O'Neill's most successful work. It transposes the events of *Aeschylus' *Oresteia* to a Puritanical family in New England, replacing the old acceptance of fate with a modern doctrine of psychological causation. A nostalgic comedy, *Ah, Wilderness!* (1933), and a somewhat barren and over-intellectualized play about faith, *Days without End* (1934; London, 1943), followed. Then for 12 years O'Neill retired from the theatre; he did an enormous amount of writing, including several plays that are parts of a series of nine interrelated plays, but refused to allow any of them to be staged. But in 1946 *The Iceman Cometh* (London, 1958), a vast play about a group of bar-room derelicts and their pipe-dreams, enjoyed a long run in New York and aroused a great deal of critical comment. Like *A Moon for the Misbegotten* (1947; London, 1960), it is partly expository drama and partly a disquisition on faith.

Some of O'Neill's plays were produced posthumously. *Long Day's Journey into Night*, written in 1941, was first seen in Stockholm in 1956; it was produced in New York the same year and in London in 1958. A largely autobiographical study, set during a single day, of a miserly actor father, his morphine-addicted wife, and their two sons, it has been several times revived, notably at the *National Theatre in 1971, with *Olivier. *A Touch of the Poet*, written in 1940, was produced in Stockholm in 1957 and New York in 1958 (London, 1988); it examines the painful conflict of an immigrant father and his American-born son. *More Stately Mansions*, written in 1938, was produced in Stockholm in 1962, in a translation which cut the playing time from 10 hours to five; the play was seen in New York in 1967 and at *Greenwich in 1974.

O'Neill was awarded the *Nobel Prize for Literature in 1936, and in 1959 the Coronet Theatre in New York was renamed in his honour. He was an introspective and troubled man, continually dogged by illness; all his talent and energy went into his work for the theatre, through which he attempted to examine the soul of modern man. Though his language is often clumsy, and some of his plays become melodramatic and absurd, he is a major dramatist whose best plays, offering strong parts for actors, continue to be revived in many countries.

The Eugene O'Neill Theatre Center, Waterford, Conn., founded in 1963, provides premises and resources for a variety of programmes and organizations, including the annual National Playwrights' Conference and the National Theatre of the Deaf.

O'Neill, James (1847–1920), Irish-born American actor, once considered the rival and successor of Edwin *Booth. Taken to America as a small child, he made his stage début in Cincinnati in 1867. In 1871 he joined the company of James V. McVicker in Chicago, and five years later was at the *Union Square Theatre in New York. He then went to San Francisco, where among other parts he played Christ in a Passion play directed by David *Belasco at the Grand Opera House. He married in 1875 Ella [Mary Ellen] Quinlan, by whom he had two sons, the younger being the playwright Eugene *O'Neill. James O'Neill was considered a good Shakespearian actor, but romantic swashbuckling roles suited him best, and in 1883 he found the perfect part for himself in Edmond Dantès, the hero of the elder *Dumas's *The Count of Monte Cristo*. This became so popular with the American public that for the next 30 years O'Neill played almost nothing else, being seen in it at least 6,000 times. The resultant frustration told heavily on him and he took to drink, his gradual decline, and that of his long-suffering wife, being tragically depicted in Eugene's autobiographical play *Long Day's Journey into Night*.

O'Neill, Maire [Molly Allgood] (1887–1952), Irish actress, sister of Sara *Allgood, and with her a member of the *Abbey Theatre company, where she created the part of Pegeen Mike in *Synge's *The Playboy of the Western World* (1907) and the title-role in *Deirdre of the Sorrows* (1910), written for her by Synge who was at that time her fiancé. She remained at the Abbey until 1911, returning occasionally to play there until 1917. She appeared subsequently at the Liverpool Repertory Theatre (see LIVERPOOL

PLAYHOUSE), and during a long and distinguished career was seen in London and New York, often in Irish plays, and especially those of Sean *O'Casey, with her sister and her second husband Arthur *Sinclair, whom she married in 1926.

Open Air Theatre, London, in Queen Mary's Garden, Regents Park. Founded by Sydney *Carroll, it opened in 1933 with a production of *Twelfth Night* directed by Robert *Atkins, who was closely associated with the theatre. Its policy was to produce mainly the plays of Shakespeare, though with a minority of plays by other writers. The theatre became a permanent fixture of the London summer season, and a marquee was erected for use in wet weather. In 1962 the New Shakespeare Company was founded to mount productions, and in 1975 a new auditorium and theatre complex was built, seating 1,178, with permanent seats replacing the earlier deck-chairs.

Open Stage, raised platform built against one wall of the auditorium, with the audience on three sides. This was sometimes known in England as an 'arena stage', a term reserved in the United States for *theatre-in-the-round. The open (or thrust) stage derives basically from the Elizabethan platform stage, and is in use today in a number of theatres, including the *Chichester Festival Theatre, the *Guthrie Theater, the *Olivier Theatre, the *Stratford (Ontario) Festival theatre, and the *West Yorkshire Playhouse. Most of the new all-purpose theatres make provision for open-stage productions, which call for different techniques in acting, staging, setting, and lighting from those used on proscenium-arch stages.

Open Theater, group established in New York in 1963 by **Joseph Chaikin** (1935–), a former member of the *Living Theatre, with the intention of exploring by means of *Collective Creation the unique powers of live theatre—which, according to Chaikin's book *The Presence of the Actor* (1972), is based upon the encounter between live performers and spectators—and to concentrate on abstraction and illusion in contrast to the realistic and psychological method of *Stanislavsky typical of the modern conventional theatre. The group's first performances were exercises and improvisations developed into short productions by Jean-Claude *Van Itallie and **Megan Terry** [Marguerite Duffy] (1932–), an avant-garde playwright whose other work included *Approaching Simone* (1970), a play about the French philosopher Simone Weil. These productions made frequent use of 'transformation', whereby one object or character becomes another, as where the performers were first parts of an aeroplane as it took off and were then 'transformed' into the people on the plane. The first long play to be developed collectively was *Viet Rock* (1966), a collage of transformations on the war in Vietnam created in a workshop conducted by Megan Terry and produced under the aegis of *La Mama Experimental Theatre Club. Later productions were *The Serpent* (1968), based on the Old Testament and including images from the assassinations of President Kennedy and Martin Luther King; *Terminal* (1969), a study of human mortality and responses to death; and *Mutation Show* (1971) which focused upon human adaptation. The last production created by the group before disbanding was *Nightwalk* (1973), an investigation into levels of sleep. Megan Terry later became playwright in residence at the Magic Theatre in Omaha.

Opera Comique, London, at the junction of the Strand and Holywell Street. This stood back to back with the old *Globe Theatre in Newcastle Street, the pair being known as the Rickety Twins, having been hastily erected in the hope of compensation—which was not forthcoming—when the area was rebuilt. It opened in 1870 with a French company under *Déjazet in *Sardou's *Les Prés St Gervais*. In 1871 actors from the *Comédie-Française under *Got, driven from home by the Franco-Prussian War, made their first appearance outside Paris. The theatre was then used mainly by visiting foreigners—including *Ristori in 1873—with long periods of inactivity, until in 1877 D'Oyly Carte took over and produced *Gilbert and Sullivan's *The Sorcerer* (1877), *HMS Pinafore* (1878), *The Pirates of Penzance* (1880), and *Patience* (1881), the last transferring to the *Savoy Theatre later in the year. In 1884 the Opera Comique closed for redecoration, reopening in 1885, but with little success. For some years it was used for special performances and try-outs, being partly rebuilt in 1895, but at the end of 1899 it closed for good, being demolished in 1902.

Opéra-Comique, Paris, see COMÉDIE-ITALIENNE.

Opera House, Manchester, opened in 1912 as the New Theatre, becoming the New Queen's in 1915 and taking its present name in 1920. Seating 2,000, it staged opera, ballet, and straight plays, but in the 1970s it was felt that the city could support only one large theatre and the choice fell on the *Palace. The Opera House became a bingo hall, but after the success of the refurbished Palace it was purchased as the latter's sister theatre. Also refurbished, it reopened in 1984, once again becoming one of the country's most important large touring theatres, staging a successful run of the musical *Barnum* followed by nine months of *Evita*, the longest provincial run of a musical.

OP Riots, see COVENT GARDEN.

Orange Moll, see MEGGS, MRS.

Orange Street Theatre, see DUBLIN.

Orange Tree Theatre, Richmond, Surrey, see FRINGE THEATRE.

Orchestra, from the Greek word for 'a dancing place'. In ancient Greek theatres the name was given to the circular area where the *chorus performed its songs and dances, as opposed to the smaller raised stage or *logeion* used by individual actors. In Roman theatres the orchestra was a semicircular space in front of the stage, with seating reserved for senators and distinguished visitors. Towards the end of the 17th century the term was revived to describe the area in front of the stage where the musicians sat, and was soon extended to the players themselves. Orchestra stalls, in a proscenium theatre, are the front rows of seats on a level with the orchestra pit.

Oregon Shakespeare Festival, see SHAKESPEARE FESTIVALS.

Orrery, Lord [Roger Boyle, first Earl of Orrery] (1621–79), Restoration nobleman and man of letters, author of some forgotten comedies and of *Mustapha* (1665) and *The Black Prince* (1667), which *Dryden, generously though perhaps not accurately, hailed as the first examples of *heroic drama. The printed versions of these and Orrery's other plays are useful on account of their detailed stage directions.

Orton, Joe (1933–67), English dramatist, whose first play, *Entertaining Mr Sloane* (1964; NY, 1965), shocked and amused its audience by the contrast between its prim-and-proper dialogue and the violence and outrageousness of its action. The same contrast was apparent in *Loot* (1966; NY, 1968), a satire on police corruption and the conventions of detective fiction. In 1967 an earlier play *The Ruffian on the Stair*, written for radio, was staged at the *Royal Court Theatre with *The Erpingham Camp*, previously televised, in a double bill, *Crimes of Passion* (NY, 1969). Orton was brutally murdered by his homosexual partner who then committed suicide—an episode as violent and bizarre as any in his plays. His last full-length work, *What the Butler Saw*, a parody of conventional farce, was produced in 1969 (NY, 1970), with Ralph *Richardson in the lead in London, and had a mixed reception. A one-act play, *Funeral Games*, was produced at the end of 1970.

Osborne, John James (1929–), English dramatist who was for some years an actor, making his London début in 1956. Although he had already had two plays produced in the provinces, he first came into prominence as a playwright when *Look Back in Anger* was produced at the *Royal Court Theatre. It was the *English Stage Company's first outstanding success, and the date of the first night, 8 May 1956, is considered a landmark in the modern theatre (see KITCHEN-SINK DRAMA); it was seen in New York in 1957. With its 'angry young man' hero, rude, eloquent, and working-class, it marked a radical departure from the traditional West End play. It did not, however, transfer to the West End, and its influence on the commercial theatre now seems negligible. Osborne's next play, also seen at the Royal Court where virtually all his plays were produced, was *The Entertainer* (1957; NY, 1958), in which Laurence *Olivier gave an outstanding performance as the seedy music-hall artist Archie Rice. *Epitaph for George Dillon* (London and NY, 1958), written in collaboration, and a satirical musical about a gossip columnist, *The World of Paul Slickey* (1959), were less successful, but *Luther* (1961; NY, 1963), with Albert *Finney in the name part, was another major achievement. After a double bill at the Royal Court, *The Blood of the Bambergs* and *Under Plain Cover* (billed as *Plays for England*, 1962), came *Inadmissible Evidence* (1964; NY, 1965), one of his best plays, about a solicitor under stress. A year later *A Patriot for Me* was refused a licence by the *Lord Chamberlain because it dealt with homosexuality, and it was therefore staged privately for members of the English Stage Society; it was produced in New York in 1969. In 1966 Osborne adapted Lope de *Vega's *La fianza satisfecha* as *A Bond Honoured* for the *National Theatre. Two new plays in 1968, *Time Present*—another tirade play, this time featuring an angry young actress—and *The Hotel in Amsterdam*, failed to enhance his reputation. They were followed by the nostalgic and seemingly conservative *West of Suez* (1971) and by *A Sense of Detachment* (1972), which has no plot and depends on the reaction between the actors on the stage and two actors pretending to be members of the audience. Osborne's adaptation of *Ibsen's *Hedda Gabler* was also produced in 1972. *The End of Me Old Cigar* (*Greenwich, 1975) and *Watch It Come Down* (NT, 1976) had poor receptions, and he wrote no more for the stage until *Déjà Vu* (1991), which shows the anti-hero of *Look Back in Anger* 35 years later.

Osborne's second wife was the actress **Mary Ure** (1933–75), who played the feminine lead in *Look Back in Anger* in both London and New York. She was also seen in Arthur *Miller's *The Crucible* and *A View from the Bridge* (both 1956), at the *Shakespeare Memorial Theatre in 1959, and in New York in *Giraudoux's *Duel of Angels* (1960) and *Pinter's *Old Times* (1971).

Jill Bennett (1931–90), his fourth wife, appeared in several of his plays, notably in *Time Present* and his adaptation of *Hedda Gabler*. Other roles were in *Anouilh's *Dinner with the Family* (1957) and as Gertrude in the Royal Court production of *Hamlet* in 1980.

Ostrovsky, Alexander Nikolayevich (1823–86), Russian dramatist, many of whose numerous plays remain in the Soviet repertory. He first attracted attention with *The Bankrupt* (1849), a study of corruption in the Moscow merchant class which cost him his job as a civil servant and condemned him to a life of constant struggle and near poverty. Banned from the stage for 13 years, it circulated freely in manuscript and was eventually staged as *It's All in the Family*. It was followed by a number of historical plays, a fairy-tale play, *The Snow Maiden* (1873), which provided the basis for an opera by Rimsky-Korsakov, and the series of realistic contemporary satires for which Ostrovsky is best known. He is difficult to translate into English, owing to the richness of his language and his use of local colour, but three of his satires—*Even a Wise Man Stumbles* (1868), also known as *Enough Stupidity in Every Wise Man*; *Easy Money* (1870), based on *The Taming of the Shrew*; and *Wolves and Sheep* (1875)—were published in 1944 in translation. Another satire, *The Forest* (1871), about two strolling players in provincial Russia, was given its British première by the *RSC in 1981, with Alan *Howard. Of the plays dealing with the position of women in Russian society which he wrote in his later years the best is usually acknowledged to be *The Poor* (or *Dowerless*) *Bride* (1879). Outside Russia his best-known play is a domestic tragedy, *The Storm* (1860). It was first seen in translation in New York in 1900, and in London in 1929, and provided the plot of Janáček's opera *Kátya Kabanová*. Most of Ostrovsky's plays were first produced at the *Maly Theatre, Moscow, sometimes known as the House of Ostrovsky, where he found a friend and ideal interpreter in the actor Prov *Sadovsky. He was deeply concerned with the position of Russian actors, and served as President of the Society of Russian Dramatists and Operatic Composers from its foundation in 1870 until his death.

Ostushev, Alexander, see MALY THEATRE, Moscow.

Other Place, The, the *RSC's studio theatre in Stratford-upon-Avon, which opened in 1974. A corrugated 'tin hut' converted from a rehearsal space, it seated 140 and presented plays in *repertory. Its programmes, which consisted of both new plays and classic works, included some notable small-scale productions of Shakespeare, among them *Macbeth* in 1976 (directed by Trevor *Nunn), *Pericles* in 1979, *Timon of Athens* in 1980, and *Othello* (Trevor Nunn), in 1989. Revivals of rarely performed plays included *Ford's *Perkin Warbeck* in 1975 and *'Tis Pity She's a Whore* in 1977; *Beaumont and *Fletcher's *A Maid's Tragedy* in 1980; and *Farquhar's *The Twin Rivals* and Ford, *Dekker, and

*Rowley's *The Witch of Edmonton* in 1981. (Productions of this type are now staged at The *Swan.) New plays such as David Edgar's *Destiny* (1976), Pam Gems's *Queen Christina* (1977) and her musical *Piaf* (1978), Christopher *Hampton's *Les Liaisons dangereuses* (1984), and Doug Lucie's *Fashion* (1987) were premièred there. The theatre closed in 1989 and a new brick-built Other Place, seating 289, opened on a different site in 1991.

Other Stage, New York, see PUBLIC THEATRE.

O'Toole, Peter (1932–), Irish-born actor, who began his career with the *Bristol Old Vic, 1955–8, and made his first appearance in London in 1956 with the company in *Shaw's *Major Barbara*. He built up a formidable reputation in Bristol, particularly with his Jimmy Porter in *Osborne's *Look Back in Anger* in 1957. He made his first outstanding success in London as the earthy, cynical trouble-maker Private Bamforth in Willis *Hall's *The Long and the Short and the Tall* (1959). In 1960 he was with the *Shakespeare Memorial Theatre company, playing Petruchio in *The Taming of the Shrew* and giving an electrifying performance as Shylock in *The Merchant of Venice*. He then moved into films and made only intermittent returns to the stage in the title-role in *Brecht's *Baal*, as Hamlet in the *National Theatre company's inaugural production (both in 1963), and as the Yorkshire novelist-hero of David *Mercer's *Ride a Cock Horse* (1965). A year later he was in Dublin to play Jack Boyle in *O'Casey's *Juno and the Paycock*, and in 1969 at the *Abbey Theatre there as Vladimir in *Beckett's *Waiting for Godot*. He returned to the Bristol Old Vic in 1973 in the title-role of *Chekhov's *Uncle Vanya*, and in 1980 he returned to the London stage to give a controversial performance as Macbeth for the Old Vic Company (see PROSPECT THEATRE COMPANY). He later appeared in the West End in several plays by Shaw: *Man and Superman* in 1982, *Pygmalion* (as Professor Higgins) in 1984 (NY, 1987), and *The Apple Cart* in 1986. In 1989 he gave a virtuoso display as a well-known contemporary columnist in Keith Waterhouse's *Jeffrey Bernard is Unwell*.

Otway, Thomas (1652–85), English dramatist, who in 1670 made his first and last appearance on the stage in Aphra *Behn's *The Forced Marriage*. He then turned to playwriting, and in 1675 produced *Alcibiades*, a successful tragedy which provided a fine part for Mrs *Barry, with whom he was madly in love. It was followed by *Don Carlos* (1676), a tragedy in rhymed verse, a comedy entitled *Friendship in Fashion* (1678), and by his two finest works, *The Orphan; or, The Unhappy Marriage* (1680) and *Venice Preserv'd; or, A Plot Discovered* (1682), tragedies in which *Betterton and Mrs Barry appeared together. Both plays were frequently revived during the 18th century. *The Orphan* was last seen in London in 1925, and *Venice Preserv'd*, which continued in favour during the 19th century, was given a splendid production in 1953 with *Gielgud and *Scofield. One of Otway's most successful plays in his own day was his translation of *Molière's *Les Fourberies de Scapin*. As *The Cheats of Scapin*, this was produced at *Dorset Garden as an *after-piece to his *Titus and Berenice* (1676) (based on *Racine). Other works were a comedy, *The Soldier's Fortune*, first seen at *Drury Lane in 1680, and its sequel, *The Atheist; or, The Second Part of the Soldier's Fortune* (1683). His best work shows great depth and sincerity.

OUDS, see OXFORD.

Ouspenskaya, Maria, see AMERICAN LABORATORY THEATRE.

Owens, John Edmond (1823–86), London-born American actor, best known as an eccentric comedian and an outstanding interpreter of Yankee characters. His most famous part was that of Solon Shingle in Joseph S. Jones's *The People's Lawyer*, in which he toured successfully throughout the United States and also appeared in London in 1865, where he was seen by *Dickens. He was also good as Dr Pangloss in the younger *Colman's *The Heir-at-Law*, as Caleb Plummer in *Dot*, *Boucicault's dramatization of Dickens's *The Cricket on the Hearth*, in the title-role of John Poole's *Paul Pry*, and as Mr Toodles in *The Toodles*, adapted by W. E. *Burton from R. J. Raymond's *The Farmer's Daughter of the Severnside*. In 1876 he appeared in New York as Perkyn Middlewick in H. J. *Byron's *Our Boys*, and later joined the *stock company at *Madison Square, where he played with Annie *Russell. He retired in 1885.

Oxberry, William (1784–1824), English actor, whose portrait hangs in the *Garrick Club. He appeared in provincial theatres and in minor roles in London ones, and is best remembered as the author of *The Actress of All Work* (1819), a farce which depicts one actress in six different roles. Oxberry also published a number of volumes of theatrical interest and gave his name to *Oxberry's Dramatic Biography*, published posthumously by his widow. Oxberry's son **William Henry** (1808–52) was also an actor and from 1843 to 1844 edited *Oxberry's Weekly Budget*, in which many melodramas otherwise unknown were first published. He adapted a number of plays from the French, of which the most successful was *Matteo Falcone; or, The Brigand and his Son* (1836). He was a quaint little man, unsuited to tragedy but good in burlesque.

Oxford, home of England's oldest university, probably had performances of *liturgical drama in medieval times, but the first play to be recorded (at Magdalen) dates from about 1490.

In the 1540s undergraduates at several colleges acted plays in Latin. There seems to have been less acting at Oxford than at *Cambridge, but in 1566 Queen Elizabeth I was present at a production in Christ Church of Edwardes's *Palaemon and Arcyte*, based on Chaucer's *The Knight's Tale*. In 1567 a comedy was produced at Merton with the intriguing title *Wylie Beguylie*. Unfortunately, as with most other plays of the time, the manuscript is lost, as happened also with the texts of the plays given before Charles I in 1636. Up to this point drama in Oxford seems to have been left entirely to amateurs, and professional visits seem to have been discouraged, though *Strange's Men played in an innyard in the city in 1590–1, the *King's Men were seen in a tennis-court in 1680, and a professional company under *Betterton was in Oxford in 1703. In the 18th century there was little beyond private theatricals, and some quasi-official performances at Commemoration.

The foundation of the Oxford University Dramatic Society (OUDS) in 1885 by a group of undergraduates which included Arthur *Bourchier, later an important actor-manager, and Cosmo Gordon Lang, a future Archbishop of Canterbury, and a cousin of the actor-manager Matheson *Lang, again made acting an acceptable extra-curricular activity in the university. The first production, given in the Town Hall, was *Henry IV, Part One*, in which Bourchier played Hotspur and Lang spoke a Prologue. Most of the plays produced in future years were comedies by Shakespeare, given in a college garden or hall.

During the First World War the OUDS was disbanded, but it started up again in 1919 with a production of Hardy's epic poem *The Dynasts*. The society then pursued its former policy of indoor and outdoor productions. One important venture was the staging in 1931 of Flecker's *Hassan*, with Peggy *Ashcroft as Pervaneh, since the company still refused to admit women members. Meanwhile a new society known as Friends of the OUDS had been formed by **Nevill Coghill** (1899–1980), later Merton Professor of English Literature, which filled the gap with 12 productions between 1940 and 1946, the last being *Ibsen's *The Pretenders*. The society was then re-formed, and women were admitted to full membership; but in 1950 it was forced by financial troubles to give up its premises, though it now has a club room and office in the Burton–Taylor Theatre (see OXFORD PLAYHOUSE). Its productions continue to reach a high standard. In 1936 Coghill had also founded a new society, the Oxford Experimental Theatre Club (ETC), which unlike OUDS was designed to leave everything connected with the production in the hands of undergraduates. It also chooses plays not within the scope of the older society, particularly those illustrating the current experimental trends in the theatre, even becoming involved in *community theatre. The university also has a Visiting Chair of Drama. The Apollo Theatre, formerly the New, which opened in 1934, is not a prime venue for touring companies, except for opera.

Oxford Music-Hall, London, at the corner of Oxford Street and Tottenham Court Road. One of London's best-known *music-halls, it opened under the management of Charles *Morton in 1861, having on its *act-drop and on the cover of its programmes a view of Magdalen Tower, Oxford. It had a successful career, both George *Robey and Harry Tate making their London débuts there, and was rebuilt in 1869, 1873, and 1893. In 1917 it was taken over by C. B. *Cochran, who renamed it the Oxford Theatre and presented there an extravaganza, *The Better 'Ole*, by Bruce Bairnsfather, the famous cartoonist and war correspondent, which ran for 811 performances. In 1920 the old hall was fully converted into a theatre, and it opened in 1921 with a spectacular long-running *revue, *The League of Notions*. During 1922 it was used briefly for films, but returned to live theatre with plays starring Sacha *Guitry and Yvonne *Printemps and a visit from Eleonora *Duse. In 1924 the *Old Vic company made its first West End appearance at the Oxford, with productions by Robert *Atkins of *As You Like It*, *The Taming of the Shrew*, *Twelfth Night*, and *Hamlet*. A series of unsuccessful ventures followed, and the theatre closed in 1926. It was demolished in the following year.

Oxford Playhouse, in Beaumont Street, the last theatre to be built in Britain before the Second World War. It replaced an earlier and very inconvenient theatre in the Woodstock Road, known as the Red Barn and formerly a Big Game Museum. It was adapted by J. B. *Fagan, who from 1923 to 1928 intermittently gave seasons there of good British and Continental plays, including *Chekhov and *Ibsen, his company including at various times, in the early stages of their careers, Tyrone *Guthrie, Flora *Robson, and John *Gielgud. The venture did not receive the support it deserved, and after Fagan left the theatre stood empty for two years until it was reopened in 1930 with a company which produced mainly comedies by, among others, Ben *Travers, Noël *Coward, and Oscar *Wilde. In 1938, after a public appeal for funds, the present Playhouse was built, opening with Fagan's comedy *And So to Bed*. It flourished during the war years, but afterwards audiences dwindled and it was not firmly established until 1956, after being closed for some months, with the support of the *Arts Council and financial assistance from Richard *Burton. It reopened with *Giraudoux's *Electra* produced by the Meadow Players, the company remaining in occupation for 17 years and establishing a high reputation, many West End actors appearing with it as guest stars. In 1961 the university bought the remaining lease of the theatre, and two years later it was redecorated and enlarged, the seating capacity being increased to 700. Richard Burton and his then wife Elizabeth Taylor appeared there in 1966 with the *Oxford University Dramatic Society in a production of *Marlowe's *Dr Faustus* to raise money for an extension, later known as the Burton–Taylor Theatre. In all these enterprises a leading part was played by the Oxford don Nevill Coghill, who until his retirement was chairman of the theatre's governing body. Because of financial difficulties the Meadow Players ceased operations in 1973, and were succeeded a year later by Anvil Productions (the Oxford Playhouse Company), which presented an enterprising programme, including several world premières and new translations of foreign plays. The theatre was also used by touring companies and by the main university dramatic societies. Though the company was highly praised and toured widely, its theatre was closed by the university in 1987, the latter withdrawing its annual subsidy. The company, however, renamed the Oxford Stage Company, continued to tour. The theatre reopened in 1991 as a touring venue.

Oxford's Men. The first mention of a company of players under this name occurs as early as 1492, and in 1547 the 'players of the Earl of Oxford' (who were to be disbanded in 1562 on the death of the 16th Earl) caused a scandal by acting in Southwark while a dirge was being sung in St Saviour's for Henry VIII who had recently died. This company played mainly in the provinces. A new company under the patronage of the 17th Earl, Edward de Vere, who was himself something of a playwright, was formed in 1580 in conjunction with the players who had been for a short time under the patronage of the Earl of Warwick, and played at the *Theatre. The actors got into trouble for brawling and were banished to the provinces until 1584, when the boys belonging to the company appeared with other *boy actors at the first *Blackfriars, in two of *Lyly's plays. They were also seen at Court in a lost play, *Agamemnon and Ulysses*, probably written for them by their patron. The company was disbanded in 1602, after some final appearances at the Boar's Head in Middlesex, where they may have been joined by the Earl of Worcester's Men.

Ozerov, Vladislav, see SEMENOVA.

P

Pabst Theater, see MILWAUKEE REPERTORY THEATER.

Page, Geraldine (1924–87), American *Method actress, who made her début in Chicago in 1940 and in New York in 1945. She then appeared for several years with stock companies and did not come into prominence until 1952, when her portrayal of Alma Winemiller in a New York revival of Tennessee *Williams's *Summer and Smoke* was widely acclaimed. In 1954 she scored a big success as Lizzie Curry, the plain spinster, in N. Richard Nash's *The Rainmaker* (London, 1956). She successfully succeeded Margaret *Leighton in the New York production of *Rattigan's double bill *Separate Tables* (1957), and in 1959 created the role of the Princess in Williams's *Sweet Bird of Youth*, giving a virtuoso performance as a fading star. A revival of *O'Neill's *Strange Interlude* in 1963 in which she played Nina Leeds was followed in 1964 by a production of *Chekhov's *Three Sisters* in which she played first Olga and then Masha. She was later seen in such plays as *Shaffer's double bill *White Lies* and *Black Comedy* (1967), *Ayckbourn's *Absurd Person Singular* (1974), and a triple bill of *Strindberg's *Creditors*, *The Stronger*, and *Miss Julie* (1977). In 1980 she starred in Williams's *Clothes for a Summer Hotel*. Much of her career was devoted to non-commercial work, and in the 1980s alone she appeared *Off-Broadway in plays by *Ibsen (as Mrs Alving in *Ghosts*), *Giraudoux, *Maugham, and Robert *Bolt, while in 1985 she was in Sam *Shepard's *A Lie of the Mind*. She had great sensitivity and a wide range, her comic gifts being well shown in her role as the eccentric Madame Arcati in a revival of *Coward's *Blithe Spirit*, during the run of which she died.

Pageant, word which has over the centuries completely changed its meaning, though retaining always its connection with the idea of 'pageantry'. In medieval times in England it was used variously to describe the 'carts', each one manned and decorated by a different guild, which carried the notables of the city in procession on civic occasions; the last vestige of this is the procession on Lord Mayor's Day, dating from the 13th century, which still takes place in London early in November. 'Pageant' was also the word applied to those fixed points within a city's walls at which an entertainment could be given to welcome important visitors. Such places were the market cross with its steps, any raised permanent structure such as a water conduit, and the space above and around the city gates, often augmented by the use of scaffolding. By extension it was then applied to the 'pageant wagon', a wheeled vehicle on which a scene from a religious play could be performed before an audience assembled at a fixed point before being drawn away to perform the same scene at another pre-arranged station, usually a crossroads, a market square, or other open space. These 'pageants' or perambulating stages, which were very popular in England during the period covered by the *liturgical drama, were usually two-storied, the lower story being curtained off to serve as a dressing-room for the actors, though it could on occasion represent Hell.

A further use of the word 'pageant' was to describe the elaborate structures of wood and painted canvas used in the Tudor *masques. These were often ingenious machines, fixed or movable, designed and built at considerable expense for the purposes of a single night's entertainment.

Early in the 20th century there arose a passion for elaborate and partly processional open-air shows celebrating the history or legends of a particular town, and these were referred to as 'pageants'. The first was produced by Louis N. *Parker at Sherborne in 1905, and it was followed by many others. The performers were usually amateurs, townspeople and school-children, directed by a professional who was responsible for the songs, dances, and short interludes of dialogue which made up the whole, including sometimes re-enactments of medieval joustings and tournaments. The performances usually took place in a large park, or in the countryside beyond the city boundaries, and it was usual for the properties and costumes to be made by the local population.

From the countryside the pageant soon spread to the theatre, and during the First World War several patriotic pageants were produced in London. In 1918 *Drury Lane celebrated its own history in *The Pageant of Drury Lane*.

Paines Plough, see FRINGE THEATRE.

Palace Theatre, London, in Cambridge Circus, Shaftesbury Avenue. This theatre, with its four-tier auditorium seating 1,697 (reduced in 1908 to 1,462), opened in 1891 as the Royal English Opera House. A year later, as the Palace Theatre of Varieties, it became a successful *music-hall under Charles *Morton. It took its present name in 1911, in which year the first Royal Command Variety Show was seen there, as was *Grock on his first visit to London. It continued to present variety bills and *revues, the first being *The Passing Show* (1914). In 1925 the musical *No, No, Nanette!* began its long run, followed by a number of other successful musicals, including *The Girl Friend* (1927) and *Frederica* (1930). After some straight plays in the 1930s the theatre reverted to revue with a series of *Cochran shows, and after appearing in *Under Your Hat* (1938) Cicely *Courtneidge and Jack Hulbert remained in the theatre for *Full Swing* (1942) and *Something in the Air* (1943). Among later successes were *Gay Rosalinda* (1945) and *Song of Norway* (1946), and in 1949 the long run of *Novello's *King's Rhapsody* began. During the 1950s the Palace was used by a number of foreign companies brought to London on the initiative of Peter *Daubeny; Laurence *Olivier was also seen there in John *Osborne's *The Entertainer* (1957). In 1960 Rodgers and *Hammerstein's musical *Flower Drum Song* opened, to be followed in 1961 by their *The Sound of Music*, which ran for six years. In 1968 Judi *Dench made a great success in *Cabaret*, and in 1970 Danny La Rue began a long season in drag. He was followed by the rock musical *Jesus Christ Superstar*, which opened in 1972 and ran for eight years. There were revivals of Rodgers and Hammerstein's *Oklahoma!* in 1980 and of Rodgers and *Hart's *On Your Toes* in 1984. The musical *Les Misérables* began a long run in 1986.

Palace Theatre, Manchester, seating 2,000, with a magnificent red and gold interior. Built in 1891 for touring companies, it quickly became the most famous theatre in the English provinces, visited in its early years by many famous stars, particularly of music-hall, variety, and musical comedy. Between 1978 and 1981 it was closed for major refurbishment, the massive extension of the stage and the technical refitting equipping it to take all major touring companies, including the Royal Opera. Since reopening with the rock musical *Jesus Christ Superstar* it has been highly successful, especially with opera, ballet, and pantomime.

Palace Theatre, New York, on Broadway, between 46th and 47th Streets. This epitome of the big American *vaudeville theatre was built by the impresario **Martin Beck** (1865–1940), who later built the *Martin Beck Theatre. Seating 1,358, it opened in 1913, but at first, partly because of its uptown location, it was not very successful. A visit by Sarah *Bernhardt brought it into the limelight, and it was acquired by a syndicate. The outstanding stars of vaudeville made their appearance at the Palace and it became the mecca for entertainers from all over the world; but with vaudeville's decline and the end of 'two-a-day' performances in 1932 the

theatre had to fight for survival. There were combination bills of films and vaudeville, appearances by such stars as Judy Garland and Danny Kaye, and even 'four-a-day' programmes, but in the end the building became a cinema. It took on a new lease of life, however, after its acquisition in 1965 by the Nederlanders, when it became the home of big musicals and also provided offices for many theatrical organizations. Famous musicals staged there have been *Sweet Charity* (1966), *George M!* (1968), *Applause* (1970), and *Lorelei* (1974). After a break in 1974 when the *RSC appeared in a revival of *Boucicault's *London Assurance* there were more musicals: a revival of *Oklahoma!* in 1979, *Woman of the Year* (1981), and *La Cage aux folles* (1983).

Palace Theatre, Watford, Hertfordshire. Built as a music-hall in 1908, this theatre began to stage plays during the First World War, becoming a repertory theatre in 1932. In 1964 the local council took it over and established it as a *civic theatre run by a Trust. Attractive and intimate, it seats 467 plus 200 in the gallery. There is no resident company, but well-known guest stars often appear. The theatre acquired a reputation for enterprising and ambitious programmes, giving the British premières of Tennessee *Williams's *Sweet Bird of Youth* (1968) and Lillian *Hellman's *The Autumn Garden* (1979). Other productions include *Pinter's *The Homecoming* (1969), directed by and starring the author, and the British première of Simon *Gray's *Molly* (1977), later seen in the West End. Dario *Fo's *Trumpets and Raspberries* (1984) and Gray's revised version of *The Common Pursuit* (1988) also had their British premières here, and there was a notable revival of Bernstein's musical *Wonderful Town* in 1986. All three later had good West End runs.

Palais-Royal, Théâtre du, Paris, originally a small private playhouse in the home of Cardinal *Richelieu, which was rebuilt at great expense in the last years of his life, with superb interior decorations and all the newest stage machinery. It held about 600, and was formally inaugurated in 1641 with a spectacular performance of Desmarets's *Mirame* in the presence of Louis XIII and his Court. After Richelieu's death in the following year the theatre became the property of the King, and was used intermittently for Court entertainments until 1660, when it was given to *Molière in place of the demolished *Petit-Bourbon. It remained in use until 1670, when it was rebuilt and enlarged and equipped with the new machinery necessary for productions of opera and spectacular musical plays. It reopened with Molière's *machine play *Psyché* (1671) and Molière played there in *Le Malade imaginaire* on the night of his death, 17 Feb. 1673. Lully, who held the monopoly of music in Paris, immediately claimed the theatre to house his new Academy of Music, and it was called by that name until it was burnt down in 1763. Rebuilt, it was again destroyed by fire in 1781. In 1784 the whole area occupied by the Palais-Royal was reconstituted by its owner, the Duc de Chartres (later Philippe-Égalité), cousin of Louis XVI, and several theatres were built there, most of which at some time called themselves the Palais-Royal. One, which opened in 1790 as the Variétés-Amusantes, was later occupied by *Talma and his pro-Revolutionary comrades as the *Théâtre de la République. It later became the *Comédie-Française. Another Palais-Royal, which opened in 1831 after having had a variety of other names, specialized in *vaudeville and *farce, and among other plays witnessed the first nights of *Labiche's famous comedy Un chapeau de paille d'Italie (1851) and of many farces. In England towards the end of the 19th century the term 'Palais-Royal farce' was used for such broadly suggestive adaptations from the French as The Pink Dominoes (1877) and *Feydeau's The Girl from Maxim's (1902). The fortunes of the theatre declined somewhat during the 20th century, until in 1958 it was restored to house *Barrault's company on its return from an extended tour abroad. It is one of the largest and best preserved 19th-century playhouses in Paris, and is now used for the presentation of light comedies and *boulevard plays.

Palitzsch, Peter (1918–), German director, who in the early 1950s was with the *Berliner Ensemble, where he co-directed *Synge's The Playboy of the Western World in 1956, *Vishnevsky's The Optimistic Tragedy in 1958, and the first production of *Brecht's Der aufhaltsame Aufstieg des Arturo Ui (1959). In 1962 he moved to the Federal Republic, where he again directed many of Brecht's plays, as well as Walser's Überlebensgroß Herr Krott (1963) and Der schwarze Schwann (1964), Dorst's Toller (1968), and a production based on *Barton and *Hall's Shakespeare cycle The Wars of the Roses (1967). He was Artistic Director at Stuttgart, 1967–72, and Frankfurt, 1972–80.

Palladio, Andrea (1518–80), Italian architect. Born Andrea di Pietro, the name by which he is now known (taken from Pallas Athene) was bestowed on him by his patron in recognition of his genius. It has in turn given its name to the Palladian style of architecture, based on the principles of classical antiquity as Palladio, under the influence of *Vitruvius, interpreted them in his buildings. Inigo *Jones, a pupil and admirer of Palladio, brought his ideas to the notice of English architects, influencing both theatre and public buildings throughout the land. One of Palladio's greatest though least fruitful achievements was the *Teatro Olimpico, which was finished after his death by his pupil **Vincenzo Scamozzi** (1552–1616).

Palladium, see LONDON PALLADIUM.

Pallenberg, Max (1877–1934), Austrian actor, who worked for some time with *Reinhardt. A versatile and subtle comedian much given to extemporization, he was excellent in broad comedy but could also play tragi-comic and serious parts, as was proved by his performance as the Barker in *Molnár's Liliom, and as Schweik in *Piscator's production of Die Abenteuer des braven Soldaten Schweik (based on a novel by the Czech author Hašek) in 1928.

Palliata, see FABULA 2: PALLIATA.

Palmer, Albert Marshman (1838–1905), American theatre manager, who controlled a number of New York theatres, among them *Wallack's 1, to which he gave his own name. Well educated, and a man of refined tastes, he sought to rival *Daly and *Wallack, and built up in each of his theatres a good company. He did not, like Daly, create stars, but chose his actors wisely and set many on the road to fame. In his early years he imported plays from Europe, but later he encouraged new American playwrights, being responsible, among other productions, for the premières of Clyde *Fitch's Beau Brummell (1890) and Augustus *Thomas's Alabama (1891). He retired in 1896 to become manager for Richard *Mansfield. He did a great deal for the American stage, his influence on staging, backstage conditions, and the fostering of native talent being extremely beneficial.

Palsgrave's Men, see ADMIRAL'S MEN.

Pantalone, one of the stock characters, or 'masks', of the *commedia dell'arte, from which is derived the Pantaloon of the English *harlequinade. In both capacities he retained his place in the scenario as the father or guardian of the young girl, opposing her marriage to the young lover because he wished (as her father) to marry her to one of his friends, or (as her guardian) to marry her himself. Invariably represented as an elderly Venetian, he wore a long black coat over a red suit, with Turkish slippers, a long pointed beard, and a skullcap. By turns avaricious, suspicious, amorous, and gullible, he was the butt of the *zanni, who finally outwitted him and brought the young couple together. The character came into the English theatre through the visits of Italian comedians to London, and in Elizabethan times 'pantaloon' became a generic term for any old man, as in Shakespeare's reference to the sixth age of man (As You Like It, II. vii): 'the lean and slipper'd pantaloon With spectacles on nose and pouch on side'; but with the development of Weaver's 'Italian Night Scenes' the name was again limited to a specific role. In the harlequinade he was the father or guardian of *Columbine, who, with the help of *Clown, finally succeeded in eloping with *Harlequin. English Pantaloons soon discarded the long coat and the skullcap, and James Barnes, one of the earliest and most famous Pantaloons of the early 19th century, played the part in short striped knee-breeches, a matching jacket with a short cape, and a fringe of beard with a stiff pigtail sticking up behind.

Pantomime, word which has drastically changed its meaning over the years. It is derived from the Latin *pantomimus*, 'player of many parts', but a misunderstanding as to the art of the Roman player of pantomime led to its adoption as the description of a story told in dancing only. A confusion in the public mind between such ballets and the story-telling dances of the *harlequinade led to the adoption of the term 'pantomime' for this offshoot of the *commedia dell'arte*, which became so popular in England when performed by John *Rich and others that it was eventually lengthened to provide a whole evening's entertainment. To lessen the burden on the dancers, the harlequinade was preceded at first by a classical fable which gave some Immortal the opportunity of handing *Harlequin his magic wand. In the 19th century this opening scene was elaborated (and incidentally gave rise to the fashionable *burlesque and *extravaganza), and the subjects gradually changed to fairy-tales, actresses being cast as the young heroes, the so-called *principal boys. At the same time the comic elderly characters, who had previously been played by men, became the prerogative of the knockabout comedian and were henceforth known as *Dames. The success of the fairy-tale openings—*Cinderella, Babes in the Wood, Aladdin, Red Riding Hood*—caused them to be spun out for so long that the harlequinade was relegated to a short scene at the end, and although it lingered on in some theatres, particularly in the provinces, it eventually disappeared. Pantomimes in their new and entirely English form were soon associated with Christmas, and most theatres produced them. They usually opened on Boxing Day (26 Dec.) and ran till March, some actors playing in nothing else. With the importation of speciality acts from the *music-halls, the show became such a hotchpotch of incongruous elements—slapstick, romance, topical songs, male and female impersonation, acrobatics, splendid settings and costumes, precision and ballet dancing, trick scenery, and transformation scenes—that for a time the phrase 'a proper pantomime' was used outside the theatre in colloquial English to signify 'a state of confusion'.

The word *pantomime* was also used in France for the wordless *Pierrot plays of *Deburau until the genre disappeared. Its most famous production was *L'Enfant prodigue* (*The Prodigal Son*), widely performed during the 1890s with a girl as Pierrot. *Fox also staged pantomimes in the USA, while another meaning was given to the word during the craze for *melodrama, when it signified the use of dumb show to convey ideas wordlessly. In this sense it is an important element in all acting and dancing, particularly ballet. Modern performers in dumb show, in order to distinguish their art from the popular idea of pantomime, describe it as *mime.

Pantomimus, name given to a performer popular in Imperial Rome, who by movement and gesture alone represented the different characters in a short scene based on classical history or mythology which was sung in Greek by the *chorus accompanied by musicians—usually flutes, pipes, cymbals, and trumpets. The *pantomimus* wore the costume of the tragic actor—a long cloak and a silken tunic—and a mask with no mouthpiece, changing it when necessary. As many as five masks could be used in one scene. The most famous *pantomimi* were Bathyllus of Alexandria, Pylades of Cilicia, and Paris, who was put to death by *Nero out of professional jealousy. The art of the *pantomimus* was considered by St Augustine more dangerous to morals than the Roman circus, since it dealt exclusively with guilty passions and by its beauty and seductiveness had a disastrous effect on female spectators, although unlike its rival the Roman *mime it was never coarse or vulgar.

Papp [Papirofsky], **Joseph** (1921–91), American theatre producer-director, who had over 10 years' theatrical experience, mostly backstage, before founding the New York Shakespeare Festival in 1954. His first productions were given in the Emanuel Presbyterian Church in East 6th Street, most of the actors giving their services. Two years later Papp presented *Julius Caesar* and *The Taming of the Shrew* in the East River Park amphitheatre, using a portable stage mounted on a truck, but in 1957 the City of New York offered him a site in Central Park, where *Romeo and Juliet, Macbeth*, and *The Two Gentlemen of Verona* were seen. The company occupied various open-air sites in the Park up to 1962, when a permanent home, the Delacorte Theatre, also in the open air, was built there for it. Financed by public and private donations, it seats 1,936, and as with all Papp's other ventures in the Park admission is free. It opened with a production, directed by Papp himself, of *The Merchant of Venice*, and achieved an enviable reputation, with the appearance of stars such as James Earl *Jones, Julie *Harris, and Colleen *Dewhurst. Among its productions a musical version of *The Two Gentlemen of Verona* in 1971 was particularly successful, later having a long run at the *St James Theatre. Meanwhile Papp had founded the *Public Theatre in 1967, and in 1973 he also became director of the *Vivian Beaumont and the Mitzi E. Newhouse theatres in the Lincoln Center. A financial crisis in 1977 forced him to relinquish the management of the last two theatres, though he still retained control of his earlier projects. In 1980 another crisis compelled the Delacorte to stage only one production, a modernized version of *Gilbert and Sullivan's *The Pirates of Penzance.* Fortunately it was an outstanding success, later having a long run at the Uris Theatre (now the *Gershwin), and Papp's difficulties were finally resolved by the award of a permanent subsidy from the City of New York. In 1981 there was once again a Shakespeare season at the Delacorte, with *Henry IV, Part One* and *The Tempest.* The Delacorte staged some of the productions in the complete Shakespeare cycle begun by Papp at the Public Theatre in 1988.

Parade, in the French theatre, a short sketch acted by fairground actors outside their booth in order to induce the spectators to pay their entrance-fees to see the play given inside. A volume of plots of such *parades* acted on the first-floor balcony of the Théâtre de la Foire Saint-Germain (see FAIRS), destroyed in 1756, shows the affinity of the genre with the scenarios of the Italian *commedia dell'arte,* but there is also evidence of direct survival from French medieval *farce.* The *parade* died out with the disappearance of the old theatres of the fairs in the mid-18th century but was revived by *Bobèche and Galimafré on the post-Revolution *boulevard du Temple. A vogue for this essentially popular form among the more sophisticated audiences of the private theatres fashionable in the second half of the 18th century was catered for by a number of dramatists of the time, among them *Beaumarchais.

Paradiso, elaborate piece of stage machinery invented by **Filippo Brunelleschi** (1377–1446) for the representation of the Annunciation which took place annually in the church of San Felice in Florence. It consisted of a group of choir boys, representing cherubim, suspended in a copper dome from the roof of the church. This was lowered by crane to a platform, and from it then emerged the actor who was to play St Gabriel. When he had finished his part, he re-entered the dome and returned to heaven. This device was the forerunner of many similar mobile chariots, which were improved by such additions as the clouds of cottonwool with which **Francesco D'Angelo** (1447–88) masked the machinery needed for Christ's Ascension. Painted canvas, mounted on battens, later took the place of cottonwool, but the basic principle of Brunelleschi's device remained in use for the transport of any supernatural being until almost the end of the 18th century.

Parallel, see ROSTRUM.

Parigi, Giulio and **Alfonso,** see SCENERY.

Parker, Louis Napoleon (1852–1944), English dramatist and pageant-master, who was for a time a music master at Sherborne School in Dorset. The success of his early plays enabled him to resign in 1892 and go to London, where he devoted the rest of his long life to the theatre. Among his numerous works the most successful were *Rosemary* (1896), written in collaboration with Murray Carson, who worked with him on a number of other plays including *Pomander Walk* and *Disraeli* (both 1911), the latter providing a fine part for George *Arliss, in which he had a long run in New York. It was seen in London in 1916. Parker was much in demand as a producer of the civic *pageants so popular in Edwardian England. During the First World War he was responsible for several patriotic pageants in London, and in 1918 he devised *The Pageant of Drury Lane.*

Park Theatre, New York, the first outstanding theatre of the United States, known as the 'Old Drury' of America. Under *Hallam and *Hodgkinson it opened in 1798 with *As You Like It,* followed by *Sheridan's *The School for Scandal.* The engagement of the brilliant young actor Thomas *Cooper, who made his first appearance as Hamlet on 28 Feb., brought prosperity to the theatre, but his success was jeopardized by constant quarrels between the managers, one of whom, *Dunlap, finally took over alone. His repertory consisted mainly of contemporary successes from London and the European stage, together with some of his own plays, but older plays were sometimes staged, and in 1804 *Twelfth Night* was seen for the first time in New York. A year later Dunlap went bankrupt and was replaced by Cooper, but it was not until a business man, Stephen *Price took control in 1808 that the theatre prospered. Under him, in 1809, the first American play on Red Indian life was produced—*The Indian Princess; or, La Belle Sauvage,* by J. N. *Barker. Price also inaugurated the policy of importing foreign stars, G. F. *Cooke appearing under him in 1810. Ten years later, just as Edmund *Kean was about to appear at the theatre, it was burnt down. Rebuilt, it reopened in 1821 and embarked on a period of prosperity which lasted until the death of Price in 1840. It then went downhill, and when in 1848 it was again destroyed by fire it was not rebuilt.

Pasadena Playhouse College of Theatre Arts, see SCHOOLS OF DRAMA.

Pashennaya, Vera Nikolayevna (1887–1962), Soviet actress, who joined the *Maly Theatre company in 1905 and remained there until 1922. She later went to the *Moscow Art Theatre, and in 1941 was appointed to the Chair of Acting in the Shchepkin Theatre School. She appeared in a wide range of parts, including *Schiller's Maria Stuart, Emilia in *Othello,* and Tanya in Leo *Tolstoy's *Fruits of Enlightenment.* Her conception of Vassya Zheleznova in *Gorky's play of that name was characterized by psychological insight, and all her roles gave proof of dramatic force and inner concentration. In 1955 she celebrated her jubilee as an actress and a year later she gave a memorable performance in Sofronov's *Money* at the Maly. In 1962 she was equally successful in the same author's *Honesty*.

Paso, term applied in 16th-century Spain to a short comic scene which later developed into the *entremés*. It depended for its appeal on a simple plot, quick-fire dialogue and repartee, and the use of a few well-known types from the *commedia dell'arte*. One of the chief characters was the *bobo* or rustic clown, who became the *gracioso* of the *entremés*. The best-known writer of *pasos* was Lope de *Rueda.

Pasquino, Pasquin, one of the minor *zanni* or servant roles of the *commedia dell'arte*. The name was adopted towards the end of the 16th century and passed into French comedy as Pasquin, being the name of the valet in the plays of *Destouches. In the French 17th-century theatre the expression 'the Pasquin of the company' designated the actor who played the satiric roles. In the 18th century the word 'pasquin' or 'pasquinade' was applied in England to a lampoon, squib, or satiric piece, often political. Henry *Fielding used it as the title of a production at the *Haymarket in 1736 and often as a journalistic pseudonym.

Pass Door, fireproof door placed in an inconspicuous part of the proscenium wall, leading from the auditorium to the side of the stage and so backstage. It is normally used only by those connected with the theatre, ordinary members of the audience penetrating behind the scenes by means of the *stage door which opens on to the street.

Passion Play, medieval religious drama in the vernacular, which dealt with the events from the Last Supper to the Crucifixion, unlike the *mystery play which presented in cyclic form Bible history from the Creation to the Second Coming. The establishment in 1313 of the feast of Corpus Christi (the Thursday after Trinity Sunday) gave a great impetus to the enactment of Passion plays in large cities throughout Europe, and also led to a tradition of open-air productions of the Good Friday story in many small towns and villages, all the actors being drawn from the local community. Most of these died out during the 15th century, but, helped by the Catholic Counter-Reformation in the following century, a few were revived in Switzerland, Austria, and Germany. The only one to have become well known is that given decennially since 1634 at Oberammergau in Bavaria. This was first performed during a visitation of plague and remains entirely amateur, the villagers dividing the parts among themselves and, as in earlier times, being responsible also for the production, music, costumes,

and scenery. There have been several revisions in recent times of the Oberammergau text, caused mainly by the desire to muffle its anti-Semitism, but it still remains controversial.

Pastor, Tony [Antonio] (1837–1908), American *vaudeville performer and manager. He made his first appearance as a child with Barnum, and later travelled with circuses and *minstrel shows as a singing clown, ring-master, or 'blackface' ballad-singer. In 1861 he made his début in *variety, which had sunk to a low level of vulgarity. Determined to clean it up, he opened a theatre on Broadway in 1865, moving 10 years later to Buckley's Minstrel Hall, later the *Olympic Theatre 3, which he renamed Tony Pastor's Opera House. He banned drinking and smoking and discouraged the more vulgar acts, appearing always in his own shows—he had a rich tenor voice well suited to the popular ballads of the day, of which he himself claimed to have composed over 2,000. In 1881 he presented the first performance of what was later called vaudeville at his newly acquired Fourteenth Street Theatre, which he ran until his death, many famous stars, including *Weber and Fields and Lillian Russell, appearing there.

Pastoral, dramatic form which evolved in Italy from pastoral poetry by way of the dramatic eclogue or shepherds' play. There were pastoral elements in early plays, but the first true pastoral was *Tasso's *Aminta* (1573), which was widely imitated. Another important work was Guarini's *Il pastor fido* (1596–8), which was seen in London in 1602, and had a great influence, particularly on *Lyly. The best of the many reworkings of inherently repetitive themes, and the representative pastoral of the next generation, was Bonarelli's *Filli di Sciro* (1607). The first English pastoral was the charming and poetic *The Faithful Shepherdess* (1608) by *Fletcher. The genre never became acclimatized in England, though it had some influence on the *masque and on Shakespeare. It was more at home in France, where it inspired a number of dramatists, among them *Hardy and above all *Mairet, whose *Sylvie* (1626) is generally regarded as the finest French pastoral, and whose *Silvanire* (1630) served as a vehicle for the introduction of the *unities into French drama. Although the vogue for the pastoral was short-lived, its introspective lovers, who analyse their feelings at great length, stimulated the type of comedy written in France in the 1630s by such authors as *Corneille.

Patent Theatres, the *Theatres Royal, *Drury Lane and *Covent Garden, which operate under Letters Patent, or Charters, given by Charles II in 1662 to Thomas *Killigrew for Drury Lane and Sir William *Davenant for *Lincoln's Inn Fields, whence it descended in 1732 via *Dorset Garden to Covent Garden. These were for the establishment of two companies to be known as 'the King's Servants' and 'the Duke of York's Servants' respectively. The company at Drury Lane was technically part of the Royal Household, and some members of it were sworn in as Grooms of the Chamber, being given an allowance of scarlet cloth and gold lace for their liveries. The charters, which in the course of the next 200 years changed hands many times, fluctuating sharply in value, are still in existence and form an integral part of the leases of the theatres. They rendered the two theatres independent of the *Lord Chamberlain as licenser of theatre buildings, though until the abolition of the theatre *censorship in 1968 they still remained accountable to him for the plays which they put on. The monopoly in 'serious' or 'legitimate' acting established by Charles II and reinforced by the Theatres Act of 1737 was finally broken in 1843.

Pätges, Johanne, see HEIBERG, JOHAN.

Pavilion, The, New York, see CHATHAM THEATRE I.

Pavilion Music-Hall, see LONDON PAVILION.

Pavilion Theatre, London, in the Whitechapel Road. The first theatre on this site opened in 1828. The building was destroyed by fire in 1856, but was rebuilt with a capacity of 3,500. Under Morris Abrahams, who became its manager in 1871, it catered with great success for the large Jewish population of the neighbourhood and was rebuilt in 1874, its three-tier auditorium then holding about 2,500. It was altered again in 1894 for Isaac Cohen, who presented a number of successful plays, including several by Sholom *Aleichem, and also added an annual Christmas *pantomime to the theatre's repertory. It closed in 1933, and stood empty for many years, being severely damaged by bombing during the Second World War and finally demolished in 1961.

Paxinou [née Konstantopoulou], **Katina** (1900–73), Greek actress, who in 1940 married as her second husband Alexis *Minotis, under whose direction she frequently acted. Trained as a singer, she made her début as an actress in 1924, and was first seen in New York in 1930, one of her roles being Clytemnestra in *Sophocles' *Electra*. In 1932 she joined the Greek National Theatre company, playing Clytemnestra again in *Aeschylus' *Agamemnon* and becoming the company's leading lady. She translated and directed several English-language plays, among them Eugene *O'Neill's *Anna Christie*, in which she played the title-role, in 1932. One of her finest parts was Mrs Alving in *Ibsen's *Ghosts*, in which she performed annually for six years beginning in 1934. She was also much admired as Phaedra in *Euripides' *Hippolytus*, Lady Windermere in *Wilde's *Lady Windermere's Fan*, both in 1937, Goneril in *King Lear*, and Mrs Chevely in Wilde's *An Ideal Husband*, both in

1938. In 1939 she made her first appearance in London as Sophocles' Electra, followed by Gertrude in *Hamlet*. In the following year she gave her first performance in English, also in London, as Mrs Alving, and in 1942 played Ibsen's Hedda Gabler in English in New York. After some years in films she returned to the stage in New York in 1951 as Bernarda in *García Lorca's *The House of Bernarda Alba*, a part in which she was seen in Greece in 1954, appearing there in 1957 in *Fry's *The Dark is Light Enough*. She was seen again in London in the 1966 *World Theatre Season, playing Jocasta in Sophocles' *Oedipus Rex* and the title-role in Euripides' *Hecuba*.

Payne, John Howard (1791–1852), American actor and dramatist, who as a precocious 14-year-old had his first play *Julia; or, The Wanderer* produced in New York. Three years later he went on the stage and made a great reputation, touring the larger American cities in such parts as Hamlet, Romeo, Young Norval in *Home's *Douglas*, and Rolla in *Kotzebue's *Pizarro*. In 1811 he appeared at the *Chestnut Street Theatre, Philadelphia, as Frederick in his own version of Kotzebue's *Das Kind der Liebe*, which, as *Lovers' Vows*, was already well known in English translations by Mrs *Inchbald and Benjamin Thompson. Two years later he sailed for England, and appeared at *Drury Lane with great success. A visit to Paris brought him the friendship of *Talma and the freedom of the *Comédie-Française, and for many years he was engaged in the translation and adaptation of current French successes for the English and American stages. Of his own plays—he has been credited with some 50 or 60—the best was *Brutus; or, The Fall of Tarquin* (1818), in which Edmund *Kean was seen at Drury Lane; he adapted several *melodramas by *Pixérécourt, of which the most successful was *Therese; or, The Orphan of Geneva* (1821), which provided him with the money to pay off the debts incurred by his unsuccessful attempt to manage *Sadler's Wells Theatre. With Washington *Irving, a lifelong friend, he also adapted Duval's *La Jeunesse de Henry V* (1806) as *Charles II; or, The Merry Monarch* (1824). In 1842 he was appointed American Consul at Tunis, where he died. He is now best remembered for the lyric of 'Home, Sweet Home', part of the libretto which he wrote for Henry Bishop's opera *Clari, the Maid of Milan*, first performed at *Covent Garden in 1823.

Peacock Theatre, see ABBEY THEATRE.

Pedrolino, see COMMEDIA DELL'ARTE and PIERROT.

Peele, George (1556–96), Elizabethan dramatist, who worked mainly for *Henslowe, often in collaboration with other playwrights of the time. It has even been suggested, without proof, that he worked with Shakespeare on *Henry VI, Parts One and Three, King John,* and *Titus Andronicus.* His first extant play appears to be *The Arraignment of Paris* (*c.*1581), one of the earliest examples in English drama of the *pastoral. It was followed, in uncertain order (the dates are conjectural), by *David and Fair Bethsabe* (*c.*1587); *The Battle of Alcazar* (*c.*1589), of which both the full text and the *platt used backstage by the actors survive; *The Old Wives' Tale* (*c.*1590), Peele's best-known work, a mixture of high romance and English folklore once dismissed by the critics as negligible nonsense, but now considered a landmark in the development of English comedy; and finally *Edward I* (*c.*1591), which survives only in a mutilated form. Peele was also the author of several texts for pageants, of which *Descensus Astraea* (1591) survives, and of a number of charming lyrical poems, some included in his play-texts.

Peking Opera, name given to the popular and traditional theatre of China, perfected in Peking in the 19th century, though its roots go back to about AD 740, when the first dramatic school, known as the Pear Garden, was founded by the T'ang Emperor Ming Huang. A flexible and harmonious combination of recitation, mime, song, dancing, and acrobatics, Peking Opera drew its material from well-known myths and folk-tales and from history, and the libretto provided little more than a framework within which highly trained actors could display their skill. Over the centuries a set of conventional symbols was developed by which an oar, for instance, represented a boat, a whip indicated a horse, blue cloth stood for the sea, and waving banners for high winds. A hat wrapped in a red cloth signified a decapitated head, a woman's tiny red shoe represented needlework, a fan was a sign of frivolity. With such symbolism, allied to the equally symbolic placing of chairs, tables, and benches, the Chinese actor was able to dispense with scenery and leave a great deal to the imagination. In the same way the costumes, sumptuous but seldom historically accurate, indicated by the predominating colour the rank and character of the wearer. Emperors wore yellow, high officials red, worthy citizens blue, rough characters black. The warrior wore a magnificent costume with tigers' heads embroidered on the wide padded shoulders, and a brilliantly coloured headdress surmounted by two pheasant plumes, often six or seven feet long, which were tossed and twirled in pride or anger. Make-up also had its significance, white indicating treachery, red courage, blue ferocity, yellow strength, gold immortality, pink and grey old age. For ceremonial occasions embroidered robes were worn with long 'rippling water' sleeves. These played an important part in the graceful, gliding ritual dances, which were accompanied by wooden instruments like castanets to beat the time; a small stringed instrument like a violin, which also accompanied the falsetto singing; a wind instrument resembling a clarinet; and brass percussion instruments. All the parts were played by men, and female impersonation, as in the Elizabethan theatre, reached a high point of perfection, as exemplified in the acting of *Mei Lanfang, the only Chinese actor to make a reputation outside his own country. In recent times women have appeared on the stage and many other innovations have been introduced. The Peking Opera has now been freed from the restrictions imposed during the Cultural Revolution.

Pélissier, Harry, see COMPTON, FAY, and REVUE.

Pembroke's Men, company under the patronage of the 1st Earl of Pembroke, first mentioned in 1592. Among the play-books which the actors sold to the booksellers a year later were *Marlowe's *Edward II* and two anonymous plays, *The Taming of a Shrew* and *The True Tragedy of Richard Duke of York.* Shakespeare's *The Taming of the Shrew* and *Richard III* may have been revisions of these last two. Pembroke's Men are also named together with *Sussex's Men on the title-page of *Titus Andronicus,* which Shakespeare refashioned from *Titus and Vespasian.* The 2nd and 3rd parts of his *Henry VI* were also in their repertory, the first part having been written for *Strange's Men. Shakespeare's connection with Pembroke's Men ceased when he joined the *Chamberlain's Men on their foundation in 1594, and the company underwent an eclipse until 1597 when they leased the newly built *Swan Theatre. They were soon in trouble over their production of *Nashe's and Ben *Jonson's satire *The Isle of Dogs.* This caused the group to disband, and some of the leading actors joined the *Admiral's Men at the *Rose Theatre under *Henslowe. The others went into the provinces and may have made an unsuccessful attempt to return to London in about 1600. No further record of them is known, but they may have joined a short-lived company belonging to the Earl of Worcester which formed the basis of *Queen Anne's Men, founded in 1603.

Penley, William Sydney (1852–1912), English actor-manager, who made his first appearance on the stage at the *Royal Court Theatre in 1871 in farce, and then toured in light and comic opera. He was for some years at the *Strand Theatre playing burlesque. The first outstanding success of his career came when he succeeded *Tree in the title-role of *Hawtrey's *The Private Secretary* (1884), a part with which he became so identified that he is usually believed to have been the first to play it. He was also closely connected with Brandon *Thomas's *Charley's Aunt* (1892), playing Lord Fancourt Babberley during its run of 1,466 performances, a record for the period. In 1900 he opened the former Novelty Theatre (later the *Kingsway) as the Great Queen Street Theatre and appeared there in revivals of his most successful parts, retiring a year later. Much of his success as a comedian lay in his dry humour, his serious, rather pathetic, face, and the solemnity of his voice and manner contrasted with the farcical lines of his part.

Penny Gaff, see GAFF.

Penny Plain, Tuppence Coloured, see TOY THEATRE.

People Show, see COLLECTIVE CREATION.

People's National Theatre, London, see PRICE, NANCY.

Pepper's Ghost, device by which a ghost can be made to appear on stage, so called because it was perfected and patented by 'Professor' J. H. Pepper, a director of the Royal Polytechnic Institution in London. It is based roughly on the principle that a sheet of glass can be both reflective and transparent, so that a reflection of a figure can appear side by side with an actual performer on the stage. The Ghost Illusion, as Pepper called it, was first demonstrated privately at the Polytechnic in 1862, and then exhibited publicly with great success. *Dickens used it in connection with his readings of *The Haunted Man.* It was first used in the theatre—though it had previously been seen in many music-halls—in 1863, at the *Britannia, and several plays there were written specially to introduce it. But it was never widely adopted, probably because it was difficult to place in position, and also because the 'ghost' could not speak. This made it useless for *Boucicault's *The Corsican Brothers,* for instance, where the ghost speaks in the last act. The device did, however, enjoy a semi-dramatic life in the Ghost shows, based on popular melodramas, which toured the provincial fairs until the early 20th century, and gave rise to the expression 'It's all done with mirrors'.

Pepys, Samuel (1633–1703), English naval administrator under Charles II and James II, whose *Diary* contains a good deal of information on the London theatre in the early days of the Restoration. In it he recorded the plays he saw and the actors in them, noted his opinion of them, and related stray items of backstage gossip retailed to him by his friends in the theatre, of whom one of the closest was Mary *Knepp. To him we owe many illuminating glimpses of the green room, of the theatre under reconstruction, and of the rowdy talkative audiences of his day.

Percy, Esmé Saville (1887–1957), English actor, particularly admired for his work in Shakespeare and *Shaw. He studied for the stage under Sarah *Bernhardt, and made his first appearance in *Benson's company in 1904, later touring with him in a number of leading Shakespearian roles. In 1908 he was with *Granville-Barker at the *Royal Court Theatre, and then joined Miss *Horniman's repertory company in Manchester. During and after the First World War he was in charge of entertainment for the troops, producing over 140 plays. He returned to London in 1923, remaining active in the theatre until his sudden death. He was a recognized authority on Shaw and from 1924 to 1928, with the Charles Macdona Players, appeared in a wide variety of Shavian parts, including the male leads in *Man and Superman, Androcles and the Lion, The Doctor's Dilemma, Pygmalion,* and *The Apple Cart.* He also directed the Hell scene (Act 3, scene 2) from *Man and Superman* as a separate production—*Don Juan in Hell*—at the *Little Theatre in 1928, playing Don Juan himself. Among his other roles were Hamlet, Gaev in *Chekhov's *The Cherry Orchard,* and Matthew Skipps in *Fry's *The Lady's not for Burning* (1949), of which he was co-director.

Peregrinus, see LITURGICAL DRAMA.

Peretz, Isaac Leib (1852–1915), Jewish writer and lawyer. Born in Poland, he wrote first in Hebrew but later changed to Yiddish. Some of his short stories dealing with life in Hassidic villages were dramatized and produced by *Ohel on its first public appearance in Tel Aviv in 1926. His dramatic poem *Night in the Old Market* was dramatized and produced in 1925 at the *Moscow State Jewish Theatre.

Performance Group, see COLLECTIVE CREATION.

Periaktoi, scenic devices used in the Roman, and perhaps earlier in the Hellenistic, theatre. According to *Vitruvius, they were triangular prisms set on each side of the stage which could be revolved on their axes to indicate a change of scene, each of their three surfaces bearing an indication of a locality, as waves for the sea, ships for a harbour, trees for a wood. The publication of Vitruvius's treatise in 1511 led to the adoption and improvement of the *periaktoi* by Renaissance theatre architects, particularly by **Bastiano da San Gallo** (1481–1551), who increased their size and their number, placing several one behind the other on each side of the stage, and providing removable painted canvas panels for each of the three faces of the prism, so as to make possible a greater number of variations in the scenery. These improved *periaktoi* were later known as *telari.*

Perkins, (James Ridley) **Osgood** (1892–1937), American actor, who, after seeing active service in the First World War, made his first appearance on the stage in *Kaufman and *Connelly's *Beggar on Horseback* (1924). His performance as Walter Burns, the archetypal cynical newspaper editor in Hecht and MacArthur's *The Front Page* (1928), confirmed his reputation as a skilled comic actor, noted for his thin, expressive face and mobile hands; but he was also a sympathetic Astrov in *Chekhov's *Uncle Vanya* in 1930. He played Sganarelle in *Molière's *The School for Husbands* (1933), supported the *Lunts in Noël *Coward's *Point Valaine* (1935), and was in S. N. *Behrman's *End of Summer* (1936). He was to have been the leading man in Rachel *Crothers's *Susan and God*, but died of a heart attack soon after its first performance in Washington.

Peruzzi, Baldassare, see SCENERY.

Petit-Bourbon, Salle du, Paris, the first Court Theatre of France, in the gallery of the palace of the dukes of Bourbon. A long, finely proportioned room, with a stage at one end, it had formerly been used for balls and ballets, and the first recorded professional company to play in it was the *commedia dell'arte* troupe the *Gelosi in 1577. In 1604 the famous Isabella *Andreini made her last appearance there. In 1645 the great Italian scene-painter and machinist *Torelli was invited to supervise the production of opera in the theatre, and in 1658, when it was again in the possession of a *commedia dell'arte* troupe under *Fiorillo, *Molière's company, fresh from the provinces, was allowed to share it with them. For this privilege Molière paid a heavy rent and was allotted the less profitable days—Mondays, Wednesdays, Thursdays, and Saturdays—the Italians keeping the more lucrative Tuesdays and Sundays for themselves. He opened in 1658 with five plays by *Corneille in quick succession, and not until the end of the month did he put on one of his own farces, *L'Étourdi*, followed by *Le Dépit amoureux*. The Petit-Bourbon also saw the first nights of *Les Précieuses ridicules* (1659) and *Sganarelle* (1660), before it was suddenly scheduled for demolition, and the company, in the full tide of their success—the Italians having already left—found itself homeless. Louis XIV gave them Richelieu's disused theatre in the *Palais-Royal and the Petit-Bourbon disappeared, Molière taking the boxes and fittings with him and Torelli's scenery and machinery being given to *Vigarani.

Phelps, Samuel (1804–78), English actor and manager, who toured the provinces for several years, making a name for himself as a tragedian, and in 1837 made his first appearances in London at the *Haymarket Theatre under Ben *Webster, playing Hamlet, Richard III, Othello, and Shylock with considerable success. Later in that year he joined the company at *Drury Lane, where he appeared with Macready in Shakespeare and also created the role of Lord Trensham in *Browning's *A Blot in the 'Scutcheon* (1843). In 1844 he took advantage of the abolition of the *Patent Theatres' monopoly in

serious drama in 1843 to assume the management of *Sadler's Wells Theatre, which he made for the first time a permanent home of Shakespeare, though other plays also figured in his repertory. With his fine and imaginative productions he did much to redeem the English stage from the triviality into which it had fallen. The plays he directed during his long tenancy were remarkable for their scenic beauty, but he always made sure that the author's text was paramount. By the time he retired in 1862 he had produced all but six of Shakespeare's plays. In 1854 *Pericles* was seen in its original form for the first time since the Restoration. Phelps appeared in most of the plays himself, his best parts being considered Lear and Othello. He was not suited to romantic comedy; he was, however, good in pathetic parts, and in spite of being mainly a tragedian gave an excellent comic performance as Bottom in *A Midsummer Night's Dream*. He was also much admired as Sir Pertinax McSycophant in *Macklin's *The Man of the World*. He remained on the stage until almost the end of his life, making his last appearance as Wolsey in *Henry VIII* on 31 Mar. 1878 at the Aquarium (later the *Imperial) Theatre.

Philadelphia, Pa., one of the first American cities to have theatrical entertainment. In 1749 *Kean and Murray's company acted *Addison's *Cato* in a converted warehouse. The same building, subsequently used for other purposes but not demolished until 1849, housed the elder Lewis *Hallam's company in 1754, two years after its arrival from England. David *Douglass brought his *American Company to Philadelphia in 1759, and returned in 1766 to build the *Southwark Theatre, considered by some to be the first permanent theatre in the United States. Later theatres were the *Chestnut Street, which opened in 1793, the *Walnut Street, which became a theatre in 1811 and is still in use, and the Arch Street, opened by William B. *Wood in 1828. In the 1830s, Philadelphia's undoubted theatrical supremacy passed to New York. The Arch Street Theatre, however, was to have its greatest artistic and financial success from 1860 to 1892 under the able management of Mrs John *Drew. The city, like Boston, cannot now be considered of major theatrical importance, although there was a renaissance in the 1980s, with some US and world premières. The Philadelphia Drama Guild has done valuable work since its foundation in 1956, and from 1971 to 1980 was housed in the old Walnut Street Theatre. It then took up residence in the Zellerbach Theatre, a new theatre in the Annenberg Center at the University of Pennsylvania.

Philipe, Gérard (1922–59), French actor, who studied at the Paris Conservatoire, and made his début in 1943 in *Giraudoux's *Sodome et Gomorrhe*, first attracting attention in 1945 in the title-role of *Camus's *Caligula*. In 1951, by which time he had made an outstanding reputation both on stage and screen, he joined the *Théâtre National Populaire (TNP) under *Vilar and gave a superb performance as the hero of *Corneille's *Le Cid*. He continued to act for the TNP until his sudden death at 37 (he was buried in the costume he wore as Le Cid), and his fame and popularity did much to attract a young audience. Among the plays in which he appeared in Paris, at the *Avignon Festival, and on tour in Russia, the USA, and Canada, were *Kleist's *Prinz Friedrich von Homburg* (1951), *Musset's *Lorenzaccio* (1952), *Hugo's *Ruy Blas*, and Shakespeare's *Richard II* (both 1954). In 1958 and 1959 he was seen in *Les Caprices de Marianne* and *On ne badine pas avec l'amour*, both by Musset. He also played Eilif in the first French production of *Brecht's *Mother Courage* (1951).

Philips, Ambrose (1674–1749), English dramatist, son of a draper in Shrewsbury, well educated, and a member of *Addison's circle. His chief claim to fame is that in *The Distrest Mother* (1712), an adaptation of *Racine's *Andromaque* (1667), he wrote one of the best pseudo-classical tragedies in English, second only to Addison's *Cato* (1713). Henry *Fielding parodied it in *The Covent Garden Tragedy* (1732), proving that it still held the stage in his day. Philips, who was nicknamed Namby-Pamby by Swift for his poor verses, wrote two other undistinguished tragedies, *The Briton* (1722) and *Humfrey, Duke of Gloucester* (1723).

Phillips, Augustine (?–1605), Elizabethan actor, who was with *Strange's Men, playing with them in *Tarleton's *Seven Deadly Sins* (1590–1), and later with the *Admiral's Men. He joined the *Chamberlain's Men on their foundation in 1594, and remained with them until his death. He was a friend of Shakespeare, with whom he acted in *Jonson's *Every Man in His Humour* (1598) and *Sejanus* (1603), and is listed among the players in Shakespeare's plays. He was one of the five original shareholders in the *Globe Theatre, and on his death left Shakespeare a 30s. gold piece.

Phillips, Robin (1942–), English director, who was first an actor, making his début as Mr Puff in *Sheridan's *The Critic* at the *Bristol Old Vic in 1959. He became an associate director there a year later, and acted at *Chichester in 1962 and the *Oxford Playhouse in 1964 before becoming an assistant director with the *RSC in 1965. He was Associate Director at the *Northcott Theatre in Exeter, 1967–8, but returned to the RSC to direct *Albee's *Tiny Alice* (1970). Later in the same year he directed Ronald *Millar's *Abelard and Heloise* (NY, 1971). Returning to Chichester, he directed *Anouilh's *Dear Antoine* in 1971 and *Fry's *The Lady's not for Burning* in 1972, in which year he played

Louis Dubedat in *Shaw's *The Doctor's Dilemma* there. After a season as Artistic Director at the *Greenwich Theatre in 1973 he took up the same position in 1974 at the *Stratford (Ontario) Festival, which he ran dynamically until 1980. Among his notable productions were *Congreve's *The Way of the World* (1976), *As You Like It* (1977), *Macbeth*, and *Coward's *Private Lives* (both 1978), all with Maggie *Smith, and *King Lear* (1979) with Peter *Ustinov. His production of Edna O'Brien's *Virginia* (1980), with Maggie Smith as Virginia Woolf, was seen in London in 1981. In 1985 he directed *Antony and Cleopatra* at Chichester, but in 1987 and 1988 he was back in Stratford running the Young Company at its Third Stage.

Phillips, Stephen (1864–1915), English poet and dramatist, who for a short time was a member of the Shakespearian company of his cousin Frank *Benson. He had had two plays, *Herod* (1900) and *Ulysses* (1902), produced by *Tree at *Her Majesty's Theatre when, later in 1902, his poetic tragedy *Paolo and Francesca* was produced by George *Alexander at the *St James's Theatre, with the young Henry *Ainley as Paolo. Its great success was believed to herald a new era of poetry in the English theatre, but his next play, *Nero* (1906), showed such a falling-off that, like the rest of Phillips's work, it was soon forgotten.

Phillpotts, Eden (1862–1960), English dramatist and novelist, author of a number of light comedies of English rural life, of which the most successful was *The Farmer's Wife*. First seen at the *Birmingham Repertory Theatre in 1916, it was revived in 1924 with Cedric *Hardwicke as Churdles Ash. Transferred to the *Royal Court Theatre, London, it had a long run and has been revived several times. Also produced at the Birmingham Repertory Theatre and in London were *Devonshire Cream* (1924), *Jane's Legacy* (1925), and *Yellow Sands* (1926), in the last of which Phillpotts collaborated with his daughter.

Phlyax, form of ancient Greek *mime play, or *farce, which bridged the gap between Athenian and Roman comedy. It was probably the model for the *fabula atellana. Much of it was improvised, and consisted of burlesques of earlier plays interspersed with scenes of daily life played by actors in ludicrously padded costumes, each male character having also a gigantic phallus. Our knowledge of the *phlyakes* derives mainly from vase paintings which portray the characters and settings of the 4th century BC. The form of stage depicted is important, for from it may have developed the salient forms of the Roman theatre, which differed markedly from the Greek. The most primitive type consisted of roughly-hewn posts supporting a wooden platform. Later the posts appear to have been joined by panels of wood with ornamental patterns, while later still the structure, though still not permanent, had a background for the actors with a practicable door and windows used in the course of the play.

Phoenix Arts Centre, Leicester, see HAYMARKET THEATRE.

Phoenix Society, London, group founded in 1919 under the auspices of the *Stage Society to present plays by early English dramatists, most of which, apart from those by Shakespeare, seemed to have fallen out of the repertory. The Stage Society began the work of revival in 1915, and continued to produce every year one *Restoration comedy, by *Farquhar, *Congreve, or *Vanbrugh, until 1919. The Phoenix Society was then constituted, and in the six years of its existence staged 26 plays, by, among others, *Marlowe, *Jonson, *Beaumont and *Fletcher, *Heywood, *Ford, *Dryden, *Otway, and *Wycherley. From the beginning, enthusiastic support was given to the Phoenix Society by actors and actresses, many of them already well established; two permanent adaptable sets were designed by Norman *Wilkinson; and all but two of the productions, for which Edith Craig (see CRAIG, GORDON) was responsible, were directed by **Allan Wade** (1881–1954). In 1923 a brilliant performance of Fletcher's *pastoral The Faithful Shepherdess was given with elaborate scenes and dresses and music arranged and conducted by Sir Thomas Beecham. The influence of these performances helped considerably to combat the indifference, and in some cases hostility, once shown to early English drama, several of the plays staged having since been frequently and successfully revived on the public stage.

Phoenix Theatre, London, in Drury Lane, see COCKPIT.

Phoenix Theatre, London, on the corner of Charing Cross Road and Phoenix Street, from which it takes its name. A three-tier house with a capacity of 1,012, it opened in 1930 with *Coward's Private Lives. Later successes by Coward were To-Night at 8.30 (1936) and Quadrille (1952). In its early years the Phoenix had a chequered history, but it later housed a number of important plays, including *Saint-Denis's productions of *Bulgakov's The White Guard and of Twelfth Night (1938); *Gielgud's revivals of *Congreve's Love for Love (1943) and of The Winter's Tale (1951) and Much Ado about Nothing (1952); revivals of *Vanbrugh's The Relapse (1948), from the *Lyric, Hammersmith, and *Farquhar's The Beaux' Stratagem (1949); and Hamlet (1955) with Paul *Scofield. Notable modern plays included *Wilder's The Skin of Our Teeth (1945) with Vivien *Leigh, *Rattigan's Playbill (1948), and *Osborne's Luther with Albert *Finney (1961) transferred from the *Royal Court. In 1965 Gielgud

appeared in his own version of *Chekhov's Ivanov, and in 1968 a musical dramatization of four of Chaucer's Canterbury Tales started a long run. Later productions were Tom *Stoppard's Night and Day (1978), *Fo's Trumpets and Raspberries (1984), Derek *Jacobi's Richard II (1988) and Richard III (in 1989), and Peter *Hall's production of The Merchant of Venice, also in 1989, with Dustin Hoffman.

Phoenix Theatre, New York, on 2nd Avenue at 12th Street, formerly the Yiddish Art Theatre, run by Maurice *Schwartz. Seating 1,100, and intended for the production of non-commercial plays, it opened as the Phoenix in 1953 with Sidney *Howard's posthumous Madam, Will You Walk? The repertory was mainly European, the first seasons including Coriolanus, *Chekhov's The Seagull, *Shaw's Saint Joan with Siobhán *McKenna, and *Turgenev's A Month in the Country. The company moved in 1961 to the East 74th Street Theatre, seating 204, renamed the Phoenix 74th Street Theatre, achieving a critical success with Conway's Who'll Save the Plowboy? and both critical and commercial success in the following year with *Kopit's Oh Dad, Poor Dad, Mamma's Hung You in the Closet and I'm Feelin' So Sad. In 1964 the company joined with the Association of Producing Artists, formed in 1960 to present quality plays in *repertory, which had alternated touring with residence at various theatres; Rosemary *Harris was its best-known performer. The APA-Phoenix company, later moving to the *Lyceum, staged such productions as *Pirandello's Right You Are, if You Think You Are in 1966 and *Ionesco's Exit the King in 1968. The company left the Lyceum in 1969, the APA being dissolved in 1970. The Phoenix company then produced plays at various theatres on and off Broadway until in 1972 the New Phoenix Repertory Company was formed, playing for the next three seasons at several university theatres and for limited engagements on Broadway. In 1975 it moved to the *Off-Broadway Playhouse Theatre in West 48th Street, and in the following year to the new 250-seat Marymount Manhattan Theatre, where, as the Phoenix Theatre, it presented the work of new playwrights. In 1982 the company moved to the Theatre at St Peter's Church in the Citicorp building, but closed almost immediately owing to the poor reception of the opening production and financial difficulties.

Piaf, Edith [Giovanna Gassion] (1915–63), French singer and entertainer, who had a hard and unhappy childhood, supporting herself at an early age by singing in the streets. Once launched into cabaret and music-halls, however, she rapidly became a popular favourite and something of a cult among a group of influential critics. Although she never appeared in England, her recordings sold there by the million, and the personal tragedy of her life was well known—her first lover (and manager) was murdered, she made several unhappy marriages, and had continually to face the problems of alcoholism, drug addiction, and ill health. Her massive yet superbly controlled voice, emerging from a tiny, shabbily dressed body, contrived to be at the same time strident and intensely moving. Her style was deceptively simple and nostalgic and highly personal. Although she was primarily a singer—she made extensive tours of Europe and North America with Les Compagnons de la Chanson, for instance—she also appeared in a number of films, and was seen on stage in Le Bel Indifférent (1941), specially written for her by *Cocteau, who was devoted to her and died on the same day as she did. She also appeared in *Achard's La P'tite Lili (1951). Among the songs she made famous were 'Je ne regrette rien', 'Mon légionnaire', 'La Vie en rose', of which she wrote both the words and the music, 'Le Voyage du pauvre nègre', and 'Pour deux sous d'amour'.

Picard, Louis-Benoît (1769–1828), French actor-manager, one of the few successful dramatists of the Napoleonic era, who with unquenchable gaiety flourished equally under the Revolution, the Empire, and the Restoration. He had the knack of amusing the public without touching on controversial subjects, which made him invaluable to Napoleon, and his caustic humour took as its target the newly rich and newly risen. He excelled in the portrayal of bourgeois and provincial interiors, as is evident in Médiocre et rampant (1797)—the title comes from a speech by *Beaumarchais's Figaro; but more amusing is the lighthearted Le Collatéral; ou, La Diligence à Joigny (1799), which with La Vieille Tante (1811) and Les Deux Philibert (1816) is among the best of his comedies. He was the originator of a new type of light satiric play with music, typified by La Petite Ville (1801) which, with music by Lehár, was produced in Vienna as Die lustige Witwe (1905) and two years later, as The Merry Widow, captivated the English-speaking world. The published texts of Picard's plays give careful directions for settings and costume, in which he aimed above all at pictorial effect, and he can be credited with the creation of one new comic character, the valet Deschamps. He was from 1816 to 1821 manager of the *Odéon, having given up acting in 1807 to qualify for the award of the Légion d'Honneur, which even Napoleon did not dare give to an actor.

Piccadilly Theatre, London, in Denman Street. This theatre, with its three-tier auditorium seating 1,193, opened in 1928 with a musical play Blue Eyes and then became a cinema. It returned to live theatre in 1929 with a revival of the operetta The Student Prince, and in 1933 housed *Bridie's The Sleeping Clergyman.

It was then used for the transfer of long runs at reduced prices. After being closed for some time at the beginning of the Second World War, it reopened with *Coward's Blithe Spirit (1941), and among later successes were *Gielgud in Macbeth (1942) and the American musical Panama Hattie (1943). The building was then damaged by flying bombs and had to be closed. It did not reopen until 1945, with Agatha *Christie's Appointment with Death. Successful productions were *Werfel's Jacobowsky and the Colonel (1945), John *Van Druten's The Voice of the Turtle (1947), and *Ustinov's Romanoff and Juliet (1956). Edward *Albee's Who's Afraid of Virginia Woolf? (1964) began its long run here, and Robert *Bolt's Vivat! Vivat Regina! transferred from *Chichester in 1970. In 1973 the musical Gypsy had a good run, as did a revival of the musical Very Good, Eddie! in 1976. The theatre later provided a West End home for several of the *RSC's productions—*O'Keeffe's Wild Oats (1977), Peter *Nichols's Privates on Parade (1978), Pam Gems's Piaf and *Kaufman and *Hart's Once in a Lifetime, which ran in repertory (1980), and Willy Russell's enormously successful Educating Rita (also 1980). The musical Mutiny! (1985) had a good run, and a revival of *Sondheim's A Little Night Music moved there from Chichester in 1989.

Piccolo Teatro della Città di Milano, the first *teatro stabile, or permanent theatre company, to be set up in Italy. Founded in 1947 by **Giorgio Strehler** (1921–) and the actor-director **Paolo Grassi** (1919–81) with a municipal grant, it later received a state subsidy, and became the model for similar ventures in Rome, Turin, and elsewhere. It opened with *Gorky's The Lower Depths, and Strehler's production of *Goldoni's Il servitore di due padroni was staged in the same season. The company quickly achieved an excellent reputation, one of Strehler's achievements being the introduction to Italian audiences of the plays of *Brecht. In 1967 the company was seen in London during the *World Theatre Season, again in Goldoni's comedy. A year later Strehler resigned over the municipality's failure to provide a new theatre for the company, but he returned as sole director in 1972. The company has staged over 200 productions encompassing revivals of the classics (especially Shakespeare and Goldoni), contemporary foreign works, and new Italian plays. It tours extensively in Italy and has visited nearly 30 foreign countries.

From 1983 to 1990 Strehler was also Artistic Director of the Théâtre de l'Europe at the *Odéon in Paris. In 1990 at the Piccolo Studio he staged the first part of a two-part adaptation of *Goethe's Faust, also playing the title-role and supervising the Italian translation.

Pickelhering, see ENGLISH COMEDIANS.

Pierrot, stock character in the French and English theatres, derived from the *commedia dell'arte* mask *Pedrolino. The transformation is usually attributed to an Italian actor named Giuseppe Giaratone or Giratoni, who joined the *Comédie-Italienne in Paris in about 1665. He accentuated the character's simplicity and awkwardness, an important feature of his later manifestations, and dressed him in the familiar costume, a loose white garment with long sleeves, ruff, and large hat whose soft brim flapped round his whitened face. This, with some modifications, has remained his distinguishing garb ever since, but his character was fundamentally altered by *Deburau, who for some 20 years played Pierrot at the *Funambules. He was succeeded in the part by his son and later by Paul Legrand, who made the character less amusing and more sentimental. This was later developed by a host of imitators until the robust country bumpkin of early days became a lackadaisical, love-sick youth, pining away through unrequited love and much addicted to singing mournful ballads under a full moon. Meanwhile the Pierrot of Deburau had become well known in London, where in 1891 he was seen as the hero of a wordless play *L'Enfant prodigue* (*The Prodigal Son*), produced in Paris in 1889. The character, which had been ousted from the *harlequinade by the English *Clown, still occasionally appeared in *pantomime and *revue, and regained much of his old vigour when he was incorporated into the new *Pierrot troupes.

Pierrot Troupes, form of entertainment which, like the Christmas *pantomime, appears to be purely English. It took the solitary, lovelorn *Pierrot of *Deburau's successors, and multiplied him by the score, turning him into a jolly, gregarious entertainer, a versatile member of an organized concert party. Dressed in a modified Pierrot costume—short white frilly frock for the girls, or Pierrettes, loose white (or black) suits for the men, with tall dunces' caps, the whole enlivened by brightly coloured (or black on white) buttons, ruffs, ruffles, and pompons—the various members of the troupe sang, danced, juggled, told funny stories, and engaged in humorous backchat. After an initial appearance at Henley Regatta, the Pierrots spread all over England, even ousting the blackfaced *minstrel shows from the beaches and pier-pavilions of the seaside towns, and eventually reaching London with Pélissier's *Follies*, the apotheosis of the genre, in the 1900s. The Pierrot troupes were later replaced by the more sophisticated *revue, but lingered on in summer resorts, and a successful revival of the old Pierrot show, staged by the Co-Optimists under **Davy Burnaby** (1881–

1949), had an unexpected success during several London seasons in the 1920s.

Pike Theatre Club, see DUBLIN.

Pilgrim Players, see BIRMINGHAM REPERTORY THEATRE and BROWNE, E. MARTIN.

Pinero [Pinheiro], **Sir Arthur Wing** (1855–1934), English dramatist. He was an actor for 10 years, making his first appearance on the stage at the Theatre Royal, Edinburgh, in 1874. In 1877 he had his first play, *£200 a Year*, produced at the *Globe Theatre. Popularity came with *The Magistrate* (1885), the first of the *Royal Court Theatre farces which became all the rage. It was followed by *The Schoolmistress* (1886), *Dandy Dick* (1887), *The Cabinet Minister* (1890), and *The Amazons* (1893), the first three having been revived many times and still ranking as some of the best farces in the English language. In 1888 a frankly sentimental play, *Sweet Lavender*, confirmed Pinero's pre-eminence in the contemporary theatre, and with *The Profligate* (1889) he made his first venture into the 'theatre of ideas', following it in 1893 with *The Second Mrs Tanqueray*, a 'problem play' in which Mrs Patrick *Campbell startled the town. In a theatre long given over to *farce, *burlesque, and *melodrama it appeared revolutionary—a 'serious' English play which made money. During the next 30 years Pinero was regularly productive. *The Notorious Mrs Ebbsmith* (1895) was Paula Tanqueray's successor; *Trelawny of the 'Wells'* (1898) was a theatrical romp; *The Gay Lord Quex* (1899) was a brilliant piece of theatricalism, containing a third act which is perhaps the author's masterpiece of contrivance. With the turn of the century came a long succession of serious plays—*Iris* (1901), *Letty* (1903), *His House in Order* (1906), and *Mid-Channel* (1909). They were progressively less successful, for Pinero had written himself out, and his last play, *A Cold June* (1932), was a pathetic failure. His reputation now rested on his farces and on *Trelawny of the 'Wells'*, whose title-role presented an irresistible challenge to spirited young actresses.

Pinter [Da Pinta], **Harold** (1930–), English dramatist, who for some years acted in the provinces under the name of David Baron. His first full-length play *The Birthday Party* (1958; NY, 1967) baffled the critics. It was followed by a double bill of *The Room* and *The Dumb-Waiter* (1960), but he came into prominence only with *The Caretaker* (also 1960; NY, 1961), an archetypal Pinter play in which the three characters, one of them a tramp, demonstrate the difficulties of communication between human beings and their consequent isolation. Next came five one-act plays—*A Night Out, A Slight Ache* (both 1961), *The Collection* (1962), produced by the *RSC, and, in a double bill, *The Lover* and *The Dwarfs* (1963), both of which

Pinter himself directed. His next full-length play, *The Homecoming* (1965; NY, 1967), was also produced by the RSC, as were another double bill, *Landscape* and *Silence* (1969; NY, 1970), and the full-length *Old Times* (London and NY, 1971). A double bill of *Tea Party* and *The Basement* was seen in New York in 1968 (London, 1970). *No Man's Land* (*National Theatre, 1975; NY, 1976) provided excellent parts for John *Gielgud and Ralph *Richardson. In *Betrayal* (NT, 1978; NY, 1980) successive acts move backwards in time. In 1980 *The Hothouse*, written over 20 years earlier, was presented in London. The triple bill *Other Places* (NT, 1982) consisted of *Family Voices*, *Victoria Station*, and *A Kind of Alaska*, the last showing a woman's awakening from a 29-year sleep; at the *Duchess (1985) *One for the Road* replaced *Family Voices*. He has directed a number of plays by other writers, including James Joyce's only play *Exiles* and several works by Simon *Gray, and has written many screenplays.

Pinter is probably the most influential of modern English playwrights, his highly individual style, when attempted by others, being labelled 'Pinteresque'. His plays are sparsely populated, and their plots slight. His characters' motives often remain obscure, their backgrounds indefinite, their fate at the end of the play indeterminate; and their language, at least in his early plays, includes the repetitions and illogicalities of ordinary speech. The scripts are as carefully balanced, with pauses and silences exactly indicated, as musical scores.

Many of the leading feminine roles in his plays were played by his first wife **Vivien Merchant** (1929–82), who also played Lady Macbeth for the RSC in 1967 and was in David *Mercer's *Flint* and Joyce's *Exiles* (both 1970).

Pioneer Players, see CRAIG.

Pipe Batten, see BARREL.

Pip Simmons Theatre Group, see COLLECTIVE CREATION.

Pirandello, Luigi (1867–1936), Italian dramatist, awarded the *Nobel Prize for Literature in 1934. Although well known as a novelist and critic before he turned to the theatre, it is as the playwright who brought Italy, after a period of stagnation, back into the mainstream of European drama that he is chiefly remembered. The success of his plays meant that the Italian straight theatre was once again able to challenge the supremacy of opera. His early plays were dramatizations of some of his own short stories, and it was not until the production of his best-known work *Sei personaggi in cerca d'autore* in 1921 that he became internationally famous. As *Six Characters in Search of an Author*, this was produced in London and in New York in 1922, and has several times been revived. The play, which is central to an understanding of

Pirandello's work, concerns a group of characters from another play who invade the stage during a rehearsal and become the arbiters of their own destiny. It forms part of a trilogy of which the other sections, less well known in English, are *Ciascuno a suo modo* (*Each in His Own Way*, 1924) and *Questa sera si recita a soggetto* (*Tonight We Improvise*, 1929). In these, as in all his plays, particularly in his finest tragedy *Enrico IV* (1922), seen in London in 1925 (NY, 1947) as *Henry IV*, Pirandello was concerned with the futility of human endeavour and the impossibility of establishing an integrated, objective personality for any human being. With consummate ability he pushed forward the frontiers of drama by bringing to the stage the mental and psychological preoccupations of his own day and becoming the chronicler of an age of decay and fragmentation. Among his other plays, those which have been successfully produced in English are *Così è* (*si vi pare*) (*Right You Are, if You Think You Are*, 1917); *Il giuoco delle parti* (*The Rules of the Game*, 1918); *Lazzaro* (*Lazarus*, 1929); and *Come tu mi vuoi* (*As You Desire Me*, 1930). Of the one-act plays, the best known is probably *L'uomo dal fiore in bocca* (1923), which as *The Man with a Flower in His Mouth* (the flower being an inoperable cancer) was produced in London in 1926.

Pirandello had an unhappy life, his wife, whom he married in 1894, becoming mentally ill in 1904. For 15 years he cared for her at home, finding his only consolation in work. After she had been consigned to a mental home he was able to devote more of his time to the theatre and in 1925 established his own company in Rome, proving himself an excellent director with an acute awareness of the technical problems of stagecraft. He toured widely in Europe and America with this company, and became well known as an actor, particularly in his own plays.

Piron, Alexis (1689–1773), French dramatist, whose first works were farces for the theatres of the Paris *fairs. He evaded the law of 1718 which forbade the use of more than one speaking actor in a fairground theatre by writing a series of monologues, of which the first was *Arlequin Deucalion* (1722). Encouraged by their success, he sent a comedy, *L'École des pères*, to the *Comédie-Française, where it was produced in 1728. It represents an interesting mixture of old and new in the theatre of the time, for it stands on the threshold of the *comédie larmoyante which was shortly to be introduced by *La Chaussée, but its author still evidently believes that comedy should amuse first and only incidentally instruct. Piron's best work (and one of the outstanding comedies of the 18th century) was *La Métromanie* (1738), which makes gentle fun of bad poets who insist on reading their effusions to their reluctant friends. Of his tragedies the best was *Gustave Wasa* (1734), which remained in the repertory of the Comédie-Française for many years.

Piscator, Erwin Friedrich Max (1893–1966), German director, who during the 1920s worked in *Berlin where he devised a form of *epic theatre, intended to reinforce the impact of his strong pacifist and Communist convictions, which encompassed the whole of society in its political and economic complexity and greatly influenced the later work of *Brecht. He anticipated the trend away from a completed playscript, and was one of the first directors to use film-strips and animated cartoons in conjunction with live actors. Among his productions at this time was *Gewitter über Gottland* (1927) by Elm Welk, which included film projections of the Russian Revolution and cost him his job at the *Volksbühne. He then opened his own theatre, where he produced *Toller's *Hoppla, wir leben!* (1927) in a revolving multi-level set with seven or eight acting areas and surfaces for film and slide projections; Alexei *Tolstoy's *Rasputin* (also 1927), staged in a huge revolving steel hemisphere symbolizing the earth; and *Die Abenteuer des braven Soldaten Schweik* (1928) adapted by Brecht and others from the novel by Hašek. This was Piscator's most successful production, and technically his most ambitious, but it was so expensive that the theatre was soon forced to close. After some desultory freelance work, Piscator left Germany in 1933, and reached New York, via Paris, in 1939. There he directed a Dramatic Workshop, where he mounted a number of teaching productions including his own adaptation of Leo *Tolstoy's *War and Peace* (1942); among his students were Arthur *Miller and Tennessee *Williams. He returned to Germany in 1951, and in 1962 he became Director of the West Berlin Freie Volksbühne, where he directed several *documentary dramas—*Hochhuth's *Der Stellvertreter* (1963), *Kipphardt's *In der Sache J. Robert Oppenheimer* (1964), and *Weiss's *Die Ermittlung* (1965). Piscator died suddenly during the rehearsals of Hochhuth's *Soldaten*. His influence was apparent in the work of Joan *Littlewood at *Theatre Workshop in England and in that of the *Living Theatre in the USA.

Pistoia, Il, see SACRA RAPPRESENTAZIONE.

Pit, name formerly given to the ground floor of the theatre auditorium, generally excavated below ground level. In the early playhouses the stage and lower boxes were approximately at ground level, and the whole space sunk between these was called the pit, from the Elizabethan cockpit used for cock-fighting. In the early 19th century the lower boxes were replaced by a raised circle, with the pit extending underneath. Shortly after, the old rows of pit seats near the orchestra were replaced by the higher-priced stalls, and the name 'pit' was applied only to the more distant rows, now called the rear stalls. Modern theatres have no pit.

Pit, The, the *RSC's second theatre in the Barbican Centre, so called because of its basement location. Successor to The Warehouse (see DONMAR WAREHOUSE), it has flexible seating for up to 200. Notable productions which originated there include Arthur *Miller's *The Archbishop's Ceiling* (1986), Vladimir Gubaryev's *Sarcophagus* (1987), Howard *Barker's *The Bite of the Night* (1988), and *Sophocles' *Electra* and Richard Nelson's *Some Americans Abroad* (both 1989).

Pitlochry Festival Theatre, Perthshire, Scotland. The idea of building a 'theatre in the hills' for the presentation of plays in repertory came to its founder, John Stewart, during a visit to the *Malvern Festival before the Second World War; but it was not until 1951 that the first theatre at Pitlochry opened. A large marquee with a fan-shaped auditorium and a very wide proscenium opening, this theatre was so well designed that its features were retained in the more permanent structure built in 1953. A new purpose-built theatre on the south bank of the River Tummel opened in 1981. It has won several architectural awards and cleverly symbolizes Pitlochry's tented origins through the use of 'king pole' pillars reaching up to a vaulted ceiling in the foyer. A particularly striking feature is the window wall stretching almost the full width of the front-of-house area which affords the visitor a spectacular view of the river and hills. Pitlochry continues to present up to six plays in six days each summer, the season running from early May to the beginning of October. Sunday concerts, various foyer events, art exhibitions, and craft demonstrations augment the six-play repertoire.

Pitoëff, Georges (1887–1939), Russian-born French actor, who for two years directed his own amateur company in St Petersburg, where he was in contact with *Stanislavsky and *Meyerhold. From 1915 to 1921 he lived in Geneva, and then settled in Paris. With his wife **Ludmilla** (1896–1951), an excellent actress, he appeared in several theatres, including that of *Copeau, until in 1924 he took a company to the Théâtre des Arts (now the Hébertot) and from there to the *Mathurins, where much of his best work was done. With very little money, but a great deal of ingenuity and hard work, he attempted to enliven the French theatre, which he considered to lack both ideas and imagination, by bringing forward the best work of foreign dramatists, as well as such innovatory French writers as *Claudel, *Cocteau, and *Anouilh. The value of his work lay not only in the plays but in the subtle and entirely personal interpretations he gave, often centred on some brilliantly simple piece of scenography or decorative technique. He was himself an excellent actor, and a complete man of the theatre, adapting, translating, directing, and acting at one and the same time. After his death his wife took the company on an extended tour of North America, her finest parts being Nora in *Ibsen's *A Doll's House*, Marthe in Claudel's *L'Échange*, and the Hostess in *Goldoni's *La locandiera*. In London in 1930 she played to perfection *Shaw's St Joan in French, together with the younger *Dumas's *La Dame aux camélias*.

Their son **Sacha** (1920–90) was an actor and director who continued his father's search for new authors. As an actor he conveyed a strange, fierce spirituality and was at his best in self-contained, solitary roles.

Pixérécourt, (René-Charles) **Guilbert de** (1773–1844), French dramatist, who alone or in collaboration wrote nearly 100 plays and for 30 years provided the staple fare of the secondary theatres. He led an unsettled life until in 1797 one of his many early plays, *Les Petits Auvergnats*, was put on. It was successful enough to persuade him to devote the rest of his life to the theatre, and in 1798 he produced *Victor; ou, L'Enfant de la forêt*, the first of the long series of plays by which he is chiefly remembered and for which he himself coined the name *melodrama. The most successful of these, which earned him the nickname of 'the *Corneille of the boulevards', was *Coelina; ou, L'Enfant du mystère* (1800), which, in a translation by Thomas *Holcroft as *A Tale of Mystery* (1802), was the first melodrama seen on the English stage. The success of his plays made Pixérécourt a wealthy man, but he was ruined when the Théâtre de la *Gaîté, of which he was a director and where many of his most successful plays were produced, was burned down in 1835. This disaster, joined to the effects of a serious illness, ended his career, and he retired to Nancy to die a lingering death. His work typifies the mixture of ferocity and idealism of the French Revolution, when blood and tears were shed with equal facility. He was aware of the literary shortcomings of his plays, but believed sincerely in their importance as a vehicle for moral teaching. He took great trouble over his productions and settings, often inventing new machinery and spectacular effects, for, as he himself said, he wrote for 'those who cannot read'—a large and enthusiastic audience. Pixérécourt also had an immense influence in England, where most of his plays were seen soon after they appeared, their main characteristics being preserved in the contemporary *toy theatre.

Place, The, see ROYAL SHAKESPEARE COMPANY.

Planché, James Robinson (1795–1880), English dramatist of Huguenot descent, a prolific writer of *burlesques, *extravaganzas, and *pantomimes, most of which were seen in the smaller theatres of London, particularly the *Adelphi and the *Olympic. The best known of Planché's works, *The Vampire; or, The Bride of the Isles* (1820), introduced to the English stage the vamp(ire) *trap, and for Mme *Vestris at the *Lyceum he wrote the spectacular extravaganza *The Island of Jewels* (1849) for which the painter Beverley produced some remarkable scenic effects. Having had his first play, a burlesque entitled *Amoroso, King of Little Britain*, produced in 1818, Planché ended his career with a pantomime, *King Christmas* (1871), and a spectacle play for Covent Garden, *Babil and Bijou; or, The Lost Regalia* (1872), written in collaboration with Dion *Boucicault. Although Planché's work was enormously successful in his own day it has no literary merit and divorced from its music and spectacular effects is quite unreadable. It depended largely on its staging, acting, and topicality, and taken as a whole provides an excellent picture of the London stage over more than 50 years. It has been claimed that many of *Gilbert's libretti for the Savoy operas were based on texts of Planché's plays. In 1823 Planché, who was a serious student of art, designed, with some approximation to historical accuracy, the costumes of Charles *Kean's *King John*. An unauthorized production of one of his own plays led him to press for the reform of the laws governing dramatic *copyright, and it was largely owing to his efforts that new legislation was passed, giving greater protection to British dramatists.

Planchon, Roger (1931–   ), French director, actor, and dramatist, whose first production, a burlesque which he mounted with his own amateur group, won a prize in Lyons in 1950. The company then turned professional and, living as a community, built their own 100-seat theatre which opened in 1952. By 1957 they had won a considerable reputation as an experimental group. Planchon was conducting investigations into various forms of stagecraft, including the Elizabethan theatre and American gangster films. Like many of his contemporaries he was strongly influenced by *Vilar; but after a meeting with *Brecht in 1954 he became the leading director of his plays in France, with translations of *Der gute Mensch von Sezuan* in 1954, *Furcht und Elend des Dritten Reiches* in 1956, and *Schweyk im Zweiten Weltkrieg* in 1962. His preoccupation with *epic theatre led him to seek a larger building, and in 1957, at the invitation of Villeurbanne, an industrial satellite-town of Lyons, he took his company to the 1,300-seat Théâtre de la Cité. There he addressed himself to factory workers and, through meetings, publications, exhibitions, and door-to-door salesmanship, created an entirely new audience. After a successful visit to Paris in 1961, the company was awarded a government subsidy, and thus became the first national theatre in the French provinces, inheriting in 1972 the name *Théâtre National Populaire after the closing of the Palais de *Chaillot. Planchon is a brilliant actor, and as D'Artagnan in his own adaptation of the elder *Dumas's *Les Trois Mousquetaires* was seen in 1960 in London and at the *Edinburgh Festival. As a director he abandoned the interpretations of French classics standardized by the *Comédie-Française, and in his productions of *Marivaux's *La Seconde Surprise de l'amour* and *Molière's *George Dandin* in 1959 extended the biting social satire underlying the buffoonery to include overt Marxist references. His iconoclasm culminated during the student unrest of 1968 in a *Mise en pièces et contestation de Cid*, which, basing itself on *Corneille's masterpiece, attacked the very foundations of French classical drama. He later applied the same methods to Shakespeare's *Henry IV, Parts One and Two*. His other notable productions include *Gogol's *Dead Souls*, adapted by *Adamov, in 1959, and *Marlowe's *Edward II*, in his own adaptation, in 1960. In 1968 he committed himself wholeheartedly to an anti-Establishment declaration by the managers of most of France's subsidized playhouses. Thereafter he played a less prominent role at Villeurbanne, laying more stress on the collective nature of the company, *Chéreau being his co-director, 1972–81. He also began to write plays himself, in a style curiously closer to *naturalism than to the critical *realism of his former master Brecht; among them are *La Remise* (1962), *Dans le vent* (1968), *Gilles de Rais* (1976), and *L'Avare* (1986).

Platform, see ROSTRUM.

Platform Stage, see FORESTAGE.

Plato (427–348 BC), Greek philosopher, member of a highly aristocratic family. In his youth he was a dramatist, but on coming under the influence of Socrates he burnt his plays and devoted his life to philosophy and mathematics. Much of his written work is in dialogue form—a development of the *mime—and although in the more abstruse works, and in most of the *Republic*, the dialogue is only nominal, elsewhere it is consistently dramatic, with occasional passages of astonishing vividness and power. The character sketch of Ion, and the opening scenes of the *Protagoras*, are good examples of Plato's dramatic skill; his mastery of ironic comedy is shown by his picture of the Sophists in the *Euthydemus*; of tragedy by the scene of Socrates' death in the *Phaedo*. His theories of literature and drama have had an immense influence. His 'inspirational' theory of poetry is the direct source of the idea of the *furor poeticus*, through a 16th-century translation of his *Ion* which greatly influenced French poets of the time. In other dialogues Plato is much less sympathetic to literature, and it is evident that he would admit poetry into his ideal society

only under a paralysing censorship. He objects in particular to drama because it appeals especially to the ignorant, debilitates the community by appealing to the emotions and not to reason, and propagates blasphemous and impossible ideas about the deity (e.g. by repeating stories of strife between gods). These criticisms are important chiefly for the reply which they drew from *Aristotle.

Platt, Elizabethan term for an outline, or 'plot', of a play, giving the main points of the action, the division into acts, and the actors' entrances and exits. Posted up behind the scenes during the performance, it was intended to help in the organization of calls and properties. It was a purely utilitarian device and not, like the scenario of the *commedia dell'arte*, a basis for improvisation. A few specimens have survived among *Henslowe's papers.

Plautus, Titus Maccius [Maccus] (c.254–184 BC), Roman dramatist, of whose life and background very little is known. He is believed to be the author of about 20 of the 150 plays attributed to him by the end of the 1st century BC. They are all free renderings of Greek *New Comedy, with complicated plots, strongly marked characters, and scenes of love-making, trickery, and debauchery, interspersed with songs, jokes, puns, and topical allusions. Although Roman dramatists were warned against political and personal satire, there seems to have been no ban on indecency, and several of the plays are set in brothels. But though Plautus often carried farce to outrageous lengths, he knew where to draw the line when the honour of a respectable woman was in question. The variety of plot found in his works is considerable, and if he lacks the subtle effects of his contemporary *Terence, he offers instead a flow of wit and a vigour of language which help to explain his success on the Roman stage—and in translation through the centuries. His *Menaechmi* was the source of Shakespeare's *The Comedy of Errors*, the *Aulularia* of *Molière's *L'Avare*. Even when they were no longer acted Plautus' plays were read and enjoyed, and their shafts of wit passed into common parlance—a process which still continues. The successful American musical *A Funny Thing Happened on the Way to the Forum* (1962; London, 1963) owed its existence to several of Plautus' plays combined.

Play, any work written to be acted, and entirely or mainly spoken. (Plays written to be read—see CLOSET DRAMA—remain outside the mainstream of the theatre.) A play may require some music (see INCIDENTAL MUSIC), but if music is paramount the work becomes an opera; if dancing is the main attraction the work becomes a ballet, or a *mime. A play may be either a *tragedy, with subdivisions into epic, historical, or neo-classical, or a *comedy, which includes farce. Hybrid forms are *ballad opera, *burlesque, *burletta, *harlequinade, *musical comedy, and *pantomime.

Although a play may be no more than a duologue on a bare platform, in its usual acceptance it covers a text, originally written by the dramatist (or playwright, which reveals his status as one worker among many, as with wheelwright), played in a suitable building by actors, together with scenery and costumes (usually specially designed) which help to interpret its inner meaning. The actors are chosen and controlled by the *director, who serves as interpreter between the author and the public. The director may function only as a director or, like such playwrights as *Molière and Shakespeare, may experience the advantages and disadvantages of actual daily participation in the life of the theatre company.

Playbill, Poster, Programme, forms of theatrical advertising. Although posters—placards hung on posts—are believed to have been used in London and Paris about the middle of the 17th century (earlier announcements were probably written on a wall, as in Pompeii), the first known English poster, which probably served also as a playbill for distribution by hand, is in the Public Record Office and dates from 1672. The earliest survival from an established theatre is dated 1687, and advertises a performance at *Drury Lane of *A King and No King*, by *Beaumont and *Fletcher. In the 19th century the poster was enlarged to contain descriptions of scenery and the names of scene-painters, such as Telbin and *Stanfield, as well as actors, but this made it too large to be manipulated with comfort, and in the 1850s the *Olympic Theatre, followed by others, began to issue small playbills folded in the middle, and initially free, for use in the theatre. Advertisements first appeared in theatre programmes in the 1860s, which also saw the beginning of the 'magazine' programme.

Large coloured posters first appeared in France in the 19th century, and such artists as Jules Chéret, beginning in 1866, and especially Toulouse-Lautrec in the 1890s made the poster into a work of art; the hoardings of Paris in 1895 reached a standard never surpassed. The Beggarstaff Brothers produced some splendid posters in England, among them work commissioned by *Irving for the *Lyceum. Posters for *melodrama, however, loomed much larger on English hoardings of the last 25 years of the 19th century and the first decade of the 20th than any other kind of theatrical advertisement. Before 1900 American posters resembled those for English melodrama, but well-known artists such as Charles Dana Gibson and Norman Rockwell later took up the work. The traditions of the artistic poster were carried on in the 20th century in England by Frank Brangwyn, Lovat

*Fraser, and others. The European poster in the early 20th century was much influenced by the French school of designers, and the influence of *Reinhardt was manifest in German theatrical publicity.

Theatre programmes gradually became more elaborate. About 1880 thin cardboard programmes came into use for a time, often printed in colour. Programmes of the 1880s varied enormously in size, and almost every theatre had its own cover design. After 1910 the 6 in. by 9 in. programme became common and there was more uniformity. Since the First World War programmes have generally been booklets containing theatre news and other articles as well as information on the particular production together with advertising. A few theatres, including the *National Theatre and those occupied by the *RSC, provide a free cast list. The advertising of theatrical productions is now fairly modest. Whereas hoardings in Paris and other cities in the 1890s mostly displayed advertisements for theatres, music-halls, and other entertainments, there is now much greater competition provided by the ampler funds of other vendors.

Playbox, see MELBOURNE THEATRE COMPANY.

Players' Club, New York, founded in 1888 on the lines of the *Garrick Club in London. Its first president was Edwin *Booth, who bought, donated, and endowed a house for it in Gramercy Park. He kept a suite of rooms there for the rest of his life, and like his successors, Joseph *Jefferson and John *Drew, died in office. The club has a large collection of theatrical relics, including jewellery and weapons owned by famous actors, paintings of American and foreign players, death-masks, and the fine Hampden–Booth Theatre Library, opened to theatre research workers in 1957 as a memorial to Walter *Hampden, the club's fourth president, who served 27 years before his retirement in 1955. He was succeeded by Howard *Lindsay, who retired in 1965, being followed by Dennis *King and others. Women were formerly admitted to the premises only for an afternoon reception on Shakespeare's birthday, but from 1946 were invited to four annual Open House nights, and from 1989 were admitted to the club, Helen *Hayes becoming the first woman member.

Players' Theatre, London, membership club which began life as Playroom Six, on the first floor of 6 New Compton Street. It opened in 1927, and five months later Peggy *Ashcroft made her London début there. In 1929 the company moved to the ground floor of the building, and the name was changed to the Players' Theatre. A further move in 1936 took the Players' to premises in King Street, Covent Garden, once occupied by *Evans's (late Joy's) 'song-and-supper-rooms'. The new venture was not a success, but in 1937 Peter Ridgeway reopened the premises as the New Players', with an entertainment in the style of the old 'song-and-supper' evenings; the programmes became famous as *Ridgeway's Late Joys*. The same mixture of songs and sketches continued to fill the bill after Ridgeway's death in 1938, and in 1940 the company moved to 13 Albemarle Street. In 1945 they were able to acquire part of the premises formerly occupied by *Gatti's-Under-the-Arches, opening there in 1946. Their popularity was enhanced by the production in 1953 of Sandy Wilson's musical pastiche *The Boy Friend*, and the theatre continued to present 'Victorian entertainment', including an annual *pantomime, to a small but devoted membership. The company occupied the *Duchess Theatre, 1987–90, while a new theatre, similar to the old, was built 'under the Arches'.

Playfair, Sir Nigel (1874–1934), English actor-manager and director, who was intended for the law, but in 1902 made his first appearance on the stage and later toured with *Benson's company. Two years later he played Hodson in the first production of *Shaw's *John Bull's Other Island*, and in 1911 appeared as Flawner Bannal in the first production of the same author's *Fanny's First Play*. In 1914 he was seen as Bottom in *Granville-Barker's production of *A Midsummer Night's Dream*. He took over the *Lyric Theatre, Hammersmith, in 1918, and made it one of the most popular and stimulating centres of theatrical activity in London, producing there *Drinkwater's *Abraham Lincoln* (1919); *Gay's *The Beggar's Opera* (1920) which ran for nearly 1,500 performances; *Congreve's *The Way of the World* (1924), with Edith *Evans; *Bickerstaffe's *Lionel and Clarissa* (1925), its first revival since the original production in 1768; *Riverside Nights* (1926), an intimate revue by A. P. Herbert and others; *Farquhar's *The Beaux' Stratagem* (1927), again with Edith Evans; *When Crummles Played* (also 1927), a burlesque of *Lillo's *The London Merchant* (1731) set against the theatrical background of *Dickens's *Nicholas Nickleby*; a stylized black-and-white revival of *Wilde's *The Importance of Being Earnest* (1930), with John *Gielgud as John Worthing; and numerous other plays, old and new, in many of which he himself appeared. He also produced the *Čapeks' *The Insect Play* (1923) at the *Regent Theatre, being part-author of the English translation, and many other plays. He remained at the Lyric Theatre until 1932, his final production being A. P. Herbert's musical *Derby Day*.

Playhouse, London, in Northumberland Avenue. As the Royal Avenue Theatre, holding about 1,500 in four tiers, this opened in 1882 with Offenbach's *Madame Favart*, the first of a series of light operas presented by various

managements. In 1890 George *Alexander began his career as an actor-manager with Hamilton Aidé's *Dr Bill*, and a year later Henry Arthur *Jones's *The Crusaders* was a success. The most important event of these early years was the production of *Shaw's *Arms and the Man* (1894). Mrs Patrick *Campbell and *Forbes-Robertson were at the Avenue in 1899, and in 1904 came Somerset *Maugham's first play *A Man of Honour*. At the beginning of 1905 Cyril *Maude took over the theatre and started to rebuild it, but when the work was almost completed part of Charing Cross station collapsed on it. Maude received £20,000 compensation and started building again. The new theatre, named the Playhouse, its three-tier auditorium having a seating capacity of 679, opened in 1907 with Clyde *Fitch's *Toddles*. With his wife Winifred *Emery as his leading lady, Maude remained until 1915, producing and playing in a number of successful plays. In 1916 Gladys *Cooper began her long association with the Playhouse. As joint manager until 1928, and then by herself, she produced and acted in many plays, her final production being Keith Winter's *The Rats of Norway* in 1933, in which Laurence *Olivier also appeared. After her departure the theatre had no settled policy until in 1938–9 Nancy *Price took it over, as the Playhouse Theatre, as a base for her People's National Theatre. It then closed for a time, to reopen in 1942 under its former name with a revival of Maugham's *Home and Beauty*. In 1943 it staged a season by the *Old Vic company, bombed out of their own theatre, which included a new Soviet play, *The Russians* by K. Simonov. The last successful production was a dramatization of Agatha *Christie's *Murder at the Vicarage* in 1949. In 1951 it became a BBC studio, housing a number of popular radio programmes until it was relinquished in 1975. After becoming derelict the theatre was handsomely restored, reopening as an 800-seat theatre in 1987. The new theatre's policy is to initiate its own productions, the first of which was the musical *Girlfriends*. The Peter *Hall Company occupied the theatre in 1991.

Playhouse Theatre, New York, on West 48th Street, between Broadway and the Avenue of the Americas. Built by William A. *Brady, this opened in 1911, and had its first success with *Broadhurst's *Bought and Paid For* the same year. In 1915–16 Brady's wife Grace *George appeared in a repertory which included Henry Arthur *Jones's *The Liars*, the first American performance of *Shaw's *Major Barbara*, and a revival of his *Captain Brassbound's Conversion*. Later successful new plays were *For the Defence* (1919) by Elmer *Rice and *The Show-Off* (1924) by George *Kelly. The next important productions were Robert *Sherwood's *The Road to Rome* (1927), and Rice's *Street Scene* (1929), which won him a *Pulitzer Prize. After a somewhat blank period the theatre had a further success with *Abbott and Holm's *Three Men on a Horse* (1935) and several revivals. In 1945 Tennessee *Williams's *The Glass Menagerie*, in which Laurette *Taylor made her last appearance, began a long run, and in 1959 Gibson's *The Miracle Worker* did well. The theatre was demolished in 1968.

Playroom Six, London, see PLAYERS' THEATRE.

Playwrights' Company, New York, see THEATRE GUILD.

Playwrights' Theatre, New York, see PROVINCETOWN PLAYERS.

Plinge, Walter, name used on English playbills to conceal a doubling of parts, particularly in a Shakespeare play, the American equivalent being George *Spelvin. There are two versions of the origin of the name. It may have been that of the landlord of a public house near the stage door of the *Lyceum in about 1900 who was popular with *Benson's company, or it may have been a fictitious name for a convivial acquaintance, the announcement by one of the company that Mr Plinge was waiting being an invitation to go for a drink. Oscar *Asche may have been the first to use it on a playbill. It still makes an occasional appearance in London and provincial cast lists. The name was never used, as Spelvin's was, for dolls or animal actors.

Plowright, Joan Anne (1929–), English actress, third wife of Laurence *Olivier. She made her first appearance on the stage in 1951, and in 1956 joined the *English Stage Company, making an outstanding success as Margery Pinchwife in *Wycherley's *The Country Wife*. She was also much admired as the Old Woman in *Ionesco's *The Chairs* (1957; NY, 1958); in *Osborne's *The Entertainer* (1957; NY, 1958); as Beatie, the working-class girl who discovers culture, in *Wesker's *Roots* (1959); and as Daisy in Ionesco's *Rhinoceros* (1960), in which year she also played Josephine in Shelagh Delaney's *A Taste of Honey* in New York. After a most interesting performance as Sonya in *Chekhov's *Uncle Vanya* during the first *Chichester Festival in 1962 she joined the *National Theatre company at the *Old Vic on its inception, playing Sonya again and also the title-role in *Shaw's *Saint Joan* (1963), Maggie in *Brighouse's *Hobson's Choice*, and Hilde in *Ibsen's *The Master Builder* (both 1964). Among other parts with the National were Beatrice in *Zeffirelli's production of *Much Ado about Nothing*, Masha in Chekhov's *Three Sisters* (both in 1967), Portia in *The Merchant of Venice* (1970), and Rosa, the harassed Neapolitan mama in *De Filippo's *Saturday, Sunday, Monday* (1973). At the *Lyric Theatre in 1975 she appeared in repertory as Arkadina in Chekhov's *The Seagull* and Alma in Ben *Travers's *The Bed before Yesterday*. In

1977 she had another long run in De Filippo's *Filumena* (NY, 1980). Later in 1980 she returned to the West End, playing a working-class old lady in Alan *Bennett's short-lived *Enjoy*. She was then seen in Chekhov's *The Cherry Orchard* in 1983, as Ranevskaya, *Congreve's *The Way of the World* (Chichester and London in 1984), as Lady Wishfort, Shaw's *Mrs Warren's Profession* at the National in 1985, and *García Lorca's *The House of Bernarda Alba* (1986). Her forte is an earthy regional realism that places her characters in a precise social context with an emphasis on down-to-earth indomitability.

Pluchek, Valentin, see SATIRE, THEATRE OF.

Plummer, (Arthur) **Christopher Orme** (1929–), gifted and versatile Canadian actor, with a flair for irony and flamboyant comedy. He gained his early experience with the Canadian Repertory Company in Ottawa and made his New York début in 1954. For several years he appeared mainly in Shakespeare, being seen as Mark Antony in *Julius Caesar* and Ferdinand in *The Tempest* in the opening season at Stratford, Conn., in 1955 (see AMERICAN SHAKESPEARE THEATRE). In 1956 he played Henry V at the *Stratford (Ontario) Festival, where his other roles included Aguecheek in *Twelfth Night* and Hamlet the following year, Leontes in *The Winter's Tale* (1958), the title-roles in *Macbeth* and *Rostand's *Cyrano de Bergerac* (1962), and Antony in *Antony and Cleopatra* (1967). In 1961 he was seen at the *RSC as Benedick in *Much Ado about Nothing* and as Richard III. He also appeared in a number of modern plays in New York, including *Fry's *The Dark is Light Enough* (1955), *MacLeish's *J.B.* (1958), *Brecht's *Arturo Ui* (1963, in the title-role), and *Shaffer's *The Royal Hunt of the Sun* (1965, as Pizarro). In London he played Henry II in *Anouilh's *Becket* (1961) and was in revivals of *Giraudoux's *Amphitryon 38* and *Büchner's *Danton's Death* for the *National Theatre company in 1971. He portrayed Anton *Chekhov in Neil *Simon's *The Good Doctor* (NY, 1973), and in 1981 returned to the American Shakespeare Theatre to play Iago in *Othello* and the title-role and Chorus in *Henry V*, repeating Iago in New York in 1982. In 1988, opposite Glenda *Jackson in New York, he again played Macbeth.

Plymouth Theatre, New York, on West 45th Street. This 1,063-seat theatre opened in 1917, and a year later saw the first production in New York in English of *Ibsen's *The Wild Duck*, starring *Nazimova, who was also seen in his *Hedda Gabler* and *A Doll's House*. Later in the year John *Barrymore appeared in a dramatization of *Tolstoy's *Redemption*, renamed *The Living Corpse*, and in 1919 he was with his brother Lionel *Barrymore in Benelli's *The Jest*. Among later successes were *What Price Glory?* (1924) by Maxwell *Anderson and Laurence Stallings, which had a long run, Elmer *Rice's

Counsellor-at-Law (1931), Robert *Sherwood's adaptation of Deval's *Tovarich* (1936), and Sherwood's own *Abe Lincoln in Illinois* (1938). Other notable productions were the musical *Lute Song* (1946), co-adapted by Sidney *Howard from a Chinese play, *Shaw's *Don Juan in Hell* (1952), from *Man and Superman*, *Giraudoux's *Tiger at the Gates* (1955), and *Wesker's *Chips with Everything* (1963). Alec *Guinness played Dylan Thomas in Sidney Michaels's *Dylan* (1964), and Neil *Simon's *The Star-Spangled Girl* (1966) and *Plaza Suite* (1969) were staged, while the 1970s brought Peter *Shaffer's *Equus* (1974) and Simon *Gray's *Otherwise Engaged* (1977). In 1981 the Plymouth housed the *RSC's production of *Piaf* by Pam Gems and its marathon adaptation of *Dickens's *Nicholas Nickleby*. Other British works were *Hare's *Plenty* (1983) and *Stoppard's *The Real Thing* (1984), while in 1987 Peter *O'Toole starred in Shaw's *Pygmalion*.

Poel [Pole], **William** (1852–1934), English actor and director, who altered his name in deference to his father's dislike of the theatre, his chosen profession. He made his first appearance on the stage in 1876, and from 1881 to 1883 was manager of the *Old Vic under Emma Cons. In 1894 he founded the Elizabethan Stage Society, which was to have an enormous influence on the staging and production of Shakespeare in the first half of the 20th century. On a stage modelled in accordance with his ideas of an Elizabethan stage, with the minimum of scenery, and with music by the Dolmetsch family, Poel produced in a variety of halls and courtyards a number of Elizabethan plays by Shakespeare, *Marlowe, *Jonson, *Beaumont and *Fletcher, *Middleton, *Rowley, and *Ford, beginning with *Twelfth Night* in 1895. He also staged the Dutch medieval *morality play *Everyman* for the first time for 400 years. His last production for the Elizabethan Stage Society was *Romeo and Juliet* in 1905. Financially the venture had not been a success, but artistically it vindicated Poel's theories, and it undoubtedly stimulated other directors to experiment with simple settings and so free Shakespeare from the cumbersome trappings of the late 19th century. Poel continued to work in the theatre until his death, and was responsible for revivals, under various auspices, of the old improvised *Hamlet* play of the *English Comedians, *Fratricide Punished*, first seen in England at the *Oxford Playhouse in 1924; of the anonymous *Arden of Feversham* (1925); and of *Peele's *David and Bethsabe* (1932) for the first time since 1599.

Poelzig, Hans, see GROSSES SCHAUSPIELHAUS.

Poetic Drama, term applied to plays written in verse or in a heightened, 'poetic' form of prose, which in the 19th and 20th centuries constituted an attempt to restore the medium of poetry to the stage. In earlier times all plays throughout Europe were in verse, and tragedy continued to be so written long after prose had become the accepted medium for comedy. Shakespeare interpolated comic scenes in prose into his great poetic plays; and by the time of *Dryden, prose comedies existed side by side with tragedies in verse. As the theatre increasingly attracted a mass audience, prose (with a greater or lesser approximation to everyday speech) became the accepted mode of expression for all plays. Works written by poets in dramatic form—Lord *Byron's *Werner* (1830), *Browning's *A Blot in the 'Scutcheon* (1843), *Shelley's *The Cenci* (1886), and *Tennyson's *Becket* (1893)—had some success, but on the whole the public preferred the rhetorical dramas of Sheridan *Knowles and *Bulwer-Lytton. Stephen *Phillips briefly revived blank verse in the commercial theatre in London, but he was the last of the poetic dramatists in the tradition of the 19th century.

About the turn of the century, poetic drama, under such diverse influences as the *nō plays of Japan, *Ibsen (whose 'realism' is fundamentally that of a poet), and the writings of the French *Symbolist poets, became more assured, and its authors were encouraged to think in terms of a theatre of their own. The leaders of the Irish literary revival, *Yeats, *Synge, and Lady *Gregory, produced plays of great poetic beauty combined with sound dramatic structure. They did not, however, establish the poetic theatre that had been hoped for, though Yeats's integrity as a poet and dramatist raised the standards of poetic drama and deeply influenced his Irish and English contemporaries. Synge's plays, though written in prose, are the work of a true poet, as are those of *O'Casey.

Among the English poetic dramas of the early 20th century John Masefield's *The Tragedy of Nan* (1908), in poetic prose, John *Drinkwater's *Rebellion* (1914), Gordon Bottomley's *Gruach*, and Flecker's *Hassan* (both 1923) were the most important. In the 1930s a number of poets broke away from the traditional verse play, employing free verse and the concepts of modern symbolism. T. S. *Eliot's *Murder in the Cathedral* (1935) was notable for its fine poetry, and has frequently been revived, while *The Dog beneath the Skin* (1936) and *The Ascent of F.6* (1937) by W. H. *Auden and Christopher Isherwood were valued in their day for their wit, satire, and social criticism.

The first important American poetic dramatist was William Vaughn *Moody. Maxwell *Anderson carried on a long and valiant fight to establish poetic drama on the American stage. Other American poets have written plays, for the most part characterized by great individuality and a considerable degree of experiment, but only a few have been produced. On the whole the production of poetic drama in the 20th-century theatre has had to depend on university theatres, drama schools, and groups specifically formed to present them. In England in the 1930s a number of plays in verse were staged at Canterbury and other cathedrals, including *The Zeal of Thy House* (1937) by Dorothy L. Sayers, which later had a London run, and *Christ's Comet* (1938) by Christopher Hassall. After the Second World War the *Mercury Theatre was for a time exclusively devoted to poetic drama, and it returned to the commercial theatre again with the production of new plays by Christopher *Fry and T. S. Eliot, while Dylan Thomas's *Under Milk Wood* (1953), written for radio, had many stage readings and performances; but these were isolated phenomena.

After the collapse of the *Federal Theatre Project in the USA in 1939, the Poets' Theatre, founded in 1951 in Cambridge, Mass., became one of the most important agencies for commissioning and producing poetic drama. Richard Eberhart, its founder and first president, wrote *The Apparition* (1951) and *The Visionary Farm* (1952), both well received; Archibald *Mac-Leish's poetic *J.B.* (1958; London, 1961) had a long run on Broadway. During the 1950s and 1960s adaptations in English of poetic plays by *Anouilh and *Giraudoux, and later those of *Beckett and *Ionesco, which combine poetry of a high order with Surrealism and the world of dreams, were commercially successful in London and New York, but poetic drama has during the past decades become increasingly identified with experimental work, and the theatre as a whole remains firmly committed to prose.

Pogodin [Stukalov], **Nikolai Fedorovich** (1900–62), prolific Soviet dramatist, writing mostly in the style of *Socialist Realism. He was originally a journalist, and his first play *Tempo* (1930) was a documentary on building. His best-known work *Aristocrats* (1934), dealing with the regeneration of a number of petty criminals engaged on the construction of a canal from the Baltic to the White Sea, was produced at the *Realistic Theatre. Both plays were published in English translations and *Aristocrats* was produced in London at *Unity Theatre in 1937. Of his other plays the most successful was *Poem about an Axe* (1931), produced by *Popov, as was *My Friend* (1932). Two war plays, *Moscow Nights* (1942) and *The Ferryboat Girl* (1943), were also well received. He wrote a trilogy based on the life of Lenin—*The Man with a Gun* (1937), *The Kremlin Chimes* (1942), and *The Third, Pathétique* (1958). A revised version of *The Kremlin Chimes* was produced in 1956 and performed in London by the *Moscow Art Theatre in 1964. His other works include *A Petrarchan Sonnet* (1957), somewhat more critical of the regime, and *The Little Student* (1959).

Poisson, Raymond (c.1630–90), founder of a family of actors who served the French stage for at least three generations. Raymond, known as **Belleroche**, joined the company at the Hôtel de *Bourgogne in the 1650s, and proved himself an excellent comic actor, who made the part of the valet *Crispin in *Scarron's *L'Écolier de Salamanque* (1654) particularly his own, introducing it into several of his own plays and reserving it for himself. His many light comedies have been forgotten, but one, *Le Baron de la Crasse* (1663), is interesting, as it depicts a strolling company whose leader may be intended as a satirical portrait of *Molière. Poisson's son **Paul** (1658–1735) carried on the family tradition, playing many of his father's parts at the *Comédie-Française from 1686 to 1724. He married an actress, and their sons **Philippe** (1682–1743) and **François-Arnould** (1696–1753) were both on the stage.

Polichinelle, French character, particularly prominent in *puppet-shows, which derived from the *commedia dell'arte* mask *Pulcinella. Like his prototype, he was humpbacked with a big hooked nose, but not at all doltish, being regarded by many as the epitome of quick wit and sardonic Gallic humour. He flourished particularly in Lyon and the surrounding district, and in 1660 was imported into England in the wake of Charles II's entourage, where as Punchinello he took fashionable society by storm and was transmogrified into the very English figure of *Punch.

Political Theatre, conscious use of the theatre to make political statements. It probably began when actors made political propaganda in factories during the Russian Revolution. With notable exceptions (such as Nazi Germany) it supported the left, and it played a dominant role in East European theatre. *Brecht and *Piscator made use of it. In the USA in the 1930s it was manifested in the work of the *Living Newspaper, which like other examples of political theatre used *documentary theatre techniques; and of the socially conscious *Group Theatre, especially the plays of Clifford *Odets. *Theatre Workshop was often political, and most of the *Fringe playwrights are ideologically committed. Rare examples from the commercial theatre include Walter Greenwood's *Love on the Dole* (1935) and *Priestley's *They Came to a City* (1943).

Politis, Fotos (1890–1934), first director of the Greek National Theatre, founded in 1930. In the four years before his death he laid the foundations of the company's excellent ensemble playing, imbuing all the actors who worked under him with his own sense of devotion and self-abnegation in the service of the theatre. Although he was open to all the influences of the modern theatre and his productions were often experimental, he was convinced that the strength of the Greek theatre lay in its past, and he always insisted on the importance of maintaining the classical repertory, a task carried on by his successors *Minotis and *Rondiris.

Poliziano, Angelo (1454–94), Italian scholar, sometimes known as Politian. Born Angelo Ambrogini, he took a new name after the death of his father. As a precocious youth he became the friend and protégé of Lorenzo de' Medici (the Magnificent), and ultimately tutor to his two sons. He was the author of the first important play to be written in the vernacular, the *Favola d'Orfeo* (1472), performed in Mantua to celebrate the betrothal of Clara Gonzaga. With no act divisions and a *multiple set, as in the *sacra rappresentazione*, it takes its subject not from the Bible but from classical mythology, and the prologue, formerly spoken by an angel, is given to the Roman god Mercury. It thus forms an interesting link between the old and new drama, and points to the imminent development of new forms of secular theatre, particularly the *pastoral.

Pomponius, Lucius, see FABULA I: ATELLANA.

Poncho Forum, see SEATTLE REPERTORY THEATRE.

Popov, Alexei Dmitrevich (1892–1961), Soviet director, who was at the *Moscow Art Theatre, 1912–18, and in 1923 was at the *Vakhtangov Theatre. In 1931 he was invited to direct the Theatre of the Revolution, later the *Mayakovsky Theatre, where he was responsible for the production of *Pogodin's *Poem about an Axe* (1931) and *My Friend* (1932), and a version of *Romeo and Juliet* (1936) which remained in the repertory for many years. In 1936 he went to the *Red Army Theatre in Moscow, where he directed two years later a production of *The Taming of the Shrew* which made theatre history, as well as a spectacular version of *A Midsummer Night's Dream* in 1940. During the war he was evacuated with his company, but continued to produce new plays, in particular a brilliant documentary on the defence of Stalingrad. Back in Moscow he made his theatre, now known as the Central Theatre of the Soviet Army, one of the most exciting in the city, winning particular acclaim for his productions in 1951 of *Gogol's *The Government Inspector* and in 1956 of the rewritten version of Pogodin's *The Kremlin Chimes*. A tall, strikingly handsome man, he had the courtly manners of an earlier generation. As a director he was the opposite of *Okhlopkov in that he first considered all the details of a production and then combined them into an integrated whole.

His son, the actor and director **Andrei Popov** (1918–83), succeeded him as Artistic Director in 1963, having already established his reputation as an actor with his performances in such roles as Petruchio in *The Taming of the Shrew* in 1956

and the Tsar in A. K. *Tolstoy's *The Death of Ivan the Terrible* in 1960.

Porta, Giambattista Della, see DELLA PORTA.

Portable Theatre, see BOOTHS and FAIRS.

Portal Opening, see FALSE PROS(CENIUM).

Porter, Cole (1891–1964), American composer and lyricist. From a wealthy background, he was intended for the law, but soon turned to music. His first complete score was the short-lived *See America First* (1916), but his heyday did not arrive until the 1930s, with *Gay Divorce* (1932; London, 1933), featuring the classic 'Night and Day', and *Anything Goes* (1934; London, 1935). After successful but lesser shows such as *Leave it to Me* (1938), *Du Barry was a Lady* (1939; London, 1942), *Panama Hattie* (1940; London, 1943), and *Something for the Boys* (1943; London, 1944), he wrote a superb score for *Kiss Me, Kate* (1948; London, 1951) about feuding husband-and-wife thespians, based on *The Taming of the Shrew*. His last two shows, *Can-Can* (1953; London, 1954) and *Silk Stockings* (1955), based on Garbo's film *Ninotchka*, were set in Paris. His stylish and sophisticated output ranged from love songs such as 'In the Still of the Night' to the witty sexual cynicism of 'My Heart Belongs to Daddy', the extended similes of 'You're the Top' and 'Let's Do It', and the Latin American 'Begin the Beguine'.

The lead in *Kiss Me, Kate* was played by **Alfred Drake** [Alfredo Capurro] (1914–), who was later to play 'straight' Shakespeare: Othello and Benedick in *Much Ado about Nothing* at Stratford, Conn., and Claudius to Richard *Burton's *Hamlet* (1964). Among other musical successes were Rodgers and *Hammerstein's *Oklahoma!* (1943), which brought him to stardom, and *Kismet* (1953; London, 1955), based on *Knoblock's play and using Borodin's music.

Porter, Eric Richard (1928–), English actor, who made his first appearance in 1945 at the *Shakespeare Memorial Theatre and eventually joined *Gielgud's company at the *Lyric Theatre, Hammersmith in 1953. He was for a time with the *Bristol Old Vic, where he played Becket in a revival of T. S. *Eliot's *Murder in the Cathedral* and Father James Browne in Graham *Greene's *The Living Room*, as well as a number of classical parts, including the title-roles in *Chekhov's *Uncle Vanya* and *Jonson's *Volpone*. In 1957 he toured England with the *Lunts in *Dürrenmatt's *Time and Again*, in which in 1958 he made his début in New York, where it was renamed *The Visit*. In 1959 he was seen at the *Royal Court Theatre as Rosmer in *Ibsen's *Rosmersholm* and he then joined the *RSC, where his roles included a fine Ulysses in *Troilus and Cressida*, Malvolio in *Twelfth Night*, and Leontes in *The Winter's Tale* (all 1960), the title-roles in *Anouilh's *Becket* (1961) and in *Macbeth* (1962), Barabas in *Marlowe's

The Jew of Malta in 1965 (in conjunction with Shakespeare's Shylock), and Ossip in *Gogol's The Government Inspector (1966). In 1968 he returned to the RSC as King Lear in *Nunn's production and also as Marlowe's Doctor Faustus, repeating the latter role on tour in the USA a year later. At Christmas 1971 he doubled the parts of Mr Darling and Captain Hook in *Barrie's Peter Pan. In 1988 he returned to the stage after a long absence as Big Daddy in Tennessee *Williams's Cat on a Hot Tin Roof at the *National Theatre. In 1989 at the *Old Vic, under Jonathan *Miller's direction, he again played Lear.

Porte-Saint-Martin, Théâtre de la, Paris, celebrated playhouse, built in 1782 to replace the Opéra, which had been burnt down. The opera company remained there until 1794, and the building was apparently not used as a theatre again until 1810, when one of the first plays to be presented was a melodrama by *Pixérécourt. In 1822 an English company appeared unsuccessfully in Othello and in 1827 *Frédérick played for the first time with Mme *Dorval, whose career was to be linked spectacularly with the Porte-Saint-Martin. The great days of the theatre were in the 1830s, when it saw the first night of the elder *Dumas's Antony and Le Tour de Nesle and *Hugo's Marion Delorme and Lucrèce Borgia; but with the decline of Romantic drama the fortunes of the theatre also declined and in 1840 it closed after the banning of Balzac's Vautrin. When it reopened it had no settled policy, but continued to present revivals and commonplace and lachrymose melodramas such as Dennery's Marie-Jeanne; ou, La Femme du peuple (1846), in which Mme Dorval made her last appearance. It was burnt down in the rioting of 1870 and rebuilt on the original plans, but somewhat smaller. It had a further moment of glory in the 1880s when it was acquired by Sarah *Bernhardt, who had appeared there 18 years earlier in the fairy-tale play La Biche au bois and now returned in a revival of Meilhac and Halévy's Frou-Frou. In 1898 the record run of *Rostand's Cyrano de Bergerac again made the theatre one of the most popular in Paris. Because of its great size it was later unable to compete with the cinema, and from 1936 to 1978 it was devoted almost entirely to musical comedy. Marcel *Marceau then took it over as a base for his École de mimodrame. It housed the *Comédie-Française when the latter was strike-ridden, and in 1989 staged an adaptation of *Camus's novel La Peste.

Portman, Eric (1903–69), English actor, who made his début in 1924 and later was at the *Old Vic Theatre, where in 1928 he was much admired as Romeo to the Juliet of Jean *Forbes-Robertson. In the following year he was seen as Stephen Undershaft in a revival of *Shaw's Major Barbara, and he also appeared as

Eben in *O'Neill's Desire under the Elms (1931). He made his début in New York in 1937 in a dramatization of Flaubert's Madame Bovary, and returned to London the following year to appear in *Priestley's I Have Been Here Before. In later years he was outstandingly successful in *Rattigan's double bill of The Browning Version and Harlequinade (1948), and gave fine performances as the Governor in the Christies' His Excellency (1950) and as Father James Browne in Graham *Greene's The Living Room (1953). He was seen later in another double bill by Rattigan, Separate Tables (1954), in which he also acted successfully in New York in 1956, following it on Broadway with several important roles, notably Cornelius Melody in O'Neill's A Touch of the Poet (1958) and Cherry in *Bolt's Flowering Cherry (1959). He returned to England in 1964, and his last role on the stage was in a revival of *Galsworthy's Justice in 1968.

Possart, Ernst von (1841–1921), German actor and theatre manager, who made his first appearance on the stage at Breslau in 1860, playing Iago in Othello. His later roles included Shylock in The Merchant of Venice, Franz Moor in *Schiller's Die Räuber, Mephistopheles in *Goethe's Faust and Carlos in his Clavigo, and the title-role in *Lessing's Nathan der Weise. In 1864 Possart was in *Munich, where he became manager of the theatre in 1875, and well known for his productions of Shakespeare and Schiller with enormous casts in opulent sets. A fine-looking man, with an alert, intelligent face and flashing eyes, he was noted for the beauty of his voice and the dignity and ease of his movements. He was seen in New York in 1910, with much success.

Posset, Jan, see ENGLISH COMEDIANS.

Poster, see PLAYBILL.

Potier des Cailletières, Charles-Gabriel (1774–1838), French actor, considered by some the greatest comedian of his day. After some years in the provinces, he made his début in Paris in 1809 at the Théâtre des *Variétés, and once he had established his reputation he never lost his hold on the affections of the public, spending his last years at the *Palais-Royal, where he had first appeared in 1831. His one fault was a weak voice, but this was offset by the subtlety and vivacity of his acting, which conveyed his meaning without the need for words.

Poulsen, Emil (1842–1911) and **Olaf** (1849–1923), Danish actors, both at the *Royal Theatre, who made their first appearance on the stage on the same day in 1867, Emil playing the title-role in *Holberg's Erasmus Montanus while Olaf played Jacob. Both were excellent actors, Emil being best remembered for his appearances in Holberg, *Molière, and Shakespeare, and taking leading roles in the Danish premières of *Ibsen's plays. Olaf was outstanding in the comic roles of Bottom in *A Midsummer Night's Dream* and Falstaff in *The Merry Wives of Windsor* and *Henry IV*, and in Holberg. Emil's elder son **Adam** (1879–1969) was for some years Director of the Dagmarteatret, Copenhagen, and his younger, **Johannes** (1881–1938), was with the Royal Theatre as actor and director from 1909 until his death, commissioning *Craig to work on a revival of Ibsen's *The Pretenders* in 1926.

Power, Nellie, see MUSIC-HALL.

Power, Tyrone (1797–1841), Irish actor, who first appeared on the stage in 1815 but made little stir until in 1826 he took to specializing in such Irish roles as Sir Lucius O'Trigger in *Sheridan's *The Rivals*. He also appeared in his own plays, among them *St Patrick's Eve* (1832), *Paddy Cary, the Boy of Clogheen* (1833), and *O'Flannigan and the Fairies* (1836). In 1840 he left London for the United States, where he was already a firm favourite after two protracted visits, and on the return journey was drowned in the sinking of SS *President*. His grandson [Frederick] **Tyrone** (1869–1931) was also an actor, as was his son, the third **Tyrone** (1913–58), who had a career in films which overshadowed his work in the theatre but was for some years on the American stage, making his début in Chicago in 1931 and playing opposite Katharine *Cornell in Christopher *Fry's *The Dark is Light Enough* (1955). In 1950 he was in London in *Mister Roberts*, by Thomas Heggen and Joshua *Logan.

Praetexta, see FABULA 3: PRAETEXTA.

Pray, Malvina, see FLORENCE.

Prehauser, Gottfried, see HANSWURST.

Preston, Thomas (*fl.* 16th cent.), English dramatist, author of a popular tragi-comedy, *Cambyses King of Persia* (*c.*1569), written with bombastic eloquence, thus giving rise to Falstaff's remark in Shakespeare's *Henry IV*, *Part One* that he would speak passionately in 'King Cambyses' vein'. This play marks the transition from the medieval *morality play to the Elizabethan historical drama. Its author, who may also have written *Sir Clyomon and Sir Clamydes* (*c.*1570), was evidently not the Thomas Preston (*c.*1537–98) who in 1592, as Master of Trinity Hall, petitioned for the banning of plays in *Cambridge.

Price, (Lilian) **Nancy Bache** (1880–1970), English actress and theatre manager, who made her first appearance on the stage in *Benson's company in 1899. She made her first success as Calypso in Stephen *Phillips's *Ulysses* (1902), and then appeared as Rosa Dartle in *Em'ly*, a dramatization of *Dickens's *David Copperfield*, and as Hilda Gunning in *Pinero's *Letty* (both 1903). She continued to appear in a wide range

of parts, but is now chiefly remembered as the founder and guiding spirit of the People's National Theatre in London, which began in 1930 and during the next few years was responsible for the production of over 50 plays, ranging from *Euripides to *Pirandello, and including Susan *Glaspell's *Alison's House* (1932), *Lady Precious Stream* (1934), a Chinese play by S. I. Hsiung, and Mazo de la Roche's *Whiteoaks* (1936), in which she played for two years the part of old Adeline Whiteoaks. These were all produced at the *Little Theatre, which she made her headquarters. During the Second World War she toured as Madame Popinot in Seymour *Hicks and Ashley *Dukes's *Vintage Wine* and was in Liverpool with the *Old Vic company, but returned to London in 1948, making her last appearance in Eden *Phillpotts's *The Orange Orchard* (1950).

Price, Stephen (1783–1840), American theatre manager, who in 1808 bought a share in the management of the *Park Theatre, New York, and gradually assumed complete control. He inaugurated the pernicious policy of importing famous European actors, beginning with G. F. *Cooke in 1810–11, which by 1840 had wrecked the old resident companies of the larger American towns. He spent a good deal of time in London between 1820 and 1839, engaging English and Continental actors and singers for America, and demonstrating his love of spectacular and freakish effects—real horses and tigers on stage, for example—during his tenancy of *Drury Lane from 1826 to 1830. Price, whose whole theatrical career was closely bound up with that of the Park Theatre, died opportunely just as its fortunes were beginning to decline: he was the first notable American manager who was not also an actor or playwright.

Priestley, J(ohn) **B**(oynton) (1894–1984), English dramatist, novelist, and critic. His first play, a dramatization of his own best-selling novel *The Good Companions* (London and NY, 1931) undertaken in collaboration with Edward *Knoblock, was followed by *Dangerous Corner* (London and NY, 1932), an ingenious play in which a chance remark at a dinner party produces a chain of revelations which lead eventually to a suicide; but the play then returns to its beginnings and the words pass unnoticed. After *Laburnum Grove* (1933; NY, 1935) and *Eden End* (1934; NY, 1935), the latter mingling gentle melancholy and rich humour in a way Priestley never again achieved, came two excellent plays influenced by Dunne's *An Experiment with Time*, *Time and the Conways* and *I Have Been Here Before* (both 1937; NY, 1938), the former being particularly effective, with its second act set 20 years later than the first and third. These 'time-plays' were followed by a rollicking farce, *When We Are Married* (1938; NY, 1939), which concerns three Yorkshire couples who find after many years that their marriages are not legal. In his next two plays, *Music At Night* (also 1938) and *Johnson over Jordan* (1939), in which Ralph *Richardson gave a fine performance, Priestley sought to give modern drama a new depth, but the technical means he employed were not to the taste of the public, though *They Came to a City* (1943), an earnest left-wing political tract, proved surprisingly popular in the West End. Another play in the style of *Dangerous Corner*, *An Inspector Calls* (1946; NY, 1947), in which Richardson again appeared, was followed in 1947 (NY, 1948) by one of Priestley's best plays, *The Linden Tree*, in which Lewis *Casson and Sybil *Thorndike played an academic and his wife confronted at a family reunion by the contrasting ideologies of their three adult children. His later plays, such as *Home is Tomorrow* (1948) and *Summer Day's Dream* (1949), proved less memorable. *Dragon's Mouth* (1952; NY, 1955) and *The White Countess* (1954) were written in collaboration with his third wife, the archaeologist Jacquetta Hawkes, while his last plays included *Mr Kettle and Mrs Moon* (1955), *The Glass Cage* (1957), and a dramatization of Iris Murdoch's novel *A Severed Head* (1963; NY, 1964) in collaboration with the author. He was the first President of the *International Theatre Institute.

Prime Minister of Mirth, see ROBEY.

Prince, Hal, see MUSICAL COMEDY and SONDHEIM.

Prince Charles's Men, company usually known as the Prince's Men formed by *Alleyn in 1616, on the death of his father-in-law *Henslowe, under the patronage of the Prince of Wales (later Charles I). They were seen at several theatres including the *Hope before settling in 1619 at the *Cockpit, where they remained until 1622. They were then ousted by the newly formed *Lady Elizabeth's Men and went first to the *Curtain and then to the *Red Bull. On the accession of Charles I in 1625 they were disbanded, many of the leading actors joining the *King's Men, but in 1631 a new company, under the same name and the patronage of the young Prince Charles (later Charles II), was formed to play at *Salisbury Court, taking in some of the actors from the *Fortune, where the new company itself played from 1640 to the closing of the theatres in 1642.

Prince Edward Theatre, London, in Old Compton Street, Soho. It opened in 1930 with a musical, *Rio Rita*, which had a short run. Except for a play on the adventures of Sexton Blake the theatre was used mainly for musicals and revues, of which the best were *Nippy* (also 1930) and *Fanfare* (1932). In 1935, after a Christmas *pantomime, *Aladdin*, it closed, to

reopen in 1936 as the London Casino, a cabaret-restaurant featuring a spectacular stage show. On the outbreak of war in 1939 it became the Queensberry All-Services Club, reopening as a theatre in 1946 with another pantomime, *Mother Goose*, followed by a revival of Ivor *Novello's *The Dancing Years*. Three editions of Robert Nesbitt's revue *Latin Quarter*, beginning in 1949, kept the theatre open, but in 1954 it became a cinema, except for a lavish *Cinderella* in 1974. In 1978 it again reverted to use as a theatre under its original name with Andrew Lloyd Webber's musical *Evita*, which ran until in 1986 it was replaced by another musical *Chess*. In 1989 the latter was succeeded by a revival of Cole *Porter's *Anything Goes*.

Prince Henry's Men, see ADMIRAL'S MEN.

Prince of Wales Theatre, London, in Coventry Street. (The name is also found on playbills and programmes as Prince of Wales' and Prince of Wales's.) It opened as the Prince's Theatre in 1884, and on 3 Mar. a free adaptation of *Ibsen's *A Doll's House* (as *Breaking a Butterfly*) was produced, but aroused little interest. It was followed by *The Private Secretary*, a German play adapted by *Hawtrey, which was not at first a success but when transferred to the *Globe had a long run. The first successful production at the Prince of Wales, as it was renamed in 1886, was the wordless play *L'Enfant prodigue* (1891), superbly mimed, which served to introduce *Pierrot to London in something other than *pantomime. *In Town*, often considered the first English *musical comedy, was presented here by George *Edwardes in 1892, and was followed by the equally successful *A Gaiety Girl* (1893). The theatre moved over to straight plays with *Forbes-Robertson and Mrs Patrick *Campbell in *Maeterlinck's *Pelléas and Mélisande* and *Martin-Harvey in Wills's adaptation of *Dickens's *A Tale of Two Cities* as *The Only Way*, transferred from the *Lyceum. Marie *Tempest appeared as Nell *Gwynn (1900) and Becky Sharp (1901), but the theatre returned to musical comedy between 1903 and 1910, including *Miss Hook of Holland* (1907) and *The King of Cadonia* (1908). For the next two decades it housed musicals, plays such as Ivor *Novello's *The Rat* (1924), and revues including *Co-Optimists* (1923) and *Charlot's Revue* (1924). During much of the 1930s the theatre was given over to non-stop *revue, and when *Encore les dames* closed in 1937 the building was demolished. A new theatre under the old name, seating 1,139 in two tiers, opened in 1937 with *Les Folies de Paris et Londres*. In 1943 Sid Field made his London début here in *Strike a New Note*, and he returned to star in *Piccadilly Hayride* (1946). Three years later he also starred in Mary Chase's comedy about an imaginary rabbit, *Harvey*, which had a long run, as did Paul Osborn's *The World of Susie Wong* (1959) and Neil *Simon's

Come Blow Your Horn (1962), while *Funny Girl* (1966) with Barbra Streisand, *Sweet Charity* (1967), and *Promises, Promises* (1969) brought to London three successful Broadway musicals. Later long-running shows were Bernard Slade's *Same Time, Next Year* (1976), *Ayckbourn's *Bedroom Farce* (1978), and a revival of the musical *Guys and Dolls* in 1985, the last two both transferred from the *National Theatre. Rodgers and *Hammerstein's musical *South Pacific* was successfully revived in 1988, and Andrew Lloyd Webber's musical *Aspects of Love* began a long run in 1989.

The *Scala, under the *Bancrofts, was also known as the Prince of Wales's.

Princess's Theatre, London, in Oxford Street. After the passing of the Licensing Act of 1843, the Princess's Theatre, which had opened in 1840 for concerts and later been used as an opera-house, was used for plays, Charlotte *Cushman making her London début there in *Macbeth*, with Edwin *Forrest, in 1845. A number of famous actors were seen briefly until Charles *Kean took over, opening in 1850 with *Twelfth Night*. Kean's management was memorable, both for his productions of Shakespeare and for his success in adapting French drama to suit English audiences. Among his most popular productions in this field were Oxenford's *Pauline* (1851), *Boucicault's *The Corsican Brothers* (1852) and *Louis XI* (1855), and Charles *Reade's *The Courier of Lyons* (1854) (better known as *The Lyons Mail*). It was during a spectacular revival of *Henry VIII* that limelight (see LIGHTING) was used for the first time in the theatre. In 1856 Ellen *Terry, then 9 years old, made her first appearance on the stage as Mamillius in *The Winter's Tale*, her elder sister Kate being already a member of the company. Kean left the Princess's in 1859 after appearing once more as Wolsey, one of his finest parts, and Augustus *Harris took over. Under him Henry *Irving made his first appearance in London in Oxenford's *Ivy Hall* and then returned to the provinces. A year later *Fechter appeared in *Hugo's *Ruy Blas*, and as Hamlet, with great success. Boucicault's melodramas, including *The Streets of London* (1864), revived many times under many names, and *Arrahna-Pogue* (1865), were followed by Charles Reade's *It's Never Too Late to Mend*, whose first night on 4 Oct. 1865 caused a riot because of the realistic flogging of a boy in the prison scene. The fortunes of the theatre then gradually declined, one of the few successes being scored by *Warner in Reade's *Drink* (1879), based on *Zola's *L'Assommoir*. Out of the profits from this production the management decided to rebuild the theatre, which was demolished in 1880. The new Royal Princess's opened six months later with Edwin *Booth as Hamlet. Wilson *Barrett later took over, and under him

Sims's *The Lights o' London* (1881) and Henry Arthur *Jones and Herman's *The Silver King* (1882) had long runs. But there were many failures, and in 1902 the theatre closed. Three years later it became a warehouse, and in 1931 it was demolished.

Prince's Theatre, Bristol, see BRISTOL.

Prince's Theatre, London, see PRINCE OF WALES THEATRE and SHAFTESBURY THEATRE 2.

Principal Boy, chief character in the English *pantomime—Aladdin, Dick Whittington, Robinson Crusoe, Prince Charming—traditionally played by a woman in a blond wig, short tunic, fleshings, and high heels. The character had its origin in the 'breeches parts' and in *burlesque, but was not firmly established until the 1880s. The tradition was carried on in the 20th century by a number of excellent actresses, but a tendency towards realism led, as early as 1938, to the appearance of an actor in a role hitherto reserved for actresses. This was not favourably received and pantomimes still usually have a woman in the part.

Printemps, Yvonne (1895–1977), French actress and singer, who made her first appearance in Paris in revue in 1908. She was for some time at the *Folies-Bergère, and also appeared at the *Palais-Royal. In 1916 she joined the company of Sacha *Guitry whom she married in 1919, appearing with him in a number of plays including his own *Nono* in which she was first seen in London in 1920. She played in English for the first time in *Coward's *Conversation Piece* (London and NY, 1934), and was also seen in London in Ben *Travers's *O Mistress Mine* (1936), both with her second husband Pierre *Fresnay. She made her first appearance in New York in 1926 with Guitry in his *Mozart*, playing the title-role, one of her most charming parts, which she had played in London earlier in that year. In 1937 with Fresnay she took over the Théâtre de la Michodière, where she appeared in a succession of musical plays and light comedies during the next 20 years.

Pritchard [*née* Vaughan], **Hannah** (1711–68), English actress, wife of the actor **William Pritchard** (1707–63). She already had some experience of acting when in 1733 she made her first appearance at *Drury Lane, where she remained for some years, spending the summer seasons from 1741 to 1747 at the Jacobs Wells Theatre in *Bristol, and being from 1743 to 1747 under *Rich at *Covent Garden. She quickly achieved an enviable reputation in light comedy, being seen as Lady Brumpton in *Steele's *The Funeral* and Lady Townly in Colley *Cibber's *The Provok'd Husband*. She was much admired as Rosalind in *As You Like It* and Beatrice in *Much Ado about Nothing*, and played Nerissa to the Portia of Kitty *Clive in *Macklin's epoch-making production of *The*

Merchant of Venice in 1741. In 1748 *Garrick, to whose Chamont in *Otway's *The Orphan* she had played Monimia in 1742, invited her to join his new company at Drury Lane, and she remained one of his leading ladies for 21 years, retiring only a few months before her death. Under him she continued her outstanding career in comedy, one of her creations being Mrs Oakly in the elder *Colman's *The Jealous Wife* (1761); but she widened her range to include tragedy, giving more prominence than usual to Gertrude in *Hamlet* and Queen Katharine in *Henry VIII*. She was also the first and only interpreter of the heroine in Dr *Johnson's *Irene* (1749); but the part in which she excelled all her contemporaries was Lady Macbeth, which she first played with Garrick in 1748, and in spite of the increasing obesity which troubled her later years chose for her farewell to the stage in 1768, after which Garrick never played Macbeth again.

Mrs Pritchard's two brothers, **Henry** (1713–79) and **William** (1715–63) **Vaughan**, were both on the stage, playing mainly low comedy, and her daughter **Hannah** (1739–81), who married an actor and became a member of Garrick's company in 1756, making her début as Juliet to her mother's Lady Capulet. She retired on the death of her husband in 1768.

Private Theatres in England. The vogue for private theatricals in English high society, which reached its peak from 1770 to 1790, resulted in the building of several private theatres. The first specially erected indoor private theatre which could vie with such Continental Court theatres as *Česky Krumlov, *Drottningholm, or *Gripsholm was erected at Wargrave in 1789 for the Earl of Barrymore. Rectangular in shape, it held 400, and contained two tiers of boxes and two stage boxes unusually placed over the orchestra well. After ruining its owner, who also had a small puppet theatre in Savile Row, London, it was demolished in 1792. In 1793 the Lady Elizabeth Craven, Margravine of Anspach, built a theatre for the production of her own plays at Brandenburgh House, Hammersmith, which remained in use until 1804. It was in castellated style, and is said to have resembled the Bastille rather than a Temple of the Muses. An engraving of the interior shows a large central box and a parterre raised on a shallow platform after the Continental model.

A theatre in a simpler style was converted from a kitchen for Sir Watkins Williams Wynn at Wynnstay, and was used for annual performances from 1771 to 1789. Other private theatres were the Duke of Marlborough's at Blenheim Palace in Oxfordshire, converted from a greenhouse and in use from 1787 to 1789, and the Duke of Richmond's at Richmond House, London, constructed out of two rooms in 1787 to hold about 150. The wardrobe book of Wynnstay and the sale catalogue of Wargrave give ample evidence of the sumptuous décors provided for performances at these theatres, where professionals frequently acted with or coached the amateurs.

The Duke of Devonshire's private theatre at Chatsworth, Derbyshire, dating from about 1830, is the oldest still in existence. Another theatre of that period was at Burton Constable, in Yorkshire, which functioned from 1830 to 1850. Victorian private theatres include Charles *Dickens's at Tavistock House, in London; a small one attached to Campden House in Kensington in the 1860s; one at Capethorne Hall, Cheshire, in use in 1870; the artist Herkomer's at Bushey, Hertfordshire; and the singer Patti's at 'Craig y Nos, Wales, still extant. In the 20th century Lord Bessborough gave annual productions for some years at his theatre at Stansted, Essex, and Lord Faringdon built a theatre, still in use, at Buscot Park, Berkshire. The late John Christie's opera house at Glyndebourne, Sussex, and Nicholas Sekers's theatre at Rosehill, Whitehaven, Cumbria, though founded by private individuals, both employ professional companies, are open to the general public, and charge for admission, and are thus not strictly to be classed as private theatres.

Proctor, Frederick Francis, see VAUDEVILLE.

Producer, American term for the man responsible for the financial side of play-production, for the buying of the play, the renting of the theatre, the engagement of actors and staff, and the handling of the receipts. In Britain most of these functions are assumed by the manager. Formerly the person who was responsible for the actual staging of the play and the conduct of rehearsals was known in Britain as the producer, but this has now been universally replaced by the American term *director.

Profile Board, see FLAT.

Programme, see PLAYBILL.

Proletkult Theatre, Moscow, see TRADES UNIONS THEATRE.

Prologue, introductory poem or speech, which originally explained or commented on the action of the play which it preceded. It was first used by *Euripides and later by the Elizabethans, who applied to it the name *Chorus. Together with the epilogue, which closed the action, the prologue was extensively used during the Restoration period, providing a good deal of incidental information on the contemporary theatre. It survived well into the 18th century, and disappeared with the crowded bills of the 19th century. It is now used only on special occasions. At their best the prologue and epilogue were witty and sometimes scurrilous commentaries on politics and social conditions, written by outstanding men of the theatre, such as *Dryden and *Garrick, and spoken by the leading actor or actress.

Promenade Productions, see THEATRE-IN-THE-ROUND.

Prompt Book, Box, Corner, Side, see STAGE DIRECTIONS.

Prompter, person responsible for supplying the necessary 'next line' to an actor who has forgotten his part—a function assumed in the Elizabethan theatre by the *book-holder. The prompter attends all rehearsals, and should note in his copy of the play all alterations in the script and, if so instructed by the *stage-manager, all the moves, which should also be in the stage-manager's copy. Prompt copies of early plays are exceedingly valuable documents, as they often contain lists of actors' names and sometimes provide the only clue to the author.

Prophet Play, see LITURGICAL DRAMA.

Props, usual term for stage properties. It covers anything essential to the action of the play which does not come under the heading of *costume, *scenery, or furniture. Hand-props are those which an actor handles—letters, documents, revolvers, newspapers, knitting, snuff boxes, and so on. These are given to him as he goes on stage, and taken from him as he comes off, and are not his personal responsibility. Other props—stuffed birds, food in general, dinner-plates, telephones—are placed on stage by the property man, who is responsible for all props under the direction of the *stage-manager. He has for storage a property room backstage, from which he is expected to produce at a moment's notice anything that may be required. He must also prevent the removal from it of oddments by members of the company.

Proscenium, word which is classical in origin, and in the later Greek theatre meant the area in front of the stage. It was used in Renaissance Italy for the draperies surrounding the stage picture, introduced into England by Inigo *Jones with his scenery for Court *masques, and in its modern meaning is applied to the permanent or semi-permanent wall dividing the *auditorium from the *stage. It has hanging in it the front *curtain, and can be made smaller by the use of a *false pros(cenium). It was once a feature of considerable architectural complexity, forming an essential link between the auditorium and the scene, but it is now generally regarded as a hindrance, particularly in the production of Elizabethan plays and modern *epic theatre. The tendency is to abolish it wherever possible in favour of the *open stage or *theatre-in-the-round, but one obstacle which confronts those who wish to experiment with these new forms in the older London theatres is the presence within the proscenium arch of the Safety Curtain demanded by fire regulations.

Proscenium Border, Pelmet, see BORDER.

Proscenium Doors, or Doors of Entrance, permanent feature of the English Restoration playhouse. Set on each side of the *forestage, they had practicable knockers and bells, and provided the usual means of exit and entrance for the actor. Leaving by one door and returning by another, he was presumed to be in another room even though the *wings and *backcloth remained unchanged. The number of doors varied from four to six not only in different theatres but also in the same theatre at different periods. By the early 18th century they were reduced to one on each side, and by the beginning of the 19th they were used only by the actor 'taking a bow' after the play. They were then known as Call Doors.

Prospect Theatre Company, founded as Prospect Productions in Oxford in 1961. After three successful summer seasons it became a touring company based on the Arts Theatre, *Cambridge, with **Toby Robertson** (1928–) as its Artistic Director, 1964–79. It soon achieved an outstanding reputation for its high standard of acting and ensemble playing and its staging of classic plays with brilliant costumes but minimal sets. From 1967 it had a unique association with the *Edinburgh Festival, where it presented numerous productions. After 1969, when the association with Cambridge ended, the lack of a base large enough for its extended activities was a severe handicap, until in 1977 a highly successful season at the *Old Vic Theatre, recently vacated by the *National Theatre company, led to its establishment there, and a change of name two years later to the Old Vic Company. Among its notable productions were *Richard II* in 1968 and *Marlowe's *Edward II* in 1969, Ian *McKellen playing the title-role in both. A production of *Hamlet* with Derek *Jacobi in 1977 was taken on an extensive overseas tour, being seen at Elsinore, Denmark, and in China, marking the first visit by a British company to the People's Republic. In 1980 Toby Robertson was succeeded by **Timothy West** (1934–), who had first joined the company in 1966, when he played Prospero in *The Tempest*, his many later roles including King Lear (1971) and Holofernes in *Love's Labour's Lost* (1972). West's first season as Artistic Director included a much criticized but commercially successful production of *Macbeth* with Peter *O'Toole and *The Merchant of Venice* with West himself as Shylock. The company appeared to be prospering when it lost its *Arts Council grant, and it had to be disbanded in 1981.

Protagonist. In ancient Greek drama the tragic poet was originally allowed only one actor and a *chorus. By the 5th century BC, when three actors were allowed, each playing if necessary more than one part, they were named the

Protagonist, Deuteragonist, and Tritagonist (from *agon*, contest). This probably did not mean that the Protagonist was the best actor of the three. He may have been allotted the best role, but in the plays of *Euripides and *Sophocles there are usually at least two major roles of equal importance. It may merely have been that he was the first to speak. There is certainly no indication that the Tritagonist was a third-rate or small-part player.

Provincetown Players [and **Playhouse**], group of American actors and playwrights founded in 1916 by Susan *Glaspell and others, whose ardent experimentalism gave Eugene *O'Neill the opportunities he needed at a vital stage in his career. Their first season, at the Wharf Theatre, Providence, RI, a converted fishing shack, included the first of his plays to be staged, the one-act *Bound East for Cardiff*, as well as new works by Susan Glaspell, Edna St Vincent Millay, Laurence *Langner, and others. Later in the year the group moved to the Playwrights' Theatre in Greenwich Village, New York (although they continued to make the Wharf Theatre their summer headquarters until 1921), leaving it in 1918 for the Provincetown Playhouse, a converted stable a few doors away. They ceased operations in 1921, but three years later the Playhouse reopened under the management of Kenneth MacGowan, Robert Edmond *Jones, and O'Neill himself, working in conjunction with the Greenwich Village Theatre, which saw the first production of O'Neill's *Desire under the Elms* (1924). After MacGowan's departure it functioned for a further year as the Irish Theatre, but then closed and was demolished in 1930. At the Provincetown Playhouse a number of contemporary European plays were staged; classical revivals included *Congreve's *Love for Love* in 1925. Some of the original members of the Provincetown Players continued to work at the Playhouse, but in 1929, after an unsuccessful move to the *Garrick Theatre on Broadway, the group disbanded.

Prowse, Philip, see CITIZENS' THEATRE.

Public Theatre, New York, on Lafayette Street, was converted by Joseph *Papp from the old Astor Library in 1967. Its auditoriums include the three-quarters arena Anspacher, seating 275, which opened in 1967 with the rock musical *Hair*, later to have a long run at the *Biltmore; the flexible Other Stage, seating 108, opened in 1968 and renamed the Susan Stein Shiva in 1984; the Newman, a proscenium-arch theatre seating 300, opened in 1970; and Martinson Hall, a flexible theatre seating 191 opened in 1971. The 150-seat South Hall opened in 1973 as a concert hall; it was eventually renamed the LuEsther Hall and used for plays and other presentations. The Public Theatre stages new American works and the American premières of foreign plays. Among the former were

Charles Gordone's *No Place to be Somebody* (1969), the first *Off-Broadway play to win a *Pulitzer Prize; David *Rabe's *Sticks and Bones* (1971); Jason Miller's *That Championship Season* (1972), another Pulitzer Prize-winner; and *A Chorus Line* (1975), which went on to become the longest running Broadway musical of all time. Foreign imports included Václav *Havel's *The Memorandum* (1968) and *Largo Desolato* (1986), and many imports from London, including plays by David *Hare and Caryl Churchill. In 1988 the theatre staged *A Midsummer Night's Dream*, the first in a series to include the entire Shakespeare canon performed only by American actors and staged either at the Public Theatre or in the summer at Papp's Delacorte Theatre. The Public Theatre's future has been assured by a permanent subsidy from the City of New York.

Pulcinella, one of the *zanni or comic servant masks of the *commedia dell'arte, a humpbacked, doltish fellow. In his Italian form he originated from Naples. Little is known of his roles or of his original dress, but he is important as the ancestor of the French *Polichinelle and the English *Punch. Some Italian critics believe that the popularity of Pulcinella and the disproportionate attention paid to his buffoonery was one of the main reasons for the decline of the *commedia dell'arte*.

Pulitzer Prize, founded under the will of Joseph Pulitzer (1847–1911), awarded annually (unless the judges decide to withhold it) for the best play, preferably one dealing with American life, produced in New York during the preceding 12 months. It has been awarded as follows:

1916–17:	No award
1917–18:	*Why Marry?* (Jesse Lynch Williams)
1918–19:	No award
1919–20:	*Beyond the Horizon* (Eugene *O'Neill)
1920–1:	*Miss Lulu Bett* (Zona Gale)
1921–2:	*Anna Christie* (Eugene *O'Neill)
1922–3:	*Icebound* (Owen Davis)
1923–4:	*Hell-Bent fer Heaven* (Hatcher Hughes)
1924–5:	*They Knew What They Wanted* (Sidney *Howard)
1925–6:	*Craig's Wife* (George *Kelly)
1926–7:	*In Abraham's Bosom* (Paul *Green)
1927–8:	*Strange Interlude* (Eugene *O'Neill)
1928–9:	*Street Scene* (Elmer *Rice)
1929–30:	*The Green Pastures* (Marc *Connelly)
1930–1:	*Alison's House* (Susan *Glaspell)
1931–2:	*Of Thee I Sing* (George S. *Kaufman, Morrie Ryskind, Ira and George *Gershwin)
1932–3:	*Both Your Houses* (Maxwell *Anderson)
1933–4:	*Men in White* (Sidney *Kingsley)
1934–5:	*The Old Maid* (Zoë *Akins)
1935–6:	*Idiot's Delight* (Robert E. *Sherwood)

1936–7: You Can't Take It With You (Moss *Hart and George S. *Kaufman)
1937–8: Our Town (Thornton *Wilder)
1938–9: Abe Lincoln in Illinois (Robert E. *Sherwood)
1939–40: The Time of Your Life (William *Saroyan)
1940–1: There Shall Be No Night (Robert E. *Sherwood)
1941–2: No award
1942–3: The Skin of Our Teeth (Thornton *Wilder)
1943–4: No award
1944–5: Harvey (Mary Chase)
1945–6: State of the Union (Howard *Lindsay and Russel Crouse)
1946–7: No award
1947–8: A Streetcar Named Desire (Tennessee *Williams)
1948–9: Death of a Salesman (Arthur *Miller)
1949–50: South Pacific (Richard Rodgers, Oscar *Hammerstein II, and Joshua *Logan)
1950–1: No award
1951–2: The Shrike (Joseph Kramm)
1952–3: Picnic (William *Inge)
1953–4: The Teahouse of the August Moon (John Patrick)
1954–5: Cat on a Hot Tin Roof (Tennessee *Williams)
1955–6: The Diary of Anne Frank (Frances Goodrich and Albert Hackett)
1956–7: Long Day's Journey Into Night (Eugene *O'Neill)
1957–8: Look Homeward, Angel (Ketti Frings)
1958–9: J.B. (Archibald *MacLeish)
1959–60: Fiorello! (Jerome Weidman, George *Abbott, Sheldon Harnick, and Jerry Bock)
1960–1: All the Way Home (Tad Mosel)
1961–2: How to Succeed in Business Without Really Trying (Abe Burrows, Willie Gilbert, jack Weinstock, and Frank Loesser)
1962–3: No award
1963–4: No award
1964–5: The Subject Was Roses (Frank D. Gilroy)
1965–6: No award
1966–7: A Delicate Balance (Edward *Albee)
1967–8: No award
1968–9: The Great White Hope (Howard Sackler)
1969–70: No Place To Be Somebody (Charles Gordone)
1970–1: The Effect of Gamma Rays on Man-in-the-Moon Marigolds (Paul Zindel)
1971–2: No award
1972–3: That Championship Season (Jason Miller)
1973–4: No award
1974–5: Seascape (Edward *Albee)
1975–6: A Chorus Line (Michael Bennett, James Kirkwood, Nicholas Dante, Marvin Hamlisch, and Edward Kleban)
1976–7: The Shadow Box (Michael Cristofer)
1977–8: The Gin Game (D. L. Coburn)
1978–9: Buried Child (Sam *Shepard)
1979–80: Talley's Folly (Lanford *Wilson)
1980–1: Crimes of the Heart (Beth Henley)
1981–2: A Soldier's Play (Charles Fuller)
1982–3: 'night, Mother (Marsha Norman)
1983–4: Glengarry Glen Ross (David *Mamet)
1984–5: Sunday in the Park With George (James Lapine and Stephen *Sondheim)
1985–6: No award
1986–7: Fences (August *Wilson)
1987–8: Driving Miss Daisy (Alfred Uhry)
1988–9: The Heidi Chronicles (Wendy Wasserstein)
1989–90: The Piano Lesson (August *Wilson)

Punch and Judy, English *puppet-show, presented on the miniature stage of a tall collapsible *booth traditionally covered with striped canvas. It was once a familiar sight in the streets of large cities and can still be seen occasionally in seaside towns. Punch, the chief character, with his humped back and hooked nose, evolved from the *Pulcinella of the *commedia dell'arte, and first appeared in London as part of the Italian marionette shows popular after the Restoration. While retaining the physical peculiarities of his Italian prototype, Punch soon became the ubiquitous English buffoon of every puppet-play of the period, equally at home with Adam and Eve, Noah, or Dick Whittington, taking over many of the characteristics of the old *Vice of the medieval *mystery play. When in the early years of the 18th century fashionable London grew tired of his antics, he migrated to the country fairs, took a wife (first called Joan, later Judy), and adopted the familiar high-pitched voice produced by introducing a 'swazzle' or squeaker into the mouth of the showman who spoke for him. Towards the end of the century he went into eclipse, but emerged again in the 19th century as a hand- or glove-puppet, a reversion to the style of the early English puppet-show which had temporarily been ousted by the Italian stringed marionettes. The change proved economically worthwhile, for one man could carry the portable booth on his back and present all the characters with his own two hands, with a mate (or wife) to 'bottle', or collect pennies from the audience. In the more or less standardized version of the play, which dates from about 1800, Punch, on the manipulator's right hand, remains on stage all the time, while the left hand provides a series of characters—baby, wife, priest, doctor, policeman, hangman—for him to nag, beat, and finally kill, until he is eaten by a crocodile, carried off by the Devil, or allowed to remain in solitary triumph, his only companion being his faithful *Toby—a live dog, usually a terrier, who sits on the ledge of the booth during the entire performance.

Puppet, inanimate figure controlled by human agency, which can be larger than life or only a few inches high. It is probably as old as the theatre itself, and it is possible that many of the wonder-working idols of pagan times were in effect immense puppets controlled by their officiating priests; but in its modern sense a puppet is a semblance of a creature—man, bird, beast, fish—given movement and the appearance of life by direct human assistance.

There are several different kinds of puppet, among them the hand- or glove-puppet, the rod-puppet, the marionette, all of which are rounded figures, and the flat puppets of the *shadow-show and the *toy theatre. Because of the popularity of the *Punch and Judy show the hand-puppet is the best known in England. The successful hand-puppet play—or *motion, as it was called in England—concentrates on broad, simple effects, humorous dialogue, and knockabout comedy. Many of the popular national puppet characters are hand-puppets, carried across Europe by wandering showmen. Apart from the English Punch there is the French *Guignol, a generous, bibulous, and witty Lyonnais silk-weaver; the German *Kasperle, a slyly astute peasant; the Russian Petrushka; and the Italian *Pulcinella, the father of them all. There are hand-puppets in China not very dissimilar from the European types; in India, where they were once very common, they survive mainly in Kerala.

An extension of the hand-puppet, still to be found in India, is the rod-puppet, a full-length rounded figure supported and controlled from below. Its movements are comparatively slow and limited, but the control is absolute, and broad gestures of rare beauty with the arms can be obtained. The most famous and beautiful rod-puppets are found in Java. Some striking effects with rod-puppets were achieved in Vienna, where the stage was seen through a convex lens which enlarged the figures, lending them an aura of enchantment and mystery. The simplest form of rod-puppet is the *Fool's *marot* or bauble, a replica of his own head with its cap and bells, fastened to a stick. The Bread and Puppet Theatre in New York, a radical political protest group founded in the 1960s, used giant puppets manipulated by both internal and external operators with rods sometimes as long as 30 feet; they were effectively used in street theatre.

The most elaborate form of puppet is probably the string puppet or marionette, originally controlled from above by rods or wires running to the centre of the head and to each limb, as in the Sicilian puppets used in productions of *Ariosto's *Orlando furioso*. Between 1770 and 1870 they were entirely manipulated by strings, thus allowing far more flexibility in limb and head movements. A standard marionette has a string to each leg and arm, two to the head, one to each shoulder (which take the weight of the body), and one to the back—nine strings (actually fine threads) in all. An elaborate figure can have two or three times this number. All the strings are gathered together on a wooden 'crutch' or control, held in one hand by the manipulator, while with the other he plucks at whatever strings are required. The figures vary in size from 12 to 18 inches for home use and up to 2 or 3 feet for public performances. The *bunraku* puppets of Japan, seen in London in 1968 during the *World Theatre Season, are about two-thirds life-size. They are sometimes strung like marionettes, but more often manipulated by as many as three operators to each figure, working in full view of the audience, and controlling their charges by means of wires and levers in their backs. Indian string puppets, now mostly used for the Tamil 'dance of the dolls', are manipulated somewhat differently.

The Italian Fantoccini puppets, who appeared in London at the Restoration, and the Puppet Theatre in the Piazza in Covent Garden between 1710 and 1713, were all marionettes, as were those used by Samuel *Foote for satirical purposes in 1733, and by Charles *Dibdin, who erected a puppet theatre at Exeter in 1775. The fortunes of the marionette then waned, but there was a revival of interest in the early 20th century, fostered by Gordon *Craig with his emphasis on the actor's role as an 'Übermarionette'. This led to an artistic flowering which bore fruit in the work of such groups as the Hogarth Puppets and John Wright's marionettes at the Little Angel Theatre in Islington. But in spite of the foundation in 1925 of a British Puppet and Model Theatre Guild, puppets in England have a limited appeal and are often thought of only as educational. In the United States, where there was no tradition of puppetry, they were also slow to establish themselves, and still attract only a minority audience. The true home of the puppet-theatre is the Far East and Eastern Europe, where it covers everything from elementary education in backward areas to sophisticated cabaret shows in the big cities.

Purim Plays, associated with the Jewish Festival of Purim on the 14th Adar (roughly the middle of March), appear to have originated in France and Germany as early as the 14th century, mainly as extemporized entertainments centring on the Old Testament story of Esther and Haman. Under the influence of the masquerades and mumming of the Italian Carnival, they developed into plays featuring racy dialogue, with interposed songs and dances, which widened their scope to include other Old

Testament figures such as Joseph and his brethren, David and Goliath, Moses and Aaron. Mostly in one act, they featured comic rabbis, apothecaries, midwives, and devils, the whole ending with a final chorus foretelling Israel's salvation. The religious authorities, who had at first opposed the acting of Purim plays, finally bowed to public demand and tolerated them as long as they did not overstep the bounds of decency. Given mainly in the countryside, they spread rapidly across Europe to the east, and remained popular in some remote districts until the end of the 19th century. They had a considerable influence on the development of *Jewish drama, and took on a literary form in the 17th century. Under the influence of the *commedia dell'arte Goliath or Haman assumed some of the characteristics of the *Capitano, Abraham or Jacob those of *Pantalone, the Devil those of *Harlequin. The *English Comedians had in their repertory a version of Esther and Haman, printed in German in 1640: if there was an original version in English it is lost. The first original German Purim play is an Ahasuerus acted at Frankfurt-am-Main in 1708. The Purim players were constantly in trouble with the authorities for the accretion of vulgar and bawdy ornamentations which gathered round the original plots, and in 1720 the reaction set in. In Germany the Haskala (Enlightenment) groups continued the fight against vulgarity and improved a number of accepted texts by relegating the comic figures to a subplot, and stressing the religious and educational elements. By the early 19th century Purim plays were covering the whole range of drama, but the founding by *Goldfaden of a permanent Yiddish theatre which embodied in its productions much of the spirit and method of the Purim plays finally led to their disappearance.

Puritan Interregnum, name given in English history to the period of Puritan domination in the mid-17th century. Opposition to the building of playhouses and the performing of plays by professional companies had been growing steadily among the Puritans since the beginning of the century, and this culminated in 1642 in a Parliamentary Ordinance which led to the closing of theatres throughout the country and the dispersal of the acting companies. Under Charles I the London theatre, its actors and its dramatists, had become increasingly attached to the Royalist cause, and the assembling of audiences in large playhouses had provided excellent opportunities for subversive activities. The Puritans maintained that they had no objection in principle to drama; under the Commonwealth plays continued to be acted in schools, with the approval of Cromwell himself, and even in private houses, and in 1656 *Davenant was allowed to produce publicly at

Rutland House his 'entertainment with music' The Siege of Rhodes, now regarded as the first English opera. But the Puritans held that there were sound political and social, as well as religious, reasons for the banning of stage-plays, which, as the Ordinance said, did not agree 'with private calamities . . . nor with the seasons of humiliation'. As a result, for 18 years actors were deprived of their livelihood and their theatres stood derelict, many of them never to be used again. Some actors joined the army, some returned to their old trades. Only the boldest, or the most desperate, tried to evade the ban. Evidence of surreptitious performances is given by court judgements imposing fines or terms of imprisonment on actors found playing, usually in the smaller theatres such as the *Cockpit or the *Red Bull. The worst consequence of the closing of the theatres was perhaps that ordinary people lost the habit of playgoing and it took a long time to win them back again: even now the bulk of the population has not been won over.

Pushkin, Alexander Sergeivich (1799–1837), Russian poet, who at the age of 8 reputedly wrote little plays in French, which he acted with his sister. It is evident from his letters and other sources that Pushkin contemplated writing a series of dramatic works of which his great drama Boris Godunov alone was completed. It is notable as being the first Russian tragedy on a political theme—the relationship between a tyrant and his people—which, though set back in time, was actually a burning contemporary problem; and it does not rely on a love-intrigue. In other respects, too, it was revolutionary; it was broken up into scenes and episodes, it mingled poetry with prose, and it made use of colloquial Russian speech. It was not published until six years after its completion in 1825, owing to trouble with the censorship, and was not seen on the stage until 1870. Four years later it was used for the libretto of an opera by Mussorgsky, in which form it is best known today. Just before his death in a duel Pushkin completed a series of one-act tragedies, one dealing with *Don Juan, one with the rivalry of Mozart and Salieri, and a third with a miser who owes something to Harpagon in Molière's L'Avare, but more to Shylock in The Merchant of Venice. With some unfinished scenes taken from Russian folklore, these made up the total of Pushkin's work for the theatre. Yet though he is primarily remembered as a poet, the Russian theatre owes him a great debt, since it was he who first made Russian a literary language. There are theatres named after him in Leningrad and Moscow, and his works are quoted by Russians much as Shakespeare's are by the English.

Pushkin Theatre, Leningrad. This theatre, renamed in 1937 in honour of Russia's greatest poet, was founded in what was then St Petersburg as the Alexandrinsky. It opened in 1824, the same year as the *Maly Theatre in Moscow. It had a fine leading actor in *Karatygin but no dramatists of the calibre of *Gogol and *Ostrovsky, and it never developed a settled policy. For many years its programmes consisted of opera and ballet, and later of patriotic melodramas, and it was not until the end of the 19th century that the first stirrings of *realism were felt with the production of such plays as *Strindberg's *The Father.* The first production of *Chekhov's *The Seagull* in 1896 was a complete failure, the company's old-fashioned technique being inadequate to the task of conveying the subtlety of the author's characterization. Just before the October Revolution *Meyerhold was working at the Alexandrinsky, his last production there being a revival of *Lermontov's *Masquerade.* Under the guidance of *Lunacharsky the theatre weathered the storms of the early 1920s, and by 1924 was ready to include Soviet plays in its repertory, one of the directors at this time being *Radlov. In 1937 Meyerhold returned to produce *Masquerade* again, and during the Second World War the company went on tour, returning to Leningrad in 1944. Interesting landmarks during its later history were a successful production of *The Seagull* in 1954, in which year it also staged *Hamlet,* and the 1955 revival by *Tovstonogov of *Vishnevsky's *The Optimistic Tragedy.* More recent productions have included Ostrovsky's *The Last Sacrifice* and Shteyn's *Night without Stars* (both 1975). The theatre has traditionally been a stronghold of fine acting, and although the death of *Cherkassov in 1966 and of several other leading actors in the early 1970s weakened the company for a time, enough good young actors remained to sustain its high standards.

Pushkin Theatre, Moscow. This theatre opened in 1951 under **Vasily Vasilyevich Vanin** (1898–1951), who shortly before his death staged there a revival of *Krechinsky's Wedding* by *Sukhovo-Kobylin. Interesting productions included revivals of *Ostrovsky's *At a Busy Place* in 1952, *Chekhov's *Ivanov* in 1960, and Sholokhov's *Virgin Soil Upturned* in 1963. The theatre also staged a new translation of *Wilde's *The Importance of Being Earnest* in 1957, a dramatization of Goncharov's *Oblomov* (1969), and *Schiller's *The Robbers* (1986).

Q

Q Theatre, Kew Bridge, London, 500-seat theatre which opened in 1924 in a converted hall with a revival of Gertrude Jennings's highly successful comedy *The Young Person in Pink*. Over 1,000 plays were presented there, half of which were new; of these over 100 were transferred to the West End, Frederick Knott's *Dial M for Murder* being probably the best remembered. Terence *Rattigan's first play *First Episode* (1933) was tried out at the Q before being transferred, with Max *Adrian, who had made his first professional appearance at the Q, in the leading role. Other well-known actors who made their débuts at this theatre were Anthony *Quayle in *Robin Hood* (1931) and Dirk Bogarde in a revival of *Priestley's *When We Are Married* in 1939. In 1955 the local authority refused to renew the theatre's licence unless it was rebuilt. The campaign to raise the necessary funds was a failure, and the building was demolished in 1958.

Quaglio, family of scenic artists extending over several generations, of whom the first were the brothers **Lorenzo** (1730–1804), important in the history of neo-classical design, and **Giuseppe** (1747–1828). Of Italian origin, they worked mainly in foreign Courts, and in the late 18th century established themselves in *Munich, where their sons, grandsons, and great-grandsons were connected with the Court Theatre. A Quaglio was also working at the Berlin Court Theatre as late as 1891.

Quayle, Sir (John) **Anthony** (1913–89), English actor and director. He made his first appearance on the stage in London in 1931, and soon gave proof of solid qualities, notably during several seasons with the *Old Vic, where he played a wide variety of parts including John Tanner in *Shaw's *Man and Superman* (1938). He was first seen in New York in 1936 in *Wycherley's *The Country Wife*. After six years in the army during the Second World War he returned to the theatre as Jack Absolute in *Sheridan's *The Rivals* (1945), also directing a dramatization of Dostoevsky's *Crime and Punishment* (1946) starring John *Gielgud and Edith *Evans. In 1948 he succeeded Barry *Jackson as Director of the *Shakespeare Memorial Theatre, where he directed a number of plays and also appeared as Petruchio in *The Taming of the Shrew* in 1948, Falstaff in both parts of *Henry IV* in 1951 and *The Merry Wives of Windsor* in 1955, Coriolanus in 1952, and Othello in 1954. In 1956 he played the title-role in *Marlowe's *Tamburlaine the Great* in New York, and shortly afterwards left Stratford to appear in London in a succession of non-classical parts which included Eddie in Arthur *Miller's *A View from the Bridge* (1956), James Tyrone in *O'Neill's *Long Day's Journey into Night* (1958), which he also directed, Marcel Blanchard in *Coward's *Look after Lulu* (1959), and Cesareo Grimaldi in *Billetdoux's *Chin-Chin* (1960). After his Sir Charles Dilke in Bradley-Dyne's *The Right Honourable Gentleman* (1964) he was seen in New York in the title-role of *Brecht's *Galileo* and in *Ustinov's *Halfway up the Tree* (both 1967), and in 1970 he returned there as Andrew Wyke in Anthony Shaffer's *Sleuth* after playing the part in London. He directed Simon *Gray's adaptation of Dostoevsky's *The Idiot* for the *National Theatre in the same year. In 1976 he partnered Peggy *Ashcroft in *Arbuzov's two-character play *Old World*, repeating the role in New York in 1978. In the same year he directed and appeared in *The Rivals* and also played Lear for the *Prospect Theatre Company at the Old Vic; and in 1983 he founded the Compass touring company, usually playing the leading roles himself, including Lear again in 1987.

Queen Anne's Men, company, usually known as the Queen's Men, formed on the accession in 1603 of James I from a combination of Worcester's and *Oxford's Men. It included Christopher *Beeston, who later became its manager, and the playwright Thomas *Heywood, and was under the direct patronage of James's wife. It had a successful career playing at the *Curtain and at the *Red Bull until 1616, when after internal dissensions Beeston took some of the company to his new theatre the *Cockpit. The venture was not a success, and on the death of Queen Anne in 1619 the company disbanded, some of the actors going back to the Red Bull or to the provinces until 1625, when several, including Beeston, joined the newly formed *Queen Henrietta's Men.

Queen Elizabeth's Men, Elizabethan company, usually known as the Queen's Men (as, somewhat confusingly, were also the later *Queen Anne's and *Queen Henrietta's Men). Formed in 1583, it included the jester Richard *Tarleton, whose death in 1588 was a great blow. The company played at several of the *innyards used as theatres, and also at the *Theatre, the *Curtain, and the *Rose up to

1594, when it was disbanded and replaced by the *Admiral's Men.

Queen Henrietta's Men, company of players, usually known as the Queen's Men, formed on the accession of Charles I in 1625 and placed under the direct patronage of his wife. It was headed by Christopher *Beeston, and included several other members of the disbanded *Queen Anne's Men, to which it was in some ways a successor. Among the successful plays given by this company at their theatre the *Cockpit were about 20 which their official dramatist James *Shirley wrote for them, either alone or in collaboration, between 1625 and 1636. They also appeared at Court in Thomas *Heywood's *masque Love's Mistress (1634), for which Inigo *Jones designed some excellent scenery. When plague closed the theatres in 1636 the company was disbanded, and replaced at the Cockpit by Beeston's Boys, but a new company under the old name was formed in 1637 and played successfully at *Salisbury Court until the closing of the theatres in 1642.

Queen's Theatre, London. 1. In Long Acre. This building, erected in 1849, was known originally as St Martin's Hall, and was used on a number of occasions by Charles *Dickens reading from his own works. In 1867 it was converted into the second largest theatre in London, and in 1868 Henry *Irving and Ellen *Terry appeared together for the first time in Katharine and Petruchio, *Garrick's version of The Taming of the Shrew. When Ellen Terry returned to the stage after some years in retirement she was seen at the Queen's as Philippa in *Reade's The Wandering Heir (1873). In spite of changing its name to the National Theatre in 1877 the theatre had little further success and finally closed in 1879.

2. In Shaftesbury Avenue. It is the sister theatre of the present *Globe Theatre, which it adjoins, the auditoriums and stages of the two playhouses being separated only by a party wall. With a seating capacity of 1,160 in three tiers it opened in 1907, and had its first success a year later with a musical The Belle of Brittany. Fashionable tango-teas were held in 1913, and in 1914 the theatre did well with Glass and *Klein's Potash and Perlmutter. Among later productions were *Fagan's And So to Bed (1926) with Yvonne *Arnaud and the *Malvern Festival production of *Shaw's The Apple Cart (1929) with Edith *Evans. In the following year John *Gielgud repeated his *Old Vic triumph as Hamlet, and other successful productions were Besier's The Barretts of Wimpole Street (1930), *Knoblock's Evensong (1932), and Robert *Morley's Short Story (1935). In 1937–8 Gielgud returned with a season of four plays. He was also in Dodie Smith's Dear Octopus (1938), which was still running when the theatres closed on the outbreak of the Second World War in 1939. The

Queen's had reopened and was occupied by Daphne Du Maurier's Rebecca when on 24 Sept. 1940 it was badly damaged by bombs. It did not reopen until 1959, after the complete restoration of the front-of-house, the first production being Gielgud's solo recital The Ages of Man. Later came The Aspern Papers, adapted from Henry *James's novel, *Bolt's The Tiger and the Horse (1960), both with Michael *Redgrave, and Anthony Newley in his own musical Stop the World—I Want to Get Off (1961). In 1964 there was a revival of *Chekhov's The Seagull with Peggy *Ashcroft and Vanessa *Redgrave. Noël *Coward made his last appearance on the stage in his Suite in Three Keys (1966), and Peter *Ustinov's Halfway up the Tree (1967) had a long run. Alan *Bennett's Getting On was seen in 1971, and Maggie *Smith in 1972 in a revival of Coward's Private Lives. The *National Theatre production of *De Filippo's Saturday, Sunday, Monday moved there in 1974, and in 1975 Alan *Bates starred in Simon *Gray's Otherwise Engaged. In 1977 Alan *Ayckbourn's Just between Ourselves had a comparatively short run, but Alan Bennett's The Old Country, starring Alec *Guinness, was more successful. Tom *Courtenay gave a brilliant performance in 1980 in the title-role of Ronald Harwood's The Dresser, and in the following year another play by Simon Gray, Quartermaine's Terms, received high praise. Later productions were Julian Mitchell's Another Country (1982), The Seagull, again with Vanessa Redgrave, in 1985, the musical Wonderful Town in 1986, and Alan Bennett's double bill Single Spies (1989). (See also DORSET GARDEN THEATRE, HER MAJESTY'S THEATRE, and SCALA THEATRE.)

Quem Quaeritis?, see LITURGICAL DRAMA.

Questors Theatre, London, in Mattock Lane, Ealing, home of one of Britain's leading amateur groups, founded in 1929, which in 1933 adapted a disused chapel for its productions. After the Second World War the company raised funds to construct a sophisticated flexible theatre, adaptable for various types of staging and seating between 325 and 450. Rehearsal rooms and other facilities are also housed in the complex. The building proceeded in stages as money became available, the theatre itself being opened in 1964. A professionally staffed part-time training course for young actors has been run since 1947 and young people's groups, from the age of 5 upwards, meet weekly.

The Questors was one of the founders of the Little Theatre Guild of Great Britain (1946) and participated in the founding of the International Amateur Theatre Association (1952). (For both see AMATEUR THEATRE.) From 1960 to 1977 it held an annual Festival of New Plays: one of its first successes was Next Time I'll Sing to You (1962) by James *Saunders, and other plays by him were given at subsequent festivals. Works by Marcel Aymé, James Joyce, William *Saroyan, Jacques *Audiberti, Friedrich *Dürrenmatt, Günter Grass, and Michel *Tremblay were first introduced to British audiences by the Questors. Six International Theatre Weeks between 1969 and 1982 featured outstanding companies from Canada, the USA, and Europe. The annual Questors Student Playwriting Competition was inaugurated in 1986.

Quick, John (1748–1831), English comedian, who made his first appearance at Covent Garden in 1767 and remained there until his retirement in 1798. Unacceptable in tragedy, he had a wide range of comic parts, from Clown in pantomime to Polonius in *Hamlet*, and was the first to play Tony Lumpkin in *Goldsmith's *She Stoops to Conquer* (1773) and Bob Acres in *Sheridan's *The Rivals* (1775). A small, impetuous, chubby-cheeked man, he was the favourite actor of George III, and his portrait by Zoffany hangs in the *Garrick Club in London.

Quick-Change Room, small, closed recess opening off the stage, where players can change their clothes when a brief absence from the stage does not allow them to return to their dressing-rooms.

Quilley, Denis Clifford (1927–   ), English actor, equally at home in straight plays and in *musical comedy. He made his first appearance at the *Birmingham Repertory Theatre in 1945 and was first seen in London in 1950. He played his first leading role in the West End in the musical *Wild Thyme* (1955), and was then in several more musicals: *Grab Me a Gondola* (1956), Bernstein's *Candide* (1959), in the title-role, *Irma la Douce* (1960; NY, 1961), Rodgers and *Hart's *The Boys from Syracuse* (1963), and *High Spirits* (1964), based on *Coward's *Blithe Spirit*. Although he was seen at *Nottingham Playhouse in 1969 as Macbeth and as Archie Rice in *Osborne's *The Entertainer*, his career as a straight actor received its biggest impetus when he joined the *National Theatre company in 1971. He remained for over five years, playing a wide variety of parts, among them Jamie in *O'Neill's *Long Day's Journey into Night* in 1971, Hildy Johnson in Hecht and MacArthur's *The Front Page* in 1972, Caliban in *The Tempest* in 1974, and Claudius in *Hamlet* in 1975. In 1977 he joined the *RSC to appear in Peter *Nichols's musical *Privates on Parade*, and was also seen at the *Albery as Morell in *Shaw's *Candida*. He starred in the London productions of the musicals *Sweeney Todd* (1980), by *Sondheim, and *La Cage aux folles* (1986), and in 1985 was Antony to Diana *Rigg's Cleopatra at *Chichester. He returned to the National in 1990 to play Sir Oliver Surface in *Sheridan's *The School for Scandal*.

Quin, James (1693–1766), English actor, the last of the declamatory school whose supremacy

was successfully challenged by *Garrick. He made his first appearance at the Smock Alley Theatre in Dublin in 1712 and two years later was playing small parts at *Drury Lane, where he made an unexpected success as Bajazet in *Rowe's *Tamerlane* when the actor billed to play the part was taken ill. In 1718 he went to *Lincoln's Inn Fields Theatre where he remained for 14 years, appearing first as Hotspur in *Henry IV, Part One*, and then in a wide range of Shakespearian parts. He later went to *Covent Garden and finally returned to Drury Lane, retiring in 1751 to live in Bath. A man of great gifts but very little education, he was obstinate and quarrelsome and a great stickler for tradition, refusing to the end to alter one detail of his original costumes. A portrait of him as the hero of Thomson's *Coriolanus* (1748) shows him equipped with plumes, peruke, and full spreading short skirt in the style of the later 17th century. There is a description of his acting in Smollett's *Humphry Clinker* (1771).

Quinault, Philippe (1635–88), French dramatist and librettist, who became valet to *Tristan l'Hermite, under whose name his first play *Les Rivales* (1653) was accepted by the actors at the Hôtel de *Bourgogne. When they discovered the truth they were reluctant to pay the fee agreed upon, and the resultant negotiations, which resulted in the author being given a fixed share in the takings from each performance, have been cited as the origin of the *royalty system, though the point is still disputable. The last of Quinault's plays to profit from the advice of Tristan, shortly before his death, was *La Comédie sans comédie* (*A Play with No Plot*, 1655), which consisted of four scenes in different genres designed to show off the talents of the actors at the *Marais. By 1666 Quinault had written a number of plays, of which the most successful were the tragedy *Astrate roi de Tyr* (1664) and the comedy *La Mère coquette* (1666). He then married a wealthy young widow who despised the theatre and made him give it up; but in 1668, being well known for his facility in light verse, he was persuaded to write the libretto for the first of Lully's French operas *La Grotte de Versailles*. He continued to write Lully's librettos until the latter's death in 1687, and such was the purifying power of music that his wife made no objections. He also collaborated with *Molière and *Corneille in the lyrics for *Psyché* (1671) and contributed to *Les Fêtes de l'Amour et de Bacchus* (1672) in which Molière also had a hand.

Quintero, José Benjamin (1924–   ), American director, born in Panama, co-founder of the *Off-Broadway group at the *Circle-in-the-Square. He remained with the company until 1964, his productions for it including an excellent revival of *O'Neill's *The Iceman Cometh* in 1956. In the same year he directed the American première of O'Neill's *Long Day's Journey into Night* at the *Helen Hayes Theatre. In 1958, at the Festival of Two Worlds in Spoleto, Italy, he directed the same author's *A Moon for the Misbegotten*, again directing it in New York in 1973 and Oslo in 1975. Quintero continued to concentrate on plays by O'Neill, including *Strange Interlude* in 1963, *Marco Millions* in 1964, *More Stately Mansions* (1967), *A Touch of the Poet*, and *Anna Christie*, both in 1977. Among his other productions were Tennessee *Williams's *Summer and Smoke* (1952) and *Clothes for a Summer Hotel* (1980), and *Cocteau's *The Human Voice* (1979). He directed *The Iceman Cometh* again in 1985 and *Long Day's Journey into Night* again for O'Neill's centenary in 1988.

Quintero, Serafín and **Joaquín Álvarez,** see ÁLVAREZ QUINTERO.

R

Rabe, David (1940–), American dramatist, whose importance was immediately recognized when his first two plays, *The Basic Training of Pavlo Hummel* and *Sticks and Bones*, were produced in New York at the *Public Theatre in 1971. Both related to the war in Vietnam, where he had served as a draftee. *Pavlo Hummel* concerns a neurotic and unhappy teenager who after undergoing his basic military training is killed in Vietnam after a quarrel with another soldier. In *Sticks and Bones* (London, 1973) a man blinded in Vietnam returns to his parents, Ozzie and Harriet, whose behaviour resembles that of the characters in a popular television show whose names they bear. *The Orphan* (1975) was loosely based on the *Oresteia* of *Aeschylus and described by the author as the third play of his Vietnam trilogy. *Boom Boom Room* (1973) was revised in the following year as *In the Boom Boom Room* (London, 1976); and *Streamers* (1976; London, 1978) is a powerful play set in an army barracks in Virginia. *Hurlyburly* (1984) savagely portrays a group of male Hollywood hedonists in the late 1970s.

Rachel [Élisa Félix] (1820–58), one of the greatest tragediennes of the French stage, child of a poor Jewish family. Befriended by an impresario who found her singing in the streets, the 13-year-old girl was sent to Saint-Aulaire's drama school in the old Théâtre Molière. Further study at the Conservatoire was cut short by her father, who was anxious to make money out of her, and in 1837 she was taken into the company of the *Gymnase and, coached by *Samson, she entered the *Comédie-Française in 1838, appearing as Camille in *Corneille's *Horace* with some success. She then appeared in other plays by Corneille which had been neglected since the death of *Talma and in the works of *Racine, whose *Phèdre* was to provide her greatest part. She also appeared in some modern plays, including the first production of *Adrienne Lecouvreur* (1849) by *Scribe and Legouvé. But it was in the great classical roles that she mainly appeared on tour—in the French provinces, in Europe, going as far as Russia, in London, where she first appeared in 1841 with outstanding success, and in America, in 1855. It was this visit which finally brought to a head the tubercular condition from which she was suffering, the result of early hardships and later overwork combined with a feverish succession of amorous intrigues. In spite of her superb acting her visit to the United States was a failure financially, mainly because of the language

barrier. An account of her last tragic journey, written by a member of her company, was later published. Her four sisters and her brother were all on the stage.

Racine, Jean (1639–99), French playwright and poet, one of the greatest tragic dramatists in the history of the theatre. Orphaned at the age of 4, he was brought up by his grandparents and his aunt Agnès, later Abbess of Port-Royal, where Racine was educated after a few years at the Collège de Beauvais. He was an excellent scholar, an enthusiastic admirer of the Greek dramatists, and at 19 already a good poet. He soon escaped from the restraining influence of Port-Royal and the Jansenists and led a free, though not particularly dissipated, life. He was quickly accepted in literary circles, where he made the acquaintance of *Molière, who in 1664 staged his first play *La Thébaïde; ou, Les Frères ennemis* at the *Palais-Royal. It was successful enough for Molière to accept his second play *Alexandre le Grand* (1665), but a fortnight after its production Racine allowed the actors at the Hôtel de *Bourgogne to stage it in direct competition with Molière's company. His excuse was that he was not satisfied with the Palais-Royal production, and it was true that the Hôtel de Bourgogne had a higher reputation in tragedy; but it is possible that feminine intrigue played a part, as **Mlle du Parc** [Marquise-Thérèse de Gorla] (1633–68), a fine tragic actress and Racine's mistress at the time, left Molière's company in order to play the lead in Racine's next play *Andromaque* (1667) at the Hôtel de Bourgogne. After this double betrayal Molière never spoke to Racine again and Racine did not long enjoy his triumph as Mlle du Parc died suddenly the following year. There was some unkind gossip about the affair at the time, and in 1680 the infamous poisoner Catherine Voisin alleged at her trial that Racine had removed Mlle du Parc to make way for Mlle *Champmeslé, who had come from the *Marais to play Hermione to du Parc's Andromaque. She certainly became Racine's mistress and was closely associated with his later work.

With the production of *Andromaque* Racine first achieved recognition as an outstanding dramatist, rival of the ageing *Corneille and in some ways superior to him. It was followed by Racine's only comedy *Les Plaideurs* (*The Litigants*, 1668), based partly on *Aristophanes' *Wasps*. It was intended for the Italian actors at the Palais-Royal, but on the departure of Tiberio

*Fiorillo (Scaramouche) from Paris it was transferred to the Hôtel de Bourgogne, where after a slow start it was extremely successful and frequently revived. It remained in the repertory, and in 1920 stood sixth in the list of plays most frequently given at the *Comédie-Française. Sadly, Racine wrote no more comedies. His next tragedy, Britannicus (1669), was not a success, in spite of its exquisite poetry, but according to *Boileau the portrait of Nero deterred Louis XIV from featuring himself further in Court ballets and entertainments. With his next play Racine found himself at odds with Corneille. Either by accident or design they were both working on the same subject, and the production of Racine's Bérénice (1670) took place only a week before Molière produced Corneille's Tite et Bérénice. Racine's version was more generally admired, and he followed it with two more tragedies, Bajazet (1672) and Mithridate (1673), both historical and oriental subjects treated in a contemporary style very different from Corneille's. Racine was often criticized for abandoning the latter's heroic mood but was now firmly established as the leading tragic dramatist of his day, and his Iphigénie (1674) was a brilliant success. But with his greatest play, Phèdre (1677), he finally terminated his career as a tragic dramatist. Several events probably contributed to the sudden cessation of his dramatic work. A mediocre dramatist, Pradon, was persuaded to present a tragedy on the same subject two days earlier than Racine's, and his enemies—he had plenty, for he was not a particularly likeable man—made sure that its reception far outdid that accorded to Racine's. There were also more sensible reasons for his withdrawal. He had recently been looked on favourably by Louis XIV, who offered him a position as historiographer royal, a Court appointment incompatible with an active career in the theatre. His relations with Port-Royal, where he had been brought up, ameliorated, he married and had seven children, and he turned his undoubted gifts to the studying and recording of French history. It was not until 1689 that at the request of Mme de *Maintenon he wrote, for performance by the young ladies at her school at Saint-Cyr, the tender and poetic Esther and the more powerful but no less poetic Athalie. Neither was seen in public for many years, though Athalie proved the more enduring. But it is by their performance in Phèdre that the greatest actresses, and not only in Paris, have proved their worth. Like Corneille, Racine had a short period of recognition in London, which is echoed in the temporary success of *Addison's Cato (1713), the only successful English play in the French neo-classical style.

RADA, see SCHOOLS OF DRAMA.

Radio City Music-Hall, New York, in the Rockefeller Center. This gigantic theatre, seating 6,200, was planned as part of the 24-acre Radio City, opening on 27 Dec. 1932 with a staff which included Robert Edmond *Jones as designer and Martha Graham as choreographer. Its companion house, the RKO Roxy cinema, opened on 29 Dec. The opening production was not a success and the Music-Hall closed, to reopen a month later as a combined film and *variety theatre. The variety shows, presented four times daily, featured the tap-dancing Rockettes and had a central theme, as in the holiday extravaganzas The Nativity and The Glory of Easter. The stage had a proscenium height of 60 ft. and a width of 100 ft., the full stage being 66 ft. deep by 140 ft. wide and equipped with a turntable with elevator sections which could be raised and lowered 40 ft. The auditorium and public areas were restored to their original art deco glory when the building was designated a landmark in 1979. The theatre is now used for spectacular shows and performances by big crowd-pullers, and the traditional Christmas and Easter shows continue.

Radlov, Sergei Ernestovich (1892–1958), Soviet actor and director, whose work was done mainly in Leningrad, where he started his theatrical career under *Meyerhold and later opened his own theatre. He did some important work in *Expressionist forms of staging during the 1920s and pioneered the staging of modern operas by Berg and Prokofiev; but although his outstanding productions included *Ostrovsky's The Dowerless Bride and *Ibsen's Ghosts, it was as a director of Shakespeare that he was pre-eminent, with productions of Romeo and Juliet, Hamlet, and Othello, the last being seen also at the Moscow *Maly. His finest work was his production of King Lear for the *Moscow State Jewish Theatre, with *Mikhoels in the title-role.

Raimund, Ferdinand [Jakob Raimann] (1790–1836), Austrian playwright and actor, a popular farce-player in *Vienna from 1813 to 1823. Having decided that his unique combination of comic gifts needed a vehicle specially created to display them to advantage, he began to write his own plays. The first, Der Barometermacher auf der Zauberinsel (The Barometer-Maker on the Magic Island, 1823), was a great success, and was followed by Das Mädchen aus der Feenwelt; oder, Der Bauer als Millionär (The Girl from Fairyland; or, The Millionaire Farmer, 1826), which preaches, with the help of magical forces and a whole host of allegorical personages, the doctrine of contentment on small means. But perhaps Raimund's best play is Der Alpenkönig und der Menschenfeind (The King of the Alps and the Misanthrope, 1828), in which a kindly mountain spirit cures a misanthropist by assuming his shape and character, while the misanthrope, disguised as his own brother-in-law, has to watch the havoc caused by his suspicions and ill-will. Encouraged by his success, Raimund set about educating himself by studying Shakespeare and—typically for a Viennese playwright—*Calderón. The result was a number of plays which were out of tune with the demands of his audiences and with his own essentially unreflective genius. It was not until he returned to his earlier style in *Der Verschwender* (*The Prodigal*, 1834) that he was again successful; but his last years were overshadowed by the rising popularity of *Nestroy and he finally committed suicide.

Rain Box, see SOUND EFFECTS.

Rake, slope of the stage floor from the back wall to the footlights. This had once a definite purpose in that it aided the illusion of scenes painted in perspective. With the passing of such scenes it ceased to have any practical use since, contrary to traditional belief, it did not give the dancer a better basis for a leap, nor did it make the actors up-stage more visible to the audience. (This is better done by grouping them on *rostrums at least a foot higher than the group in front.) The raked stage was limited in practice to a slope of 4 per cent and was often as small as $1\frac{1}{2}$ per cent. It had serious disadvantages in the setting of scenery, as pieces set diagonally could not join neatly to squarely vertical neighbours; also the side-flats of a *box-set needed wedges under the base of each to compensate for the slope, or else they had to be built out of true with sloping bottom rails and ceased to be interchangeable. Moreover, any setting of pieces on a *boat truck became dangerous, since the truck might run off on its own down the incline. The modern practice is to have a flat stage floor, and to rake the floor of the *auditorium, although a temporary rake may be used to achieve a particular effect by laying plywood sheets over joist-frames laid parallel to the front of the stage and increasing in height up-stage.

Rame, Franca, see FO.

Randall, Harry, see LENO.

Rastell, John (?–1536), brother-in-law of Sir Thomas More, and father-in-law, through the marriage of his daughter Elizabeth, of John *Heywood, some of whose *interludes were printed by Rastell's son William. John Rastell is believed to have been the author of *Calisto and Meliboea*, an adaptation of part of de Rojas's *Celestina, and of *The Dialogue of Gentleness and Nobility*, both of which were acted in his own garden at Finsbury in about 1527, and printed by him in the same year. He may also have been the author of an earlier interlude entitled *The Play of the Four Elements* (c.1517).

Rattigan, Sir Terence Mervyn (1911–77), English playwright, author of a number of well-constructed and theatrically effective works. His first play, *First Episode* (1933; NY, 1934), written in collaboration, was followed by an immensely successful light comedy, *French without Tears* (1936; NY, 1937). He later achieved a further success with *Flare Path* (London and NY, 1942), a topical war play with an RAF background. Next came two more comedies, *While the Sun Shines* (1943; NY, 1944) and *Love in Idleness* (1944), the latter providing an excellent vehicle for the *Lunts, who played it in London and in 1946 in New York as *O Mistress Mine*. Rattigan, hitherto considered no more than an astute purveyor of light entertainment, now began to show signs of a more serious purpose with *The Winslow Boy* (1946; NY, 1947), based on the true story of a father's fight to clear his young son of a charge of petty theft. *Playbill* (1948; NY, 1949) consisted of two short plays, *The Browning Version* and *Harlequinade*, the former an excellent study of a repressed schoolmaster and his ill-matched wife. *Adventure Story* (1949), with Paul *Scofield as Alexander the Great, proved an interesting failure, and with *Who is Sylvia?* (1950) Rattigan returned to his former vein of light comedy. His next play, however, *The Deep Blue Sea* (London and NY, 1952), was a deeply emotional study of a judge's wife (played successively by Peggy *Ashcroft, Celia *Johnson, and Googie *Withers) who falls in love with a feckless, drunken, ex-RAF fighter pilot and twice attempts suicide. An excursion into Ruritanian romance followed with *The Sleeping Prince* (1953; NY, 1956), in which Laurence *Olivier and Vivien *Leigh appeared. It was in the preface to his collected plays in 1953 that Rattigan first used the term 'Aunt Edna' to indicate the ordinary unsophisticated playgoer who has no use for experimental, avant-garde plays. It has since proved a useful term for drama critics and an Aunt Sally for the progressives. Rattigan's next play, *Separate Tables* (1954; NY, 1956), was a double bill (*The Window Table* and *Table Number Seven*) which portrayed with compassion the problems of a group of characters in a Bournemouth hotel. It was followed by *Variations on a Theme* (1958); *Ross* (1960; NY, 1961) with Alec *Guinness as T. E. Lawrence; *Joie de Vivre* (also 1960), a musical version of *French without Tears* which lasted only four nights; and *Man and Boy* (London and NY, 1963). After an unusually long interval came *A Bequest to the Nation* (1970), on Nelson and Lady Hamilton; another double bill, *In Praise of Love* (*Before Dawn* and *After Lydia*) (1973; a 'full version' of *After Lydia* was produced in New York as *In Praise of Love* in 1974); and *Cause Célèbre* (1977), based on a real-life murder case, which was still running when Rattigan died.

Ravenscroft, Edward (*fl.* 1671–97), English dramatist, whose career as a playwright started in 1671 with an adaptation of *Molière's Le Bourgeois gentilhomme* as *Mamamouchi; or, The Citizen Turned Gentleman.* Another adaptation of Molière, *The Careless Lovers* (1673), based on *Monsieur de Pourceaugnac*, was followed by an original comedy, *The Wrangling Lovers; or, The Invisible Mistress* (1676). An adaptation of Molière's *Les Fourberies de Scapin* was forestalled by *Otway's version, and had to be postponed until 1677, in which year there also appeared *King Edgar and Aldreda*, a tragi-comedy, and *The English Lawyer*, an English version of George Ruggle's *Ignoramus*. Ravenscroft's best work, however, was a farce, the outrageous *The London Cuckolds* (1681), which it became the tradition to produce at both *Patent Theatres on Lord Mayor's Day until *Garrick stopped it at *Drury Lane in 1751 and *Covent Garden dropped it three years later. It was revived in 1782, and then disappeared until 1979, when it was revived at the *Royal Court Theatre. A further adaptation of Molière, *The Anatomist; or, The Sham Doctor* (1696), was Ravenscroft's last comedy, and his last play, a tragedy, was *The Italian Husband* (1697).

Raymond, John T. [John O'Brien] (1836–87), American actor, who made his first appearance in 1853 and was immediately recognized as an excellent comedian. After touring in stock companies for some years he joined Laura *Keene's company in 1861, taking over from Joseph *Jefferson the part of Asa Trenchard in Tom *Taylor's *Our American Cousin*, in which he was also seen in London in 1867. Among his other parts were Tony Lumpkin in *Goldsmith's *She Stoops to Conquer*, and Crabtree in *Sheridan's *The School for Scandal*, but he was best known for his playing of Colonel Mulberry Sellers in Mark Twain's *The Gilded Age* (1874) and Ichabod Crane in *Wolfert's Roost* (1879), a dramatization of Washington *Irving's *The Legend of Sleepy Hollow*. An able and energetic actor who remained on the stage until his death, he had a long, imperturbable face and a slow seriousness of manner which made his comedy all the more appealing. He was popular with the public and with his fellow actors, and once established as a star appeared mainly in plays by and about Americans, except for *Pinero's *The Magistrate* in which he was inimitable as Posket.

Reade, Charles (1814–84), English novelist, also the author of a number of plays, of which the best known are *Masks and Faces* (1852), dealing with *Garrick and Peg *Woffington and written in collaboration with Tom *Taylor; *The Courier of Lyons* (1854), which as *The Lyons Mail* (1877) provided a marvellous vehicle for *Irving in the dual roles of Dubosc and Lesurques, parts played later by *Martin-Harvey; and

It's Never Too Late to Mend (1864), based on his own novel. He also dramatized *Tennyson's *Dora* (1867), and in 1874 persuaded Ellen *Terry to return to the theatre after her liaison with Godwin to take over the part of Philippa Chester in his play *The Wandering Heir* (1873), first produced with Mrs John *Wood in the part. Among his later plays the best was a version of *Zola's novel *L'Assommoir* (1877) as *Drink* (1879), which he wrote in collaboration with Charles *Warner who played Coupeau. Reade was essentially a novelist, and his best work for the theatre was done in collaboration with more theatrically minded men, or based on existing foreign plays.

Realism, movement in the theatre at the end of the 19th century which replaced the *well-made play and the declamatory acting of the period by dramas which approximated in speech and situation to the social and domestic problems of everyday life, played by actors who spoke and moved naturally against scenery which reproduced with fidelity the usual surroundings of the people they represented. The movement began with *Ibsen and spread rapidly across Europe, upsetting the established theatre and demanding a new type of actor to interpret the new plays. This was achieved by the system of *Stanislavsky and by the later advocates of *naturalism, the logical outcome of realism. (See also SOCIALIST REALISM.)

Realistic Theatre, Moscow, small theatre, also known as the Krasnya Presnya Theatre from the district in which it was situated. It opened in 1921 and was originally one of the studios attached to the *Moscow Art Theatre, on whose methods its early productions were based. A decisive change in its history came with the appointment in 1930 of *Okhlopkov as its Artistic Director. His experimental techniques, while interesting and valuable, had necessarily a limited appeal, and in 1938 an unsuccessful attempt was made to merge the company with that of *Taïrov at the *Kamerny Theatre. Okhlopkov then went to the Theatre of the Revolution (later the *Mayakovsky Theatre) and the Realistic was closed.

Reaney, James Crerar (1926–   ), Canadian poet and playwright. His first major play *The Killdeer* (1960, revised 1968), a flawed but haunting work set in rural Ontario, unravels a skein of emotional entanglements through a macabre murder mystery. *The Easter Egg* (1962) has a similar theme of stunted childhood and spiritual rebirth; it was followed by *The Sun and the Moon* (1965; originally *The Rules of Joy*, 1958) and *The Three Desks* (1966), which are awkwardly realistic and even melodramatic. Other plays, including a series written for children, experiment with the repetition of words and the playing of games. *Colours in the Dark* (1967) was written for performance at the

*Stratford (Ontario) Festival as part of the celebration for Canada's Centennial year. Kaleidoscopically combining scenes from the author's childhood in Ontario, the history of Canada, and even the biblical account of the Creation, it has been several times revived. *Sticks and Stones* (1973), *St Nicholas Hotel* (1974), and *Handcuffs* (1975) together form a moving and detailed account of the events leading up to the massacre in 1880 of a legendary Irish family, the Donnellys, who emigrated to Southwestern Ontario. The trilogy was created by Reaney in collaboration with the actors and the director. Subsequent plays continue his exploration of Canada's 19th-century history, though less successfully.

Red Army Theatre, Moscow, the first and better-known name of the Central Theatre of the Soviet Army. This was founded in 1919, its company being drawn from groups of actors who had for some time been entertaining army audiences, and was intended for the performance of plays about, or of interest to, the men of the Red Army. Its first director was *Zavadsky, but Alexei *Popov, who directed it from 1937 until his death in 1961, widened its scope and made it one of the outstanding theatres of the Soviet Union. Under him the company moved in 1940 to a specially designed and admirably equipped theatre shaped like a five-pointed star and capable of seating an audience of 2,000. Though directing notable productions of the classics Popov did not neglect new Soviet plays, reviving *Vishnevsky's *The Unforgettable 1919* in 1952, soon after the author's death, and giving the first production of Shteyn's *A Game Without Rules* (1962). In 1963 Popov's son Andrei (see POPOV, ALEXEI) became Artistic Director of the theatre, which includes a number of recent foreign plays in its repertory but maintains the traditional association with the armed forces, frequently staging plays with a military theme. For the 30th anniversary of the end of the Second World War it produced in 1974–5 an adaptation of Kurt Vonnegut's novel *Slaughterhouse Five* as *The Wanderings of Billy Pilgrim*. In the 1980s it staged works by *Gorky, Lope de *Vega, and Tennessee *Williams.

Red Bull Theatre, London, in Upper Street, Clerkenwell. This was originally an inn in whose yard plays were occasionally given, and was adapted as a permanent theatre in about 1605. A square structure with a tiring house adjoining the stage and galleries round, it was occupied by Queen Anne's Men until 1617 and then by other companies. It was renovated and partly rebuilt in 1625 and may have been roofed in for use in bad weather. Contemporary dramatists referred to it in slighting terms, and it appears to have been what was later known as a *gaff, or *blood-tub, specializing in strong drama with plenty of devils and red fire.

When the theatres were closed during the Commonwealth, surreptitious shows and puppet-plays were sometimes given at the Red Bull, and at the Restoration a company under Michael *Mohun appeared there. Killigrew also reappeared there before he went to the Vere Street Theatre, taking some of the best actors in the company with him. On 23 Mar. 1661 *Pepys saw those who remained behind in a poor production of *Rowley's *All's Lost by Lust*. The theatre fell into disuse soon after; it was still standing in 1663 but by 1665 it had vanished.

Rederykers, see CHAMBERS OF RHETORIC.

Redgrave, Sir Michael Scudamore (1908–85), English actor, formerly a schoolmaster. He made his first appearance on the professional stage with the *Liverpool Playhouse company, where he remained from 1934 to 1936, playing a wide variety of parts and marrying a fellow member of the company, **Rachel Kempson** (1910–). In 1936 they were together at the *Old Vic, where Redgrave made his first London appearance as Ferdinand in *Love's Labour's Lost*, his other roles including Horner in *Wycherley's *The Country Wife*, in which he displayed a gift for comedy too rarely exploited. He joined *Gielgud's repertory season at the *Queen's Theatre in 1937 and was then seen at the *Phoenix Theatre as Alexei Turbin in *Bulgakov's *The White Guard* and as a richly comic Sir Andrew Aguecheek in *Twelfth Night* (both 1938). His first notable appearance in a modern play was as Harry in T. S. *Eliot's *The Family Reunion* (1939), after which in 1940 came an interesting Macheath in *Gay's *The Beggar's Opera* and Charleston in Robert Ardrey's *Thunder Rock*. After serving in the Royal Navy he returned to the theatre in 1943 as Rakitin in *Turgenev's *A Month in the Country*, followed by the title-role in Thomas Job's *Uncle Harry* (1944) and Stjerbinsky in S. N. *Behrman's adaptation of *Werfel's *Jacobowsky and the Colonel* (1945). His first outstanding role in Shakespeare was Macbeth (1947; NY, 1948); he then played the Captain in *Strindberg's *The Father* (also 1948), subsequently rejoining the Old Vic company at the New Theatre (see ALBERY), where in 1950 he played Hamlet for the first time. In 1951 he was at the *Shakespeare Memorial Theatre, playing Richard II and Prospero in *The Tempest*, and he appeared in London a year later in *Odets's *Winter Journey* (known in America as *The Country Girl*). Returning to Stratford in 1953, he played Shylock, King Lear, and Antony, and then deserted the classics for some years, being seen in *Giraudoux's *Tiger at the Gates* (London and NY, 1955) and N. C. *Hunter's *A Touch of the Sun* (1958). In 1958 he also played Hamlet and Benedick in *Much Ado about Nothing* at Stratford. A year later he starred in his own adaptation of Henry *James's *The Aspern Papers*, which was followed by Robert *Bolt's *The Tiger and the Horse* (1960) and Graham *Greene's *The Complaisant Lover* (NY, 1961). At the first *Chichester Festival in 1962 he gave an outstanding performance as *Chekhov's Uncle Vanya, repeating it the following year during the *National Theatre's first season at the Old Vic, where he was also seen as Claudius in *Hamlet* in 1963, Hobson in Harold *Brighouse's *Hobson's Choice*, and Solness in *Ibsen's *The Master Builder* (both 1964). He took over from Alec *Guinness in John *Mortimer's *A Voyage round My Father* in 1972, and was in Simon *Gray's *Close of Play* at the National Theatre in 1979. A fine actor, with a good presence and a superb speaking voice, he was particularly successful in the portrayal of men of intellect and sensibility flawed by emotional turbulence.

His two younger children, **Corin** (1939–   ) and **Lynn** (1943–   ), are also on the stage, the latter being well known in the USA where she now resides.

Redgrave, Vanessa (1937–   ), English actress, considered by many the finest actress of her day, with a great talent for conveying ecstasy and excitement. She made her first appearance on the stage in 1957, and her London début in N. C. *Hunter's *A Touch of the Sun* (1958) with her father Michael *Redgrave. She then played Sarah Undershaft in *Shaw's *Major Barbara*, and joined the *Shakespeare Memorial Theatre company as Helena in *A Midsummer Night's Dream* and Valeria in *Coriolanus* (both 1959). A year later she was with her father again in Robert *Bolt's *The Tiger and the Horse*, and back in Stratford in 1961 and 1962 she was much admired as Rosalind in *As You Like It*, Katharina in *The Taming of the Shrew*, and Imogen in *Cymbeline*. In 1964 she was seen at the *Queen's as a touchingly vulnerable Nina in *Chekhov's *The Seagull*, and in 1966 scored a great personal triumph as the embattled Scottish spinster schoolmistress in an adaptation of Muriel Spark's novel *The Prime of Miss Jean Brodie*. Her involvement in Marxist politics and in filming somewhat curtailed her theatrical activities, but she was seen briefly in 1972 as Polly Peachum in *Brecht's *The Threepenny Opera* and Viola in *Twelfth Night*, and in 1973 returned to the West End as Gilda in *Coward's *Design for Living*. Three years later she gave a highly praised performance as Ellida in *Ibsen's *The Lady from the Sea* in New York (London, 1979). In 1984 in London she was in her father's adaptation of Henry *James's *The Aspern Papers*, and in 1985 she returned to the Queen's in *The Seagull*, this time as Arkadina. After a season playing both Katharina and Cleopatra in 1986 she gave three superb performances: as Mrs Alving in Ibsen's *Ghosts* (also in 1986); in *O'Neill's *A Touch of the Poet* in 1988; and as

Lady Torrance in Tennessee *Williams's *Orpheus Descending* (also in 1988; NY, 1989). She was formerly the wife of the director Tony *Richardson.

Regent Theatre, London, in the Euston Road. This theatre, seating 1,310 in three tiers, opened in 1900 as the Euston Music-Hall, and became a playhouse in 1922, opening with *Body and Soul* by Arnold *Bennett. It had a short but interesting life, being occupied soon after its opening by Rutland Boughton's 'music-drama' *The Immortal Hour*, with Gwen *Ffrangcon-Davies. This was directed by Barry *Jackson, whose own play for children *The Christmas Party* opened on 20 Dec. In 1923 *The Insect Play* by the *Čapeks had a short run, with *Gielgud playing Felix, and a year later Gielgud and Gwen Ffrangcon-Davies starred in *Romeo and Juliet*, Gielgud being then 19 years old. The theatre was used extensively by play-producing societies such as the Fellowship of Players, who staged a number of Shakespeare's plays for short runs; the *Phoenix Society, which revived Elizabethan and Restoration comedies; the Pioneer Players, whose production of Susan *Glaspell's *The Verge* (1925) was directed by Edith Craig (see CRAIG, GORDON); the Repertory Players, who in 1926 produced *O'Neill's *Beyond the Horizon*; the *Stage Society, with James Joyce's *Exiles* (also 1926); and the Three Hundred Club, with D. H. *Lawrence's *David* (1927). During a short repertory season in 1925 the Macdona Players were responsible for the first London production of *Shaw's *Mrs Warren's Profession*, and for *Man and Superman* in its entirety for the first time. The last play to be presented at this theatre was a Jewish comedy by Izak Goller, *Cohen and Son*, in 1932. When it closed, the building became a cinema.

Regional Theatres, see CIVIC THEATRES and REPERTORY THEATRES.

Regnard, Jean-François (1655–1709), French dramatist, who wrote for the *Comédie-Italienne from 1688 to 1696, often in collaboration, and from 1694 to 1708 for the *Comédie-Française also, ranking second to *Molière as a writer of comedy. His first successful production there was *Attendez-moi sous l'orme* (1694); of those that followed the best were *Le Joueur* (1696) and *Le Légataire universel* (1708), the latter remaining in the repertory of the Comédie-Française until the early 20th century. Though Regnard inevitably suffers by comparison with his illustrious predecessor, he had a considerable talent for comic situations and witty dialogue, and it is interesting to trace in his plays the gradual emergence as the central figure of the valet, forerunner of *Beaumarchais's Figaro.

Rehan [Crehan], Ada (1860–1916), Irish-born American actress, who was taken to the United States as a child of 5 and at 14 appeared on the stage for the first time in New Jersey, joining Mrs John *Drew's *stock company at the Arch Street Theatre in *Philadelphia a year later. There, by a printer's error, she was billed as Rehan (for Crehan), a name which she retained and made famous. In 1879 she was engaged for Augustin *Daly's New York company with which she remained until his death. She soon became one of the most popular actresses of the day in New York and in London, where she made her first appearance in 1884 at Toole's Theatre (see TOOLE). In 1891 she laid the foundation stone of Daly's own theatre in London, where she was seen in a wide range of parts, including one always connected with her, Katharina in The Taming of the Shrew. She had first played this in New York in 1887, when the Induction to the play was given for the first time there. Other parts in which she was much admired were Lady Teazle in *Sheridan's The School for Scandal and Rosalind in As You Like It. She was essentially a comedienne, and only that side of her art was developed by Daly; unfortunately the turn of the century demanded a new style of acting, and after Daly's death in 1899 she found herself outmoded. She continued on her own, but with dwindling success in spite of her attractive personality, and made her last public appearance in 1905.

Rehearsal, session during which the *director, his cast, and his technical staff work on a play, preparing it for presentation. We know nothing of rehearsals in the classical theatre, and little of those in the medieval, though players, particularly women unable to read, must have been taught their parts orally, and great skill was obviously needed to direct the large crowds demanded by the *liturgical drama. Companies in later times were no doubt rehearsed by their leader, who was also the chief actor and often, as with *Molière, the author of the play. A glimpse of Elizabethan actors in rehearsal is given in A Midsummer Night's Dream. In the days of the *stock company, in England and elsewhere, there were very few rehearsals. New plays were merely gone through to check cues and settle entrances and exits. A newcomer in an old play was left to learn his way about by trial and error, while a visiting star walked through his lines and left the company to adapt itself to his acting during the actual performance. New managements constantly began their reforms with an endeavour to institute regular rehearsals, and some degree of co-ordination was finally achieved by the *stage-manager, who eventually became the all-important stage Director of modern times. (See also AUDITION and IMPROVISATION.)

Reid, (Daphne) **Kate** (1930–), English-born Canadian actress, who trained for the theatre in New York and Toronto, making her first appearance as a student in 1948 and her London début in 1958 in Chetham-Strode's The Stepmother. In 1959 she joined the company at the *Stratford (Ontario) Festival theatre, appearing there during several seasons, notably as Lady Macbeth and Katharina in The Taming of the Shrew in 1962; in 1965 she returned as Ranevskaya in *Chekhov's The Cherry Orchard. She made her début in New York in 1962 as Martha in *Albee's Who's Afraid of Virginia Woolf? at matinées, also playing Caitlin Thomas opposite Alec *Guinness in Sidney Michaels's Dylan (1964). Two years later she starred in Tennessee *Williams's double bill Slapstick Tragedy, and in 1968 (London, 1969) she was in Arthur *Miller's The Price. She was at Stratford, Conn. (see AMERICAN SHAKESPEARE THEATRE), in 1969 (Gertrude in Hamlet) and 1974 (the Nurse in Romeo and Juliet and Big Mama in Tennessee Williams's Cat on a Hot Tin Roof). She moved with the last to New York, where earlier in the year she had been seen in Brian *Friel's The Freedom of the City. At the *Shaw Festival in Ontario in 1976 she played in Mrs Warren's Profession and The Apple Cart, and in 1979 she returned to New York as Henny in John *Guare's Bosoms and Neglect, repeating the role at Stratford (Ontario) in 1980 as well as appearing in Twelfth Night and D. L. Coburn's The Gin Game. In 1984 in New York she was Linda Loman to Dustin Hoffman's Willy in a revival of Miller's Death of a Salesman.

Reinhardt [Goldmann], **Max** (1873–1943), Austrian actor, director, and impresario, who dominated the stage in *Berlin between 1905 and 1918, and remained an influential figure in the German-speaking theatre until he left for the United States in 1938. He regarded the script of a play as a score to be interpreted, and his integrated productions, with their careful harmonization of voice, movement, music, and setting, established for many years the pre-eminence of the director, more especially in the *Symbolism and Impressionism which superseded the earlier *naturalism of *Brahm and was, after the First World War, itself superseded by the *Expressionism of *Jessner and *Piscator. As a young actor Reinhardt appeared in *Vienna and Salzburg and in 1894 joined the company at the *Deutsches Theater in Berlin, where he profited from the tuition of Brahm. Some experience of directing plays by *Strindberg, *Wedekind, and *Wilde at the Kleines Theater in 1902, and above all the success of his production in 1903 of *Gorky's The Lower Depths, in which he played Luka, led him to abandon acting and devote himself entirely to directing. At the Neues Theater am Schiffbauerdamm he produced *Hofmannsthal's Elektra (1903) against a primitive Mycenaean façade, the action taking place at night by the light of flickering torches. He went on to direct *Lessing's Minna von Barnhelm and *Schiller's

Kabale und Liebe, and in 1905 the first of his 12 versions of *A Midsummer Night's Dream*.

By this time Reinhardt had succeeded Brahm as Director of the Deutsches Theater and had bought the theatre outright, erecting next door to it the Kammerspiele for intimate productions. At the Deutsches Theater there was always a play by Shakespeare or a German classic—Schiller's *Die Räuber* in 1908 and *Don Carlos* in 1909; and *Goethe's *Clavigo* in 1908 and both parts of *Faust* in 1908 and 1911 respectively. Meanwhile at the Kammerspiele he concentrated on modern plays, including *Ibsen's *Ghosts*, the first production of Wedekind's *Frühlings Erwachen* (both 1906), and the chamber plays of Strindberg. From 1915 to 1920 he sponsored at the Berlin *Volksbühne matinée performances of new plays by young authors, while elsewhere he was exploring the potential of vast acting areas. In 1910 he produced *Sophocles' *Oedipus the King* in the Zirkus Schumann in Berlin; in 1911 he directed Vollmöller's *The Miracle* in London, converting Olympia into a vast flamboyant Gothic cathedral embracing both actors and audience; and in 1919 he opened the conversion of the Zirkus Schumann, the *Großes Schauspielhaus, with *Aeschylus' *Oresteia* and Romain *Rolland's *Danton*, in which, as in previous productions, his superb handling of crowd scenes was seen at its best. Since 1917 he had also been involved in the founding and running of the Salzburg Festival, where in 1920 he directed Hofmannsthal's *morality play *Jedermann* in an open space in front of the Cathedral. In 1924 he returned to Vienna and took over the Theater in der Josefstadt, and there and at the Komödie am Kurfürstendamm in Berlin he directed nearly 30 plays, paying particular attention to the schooling and directing of the actor, who was now his main concern. His repertory in both theatres was predominantly modern, and included not only *Shaw and *Pirandello but *Cocteau and *Molnár. In 1938 he left Austria for the United States, married Helene Thimig, who with Gertrud *Eysoldt had been one of his leading ladies for many years, and settled in Hollywood where he died.

Réjane [Gabrielle-Charlotte Réju] (1857–1920), French actress, who made her first appearance in 1875 at the Théâtre du *Vaudeville. She was soon recognized as a leading player of comedy and appeared at many Parisian theatres. She was also a frequent visitor to London, making her first appearance there in 1894. Few of her parts were memorable, with the exception of the title-role in *Sardou's *Madame Sans-Gêne* in which she was seen in New York in 1895, and she never ventured on the classics, but in her own line of light comedy she was unapproachable. She opened her own theatre in Paris in 1906 and retired in 1915.

Religious Drama, see JESUIT DRAMA; LITURGICAL DRAMA; MIRACLE PLAY; MORALITY PLAY; MYSTERY PLAY; PASSION PLAY; SAINT PLAY.

Renaissance, Théâtre de la, Paris. The first theatre of this name opened in 1826 as the Salle Ventadour, but was taken over and renamed 10 years later by Anténor Joly, who produced there plays by *Hugo and the elder *Dumas. It had a brief moment of glory in 1838, when *Frédérick made an outstanding success in the title-role of Hugo's *Ruy Blas*, but its fortunes thereafter declined and it closed in 1841.

A second Renaissance opened in 1873 on a site cleared by the burning down of the Théâtre de la *Porte-Saint-Martin two years previously. It began with strong drama, including *Zola's *Thérèse Raquin* (1873), but soon changed to lighter works by *Labiche and *Feydeau, though in 1885 *Becque's serious social problem play *La Parisienne* had a great success. In the 1890s the theatre came under the management of Sarah *Bernhardt. She appeared in a series of her former roles, both classical and modern, with great success: Phèdre, Camille, Magda, and finally *Musset's Lorenzaccio, written in 1834, but not previously produced. When she left her place was taken by *Gémier and later by Lucien Guitry (see GUITRY, SACHA). The theatre's prestige declined with the coming of the cinema, but it saw successful revivals of plays by Sacha Guitry and Feydeau, and in 1956 it was taken over by the actress Vera Korene, who restored it and brought to it a more contemporary repertoire, including plays by *Sartre and *Albee.

Renaud, Madeleine-Lucie (1900–   ), French actress, who in 1923 made her début at the *Comédie-Française and soon became known as an outstanding interpreter of classical comedy, particularly in the plays of *Molière and *Marivaux. In 1940 she married as her second husband J.-L. *Barrault, who had just joined the company, and appeared in several of his productions, including in 1943 *Claudel's *Le Soulier de satin*. In 1946 they left the Comédie-Française to establish the Compagnie Renaud–Barrault, for which she performed a wide range of parts, adding to her former repertory plays by Lope de *Vega, *Chekhov, and *Feydeau, as well as such later authors as *Anouilh, *Salacrou, *Fry, and *Beckett, in whose *Oh! les beaux jours* she was seen in London in 1965. She starred in several of Marguerite *Duras's plays, including *Des journées entières dans les arbres* (1965), *L'Éden cinéma* (1977), and *Savannah Bay* (1983). In 1973 she appeared as Maude in Colin Higgins's *Harold and Maude*, and in 1979 in *Kopit's *Wings*. The beauty of her voice and person and above all her fine, mature intelligence made her outstanding.

Repertory, Repertoire, collection of parts played by an actor or actress, or, more usually, the plays in active production at a theatre in any one season, or which can be put on at short notice, each taking its turn in a constantly changing programme. This system was at one time common to all theatres, and is still in use on the Continent, but in the commercial theatres of London and New York it has been superseded by the continuous run of one play, which may last a year or more. An effort was made in the early 20th century to reintroduce the true repertory system into Britain with the establishment of *repertory theatres, but it was not successful, and today only the *RSC and the *National Theatre company adhere to it, though it is adopted by seasonal theatres such as the *Pitlochry Festival Theatre.

Repertory Theatres. After the disappearance in the late 19th century of the old *stock companies in England and Scotland, a movement was set on foot to bring back the 'true' *repertory system as practised on the Continent. It was begun by such pioneers as Frank *Benson at Stratford-upon-Avon, Lena *Ashwell and Charles *Frohman in London, and Alfred Wareing at the *Glasgow Repertory Theatre. The strongest impetus to its development outside London, however, was given by Miss *Horniman in Manchester. Attempts to re-create 'true' repertory met with little success, theatre staffs and audiences having become accustomed to the continuous run of one play at a time, and the term 'repertory theatre' has now come to signify merely that a theatre puts on its own productions for a series of short runs instead of housing *touring companies.

The oldest surviving repertory theatre in Britain is the *Liverpool Playhouse, where repertory was introduced in 1911. Two years later came the *Birmingham Repertory Theatre. The success of these two theatres led to others, mostly presenting twice-nightly productions for a week at a time, gradually changing to once-nightly. By the early 1950s there were over 100 repertory companies, but the general standard was not high. The spread of television in the 1950s destroyed many of the weakest companies, and by 1960 only 44 were left. At its best the old repertory theatre was an excellent training ground for young actors. A weekly change of bill, with the strain of rehearsing one play while acting in another and learning lines for a third, and the added tension of frequent first nights, could do harm, and most actors remained in repertory only for a year or two. Yet it made for versatility and resourcefulness, was excellent for training the memory, and soon gave beginners self-confidence; many excellent players graduated from weekly rep.

A new era was ushered in by the opening of the *Belgrade Theatre in Coventry in 1958, in which the *civic theatre, founded by and run in close co-operation with the local authority, was to predominate. Fine new subsidized theatres, tending to be called 'regional' rather than 'repertory', were built, with bars and restaurants, spacious foyers which housed art and other exhibitions, and often with a small studio theatre. More money, better working conditions, and longer runs (three or four weeks is now usual) led to a rise in standards, and the best repertory theatres can challenge comparison with the West End, to which some of their productions have been transferred. The old bond between the local theatre and its audience, impossible in the changing world of the commercial theatre, is being forged anew, though performers are more transient now that casting is normally done for individual plays rather than the season. The number of repertory theatres has greatly increased since 1960, and now exceeds the number used by touring companies. Among them are the *Bristol Old Vic; the *Crucible, Sheffield; the *Everyman, Liverpool; the *Haymarket, Leicester; the *Library, Manchester; the *New Victoria, Newcastle-under-Lyme; the *Northcott, Exeter; *Nottingham Playhouse; the *Octagon, Bolton; the *Palace, Watford; the *Royal Exchange, Manchester; the *Stephen Joseph Theatre in the Round, Scarborough; the Theatre Royal, *York; and *West Yorkshire Playhouse, Leeds. In London are the *Greenwich Theatre, the *Lyric Theatre, Hammersmith, and the *Royal Court Theatre.

The USA saw a similar development of regional theatres in the 1960s and 1970s, though the first of them, the *Cleveland Play House, has been a professional theatre since 1921. In 1977 there were non-commercial professional theatres in 34 states plus *Washington, DC, and touring companies from these groups visited 13 more states. See entries for ACTORS THEATRE OF LOUISVILLE; ALLEY THEATRE; AMERICAN CONSERVATORY THEATRE; ARENA STAGE; ASOLO THEATRE COMPANY; CINCINNATI PLAYHOUSE IN THE PARK; DALLAS THEATER CENTER; GOODMAN THEATRE; GUTHRIE THEATER; HARVARD UNIVERSITY; LONG WHARF THEATRE; MARK TAPER FORUM; MILWAUKEE REPERTORY THEATER; SEATTLE REPERTORY THEATRE; STUDIO ARENA THEATRE; TRINITY REPERTORY COMPANY.

Republic Theatre, New York, on West 42nd Street, between 7th and 8th Avenues. Built by the first Oscar Hammerstein, this opened in 1900 with James A. *Herne in his own play *Sag Harbor*, in which the young Lionel *Barrymore made his first appearance in New York. In 1902 the theatre was taken over by David *Belasco, who named it after himself and produced there several of his own plays, including *The Darling of the Gods* (1902), *Sweet Kitty Bellairs* (1903), *Adrea*, and *The Girl of the Golden West* (both 1905). He also produced *Klein's *The Music Master* (1904), in which *Warfield made a great success, and Tully's *The Rose of the Rancho*

(1906). When in 1910 Belasco's second theatre, the Stuyvesant, was renamed the *Belasco, the Republic reverted to its original name. The last play to be seen there, before it became a *burlesque house in 1931 and then a cinema named the Victory in 1942, was John Huston's *Frankie and Johnny* (1930).

République, Théâtre de la, see COMÉDIE-FRANÇAISE.

Restoration Comedy, social comedy of manners which flourished on the Restoration stage. Artificial, cynical, licentious, and extravagant, it was a reaction against the austerity of the Puritan years, when the theatres were closed, and perhaps also reflected the House of Stuart's French exile. It is exemplified by the plays of *Congreve, its greatest exponent, *Wycherley, *Farquhar, *Vanbrugh, and *Etherege. *Dryden also wrote plays in this vein, but was better known for his *heroic dramas.

Reumert, Poul (1883–1968), Danish actor who, beginning his career in Copenhagen in 1902 in the city's minor theatres, worked from 1911 mostly with the *Royal Theatre. He became Denmark's leading actor of his generation, renowned particularly for his playing of *Holberg and *Molière (especially Alceste in *The Misanthrope* and *Tartuffe*). He was also seen in the work of more recent playwrights such as *Strindberg and *Dürrenmatt, notably as Edgar in the former's *The Dance of Death* and in the latter's *The Visit*. He normally played opposite **Bodil Ipsen** (1889–1964), a Danish actress particularly remembered for her performances as Nora in *Ibsen's *A Doll's House*, Mrs Alving in his *Ghosts*, as Strindberg's Miss Julie, and as Alice in his *The Dance of Death*.

Revels Office, see MASTER OF THE REVELS.

Revenge Tragedy, name given to those Elizabethan plays, of which *Kyd's *The Spanish Tragedy* (c.1585–9) was the first, dealing with bloody deeds demanding retribution. Their sublimity could easily turn to *melodrama; indeed, in a cruder form, the revenge motif underlay many of the famous melodramas of the 19th century. Among Shakespeare's plays, *Titus Andronicus* (c.1592) may be considered the lowest form of the Revenge tragedy and *Hamlet* (c.1600–1) its finest flowering. Under the same heading come such plays as *Chapman's *Bussy d'Ambois* (c.1604), *Tourneur's *The Revenger's Tragedy* and *The Atheist's Tragedy* (both c.1606), John *Webster's *The White Devil* (1612) and *The Duchess of Malfi* (1614), and *Middleton's *The Changeling* (1622).

Revolution, Theatre of the, Moscow, see MAYAKOVSKY THEATRE.

Revolving Stage, scenic device which originated in the *kabuki* theatre in Japan in the 17th century, and was brought into Europe at the end of the 19th century. The advantage of it was that three or even more scenes could be set in advance on the revolve and presented to the audience in turn. The scenery itself could be solid *built stuff, but considerable ingenuity was needed to fit the pieces into the various segments of a circle, and to overcome this it is now usual to combine a revolving section with laterally and vertically moving stage machinery. The *Lyttelton and *Olivier stages at the *National Theatre in London incorporate a disc and a drum revolve respectively.

Revue, French word meaning 'survey', used for an entertainment consisting of a number of short items—songs, dances, sketches, monologues—which are not normally related. Unlike the English *music-hall and American *vaudeville, in which a succession of performers appeared, revue players reappeared in various numbers throughout the programme. The first revue to be staged in London, at the *Royal Court Theatre, was *Under the Clock* (1893) by Seymour *Hicks and Charles Brookfield, but it was only part of an evening's entertainment. The first American revue, described as a 'review', was *The Passing Show* (1894). *Pot-Pourri* (1899) was the first English revue to be so described on the playbill. A specialized type of revue that became widely popular at holiday resorts was that presented by the characteristically English *pierrot troupes. One of the first to devise such entertainments was **Harry Pélissier** (1874–1913), whose company wore black-and-white *pierrot costumes against a setting of black-and-white curtains, Pélissier himself acting as compère and writing much of the material. The London success of the *Pélissier Follies*, 1908–12, finally established the popularity of modern revue with West End audiences, and the pierrot costume was widely adopted. One of the most famous concert-parties of this type was The Co-Optimists, who appeared in London during several seasons in the 1920s.

In America *Ziegfeld's *Follies*, an entertainment somewhat similar to Pélissier's, which ran through 25 editions and established revue as an important feature of the New York theatrical scene, was first seen in 1907. The star of the series from 1910 onwards was **Fanny Brice** [Fannie Borach] (1891–1951), a singer and comedienne whom Ziegfeld discovered doing impersonations in *burlesque. She later married **Billy Rose** and appeared in his *Crazy Quilt* (1931). (She was the subject of a Broadway musical, Styne's *Funny Girl*, in 1964.) The edition of 1910 introduced another outstanding newcomer, Bert *Williams, and in two later editions the musical-comedy star **Marilyn Miller** also appeared. Another entertainer, later famous in films, **Al Jolson** [Asa Yoelson] (1886–1950), appeared at the newly built *Winter Garden Theatre in the revue *La Belle Paree*

(1911), and so began a long association with that playhouse during which he starred there in many musical shows, including *Whirl of Society* (1912) and *Dancing Around* (1914).

Just before the First World War revue was at the height of its popularity in London, and was seen at the *Alhambra, the *Empire, and the London *Hippodrome, this last housing from 1912 to 1925 the revues of **Albert de Courville** (1887–1960). Much of the music, and many of the artists, came from America, de Courville's *Hullo Ragtime!* (1912) being a typical example. In 1914 **Alfred Butt** (1878–1962) inaugurated a series of revues at the *Palace Theatre, the first of which, *The Passing Show*, introduced to London the American star **Elsie Janis** [Bierbower] (1889–1956), who, originally on the stage as a child, as 'Little Elsie', went into musical comedy, and eventually became famous in such revues as *The Century Girl* (1916), *Hullo! America* (1918), *Elsie Janis and her Gang* (1919 and 1922), and *Puzzles of 1925*. She also appeared in her own play *A Star for a Night* (1911).

A new type of 'intimate' revue, which relied more on witty dialogue than on dress and dancing, was essayed by C. B. *Cochran with *Odds and Ends* (1914), but his best-known revues were those at the *London Pavilion from 1918 to 1931, which included three by Noël *Coward, Rodgers and *Hart's *One Dam Thing after Another* (1927), and Cole *Porter's *Wake up and Dream* (1929). During roughly the same period **André Charlot** (1882–1956), who came from Paris in 1912 and was associated with the pre-war productions of revue at the Alhambra, was putting on such shows as *A to Z* (1921), Coward's *London Calling* (1923), and, most notably, *Charlot's Revue* (1924 and 1925), which was popular in both London and New York. The outstanding name in revue in New York at the time, however, was **George White** [Weitz] (1890–1968), whose *Scandals of 1919* was successful enough to warrant 12 further editions. Also in 1919 came the first of the *Greenwich Village Follies*, of which seven more editions were to appear during the 1920s. The first was presented by the Canadian **John Murray Anderson** (1886–1954), known as 'the king of revue', who was involved as director, lyricist, or author with several of the later editions. He produced and staged 34 musical comedies and revues, including three of the later editions of the *Ziegfeld Follies*. **Ed Wynn** [Edwin Leopold] (1886–1966), one of the greatest American comedians of his day, had his first starring role in revue in the *Ed Wynn Carnival* (1920). He presented on stage a personality of staggering ineptitude, with a lisp, fluttering hands, and outrageous costumes. He later appeared in such revues as *The Grab Bag* (1924), *The Laugh Parade* (1931), and *Boys and Girls Together* (1940), as well as in musical comedies, usually tailored to

suit his personality, among them *Simple Simon* (1930) and *Hooray for What!* (1937). Another major star of American revue was **Eddie Cantor** [Isidore Itzkowitz or Iskowitz] (1892–1964), an eye-rolling singer and comedian who often appeared in blackface. He was in several editions of the *Ziegfeld Follies* as well as in other revues such as *Make it Snappy* (1922) and musical comedies—*Kid Boots* (1923; London, 1926) and *Whoopee* (1928).

In 1923 Earl *Carroll presented the first of 11 editions of his *Vanities*. Other successful revues were *The Band Wagon* (1931), whose score by Arthur Schwartz and Howard Dietz was perhaps the best ever written for a revue, and *As Thousands Cheer* (1933), with music by Irving *Berlin, in which **Marilyn Miller** made her last appearance on the stage.

In London, Jack Hulbert and Cicely *Courtneidge co-starred in several revues, while **Clarkson Rose** (1890–1968) presented every summer his famous seaside revue *Twinkle*, most of which he wrote himself, combining it with appearances in *pantomime at Christmas. Continuous revue was introduced at the *Windmill Theatre in 1932. From 1934 to 1939 Norman *Marshall staged intimate revue annually at the *Gate Theatre, bringing fame to one of the supreme exponents of this genre, **Hermione Gingold** (1897–1987). It also flourished at the *Little Theatre with *Nine Sharp* (1938) and *The Little Revue* (1939), written by **Herbert Farjeon** (1887–1945), the major revue librettist of the period. Grandson of the American actor Joseph *Jefferson, he was also a dramatic critic and author, and with his sister Eleanor was responsible for the musical plays *The Two Bouquets* (1936), *An Elephant in Arcady* (1938), and *The Glass Slipper* (1944). Both the Farjeon revues featured the Australian-born comedian **Cyril Ritchard** (1898–1977), who had settled in England and was later seen in Coward's revue *Sigh No More* (1945), as well as in a number of straight roles, among them Tattle in *Congreve's *Love for Love* in New York in 1947 and Sir Novelty Fashion in *Vanbrugh's *The Relapse* in London in 1948 (NY, 1950). He eventually moved to America.

The vogue for revue continued during and after the Second World War, **George Black** (1890–1943) presenting in London a series beginning with *Apple Sauce* (1940), and his sons George and Alfred continuing the tradition. On a more intimate scale, the biggest success of the 1940s was the *Sweet and Low* series with **Hermione Gingold**, for which the books were written by **Alan Melville** (1910–83). He was connected with many other revues and was the author of several plays, among them *Castle in the Air* (1949), *Dear Charles* (1952), and *Simon and Laura* (1956). Gingold also appeared in several revues with Hermione *Baddeley. **Dora Bryan** (1924–) was seen in revues such as

The Lyric Revue (1951), The Globe Revue (1952), and Living for Pleasure (1958), as well as in *musical comedy and plays such as *Shaw's Too True to be Good (1965) and Ronald *Millar's They Don't Grow on Trees (1968), in which she played nine parts. In Bamber Gascoigne's Share My Lettuce (1957), Maggie *Smith made her London début. Also in the cast was **Kenneth Williams** (1926–88), an actor well known for his 'camp' style and wide range of silly and affected voices. He was later seen in the revues Pieces of Eight (1959) and One over the Eight (1961), and in such diverse straight parts as the Dauphin in Shaw's Saint Joan (1954), Julian in Peter *Shaffer's The Public Eye (1962), and the title-role in Robert *Bolt's Gentle Jack (1963). At the Drop of a Hat (1956; NY, 1959) and At the Drop of Another Hat (1963; NY, 1966) were two-man entertainments by **Michael Flanders** (1922–75), confined to a wheel-chair by polio-myelitis, who wrote the words, and Donald Swann. Beyond the Fringe, a brilliantly clever satirical revue which originated with the *Cambridge Footlights Club and had already been seen at the *Edinburgh Festival, arrived in London in 1961 (NY, 1962); it was the joint work of Alan *Bennett, Jonathan *Miller, Peter Cook, and Dudley Moore. Thereafter topical revue was mainly confined to television, and since on economic grounds spectacular revues were impossible, the genre could not survive, though Oh! Calcutta! (NY, 1969; London, 1970), devised by the drama critic Kenneth *Tynan, in which both men and women appeared naked for much of the show, attracted curious audiences.

Riccoboni, Luigi (1676–1753), actor of the *commedia dell'arte, known as Lélio. The son of Antonio, who played *Pantalone in London in 1678/9, Luigi was a fine actor who ran his own company in Italy for many years, and in 1716 returned with an Italian company to Paris, where the *Comédie-Italienne had been closed for nearly 20 years. His attempts to re-establish the old improvised comedy, which was already dying out in Italy, met with mixed fortunes and he relied increasingly on written texts, which hastened the complete absorption of the Italians into the French theatre.

Rice [Reizenstein], Elmer (1892–1967), American lawyer and dramatist, whose On Trial (1914) was the first American play to employ the flashback technique of the cinema. Rice's first major contribution to the theatre was the expressionistic fantasy The Adding Machine (1923), which satirized the growing regimentation of modern man in the machine age through the life and death of the arid bookkeeper Mr Zero. Street Scene (1929), which followed, was awarded a *Pulitzer Prize for its realistic chronicle of life in the slums. The author later adapted it as the libretto of an opera with

music by Kurt Weill, first performed in 1947. Counsellor-at-Law (1931) drew an equally realistic picture of the legal profession. The depression of the 1930s inspired We, the People (1933), the Reichstag trial was paralleled in Judgement Day, and conflicting American and Soviet ideologies formed the subject of Between Two Worlds (both 1934). When these plays failed on Broadway Rice retired from the theatre, but returned two years later to help found and run the Playwrights' Producing Company. His later plays included American Landscape (1938), Two on an Island, Flight to the West (both 1940), the last a fervent denunciation of Nazism, and A New Life (1942). He recaptured some of the success of his early plays with the fantasy Dream Girl (1945) and presented a modern psychoanalytical variation on the Hamlet theme in Cue for Passion (1958), in which Diana *Wynyard played the Gertrude-like character Grace Nicholson.

Rice, T(homas) D(artmouth) (1808–60), American *vaudeville performer and Negro impersonator, known as 'Jim Crow' from the refrain of his most famous song-and-dance act, which he first performed in 1828 in Kentucky. It caught the public fancy and was soon being performed all over the United States. The 'trucking dance' with which Rice accompanied the song was an important element in its success, and his decision to perform it in 'blackface' led directly to the craze for *minstrel shows, though Rice himself never became part of a troupe, preferring to work alone. In 1833 he visited Washington, where the 4-year-old Joseph *Jefferson the third, later one of America's finest actors, appeared with him, being tumbled out of a sack at the conclusion of Rice's turn to mimic his song and dance. Rice himself was seen as Jim Crow in London at the *Surrey Theatre in 1836, accompanied by a full-scale publicity campaign featuring 'Jim Crow' pipes, hats, and cartoons. The name later became a term of reproach applied to subservient Negroes, and was also applied to the racial segregation laws in the American South.

Rich, John (c.1692–1761), English actor and theatre manager, son of **Christopher Rich** (?–1714), manager first of *Drury Lane and then of the old *Lincoln's Inn Fields Theatre, which John inherited from him, and where he was responsible in 1728 for the production of *Gay's The Beggar's Opera. With the profits from this highly successful venture he built the first *Covent Garden theatre, transferring to it the patent granted by Charles II to William *Davenant, and opening it in 1732 with a company headed by James *Quin. Though almost illiterate John Rich was an excellent actor in dumb-show and, developing the ideas of Weaver, he popularized *pantomime in England, playing *Harlequin himself under the name of Lun. He produced a pantomime annually from 1717 to 1760, his own masterpiece in acting being 'Harlequin Hatched from an Egg by the Sun', which he performed in *Harlequin Sorcerer* at Tottenham Court Fair in 1741.

Richard Rodgers Theatre, New York, on West 46th Street. With a seating capacity of 1,338, it opened as Chanin's Forty-Sixth Street Theatre in 1925 and became the Forty-Sixth Street Theatre in 1932. It is best known for such musicals as *Hellzapoppin* (1938), *Du Barry was a Lady* (1939), *Panama Hattie* (1940), *Finian's Rainbow* (1947), *Guys and Dolls* (1950), and *Damn Yankees* (1955). In 1958 *Gielgud occupied the theatre with his one-man recital from Shakespeare *The Ages of Man*. Later musicals included *How to Succeed in Business without Really Trying* (1961), *1776* (1969), *Chicago* (1975), *The Best Little Whorehouse in Texas* (1978), and *Nine* (1982), with a book by Arthur *Kopit. August *Wilson's *Fences* (1987) was also staged there. The theatre received its present name in 1990.

Richardson, Ian (1934–), Scottish actor, born in Edinburgh, who studied for the stage in Glasgow, and in 1958–9 was at the *Birmingham Repertory Theatre. In 1960 he joined what was to become in 1961 the *RSC, his roles including Antipholus of Ephesus in *The Comedy of Errors* in 1962. In London in 1964 he was seen as the Herald in *Weiss's *Marat/Sade* and Ford in *The Merry Wives of Windsor*, a year later playing Marat in both London and New York. In 1965 he was also seen in *Brecht's *Puntila*. With a fine, resonant voice, great bodily plasticity, and a powerful sense of controlled emotion, he was admirable as Vendice in *Tourneur's *The Revenger's Tragedy* in 1966, in the title-roles of *Coriolanus* in 1967 and *Pericles, Prince of Tyre* in 1969, and particularly as Prospero in *The Tempest* in 1970. He left the company to appear in the musical *Trelawny* (1972), based on *Pinero's *Trelawny of the 'Wells'*, but returned to alternate the roles of Richard II and Bolingbroke in *Richard II* in both London in 1973 and New York a year later; also in 1974 he was seen in London in *Gorky's *Summerfolk* and *Wedekind's *The Marquis of Keith*. After playing the title-role in *Richard III* in 1975 he starred in New York in 1976 as Higgins in a revival of the musical *My Fair Lady*. During the 1980s he worked largely for the cinema and television.

Richardson, Sir Ralph David (1902–83), English actor, surpassed only by *Gielgud and *Olivier among his contemporaries, who made his first professional appearance at Lowestoft in 1921 as Lorenzo in *The Merchant of Venice*. He joined the *Birmingham Repertory company in 1926, making his London début later the same year as the Stranger in *Sophocles' *Oedipus at Colonus*. In 1930 he went to the *Old Vic,

where his reputation was chiefly made, his roles including Caliban in *The Tempest*. His Toby Belch in *Twelfth Night* on the reopening of *Sadler's Wells in 1931 was followed by Petruchio in *The Taming of the Shrew*, Bottom in *A Midsummer Night's Dream*, and Henry V in the Vic-Wells company's 1931–2 season. After leaving the Old Vic he was seen in several important modern plays, including *Maugham's *For Services Rendered* (1932) and *Sheppey* (1933) and *Priestley's *Eden End* (1934) and *Cornelius* (1935). His New York début, as Chorus and Mercutio in *Romeo and Juliet* at the end of 1935, preceded the title-roles in *Othello* at the Old Vic (1938) and Priestley's *Johnson over Jordan* (1939). He then served in the Fleet Air Arm, returning to the Old Vic as co-director, 1944–7. His diverse roles there included *Chekhov's Uncle Vanya, Inspector Goole in Priestley's *An Inspector Calls*, and *Rostand's Cyrano de Bergerac, as well as, unforgettably, *Ibsen's Peer Gynt and Falstaff in *Henry IV*, the last also in New York in 1946. In 1949 came another much-acclaimed performance as the tyrannical Dr Sloper in Ruth and Augustus Goetz's *The Heiress*, based on Henry *James's novel *Washington Square*. After his Vershinin in an all-star production of Chekhov's *Three Sisters* in 1951 he went to the *Shakespeare Memorial Theatre to play Prospero in *The Tempest* and the title-role in *Jonson's *Volpone* (both 1952), and was then seen in N. C. *Hunter's *A Day by the Sea* (1953). At the Old Vic in 1956 he gave a fine performance in the title-role of *Timon of Athens*, and he was back in New York in *Anouilh's *The Waltz of the Toreadors* (1957). Later in the year he created in London another notable role, the self-deluding Cherry in Robert *Bolt's *Flowering Cherry*, which was followed by Graham *Greene's *The Complaisant Lover* (1959), Sir Peter Teazle in *Sheridan's *The School for Scandal* (1962; NY, 1963), and *Pirandello's *Six Characters in Search of an Author* (1963). He appeared in revivals at the *Haymarket: as William the Waiter in *Shaw's *You Never Can Tell*, Sir Antony Absolute in Sheridan's *The Rivals* (both 1966), and Shylock in *The Merchant of Venice* (1967). He then returned to modern plays: Joe *Orton's *What the Butler Saw* (1969), David *Storey's *Home* (London and NY, 1970), superbly partnering Gielgud, and *Osborne's *West of Suez* (1971), as an elderly writer turned television pundit. Joining the *National Theatre company, he played the title-role in Ibsen's *John Gabriel Borkman* (1975) and Hirst in *Pinter's *No Man's Land* (also 1975; NY, 1976), sharing the honours again with Gielgud in another virtuoso display. He returned to London as the Author in William Douglas *Home's *The Kingfisher* (1977), a part specially written for him. He later returned to the National Theatre in several productions, including David Storey's

Early Days (1980; NY, 1981). The range and variety of his work made him difficult to classify. Though not by nature a tragedian he could rivet the attention in a tragic or sinister role, was outstanding in pathos, and in comedy gave free rein to his own eccentric and amusing personality.

Richardson, Tony [Cecil Antonio] (1928–91), English director, who in 1955 joined the company at the *Royal Court Theatre, where among other new plays he directed John *Osborne's *Look Back in Anger* (1956; NY, 1957) and *The Entertainer* (1957; NY, 1958). Other outstanding productions for the Royal Court were *Ionesco's *The Chairs* and *The Lesson* (1958) in a double bill. He then went to the *Shakespeare Memorial Theatre, where he directed *Pericles* (also 1958) and *Othello* (1959) with Paul *Robeson in the title-role. He returned to the Royal Court to direct *Coward's *Look after Lulu* (1959), Osborne's *Luther* (1961), and *A Midsummer Night's Dream* (1962), and in 1962 he also directed David Turner's *Semi-Detached* in the West End, with *Olivier. After directing *Luther* in New York in 1963 he remained on Broadway, being responsible for productions of *Brecht's *Arturo Ui* (also 1963) and Tennessee *Williams's *The Milk Train Doesn't Stop Here Anymore* (1964). Back in London he directed, also in 1964, a new version of *Chekhov's *The Seagull* with Vanessa *Redgrave (to whom he was then married) as Nina. Later in the same year he was responsible for the first British production of Brecht's *St Joan of the Stockyards*. In 1969 he directed *Hamlet*, later seen in New York and on tour in the USA, and in 1972 he was responsible for productions of Brecht and Weill's *The Threepenny Opera*, again with Vanessa Redgrave, and John *Mortimer's *I, Claudius*, based on Robert Graves's novels. He also directed Vanessa in *Antony and Cleopatra* at Sam *Wanamaker's Bankside Globe in 1973, and in *Ibsen's *The Lady from the Sea* in New York in 1976. He subsequently lived in Los Angeles, where in 1979 he directed *As You Like It*. A well-known film director, he worked latterly for the cinema and television.

Richelieu, Armand-Jean du Plessis de, Cardinal (1585–1642), French statesman, for many years the virtual ruler of France. He did a great deal for the theatre, and by his patronage of *Mondory helped to establish a permanent professional playhouse in Paris and to raise the status of the actor. He had strong leanings towards dramatic authorship, and set up a company of five, including *Corneille and *Rotrou, to write plays under his direction. They were not very successful, and Corneille resigned after a disagreement over his share of the plot. Richelieu built a very well-equipped theatre in his palace which was first used in

1641 and later, as the *Palais-Royal, became famous under *Molière.

Richmond, Surrey, small town on the River Thames, not far from London, which became a fashionable resort after the opening of Richmond Wells in 1696. The first recorded theatrical performance was in 1714, when the Duke of Southampton's Servants presented an adaptation of *The Virgin Martyr* by *Dekker and *Massinger, retitled *Injured Virtue*. Four years later a temporary theatre building was set up on Richmond Hill and was regularly visited by the Prince of Wales, then residing at Richmond Lodge. The venture was a success and in 1719 Penkethman opened a permanent theatre in a converted stable with the younger *Killigrew's new comedy *Chit-Chat*, and continued to run it successfully until his death in 1725. In 1730 a new theatre opened with a season by the company from the *Lincoln's Inn Fields Theatre. Competition from a theatre on Richmond Green, which opened in 1765, forced it to close in 1767. This theatre had a three-tier auditorium and a proscenium opening 24 ft. wide. Its first production was *Bickerstaffe's *Love in a Village*. For some years the theatre provided summer employment for actors from the *Patent Theatres, but once the novelty had worn off audiences dwindled. From 1831 until his death in 1833 Edmund *Kean took over the theatre, which was then known as the King's, as manager and leading actor, living for some time in the house attached to it. The theatre finally closed in 1884. The present Richmond Theatre, which holds 875, opened on Richmond Green in 1899 with a production of *As You Like It*, Ben *Greet playing Touchstone and Dorothea Baird Rosalind. It has housed *touring companies and also functioned as a *variety theatre (being known for a time as the Richmond Hippodrome and Theatre), but for much of its history has been a *repertory theatre. Since 1973, however, it has staged touring productions, plus an annual *pantomime, and is a stopping-place for plays on pre-London runs. The building is preserved as being of architectural merit.

Richmond, Va. The first theatre in Richmond was built in the 18th century at the intersection of Twelfth and Broad Streets, and was destroyed by fire in 1811 with the loss of 72 lives. A new theatre, the Marshall, opened in 1818 and saw the first appearance in America of the elder *Booth, in 1821, and Edwin *Forrest, in 1841; Joseph *Jefferson the third managed it from 1854 to 1856, and in 1859 John Wilkes *Booth, the assassin of President Lincoln, was a member of its *stock company. The theatre burned down on 1 Jan. 1862 and was replaced in 1863 by a lavishly decorated new theatre which became known as the New Richmond. It was destined to become the leading Confederate theatre, all the outstanding theatre personalities of the South appearing there. It retained much of its glamour in the post-war years, but was finally demolished in 1896.

Richmond, Yorkshire, see GEORGIAN THEATRE ROYAL.

Rigg, Diana (1938–), English actress, who had spent some time in provincial repertory companies before joining the *Shakespeare Memorial Theatre company in 1959. She made her first appearance in London with the company, renamed the *RSC, in 1961, in *Giraudoux's *Ondine*, and remained with it until 1964, making her first appearance in New York in that year as Adriana in *The Comedy of Errors* and Cordelia in *King Lear*. She returned to the RSC in 1966 to play Viola in *Twelfth Night*. After starring opposite Keith *Michell in Ronald *Millar's *Abelard and Heloise* (1970; NY, 1971) she joined the *National Theatre company, where she displayed a gift for comedy in handling the verbal pyrotechnics of *Stoppard's *Jumpers* (1972) and showed her versatility as a striking Lady Macbeth (also 1972) and an enchantingly frivolous Célimène in *Molière's *The Misanthrope* (1973; NY, 1975). In 1974 she played Eliza Doolittle to Alec *McCowen's Professor Higgins in a revival of *Shaw's *Pygmalion*, and she then returned to the National Theatre in *Phaedra Britannica* (1975), a transposition of *Racine's *Phèdre* to 19th-century British India. In 1978 she was in another play by Stoppard, *Night and Day*, in which she played the intelligent, bored wife of a mining engineer in an African state. She was seen in Shaw's *Heartbreak House* in 1983, and in 1985 in *Ibsen's *Little Eyolf* and, at *Chichester, as Cleopatra in *Antony and Cleopatra*. In 1987 she further extended her range by appearing in *Sondheim's *Follies*.

Ring, The, London, large octagonal structure, originally a chapel, which stood in Blackfriars Road, London, about 500 yards from Blackfriars Bridge. Built in 1783, it later became a well-known boxing ring, and sprang into temporary theatrical fame when Robert *Atkins used it for some of the earliest *theatre-in-the-round productions in England, producing there in 1936–7 *Henry V*, *Much Ado about Nothing*, and *The Merry Wives of Windsor*. It then sank back into obscurity and was demolished some time after the Second World War. It has sometimes been confused with the nearby *Rotunda.

Rise-and-Sink, method of effecting a *transformation scene made possible by the provision of space above the stage with a low *grid. Though not as lofty as the later *flies, this would take the depth of half a backcloth, and a scene could therefore be changed quickly by causing the upper half to ascend and the lower half, possibly framed out and with a profiled

upper edge, to descend into the *cellar on *slotes.

Ristori, Adelaide (1822–1906), Italian actress, who made an international reputation, particularly in tragedy. She was only 14 when she gave an outstanding interpretation of the title-role in Pellico's *Francesca da Rimini* and at 18 she played Mary Queen of Scots in *Schiller's *Maria Stuart*, a part with which she was later closely associated. She then joined the Compagnia Reale Sarda and at the time of her marriage in 1847 was the leading lady of Domeniconi's company where she played opposite *Salvini. After a brief retirement she rejoined the Reale Sarda and in 1855 went with Ernesto *Rossi to Paris, where she soon became a serious rival to *Rachel in tragedy, though she was also seen in comedies by *Goldoni. She visited London and toured the main provincial towns of Great Britain a number of times between 1856 and 1882, on her last visit playing Lady Macbeth in English at *Drury Lane to a mixed reception. She made the first of four highly successful tours of the United States in 1866 and retired from the stage in 1885. A stately woman, of strong physique and commanding presence, she sometimes wasted her talents on popular romantic melodrama, but could encompass with equal brilliance parts as diverse as *Euripides' Medea and Goldoni's Mirandolina in *La locandiera*.

Ritchard, Cyril, see REVUE.

Ritterdrama, offshoot of the *Sturm und Drang drama, in which the valour of medieval knights was displayed in scenes of battle, jousting, and pageantry, often with a marked vein of Bavarian local patriotism. Written in prose and irregular in form, this 'feudal drama' had as its theme strong passions and contempt for the conventions and fostered the taste for romantic and medieval settings kindled by *Goethe's *Götz von Berlichingen* (1773) and Klinger's *Otto* (1774). Indeed, the acquisition by theatre companies of wardrobes of medieval costumes for the *Ritterdrama* led to a new awareness of 'picturesqueness' and historical realism which was carried over into the staging of *melodrama. Among the authors of such plays were **Josef August von Törring** (1753–1826), Bavarian Minister of State, with *Kasper der Thorringer* (pub. 1785) and *Agnes Bernauerin* (1780), and **Joseph Marius Babo** (1756–1822), for some time director of the Court theatre in *Munich, with *Otto von Wittelsbach* (1782). Reactionary influences caused the *Ritterdrama* to be banned from the Munich stage, but its vogue continued elsewhere, notably in Austria, where **Karl Friedrich Hensler** (1761–1825) fused this type of drama with the native operatic fairy-tale in *Das Donauweibchen* (1797).

Ritz Theater, New York, see WALTER KERR THEATER.

Riverside Studios, Hammersmith, London, arts centre housing theatre, concerts, films, dance programmes, and exhibitions. Originally a foundry, it was converted into film studios between the wars, and was at one time the largest television centre in Europe. When the BBC vacated the premises in 1974 the lease was acquired by the Hammersmith Borough Council, which in 1975 formed an independent Trust to administer the building. The centre opened part-time in 1976 and full-time in 1978. The theatre accommodates visiting British and foreign companies. It has also produced its own work, including several plays staged by Peter *Gill, the centre's first Director. Among visiting companies have been the *Fringe group Joint Stock in a dramatization of Robert Tressell's novel *The Ragged Trousered Philanthropists* (1978); the *Maly Theatre of Leningrad in Alexander Galin's *Stars in the Morning Sky* and the Asian theatre company Tamasha performing in Hindi and English (both 1988); and a *Haymarket, Leicester, double bill of *Beckett's *Krapp's Last Tape* and *Catastrophe* (1990). Artists such as Alan *Bates, Albert *Finney, Helen *Mirren, and Vanessa *Redgrave have appeared at the centre.

Rix, Sir Brian Norman Roger (1924–), English actor and manager, who made his first appearance on the stage in 1942 and a year later was seen in London with Donald *Wolfit's company in *Twelfth Night*. He served with the RAF, 1944–7, and on demobilization formed two repertory companies. He had his first outstanding success with Colin Morris's farce *Reluctant Heroes* (1950), which he presented on tour and then took to the *Whitehall Theatre in London, where it ran for nearly four years with himself as Gregory. It was succeeded by four more highly successful farces which he presented at the same theatre, appearing in them all himself: John Chapman's *Dry Rot* (1954) and *Simple Spymen* (1958); Ray Cooney and Tony Hilton's *One for the Pot* (1961); and Ray Cooney's *Chase Me, Comrade* (1964). The combined runs of these easily exceeded the record previously held by the *Aldwych Theatre of 10 years' continuous presentation of farce by one manager in the same theatre. In 1967 Rix moved to the *Garrick, where he presented and starred in a repertory of Ray Cooney and Tony Hilton's *Stand by Your Bedouin*, Anthony Marriott and Alistair Foot's *Uproar in the House*, and Harold Brooke and Kay Bannerman's *Let Sleeping Wives Lie*, the last eventually running on its own and being followed by Michael Pertwee's *She's Done It Again!* (1969) and *Don't Just Lie There, Say Something* (1971). After Pertwee's *A Bit between the Teeth* (1974) at the *Cambridge Theatre he returned in 1976 to the Whitehall

to co-present and star in *Fringe Benefits* by Peter Yeldham and Donald Churchill. He left the theatre in 1980 to work full time with a charity for the mentally handicapped.

Robards, Jason (1922–), American actor, who made his Broadway début in 1947 but spent some years in obscurity, including periods as assistant stage manager, and was still virtually unknown when in 1956 he gained enormous acclaim for his performance as Hickey in *O'Neill's *The Iceman Cometh* at the *Circle-in-the-Square. He followed it in the same year with a fine portrayal of the alcoholic elder son Jamie in the same author's *Long Day's Journey into Night*, and sealed his reputation as an interpreter of O'Neill with the roles of Erie Smith in *Hughie* in 1964, Jamie again, 10 years later, in *A Moon for the Misbegotten* in 1973, James Tyrone, father of the character he originally portrayed, in another production of *Long Day's Journey into Night* in 1976, and Cornelius Melody in *A Touch of the Poet* in 1977. He was at the *Stratford (Ontario) Festival in 1958, and in 1960 played Julian Berniers in Lillian *Hellman's *Toys in the Attic*. Later roles included Murray Burns in Herb Gardner's long-running *A Thousand Clowns* (1962), Quentin in Arthur *Miller's *After the Fall* (1964), the Vicar of St Peter's in John *Whiting's *The Devils* (1965), and Frank Elgin in a revival of *Odets's *The Country Girl* in 1972. In 1983 he starred in a revival of *Kaufman and *Hart's *You Can't Take it with You*, and in 1988, to mark O'Neill's centenary, he again played James Tyrone, together with Nat Miller in O'Neill's *Ah, Wilderness!*

Roberts, Arthur, see MUSIC-HALL.

Robert S. Marx Theatre, see CINCINNATI PLAYHOUSE IN THE PARK.

Robertson, Agnes, see BOUCICAULT, DION.

Robertson, T(homas) **W**(illiam) (1829–71), English dramatist, eldest of the 22 children of an actor. Several of his brothers and sisters were on the stage, the most famous being the youngest girl Madge who became Mrs *Kendal. Robertson himself acted as a child, and later appeared as an adult in Lincoln, where the Robertson family had for many years been in control of the theatres on the Lincoln *circuit. There he made himself generally useful, painting scenery, writing songs and adapting plays, and playing small parts. He was, in fact, trained in the old school which he was later to destroy, a process which can be studied, with reservations, in *Pinero's *Trelawny of the 'Wells'* (1898). Yet his earliest plays were in no way remarkable. He wrote them quickly and sold them cheaply to Lacy, the theatrical publisher. The first of them, *The Chevalier de St George* (1845), was produced at the *Princess's Theatre, and although they were all moderately successful, it was not until the production of *David Garrick* (1864) at the *Haymarket Theatre that he came to the attention of the public. Such plays as *Society* (1865), *Ours* (1866), *Caste* (1867), *Play* (1868), and *School* (1869), whose monosyllabic titles alone come as a refreshing change from those of earlier, and even some contemporary, plays, were all seen at the Prince of Wales, later the *Scala Theatre, where they established the reputation not only of the author but also of the newly formed *Bancroft management. With this series Robertson founded what has been called the 'cup-and-saucer drama'—the drama of the realistic, contemporary, domestic interior. His rooms were recognizable, his dialogue credible; his plots, though they now seem somewhat artificial, were true to his time, embodying serious social content, and an immense advance on anything that had gone before. *Caste* in particular still holds the stage, and some of the others would revive well. Robertson, a convivial creature with a brilliant flow of conversation, was active in the production of his own plays, fulfilling some of the functions of the modern *director. He enjoyed a few years of fame and adulation before dying at the height of his success, leaving a permanent mark on the theatre of his time and foreshadowing the work of many modern dramatists.

Robertson, Toby, see PROSPECT THEATRE COMPANY.

Robeson, Paul Bustill (1898–1976), American Negro actor and singer, who made his first appearance on the stage in 1921. He created a sensation when he appeared for the *Provincetown Players in 1924 as Jim Harris in *O'Neill's *All God's Chillun Got Wings,* and even more by his playing of Brutus Jones in the same author's *The Emperor Jones,* in which he was first seen in London a year later and in Germany in 1930. His singing of 'Ole Man River' in the London production of the musical *Show Boat* (1928) first revealed the haunting quality of his superb bass voice, and led him to devote much of his time to touring in recitals of Negro spirituals. He was, however, seen again in London in the title-role of *Othello* (1930), which when he appeared in it in New York in 1943 achieved the longest run of a Shakespeare play on Broadway up to that time. He also played the part at Stratford-upon-Avon in 1959. Another of his outstanding roles was 'Yank' in O'Neill's *The Hairy Ape* (1931). During his active career Robeson did much to further the interests of the Negro people. A highly gifted, sincere, and courageous man, he was one of the best-known Negro personalities of his time.

Robey [Wade], Sir George Edward (1869–1954), English actor and *music-hall comedian. He made his first appearance at the *Oxford Music-Hall in 1891, and was so successful that he was at once engaged to appear again 10 days later. He was soon playing at all the leading London and provincial halls, earning for himself the nickname of 'Prime Minister of Mirth'. Originally a singer of comic songs, he was equally admired in a series of humorous sketches featuring Shakespeare, Charles II, Henry VIII, The Caretaker, The Gladiator, and other historic and imaginary characters, and he made a great success in the *revue *The Bing Boys are Here* (1916), in which he played Lucius Bing, as he did in the sequel, *The Bing Boys on Broadway* (1918). He was himself the author of a revue, *Bits and Pieces* (1927), in which he appeared with great success, but it is for his music-hall turns that he is chiefly remembered. His humour is robust, and on stage he seemed to consist almost entirely of a bowler hat and two enormous black eyebrows. He showed his versatility by appearing as Dame Trot in the *pantomime *Jack and the Beanstalk* (1921); Menelaus in *Helen!* (1932), a new English version of Offenbach's *La Belle Hélène*; and Falstaff in *Henry IV, Part One* at *Her (then His) Majesty's Theatre in 1935, being the first music-hall star to appear in Shakespeare.

Robin Hood, legendary English hero, whose name first appears in *Piers Plowman* (1377). He typifies the chivalrous outlaw, champion of the poor against the tyranny of the rich. It is impossible to identify him with any historical personage, though Anthony *Munday made him the exiled Earl of Huntingdon. Since he is always dressed in green he may be a survival of the Wood-man, or Jack-in-the-Green, of the early pagan *folk festivals, or he may have been imported by minstrels from France. By the end of the 15th century he and his familiar retinue of Maid Marian, Little John, Friar Tuck and the Merry Men, with their *hobby horse and morris dance, were inseparable from May-Day revels, and the protagonists of many rustic dramas. These, however, cannot be considered folk plays, as were the *mumming play and the Plough Monday Play, since they were specially written by minstrels. In due course, the May-Day festivities found their way to Court, where they became mixed up with allegory and pseudo-classicism: Henry VIII, in particular, enjoyed many splendid Mayings, including one in which he was entertained by Robin Hood to venison in a bower. After that their popularity waned, and they were finally suppressed by the Puritans. The story of Robin Hood and his Merry Men became a favourite subject for 19th- and 20th-century Christmas *pantomime.

Robins, Elizabeth (1862–1952), American actress, who in 1885 made her first appearance on the stage with the Boston Museum stock company. She later toured with Edwin *Booth and Lawrence *Barrett, and appeared in the elder *Dumas's *The Count of Monte Cristo* with James *O'Neill. In 1889 she made her first visit to London, where the greater part of her professional life was passed, and she became prominently identified with the introduction of *Ibsen to the London stage, playing Martha Bernick in *The Pillars of Society* (1889), Mrs Linden in *A Doll's House*, the title-role in *Hedda Gabler* (both 1891), being particularly fine in the latter part, Hilda in *The Master Builder*, Rebecca West in *Rosmersholm*, Agnes in *Brand* (all 1893), Asta Allmers in *Little Eyolf* (1896), and Ella Rentheim in *John Gabriel Borkman* (1896). She held the stage rights of most of these plays, and was responsible, sometimes in conjunction with such advanced groups as Grein's Independent Theatre, for their initial productions. She was also seen in a wide variety of other parts, notably the title-role in *Echegaray's *Mariana* (1897), after which she retired, apart from a brief return to the stage as Lucrezia in Stephen *Phillips's *Paolo and Francesca* (1902), and devoted herself to novel-writing.

Robinson, Bill, see VAUDEVILLE, AMERICAN.

Robinson, (Esmé Stuart) Lennox (1886–1958), Irish dramatist, actor, director, and critic. His first play, *The Clancy Name* (1908), was staged at the *Abbey Theatre, Dublin, of which he was still a director at his death. In his early plays—*The Cross Roads* (1909), *Harvest* (1910), *Patriots* (1912), *The Dreamers* (1913)—he treated political and patriotic themes as matter for tragedy, with no weakening into sentiment; but his comedy *The White-Headed Boy* (1916; London, 1920; NY, 1934) first made him known outside his own country. Like *Crabbed Youth and Age* (1922; NY, 1932), it showed his skill as a structural craftsman and as a creator of character in the style of the comedy of manners. In *The Big House* (1926; NY, 1933; London, 1934), Robinson became the first Irishman to write a play on the changing order of Ireland's civilization, but in *The Far-Off Hills* (1928; London, 1929; NY, 1932) and *Church Street* (Dublin and NY, 1934) he returned to comedy, in which his best work was done, achieving in the latter a tragi-comedy or 'mingled drama' whose satire was genial and whose ironies were tragic. His later works included *All's Over Then* (1932; London, 1934); *Is Life Worth Living?* (also known as *Drama at Inish*, Dublin, London, and NY, 1933); *Killycreggs in Twilight* (1937; London, 1940); and *The Lucky Finger* (1948).

Robson, Dame Flora (1902–84), English actress. She made her début in Clemence *Dane's *Will Shakespeare* (1921) and then toured with Ben *Greet and was at the *Oxford Playhouse under J. B. *Fagan. After leaving the theatre for several years she returned in 1929 and was seen in London in Eugene *O'Neill's *Desire under the Elms* (1931), her performance as

Abbie Putnam, followed by roles in *Bridie's *The Anatomist* (also 1931), *Priestley's *Dangerous Corner*, and *Maugham's *For Services Rendered* (both 1932), bringing her to the head of her profession. In 1933 she joined the company at the *Old Vic, playing a wide variety of parts and showing in such roles as Gwendolen Fairfax in *Wilde's *The Importance of Being Earnest* and Mrs Foresight in *Congreve's *Love for Love* an unsuspected talent for high comedy. During the next few years she appeared in the West End in a number of modern works, including the title-roles in Bridie's *Mary Read* (1934), Wilfrid Grantham's *Mary Tudor* (1935), and O'Neill's *Anna Christie* (1937). She went to the USA in 1938, making her New York début in 1940, and appearing in Hollywood films. On returning to London she was seen as Thérèse in *Guilty* (1944), a dramatization of *Zola's novel *Thérèse Raquin*. In 1948, after playing Lady Macbeth in New York, she again displayed her talent for comedy as Lady Cicely Waynflete in *Shaw's *Captain Brassbound's Conversion*, and then achieved a great success in Lesley Storm's *Black Chiffon* (1949; NY, 1950), bringing to the role of a middle-class kleptomaniac the claustrophobic effect of controlled nervous tension in which she excelled. She was Paulina in *Gielgud's production of *The Winter's Tale* in 1951, and among her later parts were the governess in *The Innocents* (1952), based on Henry *James's *The Turn of the Screw*; Mrs Alving in *Ibsen's *Ghosts* (1958) at the Old Vic; and Miss Tina in Michael *Redgrave's adaptation of Henry James's *The Aspern Papers* (1959). She was also in revivals of Ibsen's *John Gabriel Borkman* (1963), Wilde's *The Importance of Being Earnest* (this time as Miss Prism), and *Anouilh's *Ring round the Moon* (both 1968). Her last West End appearance was in 1969 in a revival of Rodney Ackland's adaptation of Hugh Walpole's novel *The Old Ladies*.

Robson, Frederick [Thomas Robson Brownhill] (1821–64), English actor who first made his name as a singer of comic songs, appearing in 1844 at the *Grecian Theatre; he may have been seen there in his later famous character of Jem Baggs in a revival of Mayhew's *The Wandering Minstrel*, which had been produced originally at the Fitzroy Theatre (later the *Scala) in 1834. After enjoying great local popularity, Robson visited Dublin in 1850 and on his return joined the company at the *Olympic Theatre, famous for its *burlesques. It was there that he probably first sang in *The Wandering Minstrel* the popular ballad 'Villikins and his Dinah' by E. L. Blanchard. Short and ugly, and a heavy drinker, he was nevertheless a powerful actor of great charm, affectionately known as 'the great little Robson'. A number of burlesques were written specially for him, of which *Planché's *The Yellow Dwarf* (1854), exploited to the full his genius for blending the comic with the macabre. He also appeared with much success in the title-role of Palgrave Simpson's *Daddy Hardacre* (1857) and as Sampson Burr in Oxenford's drama *The Porter's Knot* (1858).

Robson, Stuart [Henry Robson Stuart] (1836–1903), American comedian who first appeared on the stage as a boy of 16 and after 10 years in various *stock companies became principal comedian in Laura *Keene's company. He also spent some years with Mrs John *Drew at the Arch Street Theatre in *Philadelphia and was for a time with the younger William *Warren at the Boston Museum. In 1873 he was seen in London, and shortly afterwards began a long association with the comedian **William H**(enry) **Crane** (1845–1928). They appeared together as the two Dromios in *The Comedy of Errors* in 1877 and parted amicably in 1889 after the outstanding success of Bronson *Howard's *The Henrietta* (1887), specially written for them. Robson then appeared as Tony Lumpkin in *Goldsmith's *She Stoops to Conquer* in 1893 and in several specially written plays, dying suddenly on tour shortly after celebrating his stage jubilee.

Rogers, Paul (1917–   ), English actor, with a high reputation in both classical and modern plays. After his début, in London, in 1938, he went to the *Shakespeare Memorial Theatre in 1939, and after serving in the Royal Navy returned to acting in 1946, joining the *Bristol Old Vic a year later. From 1949 to 1960 he was with the *Old Vic company, leaving it to appear in T. S. *Eliot's *The Confidential Clerk* (1953) and *The Elder Statesman* (1958). He made his New York début in 1956 as John of Gaunt in *Richard II*, and during his time with the Old Vic played many leading roles in Shakespeare, including Shylock in *The Merchant of Venice*, Henry VIII (both 1953), Macbeth (1954 and 1956), and King Lear (1958). After leaving the company he was seen in the London production of Archibald *MacLeish's *J.B.* (1961), followed by *Ustinov's *Photo Finish* (1962; NY, 1963) and *Chekhov's *The Seagull* (1964). A year later he became a member of the *RSC, where he gave brilliant performances as Max in *Pinter's *The Homecoming* (1965; NY, 1967) and as the muffin-faced mayor in *Gogol's *The Government Inspector* in 1966. He returned to the West End to give a display of virtuosity in three roles in Neil *Simon's *Plaza Suite* (1969), and in 1970 took over the part of the tweedy, fiction-writing bully in Anthony Shaffer's *Sleuth* (NY, 1971). He appeared with the *National Theatre company in *Shaw's *Heartbreak House* in 1975 and *Jonson's *Volpone* in 1977, and was twice seen in *Granville-Barker's plays: *The Marrying of Ann Leete* in 1975 for the RSC, and *The Madras House* for the National in 1977. Later roles included Sir in Ronald Harwood's *The Dresser*

in New York in 1981 and Hornby in Pinter's *A Kind of Alaska* (NT, 1982).

Rojas, Fernando de, see CELESTINA, LA.

Rojas Villandrando, Agustín de (*c.*1572–1625), Spanish actor and dramatist, who was for a short time a strolling player, and wrote entertainingly of his experiences in a picaresque novel *El viaje entretenido* (*The Entertaining Journey*), published in 1603. In addition to a number of his own short playlets, or *loas*, this contains also a good deal of information on the early days of the Spanish theatre, and was not without influence on other European writers, notably *Scarron (in *Le Roman comique*, 1651) and *Goethe (in *Wilhelm Meister*, 1821). No English or other translations are available.

Rojas Zorrilla, Francisco de (1607–48), Spanish dramatist, friend of *Calderón, who in his comparatively short working life produced an astonishing number of plays. The best-known is *Del rey abajo, ninguno* (*None but the King*, 1650), in which Garcia del Castañar, a nobleman living quietly in the country, suspects the king of having seduced his wife. He can do nothing about it, except plan to kill his wife to avenge his honour, but when he discovers that the seducer is merely a courtier he kills him instead. Rojas was also the first to write a true *comedia a figurón* in *Entre bobos anda el juego* (*The Sport of Fools*, pub. 1645). His plays were well known in France, where they influenced among others *Lesage and Thomas *Corneille.

Rolland, Romain Edmé Paul Émile (1866–1944), French writer, awarded the *Nobel Prize for Literature in 1915. He is best known for his 10-volume *roman-fleuve Jean Christophe* (1906–12), and he was a theorist of the theatre rather than a working playwright, his *Le Théâtre du peuple* (1903) having a considerable influence on *Gémier. Two of his plays on the French Revolution, *Danton* (1900) and *Le 14 juillet* (1902), had some success, however, and the former was given a brilliant production by *Reinhardt in Berlin in 1919. *Robespierre*, written in 1938, also had its first production in Germany, in Leipzig in 1952.

Roller, Andrei Adamovich [Andreas Leongard] (1805–91), Russian theatre designer. After studying in Vienna he worked as a stage mechanic and designer in Austria, France, and England, and from 1834 to 1879 was chief stage designer and machinist in the Imperial theatres in St Petersburg. He created a number of striking décors in the romantic tradition, devising splendid *transformation scenes and ingenious tableaux, and experimenting with the use of light for dramatic emphasis—shafts of moonlight through a window or sunlight filtering through chinks in a vaulted interior. He played an active part in the reconstruction of the scenic apparatus for the theatre at the

Tsarskoye Selo Hermitage, as well as the Bolshoi and Hermitage theatres in St Petersburg, and was responsible for much of the restoration of the Winter Palace after its destruction by fire. His work had a great influence on the development of scenic design in Russia.

Roll-out, see TRAP.

Romains, Jules [Louis-Henri-Jean Farigoule] (1885–1972), French poet, novelist, and dramatist. His first play, *L'Armée dans la ville*, was produced by *Antoine in 1911, but it was not until after the First World War that he began his close association with the theatre through his friendship with *Cocteau and *Copeau. He worked for a time at the *Vieux-Colombier, where his *Cromedeyre-le-Vieil*, a mythic portrayal of a mode of communal existence and experience, was produced in 1920. Louis *Jouvet produced and played in the three farces which followed—*M. Le Trouhadec saisi par la débauche* (1922), *Knock; ou, le Triomphe de la médecine* (1923), and *Le Mariage de M. Le Trouhadec* (1925). *Knock*, in which Jouvet played the quack doctor adept at manipulating human credulity, had an immense success and was frequently revived. It was seen with equal success in London (1926) and New York (1928). Romains's later plays included *Jean le Maufranc* (1926), initially unsuccessful but, rewritten as *Musse; ou, L'École de l'hypocrisie* (1930), offering *Dullin a fine part; *Le Dictateur* (also 1926), which had its greatest success outside France; an excellent adaptation of *Jonson's *Volpone* (1928); and *Donogoo* (1931), a stage version of a film scenario of 1920 in which the Le Trouhadec saga had its beginnings. Romains then concentrated on his 27-volume panoramic novel *Les Hommes de bonne volonté*. *Volpone* was successfully revived by *Barrault at the *Marigny in 1955, with himself as Mosca.

Roman Drama. A succession of writers adapted Greek originals to the Roman taste for rhetoric, spectacle, and sensationalism, buffoonery, homely wit, and biting repartee. Comedy almost certainly arose from a blending of the bawdy extempore joking by clowns at harvest festivals and marriage ceremonies with the performances of masked dancers and musicians who came from Etruria. From Etruria also came such theatrical terms as *histrio* for actor and *persona* for the person indicated by the mask. By the 3rd century BC Roman audiences would have been familiar with the secular entertainments of other Italian peoples—the *phlyax comedies of Tarentum, for instance, and the Oscan *fabula atellana*, which were not unlike the comedy of mainland Greece. Although *mime may well have been international from its inception, it was not until the beginning of the 3rd century that the spread of Greek drama led to the emergence of an organized Roman theatre, particularly after 250 BC, when plays

were produced at the public festivals as part of the general celebrations. The advance to true drama was made by Livius *Andronicus who was the first to produce at a public festival a *fabula palliata*, or Greek play in translation. From then until the end of the Republic adaptations of Greek tragedies and comedies predominated in the Roman theatre, together with a very few examples of the *fabula praetexta*, on a historical Roman theme, and a considerable number of the new-style *fabula togata*, or play on everyday Roman life. Among Roman dramatists there was no invention of plot or character on a grand scale. They were mainly concerned with supplying a non-reading public with live entertainment which blended a well-known story with topical allusions and a plentiful supply of comic business. Only in *Terence is it possible to discern a conscious artistic impulse to improve on his Greek model. Other dramatists following Andronicus, among them *Plautus, were content to adapt what was offered by the Greeks to the demands of a Roman audience, with the proviso that since dramatic performances were provided free by the authorities nothing should be said or indicated which might reflect on them. With moral and religious matters the censorship was less concerned, though in general the extant Latin comedies are fairly free from indecency.

At first Rome had no permanent theatre. The simple wooden buildings required for a production could be put up as and when required. Interior scenes and changes of setting within a play were unknown, and the absence of a front curtain made it necessary for every play to begin and end with an empty stage. The characters in comedy were types rather than individuals—old gentlemen, usually miserly and domineering, young gentlemen, usually extravagant and spineless, jealous wives, treacherous pimps, intriguing slaves, boastful captains—and custom prescribed the appropriate costume, wig, and mask for each type. Since respectable women were not supposed to appear in public, the unmarried heroines of Greek *New Comedy and its Latin derivatives were somewhat less than respectable—courtesans, or maidens separated by some mishap from their parents and brought up in humble circumstances. If marriage was to be the outcome of the hero's wooing, then it was necessary to show that the girl he wished to marry was not only chaste but the long-lost child of respectable parents. It is hardly surprising that in general the plots of Roman comedies turn on intrigues, attempts to raise money, and swindling and deception of many kinds.

The use of costumes and *masks enabled small companies of five or six actors to perform almost any play by doubling of parts. The leading actor was probably also the director. Such actors, such as *Roscius Gallus, could rise to fame and fortune, since the theatrical profession was not yet regarded as degrading in itself. There is no mention of slaves acting until Roscius, near the end of his career, trained a slave to appear on stage. Revivals of old plays were rare, since audiences would naturally be attracted to a new work. This would be bought from the author by a manager authorized by the festival authorities to hire the necessary actors and costumes and settle all business matters, his profit arising from what he could save on the fixed sum allotted to him.

Though tragedy seems to have attracted fewer writers than comedy, the popularity and influence of the leading tragic writers appear to have been great. In general they selected as models the more melodramatic of their Greek originals. Exciting plots, flamboyant characters, gruesome scenes, were apparently more attractive to Roman crowds than the poetic qualities for which Greek tragedies are admired today. In substance the plays underwent little alteration, the chief difference between the Greek original and its Latin adaptation being the development of rhetoric at the expense of truth and naturalness. In comedy, on the other hand, the free-and-easy methods of the early writers seem to have been succeeded by greater fidelity to the originals.

With the end of the 2nd century BC a fundamental change came over the Roman theatre. Few new plays were being written, and those mostly not for production. Theatres, which had for some time been permanent buildings, the first, in stone, dating from 55 BC, became more and more elaborate as the level of entertainment presented in them sank. Rustic farce (fabula atellana) and mime predominated. Such plays as were staged were marred by tasteless extravagance. The introduction of a curtain led to elaborate scenic effects and quick changes of settings, to the detriment of the spoken text. A divorce thus set in between the drama and the stage. It is true that under the Empire many plays were written, and many fine theatre buildings were erected in Europe, Asia, and Africa, the ruins of which still stand; but the plays and the theatres had very little to do with each other. Popular taste was all for mime and *pantomime, and the stage was given over to these and other forms of light entertainment, while the writing of plays became the prerogative of educated literary men. The tragedies ascribed to *Seneca and such works as the anonymous Octavia were *closet dramas, intended to be read among friends. Though full of clever rhetoric, they were not written with the limitations of the stage in mind, and were probably never performed. Yet as the only examples of Latin tragedy to reach the modern world they were destined to exercise an immense influence on the playwrights of the Renaissance.

It was not only the degeneration of public taste which led to the disintegration of the serious theatre in Imperial Rome. There was the added problem of the constant hostility of the Christian Church. No Christian could be an actor, under pain of excommunication, and all priests and devout persons refrained from attendance at theatrical performances of any kind. In the 6th century the theatres in Europe finally closed, the much-harassed actors being forced to rely on private hospitality or take to the road. From then until the 10th century there was nothing but an undercurrent of itinerant entertainers—mimes, acrobats, jugglers, bear-leaders, jongleurs, and minstrels—who kept alive some of the traditions of the classical theatre, while the Church quietly absorbed such pagan rites as the *folk play and the *mumming play into its own ritual, unconsciously preparing the way for the revival of the theatre it had tried to suppress.

Romashov, Boris Sergeivich (1895–1958), Soviet dramatist, whose first play, Meringue Pie (1925), was a satire on bourgeois elements in Soviet society. Among his later plays, which dealt mainly with episodes of the Civil War, were The End of Krivorilsky (1927), The Fiery Bridge (1929), and Fighters (1934). This last, perhaps his most important work, deals with the Red Army in peacetime, and the clash between private and public interests among the officers. In 1942 a play dealing with the defence of Moscow, Shine, Stars!, was put on less than a year after the events it deals with took place. Its theme is the courage of a young student galvanized by war into becoming a man of action, and it proved immensely popular. In 1947 A Great Force, which depicts the conflict between conservative and progressive scientists, was seen at the *Maly Theatre, Moscow.

Rondiris, Dimitrios (1899–1981), Greek director, who made his first appearance on the stage in 1919, playing Florizel in a translation of The Winter's Tale. During the next 10 years he was seen with a number of Greek companies, and then went to Vienna, where for three years he studied the history of art and attended Max *Reinhardt's seminars. On his return to Athens he was appointed a director of the Greek National Theatre and in 1936 inaugurated the first modern festival of Greek tragedy with a production of *Sophocles' Electra in the Theatre of Herod Atticus in Athens. This festival has now become an annual event, as has the festival of Greek plays at *Epidauros which he also organized. In 1939 he directed a production of Hamlet with Alexis *Minotis which was seen in London. He became Director-General of the National Theatre in 1946, and his production of Electra was seen in New York in 1952, with Katina *Paxinou. Five years later he founded the Piraikon Theatre in the Piraeus, where for some years before his retirement he continued staging classical tragedies in a modern style, taking his company on tour as far afield as Russia and Israel.

Rope House, see HAND WORKING.

Roscius Gallus, Quintus (*c.*120–62 BC), Roman actor, the most famous of his day. He was a friend of Cicero, who delivered on his behalf during a lawsuit the speech *Pro Roscio Comoedo*, and though of middle-class origins was raised to equestrian rank by the dictator Sulla. He appeared in the plays of *Plautus—one of his best parts being Pseudolus—and *Terence, as well as of lesser dramatists, and by his example did much to raise the status of actors in the Roman world. Much of his success was due to careful study of his parts; he thought out and rehearsed every gesture before using it on stage, and his name became synonymous with all that was best in acting. Shakespeare referred to him in *Hamlet*, and many outstanding actors were called after him.

(For the Dublin, or Hibernian, Roscius, see BROOKE; Young Roscius, see BETTY.)

Rose, Billy, see NEDERLANDER THEATRE, REVUE, and ZIEGFELD THEATRE.

Rose, Clarkson, see REVUE.

Rosenplüt, Hans, see FASTNACHTSSPIEL and MASTERSINGERS.

Rose Theatre, London, in Rose Alley, off Southwark Bridge Road. This theatre, 94 ft. square, constructed of wood and plaster on a brick foundation, octagonal in shape, and partly thatched, opened in 1587. It was built for Philip *Henslowe and his partner and stood in an area known as the Liberty of the Clink on the site of a former rose garden, about halfway between the later *Globe and *Hope theatres. It is not known which company first played there, but in 1592, after extensive repairs had been done and alterations made, *Strange's Men were there, and from 1594 the *Admiral's Men, under Henslowe's son-in-law *Alleyn, used it almost exclusively until they moved to the *Fortune in 1600. Several of *Marlowe's plays were first performed there, as probably was Shakespeare's *Henry VI*, according to Henslowe's accounts a highly profitable venture. After the departure of Alleyn, several companies used it, notably *Pembroke's Men and Worcester's Men, the latter (as *Queen Anne's Men) being still in residence in 1604, just before Henslowe's lease expired. The building may then have been used for some kind of amusement, but not for plays, and it was pulled down the following year. Parts of the site were exposed during development work in 1989.

Rossi, Ernesto Fortunato Giovanni Maria (1827–96), Italian actor, who excelled in tragedy. As a young boy he acted with a children's company which played in private houses and made his professional début at the age of 18 in his birthplace, Livorno (Leghorn). He later joined the company of Gustavo *Modena, and in 1852 played opposite Adelaide *Ristori, with whom he went to Paris in 1855. Although his Othello was later eclipsed by that of *Salvini, he was the first Italian actor to play the part, in Milan in 1856; he also played Hamlet and was considered good as Lear, though his interpretations of Shakespeare did not meet with approval in England or America.

Rosso di San Secondo, see GROTTESCO, TEATRO.

Rostand, Edmond Eugène Alexis (1868–1918), French romantic dramatist, whose colourful poetic plays came as a relief after the drab realities of the naturalistic school. His first play *Les Romanesques* (1894; see RUN), a delicious satire on young lovers, was followed by the more serious but still tender and lyrical *La Princesse lointaine* (1895). A biblical play, *La Samaritaine* (*The Woman of Samaria*, 1897), was less successful, but in *Cyrano de Bergerac* (1898), written for *Coquelin aîné, he achieved a marvellous fusion of romantic bravura, lyric love, and theatrical craftsmanship, and its success was overwhelming. It became a perennial favourite, not only in France but in England and America, where something of its quality was apparent even through a pedestrian translation. *L'Aiglon* (*The Eaglet*, 1900), in which *Bernhardt played the ill-fated son of Napoleon, had less vigour, but appealed by its pathetic evocation of fallen grandeur and the frank sentimentality of its theme. Rostand's last play was *Chantecler* (1910), by some critics accounted his best, as it is certainly his most profound, work. The verse is masterly and the allegory unfolds effortlessly on two planes of consciousness, the beast's and the man's. *La Dernière Nuit de Don Juan* was left unfinished, but indicates how Rostand's undoubted talent might have matured.

Rostrum, any platform, from a small dais for a throne to a large battlement on which actors can assemble. Known in America as a Parallel, it is usually made with a removable top and hinged side-frames, to fold flat for packing, though permanent 'stock' or Rigid Rostrums are also used. A Rostrum is usually approached by steps or a ramp, and quitted off-stage by 'lead-off' steps. When necessary a canvas-covered *flat, known as a rostrum-front, is placed to hide the front of the platform from the audience.

Roswitha, see HROSWITHA.

Rotrou, Jean de (1609–50). French dramatist, next to *Corneille the most important of his day. He was only 19 when he had two plays produced at the Hôtel de *Bourgogne, where he may have succeeded *Hardy as official dramatist to the troupe. His popularity may be gauged from the fact that he had four plays produced in Paris in 1636. More than 30 of his works survive, some of them the best extant examples of the *tragi-comedy of the time, though he was also instrumental with *Mairet and Corneille in establishing neo-classical tragedy, of which his *Hercule mourant* (1634) is an early example. Rotrou, like Corneille, was interested in Spanish literature, and translated one of Lope de *Vega's plays as *La Bague de l'oubli* (1629), the first extant French play to be based on a Spanish source and the first notable French comedy, as distinct from farce. He was also the author of one of the many versions of the story of Amphitryon, *Les Sosies* (1637), considered one of his best plays. Of his later works a tragedy, *Cosroës* (1649), remained in the repertory until the early 18th century, while *Venceslas* (1647), a tragedy based on a play by *Rojas Zorrilla, was still being played up to 1857. In many ways the most appealing of Rotrou's later plays, and a masterpiece of baroque tragedy, is *Le Véritable Saint-Genest* (1645). Again based on a play by Lope de Vega, this portrays the conversion of the actor Genest while playing the part of the martyr St Adrian, with the result that he is himself taken away to suffer martyrdom. Rotrou, a man of great charm and nobility of character, held important municipal offices in his native town of Dreux, and died there during a plague, having refused to abandon his official post and seek shelter elsewhere.

Rotunda, The, London, hall in Blackfriars Road, at the corner of Stamford Street, sometimes confused with the *Ring, which was on the opposite side of the road about 500 yards away. It opened in 1790 as a museum, but was used for musical and dramatic entertainments as early as 1829 and from 1833 to 1838 was known as the Globe Theatre. It became the Rotunda under the management of the composer John Blewitt, who called it a 'musick hall', in the old sense. Although it held only 150, it enjoyed much local popularity until, having changed its name again to the new-style Britannia Music-Hall, it lost its licence in 1886, after an illegal cockfight. It then became a warehouse and was finally demolished in 1945.

Round House, London, in Chalk Farm Road, Camden Town, disused locomotive shed, dating from the mid-19th century, taken over in 1961 by Arnold *Wesker to serve as the home of his Centre 42, which aimed to make the arts more widely available, primarily through trade union involvement. The venture was not a success, but the theatre has housed some interesting productions 'in-the-round' including in 1968 an experimental production of *The Tempest* by Peter *Brook. Later in the same year a play about Nelson, *The Hero Rises Up*, by John

*Arden and his wife, had a short run, and 1969 saw the production of Kafka's *Metamorphosis* and *In the Penal Colony*, a visit from the *Living Theatre, and Tony *Richardson's production of *Hamlet*. In 1970 the erotic revue *Oh! Calcutta!* had a short run before moving on to the West End, and in 1971 the American musical *Godspell* opened, transferring later to *Wyndham's. The *Open Theater from New York appeared in 1973. The theatre continued to be used for productions that might not otherwise find a home in London, including *Brenton's *Epsom Downs*, Tennessee *Williams's *The Red Devil Battery Sign* (both 1977), and David *Rabe's *Streamers* (1978). After a complete reconstruction in 1979 the theatre reopened with Ronald Harwood's adaptation of Evelyn Waugh's novel *The Ordeal of Gilbert Pinfold*, originally seen at the *Royal Exchange Theatre, Manchester; other productions from the same theatre followed. The theatre also staged *Ayckbourn's *Season's Greetings* (1980) and *Suburban Strains* (1981). It closed in 1984 but the Round House Arts Centre Trust hopes to reopen it.

Roussin, André, see MORLEY.

Rowe, Nicholas (1674–1718), English dramatist, one of the few of the Augustan age to display any real dramatic power. Of his seven tragedies the early ones, *The Ambitious Step-Mother* (1700) and *Tamerlane* (1701), in which *Betterton was outstanding in the title-role, were written in a somewhat frigid neo-classical style. His later masterpieces, however, which include *The Fair Penitent* (1703), based on *Massinger's *The Fatal Dowry*, and *The Tragedy of Jane Shore* (1714), have genuinely moving and poetic passages, and both were frequently revived, Mrs *Siddons being particularly admired in the leading roles.

Rowley, Samuel (c.1575–1624), English actor and dramatist, a member of the *Admiral's Men, whom he joined in 1597. He is credited with having a hand in writing a number of lost plays which preceded, and possibly provided material for, some of Shakespeare's. Rowley's only extant play, a chronicle drama on the life of Henry VIII, is entitled *When You See Me, You Know Me* (1603). He is believed to have altered *Marlowe's *Dr Faustus* for *Henslowe, mainly by adding some inferior comic passages for a revival in 1601.

Rowley, William (c.1585–1626), English actor and dramatist, known to have collaborated with several contemporary playwrights, notably *Middleton, with whom he wrote an outstanding play, *The Changeling* (1622). He also worked with *Dekker and *Ford on *The Witch of Edmonton* (1621), with *Fletcher on *A Maid in the Mill* (1623), and with *Webster on *A New Wonder, a Woman Never Vexed* (1624) and *A Cure for a Cuckold* (1625). He is known to have played Plumporridge, the fat clown, in *The Inner Temple Masque* in 1619, Jacques, a fat fool, in his own play *All's Lost By Lust* (c.1622), and the fat bishop in Middleton's *A Game at Chess* (1624).

Royal Academy of Dramatic Art, see SCHOOLS OF DRAMA.

Royal Amphitheatre, London, see ASTLEY'S AMPHITHEATRE.

Royal Arctic Theatre, Canada, unusual form of entertainment which flourished intermittently in the 19th century. During the Royal Navy's search for a north-west passage to the Pacific British ships had to winter in the Arctic circle and to keep up morale plays were regularly performed, either in the ships that lay frozen in the ice or in snow-houses built alongside. Although known by various names the enterprise has come to be known as Arctic Theatre. There were three main periods of activity, the first in 1819, when *Garrick's Miss in her Teens* is known to have been acted on the quarter-deck of HMS *Hecla*. This tradition of play-acting was revived between 1850 and 1853, in the wake of the activity which followed the disappearance of Sir John Franklin. A series of theatrical 'seasons' was advertised in playbills and in 'The Illustrated Arctic News' published on board ship. It was reported that one stage made of ice was so engulfed in clouds of vapour caused by the breath of the audience that the actors were virtually invisible. It was in such conditions that what was probably the most northerly production of Shakespeare ever given was seen, when in 1852 the first act of *Hamlet* was performed, with, improbably, the roles of Hamlet and Ophelia doubled by the same actor. The Arctic Theatre came to an end in 1875–6, when British ships were attempting to reach the North Pole. A *pantomime, *Aladdin and the Wonderful Scamp*, was performed below decks on HMS *Alert* in 1875, and the officers and crew of HMS *Discovery*, anchored further south, alternated productions every two weeks in a snow-theatre on land. From the surviving illustrations and accounts it appears that the scenery, props, and costumes were often skilfully made and extremely elaborate.

Royal Artillery Theatre, London, at Woolwich. This was built by public subscription, with the help of the War Office, to take the place of a theatre constructed in the old garrison church in 1863 and used for amateur dramatics. The new theatre was fully professional and from 1909 to 1940 was controlled by Mrs Agnes Littler who, with her husband, was injured when a bomb dropped outside the building in 1918. Among the famous actors who appeared here were Lillie *Langtry, Violet *Vanbrugh, George *Robey, and Albert *Chevalier. During the Second World War the theatre, which was used by various Service units for entertaining the troops, was again damaged by bombing, and, attempts to reopen it having failed, it closed down finally in 1956.

Royal Avenue Theatre, London, see PLAYHOUSE.

Royal Brunswick Theatre, London, see ROYALTY THEATRE 1.

Royal Circus, London, see SURREY THEATRE.

Royal Coburg Theatre, London, see OLD VIC.

Royal Court Theatre, London. **1.** In Lower George Street, Chelsea. A badly adapted Nonconformist chapel, it opened as the New Chelsea in 1870 and had no success at all, even when it changed its name to the Belgravia, until after reconstruction and redecoration it opened as the Royal Court in 1871 with *Randall's Thumb* by W. S. *Gilbert, who provided further successes with *The Wedding March* and *The Happy Land* (both 1873). In the latter he burlesqued contemporary politicians so mercilessly that the *Lord Chamberlain intervened, and the actors' make-up had to be altered. In 1875 *Hare took over with a good company led by the *Kendals, and produced a number of successful plays. He was followed in 1879 by Wilson *Barrett, and with him as her leading man *Modjeska made her first appearance in London a year later. In 1885 a company which included Marion *Terry, Mrs John *Wood, and Brandon *Thomas inaugurated a successful series of farces by *Pinero, beginning with *The Magistrate* and followed by *The Schoolmistress* (1886) and *Dandy Dick* (1887). The theatre finally closed in 1887 to make way for street improvements and was completely demolished.

2. On the east side of Sloane Square. This theatre was built to replace the above, and had a three-tier auditorium with a seating capacity of 642. It opened in 1888 with Grundy's *Mamma!*, and was for a time less successful than the old theatre, in spite of the popularity of Pinero's new farce *The Cabinet Minister* (1890). His record-breaking *Trelawny of the 'Wells'* (1898) first brought the Royal Court back into the limelight, its next outstanding success being Charles Hannon's *A Cigarette-Maker's Romance* (1901) starring *Martin-Harvey. In 1904 J. E. Vedrenne and Harley *Granville-Barker took over the theatre and produced a remarkable series of plays, both new and old, which ranged from Shakespeare to *Shaw, *Galsworthy, and Barker himself. After Barker and Vedrenne had moved to the *Savoy, Somerset *Maugham's *Lady Frederick* (1907) filled the house to capacity; but then its fortunes declined until after the First World War, when J. B. *Fagan took over, one of his first productions being the British première of Shaw's *Heartbreak House* in 1921. In 1924 Barry *Jackson brought his *Birmingham Repertory Theatre company to the Court, opening with the five parts of Shaw's *Back to Methuselah*, followed by Eden *Phillpotts's popular comedy *The Farmer's Wife*. In 1928 Jackson returned with controversial productions of *Macbeth* and *The Taming of the Shrew*, both in modern dress, and in 1932, after three seasons of Shaw's plays by the Macdona Players, the theatre became a cinema. It was badly damaged in 1940, but after extensive renovation it reopened as a theatre in 1952, the only productions of note being Frank Baker's *Miss Hargreaves*, starring Margaret *Rutherford, the revue *Airs on a Shoestring* (1954), and *Brecht's *The Threepenny Opera* (1956). In April 1956 it was taken over by the *English Stage Company under George *Devine. A policy of presenting new playwrights was justified by the success of John *Osborne's *Look Back in Anger*, followed in 1958 by Arnold *Wesker's *Chicken Soup with Barley* and in the following year the remaining plays of the trilogy, *Roots* and *I'm Talking about Jerusalem*. In 1957 came Osborne's *The Entertainer*, with Laurence *Olivier, and in 1959 *Serjeant Musgrave's Dance* by John *Arden, who, like Osborne and Wesker, remained closely associated with this theatre. Before the building closed in 1964 for reconstruction it had seen a number of important productions, including N. F. Simpson's *One-Way Pendulum* (1959) and *Ionesco's *Rhinoceros* (1960). After the reopening came two plays by Osborne, *Inadmissible Evidence* (1964) and *A Patriot for Me* (1965). Following the death of Devine, William *Gaskill took over, 1965–72. Under him David *Storey's first play *The Restoration of Arnold Middleton* was seen in 1967, followed by Osborne's *Time Remembered* and *The Hotel in Amsterdam* (both 1968). Christopher *Hampton was resident dramatist, 1968–70. In 1969 the theatre celebrated the end of stage *censorship by producing three plays by Edward *Bond, and in the same year Storey's *In Celebration* and *The Contractor* were well received. An outstanding success was also scored by his *Home*, which, with John *Gielgud and Ralph *Richardson, went on eventually to New York. Gielgud was seen again in Charles *Wood's *Veterans* (1972). Howard *Barker's *Stripwell* (1975) received a major production, and David *Hare's *Teeth 'n' Smiles* (also 1975) and Mary O'Malley's *Once a Catholic* (1977) both transferred to *Wyndham's, the latter with enormous success. Notable later productions included Martin Sherman's *Bent* (1979) with Ian *McKellen and *Hamlet* (1980) directed by Richard *Eyre. The 1980s, though financial constraints caused a severe diminution of activity, produced such notable works as Michael Hastings's *Tom and Viv* (1984), David *Mamet's *Edmond* (1985), Larry Kramer's *The Normal Heart* and Alan *Bennett's *Kafka's Dick* (both 1986), and Timberlake Wertenbaker's *Our Country's Good* (1988).

The 1980s was also the decade of the prolific English dramatist **Caryl Churchill** (1938–), though she began writing radio plays in the 1960s and her first full-length play *Owners* was produced at the Royal Court in 1972 (NY, 1973). Many of her plays, which cover socialist and feminist themes, have been produced there, including *Light Shining in Buckinghamshire* (1976), set during the English Civil War (one of several composed by *Collective Creation), and *Cloud Nine* (1979; NY, 1981), which contrasts sexual attitudes under Victorian colonialism and in 1979. Among her works staged in the 1980s were *Top Girls* (1982; NY, 1982), which featured a dinner party of female historical characters, the actresses who played them returning as modern women; and *Fen* (1983; NY, 1983), set on an East Anglian farm. *Serious Money* (1987; NY, 1987), a satire on the City, had a long run in the West End.

In 1969 the Royal Court added to its amenities the Theatre Upstairs, an adaptable theatre housed in a former rehearsal room and intended for experimental and low-budget works. Its productions have included several of Sam *Shepard's plays. The theatre also runs a Young People's Theatre School for young playwrights, who it is hoped will contribute to the theatre's reputation as the 'National Theatre of New Writing'.

Royal Dramatic Theatre (Kungliga Dramatiska Teatern), Stockholm, can trace its origins back to 1737, when the Swedish Dramatic Theatre was established. Over the years it underwent many changes of name and played in a succession of different buildings under various administrations, until in 1788, when it was housed in the Stora Bollhuset where it remained until 1792, it was brought into association with the Swedish Royal Opera, founded in 1773, the two being placed under a single controlling body in 1813. In 1794, after a season in the opera house, the theatre company moved to the Kungliga Mindre (Small Royal) Teatern where it remained until the building burned down in 1825, when it returned to the opera house. In 1863 it took over a theatre built in 1842 called successively the Nye (New) and then the Mindre, finally becoming known simply as Dramaten. The present building on Nybroplan was built in 1904–7 and is a fine example of Swedish architecture; it opened in 1908 with a performance of *Strindberg's *Mäster Olof*. Extensive alterations were carried out in 1936 and 1957–60. The theatre is run as a limited company, with financial support from state lotteries. It has several studio theatres.

Olof Molander (1892–1966) greatly influenced the theatre, becoming its head in the 1930s. In the 1920s he mounted productions of classics in the *Reinhardt tradition which, with his later productions of Strindberg, made him the principal architect of Sweden's post-naturalistic style of theatre. Alf *Sjöberg, Anders *Ek, and above all Ingmar *Bergman were also important in the theatre's history. Three of Eugene *O'Neill's plays had their premières there posthumously.

Royal English Opera House, London, see PALACE THEATRE.

Royale Theatre, New York, on West 45th Street, between Broadway and 5th Avenue. Seating 1,059, it opened in 1927 with *Piggy*, a *musical comedy, and later housed Winthrop *Ames's productions of *Gilbert and Sullivan. Mae *West was seen here in her own play *Diamond Lil* (1928), which ran for nearly a year, and the *Theatre Guild put on Maxwell *Anderson's *Both Your Houses* (1933). In 1934 the theatre was renamed the John Golden, presenting a series of moderately successful comedies, but from 1936 to 1940 it was used for broadcasting. It returned to drama under its old name in 1940, successful productions including *Fry's *The Lady's not for Burning* (1950), Thornton *Wilder's *The Matchmaker* (1955), a *revue entitled *La Plume de ma tante* (1958), and Tennessee *Williams's *The Night of the Iguana* (1961). Robert Shaw's *The Man in the Glass Booth* (1968) had a good run, but the biggest hit in the theatre's history was the musical *Grease*, which moved there in 1972 and ran until 1980. A later musical was Andrew Lloyd Webber's *Song and Dance* (1985).

Royal Exchange Theatre, Manchester. In 1968 the 69 Theatre Company, an offshoot of the 59 Theatre Company which had successfully played for a season at the *Lyric Theatre, Hammersmith, took possession of the university theatre in *Manchester and achieved a great success, seven of its 21 productions being transferred to London. The company, which included actors, directors, and designers trained under Michel *Saint-Denis at the *Old Vic Theatre School, was looking for a permanent home and decided to stay in Manchester. In 1972 it leased the Royal Exchange, formerly used for cotton trading, within which a temporary theatre was erected in 1973, followed in 1976 by a permanent one. This theatre is a module enclosed in clear glass and suspended from four of the pillars in the hall of the Exchange. It is a *theatre-in-the-round, based on a seven-sided figure, and no seat is more than about 30 ft. from the stage. The ground floor seats 400 and the two balconies 150 each. The auditorium can be converted for thrust-stage productions, and the foyer, one of the largest in the world, consists of the whole of the Royal Exchange hall outside the module. The venture was made possible by grants from two local authorities, the *Arts Council, and public subscription. The company, which is a *repertory company, retained its high reputation in its new home

and attracted a large number of outstanding players, including Albert *Finney in *Chekhov's *Uncle Vanya* and *Coward's *Present Laughter* (both 1977), Vanessa *Redgrave in *Ibsen's *The Lady from the Sea* (1978), and Helen *Mirren in *Webster's *The Duchess of Malfi* (1980). The last two were transferred to London, as were three plays by **Ronald Harwood** (1934–) which had their world premières at Manchester—*The Ordeal of Gilbert Pinfold* (1977), based on Evelyn Waugh's novel, with Michael *Hordern; *A Family* (1978) with Paul *Scofield; and *The Dresser* (1980) with Tom *Courtenay, also seen in New York, about the relationship between an elderly actor and his dresser. The repertoire combines classics with new work, mostly commissioned. The theatre is also used for talks, concerts, children's shows, and workshops, and the company has its own mobile theatre-in-the-round for touring.

The co-founder and joint Artistic Director of the 69 Theatre Company, and one of the resident artistic directors of the Royal Exchange Theatre Company, was **Michael Elliot** (1931–84), whose first London stage production was Ibsen's *Brand* in 1959 for the 59 Theatre Company. He directed the *RSC's famous production of *As You Like It* in 1961–2 with Vanessa Redgrave, and was Artistic Director at the Old Vic, 1962–3. In Manchester he directed such productions as *The Tempest* (1969), *Uncle Vanya*, *The Ordeal of Gilbert Pinfold*, an adaptation of Dostoevsky's *Crime and Punishment* (1978), and *The Dresser*. He twice directed Ibsen's *Peer Gynt*, at the Old Vic in 1962 and in Manchester in 1970.

Royal (Holborn) Music-Hall, see HOLBORN EMPIRE.

Royal Lyceum Theatre, Edinburgh. Opened in 1883 to accommodate touring companies, mainly from London, this theatre, seating 904, was named after the London theatrical home of Henry *Irving, whose company was the first to appear in it, its repertory including *Much Ado about Nothing*, *Hamlet*, and Lewis's *The Bells*. Built at a cost of £17,000 by C. J. Phipps, it remains one of the most beautiful of Phipps's buildings. It played host to most of the prominent touring companies, and between the late 1930s and the mid 1950s the company formed by the grandson of Wilson *Barrett played regular seasons there. From 1947 the theatre was used as one of the principal stages for the *Edinburgh Festival. In 1965 it was purchased by the Edinburgh Corporation to house the Royal Lyceum Theatre Company, to be administered by the Edinburgh Civic Theatre Trust. Opening with *Servant o' Twa Maisters*, a Scots version of the play by *Goldoni, the newly formed company flourished and developed a national reputation. It toured abroad and created a successful offshoot, the Young Lyceum Company, based nearby in a converted Gaelic church. From the mid-1970s, however, there was continuing controversy about proposed developments to the building and the funding needed to enable the company to play continuously. The policy of staging contemporary and classic international drama continued, but further problems with funding in the 1980s caused reductions in the company's programme and the closure of the studio theatre.

Royal National Theatre of Great Britain, London, see NATIONAL THEATRE.

Royal Opera House, London, see COVENT GARDEN.

Royal Princess's Theatre, Glasgow, see CITIZENS' THEATRE.

Royal Shakespeare Company. In 1961 the company playing at the *Shakespeare Memorial Theatre was reorganized and given its present name, the theatre being at the same time renamed the *Royal Shakespeare. Its first director was Peter *Hall, under whom the company embarked on an ambitious programme which included establishing a permanent base in London at the *Aldwych Theatre to stage modern plays and non-Shakespearian classics as well as transfers of the Shakespeare productions from Stratford. It also housed the annual *World Theatre Seasons. The plays of Shakespeare remained the preoccupation of the Stratford-based company, and RSC companies toured all over the world. In 1970, two years after Hall had been succeeded by Trevor *Nunn as Artistic Director, a Stratford company became the first British theatrical ensemble to visit Japan, appearing there with great success in *The Winter's Tale* and *The Merry Wives of Windsor*. In 1965 the company established Theatregoround, a travelling troupe which with more flexible staging and smaller casts than usual was able to perform in schools, factories, and church halls. For a time it was mainly through this organization that the work of the RSC was seen in the provinces, but unfortunately the project had to be abandoned in 1971. From 1978, however, the company has undertaken small-scale tours to similar venues, as well as major tours to regional theatres and an annual season in Newcastle-upon-Tyne. In 1974, after playing three seasons of experimental work at The Place in London, the RSC opened its own studio theatre in Stratford, The *Other Place. This in turn found its London counterpart in 1977 in The Warehouse (see DONMAR WAREHOUSE). These small theatres housed much of the RSC's most exciting small-scale work. In 1982 the company moved its London base to the Barbican Centre, which houses both the *Barbican Theatre and the successor to the Warehouse, The *Pit. A third theatre, The

*Swan, opened in Stratford in 1986. In 1987 Terry *Hands, who had been appointed Joint Artistic Director with Trevor Nunn in 1978, succeeded him as sole Artistic Director, and from 1991 the company was governed by a triumvirate including Adrian *Noble as Artistic Director.

Royal Shakespeare Theatre, name given to the *Shakespeare Memorial Theatre, Stratford-upon-Avon, on the formation of the *RSC in 1961. The first season included Vanessa *Redgrave's enchanting Rosalind in *As You Like It* and *Zeffirelli's disastrous *Othello*, while 1962 provided outstanding productions of *The Comedy of Errors* and *King Lear*, the latter directed by Peter *Brook with Paul *Scofield in the title-role. In 1963 the theatre mounted John *Barton's three-play adaptation of the *Henry VI* trilogy with *Richard III* as *The Wars of the Roses*, followed a year later by *Richard II* and *Henry IV, Parts One and Two*. In 1968 there was Eric *Porter's *King Lear*, and in 1970 another production by Peter Brook, an idiosyncratic reading of *A Midsummer Night's Dream*. Four Roman plays, *Coriolanus*, *Julius Caesar*, *Antony and Cleopatra*, and *Titus Andronicus*, under the joint title of *The Romans*, were seen in 1972, and in 1974 the season, which had previously begun in March and ended in December, was extended into January, an innovation which proved so successful that it was adopted permanently. A highly acclaimed production of *Henry V* in 1975, with Alan *Howard, established a record by playing for over 100 performances at Stratford during the year; it was revived three times and widely seen in Britain and Europe, as well as in New York. In 1976 there was a popular musical version of *The Comedy of Errors*, and a year later the first production since Shakespeare's lifetime of the unadapted texts of the *Henry VI* trilogy. Terry *Hands directed Derek *Jacobi in *Much Ado about Nothing* in 1982, which went to New York. Antony *Sher's memorable *Richard III* came in 1984 and John *Wood's Prospero in *The Tempest* in 1988, which also saw the production of *The Plantagenets*, another three-play adaptation of *Henry VI* and *Richard III*.

Royal Standard Music-Hall, London, see VICTORIA PALACE.

Royal Theatre (Kongelige Teater), Copenhagen, Denmark's national theatre. The original building, designed for the *Danske Skueplads and completed in 1748, was extensively altered and enlarged in 1773. The present building, close to the original site in Kongens Nytorv, dates from 1874. In the late 19th century it achieved a European reputation as the theatre most closely associated with *Ibsen. Poul *Reumert and Mogens *Wieth were closely associated with the theatre. There are several annexe theatres, and a travelling company regularly tours the provinces. In 1987 the theatre staged a production of *Wilde's *The Importance of Being Earnest* directed by Clifford *Williams and designed by Carl *Toms.

Royalty, in the modern commercial theatre the payment made to a dramatist—usually a fixed percentage of the takings—every time his play is performed. Amateur societies generally pay a fixed fee for each night or group of nights of a particular production. The payment of royalties is of comparatively recent origin. In Elizabethan times plays were either bought outright by a manager, as was done by *Henslowe, or formed part of the stock-in-trade of the company to which the actor-author belonged; this was the system under which Shakespeare worked. While a play remained in manuscript no other manager or company could perform it, but once it was printed anyone could stage it on payment of a small fee. This probably helps to account for the number of early plays which have failed to survive, since they were never printed. Some authors, particularly if a play had been successful, would recover the original prompt-copy for the sake of the small fee they might receive when it was printed. The first to do this systematically was Ben *Jonson, whose collected plays and *masques were printed in 1616, and it is to his pioneer work in this field that we owe the publication of the plays of Shakespeare by *Condell and *Heminge in the First Folio of 1623. From then onwards dramatists began to bear in mind the importance of a reading public which was to increase considerably, particularly when the rise of Puritanism made it safer to read plays than to see them. As early as 1602 the custom arose of giving the author the profits of the third, sixth, and ninth performances of a play, the so-called 'author's nights' (see BENE-FIT), but with few theatres, a potentially small audience, and a constant change of bill, many plays failed to achieve even a third night. Under Restoration conditions dramatists continued to receive little monetary reward for their work; most of them, being men of substance, had little need of it and the others managed as best they could, sometimes with the help of a patron. Conditions were little better in the 18th century. One of the most successful plays of the time, *The Beggar's Opera* (1728), brought *Gay less than £700, which was considered enough to make him rich. At the end of the century a return was made to the old practice of buying plays outright. Thomas *Morton is said to have received £1,000 for one of his comedies, and Mrs *Inchbald £800 for *Wives as They Were and Maids as They Are* (1797). Unfortunately in the early 19th century the prestige, and consequently the monetary value, of dramatic works declined sharply. Farces and musical pieces earned more than straight plays, and even

they were priced very low: it was possible to buy a *burletta outright for two guineas. Consequently the theatre was inundated with a stream of thefts, plagiarisms, and adaptations of French and German plays turned out quickly by poorly paid hacks.

The first movement to secure proper recognition of dramatic authorship in England was made by *Planché, and it was mainly owing to his efforts, and those of *Bulwer-Lytton, that adequate *copyright protection was established for plays written after 1833, though the law was sometimes difficult to enforce. The first dramatic copyright law in the USA was not passed until 1856. In France, where *Beaumarchais had made great efforts to establish a royalty system earlier, *Scribe secured for himself the payment of a royalty after the success of *Le Solliciteur* (1817), and by 1832 a payment to the author of 12 per cent of the nightly receipts was usual. One of the first to profit by the new system in England was *Boucicault. For his highly successful *London Assurance* in 1841 he received only £300, whereas in 1860 he netted £10,000 for *The Colleen Bawn*, which seems to have been the first play in London to be paid for on the royalty system. Other authors were quick to follow Boucicault's example, and by degrees the standard rate of royalty became 5–10 per cent, rising perhaps to 20 per cent for established authors. During the early part of the 20th century, with the theatre largely dependent upon successful West End runs and subsequent provincial tours, dramatists tended to make either a great deal of money or very little. Since the Second World War, however, there have been increasing opportunities for them to supplement their incomes by writing for the radio, the cinema, and television. Limited subsidies have also been secured through the *Arts Council, either indirectly, because of the increase in the number of companies whose grants enable them to take risks with new works, or, less often, through bursary schemes, personal grants, and the attachment of resident dramatists to particular theatres. (See also SHARING SYSTEM.)

Royalty Theatre, London. **1.** In Well Street, Wellclose Square. In 1787 John Palmer and John *Bannister from *Drury Lane opened a theatre for which they had obtained a licence from the Governor of the Tower of London, in whose precincts the building stood, and the magistrates of the Tower Hamlets. Under the provisions of the Licensing Act of 1737 this made it technically an 'unlicensed' theatre, and when it opened with *As You Like It* and *Garrick's *Miss in Her Teens* the *Patent Theatres objected; Palmer was arrested and the theatre closed. Later it reopened with *burlesque and *pantomime only but led a precarious existence. In 1813 another Palmer took it over and changed its name to the East London, but it was still unsuccessful and in 1826 it was burned down. Rebuilt as the Royal Brunswick, it became the shortest-lived playhouse in London's theatre history. Opening on 25 Feb. 1828, it collapsed three days later owing to the weight of the roof. A rehearsal of Scott's *Guy Mannering* was in progress at the time; 15 people were killed and 20 injured.

2. In Dean Street, Soho. This small theatre was built for Fanny *Kelly, and was used by her in conjunction with a school of acting. It opened in 1840, but the newly installed stage machinery, worked by a horse, proved so noisy that it had to be removed. The theatre was not a success and closed in 1849, reopening in 1850 as the Royal Soho. After a further period as the New English Opera House it again became a school of acting, and in 1861 reopened as the New Royalty with a revival of a *melodrama in which the 13-year-old Ellen *Terry appeared. In 1875 a one-act musical farce *Trial by Jury*, put on as a stop-gap, brought *Gilbert and Sullivan together for the first time. In 1882 the theatre was partially reconstructed, reopening in 1883. The Independent Theatre Society produced *Ibsen's *Ghosts* there in 1891, and Bernard *Shaw's first play, *Widowers' Houses*, in 1892. At the end of that year Brandon *Thomas's farce *Charley's Aunt* opened there before moving to the *Globe Theatre. Ibsen's *A Doll's House* (1893) and *The Wild Duck* (1894) maintained the theatre's reputation as a home of modern drama, and between these two productions William *Poel mounted his seminal 'Elizabethan' production of *Measure for Measure*. The theatre was reconstructed again in 1895, 1906, and 1911, and many well-known managements came and went. In 1900 Mrs Patrick *Campbell revived *Sudermann's *Magda* and *Maeterlinck's *Pelléas and Mélisande*. She also gave the first English production of *Bjørnson's *Beyond Human Power* (1901). In 1912 came *Galsworthy's *The Pigeon* and *Bennett and *Knoblock's *Milestones*, in 1919 the first performance of *Maugham's *Caesar's Wife*, with Fay *Compton, and two years later The Co-Optimists concert-party. In 1924 Noël *Coward's *The Vortex* was transferred there, and in 1925 Sean *O'Casey had his first London success with *Juno and the Paycock*. John *Drinkwater's *Bird in Hand* (1928) had a long run, and two comedies, *While Parents Sleep* (1932) by Anthony Kimmins and *Bridie's *Storm in a Teacup* (1936), opened here. In 1937 *Priestley's *I Have Been Here Before* gave the theatre its last success before it closed in 1938. It was badly damaged by bombing during the Second World War, and demolished in 1955.

3. In Portugal Street, Kingsway. Built on the site of the *Stoll Theatre, this theatre has a two-tier auditorium holding 997. It opened in 1960 with *Dürrenmatt's *The Visit* starring the *Lunts. After William Gibson's *The Miracle*

Worker in 1961 it became a cinema, but reopened as a live theatre in 1970, the erotic revue *Oh! Calcutta!* being transferred there. It occupied the theatre until 1974 before moving. In 1977 the Royalty housed the successful musical *Bubbling Brown Sugar*, but it later had little success and in 1981 became a television studio, though since changing hands a few years later it has been used for theatre, concerts, and trade shows as well as television.

Royal Victoria Hall, Royal Victoria Theatre, see OLD VIC.

Rozov, Victor Sergeyevich (1913–　), Soviet dramatist, who in 1938 became an actor and director at the *Mayakovsky Theatre in Moscow. He began writing plays after the Second World War, mainly on the problems confronting young people and domestic issues, the first being *Her Friends* (1949). He soon became one of the most prolific and popular playwrights of the USSR. His works also found a responsive public abroad, especially in socialist countries. One of his best-known plays, *Alive for Ever* (1957), a revision of an earlier work, has a wartime setting and provided the scenario for the film *The Cranes Are Flying*; it was translated into English. Some of his plays deal with family events (*On the Day of the Wedding*, 1964, *The Reunion*, 1967). Other plays include *Brother Alyosha* (1972), an adaptation of Dostoevsky's novel *The Brothers Karamazov*, and *The Nest of the Wood Grouse* (1981; NY, 1984), a comedy about conflict in a government official's family. *The Little Cabin* (1982), originally banned, was performed in Moscow in 1987.

RSC, see ROYAL SHAKESPEARE COMPANY.

Rudman, Michael Edward (1939–　), American director and producer, who took an MA at Oxford and then worked mainly in Britain. He was with *Nottingham Playhouse, *Newcastle Playhouse, and the *RSC before becoming Artistic Director of the *Traverse Theatre, 1970–3, and *Hampstead Theatre, 1973–8. His productions at Hampstead included *Handke's *The Ride across Lake Constance* (1973), C. P. *Taylor's *The Black and White Minstrels* (1974), and *Frayn's *Alphabetical Order* (1975) and *Clouds* (1977), both of which he later directed in the West End. He was an associate director of the *National Theatre, 1979–88, being also Director of the *Lyttelton, 1979–81. Notable productions at the National were *Maugham's *For Services Rendered* (1979), Neil *Simon's *Brighton Beach Memoirs* (1986), which moved to the *Aldwych, *Pirandello's *Six Characters in Search of an Author*, and *Beckett's *Waiting for Godot* (both 1987). He directed Arthur *Miller's *Death of a Salesman* at Nottingham (1966), the National (1979), and, with Dustin Hoffman, in 1984 in New York, where he also directed

*Storey's *The Changing Room* (1973) and in 1975 *Hamlet*. He was Director of *Chichester Festival Theatre, 1989–91.

Rueda, Lope de (c.1505–65), Spain's first actor-manager and popular dramatist, whose plays were performed by his own company in squares and courtyards throughout the country, as well as in palaces and great houses. He was much admired by *Cervantes, particularly for his playing of comic rascals and fools, and his reputation led the great *commedia dell' arte* actor *Ganassa to visit Spain with his troupe in 1574. Rueda's dialogue, mainly in prose, is natural, easy, and idiomatic, with a strong sense of the ridiculous and a happy satirizing of the manners of his day. He was the originator of the *paso*, or comic interlude, of which the best is *Las aceitunas* (*The Olives*). Two of his plays are based on Italian originals which were also used by Shakespeare—*Eufemia*, like *Cymbeline*, deriving from Boccaccio, and *Los engañados*, like *Twelfth Night*, from the anonymous *Gl'ingannti* (1531).

Ruggeri, Ruggero (1871–1953), Italian actor, who made his first appearance in 1888 and from 1891 to 1899 was with *Novelli's company, where he laid the foundations of his future career, retaining until the end the taste he then acquired for strong dramatic and romantic parts. He was a fine if somewhat old-fashioned Hamlet and an excellent Iago in *Othello*, and in 1904 he scored a great success as Aligi in *D'Annunzio's *La figlia di Jorio*. He was for some years leading man in the company run by *Pirandello, in whose *Enrico IV* he appeared in London shortly before his death and in whose *Sei personaggi in cerca d'autore* he was much admired. He played opposite most of the outstanding actresses of his day, particularly Emma *Gramatica. In his later years he was hostile to modern trends, disliking the plays of *Brecht and considering T. S. *Eliot the best contemporary dramatist.

Ruiz de Alarcón y Mendoza, Juan (1580–1639), dramatist of Spain's Golden Age, rival and enemy of Lope de *Vega Carpio. Born in Mexico City, he went to Spain in 1600 and attained eminence in the law and politics. A hunchback, embittered by the ridicule attracted by his deformity in an age without pity, he retaliated by making the admirable hero of one of his best plays, *Las paredes oyen* (*Walls Have Ears*, 1617), similarly afflicted. Much of his work, satirical, well constructed, and with some excellent character-drawing, was more popular abroad than in his own country; *La verdad sospechosa* (*Truth Itself Suspect*), probably his best-known play, about a young man so given to lying that when he speaks the truth he is not believed, was the source of *Corneille's *Le Menteur* (1644) and *Goldoni's *Il Bugiardo* (*The Liar*, 1750). The former provided the plot

for *Steele's *The Lying Lover; or, The Ladies' Friendship* (1703), on which *Foote based his comedy *The Liar* (1762).

Run, total number of consecutive performances of one play, usually at one theatre. When audiences were small and mainly local all theatres worked on the *repertory system and changed their bill practically every night. A play which was successful on its first appearance would be retained in the repertory; an unsuccessful one might be given again once or twice and then dropped. In the 19th century, with the vast increase in the number of potential theatregoers and the improvement in transport, plays began to remain in a theatre for as long as the actor-managers were willing to appear in them, and later for as long as the audiences would pay to see them, with the result that one successful production could tie up a theatre and a number of actors over a period of years.

In London Agatha *Christie's *The Mousetrap* opened in 1952 and was still running 40 years later. The *Whitehall, with a series of successful farces from 1954 to 1966, had only four plays in 12 years; and five American musicals, from *Oklahoma!* in 1947 to *My Fair Lady* in 1958, accounted for 16 years in the life of *Drury Lane. In New York the musical *The Fantasticks*, based on *Rostand's *Les Romanesques*, which opened at the *Off-Broadway Sullivan Street Playhouse in 1960, was still running in the 1990s. The musical *A Chorus Line* ran for 15 years and the musicals *Fiddler on the Roof* and *Grease* each ran for eight years.

Rundhorizont, see CYCLORAMA.

Russell, Annie (1864–1936), English-born American actress, particularly effective in *ingénue* roles. She was on the stage as a child, and in 1881 appeared with immense success in New York in the title-role of *Esmeralda*, written by William *Gillette in collaboration with Frances Hodgson Burnett. In 1891 she was forced to retire through illness, but in 1894 she returned and was seen in a number of new plays, among them Bret Harte's *Sue*, in which she was seen in London for the first time in 1898. On a subsequent visit in 1905 she created the heroine of *Shaw's *Major Barbara*. She had essayed several Shakespearian parts, including Viola in *Twelfth Night* in 1909, before she organized in 1912 an Old English Comedy company, for which she played Beatrice in *Much Ado about Nothing*, Kate Hardcastle in *Goldsmith's *She Stoops to Conquer*, and Lydia Languish and Lady Teazle in *Sheridan's *The Rivals* and *The School for Scandal*. She retired in 1918.

Russell, Fred, see MUSIC-HALL.

Russell, Lillian, see BURLESQUE, AMERICAN, and VAUDEVILLE, AMERICAN.

Russell, Willy, see EVERYMAN THEATRE, Liverpool.

Russell Street Theatre, see MELBOURNE THEATRE COMPANY.

Rutherford, Dame Margaret (1892–1972), English actress. She was late in starting her stage career, being in her early thirties when a small legacy enabled her to give up teaching elocution and the piano. She made her first appearance in pantomime at the *Old Vic in 1925, under Robert *Atkins, subsequently playing several small parts without much success. The turning-point in her career came when she played the formidable and extremely eccentric Bijou Furze in *Spring Meeting* (1938) by M. J. Farrell and John Perry. She was immediately recognized as an outstanding comedienne, and in the following year played Miss Prism in *Wilde's *The Importance of Being Earnest*. After a serious role as Mrs Danvers, the housekeeper in Daphne Du Maurier's *Rebecca* (1940), she achieved a great success as the medium Madame Arcati in *Coward's *Blithe Spirit* (1941). She then made her New York début as Lady Bracknell in *The Importance of Being Earnest* (1947), and returned to London to appear as Miss Whitchurch in John Dighton's farce *The Happiest Days of Your Life* (1948); Madame Desmortes (in a bath-chair) in *Anouilh's *Ring round the Moon* (1950); and Lady Wishfort in *Congreve's *The Way of the World* in 1953, in which she gave a masterly impression of an 'old peeled wall'. She was then seen in Anouilh's *Time Remembered* (1954) and, after an Australian tour, in Rodney Ackland's adaptation of John Vari's *Farewell, Farewell, Eugene* (1959; NY, 1960). She played Bijou again in *Dazzling Prospect* (1961), the sequel to *Spring Meeting*, and was then seen as two of *Sheridan's ladies, Mrs Candour in *The School for Scandal* in 1962 and Mrs Malaprop in *The Rivals* (her last appearance on the stage) in 1966. At her best in comedy, she could convey pathos and also inject a certain sinister element into some of her personifications of unpredictable old ladies.

Ruzzante, Il, see BEOLCO.

Rylands, George, see CAMBRIDGE.

S

69 Theatre Company, see ROYAL EXCHANGE THEATRE.

Sabbattini, Nicola, see LIGHTING.

Sachs, Hans (1494–1576), German dramatist, best known as the hero of Wagner's opera *Die Meistersinger von Nürnberg* (1868). He was a cobbler by trade, but from about 1518 poured out a stream of plays on subjects taken from the Bible, legend, history, the classics, and popular folklore, all treated in a prosaic and unhistorical way and employed for the purpose of moral instruction—Sachs was one of the first writers of his day to support Luther. Most of the tragedies are travesties, but the comedies—plays with happy endings but not necessarily humorous—are somewhat better. His fame rests chiefly on his Carnival plays, of which he wrote about 200, notable for their homespun humour and lively pictures of daily life. In them he turned the horseplay of the *Fastnachtsspiel* into a simple, amusing folk-play. The dialogue is natural and unforced and the rough-and-ready verse not unpleasing. Sachs trained his actors and directed his plays himself, adapting for the purpose a disused church, which thus became Germany's first theatre building.

Sackville, Thomas, see ENGLISH COMEDIANS.

Sackville, Thomas, 1st Earl of Dorset, see NORTON.

Sacra Rappresentazione (pl. *sacre rappresentazioni*), religious play of 15th-century Italy, which, like the Spanish *auto sacramental* and the English and French *mystery play*, developed from *liturgical drama. It was chiefly the product of Florence, where it was produced not, as in England, by craft guilds or, as in France, by specially organized adult lay bodies, but by groups of young people from religious or educational centres, who enacted scenes from the Bible or Church history either indoors, in a hall, or in the open square or market-place, in a *multiple setting. One play consisting of a cycle of episodes from the Old and New Testaments required no fewer than 22 booths or *edifizi*, indicating different localities; and a play on the life of St Olive, based on a popular legend, took two days to perform. As elsewhere, too many comic interpolations and the gradual introduction of secular and often impious elements led to a decline in the quality of new productions, but the traditions which had governed the development of the liturgical play, particularly its free-flowing style of composition

and its mingling of comedy and tragedy, were still strongly felt even by those late 15th-century dramatists whose plays were being written under the influence of the tragedies of *Seneca, as can be seen in the *Timon* (1497) and *Sofonisba* (1502) of **Galleotto del Carretto** (?–*c*.1530) and the *Filostrato e Panfile* (1499) of **Antonio Caminelli** (1436–1502), known as 'Il Pistoia'.

Sadler's Wells Theatre, London, in Rosebery Avenue, Finsbury. In 1683 Thomas Sadler discovered a medicinal spring in his garden and established there a popular pleasure-garden which became known as Sadler's Wells. Two years later he built a wooden music-room to house concerts, which in 1699 was run by a Mr Miles who renamed it Miles's Musick House. Miles died in 1724, and although the Wells continued to function it rather lost its reputation until in 1746 a local builder took it over. In 1753 he engaged a regular resident company of actors, and the old 'musick house' became a theatre. A version of *The Tempest*—probably *Dryden's—was performed there in 1764, and in the following year the old wooden building was replaced with a stone one. Under Tom *King of *Drury Lane from 1772 the theatre prospered, attracting fashionable audiences. In 1781 *Grimaldi made his first appearance there at the age of 3, and in 1801 Edmund *Kean, as 'Master Carey', aged about 13, was also in the company. In 1804 Charles *Dibdin, who took over when King died, yielded to the fashionable craze for *aquatic drama and installed a large tank on stage, filled with water from the New River, for the production of mimic sea battles. As the Aquatic Theatre, the renovated Wells opened with *The Siege of Gibraltar*, complete with naval bombardment, but soon returned to its former name. A false alarm of fire in 1807 caused a panic which resulted in the death of 20 people. In 1828 Grimaldi, who had been closely associated with the theatre for many years, returned to make his farewell appearance. Following the breaking of the monopoly of the *Patent Theatres in 1843, the theatre was let to Samuel *Phelps, who in 1844 inaugurated with *Macbeth* a series of productions of plays by Shakespeare. After his retirement in 1862 the Wells reverted to mixed popular entertainment, being used as a skating rink and for prize-fighting before being closed in 1878 as a dangerous structure. A year later *Bateman's widow, who had taken over the *Lyceum on her husband's death, moved to Sadler's Wells, had the interior reconstructed, and reopened it with her daughter Isabel as the leading lady of a good company. Attempts to revive the theatre's reputation were only partially successful, however, and Mrs Bateman died in 1881 heavily in debt. The theatre then became a local home for *melodrama, and in 1893 a music-hall, before finally closing down in 1906. Twenty-one years later Lilian *Baylis took over the derelict building and erected a new theatre intended as a North London counterpart to the *Old Vic in South London. This opened in 1931 with *Twelfth Night*, in which *Gielgud played Malvolio. It had a seating capacity of 1,650 (later reduced to 1,499) in three tiers, and a proscenium width and stage depth of 30 ft. each. The original policy of alternating productions between Sadler's Wells and the Old Vic proving uneconomic, it was decided in 1934 to make the latter the permanent home of drama while the Wells became the home of the opera and ballet companies. It closed in 1940 and a year later suffered damage by enemy action, not reopening until 1945. After the departure of the ballet company to *Covent Garden the opera company carried on alone, though the theatre was sometimes used by visiting companies, including in 1958 the *Moscow Art Theatre. In 1968 the Sadler's Wells opera company moved to the *Coliseum, becoming the English National Opera, and the theatre has since been used by a number of distinguished visiting dance companies, both from Britain and overseas; it also houses opera and children's theatre.

The **Lilian Baylis Theatre**, seating 200, associated with Sadler's Wells and adjacent to it, opened in 1988 with John *Guare's *House of Blue Leaves*.

Sadovsky [Yermilov], **Prov Mikhailovich** (1818–72), Russian actor, who first appeared on the provincial stage in Tula at the age of 14 and in 1839 joined the company at the *Maly Theatre in Moscow, coming to the fore in *Ostrovsky's plays, of whose comic roles he proved to be the ideal interpreter. Unlike *Karatygin and *Mochalov, he followed *Shchepkin in a realistic approach to his art and was largely responsible for keeping Ostrovsky's plays in the repertory until they won public acceptance.

Safety Curtain, see CURTAIN.

Sainete, name given in 17th-century Spain to a short comic scene played between the acts of a serious full-length play. Originally it resembled the *paso and the *entremés, but later differed from them by its greater use of music, its diversity of scenes and characters from city life, and a hint of social criticism in its racy dialogue. The great writer of *sainetes* in the 18th century was Ramón de la Cruz, but the genre lives on, and can be found in the works of such modern writers as Carlos Arniches and the *Álvarez Quintero brothers. With the one-act

zarzuela, the *sainete* forms part of the *género chico*.

St Albans, Hertfordshire, see VERULAMIUM.

Saint-Denis, Michel Jacques (1897–1971), French actor and director, who began his career under his uncle Jacques *Copeau at the *Vieux-Colombier. In 1930 he founded the Compagnie des Quinze, for which he directed *Noé*, *Le Viol de Lucrèce*, and *La Bataille de la Marne*, all by André *Obey, as well as a number of other plays, in all of which he himself appeared. The company achieved a great reputation, but was eventually forced to disband, and Saint-Denis settled in London, directing Obey's *Noé* in translation, with John *Gielgud as Noah, in 1935 and the Elizabethan tragi-comedy *The Witch of Edmonton* at the *Old Vic in 1936. He then founded the London Theatre Studio for the training of young actors. It had already made several interesting contributions to the English theatre, one of its graduates being Peter *Ustinov, when the outbreak of the Second World War caused it to close down. After working with the French section of the BBC during the war, as Jacques Duchesne, Saint-Denis returned to the theatre, becoming head of the short-lived drama school at the Old Vic. On its demise he returned to France to run the Centre Dramatique de l'Est based on Strasbourg. In 1957 he was appointed artistic adviser to the *Vivian Beaumont Theatre in New York in connection with the Lincoln Center, and in 1962 he became general artistic adviser to the *RSC, for whom he directed at the *Aldwych Theatre in 1965 *Brecht's *Squire Puntila and his Servant Matti*. As a frequent adjudicator of the Dominion Drama Festival in Canada, he also helped in the foundation of the National Theatre School of Canada in Montreal in 1960, advising particularly on its basic curriculum and method of training.

Sainthill, Loudon (1919–69), theatre designer, born in Tasmania, who worked for some years in Australia before coming to England. In 1951 he achieved instant success with his designs for *The Tempest* at *Stratford-upon-Avon, where he was subsequently responsible for the décor for *Pericles* (1958) and *Othello* (1959). In London he worked for the *Old Vic and also designed the sets for such diverse productions as Errol Johns's *Moon on a Rainbow Shawl* (1958), Tennessee *Williams's *Orpheus Descending* (1959), *Half a Sixpence*, the musical version of H. G. Wells's *Kipps*, and *The Wings of the Dove*, based on Henry *James's novel (both 1963). Shortly before his death he was given a Tony award for his costumes for a New York musical based on Chaucer's *Canterbury Tales* (1968). He also designed for the ballet, opera, and pantomime.

St James's Theatre, London, in King Street, Piccadilly. Designed by Samuel *Beazley, this theatre opened with a seasonal *burletta licence in 1835. His venture was a failure, but among the plays he staged were the first dramatic works of Charles *Dickens, the two-act farce *The Strange Gentleman*, and a ballad opera *The Village Coquettes* (both 1836). The theatre was renamed the Prince's in honour of Queen Victoria's consort, but in 1842 returned to its original name. For a time it became a home of French drama, *Rachel appearing there on several occasions. In 1866 *Irving enjoyed his first London success as Rawdon Scudamore in *Boucicault's *Hunted Down*; a month later W. S. *Gilbert's first burlesque, *Dulcamara; or, The Little Duck and the Great Quack*, was produced. In 1869 Mrs John *Wood took the theatre and had it partly reconstructed, remaining in nominal control until 1876 when she left for America. John *Hare and W. H. *Kendal took over in 1879 and again reconstructed the building as a four-tier theatre with a seating capacity of 1,200, reopening it on 4 Oct. They enjoyed a steady success and established a tradition of good stagecraft, much attention being paid to scenery, furniture, and costumes. Among the productions by their excellent company were several early plays by *Pinero, including *The Money Spinner* and *The Squire* (both 1881), and they also engaged the young George *Alexander who later became manager of the theatre, inaugurating its most brilliant period in 1891 with the transfer of R. C. Carton's *Sunlight and Shadow*. In 1899 the theatre closed for the reconstruction of the interior, to reopen in 1900 with Anthony Hope's *Rupert of Hentzau*. Alexander remained until 1917. He died a year later, and Gilbert Miller, the American impresario, son of Henry *Miller, took over the theatre, remaining in control until it was demolished. During the next few years many famous players appeared at the St James's, including Sybil *Thorndike and Lewis *Casson, Edith *Evans, and Noël *Coward, but there were few memorable plays until 1923, when George *Arliss starred in William *Archer's long-running thriller *The Green Goddess*. In 1925 Gladys *Cooper and Gerald *Du Maurier had great success with *Lonsdale's *The Last of Mrs Cheyney*. The 1930s saw A. A. Milne's *Michael and Mary* (1930), *The Late Christopher Bean* (1933), adapted by Emlyn *Williams from the French, an excellent adaptation of Jane Austen's *Pride and Prejudice* (1936), and Clifford *Odets's *Golden Boy* (1938). The biggest success of the war years was *Turgenev's *A Month in the Country* in 1943, with Michael *Redgrave as Rakitin. The run of Agatha *Christie's *Ten Little Niggers* was interrupted when the roof was damaged by bombing in 1944, but in 1945 Emlyn Williams appeared in his own play *The Wind of Heaven*, and in 1949 Paul *Scofield had his first leading role in the West End as Alexander the Great in *Rattigan's *Adventure Story*. In 1950 Laurence *Olivier took over the

theatre, opening with *Fry's *Venus Observed* and scoring a great success with Vivien *Leigh alternating *Shaw's *Caesar and Cleopatra* with Shakespeare's *Antony and Cleopatra* during 1951 for the Festival of Britain. In the same year the St James's returned to an old tradition when it housed the *Renaud–*Barrault company in a season which included *Salacrou's *Les Nuits de la colère*, *Claudel's *Partage de midi* with Edwige *Feuillère, and *Marivaux's *Les Fausses Confidences* with Madeleine Renaud. The only later productions of note were Odets's *Winter Journey* (1952) and Rattigan's *Separate Tables* (1954), which gave the theatre its longest run—726 performances—with Margaret *Leighton and Eric *Portman playing the leading roles in both parts. The theatre closed in 1957 and, after an unsuccessful campaign to preserve it, was pulled down and replaced by offices.

St James Theatre, New York, on West 44th Street, between Broadway and 8th Avenue. This large theatre, seating 1,583, opened as Erlanger's in 1927 for spectacular and musical shows, and received its present name in 1932. Among its early productions were revivals of *Goldsmith's *She Stoops to Conquer* and *Sardou's *Diplomacy* (both 1928), and Lion Feuchtwanger's *Jew Süss* (1930) dramatized by Ashley *Dukes. The theatre then housed light opera and musical shows, returning to straight drama with a dramatization of James Hilton's *Lost Horizon* in 1934. In 1937 Margaret *Webster's highly acclaimed production of *Richard II* starred Maurice *Evans, who also appeared in her production of *Hamlet* in 1938 and *Henry IV, Part One* in 1939. A year later he was seen with Helen *Hayes in *Twelfth Night*, and in 1941 Canada *Lee gave a fine performance in Paul *Green's *Native Son*. The theatre was then occupied for several years by Rodgers and *Hammerstein's *Oklahoma!* (1943); their *The King and I* (1951), with Gertrude *Lawrence, was also successful. *Anouilh's *Becket* (1960) did well, as did *Osborne's *Luther* (1963), in which Albert *Finney made a successful Broadway début, the latter moving in 1964 to make way for another long-running musical *Hello, Dolly!* Later hits were *Papp's musical version of *Two Gentlemen of Verona* in 1971; the London *National Theatre's production of *Molière's *The Misanthrope* in 1975; two notable new musicals, *On the Twentieth Century* (1978) and *Barnum* (1980); and an evening of music and lyrics by the *Gershwins, *My One and Only* (1983).

St Martin's Hall, London, see QUEEN'S THEATRE 1.

St Martin's Theatre, London, in West Street, St Martin's Lane, small theatre seating 550 which opened in 1916 with C. B. *Cochran's production of a 'comedy with music' *Houp La!* It was followed by the first public performance of *Brieux's sensational *Damaged Goods* (1917). Later in the same year Seymour *Hicks appeared in his own play *Sleeping Partners* and in 1920 Alec L. Rea and Basil *Dean took over, producing a number of notable new plays, among them *Galsworthy's *The Skin Game* (1920) and *Loyalties* (1922), Clemence *Dane's *A Bill of Divorcement* (1921), Charles McEvoy's *The Likes of Her* (1923) with Hermione *Baddeley, and Frederick *Lonsdale's *Spring Cleaning* (1925). Dean then withdrew, but Rea continued in management until 1937 with such productions as Arnold Ridley's *The Ghost Train* (1925), Reginald Berkeley's *The White Château* (1927), considered by some critics the best play yet written about the First World War, Rodney Ackland's *Strange Orchestra*, the first modern play to be directed by John *Gielgud, and Merton Hodge's *The Wind and the Rain* (both 1932). After J. B. *Priestley's *When We Are Married* (1938) the theatre had no settled policy, and plays succeeded one another quickly, only Kenneth Horne's *Love in a Mist* (1941), Rose Franken's *Claudia* (1942), and Edward Percy's *The Shop at Sly Corner* (1945) achieving good runs. Later productions included the revue *Penny Plain* (1951) with Joyce *Grenfell and two plays by Hugh and Margaret Williams, *Plaintiff in a Pretty Hat* (1957) and *The Grass is Greener* (1958). The 1960s saw a revival of *Synge's *The Playboy of the Western World* (1960) with Siobhán *McKenna, Hugh *Leonard's *Stephen D* (1963), Donald Howarth's *A Lily in Little India* (1966), and Robert Shaw's *The Man in the Glass Booth* (1967). Anthony Shaffer's thriller *Sleuth* opened in 1970 with Anthony *Quayle. Several short runs followed before the theatre was taken over in 1974 by Agatha *Christie's *The Mousetrap*, then in its 22nd year, which looks set to run indefinitely.

St Martin's Theatre, Melbourne, see MELBOURNE THEATRE COMPANY.

St Nicholas's Theatre Company, see CHICAGO and MAMET.

Saint Play, in early religious drama a play based on the life and legends of a Christian saint or martyr. One of the earliest extant plays, in a form similar to *liturgical drama but not connected with the liturgy, is a short work by Hilarius, an English 'wandering scholar' resident in France, dramatizing one of the miracles performed by St Nicholas. Another deals with an episode in the life of St Paul. As the patron saint of children St Nicholas became a favourite subject of plays intended to be performed at or near Christmas—his feast being on 6 Dec.—by choirboys and schoolboys, and perhaps linked with the tradition of the *boy bishop. The celebration of the name-day of other saints chosen as patrons of churches, colleges, and guilds offered ample opportunity for the dramatization of episodes in their lives which

employed, instructed, and entertained both the local clergy and the laity. In France similar plays, and those specifically devoted to incidents in the life of the Virgin Mary, were known as *miracle plays.

Salacrou, Armand (1899–1989), French dramatist whose plays, though often comedies or farces, were imbued with social and metaphysical observation. The finest but least characteristic of them is *Les Nuits de la colère*, which deals in a chronologically free-flowing form with the German occupation of France during the Second World War. Presented by the *Renaud–*Barrault company, it opened in 1946 and was seen in London during the company's visit in 1951. It was produced in New York in 1947 in an English translation as *Nights of Wrath*. Unlike his younger contemporaries *Anouilh and *Sartre Salacrou made little impact upon the English-speaking theatre; his only other plays to have been performed in translation appear to be *L'Inconnue d'Arras* (1935), seen in 1948 and 1954 as *The Unknown Woman of Arras*, in which a man's entire life is relived in the instant of his suicide; the farcical *Histoire de rire* (1939), produced in London in 1957 as *No Laughing Matter*; and *L'Archipel Lenoir* (1947), produced as *Never Say Die* in London in 1966. Salacrou's plays also include two early experiments in Surrealism, *Tour à terre* and *Le Pont de l'Europe* (both 1925); his first success, the comedy *Atlas-Hotel* (1931); *Une femme libre* (1934), a study of differing kinds of male domination; a historical drama, *La Terre est ronde* (1938), set in Florence in the time of Savonarola; and *La Rue noire* (1967), his last play.

Salisbury Court Theatre, London, last playhouse to be built in London before the Civil War. A private, roofed theatre, built of brick, it was erected in 1629 on part of the site of Dorset House, between Fleet Street and the Thames, the area being 140 ft. by 42 ft., and cost £1,000. It was used by the *King's Men from 1629 to 1631, one of their productions being *Ford's *The Broken Heart*, by *Prince Charles's Men from 1631 to 1635, and by *Queen Henrietta's Men from 1637 to 1642, when the theatres closed. During the Commonwealth surreptitious performances were given there, and the interior fittings were destroyed by soldiers in 1649, during a raid. William *Beeston restored it in 1660, and *Davenant's company played there before he went to his *Lincoln's Inn Fields Theatre. In 1661 George *Jolly was in possession, and Beeston himself ran a company there from 1663 to 1664. The building was burnt down in the Great Fire of London in 1666. It was sometimes confused with the *Whitefriars Theatre, which it replaced.

Salle des Machines, Paris, large, lavish, and well-equipped theatre in the Tuileries, built by *Vigarani in 1660 to house the spectacular shows given in honour of the marriage of Louis XIV. It continued in use for many years for Court entertainment, and was later under the control of the artist and scenic designer Jean *Bérain. It was, however, under *Servandony that it reached the height of its splendour, many magnificent spectacles being given there with his designs and machinery, some of them being wordless displays of lighting and décor well suited to a house whose acoustics were reputedly abominable. For a few years after 1770 it was the home of the *Comédie-Française.

Salvini, Tommaso (1829–1915), Italian actor, who made an international reputation, particularly in the tragic heroes of Shakespeare. He began his career at 14 and five years later made an outstanding success in the title-role of *Alfieri's *Oreste*, playing opposite *Ristori. His political activities caused him to leave Italy for a time and on his return he joined Dondini's company, playing Hamlet and Othello. In 1861 he added to his repertory one of his finest modern parts, that of the ex-convict Corrado in *Giacometti's *La morte civile*. His international career began effectively in 1869, when he toured Spain and Portugal; in 1871 he progressed in triumph through the major cities of Latin America, and in 1873 he made his first appearance in the USA, where in 1886 he played the Ghost and Othello to the Hamlet and Iago of Edwin *Booth. He was first seen in London in 1875, as Othello at *Drury Lane, creating a sensation by depicting the savagery beneath apparent urbanity. His Hamlet inevitably invited comparison with *Irving's but was well received. On a visit to Scotland in 1876 he was seen in *Macbeth*, which he considered Shakespeare's finest play; this was not seen in London until 1884, when he also played Lear to a British audience for the first time. Possessed of a powerful physique, graceful in movement and gesture, and above all equipped with a superbly flexible and musical voice, he studied his roles long and carefully, and his interpretations were rich in imaginative insight.

Sam H. Harris Theatre, New York, on West 42nd Street, between Broadway and 8th Avenue. This opened as a cinema in 1914 but as the Candler Theatre was sometimes used for plays, Elmer *Rice's *On Trial* being seen there in 1914 and *Galsworthy's *Justice* in 1916. Later that year it was renamed the Cohan and Harris Theatre, receiving its present name in 1921 after *Cohan relinquished his share in the management. In 1922 theatrical history was made there when John *Barrymore played Hamlet 101 times. The building again became a cinema in 1932.

The American producer **Sam H**(enry) **Harris** (1872–1941) was in partnership with Cohan,

1904–20, and then continued independently. He built the *Music Box, with Irving *Berlin, in 1921, his productions there including the Gershwins' musical *Of Thee I Sing* (1931), *Kaufman and Edna Ferber's *Dinner at Eight* (1932), *Steinbeck's *Of Mice and Men* (1937), and Kaufman and Moss *Hart's *The Man Who Came to Dinner* (1939).

Samson, Joseph-Isidore (1793–1871), French actor, who entered the Conservatoire at 16, and subsequently spent several years in the provinces. In 1819 he appeared in Paris at the opening of the new *Odéon theatre, and made such a good impression that he was retained as leading man until 1826, when the *Comédie-Française claimed him. Apart from a few years at the *Palais-Royal, he remained there until his retirement in 1863. A handsome man, with a fine profile and a mass of curly hair, he was accounted a good actor, and also one of the finest teachers of acting at the Conservatoire where he remained on the staff until his death; it is as the teacher of *Rachel that he is now chiefly remembered. By instructing her in the classical tradition, which he himself had inherited from *Talma, he contributed to the revival of French tragedy with which Rachel was associated.

Sam S. Shubert Theatre, New York, see SHUBERT THEATRE.

San Diego National Shakespeare Festival, see SHAKESPEARE FESTIVALS.

San Francisco, see AMERICAN CONSERVATORY THEATRE.

San Francisco Mime Troupe, see COLLECTIVE CREATION.

San Gallo, Bastiano da, see PERIAKTOI.

Sans Pareil, London, see ADELPHI THEATRE I.

Sans Souci, New York, see NIBLO'S GARDEN.

Sarah-Bernhardt, Théâtre, see BERNHARDT.

Sarat, Agnan (?–1613), French actor, with an ugly, squashed nose, who in 1578 took a company to Paris and leased the theatre of the Hôtel de *Bourgogne from the *Confrérie de la Passion. After a short stay he disappeared again into the provinces, but in 1600 returned to Paris as chief comedian in the company of *Valleran-Lecomte, with whom he remained until his death. Like some of his fellow comedians, he always put flour on his face when playing in farce.

Sardou, Victorien (1831–1908), French dramatist, one of the most consistently successful of his day. Like *Scribe, whose successor he was, he wrote copiously on many subjects in many styles, with expert craftsmanship and superficial brilliance. His first outstanding success was a comedy, *Les Pattes de mouche* (1860), seen in London in 1861 as *A Scrap of Paper*. Of his historical plays, the best known is *Madame Sans-Gêne* (1893). The part of the Duchess of Dantzig, a washerwoman whose husband became one of Napoleon's marshals, was created by *Réjane and played in an English adaptation by Ellen *Terry in 1897. The best of the romantic melodramas were *Fédora* (1882) and *La Tosca* (1887), the latter providing the libretto for Puccini's opera. Both plays were written for *Bernhardt, who revived them many times, and appeared also in a number of Sardou's other plays. Two social dramas which are typical of their kind were *Dora* (1877) and *Divorçons* (1880). The former, in a translation by Clement *Scott as *Diplomacy*, had a great success in London in 1878; the latter, under its original title, was seen in London in 1907. Sardou, who was always ready to exploit whatever dramatic form seemed assured of popular success, was a favourite target of contemporary critics, and *Shaw, who disliked everything he stood for, coined the word 'sardoodledom' to epitomize his *well-made plays.

Saroyan, William (1908–81), American novelist and dramatist, born in Fresno, Calif., of Armenian parents. His first play, a long one-acter produced under the auspices of the *Theatre Guild, was *My Heart's in the Highlands* (1939), which dealt with a poet's struggle to maintain his integrity in a materialistic world. His next play, *The Time of Your Life* (also 1939), a fresh and invigorating work about a group of characters in a saloon, was awarded a *Pulitzer Prize. Not as successful, but still interesting, were *Love's Old Sweet Song* (1940) and *The Beautiful People* (1941). A less optimistic note was struck with *Hello, Out There* (1942), a one-acter depicting the lynching of an innocent tramp, and *Get Away, Old Man* (1943), which dealt with a young writer's conflict with a ruthless Hollywood mogul. After the comparative failure of several other plays not seen on Broadway, Saroyan recaptured some of his early success with *The Cave Dwellers* (1957), about some decrepit performers living in an abandoned theatre. *Sam, the Highest Jumper of Them All* (1960) was created with and for Joan *Littlewood's *Theatre Workshop in London, after which Saroyan wrote little further for the stage, his last work being *The Rebirth Celebration of the Human Race at Artie Zabala's Off-Broadway Theater* in 1975. His plays have a strong vein of fantasy, and avoid conventional dramatic structure.

Sartre, Jean-Paul-Charles-Aymard (1905–80), French dramatist and philosopher, probably the best known of the post-war French playwrights outside his own country. His philosophy of Existentialism—the responsibility of each man for his own acts and for the consequences of those acts—is implicit in his dramatic works. His first play, *Les Mouches* (1942), was a modern interpretation of the story of Orestes; as *The Flies* it was seen in New York in 1947 (London, 1951). *Huis-Clos* (1944) was produced in London as *Vicious Circle* and in New York as *No Exit*, both in 1946, the year in which *Morts sans sépultures* and *La Putain respectueuse* were first seen in Paris. As *Men without Shadows* and *The Respectable Prostitute* these were seen in London in 1947. The second was produced in America in 1948 as *The Respectful Prostitute*, a more appropriate title, since the prostitute betrays her coloured lover out of a craven respect for the dictates of society. Sartre's plays have often been given different titles in translation; *Morts sans sépultures* was seen in New York in 1948 as *The Victors; Les Mains sales* (1948) was produced in London as *Crime passionel* and in New York as *Red Gloves*, both in 1948. Sartre's next play, *Le Diable et le bon Dieu*, based on the same story as *Goethe's *Götz von Berlichingen*, has been variously translated as *The Devil and the Good Lord* and *Lucifer and the Lord*. In 1956 *Unity Theatre gave the first English production of *Nekrassov* (1955), later seen at the *Royal Court, where in 1961 *Les Séquestrés d'Altona* was produced as *Altona* (NY, as *The Condemned of Altona*, 1965). In 1969 one of Sartre's many film scripts, *L'Engrenage*, was successfully adapted for the theatre, and in 1971 his adaptation of the elder *Dumas's *Kean* (1836), first seen in Paris in 1853 with Pierre *Brasseur, was produced in London.

Satire, Theatre of, Moscow. This theatre, which opened in 1924, was intended for the production of revues on domestic and political topics, but soon began staging satirical comedies such as *Katayev's *Squaring the Circle* (1926). This policy was maintained, though its tone was muted during the last years of Stalin's rule; after his death the theatre flourished again and there were revivals of *Mayakovsky's three major plays. Under the directorship of **Valentin Pluchek** (1909–), appointed in 1957, it mounted a series of successful productions, outstanding among which were *Shaw's *Heartbreak House*, *Gogol's *The Government Inspector*, *Sartre's *Nekrassov*, and *Beaumarchais's *The Marriage of Figaro*. During the late 1960s the company, originally housed on Mayakovsky Square, moved into a magnificent new theatre next door to the Tchaikovsky Concert Hall. *Rozov's *Nest of the Wood Grouse* was staged in 1981.

Satyr-Drama, in ancient Athens, burlesque plays which followed, and served as ribald comments on, the statutory tragic trilogy (see TETRALOGY) in the annual dramatic contest in connection with the *Dionysia. In the satyr-plays a heroic figure, sometimes the chief character of the preceding trilogy but very often Hercules, was shown in a farcical situation, always with a chorus of Sileni, or satyrs. These were the legendary companions of *Dionysus, and were portrayed as being half-human, half-animal, with the ears and tail of a horse. The characteristics of satyr-drama were swift action, vigorous dancing, boisterous fun, and much indecency in speech and gesture. Although *Aristotle said that Greek tragedy 'developed from the satyr-play', the connection between them is not clear and must date from long before the time of the first official festival in 534 BC. Only one satyr-play has survived in its entirety, the *Cyclops* of *Euripides, though there are also fragments of an *Ichneutae* by *Sophocles. The popularity of the satyr-play declined during the second half of the 5th century.

There is no connection between satyr-drama and modern satire, or between satyr-drama and any form of Greek comedy.

Saunders, James (1925–   ), English dramatist, most of whose plays were first produced either at the *Questors or the Orange Tree Theatre, Richmond, Surrey. After a number of one-act plays vaguely reminiscent of *Ionesco, among them *Alas, Poor Fred* and *The Ark* (both 1959), he created a big impression with *Next Time I'll Sing to You* (1962), based on the life and death of an actual Essex hermit, which in a revised form was successfully produced in the West End and New York in 1963. More one-act plays followed, and in 1964 came another unusual full-length play, *A Scent of Flowers*, dealing with the period immediately following a young girl's suicide, the girl herself appearing as the chief character; it was produced at the *Duke of York's (NY, 1969). In 1967, in collaboration with the author, Saunders dramatized for the *Bristol Old Vic Iris Murdoch's novel *The Italian Girl*, which later had a long run at *Wyndham's. It was followed by *The Borage Pigeon Affair* and *The Travails of Sancho Panza* (both 1969), the latter a children's play specially written for the *National Theatre company. Other plays included *Hans Kohlhaas* (1972), based on a play by *Kleist; *Bodies* (1977), about two couples who meet again after changing partners 10 years earlier, which had a long West End run in 1979; *The Girl in Melanie Klein* (1980), set in a private asylum and based on a novel by Ronald Harwood; and *Fall* (1981), in which three contrasting sisters gather just before their father's death. *Nothing to Declare* (1983) contains the ruminations of a 65-year-old novelist.

Saunderson, Mary, see BETTERTON.

Savary, Jérôme, see CHAILLOT, PALAIS DE.

Saville Theatre, London, in Shaftesbury Avenue, seating 1,250 on three tiers. It opened in 1931 with a series of musical plays, and the theatre continued to house musical comedy and revue until the end of 1938, when it went over to straight plays with *Shaw's *Geneva*. In the following year, *Priestley's *Johnson over Jordan*

transferred there. The Saville was damaged by enemy action in 1940, but was hastily repaired and continued in use. In 1951 Ivor *Novello's last work *Gay's the Word* was seen, with Cicely *Courtneidge, which was followed by another musical, *Love from Judy*, which ran for 594 performances. John *Clements took over in 1955 and inaugurated a fine series of revivals which continued for two years and included *Ibsen's *The Wild Duck*, *Chekhov's *The Seagull*, and *Congreve's *The Way of the World*. Later productions of note were N. C. *Hunter's *A Touch of the Sun* (1958), the musical *Expresso Bongo* (also 1958) with Paul *Scofield, and Anthony Kimmins's *The Amorous Prawn* (1959). The last 10 years of the theatre's life were dogged by uncertainty, but among the interesting productions were *Ustinov's *Photo Finish* and David Turner's *Semi-Detached* with Laurence *Olivier (both 1962) and a musical version of *Dickens's *Pickwick Papers* (1963) which ran for 694 performances. In 1967 the theatre was occupied by dance companies, and in 1968 there were revivals of Novello's *The Dancing Years* and the *Gershwins' *Lady, Be Good!* The last notable production, before the theatre closed in 1970 and was converted into twin cinemas, was *Brecht's *The Resistible Rise of Arturo Ui* (1969).

Savoy Theatre, London, in Beaufort Buildings in the Strand. Built to provide a permanent new home for the operettas of *Gilbert and Sullivan, this theatre seated about 1,000 in four tiers, and with its delicate colouring and new electric lighting was considered revolutionary. It opened in 1881 with *Patience*, transferred from the *Opera Comique, and gave its name to the whole series of Gilbert and Sullivan collaborations, most of which were seen there, as were *Merrie England* (1901) and *The Princess of Kensington* (1903), both with music by Edward German. In 1907 *Granville-Barker and J. E. Vedrenne, fresh from their successful seasons at the *Royal Court, staged the first London production of *Shaw's *Caesar and Cleopatra*, and from 1912 to 1914 a number of Shakespeare revivals. The popular children's play *Where the Rainbow Ends* had its first performance there in 1911. Later productions included *Coward's *The Young Idea* (1923), *Van Druten's *Young Woodley* (1928), and *Sherriff's *Journey's End* (1929). The building was then closed for reconstruction. The new theatre, seating 1,121 in three tiers, opened in 1929 with a revival of *The Gondoliers*, and in 1930 staged *Othello* with Paul *Robeson. During the next decade it was used mainly for transfers, but *The Man Who Came to Dinner* (1941) by George S. *Kaufman and Moss *Hart had a long run, as did two American comedies based on volumes of short stories, Ruth McKenney's *My Sister Eileen* (1943) and Clarence Day's *Life with Father* (1947), Coward's *Relative Values* (1951) with Gladys

*Cooper, and Agatha *Christie's *Spider's Web* (1954). Other successes included Alec Coppel's *The Gazebo* (1960), Coward's musical *Sail Away!* (1962), Ronald *Millar's *The Masters* (1963), and *The Secretary Bird* (1968) by William Douglas *Home, who had a further success in 1972 with *Lloyd George Knew My Father* starring Peggy *Ashcroft and Ralph *Richardson. Agatha Christie's *Murder at the Vicarage* began a long run in 1975. Robert *Morley starred in a revival of Ben *Travers's *Banana Ridge* in 1976, the *RSC's production of Shaw's *Man and Superman* was seen in 1977, and in 1979 a revival of Ray Cooney and John Chapman's *Not Now, Darling* was very popular. Notable productions in the 1980s were Michael *Frayn's *Noises Off* (1982) and the RSC's presentation of the musical *Kiss Me, Kate* in 1988. The theatre was badly damaged by fire in 1990, but is being restored.

Scala, Flamineo, see CONFIDENTI.

Scala Theatre, London, in Tottenham Street, Tottenham Court Road. This opened as the King's Concert Rooms in 1772 and in 1802, as the Cognoscenti Theatre, became the headquarters of a private theatrical club The Pic-nics, which was successful enough to attract the hostility of the *Patent Theatres. Closed in 1808, it reopened as the Tottenham Street Theatre in 1810, and in 1814 was sold to the father of the scene painter William *Beverley. He renamed it the Regency Theatre, but it had little success, and in 1820 was reopened by Brunton as the West London, his daughter Elizabeth, who later married the actor Frederick *Yates, starring in many of his productions. It was constantly in trouble with the Patent Theatres and was closed for several years, reopening in 1831 as the Queen's or alternatively the Fitzroy, the names being interchangeable. It then became a home for lurid melodrama, and was nicknamed the 'Dust Hole'. In 1865, taken over by Marie *Wilton, it was completely redecorated, renamed (by royal permission) the Prince of Wales, and reopened in the presence of the future Edward VII with immediate success. Marie Wilton's leading man was Squire *Bancroft, whom she later married, and together they built up an excellent company, including Ellen *Terry, Mr and Mrs *Kendal, and John *Hare. Under their management the epoch-making 'cup-and-saucer' dramas of T. W. *Robertson were first produced, beginning with *Society* in 1865. *Caste,* the only one to have been revived in recent times, was seen in 1867. Other important productions were *Masks and Faces* (1875) by Tom *Taylor and Charles *Reade and *Sardou's *Diplomacy* (1878). By 1880, when the Bancrofts left to go to the *Haymarket, the despised 'Dust Hole' had become a fashionable theatre, but in 1882 it was condemned as structurally unsound and closed for repairs. Owing to a long-drawn-out dispute

it was not used again except as a Salvation Army hostel. In 1903 it was demolished, only the original portico remaining to serve as the stage-door entrance of a new theatre which seated 1,193 in a three-tier auditorium. Renamed the Scala, it opened in 1905 under the management of *Forbes-Robertson, but the venture was not a success and the theatre often stood empty or was used for films, puppet-shows, and amateur productions. In 1926 Ralph *Richardson made his first London appearance at the Scala for the Greek Play Society as the Stranger in *Sophocles' *Oedipus at Colonus.* Various touring companies used the theatre, among them Donald *Wolfit's and the D'Oyly Carte in *Gilbert and Sullivan, and from 1945 there was an annual Christmas revival of *Barrie's *Peter Pan,* until in 1969 the theatre closed, being demolished in 1972.

Scamozzi, Vincenzo, see PALLADIO.

Scapino, one of the *zanni or servant roles of the *commedia dell'arte. Like *Brighella, Scapino was crafty and unprincipled, and when in danger lived up to his name, which means 'to run off' or 'to escape'. The first actor to play him was **Francesco Gabrielli** (?–1654), who was in the company which went with the younger *Andreini to Paris in 1624. Through him and later actors of the part it passed into French comedy to end up as the quick-witted and unscrupulous valet of *Molière's *Les Fourberies de Scapin* (1671).

Scaramuccia, literally, 'a little skirmisher', a character of the *commedia dell'arte generally classed with the *zanni roles, but considered by some to have approximated more to the blustering braggart *Capitano. The actor most closely associated with the part, though he may not actually have created it, was Tiberio *Fiorillo, who, abandoning the use of a mask, played the role with white powdered face, small beard, and long moustache, in a costume predominantly black but set off by a white ruff. In France he became Scaramouche.

Scarborough, Yorkshire, see STEPHEN JOSEPH THEATRE IN THE ROUND.

Scarron, Paul (1610–60), French dramatist and novelist, crippled by rheumatism at the age of 30 and forced to rely on his pen for a livelihood. He wrote a number of witty though slightly scabrous farces, of which the first two, *Jodelet; ou, Le Maître-valet* (1643) and *Jodelet soufflété* (1645), were produced at the *Marais with the comedian *Jodelet himself in the title-roles. In 1652 Scarron married the beautiful but penniless orphan Françoise d'Aubigné, who as Mme de *Maintenon was to become the second wife of Louis XIV and the virtual ruler of France. Meanwhile Scarron continued to write for the theatre. The best example of his burlesque comedy, in which he obtained his comic effects by ingenious word-play and by the incongruity between subject-matter and style, is *Don Japhet d'Arménie* (1647), frequently revived in later years by *Molière. *L'Écolier de Salamanque* (1654) is notable for the character of the valet *Crispin, long played by successive members of the *Poisson family. Scarron, whose interest in Spanish literature had led him to translate a number of Spanish plays, modernizing them and adding much material of his own, may have taken from Agustín de *Rojas the idea of his most important work *Le Roman comique* (1651), a novel which depicts the adventures of an itinerant provincial theatre company. It has considerable documentary value.

Scenario, skeleton plot of the *commedia dell'arte* play, the term replacing some time in the early 18th century the older word *soggetto*. These are not such synopses as might be drawn up by an author for his written drama, nor are they identical with the Elizabethan *platt; they were theatrical documents prepared for the use of a professional company by its leader, or by an enthusiastic amateur admirer of the *commedia dell'arte* style, and consisted of a scene-by-scene résumé of the action, together with notes on locality and special effects. Their informal elasticity allowed the insertion of extraneous business (the *burla or *lazzo) at the discretion—or according to the ability—of the actors. The term, whose plural in English is now scenarios, is commonly used today for the script of a film or the synopsis of a musical play.

Scene of Releave (Relief), see MASQUE.

Scene Room, see GREEN ROOM.

Scenery, term covering everything used on stage to represent the place in which an action is performed, including hangings, cut-outs, painted *flats, *box-sets, *built stuff, etc., but not usually movable furniture and *props.

Scenery is a comparatively recent innovation in the history of the theatre. Greek plays were acted against a stage wall, which by Roman times had become a grandiose architectural façade, the *scenae frons*, and the use of *stage machinery is indicated by mentions of the *ekkyklema and the *mechane, but except for the *periaktoi of Hellenistic and Roman times the classical stage had no scenery as we know it. In the early medieval period the interior of a church provided the setting for *liturgical drama. As the drama moved out of doors the *multiple setting began to evolve, with 'mansions' representing different localities placed around an open *platea*, or acting area, the outside wall of the church forming a backdrop; this style was to have a long-lasting influence on the theatres of France. In England, by contrast, *mystery plays were presented on mobile wagons, scene by scene, and the public playhouses of the 16th–17th centuries owed much

to their portable precursors. Theatres such as Shakespeare's *Globe were probably gaily decorated, but apart from movable properties on the platform stage (see FORESTAGE) and in the 'discovery space' (see INNER STAGE) scenery in the modern sense was not used. It is first found in the courts of Renaissance Italy where entertainments were presented with all possible splendour. Perspective painting, developed in the mid-15th century, was used to make enclosed spaces seem larger; these principles were applied to theatrical illusion by **Baldassare Peruzzi** (1481–1537), who was also influenced by the newly discovered architectural treatise by *Vitruvius. Peruzzi's pupil *Serlio published in 1545 descriptions of perspective stage settings for tragic, comic, and satyric plays. In the second half of the century **Bernardo Buontalenti** (1536–1608) was using a painted backcloth with *telari*, three-sided prisms in imitation of the classical *periaktoi*, forming side-wings. His innovations and those of his pupils (notably **Giulio**, 1590–1636, and **Alfonso**, ?–1656, **Parigi**, father and son) spread all over Europe and were introduced into England by Inigo *Jones for the elaborate Court *masques of 1600–40.

The scenic system of the masque—the origin of modern scenery—used, in its final form, a decorative *proscenium arch behind which sets of *wings, called side scenes or side shutters, framed the back scene on either side. This was the flat scene, from which the modern term 'flat' is derived. The back scene consisted of pairs of painted shutters, from two to four in number, centrally divided and sliding in *grooves placed about half-way upstage. The wings, too, could be in groups sliding in grooves, so that they could be changed according to the changes of the back scenes.

These Court performances of masques had no influence on the English public theatres, which still used the bare apron stage derived from the medieval tradition. Development along Continental lines was prevented by the outbreak of the Civil War and the proscription of playhouses until the monarchy was restored in 1660. In 17th-century Italy the widespread passion for opera gave stimulus to the work of many scene designers, including the great sculptor *Bernini, Filippo *Juvarra, and above all the *Bibienas, whose diagonal perspective, and the growing popularity of landscape painting, gradually changed the whole character of theatre decoration. In Paris, where *Mahelot had begun to oust the old-fashioned multiple setting from the Hôtel de *Bourgogne, *Torelli created a rage for *machine plays with his inventions, including the *carriage-and-frame method of rapid scene-changing. French taste was moving towards lightness, and fantasy progressing into neo-classicism, as can be seen in the succession of the *Vigaranis, Bérain, and

*Servandony at the *Salle des Machines; the *Quaglio family was to continue the trend in Germany, as did **Pietro Gonzaga** (1751–1831) in Russia. The increasing elaboration of the *Jesuit drama both echoed and influenced developments all over Europe.

Meanwhile in England the Restoration playhouse set the basic scenic style for over a century: an apron stage, with up to three pairs of *proscenium doors opening on to it, drew on the medieval tradition, while scenes composed of sets of wings and *backcloths were a direct inheritance from the flat scene of the masque. England was untouched by the aesthetic movements of France and Italy, and scene design changed little until *Garrick brought Philip de *Loutherbourg to *Drury Lane. Gradually the old architectural setting was abandoned in favour of romantic landscapes, with transparencies and elaborate cut-outs helping to create an attractive stage picture which was still in use 100 years later and lasted even longer in *pantomime.

The 19th century saw also an enthusiasm for neo-Gothic design and a growing insistence on painstaking architectural detail. This passion for authenticity had begun with the designs of *Capon for *Kemble's Shakespeare revivals at Drury Lane, 1794–1802, and at *Covent Garden in the 1810s, and was continued by *Kean in the 1850s at the *Princess's and Hawes *Craven working for *Irving at the *Lyceum, reaching its peak of elaboration with Beerbohm *Tree's productions of Shakespeare. This was the great era of stage illusion, of *traps, gauzes, and *transformation scenes, and of trompe-l'œil scene painting, with every detail painted on stretched canvas—doors, windows, draperies, even furniture, as well as outdoor vistas: William *Beverley was one of the period's most successful practitioners. For spectacular pieces such as pantomimes, cut-cloths developed to such excess that the stage picture resembled a lacy valentine. As drama began to follow the rise of realism in literature and painting, these theatrical conventions became unacceptable. The box-set, in use since the 1830s, became the normal means of presenting interior scenes; still constructed of flats lashed together, it was now supplied with real furniture and accessories and practicable doors and windows. When André *Antoine founded the *Théâtre Libre in 1887 he insisted on complete verisimilitude for his productions of naturalistic contemporary plays (even real food was used, and real fountains played, on stage); but he used scenery and properties to reinforce the mood of a play in a totally new manner.

The French *Symbolists attacked the methods of the Théâtre Libre for ignoring imagination and fantasy in the search for the exact. The French poet **Paul Fort** (1872–1960) founded in 1891 the Théâtre Mixte and in his manifesto enunciated many of the principles later adopted by the modernist school. Scenery was to be simplified, evocative rather than descriptive; there was to be frank stylization, complete harmony between scenery and costume, and the absolute abandonment of the perspective backcloth. Norman *Wilkinson and Charles Ricketts, working for *Granville-Barker, were to be the main followers in England of the Symbolist approach.

The works of some new dramatists, especially *Maeterlinck, were sufficiently imaginative to give scope to the new method. In 1893, when the Théâtre d'Art had become the Théâtre de l'Œuvre, *Lugné-Poë presented Pelléas et Mélisande at the Bouffes-Parisiens; *Stanislavsky saw it and afterwards admitted how much he owed to the experimental work being carried out in Paris. The *Moscow Art Theatre, founded by Stanislavsky in 1898, adopted Antoine's naturalism and the realistic effects of the *Meininger company, which had visited Moscow in 1885 and 1890: this low-key manner was in harmony with Stanislavsky's aim of presenting actual conditions of life through drama.

In 1899 Adolphe *Appia published his epoch-making work on the reform of staging, Die Musik und die Inscenierung, which stressed in particular the illogicality of placing three-dimensional actors against flat scenery. His proposals for solid settings of extreme simplicity, lit so as to emphasize instead of flattening the human form, were to have an immense influence on 20th-century stage décor. Electric lighting, applied in the theatre from the 1890s, made Appia's ideas practicable and opened the way to other developments in stage design. One of the most widespread was the *cyclorama, which evolved from Mariano Fortuny's Kuppelhorizont or sky-dome (see LIGHTING). This solid, curved rear wall could, with shadowless lighting, represent indefinite open space; it was latter used for projected clouds and various effects of light.

The trend towards greater simplicity continued. Gordon *Craig was more of a theorist than a practitioner and the most influential of Appia's successors. He evolved a system of large screens, with a few movable features such as flights of steps, to build up an imaginative stage picture with no concessions to realism: his production of Hamlet at the Moscow Art Theatre in 1912 is still controversial. He also urged a theatre completely created by a single man—author, director, designer, costumier—with actors (Übermarionette) under his dictatorial control. However extreme, this theory was in tune with the increasing importance of the theatre director during the 20th century. Max *Reinhardt came near to filling the role proposed by Craig. His designers provided scenery as eclectic as his choice of play: he used semi-permanent settings, *composite settings, proscenium, apron, and arena stages; mounted productions in theatres, circuses, exhibition halls, ballrooms; and in his historic production of *Hofmannsthal's *Jedermann* (1911) the streets of Salzburg became his stage and the façade of its cathedral his back-drop. In the years 1910–33 Reinhardt dominated the stage of central Europe with his grand theatrical enterprises; his more intimate productions showed great subtlety and individuality as well as an awareness of current trends in the visual arts. The American director David *Belasco, in contrast, opted for a kind of spectacular naturalism for the romantic dramas produced at his own theatre in New York, 1907–31. He was adventurous in his use of lighting, and took advantage of developments in stage machinery including the *revolving stage.

In the years immediately following the First World War, the European stage saw the brief flowering of *Expressionism, taking the form on the one hand of an extreme simplification of scenery and on the other the distortion of inanimate objects to reflect the moods of a play. *Komisarjevsky mounted productions of Shakespeare in Expressionist settings at Stratford-upon-Avon during the 1920s and 1930s. Expressionism appeared on the American stage with Robert Edmond *Jones's designs for *Macbeth* in 1921, lop-sided cardboard arches that emphasized the insecurity of the hero's moods and fortunes. Other American scene designers who came to the fore at this time included Norman *Bel Geddes and Lee *Simonson.

Britain remained for the most part indifferent to Continental and American developments. Lovat *Fraser, the only artist who might have inaugurated a movement of far-reaching significance, died young after producing his admirably simple semi-permanent set for a revival of *Gay's *The Beggar's Opera* in 1920. In general Britain remained faithful to realism and the box-set. In France the best designers still tried to get away from the tyranny of the painted scene. *Copeau indeed dispensed with scenery entirely, and the sets of *Dullin were based on Craig's idea of movable screens. In Russia the Revolution had swept away the conventional forms of theatre décor, reducing the set to bare scaffolding or the sparse clean lines of metal machinery. Décor became symbolic, simplified to the point of abstraction in the then dominant mode of Constructivism. The movement's chief exponents in Soviet theatre, Vsevolod *Meyerhold and Alexander *Taïrov, adoped a frank theatricality, rejecting all kinds of realism and taking the action among the spectators, who were in turn drawn into the action. Even the Moscow Art Theatre was affected by the artistic ferment of the 1920s; but official reaction against *Formalism in all the arts put an end to such experiments and from the early thirties *Socialist Realism became the only acceptable manner, in the theatre as elsewhere, for more than two decades. Throughout Europe, the period of fundamental innovation ended in about 1930,

the growing threat of war inhibiting the expansion of new ideas; the British and American theatres now began to catch up with Continental developments, helped by an influx of émigré artists and designers.

The popular revues of C. B. *Cochran used designers of the calibre of Rex Whistler and Oliver *Messel, whose witty, stylized settings owed much to earlier European experiments. The commercial theatre of the pre-war decade was marked by elegance and decorativeness, taken up in the years immediately after the Second World War in a revival of opulent romanticism. But the preoccupations of post-war playwrights—*Osborne, *Wesker, *Behan in Britain, Arthur *Miller and Tennessee *Williams in America—forced designers into a visual austerity that extended into the classical repertory as post-war euphoria declined. Monochrome designs, sometimes with minimal colour accents, came into vogue, perhaps in reaction to the pictorial splendour of musicals, which reached a peak in Cecil *Beaton's scenery and costumes for *My Fair Lady* (1956). The increasing use of *theatre-in-the-round and *flexible staging placed new difficulties in the way of designers. Scene-painting in particular languished. Materials produced by new industries almost ousted the wood, canvas, and papier mâché of traditional scenery: the vast range of plastics, especially, provided novel textural effects and made possible the building of structures at once massive and lightweight. Modern metal alloys, fibreglass, fibre board, plywood, and newly developed adhesives, all helped to widen the designer's range, although audiences tended to resist their more brutal evocations of contemporary artistic trends. Projected scenery had been used by *Piscator as early as 1924 and is claimed to have been first used in England in *Strindberg's *The Road to Damascus* in 1937; the technique benefited from post-war developments in optical technology, and was used with particular success by Josef *Svoboda, who took up the European tradition (inherited from the Symbolists and the Constructivists by way of the *epic theatre of *Brecht and Piscator) to achieve a heightened realism that proved applicable to a wide range of subjects. Although the naturalistic box-set is still used for single-setting plays, the great range of technical choice has tended to be applied to a deliberate anti-illusionism, with stage mechanisms and lighting equipment exposed to view. Within this general consensus, styles of interpretation vary widely. (See also COSTUME; DETAIL SCENERY; FALLING FLAPS; FULL SCENERY.)

Schechner, Richard, see COLLECTIVE CREATION.

Schicksalstragödie, see FATE DRAMA.

Schiller, (Johann Christoph) **Friedrich von** (1759–1805), outstanding German poet and dramatist. He was only 22 when his first play

Die Räuber, about the hostility between two brothers, was accepted for the theatre at *Mannheim, where it was produced in 1782 with immediate success. Schiller was appointed official dramatist to the theatre, writing for it *Fiesco* (1783) and *Kabale und Liebe* (1784). He was heavily in debt, however, having taken refuge in Mannheim, where he was living under an assumed name, from his duties as an army doctor. After the production of his first historical tragedy, *Don Carlos* (1787), he published two books, one of which, on the Thirty Years War, provided him with the material for his great dramatic trilogy *Wallenstein*, completed in 1799 and translated into English by *Coleridge. Schiller's last years, before his early death from tuberculosis, were spent in *Weimar, where he enjoyed the friendship and collaboration of *Goethe, who staged some of his best works, notably *Maria Stuart* (1800); *Die Jungfrau von Orleans* (1801), his most operatic play; *Die Braut von Messina* (1803), written in the form of ancient drama in chiselled verse, with lyrically beautiful choruses; and his last play *Wilhelm Tell* (1804). All Schiller's plays were translated into English, at first for reading rather than for the stage. The most influential was *Die Räuber*, which reinforced the *Sturm und Drang* movement unleashed by Goethe's *Götz von Berlichingen* (1773). As *The Red-Cross Knights* it was seen at the *Haymarket in 1799 and as *The Robbers* at *Drury Lane in 1851. *Kabale und Liebe* as *The Harper's Daughter* was seen at *Covent Garden in 1803 and as *Power and Principle* at the *Strand in 1850. It was seen in German in London during the *World Theatre Season of 1964. *Maria Stuart*, which brings together Elizabeth I and Mary Queen of Scots, who in real life never met, was staged at Covent Garden in 1819, the Court Theatre in 1880, and the *Old Vic in 1958.

Schiller [de Schildenfeld], **Leon** (1887–1954), Polish director and designer, an innovator in the style of *Wyspiański, who, with *Craig and *Stanislavsky, constituted the main influences on his work. An exponent of 'total theatre', he was also a champion of both romantic and realistic drama, and became one of the outstanding figures in Polish theatre history. After the First World War he worked at the Polski Teatr in Warsaw and at the Reduta Theatre before founding his own Bogusławski Theatre in 1924, where he directed, among other classics, a revival of Krasiński's *The Undivine Comedy* in 1926. His political affiliations with the left, and a controversial production of *Brecht's *Die Dreigroschenoper*, led to the closing of his theatre in 1930, and he went to Lwów, where his outstanding productions were of *Tretyakov's *Roar, China!* and *Mickiewicz's *Forefathers' Eve*; he also directed a production of the latter in

Bulgarian in Sofia in 1937. Interned at Auschwitz, he was released in 1941 and, since the Germans had closed all the Polish theatres, set to work secretly to produce, in a convent near Warsaw, some of the old Polish *liturgical dramas. In 1946 he became director of the theatre in Łódź, where his last important production was *The Tempest* in 1947.

Schlegel, August Wilhelm von (1767–1845), German dramatist who wrote 16 plays, now forgotten, and is mainly remembered as the translator, with his friend *Tieck, of 17 of Shakespeare's plays. These versions, in spite of their romanticism, are still the ones most often performed in Germany. He also translated Dante, and some of the plays of *Cervantes and *Calderón.

Schneider, Alan [Abram Leopoldovich] (1917–84), American director, born in Russia. He made his Broadway début in 1948, but though he directed Broadway hits he spent most of his career promoting works of less obvious commercial appeal, often outside New York and notably at the *Arena Stage, Washington. His pioneering work on *Beckett's plays was particularly outstanding, including *Waiting for Godot* (Miami, 1956), *Endgame* (1958), and *Happy Days* (1961). He also promoted *Pinter in the USA (*The Pinter Plays*, 1962, consisting of *The Collection* and *The Dumb-Waiter*; *The Lover*, 1964; *The Birthday Party*, 1967). He directed many plays by *Albee such as *Who's Afraid of Virginia Woolf?* (1962; London, 1964), *Tiny Alice* (1964), and *A Delicate Balance* (1966). Other American playwrights whose work he staged were Tennessee *Williams, Thornton *Wilder, and Robert *Anderson. He was killed in a road accident while working in England.

Schnitzler, Arthur (1862–1931), Austrian dramatist, a doctor by profession, who brought to his plays something of the dispassionate attitude of the consulting-room. His first work for the theatre was *Anatol* (1893), a series of sketches depicting the adventures of a young Viennese philanderer. This was followed by *Liebelei* (1896), in which a working-class girl kills herself on learning of the death of the young aristocrat who had been merely trifling with her affections. It was staged at the *National Theatre in London in 1986 as *Dalliance*, adapted by *Stoppard. *Der grüne Kakadu* (*The Green Cockatoo*, 1899) is a one-act play on an incident of the French Revolution in which Schnitzler handles with a sure touch the change from irresponsible make-believe to grim reality. *Reigen* (*The Round Dance*) was performed in Magyar in Budapest in 1912 and in the original German in Berlin and Vienna in 1921. The case brought against this linked sequence of 10 loveless sexual encounters on the grounds of obscenity was unsuccessful. Schnitzler however forbade all performances and it became well known only through a French film, *La Ronde* (1950). Immediately on the expiry of *copyright the play was given a number of productions, notably by the *RSC in London in 1982. Among Schnitzler's later plays are *Der einsame Weg* (*The Lonely Road*, 1904), a sensitive play of delicate half-lights staged at the *Old Vic in 1985; *Das weite Land* (1911), seen at the National Theatre in 1979 as *Undiscovered Country*, again adapted by Stoppard; *Der Ruf des Lebens* (*The Call of Life*, 1906); and *Professor Bernhardi* (1912), Schnitzler's only problem-play, in which he views from all angles the repercussions of an anti-Semitic incident in a Viennese hospital. Schnitzler's world vanished in the First World War, and the plays he wrote after 1918— *Komödie der Verführung* (*A Comedy of Seduction*, 1924) and *Der Gang zum Weiher* (*The Walk to the Lake*, 1925) are little more than nostalgic echoes of the past.

Schönherr, Karl (1867–1943), Austrian dramatist, whose powerful and realistic dialect dramas of Tyrolean peasant life won him widespread recognition. Writing under the influence of *Anzengrüber but with a closer affinity to *naturalism, his themes are the peasant's clinging to the soil, as in *Erde* (*Earth*, 1907), his bewilderment in the face of religious conflict, as in *Glaube und Heimat* (*Faith and Homeland*, 1910), and his defence of the Tyrol against Napoleon's armies, as in *Volk in Not* (*People in Need*, 1915). A doctor himself, Schönherr wrote about the medical profession in *Der Kampf* (*The Struggle*, 1920) and the psychology of pretence in *Der Komödiant* (*The Comedian*, 1924). A sound instinct for the theatre which did not hesitate to use the resources of melodrama where necessary was the basic element in his success.

School Drama, term applied to the academic, educational plays which appeared in all European countries during the Renaissance. Written under the influence of the humanists by scholars for performance by schoolboys, they were originally in Latin, a tradition which persisted longest in *Jesuit drama. Elsewhere they tended to slip quickly into the vernacular, and had some influence on the development of the non-academic, popular, and later professional drama.

There was a good deal of dramatic activity in English schools and colleges in the first half of the 16th century, and the first two regular English comedies—*Ralph Roister Doister* by Nicholas *Udall and *Gammer Gurton's Needle* probably by William *Stevenson—were given at Eton (or perhaps Westminster) and Christ's College, Cambridge, respectively. The tradition of an annual school production in English has survived in many schools, but the only one in Latin today is the Westminster play.

Schools of Drama. Until the present century entry into the theatrical profession was haphazard, the beginner usually joining an established company in the provinces. Some actors still join the profession without any formal training, but most go through a three-year course at a recognized drama school, of which there are some 30 in Great Britain, or study in a *university department of drama. The leading London schools include the London Academy of Music and Dramatic Art (LAMDA), founded in 1861 under the auspices of the Academy of Music; the Mountview Theatre School, established as an amateur theatre group in 1947; the Royal Academy of Dramatic Art (RADA), founded in 1904 at *Her (then His) Majesty's Theatre by Beerbohm *Tree; the Guildhall School of Music and Drama, now located at the Barbican Centre, founded in 1880 as the Guildhall School of Music and extending its coverage to drama in 1935; and the Central School of Speech and Drama, founded in 1906 in the Albert Hall by the actress **Elsie Fogerty** (1866–1945), mainly for the teaching of poetic speech, and now housed in the Embassy Theatre. In Scotland the most important school of drama is at the Royal Scottish Academy of Music and Drama, founded in 1950 mainly through the efforts of James *Bridie.

Apart from the recognized drama schools, for which the local authorities will give grants, there are a number of stage schools for children of which the best known was founded by Italia *Conti. These are privately operated and give stage training as well as a standard education. Of recent years many polytechnics and colleges of education have introduced drama courses, but in most cases they lack the expertise and facilities of the specialized schools. Training at any type of drama school does not of itself entitle a person to membership of British Actors' *Equity, a prerequisite of professional employment, which has its own methods of selection.

In the USA most actor training has now fallen within the sphere of the universities, sometimes, as at *Yale, in conjunction with fully professional theatre activities. A notable early school was the American Academy of Dramatic Arts, New York, founded in 1884 as the Lyceum Theatre School of Acting and receiving charters from the University of the State of New York in 1899 and 1952. The *Goodman Theatre, like many others, combined a resident professional company with a student training programme, though the drama school is now affiliated to a university. The Neighborhood Playhouse School of the Theatre in New York City, formally founded in 1928, grew out of efforts to enrich the lives of neighbourhood children. It offers a two-year apprenticeship under teachers who are theatre professionals. On the West Coast, the Pasadena Playhouse College of Theatre Arts was also opened in 1928, and eventually was allowed to grant academic degrees in theatre arts. The Actors' Studio, founded in 1947 by Elia *Kazan and others, is another important training ground for theatre professionals. Many actors have set up their own schools to teach acting and directing.

The leading and longest established French school is the Paris Conservatoire, which began in 1786 as the École de Déclamation, taking its present name in 1793. Among its first pupils was *Talma and a later pupil was *Samson, who returned to become one of its finest teachers. The school was reorganized after the student unrest of May 1968. The oldest and most important training establishment in the Soviet Union is the *Lunacharsky State Institute of Theatre Art (GITIS), founded in 1878. Originally both a music and drama school, it became a Conservatoire in 1886, and counted among its teachers *Nemirovich-Danchenko. In 1934, after many changes of name and status, it was given the name of the Soviet Union's first Minister for Education. Its pupils are drawn from some 40 different nationalities, half of them being external students already working in the theatre. In 1958 the Institute opened its own theatre building for practical work.

Schreyvogel, Josef, see VIENNA.

Schröder, Friedrich Ludwig (1744–1816), German actor, son of the actress **Sophia Schröder** [née Biereichel] (1714–92). He played as a child in the travelling company of his stepfather *Ackermann but in 1756, on the outbreak of war, became separated from it and for the next few years lived by his wits, becoming in the process an expert acrobat and rope-dancer. He eventually rejoined the company in Switzerland and resumed his acting career, but considered it very inferior to that of a tumbler. It was the advent of *Ekhof, then at the height of his powers, that made him realize what acting could be. During the next few years he studied and practised his art to such good effect that he gradually took over most of Ekhof's parts, leaving the older man no option but to withdraw. Following Ackermann's death in 1771, Schröder assumed artistic control of the company, his mother retaining financial control. At the head of an admirable company, which included his half-sisters Dorothea and Charlotte Ackermann, he was at the forefront of the new movement in Germany, responsible for the production of several important new plays, including *Emilia Galotti* (1772) by *Lessing and *Götz von Berlichingen* (1773) by *Goethe. His chief enterprise was the introduction of Shakespeare in action to young Germans who had previously met him only on the printed page. The adaptations of Shakespeare's tragedies, in which Romeo, Juliet, Cordelia, Ophelia, and even Hamlet survived, were made by Schröder

himself. He began with *Hamlet* in 1776, in which *Brockmann played the title-role and Schröder the Ghost (he later played Laertes, the Grave-digger, and Hamlet), and by 1780 11 of Shakespeare's plays had been performed, of which *Othello* was a failure and *King Lear* an outstanding success. To offset all this pioneer work Schröder continued to give his audiences a more conservative repertory of now-forgotten plays and monodramas, ballets and light musical pieces. The first glorious phase of his Hamburg management ended when, tiring at last of the constant friction with his mother over money matters, he went in 1782 with his wife **Christine Hart** (?–1829) as guest-artist to the Burgtheater in *Vienna, where he remained for four years. He exercised a salutary influence on his fellow actors, modifying their pompous ranting in tragedy and old-fashioned fooling in comedy, and may be said to have laid the foundations of the subtle ensemble playing which was later a distinguishing feature of the Burgtheater's productions. In 1786 he had once more to take over the company which his mother, at the age of 72, could no longer control. He again gave it an important position in German theatrical life, but the fire and enthusiasm of the earlier period were lacking and the energy that should have gone to the production of the last great plays of *Schiller and Goethe was dissipated on the trivialities of *Iffland and *Kotzebue. Nevertheless, once given a free hand financially, Schröder prospered, and in 1798 he was able to buy and retire to a country estate, where he died.

Schröder, Sophie (1781–1868), Austrian actress, who played leading parts at the *Vienna Burgtheater. She created the part of Bera in *Grillparzer's *Die Ahnfrau* (1817) and of Medea in his *Das goldene Vliess* (1821), and it was said that her playing of the title-role in his *Sappho* (1818) established the 'noble simplicity' which remained the hallmark of the Burgtheater's style of acting for many years. She was much admired in the plays of *Goethe, *Schiller, *Kleist, and particularly Shakespeare, her Lady Macbeth in 1821 eschewing violent outbursts and impressing the spectators with her tenacity of purpose and unshakeable resolution: the contemporary critic Heinrich Laube speaks of her 'ideal form' enlivened by a passionate temperament, her grace of movement and gesture, and the moving purity of her diction.

Schuster, Ignaz (1779–1835), Austrian actor, one of the most popular comic figures in the Viennese popular theatre. After a period with Karl *Marinelli's Baden company, he joined the Leopoldstädter company in *Vienna in 1801 and remained with it until his death. From 1805 onwards he appeared in local *Singspiele*. Schuster, who was small and deformed, did not achieve fame until 1813 when he created the part of *Staberl the umbrella-maker in Bäuerle's *Die Bürger in Wien*. Thereafter he was the hero of innumerable *Staberliaden*, many of them written by Karl Carl, and gradually replaced Anton Hasenhut in popular favour.

Schwartz, Evgenyi, see SHWARTZ.

Schwartz, Maurice (1889–1960), Jewish actor and director. Born in the Ukraine, he went to the United States as a child, and later appeared in Yiddish plays in several towns before joining a company in New York. In 1918 he was at the Irving Place Theatre, where in 1919 he directed a performance of a play by Peretz *Hirschbein which proved a turning-point in his career. His greatest achievement was his discovery of the work of Sholom *Aleichem, the quintessence of Jewish folk humour and characterization. Schwartz also introduced to the stage the works of a number of other Jewish writers, including Halper Leivick. In 1924 he undertook a tour of Europe which proved successful, and he returned to New York to open a theatre on Broadway for the production of European classics in Yiddish. This, however, failed, and in 1926 he opened a Yiddish Art Theatre on 2nd Avenue, the traditional home of New York Yiddish drama. He later toured South America and also visited Palestine where he worked with *Ohel. His repertory was extensive, comprising some 150 roles, and under his influence a number of Yiddish Art Theatres were founded in New York; but the widespread adoption of English by Yiddish families, and a slackening of Jewish immigration to the States, caused a decline in their audiences. In 1959 Schwartz went to Israel hoping to establish a Yiddish theatre there, but died after producing only one play.

Science Fiction Theatre of Liverpool, see EVERYMAN THEATRE, Liverpool.

Scioptikon, see LIGHTING.

Scissor Stage, see BOAT TRUCK.

Scofield, (David) **Paul** (1922–), English actor, perhaps the greatest of the post-*Olivier generation, with a particular gift for conveying moral worth without seeming priggish. He made his first appearance on the stage in 1940, and had played a number of parts when he first came into prominence with his portrayal of the Bastard in *King John* at the *Birmingham Repertory Theatre in 1945. He then went to the *Shakespeare Memorial Theatre for two seasons, his wide range of parts including Mephistophilis in *Marlowe's *Doctor Faustus* and the title-roles in *Henry V* and *Pericles*. His Young Fashion in *Vanbrugh's *The Relapse* in 1948 was followed by a return to Stratford, where he played Hamlet and Troilus in *Troilus and Cressida*. In 1949 he appeared as Alexander the Great in *Rattigan's *Adventure Story* and Konstantin in *Chekhov's *The Seagull*, and a

year later he was seen as the twin brothers Hugo and Frederick in *Anouilh's *Ring round the Moon*. After starring in Charles *Morgan's *The River Line* (1952), he gave a fine performance in 1953 as Pierre to *Gielgud's Jaffier in *Otway's *Venice Preserv'd*, offering in the same season at the *Lyric Theatre, Hammersmith, his Richard II and his Witwoud in *Congreve's *The Way of the World*. After two modern plays, Wynyard Browne's *A Question of Fact* (also 1953) and Anouilh's *Time Remembered* (1954), he went to Russia in 1955 to play Hamlet at the *Moscow Art Theatre with the Stratford company. The production was by Peter *Brook, who also directed him as the drunken priest in *The Power and the Glory*, based on Graham *Greene's novel, and as Harry in a revival of T. S. *Eliot's *The Family Reunion* (both 1956). In 1958 he made his début in a musical, *Expresso Bongo*, and after Greene's *The Complaisant Lover* (1959) was seen in probably his most famous role, Sir Thomas More in Robert *Bolt's *A Man for All Seasons* (1960; NY, 1961). Peter Brook directed him in *King Lear* in 1962 for the *RSC in a production which went from Stratford to London, several European countries, including Russia again, and New York. Later roles for the company included Timon in *Timon of Athens* at Stratford in 1965, Khlestakov in *Gogol's *The Government Inspector* and the homosexual barber in Charles Dyer's *Staircase* at the *Aldwych in 1966, and Macbeth at Stratford in 1967. He was seen in *Osborne's *The Hotel in Amsterdam* (1968), and gave a superb performance as Chekhov's Uncle Vanya at the *Royal Court in 1970. He was with the *National Theatre company in 1971 in *Zuckmayer's *The Captain of Köpenick* and *Pirandello's *The Rules of the Game*, and then took the leading role in Christopher *Hampton's *Savages* (1973), following it with Prospero in *The Tempest* in 1974 and Athol *Fugard's *Dimetos* (1976). After returning to the National Theatre in 1977 in *Jonson's *Volpone* and *Granville-Barker's *The Madras House*, he reappeared there in 1979 in *Shaffer's *Amadeus* as Salieri, in 1980 as Othello, and in 1982 as Oberon in *A Midsummer Night's Dream*. Nat in Herb Gardner's *I'm Not Rappaport* (1986) was his only other notable part in the 1980s.

Scott, Clement William (1841–1904), English dramatist and dramatic critic, who for nearly 30 years from 1872 reviewed plays for the *Daily Telegraph*, putting up a determined resistance to the new drama as typified by *Ibsen. His attack on *Ghosts* (1891) as 'a wretched, deplorable, loathsome history' was the outcome of his obstinate refusal to consider anything outside his own range of conventional morality. He was for many years editor of the *Theatre*, which he founded in 1877. After he withdrew in 1890 it distressed him by supporting his rival William

*Archer in his campaign in favour of modern drama. Scott was the author of a number of plays based on French originals and usually written in collaboration. The only one to survive is *Diplomacy* (1878), based by Scott and B. C. Stephenson on *Sardou's *Dora*, last revived in London in 1933.

Scottish Comedians, see LAUDER.

Scottish Community Drama Association, see AMATEUR THEATRE.

Scottish National Players, company which made its first appearance in Jan. 1921 in a Glasgow hall under the auspices of the St Andrew Society of Glasgow, presenting three new Scottish one-act plays. By the end of that year a further five one-act plays and two full-length ones had been produced. The company, which became independent in 1922, was wound up in 1934, two years after a proposal to turn professional had been rejected; but the Players continued independently, their last production (after a war-time interruption) being in 1947. A notable feature of their work was the summer camping tours, which brought drama to relatively remote parts of the country. Although the movement did not succeed in its aim of creating a Scottish National Theatre similar to the *Abbey Theatre, Dublin, it did call into being a body of plays which sought to reflect the variety of Scottish life, with the emphasis on the rural rather than the urban scene; among the 30 or so authors who wrote for the company were James *Bridie and Robert *Kemp. During the years 1921–31 the Players produced 62 new plays, their director in 1926 and 1927 being Tyrone *Guthrie.

Scribe, (Augustin) **Eugène** (1791–1861), French dramatist, originator and exploiter of the *well-made play. A prolific writer, he was responsible, alone or in collaboration, for more than 400 works, comprising tragedies, comedies, vaudevilles, and librettos for light opera. His early plays were failures and it was not until 1815 that he achieved fame with *Une nuit de la Garde Nationale*. Even more successful was *Un verre d'eau* (1850), translated into English as *A Glass of Water; or, Great Events from Trifling Causes Spring* (1863) and also as *The Queen's Favourite* (1883). The most successful of Scribe's plays, however, and the only one now remembered, was *Adrienne Lecouvreur* (1849), written in collaboration with Legouvé. The play, though historically incorrect, provided a fine part for *Rachel and later for *Bernhardt. In translation it was played by *Ristori, *Modjeska, and Helen *Faucit, among others. Scribe's plays, skilfully constructed with the utmost neatness, economy, and banality, came as a relief to a middle-class audience surfeited with the incoherence of the Revolution and the excess of the Romantics. Though his collaborators contributed much to the common stock, the stagecraft was his alone; with no depth or delicacy of perception, he had an uncanny flair for knowing what the public wanted, and how to give it to them with the maximum dramatic effect. Although he was to some extent unfairly blamed for all the shortcomings of the dramatists who succeeded him, he had an immense influence on them, particularly on *Labiche and *Sardou, and there is no doubt that his much-sought-after librettos, written for such musicians as Meyerbeer, Offenbach, and even Verdi (*Les Vêpres siciliennes*, 1855), helped to make French romantic opera a model of theatrical effectiveness.

Scrim Drop, see CLOTH.

Scruto, see TRANSFORMATION SCENE.

Scudéry, Georges de (1601–67), French dramatist, brother of the celebrated Madeleine de Scudéry, in whose novels he had a hand. He wrote in all 16 plays, of which the first, *Ligdamon et Licias; ou, La Ressemblance* (1630), which turns on the unlikely coincidence of two young men, not related, being as alike as identical twins, shows already the lively imagination, love of excessive rhetoric, and predilection for the unusual which was to be apparent in all his later works. The best of these were the tragedy *La Mort de César* (1635); the tragi-comedy *Le Trompeur puni* (1631), which *Floridor's company played at the *Cockpit in London in 1635; and the comedy *La Comédie des comédiens* (1635), which portrays on stage the company of which *Montdory was the leader, several parts being given the names of the actors playing them. Scudéry, who had been resentful at not being one of the five dramatists chosen to write *Richelieu's plays for him, was consoled a little by his election as a founder-member of the French Academy, an honour he had done little to deserve. He took an active part in the literary quarrel over *Corneille's *Le Cid* (1637) and had the pleasure of finding his own tragi-comedy *L'Amour tyrannique* (1638) praised above that of Corneille by Richelieu; but it is now forgotten.

Sea Rows, see GROUNDROW.

Seattle Repertory Theatre, Seattle, Wash., opened at the 895-seat Seattle Center Play House in 1963, its first season including *King Lear*, Max *Frisch's *The Firebugs*, Christopher *Fry's *The Lady's not for Burning*, and Arthur *Miller's *Death of a Salesman*. A financial crisis in 1970 (overcome by 1975) led to reduced emphasis on the classics. In 1983, the Bagley Wright Theatre, consisting of the 868-seat Mainstage and the 142-seat Poncho Forum, was erected at the Seattle Center to house the company. Three production schedules have been established: the Main Season, Oct.–May, on the Mainstage; the more intimate Stage Two; and the Other Season. The emphasis is on new works, and many plays originating at the Seattle Rep. have moved to Broadway and other venues. The company also tours extensively along the west coast. The Seattle Children's Theatre uses the Poncho theatre for its productions. A Contemporary Theatre, also in Seattle, was founded in 1965, and operates a 449-seat theatre with a thrust stage. At one time it worked in close co-operation with the Seattle Rep., but it now produces independently and also operates a Young ACT, founded in 1966, for younger audiences.

Sedley, Sir Charles (c.1639–1701), Restoration dramatist, wit, and man of letters, friend of Rochester and *Etherege. He wrote several plays, of which the best are *The Mulberry-Garden* (1668), a comedy of contemporary manners which owes something to *Molière's *L'École des maris* (1661) and something to Etherege's *Comical Revenge* (1664), and the lively but licentious *Bellamira; or, The Mistress* (1687), based on the *Eunuchus* of *Terence. Sedley was also the author of a dull tragedy on the subject of Antony and Cleopatra, written in imitation of *Dryden's heroic drama *All for Love* (1677).

Sedley-Smith, William Henry (1806–72), American actor, stage manager, and dramatist, who in his day was considered one of the best light comedians in the United States. Born in Wales, he left home at 14 to become an actor, adding Sedley to his own name of Smith. After touring the English provinces for some years he went to America in 1827, making his first appearance there at the *Walnut Street Theatre, Philadelphia. For many years he toured the larger American cities, and he was stage-manager of the Boston Museum, 1843–60. He is chiefly remembered today for his famous temperance drama *The Drunkard; or, The Fallen Saved* (1844), a melodramatic tract which achieved an astonishing success. After its first performance at the Boston Museum it was revived by Barnum at his *American Museum in New York in 1850, and became the first American play to run for nearly 200 performances. Frequently revived, it was put on again in Los Angeles in 1933 (at about the time prohibition ended) and ran for 26 years, notching up 9,477 performances.

Segelcke, Tore (1901–79), Norwegian actress, one of the leading Scandinavian players of her generation. After making her début at Det *Norske Theatret in Oslo in 1921, she went to Bergen in 1924, staying there for four years before returning to Oslo to join the *Nationaltheatret, where she remained. Her *Ibsen roles included Aase in *Peer Gynt*, Mrs Alving in *Ghosts*, Hilde Wangel in *The Master Builder*, and, supremely, Nora in *A Doll's House*, a performance which won her international fame. She was also powerful in Shakespeare and such *O'Neill roles as Nina Leeds in *Strange Interlude*

and Lavinia Mannon in *Mourning Becomes Electra*.

Semenova, Ekaterina Semenovna (1786–1849), Russian actress, whose main triumphs were in the tragedies of Shakespeare, *Racine (particularly *Phèdre*), and *Schiller. The daughter of a serf, she was sent at 10 years of age to the Theatre School in St Petersburg (Leningrad), and seven years later made her formal début, attracting attention by her acting in the somewhat frigid tragedies of **Vladislav Ozerov** (1770–1816). She made a great impression on her contemporaries by her beauty and her superb contralto voice, and *Pushkin dedicated some of his poems to her.

Seneca, Lucius Annaeus (c.4 BC–AD 65), Roman dramatist, philosopher, satirist, and statesman, the tutor, and later the victim, of *Nero. Nine tragedies adapted from the Greek are attributed to him—the *Hercules Furens*, *Medea*, *Phaedra* (or *Hippolytus*), and *Troades* (all possibly based on *Euripides), the *Agamemnon* (on *Aeschylus), the *Oedipus*, *Phoenissae* (or *Thebais*), and *Hercules Oetaeus* (on *Sophocles), and the *Thyestes* (on an unknown original). *Octavia*, based on the life of Nero's unhappy wife, was formerly attributed to Seneca, but is now considered not to be by him, though its author is still unknown.

As the only extant dramas from the Roman empire Seneca's tragedies are important historically, and their influence on the development of drama in modern times has been profound, in spite of the fact that they were *closet dramas. In spite of the fact that Seneca's alterations of his Greek models are usually for the worse, it would be unfair to deny the dramatic power of many of his scenes or the beauty of some of his choral passages, which offset the atmosphere of gloom and horror, brutality and treachery, which pervades these plays and reflects that of contemporary life. When Seneca writes of the intrigues of countries, the instability of princes, the crimes of tyrants, the courage of men in peril of death, there is something more than literary artifice and imagination: he had experienced it all. Perhaps that is why, for the Renaissance, Seneca was the model writer of tragedy. His Latin was easily understood, his plays were divided into the five acts demanded by Horace, their plots, however melodramatic, were universally intelligible, and even his rant and rhetoric appealed to the taste of the time. His line-by-line exchange of dialogue, his chorus, his tyrants, ghosts, and witches, his corpse-strewn stage, all reappear in Elizabethan drama, and even if such effects had already been used by dramatists before Shakespeare, they were reinforced by the reading, and possibly the acting, of translations of the tragedies as early as the 1550s, long before the publication of Seneca's 'Tenne Tragedies' in 1581. Their influence is already apparent in the earliest English tragedy, *Gorboduc* (1562) by *Norton and Sackville; in *Gascoigne's *Jocasta* (1566); in Shakespeare's early plays, particularly *Richard III* (*c.*1593) (which shows what splendid results can be achieved when Senecan material is used by a man of genius) and *Titus Andronicus* (*c.*1594); and in *Jonson's two tragedies *Sejanus* (1603) and *Catiline* (1611).

Serlio, Sebastiano (1475–1554), Italian painter and architect. He published a treatise on architecture, largely based on Peruzzi's notes and drawings, of which the second part, dealing with perspective in the theatre, appeared in 1545. It was published in an English translation as *The Second Book of Architecture* in 1611. Writing with temporary theatres set up in princely or ducal banqueting-halls in mind, Serlio described and illustrated three basic permanent sets, the tragic, the comic, and the satyric. These, with their symmetrical arrangement of houses or trees in perspective on either side of a central avenue, had an immense influence on scene design everywhere. They survived the introduction of the *scena d'angolo*, or diagonal perspective, by the *Bibienas, and traces of them can still be seen in the scenery of 19th-century *melodrama. Serlio was also a pioneer of *lighting, one section of his book dealing with the general illumination of the stage and theatre and the imitation of such natural phenomena as sunshine and moonlight.

Servandony, Jean-Nicolas (1695–1766), French scenic artist, born in Lyons, who in an effort to appear fashionably Italian changed the spelling of his name to Servandoni. After studying in Italy, he worked there and in other European countries, and was one of the first to adopt the neo-classic style in reaction against the universally popular baroque. He then settled in Paris, where he assumed control of the *Salle des Machines. The influence of the work he did there in a series of spectacular productions was soon apparent in Germany and even in Italy. Later he worked in Dresden and Vienna, and in 1749 was in London, where he married.

Set, the surroundings, visible to the audience, in which a play develops. Originally the phrase was 'set scene'—that is, an arrangement of painted and built components prepared or 'set up' in advance and revealed by the opening of a front scene, as opposed to a 'flat scene', where the *flats slid on and off stage in full view of the audience. An alternation of set and flat scenes was common in the English theatre until almost the end of the 19th century, and with the elaboration of *built stuff in Victorian times, a specially written front scene, the *carpenter's scene, was often provided to allow time for its erection. The word 'set' now covers everything arranged on the stage, ranging from the simplicity of a *curtain set to the detailed naturalism of a *box-set, and its derivative, 'setting', has

become the general term for the whole theatrical art of designing and staging the *scenery of a play. (See also MULTIPLE SETTING.)

Set Piece, *flat cut to the silhouette of, for example, a house or a fountain; also the name given to the solid three-dimensional elements of full scenery, also known as *built stuff.

Settle, Elkanah (1648–1724), Restoration dramatist, who began his theatrical career by staging *drolls at the London *fairs. He then turned to playwriting, his first play, *Cambyses, King of Persia* (1671), being put on at *Lincoln's Inn Fields Theatre. A second, *The Empress of Morocco* (also 1671), was a *heroic drama first presented at Court and revived in 1673 at *Dorset Garden with *Betterton in the leading role. It was the first English play to be published with scenic illustrations, which have provided valuable evidence on the theatre of the time. It was parodied in a farce produced at *Drury Lane towards the end of 1673. Settle wrote a number of other plays, mainly tragedies, now forgotten but successful enough in their day to enrage *Dryden, who considered his own popularity at Court endangered by that of Settle and satirized him as Doeg in *Absalom and Achitophel*. Towards the end of his life Settle returned to Bartholomew Fair, writing and acting in drolls at Mrs Minn's (or Myn's) booth, where he is also recorded as having played a 'dragon in green leather of his own invention'.

Set Waters, see GROUNDROW.

Seyler, Athene (1889–1990), English actress, outstanding in comedy. After training at the Royal Academy of Dramatic Art, where she won the Gold Medal in 1908, she appeared in the West End from 1909 in a wide variety of mainly modern roles before making her first venture into *Restoration comedy, at which she was to excel, playing Cynthia in a revival of *Congreve's *The Double Dealer* in 1916. It was followed by Melantha in *Dryden's *Marriage à la Mode* in 1920, Mrs Frail in Congreve's *Love for Love* in 1921, and Lady Fidget in *Wycherley's *The Country Wife* in 1924. In 1920 she also made her first appearance in Shakespeare, playing Rosalind in *As You Like It*. Her later roles in his plays included Titania and Hermia in *A Midsummer Night's Dream*, Beatrice in *Much Ado about Nothing*, Emilia in *Othello*, and the Nurse in *Romeo and Juliet*. She was also seen in plays by Oscar *Wilde (*The Importance of Being Earnest*, as Lady Bracknell and Miss Prism, *Lady Windermere's Fan*, and *A Woman of No Importance*); *Sheridan (Mrs Candour in *The School for Scandal* and Mrs Malaprop in *The Rivals*); *Shaw (*Candida*, Mrs Higgins in *Pygmalion*, and Mrs Mopply in *Too True to be Good*); and *Chekhov (Ranevskaya in *The Cherry Orchard*). She also featured in such modern plays as Lillian *Hellman's *Watch* on the Rhine (1942), Mary Chase's *Harvey* (1949), *Rattigan's *Who is Sylvia?* (1950), and John *Van Druten's *Bell, Book and Candle* (1954). In 1958 she was seen in Peter Coke's long-running *Breath of Spring*, and in 1962 she played an Old Bawd in John *Fletcher's *The Chances* at the first *Chichester Festival with unimpaired wit and vitality. She was last on the stage in 1966, as Martha Brewster in a revival of Joseph Kesselring's *Arsenic and Old Lace*.

She married as her second husband the actor **Nicholas Hannen** (1881–1972), who appeared with her in a number of productions, playing Oberon to her Titania and Benedick to her Beatrice. His first London success was as Nelson in *Granville-Barker's dramatization of Hardy's *The Dynasts* (1914), and his many roles ranged from Greek tragedy (Menelaus in *Euripides' *The Trojan Women*) to modern comedy (Samson Raphaelson's *Accent on Youth*). He made a big impression in Allan Monkhouse's *The Conquering Hero* (1924), which was followed by Philip Madras in Granville-Barker's *The Madras House* (1925). He was with the *Old Vic company on several occasions, including 1944–7, when he played the title-role in *Henry IV, Parts One and Two*.

Seyrig, Delphine (1932–90), Lebanese-born French actress, who became well known in the early 1960s through Resnais's film *L'Année dernière à Marienbad* and her performance as Nina in *Chekhov's *The Seagull*. The mysterious resonance and slightly foreign intonation of her voice, and the supple exoticism of her gestures, as well as her extremely sophisticated presence on stage, made her much sought after by directors for such various plays as *Pirandello's *Enrico IV* and *Non si sa come*, *Turgenev's *A Month in the Country* (1964), *Arrabal's *Le Jardin des délices* (1969), and *Handke's *Der Ritt über den Bodensee* (1973). She played Shakespeare's Cleopatra in 1976 and appeared in new British plays by *Pinter (*The Collection* and *The Lover*, 1965; *Old Times*, 1971), James *Saunders (*Next Time I'll Sing to You*, 1966), *Stoppard (*Rosencrantz and Guildenstern are Dead*, 1967), and *Ayckbourn (*Woman in Mind*, 1987).

Shadow-Show, form of puppetry in which flat, jointed figures are passed between a translucent screen and lighted candles or, nowadays, electric light bulbs, so that the audience, seated in front of the screen, sees only their shadows. It originated in the Far East, particularly in Java and India, and in an increasingly crude form spread to Turkey and so to Greece, where it gave rise to plays centred on the comic character Karaghiozis (in Turkish, Karagöz) which can still be seen in a rudimentary form in some Greek villages and in the back streets of Athens. As 'les Ombres Chinoises', shadow-shows were popular in Paris for about 100 years. In 1774 Dominique Séraphin opened a theatre devoted to them in Versailles, moving in 1784 to the Palais Royal, where his nephew continued his work until 1859. It was Séraphin who first introduced to Paris the classic shadow-play *The Broken Bridge*, in which a frustrated traveller indulges in an impassioned but silent argument with a workman on the other side of the river. This was well known in the streets of London, where, as the Galanty Show, shadow-plays continued to be given up to the end of the 19th century, usually in *Punch and Judy booths with a thin sheet stretched across the opening and candles behind. There was a literary revival of the shadow-show at the Chat Noir in Paris in the 1880s, and in the 1930s Lotte Reiniger employed the technique of the shadow-show for her animated films. Her puppets were made of tin, as were those used at the Chat Noir and in the English Galanty Show, but in Java and Bali, where the shadow-play survives in its traditional form, they are cut from leather. Manipulation is by bamboo rods or concealed wires running up the centre of the figure and operated from below the screen, except in Turkey and Greece, where the rod is held at right angles to it, and fastened to the flat figure in the centre of the back.

Shadwell, Thomas (c.1642–92), Restoration dramatist, whose first play *The Sullen Lovers; or, The Impertinents* (1668) was based on *Molière's *Les Fâcheux* (1661). The comedies which followed it seem, however, more indebted to *Jonson, whom Shadwell much admired. The best known of them is *Epsom Wells* (1672). Shadwell has been much criticized for his adaptation of *The Tempest* as an opera, *The Enchanted Island* (1674), in which, following the examples of *Davenant and *Dryden, everything was subordinated to the stage machinery and scenery. He then returned to comedy with *The Libertine* (1675) and *The Virtuoso* (1676) before rewriting *Timon of Athens* as *The Man-Hater* (1678), and in his last years produced two of his best comedies, *The Squire of Alsatia* (1688) and *Bury Fair* (1689), which give interesting though somewhat scurrilous pictures of contemporary manners. His last play, *The Volunteers; or, The Stock Jobbers*, was produced posthumously. It was ironic that Shadwell, who was mercilessly satirized by Dryden in *MacFlecknoe; or, A Satire upon the True-Blue Protestant Poet T.S.* (1682), should have succeeded him on political grounds as Poet Laureate in 1688.

Shaffer, Peter Levin (1926–   ), English dramatist, whose first play *Five Finger Exercise* (1958), a naturalistic study of family tensions directed by *Gielgud, had a great success in London (NY, 1959). It was followed by a double bill, *The Private Ear* and *The Public Eye* (1962; NY, 1963), with Maggie *Smith in London, which also did well, and *The Merry Roosters Panto*

(1963), written for Joan *Littlewood. Shaffer's next play, *The Royal Hunt of the Sun* (1964), an epic tragedy about the murder of the Aztec king Atahualpa by Pizarro, was first performed at *Chichester and then by the *National Theatre company and in New York, as was the one-act *Black Comedy* (1965), an ingenious farce in which most of the action is supposed to take place in the dark, although the stage is lit. In New York, and later in London, the latter was joined in a double bill with another one-acter, *White Lies* (NY, 1967; London, as *The White Liars*, 1968). In *The Battle of Shrivings* (1970) an elderly pacifist reminiscent of Bertrand Russell (played by Gielgud) was shown in conflict with an anti-liberal poet. Two later plays received high praise: *Equus* (1973; NY, 1974) deals with a stable-boy who blinds horses and his subsequent treatment by a psychiatrist; *Amadeus* (1979; NY, 1980) portrays the bitterness aroused in the composer Salieri by the genius of Mozart. After *Yonadab* (NT, 1985), based on an episode about two half-brothers taken from the Book of Samuel, he had another enormous hit with the comedy *Lettice and Lovage* (1987; NY, 1990), in which Maggie Smith played the fantasizing guide to a country house.

Shaffer's twin brother **Anthony** achieved an unexpected success with his first play *Sleuth* (London and NY, 1970) which his later plays, including *Murderer* (1975), did not equal.

Shaftesbury Theatre, London. 1. In Shaftesbury Avenue. This four-tiered theatre, seating 1,196, was the first to be built in the new Shaftesbury Avenue, and opened in 1888 with *Forbes-Robertson as Orlando in *As You Like It*, which was not a success; but E. S. *Willard did well in Henry Arthur *Jones's *The Middleman* (1889) and *Judah* (1890). In 1898 came the long-running musical *The Belle of New York*, and the first Negro musical *In Dahomey*, starring Bert *Williams, was seen in 1903, and ran for 251 performances. Seasons of Grand *Guignol and revivals followed, and in 1909 Cicely *Courtneidge made her London début in Lionel Monckton's *The Arcadians* under the management of her father Robert Courtneidge. In 1921 Clemence *Dane's *Will Shakespeare* was well received, and the theatre scored an immense success with *Tons of Money* (1922) by Will *Evans and Valentine, which brought together Ralph *Lynn, Robertson *Hare, Tom *Walls, and Mary Brough (see BROUGH, LIONEL), so laying the foundation of the future *Aldwych farces. The next year saw the London début of Fred Astaire and his sister Adèle in the musical farce *Stop Flirting*, but future productions were less successful; the last was a revival of Oscar Straus's *The Chocolate Soldier* (1940), and in 1941 the theatre was destroyed by bombs.

2. At the Holborn end of Shaftesbury Avenue. This theatre, seating 1,300 in three tiers, was built by the *Melville brothers to house their own *melodramas. It opened on 26 Dec. 1911 as the New Prince's Theatre, but the 'New' was soon dropped. After 1916 it had no settled policy, and its productions ranged from straight plays to ballet, pantomime, and opera. There were also revivals of *Gilbert and Sullivan during the 1920s, and distinguished foreign visitors included Sarah *Bernhardt in 1921, on her last visit to London; Diaghilev's Ballets Russes in 1921 and 1927; and Sacha *Guitry with Yvonne *Printemps in 1922. In 1924 *Darlington's *Alf's Button* had an unexpected success. Sybil *Thorndike and Henry *Ainley were seen in *Macbeth* in 1926, and in 1927 George *Robey appeared in the revue *Bits and Pieces*. A year later the *Gershwins' musical comedy *Funny Face* began a long run. Two dramatizations of stories by Edgar *Wallace, *The Frog* (1936) and *The Gusher* (1937), were popular, as were two musicals, *Wild Oats* (1938) and *Sitting Pretty* (1939). The theatre was badly blasted in 1940–1 but managed to stay open, and for a time housed the *Sadler's Wells ballet and opera companies. After the war the main successes were again the Gilbert and Sullivan seasons, but there were long runs of *His Excellency* (1950) by Dorothy and Campbell Christie and of two American musicals, *Pal Joey* (1954) and *Wonderful Town* (1955). In 1962 the theatre closed for renovation, and it reopened under its present name in 1963 with another American musical, *How to Succeed in Business without Really Trying*. In 1966 a farce by Philip King and Falkland Cary, *Big Bad Mouse*, began a long run, and it was followed by the epoch-making American rock musical *Hair* (1968), which just failed to complete its 2,000th performance in 1973 owing to the collapse of the auditorium ceiling. The theatre reopened in 1974 with a revival of the musical *West Side Story*. It was followed by a series of short runs which left the theatre in the doldrums until in 1980 the musical *They're Playing Our Song*, written by Neil *Simon, attracted large audiences.

In 1983 the theatre was taken over by the Theatre of Comedy, which still owns it. Its first production was Ray Cooney's *Run for Your Wife*, which continued its run at other theatres. The Shaftesbury later staged *Sondheim's *Follies* (1987) and David Henry Hwang's *M. Butterfly* (1989).

Shakespeare. 1. Life and Works, William Shakespeare (1564–1616), first son and third child of John Shakespeare, a glover yeoman of Stratford-upon-Avon, and his wife Mary Arden, was christened on 26 Apr. 1564; tradition asserts that his birthday was 23 Apr., St George's Day. John Shakespeare became an alderman in 1565 and in 1568 bailiff (that is, mayor). His position would have qualified William to attend the grammar school of his native town, but its archives have not survived and the first record of his activities is that of his marriage, evidently a hasty one, to a lady whom extant documents almost certainly (but not positively) identify as Anne Hathaway of Shottery. The wedding took place in 1582; a daughter was born in 1583, and twins followed in 1585. Another gap in our knowledge extends from this time until 1592, when a pamphlet written by the dying Robert *Greene shows Shakespeare evidently well established in London as actor and dramatist. In 1593 and 1594 he dedicated his poems *Venus and Adonis* and *The Rape of Lucrece* to the Earl of Southampton, in terms that suggest familiarity, and by the beginning of 1595 he had evidently become a *sharer in the company known as the *Chamberlain's Men. Evidence of his rise in the world appears in his father's successful application in 1596 to the Heralds' College for a coat of arms, and by the poet's purchase in 1597 of the large house known as New Place in Stratford, to which he was to retire in 1610. His will, signed on 25 Mar. 1616, preceded his death (23 Apr., his birthday) by about a month; tradition says he died after a too convivial evening with *Drayton and *Jonson. He was buried in the chancel of Stratford church.

Shakespeare came just at the right moment to make full and fresh use of the teeming drama of his time, finding a novel and flexible stage apt for his purposes and an eager audience representative of all classes to encourage and inspire. The man and the time were in harmonious conjunction. Starting to write probably about 1590, he contributed at least 36 plays to the theatre. Of these, 16 were printed in quarto during his lifetime, but, apart from the fact that some are obviously bad texts, surreptitiously obtained, the publishing conditions of the age make it probable that he himself did not read the proofs. In 1623 *Heminge and *Condell of the *King's Men, as the Chamberlain's Men were called after 1603, issued the entire body of his dramatic work in folio form; this volume presents the only texts of another 20 plays, and is probably the most important single volume in the entire history of literature. Arranging the contents under the headings of Comedies, Histories, and Tragedies, the editors of the Folio give no indication of the dates of composition of the separate items, but from a careful scrutiny of such external evidence as exists, and from 'internal' tests (the quality of the blank verse, use of prose and rhyme, etc.), most scholars are agreed, at least in general terms, concerning their chronology. The prefatory matter shows how highly Shakespeare was esteemed by his fellow actors and his great contemporary Jonson.

Shakespeare probably started his career by writing, unaided or in collaboration, a historical tetralogy consisting of the three parts of *Henry VI* (1591–2) and *Richard III* (1593). Richard

*Burbage won fame in the role of Richard, and the play was still being presented in 1633. The success of these plays no doubt encouraged Shakespeare to write *King John* (1594), based on an older two-part dramatization of that monarch's reign. This tragedy stands alone, but shortly afterwards, about 1595, another historical tetralogy was started with *Richard II*, which, during the Essex conspiracy in 1601, won notoriety because of its abdication scene; it was still in the repertory of the *Globe Theatre in 1631, and from the year 1607 comes an interesting record of its popularity, when it was produced on the high seas by sailors. The two parts of *Henry IV* (probably about 1597 or 1598) carry on the story of Henry Bolingbroke and introduce a richly contrasting comic element with the character of Falstaff (originally named after the historical Sir John Oldcastle), while the general theme is rounded off with *Henry V* (probably 1599). With this play Shakespeare closed his career as a writer of histories, save for the late *Henry VIII*, produced in 1613, in which he is believed by many scholars to have collaborated with the young *Fletcher.

Early in his career he tried his hand at comedy, experimenting with the courtly, satirical *Love's Labour's Lost* (1592), *The Comedy of Errors* (1593) in the style of *Plautus, and the more robust *The Taming of the Shrew* (1593/4). These plays convey the impression of a young dramatist unsure of his orientation, yet all are skilful and succeeded in holding the stage. *The Two Gentlemen of Verona* (c.1594) shows perhaps even less assurance. Little is known about the early stage history of those rich and lyrical plays written between 1595 and 1599, *A Midsummer Night's Dream*, *Much Ado about Nothing*, *As You Like It*, and *Twelfth Night*. Following the production of *Henry IV* (1597–8) comes *The Merry Wives of Windsor* (1600–1), an aberration in this series; probably tradition is right in saying it was written at the command of Queen Elizabeth, who wanted to see Falstaff in love. In *The Merchant of Venice* (1596–7) a break in the almost perfect balance observable in the other comedies is patent, and this leads to a couple of so-called 'dark comedies'—*All's Well that Ends Well* (1602) and *Measure for Measure* (given at Court in 1604), in which the romantic material is strained almost to breaking. With these may be associated the cynically bitter *Troilus and Cressida* (1602), which possibly was acted not on the public stage but privately.

These were composed at the same time as Shakespeare was reaching towards the deepest expression of tragic concepts. Already at the very beginning of his career he wrote (possibly in collaboration) *Titus Andronicus* (c.1592), a play which despite or because of its bloodiness remained popular. Again, in the midst of his lyrical comedies he made a second attempt at tragedy in *Romeo and Juliet* (about 1595). Then came the great series of tragedies and Roman plays. *Julius Caesar*, seen by a visitor to London in 1599, was probably the first, but *Hamlet* must have come very soon after. *Othello* may have been new when it was presented at Court in 1604; it was evidently very popular. *King Lear* followed not many months later; it appeared at Court in 1606, and about the same time came *Macbeth*, which, linked in theme with *Julius Caesar*, clearly addresses itself to a Jacobean Court. The classical subject-matter of *Julius Caesar* is paralleled in *Antony and Cleopatra*, in *Coriolanus*, and in *Timon of Athens*, which seems to have survived only in a draft (all probably c.1607 or 1608).

In *King Lear* Shakespeare had turned to ancient British history, and the atmosphere of this play is reproduced, albeit with a changed tone, in *Cymbeline*. Seen in 1611, it was written probably about 1610; there was a revival at Court in 1634. Another play seen in 1611, *The Winter's Tale*, is similar in spirit, darker than the early comedies of humour and including incidents reminiscent of the tragedies, yet ending with solemn happiness. Evidently popular, it had several Court productions. *The Tempest*, gravest and serenest of all the dramas, was presented at Court in 1611 and in the following year.

To Shakespeare have been attributed, in whole or in part, several other dramas. His hand in *The Two Noble Kinsmen*, which was printed in 1634 as by him and Fletcher, and in *Pericles*, printed as his in 1609 and added to the Third Folio of 1664, has generally been accepted. There may be some pages of his own writing in the manuscript of *Sir Thomas More*, dating probably from the mid-1590s. Less likely, although still possible, is his participation in *Edward III*, printed in 1596.

2. Production in English. Although some of Shakespeare's plays remained continuously in the repertory of the British theatre, from the Restoration until the end of the 19th century few people had an opportunity of seeing them in their original form. For this the change in *theatre buildings and theatrical technique was partly responsible, but the main onus lay on those who, while professing their admiration for Shakespeare, deliberately altered his texts to make them conform to the requirements of a new age. During the Commonwealth his comedies were pillaged to provide short entertainments or *drolls, such as that of 'Bottom the Weaver', taken from *A Midsummer Night's Dream*. *Macbeth* was revised and embellished with singing and dancing by *Davenant; his 'singing witches' were to last until 1847. *Romeo and Juliet* was sometimes played with a happy ending. Neither play was revived in its proper form until 1744. *The Tempest*, adapted by Davenant and *Dryden, was then made into an opera by *Shadwell. *A Midsummer Night's Dream* was combined with *masque-like episodes to music by Henry Purcell to become *The Fairy Queen*, and *Lacy made a new version of *The Taming of the Shrew*. In 1681 Nahum *Tate rewrote *King Lear*, omitting the Fool (who was not seen again until 1838), sending Lear, Gloucester, and Kent into peaceful retirement, and keeping Cordelia alive to marry her lover Edgar. He also tackled *Richard II* and *Coriolanus*, but with less success. An adaptation which survived even longer than Tate's *Lear* was the *Richard III* of Colley *Cibber. First given in 1700, and containing passages from *Henry IV*, *Henry V*, *Henry VI*, and *Richard II*, as well as a good deal of Cibber's own invention, it proved immensely popular and provided an excellent part for a tragic actor. Also popular were versions of other plays by Cibber, Shadwell, and later *Garrick. The last, though in some ways he tried to prune the excrescences of the Restoration texts and gave the first recorded performances of *Antony and Cleopatra*, retained Cibber's *Richard III*, gave Macbeth a dying speech, and caused Juliet to awake before the death of Romeo, giving them a touching final conversation. He even made short versions of four of the comedies, and his *Katharine and Petruchio* (1756) remained popular until well into the 19th century.

Garrick's tampering with *Hamlet*, including his omission of the Grave-diggers, marks the end of this phase in the treatment of Shakespeare's texts. In 1741 *Macklin had rescued Shylock, in the so-called *Jew of Venice*, from the hands of the low comedian. John Philip *Kemble reformed the costuming of Shakespeare's plays. His Othello was still a scarlet-coated general, his Richard III wore silk knee-breeches, and Lear defied the storm in a flowered dressing-gown. But, helped by his sister Sarah *Siddons, who was the first to discard the hoops, flounces, and enormous headgear of earlier tragic heroines, he made an effort to combine picturesqueness with accuracy. Edmund *Kean restored the original ending to *King Lear*, and the original plays gradually emerged. Mme *Vestris and the younger *Mathews revived *Love's Labour's Lost* and *A Midsummer Night's Dream* in 1839–40 with the original text, and the freedom of the theatres in 1843 enabled Samuel *Phelps to embark on his fine series of productions at *Sadler's Wells from 1844 to 1862, while Charles *Kean staged his equally remarkable Shakespeare seasons at the *Princess's.

Shakespeare was now presented in a reasonably correct form. *Macready should be credited with having restored the Fool to Lear, though played by a woman, and with having revived *The Tempest* without Dryden's interpolations, which included a male counterpart of Miranda. Phelps, who in 1845 restored a male Fool, and in 1847 put on *Macbeth* without the singing witches, as well as reviving *The Winter's Tale*, was more concerned with the text than with the scenery, which was pleasantly sober and unobtrusive. Elsewhere the newly restored texts were in danger of disappearing under the elaboration of detail, while the action of the plays, designed for an untrammelled stage, was constantly rearranged and held up because of the necessity for elaborate scene changes. Still in this tradition were Henry *Irving's productions at the *Lyceum (1878–1902) and Beerbohm *Tree's at the *Haymarket (1887–97) and later at *Her Majesty's.

The publication in 1888 of de Witt's drawing of the *Swan Theatre encouraged attempts to reproduce not only the text of Shakespeare's plays but also the physical conditions in which they were first seen. The main interest had already switched from the problem of the text to the problem of interpretation. This became even more important as the director gained the upper hand. To this was added the problem of providing a building suitable for Shakespeare, an approximation to the original Elizabethan stage. This was first tackled by William *Poel. Robert *Atkins's productions at the *Ring, Blackfriars, made an effort to solve the difficulties by presenting Shakespeare-in-the-round. After the Second World War the stages at the *Shakespeare Memorial Theatre, Stratford-upon-Avon, and the *Old Vic crept out beyond the proscenium arch, which Stratford finally abolished altogether. These and other developments showed that Shakespeare cannot adequately be presented in the proscenium-arch theatre, though even there efforts have been made to present the plays more simply and coherently, allowing the action to flow unchecked for as long as possible. One of the landmarks in the history of Shakespearian production was undoubtedly *Granville-Barker's first season at the *Savoy in 1912; another was the introduction in 1914 of the first season of Shakespeare at the Old Vic. Experiments in presenting the plays have been frequent, from Barry *Jackson's modern-dress *Hamlet* to the fantastications of *Komisarjevsky, the elaborations of *Reinhardt, and the challenges to tradition of Peter *Brook. There have been a *Lear* with Japanese décor, a *Hamlet* set in Victorian times, a *Romeo and Juliet* in modern Italian style by *Zeffirelli. But, with all its divergencies and aberrations, the main trend since 1900 has been the simplification of the background, by the use of a permanent set, a bare stage, or symbolic settings, and a consequent insistence on the importance of the text, the free flow of the verse, and the unhampered action of the plot. These ideals have been discernible behind most of the productions of the *RSC, diverse in style though these have been.

In America the first productions of Shakespeare were given by visiting British actors, and the situation was therefore much the same as in contemporary Britain. For instance, the first recorded play, *Richard III*, acted in New York City in 1750 by Thomas Kean and William Murray's company, was in Colley Cibber's version. Little information is available on the texts of subsequent productions, but it is reasonable to suppose that they were substantially those current in the English theatre at the time. Although there was not, as in Europe, a language barrier or a preconceived notion of classical writing to be overcome, there were moral difficulties. Even Shakespeare's reputation was not always sufficient to overcome deep-rooted prejudices against play-going, and *Othello* was first introduced to Boston as 'a Moral Dialogue against the Sin of Jealousy'. Probably more people in the 18th century were reading Shakespeare as a poet than seeing him as a playwright. Although the eastern cities had opportunities of seeing the full-length plays in theatres modelled on contemporary British lines, in the West it was the lecturers, elocutionists, entertainers, and showboat companies who first popularized Shakespeare, in isolated scenes and speeches. It may be said that from the earliest times Shakespeare played a large part in the emergent culture of the pioneer peoples.

After the first 50 years it is difficult to disentangle the imported productions of Shakespeare from those of the young but vigorous American theatre. Thomas Abthorpe *Cooper and Henry *Wallack, who both appeared in Shakespeare early in their careers, were typical of the new generation of actors who, though born in Britain, spent the latter and greater part of their working lives in America. The honour of being the first native-born actor to play leading roles in Shakespeare must probably go to John Howard *Payne, who as a youth of about 17 played Hamlet and Romeo in 1809, but the greatest was undoubtedly Edwin *Booth, whose Hamlet in 1864 was generally admired. The 19th-century personality cult of the 'star' actor and the insistence on elaborate trappings was as prevalent in America as in Britain, and was reinforced by the many tours undertaken by such London companies as Charles Kean's and later Irving's.

The 20th century saw, as in Britain, the gradual liberation of the play from over-decoration, with the simplified settings of Robert Edmond *Jones and Norman *Bel Geddes, and the new approach to a purified text. This was reinforced by a phenomenon peculiar to America—the invasion of the world of the theatre by the universities, which did not take place in Britain until much later. The proliferation of *university departments of drama whose syllabuses led to a degree in drama has not been without its dangers, but it has led to much scholarly work on the problems of Shakespeare and the Elizabethan theatre in general, and to a wider spread of interest in and productions of Shakespeare. This is particularly fortunate since the commercial theatre on Broadway has not on the whole been enthusiastic about Shakespeare's plays. With no tradition of Shakespearian acting, with the same handicap as in London of unsuitable theatres, and with no pressing demand from the audience, managers have found them expensive to stage and uncertain in their box-office returns. The exceptions have mostly been due to the efforts of an individual—John *Barrymore, Eva *Le Gallienne, Orson *Welles, Maurice *Evans, Margaret *Webster, Joseph *Papp. It is therefore the universities who have mainly kept Shakespeare alive on the American stage, either by incorporating his plays in the repertory of a community theatre, or by organizing festivals devoted solely to his works. The main dangers of academic Shakespeare are pedantry in the presentation and immaturity in the actors, but a healthy spirit of experiment may do much to redress the balance. It certainly seems as if the future of Shakespeare in the USA lies rather with the community and the *Off-Broadway theatres, and with the *Shakespeare Festivals, than with the commercial theatre.

Shakespeare Festivals. Festivals of Shakespeare's plays by professional companies are given in the two Stratfords—in England at the *Royal Shakespeare Theatre, and in Canada at the *Stratford (Ontario) Festival theatre. (There was a festival at Stratford, Conn., 1955–81; see AMERICAN SHAKESPEARE THEATRE.) Professional open-air festivals are also held in Central Park, New York City, under the auspices of Joseph *Papp, and in Regent's Park, London (see OPEN-AIR THEATRE). A new *Globe Theatre near the site of the original is due to open in 1992. (See WANAMAKER.)

The oldest Shakespeare festival in America, founded in 1935, is that held out of doors during the summer at Ashland, Oregon. Plays were first given by students in a roofless structure, which was damaged by fire in 1940. The festival was then suspended until 1947, when a new stage was built and professional actors were engaged. An Elizabethan-type theatre, the Old Globe, was designed for the 1935–6 California Pacific International Exposition, Balboa Park, San Diego, where a summer Shakespeare festival was inaugurated in 1949. The first actors were local amateurs, but in 1954 student actors and technicians from colleges and drama schools throughout the country were enrolled, and since 1959 the major roles have been played by professionals. The theatre, originally roofless like its predecessor, was later roofed over. After it was destroyed by fire in 1978 plays were performed on the Lowell Davies Festival

Theatre stage adjacent to the Old Globe, but a new Old Globe Theatre seating 581 was opened in 1982. The Shakespeare Theater at the Folger (see WASHINGTON) stages an Oct.–June season. There are about 30 Shakespeare festivals in the United States.

A Danish festival, devoted entirely to productions of *Hamlet*, was held in the courtyard of Kronborg Castle at Elsinore. It was inaugurated in 1937 with a performance by Laurence *Olivier and the *Old Vic company. In 1938 there was a German company headed by *Gründgens, and in 1939 another British company under John *Gielgud. After the Second World War efforts were made to revive it, and Swedish, Norwegian, Finnish. American, and Irish Hamlets were seen, Michael *Redgrave heading a third British company in 1950. After 1954, however, there were no further productions until the *Prospect Theatre Company's visit in 1978, with Derek *Jacobi in the title-role.

There are a number of amateur festivals in England, of which the most important is probably that held in London in the week nearest to Shakespeare's birthday (23 Apr.), which includes a production in the yard of the George Inn, Southwark, a site which more closely than any other appears to resemble the *innyards in which London companies appeared before the erection of the *Theatre in 1576.

Shakespeare   Globe   Museum,   see WANAMAKER.

Shakespeare in Translation. During the first quarter of the 17th century the *English Comedians, travelling on the Continent, included in their repertory cut versions of several of Shakespeare's plays. Otherwise he remained virtually unknown outside England until well into the 18th century, when *Voltaire first drew attention to him in his *Lettres philosophiques* (1734). Early translations of his plays, mostly incomplete and bowdlerized, did little to help. The first versions to be widely read, though not acted, were French prose versions by **Pierre Le Tourneur** (1736–88). These were the only ones known in Italy also until the publication in 1819–22 of Italian versions by Michele Leoni, which, together with those of La Place, were read by the novelist Manzoni, the first Italian to be influenced by Shakespeare. They also provided *Ducis, who knew no English, with the basis of some of the first French stage versions used by *Molé. These were tailored to fit the *unities, so that Desdemona in *Othello*, for instance, was wooed, wedded, and murdered in the space of 24 hours, while *Hamlet, Romeo and Juliet*, and *King Lear* had the happy endings being given to them in contemporary productions in London. In Germany it was the great actor *Schröder who first put Shakespeare on the stage with a production of *Hamlet* at Hamburg in 1776. It was successful enough to encourage Schröder to put on *Othello*, *The Merchant of Venice*, and *Measure for Measure*, which were all failures, and finally *King Lear*, which was an unqualified success. Good acting versions of other plays followed, including some by Eschenburg (the Mannheim Shakespeare) and by A. W. von *Schlegel, completed by *Tieck and others. Between 1869 and 1871 good acting versions were produced by the actor Emil *Devrient, and Shakespeare became so popular that he was acclaimed as *Unser* ('Our') Shakespeare, and his genius was thought to be akin to that of modern Germans. Romanticism brought him into favour in France, as did the publication of his complete works translated by François Victor Hugo, younger son of Victor *Hugo, from 1856 to 1867.

In Italy Shakespeare established himself more slowly, but it was Italian actors and actresses who proved themselves the finest players of his great tragic roles. The first Othello was Ernesto *Rossi, who went on to play Macbeth, Lear, Coriolanus, Hamlet, and Shylock, while Adelaide *Ristori scored a triumph as Lady Macbeth. In spite of fine productions of *The Taming of the Shrew*, with Ermete *Novelli, and an outstanding *A Midsummer Night's Dream* with Mendelssohn's music in Rome in 1910, the predilection of the Latin genius for his tragic side was shown again in the triumph of *Gide's translation of *Hamlet* in 1946, ably interpreted by *Barrault.

Spain was slow to react to Shakespeare; even now there are only a couple of complete translations. These are used also in South America, except in Brazil, where Portuguese translations are current. The first Greek translator was Demetrios Bikelas, some of whose versions, made between 1876 and 1884, were acted in Athens.

There is probably no country in the world today where some knowledge and appreciation of Shakespeare is not to be found. The first translations in central Europe were probably based on German originals, but it was not long before the desire to read him in the original, and so fathom the mystery of his universal appeal, caused many admirers to learn English for his sake. In Russia *The Merry Wives of Windsor* was translated by Catherine the Great in 1786 as *What it is Like to have Linen in a Basket*. The main Russian theatres have some Shakespeare plays in their permanent repertory, the most popular being *Hamlet*. Shakespeare's plays are no longer acted in India by Europeans for Europeans, as they once were, but there are indigenous versions of some of the plays. Japan has also adapted Shakespeare's plays to the conventions of its own theatre. Some of the local re-creations of well-known Shakespearian plots in other settings, as in the Zulu *Macbeth*, emphasize that with Shakespeare nothing is impossible!

Shakespeare Memorial Theatre, Stratford-upon-Avon, Warwickshire. This theatre, devoted to the production of plays by Shakespeare, stood on a riverside site in his birthplace, donated by Charles Edward Flower, member of a local family of brewers. A bright-red brick building in a pseudo-Gothic style, it opened in 1879 on 23 Apr. (Shakespeare's birthday), and attracted a good deal of adverse criticism on account of its gabled and turreted exterior, bare interior, and inadequate stage. It was, however, destined to house many fine productions with outstanding actors during the annual festival of Shakespeare's plays, which from 1886 to 1919 were directed mainly by Frank *Benson, and afterwards by W. *Bridges-Adams, and even to gain the affection of some of those who visited it regularly, until in 1926 it was destroyed by fire, leaving the library and picture gallery, added in 1883, still standing, though badly damaged. The company moved to a local converted cinema while plans were put in hand for a new theatre, on the same site but with an extension into the adjoining Bancroft gardens. The shell of the old theatre was converted into a conference hall, now used for rehearsals. Much of the money needed to build a new theatre came from the USA, and the moving spirit of the appeal was again a Flower—Sir Archibald.

The new building, designed by Elizabeth Scott, grandniece of the architect Sir Gilbert Scott, opened on 23 Apr. 1932. It was purely functional both inside and out, with high windowless walls, a fan-shaped auditorium seating about 1,500, and a wide stage. Again it caused widespread controversy; just as the first theatre had been dubbed 'a wedding cake', so the second was dismissed as 'a factory' or 'a tomb'. The actors suffered from cramped conditions backstage and from the distancing effect on their performances of the large orchestra pit.

Two years later Bridges-Adams retired, after extending the annual season from three or four weeks to five months and inviting *Komisarjevsky to direct several plays, including a controversial production of *Macbeth* with aluminium screens and vaguely modern uniforms. He returned under Bridges-Adams's successor Ben Iden Payne, who introduced dramatists other than Shakespeare into the programme—*Jonson for the tercentenary of his death in 1937, *Goldsmith in 1940, *Sheridan in 1941. This policy continued until 1946, since when Shakespeare has reigned virtually supreme. A full programme was maintained during the Second World War, Payne being succeeded by Milton Rosmer in 1943 and Robert *Atkins in 1944; under the latter the forestage was carried out over the orchestra pit, with a welcome gain in contact between actors and audience. In 1945 Barry *Jackson took over and initiated a number of reforms, including the spacing out of first nights over the whole season instead of crowding them all into the first fortnight and the appointment of a different director for each play instead of a resident director for the season. Improvements were made both in the auditorium and backstage, including the enlargement and refitting of the workshops. Jackson also invited promising youngsters to join the company, including Paul *Scofield and Peter *Brook, and in 1948 Robert *Helpmann appeared as King John, Shylock, and Hamlet. In the autumn of that year Anthony *Quayle took over as director, and under him a number of leading players, including Peggy *Ashcroft, John *Gielgud, Diana *Wynyard, and Michael *Redgrave, appeared in a series of brilliant productions. The front curtain was removed, thus further integrating stage and auditorium. Glen *Byam Shaw succeeded Quayle in 1956, after being co-director for some years. Overseas touring, which began with tentative visits to North America and Australia before 1939, increased after the Second World War, and there were visits to Moscow in 1955 and Leningrad in 1958.

The formation of the *RSC in 1961, with Peter *Hall as director, began a new era, with the company appearing not only at Stratford, where the theatre was renamed the *Royal Shakespeare Theatre, but also in London, at the *Aldwych Theatre and, from 1982, the *Barbican Theatre.

Shakespeare Theater at the Folger, see WASHINGTON.

Shank, John (?–1636), English actor, who appears in the actor-list of Shakespeare's plays, and whose name is found in many different spellings. He apparently began his career with *Queen Anne's Men, though no proof of this has yet been found. He is known to have been with the *King's Men, joining them either shortly before, or at the time of, the death in c.1611 of *Armin, whose position as chief clown he inherited. He was a comedian, well thought of as a singer and dancer of *jigs, and appears to have been very popular with his audience; though the written lines of his roles are often few, it seems that he was allowed considerable licence in gagging. From the number of boy-apprentices with whom his name is connected, and from the fact that so many young men are recorded in the parish burial register as having died at his house in Cripplegate, it has been inferred that he undertook the training of apprentices who lodged with him.

Shared Experience, see FRINGE THEATRE.

Sharers, the name given in the Elizabethan theatre to those members of a company who owned part of the wardrobe and playbooks, as distinct from the apprentice, who was not paid, or the hired man, who was on a fixed wage.

Those who also owned shares in the actual building were known as *housekeepers.

Sharing System. All the English provincial companies in the first half of the 18th century worked on shares, though somewhat differently from the Elizabethans. After the expenses of the night had been paid the profits, even what remained of the candles, were shared among all the members of the company equally, except that the manager took four extra parts known as dead shares for expenses such as scenery and wardrobe. Every night an account was made up and the money paid out, the younger and less experienced members getting the same as the leading actors. Although this 'commonwealth' system was good in many ways it was open to abuse by unscrupulous managers, for the accumulation of important bills which the management owed, known as the stock debt, sometimes figured among the expenses long after it had been paid off. Such abuses led to the system falling out of favour in the mid-century and being replaced by the payment of salaries, which varied in proportion to the importance of the player and the affluence of the manager but were always very low in comparison to London salaries. The actors often depended more on their *benefit performances, of which they had at least one every year, than on their shares or their salaries.

Shatrov, Mikhail, see VAKHTANGOV THEATRE.

Shaw, George Bernard (1856–1950), Irish dramatist, critic, and social reformer, born in Dublin. His mother, a singer and teacher of singing, early imbued him with a knowledge and love of music which qualified him, after migrating to London in 1876, to become music critic of the *Star* (under the pseudonym of Corno di Bassetto) from 1888 to 1890 and of the *World* from 1890 to 1894. Admirable though his music articles were, their quality was undoubtedly surpassed by that of the dramatic criticism he contributed to the *Saturday Review*—where his successor was Max *Beerbohm—between Jan. 1895 and Dec. 1898. His interest in social and political reform had led him in 1884 to join the Fabian Society, on whose behalf he soon became a fluent and effective speaker. Though he had no particular love for the theatre of his own time, he was not slow to recognize its value as a platform. This, and his admiration for the 'new drama' of *Ibsen, which in translations by Shaw's friend William *Archer was becoming known in London, led him to the writing of plays. The production in 1892, under the auspices of Grein's Independent Theatre Club, of *Widowers' Houses* (begun in 1885) inaugurated his long career as the foremost dramatist of his day. This was a private production for Club members, as were the productions in 1902 and 1905 of *Mrs Warren's Profession*, about prostitution, and *The*

Philanderer (both written in 1893). The first of Shaw's plays to be presented publicly was *Arms and the Man*, a satire on militarism, produced at the Avenue (later *Playhouse) Theatre in 1894 as part of a repertory season run by Florence *Farr, who had played in the first production of *Widowers' Houses*. The others ran into trouble with the censor and, taken in conjunction with Shaw's lectures and writings on behalf of the Fabians, caused him to be regarded in many quarters as a subversive influence. Deliberately disregarding the current conventions of the *well-made play, he set out to appeal to the intellect and not the emotions of his audiences, and introduced on stage subjects such as slum landlordism, prostitution, war, religion, family quarrels, health, economics, and the position of women. Thought, not action, was the mainspring of the Shavian play; but, as audiences were eventually to realize, it was thought seasoned by wit and enlivened by eloquence. The popularity of Shaw's plays dates from the seasons of 1904–7 at the *Royal Court Theatre, run by *Granville-Barker and J. E. Vedrenne, when 10 were performed in repertory—*Candida, John Bull's Other Island, How He Lied to Her Husband, You Never Can Tell* (first publicly produced at the *Strand in 1900), *Man and Superman* (without Act 3, produced separately in 1907 as *Don Juan in Hell*), *Major Barbara, The Doctor's Dilemma*, about medical ethics, *Captain Brassbound's Conversion*, with Ellen *Terry, *The Philanderer*, and the one-act *The Man of Destiny*. When Vedrenne and Barker moved to the *Savoy Theatre at the end of 1907, they produced there *The Devil's Disciple*, first seen in New York with Richard *Mansfield as early as 1897, and *Caesar and Cleopatra*, first performed in 1906 in German in Berlin under Max *Reinhardt and, in English, in New York the same year.

The chronology of Shaw's plays is complicated by the fact that many of them had *copyright, private, amateur, or foreign professional productions before being seen in London. His most important works after *Misalliance* (1910) are *Fanny's First Play* (1911; NY, 1912); *Androcles and the Lion* (1913; NY, 1915); *Pygmalion*, first seen in Vienna in 1913, in which Mrs Patrick *Campbell played Eliza Doolittle in 1914 in both London and New York; and *Heartbreak House* (NY, 1920; London, 1921), a 'fantasia' in the Russian (i.e., Chekhovian) manner first produced by the *Theatre Guild, as was the 'metabiological Pentateuch' *Back to Methuselah* (1922; London, 1924). *Saint Joan* (NY, 1923), Shaw's finest play in the opinion of most critics, was seen in London in 1924 with Sybil *Thorndike as an unforgettable Joan, and *The Apple Cart* (1929) was brought to London from the *Malvern Festival, as were *Too True to be Good* (1932), *Geneva* (1938), and Shaw's last memorable full-length play *In Good

King Charles's Golden Days* (1939). Subtitled by Shaw 'a true history that never happened', this brings together Charles II, Isaac Newton, and other historical characters.

It would be impossible to list all the revivals and translations of Shaw's major plays. *Pygmalion, Major Barbara*, and *Saint Joan* are the most popular, followed by *Candida, Man and Superman, Arms and the Man, The Doctor's Dilemma*, and *You Never Can Tell*. Many of the first productions, particularly of the earlier plays, were directed by Shaw himself, who earned the respect and admiration of his actors and might have been a successful actor himself. He also supervised the printing of his plays with meticulous attention to layout and typography, his detailed stage directions forming a running commentary which helps the reader to visualize the scene; his prefaces to the published plays enlarge on their arguments.

Shaw, Glen Byam, see BYAM SHAW.

Shaw Festival, at Niagara-on-the-Lake, Ontario, Canada, annual event inaugurated in 1962. Originally amateur, it became professional in 1965, and although intended for the production of plays by Shaw only—the opening productions being *Candida* and the 'Don Juan in Hell' scene from *Man and Superman*—it now presents also plays by Shaw's contemporaries around the world. A new proscenium theatre seating 861 opened in 1973, and the season, which at first lasted for eight nights only and later in the 1960s for 3–6 weeks, now extends from mid-Apr. to mid-Oct. Currently the company is predominantly Canadian, but in the 1970s guest actors included Jessica *Tandy, Micheál *MacLiammóir, and Ian *Richardson.

Shaw Theatre, see NATIONAL YOUTH THEATRE OF GREAT BRITAIN.

Shchepkin, Mikhail Semenovich (1788–1863), Russian actor, son of a serf, who in his youth played small parts in private amateur theatricals in Kursk, his birthplace, and also appeared at the Kursk theatre during several summer seasons. In 1808 he took over a number of leading comic roles there, which he continued to play until the company was disbanded in 1816. A movement was then set on foot by Prince Repin, one of his admirers and patrons, to purchase his freedom, and in 1822 he was invited to join the company at the *Maly Theatre in Moscow, where he made his début in Nov. 1822 in Zagoskin's comedy *Gospodin Bogatonov; or, A Provincial in the Capital*. From 1825 to 1828 he was in St Petersburg, but it was in Moscow that he did his best work, particularly during the 1840s, when *Griboyedov's *Wit Works Woe* and *Gogol's *The Government Inspector* provided him with two of his best parts. He was also noted for playing a number of minor comic characters in *Molière and Shakespeare, though sometimes, particularly in his later years, he was criticized for too boisterous buffoonery. Nevertheless his influence was far-reaching, and he was recognized by *Stanislavsky as the founder of the tradition of realistic acting on the Russian stage, his synthesis of *Mochalov's passion and *Karatygin's technical perfection creating a style well suited to the plays of *Ostrovsky, *Chekhov, and *Gorky.

Sheffield, Yorkshire, see CRUCIBLE THEATRE.

Sheldon, Edward Brewster (1886–1946), American dramatist, who trained under George Pierce *Baker and had his first play *Salvation Nell* (1908) produced when he was only 22, with Mrs *Fiske as the heroine. It was a great success and Sheldon was immediately hailed as the rising hope of the new American school of realistic dramatists, a position he appeared to consolidate with *The Nigger* (1909), a courageous handling of the Negro problem, and *The Boss* (1911), a study of modern industrial conditions. But all his serious work was overshadowed by the immense popular success of his romantic play *Romance* (1913), which with **Doris Keane** (1881–1945) as an Italian opera singer had long runs in London and New York. It made an immense emotional appeal to audiences all over the world, and was translated into French and other languages. Handicapped by serious illness, Sheldon did most of his later work in collaboration, translating and adapting a number of popular successes, one of the best being a version of Benelli's *La cena delle beffe* (1909) as *The Jest* for John and Lionel *Barrymore in 1919.

Shelley, Percy Bysshe (1792–1822), English Romantic poet, and author of several plays in verse, of which the best known is *The Cenci*. Published in 1819, this was first acted by the Shelley Society in 1886. It has been several times revived, notably in 1922 and 1926 with Sybil *Thorndike, and in 1959 at the *Old Vic with Barbara *Jefford. Though pure poetry, it is poor drama, being confused in action and in style somewhat too indebted to Shakespeare.

Shepard, Sam [Steve Shepard Rogers] (1943–), prolific American dramatist, whose plays, influenced by popular culture, reflect contemporary America, including the violent conflict between the dreams of the early American pioneers and modern industrialization. His early one-act plays, however, such as *Cowboys* (1964) and *Icarus's Mother* (NY, 1965; London, 1970), were poetic images lacking formal structure, his first full-length play being *La Turista* (NY, 1966; London, 1969). In the early 1970s he lived for some years in London, where his work included one of his best plays *The Tooth of Crime* (London, 1972; NY, 1973) about an ageing rock star who is displaced by a rival and

commits suicide. *The Curse of the Starving Class* (London, 1977; NY, 1978) portrays the selfishness and indifference within a poor farming family in California; while the *Pulitzer Prize-winning *Buried Child* (NY, 1978; London, 1980), also set on a farm, depicts a family in an even more advanced state of disintegration. *Seduced* (NY, 1979; London, 1980) features a dying recluse based on Howard Hughes, and *True West* (NY, 1980; *Cottesloe, 1982) is about rival Californian brothers, one a successful Hollywood script writer and the other a petty crook. Both *Fool for Love* (NY, 1983; Cottesloe, 1984) and *A Lie of the Mind* (NY, 1985; London, 1987) portray sexual conflict; the first a stormy encounter in a desert motel room between a man and his half-sister and former mistress, the second the aftermath of an attempted wife-murder. He is also a film actor and script writer.

Sher, Antony (1949–), South African-born actor, who trained in London and then worked at the *Everyman, Liverpool, *Nottingham Playhouse, and the *Royal Lyceum Theatre, Edinburgh; in the West End; and at the *National Theatre (Sam *Shepard's *True West*, 1982). After joining the *RSC as an associate artist he achieved a high reputation with a series of dangerous and unorthodox interpretations of Shakespeare: the Fool in *King Lear* (1982), Richard III (1984), Malvolio in *Twelfth Night*, and Shylock in *The Merchant of Venice* (both 1987). Other roles for the RSC included the friar in Peter Barnes's *Red Noses* (1985), set during the Black Death, and the title-role in Peter Flannery's *Singer* (1989), about a Polish Jew who survives Auschwitz to become a slum landlord in post-war Britain. He scored a West End success in 1985 as a drag queen in Harvey Fierstein's three linked plays *Torch Song Trilogy*.

Sheridan, Mark, see MUSIC-HALL.

Sheridan, Richard Brinsley (1751–1816), English dramatist, theatre manager, and politician, whose best play, *The School for Scandal*, stands as the masterpiece of the English comedy of manners, with all the wit but none of the licentiousness of the *Restoration comedy from which it derives. Sheridan, the son of an actor, **Thomas Sheridan** (1721–88), and a writer, **Frances** [née Chamberlaine] (1724–66), whose plays *The Discovery* and *The Dupe* were produced by *Garrick (both in 1763), was born in Dublin, educated at Harrow, and intended for the law. But after contracting a romantic marriage with a singer he settled in London and had his first play, *The Rivals*, produced at *Covent Garden in 1775. This was followed later in the same year by a farce, *St Patrick's Day; or, The Scheming Lieutenant*, and a comic opera, *The Duenna*. In 1776 Sheridan bought Garrick's share in *Drury Lane Theatre, which he rebuilt in 1794, remaining in charge there until its destruction by fire in 1809, always in financial difficulties. His later plays, all produced at Drury Lane, include *A Trip to Scarborough*, altered from *Vanbrugh's *The Relapse*, and *The School for Scandal* (both 1777); *The Critic; or, A Tragedy Rehearsed* (1779), the best of the many burlesques stemming from Buckingham's *The Rehearsal* and the only one to have been constantly revived; and in 1799, when he had practically deserted the theatre for politics, *Pizarro*, an adaptation of a popular drama by *Kotzebue. Sheridan was also part-author of several entertainments and wrote the pantomime *Robinson Crusoe* for Drury Lane in 1781 as an *after-piece to *The Winter's Tale*. He exploited to the full the popular taste for spectacle and pantomime, helped by the scenic artist Philip de *Loutherbourg, and all his plays were produced with remarkable scenic effects and lavish costumes. His management of Drury Lane was marked by a succession of quarrels with the managers of the smaller theatres—*Astley's, *Sadler's Wells, the new *Royalty—whose success alarmed him, and he was often instrumental in embroiling his rivals with the authorities. His last years were unhappy, and he never recovered from the destruction of Drury Lane.

Sheridan Square Playhouse, see CIRCLE REPERTORY COMPANY.

Sherman Theatre, see CARDIFF.

Sherriff, R(obert) **C**(edric) (1896–1975), English dramatist and novelist, who became widely known for his realistic and moving play *Journey's End* (1928), the first to deal successfully in the theatre with the First World War. Originally produced on a Sunday evening by the *Stage Society, it was taken into the commercial theatre by Maurice *Browne, who played Lieutenant Raleigh, the part of Captain Stanhope, first played by Laurence *Olivier, being taken over by Colin Clive; also in the all-male cast was Robert *Speaight. Dealing with the reactions of a small group of men in a dug-out just before an attack, it made an immediate impact, and was subsequently translated and played all over the world, being first seen in New York in 1929. Among Sherriff's other plays are *Badger's Green* (1930), on village cricket; *St Helena* (1935; NY, 1936), written in collaboration and dealing with the last years of Napoleon; *Miss Mabel* (1948); and two plays in which Ralph *Richardson gave impressive performances—*Home at Seven* (1950), a study of amnesia, and *The White Carnation* (1953). *The Long Sunset* (1955) was set towards the end of the Roman occupation of Britain; and *The Telescope* (1957) provided the basis for a musical, *Johnny the Priest*, seen in 1960, in which year Sherriff's last play *A Shred of Evidence* was also produced.

Sherwood, Robert Emmet (1896–1955), American dramatist, who scored a success with his first play *The Road to Rome* (1927; London,

1928), a satirical treatment of Hannibal's march across the Alps which deflated military glory. It was followed by *The Love Nest* (also 1927), based on a short story by Ring Lardner, and *The Queen's Husband* (1928; London, 1931), which drew an amusing portrait of a henpecked king. *Waterloo Bridge* and *This is New York* (both 1930) were failures, but with his next play, *Reunion in Vienna* (1931; London, 1934), brilliantly interpreted by the *Lunts, Sherwood again achieved success. With *The Petrified Forest* (1935; London, 1942), about an idealist's virtual suicide, he began to take cognizance of the rapidly deteriorating world situation, and although there was a light-hearted interval with *Tovarich* (London, 1935; NY, 1936), based on a play by Jacques Deval, in the *Pulitzer Prize-winner *Idiot's Delight* (1936; London, 1938) his ironic pessimism grew darker; in it he foretold a Second World War (he had fought in the first one), and the intellectual bankruptcy of Western civilization. *Abe Lincoln in Illinois* (1938) showed Lincoln as a man of peace who entered the political arena reluctantly, and paralleled contemporary political struggles with those of his day. *There Shall be no Night* (1940; London, 1943), written in response to the invasion of Finland, showed a pacifist scientist choosing war as preferable to slavery. After a few years of political activity, during which he wrote no new plays, Sherwood returned to the theatre with *The Rugged Path* (1945). His last play, *Small War on Murray Hill*, a mildly romantic comedy about the American Revolution, was produced posthumously in 1957.

Shields, Ella, see MUSIC-HALL.

Shiels [Morshiel], George (1881–1949), Irish dramatist, who drew upon a wide knowledge of character and social relations in both parts of Ireland and in America for his realistic comedies. He made his name with the one-act *Bedmates* (1921), his later sympathetic satires of contemporary life including *Paul Twyning* (1922; London, 1933), *Professor Tim* (1925; London, 1927), and *The New Gossoon* (1930), seen in London in 1931, the title being subsequently altered to *The Girl on the Pillion*. Shiels's serious plays, such as *The Passing Day* (1936; London, 1951), *The Rugged Path* (1940), and *The Summit* (1941), showed that his dramatic gifts were not limited to comedy. *The Rugged Path*, in particular, had an unprecedentedly long run at the *Abbey Theatre, where 22 of his plays were produced, and was also seen in England in 1953.

Shirley, James (1596–1666), leading dramatist of London when the Puritans shut the playhouses in 1642. He wrote about 40 plays, most of which have survived in print though not on the stage. These include tragedies such as *The Maid's Revenge* (1626); *The Traitor* (1631), Shirley's most powerful play, a *revenge tragedy into which he imported a *masque of the

Lusts and Furies; *Love's Cruelty* (1631); and *The Cardinal* (1641). His best work, however, is found in his comedies, which provide a link between those of Ben *Jonson and the Restoration playwrights. The most successful were *The Witty Fair One* (1628), *Hyde Park* (1632), *The Gamester* (1633), later adapted by *Garrick, *The Lady of Pleasure* (1635), and *The Sisters* (1642). A prompt-book of this last, dating from the early years of the Restoration, supplies some interesting stage directions, and is now in the library of Sion College. Shirley survived the Commonwealth and was popular in the early days of the Restoration, no less than eight of his plays being revived, including *The Cardinal*, which *Pepys saw in 1667, the author having died of exposure during the Great Fire of London.

Showboat, name given to the floating theatres of the great North American rivers of the West, particularly on the Mississippi and the Ohio, which represented an early and most successful attempt to bring drama to the pioneer settlements. It is not known who first built a showboat or at what date. The first record of actors travelling by boat dates from 1817, when Noah *Ludlow took a company of players along the Cumberland river to the Mississippi in his 'Noah's Ark'. But they acted on land, and it was apparently William *Chapman, formerly an actor in London and New York, who first commissioned the building of a true showboat. The interior was long and narrow, with a shallow stage at one end and benches across the width of the boat, the whole being lighted by candles. Here the Chapmans with their five children played one-night stands along the rivers wherever enough people could be found to provide an audience. The entrance fee was about 50 cents, and the staple fare strong melodrama or fairy-tale plays, ranging from *Kotzebue's *The Stranger* to *Cinderella*. Starting in the autumn from as far upstream as possible, usually Pittsburgh, the showboat made its way downstream to New Orleans, where it was abandoned, the company returning to their starting-point to descend the river again in a new boat. Steam-tugs were later used to take the showboat, renamed the 'Steamboat Theatre', back to its point of departure, some managers even owning their own tugs. By the time Chapman died in 1839 there were a great number of showboats on the rivers, but his widow, with her two sons, continued to operate under the name of 'Chapman's Floating Palace' until 1847, when they sold the boat to Sol *Smith, who lost it a year later in a collision. Another showboat captain of these early days was Henry Butler, an old theatre manager who took a combined museum and playhouse up and down the Erie Canal for many years, showing stuffed animals and waxworks by day,

and at night producing *nautical dramas such as *Jerrold's *Black-Ey'd Susan*. A more elaborate boat was the 'Floating Circus Palace'; built in Cincinnati in 1851, it was intended for spectacular equestrian shows, having living quarters, dressing rooms, stables, a museum, and a ring capable of holding 45 horses. The steamboat which accompanied it, the *James Raymond*, had a theatre used for straight dramatic performances. Showboats increased in numbers and popularity until the Civil War in 1861 drove them off the rivers.

It was not until 1878, when Captain A. B. French took his *New Sensation*, the first of five successive boats, along the Mississippi that showboats again became familiar sights. French had to live down a good deal of prejudice, but the high moral tone of his productions and the good behaviour of his small company soon made his entertainments popular. At one time French and his wife, the first woman to hold a pilot's licence and master's papers on the Mississippi, ran two showboats, piloting one each. Like other showboat captains of the time, they usually avoided the cities and larger towns, not wishing to risk comparison with the theatres springing up everywhere. A formidable rival to the Frenches was Captain E. A. Price, owner of the *Water Queen*, built in 1885. This had a stage 19 ft. wide, lit by oil, a good stock of scenery, a company of about 50, and, like all showboats, a steam calliope. Another well known manager was Captain E. E. Eisenbarth, owner of the first boat to bear the name *Cotton Blossom*. This was capable of seating a large audience, and on a stage 20 ft. by 11 ft. presented a three-hour entertainment of straight solid drama, usually popular melodrama. The *Cotton Blossom* was one of the first showboats to be lit by electricity. In 1907 Captain Billy Bryant, author of *Children of Ol' Man River* (1936), began his career as a showboat actor when his father launched the *Princess*. By 1918 the Bryants were able to build their own boats, on which they gave successful revivals of many good old melodramas, the most popular being *Ten Nights in a Bar Room*, one of the many dramatizations of a temperance novel by William Pratt. Other well-known showboat personalities were the Menkes, four brothers who in 1917 bought French's fifth and last *New Sensation*, built in 1901, from Price, who had purchased it from French's widow in 1902. They kept the old name and in 1922-3 took the boat on a trip which lasted a year and covered 5,000 miles. The plays given in her, as in the other boats owned by the Menkes, were still mainly melodramas and typical Victorian entertainments, though in the late 1920s they took to presenting musical comedies. In the years before the slump of 1929 a number of new managements made their appearance on the water, mostly offering melodrama and variety. But they were hard hit by the economic depression, and for some time the Menkes' *Golden Rod* and Bryant's *Showboat* were the only ones still functioning. The last to be built was the *Dixie Queen*, launched in 1939, and converted into an excursion boat in 1943. Found nowhere but in the USA, the showboat represents a survival of the colourful pioneering days of the Golden West.

Show Portal, see FALSE PROS(CENIUM).

Shubert, Lee (1875–1953), **Sam S.** (1876–1905), and **Jacob J.** (1880–1963), three brothers, American theatre owners and producers. Sam died in a railway accident, but the others continued to prosper and by 1916, having broken the monopoly of the *Theatrical Syndicate, controlled most of the theatres in New York and the principal cities of the United States. They also established themselves in London in the 1920s, but withdrew in 1931. In the 1950s the US government took legal action against them because of their monopoly, and in 1956 they were forced to relinquish some of their theatres. The Shubert Organization is now responsible for the operation and management of 17 theatres in New York, including the *Shubert, the *Winter Garden, the *Broadway, the *Ethel Barrymore, the *Plymouth, and the *Lyceum, and several theatres in other cities. Whilst most of its productions have been musicals, it also stages straight productions such as *Shaffer's *Amadeus* (1980), the *RSC's *Nicholas Nickleby* (1981), David *Mamet's *Glengarry Glen Ross* (1984), and *Hampton's *Les Liaisons dangereuses* (1987).

Shubert Theatre, New York, on West 44th Street, seating 1,469, opened in 1913 under the control of Lee and J. J. *Shubert. They originally named it the Sam S. Shubert after their dead brother, but it is now usually called the Shubert. It opened with *Forbes-Robertson in *Hamlet*, and the first American play to be seen there was *A Thousand Years Ago* (1914) by Percy Mackaye. The theatre was intended mainly for musicals, but among its straight successes were a dramatization of Harold Frederic's novel *The Copperhead* (1918), with Lionel *Barrymore, *Fagan's *And So to Bed* (1927), with Yvonne *Arnaud, and Sinclair Lewis's *Dodsworth* (1934). Elisabeth *Bergner made her Broadway début there in Margaret Kennedy's *Escape Me Never* (1935), and the *Lunts were seen in *Sherwood's *Idiot's Delight* (1936) and *Giraudoux's *Amphitryon 38* (1937). Successful musicals included *Paint Your Wagon* (1951), *Stop the World—I Want to Get Off* (1962), *Promises, Promises* (1968), and *Sondheim's *A Little Night Music* (1973). *A Chorus Line*, which opened in 1975, became the longest running Broadway musical of all time in 1989, closing in 1990.

Shusherin, Yakov Emelyanovich (1753–1813), Russian actor, who joined the Moscow troupe of M. E. *Meddoks about 1770. Later

he studied for a time under *Dmitrevsky and in 1785 had a great success as Iarbas in *Dido* by Y. B. Knyazhin. He went to St Petersburg in 1786, where he played the title-role in *Sumarokov's *Dmitri the Impostor* and Count Appiani in *Lessing's *Emilia Galotti*, two of his best roles, becoming a great favourite with the Empress Catherine the Great after appearing successfully in her comedy *The Discontented Family* and her chronicle play *The Early Rule of Oleg*. He survived the transition from the classical to the sentimental style and was as popular in the fashionable *comédie larmoyante* as in *King Lear*, Ozerov's neo-classical *Oedipus in Athens*, and *Sophocles' *Philoctetes*.

Shwartz [Schwartz, Shvarts], Evgenyi Lvovich (1896–1958), Soviet dramatist, who in 1925 abandoned acting to concentrate on writing plays for young people. Several of the most successful were based on fairy-tales—*The Naked King* (1934) (based on Andersen's *The Emperor's New Clothes*), *Red Riding Hood* (1937), *The Snow Queen* (1938)—but his best-known work, *The Dragon*, was entirely original. First seen in Leningrad in 1944, it was produced in an English translation at the *Royal Court Theatre in 1967, and proved to be an outspoken and hilarious debunking of political tyranny, with a daring hero, a beautiful heroine, a talking cat, villains, clowns, and a three-headed dragon. Among his other plays, *The Shadow* (1940) was directed by *Akimov, who was also responsible for the production of his last play, a satirical fairy-tale entitled *An Ordinary Miracle* (1956).

Siddons [née Kemble], Sarah (1755–1831), English actress. The eldest of the 12 children of an actor-manager in the provinces, she spent her childhood travelling with his company, and at the age of 18 married **William Siddons** (1744–1808), also a member of the company. They played together in the provinces, returning there in 1775 after Mrs Siddons had made a first and unsuccessful appearance at *Drury Lane under *Garrick, and were seen in *York with Tate *Wilkinson and *Bath. A second appearance by Mrs Siddons in London in 1782 was more successful, and she was soon acclaimed as a tragic actress without equal, a position she maintained until the end of her career. She began, however, at the zenith of her powers, and unlike her brother John Philip *Kemble did not improve with age. Among her early parts were Isabella in *Southerne's *The Fatal Marriage*, Belvidera in *Otway's *Venice Preserv'd*, and the title-role in *Rowe's *Jane Shore*. Later she proved outstanding as Constance in *King John*, Zara in *Congreve's *The Mourning Bride*, and above all as Lady Macbeth, the part in which she made her farewell appearance in 1812. She returned in 1819 to play Lady Randolph in *Home's *Douglas* for the *benefit of her younger brother Charles *Kemble, but was only the shadow of her former self. In her heyday critics were unanimous in their praise of her beauty, tenderness, and nobility. A superbly built and extremely dignified woman, with a rich, resonant voice and great amplitude of gesture, she wisely refused to appear in comedy. Her intelligence and good judgement made her the friend of such men as Dr *Johnson. Painters, including Gainsborough, delighted in painting her, Reynolds immortalizing her in 1784 as 'The Tragic Muse'. She had seven children, four girls dying in infancy.

Sierra, Gregorio Martínez, see MARTÍNEZ SIERRA.

Sigurjónsson, Jóhan (1880–1919), the first fully professional Icelandic dramatist, although after *The Hraun Farm* (1908) he wrote his plays in both Danish and Icelandic and they were premièred at the *Royal Theatre, Copenhagen. His best-known play, *Bjærg-Ejvind og hans hustru* (*Eyvind of the Hills and His Wife*, 1911), a tragedy of misplaced love, brought him international recognition. *Ønsket* (*Loft's Wish*, 1915) told of a man who, like *Faust, attempted to harness the powers of evil. His last play was *The Liar* (1917).

Silence, Theatre of, term used for plays which, like those of *Maeterlinck, who is regarded as the founder of the genre, and particularly of J.-J. *Bernard, are important as much for what they omit from their dialogue as for what they actually say—a theatre, in fact, of pregnant pauses, during which the imagination of the audience supplies the missing ingredient, which is not only unexpressed but perhaps cannot be expressed in words. Hence the French term for this type of play, *le théâtre de l'inexprimé*. It is also known as the theatre of *l'école intimiste*.

Silva, Nina de, see MARTIN-HARVEY.

Sim, Alastair (1900–76), outstanding Scottish actor, who could combine a roguish conspiratorial glee with prim old-maidish fastidiousness. He made his first appearance on the stage in 1930, and in 1932–3 was at the *Old Vic, appearing later in Edgar *Wallace's *The Squeaker* (1937). In 1941 he first played Captain Hook in *Barrie's *Peter Pan*, a role he was to repeat many times; but he first made a decided impact on the theatrical scene with his Mr McCrimmon in James *Bridie's *Mr Bolfry* (1943), the first of a series of roles in plays (which he also directed) by the same author: *It Depends What You Mean* (1944), *The Forrigan Reel* (1945), *Dr Angelus* (1947), *The Anatomist* (1948), and *Mr Gillie* (1950). After playing the title-role in a revival of *Mr Bolfry* in 1956 and the Emperor in William Golding's *The Brass Butterfly* (1958), he was seen again at the Old Vic in 1962 as a somewhat improbable Prospero in *The Tempest*. He found himself on more familiar ground as Colonel Tallboys in *Shaw's

Too True to be Good in 1965, and as a singularly decrepit Lord Ogleby in *Colman the elder's *The Clandestine Marriage* at *Chichester in 1966, where in 1969 he gave a masterly performance (seen also in London) as Mr Posket in *Pinero's *The Magistrate*. It was followed by an appearance in William Douglas *Home's *The Jockey Club Stakes* (1970), and by another excellent Pinero performance as the sporting Dean in *Dandy Dick* (Chichester and London, 1973). He made his last appearance on the stage as Lord Ogleby in London in 1975. He was also a famous film star.

Simon, (Marvin) Neil (1927–), American dramatist, a witty, prolific, and accurate observer of American marital and family relationships who once had four plays running simultaneously on Broadway. He began by writing sketches for *revue, his first show for Broadway being *Catch a Star!* (1955); he collaborated on the sketches with his brother Daniel, with whom he also wrote his first play *Come Blow Your Horn* (1961; London, 1962). After the book of the musical *Little Me* (1962; London, 1964) came a string of successful plays: *Barefoot in the Park* (1963; London, 1965); *The Odd Couple* (1965; London, 1966), in which two heterosexual males—one fussy, one not—share accommodation; *The Star-Spangled Girl* (1966); and *Plaza Suite* (1968; London, 1969), three one-act plays set in the same hotel suite. He also wrote the book for two more musicals, *Sweet Charity* (1966; London, 1967) and *Promises, Promises* (1968; London, 1969). His last play of the 1960s was *The Last of the Red Hot Lovers* (1969; London, 1979) in which the three acts are concerned with a man's three unsuccessful attempts to commit adultery. Simon's success continued into the 1970s with *The Gingerbread Lady* (1970; London, 1974), a more serious play about an alcoholic cabaret star; *The Prisoner of Second Avenue* (1971), about the stresses of redundancy and New York apartment life; and *The Sunshine Boys* (1972; London, 1975), about a *vaudeville team. He broke new ground in 1973 with *The Good Doctor*, a collection of sketches based on *Chekhov's stories, and again in 1974 with *God's Favorite*, based on the Book of Job. Neither these works nor *The Trouble with People . . . and Other Things*, which he wrote in the same year with Daniel, were successful, but his popularity returned with *California Suite* (1976), constructed on the same principle as *Plaza Suite*, which was followed by the semi-autobiographical *Chapter Two* (1977; London, 1981), concerning a widower's second marriage, the book of the musical *They're Playing Our Song* (1978; London, 1980), and *I Ought to be in Pictures* (1980). The 1980s produced a trilogy about his youth: *Brighton Beach Memoirs* (1983; *National Theatre, 1986), *Biloxi Blues* (1985), and *Broadway Bound* (1986); a female version of *The Odd Couple* (1985); and *Rumors* (1988), more farce-like than his usual work.

Simonov, Reuben Nikolaivich (1899–1968), Soviet actor and director, who in 1920 joined the *Moscow Art Theatre's Third Studio (later the *Vakhtangov Theatre). He was an excellent actor, his best parts being Cyrano de Bergerac in *Rostand's play and Kostya Kapitan in *Pogodin's *Aristocrats*; but after a brilliant performance by students which he himself had trained in the dramatization of Sholokhov's *Virgin Soil Upturned* in 1931 he devoted himself almost entirely to production. He was Chief Director of the Vakhtangov from 1939 until his death, being succeeded by his son **Yevgenyi** (1925–), who had been Director of the *Maly. Among the elder Simonov's outstanding productions were Pogodin's *The Man with the Gun* (1937), *Korneichuk's *The Front* (1943), and his own adaptation of *Gorky's *Foma Gordeyev* (1956). His son's productions included *Antony and Cleopatra* in 1972 and *Little Tragedies*, based on *Pushkin in 1975.

Simonson, Lee (1888–1967), American designer, whose first work for the stage was done in connection with the *Washington Square Players. He later became one of the founders and directors of the *Theatre Guild, for which many of his finest sets were done, among them those for Leo *Tolstoy's *The Power of Darkness*, *Strindberg's *The Dance of Death* (both 1920), *Molnár's *Liliom* (1921), and *Ibsen's *Peer Gynt* (1923). His sets for Elmer *Rice's *The Adding Machine* (also 1923) were *Expressionistic. He also designed the sets for the first productions of two plays by *O'Neill, *Marco Millions* (1928) and *Dynamo* (1929)—the latter a constructivist set based on the interior of a power house. He was responsible for Robert *Sherwood's *Idiot's Delight* (1936) and Maxwell *Anderson's *The Masque of Kings* (1937), and did the settings for the world premières of *Shaw's *Heartbreak House* (1920), *Back to Methuselah* (1922), and *The Simpleton of the Unexpected Isles* (1935). For other managements he designed mainly for new American plays, among them Sherwood's *The Road to Rome* (1927), O'Neill's *Days without End* (1934), and Maxwell Anderson's *Joan of Lorraine* (1946).

Simov, Victor Andreyevich (1858–1935), Russian stage designer, who joined the *Moscow Art Theatre on its foundation in 1898, and except for a short interval after the First World War, when he worked as a freelance, spent the rest of his life there. He was responsible for the settings of the theatre's first production, Alexei *Tolstoy's *Tsar Feodor Ivanovich*, and of *Chekhov's plays as they joined the repertory, and among his other pre-war designs were those for *Gorky's *The Lower Depths* (1902), Shakespeare's *Julius Caesar* (1903), and Leo *Tolstoy's *The Living Corpse* (1911). After his return to the Moscow Art Theatre in 1925 he designed the settings for, among other productions, *Ivanov's *Armoured Train 14-69* (1927) and *Gogol's *Dead Souls* (1932).

Simpson, N(orman) **F**(rederick), see ABSURD, THEATRE OF THE.

Sinclair, Arthur [Francis Quinton McDonnell] (1883–1951), Irish actor, second husband of Maire *O'Neill, who made his first appearance on the stage in *Yeats's *On Baile's Strand* (1904) with the *Irish National Dramatic Society and was with the *Abbey Theatre until 1916, when he led a mass resignation of the players over St John *Ervine's management of the theatre. He then formed his own company and toured Ireland and England, subsequently appearing in variety theatres in Irish sketches. He built up a great reputation as an Irish comedian, among his finest parts being Old Mahon in *Synge's *The Playboy of the Western World* and John Duffy in Lennox *Robinson's *The White-Headed Boy*. He was much admired in *O'Casey's plays, especially as Captain Boyle in *Juno and the Paycock*, Fluther Good in *The Plough and the Stars*, and Seumas Shields in *The Shadow of a Gunman*.

Sinden, Donald Alfred (1923–), English actor with a rich, resonant voice, equally at home in the classics and in popular commercial theatre. He joined the *Shakespeare Memorial Theatre company in 1946, and was later with the *Bristol Old Vic. After becoming a leading man in films he resumed his stage career in 1957, but his roles were at first unremarkable except for his doubling of Captain Hook and Mr Darling in *Barrie's *Peter Pan* in 1960 and the title-role in Archibald *MacLeish's *J.B.* in 1961. In 1963 he joined the *RSC, where he played Richard Plantagenet in John *Barton's adaptation *The Wars of the Roses* and was also seen in Henry *Livings's *Eh?* (1964). For some years he alternated between the RSC and modern comedy: Terence Frisby's *There's a Girl in My Soup* (1966); *Vanbrugh's *The Relapse* in 1967, in which he gave a brilliant performance as Lord Foppington; Ray Cooney and John Chapman's *Not Now, Darling* (1968); Malvolio in *Twelfth Night* and the title-role in *Henry VIII* (both 1969). In 1970 he played Sir William Harcourt Courtly in the RSC's revival of *Boucicault's *London Assurance* (NY, 1974). He also appeared in *Rattigan's double bill *In Praise of Love* (1973) and as Dr Stockman in *Ibsen's *An Enemy of the People* at *Chichester in 1975. Later appearances confirmed his versatility, ranging from Lear, Benedick (both 1976), and Othello (1979) to farces by Alan *Bennett (*Habeas Corpus*, NY, 1975) and John Chapman and Anthony Marriott (*Shut Your Eyes and Think of England*, 1977). The pattern continued in the 1980s, with *Chekhov's *Uncle Vanya* in 1982, *Sheridan's Sir Peter Teazle in *The School* *for Scandal* in 1983, and *Shaw's Undershaft in *Major Barbara* (Chichester in 1988) interspersed with *Coward's *Present Laughter* in 1981 and Ray Cooney's *Two into One* (1984). In 1990 he gave a solo performance as Oscar *Wilde.

Singspiele, see BALLAD OPERA.

Sjöberg, Alf (1903–80), Swedish director and manager attached to the *Royal Dramatic Theatre from 1925, first as an actor and from 1929 as director. His imaginative direction and sensitive psychological interpretations left a deep mark on the Swedish theatre. His major productions included *As You Like It* (1938), *The Merchant of Venice* (1944), *Twelfth Night* (1946), *Richard III* (1947), and *Hamlet* (1960), as well as *Brecht (whom he introduced to Sweden), *Ibsen (*Brand*, 1950; *Rosmersholm*, 1959), *García Lorca (*Blood Wedding*, 1944), *Sartre (*Les Mouches*, 1945), and Arthur *Miller (*Death of a Salesman*, 1949). He was also an outstanding director of *Strindberg, his screen adaptations of *Miss Julie* (1951) and *The Father* (1969) being internationally famous.

Skelton, John (c.1460–1529), English poet and satirist, tutor to the future Henry VIII. He is believed to have written an interlude played before the Court at Woodstock, a comedy, and three *morality plays; the only one to survive is *Magnyfycence*, acted probably between 1515 and 1523. First printed in 1530 by John *Rastell, it was reprinted by the Early English Text Society in 1906. In this play Magnyfycence, a benevolent ruler, is corrupted by bad counsellors (Folly, Mischief, etc.), but restored by good ones (Good Hope, Perseverance, etc.).

Skinner, Otis (1858–1942), American actor, who made his first appearance on the stage in Philadelphia in 1877 and was seen in New York, at *Niblo's Garden, in 1879. After some years with Edwin *Booth and Lawrence *Barrett he joined Augustin *Daly and with him made his first appearance in London, as Romeo. For two years he toured with *Modjeska, and later played Young Absolute in *Sheridan's *The Rivals* to the Sir Anthony of Joseph *Jefferson. Among his later successes were *Your Humble Servant* (1910) and *Mr Antonio* (1916), both specially written for him by Booth *Tarkington; but he made his greatest and most lasting success as Hajj in *Knoblock's *Kismet* (1911). In 1926 he appeared as Falstaff in *Henry IV, Part One*, and in 1928 played the same character in *The Merry Wives of Windsor*. Among his last appearances were Shylock in *The Merchant of Venice* in 1931 and Thersites in *Troilus and Cressida* in 1933.

His daughter, **Cornelia Otis Skinner** (1902–79), first appeared with her father's company in 1921, and became celebrated in America as a *diseuse*, in which capacity she was seen in London

in 1929 in *The Wives of Henry VIII* and other sketches.

Sky-Dome, see LIGHTING.

Slapstick, American term for the bat, or flexible divided lath, used by a *Harlequin. With it he gave the signal for the *transformation scene and other changes of scenery in the *harlequinade and belaboured the backsides of his enemies, making a great noise under pretence of striking heavy blows. The word became current in England in about 1900, and was later applied to the boisterous comedy of the *clown and the low comedian, particularly to the knockabout farces and sketches of the *music-hall.

Slips, name given in the late 18th and early 19th centuries to the ends or near-stage extremities of the upper tiers of seats in the theatre. It is still in use at *Covent Garden.

Sloat, see SLOTE.

Sloman, Charles (1808–70), performer in the 'song-and-supper rooms' which preceded the early *music-halls, and the original of Young Nadab in Thackeray's novel *The Newcomes*. He was best known for his extempore doggerel verses on subjects suggested by members of the audience or on the dress and appearance of those seated in the hall; but he was also a writer of songs, both comic and serious, for himself and his contemporaries. He was at the *Rotunda in 1829, and probably appeared at *Evans's, being at the height of his fame in the 1840s. He later fell on hard times, and one of his last engagements was as *Chairman of the *Middlesex Music-Hall, a position he had previously filled when it was the Mogul Tavern or 'Old Mo'. He died soon after, a pauper, in the Strand workhouse.

Slote or **Sloat** (Hoist in America, *Cassette* in France), form of *trap for raising a long narrow piece of scenery—a *groundrow, for example—from below to stage level, or for carrying an actor up or down from the stage. It appears that the slote was already standard equipment at *Drury Lane as early as 1843, and in *The Orange Girl*, by H. T. Leslie and N. Rowe, produced at the *Surrey Theatre in 1864, there are elaborate stage directions for the conveyance of the heroine by a slote from a height downwards to a trap. Since the slote is also referred to as a 'slide' one of its variants may have been the Corsican Trap (see TRAPS). To allow free passage to the slote the floor-board above it was divided vertically, each half dropping and being pulled to the side of the stage by ropes, leaving an opening in the stage floor. On occasion a *cloth might be attached to the base of a groundrow and be drawn up from its rolled position in a Slote Box below by means of lines from the *grid. It would thus unroll as the groundrow rose in the air, for instance during a *transformation scene. With two slotes set in each of three sections of the stage, the elements of six separate scenes could be raised from below to stage level.

Słowacki, Juliusz (1809–49), Polish Romantic poet and playwright, who was exiled in 1831 and thereafter lived mainly in Paris. He wrote over 20 plays, of which only one was staged during his lifetime. This was *Mazeppa*, in a Hungarian version seen in Budapest in 1847. The Polish original was seen in Cracow soon after Słowacki's death, but it was not until 1899 that his finest play, *Kordian*, written in 1833, was finally produced. This, like the unfinished *Horsztyński* (written in 1835), shows very clearly the influence of Shakespeare, who first became known to Słowacki in 1831 when on a visit to London he saw Edmund *Kean in *Richard III*. All Słowacki's plays had been seen on the Polish stage by 1905, but it was not until Leon *Schiller's rediscovery of the Romantic playwrights that his plays became part of the permanent repertory. The revival of *Kordian* by *Axer in 1954 marked the end of *Socialist Realism in the Polish theatre, and led to further productions of such works as *Beniowski* in 1971 and *Balladyna* in 1974. Other important plays by Słowacki are *Lilla Weneda*, based on an early legend, and two late tragedies, *The Silver Dream of Salomé* and *Samuel Zborowski*. His only comedy, a satire on the excesses of Romanticism entitled *Fantazy*, was written in about 1841 and first performed in 1867. His translation of *Calderón's *Il principe constante* was produced by *Grotowski's Laboratory Theatre and was seen in London in 1969.

Sly, William (?–1608), English actor, who appears in the actor-list of Shakespeare's plays. He was connected with the theatre from about 1590, when his name appears in the cast of *Seven Deadly Sins, Part II*, a lost play, probably by *Tarleton, of which the *platt is preserved among *Henslowe's papers at Dulwich College, where there is also a portrait of Sly by an unknown artist. He joined the *Chamberlain's Men on their formation in 1594, appearing in four plays by Ben *Jonson, *Every Man in His Humour* (1598), *Every Man out of His Humour* (1599), *Sejanus* (1603), and *Volpone* (1606). Though he was not one of the original shareholders of the *Globe Theatre, he became one, since he mentions it in his will. He also had a seventh share in the *Blackfriars, later taken over by Richard *Burbage.

Smith, Albert (1816–60), interesting but forgotten figure of literary London, who dramatized several of *Dickens's novels for the stage and also produced some ephemeral plays and novels of his own. His chief enterprise was a series of one-man entertainments, of which the first was *The Overland Mail* (1850). This was an amusing account of a recent trip to India,

interspersed with topical songs and stories and illustrated by scenery specially painted for the occasion by William *Beverley. It proved such a success that Smith followed it with *The Ascent of Mont Blanc*, given at the Egyptian Hall, Piccadilly, again with scenery by Beverley, and then by a similar programme on China. He married in 1859 Mary, the actress–daughter of Robert *Keeley, and died at the height of his popularity. His simple entertainment, whose charm lay as much in the spontaneity and wit of its presentation as in the actual material, was frequently patronized by Queen Victoria and the royal children.

Smith, Dodie, see TEMPEST.

Smith, Dame Maggie [Margaret] **Natalie** (1934–   ), English actress, with a flair for high comedy but also outstanding in dramatic roles. She made her stage début in *Oxford in 1952, playing Viola in an OUDS production of *Twelfth Night*, and after training at the *Oxford Playhouse Drama School went to New York to appear in the revue *New Faces of '56*. Back in London she scored a big success as the leading comedienne in another revue, *Share My Lettuce* (1957), and then went to the *Old Vic, where she was admirable as Lady Plyant in *Congreve's *The Double Dealer* in 1959 and played Maggie Wylie in *Barrie's *What Every Woman Knows* in 1960, in which year she took over from Joan *Plowright the part of Daisy in Ionesco's *Rhinoceros*. Her growing reputation was enhanced by performances in *Anouilh's *The Rehearsal* (1961), Peter *Shaffer's double bill *The Private Ear* and *The Public Eye* (1962), and Jean Kerr's *Mary, Mary* (1963). She next joined the *National Theatre company at the Old Vic, being a superb Silvia in *Farquhar's *The Recruiting Officer* and a year later showing unsuspected depths of passion as Desdemona to the Othello of Laurence *Olivier. Among her later parts with this company were Hilde Wangel in *Ibsen's *The Master Builder*, Myra in a revival of *Coward's *Hay Fever* (both 1964), Beatrice in *Much Ado about Nothing*, and the title-role in *Strindberg's *Miss Julie* (both 1965). In 1970 she excelled as Mrs Sullen in Farquhar's *The Beaux' Stratagem* and Hedda Gabler in Ingmar *Bergman's production of Ibsen's play, in which she carried a strong charge of sexual frustration. On leaving the National Theatre she was a much-acclaimed Amanda in Coward's *Private Lives* (1972; NY, 1975). She then spent several seasons at the *Stratford (Ontario) Festival, where she was seen in many important roles including Cleopatra in *Antony and Cleopatra*, Millamant in Congreve's *The Way of the World*, and Masha in *Chekhov's *Three Sisters* (all 1976), and Lady Macbeth (1978). In 1979 she took over from Diana *Rigg in the London production of Tom *Stoppard's *Night and Day* and then played the role in New York, after which she returned to Stratford, Ontario, where she starred as Virginia Woolf in Edna O'Brien's *Virginia* in 1980 (London, 1981). At *Chichester and in London in 1984 she again played Millamant, and in 1987 (NY, 1990) she scored one of her biggest successes in a made-to-measure role in Shaffer's comedy *Lettice and Lovage*.

Smith, Richard Penn (1799–1854), American dramatist, born in Philadelphia, a lawyer by profession, and the author of some 20 plays, of which 15 were acted. They are of all types, ranging from farce to romantic tragedy, and represent the transition in the American theatre from the play imported or inspired by Europe to the true native production of later years. Most of Smith's comedies were adaptations from the French, while his romantic plays were based mainly on incidents from American history. What is believed to have been his best work, a tragedy entitled *Caius Marius*, has not survived. It was produced in 1831 by Edwin *Forrest with himself in the title-role, and proved extremely successful. It was possibly Forrest's aversion to the printing of plays in which he appeared that caused it to be lost. Another interesting play, also lost, was *The Actress of Padua* (1836), based on Victor *Hugo's *Angelo*, which marks the first appearance in American theatrical history of French Romanticism. It was revived by Charlotte *Cushman in the early 1850s, and was seen in New York as late as 1873, probably with some alterations by John *Brougham, to whom it has been attributed.

Smith, Sol(omon) **Franklin** (1801–69), pioneer of the American theatre on the frontier. After a hard childhood he ran away from home to become an actor, travelling with the *Drakes and other itinerant companies. In 1835 Smith went into partnership with Noah *Ludlow, and together they prospered, dominating theatrical activity in the South and West and building in St Louis the first permanent theatre west of the Mississippi. Smith, who later appeared as a guest artist at the *Park Theatre in New York and also in Philadelphia, was an excellent low comedian, excelling in such parts as Mawworm in *Bickerstaffe's *The Hypocrite*. His son **Marc**(us) (1829–74) was also on the stage, specializing in Old Comedy with Lester *Wallack and Mrs John *Wood, and being well received in London. He made his last appearance at the *Union Square Theatre in Hart Jackson's adaptation of a French play, *One Hundred Years Old*, in 1873.

Smith, William (1730–1819), English actor, known as 'Gentleman' Smith on account of his elegant figure, fine manners, and handsome face. He made his first appearance at *Covent Garden in 1753, and remained there until 1774, then going to *Drury Lane, where he created the part of Charles Surface in *Sheridan's *The

School for Scandal (1777). He also played Macbeth to Mrs *Siddons's Lady Macbeth when she first appeared in the part in 1785, and alternated Hamlet and Richard III with *Garrick, whom he greatly admired, though his own style of acting approximated more to that of *Quin. He was playing most of the big tragic roles when John Philip *Kemble first went to Drury Lane in 1783, and continued to appear in them until his retirement in 1788. In an age which expected its actors to turn their hands to anything from tragedy to pantomime—even Garrick is said to have once played *Harlequin— Smith's proudest boast was that he had never blacked his face, never played in a farce, and never ascended through a trap-door. He would also never consent to appear at the theatre on a Monday during the hunting season, as he was a zealous rider to hounds.

Smith, William Henry Sedley-, see SEDLEY-SMITH.

Smithson, Harriet [Henrietta] **Constance** (1800–54), English actress of Irish descent, who made her first appearance at *Drury Lane in 1818 as Letitia Hardy in Hannah *Cowley's *The Belle's Stratagem* and then played Lady Anne and Desdemona to the Richard III and Othello of Edmund *Kean. She spent some time at *Covent Garden and the *Haymarket Theatre and then returned to Drury Lane to create the part of the Countess Wilhelm in John Howard *Payne's adaptation of *Pixérécourt's melodrama *Valentine* as *Adeline; or, The Victim of Seduction* (1822). In 1827 she went with Charles *Kemble to Paris, playing Ophelia to his Hamlet, and returned there a year later with *Macready, as Desdemona in *Othello*. To the amazement of the London critics, who had not thought highly of her acting, she was received on both occasions with acclamation, and aroused enormous enthusiasm among the young Romantics of the day, the critic Janin declaring that she had revealed Shakespeare to France and made his plays the prerogative of the tragic actress, thus forestalling *Rachel, who many years later mentioned her as 'a poor woman to whom I am much indebted'. Her fame was short-lived, as she retired in 1833 to make an ill-judged and unhappy marriage with the composer Berlioz, who had fallen madly in love with her the first time he saw her on stage and after a period of doubt and despair again fell temporarily under her spell when they finally met off stage in 1832.

Smock Alley Theatre, see DUBLIN.

Smoktunovsky, Innokenti Mikhailovich (1925–), Soviet actor, who served a long apprenticeship in the provinces before joining the Leningrad Grand *Gorky Theatre company in 1957. He remained there until 1960, scoring a great success as Prince Myshkin in a dramatization by *Tovstonogov of Dostoevsky's *The Idiot* in 1957; the production was seen during the 1966 *World Theatre Season in London. After some years in films, during which he became internationally known for his Hamlet, Smoktunovsky returned to the theatre in 1973, appearing at the *Maly Theatre, Moscow, in the title-role of A. K. *Tolstoy's *Tsar Feodor Ivanovich*. Other roles included Sergei in *Arbuzov's *It Happened in Irkutsk*, Gaev in *Chekhov's *The Cherry Orchard*, and the title-roles in his *Ivanov* and *Uncle Vanya*. He returned to London in the last role with the *Moscow Art Theatre in 1989.

Snow, Sophia, see BADDELEY, ROBERT.

Socialist Realism, term applied, in all artistic fields, to the sober style that succeeded the wave of experiment following the Russian Revolution. As expounded by *Lunacharsky, it was intended to make the theatre an instrument for the education of the masses in Communism. Apparently the term was first used in 1932, during a period of protest against the *Formalism of such directors as *Meyerhold and *Taïrov, whose work at the time was considered too abstract for the new audiences, and useless as social propaganda. While admitting that a production must not be untrue, either to present-day facts or to knowledge of the past, it nevertheless entailed the depiction of the truth in terms that a worker-audience could understand, and the interpretation of the classics in the light of present-day trends. Everything in the theatre, even the writing of new plays, was therefore bound up with the approach to, or the exposition of, the upheavals which led to the Revolution of 1917. Although the criteria of Socialist Realism, of which *Gorky (with *The Mother* and *Enemies*) is considered the founder and *Mayakovsky the first brilliant exponent, have shifted with changing conditions over the years its basic policy on the problems of theatrical creation and interpretation apparently remains the same.

Society for Theatre Research, founded in 1948 after a meeting held in the *Old Vic Theatre. The society, whose first president was Gabrielle Enthoven, arranges a winter programme of lectures and publishes a journal, *Theatre Notebook*, three times a year and one or more annual publications as well as occasional pamphlets. It encourages the preservation of theatre buildings, of source material on the theatre, and of photographic records. It was instrumental in founding the British *Theatre Museum, to which it presented its library, and in 1955 held in London the first international conference on theatre history, which resulted in the foundation of the *International Federation for Theatre Research. It makes annual study grants towards research on the British theatre

and organizes an annual William *Poel Memorial Festival involving students from the principal acting schools. It also sponsors lectures in memory of Gordon *Craig.

In 1956 an American Society for Theatre Research was founded. Unlike most national theatre societies, it is concerned not only with American theatre but with any aspects of theatre studies which may interest its members. It holds an annual meeting in New York, and publishes an occasional newsletter and an annual volume of essays, *Theatre Survey*.

Sock, see BUSKER.

Soggetto, Commedia a, see COMMEDIA DELL'ARTE.

Soho Poly, London, see FRINGE THEATRE.

Soldene, Emily, see MUSIC-HALL.

Soleil, Théâtre du, French workers' co-operative drama group, formed in Paris in 1964 under the direction of **Ariane Mnouchkine** (1934–), whose long association with it made her one of France's foremost theatrical figures. Their first productions, not particularly successful, established their method of work—months of rehearsal, participation of all the actors on the technical side, and experiments with various levels of illusion and acting conventions. The company became famous in 1967 with a production of *Wesker's *The Kitchen* in the disused Cirque Medrano, and in 1969 produced, by *Collective Creation, a play of their own entitled *Les Clowns*, which was followed by *1789* (1970), a view of the French Revolution through ordinary people's eyes, and *1793* (1972). In 1972, being homeless, they were offered cheaply, by the town council of Paris, the lease of a shed in the abandoned Cartoucherie de Vincennes, a former army training ground and ammunition dump where in 1975 they staged *L'Âge d'or*. Their presence attracted a large audience of supporters of popular theatre, and adjacent sheds soon housed other experimental companies. In 1979 the company staged Mnouchkine's own adaptation of Klaus Mann's *Mephisto*. They then turned to Shakespeare with productions of *Richard II* and *Twelfth Night* in 1982 (the latter originating at *Avignon) and *Henry IV, Part One* in 1984. All three productions showed the influence of oriental theatre, which was again apparent in two epic plays by Hélène Cixous: *Norodom Sihanouk, King of Cambodia* (1986), about contemporary Cambodia, and *The India of Their Dreams* (1987).

Somi, Leone di, see LIGHTING.

Sondheim, Stephen Joshua (1930–), American composer and lyricist, whose first music for the theatre was the incidental music for N. Richard Nash's *Girls of Summer* (1956). He then wrote the lyrics for two major musicals, *West Side Story* (1957; London, 1958) and *Gypsy*

(1959; London, 1973). The first shows for which he wrote both music and lyrics were *A Funny Thing Happened on the Way to the Forum* (1962; London, 1963), drawn from the comedies of *Plautus, and the short-lived *Anyone Can Whistle* (1964). He wrote only the lyrics for *Do I Hear a Waltz?* (1965), but thereafter he was responsible for both words and music. In the 1970s he firmly established his pre-eminence with a series of brilliant and innovative musicals which extended the boundaries of the genre, the music and the densely packed lyrics being ever more closely integrated with the text. *Company* (1970; London, 1972), about the problems of marriage and bachelorhood, was followed by *Follies* (1971; London, 1987), in which a group of middle-aged ex-showgirls hold a reunion in a soon-to-be-demolished theatre where they had earlier performed. The operetta-like *A Little Night Music* (1973; London, 1975), which contains some of Sondheim's most hauntingly beautiful music, was based on Ingmar *Bergman's film *Smiles of a Summer Night*. *Pacific Overtures* (1976) looks from the Japanese point of view at the opening up of Japan to American trade in the 19th century, the music making use of oriental instruments. Victorian London is the setting for *Sweeney Todd* (1979; London, 1980), which has more singing than spoken dialogue. His adaptation of *Kaufman and *Hart's play *Merrily We Roll Along* (1981), which moves back in time as the evening progresses, was a failure; but *Sunday in the Park with George* (1984; *National Theatre, 1990), inspired by the French painter Georges Seurat, won the *Pulitzer Prize. *Into the Woods* (1987; London, 1990) imagines the subsequent lives of Grimms' fairy-tale characters. From *Company* onwards his works (apart from the last two) were directed by Hal Prince.

Son et Lumière, outdoor spectacle presented at night at historic buildings or in natural settings. It originated in France, the French name meaning 'sound and light'. Recorded sound (words and music together with gunfire, thunder, and other special effects) is combined with light from floodlights, searchlights, and laser beams to provide an appropriate historical narrative.

'Song-and-Supper Rooms', see EVANS'S.

Sonnenfels, Josef von, see VIENNA.

Sophocles (496–406 BC), Greek tragic dramatist, born of good family at Colonus near Athens. As a boy he was celebrated for the beauty of his voice and figure, and took part in a boys' dance which celebrated the victory of Salamis in 480 BC. (Athenian tradition linked its three great tragic poets to this battle; *Aeschylus fought in it, and *Euripides was said, inaccurately, to have been born while it was in progress.) Sophocles is said to have written over 100 plays, of which seven tragedies are extant, as well as substantial parts of a *satyr-drama, the *Ichneutae* (*The Trackers*), dealing with the theft by Hermes of Apollo's cattle. He held important civic and military offices, and seems always to have enjoyed the respect and esteem of his fellow-Athenians. The extant plays are: *Ajax* and *Antigone* (perhaps written closely together, c.442–441), the *Trachiniae*, *Oedipus the King* (c.429), *Electra*, *Philoctetes* (409), and *Oedipus at Colonus* (written at the end of his life, and produced posthumously by his son).

Sophocles' life embraced the most vital and crucial period of Athenian history, from the defeat of the foreign menace in the Persian wars, through the subsequent economic and cultural expansion, to the years of decline; he died just before the final defeat of Athens by Sparta in the Peloponnesian War. Although his plays were performed side by side with those of Euripides they reveal a serenity which his younger contemporary lacked—a serenity that comes from triumph over suffering, not avoidance of it. Few moments in drama are more poignant than Sophocles' tragic climaxes. His language is clearer and more incisive than that of Aeschylus; his characters are more fully rounded; but he inherited his predecessor's concern with questions of moral law, though he set those questions in a framework with which his audience could more immediately identify. His technique is often to isolate powerful, resourceful individuals against a background of crisis—*Oedipus the King* is set in a city ravaged by plague, *Antigone* in the same city decimated by war; the chief character in *Philoctetes* is a desperate castaway on a desert island—and to show their response to the various demands upon them.

Writing in an age when many hailed expediency as the only guiding principle, he constantly reaffirmed the necessity to respond to a higher moral imperative; though he is always ready to pay tribute to purely human attributes, contriving to hold a balance between the old religion and the new morality. His final play exemplifies his serenity—Oedipus, after a life of torment, goes at last to a tranquil rest—but also foreshadows the closing of an age; although great spirits such as Oedipus may continue to live in legend and tradition, the living world is left to lesser men.

The analysis of tragedy made by *Aristotle in his *Poetics* is based in the main on Sophoclean drama, which he regarded as the mature form of tragedy, therefore neglecting the earlier Aeschylean form.

Sorano, Daniel (1920–62), French actor, one of the finest of his day, who began his career with the Grenier de Toulouse, one of the Centres Dramatiques, which in 1965 celebrated its 20th anniversary by taking possession of a new theatre named after him, and with which he was seen on a visit to Paris, where he made a great impression as Scapin in *Molière's *Les Fourberies de Scapin*. He also played Biondello in their excellent production of *The Taming of the Shrew* as *La Mégère apprivoisée*. In 1952 Sorano joined the company of the *Théâtre National Populaire, where his most notable roles were Sganarelle in *Don Juan*, Mascarille in *L'Étourdi*, and Argan in *Le Malade imaginaire*, all by Molière; Arlequin in *Marivaux's *Le Triomphe de l'amour*, Figaro in *Beaumarchais's *Le Mariage de Figaro*, Don César de Bazan in *Hugo's *Ruy Blas*, and the Chaplain in *Brecht's *Mutter Courage und ihre Kinder*. He was an excellent interpreter of Shakespeare and other Elizabethan dramatists, appearing in 1961 in a French version of *Ford's *'Tis Pity She's a Whore* under the direction of *Visconti, and being particularly admired as Richard III and as the Porter in *Macbeth*. His last role was Shylock, with *Barrault's company, a powerful and sober interpretation. He died at the height of his powers.

Sorge, Reinhard Johannes (1892–1916), German poet, who began as a disciple of Nietzsche and fell, a devout Catholic, in the First World War. His most important play *Der Bettler* (*The Beggar*), written in 1912 but not performed until 1917 in a production by *Reinhardt, was a drama of social protest which foreshadowed the revolt of youth against the older generation and the striving for a higher spiritual orientation, two of the most insistent themes of *Expressionism. The structure of the play is determined by the subject-matter: the main character's inner development provides the sole link in a series of loosely connected episodes. The script specifies the use of sophisticated spot lighting, then in its infancy, to effect changes within *composite settings. The plays he wrote after his conversion to Catholicism are on mystical and religious themes—*König David* (pub. 1916) and *Der Sieg des Christos* (pub. 1924).

Sorma [Zaremba], **Agnes** (1865–1927), German actress, who in 1883 was engaged for the newly founded *Deutsches Theater in *Berlin, where she soon became popular in young girls' parts. Some of her first mature successes were scored in revivals of *Grillparzer's works, particularly *Weh dem, der lügt!* in which she played opposite Joseph *Kainz. She was also seen as Juliet in *Romeo and Juliet*, Ophelia in *Hamlet*, and Desdemona in *Othello*, and later as Nora in *Ibsen's *A Doll's House*, a part she continued to play for many years, notably on an extended tour of Europe and on her first visit to New York in 1897. She was a distinguished interpreter of the heroines of *Sudermann and *Hauptmann and, in a lighter vein, of the Hostess in *Goldoni's *La locandiera*. From 1904 to 1908 she worked under Max *Reinhardt.

Sothern, Edward Askew (1826–81), English actor, who first appeared in the provinces under the name of Douglas Stewart and then went to America, where he played unsuccessfully in Boston, being considered by a contemporary critic 'under-taught and over-praised'. After some years on tour in North America he joined Lester *Wallack as Sothern, by which name he was known thereafter, and in 1858 was with Laura *Keene, making an immense success in Tom *Taylor's *Our American Cousin* as Lord Dundreary, a part ever after associated with him. Joseph *Jefferson, who played Asa Trenchard in the same production, describes in his autobiography how Sothern, who was at first dismayed by the few lines offered him, began to introduce 'extravagant business' into his part. By the end of the month he was the equal of any other character; by the end of the run he was the whole play. He was equally successful in London in 1861, where long side-whiskers, as he wore them, became fashionable as 'dundrearies'. The play became practically a series of monologues, and several other sketches were written round Sothern's creation. Another of his great parts was the title-role in *Brother Sam* (1865), which he wrote in collaboration with *Buckstone and Oxenford, some people preferring it to Dundreary. He also appeared in T. W. *Robertson's *David Garrick* (1864), and in H. J. *Byron's comedy of an old actor, *A Crushed Tragedian* (1878), which after a poor reception in London was a great success in New York. It was as an eccentric comedian that Sothern did his best work.

Sothern, Edward Hugh (1859–1933), one of three actor-sons of Edward Askew *Sothern, who inherited much of his father's charm and talent. Educated in England, he intended to become a painter but took to the stage, becoming in 1904 Daniel *Frohman's leading man at the *Lyceum Theatre in New York and remaining there until 1907. A light comedian and an excellent romantic hero in such popular plays as Hope's *The Prisoner of Zenda* and Justin McCarthy's *If I Were King*, he later formed his own company for the production of plays by Shakespeare with his second wife Julia *Marlowe, whom he married in 1911 and took with him on his return to England after an absence of 25 years. Their joint appearances in such plays as *Romeo and Juliet* were very well received. On Julia's retirement in 1924, after an accident, she and Sothern presented the scenery, costumes, and properties for 10 of his Shakespeare productions to the *Shakespeare Memorial Theatre. He continued to act intermittently until 1927 and devoted much of his later years to public readings and lectures. A small but dignified man, with a handsome, sensitive face, he was the ideal romantic hero of the late 19th century, and although by hard work he achieved some success in tragedy he was at his best in high comedy, his curious combination of gifts making him an excellent Malvolio in *Twelfth Night*. He also revived on several occasions his father's old part of Lord Dundreary in Tom *Taylor's *Our American Cousin*.

Sotie, topical and satirical play of medieval France, whose best-known author is **Pierre Gringore** (*c.*1475–1538). The *sotie* was not a farce, though the two forms had elements in common, and it was often inspired by political or religious intrigue. It was intended for amusement only and had no moral purpose, though it often served as a prelude to a *mystery or *morality play. The actors, or *sots* (fools), wore the traditional fool's costume—dunce's cap, short jacket, tights, and bells on their legs. *Soties* were acted not only by amateur companies of students and law clerks but by semi-professional and more or less permanent companies, somewhat in the tradition of the *commedia dell'arte, each with its own repertory. There are a number of extant texts, of which the *Recueil Trepperel* is the most representative.

Sound, see ACOUSTICS.

Sound Effects. Until the introduction of disc recordings into the theatre in the 1950s, all sound effects were produced live offstage, many by very old methods that still survived unchanged. It is widely held that the spontaneity of 'live' effects is superior to the recorded reproduction of reality, and some effects, such as weather and bells, are better when produced live. Some examples of live sound effects are:

Wind, produced by a wind machine, a hand-driven drum on to which is mounted a series of strips which rub against a length of sail canvas draped over the drum and fixed at one end. This is particularly effective when operated sensitively, but the machine needs regular maintenance.

Rain, *surf*, and *hail*, usually represented by dried peas or small lead shot being shaken in a shallow box containing some fixed obstruction, such as nails. A cylinder, which is easier to operate by hand, can be used instead of a box, and the substitution of larger shot or marbles produces an effective hailstorm.

Thunder is still produced most effectively by means of a thunder-sheet. This is an iron sheet, as large as practicable, at least 6′ by 2′6″, suspended in the flies and vibrated by jerking a handle at the bottom. Less resonant thunder can be produced in the same way by using large sheets of plywood. The thunder-sheet replaced the 18th-century thunder-run, a series of wooden troughs built into the theatre's back wall, each separated by a gap so that cannon balls rumbled down and fell with a considerable crash. This method lost favour because once the cannon balls had been released there was no way of controlling the effect. A thunder-run

still survives at the Theatre Royal, Bristol (see BRISTOL OLD VIC), though it is seldom used.

Door slams are only effectively produced by slamming an actual door. Should one not be available offstage, then a small door is used, 3′ by 2′ by 1″, fixed to a portable base and heavily constructed with all appropriate locks, etc.

Horses' hooves are still simulated most effectively by coconut shells, although much skill is needed in the operator. The shells can be used against each other, or against slate or carpet, to simulate cobbles, tarmac, or turf.

Explosions and gunfire. Individual explosions are best obtained by maroons placed in appropriate bombtanks and fired electrically. For shots, other than those obtained by striking wood against leather, a starting pistol should be used. (There are strict regulations governing explosions and the use of firearms in the theatre.) Recordings are the best way to convey the sound of continuous bombardment.

Bells. Doorbells and telephones best utilize their real components. It is desirable for actors to be able to press the doorbells themselves, and also that the operation of lifting the telephone handset cuts off the bell, although it is actuated from offstage. For chimes, a set of tubular bells is preferable to recordings.

Breaking glass is simulated by a quantity of broken glass and china flung from one bucket into another.

Effects recorded on 78 rpm discs persisted well into the 1960s, usually alongside live effects. A theatre would typically have two independent turntables and amplifiers, as an insurance against breakdowns, each with access to several loudspeakers. The records were generally hired from specialist companies.

Tape decks are also installed in pairs, for back-up purposes but also because one may be taken up by tapes of house-music or by loops—small continuous tapes of repetitive sounds such as birdsong. Other effects are combined on a conventional tape in the order in which they will be used. Many sound effects are now recorded on cartridge machines similar to those used for advertising 'jingles'.

Southerne, Thomas (1660–1746), English dramatist, a friend of *Dryden, for whose plays he wrote a number of prologues and epilogues, and of Aphra *Behn, two of his plays, *The Fatal Marriage; or, The Innocent Adultery* (1694) and *Oroonoko* (1695), being based on her novels. His tragedies, which include his first play *The Loyal Brother; or, The Persian Prince* (1682), *The Fate of Capua* (1700), and *The Spartan Dame* (1719), show a mingling of heroic and sentimental drama which had some influence on the development of 18th-century tragedy as exemplified by Nicholas *Rowe. He was also the author of three comedies which enjoyed some success when first produced: *Sir Anthony*

Love; or, The Rambling Lady (1690), The Wives' Excuse; or, Cuckolds Make Themselves (1691), and The Maid's Last Prayer; or, Any, Rather than Fail (1693). They contain some witty scenes, but are weak in construction and overloaded with detail and have not been revived. It was on Southerne's recommendation that Colley *Cibber's first play Love's Last Shift; or, The Fool in Fashion was produced at *Drury Lane in 1696.

South Street Theatre, Philadelphia, see SOUTH-WARK THEATRE.

Southwark Theatre, the first permanent play-house to be erected in *Philadelphia, and pos-sibly the first in America. A rough brick and wood structure painted red, its stage lit by oil lamps, it was built in 1766 by David *Douglass, manager of the *American Company, and opened on 12 Nov. with *Vanbrugh's The Provoked Wife and Isaac *Bickerstaffe's Thomas and Sally. Early in 1767 it saw the production for one night of *Godfrey's The Prince of Parthia, the first American play to be staged professionally. During the War of Independence the building, which from its position was sometimes known as the South Street Theatre, was closed, but after the departure of the British it reopened for a short time in the autumn of 1778. In 1784 the younger *Hallam and John *Henry, who had assumed the management of the American Company, stopped in Philadelphia on their way back to New York from the West Indies, and reopened the theatre for a while. It continued to be used for plays after their departure until in 1821 it was partly destroyed by fire. It was finally demolished in 1912.

Soviet Army, Central Theatre of the, MOSCOW, see RED ARMY THEATRE.

Sovremennik Theatre, Moscow, founded in 1957, after a group of young actors from various other playhouses in the city had come together under the director and actor **Oleg Efremov** (1927–) to stage the first production of *Rozov's Alive for Ever (1957). The success of their venture led them to remain together, and the building in which they had appeared, lent to them by the *Moscow Art Theatre, was given the name Sovremennik or Contemporary. A strong humanist and lyrical element was characteristic of their best work in their early years. Apart from the work of Soviet dramatists, including a trilogy to mark the 50th anniversary of the Russian Revolution in 1967, the company staged *Osborne's Look Back in Anger in 1966 and *Albee's The Ballad of the Sad Café in 1967. After Efremov moved to the Moscow Art Theatre in 1971 the company came under the control of its leading actress **Galina Volchek** (1934–), who in 1975 invited the Artistic Director of the *Crucible Theatre, Sheffield, to direct Twelfth Night, the Sovremennik's first

Shakespeare production. In 1985 she played Martha in Albee's Who's Afraid of Virginia Woolf? Sovremennik Two opened in 1987.

Soyinka, Wole [Akinwande Oluwole] (1934–), Nigerian writer, the first to become known for his dramatic works outside his own country. Born at Abeokuta in Western Nigeria, he attended University College, Ibadan, before taking a degree in English at Leeds University, where he wrote several plays including The Lion and the Jewel (1957; *Royal Court, 1966), which satirically contrasts old and new Nigeria. In 1958 he joined a writers' group conducted by George *Devine and William *Gaskill at the Royal Court, where his one-act play The Invention (1959) was given a Sunday night performance. He returned to the University of Ibadan in 1960 as a research fellow in drama and wrote The Dance of the Forests for the Nigerian independence celebrations. He then produced two satirical revues, The Republican (1964) and Before the Blackout (1965), and also in 1965 had a play, The Road, produced in London. This was followed by Kongi's Harvest (1966), about a tyrannical modern African state, produced at the Dakar World Festival of Negro Arts and, in 1968, in New York. Active in politics, he was arrested in 1965 and again in 1967, when his attempts to negotiate a truce in the Biafran War led to two years in jail. He returned to the School of Drama at Ibadan of which he had been appointed Director in 1967, but resigned in 1972 after disturbances on the campus. His next play, Madmen and Specialists (NY, 1970), expressed his horror of war and destruction. In 1973 the *National Theatre in London commissioned his adaptation of *Euripides' Bacchae, and he went home in 1975 to occupy the Chair of Comparative Literature at the University of Ifé. Death and the King's Horseman, written in England in 1975 and staged in Ifé in 1976 (NY, 1987), deals with an incident in Yoruba history in 1946, when the British opposed the ritual suicide of the King's Horse-man after his King's death. Among his later works are Opera Wonyosi (1977), based on *Gay's The Beggar's Opera and *Brecht's The Threepenny Opera, and A Play of Giants (1984), based on *Genet's The Balcony, both of which again attack modern African tyrants. Essentially a political satirist, Soyinka fully deserves his international reputation, being a sophisticated craftsman with a fine command of the English language and an innovative approach to stage-craft. He was awarded the *Nobel Prize for Literature in 1986.

Speaight, Robert William (1904–76), English actor and author, who played with the OUDS while at *Oxford. A year later he was in London, and before joining the *Old Vic company in 1931, where his roles included King John and Hamlet, had been seen in a number of plays, including *Sherriff's *Journey's End* (1928) and Oscar *Wilde's *Salome* (1931). He came into prominence when he appeared in the Chapter House of Canterbury Cathedral as Becket in T. S. *Eliot's *Murder in the Cathedral* (1935), playing the same part later in London and on tour and in 1938 in New York, where he was also seen as Chorus in *Five Kings* (1939), an adaptation by Orson *Welles of Shakespeare's history plays. Back in London he was in Ronald Duncan's *This Way to the Tomb* (1946) and made his last appearance in the West End in Charles *Morgan's *The Burning Glass* (1954). He also acted in plays and pageants connected with religious foundations.

Spectacle Theatres, name given to the early ornate playhouses of which only Italy has been able to preserve any 16th- and 17th-century examples. These are *Palladio's *Teatro Olimpico in Vicenza, Scamozzi's Court Theatre at Sabbioneta, and Aleotti's *Teatro Farnese at Parma. All these theatres, and many like them which no longer exist, were sumptuously decorated and had finely equipped stages, with machinery capable of dealing with the most elaborate settings. (See THEATRE BUILDINGS.)

Spectatory, see AUDITORIUM.

Spelvin, George, fictitious stage name, the American equivalent of Walter *Plinge, used to cover doubling. It is first found in New York in 1886 in the cast-list of Charles A. Gardiner's *Karl the Peddler*, and in a comedy entitled *Hoss and Hoss* (1895), by William Collier, Sr., and Charles Reed, is even given credit for supplying some of the gags. It is estimated that George Spelvin or his relatives (several variations of the Christian name have been used) have figured in more than 10,000 Broadway performances since George's début. The name has also been applied, theatrically, to dead bodies, dolls substituting for babes in arms, and animal actors.

Spencer, John, see ENGLISH COMEDIANS.

Spieltreppe, see JESSNER.

Spot Bar, see BARREL.

Staberl, Austrian stock comic character, the umbrella-maker Chrysostomos Staberl, played by Ignaz *Schuster in *Bürger von Wien* (1813) by **Adolf Bäuerle** (1786–1859) with such success that he became a household word and appeared in some 25 sequels by various hands, including four more by Bäuerle. He marks the transition from such imported clowns as *Hanswurst and *Kasperle to native comedy, and was instantly recognizable as an archetypal Viennese working-class character, genial and shrewd, but something of a grouser and layabout.

Stackner Cabaret, see MILWAUKEE REPERTORY THEATER.

Stage, space in which the actors appear before the public. In its simplest form this is a cleared area with the audience sitting or standing all round, indoors or out, with or without a raised platform. At its most complicated, as in the picture-frame or *proscenium-arch theatre, it is an elaborate structure, having a stage floor with a *rake, surmounted by *flies and a *grid, with a cellar underneath to house the machinery needed for working the *traps and for mechanical scene-changing. In this case the *boards will be removable to permit the passage of actors and props from the cellar. An intermediate form of stage, found in the Elizabethan playhouse, is a high platform built against a permanent wall structure with a cellar below. This may or may not have an *inner stage for the provision of simple scenic effects. The part of the stage floor in front of the proscenium arch is known as the *forestage or apron stage. (See also OPEN, REVOLVING, and TERENCE-STAGE.)

The word 'stage' is also used of the whole ensemble of acting and theatre production, excluding the texts of the plays, which belong to dramatic literature. To be 'on the stage' is to be an actor; to be 'on stage' is to appear in the scene then being played; 'off-stage' refers to a position close to the stage where the actor is invisible, though not inaudible, to the audience; *back stage includes the area around and behind the stage used by theatre workers, who enter the building through the *stage door.

Stage, Audience on the, see AUDIENCE ON THE STAGE.

Stage Brace, see FLAT.

Stage Directions, notes added to the script of a play to convey information about its performance not already explicit in the dialogue. Stage directions in the English theatre are based on two important peculiarities: they are all relative to the position of an actor facing the audience—right and left are therefore reversed from the spectators' point of view—and they all date from the time when the stage was raked, or sloped upwards towards the back. Thus movement towards the audience is said to be 'down stage', movement away from the audience 'up stage'. (When one actor moves upstage of another, the latter must turn away from the audience to face him; hence the expression 'to upstage', meaning to focus attention on oneself at the expense of someone else.) 'To go off' is to leave the stage; 'to go off a little' is to withdraw to one side while remaining in view of the audience. Movement round an object on stage is expressed as 'above' or 'below' that object, and not as 'in front' or 'behind'. This too dates back to the steeply raked stage of earlier times, and is also useful in avoiding confusion between the front or back of the object and the front or back of the stage.

The simplest examples of stage directions are such single words as 'enters' or 'turns', to which may be added the Latin word *exit* ('he goes out'), now inflected as a normal English verb: 'you exit, he exits, they exit'. The plural form *exeunt* is obsolete except in the conventional phrase *exeunt omnes* ('all go out'). The antithesis *manet* ('he remains') has disappeared, but as early as 1698 it was used to indicate that a character remained on stage at the end of a scene to take part in the next scene, even if the scenery had to be changed (as was done in full view of the audience), thus ensuring that the action of the play carried on without a break.

In printed copies of old plays, particularly *melodramas, such terms as RUE ('right upper entrance') and L. 2. E. ('left second entrance') are found. These relate to the time when the side-walls of the stage were composed of separate *wings, the entrance being the passage between one wing and its neighbour. An earlier variant of this terminology (found as far back as 1748) was to indicate an entrance as 'in the second (or third) *grooves', signifying that the actor entered behind the wing running in the second (or third) groove in the stage floor.

Two other terms, which apply only to the English-speaking theatre, are 'prompt side' (on the stage left), which houses the *prompt corner, and 'opposite prompt' (on the stage right), usually spoken and written as OP.

Stage Door, usual means of access to the area back stage for actors and stage-hands, situated at the back or side of the theatre. Immediately inside it is the cubicle of the stage-door keeper (formerly known as the hall keeper) who checks the arrival and departure of staff, prevents the entry of unauthorized persons, and transmits messages. Nearby is the Call Board, on which schedules of rehearsals and other items of information needed by the actors are posted, including the 'Notice' informing them of the end of the play's run. The stage door is never used by the audience, who enter by the main front and subsidiary side doors, but in Elizabethan times the stage (or tiring-house) door was used by those members of the audience who had seats on the stage.

Stage Floor, see BOARDS.

Stage-Keeper, functionary of the Elizabethan theatre responsible for the sweeping and clearing of the acting space, and probably for other odd jobs.

Stage Machinery, mechanical devices used to make quick scene changes and increase realism. The Greeks employed the *mechane for propulsion through the air, and the *ekkyklema for movement forward. Machinery was also used by the Romans. The *periaktoi which provided

scene changes in Renaissance Italy were rotated mechanically, and 'flying angels' and the *Paradiso could also be seen at this time. Giacomo *Torelli invented the *carriage-and-frame devices which moved scenery by means of carriages beneath the stage; he also pioneered the *machine plays which specialized in elaborate mechanical effects. Philip de *Loutherbourg, who worked for *Garrick, was especially good at reproducing climatic effects and natural phenomena such as fire and volcanoes. The 19th century saw the introduction of *traps below the stage, through which actors could emerge; the *grid above, permitting scenery to be 'flown'; and the *revolving stage. Whole stages can now be moved into the wings and instantly replaced by new sets; individual sections of a stage can be raised or lowered. There is almost no illusion which modern stage machinery, together with modern *lighting, cannot achieve.

Stage-Manager, member of theatre staff responsible for the overall management of everything connected with the stage and backstage. In the days before the pre-eminence of the *director, the stage-manager also conducted rehearsals, which were often perfunctory and were called usually for the instruction of new members of the company and to co-ordinate the movements of the resident actors with those of a guest-star. In the 19th century men such as *Belasco, though called stage-managers, occupied virtually the position held by the director today, often combining it with the office of *prompter.

Stage Society, London, organization founded in 1899 to produce plays of artistic merit not likely to be performed in the commercial theatre. In order to make use of professional actors, performances were given in selected theatres on Sunday nights, when they were normally closed. This led to a police raid in 1899 on the *Royalty Theatre, where the society was giving its first production, *Shaw's *You Never Can Tell.* It was argued, successfully, that the theatre was being used as a private place and was therefore not subject to the ban on Sunday opening. Among the plays produced at similar Sunday performances was Shaw's *Mrs Warren's Profession* in 1902, which firmly established the society's right to perform plays which had been refused a licence by the *Lord Chamberlain. Other unlicensed plays were *Maeterlinck's *Monna Vanna* (also 1902), Leo *Tolstoy's *The Power of Darkness* (1904), *Granville-Barker's *Waste* (1907), *Pirandello's *Six Characters in Search of an Author* (1922), and James Joyce's *Exiles* (1926). These were all seen later in the commercial theatre, where the most successful of the society's productions, R. C. *Sherriff's *Journey's End* (1928), was seen only a few weeks after its original production.

The Stage Society functioned for 40 years,

during which it staged over 200 plays, many of them first performances of American and foreign plays in England. During the First World War it experimented with the revival of classic plays (in the absence of suitable modern material) and so aroused an interest in Jacobean and Restoration plays which was to lead to the establishment of the *Phoenix Society and be a marked feature of the 1920s. In 1926 in face of rising costs and a declining membership, which had reached its peak just before the war, the society merged with Phyllis Whitworth's Three Hundred Club, whose productions since its foundation in 1923 had included *A Comedy of Good and Evil* (1924), by Richard Hughes, J. R. Ackerley's *The Prisoners of War* (1925), and J. E. Flecker's *Don Juan* (1926). Important productions after the merger, which lasted until 1931, were *The Widowing of Mrs Holroyd* (1926) and *David* (1927), both by D. H. *Lawrence, and John *Van Druten's *Young Woodley* (1928) and *After All* (1929). The society's last production was *García Lorca's *Bodas de sangre* as *Marriage of Blood* (later known as *Blood Wedding*) in 1939 at the *Savoy Theatre. The Second World War created conditions in which the society could not survive, and an attempt to revive it after the war was not successful.

Stagg, Charles and **Mary,** see WILLIAMSBURG.

Stalls, individual seats between the stage front and the *pit, those nearest the stage being known as Orchestra Stalls. They first appeared in the 1830s to 1840s, after the raising of the first circle had allowed the pit to extend further back. With the exception of *boxes and the front of the Dress Circle they are the most expensive seats in the theatre. They were at one time called by their French name of *fauteuils.* In some theatres the term Balcony Stalls was applied to the front rows of the Dress Circle.

Standard Theatre, London, in Shoreditch, originally a pleasure-garden attached to the Royal Standard public-house, was already in use in 1837. The first substantial building on the site, which held over 3,000, had a circus-ring which could be boarded over for concerts and dramatic performances. Burnt down in 1866, it was rebuilt on a larger scale, and reopened in 1867 as the Standard Theatre under the management of John Douglass, who remained there until his death in 1888. He maintained a good stock company, and his pantomimes rivalled those of the *Britannia in Hoxton and even those of *Drury Lane. It was the first of the north-eastern suburban theatres to attract visiting stars from the West End, Henry *Irving being seen there in 1869. In 1876 it was again rebuilt, with a seating capacity of 2,463 in four tiers, and reopened as the New National Standard Theatre, but was always known by its old name. In 1889 it was taken over by the *Melville brothers, who ran it successfully until

1907, after which, as the Olympia, Shoreditch, it became a *music-hall, and in 1926 a cinema. It was badly damaged by bombs in 1940 and demolished soon after.

Standard Theatre, New York, on the west side of Broadway, between 32nd and 33rd Streets. This opened as the Eagle in 1875, and had a prosperous career as a home of variety and light entertainment. Renamed the Standard, it witnessed in 1879 the first production in New York of *Gilbert and Sullivan's *HMS Pinafore,* which ran for 175 nights. In 1881 *Patience* ran through the whole season, as did *Iolanthe* in the following year. Destroyed by fire in 1883, the theatre was rebuilt as a home for light opera. In 1896, as the Manhattan, it again became a playhouse under William A. *Brady with *What Happened to Jones* by George *Broadhurst, and from 1901 to 1906 it housed a company run by Mrs *Fiske. It was demolished in 1909 to make way for Gimbel's department store.

Stanfield, Clarkson (1793–1867), English artist, who was for a few years a scene-painter, by many considered the most gifted to work in the London theatre. His first scenes were painted for the *Royalty Theatre in Wellclose Square; in 1831 he was at the Theatre Royal in Edinburgh, and he then returned to London to become scenic director at the Coburg Theatre (later the *Old Vic) and eventually at *Drury Lane. By 1834 he had achieved sufficient eminence as an artist to give up scene-painting as a profession, though he occasionally did some work for his friends, assisting *Macready with scenery for his *pantomimes in 1837 and 1842 and painting a backdrop in 1857 for *Dickens's production at his private theatre in Tavistock House of Wilkie Collins's *The Frozen Deep.* His last theatrical work was a drop scene for the *Adelphi Theatre, which he painted for Ben *Webster in 1858.

Stanislavsky [Alexeyev], **Konstantin Sergeivich** (1863–1938), Russian actor, director, and teacher of acting. He had already had some experience in the theatre when in 1898 he founded with *Nemirovich-Danchenko the *Moscow Art Theatre. In its first productions—Alexei *Tolstoy's *Tsar Feodor Ivanovich* (1898), *Ostrovsky's *The Snow Maiden* (1900), Leo *Tolstoy's *The Power of Darkness* (1902), *Julius Caesar* (1903)—Stanislavsky put into practice the theories he had formulated under the influence of the *Meininger company, and out of his own experience. Rejecting the current declamatory style of acting, he sought for a simplicity and truth which would give a complete illusion of reality. His ideas were particularly acceptable in America, where in the 1930s the *Group Theatre made them the basis of a system of actor-training known as the *Method. One of Stanislavsky's greatest achievements was his staging of *Chekhov's plays, which he produced as lyric dramas, underlining the emotional moments with music and showing how Chekhov's apparently passive dialogue demands great subtlety and a psychologically orientated internal development of the role, with great simplicity of external expression. During the years of upheaval leading to the uprising of 1905 Stanislavsky produced the plays of Maxim *Gorky—'stormy petrel of the Revolution'—including The Lower Depths (1902). In the years of reaction (1905–16) he turned to *Symbolism with *Maeterlinck and *Andreyev in aestheticized, stylized productions. Under the Soviet regime, after an initial period of adjustment, he continued his work, but in later years, because of ill health, he gave up acting and concentrated on production and teaching. Among his best roles were Astrov in Uncle Vanya, Vershinin in Three Sisters, Gaev in The Cherry Orchard, Dr Stockmann in *Ibsen's An Enemy of the People, and Rakitin in *Turgenev's A Month in the Country.

Stapleton, (Lois) **Maureen** (1925–), American *Method actress of great power and subtlety, excelling in character roles. She made her New York début in *Synge's The Playboy of the Western World in 1946, but achieved stardom as the tempestuous Serafina in Tennessee *Williams's The Rose Tattoo in 1951. She then appeared in revivals (Lady Anne in Richard III in 1953, Masha in *Chekhov's The Seagull in 1954) before starring in two more plays by Williams—as Flora in Twenty-Seven Wagons Full of Cotton (1955) and Lady Torrance in Orpheus Descending (1957). Her Ida in S. N. *Behrman's The Cold Wind and the Warm (1958) and Carrie in Lillian *Hellman's Toys in the Attic (1960) were followed in 1965 by another Williams role, Amanda Wingfield in a revival of The Glass Menagerie, while in 1966 she played Serafina again. She was seen to advantage in two plays by Neil *Simon: in three separate roles in Plaza Suite (1968) and as the alcoholic heroine of The Gingerbread Lady (1970). In 1972 she was an effective Georgie Elgin opposite Jason *Robards in a revival of *Odets's The Country Girl. In 1975 she again played Amanda, and in 1981 was Birdie Hubbard to Elizabeth Taylor's Regina in a revival of Lillian Hellman's The Little Foxes.

Star Theatre, New York, at the north-east corner of Broadway and 13th Street. It was opened by the elder James *Wallack in 1861, though he never himself appeared there, and after his death in 1864 it was managed by his son Lester *Wallack, who in 1881 left it to open his own theatre on Broadway. In 1882 the old theatre, renamed the Star, and under the management of H. E. *Abbey, housed Ellen *Terry and Henry *Irving with their company from the London *Lyceum on their first visit to the USA. Not wishing to challenge Edwin

*Booth in Shakespeare, Irving appeared in modern plays, of which the staging, lighting, and acting proved a revelation to New York audiences. Later visitors to the Star were Booth himself, Mary *Anderson in 1885, returning after two years in London, *Modjeska in Maurice *Barrymore's Nadjezda (1886), Sarah *Bernhardt in some of her most popular parts, and Wilson *Barrett making his American début, also in 1886. The theatre was demolished in 1901.

Steele, Sir Richard (1672–1729), English soldier, politician, essayist, pamphleteer, and incidentally dramatist, in which capacity he was one of the first to temper the licentiousness of Restoration drama with sentimental moralizing. His three early comedies—The Funeral; or, Grief à-la-mode (1701), The Lying Lover; or, The Ladies' Friendship (1703), and The Tender Husband; or, The Accomplish'd Fools (1705)—had only a moderate success, and he turned his attention to founding and editing, with *Addison, the Tatler (1709–11) and the Spectator (1711–12), and also the first English theatrical periodical, the Theatre, which appeared twice a week from 1719 to 1720. It was not until 1722 that he produced his last and most important play, The Conscious Lovers, a sentimentalized adaptation of *Terence's Andria, marked throughout by a high moral tone and given at *Drury Lane under Colley *Cibber's direction. It was a great success, and was immediately translated into German and French, exercising an immense influence on the current European drift towards *comédie larmoyante.

Steele, Tommy, see LONDON PALLADIUM and MUSICAL COMEDY.

Stein, Peter (1937–), German director of world renown. He worked initially in Munich, Bremen, and Zürich, his early work including Edward *Bond's Saved (1967) and a politically provocative staging of *Weiss's Vietnam Discourse (1968). He moved to the Schaubühne, Berlin, as Artistic Director, 1970–85, making his début there with *Brecht's adaptation of *Gorky's novel The Mother, which was followed by such notable productions as *Ibsen's Peer Gynt (1971), Gorky's Summerfolk (1973), and *Aeschylus' Oresteia (1980). A realist with a preference for ensemble acting, he was especially praised for his interpretations of *Chekhov's Three Sisters (1984) and The Cherry Orchard (1989) and *Racine's Phèdre (1987). His *Expressionist production of *O'Neill's The Hairy Ape was seen at the *National Theatre in London in 1987. He makes an annual return to the Schaubühne, and also directs opera.

Steinbeck, John Ernst (1902–68), American novelist, awarded the Nobel Prize for Literature in 1962. He adapted for the stage two of his own novels, already dramatic both in structure and dialogue—*Of Mice and Men* (1937; London, 1939), a realistic picture of itinerant farm labour and of a feeble-minded farmhand; and *The Moon is Down* (1942; London, 1943), portraying the occupation by the Germans of a peaceful town, and its resistance movement, which was more favourably received in Europe than in America. Dramatized by other hands were *Tortilla Flat* in 1938 and *Burning Bright* in 1950.

Stephen Joseph Theatre in the Round, Scarborough, North Yorkshire. **Stephen Joseph** (1921–67), son of *revue star Hermione Gingold, set up his first experimental *theatre-in-the-round in the public library of this seaside resort for a summer season in 1955. His aim was to create an ambience in which young playwrights could work closely with a small group of actors. The season became an annual event, and Alan *Ayckbourn's first play was produced there in 1959, as was James *Saunders's *Alas, Poor Fred*, the first of his plays to be produced professionally. The theatre survived Joseph's death through the perseverance of its manager, who invited Ayckbourn to become Director of Productions in 1970. In 1976 the company moved to new premises on the ground floor of a school and the theatre took its present name. The new auditorium seats 300 and there is a studio theatre; the company now operates throughout the year. Most of Ayckbourn's plays have been launched from Scarborough and many other writers have benefited from the policy of presenting new (often first) plays.

Stern, Ernst (1876–1954), scene designer, born in Bucharest, who worked for many years in Germany, making his début in 1904 with designs for Emanuel von Bodman's *Die heimliche Krone*. From 1906 to 1921 he worked for Max *Reinhardt, solving the problems set by the director's predilection for the revolving stage, and the consequent need to think in terms of ground-plans and models rather than pictures. His sets for Reinhardt's Shakespearian productions included *Twelfth Night* in 1907, *Hamlet* in 1909, and *A Midsummer Night's Dream* in 1913; of his other designs, the most important were probably those for *Goethe's *Faust II*, *Aeschylus' *Oresteia*, Vollmöller's *The Miracle* at Olympia in London (all 1911), and *Ibsen's *John Gabriel Borkman* in 1917. From the late 1920s he worked mainly in London, designing the sets for *Coward's *Bitter Sweet* (1929), Rodgers and *Hart's *Ever Green* (1930), and for another musical *White Horse Inn* (1931), reproducing the designs he had made for an earlier version in Germany. Between 1943 and 1945 he also designed a number of sets for Shakespeare productions by Donald *Wolfit.

Sternheim, Carl (1878–1942), German dramatist, at the height of his popularity in the 1920s. Banned under the Nazis, his plays have since been revived many times, and he is considered one of the few 20th-century German playwrights to have excelled in comedy. At first his clipped dialogue and grotesque situations seemed to show his affinity with *Expressionism, but in revival he appears more closely related to *realism. Several of his bitter anti-bourgeois satires follow the fortunes of one family in its rise to social eminence under the collective title of *Aus dem bürgerlichen Heldenleben* (Scenes from the Heroic Life of the Middle Classes). The first of these was *Die Hose* (1911); produced by *Reinhardt in *Berlin, this caused such a scandal because of its 'indelicate subject'—the loss of a lady's knickers in embarrassing circumstances—that it was banned, and its author left Germany to reside permanently in Brussels. In a translation by Eric *Bentley, it was seen in London in 1963 as *The Knickers*. It was followed by *Die Kassette* (*The Money Box*, 1911) and *Bürger Schippel* (1913), which shows the chief character, a plumber, rising to middle-class status; it was seen in London in an adaptation by C. P. *Taylor in 1975. *Der Snob* (1914) deals with the next generation's adaptation to high society, and was the first play to be revived in Berlin in 1945. It was followed by *Der Kandidat* (1915), *Tabula rasa* (1919), and *Das Fossil* (1925). Among Sternheim's other plays *Die Marquise von Arcis* (1919) was adapted by Ashley *Dukes in 1935 as *The Mask of Virtue*, in which Vivien *Leigh made her first outstanding success in London.

Stevenson, William (?–1575), Fellow of Christ's College, Cambridge, believed to have been the author of *Gammer Gurton's Needle*, a play which, with *Ralph Roister Doister* by *Udall, stands at the beginning of English comedy. No definite date can be assigned for its performance but it was probably given in Cambridge between 1552 and 1563. It was printed in 1575, the year of Stevenson's death, and may be identical with a play referred to as *Diccon the Bedlam*, the name of the chief character in *Gammer Gurton's Needle*. This play has also been attributed, with little likelihood, to Dr John Still, Bishop of Bath and Wells in 1593, and to Dr John Bridges, mentioned as its author in the *Martin Marprelate* tracts. Although structurally it conforms to the classic type, its material is native English, and one of its characters, Hodge, has given his name to the conventional English farm labourer.

Stichomythia, type of dialogue employed occasionally in classical Greek verse drama, in which during passages of great emotional tension two characters speak single lines alternately. The device, which has been likened to alternate strokes of hammers on the anvil, was used by Shakespeare, notably in *Richard III* (see I. ii. 193–203; IV. iv. 344–70). Echoes of it can be found in modern prose plays, particularly in the clipped dialogue of the 1920s, as in some of *Coward's early works, and it is found even in

*Beckett's *Waiting for Godot* (1955), where, though outwardly comic, it engenders a mood of almost hysterical excitement. To be effective, the device must be used sparingly. It is sometimes referred to as 'cat-and-mouse', 'cut-and-parry', or 'cut-and-thrust' dialogue.

Stiemke Theater, see MILWAUKEE REPERTORY THEATER.

Stock Company, name applied in the English theatre of the 1850s to a permanent troupe of actors attached to one theatre or group of theatres and operating on a true *repertory system, with a nightly change of bill, to distinguish it from the newly emergent *touring companies. The system was, however, in use long before, and the term could have been applied to the companies at *Covent Garden and *Drury Lane from the Restoration onwards, to the 18th-century *circuit companies in the English provinces, and to the resident companies of the early 19th century in the larger towns of the USA. In both countries the stock company found itself threatened by the establishment of the long *run and the touring company, which in the 1880s, helped by cheap and easy railway transport, finally triumphed. The stock company then ceased to exist, the last in London being that of Henry *Irving at the *Lyceum Theatre. It had been an excellent training-ground for young actors, combining a variety of stage experience with some element of security, functions that were to some extent taken over by the *repertory theatres.

Each player in the old stock company undertook some special 'line of business', though usually ready to play anything else when required. The recognized leader of the troupe was the Tragedian, who played Hamlet and Macbeth and might appear also in serious comedy. The Juvenile Tragedian played Laertes or Macduff, combining such roles with light comedy. The Juvenile Leads played the young lovers and the youthful heroes and heroines. The Old Man and the Old Woman appeared in such parts as Sir Anthony Absolute in *Sheridan's *The Rivals* and the Nurse in *Romeo and Juliet*, while the Heavy Father (or Heavy Lead) played tyrants in tragedy and from the 1830s onwards villains in *melodrama. The Heavy Woman played Lady Macbeth or Emilia in *Othello*. The Low Comedian, who ranked next in importance to the Tragedian, played leading comic parts of a broad, farcical, or clownish type, together with minor roles in tragedy, and the Walking Lady and Gentleman played the secondary parts, such as Careless in Sheridan's *The School for Scandal*. They were usually beginners, and were poorly paid. General Utility, or simply Utility, also poorly paid, played minor roles in every type of play; the Supernumerary, or Super, was engaged merely to walk on, had nothing to say, and was not paid at all. The term Super is still used in the theatre for members of a crowd; a small, individualized part with few or no lines is now called a Walk-on; the term Juvenile Lead persists only in an ironic or pejorative sense; but Heavy Father has become part of the language in general. Apart from the above, the company might contain such specialized players as the First Singer, the First Dancer, the Countryman, or the Singing Chambermaid. The term 'summer stock' is still used in the USA for summer theatres.

Stockfisch, Hans, see ENGLISH COMEDIANS.

Stoll Theatre, London, in Kingsway. This theatre, which held 2,430 persons on four tiers, was built for Oscar Hammerstein I who, intending it as a rival to *Covent Garden, opened it as the London Opera House in 1911. The venture proved a failure and Hammerstein gave up in 1912. Revue and variety were then staged there under a succession of managers, including C. B. *Cochran, until in 1916 it was taken over by **Oswald Stoll** [Gray] (1866–1942), who already controlled the *Coliseum and a chain of variety theatres, and became a cinema under the name of the Stoll Picture Theatre. It returned to live theatre, still as the Stoll, in 1941, being used for revivals of popular musicals. In 1947 the huge stage was used for ice shows, and in 1951 the theatre became the headquarters of the Festival Ballet. The *Gershwins' *Porgy and Bess* had its first London production there in 1952, and two years later Ingrid Bergman made her first appearance in London in *Joan of Arc at the Stake* by *Claudel and Arthur Honegger. A musical version of *Knoblock's *Kismet* opened in 1955, and with 648 performances gave the theatre its longest run to date. In 1957 Laurence *Olivier and Vivien *Leigh began a five-week season of *Titus Andronicus*, directed by Peter *Brook. This proved to be the last play seen at the Stoll, the new *Royalty Theatre being built on part of the site.

Stoppard [Straussler], **Tom** (1937–), Czech-born British dramatist, who first came into prominence with the production by the *National Theatre company of *Rosencrantz and Guildenstern are Dead* (1967), a play in the style of the Theatre of the *Absurd in which two minor characters from *Hamlet* are shown as powerless spectators at the play's events. Its unexpected success, both in England and abroad—it began a long run in New York in the same year—led to the production in London of a play written earlier, *Enter a Free Man* (1968; NY, 1974), first seen in Hamburg in 1964. Stoppard's next play, *The Real Inspector Hound* (also 1968; NY, 1972), is a one-acter about two dramatic critics who get drawn into the action of a play. It was followed by two more one-act plays, *After Magritte* (1970; NY, 1972) and

Dogg's Our Pet (1972). In *Jumpers* (also 1972; NY, 1974) a Professor of Moral Philosophy defends intuitive values against contemporary rationalism; while *Travesties* (1974; NY, 1975) is a lively debate, involving James Joyce and Lenin, on the justification for art. Produced in London by the National Theatre company and the *RSC respectively, they provided supreme examples of Stoppard's astonishing verbal dexterity. A double bill, *Dirty Linen* and *New Found Land* (1976; NY, 1977), ran for four years at the *Arts Theatre, and after a 'music-drama', *Every Good Boy Deserves Favour* (1977; NY, 1979), came *Night and Day* (1978; NY, 1979), his most naturalistic play to date, a debate on the freedom of the press set in a fictitious African country. He had a big success with *The Real Thing* (London, 1982; NY, 1984), in which an opening scene about marital infidelity turns out to be from a play, and is followed by a witty portrayal of the playwright's own relationships. *Hapgood* (1988), a spy thriller about a defecting Russian nuclear physicist, includes expositions on various principles of physics; while *Artist Descending a Staircase* (also 1988; NY, 1989) began as a radio play. He is also a notable adaptor: *Mrożek's *Tango* (RSC, 1966); *García Lorca's *The House of Bernarda Alba* (1973); *Havel's *Largo Desolato* (1987); and, all at the National Theatre, *Schnitzler's *Das weite Land* as *Undiscovered Country* (1979), his *Liebelei* as *Dalliance* (1986), *Nestroy's *Einen Jux will er sich machen* as *On the Razzle* (1981), and *Molnár's *Play at the Castle* as *Rough Crossing* (1984).

Storey, David Malcolm (1933–), English novelist and playwright, who was already well known through his books, particularly *This Sporting Life* (1960), when his first play, *The Restoration of Arnold Middleton* (1967), was staged at the *Royal Court Theatre, where most of his later plays were seen. It was followed by *In Celebration* (1969), centring on a 40th wedding anniversary and starring Alan *Bates, and the extremely successful *The Contractor* (1970; NY, 1973), which deals with the erection and dismantling of a marquee for a wedding reception. Both these plays were directed by Lindsay *Anderson, who was also responsible for the production of several of Storey's later works. His next plays were *Home* (London and NY, 1970), an impressionistic play in which John *Gielgud and Ralph *Richardson gave fine performances as two inmates of a mental home, and *The Changing Room* (1971; NY, 1973), about a Rugby football team. After a symbolic drama entitled *Cromwell* and *The Farm* (both 1973), seen in New York in 1978 and 1976 respectively, came *Life Class* (1974; NY, 1975), again with Alan Bates in London, in which an art class provides a background for a discussion of life and art. In *Sisters* (1978; London, 1989)

a council house is used as a brothel, and in *Early Days* (1980) Ralph Richardson played an elderly politician reviewing the course of his life. The last was produced at the *National Theatre, as was *The March on Russia* (1989), later renamed *Jubilee*, which shows the characters from *In Celebration* 20 years on.

Storm and Stress, see STURM UND DRANG.

Strand Musick Hall, London, see GAIETY THEATRE.

Strand Theatre, London. **1.** In the Strand. It opened in 1832 as Rayner's New Strand Subscription Theatre. The last battles between the unlicensed houses and the *Patent Theatres were being waged, and the opening attraction was a *burlesque on the current situation entitled *Professionals Puzzled; or, Struggles at Starting*. The enterprise was not a success and the theatre closed. It reopened in 1833, closed again, but in 1836 Douglas *Jerrold reopened it, adding a gallery to the auditorium. He enjoyed some success with dramatized versions of novels by Charles *Dickens, *The Pickwick Papers* being retitled *Sam Weller*. After a succession of managers William *Farren took over in 1848, staging an adaptation of *Goldsmith's novel *The Vicar of Wakefield*. After he left the theatre was renamed Punch's Playhouse and sank into obscurity, but reopened in 1858 as the Strand with a season of burlesques by H. J. *Byron. The unexpected success of these productions enabled the theatre to be largely reconstructed in 1865. New safety regulations, however, forced it to close in 1882, and it was again largely rebuilt. In 1884 a revival of H. J. Byron's *Our Boys* did well, and it was followed by other successful revivals. In 1901 a musical play *The Chinese Honeymoon* began a long run before the theatre finally closed in 1905, the site now being occupied by the Aldwych underground station.

2. In the Aldwych. This theatre was built for the American impresarios Sam and Lee *Shubert, its exterior being identical to that of the *Aldwych Theatre at the other end of the block. It has a four-tier auditorium seating 1,084 and opened in 1905 as the Waldorf. After the destruction of his own theatre in 1905 Cyril *Maude made it his headquarters, remaining until 1907, when Julia *Marlowe and E. H. *Sothern appeared in a series of plays which included *Kester's *When Knighthood was in Flower*. In 1909 the name of the theatre was changed to the Strand and in 1911 it became the Whitney after its American manager, reverting to the Strand in 1913 when he left, and scoring a success at last with Matheson *Lang in *Mr Wu* by Harry Vernon and Harold Owen. In 1915 the building, which was then occupied by Fred *Terry and Julia *Neilson, was slightly damaged during a Zeppelin raid. Arthur *Bourchier took over in 1919 and successfully produced A. E. W. Mason's *At the Villa Rose* (1920), Ian Hay's *A Safety Match* (1921), and a dramatization of R. L. Stevenson's *Treasure Island* (1922). Eugene *O'Neill's *Anna Christie* had its London première there in 1923, and George *Abbott and Philip Dunning's *Broadway* (1926) was another successful American play. Later hits included a farce by Austin Melford, *It's a Boy* (1930), *1066 and All That* (1935), Vernon Sylvaine's *Aren't Men Beasts!* (1936), and Ben *Travers's *Banana Ridge* (1938). In 1940, at the height of the blitz, Donald *Wolfit gave midday productions of Shakespeare. During one of them the building was badly blasted, but it was soon repaired, and did outstandingly well with Kesselring's *Arsenic and Old Lace* (1942). Post-war successes included Vernon Sylvaine's farce *Will Any Gentleman?* (1950); *Sailor, Beware!* (1955), a farce by Philip King and Falkland Cary; the revue *For Adults Only* (1958); and two plays by Ronald *Millar based on novels by C. P. Snow, *The Affair* (1961) and *The New Men* (1962). *Sondheim's *A Funny Thing Happened on the Way to the Forum* (1963) was followed by the thriller *Wait until Dark* (1966) by Frederick Knott. In 1968 *Not Now, Darling* by Ray Cooney and John Chapman began a long run. The comedy *No Sex Please— We're British* by Anthony Marriott and Alistair Foot opened in 1971 and despite poor reviews ran there until moved to the *Garrick Theatre in 1982. *Stoppard's *The Real Thing* opened in that year, and in 1989 *Much Ado about Nothing* and *Chekhov's *Ivanov* ran in *repertory.

Strange's Men, company which after playing in the provinces first appeared at Court in 1582, at the same time as the company of Lord Strange's father, the 4th Earl of Derby, a duplication which has caused some confusion in the records. This company, which appears to have amalgamated intermittently with the *Admiral's Men, was at the *Theatre in 1590–1 and may have been the first to employ Shakespeare, either as actor and playwright or as playwright only. They were also at the *Rose under *Henslowe in 1592–3, when they may have produced the first part of Shakespeare's *Henry VI* and perhaps *The Comedy of Errors.* They also appeared in the lost *Titus and Vespasian* (1592), which may have been an early draft of *Titus Andronicus* (1594), and in *A Jealous Comedy* (1593), a possible source of *The Merry Wives of Windsor* (1600). The company's repertory included the anonymous *A Knack to Know a Knave,* *Greene's *Orlando furioso* and *Friar Bacon and Friar Bungay,* and *Marlowe's *The Jew of Malta,* as well as a number of plays now lost. They separated themselves from the Admiral's Men in 1594, and on the death of their patron later in that year went into the provinces, some of their actors remaining in London to join the newly formed *Chamberlain's Men.

Stranitzky, Joseph Anton (1676–1726), Austrian actor, associated with the comic figure *Hanswurst. He began his theatrical career with a travelling puppet-show and in about 1705 arrived in *Vienna, where he headed a company of 'German Comedians' modelled on the lines of the *English Comedians. They played at the Ballhaus from 1707 and in 1711 he took over the newly built Kärntnertortheater, making it the first permanent home of German-language comedy. For this company Stranitzky adapted old plays and opera librettos, making Hanswurst (a role which he bequeathed to Prehauser) the central comic character.

Strasberg, Lee (1901–82), American theatre director, who had had some acting experience with the *Theatre Guild before, in 1931, he helped to found its offshoot the *Group Theatre. With Cheryl *Crawford he directed its first production, Paul *Green's *The House of Connelly,* and later was responsible for a number of other productions, including Maxwell *Anderson's *Night over Taos* (1932), Sidney *Kingsley's *Men in White* (1933), and Paul Green's *Johnny Johnson* (1936). When the Group Theatre ceased production he directed such plays as *Odets's *Clash by Night* (1941) and *The Big Knife* (1949) for other managements. A firm believer in the *Method for the formation of actors, he became in 1950 a director of the Actors' Studio, where in 1965 he was responsible for the production of *Chekhov's *Three Sisters* seen during the *World Theatre Season in London. He also directed for the Studio the première of Odets's *The Silent Partner* (1972), Paul Zindel's *The Effect of Gamma Rays on Man-in-the-Moon Marigolds,* and *O'Neill's *Long Day's Journey into Night* (both 1973).

Stratford, Conn., see AMERICAN SHAKESPEARE THEATRE.

Stratford (Ontario) Festival, Canada, annual drama and music festival which lasts from mid-May to the end of Oct. It was founded for the production of plays by Shakespeare, along the lines of the festival at Stratford-upon-Avon. Its first theatre was an immense tent housing an approximation to an Elizabethan *open stage, thus fulfilling a long-standing dream of Tyrone *Guthrie, its first Director, and his designer Tanya *Moiseiwitsch, both from England. The opening production in 1953 was *Richard III* with Alec *Guinness, followed by *All's Well that Ends Well.* Concerts, opera, and a film festival were later added to the programme, and in 1957 a permanent theatre was erected, which retained the spirit of the tent and the original stage. The conical roof, locked by 34 girders which meet at the centre like the spokes of a wheel, and the cantilevered balcony are without visible means of support. The auditorium sweeps in a full semi-circle round the stage, from which no seat is more than 65 ft. away, although the building seats 2,262. In 1956 Guthrie was succeeded by another English director, Michael Langham, who continued to maintain a permanent Canadian company with visiting stars, though it was not long before the stars themselves were Canadian. In 1956 the company appeared at the *Edinburgh Festival in a distinctively Canadian production of *Henry V* in which the French king was played by Gratien *Gélinas, with actors from the Théâtre du *Nouveau Monde as his courtiers, Henry V being played by the Canadian actor Christopher *Plummer, supported by an English-language cast. A year later Plummer inaugurated the new theatre building with his performance of Hamlet. The company has also visited the USA, London, Continental Europe, Australia, and *Chichester. In 1963 a second theatre, the Avon, a proscenium house seating 1,100, was opened, and in 1971 an experimental Third Stage was added.

Two Canadian directors, Jean *Gascon and John *Hirsch, succeeded Langham in 1967, and after Hirsch resigned in 1969 Gascon remained as sole director until 1974. The repertoire was expanded to include Jacobean authors and *Molière, as well as European classics ranging from the elder *Dumas's *The Three Musketeers* to Samuel *Beckett's *Waiting for Godot*. The Shakespeare canon was completed with *Titus Andronicus* in 1978. Gascon was succeeded in 1975 by an English director, Robin *Phillips, amid a storm of nationalistic controversy. He pioneered a Young Company, offering stardom to young Canadian actors, and invited many Canadian directors to share productions with him. He also experimented with double and triple casting of certain leading roles, and astutely utilized both the Avon Theatre and the Third Stage for Shakespeare productions, hitherto reserved for the main festival stage. The festival was given a more truly festive character by opening several seasons with a casual 'Gala' or 'Revel'. Although stars had been imported over the years, among them Irene *Worth, Paul *Scofield, and Julie *Harris, with Phillips the idea became central. In addition to engaging the Canadian actor Hume *Cronyn and his wife Jessica *Tandy, Phillips essentially built four of his six seasons around Maggie *Smith and Brian Bedford, and in 1979 and 1980 Peter *Ustinov was invited to play King Lear. After a financial crisis John Hirsch returned as Director, 1981–5, his varied programme including most of Shakespeare's comedies, a *Gilbert and Sullivan cycle, and three modern American classics. John *Neville's tenure as Director, 1986–9, was marked by the introduction of musicals on the main stage, starting with Rodgers and *Hart's *The Boys from Syracuse*, and the revival of rarely

seen plays such as *Dekker's *The Shoemaker's Holiday* and *Brecht's *Mother Courage*. Only one Canadian play, *The Canvas Barricade* (1963) by Donald Jack, has been shown on the festival stage, but a number have had major productions at the Avon, among them James *Reaney's *Colours in the Dark* (1967) and *Foxfire* (1980) by Susan Cooper and Hume Cronyn. Others have been seen on the Third Stage. Long-standing Canadian actors at Stratford have included William *Hutt, Martha Henry, Douglas Rain, and Kate *Reid.

Stratford-upon-Avon, Warwickshire, birthplace of Shakespeare, and the scene of an annual festival of his plays. The first festival, organized by *Garrick in 1769, was chiefly remarkable for the fact that none of Shakespeare's plays was performed. After that there were only sporadic performances by strolling companies. In 1864, on the initiative of the then mayor, Edward Fordham Flower, member of a local family of brewers, actors from London were invited to appear in a Grand Pavilion, specially built for the occasion, to celebrate the 300th anniversary of Shakespeare's birth, the plays performed being *Twelfth Night*, *The Comedy of Errors*, *Romeo and Juliet*, *As You Like it*, *Othello*, and *Much Ado about Nothing*. Some years later Flower's son, Charles Edward Flower, donated the site and provided most of the money for the building of the *Shakespeare Memorial Theatre, which later became the *Royal Shakespeare Theatre. The *Other Place opened in 1974, and The *Swan in 1986.

Apart from the Royal Shakespeare Theatre's library and picture gallery, Stratford-upon-Avon has a library and conference centre attached to Shakespeare's birthplace. There is also an Institute of Shakespeare Studies sponsored by the University of Birmingham, which grew out of the lectures arranged by Professor Allardyce *Nicoll for the *British Council at Mason Croft, the former home of the novelist Marie Corelli; it provides a centre for research work all the year round and a meeting-place for Shakespeare scholars during the summer conference. Hall Croft, the home of Shakespeare's daughter Susanna after her marriage in 1607, also provides a useful centre of activity during the Shakespeare season.

Stratton, Eugene, see MUSIC-HALL.

Strauss, Botho (1944–), German playwright, novelist, and poet, whose non-naturalistic plays scathingly attack modern materialism. After co-editing *Theatre Today*, 1967–70, he wrote *The Hypochondriacs* (1972) and *Familiar Faces* (1975). He then won international praise with *Three Acts of Recognition* (1977; NY, 1982), about relationships between people at an art gallery the day before an exhibition begins. His next play, *Big and Little* (1978), deals in 10 episodes with the experiences of a middle-aged bag-lady, filled with unshakeable optimism, in modern society. It was performed in New York in 1979 and in London, as *Great and Small*, with Glenda *Jackson, in 1983. Other plays include *The Park* (1984; NY, 1988), in which Titania and Oberon from *A Midsummer Night's Dream* find themselves in a modern park; and *The Tourist Guide* (1986; London, 1987), set in Greece, about an affair between a female tourist guide and a middle-aged German teacher. Strauss worked with *Stein at the Schaubühne, 1970–5, adapting for production such plays as *Ibsen's *Peer Gynt* (1971) and *Gorky's *Summerfolk* (1974); Stein also directed the world première of *Big and Little* there.

Street Theatre, see BUSKER and COMMUNITY THEATRE.

Strehler, Giorgio, see ODÉON and PICCOLO TEATRO DELLA CITTÀ DI MILANO.

Strindberg, (Johann) **August** (1849–1912), Swedish dramatist who also wrote prolifically in the fields of fiction, criticism, and social commentary. After an unhappy childhood, his university career was cut short for financial reasons, and these early miseries set the tone both of his later life and of his writings. His first major play, *Mäster Olof* (1872), was a study of the humanist Olaus Petri, a prose work later rewritten in verse. In 1877 Strindberg married an actress; the first of his three disastrous experiences of marriage, it ended in divorce in 1891. Discouraged by the lack of success of his early dramas, he spent the greater part of the years 1883 to 1896 abroad, though he had to return to Sweden during this time to stand trial for blasphemy in his collection of stories *Giftas* (*Married Life*); he was acquitted. Misogyny and despair mark the plays of this period, notably *Fadren* (*The Father*, 1887), *Fröken Julie* (*Miss Julie*), and *Fordringsägare* (*The Creditors*), both 1888; and the one-act *Den starkare* (*The Stronger*, 1889). In them, as well as in the important Preface to *Fröken Julie*, Strindberg denounces the moral corruption of human nature, and concentrates on sin, crime, abnormality, and the pitiless battle between the sexes as he saw it. In 1895–7, following his divorce from his second wife, Strindberg suffered a severe mental breakdown, his 'Inferno crisis'; he became interested in alchemy and occultism and, abandoning his previous allegiance to the philosophy of Nietzsche, came under the influence of Swedenborg. The *naturalism of his earliest dramatic works developed into a highly charged *realism (he preferred the term 'neo-naturalism') which has been seen as an anticipation of *Sartre. In this mode he explored his conception of spiritual reality in *Advent* (1898); in five plays completed in 1899—in the historical dramas *Folkungasagan* (*The Saga of the Folkungs*), *Gustaf Vasa*, *Erik XIV*, and *Gustaf Adolf*, and in *Brott och brott* (*Crimes and Crimes*);

in *Påsk* (*Easter*, 1900); and supremely in *Dodsdansen I–II* (*The Dance of Death*, 1901). His work thereafter became progressively more experimental, symbolic, and dream-like, with *Till Damaskus I–II* (1898–1904) and (in 1901) *Kronbruden* (*The Crown Bride*, seen in New York as *The Bridal Crown*), *Svanehvit* (*Swanwhite*), and *Ett drömspel* (*A Dream Play*), all of them containing elements in which the major movements of 20th-century drama are rooted, notably *Expressionism, Surrealism, and the Theatre of the *Absurd. Also in 1901 Strindberg married his third wife, another actress; they were divorced in 1904.

Strindberg was always closely concerned with the practical presentation of his plays, and in 1907, with the director August Falcke, he established the Intima Teatern in Stockholm, writing his last group of plays—*Spöksonaten* (*The Ghost Sonata*), *Oväder* (*The Storm*), *Brända tomten* (*The Burnt House*), and *Pelikanen* (*The Pelican*)—for performance there. His last drama, *Stora landsvägen* (*The Great Highway*, 1909), contains many autobiographical elements.

Although many of Strindberg's plays have been translated and acted in English, only a few are regularly performed in English-speaking countries. The best known are probably *The Father* (first seen in New York in 1912 and in London in 1927) and *Miss Julie* (produced privately in London in 1927 by the *Stage Society and publicly in 1935; not seen in New York until 1956). In America the *Provincetown Players first performed *The Ghost Sonata* in 1924 (as *The Spook Sonata*, under which title it was seen in London in 1927) and *A Dream Play* in 1926 (London, 1933), and both productions had a great influence on the development of *O'Neill. Among other productions were *Advent* (London, 1921); *The Creditors* (NY, 1922; London, 1952); *Easter* (NY, 1926; London, 1928); *The Dance of Death* (NY, 1923; London, 1928—superbly revived in 1967 by the *National Theatre, with Laurence *Olivier); *The Road to Damascus* (London, 1937; NY, as *To Damascus*, 1961); *The Stronger* (NY, 1937); *There are Crimes and Crimes* (London, 1946).

Strip-Tease, see BURLESQUE, AMERICAN.

Studio Arena Theatre, Buffalo, NY, non-profit-making organization which serves the 1.5 million people of the western part of the state. It grew out of a community theatre group founded in 1927, and since 1965, when it took over an adapted nightclub, has been fully professional. In 1978 it moved across the street to a new 637-seat thrust-stage theatre converted from a *burlesque house. A subscription series of seven plays is offered in a season which runs from Sept. to May, productions including a wide variety of old and new plays and musicals. There is a decided emphasis, however, on contemporary theatre, and over 30 world or

American premières have taken place. The theatre school, founded in 1927, offers an integrated training programme and plays for children.

Sturm und Drang (Storm and Stress), name given to one aspect of 18th-century German Romanticism which carried to excess the doctrine of the rights of man and Rousseau's plea for a return to nature. It took its name from the title of a play by Klinger, produced in 1776, and was much influenced by Shakespeare. Among the frequently recurring themes in the plays which characterize the movement are the tragedy of the unmarried mother executed for infanticide while her seducer goes free, treated by *Goethe and Klinger; the conflict between hostile brothers in love with the same woman, as in *Julius von Tarent* (1776) by Leisewitz and *Schiller's *Die Räuber* (1782); and the overmastering power of love, hurling even honourable natures into crime, as in *Golo und Genoveva* (c.1780) by 'Maler' Müller. The movement had repercussions all over Europe, and its influence can clearly be seen in early 19th-century English *melodrama. A native German offshoot of it was the *Ritterdrama*.

Stuyvesant Theatre, New York, see BELASCO THEATRE.

Sudakov, Ilya Yakovleivich (1890–1969), Soviet actor and director, who trained at the *Moscow Art Theatre where, under *Stanislavsky's supervision, he directed *Bulgakov's *Days of the Turbins* (1926) and *Ivanov's *Armoured Train 14-69* (1927) and together with *Nemirovich-Danchenko staged Ivanov's *Blockade* (1929), *Afinogenov's *Fear* (1935), and *Trenev's *Lyubov Yarovaya* (1936). In 1937 he became Artistic Director of the *Maly Theatre, where some of his best productions were *Gutzkow's *Uriel Acosta* (1940), *Invasion* (1942) by *Leonov, and *Front* (1943) by *Korneichuk. Among his post-war productions were revivals of *Chekhov's *Uncle Vanya* (1947), Afinogenov's *Mother of Her Children* (1954), and Alexei *Tolstoy's *The Road to Calvary* (1957). In 1959 he became Director of the Gogol Theatre (formerly the Moscow Transport Theatre), where he remained until his death.

Sudermann, Hermann (1857–1928), German novelist and dramatist, whose first play, *Die Ehre* (*Honour*), was produced with great success in 1889. He became the main exponent of the new theatre of *realism in Germany, much influenced by *Ibsen, as can be seen in *Das Glück im Winkel* (*Happiness in a Quiet Corner*, 1896) and *Johannsfeuer* (*The St John's Eve Fire*, 1900). The play by which he is chiefly remembered is *Heimat* (1893), which provided a popular melodramatic vehicle for many famous actresses, including *Duse and *Bernhardt, who both played it in London at the same time in 1895. As *Magda*, in a translation by Louis N. *Parker, it was first seen in English in 1896 at the *Lyceum Theatre with Mrs Patrick *Campbell in the title-role, a part played in 1923 by Gladys *Cooper and in 1930 by Gwen *Ffrangcon-Davies. It was produced in New York in 1904 and revived in 1926.

Sukhovo-Kobylin, Alexander Vasilievich (1817–1903), Russian dramatist, whose whole life was overshadowed by the death of his mistress, whom he was falsely suspected of having murdered. He had already started to write a play, *Krechinsky's Wedding*, before this tragic event, and while in jail he finished it. It was finally staged at the Moscow *Maly Theatre in 1855. Its subject, like that of his two later plays—*The Case* (written in 1857 but not performed until 1881) and *Tarelkin's Death* (written in 1868 and staged in 1900), with which it forms a trilogy whose characters interlock—was the decay of the patriarchal life of old Russia and the growing power of a corrupt bureaucracy. Although Sukhovo-Kobylin always asserted that he was not a revolutionary, he was regarded as a dangerous man and his plays were banned by the censor. He was a great friend and admirer of *Gogol, whose influence is seen in the character of Krechinsky. Worn out by his struggles with the censorship, he finally gave up the theatre and in 1857 settled permanently in France. A musical version of *Tarelkin's Death* was presented by *Tovstonogov in 1984.

Sumarokov, Alexei Petrovich (1718–77), Russian dramatist, who wrote in the neo-classical style imported from Germany and France, but took his subjects from Russian history and in his dialogue endeavoured to refine and purify the Russian language, writing with great intensity and economy. His plays, of which the first, *Khorev*, was produced in 1749, were acted by students from the Cadet College for sons of the nobility, of which Sumarokov had been one of the first pupils on its foundation in 1732. It was largely owing to his efforts that the first professional Russian company was formed, and in 1756 he was appointed head of the Russian (as distinct from the Italian and French) Theatre in St Petersburg, remaining there until 1761, when his liberal outspokenness caused his plays to be banned and he himself to be dismissed from office.

Summer Stock, see STOCK COMPANY.

Super(numerary), see STOCK COMPANY.

Surrey Music-Hall, Surrey Gardens Music-Hall, London, see MUSIC-HALL.

Surrey Theatre, London, in Blackfriars Road, Lambeth. This stood on the site of the Royal Circus which, rebuilt after a fire in 1806, was converted into a theatre by Robert *Elliston, who gave it the name by which it was thereafter known. To avoid trouble with the *Patent Theatres, Elliston introduced a ballet into every production, including *Macbeth*, *Hamlet*, and *Farquhar's *The Beaux' Stratagem*. He left in 1814, and the Surrey became a circus again until it reopened as a theatre in 1816, with little success. Not until Elliston returned did its fortunes change, with the production in 1829 of Douglas *Jerrold's *Black-Ey'd Susan*, which with T. P. *Cooke as William, the nautical hero, had a long run. Osbaldiston took over when Elliston died, and among other plays produced Edward *Fitzball's *Jonathan Bradford; or, The Murder at the Roadside Inn*, which ran for 260 nights. But it was Richard Shepherd (who succeeded Alfred Bunn in 1848 and remained at the theatre until 1869) who established the theatre's reputation for rough-and-tumble *transpontine melodrama. In 1865 the theatre was again burnt down, and a new theatre, seating 2,161 in four tiers, opened in the same year. Little of note took place until 1881, when George *Conquest took over, staging sensational dramas, many of them written by himself, which proved extremely popular, and each Christmas he put on an excellent *pantomime. The Surrey prospered until his death in 1901, but thereafter went rapidly downhill until in 1920 it became a cinema. It finally closed in 1924 and the building was demolished in 1934.

Susan Stein Shiva, see PUBLIC THEATRE.

Susie, see TOBY 2.

Sussex's Men, company of players founded in about 1569 by the 3rd Earl of Sussex. They were seen at Court in 1572, but then played chiefly in the provinces until in 1593–4 they appeared for a six-week season at the *Rose under the management of *Henslowe. The one new play in their repertory was *Titus Andronicus*, which Shakespeare may have refashioned for them from an earlier play, *Titus and Vespasian*, in which *Pembroke's Men had appeared in 1592. Having lost their patron, the 4th Earl who died in 1593, they came under his son but were disbanded soon after they left the Rose.

Suzman, Janet (1939–), South African-born actress who has made an outstanding reputation for herself in England, mainly with the *RSC, which she joined in 1962 after making her first appearance on the stage earlier the same year in Ipswich in Keith Waterhouse and Willis *Hall's *Billy Liar*. She was first seen in London in *The Comedy of Errors*, and came into prominence with her embattled Joan La Pucelle in *Henry VI* in 1963. A modern role in *Pinter's *The Birthday Party* (1964) was followed by an impressive gallery of Shakespeare's heroines— Portia in *The Merchant of Venice* and Ophelia in *Hamlet* (both 1965), Katharina in *The Taming of the Shrew* (1967), Rosalind in *As You Like It*,

and Beatrice in *Much Ado about Nothing* (both 1968), her Katharina and Beatrice both being seen also in the USA. In 1972 she scored a striking success as Cleopatra in *Antony and Cleopatra*, endowing the character with vocal strength, voluptuousness, and a mettlesome intellect; in the same year she also played Lavinia in *Titus Andronicus*. She gave a sensitive performance as a dilapidated South African whore in Athol *Fugard's *Hello and Goodbye* in 1973. Outside the RSC, she played Masha in Jonathan *Miller's production of *Chekhov's *Three Sisters* in 1976, and a year later was seen in *Brecht's *The Good Woman of Setzuan* at the *Royal Court and in *Ibsen's *Hedda Gabler*. She returned to the RSC in 1980 as Clytemnestra and Helen in *The Greeks*, based mainly on *Euripides' Trojan plays. She appeared in Fugard's *Boesman and Lena* at *Hampstead in 1984, as *Racine's Andromache at the *Old Vic in 1988, and in Ronald Harwood's *Another Time* (1989). She was formerly married to Trevor *Nunn.

Svoboda, Josef (1920–), Czech scene designer, associated with over 500 productions, who became chief designer at the Czech National Theatre in 1948 but has also worked for playhouses and opera houses all over the world. In Prague some of his best work was done for *Gogol's *The Government Inspector* in 1948, *Hamlet* in 1959, *Chekhov's *The Seagull* in 1960, and *Sophocles' *Oedipus Rex* in 1963. He also designed the production of the *Čapeks' *The Insect Play* which the Czech National Theatre took to London in the *World Theatre Season of 1966, as a result of which he was commissioned by the *National Theatre company to design productions of *Ostrovsky's *The Storm* in 1966 and Chekhov's *Three Sisters* in 1967. He returned to the Old Vic in 1970 to design Simon *Gray's adaptation of Dostoevsky's *The Idiot*, and later work included *Claudel's *Partage de midi* at Louvain in 1984. His originality and technical ingenuity are displayed in his use of machinery, lighting, and electronic devices. He is also well known for his Laterna Magika, a *revue-type entertainment combining live action with projected images which arose out of the work he did for the Brussels World's Fair in 1958. It is now regularly presented in Prague in association with the Czech National Theatre.

Swan, The, the *RSC's third theatre in Stratford-upon-Avon, opened in 1986 in the shell of the *Shakespeare Memorial Theatre destroyed by fire in 1926. A Jacobean-style playhouse seating 430 in three galleries curving round an apron stage, it stages the work of Shakespeare's contemporaries and successors, once highly popular but now rarely performed. The opening season consisted of *The Two Noble Kinsmen*, possibly the joint work of *Fletcher and Shakespeare, *Jonson's *Every Man in His Humour*, Aphra *Behn's *The Rover*, and Thomas *Heywood's *The Fair Maid of the West*. It occasionally stages some of Shakespeare's own plays, such as *Titus Andronicus* in 1987, and in 1989 premièred a new play, Peter Flannery's *Singer*, with Anthony *Sher.

Swanston, Eliard [Eyllaerdt, Hilliard] (?–1651), prominent member of the *King's Men from 1624 to the closing of the theatres in 1642. He appears to have been on the stage for at least two years before joining the company, and was not only one of its leading actors, but also important on the business side, together with *Lowin and *Taylor. He played a variety of roles, many of them in revivals, including Othello and Richard III and *Chapman's Bussy d'Ambois. During the Civil War he became a Parliamentarian, in contrast to most of the other actors, who remained staunchly Royalist, and earned his living as a jeweller.

Swan Theatre, London, on Bankside in Southwark, in Paris Garden. The fourth theatre to be built in London, it was a circular building constructed in about 1595. Although it did not have a very exciting history, a good deal is known about it, as Johannes de Witt, a Dutch visitor to London in 1596, sent a sketch of its interior, probably made from memory, to a friend in Utrecht, who copied it into his commonplace book. This is the only known drawing of an Elizabethan theatre interior and presents many interesting and puzzling features. According to de Witt's account of it, it held about 3,000, and was of wood on a brick foundation with flint and mortar work between wooden pillars painted to resemble marble. It served *Henslowe as a model for the *Hope, and is frequently mentioned in his diary, but it had no regular company, and was used as often for sports and fencing as for plays. In 1597 trouble arose over the presentation by *Pembroke's Men of *The Isle of Dogs*, a 'seditious' comedy by *Jonson and *Nashe, which landed Jonson and others in prison. The company was broken up, the leading actors joined the *Admiral's Men at the *Rose, and the Swan was used only intermittently. In 1602 an infuriated audience wrecked the interior after paying for an entertainment which never materialized. The last play to be seen there was *Middleton's *A Chaste Maid in Cheapside* (1611), performed by *Lady Elizabeth's Men. When the Hope Theatre opened in 1614 the Swan fell into disuse, though it was still being used in 1620 for prize-fights.

Swazzle, see PUNCH AND JUDY.

Sydney Opera House, unusual and controversial building, designed by Jorn Utzon, which took nearly 15 years to erect. About 8,000 helped in the construction, the largest number on site at any one time being 1,500.

The building opened in 1973 with a permanent staff of about 350. It has four major auditoriums, two of which are devoted to drama. The Drama Theatre, seating 544, was first occupied by the *Old Tote Theatre Company, which in Jan. 1979 was replaced by the newly formed Sydney Theatre Company. During 1979 the STC held a season of plays performed by other companies, and in 1980 its permanent occupation of the Drama Theatre began with Darrell's *The Sunny South*. The Playhouse, seating 398, has been operating as a venue for drama since 1983, opening with Arthur *Miller's *All My Sons*. Other auditoriums, including the Opera Theatre and the Broadwalk Studio, have also been used for drama.

Symbolism. Symbols have been used on the stage since the earliest times. Much of Elizabethan 'stage furniture' was symbolic, a throne standing for a Court, a tent for a battlefield, a tree for a forest. Symbolic elements are found in *Chekhov and in the later plays of *Ibsen and *Strindberg. But Symbolism as a conscious art-form, conceived as a reaction against *realism, came into the theatre with *Maeterlinck, writing under the influence of Mallarmé and Verlaine. His characters have no personality of their own, but are symbols of the poet's inner life. This aspect was intensified in the early plays of *Yeats. Other dramatists to come under the influence of Symbolism include *Andreyev and *Evreinov in Russia, Hugo von *Hofmannsthal and the later *Hauptmann (with *Die versunkene Glocke*) in Germany, *Synge (*The Well of the Saints*) and *O'Casey (*Within the Gates*) in Ireland, and *O'Neill in the United States.

Syndicate, see THEATRICAL SYNDICATE.

Synge, (Edmund) **John Millington** (1871–1909), Irish dramatist, with *Yeats a leading figure in the Irish dramatic movement. Although he died prematurely, his six completed plays establish him as the greatest of modern Irish dramatists. His control of structure, whether in comedy or tragedy, is assured; his revelation of the characters and thought-processes of a subtle and imaginative peasantry is penetrating; his language, and especially his imagery, is rich, live, and essentially poetic. An early play, *When the Moon Has Set*, apparently written in 1901 and rejected for production by Yeats and Lady *Gregory, was found among his papers after his death and published with his other works in 1962–8; but the first of his plays to be performed, by the *Irish National Dramatic Society, was *In the Shadow of the Glen* (1903), the first of a series of grave, original studies of Irish thought and character which drew upon the author the hostility of some of his early audiences. It was first seen in London in 1904, and was followed by *Riders to the Sea*, produced by the *Abbey Theatre in the same year, a one-act tragedy whose brevity and intensity make it one of the best of modern short plays. It was first seen in London in 1904, in a double bill with *In the Shadow of the Glen*, and in New York in 1920. A third play, *The Tinker's Wedding*, begun as early as 1902 but revised several times before its publication in 1908, was given its first production in London in 1909, being considered 'too dangerous' for an Abbey audience; but its comedy, drawn from the life of the Irish roads, is richer and more jovial than any other that Synge wrote. It was seen in New York, with *Riders to the Sea* and *In the Shadow of the Glen*, in 1957. *The Well of the Saints* (Dublin and London, 1905; NY, 1931) is another comedy in which poetic beauty is mingled with an underlying irony that is potentially tragic.

The climax of Synge's achievements in the theatre is the comedy of bitter, ironic, yet imaginative realism, *The Playboy of the Western World* (1907). The unsparing though sympathetic portraiture in this play caused riots in the Abbey Theatre on its first production, and disturbances led by certain Irish patriots when it was first produced in New York in 1911. It had its first London production in 1907. It has long been accepted as Synge's finest work; his power is here seen at its fullest, as it could not be in the unfinished and unrevised *Deirdre of the Sorrows*, his last play, in which he turned back, as Yeats had done before him, to the ancient legends of Ireland. This was first produced posthumously at the Abbey in 1910 and was seen in London in the same year.

T

Tabarin [Antoine Girard] (?–1626), French actor, who in about 1618 set up a booth on the Pont Neuf with his brother, the quack-doctor Mondor. Here, and in the Place Dauphine, Tabarin would mount a trestle platform and, donning his famous floppy and polymorphic hat, would put the holiday crowd into a good humour before Mondor began the serious business of the day, selling his nostrums and boluses. Most of his material Tabarin wrote himself, or rather sketched out in the style of the *commedia dell'arte* scenarios; and when a person unknown brought out in 1622 a collection of *œuvres tabariniques*, he himself published in the same year *L'Inventaire universel des œuvres de Tabarin*, containing his farces, puns, jokes, and monologues. He was probably much influenced by the Italian actors who often played in Paris, and from them he took the 'sack-beating' scene which *Molière borrowed for *Les Fourberies de Scapin* (1671). Tabarin himself never appeared in a theatre, but he was remembered long after his death, and his name passed into everyday speech (*faire le tabarin* = play the fool).

Tabernaria, see FABULA 4: TOGATA.

Tableau Curtain, Tabs, see CURTAIN.

Taganka Theatre, Moscow. This theatre, whose official name is the Theatre of Drama and Comedy, is situated on Taganka Square, at some distance from the centre of the capital, and seats about 600. It was founded in 1946 and during the next 18 years presented more than 50 Soviet plays. In 1964 *Lyubimov took it over and made it a centre for avant-garde work, presenting a wide range of innovatory productions including an adaptation of John Reed's book *Ten Days That Shook the World* in 1967, *Molière's *Tartuffe* in 1969, and in 1973 Peter *Weiss's *How Mr Mockinpott was Relieved of His Suffering* and *Gorky's *The Mother*. There was a modern-dress *Hamlet* in 1974 and in 1977 a brilliant staging of *Bulgakov's novel *The Master and Margarita*. There have also been 'montage' versions of poems by *Mayakovsky, *Pushkin, and others. Lyubimov went into exile in 1984, but was reinstated in 1989 and restored some of his banned productions.

TAG Theatre Company, see CITIZENS' THEATRE.

Taïrov, Alexander Yakovlevich (1885–1950), Soviet director, who after some experience with *Meyerhold in St Petersburg joined Gaideburov's Mobile Theatre in 1908. Five years later, while working for *Mardzhanov, he met his future wife **Alisa Koonen** (1889–1974), and together they opened the *Kamerny Theatre in 1914. The first productions there, which included *Wilde's *Salome*, showed clearly Taïrov's preoccupation with new spatial possibilities in staging, and the Cubist-influenced designs of his scenic artists perfectly matched his intentions. In common with Meyerhold he was antipathetic to *naturalism, but unlike him held the importance of the actor to be fundamental, demanding from his company all-round ability, including virtuoso pantomimic and acrobatic techniques. The position of his wife helped to emphasize the place of the actor in his productions, and almost every one of them became as much a medium for her remarkable talents as for his own experiments. Combining great beauty with exceptional plasticity of movement, she was outstanding as Adrienne Lecouvreur in *Scribe's play, as Juliet, *Racine's Phèdre, and *Shaw's St Joan.

The 1917 Revolution led to the abandonment of Taïrov's aesthetic excesses and to a sharpening of his social consciousness, as reflected in his productions, much influenced by *Expressionism, of *Ostrovsky's *The Storm* in 1924 and of three plays by *O'Neill—*The Hairy Ape* and *Desire under the Elms* in 1926 and *All God's Chillun Got Wings* in 1929. In the late 1920s Taïrov was severely criticized for his *Formalism and for his choice of plays, which included *Bulgakov's *The Red Island*. He replied to his critics by staging in 1934 *Vishnevsky's *An Optimistic Tragedy*, now considered one of the finest products of *Socialist Realism. This was followed in 1938 by a dramatization of Flaubert's novel *Madame Bovary* with Alisa Koonen. Although Taïrov's theatre was put by the authorities under the control of a committee he was not dismissed, as Meyerhold had been in similar circumstances, but continued to work until a year before his death.

Talk of the Town, London, see HIPPODROME.

Talma, François-Joseph (1763–1826), French actor, brought up in England, where he was about to join a London company when his father sent him back to Paris. There he entered the newly founded École de Déclamation, shortly to become the Conservatoire, and after tuition from *Molé and others made his début at the *Comédie-Française in 1787 in *Voltaire's *Mahomet*. Although handsome, with a fine presence and a resonant voice which made him an excellent speaker of verse, he had played only small parts when in 1789 he appeared in the title-role of Chénier's *Charles IX*, which all the older actors had refused because of its political implications. Talma declaimed the revolutionary speeches with such fervour that the theatre was in an uproar, and was eventually closed. Supported by some of the younger members of the company, he moved to the Théâtre de la Révolution (the present Comédie-Française), where he appeared in *Corneille's *Le Cid* and other classical revivals, as well as in some of Shakespeare's tragic parts in translations by *Ducis. In 1799, when Napoleon, who had met and become friends with Talma at Mlle Montansier's *salon*, reconstituted the Comédie-Française, Talma rejoined the company and put in hand many reforms, notably in the costuming of plays. He was the first French actor to play Roman parts in a toga instead of contemporary dress, and he also reformed theatrical speech, suppressing the exaggerations of the declamatory style and allowing the sense rather than the metre to dictate the pauses. One of his great successes at this time was in a revival of *La Fosse's *Manlius Capitolinus* in 1805. In 1817 he visited London, appearing at *Covent Garden in extracts from his best parts, and he attended John Philip *Kemble's farewell performance and banquet. He remained on the stage until four months before his death with no lessening of his great powers.

Tamayo y Baus, Manuel (1829–98), Spanish dramatist, who represents the transition from Romanticism to Realism in the Spanish theatre. His early plays were first performed by his parents, who were both on the stage, and were mainly translations from the French and German. Of his own plays, *La locura de amor* (*The Insanity of Love*, 1855) and *La bola de nieve* (*The Snowball*, 1856), dealing with jealousy in historical and modern settings respectively, were the first to succeed. The only one to be remembered now is *Un drama nuevo* (*A New Play*, 1867), written under the pseudonym Joaquín Estebánez. It is set in Elizabethan times, with Shakespeare as one of the characters. An actor who is playing Yorick (the 'king's jester' referred to in *Hamlet*) kills on stage during a mock fight the young actor with whom he suspects his wife is in love, egged on by the Iago-like character called Walton.

Tandy, Jessica (1909–   ), American actress, born in London, who began her stage career at the *Birmingham Repertory Theatre in 1928, being seen briefly in London a year later before going to New York in 1930 to play in G. B. Stern's *The Matriarch*. On her return to London she appeared in a wide variety of plays, including Dodie Smith's *Autumn Crocus* (1931), *Hamlet* in 1934 as Ophelia opposite John *Gielgud, and

*Rattigan's *French without Tears* (1936). After a season with the *Old Vic company in 1937 she returned to New York to play Kay in the Broadway production of J. B. *Priestley's *Time and the Conways* (1938), and in the 1940s she settled permanently in the USA. The role of Blanche Du Bois in Tennessee *Williams's *A Streetcar Named Desire* (1947), which she played on Broadway for over two years, established her as one of America's leading actresses.

With her second husband Hume *Cronyn she formed a notable stage partnership. Their early plays together included Jan de Hartog's two-character study of a marriage *The Fourposter* (1951) and N. C. *Hunter's *A Day by the Sea* (1955). She appeared in Peter *Shaffer's *Five Finger Exercise* (1959) and at the *American Shakespeare Theatre in 1961, where she played Lady Macbeth, before joining her husband again in the London production of Hugh Wheeler's *Big Fish, Little Fish* (1962), and in the opening season at the *Guthrie Theater, Minneapolis (1963), when she played Gertrude in *Hamlet*, Olga in *Chekhov's *Three Sisters*, and Linda Loman in Arthur *Miller's *Death of a Salesman*. They returned to the Guthrie in 1965 and were also seen in the New York productions of *Dürrenmatt's *The Physicists* (1964) and *Albee's *A Delicate Balance* (1966). She starred independently in Tennessee Williams's *Camino Real* (1970) and Albee's *All Over* (1971), and continued her partnership with Cronyn in *Beckett's *Happy Days* (1972) and the double bill *Noël Coward in Two Keys* (1974). They were both at the *Stratford (Ontario) Festival in 1976, where she played Lady Wishfort in *Congreve's *The Way of the World* and Hippolyta/Titania in *A Midsummer Night's Dream*. In New York (1977; London, 1979) they starred in another two-character play, D. L. Coburn's *The Gin Game*, set in an old people's home, and at Stratford (Ontario) she played Mary Tyrone in *O'Neill's *Long Day's Journey into Night* in 1980. She was with Cronyn in *Foxfire* (1982), which he jointly adapted (with Susan Cooper) from books of folklore. She starred in a revival of Tennessee Williams's *The Glass Menagerie* in 1983, and in 1986 was with her husband in Brian Clark's *The Petition*, which again had only two characters.

Tanguay, Eva, see VAUDEVILLE, AMERICAN.

Tarkington, (Newton) **Booth** (1869–1946), American novelist, and author of a number of plays, of which the best known is the romantic costume drama *Monsieur Beaucaire* (1901). Based in collaboration on his novel of the same title, this was spoilt artistically by the substitution of a conventionally happy ending for the ironic ending implicit in the book. It was nevertheless very successful in America, with Richard *Mansfield in the name part, and also in London, where it later provided the libretto for a musical play with a score by Messager. Among Tarkington's other plays were two for Otis *Skinner, *Your Humble Servant* (1910) and *Mr Antonio* (1916); a charming comedy of youth entitled *Clarence* (1919); and a social drama on the theme of snobbery, *Tweedles* (1923), which failed in production.

Tarleton [Tarlton], **Richard** (?–1588), most famous of Elizabethan clowns, probably the 'Yorick' referred to by Hamlet. A drawing of him in a manuscript now in the British Museum, reproduced in *Tarleton's Jests* (a posthumous work), shows that he was short and broad, with a large, flat face, curly hair, a wavy moustache, and a starveling beard. His usual dress was a russet suit and buttoned cap, with short boots strapped at the ankle, as commonly worn by rustics at the time. A money-bag hangs from a belt at his waist, and he is shown playing on a tabor and pipe. He was one of *Queen Elizabeth's Men, and much of his clowning is believed to have been extempore. Both *Marlowe and Shakespeare may have had him in mind when they railed at 'clownage'. It may have been Shakespeare's desire to confine Tarleton's gagging within reasonable limits that led him to write out in full such richly comic parts as Launce and Speed in *The Two Gentlemen of Verona*, Bottom in *A Midsummer Night's Dream*, Dogberry in *Much Ado about Nothing*, and the First Grave-digger in *Hamlet*. There can be no doubt that the genius of Tarleton was responsible for much of the mingling of tragedy and comedy in early English plays, but his greatest achievements lay in the *jig. The music for some of these has survived, but the only known libretto, *Tarltons Jigge of a horse loade of Fooles* (c.1579), is now considered to be a forgery. Tarleton is, however, known to have written for the Queen's Men a composite play, now lost, entitled *The Seven Deadly Sins*, the first part containing five short plays, the second three. The outline, or *platt, of the second part has been preserved in manuscript. His enduring popularity may be judged by the number of taverns named after him; one, The Tabour and Pipe Man, with a sign-board taken from the frontispiece to *Tarleton's Jests*, still stood in the Borough in London 200 years after his death, while the action of William Percy's *Cuckqueans and Cuckolds Errant* (1601) is said to take place in the Tarlton Inn, Colchester. Tarleton himself for some time ran an eating-house in the City, in Paternoster Row.

Tasso, Torquato (1544–95), Italian poet and playwright, whose *L'Aminta* (1573) was the first true *pastoral and the pattern for many that followed in Italy and France. It was also the source of Browne's speech in *Love's Labour's Lost*, Act IV, Scene iii: 'From women's eyes this doctrine I derive.' This tale of rustic life—rustic in an artificial sense, since it deals with the loves of idealized shepherds and shepherdesses—was first translated into English in 1591 as *Phillis and Amyntas*, and in the following century at least four separate translations were made. The latest version is apparently that prepared by Leigh *Hunt in 1820. Tasso, best remembered for his great epic poem *Gerusalemme liberata* (1581), which may have influenced the writing of *Cymbeline*, was also the author of *Torrismondo* (1587), which shows an early mingling of tragedy and romance. Though classic in form, it deals in romantic fashion with the love of King Torrismondo for Rosmonda, whom he marries. When she is discovered to be his sister, he commits suicide.

Tate, Harry, see MUSIC-HALL.

Tate, Nahum (1652–1715), a poor poet and worse playwright, who collaborated with *Dryden in the second part of *Absalom and Achitophel* (1682) and with Brady in a metrical version of the Psalms, published in 1696, which long remained popular in the Church of England. In 1692 he succeeded Thomas *Shadwell as Poet Laureate, and he was pilloried by Pope in *The Dunciad* (1728). Otherwise he is mainly remembered for his extremely odd versions of some of the plays of Shakespeare. His *History of King Richard II* (1680) was intended to blacken the character of Bolingbroke and render Richard II wholly sympathetic. In *King Lear* (1681) the character of the Fool is omitted, and Cordelia survives to marry Edgar. This remained the standard acting text until *Macready's production of the original version in 1838. *Coriolanus*, as *The Ingratitude of a Common-Wealth; or, The Fall of Caius Martius Coriolanus* (also 1681), was not so badly mangled except in the last act, which incorporates the worst features of *Titus Andronicus*, probably in an unsuccessful bid for popularity.

Taylor, C(ecil) P(hilip) (1929–81), Scottish dramatist, formerly an electrician, who settled in Northumberland. A prolific writer who wrote well over 60 stage plays in 20 years, as well as many for television and radio, he had a strong commitment to regional theatre. His first play *Aa Went to Blaydon Races* was produced in Newcastle, and the *Traverse Theatre, Edinburgh, gave the first performances of *Happy Days are Here Again* and *Of Hope and Glory* (both 1965); *Allergy* (1966), *Lies about Vietnam and Truth about Sarajevo* (both 1969); *Passion Play* (1971); *The Black and White Minstrels* (1972), with Alan *Howard, about two Glasgow couples whose left-wing convictions are strained by a difficult Black tenant; *Next Year in Tel Aviv* and *Columba* (both 1973); *Schippel* (1974), adapted from *Sternheim; *Gynt!* (1975), adapted from *Ibsen; and the two-part *Walter* (1977). Although Taylor's primary concern was political, political ideas do not dominate his characters, but rather act through them: thus in *Bread* *and Butter* (1966) the defeats and compromises of the left are seen through the changing circumstances of two ordinary Glasgow-Jewish couples. Taylor's most highly praised play was his penultimate one, *Good*, which the *RSC produced at The Warehouse (see DONMAR WARE-HOUSE) in 1981 and in the West End and New York in 1982. Tracing the self-deception and moral inertia which enable a liberal German professor, played by Howard, to work in Auschwitz, it embodies Taylor's constant theme of the need for total integrity. Apart from *Good*, only *Schippel* (as *The Plumber's Progress*, 1975) and *And a Nightingale Sang . . .* (1979; NY, 1983), portraying a British working-class family during the Second World War, had mainstream London productions. *Bandits*, about racketeering in 1960s Newcastle, was seen at The Warehouse in 1977 and in New York in 1979, and *The Black and White Minstrels* (1974), again with Howard, and *Goldberg* (1976) were produced at the *Hampstead Theatre. The *Newcastle Playhouse's posthumous production in 1982 of his last play, *Bring Me Sunshine, Bring Me Smiles*, was also seen in London. Of all modern Scottish dramatists, Taylor most successfully learned the lesson of *Bridie's career, that an honest presentation of what one knows best can be of more than local significance.

Taylor, Joseph (*c.*1585–1652), English actor, who joined the *King's Men in 1619, by which time he was already well known, and took over many of Richard *Burbage's parts. He also appeared as the handsome young lovers or dashing villains of *Beaumont and *Fletcher. It is unlikely that, as was once thought, he was coached by Shakespeare himself in the part of Hamlet, but he may well have seen Burbage act it, and he is certainly said to have appeared to advantage in the part. Other roles in which he gave excellent performances in revivals appear to have been Iago in *Othello*, Truewit and Face in Ben *Jonson's *Epicoene* and *The Alchemist*, and Ferdinand in *Webster's *The Duchess of Malfi*. With *Lowin he became one of the chief business managers of the King's Men after the deaths of *Condell and *Heminge.

Taylor (*née* Cooney), **Laurette** (1884–1946), American actress. She had already had a long career, being on the stage from childhood as 'La Belle Laurette' in *vaudeville, when in 1912 she appeared at the *Cort Theatre, New York, as Peg in *Peg o' My Heart*, by her second husband **John Hartley Manners** (1870–1928). She played the part in London in 1914. It was always thereafter associated with her, and she also starred in many of his other 30 plays, though they were not worthy of her considerable talent. After Manners's death she retired from the stage and had alcohol problems, returning occasionally in such roles as Mrs Midgit in a revival of Sutton Vane's *Outward Bound* (1938). In 1945 she gave an outstanding performance as the mother in Tennessee *Williams's *The Glass Menagerie*, her last appearance.

Taylor, Tom (1817–80), English dramatist, whose first play, *A Trip to Kissingen* (1844), was produced by the *Keeleys at the *Lyceum. A prolific writer, and for some time editor of *Punch*, he continued his output of plays until about two years before his death. The best known of his works are probably *To Parents and Guardians* (1846); *Masks and Faces* (1852), a comedy on the life of Peg *Woffington written in collaboration with Charles *Reade and frequently revived; *Still Waters Run Deep* (1855), based on a French novel and remarkable in its time for its frank discussion of sex; *Our American Cousin* (1858), first produced in New York and noteworthy because of the appearance in it of E. A. *Sothern as Lord Dundreary; *The Overland Route* (1860); *The Ticket-of-Leave Man* (1863), a melodrama on a contemporary theme of low life which had much influence on such later works as H. A. *Jones and Herman's *The Silver King*; and two plays written in collaboration, *New Men and Old Acres* (1869) and *Arkwright's Wife* (1873). Taylor was himself an enthusiastic amateur actor, playing at *Dickens's private theatre in Tavistock House and being one of the leading members of the Canterbury Old Stagers. He had little originality, borrowing his material freely from many sources, but his excellent stagecraft and skilful handling of contemporary themes make him interesting as a forerunner of T. W. *Robertson.

Tearle, Sir Godfrey Seymour (1884–1953), English actor. He made his first appearance at the age of 9 in the company run by his father Osmond *Tearle and rejoined it six years later, remaining until his father's death. In spite of his fine voice and natural authority he was slow to achieve recognition, but he was an ideal romantic hero and in 1920 scored a popular success in the dramatization of Robert Hichens's novel *The Garden of Allah*. His finest part was Commander Edward Ferrars in Charles *Morgan's *The Flashing Stream* (1938; NY, 1939), though he was much admired as Maddoc Thomas in Emlyn *Williams's *The Light of Heart* (1940). He achieved belated recognition as a Shakespearian actor with his Antony to the Cleopatra of Edith *Evans in 1946, repeating the role in New York in 1947, and with Othello and Macbeth at the *Shakespeare Memorial Theatre in the 1948–9 season. Other later roles included Hilary Jesson in *Pinero's *His House in Order* in 1951 and the title-role in Raymond *Massey's *The Hanging Judge* (1952).

Tearle, (George) Osmond (1852–1901), English actor, who made his first appearance in Liverpool in 1869, and two years later was seen in the provinces as Hamlet, a part he subsequently played many times. After a further six years' experience he appeared in London and soon formed his own company, with which he toured. In 1880 he joined the *stock company at Lester *Wallack's theatre in New York, making his first appearance as Jaques in *As You Like It*. He then alternated between London and New York until in 1888 he organized a stock company to play Shakespeare at Stratford-upon-Avon, which proved an admirable training-ground for young actors. He was himself a fine Shakespearian actor, combining excellent elocution with a natural elegance and dignity. He had a high reputation in the English provinces, and made his last appearance at Carlisle, dying a week later. He married the granddaughter of W. A. *Conway, and their son Godfrey (above) was also an actor.

Teaser, see FALSE PROS(CENIUM).

Teatro Campesino, Chicano, de la Esperanza, de la Gente, de los Barrios, del Piojo, see CHICANO THEATRE.

Teatro Eslava, Madrid, see MARTÍNEZ SIERRA, GREGORIO.

Teatro Farnese, Parma, begun in 1619 and opened in 1628, was designed by the Italian theatre architect **Giovanni Battista Aleotti** (1546–1636). With the rising passion for operatic spectacles in mind, it consisted of a single-storey auditorium with parallel sides and a rounded end enclosing an open space, suitable for processions, tournaments, or sea-fights, behind which rose a high stage with a proscenium opening. On stage was a complete system of *wings, probably invented and built by Aleotti himself. The building was partly destroyed in the Second World War but was reconstructed in plain timbers and weathered brick. It is not normally used for dramatic performances.

Teatro Olimpico, Vicenza, outstanding example of academic Italian Renaissance theatre architecture. Designed by *Palladio and completed after his death by his pupil Scamozzi, it opened in 1585 with a production of *Sophocles' *Oedipus the King* in a new translation. The building, which still stands, was based strictly on the contemporary neo-classical idea of a Greek theatre, formulated by *Vitruvius, and in spite of its splendour it had little or no influence on the development of later theatre buildings in Italy.

Teatro Popolare Italiano, see GASSMAN.

Teatro por Horas, see GÉNERO CHICO.

Teatro Stabile (pl. *teatri stabili*), name given to the permanent professional companies established in the major cities of Italy since the Second World War. Although attempts had been made earlier to set up stable troupes in some of the larger towns, these had proved short-lived, whereas some of the *teatri stabili* have survived long enough to become an important part of the Italian theatrical scene, among them the *Piccolo Teatro della Città di Milano and those in Genoa, Rome, and Turin.

Telari, see PERIAKTOI.

Téllez, Fray Gabriel, see TIRSO DE MOLINA.

Tempest, Dame Marie [Mary Susan Etherington] (1864–1942), English actress. Trained as a singer, she was seen in operetta and musical comedy and in 1890 went to New York, later touring Canada and the USA in light opera. Returning to London in 1895 she appeared under George *Edwardes at *Daly's Theatre, being seen in *The Geisha* (1896), *San Toy* (1899), and other musical comedies. She then deserted the musical stage and from the time of her successful appearance as Nell *Gwynn in Anthony Hope's *English Nell* (1900) appeared only in comedy. She was much admired in two plays by her second husband **Cosmo Gordon-Lennox** (1869–1921), *Becky Sharp* (1901), based on Thackeray's *Vanity Fair*, and *The Marriage of Kitty* (1902), which she subsequently revived many times. She toured America, Australia, and elsewhere from 1914 to 1922 with a varied repertory of light comedies. On her return she became noted for playing charming and elegant middle-aged women—Judith Bliss in Noël *Coward's *Hay Fever* (1925), the title-role in St John *Ervine's *The First Mrs Fraser* (1929), and Fanny Cavendish in *Theatre Royal* (1934) by Edna Ferber and George S. *Kaufman. In 1935 she celebrated her stage jubilee with a matinée at *Drury Lane, but she continued to act, being seen as Georgina Leigh in Robert *Morley's *Short Story* (1935) with her third husband, **W. Graham Browne** (1870–1937), who had appeared with her in many previous productions, some of which he directed. She made her last appearance as Dora Randolph in *Dear Octopus* (1938).

This play, in which a matriarch presides over a reunion of her family—the 'octopus' of the title—was the best-known work of **Dodie** [Dorothy] **Smith** (1896–1990), the author of several other well-written escapist plays, beginning with *Autumn Crocus* (1931) and including *Call it a Day* (1935).

Templeton, Fay, see VAUDEVILLE, AMERICAN.

Tennent, H. M., Ltd., see BEAUMONT, HUGH.

Tennis-Court Theatres. Both in Paris and in London in the mid-17th century, tennis-courts were converted into theatres. Among the most famous were the *Illustre-Théâtre (1644), where *Molière first acted in Paris, and, in London, *Killigrew's Vere Street Theatre (1660) and *Davenant's *Lincoln's Inn Fields Theatre (1661). They were used only for a short time, and their conversion into theatres must have entailed no more alteration than could have been easily removed to allow them to revert to their original use. The rectangular shape of the tennis-court auditorium, which approximated more to the private than the public Elizabethan theatre auditorium, had an important influence on the development of the French playhouse, and may also have influenced the eventual shape of the English Restoration theatre, just as the boxes round the pit may have developed partly from the 'pent-house' or covered way which ran along one side and one end of the court itself.

Tennyson, Alfred, Lord (1809–92), English poet, who succeeded Wordsworth as Poet Laureate. Though not much in touch with the contemporary stage, he nevertheless contributed to the *poetic drama of his day, his first play, *Queen Mary* (1876), a frigid tragedy in blank verse on Elizabethan lines, *The Cup* (1881), and *Becket* (1893) being produced at the *Lyceum by Henry *Irving, whose Becket was considered by many the finest achievement of his career. Of Tennyson's other plays in verse *The Falcon*, based on an episode in the *Decameron*, was produced at the *St James's Theatre by the *Kendals in 1879; *Harold* was published in Tennyson's lifetime but not performed until 1928, when it was given by the *Birmingham Repertory company at the *Royal Court Theatre, London, with Laurence *Olivier as King Harold. Like many other writers of his day, Tennyson failed to amalgamate fine poetry and good theatre. None of his plays has been revived, and the success of *Becket* was mainly due to the beauty and compelling power of Irving's interpretation.

Terence [Publius Terentius Afer] (c.190–159 BC), Roman dramatist, a freed slave, probably of African parentage. Six of his plays in the *fabula palliata form are extant—the *Andria* (*The Girl from Andros*, 166 BC), the *Hecyra* (*The Mother-in-Law*, 165), the *Heauton Timorumenos* (*The Self-Punisher*, 163), the *Eunuchus*, the *Phormio* (both 161), and the *Adelphi* (*The Brothers*, 160). Although they are all based on Greek comedies (four of them on plays by *Menander), their originality is not at all comparable with that of *Plautus. Where Plautus is topical, rough, farcical, and anti-realistic, Terence's plays are distinguished by their urbanity, elegance, and smooth construction. In all there are contrasts of character—in the *Adelphi*, for instance, between the strict father and the genial uncle and between the two brothers, one rash, one timid. In the *Andria* there is a situation not found in Greek comedy—a young man of good family in love with a young lady of his own station, an episode only possible because of the greater freedom of women in Rome than in Athens. The background of all the plays is neither Greek nor Roman but independent of time and place, and perhaps because of this Terence's works later had a universal appeal. In his own day, though he achieved some measure of success in the

theatre, his audiences were not uncritical, and the *Hecyra* twice failed in production. Even Julius Caesar found Terence lacking in comic force, though many of his pithy sayings were quoted by contemporary writers. But his interest in humanity, summed up in the famous remark by a character in the *Heauton Timorumenos*: 'I am a man; and all human affairs concern me', gave his work an abiding appeal. In the schools of the Middle Ages his plays were not only read but acted; in the 10th century the Abbess *Hroswitha even made adaptations of them for her nuns at Gandersheim. With the coming of the Renaissance they were translated into several languages; their influence spread to France, where it reached as far as *Molière, and to England, where it can be traced in the first English comedy, *Udall's *Ralph Roister Doister* (c.1553), through *Lyly and Shakespeare to *Steele, whose *The Conscious Lovers* (1727) is an adaptation of the *Andria*. Several of the manuscripts and early printed editions of Terence have illustrations which provide useful information about the staging of Renaissance plays, notably the *Terence-stage.

Terence-Stage, name given to the setting shown in Renaissance editions of the plays of *Terence. In the Trechsel edition of 1493 a woodcut shows a two-storey structure with arches (*fornices*) below and an auditorium (*theatrum*) above. Three tiers of spectators are seated in front of a stage-wall (*proscenium*) divided into sections by columns, with curtains hanging between them. There is a large forestage on which a musician is seated playing a wind instrument. In a box on the side wall between the spectators and the stage are two officials (*aediles*). The appearance of this stage is vouched for by another woodcut in the same volume, showing a scene from the *Eunuchus* with four curtained arches, each labelled with the name of a character in the play. Although this is reminiscent of the 'houses' of the *liturgical drama the pillared façade is more like the *scaenae frons* of the late classical theatre building. There were evidently alternative forms of the *Eunuchus* stage.

Terrace Theater, see JOHN F. KENNEDY CENTER FOR THE PERFORMING ARTS.

Terriss, Ellaline (1871–1971), English actress, daughter of William *Terriss. Born in the Falkland Islands while her father was sheep-farming there, she returned to England with her parents at an early age, and in 1888 made her first appearance on the stage at the *Haymarket Theatre under *Tree. She was also with *Wyndham at the *Criterion Theatre for three years. She made a great success in the title-role of *Bluebell in Fairyland* (1901), a children's play by her husband Seymour *Hicks revived annually at Christmas for many years, and also appeared with him in a number of his other plays, including *The Beauty of Bath* (1906) and *The Gay Gordons* (1907), though one of her best performances had been given somewhat earlier as Phoebe Throssel in *Barrie's *Quality Street* (1902). She accompanied Hicks on tour, both in straight plays and in *music-hall sketches, and went with him to France in the First World War to play to the troops, afterwards travelling with him in Australia and Canada. She left the stage in 1929, after appearing in a revival of Hicks's *The Man in Dress Clothes*, but returned briefly to play Mrs Thornton in his *The Miracle Man* (1935).

Terriss, William [William Charles James Lewin] (1847–97), English actor, formerly a sailor, known affectionately as Breezy Bill, or No. 1, Adelphi Terrace, as his best work was done at the *Adelphi Theatre. He first appeared on the stage, unsuccessfully, in 1867, and then left England for the Falkland Islands, where his daughter Ellaline (above) was born. Returning to England in 1873 he again entered the theatre—as Doricourt in a revival of Mrs *Cowley's *The Belle's Stratagem*. One of his first outstanding appearances was as *Dickens's Nicholas Nickleby in 1875, but he finally made his name as Squire Thornhill in *Olivia* (based on *Goldsmith's novel *The Vicar of Wakefield*) in 1878, playing opposite Ellen *Terry. He later played Romeo to the Juliet of Adelaide *Neilson, and in 1880 joined *Irving at the Lyceum. But he is best remembered as the hero of a series of famous melodramas at the Adelphi, among them Sims's *Harbour Lights* (1885), *Belasco's *The Girl I Left behind Me* (1895), and Seymour *Hicks's *One of the Best* (also 1895), which Bernard *Shaw reviewed as 'One of the Worst'. The success of these and many similar plays owed much to his handsome debonair presence and vigorous acting. He was one of the most popular actors of his day, and his assassination by a madman as he was entering the Adelphi Theatre was felt as a personal loss by his many admirers.

Terry, Benjamin (1818–96), English actor, son of an innkeeper at Portsmouth, who with his wife, **Sarah Ballard** (1817–92), went on the stage, touring extensively in the provinces and later playing small parts with Charles *Kean at the *Princess's, where his young children were in the company and his wife helped in the wardrobe. He had 11 children, of whom five went on the stage—Ellen, Fred, Kate, and Marion (below), and **Florence** (1855–96)— while two sons, **George** (1850–1928) and **Charles** (1857–1933), were connected with theatre management.

Terry, Edward O'Connor (1844–1912), English actor and manager, not, as far as is known, connected with the family of Benjamin *Terry. After some years in the provinces, when he was briefly in the same company as the young Henry

*Irving and played such parts as Touchstone and Dogberry in Manchester, he made his first appearance in London in 1867, at the *Surrey Theatre, and then played in *burlesque and light comedy for several years at the *Strand Theatre. In 1876 he was engaged by *Hollingshead for the *Gaiety Theatre, where he became one of the famous 'quartet'. In 1887 he opened a theatre under his own name in the Strand, on the site of a famous 'song-and-supper room', the Coal Hole, the first production being his own adaptation of a German play as *The Churchwarden*. A year later he scored his first outstanding success as Dick Phenyl in *Pinero's *Sweet Lavender*. In 1902 Terry appeared for the first time in musical comedy with *My Pretty Maid*, followed in 1903 by *My Lady Molly*, which had a long run. After this his only commercial success was *Mrs Wiggs of the Cabbage Patch* (1907), dramatized by Alice Hegan Rice from her book. In 1910 he gave up his theatre, which then became a cinema and was demolished in 1923, and toured extensively in Australia, South Africa, and the USA. Though not a good straight actor, Terry was an excellent 'eccentric comedian' and a careful and conscientious manager.

Terry, Dame Ellen Alice (1847–1928), English actress, the second daughter of Benjamin *Terry. She made her first appearance on the stage at the age of 9, playing Mamillius in *The Winter's Tale* in Charles *Kean's company at the *Princess's Theatre. She remained with the Keans until their retirement in 1859, and in the summer of that and succeeding years toured with her elder sister Kate (below) in *A Drawing-Room Entertainment*, in which they played together in short sketches. In 1861 Ellen joined the company at the *Haymarket Theatre, leaving in 1864 to marry the painter G. F. Watts, an ill-judged union with a man twice her age which soon came to an end. She returned to the theatre for a time, leaving it again shortly afterwards to live with the archaeologist, architect, and theatrical designer **Edward Godwin** (1833–86), by whom she had two children, Edith and Edward Gordon *Craig. When, on the insistence of Charles *Reade, she reappeared on the London stage in 1874 as Philippa in his drama *The Wandering Heir*, taking over the part from Mrs John *Wood, she was as brilliant as ever, and her long absence from the theatre seemed only to have increased the excellence of her acting. After playing for a year with the *Bancrofts, she went to the *Royal Court Theatre under *Hare, playing for him one of her most successful parts, the title-role in *Olivia* (1878), an adaptation of *Goldsmith's novel *The Vicar of Wakefield*. Later in 1878 Henry *Irving, who had recently begun his tenancy of the *Lyceum, engaged Ellen Terry as his leading lady, and so inaugurated a partnership which was to become one of the outstanding features of the London theatrical scene for the next 25 years. She appeared with him in a wide variety of parts, including a good deal of Shakespeare—notably Ophelia, Beatrice, Desdemona, Juliet, Viola, Lady Macbeth, and Imogen—in revivals of contemporary plays—*Bulwer-Lytton's *The Lady of Lyons* and Selby's *Robert Macaire*—and in a few plays specially written for him—Wills's *Charles I* (1879) and *Faust* (1885), Merivale's *Ravenswood* (1890), based on Scott's novel *The Bride of Lammermoor*, *Tennyson's *The Cup* (1881) and *Becket* (1893), and Comyns Carr's *King Arthur*. After leaving the Lyceum Ellen Terry became manager of the *Imperial Theatre, where in 1903 she appeared in *Much Ado about Nothing* and *Ibsen's *The Vikings*, being seen in the same year in *Heijermans's *The Good Hope* (for the *Stage Society), and two years later in *Barrie's *Alice Sit-By-The-Fire*. In 1906 she celebrated her stage jubilee with a mammoth matinée at *Drury Lane at which 22 members of the Terry family assisted. She was at the same time appearing as Lady Cicely Waynflete, a part specially written for her by *Shaw in *Captain Brassbound's Conversion*. She seldom acted afterwards, but toured America and Australia, giving readings of and lectures on Shakespeare. Throughout her career she was an inspiration to those who played with her. She was not at her best in tragedy, though some critics admired her Lady Macbeth, the role in which she was painted by Sargent, and she unfortunately never played Rosalind in *As You Like It* which seemed, above all other parts, to have been written for her, but to a hundred other roles she imparted a freshness and vitality which was never forgotten by those who saw her.

Terry, Fred (1864–1932), English actor, the youngest child of Benjamin *Terry. He made his first appearance on the stage at the *Haymarket Theatre under the *Bancrofts, and four years later, at the *Lyceum, played Sebastian in *Twelfth Night* to the Viola of his elder sister Ellen (above). A handsome, romantic actor, he became best known for his performance as Sir Percy Blakeney in Baroness Orczy's *The Scarlet Pimpernel*, which he produced under his own management in 1905 and frequently revived in London and on tour. He was also much admired as Charles II in *Sweet Nell of Old Drury* (1900) and as Sir John Manners in *Dorothy o' the Hall* (1906), both by Paul *Kester, and in the title-roles of *Matt o' Merrymount* (1908), by Beulah Dix and Mrs Sutherland, and *Henry of Navarre* (1909), by William Devereux. In these and many other productions he was partnered by his wife Julia *Neilson until his retirement in 1929. His son **Dennis** (1895–1932) and daughter **Phyllis Neilson-Terry** (1892–1977) were both on the stage.

Terry, Kate (1844–1924), English actress, the eldest daughter of Benjamin *Terry. Trained for the stage by her father, she made her first appearance in London as Prince Arthur in Charles *Kean's production of *King John* at the *Princess's Theatre in 1852, and remained with the Keans until 1859, playing Ariel in *The Tempest* in 1857 and Cordelia in *King Lear* in 1858. She then joined the Bristol *stock company, but returned to London in 1861 to play Ophelia to the Hamlet of Charles *Fechter. She appeared in several plays by Tom *Taylor, and seemed to be heading for a brilliant career when in 1867 she left the stage to marry Arthur Lewis, by whom she had four daughters, the eldest, also Kate, becoming the mother of John *Gielgud, and the youngest, Mabel *Terry-Lewis, an outstanding actress. In 1898 Kate Lewis, who at the height of her career had been considered by some critics to be a better actress than her more famous sister Ellen, returned to the stage to play opposite *Hare in Stuart Ogilvie's *The Master*, but the venture was not a success, and she never again appeared on stage.

Terry, Marion (1852–1930), English actress, younger sister of Ellen (above). She made her first appearance on the stage in 1873, and three years later scored a success as Dorothy in *Gilbert's *Dan'l Druce, Blacksmith*. In 1879 she was engaged by the *Bancrofts for the Prince of Wales (see SCALA) where she made her first appearance as Mabel in a new play by James *Albery entitled *Duty*. An able and attractive actress, she was later with *Alexander at the *St James's, where she created the part of Mrs Erlynne in *Wilde's *Lady Windermere's Fan* (1892); with *Forbes-Robertson at the *Lyceum, where she played Andrie Lesden in Henry Arthur *Jones's *Michael and His Lost Angel* (1896); and at the *Vaudeville, where she appeared as Susan Throssel in the first production of *Barrie's *Quality Street* (1902). She continued to play elegant aristocratic parts until increasing arthritis forced her to retire, her last part being the Principessa della Cercola in *Maugham's *Our Betters* (1923).

Terry, Megan, see OPEN THEATER.

Terry-Lewis, Mabel Gwynedd (1872–1957), English actress, daughter of Kate *Terry. She made her first appearance in 1895 with John *Hare in Grundy's *A Pair of Spectacles*, and in 1898 was seen with her mother, who had been absent from the stage for 22 years, in G. Stuart Ogilvie's *The Master*. She then appeared in revivals of several of T. W. *Robertson's plays, played Muriel Eden in *Pinero's *The Gay Lord Quex* (1899), and Gloria Clandon in the first public performance of *Shaw's *You Never Can Tell* (1900). Among her later parts were Madeleine Orchard in *Martin-Harvey's production of *After All* (1902), based on *Bulwer-Lytton's novel *Eugene Aram*, and Isabel Kirke in H. H.

Davies's *Mrs Gorringe's Necklace* (1903). She retired on her marriage in 1904, and did not return to the stage until after the death of her husband, when she was seen in H. M. Harwood's *The Grain of Mustard Seed* (1920). She continued her successful career, playing in a number of revivals and creating some 30 parts, mainly in plays now forgotten. Among the best of her elegant and aristocratic characters were Dona Filomena in the *Alvarez Quinteros' *A Hundred Years Old* (1928), Lady Bracknell in *Wilde's *The Importance of Being Earnest* (1930), and her last part, Lady Damaris in Pryce's *Frolic Wind* (1935). She was extremely popular in New York, where she appeared several times from 1923 onwards.

Terson, Peter (1932–), English dramatist, whose first play to be produced, *A Night to Make Angels Weep* (1964), was performed at the Victoria Theatre, Stoke-on-Trent (see NEW VICTORIA THEATRE), where he was resident dramatist for 18 months, and for which he was ultimately to write 22 plays. It was followed by *The Mighty Reservoy* (also 1964), *All Honour Mr Todd*, and *I'm in Charge of These Ruins* (both 1966). Other works for Stoke were *Mooney and his Caravans* (1967) and a dramatization of *Clayhanger* staged in honour of Arnold *Bennett's centenary (also in 1967), in which Terson collaborated with Joyce Cheeseman. In the same year he began a long association with the *National Youth Theatre with *Zigger-Zagger*, his best-known play, about a football fan. From then onwards he wrote a play for the NYT almost every year, all of them relating to young people and their problems. They include *The Apprentices* (1968), *Fuzz!* (1969), *Good Lads at Heart* (1971; staged by the NYT in New York in 1979), and *Geordie's March* (1973). Among later subjects were life in a hotel (*The Bread and Butter Trade*, 1976), relationships between parents and children (*Family Ties*, 1977), an unemployed youth who joins the National Front (*England My Own*, 1978), and a young soldier who tries to get out of the army (*The Ticket*, 1980). Terson continued to write plays for the Victoria Theatre, such as *The Adventures of Gervaise Becket; or, The Man Who Changed Places* (1969), and *The Pied Piper* (1980), both for children; the semi-documentary *The 1861 Whitby Lifeboat Disaster* (1970); an adaptation of Herman Melville's *Moby Dick* (1972); and *Aesop's Fables* (1983), with music. He also wrote plays for other theatres, and *Strippers* (1985), in which a woman becomes a stripper when her husband loses his job, was staged in the West End. Terson has been widely praised for his honesty and directness, and has even been compared to *Chekhov. His plays for the NYT were thematic and extensively reworked in rehearsal, whereas his plays for the Victoria Theatre and elsewhere were more personal and

subject only to the normal rehearsal adjustments. He works at great speed, writing a play in two weeks and sometimes in as little as three days.

Tetralogy. In the ancient Greek *Dionysia each dramatic poet was originally supposed to submit for production and competition four plays on the same theme, character, or family, consisting of a trilogy of tragic plays followed by a burlesque of the main theme in a *satyr-drama. This rule was followed by *Aeschylus, whose *Oresteia* is the only true trilogy to survive, and it is reasonable to suppose that the satyr-drama which followed it dealt also with the House of Atreus; also that the surviving play *Prometheus Bound* was the first of a trilogy on Prometheus. Later dramatists ignored the unity of subject, and submitted 'pseudo-tetralogies', in which all four plays dealt with different subjects.

Thaddädl, last Viennese comic figure in the tradition of *Hanswurst, created by the Austrian actor **Anton Hasenhut** (1766–1841). Based on Taddeo, a German importation into the *commedia dell'arte*, Thaddädl was a clumsy youth with a falsetto voice, a perpetually infatuated, idiotically infantile booby. Although much admired by *Grillparzer, among others, the character never achieved universal popularity, and was soon forgotten. The role at its best is found in Kringsterner's *Der Zwirnhändler* (*The String Merchant*, 1801).

Thalia, see MUSES.

Theater an der Wien, see VIENNA.

Theatre, The, the first—and most appropriately named—playhouse to be erected in London, built by James *Burbage. Because of opposition from the Lord Mayor of London to actors appearing in *innyards, it had to be erected outside the City boundary, and was situated between Finsbury Fields and the public road from Bishopsgate to Shoreditch Church. Some of the money for its construction came from Burbage's father-in-law, probably an actor. It was a circular wooden building, without a roof, and cost between £600 and £700. The actual dimensions are not known, but the building was apparently commodious, with three galleries and what would now be described as boxes. It opened in 1576. Admission was one penny for standing room on the ground and a second penny for admission to the galleries; for a further penny one could obtain a stool, or what was described as a 'quiet standing'. Although the authorities disapproved, the public appear to have flocked to the new theatre, but its career, though stormy, was not very distinguished theatrically. It was used for competitions of sword-play, fencing, quarterstaff, and athletic exercises, and by a number of different companies until in 1594 the *Chamberlain's Men took over. Plays performed there may have included the original *Hamlet*, on which Shakespeare based his play, and *Marlowe's *Doctor Faustus*. *Tarleton and *Kempe also performed 'jigges and drolls' there. In spite of good audiences, there was little profit on the Theatre and continual harassment from the authorities, so when in 1597, just after James's death, the lease ran out, his son Cuthbert pulled the building down, transported the timber and other materials across the river to Bankside, and used them to build a new theatre (see GLOBE 1).

Théâtre Alfred Jarry, see ARTAUD.

Théâtre Antoine, Paris, see ANTOINE.

Theatre Behind the Gate, see KREJČA.

Theatre Buildings. The provision of permanent roofed buildings specially erected for the performances of plays came comparatively late in theatrical history. Greek open-air theatres evolved from the ritual *dithyramb performed round the altar of *Dionysus, which took place in front of the temple, and later on a site cut out of a neighbouring hillside. This provided a natural auditorium of rising tiers of seats which extended a little more than halfway round a circular *orchestra, or playing-place, backed by a low stage with a stage-wall (*skênê*) behind. This formed one wall of the dressing-rooms and storage rooms and was pierced by doors through which the actors came on stage. It also housed the machinery which worked the crane by which the god (the *deus ex machina*) finally appeared from heaven to resolve the complications of the plot. All that is known of Greek theatres in the 5th century BC, the age of the great classical tragedies and comedies, has had to be inferred from the ruins of those that still exist, many of which have been subsequently altered, rebuilt, and finally abandoned. Some, such as that at *Epidauros, have been refurbished and are used for annual festivals of Greek classical plays. The only things which seem certain about the early theatres are that the audience sat first on wooden benches and then on stone; that the *chorus occupied the circular orchestra; that a raised stage was provided for the actors, of which, in classical times, there were never more than three; and that the acoustics of these early theatres were perfect, as can be verified by anyone visiting them today: the slightest whisper from the orchestra can be heard clearly by people in the topmost seats.

In the great Hellenistic theatres which replaced the simpler ones of early times, the stage was raised, often to a height of several feet, and the stage-wall became more elaborate. Columns supported a stage-roof, and the ramps which led up to the stage were built over a colonnade (the *proskênion* = *proscenium).

The Roman theatre, unlike the Greek, was built on the flat. The early ones were of wood and have disappeared. Later ones, in stone, still exist. The much-diminished orchestra was little more than a semicircle terminated by an elaborate stage-wall (*scaenae frons*) often three storeys high, in front of which was the stage, usually about 5 ft. above ground level. This was separated from the auditorium by a curtain which descended into a trough. The exterior of the theatre, which rose in a series of colonnades to a great height, was solidly constructed, and, in the case of *amphitheatres used for chariot-races and gladiatorial combats, which had a circular arena, was also completely circular. The destruction of the Roman Empire saw the collapse of organized theatre. When it was reborn in *liturgical drama, plays were first given in churches and later in the open air, either in front of the church door, which provided an excellent stage-wall, or on raised platforms erected in the market-place. In England biblical plays were often acted on *pageants. The Renaissance, which was in full flower in Italy while other countries still clung to their medieval traditions, brought about a great change in the design of theatres. For the first time plays were produced indoors, often on stages temporarily set up in a nobleman's hall or palace. The illustrations in late 15th-century editions of *Terence's plays show the Renaissance stages on which they were acted. They combined elements of medieval staging with what had been learned of classical staging from the newly discovered works of *Vitruvius, and provided models which, in various combinations, developed into the theatres which we know today (see TERENCE-STAGE). The main innovation in 16th-century Italy was the proscenium arch, which framed the elaborate stage-picture provided for a courtly entertainment. This is still a permanent fixture in many theatres, and it is only in recent times that theatres have been built without a proscenium wall. The rise of opera and ballet in Italy led to the evolution of the horseshoe-shaped auditorium characteristic of opera-houses all over the world and typified by the 1589 theatre at Sabbionetta and the 1619 *Teatro Farnese at Parma, while the academic tradition of the classical play under the influence of Vitruvius culminated in the great *Teatro Olimpico at Vicenza, first used in 1585, with its superb *scaenae frons*. During the 16th century new theatres were built all over Europe. At first each country had its own style. The early French theatres, such as the Hôtel de *Bourgogne (1548), were long and narrow, with a space in front of the high stage—originally intended for the ball which followed the spectacle—rising tiers of seats, and galleries at the side. Many of the early theatres, up to the time of *Molière, were adapted *tennis-courts, but the Court theatres followed the Italian pattern, with a centrally placed dais—later a Royal Box—for the accommodation of the king and queen. The *Vigaranis' 1660 *Salle des Machines showed the influence of *Palladio.

In Spain the early theatres followed the pattern of the open-air stages erected in the public squares, with a stage raised on scaffolding and spectators at the windows and on the balconies of the houses all round. This was somewhat similar to the open-air *Elizabethan playhouse, which at the Restoration, and even earlier, gave way to the indoor theatre on Italian lines, though retaining in the small theatres, even in Georgian times, some purely English characteristics, with rows of *boxes behind and on both sides of a central *pit with benches, and a large *forestage with *proscenium doors on each side. On the Continent the constant moving around of Italian architects and stage designers, notably the *Bibienas, led to the adoption everywhere of the operatic tradition, with baroque and rococo decorations which lingered on until the 19th century. In 1876 Wagner introduced a new concept into his opera-house at Bayreuth, doing away with ornate decorations and replacing the hierarchy of pit, boxes, and galleries by a single fan-shaped auditorium with a steep rake. This, particularly in later adaptations, was not always successful, but something of its influence lingered on into the 20th century and was apparent in the buildings which proliferated throughout the USA, particularly the Chicago Opera House, built in 1929. Even more radical ideas were developed by *Reinhardt, for whom Hans Poelzig converted a circus into the *Grosses Schauspielhaus, Berlin. Reinhardt's concept of Theatre for the Masses survived chiefly in the Soviet Union where a number of vast indoor amphitheatres were built. The next step was inevitably towards the complete arena, as visualized in 1926 by Walter Gropius. His unrealized Totaltheater, intended for *Piscator, was oval in plan with a steep, 2,000-seat auditorium wrapped around a forestage backed by a proscenium stage.

The Grosses Schauspielhaus and the Totaltheater embodied ideas that were a crucial influence on later experiments with *theatre-in-the-round and *flexible staging, as were Norman *Bel Geddes's schemes of the 1920s and 1930s. Another theme has been the *open or thrust stage. A development of the 1970s was the small 'workshop' theatre, either added to an existing theatre (though not necessarily in the same building) or incorporated in a new theatre complex. Britain's *National Theatre (1976) incorporates such a workshop in the *Cottesloe Theatre.

The search for the perfect theatre building still continues. A theatre, once built, is difficult to get rid of, and many out-of-date theatres remain which resist adaptation and are the despair of their directors.

(See also BOOTHS; CORNISH ROUNDS; FAIRS; INNYARDS USED AS THEATRES; PRIVATE THEATRES.)

Théâtre d'Art, Paris, see LUGNÉ-POË.

Théâtre de France, Paris, see ODÉON.

Théâtre Déjazet, Paris, see DÉJAZET.

Théâtre de l'Atelier, Paris, see DULLIN.

Théâtre de l'Europe, Paris, see ODÉON.

Théâtre de l'Hôtel d'Argent, Paris, see ARGENT.

Théâtre de l'Hôtel de Bourgogne, Paris, see BOURGOGNE.

Théâtre de l'Œuvre, Paris, see LUGNÉ-POË.

Theatre de Lys, New York, see LUCILLE LORTEL THEATRE.

Théâtre des Amandiers, Nanterre, see CHÉREAU.

Théâtre des Nations, international festival of drama founded under the auspices of UNESCO, which began in 1954 when a number of foreign companies were invited to perform in Paris at the Théâtre Sarah-Bernhardt, now the Théâtre de la Ville. The festival was officially established in 1955, and from 1957 to 1965 a two-month season of foreign plays was organized annually by the manager of the Théâtre Sarah-Bernhardt, with state and municipal support. During this period some 50 countries were represented by about 150 companies with 180 plays and 150 other entertainments, and the festival played an important part in bringing groups such as the *Berliner Ensemble, the *RSC, the *Piccolo Teatro della Città di Milano, the *Living Theatre, and the *Peking Opera to the attention of Parisians, and indeed to European audiences in general, since the prestige and publicity of their Paris season brought the companies invitations to visit other festivals. In 1966 *Barrault took over the organization of the festival and transferred it to the *Odéon, which he was then managing. He was in the process of giving it a firmer international basis when the 1968 riots caused his own dismissal from state employment. After that the Théâtre des Nations led a sporadic and vestigial existence, until in 1973 the *International Theatre Institute, which had been instrumental in founding it, agreed to its becoming a peripatetic event.

Théâtre du Nouveau Monde, Montreal, see NOUVEAU MONDE, THÉÂTRE DU.

Théâtre du Soleil, Paris, see SOLEIL.

Théâtre du Vieux-Colombier, Paris, see VIEUX-COLOMBIER.

Théâtre-Français, Paris, see COMÉDIE-FRANÇAISE.

Theatregoround, see ROYAL SHAKESPEARE COMPANY.

Theatre Group, Los Angeles, see MARK TAPER FORUM.

Theatre Guild, New York, theatre production company, evolved out of the work of the *Washington Square Players for the presentation of non-commercial plays, American and foreign, to a subscription audience. It occupied the *Garrick Theatre, where its first production, in 1919, was *Benavente's *The Bonds of Interest*. Among the plays which followed were *Molnár's *Liliom* (1921), *Andreyev's *He Who Gets Slapped*, and *Čapek's *RUR* (both 1922). The first new American play to be staged was Elmer *Rice's *The Adding Machine* (1923), but on the whole productions of foreign plays predominated, with *Shaw's works very much to the fore. As early as 1920 *Heartbreak House* was seen, followed by *Saint Joan* in 1923, and in 1925 *Caesar and Cleopatra* was chosen as the first play presented by the Guild in its own Guild Theatre (see VIRGINIA THEATRE). Later Shaw productions included *Arms and the Man* (1925), *Pygmalion* (1926), *The Doctor's Dilemma* (1927), *Major Barbara* (1928), and *The Apple Cart* (1930). There were also productions of several of *O'Neill's plays including *Strange Interlude* (1928) and *Mourning Becomes Electra* (1931), the *Gershwins' folk opera *Porgy and Bess* (1935), and *Sherwood's *Idiot's Delight* (1936). Much of its work was mounted at theatres other than the Guild. The enterprise became vulnerable during the financial recession of the 1930s and early 1940s, but was rescued by the success of Rodgers and *Hammerstein's *Oklahoma!*, presented under the Guild's auspices at the *St James Theatre in 1943. It staged the same team's *Carousel* (1945) and such notable later productions as O'Neill's *The Iceman Cometh* (1946); but it never regained its former eminence. The Guild Theatre was taken over by the *American National Theatre and Academy in 1950; the Guild itself continued for a time to mount new plays, revivals, and musicals within a commercial framework.

Two important breakaway organizations were the *Group Theatre and the Playwrights' Company. The latter, founded in 1938, included playwrights such as Maxwell *Anderson, Sidney *Howard, Elmer Rice, and Robert E. Sherwood, all of whom had had plays staged by the Theatre Guild but were unhappy with it. Many of the Playwrights' Company's productions were written by its members, and it survived until 1960.

Théâtre Historique, Paris, see DUMAS *père*.

Theatre-in-the-Round, form of play presentation in which the audience is seated all round the acting area. One of the earliest forms of theatre, it was probably used for open-air performances, street theatres, and such rustic sports as the Mayday games and the Christmas *mumming play, and was revived in the 20th century by those who rebelled against the so-called 'tyranny' of the *proscenium arch.

Modern theatre-in-the-round first came into prominence in the Soviet Union, where in the 1930s *Okhlopkov in his *Realistic Theatre produced a number of plays on stages with the audience on all sides. At the same time, in England, Robert *Atkins was producing Shakespeare in the *Ring in Blackfriars, an interesting experiment which seemed to have no immediate impact. It was in America, where the idea had already been mooted by Robert Edmond *Jones in 1920, that the first theatre-in-the-round was erected at the University of Washington in Seattle in 1940. It had an elliptical acting area and auditorium contained within a circular foyer, the space between them on two sides being used for prop rooms and a lighting control booth. In the rapidly expanding world of American university drama, theatre-in-the-round flourished. Outside the universities, the most important exponent of the new method was Margo *Jones, working in Dallas. The *Circle-in-the-Square (1951) made this type of production familiar to *Off-Broadway playgoers, and in 1961 the *Arena Stage in Washington, DC, was opened.

In England the *New Victoria Theatre, Newcastle-under-Lyme, the *Stephen Joseph Theatre, Scarborough, and the *Royal Exchange Theatre, Manchester are permanent professional theatres-in-the-round. A number of new theatres, both professional and amateur, have been built with *flexible staging, in which theatre-in-the-round is an option. Inherent technical problems include special demands on lighting and set designers, as well as the need for frequent movement by the actors in order to give all sections of the audience a view of each character's face; there are also audibility problems.

A fertile development of theatre-in-the-round is the 'promenade' production, of which examples have been seen at the *Round House and at the *Cottesloe Theatre, in which the players move around the auditorium for the different episodes in the play, with the standing audience clustering about them.

Théâtre-Italien, Paris, see COMÉDIE-ITALIENNE.

Theatre Library Association, see MUSEUMS AND COLLECTIONS, THEATRICAL.

Théâtre Libre, Paris, theatre club founded in 1887 by André *Antoine for the production of plays by new *naturalistic French and foreign playwrights. Before it finally closed in 1896, mainly because of financial difficulties, it had staged 184 plays, among them those of such outstanding dramatists as *Becque, *Brieux, *Hauptmann, *Ibsen, and *Strindberg. Its innovations in playwriting, direction, and acting had a great influence on the contemporary French theatre, and its settings, scrupulously exact reproductions of real life, helped to liberate the stage from the artificiality of an earlier epoch.

Théâtre Montparnasse, see BATY.

Theatre Museum, London. This came into being in 1974 when the collections of the British Theatre Museum Association were transferred to the Victoria and Albert Museum. They joined theatrical material already in the museum, notably the collection of the English theatre historian **Gabrielle Enthoven** (1868–1950), given in 1924, consisting of programmes, play-bills, engravings, and other material relating to the London stage from the 18th century onwards. In 1987 the Theatre Museum opened its own premises in Covent Garden, which provide galleries for permanent and temporary exhibitions, a 70-seat theatre, and a research area with public reading room. The scope of the collections includes theatre, ballet, opera, music-hall, puppetry, circus, and pop music. Special collections include the Harry R. Beard collection; the Houston Rogers collection of theatrical photographs; the Anthony Hippisley Coxe circus collection; the Arnold Rood collection of material relating to Gordon *Craig; the Hinkins collection of *toy theatre sheets and play texts; and the *Royal Court Archive. The museum also holds important collections of designs, including those of the *Arts Council. Costumes in the possession of the museum include many from the Diaghilev Ballet.

Theatre Museums and Collections, see MUSEUMS AND COLLECTIONS, THEATRICAL.

Théâtre National Populaire, Paris. The first TNP was founded by *Gémier, who from 1920 to 1933 obtained Government grants to defray the cost of inviting companies to play to worker-audiences at the Palais de *Chaillot in the Trocadéro. He also organized provincial tours, pioneering the use of motorized transport for stage equipment. His work came to an end when in 1934 the Trocadéro was reconstructed to house the International Exhibition of 1937.

The second TNP was directed by *Vilar, 1951–63, and was again housed in the Palais de Chaillot, which since 1945 had been used as an assembly hall by the United Nations. The enterprise was able to support its own company, which developed in conjunction with the activities of Vilar at the *Avignon Festival. For the first five years productions were staged at Chaillot and then sent on tour, especially in the working-class suburbs of Paris; efforts were later directed rather to attracting spectators from those areas through organized trips and audience associations. An entirely new approach to public relations gave the TNP enormous popularity in the 1950s, though even then there were criticisms of the low proportion of manual workers in its audiences, and the preponderance of well-known plays in its repertory. Certainly Vilar based his work mainly on the classics, both French and foreign, but he also produced a number of contemporary plays—*Brecht's

Mutter Courage und ihre Kinder in 1953, *Jarry's *Ubu-roi* in 1958, and Brecht again with *Der aufhaltsame Aufstieg des Arturo Ui* in 1960. He overcame problems posed by the enormous auditorium (seating 1,800, with a stage 34 metres wide) by adopting reforms still at that time considered experimental: sparse use of scenery, abolition of the footlights and proscenium arch, complex lighting effects, and the use of large acting areas defined by lifts and revolving platforms. Playing in many of the productions himself, and ably seconded by Gérard *Philipe in such classic roles as *Corneille's Le Cid and Maria *Casarès in the tragic heroines of the Romantics, Vilar established a 'TNP style', which, unlike that of the *Comédie-Française, was based on economy of gesture and simplicity of speech.

The TNP was directed, 1963–72, by the actor-director **Georges Wilson** (1921–   ), who had joined it in 1952 and attracted critical acclaim in such roles as Ubu-roi. Under Wilson the repertory became more modern and international, and he introduced Brendan *Behan and Edward *Bond to the French stage; but the loss of Vilar and Philipe, together with the development of other popular theatres in the Parisian suburbs and provinces under the policy of *Décentralisation Dramatique, led to a decline in popularity. Whereas the average number of seats sold had remained at over 90 per cent until 1966, it dropped in 1970–1 to 37 per cent and in 1972 the theatre was closed by the Government, its subsidy and title passing to the group controlled by *Planchon at Villeurbanne.

Wilson continued to work independently, his later roles including James Tyrone in *O'Neill's Long Day's Journey into Night in 1973 and Othello at the Avignon Festival in 1975. He directed *Sartre's Huis-Clos in 1981 and appeared in *Beckett's Waiting for Godot in 1985.

Theatre of Comedy, London, see SHAFTESBURY THEATRE 2.

Theatre of Cruelty, see ARTAUD.

Theatre of Drama and Comedy, Moscow, see TAGANKA THEATRE.

Theatre of Fact, see DOCUMENTARY THEATRE.

Theatre of Satire, Moscow, see SATIRE, THEATRE OF.

Theatre of Silence, see SILENCE, THEATRE OF.

Theatre of the Absurd, see ABSURD, THEATRE OF THE.

Theatre of the Revolution, Moscow, see MAYAKOVSKY THEATRE.

Theatre Royal, term applied to the two London theatres, *Covent Garden and *Drury Lane, which operate under Letters Patent granted by Charles II in 1662. These *Patent Theatres originally had a monopoly of 'legitimate' drama which lasted until 1843. The *Haymarket Theatre was given the courtesy title of Theatre Royal by virtue of a patent for the summer months only obtained in 1766 by Samuel *Foote from George III. During the 19th century a proliferation of Theatres Royal all over the country—none of them having any right to the title, since these theatres were licensed by the local magistrates and not directly by the Crown—caused the term to become nothing more than a generic title for a playhouse.

For individual theatres bearing the name see also below and under ADELPHI THEATRE; ASTLEY'S AMPHITHEATRE; BATH; BRIGHTON; BRISTOL; BRISTOL OLD VIC; BURY ST EDMUNDS; EXETER; JOHN STREET THEATRE; MANCHESTER; NEWCASTLE-UPON-TYNE; and YORK.

Theatre Royal, Hobart, Tasmania, the oldest playhouse in Australia. It opened officially in 1837 with *Morton's Speed the Plough. Little of the original building is now to be seen except for the external walls, which show evidence of successive changes. The classical front façade was added in 1860 and in 1911 the auditorium was reconstructed in Edwardian style. Additions to the theatre in the 1970s provided new dressing rooms and office accommodation. A fire in 1984 seriously damaged the building, which was restored and its backstage facilities upgraded through public donation and Government funds. The theatre is now owned by the Tasmanian Government and managed by the Theatre Royal Management Board, which also brings international and interstate attractions into Tasmania.

Theatre Royal, Stratford East, London, in Gerry Raffles Square, named after the theatre's most famous manager. It seats 467 and opened in 1884 with *Bulwer-Lytton's Richelieu, being used for a time for revivals of other old plays. For many years it was run by the Fredericks family, who also controlled the Borough Theatre in Stratford High Street, used mainly by touring companies. The Royal had a somewhat undistinguished career, relying on melodrama and an annual pantomime. In 1902 it was badly damaged by fire; the stage was then enlarged and the auditorium redecorated. Between 1927 and 1933 the theatre was given over to twice-nightly variety, and from 1947 to 1949 it returned to straight plays; but the venture was not a success and in 1950 Tod Slaughter's Spring-Heeled Jack, the Terror of Epping Forest heralded the return to more popular fare. The Christmas play that year was Stevenson's Treasure Island. In 1953 Joan *Littlewood took over the theatre, making it the home of her *Theatre Workshop until 1963. A new management opened in 1964 with a translation of Max *Frisch's Graf Öderland as Edge of Reason; but Theatre Workshop returned in 1967, and did not finally leave until the end of 1973. After

the Workshop era the theatre went through a very difficult period: attendances were low and there was a danger that its subsidy from the *Arts Council would be reduced. In 1980, however, under a new manager and with a programme tailored to the needs of the area, there were signs of recovery. Productions included Howard *Brenton and Tony Howard's A Short, Sharp Shock, a satire on Mrs Thatcher's Government, and Henry *Livings's This Jockey Drives Late Nights, and in 1981 the theatre had its biggest success since Joan Littlewood's departure with Nell Dunn's Steaming, which transferred to the West End. Although the theatre mainly presents new plays it stages the occasional classic, its production of *Boucicault's The Streets of London in 1980 also moving to the West End. The theatre provides an entertainment nearly every Sunday, and is strongly involved in educational and community activities.

Theatre Union, see LITTLEWOOD.

Theatre Upstairs, see ROYAL COURT THEATRE 2.

Theatre Workshop, company founded in Kendal in 1945 by a group of actors who were dissatisfied with the commercial theatre 'on artistic, social, and political grounds'. Joan *Littlewood, who had worked with some of the members of the group earlier in the North of England, became Artistic Director and Gerald Raffles General Manager. After seven years spent touring in England and Europe, the company took over the *Theatre Royal, Stratford East, London (Chaucer's Stratford-atte-Bowe), repairing and redecorating the building themselves before opening in 1953 with Twelfth Night. The company quickly made a name for itself, and was invited to represent Great Britain at the Paris *Théâtre des Nations in 1955 with *Lillo's version of Arden of Feversham and *Jonson's Volpone and in 1956 with an adaptation of Hašek's The Good Soldier Schweik. It also visited Zürich, Belgrade, and Moscow, where it appeared at the *Moscow Art Theatre, filling it to capacity. Left-wing and indeed almost Communist in its ideology, it sought always to revivify the English theatre by a fresh approach to established plays—two of its most interesting and controversial productions being Lope de *Vega Carpio's Fuenteovejuna as The Sheep Well and Shakespeare's Richard II (both 1955)—or by producing working-class plays, some of which were subsequently transferred to West End theatres, among them Brendan *Behan's The Quare Fellow (1956) and The Hostage (1958), Shelagh Delaney's A Taste of Honey (also 1958), Frank Norman and Lionel *Bart's musical Fings Ain't Wot They Used T'Be (1959), and Stephen Lewis's Sparrers Can't Sing (1960). When Joan Littlewood left to work elsewhere the company carried on under Gerald

Raffles, until in 1963 she returned to produce *Oh, What a Lovely War!*, a 'musical entertainment' satirizing the First World War. This also transferred to the West End, after which the company dispersed. It was back in residence, however, in 1967, when once more under Joan Littlewood's direction it presented a succession of plays including Barbara Garson's *MacBird!* (a reworking of Shakespeare's *Macbeth* first seen in New York) and *Intrigues and Amours* (based on *Vanbrugh's *The Provoked Wife*). After directing Peter Rankin's *So You Want to be in Pictures?* (1973) Joan Littlewood went abroad and the group finally dispersed.

Theatrical Commonwealth, group of actors who seceded from the *Park Theatre, New York, in 1813, after disputes with the management, and set up for themselves in a converted circus on Broadway. Although none of them was outstanding, they reputedly gave excellent productions of *As You Like It* and *Sheridan's *The School for Scandal* and *The Rivals*. Emboldened by their success, the rebels even had the temerity to put on Frederick Reynolds's *The Virgin of the Sun* on the same night as the Park. They disbanded after the death of Mrs Twaits, a member of the company, at the end of 1813, several of them returning to the Park.

Theatrical Syndicate, association of American businessmen in the theatre, formed in 1896, which included Charles *Frohman. For about 16 years they controlled most of the theatres of New York and many of those in other big towns, and gradually they exerted a stranglehold over the whole entertainment life of the USA. The Syndicate's original intentions—to rationalize theatre organization and prevent exploitation—developed into a determined commercialism which suppressed competition and depressed aesthetic standards. It was powerful enough to harm those who opposed its monopoly, forcing Mrs *Fiske to play in second-rate theatres on tour and Sarah *Bernhardt to appear in a tent. Both these players, with *Belasco and above all the *Shuberts, helped in the end to break it.

Théophile de Viau (1590–1626), French poet and dramatist, who marks the incursion into the French theatre of the young nobleman. As a Huguenot and a free-thinker, he was suspected of being part-author of *Le Parnasse satirique* (1622), a collection of licentious and blasphemous verse, and was first exiled and then condemned to be burned at the stake; he escaped, was again imprisoned, and died in exile a year after being released. His only play, *Pyrame et Thisbé*, a tragedy based on the episode in Ovid's *Metamorphoses* which provided the source of the mechanicals' play in *A Midsummer Night's Dream*, was published in 1623 and perhaps acted before that. It was certainly seen at Court in 1625, and was in the repertory of the Hôtel de *Bourgogne in 1633–4, being one of the oldest plays mentioned in the Mémoire de *Mahelot. It is now recognized as a masterpiece of baroque tragedy, though its elaborate conceits caused it to be looked on unfavourably by *Boileau and other neo-classical critics. Regular in construction, according to the interpretation of the *unities current at the time, it is an excellent illustration of the final period of the *multiple stage, before this gave way to the single set characteristic of French classical drama.

Thespis, Greek poet from Icaria in Attica, usually considered the founder of drama, since he was the first to use an actor in his plays in addition to the *chorus and its leader. He won the prize at the first tragic contest in Athens, *c*.534 BC. Only the titles of his plays have survived and even these may not be authentic. Tradition has it that Thespis took his actors round in a cart, which formed their stage. In the 19th century the adjective Thespian was used of actors and acting in general, and often figured in the names of amateur companies, while 'the Thespian art' was journalese for the art of acting.

Third Stage, see STRATFORD (ONTARIO) FESTIVAL.

Thomas, Augustus (1857–1934), American dramatist, and one of the first to make use in his plays of American material. He succeeded *Boucicault as adapter of foreign plays at the *Madison Square Theatre under *Palmer, but his first popular success, an original drama entitled *Alabama* (1891), enabled him to resign and devote all his time to his own work. Among his later plays were several others based on a definite locality—*In Mizzoura* (1893), *Arizona* (1899), *Colorado* (1901), and *Rio Grande* (1916). His most successful play was *The Copperhead* (1918), in which Lionel *Barrymore made a hit. An interest in hypnotism and faith-healing was shown in *The Witching Hour* (1907), *Harvest Moon* (1909), and *As a Man Thinks* (1911), but on the whole Thomas's plays were not profound, and provided entertainment of a kind acceptable to his audiences.

Thomas, (Walter) Brandon (1856–1914), English actor, playwright, and writer of coon songs, which he sang in *music-halls himself. After some amateur experience he made his professional début with the *Kendals in 1879, remaining with them until 1885. He wrote about a dozen plays, but only one is now remembered, the farce *Charley's Aunt*, seen in London a number of times since its first production at the *Royalty Theatre in 1892, when it ran for four years with W. S. *Penley in the title-role. The play has been filmed, made into a musical, *Where's Charley?* (NY, 1948; London, 1958), and has figured in the repertory of almost every amateur and provincial theatre, as well as being played all over the world in English and in innumerable translations. At one time it was running simultaneously in 48 theatres in 22 languages, including Afrikaans, Chinese, Esperanto, Gaelic, Russian, and Zulu.

Thomás, Cornelia, see JEFFERSON, JOSEPH (1774–1832).

Thompson Shelterhouse, see CINCINNATI PLAYHOUSE IN THE PARK.

Thorndike, Dame (Agnes) **Sybil** (1882–1976), English actress. She began her long and distinguished career under Ben *Greet in 1904 in a wide variety of Shakespeare parts on tour in England and the USA, and in 1908 joined Miss *Horniman's repertory company in Manchester, where in 1909 she married the actor and director Lewis *Casson, with whom she was associated in much of her later work. During the next five years she divided her time between London—where she was seen, among other parts, as Emma Huxtable in *Granville-Barker's The Madras House (1910) and Beatrice Farrar in *Houghton's Hindle Wakes (1912)—and Manchester, where she gave an excellent performance in the title-role of St John *Ervine's Jane Clegg (1913). She made her début in New York in Somerset *Maugham's Smith (1910) opposite John *Drew. She was at the London *Old Vic under Lilian *Baylis, 1914–18, playing not only most of Shakespeare's young heroines but also, owing to the absence of young actors on war service, various supporting male roles in Shakespeare. On leaving the Old Vic she was seen as Synge de Coûfontaine in *Claudel's The Hostage and as Hecuba in Gilbert *Murray's translation of *Euripides' Trojan Women (both 1919), as well as in a number of short-lived modern plays, and from 1920 to 1922 was in Grand *Guignol seasons at the *Little Theatre with her husband and brother. One of her finest roles was St Joan (1924) in the first London production of *Shaw's play; she portrayed another Shaw heroine, Barbara Undershaft, in a revival of Major Barbara in 1929. She appeared in *Van Druten's The Distaff Side in 1933, repeating the role a year later in New York, where she also played Mrs Conway in *Priestley's Time and the Conways in 1938. Later in the same year London audiences saw one of her most memorable performances, as the elderly schoolmistress Miss Moffat in Emlyn *Williams's The Corn is Green. During the Second World War she toured with the Old Vic company for ENSA to mining towns and villages, playing Shaw's Candida, Lady Macbeth, and Euripides' Medea. She was at the New Theatre (now the *Albery) with the Old Vic company as Aase in *Ibsen's Peer Gynt (1944) and Jocasta in *Sophocles' Oedipus the King (1945). In 1947 she and her husband were in one of Priestley's best plays, The Linden Tree, and in 1949 she began a long run in the comedy

Treasure Hunt by M. J. Farrell and John Perry. She was in two long-running plays by N. C. *Hunter, Waters of the Moon (1951) and A Day by the Sea (1953), and made three overseas tours before returning to London in a revival of T. S. *Eliot's The Family Reunion in 1956 and visiting New York in Graham *Greene's The Potting Shed in 1957. She and Lewis Casson celebrated their golden wedding in 1959 by starring in Eighty in the Shade, specially written for them by Clemence *Dane, and were together in *Coward's Waiting in the Wings (1960) and as the Nurse and 'Woffles' in *Chekhov's Uncle Vanya at the first *Chichester Festival in 1962. After starring in a new comedy by William Douglas *Home, The Reluctant Peer (1964), she made her last appearance on the London stage in 1966 in a revival of Kesselring's Arsenic and Old Lace with Casson and Athene *Seyler.

Her brother (Arthur) **Russell Thorndike** (1885–1972) was an actor who also wrote the play Dr Syn (1925), in which he played the title-role. Their younger sister **Eileen Thorndike** (1891–1953) was also an accomplished actress.

Three-fold, see FLAT.

Three Hundred Club, London, see STAGE SOCIETY.

Thrust Stage, see OPEN STAGE.

Thunder Run, Thunder Sheet, see SOUND EFFECTS.

Tieck, (Johann) **Ludwig** (1773–1854), German Romantic poet and dramatist, whose early plays, satirical fairy-tales—Ritter Blaubart (Bluebeard, 1796), Der gestiefelte Kater (Puss-in-Boots, 1797), and Die verkehrte Welt (The World Upside-down, 1798)—were followed by verse dramas, Leben und Tod der heiligen Genoveva (1799) and Kaiser Oktavianus (1804). In 1824 Tieck became Director of the Court Theatre in *Dresden, where he insisted on clear diction and simplified staging, though a performance of A Midsummer Night's Dream on a specially constructed Elizabethan stage, as Tieck imagined it to have been, remained an isolated experiment. He became an influential critic, and his description of work at Dresden, collected as Dramaturgische Blätter (1826), reveals him as a man of insight and good taste. His interest in the Elizabethan theatre led him to translate several plays by Ben *Jonson, and with his daughter he completed *Schlegel's translations of Shakespeare, whose reputation in Germany he did much to further.

Tilley, Vesta [Matilda Alice Powles] (1864–1952), one of the outstanding stars of the British *music-hall, famous for her *male impersonations. She made her first appearance at the age of 4 at the St George's Hall in Nottingham, where her father William Powles, known as Harry Ball, was *Chairman, and soon after was seen in Birmingham, giving an impression of the popular contemporary tenor Sims Reeves. As The Great Little Tilley, she made her first appearance in London in 1878, at the Royal, Holborn, singing 'Near the Workhouse Door' in the character of 'Poor Jo'. It was then that she added Vesta to her pet name of Tilley, retaining it throughout her career. She was seen as *principal boy in a number of *pantomimes, including several at *Drury Lane, where she first appeared at Christmas 1892. But her true home was the music-hall, where her immaculately cut suits and elegant gestures provided a pattern followed everywhere by suburban and provincial young-men-about-town. She was not the first music-hall star to wear male attire, but she was undoubtedly the greatest. She was particularly admired for her impersonations of soldiers, and could reproduce to the life, with a faint edge of satire, the 'strut' of the guardsman or the young officer's 'glide'. Although she presented a completely convincing male image, she made no attempt to sing like a man, her voice, like her songs, many of which were written by her husband, presenting the woman's view of the men she represented. Two of her most famous numbers were 'Following in Father's Footsteps' and 'Jolly Good Luck to the Girl who Loves a Soldier'.

Times Square Theatre, New York, on West 42nd Street. This opened in 1920 with Vajda's Fata Morgana, translated as The Mirage, which ran for six months. Several musical plays were followed by Channing Pollock's The Fool (1922), more successful outside New York than on Broadway, and *Maeterlinck's Pelléas and Mélisande (1923) which, in spite of the performance of Jane *Cowl, had only a short run. Charlot's Revue of 1924, with a large cast of London favourites, was, however, a success, as was Anita Loos's Gentlemen Prefer Blondes (1926), while The Front Page, a play about journalism by Ben Hecht and Charles MacArthur, ran for 281 performances in 1928–9. In 1931 came a long run of *Coward's Private Lives, with the author, Gertrude *Lawrence, Jill Esmond, and Laurence *Olivier. The last play at this theatre before it became a cinema was Edward Roberts and Frank Cavett's Forsaking All Others (1933) which reintroduced Tallulah *Bankhead to Broadway.

Tireman, in the Elizabethan theatre the man in charge of the wardrobe, who was kept in the tiring-house, or dressing-room. He also saw to the provision of stools for those members of the audience who sat on the stage, and in the private roofed theatres he looked after the lights.

Tirso de Molina [Fray Gabriel Téllez] (c.1571–1648), Spanish ecclesiastic, whose secular works include a number of plays, of which more than 80 are extant, though he claimed to have written 400. Their technique derives from that of his near contemporary Lope de *Vega Carpio, whom he much admired, but it is modified in his case by his greater interest in the psychology of his characters. He was particularly good at drawing women at their wittiest and most intelligent in such comedies as Don Gil de la calzas verdes (The Man in Green Breeches, c.1611), in which he employs his favourite comic device of women disguised as men, and El vergonzoso en palacio (The Bashful Man at Court, c.1612). Among his historical plays, the best is undoubtedly La prudencia en la mujer (Prudence in Woman, c.1622), in which he draws an excellent picture of the heroic Queen Maria and points a moral for the Spanish politicians of his own time. Tirso was also the author of four *autos sacramentales, and of a number of religious plays of which an excellent example is El condenado por desconfiado (Damned for Lack of Faith, c.1624); but his masterpiece is El burlador de Sevilla y convidado de piedra (The Trickster of Seville and the Stone Guest, c.1630), the first of many plays on the subject of *Don Juan, which was staged by the *RSC in 1990.

Tivoli Music-Hall, London, in the Strand. Erected on the site of a beer-hall, whose name it took, this three-tier building holding 1,500 opened in 1890 as a theatre, but had no success until in 1893 it was taken over by Charles *Morton who made the 'Tiv' one of London's most popular *music-halls. Its shows lasted for anything up to four hours, and the bill often included as many as 25 turns, featuring all the great names of the music-hall. In the years immediately before the First World War its popularity began to decline and it closed in 1914, being demolished and replaced by a cinema in 1923.

Toby. 1. The dog of the *Punch and Judy show, who is a purely English character, like *Clown, having no connection with the *commedia dell'arte or the *harlequinade. The name seems to have come into use with the introduction of a live dog somewhere between 1820 and 1850, either because the first dog to be employed was already so called, or from association with the biblical Tobias, a favourite subject for a puppet-play, in which Tobias and the angel were accompanied by a dog. Toby is usually a small, quick-witted mongrel terrier. Wearing a ruff round his neck, he sits on the sill of the puppet-booth window and takes no part in the action, unless Punch pets him or, alternatively, urges him to bite the other characters. After the show he goes round among the audience, with whom he is a firm favourite, collecting pennies in a little bag which he holds in his mouth.

2. A stock character in the folk theatre of the United States, who represents the country bumpkin triumphing over the 'city slickers'. Deriving from a mixture of the commedia dell'arte, the Shakespearian clown, the conventional stage 'silly boy', and the Yankee comedian, he has a freckled face with a blacked-out front tooth, and wears a rumpled red wig, battered hat, calico shirt, baggy jeans, and large ill-fitting boots or shoes. He first emerged in the 1900s, and Frederick R. Wilson from Horace Murphy's Comedians was the first of a long line of actors to specialize in Toby roles, which include generous use of the topical 'ad-lib', and of such theatrical gymnastics as the pratfall, glides, splits, and rubber-legs. It is traditional in revivals of Uncle Tom's Cabin for the Toby-comedian of the troupe to don blackface and a gunnysack costume to impersonate Topsy. Toby sometimes has a feminine counterpart named Susie. Although there are still a number of small-time Toby shows in remote areas, the last of any importance closed in Wapello, Iowa, in 1962, after nearly 50 years under the same managers.

Togata, see FABULA 4: TOGATA.

Tokyo Globe, salmon-pink theatre modelled on the first *Globe, and privately financed, which opened in 1988. Located in an outer suburb, it re-creates the original's layout and façade, though much of the interior shows Japanese influence. Its first occupant was the English Shakespeare Company (see BOGDANOV) with its 'Wars of the Roses' cycle. It was soon followed by the *National Theatre (in Cymbeline, The Tempest, and The Winter's Tale), the *Royal Dramatic Theatre (in Hamlet), and the *Maly Theatre, Leningrad.

Toller, Ernst (1893–1939), German poet and dramatist, one of the best and most mature exponents of *Expressionism. His first play, Die Wandlung (Transfiguration, 1919), written during his imprisonment as a pacifist after being invalided out of the trenches in 1916, is a plea for tolerance and the abolition of war. It was followed by Masse-Mensch (1920) and Die Maschinenstürmer (1922). The latter, as The Machine Wreckers, in a translation by Ashley *Dukes, was seen in London in 1923. It is less Expressionist in technique than Masse-Mensch, which as Man and the Masses was produced in New York in 1924 by Lee *Simonson for the *Theatre Guild, and also less pessimistic, since Toller, in the person of his hero Jim Cobbett, foresees the day when the rebellious workers will be an organized and stable body of intelligent men. But Toller's later plays, of which Hoppla, wir leben! (1927), first produced by *Piscator, was staged in London in 1929 as Hoppla!, became progressively less hopeful as he watched the decline of freedom in Germany, and in 1933 he left for England and the USA, where he committed suicide on hearing of the outbreak of the Second World War. Among Toller's other plays were Wunder in Amerika (on Mary Baker Eddy) and Feuer aus den Kesseln (on a naval mutiny in 1917),

both first produced in 1930. The former, as Miracle in America, was seen in London in 1934, the latter, as Draw the Fires!, in 1935. Die blinde Göttin (1932), on a miscarriage of justice, was translated in 1953 as Blind Man's Buff.

Tolstoy, Alexei Konstantinovich (1817–75), Russian diplomat and poet, author of a fine historical trilogy, containing excellent crowd scenes and written with much semi-oriental imagery, in which he idealized old feudal Russia. All three plays—The Death of Ivan the Terrible, Tsar Feodor Ivanovich, and Tsar Boris—were written between 1866 and 1870, but were banned by the censor, who finally allowed the second to be put on as the opening production of the *Moscow Art Theatre in 1898, two days after it had been seen in St Petersburg. It has frequently been revived, and there have also been productions of the complete trilogy.

Tolstoy, Alexei Nikolaivich (1882–1945), Soviet novelist and dramatist, whose first play, written after the October Revolution, shows the influence of the uprising reaching out as far as the planet Mars. His later works include a historical drama on Peter the Great and a two-part play on Ivan the Terrible; the first part, dealing with Ivan's youth and marriage, was produced at the *Maly Theatre, Moscow, in 1943, the second, dealing with Ivan's struggles to unite Russia, at the *Moscow Art Theatre later the same year. Of his plays on modern themes the best is The Road to Victory (1939), dealing with an episode of the Revolution in which both Lenin and Stalin appear. His Rasputin was staged by *Piscator in 1927.

Tolstoy, Leo Nikolaivich, Count (1828–1910), Russian novelist, critic, and dramatist. Under the influence of *Turgenev and *Ostrovsky, he started some comedies in the 1850s, which, however, remained unfinished. It was not until 1886 that he once more turned to the theatre, and by then his whole philosophy of life had changed. Under the influence of M. V. Lentovsky, Director of one of the Moscow People's Theatres, he wrote The Power of Darkness, possibly the most forceful peasant play ever written. Its main outline was taken from a criminal case heard at Tula, but in Tolstoy's hands it became a stark naturalistic document which for many years banned by the censor and first acted abroad, in Paris in 1888 under *Antoine, and in Berlin by the *Freie Bühne in 1890. It was not seen in Russia until 1895, when it was staged both at the Alexandrinsky Theatre in St Petersburg (now the *Pushkin Theatre in Leningrad) and at the Moscow *Maly. It was first seen in London in 1904, and in New York in 1920. Tolstoy's next play, a short comedy entitled The First Distiller (1887) which attacked alcoholism, was followed by The Fruits of Enlightenment, a comedy which satirizes the parasitic life of the country gentry.

Published in 1891, it was produced in the same year by *Stanislavsky, and the following year was seen at the Maly, with little success. Even the actors at the *Moscow Art Theatre, where the play was revived some years later, could not at first tackle the peasant characters successfully, and many years of experiment and experience were needed before Tolstoy's famous drama could be adequately portrayed. It was first seen in London in 1928. Tolstoy's last two plays, published in 1912, were left unfinished at his death. *Redemption* (or *The Living Corpse*), an attack on the evils of contemporary Russian marriage laws, was produced by the Moscow Art Theatre in 1911; John *Barrymore played the hero, Fedya, in New York in 1918, Donald *Wolfit in London in 1946. *The Light that Shines in Darkness*, in which the useless life of a wealthy family is contrasted with that of the poverty-stricken and overworked peasants, does not appear to have been staged. Three of Tolstoy's novels, *Resurrection*, *Anna Karenina*, and *War and Peace*, were dramatized and produced at the Moscow Art Theatre, and the last two have also been staged in English.

Tom Bradley Theatre, Los Angeles, see LOS ANGELES THEATRE CENTER.

Toms, Carl (1927–), English designer, whose work was first seen in London in 1957, and who first became known in the field of opera. In 1968 he designed the *National Theatre's productions of *Love's Labour's Lost* and *Marlowe's *Edward II*; he later worked for the company on such plays as *Vanbrugh's *The Provoked Wife* (1980), Neil *Simon's *Brighton Beach Memoirs* (1986), and Brian *Friel's adaptation of *Turgenev's novel *Fathers and Sons* (1987). He has been particularly associated with *Stoppard: *Travesties* (1974; NY, 1975) for the *RSC, *Night and Day* (1978; NY, 1979), *The Real Thing* (1982), *Jumpers* (1985), and *Hapgood* (1988); and Stoppard's translations produced at the National Theatre as *On the Razzle* (1981), *Rough Crossing* (1984), and *Dalliance* (1986). Toms's designs for *Bolt's *Vivat! Vivat Regina!* (1970; NY, 1972), *Coward's *Look after Lulu* (1978), and *Osborne's *A Patriot for Me* (1983) were seen in both *Chichester and London. As resident designer with the *Young Vic, 1970–83, he was associated with many of Frank *Dunlop's productions, and in 1989 he worked on the RSC's production of *Kaufman and *Hart's *The Man Who Came to Dinner* and on Ben *Travers's *Thark*. He is well known also in opera, ballet, and films, and has been very active outside England, especially in New York and Vienna.

Toole, John Laurence (1830–1906), English actor and theatre manager, who was for a short time, like *Garrick, clerk to a wine merchant. Success in amateur theatricals turned his thoughts to the stage, and he made his first appearance as a professional in Dublin in 1852. After further experience in the English provinces he established himself in London in 1856 as a low comedian, playing Fanfaronade in Charles Webb's *Belphegor the Mountebank*, in which Marie Wilton, later the wife of Squire *Bancroft, also made her London début. In 1857 Toole was with Henry *Irving, who remained a lifelong friend, and a year later, on the recommendation of *Dickens, another close friend, was engaged by Ben *Webster for the *Adelphi Theatre, where he remained for nine years. Among his best parts were two of Dickens's characters—Bob Cratchit in *A Christmas Carol* (1859) and Caleb Plummer in *Dot* (1862), a dramatization of *The Cricket on the Hearth* by *Boucicault. In this latter part Toole combined humour with pathos in a way which showed how well he might have played serious parts; but the public preferred him in farce, and in 1869 he began a five-year association with *Hollingshead at the *Gaiety Theatre, where he proved admirable in *burlesque. He made an appearance in New York, in 1874, but his humour was too cockneyfied for the Americans, and he never returned there. In 1879 he went into management at the Charing Cross Theatre, opened in 1869, to which he gave his own name three years later. His most important productions there were *Pinero's early comedy *Imprudence* (1881) and *Barrie's first play *Walker, London* (1892). He retired in 1895, and the theatre closed, being demolished the next year.

Topol, Chaim (1934–), Israeli actor, who had his first experience of acting while in the army, and afterwards founded a theatre of satire in Tel Aviv which achieved widespread popularity. In 1961 he went to Haifa to help establish the Municipal Theatre there, acting as assistant to the director and appearing with the company at the Venice Biennale as Azdak in *Brecht's *The Caucasian Chalk Circle*, repeating the role at *Chichester in 1969. He also played such parts as Petruchio in *The Taming of the Shrew*, Pat in Brendan *Behan's *The Hostage*, John in *Ionesco's *Rhinoceros*, and the Soldier in *Frisch's *Andorra*. In 1965 he was with the *Cameri Theatre in Tel Aviv, and in 1967 he appeared in London as Tevye in the musical *Fiddler on the Roof* in which he scored an immense success, returning in the role in 1983. In 1988 he took over the title-role in the musical *Ziegfeld* at the *London Palladium, but failed to revive its flagging fortunes.

Torelli, Giacomo (1608–78), Italian machinist and scene-painter. A practical man of the theatre rather than an artist, he made many important innovations in the designing and setting of scenery, being the inventor of the *carriage-and-frame device for moving several sets of *wings on and off stage simultaneously. He was a pupil of Aleotti, and may have worked with him on the building of the *Teatro Farnese, which was not completed until 1628. In 1640 he was responsible for the building of the Teatro Novissimo in Venice, where the magical effects of his stage mechanisms won for him the nickname of 'the great magician'. Five years later he went to Paris and inaugurated the fashion for *machine plays of which the finest was *Corneille's *Andromède* (1650). This was produced at *Molière's first theatre, the *Petit-Bourbon, which Torelli had completely refurbished backstage, painting new scenery and installing machinery of his own invention. It was all destroyed by his rival Gaspare *Vigarani in 1660, but the designs were preserved, and over a hundred years later the complete survey of contemporary theatre machinery given under 'Machines du Théâtre' in *Diderot's *Encyclopédie* (in 1772) shows that it was almost all based on Torelli's ideas.

Tormentor, see FALSE PROS(CENIUM).

Torres Naharro, Bartolomé de (c.1485–c.1524), leading Spanish dramatist of the 16th century, contemporary with *Encina and Gil *Vicente. Little is known of his life, but from 1513 he was in Rome, where most of his plays appear to have been written. His collected works appeared in 1517 under the title of *Propalladia*. The important prologue distinguishes between two types of play, the *comedia a noticia* and the *comedia a fantasía*, terms which may be said to correspond to realistic and novelesque genres. Of the former type two examples have survived, *Soldadesca*, dealing with a braggart Spanish captain, and *Tinellaria*, depicting life in the kitchen of an Italian palace. The best-known example of the novelesque plays is *Himenea*, based on part of the *Celestina*, where the relationship between master and man and the theme of the conflict of love and honour both foreshadow the plays of Lope de *Vega Carpio.

Törring, Josef August von, see RITTERDRAMA.

Totaltheater, see THEATRE BUILDINGS.

Tottenham Street Theatre, London, see SCALA THEATRE.

Touring Companies. Since many touring companies present opera, ballet, or large-scale musicals, they need larger theatre buildings than the *repertory theatres, which usually limit themselves to straight plays or modest musicals; on the other hand they need less storage and wardrobe space and fewer workshops. In the 1960s there were about 35 theatres in Britain graded as No. 1 for the use of touring companies, but soaring costs and the lack of suitable productions made it difficult to keep them all open throughout the year; the sites also proved increasingly valuable for development purposes, and their owners were often induced to sell them. After a number had been lost urgent action was taken to preserve those that remained

through purchase by the local authority or by subsidy, and there is now a touring circuit of over a dozen well-equipped theatres which can accommodate the large subsidized opera, ballet, and drama companies, while a further 20 theatres can receive less spectacular productions. Subsidized productions, however, can occupy only a part of a theatre's programme, the rest of which must be supplied by commercial shows, such as appearances by pop groups and other performers and touring productions of West End hits and plays from the repertoire. Pre-London tours are becoming rare because of cost and because West End theatres increasingly take their productions direct from the subsidized and *Fringe theatres; post-London tours with the original casts are now also rare. Some touring is undertaken by companies such as the *Cambridge Theatre Company and the English Shakespeare Company (see BOGDANOV), and Fringe theatre groups tour studio theatres, arts centres, and other more makeshift venues.

Tourneur, Cyril (1575–1626), English dramatist, of whom little is known, though he was connected with the Cecils, and may have been sent by them on secret missions abroad. He was certainly in Cadiz with Sir Edward Cecil in a secretarial capacity the year before his death. Of the three tragedies doubtfully ascribed to him, *The Nobleman* (1607) is lost, the manuscript having been destroyed by *Warburton's cook. *The Revenger's Tragedy*, which has also been credited to *Middleton, was published anonymously in 1607 and probably acted a year previously by the *King's Men. It was revived under Tourneur's name by the *RSC at Stratford-upon-Avon in 1966 and at the *Aldwych Theatre in London three years later. *The Atheist's Tragedy; or, The Honest Man's Revenge* was probably written in 1606, since echoes of *King Lear* (1605) have been noted in it.

Tovstonogov, Georgyi Alexandrovich (1915–89), Soviet Russian theatre director, one of the best known and most successful of his time. He was born in Tbilisi, where he first worked in the theatre and where he returned after graduating from the Moscow State Theatre Institute, becoming manager of the Griboyedov Theatre, 1938–46. He soon became known as a highly original director of both Russian and English-language plays, among the latter being *Sheridan's *The School for Scandal* and Lillian *Hellman's *The Little Foxes*. From 1946 to 1949 he was Director of the Moscow Children's Theatre, and from 1950 to 1956 of the Lenkom Theatre in Leningrad, opening his first season with *Fletcher's *The Spanish Curate*. In 1955 he revived *Vishnevsky's *The Optimistic Tragedy* at the *Pushkin Theatre there, and in 1956 he took over the *Gorky Theatre, making it one of the most talked-of and widely travelled of all contemporary Russian theatre companies.

Tower Theatre, London, see AMATEUR THEATRE.

Toy Theatre, or Juvenile Drama, collections of theatrical material, popular in the 19th century, which consisted of drawings of actors, scenery, and properties in a successful contemporary play, suitable for cutting out and mounting on cardboard for a performance in which they were drawn on metal long-handled slides across a small model stage, while an unseen assistant recited an extremely condensed version of the text. The sheets which made up the complete set, usually eight to twelve in all, could be bought for a 'penny plain, twopence coloured', the colouring being done by hand in bold, vivid hues that are as fresh today as when first applied. These sheets, which may have originated in those sold by theatrical agencies in Paris for the benefit of provincial and foreign managements, were probably first intended, in England at any rate, as theatrical souvenirs. They capture with astonishing fidelity the theatre of *Grimaldi, *Kean, *Kemble, *Liston, and *Vestris, and the productions of *Astley's, *Covent Garden, *Drury Lane, the *Olympic, and the *Surrey in the early 19th century. The total repertory of some 300 plays includes melodramas such as Pocock's *The Miller and His Men*, *Lillo's *George Barnwell*, and *Boucicault's *The Corsican Brothers*, ballad operas such as *Dibdin's *The Waterman*, contemporary versions of Shakespeare, and many long-forgotten pantomimes. At first considerable care was taken to reproduce the costumes, attitudes, and even the features of the actors, as well as the details of wings, backcloth, and scenic accessories; but as the toy theatre increased in popularity the quality of the drawings and reproduction fell off, though trade continued brisk until the 1850s and beyond. The old-style Juvenile Drama never quite disappeared, and as late as 1932 two shops in Hoxton still printed the sheets of the old plays from the original blocks.

Similar toy theatres were popular on the Continent in the 19th century, particularly in Germany, Denmark, and Spain, where the characters moved in grooves rather than on slides, but the plays were usually specially written for children, and not, as in England, taken from the adult repertory.

Trades Unions Theatre, Moscow. This theatre, originally known as the First Workers' Theatre, was one of several established during the early years of the Revolution which were affiliated to Proletkult (Organization for the Cultural Enlightenment of the Proletariat), by which name it was also known in its early days. Its work was much influenced by the theories of Bogdanov relating to the independent development of a specifically proletarian culture, and

it is chiefly memorable for the work done there between 1921 and 1923 by Sergei Eisenstein, especially his 'agit-montage' productions of *Ostrovsky's *Enough Stupidity in Every Wise Man* and *Tretyakov's *Gas Masks*. After Eisenstein left to devote the rest of his life to the cinema, the theatre passed through many vicissitudes until in 1932 it was given a new name as well as a new director, Alexei Dikie, whose first production was Wolf's *Sailors of Cattaro*. This was well received, but later productions were less successful and in 1936 the theatre finally closed.

Trafalgar Square Theatre, London, see DUKE OF YORK'S THEATRE.

Trafalgar Theatre, New York, see NEDERLANDER THEATRE.

Tragedian, see STOCK COMPANY.

Tragedy, play dealing in an elevated, poetic style with events which depict man as the victim of destiny yet superior to it, both in grandeur and in misery. The word is of Greek origin and means 'goat-song', possibly because a goat was originally given as a prize for a play at the *Dionysia. The classic Athenian tragedies of *Aeschylus, *Euripides, and *Sophocles developed from the choral lyric, an art which reached its height among the Dorian peoples of the Peloponnese during the 6th century BC. The earliest plays began with the *parados*, or entrance of the *chorus, which was soon preceded by a *prologos* for the actor or actors. Each formal ode, or *stasimon*, for the chorus alternated with a dramatic scene, or *episode*; lyrical dialogue between an actor and the chorus was called a *kommos*; and all that followed the final stasimon was the *exodus*. The chorus sang, or chanted, in unison, but probably spoke through its leader. As nothing is known about the music and dancing of the chorus, and the music-rhythms of the odes cannot be translated into speech-rhythms, it is impossible to dogmatize about the original productions of the great texts which have come down to us, and all translations and revivals can only be approximations. It was the subject-matter of the plays which exercised the greatest influence on the drama of the future. Taken from the myths of gods and heroes, it retained a link with its religious origins by the beneficent intervention, usually at the end of the play, of a god—the *deus ex machina*—who descended from above the stage by means of a crane or pulley. The Roman theatre produced excellent writers of comedy in *Plautus and *Terence, but no tragedies for the stage have survived; those by *Seneca, which had an immense influence on later European drama, were *closet plays.

Tragedy in Renaissance Italy, more under the direct influence of the Greeks than of Seneca, developed early, but did not produce any outstanding playwright until the 18th century, with *Alfieri. In France tragedy developed under the influence of Seneca, modified by the contemporary interpretation of *Aristotle which gave rise to the theory of the *unities of time, place, and action, though only the last was consistently observed by Greek dramatists, the unities of time and place being imposed on the play by the continuous presence of the chorus. The greatest exponents of French classical tragedy were *Corneille and *Racine, whose successors up to the end of the 18th century continued to employ their outward forms but without their inward excellence.

In England, where the influence of Seneca was paramount, *Marlowe and Shakespeare evolved a form of tragedy mingled with comedy which was *sui generis*. Because of its powerful appeal to English audiences, the English theatre remained impervious to the influence of French classical tragedy, even after the Restoration, when such plays as *Addison's Cato (1713) brought the letter but not the spirit of Corneille and Racine briefly on the English stage. Spain, too, had her native tragedy, formulated by *Calderón, and efforts to import French tragedy failed, as did the attempts of *Gottsched and Carolina *Neuber in Germany. The German theatre later produced its own writers of tragedy in *Goethe and *Schiller; but it was the melodramatic aspect of their tragedies which had the greatest appeal, and this, added to the influence of Shakespeare all over Europe at the end of the 18th century, produced the highly coloured *melodrama which in the 19th century replaced true tragedy everywhere. Meanwhile, in the 18th century, in the plays of *Lillo, *Lessing, and *Mercier, efforts had been made to apply the formula of classical tragedy to middle-class existence, resulting in 'domestic tragedy' or *tragédie bourgeoise*. It was not a success. Tragedy in the narrow theatrical sense demands a cast of heroes or demi-gods, an unfamiliar background—exotic, romantic, or imaginary—and a sense of detachment heightened by the use of verse or rhetorical prose. Even the plays of *Ibsen and his successors, though often tragic in their implications, are dramas rather than tragedies in the Greek sense. In modern times efforts have again been made to tame tragedy and bring it within the family circle. But it is interesting to note that *Murder in the Cathedral* (1935) by T. S. *Eliot, which has as protagonists a king and an archbishop, was a success, unlike his *The Family Reunion* (1939) which, though based on a Greek myth, was firmly rooted in suburbia.

Tragic Carpet, green baize stage *cloth which by the end of the 17th century was invariably spread over the stage before the performance of a tragedy, to prevent the actors from soiling their costumes when collapsing on the dusty boards. It continued in use until well into the 19th century, and is often referred to in theatrical memoirs and letters. Earlier mentions of permanent green cloth coverings, once thought to refer to the tragic carpet, are now believed to refer to the benches in the auditorium.

Tragi-comedy, bastard form of tragedy, being a play dealing with a tragic story which ends unhappily, but which contains certain elements of comedy and the remote possibility of a happy ending. Some critics have classified *Hamlet* (c.1600–1) as a tragi-comedy, but the perfect example of the type is *Corneille's *Le Cid* (1637). In France tragi-comedy was important until the mid-17th century, *Garnier and *Hardy being especially associated with the genre.

Tramway, see GLASGOW.

Transformation Scene, important element in the English *pantomime. An instantaneous change of part of a scene, such as a shop-window or a house-front, was usually done by the use of *falling flaps, a variant of which was the *chassis à développement*. By the use of the earlier *carriage-and-frame or *drum-and-shaft methods, the whole scene could be changed, the *backcloth and side *wings being drawn off simultaneously to reveal new ones behind. For the swift changes needed for the pantomime a more spectacular method, made possible by the *flies found in newer theatre buildings, was the *rise-and-sink. A quick change could also be achieved by the use of scruto (thin strips of wood fastened to a canvas backing so as to form a continuous flexible sheet); by the *Fan Effect, in which sectors of the back scene, pivoting centrally at the foot of the scene, sank sideways upon each other like collapsing fans, revealing a new scene behind; or by placing rolls of painted canvas like columns across the stage, with lines top and bottom, which when pulled drew the new scene across the old. Various *traps were used to enhance the illusion. With the advent of more sophisticated *lighting the usual method of effecting a transformation scene became the transparency, which when lit from the front was as opaque as canvas, but faded from sight when lit from behind.

Transparency, see TRANSFORMATION SCENE.

Transpontine Melodrama, term applied, usually in genial derision, to a type of crude and extravagantly sensational play staged in the mid-19th century in London theatres 'across the bridges' (i.e. on the south side of the Thames) such as the *Surrey Theatre and the *Old Vic. By extension the term was later attached to such *melodramas wherever performed.

Traps, devices by which scenery and actors can be raised to stage level from below. They are now used mainly in *pantomime, but were formerly essential for the acrobatic *trickwork of the dumb ballet, for ghostly apparitions, and for the sudden metamorphosis of one character into another. Although the modern theatre makes very little use of traps, they can still be found in some older theatres, arranged in a traditional pattern and capable of being used for their original purpose. To accommodate them the joists of the stage floor are cut and specially framed, and the opening is concealed in a variety of ways. Corner traps are used for conveying a single standing figure. Nearer the centre of the stage is the star trap, which projects an actor on stage at great speed, and is used in pantomime for the arrival of the Demon King. Almost in the centre of the stage is the large rectangular grave trap, which measures roughly 6 ft. long and 3 ft. wide. It has a platform below, which can be raised and lowered, and takes its name from its use for the graveyard scene in *Hamlet*, just as the less commonly found cauldron trap, usually square and placed further upstage, is named from the witches' scene in *Macbeth*. One of the most ingenious trap mechanisms was that devised for the apparition in *Boucicault's *The Corsican Brothers* (1852), which had to rise slowly out of the earth while gliding across the full width of the stage. This Corsican trap, or ghost glide, consisted of two long sliders, the first of which drew the trap across the stage while the second closed the aperture behind it. An inclined rail track below the stage carried a small *boat truck on which stood the actor playing the ghost. Two other traps were the *slote, which could be used for people or scenery, and the vamp trap, consisting of two spring-leaves in the backcloth or stage floor so arranged that the actor's body appeared to pass through a solid wall. It is said to have taken its name from *Planché's *The Vampire; or, The Bride of the Isles* (1820), in which it was first used. For the swift substitution of one character for another a pair of traps, similar to the corner traps, was placed side by side, so geared that one rose as the other fell. When the actor on stage took his place on the raised trap, his partner stood ready on the lower trap; at the crucial moment the traps were released, and in a second a different figure stepped from a part of the stage floor so near the original spot as to be indistinguishable from it. (See also FOOTLIGHTS TRAP.)

Traveller, see CURTAIN.

Travers, Ben (1886–1980), English dramatist, an outstanding writer of farce, who first came into prominence with *A Cuckoo in the Nest* (1925), which inaugurated a series of 'Aldwych farces', so called because they were all presented at the *Aldwych Theatre. Their casts included Robertson *Hare, Ralph *Lynn, and Tom *Walls. The best known were *Rookery Nook* (1926), *Thark* (1927), *Plunder* (1928), and *Turkey Time* (1931), the last being *A Bit of a Test* (1933).

Travers also wrote a number of plays for other theatres, including a comedy, *O Mistress Mine* (1936), and the farces *Banana Ridge* (1939), *Spotted Dick* (1940), and *She Follows Me About* (1945), in all of which Robertson Hare again appeared, being joined in *Outrageous Fortune* (1947) and *Wild Horses* (1952) by Ralph Lynn. In 1975 *The Bed before Yesterday*, starring Joan *Plowright, began a long run in London. His plays stand up reasonably well and are sometimes revived.

Traverse Theatre, Edinburgh, lively experimental theatre which opened in 1963. The original auditorium held 60 seated on either side of the acting area, but in 1969 the company moved to new premises in the Grassmarket with a multi-purpose acting area and a seating capacity of up to 120, to which a second studio space was later added. The Traverse develops new work in all areas of theatre. It has given the first productions in English of many foreign plays, including works by *Arrabal, *Jarry, *Mrożek, and Peter *Weiss. It has also had a particularly close relationship with Scottish dramatists such as C. P. *Taylor. The contributions of the Traverse to the *Edinburgh Festival have included numerous British and world premières. The theatre also plays host to *Fringe groups from all over the world, and its own productions have been seen in London, Amsterdam, the USA, Canada, Israel, and Australia. In 1992 the Traverse moved to a new location, Britain's first purpose-built theatre for new writing, with performance spaces seating 250 and 100.

Tree, Ellen, see KEAN, CHARLES.

Tree, Sir Herbert Draper Beerbohm (1853–1917), English actor-manager. The half-brother of Max *Beerbohm, he was working in the city office of his father, a grain merchant, when some successful appearances in amateur dramatics decided him to go on the stage. He made his professional début in 1878, appeared with Geneviève *Ward as Prince Maleotti in Herman Merivale's *Forget-Me-Not* (1879), and was the first to play the Revd Robert Spalding in *Hawtrey's *The Private Secretary* (1884), a part later closely associated with W. S. *Penley. Early in 1887 he became manager of the *Comedy Theatre, where his most successful production was W. O. Tristram's comedy *The Red Lamp*, with which he inaugurated his management of the *Haymarket Theatre later the same year. Among his productions there were *The Merry Wives of Windsor* (1889), Henry Arthur *Jones's *The Dancing Girl* (1891), in which he played the Duke of Guisebury to the Drusilla Ives of Julia *Neilson, *Hamlet* (1892), and Oscar *Wilde's *A Woman of No Importance* (1893). The most successful of his productions was *Trilby* (1895), based by Paul Potter on George Du Maurier's novel. Trilby was played by Dorothea Baird, and Tree himself appeared as Svengali, a part which he revived many times. The success of his tenancy of the Haymarket enabled him to build *Her Majesty's Theatre, which he opened with Gilbert Parker's *The Seats of the Mighty* (1897). There he carried on *Irving's tradition of lavishly spectacular productions of Shakespeare, staging between 1888 and 1914 18 of his plays with a magnificence much to the taste of the time, and achieving at least once, in *Richard II*, a remarkable synthesis of style and setting. He also produced a number of new plays, among them Stephen *Phillips's *Herod* (1900) and *Ulysses* (1902), and American importations such as Clyde *Fitch's *The Last of the Dandies* (1901) and *Belasco's *The Darling of the Gods* (1903). He was also responsible for the first production in English of *Shaw's *Pygmalion* (1914), in which he played Higgins with Mrs Patrick *Campbell as Eliza Doolittle. The play was directed by Shaw, but in general Tree was his own director. A firm disciplinarian, and the founder of the Royal Academy of Dramatic Art (see SCHOOLS OF DRAMA), he was in essence a romantic actor, delighting in grandiose effects and in the representation of eccentric characters which allowed his imagination free play. He ran his company well, helped by his wife **Maud Holt** (1863–1937), an excellent actress who was an active and intelligent partner in all her husband's enterprises.

Tremblay, Michel (1942–   ), French Canada's most renowned playwright. He first attracted attention when his play *Les Belles-Sœurs* (*The Sisters-in-Law*, 1968) was put on by the Théâtre du Rideau Vert in Montreal. This rowdy comedy pioneered the use on stage of *joual* (from *cheval*), the Québécois *argot* that mixes broken French, swear-words, slang, and idiomatic English. This innovation caused considerable controversy, both in Quebec and in Paris when the play was performed there in 1973. His early plays formed a largely autobiographical cycle juxtaposing the restrictive life of the lower middle class with various states of sexual emancipation; they included *La Duchesse de Langeais* (1969), a monologue by an ageing drag queen; *Hosanna* (1973), about the humiliation of a transvestite; and *Sainte Carmen de la Main* (1976), in which the heroine escapes a stifling household for cabaret martyrdom on Main Street. Tremblay at first refused to allow English-language productions of his plays in Quebec, but relented in 1976 with the election of a separatist government and the growth of his international reputation. Among his later plays are *L'Impromptu d'Outremont* (1980), a satire on the impotence of upper-class values; *Albertine, en cinq temps* (1984), a woman looking back on five stages of her life; and *Le Vrai Monde?* (*The Real World?*, 1987), which discusses the moral

dilemma of an artist interpreting his own family. He has also made several translations into French, such as *Aristophanes' *Lysistrata* (1969) and *Chekhov's *Uncle Vanya* (1984).

Trenev, Konstantin Andreivich (1884–1945), Soviet dramatist and short-story writer, whose first play, dealing with an 18th-century peasant insurrection, was staged by the *Moscow Art Theatre in 1925. In 1926 the *Maly Theatre, Moscow, produced the first version of his *Lyubov Yarovaya*, which, after intensive alterations, became one of the outstanding productions of the Moscow Art Theatre in 1937. The story of a school teacher in the Revolution, caught between love for her White Russian husband and loyalty to her ideals, it became popular all over the USSR and is still acted today. Trenev also wrote *On the Banks of the Neva* (1937), one of the first Soviet plays to bring Lenin on to the stage. His later plays include *Anna Luchinina* (1941), *Meeting Halfway* (1942), and *The General* (1944), in which the chief character is the famous Russian Field-Marshal Kutuzov (1745–1813).

Tretyakov, Sergei Mikhailovich (1892–1939), Soviet poet and dramatist, who worked at the Proletkult (later the *Trades Unions) theatre in 1923–4 with Sergei Eisenstein for whom he wrote two propaganda plays—*Are You Listening, Moscow?* and *Gas Masks*. His best-known play, *Roar, China!*, was produced at *Meyerhold's theatre in 1926. Although melodramatic and written avowedly for propaganda purposes, it is a vivid historical document, presenting with intense conviction a conflict between Chinese coolies and British imperialists which had taken place only two years previously. It was seen in translation in New York in 1930, and also in England. During the 1930s, Tretyakov, whose later plays are negligible, was the first to champion *Brecht in the Soviet Union and translated several of his plays.

Trickwork, outstanding feature of 19th-century British theatre, particularly in *pantomime, which was much admired abroad. Stage tricks depending mainly upon the working of ingenious mechanisms are dealt with under *transformation scene and *traps; but the apotheosis of the trick came when the brilliant timing and acrobatic skill of the actor were combined with the expertise of the experienced stage-carpenter. The humblest example was the roll-out, in which a flap of loose canvas was left at the bottom of a piece of scenery, through which a player could suddenly roll from behind and leap to his feet on stage. The leap, in fact, was the supreme test of the trick player. In essence no more than an acrobat's jumping through a trap in the scenery, as in the vamp trap, it could, by the skilful interplay of a group of highly trained actors and the clever multiplication and placing of different types of trap, become an entertainment fit to stand on its own. Such an entertainment was the dumb ballet, of which the best-known exponents were the *Hanlon-Lees. London audiences had a chance to see something resembling the old trickwork (which is still employed in the *commedia dell'arte theatre in the Tivoli Gardens, Copenhagen) when Peppino *De Filippo brought his company to the *World Theatre Season of 1964 in Metamorphoses of a Wandering Minstrel. (See also FALLING FLAPS.)

Trilogy, see TETRALOGY.

Trinity Repertory Company, Providence, RI, one of the best and most adventurous companies in the USA, known until 1986 as Trinity Square. It was founded in 1964 by a number of local citizens who played in a rented theatre in a church building, and became fully professional later the same year. In 1973 it moved to a permanent home in the Lederer Theatre Complex, a renovated cinema which contains two auditoriums. The downstairs theatre, seating 297, re-creates the intimate atmosphere of the original theatre; the upstairs, seating up to 500, is flexible, allowing for continual readjustment of audience and performing areas. The year-round season offers revivals of classic and contemporary plays, and the theatre is especially noted for its encouragement of living American playwrights such as Sam *Shepard and John *Guare; many world premières have been staged. Trinity Summer Rep was initiated in 1978 in the downstairs theatre, and the same season marked the beginning of the Trinity Rep Conservatory, a training programme for actors, directors, and playwrights. Project Discovery, for high school students, was started in 1966.

Tristan l'Hermite, François (c.1602–55), French dramatist and man of letters. He was poor and a compulsive gambler, and after a stormy youth settled in Paris to make a living by his pen. His first play, a tragedy entitled La Mariane (1636) on the subject of Herod's jealous love for his wife, was produced at the *Marais with *Montdory. It had an unprecedented success, competing for public favour with Le Cid, and for some time Tristan was considered a serious rival to *Corneille. La Mariane remained in the repertory until 1704, Herod providing an excellent part for a passionate and fiery actor. Of Tristan's other plays, La Mort de Sénèque (1644) is one of the best of the Roman tragedies after those of Corneille and *Racine. Le Parasite (1654), a comedy, is interesting as an attempt to introduce into France the stock figure of the sponger from Latin comedy. Two of Tristan's plays were in the repertory of *Molière's short-lived *Illustre-Théâtre.

Tritagonist, see PROTAGONIST.

Tron Theatre, Glasgow, seating 215, opened in 1978. A major home for new writing in Scotland, it mounts five productions each year. It also plays host to small-scale touring companies, national and international, and has regular jazz, folk, and cabaret, being a centre for stand-up comedy and cabaret in the west of Scotland. Originally a club, the theatre went public in 1989.

Troupes Permanentes, see DÉCENTRALISATION DRAMATIQUE.

Trouvères, see JONGLEURS.

Trucks, see BOAT TRUCKS.

Tuccio, Stefano, see JESUIT DRAMA.

Tucker, Sophie, see VAUDEVILLE, AMERICAN.

Tumbler, Tumbling, see DROP.

Turgenev, Ivan Sergeivich (1818–83), Russian novelist and dramatist, who in 1843, while studying at Berlin University, published his first play, a romantic swashbuckling drama set in Spain. His second play, a satirical comedy in the style of *Gogol entitled Penniless; or, Scenes from the Life of a Young Nobleman, was published in 1846. He later wrote several short plays, in the style of *Musset's Comédies et proverbes, among them Where It's Thin, It Breaks; The Bachelor (1849), written for *Shchepkin; and The Boarder (1850). In 1850 Turgenev wrote his dramatic masterpiece, A Month in the Country (originally entitled The Student), it was published in 1869, but not staged until 1872. It is important as the first psychological drama in the Russian theatre, and Turgenev proved himself the forerunner of *Chekhov in shifting the dramatic action from external to internal conflict. Battles with the censorship, imprisonment, and exile then led Turgenev to give up the theatre, and A Month in the Country is his only play to be known outside Russia. It was first seen in London in 1926 and New York in 1930, and has been revived several times in both cities. Brian *Friel's adaptation of Turgenev's novel Fathers and Sons was staged at the *National Theatre in 1987.

Turlupin [Henri Legrand] (c.1587–1637), French actor, who as Belleville played serious parts in tragedy, but excelled in the parts of roguish valets and is best remembered for the Turlupinades, farces played at the Hôtel de *Bourgogne in partnership with *Gaultier-Garguille and *Gros-Guillaume. He probably played at the Paris *fairs before joining a professional company in about 1615. He figures as himself, with the rest of the company, in Gougenot's La Comédie des comédiens (1631).

Turnham's Music-Hall, see METROPOLITAN MUSIC-HALL.

Tutin, Dorothy (1930–), English actress, who made her first appearance as the young Princess Margaret in William Douglas *Home's *The Thistle and the Rose* (1949), and was then with the *Bristol Old Vic and the *Old Vic in London. She first attracted attention by two brilliant performances: as Rose, a young Catholic orphan in love with a middle-aged man, in Graham *Greene's *The Living Room* (1953), and as Sally Bowles in *Van Druten's *I am a Camera* (1954). A year later she was a great success as St Joan in *Anouilh's *The Lark* and as Hedvig in *Ibsen's *The Wild Duck*. In 1958 she went to Stratford-upon-Avon, where she was an outstanding Viola in *Twelfth Night* and also played Juliet in *Romeo and Juliet* and Ophelia in *Hamlet* (all 1958), Portia in *The Merchant of Venice*, and Cressida in *Troilus and Cressida* (both 1960). Remaining with the company when it became the *RSC, she gave an absorbing study of sexual frustration as Sister Jeanne in John *Whiting's *The Devils* and added Desdemona in *Othello* to her gallery of Shakespearian heroines. She made her début in New York in John *Barton's anthology *The Hollow Crown* (1963). At the Bristol Old Vic and then in London she gave a touching account of the young Queen Victoria's resilience and humour in William Francis's *Portrait of a Queen* (1965; NY, 1968). After Rosalind in *As You Like It* (1967) and Kate in *Pinter's *Old Times* (1971) for the RSC she played Peter Pan in *Barrie's play (1971 and 1972)—to which her frail, elfin looks were admirably suited—and Maggie in his *What Every Woman Knows* in 1974. She starred in *Turgenev's *A Month in the Country* at *Chichester in 1974 and in London for *Prospect in 1975, and was Cleopatra opposite Alec *McCowen's Antony for Prospect in 1977. Her roles with the *National Theatre company, which she then joined, included Ranevskaya in *The Cherry Orchard*, Lady Macbeth, and Lady Plyant in *Congreve's *The Double Dealer* (all 1978), and Lady Fanciful in *Vanbrugh's *The Provoked Wife* (1980). She was later seen to advantage in Pinter's *A Kind of Alaska* (1985), Neil *Simon's *Brighton Beach Memoirs* (1986), and *Sondheim's *A Little Night Music* (Chichester and London, 1989). A sensitive actress of great charm and versatility, she suffers from the dearth of worthwhile parts for women in modern plays.

Two-fold, see FLAT.

Twopenny Gaff, see GAFF.

Tyl, Josef Kajetán (1808–50), Czech playwright and actor, virtually the founder of the modern Czech theatre. His first play *The Fair* (1834) was a patriotic folk-comedy with music; one of its songs, 'Where is My Home?', has become the Czech national anthem. It was produced by Tyl's own company, the Kajetán, which he continued to direct until 1837. During the troubled years of 1845 to 1851 his plays reflected the revolutionary struggle in town and country, the fairy-tale atmosphere of *The Bagpiper of Strakonice* (1847)—the source of Weinberger's opera *Schwanda the Bagpiper* (1927)—contrasting strongly with the more realistically portrayed characters of *The Bloody Trial; or, The Miners of Kutná Hora* (1848) and *A Stubborn Woman* (1849). All these plays have been revived in recent years, as well as the historical plays *Jan Hus* (1848), a verse-drama which presents Hus as fighting for the national and social rights of the Czech nation, and *The Bloody Christening; or, Drahomíra and her Sons* (1849).

Tyler, Royall (1757–1826), American dramatist, author of *The Contrast*, a light comedy in the style of *Sheridan's *The School for Scandal* and the first indigenous play to be produced in America. Tyler was a friend of Thomas *Wignell, the leading comedian of the *American Company, and it was probably owing to his influence that the play was given by them at the *John Street Theatre on 16 Apr. 1787. In return for his help Tyler gave him the *copyright. It was a success and was several times revived, though when given in Boston, Tyler's birthplace, it had, like *Othello*, to be disguised as a 'Moral Lecture in Five Parts'. It was published in 1790, George Washington heading the list of subscribers, and in 1917 was revived by Otis *Skinner. Tyler wrote several other plays, but none was as successful as *The Contrast*.

Tynan, Kenneth Peacock (1927–80), English dramatic critic, who worked for the *Observer*, 1954–8, and again, 1960–3, and for the *New Yorker*, 1958–60. He was literary manager of the *National Theatre, 1963–9 (after which he became resident in the United States): as a strong opponent of *censorship he fought, and lost, a battle for the production of *Hochhuth's controversial play *Soldiers* at the National, and he was one of the play's co-producers when it was finally staged at the New Theatre (now the *Albery) in 1968. Equally controversial, but a good deal more successful, was the 'evening of elegant erotica' *Oh! Calcutta!* which he devised. It opened in New York in 1969 and in London a year later, having long runs in both cities.

Tyne Theatre and Opera House, see NEWCASTLE-UPON-TYNE.

Tyne, Tyneside, Tynewear Theatre Company, see NEWCASTLE PLAYHOUSE.

Tyrone Guthrie Theater, Minneapolis, see GUTHRIE THEATER.

U

Udall, Nicholas (1505–56), English scholar, headmaster in turn of Eton and Westminster, and the author of *Ralph Roister Doister*. Written while he was at Eton for performance by the boys in place of the usual Latin comedy, it was probably performed there between 1534 and 1541, though efforts have been made to connect it with Udall's headmastership at Westminster and date it 1552. It was not printed until about 1566–7. This comedy, the first play in English to deserve that name, is much influenced by *Terence and *Plautus, and turns on the efforts of a vainglorious fool to win the heart and hand of a wealthy London widow. Although Udall is known to have written several other plays, they are lost, or survive only in fragmentary form. He was once credited with the authorship of *Thersites*, an interlude acted at Court in 1537 which is now considered to be the work of John *Heywood.

Ulster Group Theatre, Belfast, see GROUP THEATRE.

Unamuno y Jugo, Miguel de (1846–1936), Spanish philosopher, essayist, novelist, and playwright. The best known of his plays is *Sombras de sueño* (*Dream Shadows*, 1930). He also dramatized his own novel *Abel Sánchez* (1917), a modern version of the story of Cain and Abel, as *El otro* (*The Other*, 1926), and in *El hermano Juan* (*Brother Juan*, 1929) made a typically personal contribution to the development of the *Don Juan theme, stressing the relationship between character and author in a manner reminiscent of *Pirandello. Of Unamuno's 11 plays, nine were staged, but they were not on the whole as successful in performance as the adaptation of one of his short stories, *Nada menos que todo un hombre* (*Nothing Less than a Total Man*), made by Julio de Hoyos.

Unicorn Theatre for Children, see ARTS THEATRE, London, and CHILDREN'S THEATRE.

Union Square Theatre, New York, on the south side of Union Square, between Broadway and 4th Avenue. This opened as a *variety hall in 1871. In 1872 A. M. *Palmer took over, and for 10 years made the theatre one of the finest in New York, with an admirable stock company and visits from all the best players in the United States. *The Two Orphans* by John Oxenford, produced in 1874, made a fortune for Palmer. Other outstanding successes were *Camille*, based on the younger *Dumas's *La Dame aux camélias*, with Clara *Morris, and a dramatization of Charlotte Brontë's *Jane Eyre*. W. H. *Gillette

made his first appearance in New York under Palmer, as did Richard *Mansfield, in Feuillet's *A Parisian Romance* (1883), playing a small part which made him famous overnight. This was the last play to be put on under the management of Palmer, who in 1885 disbanded the stock company and went to the *Madison Square Theatre. The Union Square was then used by travelling companies until in 1888 it was burnt down. It was rebuilt, but never regained its former brilliance, and under various names was mostly devoted to continuous *vaudeville. It later became a *burlesque house and then a cinema, and in 1936 it was demolished.

Union Theatre, Melbourne, see MELBOURNE THEATRE COMPANY.

Union Theatre, New York, see CHATHAM THEATRE 2.

Uniti, company of the *commedia dell'arte* which has caused much discussion, since it is possible that the name refers not to a distinct company, as did *Gelosi and *Confidenti, for example, but to a combined troupe of actors from different companies formed for a special occasion. Because of this some critics consider the Uniti to be part of the history of the Gelosi, some of the Confidenti. It remains certain that a company under that name was directed by Drusiano *Martinelli and his wife Angelica in 1584 and for some years afterwards. A company calling itself the Uniti, noted in 1614, may have been a temporary amalgamation. Among its members were Silvio *Fiorillo and his son.

Unities, of time, place, and action, the three elements of drama introduced into French dramatic literature by Jean *Mairet, based on a misinterpretation of comments in *Aristotle's *Poetics*. They demanded that a play should consist of one action, represented as occurring in not more than 24 hours, and always in the same place. Aristotle insists only on the unity of action, merely mentions the unity of time, and says nothing about the unity of place, though this was to a certain extent imposed on Greek dramatists by the presence of the *chorus. The observance of the unities, defined by *Boileau in his *Art poétique* (1674), became an essential characteristic of French classical tragedy (though both *Corneille and *Racine ignored them when they wished) and found its way, with neo-classicism, into Spain and Italy. In England the influence of Shakespeare, who certainly had no regard for the unities of time and place and very little for that of action, was strong enough to counteract the efforts of Restoration writers of tragedy, of whose works only *Addison's *Cato* (1713) is now remembered.

Unit Setting, see DETAIL SCENERY.

Unity Theatre, Glasgow, left-wing theatre group formed in 1941 by the war-time amalgamation of four amateur companies, the Transport, the Clarion, the Jewish Institute, and the Workers' Theatre. It rapidly achieved success, and in the summer of 1945 presented in London a highly praised production of *Gorky's *The Lower Depths*. In the following year a full-time professional group was formed, which, like the *Scottish National Players, toured throughout Scotland as well as playing in Glasgow. Its biggest success came with *The Gorbals Story* (1948), a study of life in an overcrowded slum written by Robert MacLeish, a mixture of broad comedy and pathos reminiscent of the early plays of *O'Casey. This was revived many times, toured all over Britain, was taken into the West End briefly, and eventually filmed. Financial insecurity and the lack of a permanent home led to Unity's eventual downfall, and by the early 1950s it had ceased to be effective, although a number of its actors continued to make a vital contribution to the Scottish theatre.

Unity Theatre, London, amateur left-wing theatre group which acquired a converted church hall in Britannia Street, King's Cross, and opened there in 1936. Noteworthy productions were Ben Bengal's *Plant in the Sun*, in which Paul *Robeson played the lead, and *Odets's *Waiting for Lefty*. In 1937 the group moved to a small theatre holding 200 in Goldington Street, St Pancras, having a proscenium stage with a width of 30 ft. and a stage depth of 12 ft. A year later it presented not only the first *Brecht play to be seen in London—*Señora Carrer's Rifles*—but also the first English *Living Newspaper, a documentary on a London bus-strike. Other notable first productions by Unity, which at one time achieved professional status, were *O'Casey's *The Star Turns Red* (1940), *Sartre's *Nekrassov* (1956), and *Adamov's *Spring '71* (1962). A scheme was then put in hand to develop the theatre as a cultural centre for the Labour movement in London, and rebuilding had begun when in 1975 the premises were destroyed by fire.

University Departments of Drama. The first attempt to present theatre history and practice as an academic study leading to a university degree was made in the USA, when the Carnegie Institute of Technology, Pittsburgh, established in 1914 a Department of Drama offering a degree in theatre arts. More influential, however, was the establishment in 1925 of a postgraduate Department of Drama at *Yale, headed by George Pierce *Baker. The movement grew

so quickly that when in 1936 the American Educational Theatre Association (AETA) was founded, it had 80 members. Now, as the American Theatre Association (ATA) dealing with educational and non-commercial theatre nation-wide, it covers 1,600 drama departments in US colleges and universities.

In 1960 the McCarter Theatre at Princeton became the first professional theatre in the United States to be entirely under the control of a university. There was an expansion of university and college theatre building during the next 20 years, at the end of which virtually every educational institution with a theatre department had its own theatre, its facilities often comparing favourably with those of a Broadway or regional theatre. Some of the university theatres also function as civic theatres, serving the surrounding districts as well as the student body, often under a professional director. In some cases professional actors have been engaged to play in university theatres, in addition to student groups.

In Britain the gradual acceptance of the theatre as a subject for academic study resulted in the establishment in 1946–7 of the first Department of Drama, at Bristol University, which works closely with the *Bristol Old Vic; it was followed after 13 years by the creation at Bristol of the first Chair of Drama at a British university. A Department of Drama was established in 1961–2 at *Manchester, with Hugh *Hunt as its first Professor. Similar departments have now been set up in other universities, and there are two within the University of London. Many of these universities also have Chairs of Drama, and *Oxford has a Visiting Professor. By the end of the 1970s almost 20 British universities, and a similar number of other institutions of higher education, offered drama within a BA degree, many more offering drama in a B.Ed. degree. Some universities now have a theatre open to the public, among them the *Northcott at Exeter, the Sherman at *Cardiff, the Nuffield at Southampton, and the university theatre at Manchester.

There has also been a great expansion of university theatre studies throughout the world, particularly in Canada, France, Germany, and Italy. Theatre is studied at universities in the USSR and in places as diverse as India, Japan, Korea, and Nigeria.

University Wits, name given to a group of somewhat dissolute Elizabethan playwrights educated at Oxford or Cambridge and therefore contemptuous of such 'uneducated' writers as Shakespeare and Ben *Jonson. Among them were *Greene, *Nashe, *Peele, and, the most important in the development of the English theatre, *Marlowe. A lesser member of the group was Thomas Lodge (*c.*1557/8–1625), on whose pastoral romance *Rosalind* (1590) Shakespeare based *As You Like It*.

Up Stage, see STAGE DIRECTIONS.

Urban, Joseph (1872–1933), architect and stage designer, born in Vienna, where he worked for many years before going to Boston in 1911–12 to design sets for the opera. He later moved to New York, where he designed settings for all the *Ziegfeld Follies* from 1915 to 1932; musicals such as *Sally* (1920), *Sunny* (1925), and *Show Boat* (1927); the Metropolitan Opera; and James K. *Hackett's Shakespearian productions. He built the *Ziegfeld Theatre in New York, with its egg-shaped interior, and introduced much of the new Continental stagecraft to American theatre audiences, making use of broad masses of colour (especially his 'Urban blue') and novel lighting effects on costume and scenery.

Ure, Mary, see OSBORNE.

Uris Theatre, New York, see GERSHWIN THEATRE.

Ustinov, Sir Peter Alexander (1921–), English actor, dramatist, and director, of Russian-French descent. Trained under *Saint-Denis at the London Theatre Studio, he made his first appearance in London in 1939 with the *Players' Theatre *Late Joys* in his own sketches, and his memorable impersonation of the ageing opera-singer Madame Liselotte Beethoven-Finck admirably displayed his great gift for mimicry. His first play *Fishing for Shadows* (1940), in which he himself appeared, was based on Sarment's *Le Pêcheur d'ombres* (1921), and was followed in quick succession by *House of Regrets* (1942), *Blow Your Own Trumpet* (1943), *The Banbury Nose* (1944), and *The Tragedy of Good Intentions* (1945). Ustinov returned to the London stage in 1946, after serving in the army, to play Petrovitch in an adaptation of

Dostoevsky's *Crime and Punishment*. His next play *The Indifferent Shepherd* (1948), about a clergyman who has problems with both his beliefs and his marriage, starred Gladys *Cooper; in the same year he appeared in his adaptation of Ingmar *Bergman's *Hets* (*Frenzy*). In 1949 he acted in Linklater's *Love in Albania*, after which he appeared only in his own plays for the next 30 years. His first popular success came with *The Love of Four Colonels* (1951; NY, 1953), a fantasy in which four officers from the occupying forces, American, British, French, and Russian, woo a Sleeping Beauty in a German castle; he played a leading role in it in London himself. After *The Moment of Truth* (also 1951) and *No Sign of the Dove* (1953) came another popular success, *Romanoff and Juliet* (1956; NY, 1957), in which he also appeared, about the romance between the son of the Russian ambassador and the daughter of the American ambassador in a Ruritanian country. His play *Photo Finish* (1962; NY, 1963) was memorable for his own performance as a bedridden 80-year-old confronted with himself at the ages of 20, 40, and 60. The comedy *Halfway up the Tree* (London and NY, 1967) provided a good role for Robert *Morley in London. Ustinov then played the Archbishop in his own play *The Unknown Soldier and His Wife* (1968) at *Chichester, repeating the role in London in 1973, in which year *R Loves J*, a musical based on *Romanoff and Juliet* for which he wrote the book, was also seen at Chichester. In 1979 he was seen at the *Stratford (Ontario) Festival as King Lear; his first Shakespearian role, and in 1983 (NY, 1984) he played Ludwig in his own *Beethoven's Tenth*. He has also directed a number of plays, including several of his own, and is a notable film actor and opera director.

Utility, see STOCK COMPANY.

V

Vakhtangov Theatre, Moscow, *Moscow Art Theatre's Third Studio founded in 1921 under the direction of the Soviet actor and director **Eugene Vakhtangov** (1883–1922), who joined the MAT as an actor in 1911. During his short career he made the widest possible use of theatre forms including an unsurpassed use of the grotesque. Under his dynamic leadership the studio began to break away from the parent theatre, staging two of his best productions—*Maeterlinck's *Le Miracle de Saint Antoine* in 1921 and *Gozzi's *Turandot* in 1922. In 1926 it was officially given its present name, and many important Soviet directors worked there, among them *Meyerhold, *Popov, *Zavadsky, and *Akimov, whose somewhat unorthodox production of *Hamlet* in 1932, with music by Shostakovich, caused controversy and drew the public's attention to the company. The first two parts of *Gorky's uncompleted trilogy, *Yegor Bulychov and Others* (1932) and *Dostigayev and Others* (1933), were first seen at the Vakhtangov. *Okhlopkov worked there, and several of *Pogodin's plays were first seen at this theatre. After the Second World War the Vakhtangov again produced a number of new Soviet plays, including *Arbuzov's *City at Dawn* (1957), *It Happened in Irkutsk*, and *The Twelfth Hour* (both 1959). The high artistic standards of its founder were sustained by Reuben *Simonov from 1939 until his death in 1968, and maintained under his son Yevgenyi who succeeded him. An adaptation of Leo *Tolstoy's novel *Anna Karenina* was staged to mark Vakhtangov's centenary in 1983. In 1987 the theatre staged Mikhail Shatrov's *The Peace of Brest-Litovsk*, banned for 20 years because it showed the Bolsheviks' internal arguments about signing the peace treaty with Germany in 1917; the production was staged in London in 1989.

Valleran-Lecomte (*fl.* 1590–*c*.1615), French actor-manager, first heard of at Bordeaux in 1592. A year later he was asking permission in Frankfurt to present biblical tragedies, claiming that he had already played these and the plays of *Jodelle in Rouen and Strasbourg. Soon afterwards he had his own company, with **Marie Venier [Vernier]** (*fl.* 1590–1627), the first Parisian actress to be known by name, as his leading lady and the prolific Alexandre *Hardy as his salaried author. By the early years of the 17th century he was established intermittently at the Hôtel de *Bourgogne as the tenant of the *Confrérie de la Passion, with occasional tours of the provinces and the Low

Countries. The old actor Agnan *Sarat, who more than 20 years before had brought his own company to Paris, was Valleran-Lecomte's chief comedian, and remained with him until his death. Valleran-Lecomte himself appears to have been well received in his early years as the young lover in *commedia dell'arte*-type comedies, and was always acceptable in farce. When his company first played in Paris, according to Tallemant des Réaux in his *Historiettes*, he took the entrance money at the door himself, and he must have been one of the first managers to establish a school of acting for young players.

Vanbrugh [Barnes], **Dame Irene** (1872–1949), English actress, sister of Violet (below) and wife of the younger Dion *Boucicault. She made her first appearance in London in 1888, and was later with *Alexander at the *St James's Theatre, where she played Gwendolen Fairfax in the first production of *Wilde's *The Importance of Being Earnest* (1895). She made her first outstanding success as Sophie Fullgarney in *The Gay Lord Quex* (1899) by *Pinero, of whose heroines she became the leading exponent, having already played Ellean in *The Second Mrs Tanqueray* on tour in 1894 and Rose in *Trelawny of the 'Wells'* on its first production in 1898. She subsequently created the title-role in *Letty* (1903) and Nina Jesson in *His House in Order* (1906). She was also much admired in *Barrie's plays, particularly *The Admirable Crichton* (1902), *Alice Sit-By-The-Fire* (1905), and *Rosalind* (1912). In 1907 she appeared in the first public performance of *Shaw's *The Man of Destiny*, playing opposite the Napoleon of her husband. She was again seen in Shaw when she played Catherine of Braganza in *In Good King Charles's Golden Days* at the *Malvern Festival in 1939, having celebrated her stage jubilee the previous year. The theatre at the Royal Academy of Dramatic Art, one of London's outstanding *schools of drama, was named after her and her sister by their brother Sir Kenneth Barnes, who was Principal of the school for many years.

Vanbrugh, Sir John (1664–1726), English dramatist and architect, in which latter capacity he was responsible for the design of the first Queen's Theatre on the site of the present *Her Majesty's. This was built to house a company led by *Betterton, who appeared there in Vanbrugh's own play *The Confederacy* (1705), based on *Dancourt's *Les Bourgeoises à la mode* (1692) and often billed as *The City Wives'*

Confederacy. Vanbrugh's best plays are undoubtedly *The Relapse; or, Virtue in Danger* (1696), a sequel to and parody of *Love's Last Shift* (also 1696) by *Cibber, and *The Provoked Wife* (1697). His last play, originally called *A Journey to London*, was left unfinished at his death. It was completed and produced by Cibber in 1728 as *The Provoked Husband*. *The Relapse* was later rewritten for a more prudish stage by *Sheridan as *A Trip to Scarborough* (1777), but in its original form it had a long run in London in 1947–8, and was also revived by the *RSC at the *Aldwych in 1967.

Vanbrugh [Barnes], **Violet Augusta Mary** (1867–1942), English actress, sister of Irene (above), made her first appearance in London in 1886 and had a consistently successful career, celebrating her stage jubilee in 1937 by playing Mistress Ford in *The Merry Wives of Windsor*, a part which she first played in 1911 and frequently revived. She married Arthur *Bourchier in 1894 and for some time appeared mainly under his management, but in 1910 was with *Tree, giving a magnificent performance as Queen Katharine in *Henry VIII*, probably her finest part. She was also excellent as Lady Macbeth and Portia. Although her greatest successes were scored in Shakespeare she appeared in many modern plays, including several by Henry Arthur *Jones, and was much admired as Lady Tonbridge in Gertrude Jennings's *The Young Person in Pink* (1920), as Mrs Vexted in Millar's *Thunder in the Air* (1928), and as Princess Stephanie in Beverley Nichols and Edward *Knoblock's *Evensong* (1932).

Vance, The Great, see MUSIC-HALL.

Vandenhoff, George (1820–84), British-born American actor, son of **John Vandenhoff** (1790–1861), a good actor who nevertheless failed to reach a commanding position. George first appeared at *Covent Garden in 1839, playing Mercutio in Mme *Vestris's production of *Romeo and Juliet*. In 1842 he went to New York, and after making a successful début at the *Park Theatre as Hamlet decided to remain there. He was later leading man at both the *Chestnut Street Theatre in Philadelphia and at Palmo's Opera House, later the *Chambers Street Theatre, in New York. He was admired as Benedick in *Much Ado about Nothing* and Claude Melnotte in *Bulwer-Lytton's *The Lady of Lyons*. In 1845 he staged an English translation of *Sophocles' *Antigone*, with music by Mendelssohn, on a stage approximating to the contemporary idea of a Greek theatre. He returned to London in 1853, and after a brief appearance as Hamlet went into retirement, having little liking for the profession he had so successfully adopted. A tall, scholarly man, somewhat aloof, he made a final appearance on the stage in 1878, playing Wolsey in *Henry VIII* to the Katharine of Geneviève *Ward.

Van Druten, John (1901–57), dramatist, of Dutch extraction, born in London, but later an American citizen. He first came into prominence with *Young Woodley*, a slight but charming study of adolescence which was produced in New York in 1925. Unaccountably banned by the censor in England, it was first produced privately by the Three Hundred Club (see STAGE SOCIETY) in 1928, and when the ban was removed had a successful run at the *Savoy later that year. Van Druten's later plays, mainly light comedies, include *After All* (1929; NY, 1931), a study of family relationships; *London Wall*, a comedy of office life, and *There's Always Juliet* (both 1931, the latter seen in New York in 1932); *The Distaff Side* (1933; NY, 1934), in which Sybil *Thorndike gave a moving performance as the mother; and *Old Acquaintance* (1940; NY, 1941), a study of two women writers. *The Voice of the Turtle* (1943), a wartime comedy which had a long run in New York, was coldly received in London in 1947, but Van Druten achieved a success in both capitals with *Bell, Book and Candle* (NY, 1950; London, 1954), an amusing comedy on witchcraft in which Rex *Harrison starred. Van Druten's adaptation of stories by Isherwood as *I am a Camera* (NY, 1951; London, 1954) was later used as the basis of the musical *Cabaret* (NY, 1966; London, 1968).

Vanin, Vasily Vasilyevich, see PUSHKIN THEATRE, Moscow.

Van Itallie, Jean-Claude (1936–   ), Belgian-born American dramatist, whose early works, including *War* (1963; London, 1969), *The Hunter and the Bird* (1964), and *Where is de Queen?* (1965), revealed him as an experimental writer of considerable imaginative powers. His first critical success was *America Hurrah!* (1966; London, 1967), a trilogy of short plays which exposed his vision of the dehumanizing nature of modern society in a series of striking metaphors. It was produced in both New York and London by the *Open Theater, and Van Itallie collaborated with the group to create *The Serpent* (Rome, 1968; NY, 1970), a confrontation of innocence and corruption which was the embodiment of Van Itallie's conviction that the theatre should rediscover something of its original function as a communal rite and spiritual exemplar. His later works include *The King of the United States* (1972), revised as *Mystery Play* (1973); another trilogy, *Early Warnings* (1983), consisting of *Bag Lady*, *Sunset Freeway*, and *Final Orders*; and a theatrical version, with music, of *The Tibetan Book of the Dead* (also 1983), stylistically reminiscent of *The Serpent*. His adaptations of *Chekhov's four major plays were all seen in New York.

Variétés, Théâtre des. The first playhouse of this name opened on the *boulevard du Temple in 1779, its company moving to several other theatres before being expelled by Napoleon from the present *Palais-Royal. The name was then given to a theatre built in 1807, and saved from redevelopment in 1937 by a conservation order. Except for a partial reconstruction in 1823 and the replacement of the balcony boxes by seats in 1900, it remains in its original state, and is now mainly used for comedies, revues, and popular musical shows. Among the plays seen there was the elder *Dumas's *Kean* (1836), but its most successful period was between 1860 and 1900, when it housed a number of well-known comic operas and *vaudevilles. It is described in *Zola's novel *Nana*. The theatre was renovated in 1975, and in 1978 staged three one-act plays of *Feydeau as *Boulevard Feydeau*; it was followed by Poiret's *La Cage aux Folles*. Sacha *Guitry's *Don't Listen, Ladies!*, in which Guitry himself had acted there in 1952, was revived to celebrate the centenary of his birth in 1985.

Variety. When in the late 19th century, the era of their greatest prosperity, the *music-halls were rebuilt so as to abolish the individual supper-tables in favour of normal theatre seating, they lost much of the boisterous element, both on stage and among the audience, of the earlier days, as well as a good deal of the original 'free-and-easy' sparkle. They were then renamed 'palaces of variety' and the turns were billed as 'variety', a word already in use as the equivalent of 'music-hall', the former unbroken succession of single acts being replaced by the twice-nightly programme first instituted by Maurice de Frece at the Alhambra, Liverpool; ballets and spectacular shows were imported, as well as short plays from the legitimate stage, and the whole elaborate set-up was far removed from the simple robust humour of the old music-hall. The old name remained in use, however, side by side with the new, and the history of Variety will be found under *music-hall for England and *vaudeville for the USA.

Variety Artists' Federation, see EQUITY.

Vaudeville, French word, possibly a corruption of Vau (or Val) de Vire, meaning 'songs from the Valley of Vire' (in Normandy), where in the 15th century Olivier Basselin composed satirical couplets, sung to popular airs, against the English invaders; or, alternatively, 'voix des villes' ('songs of the city streets'). In 1674 *Boileau, in his *Art poétique*, used it in its present form to describe a satirical, often political ballad. It acquired its later meaning of 'a play of a light or satiric nature, interspersed with songs' by way of the little theatres in the Paris *fairs. Owing to the monopoly of the *Comédie-Française, plays in such theatres (sometimes no more than *booths) could only be given in dumb-show, with interpolated choruses on well-known tunes, often parodying the productions at the legitimate theatre. These *pièces en vaudevilles* were the staple fare of the *Opéra-Comique, and were written by many well-known dramatists, including *Lesage. Their popularity paved the way for the immense vogue of light opera and operetta in mid-19th-century Paris, with librettos written by such men as the prolific *Scribe and his numerous collaborators. When this particular kind of light comedy lost its popularity, the use of the term was extended to sketches on the *variety stage, whence its present-day meaning in the USA (see below).

Vaudeville, American, name adopted in the USA in the late 19th century for the type of respectable family entertainment pioneered by Tony *Pastor in the 1860s, and later developed by **Benjamin Keith** (1846–1914) and his partners, the elder Edward *Albee and **Frederick Proctor** (1851–1929). It replaced *variety, which was at a low ebb, catering mainly for the alcoholics and prostitutes who frequented the local beerhalls. After a long struggle to provide 'clean' entertainment in several New York theatres, Pastor finally settled in Fourteenth Street, where in 1881 he presented the first programme of what was later to be known as vaudeville. It consisted of eight contrasting acts—comedy, acrobatics, singing, and dancing—the cast being headed by Ella Wesner, whose *male impersonations featured several typical 'dandies' of the period and a number of English *music-hall songs. The theatre became popular with respectable family groups, including women and young girls. The transition from the lusty and often lewd acts of earlier times to the simpler and more refined bills of the turn of the century was welcomed also by the performers, who enjoyed working in the spacious theatres, of which the Keith in Boston, built by Albee in 1894, was typical. The response of the new audiences led to a new type of theatrical presentation, with sketches never lasting more than 20 minutes and written to a specific 'punch' formula; animal acts with trainers in hunting or riding kit; magic, musical, or spectacular numbers; and true humour blended with pathos replacing the excessive *slapstick of the earlier comics. One of the finest turns of the mid-1890s was that of **James J. Morton** (1862–1933) and Maude Ravel, in which Morton, with a cadaverous, expressionless countenance, made strenuous efforts to help Maude with her songs and failed every time.

The new conditions developed the 'headline' system, which led to the appearance in vaudeville of star names from the theatre—Sarah *Bernhardt, Lillie *Langtry, Mrs Patrick *Campbell among others—in short scenes and sketches. They were engaged purely for their publicity value; but by this time audiences were creating their own vaudeville stars. 'Nut' acts, or *burlesque, of which Collins and Hart, or Duffy and Sweeney, were incomparable examples, became very popular, and many new artistes, both Black and White, graced the heyday of vaudeville, which lasted roughly from 1881 to the closing of Broadway's Palace Theatre as a 'two-a-day' in 1932, by which time films and radio had driven it into decline. Among them were W. C. *Fields, Bert *Williams, and **Lew Dockstader** [George Alfred Clapp] (1856–1924), originally a leading 'blackface' performer in the *minstrel show, who was famous for his political and social satire; **Charlie Case** (1858–1916), who pioneered the quiet, slyly humorous monologue, delivered while fiddling with a short length of string, his only prop; **Joe Jackson** (?–1942), whose bicycling tramp, in 'fright' wig and red nose, created by the look in his eyes, by his gestures, and by the furrowing of his quizzical brow the puzzlement and poignancy that Case conveyed in words; and **Bill Robinson** (1878–1949), a famous tap-dancer and Negro comedian affectionately known as 'Bojangles', who was seen in the *revue *Blackbirds* (1927) and, with outstanding success, in *The Hot Mikado* (1939). These were solo turns; but there were many fine groups and double acts, one of the best of the latter being *Weber and Fields; later came the **Howard** brothers, **Eugene** (1880–1965) and **Willie** (1883–1949), who became vaudeville 'headliners' in 1912, and were later seen in the *Ziegfeld Follies* and *George White's Scandals*. A double singing act just after the turn of the century was that of **Jack Norworth** (1879–1959) and his second wife **Nora Bayes** (1880–1928). Other women singers who are still remembered are **Lillian Russell** [Helen Louise Leonard], a beautiful woman with a vivid and flamboyant personality who first appeared under Pastor in 1881; **Marie Dressler** [Leila Koeber] (1871–1934), remembered for 'Heaven Will Protect the Working Girl', which she sang in a musical, *Tilly's Nightmare* (1894), who left the stage for films in 1914; **Marie Cahill** (1870–1933), who, like **Elsie Janis**, was equally well known in *musical comedy; **Eva Tanguay** (1878–1947), singer of 'I Don't Care!', and **Fay Templeton** (1865–1939), who both had successful careers in the theatre before going into vaudeville, the first in 1907, the latter in 1916. The buxom, raunchy, brassy-voiced **Sophie Tucker** [Sonia Kalish] (1884–1966), 'the Last of the Red-Hot Mammas', became famous for songs such as 'After You've Gone' and 'Some of these Days'.

Many of the big names of vaudeville survived to become even better known in films and on the radio, among them **Ed Wynn**, who started in vaudeville at 15 and later starred in *The Perfect Fool* (1921)—which became his nickname—and other shows which he wrote and directed himself; **Eddie Cantor**, known as 'Banjo-Eyes',

first seen at the Clinton Music-Hall in New York in 1907, who later appeared in revue and musical comedy, scoring a great success as an eccentric comedian in *Kid Boots* (1923); and that other eccentric comedian, his exact contemporary, **Jimmy Durante** (1893–1980), whose big nose earned for him the sobriquet 'Schnozzle'; not to mention Bert *Lahr, who spent many years in vaudeville and burlesque.

These are only a few of the practitioners who benefited from Tony Pastor's enterprise and helped to build up the glittering world of American vaudeville. Their counterparts still exist, providing the staple fare of popular television.

Vaudeville, Théâtre du, Paris, playhouse which opened in 1792 in the rue de Chartres to house the actors forced to leave the *Comédie-Italienne when its licence was renewed for musical plays only. The company was frequently in trouble for the topical and political allusions in its productions. It eventually fell back on semi-historical pieces, based on anecdotes of heroic figures. Closed in 1838, it reopened two years later on the place de la Bourse; closed again by town planners in 1869, it moved to the chaussée d'Antin, where it operated until 1927 and then became a cinema. In its heyday this third theatre housed productions of comedies by *Labiche and others, as well as the social dramas of *Brieux and *Ibsen.

Vaudeville Theatre, London, in the Strand. Built in 1870 for **David James** (1839–93) and two other actors, it held over 1,000, opening with Andrew Halliday's comedy *For Love or Money*. It had its first success in 1871 with *Albery's *Two Roses*, which introduced Henry *Irving to London audiences, and in 1875 H. J. *Byron's *Our Boys*, in which James began a four-year run. In 1884 the theatre scored another success with Henry Arthur *Jones's *Saints and Sinners*. In 1891 it reopened after reconstruction, its seating capacity reduced to 740 on four tiers, with Jerome K. *Jerome's comedy *Woodbarrow Farm*, and later in the year came the first performances in England (at matinées only) of *Ibsen's *Rosmersholm* and *Hedda Gabler*, both with Elizabeth *Robins. The *Gatti brothers bought the theatre in 1892 and from 1900 to 1906 Seymour *Hicks and his wife Ellaline *Terriss appeared in a series of long runs under the direction of Charles *Frohman, Hicks's popular Christmas entertainment *Bluebell in Fairyland* being first seen in 1901. Among other successes were *Barrie's *Quality Street* (1902) and several musical comedies including *The Belle of Mayfair* (1906). Charles *Hawtrey then appeared in a series of comedies, and in 1915 the theatre became the home of Charlot's *revues, which continued until 1925, when the theatre again closed for rebuilding, though retaining its original façade. It reopened in 1926, continuing mainly with revues until 1937, its seating capacity having been reduced to 659 on three tiers. In 1938 Robert *Morley's *Goodness, How Sad!* was a hit. Wartime productions included Esther McCracken's *No Medals* (1944), which ran for two years. In 1947 William Douglas *Home enjoyed two good runs with *Now Barabbas . . .* and *The Chiltern Hundreds*, the latter with A. E. *Matthews. The 1950s were dominated by the record-breaking musical by Dorothy Reynolds and Julian Slade, *Salad Days*, which ran from 1954 to 1960. Ronald *Millar's *The Bride Comes Back* (1960), *Wesker's *Chips with Everything* (1962), and Joyce Rayburn's *The Man Most Likely to . . .* (1968) also did well. In 1970 the theatre changed hands and was extensively refurbished, its first outstanding success after reopening being the farce *Move Over, Mrs Markham* by John Chapman and Ray Cooney. In 1977 Agatha *Christie's *A Murder is Announced* began a long run, and notable later productions were *Frayn's *Benefactors* (1984) and *Ayckbourn's *Woman in Mind* (1986) and *Henceforward . . .* (1988).

Vaughan, Hannah, Henry, William, see PRITCHARD.

Vedrenne, John Eugene, see GRANVILLE-BARKER.

Vega Carpio, Lope Félix de (1562–1635), Spanish playwright, and the most prolific dramatist of all time; he himself claimed to have written 1,500 plays, though the actual number of extant texts is from 400 to 500, mostly written in neat, ingenious, and superbly lyrical verse. His first plays date from the period of the early open-air theatres of Madrid. When he began to write, the professional theatre in Spain was in its infancy, and he can rightly be considered the consolidator, if not the founder, of the commercial theatre in Spain. He played a large part in the development of the Spanish *comedia, and the formula he evolved for its construction in the last decade of the 16th century remained largely unchanged for 100 years.

Critics have endeavoured to cope with Lope's enormous output by dividing his plays into various categories, but it is really more important to stress the unity of his dramatic production. His world is that of a Spanish Catholic of his day—he took orders in 1614. Society is viewed in an idealized light, and its multiple social distinctions are reduced to three: king, nobles, and commoners, the last usually peasants. It is against this background that such plays as *Peribáñez y el Comendador de Ocaña* (c.1608), *Fuenteovejuna* (*The Sheep-Well*, c.1612), and *El mejor alcalde el rey* (*The King the Best Magistrate*, c.1620) should be considered. *Fuenteovejuna* deals with the rising of a village against its brutal lord; shorn of its subplot—the rising of that

same lord against his king—the play has been interpreted in terms of the class struggle. The two plots are, however, interdependent, and both are essential to the true understanding of the play. Here and elsewhere Lope reveals himself clearly as conservative rather than revolutionary: the brutal overlord is removed because he disrupts the harmony of society, failing in his duties to the king above him as well as to the commoners beneath him.

This and other historical dramas of Lope have suffered from an attempt to interpret them according to the ideas of a later age, as have those concerned with the stylized code of honour. Lope's handling of this latter theme is diverse. In the early play *Los comendadores de Córdoba*, the outraged husband, discovering that he is openly and publicly dishonoured, takes open, public, and bloody revenge. In the mature *El castigo sin venganza*, the duke secures his public position but revenges himself privately upon the bastard son and adulterous wife who have caused the loss of his honour. His public reputation is saved, since his revenge is secret, and with it the honour of the state, but this secret revenge entails the killing of the only two persons he has loved. The subtlety of the play consists in Lope's demonstration that this state of affairs is the direct consequence of the duke's own licentiousness. The theme of honour is more than a barbaric convention.

Lope also wrote for the entertainment of the Court, usually plays with a mythological or pastoral plot which made use of the most up-to-date machinery and scenic effects. *La selva sin amor* (*The Loveless Forest*, 1629) is a particularly interesting work, being partly set to music and thus a forerunner of the later *zarzuela. Equally spectacular were his *autos sacramentales* played on three carts, the centre being the main playing-space, and the outer carts bearing complicated machinery. They do not have the lyrical beauty or the subtle symbolism of those by *Calderón, but they present their theme simply, forcibly, and dramatically, and show an assured grasp of technique.

Venier, Marie, see VALLERAN-LECOMTE.

Verfremdungseffekt, see ALIENATION.

Verga, Giovanni (1840–1922), Italian playwright, born in Sicily, better known as a novelist and short-story writer. He was, however, the only completely successful writer of tragedy in the Italian theatre between *Alfieri and *Pirandello, employing *verismo not by formula but by conviction, his portraits of Sicilian life being unflinching confrontations of grey desperation. *Cavalleria rusticana* (*Rustic Chivalry*, 1884), the first and most famous example of this movement, a dramatization of one of his own short stories which later provided the libretto for Mascagni's opera, is a sparse and swiftly moving example of Verga's power to fuse two interlinked tragedies, that of a community and that of a religious man, into a coherent whole. Comparable achievements are *La lupa* (*The She-Wolf*, 1896), a most perceptive study of female sexuality and man's rank fear of it, and *La caccia al lupo* (*The Wolf-Hunt*, 1901), where he succeeds in transmuting the melodrama of jealousy into the poignancy of inescapable aloneness. Only recently have critics come to discern the unequivocal poetry informing Verga's dramas, notably the overlooked excellence of *In portineria* (*In the Porter's Lodge*, 1885), a play set, untypically, outside Sicily, and *Dal tuo al mio* (*From Yours to Mine*, 1903), a terse and evocative analysis of the clash between two classes of society.

Verismo, Italian cultural movement of the later 19th century, having affinities with French *naturalism. Subject to multiple critical interpretations, the term evades precise definition, but in general the movement constituted a reaction against the Italian academic and rhetorical tradition in favour of an objective exploration of the immediate problems of urban bourgeois and provincial society. Such an exploration was inextricably linked to an awareness of the political and social realities that came in the aftermath of the Risorgimento. One notable exponent of *verismo* in the theatre was Giuseppe *Giacosa, and later dramatists, among them *Pirandello and *D'Annunzio, were veristic in their early writings. But perhaps the richest achievements of *verismo* were in the field of regional literature, and are to be found in the work of such writers as *Verga.

Versailles. Although there was a good deal of theatrical entertainment at the palace of Versailles under Louis XIV, there was no permanent theatre there and plays were given on temporary stages erected indoors or in the gardens. It was not until 1768 that Louis XV instructed his chief architect to build a theatre in the north wing of the château. Oval in design, not rectangular as earlier French theatres had been, it was built of wood, much of it painted to resemble marble. The stage, almost as large as that of the Paris Opéra, was well supplied with machinery, and the floor of the auditorium could be raised to stage level to form a large room for balls and banquets. Lighting was provided by crystal chandeliers. The theatre was first used in 1770 for a banquet in honour of the marriage of the future Louis XVI to Marie-Antoinette. The first plays to be given there were *Racine's *Athalie* on 23 May, with Mlle *Clairon in the title-role, and on 20 June *Voltaire's *Tancrède*. When in 1837 Louis-Philippe made Versailles a museum of French military history, the opening ceremony was followed on 10 June by a gala performance of *Molière's *Le Misanthrope*. The theatre was then used occasionally for concerts and in 1855 for a banquet in honour of Queen

Victoria. In 1871 it was taken over by the Assembly, who met there during the Commune. A floor was laid over the pit, and everything above it was painted brown. This fortunately preserved the decorations below, and when in 1952 restoration began on the château, the theatre too was restored to its original colours of dark blue, pale blue, and gold. It was even found possible to replace the original material on the seats, made by the firm which had supplied it in 1768. The restoration was completed in time for an official visit by Queen Elizabeth II of England in 1957, when a theatrical and musical entertainment was given. The theatre is still occasionally used for concerts, operas, and plays.

A theatre built by Mlle *Montansier in 1777 on a site near the palace of Versailles remained in use until 1886.

Verulamium. The Roman theatre in this 2nd-century city (now St Albans) was for a long time the only one known in Britain. (Traces of others have now been found at Canterbury and near Colchester.) It was probably built between AD 140 and 150 and used mainly for sport, particularly cock-fighting. The *orchestra, which was completely circular, with seating round two-thirds of it and a small stage-building in the remaining space, could have been used for *mimes and dancing, and the stage for small-scale entertainments. In AD 200 the stage was enlarged, possibly to allow for the positioning of a slot for a curtain, and at the end of the 3rd century, after a period of disuse, the theatre was rebuilt. The auditorium was extended over part of the orchestra, the floor levels were raised, and a triumphal arch was built spanning Watling Street. The building was finally abandoned at the end of the 4th century and the site used as a municipal rubbish-dump. It was rediscovered in 1847 and excavated in 1934.

Vestris, Mme [*née* Lucia Elizabetta (or Lucy Elizabeth) Bartolozzi] (1797–1856), English actress, the wife of the French ballet-dancer Armand Vestris, who deserted her in 1820. As Mme Vestris, she had a distinguished career on the London stage. Although an excellent singer, she preferred to appear in light entertainment rather than grand opera, and was at her best in *burlesque or as the fashionable ladies of high comedy. She played in Paris for several years with great success and then appeared in London, playing alternately at *Covent Garden and *Drury Lane. In 1830 she took over the *Olympic Theatre, opening with *Olympic Revels*, by *Planché, who furnished her with a succession of farces and burlesques both at this theatre and later at the *Lyceum. During her tenancy of the Olympic she took into her company the younger Charles *Mathews, marrying him in 1838, and the rest of her career ran parallel to his. She was an excellent manageress, effecting a number of reforms in theatre management and many improvements in scenery and effects. She was one of the first to use historically correct details in costume, anticipating the reforms of Charles *Kean, was responsible for the introduction of real properties instead of fakes, and in 1832 introduced the *box-set, complete with ceiling, on to the London stage.

Vezin, Hermann (1829–1910), English actor, born in the USA, who moved to England in 1850 and remained there until his death. After working for a time in the provinces, he took over the *Surrey Theatre and appeared in a number of classic parts. Samuel *Phelps then engaged him for *Sadler's Wells, and by 1861 he was recognized as an outstanding actor in both comedy and tragedy, his best parts being Macbeth, Othello, Jaques in As You Like It, and Sir Peter Teazle in *Sheridan's The School for Scandal. He played opposite Ellen *Terry in Olivia (1878), a dramatization of *Goldsmith's novel The Vicar of Wakefield, and was for a time in *Irving's company at the *Lyceum. He made his last appearance under *Tree in 1909, playing Rowley in The School for Scandal. A scholarly man, of small stature, with clear-cut features and a dignified bearing, he was remarkable for the beauty and clarity of his diction, and lacked only warmth and personal magnetism.

Viau, Théophile de, see THÉOPHILE.

Vice, character in the English *morality play. Originally an attendant on the Devil, whom he helped in his attacks on mankind, he eventually became a figure of fun and a cynical and sardonic commentator on the action of the play. In Twelfth Night the Clown, who took over many of the Vice's attributes, refers to him as 'the old Vice . . . with dagger of lath'.

Vicente, Gil (c.1465–c.1536), Portuguese dramatist, and as Court poet from 1502 to 1536 primarily a deviser of courtly entertainments. He also wrote prolifically in a number of dramatic forms—eclogues, moralities, farces, romantic plays, and allegorical spectacles. Of his 44 extant works, 17 are in Portuguese, 11 in Spanish, and 16 use both languages, though all have Portuguese titles. The eclogues make good use of dialogue, but they are obviously dependent on *Encina. Gradually, however, a new form grew out of them: the *morality play, sharing some features with medieval European tradition, but owing most to Vicente's own powers of invention. His masterpieces in this form (1517–c.1519) were the Auto da alma and the trilogy of the Barcas (boats): Barca do Inferno, Barca do Purgatório, and Barca do Glória. These combined morality plays with outspoken social comment, possible only because of Vicente's privileged position at Court. Vicente is unquestionably the greatest religious dramatist of this period in the Peninsula, and his true successors are to be found in the Spanish writers of *autos sacramentales.

Whereas in the morality Vicente had to rely on a Spanish tradition which he then transformed, his farces seem to have had behind them indigenous popular entertainments. He was thus able to find his feet at once, and the Auto da Índia (1509) has great technical assurance. In this and in later works—of which Farsa de Inês Pereira (1523) is among the best—Vicente combined knockabout humour and pungent social satire. Vicente went on writing farces all his life, but from about 1520 his attention was mainly devoted to other forms. For a few years he experimented with romantic comedy. Comédia Rubena (1521) shows the amatory misfortunes and final triumph of the heroine; Dom Duardos (1522) and Amadis de Gaula (1523?) put chivalresque fiction on the stage; and Comédia do Viuvo (1524) leans heavily on the figure of the disguised nobleman.

The morality and the romantic comedy were theatrically the most promising forms evolved by Vicente, but in the last 12 years of his life he turned chiefly to secular allegorical fantasies. The characteristic note of the last period is a blend of allegory, lyricism, uninhibited satire, and lavish staging. Sometimes there is an organic plot, as in Auto da Feira (c.1526–8), sometimes the work is held together only by the visually dazzling allegorical framework, as in the Frágua do Amor (1524). The splendour and complexity of the stage devices used in these works are a direct outcome of the 15th-century Court mummings.

Victoria Palace, London, in Victoria Street, opposite Victoria Station. This stands on the site of the Royal Standard Hotel, whose manager John Moy obtained a licence in 1840 for the presentation of entertainments. The room in which they were given became known as Moy's Music Hall. In 1863 the hall was renovated and reopened as the Royal Standard Music Hall, which was again rebuilt in 1886 and continued its successful career until in 1910 it was acquired by Sir Alfred Butt, who built on the site the present theatre. It opened in 1911, its four-tier auditorium holding 1,565. From 1921 to 1929 and again in 1931 it housed an annual Christmas play, Frederick Bowyer's The Windmill Man. Otherwise it staged music-hall and revue, Gracie *Fields appearing there in The Show's the Thing in 1929. Walter Reynolds's patriotic play Young England (1934) was greeted with hilarity and drew vast audiences which came to cheer, jeer, and join in the dialogue. A year later Seymour *Hicks took over and produced his own play The Miracle Man, following it with revivals of some of his former successes. Lupino *Lane succeeded him in 1937, presenting a musical play Me and My Girl which ran until the outbreak of the Second World War in 1939. From 1947 the Victoria Palace was almost exclusively the home of the *Crazy Gang, each of their shows achieving more than 800 performances. The popular BBC television show *The Black and White Minstrels* took over in 1962, followed after a long run by *Magic of the Minstrels* and *Carry On, London!*, a stage version of the popular 'Carry On' film comedies. The American musical *Annie* ran from 1978 to 1981, and in 1982 Elizabeth Taylor starred in a revival of Lillian *Hellman's *The Little Foxes*. The theatre reverted to musicals with revivals of *Barnum* (1985) and *Brigadoon* (1988) and the new musicals *High Society* (1987) and *Buddy* (1989).

Victoria Theatre, London, see OLD VIC THEATRE.

Victoria Theatre, Stoke-on-Trent, Staffordshire, see NEW VICTORIA THEATRE.

Vienna, capital up to 1918 of the former Austro-Hungarian empire, and until the mid-19th century the focal point of the German-speaking theatre. Its theatrical history begins at the end of the 15th century, when the humanists encouraged their students at the university to perform the works of *Plautus and *Terence as well as new plays in Latin. By the mid-16th century plays in the vernacular were incorporating elements of indigenous folk-comedy which were to be important later, but these had for a time to give way to the splendours of *Jesuit drama. The influence of the *English Comedians, who first appeared in the neighbourhood in 1608, combined with indigenous drama to give rise to the *Haupt- und Staats-aktion, with its heroic background and its all-pervading *Hanswurst, who with his companions under *Stranitzky occupied in 1711 the Kärntnertor, the first permanent theatre building in Vienna. The Altwiener Volkstheater, or Folk-theatre of Old Vienna, reigned supreme. In 1776, however, the Emperor Josef II made the Burgtheater, built in 1741 in the Imperial palace, the home of serious drama in German, banishing the light-hearted comedies with their songs and dances to the remote suburbs. His reforms were ably implemented by the Austrian **Josef von Sonnenfels** (1733–1817), the theatre's Director from 1776, an exponent of 'regular drama' in the manner of *Gottsched. The Burgtheater's pre-eminence was established by **Josef Schreyvogel** (1768–1832), who took over in 1814 and built up a repertory which included not only the classical plays of Europe and the works of *Goethe and *Schiller but also the first plays of *Grillparzer and such lighter fare as *Kotzebue. He recruited from many parts of the German-speaking theatre a company of distinction with a polished but natural diction, the Viennese *parlando*. Meanwhile, contrary to all expectations, the old

folk-comedy flourished in the suburbs, at the Leopoldstädter Theater (1781), the Josefstädter Theater (1788), and the Theater an der Wien (1801), the first manager of the last being Mozart's librettist Emanuel Schikaneder. Two comic characters typify this last phase of the old extempore farce—*Kasperle and *Thaddädl. Their brief but glorious reign gave way in the later 19th century to the unique mixture of farce, fairy-tale magic, and parody which culminated in the great but practically untranslatable plays of *Raimund and *Nestroy and finally disappeared in the vogue for Viennese operettas. The Burgtheater, after a period of decline following the departure of Schreyvogel, was revivified by the German playwright **Heinrich Laube** (1806–84), its Director from 1849, who retained the European classics in the repertoire if they had a contemporary appeal, but also introduced the *well-made plays of France, partly because they were such admirable vehicles for the Burgtheater's intimate casual style. Its world-famous acoustics had been reinforced by the introduction of the *box-set, which Laube pioneered in the German-speaking theatre. By the time of his retirement in 1867 pre-eminence had passed to *Berlin. In 1888 the Burgtheater company moved to a new theatre which, destroyed in 1945 and rebuilt in 1955, still stands on the Ring facing the Town Hall. With the founding of the Austrian Republic in 1918, responsibility for the Burgtheater passed from the Court to the state. Heavily subsidized, it is now too unwieldy for one man to influence creatively, and its somewhat conservative repertory absorbs the theatre of the avant-garde only when it has become established. But its brilliant company, in beautifully set and costumed productions directed by outstanding men of the theatre, still makes it one of the great theatres of Europe. Of the other Vienna theatres, the Leopoldstädter, rebuilt in 1847, was demolished in 1945; the Josefstädter has a widely based repertory in which the tradition of fine acting in an intimate style, initiated by *Reinhardt when he took over in 1924, still survives; the Theater an der Wien is mainly notable for its operettas and musical shows. The Volkstheater, which presents classical and modern plays with a political flavour, was reconstituted from the earlier Deutsches Volkstheater in 1948. There are also the Raimundtheater, built in 1893 for folk-plays; the Akademietheater, seating about 500, opened in 1922; the Kammerspiele, the home of light comedy; and a number of experimental 'cellar' theatres. In summer the finely preserved rococo theatre in the Schönbrunn Palace, built in 1747, offers a season of plays mainly for visitors.

Viertel, Berthold (1885–1955), Austrian poet, actor, and theatre director, who figures as Friedrich Bergmann in Christopher Isherwood's novel *Prater Violet* (1945). After working in Vienna and Dresden, he went to *Berlin in 1922, where he directed *Hebbel's *Judith* at the *Deutsches Theater for *Reinhardt and achieved a great success with a production of Bronnen's *Vatermord* (*Patricide*). He then founded his own company, Die Truppe, but in 1927 left Germany, being in London from 1933 to 1937 and then in the USA where he worked in Hollywood and in 1945 produced in New York *Brecht's *Furcht und Elend des Drittes Reiches* as *The Private Life of the Master Race*. In 1949 he returned to Berlin, producing *Gorky's *Vassa Zheleznova* for the *Berliner Ensemble, and in 1951 became a director at the *Vienna Burgtheater, where among his productions was Tennessee *Williams's *A Streetcar Named Desire*.

Vieux-Colombier, New York, see GARRICK THEATRE.

Vieux-Colombier, Théâtre du, Paris, originally the Athénée, built in the early 19th century close to the dovecote of the former Abbey of St Germain. Until 1913 it was no more than a suburban playhouse for travelling companies; but it was then taken over by Jacques *Copeau as an experimental playhouse for the production of new and serious plays. With the help of his stage-manager Louis *Jouvet, Copeau redesigned the auditorium. Originally an oblong with a stage at one end, it was adapted so as to do away with the curtain (for the first time in the modern French theatre) and integrate the proscenium arch into a series of platforms and openings surrounding the audience on three sides. On a small scale—the theatre held only 200, and the stage mostly remained bare— Copeau tried to use features of the Elizabethan stage to involve the audience in a total experience of theatre. He opened with a translation of *Heywood's *A Woman Killed with Kindness*, and from 1913 to 1914 and again from 1919 to 1926, when the theatre became a cinema, he presented memorable productions of Shakespeare, *Molière, *Goldoni, and *Musset, and also introduced his predominantly intellectual audiences to *Claudel, *Ghéon, *Romains, and *Gide. In 1930 the Vieux-Colombier again became a theatre, when a group of Copeau's disciples, the Compagnie des Quinze (see SAINT-DENIS), re-established the tradition of intellectual experiment. After the departure of the Compagnie several notable events took place, including the production of *Sartre's *Huis-Clos*, *Vilar's production of T. S. *Eliot's *Murder in the Cathedral*, and first performances of plays by *García Lorca and Claudel; but successive managements found the theatre difficult to run successfully, and it finally closed in 1972.

Vigarani, Gaspare (1586–1663), Italian stage designer and machinist, inventor of many new stage effects. He was working in Modena when in 1659 he was called to Paris to supervise the entertainments to be given in honour of Louis XIV's approaching marriage. For these he built the *Salle des Machines in the Tuileries, and designed its *stage machinery. As the *Petit-Bourbon theatre was then being demolished, Vigarani claimed the scenery and machinery for use in his new theatre and burnt it all, thus destroying the work of a rival. After his death his son **Carlo** (?–*c*.1693), who had been working with him, was responsible for the mechanics of *Les Plaisirs de l'île enchantée* (1664), an entertainment given at Versailles with the collaboration of *Molière and Lully.

Vilar, Jean (1912–71), French actor, manager, and director, who worked under *Dullin at the Atelier for some time before the Second World War, and during the occupation of France by the Germans joined a band of young actors, La Roulette, which toured the provinces. Back in Paris in 1943 he directed *Strindberg's *The Dance of Death* and *Molière's *Don Juan* and made a brief appearance in *Synge's *The Well of the Saints* before taking over the 100-seat Théâtre du Poche. He first came into prominence in 1945 with his production, first at the *Vieux-Colombier and later in front of the Abbey of Bec-Hellouin, of T. S. *Eliot's *Murder in the Cathedral*, in which he played Becket. As a result of this he was invited in the summer of 1947 to organize the open-air *Avignon Festival, which he ran until his death. Its growing importance established him as the leading director of the French theatre, a position that was confirmed in 1951 when he became head of the *Théâtre National Populaire at the Palais de Chaillot. His productions, vast in scope and bold in conception as the great stage demanded, aroused both enthusiasm and controversy, and contributed to the simplification and deromanticizing of scenery and to the increased attention paid to the quality of lighting and costume. Vilar, an excellent actor as well as an inspired director, appeared in many of his own productions, among them the title-roles in *Macbeth* and *Richard II*, Molière's *Don Juan*, and *Pirandello's *Enrico IV*; he also played Harpagon in Molière's *L'Avare* and the Gangster in *Brecht's *Arturo Ui*, and in 1951 was seen as Heinrich in *Sartre's *Le Diable et le bon Dieu*. In the same year he played *Sophocles' Oedipus for the *Renaud–*Barrault company. Among the classics he revived for the TNP were *Lesage's *Turcaret* and *Hugo's *Ruy Blas* and *Marie-Tudor*. He also introduced his audiences to *Goldoni's *I rusteghi*, *Büchner's *Dantons Tod*, *Chekhov's *Platonov*, and *Bolt's *A Man for All Seasons*. A spare, ascetic-looking man, Vilar could on stage appear amazingly handsome, and his voice had great sonority and emotional overtones. In 1963 he resigned from the TNP in protest against the inadequate support given to it by the French Government and engaged in freelance activities until his death.

Ville, Théâtre de la, Paris, see BERNHARDT and THÉÂTRE DES NATIONS.

Villiers, George, second Duke of Buckingham, see DRYDEN and FLETCHER.

Vilna Troupe, company founded in Vilna, Lithuania, in 1916 to further the reform of the Yiddish stage inaugurated by *Hirschbein. The troupe's first (amateur) production was Sholom *Asch's *Landsleute*. A year later it moved to Warsaw and became professional, soon being recognized as one of the outstanding Yiddish companies of its day, mainly because of the excellence of its ensemble playing. Among its directors was **David Hermann** (1876–1930), who was responsible for the first production of *Peretz's dramatic poem *Night in the Old Market*. His finest work, however, was his production in Yiddish, a month after the author's death in 1920, of *Ansky's *The Dybbuk*, with which the troupe toured France, England, and America, winning international renown. Returning to Warsaw in 1924, the troupe found it had lost its theatre, and shortly afterwards moved to Vienna, where it split in two, some of the actors going to join Maurice *Schwartz in New York, others remaining active in Europe, particularly in Romania and Poland, until the 1930s. The repertory of the Vilna Troupe was at first fairly extensive and included a number of Yiddish classics; but later it became exclusively associated with and dependent on *The Dybbuk*, which probably accounted for its eventual demise.

Vining, Fanny Elizabeth, see DAVENPORT, EDWARD.

Violetta, Mlle, see GARRICK.

Virginia Theatre, New York, on West 52nd Street, between Broadway and 8th Avenue. Built by the *Theatre Guild to house its own productions, this opened in 1925 as the Guild Theatre, with Helen *Hayes in *Shaw's *Caesar and Cleopatra*. During subsequent seasons further plays by Shaw were given, as well as plays by European and new American playwrights. In 1950, after seven years as a radio playhouse, the theatre was taken over by the *American National Theatre and Academy, and as the Anta Playhouse was used for experimental productions. Renamed the Anta Theatre in 1954, it reverted to commercial use in 1957, among its productions being Archibald *MacLeish's *J.B.* (1959), James Baldwin's *Blues for Mr Charlie* (1964), seen in London during the *World Theatre Season of 1965, Peter *Shaffer's *The Royal Hunt of the Sun* (1965), and a documentary on the assassination of John Kennedy, *The Trial of Lee Harvey Oswald* (1967). Several famous visiting companies appeared there, and other productions included Charles

Gordone's *No Place to be Somebody* (1969) and William *Inge's *Summer Brave* (1975). The very successful all-Black musical *Bubbling Brown Sugar* was staged there in 1976, and *Stoppard's *Night and Day*, with Maggie *Smith, in 1979. In 1981 the theatre was sold and renamed, its capacity being enlarged to 1,220. Notable later productions were *Wild Honey* (1986), a translation by Michael *Frayn of *Chekhov's *Platonov*, and the musical *City of Angels* (1989).

Visconti, Luchino (1906–76), Italian director and scene designer, who as a young man went to Paris and became involved in film production. Although much of his early work was in films, he directed plays in Milan in 1937, and in 1945 introduced plays by a number of European dramatists, particularly *Cocteau, into the repertory of the Teatro Eliseo in Rome, where in 1948 he staged Shakespeare's *As You Like It* in Italian as *Rosalinda*. In 1946, in which year he directed *Beaumarchais's *Le Mariage de Figaro*, he joined the Paolo Stoppa–Rina Morelli company, for which he staged such modern works as Tennessee *Williams's *The Glass Menagerie* in 1946, *Anouilh's *Eurydice* in 1947, *Miller's *Death of a Salesman* in 1951, and his *A View from the Bridge* in 1958. Among his other productions were *Chekhov's *Three Sisters* (1952) and *Uncle Vanya* (1955) and, in Paris in 1961, a sumptuous version of *Ford's *'Tis Pity She's a Whore*, as *Dommage qu'elle soit putain*. He worked on *Troilus and Cressida* in Florence, and also on Tennessee Williams's *A Streetcar Named Desire*. His later work was mainly in films and in opera, but he returned to live theatre in 1973 to direct *Pinter's *Old Times* in Rome, a production disowned by the author.

Vishnevsky, Vsevolod Vitalevich (1900–51), Soviet dramatist, whose epic plays were enhanced by his own extensive active service experience. After *The First Horse Army* (1929) came his most successful play *The Optimistic Tragedy* (1934), produced at the *Kamerny Theatre by *Taïrov, whose wife Alisa Koonen played the leading role of a woman commissar with the Red Fleet who is killed in battle during the early days of the Soviet régime. It was published in an English translation in *Four Soviet Plays* (1937). In 1943, after several less successful productions, Vishnevsky collaborated on a musical, *Wide Spreads the Sea*, and in the same year wrote *At the Walls of Leningrad*. Both deal with the siege of Leningrad and were well received, as was Vishnevsky's last play, *The Unforgettable Year 1919* (1949), of which Stalin is the hero. *The Optimistic Tragedy* was revived by *Tovstonogov in Leningrad in 1955.

Visor, see MASKS, THEATRICAL.

Vitez, Antoine (1930–90), French actor and director, probably the most intellectual and uncompromising of his generation as well as the most deeply marked by Communism and the theories of *Brecht. After working in Marseilles and at the Maison de la Culture in Caen in the 1960s, he became one of the moving spirits of the 'red belt' of popular playhouses in the industrial suburbs of Paris, taking productions to halls and schools, first at Nanterre and then at Ivry, where he took over the Théâtre des Quartiers. His versions of *Sophocles' Electra (1971) and Brecht's Mother Courage (1973) established him as an original though perhaps overambitious director, able to handle actors well. As part of the modernization following the student unrest of 1968 he was appointed director and teacher at the Conservatoire, chosen not only because of his expertise in *mime—a discipline traditionally neglected by the official body of French drama teachers—but because of his firm Marxist beliefs. He was Joint Artistic Director of the Théâtre National de *Chaillot, 1973–4, becoming sole Director, 1981–8, and mounting in 1987, originally for the *Avignon Festival, the complete 12-hour version of *Claudel's Le Soulier de satin. In 1988 he was appointed Administrator of the *Comédie-Française, directing La *Celestina in Avignon and Paris in 1989. His productions of *Molière and *Racine were famous.

Vitrac, Roger (1899–1952), French poet and playwright, one of the leaders of *Dada, who in 1927 was associated with Antonin *Artaud in the foundation of the Théâtre Alfred *Jarry, where his first two plays were performed under Artaud's direction—Les Mystères de l'amour as part of the opening programme and Victor; ou, Les Enfants au pouvoir as its fourth production in 1928. Les Mystères de l'amour, which evokes the amorous and sadistic fantasies of a pair of lovers, is probably one of the most successful attempts to write a play on the Surrealist principle of automatic writing and foreshadowed the Theatre of *Cruelty, just as Victor, with its monstrous characters and tragi-comic denouement, foreshadowed the Theatre of the *Absurd. It was successfully revived in Paris in 1946 and again in 1962, the second revival being directed by *Anouilh, who considered that Vitrac had an important influence on his own work. It was seen in translation in London in 1964. Vitrac's other plays include Le Coup de Trafalgar (1934), the uncharacteristically tragic Les Demoiselles du large (1938), Le Loup-garou (The Werewolf, 1940), and Le Sabre de mon père (1951), a sexual extravaganza set in a fashionable nursing home.

Vitruvius Pollio, Marcus (fl. 70–15 BC), Roman author of a treatise in 10 books, De architectura, of which Book V deals with theatre construction, illustrated by diagrams. Discovered in manuscript at St Gallen in 1414, this was printed in 1484. The first edition with illustrations was published in 1511, and an Italian

translation appeared in 1531. This work had a great influence on the building of Renaissance theatres, and from it the new generation of theatre designers took—though not always accurately—the idea of such devices as the *periaktoi and in general the proportions and acoustic properties of the later Hellenistic and Roman theatres (see THEATRE BUILDINGS).

Vivian Beaumont Theatre, New York, on West 65th Street. Designed by Eero Saarinen with the collaboration of Jo *Mielziner and seating 1,140, this theatre opened in 1965. It forms part of the Lincoln Center for the Performing Arts, and housed the Lincoln Center Repertory Company from the *Washington Square Theatre. Its opening production was *Büchner's Danton's Death, later presentations including *Sartre's The Condemned of Altona, *Brecht's The Caucasian Chalk Circle (both 1966), Tennessee *Williams's Camino Real, and Sam *Shepard's Operation Sidewinder (both 1970). In 1973 the theatre was taken over by Joseph *Papp, reopening with David *Rabe's Boom Boom Room and presenting revivals of *Strindberg's The Dance of Death in 1974, Hamlet in 1975, Brecht and Weill's The Threepenny Opera in 1976, and *Chekhov's The Cherry Orchard, with Irene *Worth, in 1977. A smaller downstairs theatre seating 299 in the same building, which opened in 1966 as the Forum, was also taken over by Papp in 1973 and renamed the Mitzi E. Newhouse Theatre; it staged plays by Shakespeare, opening with Troilus and Cressida, and new plays such as David Rabe's Streamers (1976). In 1977 a financial deficit forced Papp to give up the theatres, which were then used only intermittently for such productions as Peter *Brook's La Tragédie de Carmen (1983), adapted from Bizet's opera, and C. P. *Taylor's And a Nightingale Sang . . . (also 1983). In 1985 however the Mitzi E. Newhouse staged *Mamet's one-acters Prairie du Chien and The Shawl followed by John *Guare's The House of Blue Leaves (1986), the last transferring to the Vivian Beaumont, which later staged other productions including *Soyinka's Death and the King's Horseman and Cole *Porter's musical Anything Goes (both 1987); the Mitzi Newhouse housed the South African musical Sarafina! (also 1987), which moved to Broadway.

Volchek, Galina, see SOVREMENNIK THEATRE.

Volkov, Feodor Grigoryevich (1729–63), Russian actor, who with his brother Grigori was running an amateur theatre in Yaroslavl when in 1752 he was commanded to give a performance in St Petersburg before the Tsar and his Court. This led to his being sent with *Dmitrevsky and other members of the company to the Cadet College for the sons of the nobility, where he was trained as an actor for the Court theatre and also given a good general education. He appeared at Court again in 1755 and in the following year joined the first professional Russian company under *Sumarokov, soon becoming his chief assistant and leading actor. Attached as they were to the party responsible for the overthrow of Peter III and the accession of Catherine the Great, Volkov and his brother were rewarded with Court offices and put in charge of the celebrations in honour of Catherine's coronation in Moscow. It was while directing a street masquerade there that Feodor caught cold and died.

Volksbühne, or People's Theatre, organization which originated in *Berlin in 1890 with the Freie Volksbühne, formed to bring good plays, preferably with a pronounced social content, to the working class at prices they could afford. In 1914 the organization opened its own theatre on the Bülowplatz, which soon became one of the leading Berlin playhouses, and by 1930 the movement had spread all over Germany, having over 300 branches with nearly half a million members, tickets being allocated by lot as they had been from the beginning. Dissolved by the Nazis in 1933, the movement was revived in 1948 and by the end of the 1950s had about 90 local centres in West Germany, the corresponding East German groups being absorbed into the state system which embraced all theatrical enterprises. From 1949 the Freie Volksbühne had its own theatre in West Berlin (the original site then being in East Berlin), and in 1962 it moved into new premises under *Piscator. The Volksbühne in East Berlin reopened after rebuilding in 1954, and was revitalized under Benno *Besson's directorship from 1969 to 1978. Heiner *Müller directed his adaptation of *Macbeth* there in 1982.

Voltaire [François-Marie Arouet] (1694–1778), French man of letters, who took as his pseudonym an anagram of 'Arouet l(e) i(eune)'. Writer, philosopher, and historian, he is important in theatre history as a playwright and critic. He was passionately addicted to the theatre throughout his long life, befriending many actors and actresses, including *Lekain and Adrienne *Lecouvreur; himself a keen actor, he built several private theatres (notably at Ferney, his last home) where he could indulge his taste for private theatricals. His contemporaries considered his tragedies as good as those of *Corneille and *Racine, but despite certain superficial innovations his plays for the most part conformed to the models established in the previous century and hardly any of them survived in the repertory even into the 19th century. However, when he was elected to the French Academy in 1746 his tragedies constituted his most substantial qualification.

His first play was a tragedy, *Œdipe* (1718), written in the Bastille, where he had been imprisoned as the author of a political lampoon. Its success, and that of other writings, brought him fame, social advancement, and a Court pension; he speculated to good effect and became a very wealthy man. Exiled in 1726 after a quarrel with the Marquis de Rohan, he went to London, where he remained until 1729. This visit was of the greatest importance to Voltaire's intellectual development (its general effect can be studied in his *Lettres philosophiques* of 1734), and to his career as a dramatist, for he learned English, frequented the London playhouses, and read Shakespeare and the Restoration dramatists in the original. Some of the plays he wrote on his return to Paris—*Eriphyle*, *Zaïre* (both 1732, the latter, his masterpiece, being based on *Othello*), *La Mort de César* (1735)—show traces of Shakespearian influence. But the differences far outweigh any similarities of detail, and it was not long before Voltaire's appreciation of Shakespeare gave way to harsh criticism. Conditioned by an entirely rationalistic approach to dramatic language, Voltaire wrote in verse, not poetry, and his plays show little creative imagination. Among his innovations in drama—and they are not fundamental changes—are the adoption in tragedy of subjects from French national history, alongside those from mythology or ancient history, and the use of the tragic form as a vehicle for the expression of controversial ideas. This latter feature makes plays such as *Alzire* (1736), *Mahomet; ou, Le Fanatism* (1741), and *L'Orphelin de la Chine* (1755) still interesting to read today. They stand beside the philosophical tales and pamphlets as examples of Voltaire's unceasing attack on religious bigotry and intolerance, and on tyrannical oppression in all its forms. As a critic he left a substantial volume of writings on the theatre, the most important, apart from the prefaces to his own plays, being his *Commentaire sur Corneille* (1764); though he admired Corneille, his preference was for Racine. His comedy *L'Écossaise* (1760) was a biting personal satire on his literary enemy Fréron, and he was critical both of *Marivaux's idiosyncratic comedy and of the *comédie larmoyante* established by *La Chaussée, describing the *drame bourgeois* which developed from it as 'a kind of tragedy for chambermaids'. This did not prevent him from writing a number of such *drames* himself— *L'Enfant prodigue* (1736), for instance, and *Nanine* (1749), based on Richardson's novel *Pamela* (1740).

Voltaire was not a great dramatist, but his plays show a breadth of treatment and force of description hardly surpassed in his own day. If his plays are never revived now the fault lies in his facility, which led him to write too much and too carelessly, and in the fact that he lived in an age of transition and reflected its momentary preoccupations. By slackening the rigid form of tragedy to something more acceptable to a larger but less educated audience, he drove it a step further on the road to *melodrama, and to him, rather than to the Romantic dramatists, goes the honour of having first introduced local colour into the theatre. Many of his later works were marred by philosophical propaganda, as in Les Guèbres which was never acted. The only play in which he achieved the impact of true tragedy was Zaïre, and there are innumerable tributes to the genuine emotion it aroused, both when it was first performed in 1732 with Dufresne and Mlle *Gaussin in the leading roles and again later in the century, when Lekain took the role of Orosmane. When put on in London, at *Drury Lane, in Aaron *Hill's translation as Zara (1736), it was the occasion of a memorable début by Susanna, wife of Theophilus *Cibber. It was also played with great success in Milan, and there were no less than seven translations of it into Italian in Voltaire's lifetime. One reason for actors and audiences alike to revere the memory of Voltaire is that he was instrumental in doing away with the intolerable nuisance of the *audience on the stage, after which dramatists were able to use the larger stage area for the introduction of more spectacle and greater

freedom of action, as Voltaire himself did in Tancrède (1760).

Vondel, Joost van den (1587–1679), Flemish dramatist, who began to write plays late in life, and produced some of his best work after 60. A humanist and a deeply religious man, he translated a number of classical plays, but his own works, while retaining the form and technique of Greek tragedy, are imbued with the Christian spirit. Many of his plays were on subjects taken from the Old Testament—Adam, Noah, Joseph, Saul, Samson—and these had a great influence on the Jewish drama which developed in Amsterdam during the first half of the 17th century, many of the Hebrew plays written at the time being based on his. Among his later plays was one on Mary Queen of Scots, Maria Stuart (1646); his masterpiece Lucifer, banned by the authorities after two performances, was published in 1654 and was known to Milton, on whose epic poem Paradise Lost (1667) it had some influence. Vondel also influenced the development of baroque drama in Germany in the 17th century, mainly through the translation and adaptation of some of his plays by *Gryphius.

W

Wade, Allan, see PHOENIX SOCIETY.

Waggon Stage, see BOAT TRUCK.

Wakefield Cycle, see MYSTERY PLAY.

Waldorf Theatre, London, see STRAND THEATRE 2.

Walking Gentleman, Lady, see STOCK COMPANY.

Walkley, A(lfred) **B**(ingham) (1855–1926), English dramatic critic, whom *Shaw satirized as Mr Trotter in *Fanny's First Play* (1911). He wrote for a number of papers before becoming dramatic critic of *The Times* in 1900, retaining this position until 1926. A cultured and conscientious writer who took himself and his duties very seriously, he wrote rather as a literary than a dramatic critic, devoting more space to the play than to the actors.

Wallace, (Richard) Edgar Horatio (1875–1932), English journalist, novelist, and playwright. He was the first to make a speciality of detective drama, many of his plays being based on his own books. Among the most successful were *The Ringer* (1926), *The Terror* (1927), *The Squeaker* (1928), *The Flying Squad* (1929), *On the Spot*, *Smoky Cell* (both 1930), and *The Case of the Frightened Lady* (1931), in all of which Wallace showed extraordinary perfection of detail, narrative skill, and knowledge of police methods and criminal psychology, fruits of his apprenticeship as a crime reporter.

Wallace, Nellie [Eleanor] **Jane** (1870–1948), one of the great stars of the English *music-hall. Born in Glasgow, she made her first appearance as a clog-dancer in Birmingham in 1888. After touring the 'halls' as one of the Three Wallace Sisters, she was seen in the provinces in a series of straight comedies, but returned to the music-halls as a solo turn in the early 1890s, billed as 'the Essence of Eccentricity'. In her songs and sketches she was invariably the ever-hopeful spinster, seemingly unaware of her shabby clothes and her plain face, plastered with badly-applied make-up, surmounted by a battered hat, and with a moth-eaten feather boa—'Me Furs'—twisted round her throat. She would rattle on, equally unaware, it appeared, of the ambiguity of her remarks, until a sudden sideways leer proved the contrary. Her best-known song was 'I lost Georgie in Trafalgar Square / Lost 'im on me 'oneymoon, but I don't care', in which she was the epitome of the rapacious female on the rampage. A mistress of the grotesque, she was probably the only woman to make a success of the *Dame in *pantomime, playing the Widow Twankey in *Aladdin*, the Cook in *Dick Whittington*, and Dame Durden in *Jack and the Beanstalk*. In 1935 she gave a superb performance as the Wicked Witch Carabosse in *The Sleeping Beauty; or, What a Witch*. She made her last appearance in 1945.

Wallack, James William the elder (1794–1864), London-born actor who became equally well known in England and the USA, where he first appeared in 1818, playing Macbeth at the *Park Theatre, having previously played leading roles at *Drury Lane. A romantic and tragic actor in the style of John Philip *Kemble, he also excelled in comedy. He had a most successful season at the *National Theatre in New York in 1837–8, with his elder brother **Henry** (1790–1870), also an actor, as his stage-manager, and then went to *Niblo's Garden and on tour. He made his last appearance in London in 1851, and a year later opened *Brougham's former theatre in New York as Wallack's Lyceum. Elegantly redecorated, well equipped, and furnished with a good *stock company in a repertory of Shakespeare and standard comedies, with some modern plays, it flourished for nine years, Wallack himself making his last appearance on the stage there in 1859. Two years later he opened the *Star Theatre on Broadway, but his inaugural speech marked his last public appearance and he retired, leaving the traditions he had established to be carried on by his son Lester *Wallack and his nephew, James *Wallack the younger. Wallack's Lyceum survived, under various names, until 1869, when it was demolished.

Wallack, James William the younger (1818–73), English-born American actor, son of **Henry**, under whom he served his apprenticeship to the stage. In 1837 he joined the company of his uncle James *Wallack the elder at the *National Theatre in New York, rising quickly from 'walking gentleman' to juvenile lead. A man of rugged physique, he was at his best in such parts as Macbeth, Othello, Iago, and Richard III. In 1865 he became a member of the stock company at the *Star Theatre under his cousin Lester *Wallack. He was excellent as Fagin in a dramatization of *Dickens's *Oliver Twist* in 1867 and in 1872 played, most terrifyingly, Mathias in Lewis's *The Bells* at *Booth's Theatre. He was not so successful in comedy, except for Jaques in *As You Like It* and Mercutio in *Romeo and Juliet*.

Wallack, Lester [John Johnstone] (1820–88), American actor, son of James *Wallack the elder, with whom he first appeared on the stage in the English provinces. In 1845 he was in *Manchester with Charlotte *Cushman and Helen *Faucit, and soon after made his début in New York at the *Broadway Theatre. In 1850, as a member of *Burton's company, he gave excellent performances as Aguecheek in *Twelfth Night* and Charles Surface in *Sheridan's *The School for Scandal*, and then became stage-manager for his father at Wallack's Lyceum, where he also played a wide range of comic and romantic parts. Under his management the *Star Theatre, which the elder James Wallack opened in 1861 and immediately relinquished, flourished until 1881, when Lester moved to a new *Wallack's Theatre, remaining there until the year before his death. He concentrated on the production of plays by English dramatists, particularly T. W. *Robertson, and one of his finest parts was Elliot Grey in his own adaptation of a novel, *Rosedale* (1863), in which he appeared for many years. He was also good as Benedick in an 1867 production of *Much Ado about Nothing*, after which he staged no more of Shakespeare's plays until 1880, when he put on *As You Like It*.

Wallack's Lyceum, see WALLACK, JAMES WILLIAM the elder.

Wallack's Theatre. 1. At Broadway and 30th Street. This was opened by Lester *Wallack in 1882 with a revival of *Sheridan's *The School for Scandal*. Lillie *Langtry, who was to have made her New York début at the second *Park Theatre on 30 Oct. that year—the day it was burnt down—first appeared at Wallack's instead. The theatre was not a success, and in 1887 Lester transferred the lease to other hands, allowing them to retain the old name. The last stock season was given under Henry *Abbey in 1888, and the theatre was then leased by *Palmer, who opened it under his own name with a French company. It reverted to its original name in 1896, and finally closed in 1915, after a season of plays directed by *Granville-Barker which included the first American production of *Shaw's *Androcles and the Lion*. It was then demolished.

2. A theatre at 254 West 42nd Street, between 7th and 8th Avenues, which in 1924 was named the Wallack after the famous theatrical family, opened originally in 1904 as the Lew Fields Theatre. Two years later it was taken over by James K. *Hackett, who named it after himself and appeared there with his own company in a number of successful productions, including Anthony Hope's *The Prisoner of Zenda* in 1908. The same year saw the first production of *Sheldon's *Salvation Nell*, with Mrs *Fiske. In 1911 and 1920 the theatre changed its name with its managers, to the Harris and the Frazee respectively, and in 1921 it housed a successful production of *Kaufman and *Connelly's *Dulcy*, with Lynn *Fontanne. It became a cinema in 1931.

The *Star Theatre was known as Wallack's from its opening in 1861 until 1882.

Waller, Emma (1820–99), British-born American actress, who emigrated to the USA with her husband, **Daniel Wilmarth Waller** (1824–82), and made her first appearance at the *Walnut Street Theatre in Philadelphia in 1857 as Ophelia to his Hamlet. She was seen in New York a year later, and from then until her retirement in 1878 played leading roles there and on tour throughout the country. Among her finest parts were Lady Macbeth; Queen Margaret in Colley *Cibber's version of *Richard III*, which she played with Edwin *Booth; Meg Merrilies in a dramatization of Scott's *Guy Mannering*, in which she was seen many times; and Julia in Sheridan *Knowles's *The Hunchback*. In the fashion of the time she also played Hamlet and Iago. After her retirement she continued to give public readings and taught elocution.

Waller, Lewis [William Waller Lewis] (1860–1915), English actor-manager, born in Spain of English parents, who made his first appearance as a professional at *Toole's Theatre in 1883. A robust and dynamic actor, with a magnificent voice, he was at his best in costume parts and particularly in Shakespeare. His Brutus in *Julius Caesar*, Faulconbridge in *King John*, and above all his Henry V were memorable. In modern-dress comedy, such as *Wilde's *An Ideal Husband* with which he opened his management at the *Haymarket in 1895, he did not appear to such advantage. He was much admired as D'Artagnan in one of the many dramatizations of the elder *Dumas's *The Three Musketeers* (1898) and in the title-role of *Monsieur Beaucaire* (1902), based by Booth *Tarkington on his own novel. The latter role was perhaps the supreme example of Waller's talent. He visited the USA for the first time in 1911, where he was successful in the New York production of *The Garden of Allah*, adapted by Robert Hichens and Mary Anderson from the former's book. He was one of the outstanding romantic actors of his day, one of the first so-called 'matinée idols', but entirely without conceit, and the hysteria of his more fervid supporters caused him much embarrassment. He married an actress, **Florence West** (1862–1912), sister-in-law of the critic Clement *Scott, who appeared with her husband in many of his outstanding successes, notably as Miladi in *The Three Musketeers*.

Walls, Tom (1883–1949), English actor, director, and theatre manager. He first appeared on the stage in Glasgow in 1905, and subsequently toured in the United States and

Canada, returning to London in 1907. He had a successful career in musical comedy, and in 1922 went into management, his first venture being the immensely successful *Tons of Money*, a farce by Will *Evans which when transferred to the *Aldwych inaugurated the long succession of 'Aldwych farces' by Ben *Travers. Walls appeared in all of them, up to and including *Turkey Time* in 1931, playing the flashy opportunist. He then went into films, but reappeared on the stage in 1938, continuing his successful career in light comedy and farce. He was for a time manager of the *Fortune Theatre, opening with *Lonsdale's *On Approval* (1927), which he directed himself. He controlled a number of touring companies in the 1930s, and in 1939 took over the Alexandra in Stoke Newington, which he ran as a repertory theatre.

Walnut Street Theatre, Philadelphia, oldest playhouse in the USA. Originally built as a circus in 1809, it was first used as a theatre two years later, competing with the *Chestnut Street Theatre and, from 1828, with the Arch Street Theatre. After the financial crisis experienced by the theatres of *Philadelphia in 1829 the Walnut Street continued to operate with a good *stock company which supported visiting guest players. Seating 1,052, it was remodelled in 1970, and from 1971 to 1980 provided a home for the Philadelphia Drama Guild.

Walter, Eugene (1874–1941), American dramatist, author of 20 plays, the best of which, though somewhat melodramatic, seemed to point the way towards a more realistic and sober approach to social problems by contemporary American playwrights. They included *Paid in Full* (1908), in which a wife gets her husband cleared of embezzlement, but in doing so comes to despise him; and *The Easiest Way* (1909), about a kept woman who loses a chance to break free. Walter failed to live up to his early promise, and his later plays are negligible. Among them were dramatizations of two popular novels by John Fox, Jr, *The Trail of the Lonesome Pine* (1912) and *The Little Shepherd of Kingdom Come* (1916).

Walter Kerr Theater, New York, on West 48th Street, between Broadway and 8th Avenue. Seating 950, this opened as the Ritz in 1921, but success did not come until 1924 with the production of Sutton Vane's *Outward Bound* and *Galsworthy's *Old English*, with George *Arliss as Sylvanus Heythorp. In the following year Ashley *Dukes's *The Man with a Load of Mischief* had a short run. A further series of failures was broken in 1927 by the long run of John McGowan's *Excess Baggage*, a comedy on the heartbreaks of vaudeville. In 1937 the theatre was taken over by the *Federal Theatre Project, which presented there the *Living Newspaper *Power*. A year later T. S. *Eliot's *Murder in the Cathedral* had a short run, and after a Federal

Theatre production in 1939 the theatre was taken over for radio and television. Unsuccessful attempts to reopen it were made in 1970 and 1972, and in 1973 it was taken over by the Robert F. Kennedy Theatre for Children, which remained there until 1976 when a financial crisis caused it to leave. The theatre fell into disrepair, but reopened in 1990, renovated and renamed, with August *Wilson's *The Piano Lesson*.

Wanamaker, Sam (1919–), American actor and director, who studied for the stage at the *Goodman Theatre in Chicago, where in 1964 he returned to play Macbeth, and made his first appearance in New York in 1942. After serving with the US armed forces he returned to the stage in 1946 in Maxwell *Anderson's *Joan of Lorraine*. He made his first appearance in London in 1952 as Bernie Dodd in *Odets's *Winter Journey*, which he also directed, and then remained in England, directing and acting in a number of plays, including Odets's *The Big Knife* (1954) and N. Richard Nash's *The Rainmaker* (1956). In 1959 he was with the company at Stratford-upon-Avon, playing Iago to the Othello of Paul *Robeson.

He directed several plays in America in the early 1960s, but his work in the theatre has since been concentrated on his efforts to rebuild Shakespeare's *Globe Theatre on Bankside near the original site. In the early 1970s he founded a Trust for this purpose, and a temporary theatre tent opened in 1972 for a summer season which included a modern-dress production of *Hamlet* with Keith *Michell. The second season, in which Vanessa *Redgrave played Cleopatra in *Antony and Cleopatra*, came to an abrupt end when the tent collapsed in a storm. A second temporary theatre of tubular steel opened in 1975, and three summer seasons were given there. After lengthy legal wrangles a permanent theatre is scheduled to open in 1992. The site will also contain a small covered theatre (based on an Inigo *Jones design) and a new museum, the whole complex, the International Shakespeare Globe Centre, forming a centre for Shakespearian studies. The nearby Shakespeare Globe Museum, originally called the Bear Gardens Museum, was founded by Wanamaker in 1972 and covers the English theatre in the 16th and 17th centuries.

Warburton, John (1682–1759), English antiquarian and book collector, who had at one time in his possession some 60 Elizabethan and Jacobean plays in manuscript, many of them unique copies of works which had never been printed. Unfortunately, through his own carelessness and the ignorance of his servant Betsy Baker, who somehow got possession of them, all but three were, as Warburton himself says, 'unluckely burnd or put under Pye Bottoms'. He left a list of the titles of these lost plays, from which it appears that the chief sufferers from Betsy's depredations were Thomas *Dekker, John *Ford, and Philip *Massinger.

Ward, Dame (Lucy) **Geneviève Teresa** (1838–1922), actress equally well known in the USA and England, and the first to be created DBE (in 1921). Originally an opera-singer (as Madame Ginevra Guerrabella), she turned to straight acting after losing her singing voice through illness and overwork, and made her début in 1873, in Manchester, as Lady Macbeth, a part which she later played in French at the *Porte-Saint-Martin in Paris. She achieved success in a variety of roles, including Julia in Sheridan *Knowles's The Hunchback and Portia in The Merchant of Venice (both 1874), *Sophocles' Antigone, Mrs Haller in *Kotzebue's The Stranger, and Belvidera in *Otway's Venice Preserv'd (all 1875). She was first seen in New York, her birthplace, in 1878, as Jane Shore in W. G. Wills's tragedy of that name, and a year later, under her own management in London, she produced Forget-Me-Not, by Herman Merivale and F. C. Grove, which proved such a success that she toured in it all over the world. Playing opposite her in the original production at the *Lyceum was the young Johnston *Forbes-Robertson. After her last visit to America in 1891 she joined *Irving at the Lyceum, her first appearance with him being as Queen Eleanor in *Tennyson's Becket. From 1900 she appeared rarely on the stage, but was occasionally seen with *Benson's company, with which she made her last appearance in London, as Volumnia in Coriolanus, at the *Old Vic in 1920.

Warehouse, The, see DONMAR WAREHOUSE.

Warfield [Wollfeld or Wohlfelt], **David** (1866–1951), American actor, who began as a programme seller and later an usher at a San Francisco theatre and in 1888 joined a travelling company, playing Melter Moss in Tom *Taylor's The Ticket-of-Leave Man. This failed after a week, and he went into variety, appearing in New York in 1890 in vaudeville and musical comedy. He was Karl in the original production of the popular musical comedy The Belle of New York (1897) and later spent three years in a burlesque company. He was adept at presenting the New York East Side Jew of his day, and had already given proof of fine qualities in his acting when in 1901 *Belasco starred him in *Klein's The Auctioneer. This was an instantaneous success and had a long run, but it was as the gentle, pathetic, self-sacrificing Anton von Barwig in The Music Master (1904), also by Klein, that Warfield set the seal on his growing reputation. He played nothing else, in New York and on tour, for three years. Among his later successes were Wes Bigelow in Belasco's A Grand Army Man (1907), and the title-roles in The Return of Peter Grimm (1911), also by Belasco, and Wills's Vanderdecken (1915). He

was also seen as Shylock in Belasco's 1922 production of The Merchant of Venice, retiring from the stage two years later.

Warner [Lickfold], **Charles** (1846–1909), English actor, who made his first appearance as a boy of 15 at *Sadler's Wells, where his father, James Lickfold, was a member of the company. Forced against his will to study architecture, he ran away from home and returned to the theatre under the name of Warner, which he afterwards retained. He first appeared in London in 1864, playing Paris in Romeo and Juliet. A year later he played Romeo, and Iago in Othello. He scored a great success as Steerforth in Little Em'ly (1869), an adaptation of *Dickens's David Copperfield, and was the first to play Charles Middlewick in H. J. *Byron's Our Boys (1875); but though good in comedy, he was at his best in *melodrama, in which he played at the *Adelphi Theatre for many years. His finest and most memorable part was Coupeau in Drink (1879), an adaptation of *Zola's novel L'Assommoir in which he collaborated with Charles *Reade. He made his last appearance in 1906 in The Winter's Tale, playing Leontes to the Hermione of Ellen *Terry, before going to America where he committed suicide.

Warren, William the elder (1767–1832), British-born American actor, who made his early appearances in the English provinces, and in 1788 was with Tate *Wilkinson, playing in support of Sarah *Siddons. In 1796 he was invited by Thomas *Wignell to join his company in *Philadelphia, and there, except for a number of short visits to New York, he spent the rest of his life. He succeeded Wignell as manager of the *Chestnut Street Theatre, in partnership with William *Wood, and under them in 1820 Edwin *Forrest made his first appearance, playing Young Norval in *Home's Douglas. Warren was a fine actor of the old school, among his best parts being Sir Toby Belch in Twelfth Night, Sir Anthony Absolute and Sir Peter Teazle in *Sheridan's The Rivals and The School for Scandal, and Old Dornton in *Holcroft's The Road to Ruin. He retired in 1829, making a final farewell appearance the year before his death. In 1806 he married as his second wife the actress Mrs *Merry, and after her death two years later married the sister-in-law of the first Joseph *Jefferson, with whose father Thomas Jefferson he had appeared in England as a young man. There were six children of this marriage, all connected with the stage. The best known was William *Warren the younger; his brother became a theatre manager and, by the marriage of his daughter Sarah Isabel, the father-in-law of the third Joseph *Jefferson; his four sisters all married actors or theatre managers.

Warren, William the younger (1812–88), American actor, son of the above, who spent most of his professional life at the *Boston Museum. He made his first appearance at the Arch Street Theatre in *Philadelphia, shortly after his father's death, playing Young Norval in *Home's *Douglas*, and then toured for several years and was a member of several good *stock companies. He was seldom seen in New York and only once, in 1845, in England. He began his long association with the Museum in 1847, and remained there until his retirement in 1883. In his early days he acted a wide variety of parts, but later specialized in comedy, being particularly admired as Touchstone in *As You Like It*, Polonius in *Hamlet*, and Bob Acres and Sir Peter Teazle in *Sheridan's *The Rivals* and *The School for Scandal*. He appeared also in a number of new plays, which his excellent acting often redeemed from mediocrity, and in 1882 celebrated his stage jubilee by playing Dr Pangloss in the younger *Colman's *The Heir-at-Law*. He was last seen on the stage as Old Eccles in T. W. *Robertson's *Caste*.

Warwick's Men, see OXFORD'S MEN.

Washington, DC. The first theatre in the federal capital, known as the United States or National Theatre, was adapted from a building already in existence. Thomas *Wignell and his company played there for some months in 1800, after which it became a post and patent office. The first purpose-built theatre was the Washington, which opened in 1804 and was later rebuilt and renamed the Washington City Assembly Rooms. A New Washington Theatre was opened in 1821 by a company under *Warren and *Wood, which used it during the summer months only. Joe *Cowell appeared there in 1828. A second National Theatre, which opened in 1835, played an important part in the social life of the city, one innovation being the turning of the old pit into a 'parquette area', with comfortable seating, which made it more acceptable to the better-class citizens. The theatre had to be rebuilt several times after fires in the 19th century; the present auditorium dates from 1922. *Ford's Theatre, which opened in 1862, became famous as the scene of Abraham Lincoln's assassination. Among other present-day theatres in Washington are the *Arena Stage and the theatres in the *John F. Kennedy Center. The Folger Theatre was founded in 1970 under the auspices of the Folger Shakespeare Library (founded by **Henry Folger**, 1857–1930) to present plays in a reconstruction of an Elizabethan playhouse seating 253. In 1985 the Library trustees withdrew financial support and the theatre is now a separate non-profit entity, the Shakespeare Theater at the Folger, with a season running Oct.–June.

Washington Square Players, New York, play-producing society founded in 1914 by a group which included Lawrence *Langner. Their first programme, in 1915, consisted of a number of one-act plays, some of them specially written for the occasion. In 1916 they began to present full-length plays, among them *Chekhov's *The Seagull*, *Ibsen's *Ghosts*, and *Shaw's *Mrs Warren's Profession*. In the same year they moved to the *Comedy Theatre, where their productions included *O'Neill's one-act play *In the Zone* (1917) and Elmer *Rice's *Home of the Free* (1918). Katharine *Cornell made her first professional appearances with them, and Lee *Simonson also began his career there. They disbanded in 1918, but it was out of their work that the *Theatre Guild evolved.

Washington Square Theatre, New York, on West 4th Street, shed-like corrugated-steel building designed by Jo *Mielziner, built as the temporary home of the Lincoln Center repertory company pending the completion of the *Vivian Beaumont Theatre. It opened in 1964 with Arthur *Miller's *After the Fall* directed by Elia *Kazan, who had been appointed co-director of the new company. It was followed by a revival of *O'Neill's *Marco Millions* and by a new play, *But for Whom Charlie*, by S. N. *Behrman. The venture was not a success, and in 1965 the company moved to its permanent home at the Lincoln Center. From Nov. 1965 to Mar. 1968 the Washington Square Theatre was occupied by a musical, *Man of La Mancha*, based on *Cervantes' *Don Quixote*. It was demolished in 1968.

Waterhouse, Keith, see HALL, WILLIS, and O'TOOLE.

Water Rows, see GROUNDROW.

Waters, Ethel (1896–1977), American Negro actress and singer, who began her stage career in 1917, singing in night-clubs and vaudeville. She made her début on Broadway in 1927 in *Africana*, an all-Negro revue, and in 1933 appeared in Irving *Berlin's revue *As Thousands Cheer*. Her first dramatic role was Hagar in the DuBose Heywards' *Mamba's Daughters* (1939), in which she proved herself an excellent actress. Subsequent Broadway roles as Petunia in the musical *Cabin in the Sky* (1940) and as Berenice in Carson McCullers's *The Member of the Wedding* (1950) enhanced her reputation. She also gave many solo performances, among them *At Home with Ethel Waters* (1953), in which she sang songs she had helped to make famous, including 'Dinah', 'Stormy Weather', and 'Am I Blue?'

Watford, Hertfordshire, see PALACE THEATRE, Watford.

Weaver, John, see HARLEQUINADE.

Webb, John (1611–72), English artist and scenic designer, a pupil of Inigo *Jones. He was employed by *Davenant to design and paint scenery for his productions, beginning with *The Siege of Rhodes* (1656), for which Webb prepared landscapes showing the general layout of the town and harbour, based probably on actual engravings of the time. In the epilogue to *The World in the Moon* (1697), the author, Elkanah *Settle, emphasizes the local origin of Webb's elaborate scenery, in marked contrast to the French fashion of the moment, with a play on the scene-painter's name: "Tis home-spun Cloth; All from an English Web.'

Weber and Fields, American comedy duo consisting of **Joseph Weber** (1867–1942) and **Lew Fields** (1867–1941), the sons of poor Jewish immigrants, who began acting as children in dime museums and beer gardens in and around New York and later appeared in 'burnt-cork' minstrelsy. They soon discarded this for a 'knockabout' act which combined slapstick clowning with the immigrant's difficulties with the English language, Fields, tall, thin, and tricky, appearing as Myer, and Weber, short, squat, and guileless, playing his stooge Mike. In 1885 they established their own company, writing *burlesques of the serious dramas of the day, and 10 years later opened the former Broadway Music-Hall under their own names, remaining there until 1904, when their careers diverged, though they were seen together again in *Hokey-Pokey* (1912). Weber ceased to act in 1918, but continued to direct plays for another 10 years, while Fields directed and appeared in a number of *musical comedies, remaining on the stage until 1929. He was the father of the dramatist Joseph *Fields.

Webster, Ben(jamin) **Nottingham** (1797–1882), English actor, manager, and dramatist. He was first a dancer, appearing as *Harlequin and *Pantaloon in *pantomime in the English provinces and at *Drury Lane. He then became an actor, proving himself an excellent broad comedian, and was with Mme *Vestris at the *Olympic Theatre. In 1837 he became lessee of the *Haymarket Theatre, where he remained for 16 years, engaging all the best actors of the day and putting on many notable productions, in many of which he himself appeared. In 1844 he took over the *Adelphi Theatre as well, and was associated there with Mme *Celeste and Dion *Boucicault, collaborating with the latter in two plays. Webster also made in 1846 an adaptation of *Dickens's *The Cricket on the Hearth* in which he appeared as John Peerybingle with great success; but his finest part was Triplet in Tom *Taylor and Charles *Reade's *Masks and Faces* (1852). He opened the New Adelphi in 1859 and continued in management until 1874, when he retired. As a character actor he was unsurpassed in his own day, but he had little liking for farce, in spite of having written several successful farces himself, and he tempered his comedy with a somewhat grim humour.

His grandson Ben Webster became a well-known actor in London and New York and was the father of Margaret *Webster.

Webster, John (*c.*1580–1634), English dramatist, whose fame rests almost entirely on two plays, *The White Devil* (1612) and *The Duchess of Malfi* (1614). Both are founded on Italian *novelle* and are passionate dramas of love and political intrigue in Renaissance Italy, compounded of crude horror and sublime poetry. Indeed, in the latter respect Webster approached Shakespeare more closely than any of his contemporaries. Both plays have held the stage, providing scope for great acting and fine settings, and as *poetic drama remained unsurpassed by any later work, except that of *Otway, until a new conception of tragedy was imported into European literature by *Ibsen. Webster's other plays, some of which were written in collaboration, were of little importance. Practically nothing is known of his life, but he may have been an actor, perhaps the John Webster who appeared with the *English Comedians in Germany in 1596.

Webster, Margaret (1905–72), English actress and director, the daughter of the actor **Ben**-(jamin) **Webster** (1864–1947), himself the grandson of B. N. *Webster, and **Dame May Whitty** (1865–1948), an actress appointed DBE for charitable work during the First World War. They were together in *Irving's company at the *Lyceum from 1895 and later became very popular in the USA, Margaret being born in New York during one of their visits. She was on the stage for some time as a child, making her first appearance in 1917, and her adult début in the chorus of *Euripides' *Trojan Women* with Sybil *Thorndike in 1924. She was a member of *Fagan's repertory company in *Oxford in 1927, toured with Ben *Greet in Shakespeare, and in 1929 was at the *Old Vic. After playing a wide variety of parts in London she went to New York in 1936, and, while continuing to act, made an outstanding reputation as a director. Maurice *Evans appeared in several of her productions of Shakespeare, and her production of *Othello* in 1943 with Paul *Robeson, in which she played Emilia, broke all records for a Shakespearian play on Broadway. With Eva *Le Gallienne and Cheryl *Crawford she founded the American Repertory Theatre, playing Mrs Borkman in *Ibsen's *John Gabriel Borkman* for it the same year, and the Cheshire Cat and the Red Queen in Lewis Carroll's *Alice in Wonderland* in 1947. She also organized a Shakespeare company which toured the United States in 1948–50 by bus and truck, and directed and appeared in *An Evening with Will Shakespeare* (1952) on tour. Among her other post-war productions were Eva Le Gallienne's translation of *Hochwälder's *Das heilige Experiment* (1943) as *The Strong are*

Lonely (NY, 1953; London, 1955), *Shaw's *Back to Methuselah* in New York in 1958, and Noël *Coward's *Waiting in the Wings* (London, 1960). In 1964 she was seen in London in a solo performance based on the life and works of the Brontë sisters.

Wedekind, Frank [Benjamin Franklin] (1864–1918), German dramatist and actor, who first appeared in cabaret, singing his own songs, and later acted in his own plays with his young wife. The first of these was *Die junge Welt* (written in 1889, prod. 1908); the best known is *Frühlings Erwachen*, written in 1891, which harks back to *Büchner in its staccato structure and intensified realism, but looks forward to *Expressionism and *Symbolism in its graveyard and schoolroom scenes in which it analyses the situation of two 14-year-old lovers who pay with their lives for the moral dishonesty of their tyrannical parents. It was produced by *Reinhardt at the *Berlin Kammerspiele in 1906, but its frank depiction of the results of sexual repression prevented its public performance in England until 1963, when as *Spring Awakening* it was seen at the *Royal Court Theatre. In 1974 a new translation by Edward *Bond, produced by the *National Theatre (NY, 1978), proved that though dated the play still retained its hold. Two of Wedekind's later plays, *Erdgeist* (written in 1893, staged in 1902) and *Die Büchse der Pandora* (written in 1894, staged 1905), served to reinforce his main thesis—that the repression of sexuality results in perversion and tragedy. As *Lulu* the two plays were staged in London and New York in 1970. Among Wedekind's other plays, *Der Marquis von Keith* (1901) was revived by *Jessner in a famous Expressionist production in 1920, and was seen at the Schillertheater in Berlin in 1963. It was not seen in London until 1974 (NY, 1979). Other plays sometimes revived are *König Nicolò; oder, So ist das Leben* (1902) and *Schloß Wetterstein*, first produced in 1917 with Elisabeth *Bergner playing opposite the author. All Wedekind's plays, with their sex-ridden men, women, and children, their gentlemen crooks, and their grotesque yet vivid cranks, typify the feverish spirit of the years before 1914.

Wehr Theater, see MILWAUKEE REPERTORY THEATER.

Weigel, Helene (1900–71), Austrian-born actress and theatre manager, who began her career in Frankfurt playing such powerful low-life characters as Marie in *Büchner's *Woyzeck* and the title-role in *Schönherr's *Der Weibsteufel*. In 1923 she went to *Berlin and under *Jessner played minor roles in which she was admired by some critics, though others found her too strident and uncontrolled. She made her first big success as Klara in *Hebbel's *Maria Magdalene* in 1925 and, after appearing as Grete in *Toller's *Der deutsche Hinkemann* in 1927, the Widow

Begbick in *Brecht's *Man ist Mann* (also 1927), and the title-role in his adaptation of *Gorky's *The Mother* in 1932, was considered unrivalled as the exponent of working-class women. In 1928 she married Brecht, accompanying him on his wanderings from 1933 to 1948, and appearing during this time only in his *Die Gewehre der Frau Carrar* in Paris in 1937 and *Furcht und Elend des Dritten Reiches* in the following year. Returning to East Berlin with him in 1948, she helped to found and run the *Berliner Ensemble, playing with the company the title-role of her husband's *Mutter Courage und ihre Kinder*, Natella in his *Die kaukasische Kreidekreis*, and Volumnia in his adaptation of *Coriolanus*. After Brecht died she took control of the company, the most important part she played after his death being Frau Luckerniddle in his *Die heilige Johanna der Schlächthofe* in 1968. Her acting in later life had an effortless simplicity, and in *Mutter Courage* she illustrated to perfection Brecht's theory of acting (see ALIENATION), seeming to stand beside the figure without identifying with it.

Weimar, town in Thuringia, Germany. The first theatre in the town was built in 1696 and used by visiting professionals, including *Ekhof in 1772. Three years later this was destroyed and replaced by a temporary theatre in the palace of the Duchess Anna Amalia, used by professionals and amateurs. There from 1775 to 1783 *Goethe, with a group of courtiers, produced plays for royal occasions. In 1784 a new Court theatre opened with a resident professional company of which Goethe became Artistic Director in 1791. He established a repertory which included plays by himself, *Schiller, his co-director from 1799 to 1805, *Lessing, Shakespeare, *Calderón, and *Voltaire. Guest artists such as *Schröder and *Iffland were imported to strengthen the resident company, and the theatre soon became famous, particularly for Goethe's fine handling of crowd scenes. After Schiller's death in 1805 Goethe continued to direct the theatre alone until 1817, by which time it had already begun to decline in popularity. The building, renovated in 1798, was burnt down in 1826 and rebuilt, but it failed to regain its audience and had a chequered career until in 1848 it came under the direction of Liszt, who remained there until 1858. After damage in the Second World War the theatre reopened in 1948, since when it has continued to serve the town as a repertory theatre.

Weise, Christian (1642–1702), German dramatist, headmaster of a school at Zittau, where he wrote and produced a number of long plays which were purely academic. The chronicle plays and tragedies of his earlier days have little to recommend them, but the later comedies, in which Weise portrays the middle class on stage for the first time in the German theatre, foreshadow to some extent the work of *Lessing a century later. His *Komödie der bösen Catharina* (1702) was based on the same plot as *The Taming of the Shrew*. He used a simple stage and, although he admitted the *Fool in the shape of Pickelhering into his plays, he demanded clear delivery in place of declamation and attempted individual characterization.

Weiss, Peter Ulrich (1916–82), German dramatist and novelist, resident after 1939 in Sweden. First known as a graphic artist and novelist, he gained international renown with his first play, *Die Verfolgung und Ermordung Jean Paul Marats dargestellt durch die Schauspielgruppe des Hospizes zu Charenton unter Anleitung des Herrn de Sade* (*The Persecution and Assassination of Marat as Performed by the Inmates of the Charenton Asylum under the Direction of the Marquis de Sade*, 1964). Known briefly as the *Marat/Sade*, it shows Sade and Marat in the asylum's bath-house, debating their contrasting philosophies of absolute individualism and unswerving dedication to social revolution, while the historical events leading up to Marat's assassination are acted out by the increasingly unruly lunatics. Peter *Brook produced it later the same year in London with the *RSC (NY, 1965). Its success led to the simultaneous production in 14 German theatres of Weiss's next play *Die Ermittlung* (1965), which, as *The Investigation*, was given a rehearsed reading in London, again under Brook, and was produced in New York in 1966. One of a wave of 'documentary dramas' (see DOCUMENTARY THEATRE), it was based on the transcript of the 1964 Frankfurt War Crimes Trial, and attempts to apportion the blame for the Auschwitz atrocities. This, like the *Marat/Sade* and *Der Turm* (*The Tower*, 1967), was published in English, as was *Die Versicherung* (*The Insurance*, 1969), a Surrealist allegory in which men and beasts intermingle. An earlier play, *Gesang vom lusitanischen Popanz* (1967), an account of the uprising in Angola and its suppression by the Portuguese, was staged, as *Song of the Lusitanian Bogey*, by the Negro Ensemble Company in New York in 1968 and London in 1969. A documentary on the war in Vietnam followed, its cumbersome title being shortened to *Viet Nam Diskurs* (*Vietnam Discourse*, 1968). After an excursion into more popular entertainment with *Wie dem Herrn Mockingpott das Leiden ausgetrieben wird* (*How Mr Mockingpott was Relieved of his Sufferings*, also 1968), in which the chief characters are two clowns, Mockingpott and Wurst, Weiss reverted to documentaries with *Trotzki in Exil* (1970), which portrayed Trotsky not only as literally an exile from his own country but as an exile of the mind, and *Hölderlin* (1971), presenting the poet as the archetypal revolutionary writer.

Wekwerth, Manfred, see BERLINER ENSEMBLE.

Welfare State International, see COLLECTIVE CREATION.

Welles, (George) **Orson** (1915–85), American actor and director, a forceful and idiosyncratic personality who had a stormy and spectacular career in both the theatre and the cinema. He made his first professional appearance at the *Gate Theatre, Dublin, in 1931, playing the Duke of Württemberg in Feuchtwanger's *Jew Süss*, and in 1933–4 was with Katharine *Cornell, playing Mercutio and later Tybalt in *Romeo and Juliet*, Marchbanks in *Shaw's *Candida*, and Octavius in Besier's *The Barretts of Wimpole Street*. He next collaborated with John *Houseman, first as an actor in Archibald *MacLeish's *Panic* (1935) and then in the *Federal Theatre Project, for which he directed several controversial productions including the 'voodoo' *Macbeth* in 1936 and *Marlowe's *Dr Faustus* in 1937, in which he also appeared. Finally he partnered Houseman in the running of the Mercury Theatre (see COMEDY THEATRE, New York). In 1939 he was seen as Falstaff in his own adaptation of Shakespeare's historical plays under the title *Five Kings*. He then turned his attention to the cinema, but returned to New York in 1941 to direct Paul *Green's *Native Son* and in 1946 to direct and play in *Around the World*, his own adaptation of Jules Verne's *Around the World in 80 Days*, with songs by Cole *Porter. He made his first appearance in London as Othello in 1951, and was seen there again in 1955 in Melville's *Moby Dick*, which he adapted, directed, and designed. In 1956 he directed and played the title-role in *King Lear* in New York, and in Dublin in 1960 re-created the role of Falstaff in *Chimes at Midnight*, his own adaptation of *Henry IV* and *Henry V*. He also designed and directed the London production of *Ionesco's *Rhinoceros*, with *Olivier, in the same year, and in 1962 his adaptation of *Moby Dick* was seen in New York. He then concentrated on films.

Well-Made Play, English version of *une pièce bien faite*, a label applied in early 19th-century France, at first in a complimentary sense, to plays written by dramatists skilled above all in putting together a plot. It soon took on a pejorative meaning, and came to be used ironically of all plays in which the action develops artificially, according to the strict laws of logic and not to the unpredictable demands of human nature; and in which the plot, to which the characters are completely subordinated, is conceived in terms of exposition, knot, and denouement, with a series of contrived climaxes to create suspense. It is commonly used in France of the works of *Scribe and *Sardou, and in Britain of such playwrights as *Robertson, *Jones, *Pinero, *Coward, and *Rattigan. The well-made play was the chief target of *Zola and other proponents of *naturalism in France, and of *Shaw's hostile criticism of productions in London from 1895 to 1898.

Werfel, Franz (1890–1945), Austrian dramatist and novelist, who first became known as one of the early poets of *Expressionism. His first theatrical success was *Die Troerinnen* (1916), an adaptation of *Euripides' *Trojan Women* acclaimed as a disguised protest against war, but his main contribution to Expressionist drama was the 'magic trilogy' *Der Spiegelmensch* (*The Mirror Man*, 1921), a modern version of the *Faust–Mephistopheles theme, showing in symbolic images man in constant conflict with his *alter ego*. This was followed by *Bocksgesang* (also 1921), produced in New York as *Goat Song* by the *Theatre Guild in 1926. In it man's rebellion against the established order is symbolized by a monster, half-goat half-man, who leads a peasants' revolt in the 18th century. In his later plays Werfel turned to historical themes presented in a more realistic manner, his greatest theatrical success being *Juarez und Maximilian* (1924) on the tragedy of the Habsburg Emperor of Mexico. Two further historical plays proved less successful: *Paulus unter den Juden* (1926), which centres on the conflict between inspired prophecy and established religion among early Christians, and *Das Reich Gottes in Böhmen* (1930). Roused by the Nazi persecution of the Jews, Werfel then wrote a verse play, *Der Weg der Verheißung*, illustrating the tragic history of Judaism through the ages. It was staged in New York in 1936 in a spectacular production by *Reinhardt. Werfel's last play, written after he had left Germany for the USA, was, unexpectedly, a comedy, *Jacobowsky und der Oberst* (set in 1940, after the fall of France), in which a Jewish refugee contrives to smuggle an anti-Semitic Polish officer through the German lines to safety. Translated by S. N. *Behrman, this was successfully staged in New York in 1944 as *Jacobowsky and the Colonel* (London, with Michael *Redgrave, 1945).

Werner, (Friedrich Ludwig) **Zacharias** (1768–1823), author of the one-act tragedy *Der 24 Februar* (*The 24th of February*, 1810), which established the so-called *fate drama in Germany. Werner was a poet and mystic who became a Roman Catholic priest in Vienna, where he was renowned for his preaching. His other plays, which include *Das Kreuz an der Ostsee* (1806), *Martin Luther* (1807), and *Kunigunde, die Heilige* (1815), bear a strong religious imprint, and mingle scenes of telling realism with weird fantasy. They were among the few German Romantic verse dramas to achieve success on the contemporary stage.

Wesker, Arnold (1932–   ), English dramatist, who left school at 16 and had a variety of jobs, including carpenter's and plumber's mate and farm labourer, before his first play *Chicken Soup with Barley* (1958) was produced at the *Belgrade Theatre in Coventry and subsequently transferred to the *Royal Court in London. It formed the first part of a trilogy about a family of Jewish Communists, the later sections being *Roots* (1959; NY, 1961), in which Joan *Plowright scored an outstanding success in London as a working-class girl discovering education, and *I'm Talking about Jerusalem* (1960). *The Kitchen*, based on Wesker's experiences as a pastrycook, was first seen at the Royal Court in 1959 (NY, 1966). In 1961 Wesker accepted the directorship of Centre 42 (see ROUND HOUSE), which he retained until he disbanded it in 1970. His next play, and only West End success, was an anti-Establishment study of RAF conscripts, *Chips with Everything* (1962; NY, 1963). It was followed by *The Four Seasons* (1965; NY, 1968), about the year two lovers spend together. Wesker's work, though seldom without interest, has since found little critical favour. *Their Very Own and Golden City* (Royal Court, 1966) describes an architect's frustrated attempts to build Utopian cities. In 1970 he directed the première of his play *The Friends* in Stockholm, and later that year it was seen in London. *The Wedding Feast* (1974), an adaptation of a story by Dostoevsky, and *The Merchant* (1976), a reworking of *The Merchant of Venice* in which Shylock receives more sympathetic handling, were also first seen in Stockholm, *The Merchant* receiving its English-speaking première in New York in 1977. *The Old Ones* (Royal Court, 1972; NY, 1974) is a study of old age in a London Jewish family. *Love Letters on Blue Paper* (1978; NY, 1984), an adaptation of his own story in which a trade union leader is dying of leukaemia, and *Caritas* (1981) were produced at the *National Theatre. The three short plays *Annie Wobbler* (1983; NY, 1986) and the double bill of *Yardsale* and *Whatever Happened to Betty Lemon?* (1987) were written for a solo actress.

West, Florence, see WALLER, LEWIS.

West, Mae (1892–1980), American actress and entertainer, and a significant figure in the history of stage *censorship. She was on the stage as a child, among her many juvenile roles being Little Willie in Mrs Henry Wood's *East Lynne*. She retired at the age of 11, returning as an adult actress in *musical comedy, her first major role on Broadway being in *Winsome Widow* (1912). She was then seen in *revue, and soon became renowned for her mastery of sexual innuendo expressed through the character of a curvaceous, Edwardian-style vamp. She first tangled with the censorship over *Sex* (1926), at *Daly's Theatre, and was charged with corruption of morals, fined, and sent to Welfare Island for 10 days. Two years later she played the title-role in her own production of *Diamond Lil*, probably her best-known vehicle. After some years in films she returned to Broadway in her own play *Catherine was Great* (1944). In 1948 she was seen in London in *Diamond Lil*, which she then took on tour until 1951. She was a glamorous and extrovert personality, whose fame can be judged by the fact that the RAF in 1940 named their inflatable life-jacket a 'Mae West'.

West, Timothy, see PROSPECT THEATRE COMPANY.

Western, (Pauline) **Lucille** (1843–77) and **Helen** (1844–68), American actresses, daughters of a comedian and his actress wife who, by a second marriage after his early death, became Mrs Jane English, the name by which she is usually known. The children toured with their mother and stepfather in a mixed entertainment which gave plenty of scope for their precocious talents in acting and dancing. Helen died before she could become well known as an adult actress but Lucille was noted for her playing of strong emotional parts in such plays as Mrs Henry Wood's *East Lynne*, the younger *Dumas's *Camille* (*The Lady of the Camellias*), *Taylor and *Reade's *Masks and Faces*, and *Kotzebue's *The Stranger*. One of her finest performances was as Nancy in a dramatization of *Dickens's *Oliver Twist*. She died at the height of her success.

Westminster Theatre, London. **1.** This opened in 1832, and during the four years of its existence was never able to obtain a licence. The site was later occupied by part of the *Imperial Theatre and subsequently by the Central Hall, Westminster.

2. In Palace Street, near Victoria Station. It opened in 1931, having originally been a chapel converted into a cinema in 1924, a new frontage being added. The two-tier auditorium seats 580. The crypt of the old building was converted into dressing-rooms and a bar, and the theatre opened with Henry *Ainley in *Bridie's *The Anatomist*. A wide variety of English and foreign plays was staged, among them revivals of *Pirandello's *Six Characters in Search of an Author* in 1932 and of *Granville-Barker's *Waste* in 1936. The *Group Theatre had its headquarters, and staged most of its productions, there. Several of Denis *Johnston's plays were first seen there, including *The Moon in the Yellow River* (1934), *The Old Lady Says 'No!'* (1935), and *A Bride for the Unicorn* (1936), and *O'Neill's *Ah, Wilderness!* (also 1936) was followed by his *Mourning Becomes Electra* (1937) and *Marco Millions* (1938). Other productions were *The Zeal of Thy House* (also 1938) by Dorothy L. Sayers, transferred from the Chapter House of Canterbury Cathedral, and T. S. *Eliot's *The Family Reunion* (1939). The Westminster was the first theatre to reopen after the compulsory closure in Sept. 1939, the production being *Priestley's *Music at Night*. Paul *Scofield made his début there in 1940 in O'Neill's *Desire under the

Elms, and other successful productions included *Afinogenov's *Distant Point* (1941) and Bridie's *Mr Bolfry* (1943) and *It Depends What You Mean* (1944). The theatre was bought in 1946 by the evangelical Oxford Group, which used it mainly for productions promoting its 'Moral Rearmament' campaign such as Malcolm Muggeridge's *Sentenced to Life* (1978). Occasional returns to the commercial theatre included revivals of *Coward's *Relative Values* (1973) and the *Gershwins' *Oh, Kay!* (1974). The theatre however was often dark until in the late 1980s it came under ecumenical management (though the site itself did not change hands). *Wilde's *An Ideal Husband* had a long run in 1989 and was followed over Christmas by adaptations of two of C. S. Lewis's children's stories.

A studio theatre, the First Floor Theatre (FFT), seating 107, was opened in 1989.

Weston's Music-Hall, see HOLBORN EMPIRE.

West Yorkshire Playhouse, Leeds, regional theatre opened in 1990 on a prime site in the city centre. It replaced Leeds Playhouse, which opened in 1970 on a temporary site provided by the university. The new theatre presents classics and new writing, the opening season including two world premières and a British première, as well as Rodgers and *Hammerstein's *Carousel*, *Shaw's *You Never Can Tell*, and Neil *Simon's *Brighton Beach Memoirs*. There are two auditoriums: the Quarry Theatre, seating 750, with an open thrust stage and a steep rake, which opened with *O'Keeffe's *Wild Oats*; and the flexible Courtyard Theatre, seating up to 350. The theatres also house touring companies, and there are two other performance spaces. There is a strong emphasis on young people's theatre, including an already established Theatre-in-Education scheme, and on forging links with the community.

Wharf Theatre, Provincetown, RI, see PROVINCETOWN PLAYERS.

Whelan, Albert, see MUSIC-HALL.

White, George, see REVUE.

White, Patrick Victor Martindale (1913–90), Australian novelist and playwright, Nobel Laureate 1973. Born in London of Australian parents, he was brought up in England and Australia, where he settled after the Second World War, becoming internationally known after the publication of his novel *Voss* in 1957. While living in London in the 1930s he wrote a number of sketches and lyrics for revues. Two of his comedies were produced in Sydney in the 1930s, but as a dramatist he is best known for later plays which, though not internationally recognized, have had considerable influence on Australian theatre, and which strongly criticize Australian society. The almost *Expressionist *The Ham Funeral* (1961), written in London in 1947, was followed by *The Season at Sarsaparilla* (also 1961), a study of middle-class suburbia with characters consciously drawn as cartoon-like stereotypes; *A Cheery Soul* (1962), based on his own short story, about a disastrous female do-gooder; and *Night on Bald Mountain* (1964). He returned to playwriting with *Big Toys* (1978), an attack on materialism and greed; *The Night of the Prowler* (also 1978); *Signal Driver* (1983), in which a couple make a symbolic 60-year attempt to signal a bus; and *Netherwood* (also 1983), about the pariah-like treatment of the mentally ill.

White Barn Theatre, Conn., see LUCILLE LORTEL THEATRE.

White-eyed Kaffir, see MUSIC-HALL.

Whitefriars Theatre, London, 'private' roofed theatre in the refectory hall of the old Whitefriars monastery, which was situated in the area around the present Bouverie Street. The hall, adapted in about 1605 on the lines of the second *Blackfriars Theatre, its dimensions being about 35 ft. by 85 ft., was used by the Children of the King's Revels, 1608–9; by the Queen's Revels, 1609–13, when the plays performed included *Field's *A Woman is a Weathercock*, *Marston's *The Insatiate Countess*, and Ben *Jonson's *Epicoene; or, The Silent Woman*; and from 1613 to 1614 by *Lady Elizabeth's Men, who appeared there in Jonson's *Bartholomew Fair*. Its later history is obscure, but it was still in use in 1621, being replaced by the *Salisbury Court Theatre in 1629. The two were often confused, *Pepys in his diary for 1660 noting that he saw *Massinger's *The Bondman* at Whitefriars, in mistake for Salisbury Court.

Whitehall Theatre, London, in Whitehall, near Trafalgar Square. An intimate modern playhouse holding 628 on two tiers, it opened in 1930, its first production being Walter Hackett's *The Way to Treat a Woman*, transferred from the *Duke of York's. Hackett and his wife Marian Lorne controlled the theatre until 1934, he writing the plays and she appearing in them. Subsequent successes included Norman Ginsbury's *Viceroy Sarah*, St John *Ervine's *Anthony and Anna* (both 1935), Alec Coppel's *I Killed the Count* (1937), and Philip King's *Without the Prince* (1940). During the Second World War the theatre mainly housed non-stop revue with Phyllis Dixey, and in 1945 *Worm's Eye View* by R. F. Delderfield began a long run, returning after a short tour in 1947 for a further 1,745 performances. In 1950 *Reluctant Heroes* by Colin Morris was the first of a series of farces produced by Brian *Rix that ran until the mid-1960s. Bryan Blackburn's musical *Come Spy with Me* (1966) was also co-presented by Rix. In 1969 *Pyjama Tops*, a comedy by Paul Raymond based on a French farce *Moumou*, began a run of 5½ years and prompted a sequel in the same style, *Come into My Bed* (1976) by

André Launay. The Black African musical *Ipi Tombi* ran there from 1978 to 1980. In the 1980s the theatre was used for a time as a Museum of War, and in 1990 it staged a revival of *Ayckbourn's *Absurd Person Singular*.

Whitelaw, Billie (1932–), English actress, who was sent by her mother for theatrical training at her local theatre to cure a stutter. She made her first appearance in London in *Feydeau's *Hotel Paradiso* (1956) and was in the *Theatre Workshop production of Alun Owen's *Progress to the Park* (1960), moving with it to the West End in 1961. In 1962 she starred in Willis *Hall and Keith Waterhouse's revue *England, Our England* and in 1964 she joined the *National Theatre company, being seen in the company's productions of *Marston's *The Dutch Courtesan* (1964) and *Pinero's *Trelawny of the 'Wells'* (1965) at *Chichester and as Maggie in *Brighouse's *Hobson's Choice* at the *Old Vic (also 1965). She joined the *RSC to appear in David *Mercer's *After Haggerty* (1971), and in 1975 gave a brilliant performance as the inefficient librarian of a newspaper office in Michael *Frayn's *Alphabetical Order*. In 1980, with the RSC again, she played Andromache and Athena in *The Greeks*, a trilogy based mainly on *Euripides, and in 1981 she returned there to star in *Passion Play*, Peter *Nichols's study of adultery. She was also in *Hampton's *Tales from Hollywood* (NT, 1983).

To this already notable career she adds the distinction of being the leading English exponent of *Beckett. She was seen, at the *Royal Court or the National Theatre, in *Play* (1964), *Not I* (1973 and 1975), *Footfalls* (1976), *Happy Days* (1979), and *Rockaby* (1982), and appeared in New York in 1984 (*Riverside, 1986) in a triple bill of *Rockaby, Enough*, and *Footfalls*.

Whiting, John (1917–63), English dramatist, who first came into prominence when his play *A Penny for a Song* (1951) was produced at the *Haymarket Theatre. A light-hearted comedy set in the time of the Napoleonic wars, it was well received by the critics but failed with the public. Also in 1951 Whiting won a playwriting competition to mark the Festival of Britain with *Saint's Day*, produced at the *Arts Theatre; it was condemned by the critics for obscurity but publicly defended by theatre people such as *Gielgud, *Guthrie, and Peggy *Ashcroft who considered it intensely poetic and theatrically exciting. Whiting's other plays were *Marching Song* (1954), a darkly symbolic drama of conscience; *The Gates of Summer*, seen on tour in 1956 and in London in 1970; and *The Devils* (1961), based on a historical work, *The Devils of Loudun* by Aldous Huxley, which was staged with great success by the *RSC. In 1962 the RSC presented a rewritten and much improved version of *A Penny for a Song*, which in 1967 was used as the libretto of an opera by Richard

Rodney Bennett. Whiting also translated *Anouilh's *Madame de . . .* and *Le Voyageur sans bagage* (as *The Traveller without Luggage*), both seen in 1959, and in 1965 the *Bristol Old Vic put on *Conditions of Agreement*, found among his papers after his death, which was also produced in New York in 1972. Whiting is generally agreed to have been in advance of his time: he died just as his reputation in England was beginning to equal that he achieved on the Continent, especially in Germany.

Whitlock, Mrs, see KEMBLE, JOHN PHILIP.

Whittle, Charles, see MUSIC-HALL.

Whitty, Dame May, see WEBSTER, MARGARET.

Whitworth, Geoffrey, see AMATEUR THEATRE.

Wieth, Mogens (1919–62), Danish actor, technically and artistically one of the finest of the century. He made his début in 1939 as a member of the *Royal Theatre, Copenhagen, where he quickly established himself as a classical actor. Among the important roles he played were Orpheus in *Anouilh's *Eurydice*, the Narrator in Stravinsky's *The Soldier's Tale*, John Worthing in *Wilde's *The Importance of Being Earnest*, and Higgins in the musical *My Fair Lady*. In 1948, during the celebrations held for the 200th anniversary of the theatre, he played 10 leading parts in 18 days. In 1953 he played Helmer in English in a production of *Ibsen's *A Doll's House* in London, and in 1962 he was engaged to play Othello and Ibsen's Peer Gynt with the *Old Vic company. He had just started rehearsals when he died suddenly.

Wignell, Thomas (1753–1803), American actor of English extraction, a cousin of the younger Lewis *Hallam, by whom he was persuaded in 1774 to join the *American Company, of which he soon became the leading man. In 1787 he was instrumental in arranging the production in New York of the first American comedy, Royall *Tyler's *The Contrast*, and in 1789, in which year George Washington, who much admired him, attended his benefit night, he spoke the prologue to *Dunlap's *The Father; or, American Shandyism*, in which he played the comic doctor. Shortly afterwards he left for England, where he recruited a fine company to play at the newly opened *Chestnut Street Theatre in Philadelphia. Among its members were James *Fennell, with whom Wignell had acted in New York, and Mrs *Merry, whom he married seven weeks before his death. The company soon achieved an enviable reputation, and on a visit to New York in 1797 was considered superior even to the American Company. Wignell, however, took it back to Philadelphia, where he remained until his death.

Wilde, Oscar Fingal O'Flahertie Wills (1854–1900), Irish wit and writer, best known as a dramatist. Educated in Dublin, he then went to Oxford, where he won the Newdigate Prize for English Verse and, in spite of a reputation for idleness, took a First Class degree in Classics. By 1881 he was well enough known as a poet and aesthete to be satirized in *Gilbert and Sullivan's operetta *Patience*, and a year later he went to New York, where his first plays, *Vera; or, The Nihilists* (1882) and *The Duchess of Padua* (1891) (as *Guido Ferrandi*), a blank-verse tragedy, were produced, without much success. It was with a succession of comedies, beginning with *Lady Windermere's Fan* (1892), produced at the *St James's Theatre by George *Alexander, that he ultimately achieved fame, following up his initial success with *A Woman of No Importance* (1893) and *An Ideal Husband* (1895), both seen at the *Haymarket Theatre. His most characteristic play, *The Importance of Being Earnest* (also 1895), with Alexander as John Worthing, was seen at the St James's. In it Wilde discarded his former vein of sentiment, which he knew to be false and a concession to the taste of the times, and returned to the pure comedy of *Congreve. Though all his comedies have been revived, *The Importance of Being Earnest* wears best, and has proved the most successful. Few comedies of the English stage have such wit, elegance, and theatrical dexterity. In the year of its first production Wilde, who was married with two young sons, was sentenced to two years' imprisonment with hard labour for homosexuality, and afterwards went to Paris, where, broken in health and fortune, he soon died. His last play, the poetic one-act *Salomé*, written in French, was banned by the censor in England, but was produced in Paris in 1896 by Sarah *Bernhardt, and later used as the libretto of an opera by Richard Strauss. It was first seen privately in London in 1905, and has since been revived occasionally. Wilde left an unfinished one-act play, *A Florentine Tragedy*, completed by Sturge Moore and produced in London in 1906. In 1960 Micheál *MacLiammóir toured in a one-man entertainment based on his work, *The Importance of Being Oscar*.

Wilder, Thornton Niven (1897–1975), American novelist and dramatist. Among his earlier works for the theatre were *The Trumpet Shall Sound* (1927), about the American Civil War; four one-act plays (1931), one of which, *The Long Christmas Dinner*, provided the libretto for a short opera by Hindemith (1961); a translation of *Obey's *Le Viol de Lucrèce*, as *Lucrece* (1932), for Katharine *Cornell; and a new version of *Ibsen's *A Doll's House* (1937). His most important works for the theatre were, however, his two stimulating and provocative plays, *Our Town* (1938; London, 1946), a picture of a small American community, in which scenery was reduced to a minimum and the characters mimed instead of using props; and *The Skin of*

Our Teeth (1942; London, 1945), a survey of man's hairsbreadth escape from disaster through the ages, both of which were awarded a *Pulitzer Prize. The latter provided an excellent part for Tallulah *Bankhead in New York and Vivien *Leigh in London. Perhaps Wilder's most popular work was his adaptation of one of *Nestroy's farces, *Einen Jux will er sich machen*. Originally produced as *The Merchant of Yonkers* (1938), it was rewritten and retitled *The Matchmaker*, under which title it was seen at the *Edinburgh Festival in 1954, later that year in London, and in New York in 1955. *A Life in the Sun* (1955), on the legend of Alcestis, was commissioned by Tyrone *Guthrie for the Edinburgh Festival, and a trilogy of one-act plays, *Three Plays for Bleecker Street* (1962), was produced in New York.

Wilkinson, Norman (1882–1934), English artist, designer of some outstanding settings and costumes for the theatre. He was usually referred to as 'Norman Wilkinson of Four Oaks', to distinguish him from a contemporary marine artist of the same name. He first worked for Charles *Frohman at the *Duke of York's Theatre in 1910, but came into prominence with his designs for *Granville-Barker's seasons at the *Savoy, his permanent set for *Twelfth Night* in 1912 being much admired. Particularly memorable were his designs for *A Midsummer Night's Dream* in 1914, with its gilded fairies and iridescent forest. He was later with Nigel *Playfair at the *Lyric Theatre, Hammersmith, worked for C. B. *Cochran and for the *Phoenix and *Stage Societies, and in 1932 designed *A Midsummer Night's Dream* for the *Shakespeare Memorial Theatre, of which he was a governor.

Wilkinson, Tate (1739–1803), English actor and provincial theatre manager, a man of good family whose passion for the theatre led him to adopt the stage as a profession. He was recommended to *Garrick who engaged him for *Drury Lane, where his imitations of contemporary players were much admired, particularly by Samuel *Foote, who took him to Dublin where he was a great success. He was less well received, however, in London on his return, and eventually went into the provinces. There he took over the *York circuit, which included Hull and Leeds, and ran it for about 30 years with marked success, many actors and actresses, among them Mrs *Jordan, John Philip *Kemble, and the elder Charles *Mathews, gaining their first experience under him. As he grew older he became very eccentric, and Foote, after a violent but short-lived quarrel, satirized him as Shift in *The Minor* (1760). In spite of his many foibles he was much loved and respected and did a great deal to keep the theatre alive in provincial towns in the North of England.

Wilks, Robert (1665–1732), English actor, engaged for *Drury Lane in 1692 to take the place of William *Mountfort, who had just been assassinated. He soon became popular with the public and his fellow actors, being conscientious and hard-working, and retaining to the end the ability to play high comedy. He was the first to appear as the fine gentlemen of Colley *Cibber's comedies, and one of his best parts was Sir Harry Wildair in *Farquhar's *The Constant Couple* (1699). He was less good in tragedy, but could always move an audience to tears in parts which, like Macduff in *Macbeth*, demanded pathos. He was also much admired as Hamlet. In 1709 Wilks took over the management with Cibber and *Doggett but, although the theatre prospered, he found himself constantly at odds with the latter, and harmony was not restored until Doggett was replaced by Barton *Booth. Even then Wilks's imperious temper often caused trouble and led several actors to migrate to *Lincoln's Inn Fields.

Willard, Edward Smith (1853–1915), English actor, who was particularly admired for his villains in contemporary *melodrama. He made his first appearance on the stage in Weymouth in 1869, and after some years in the provinces was in London, where he made a great success in Sims's *The Lights o' London* (1881) and in the title-role in *Tom Pinch*, a domestic comedy based on *Dickens's *Martin Chuzzlewit*. He enhanced his reputation with his Captain Skinner in H. A. *Jones and Herman's *The Silver King* (1882), James Ralston in Charles Young's *Jim the Penman* (1886), and Jem Dalton in a revival of Tom *Taylor's *The Ticket-of-Leave Man* in 1888. In 1889 he went into management at the *Shaftesbury Theatre, where he played among other parts Cyrus Blenkarn in H. A. Jones's *The Middleman*, and then went to America, where he appeared with such success that he returned there annually for several years, touring the USA and Canada with a repertory of his most successful parts. Among his later roles were the title-roles in *Barrie's *The Professor's Love Story* (1894), Louis N. *Parker's *The Cardinal* (1903), and a dramatization of Thackeray's *The Newcomes* as Colonel Newcome (1906), in which he made his last appearance on the stage.

His nephew **Edmund** (1884–1956) was also an actor, at his best in strong dramatic parts such as Macbeth, Othello, Jones in *Galsworthy's *The Silver Box*, and Lopakhin in *Chekhov's *The Cherry Orchard*.

Williams, Bert [Egbert Austin] (c.1876–1922), American Negro entertainer and song writer, born in the Bahamas, who joined a *minstrel show and in 1895 teamed up with George Walker, also a Negro comedian. Together they perfected a double act on tour, and later appeared in it with outstanding success at

*Koster and Bial's Music Hall in New York. In 1903 Williams produced In Dahomey, the first full-length musical written and performed by Negroes to be seen in a major Broadway theatre. It was seen in London in the same year. After Walker's death in 1909 Williams gave up management and appeared in musical *revues, chiefly the Ziegfeld Follies, usually writing his own material. A gifted actor, with a rich bass singing voice, he was a serious scholarly man off stage, but for his performances blacked himself and portrayed the shuffling, shiftless Negro of tradition. He was the first to receive equal billing with Whites on Broadway, and his pioneering work for his race included the formation of an all-Negro actors' society in 1906.

Williams, Bransby, see MUSIC-HALL.

Williams, Clifford (1926–), Welsh director. After working as an actor he became a director at various theatres, including the *Arts Theatre, London. He joined the *RSC in 1961, becoming an associate director in 1963. Among his productions there were The Comedy of Errors (1963 and 1983), *Hochhuth's The Representative (1963), *Marlowe's The Jew of Malta (1964) and Dr Faustus (1968), and *O'Keeffe's Wild Oats (1976 and 1979). He also directed As You Like It (1967) and *Shaw's Back to Methuselah (1969) for the *National Theatre. The most notable of his many productions in the commercial theatre were Hochhuth's Soldiers (NY and London, 1968), Anthony Shaffer's Sleuth (London, NY, and Paris, 1970), the nude revue Oh! Calcutta! (London and Paris, 1970), and *Pirandello's Henry IV (NY, 1973; London, 1974) with Rex *Harrison. He directed Derek *Jacobi in Hugh Whitemore's Breaking the Code (1986; NY, 1987) and in Richard II (1988) and Richard III (1989). He has also written and translated several plays.

Williams, (George) Emlyn (1905–87), Welsh actor, dramatist, and theatre director, who first appeared on the stage in *Fagan's And So to Bed (1927) in London and New York, and became well known through his roles in two plays by Edgar *Wallace, On the Spot (1930) and The Case of the Frightened Lady (1931). He had already written several plays before the long run of The Late Christopher Bean (1933), his adaptation for London of Sidney *Howard's translation of Prenez garde à la peinture by René Fauchois. His Spring 1600 (1934), depicting life backstage in Shakespeare's theatre, failed, but he achieved an enormous success, as both author and actor, with Night Must Fall (1935), in which he played the homicidal Danny. He repeated it in New York the following year, the only occasion on which he acted there in one of his own plays. He was also at the *Old Vic in 1937 as Angelo in Measure for Measure, but for the next few years was seen only in his own work.

Another failure, He was Born Gay (1937), was followed by The Corn is Green (1938; NY, 1940), his best and most popular play, in which he played a young Welsh miner encouraged by a sympathetic teacher to study for university. Later plays included The Light of Heart (1940; NY, as Yesterday's Magic, 1942), about an alcoholic actor; The Wind of Heaven (1945), which postulates Christ's reincarnation as a Welsh boy in a mining village; Accolade (1950); and Someone Waiting (1953; NY, 1956). He gave a fine performance as Sir Robert Morton in *Rattigan's The Winslow Boy (1946). In 1951 he began a solo performance as Charles *Dickens, becoming famous for his impersonation of the author reading from his own works, touring the world with it and reviving it several times. In 1955 he began another solo performance, A Boy Growing Up, in which he read from the works of his fellow Welshman Dylan Thomas. It was also given in New York, and later revived in London and 59 American cities. Williams then had a series of good roles in plays by other authors: Hjalmar Ekdal in *Ibsen's The Wild Duck in 1955; Shylock in The Merchant of Venice and Iago in Othello at Stratford-upon-Avon in 1956; and the Author in Robert Ardrey's Shadow of Heroes in 1958. In 1962 he successfully took over the part of Sir Thomas More in *Bolt's A Man for All Seasons in New York, where he also played Pope Pius XII in *Hochhuth's The Deputy in 1964; in the same year his adaptation of Ibsen's The Master Builder was produced by the *National Theatre company in London. He was in his own adaptation of *Turgenev's A Month in the Country in 1965, took over from *Gielgud in Alan *Bennett's Forty Years On in 1969, and in 1977 gave another solo performance, reading the short stories of Saki (H. H. Munro).

Williams, (Ernest George) Harcourt (1880–1957), English actor and director, who made his first appearance in Belfast in 1898 with Frank *Benson, appearing also with him in London in 1900. He was with most of the leading players of the day, touring with Ellen *Terry and George *Alexander and visiting America with Henry *Irving. In 1907 he was again in London, where he was seen as Valentine in a revival of *Shaw's You Never Can Tell, and played Count O'Dowda in the first production of the same author's Fanny's First Play (1911). After service in the First World War he returned to the stage in 1919, as The Chronicler and General Lee in *Drinkwater's Abraham Lincoln. Among his later parts were the Stranger in Chesterton's Magic (1923), the Player King to John *Barrymore's Hamlet (in 1925), and the Chancellor in Ashley *Dukes's adaptation of Neumann's Der Patriot as Such Men are Dangerous (1928). He directed at the *Old Vic Theatre, 1929–34, where his innovations in Shakespearian production were first criticized and later hailed as epoch-making, based as they were upon the theories of *Granville-Barker and Williams's own feeling for the swiftness and splendour of Elizabethan verse. He also introduced Shaw into the repertory with revivals of Androcles and the Lion and The Dark Lady of the Sonnets in 1930. In 1937 he gave a fine performance as William of Sens in Dorothy L. Sayers's The Zeal of Thy House, which he also directed, and some years later returned to the Old Vic company, with which he visited New York in 1946. He celebrated his stage jubilee in 1948 while playing again in Shaw's You Never Can Tell, this time as William the Waiter.

His wife, the actress **Jean Sterling Mackinlay** (1882–1958), was prominently associated with the movement for *Children's Theatre in England.

Williams, Kenneth, see REVUE.

Williams, Tennessee [Thomas Lanier] (1911–83), American dramatist, who after a hard youth began to write plays while still a student. His main concern was the plight of the romantic soul in an unromantic world, and the plays—whose outspokenness caused much controversy—show compassion for those who find themselves unable to function in the clear light of American reality.

In 1939 Williams was awarded a prize by the *Theatre Guild for a group of four one-act plays American Blues, but after this auspicious start his first full-length play Battle of Angels (1940) was a failure and did not reach Broadway. It was later revised as Orpheus Descending (1957; London, 1959) with some success. He finally achieved recognition with The Glass Menagerie (1945; London, 1948), a sensitive study of his own mentally afflicted sister, though in the play her handicap was transmuted into a club-foot which became the image of the arbitrary crippling of individual desires. It was followed by A Streetcar Named Desire (1947; London, 1949), a powerful study of the clash between the old and new America—symbolized by the pathetic, self-deceiving Blanche Du Bois and her brutish brother-in-law—which was awarded a *Pulitzer Prize. His next plays were Summer and Smoke (1948; London, 1951), about a prim spinster's fall from grace, and The Rose Tattoo (1951; London, 1959), about a spirited Sicilian-American widow who idealizes her married life. Camino Real (1953; London, 1957) was an unsuccessful essay in *Symbolism based on Don Quixote, and Cat on a Hot Tin Roof (1955; London, 1958) an intense drama of family relationships set on a plantation in the Mississippi Delta; the latter received another Pulitzer Prize. Suddenly Last Summer, produced with Something Unspoken in a double bill entitled Garden District (NY and London, 1958), featured cannibalism; and Sweet Bird of Youth (1959), a study of a fading film star, involved castration. Period of Adjustment (1960; London, 1962) was a marital comedy which of all Williams's works probably most resembles an orthodox Broadway offering; but The Night of the Iguana (1961; London, 1965), set in a Mexican hotel, returned to a more exotic setting. The following decade was not a happy one for Williams, whose talent seemed to be slipping away as he battled through a series of personal crises. The Milk Train Doesn't Stop Here Anymore (Spoleto, 1962; NY, 1963), the double-bill Slapstick Tragedy (1966), The Seven Descents of Myrtle (1968), and In the Bar of a Tokyo Hotel (1969) had only short runs in New York. The pungent Small Craft Warnings (1972; London, 1973) seemed to indicate the return of dramatic vigour; but the production of The Red Devil Battery Sign destined for New York in 1975 failed to reach there, though the play was seen in London in 1977. Williams's last works were The Eccentricities of a Nightingale (1976)—a revision of Summer and Smoke—Vieux Carré (1977; London, 1978), set in a New Orleans boarding house, A Lovely Sunday for Creve Coeur (1979), and Clothes for a Summer Hotel (1980), a portrayal of Scott and Zelda Fitzgerald.

Williamsburg, Va., probably the first English settlement in the New World to have a theatre. In 1716 a merchant named William Levinston, assisted by his indentured servant Charles Stagg and Stagg's wife Mary, formerly actors and teachers of dancing and elocution in England, produced some plays, one of which was possibly given in 1718 before the Governor. Little is known of the building in which they were acted. If it was specially erected it would be the first theatre in America, a distinction claimed also by the *Southwark Theatre in Philadelphia; but it may have been adapted from an existing building. From the evidence that remains it appears that the auditorium was only 30 ft. wide by 86 ft. long, with two tiers of boxes on either side and a gallery at the back. It continued in occasional use as a playhouse until 1745, and was demolished about 1769. A second playhouse, located at the back of the Capitol building, was opened in 1751 by Walter Murray and Thomas *Kean for their theatrical company. They met with little success, perhaps because of the poor quality of their productions, and left Williamsburg after two seasons. Their theatre was later remodelled by the elder Lewis *Hallam, who opened there in 1752 with The Merchant of Venice and a farcical *after-piece by *Ravenscroft, The Anatomist; or, The Sham Doctor. It is not known how long they stayed, or when the building was demolished.

Williamson, David (1942–), Australian dramatist, leader of the Australian dramatic renaissance of the 1970s, whose work is noted for its naturalistic, witty (and bawdy) dialogue.

His first play to be produced was *The Removalists* (1971; London, 1973; NY, 1974), a fierce attack on the Australian police, after which came *Don's Party* (1972; London, 1973), a study of a group of middle-class liberals at a party on the eve of an Australian election; *Jugglers Three* (also 1972); *What If You Died Tomorrow?* (1973; London, 1974), in which a man gives up medicine for novel-writing; and *The Department* (1974). *The Club* (1977; NY, as *Players*, 1978; London, 1980) is about the battle for control of a rugby club, while *Travelling North* (1979; London, 1980) portrays a love affair between a man in his seventies and a woman 20 years younger. Prolific and successful, he continued to write with conviction on various aspects of Australian life. In *The Perfectionist* (1982; London, 1983) an academic becomes a house-husband; in *Sons of Cain* (1985; London, 1986) a newspaper editor attempts to expose government corruption; and in *Emerald City* (1987; London, 1988) a screenwriter loses his idealism.

Williamson, James Cassius (1845–1913), American-born actor-manager, founder of an important Australian theatrical management. With his wife **Maggie Moore** [Margaret Sullivan] (1851–1926) he was appearing at the California Theatre, San Francisco, when in 1874 George *Coppin engaged them both for a tour of Australia. They made their début there in *Struck Oil; or, The Pennsylvania Dutchman*, expanded from an original short play by Sam W. Smith, and its immense popularity forced them to revive it constantly, much of it being rewritten by Williamson in performance. It is said to have made a profit of £10,000 in Australia, and was also staged elsewhere, having a run of 100 performances at the *Adelphi in London in 1876. Williamson returned to Australia in 1879 with the Australasian rights to *Gilbert and Sullivan's *HMS Pinafore*, in which he played Sir Joseph Porter, and he later obtained the Australasian rights to all the Savoy operas. In 1882 he founded J. C. Williamson Theatres Ltd., and eventually concentrated almost exclusively on its administration, acting only on rare occasions. The company attained a dominant position in Australian theatre which it retained until 1976.

Will's Coffee House, also known as the Rose Tavern. This was situated at No. 1, Bow Street, Covent Garden, on the west side, at the corner of Russell Street. It was kept by Will Unwin, and the Restoration wits resorted to it after the play. They mainly congregated in a room on the first floor. There are many references in contemporary literature to this famous tavern, which was frequented by Dr *Johnson and often mentioned in the *Spectator*.

Wilson, August (1945–), major American playwright, son of a White father and a Black mother, who identifies wholly as Black and writes about the Black experience in America. His *Jitney* was staged in 1978 by Black Horizons Theatre which he founded himself in 1968. He achieved prominence however with *Ma Rainey's Black Bottom* (1984; *National Theatre, 1989), in which a session in a Chicago recording studio in 1927 by the Black blues singer Ma Rainey and her back-up band is used as a symbol of White exploitation. The setting of *Fences* (1987; London, 1990), which won the *Pulitzer Prize, is a run-down house in Pittsburgh (Wilson's home town), the action covering 1957–65. The leading character, played originally by James Earl *Jones, is a Black former baseball player, prevented by racial prejudice from gaining the recognition he deserves, who works as a garbage collector and vents his frustration on his family. *Joe Turner's Come and Gone* (1988; London, 1990) is set in 1911 in a Pittsburgh boarding-house used by Blacks in transit from the South. In 1990 Wilson won a second Pulitzer Prize for *The Piano Lesson*, set in 1936, about a family dispute over a piano. He intends eventually to write a play set in each decade of the 20th century. He is also a poet.

Wilson, Georges, see THÉÂTRE NATIONAL POPULAIRE.

Wilson, Lanford Eugene (1937–), prolific American playwright, who began writing plays while at the University of Chicago, 1957–8. His early works, many of them one-acters, were seen at the *Caffè Cino and Café *La Mama, his first full-length work being *Balm in Gilead* (1964), set in a seedy all-night coffee shop. The short-lived *The Gingham Dog* (1969) was his first Broadway production. He was one of the founders of the *Circle Repertory Company, which premièred several of his later plays, including the highly successful *The Hot l Baltimore* (1973; London, 1976), a sympathetic study of characters in the lobby of a shabby hotel (symbolized by the absence of the 'e' from its electric sign). *Serenading Louie* (1976) is a study of two failing marriages. In *The 5th of July* (1978) a group who were radical students in the 1960s come together in the 1970s at the home of a legless Vietnam veteran. *Talley's Folly* (1979; London, 1982), a *Pulitzer Prize-winner, and *A Tale Told* (1981; revised as *Talley and Son*, 1985) have character linkages with the previous play, but take place (on the same evening) over 30 years earlier. *Angels Fall* (1982) depicts six people taking shelter during a nuclear scare in New Mexico; and in *Burn This* (1987; London, 1990) a tough restaurant manager collects the belongings of his dead homosexual brother, a dancer, and falls for his dancing partner.

Wilton, Marie, see BANCROFT.

Wind Machine, see SOUND EFFECTS.

Windmill Theatre, London, in Great Windmill Street, near Piccadilly Circus. This intimate theatre, which seated 326 on two levels, opened in 1931 in a converted cinema which dated from 1910, its first production being Michael Barrington's *Inquest*. In 1932 Laura Henderson and her manager Vivian Van Damm introduced a policy of continuous variety entitled *Revudeville*, with performances running from 2.30 p.m. to 11 p.m. The Windmill Girls, who appeared in nude tableaux, became a part of London life, and many famous comedians, among them Jimmy Edwards, Harry Secombe, and Tony Hancock, served their apprenticeship there. The Windmill was the only theatre to remain open throughout the Second World War, and became renowned for its slogan 'We Never Closed'. After the death of Vivian Van Damm in 1960 *Revudeville* continued under the management of his daughter Sheila until it finally ended in 1964. The building then reverted to its former role as a cinema until 1974, when it was taken over by Paul Raymond and reopened as a theatre with a comedy entitled *Let's Get Laid* which ran into 1976. It was followed by a nude revue *Rip-Off* which ran until, in 1981, the theatre closed for conversion to a theatre-restaurant.

Wings, pairs of *flats placed at each side of the stage, either facing or obliquely towards the audience. As many as eight pairs could be used at any one time, though three or four was the usual number. A form of them was used in Italy in the late 16th century by Bernardo Buontalenti, who designed the *scenery for the Florentine *intermezzi* of 1589. Inigo *Jones, on his return from the Continent, introduced them, as Side Scenes or Side Shutters, into his settings for Court *masques. From there they descended to the Restoration theatre, and, used in conjunction with a *backcloth and *borders, remained the basic elements of scenery until the introduction in 1832 of the *box-set. They are still used for *pantomime, big spectacular shows, opera, and ballet. On the Continent they were moved by the *carriage-and-frame method, but in England they usually ran in *grooves, though in some early Victorian theatres *book wings were used.

Winter, William (1836–1917), American dramatic critic, whose conservatism made him the transatlantic counterpart of Clement *Scott. He was the most powerful dramatic critic of his time in the United States, probably because he rarely committed himself to an opinion at variance with that of the great majority of his readers. Although he admired English actors, he was antagonistic to such foreign visitors as *Duse, *Bernhardt, and *Réjane, dragging irrelevancies concerning their private lives into his notices. He also denounced *Ibsen ('slimy mush'), *Maeterlinck ('lunacy'), and *Shaw.

Winter Garden Theatre, London, in Drury Lane, originally the *Middlesex Music Hall,

later adapted as a theatre with a three-tier auditorium seating 1,581. It opened in 1919 with *Kissing Time*, starring Leslie *Henson, who appeared in most of the subsequent productions up to 1926, including *Sally* (1921), *Primrose* (1924), *Tell Me More* (1925), the last two having music by George *Gershwin, and *Kid Boots* (1926). In 1927 a musical play *The Vagabond King* began a long run. The few later productions of note included *It's Time to Dance* (1943) with Jack *Buchanan and Joan Temple's *No Room at the Inn* (1946) about evacuee children. Agatha *Christie's *Witness for the Prosecution* (1953) gave the theatre another long run. Among later productions were *Feydeau's *Hotel Paradiso* with Alec *Guinness, *Shaw's *The Devil's Disciple* (both 1956), and *O'Neill's *The Iceman Cometh* (1958). After a Christmas production of a musical version of Lewis Carroll's *Alice in Wonderland* (1959) the theatre closed, standing empty until it was demolished in 1965. The *New London Theatre now occupies the site.

Winter Garden Theatre, New York, on Broadway, between 50th and 51st Streets, on the site of the American Horse Exchange. This theatre, seating 1,479, was opened by the *Shuberts in 1911 with *La Belle Paree*, which began a 14-year association with Al Jolson (see REVUE). It was used mainly for musicals and revues and then as a cinema, but returned to live theatre in 1933. It has since housed several successful musicals, among them *West Side Story* (1957), *Funny Girl* (1964), and *Mame* (1966). *Sondheim's *Follies* was seen in 1971, and in 1974 the theatre staged one of its rare straight plays when Zero *Mostel starred in *Ulysses in Nighttown*, based on James Joyce's *Ulysses*. It returned to musicals later in the year with a revival of *Gypsy*. Sondheim's *Pacific Overtures* (1976) had a disappointingly short run, but *42nd Street* (1980) and *Cats* (1982) proved enormously popular.

The *Metropolitan Theatre under *Boucicault in 1858 was renamed the Winter Garden, retaining the name until burnt down in 1867.

Winthrop Ames Theatre, New York, see HELEN HAYES THEATRE 2.

Withers, Googie [Georgette Lizette] (1917–), English actress, born in India, who studied for the stage under Italia *Conti and made her début in the children's play *The Windmill Man* (1929) by Frederick Bowyer. She was then seen in the chorus of the musical comedy *Nice Goings On* (1933) before appearing in J. B. *Priestley's *Duet in Floodlight* (1935); after N. C. *Hunter's *Ladies and Gentlemen* (1937) she was in another Priestley play *They Came to a City* (1943). In 1945 she took over the part of Amanda in *Coward's *Private Lives*, following it with Lee in Ronald *Millar's farcical comedy *Champagne for Delilah* (1949) and Georgie Elgin in *Odets's *Winter Journey* (1952), giving one of her finest performances as the supportive wife of an alcoholic actor. Later in 1952 she succeeded Peggy *Ashcroft as Hester Collyer in *Rattigan's The Deep Blue Sea, a part in which in 1955 she toured Australia and New Zealand with her Australian-born actor husband John McCallum (1914–), who partnered her in several later productions. In 1958 she played Gertrude in Hamlet and Beatrice in Much Ado about Nothing at the *Shakespeare Memorial Theatre. In the same year her husband entered theatrical management with J. C. *Williamson Theatres Ltd. in Australia, and much of her later acting has been done in Australia and New Zealand. She made her début in New York in 1961 in Graham *Greene's The Complaisant Lover and was seen in Edinburgh and London in 1963 in *Ionesco's Exit the King, and at the *Strand Theatre in 1967 in *Shaw's Getting Married. She then returned to Australia for some years, touring and playing Ranevskaya in *Chekhov's The Cherry Orchard (1972) for the *Melbourne Theatre Company. In 1976 she was back in England, appearing at *Chichester and in London as Lady Kitty in *Maugham's The Circle. She was again at Chichester as Lady Bracknell in *Wilde's The Importance of Being Earnest (1979) and in such productions as Priestley's Time and the Conways (1983) and *Anouilh's Ring round the Moon (1988).

Witkiewicz [Witkacy], **Stanisław Ignacy** (1885–1939), Polish artist, novelist, and dramatist, who led a stormy and unhappy life, mainly in exile, travelling widely in Russia and visiting Australia. He committed suicide when the Nazis invaded Poland. His plays, written under the influence of *Expressionism, were performed only in small experimental theatres during his lifetime and made very little impact. Rediscovered after the Second World War, they were recognized as Surrealistic, foreshadowing not only the work of *Brecht but containing also the seeds of *Beckett's and *Ionesco's Theatre of the *Absurd. Most of them were then revived or staged for the first time, and they were widely translated, being hailed as masterpieces. A volume published in America in 1968, six years after the appearance of his complete works—30 plays written between 1918 and 1926—in a two-volume Polish edition, contained six plays: The Shoemakers, first seen in 1957, revived in 1971; The Crazy Locomotive and The Mother, both seen in 1964, the former being produced in New York in 1977 and the latter revived by *Axer in 1970; They, seen in 1965; The Madman and the Nun, which had its first performance in German in Vienna in 1962 and was also staged in New York in 1979; and The Water Hen, revived in 1967. Among his other works are The Cuttlefish, staged in 1957; Gyubal Wahazar, revived in Warsaw in 1972; Lovelies and Dowdies, staged in 1973; and a

five-act tragedy The Pragmatists, written in 1918 and perhaps the most frequently revived.

Wodehouse, P. G., see MUSICAL COMEDY.

Woffington, Peg [Margaret] (c.1718–60), Dublin-born actress. She was engaged at the age of 12 by Mme Violante, a famous rope-dancer, to play in a children's company, being seen among other parts as Polly in *Gay's The Beggar's Opera. In 1732 she made her first appearance in London in the same work, but as Macheath, and then returned to Dublin, where she eventually joined the company at the Smock Alley Theatre. There she played a wide range of parts, including the one in which she was to become famous, Sir Harry Wildair in *Farquhar's The Constant Couple, in which she was first seen in 1740. Later the same year she was engaged by John *Rich for *Covent Garden, where she played Sir Harry with such spirit and elegance that for a long time no male actor dared attempt the part. A year later she was at *Drury Lane, making her first appearance in another *male impersonation, Silvia in Farquhar's The Recruiting Officer, and from then on divided most of her time between the two theatres. Her naturally harsh voice rendered her unfit for tragedy, but in comedy she was outstanding, being much admired as Millamant in *Congreve's The Way of the World and as Lady Townly and Lady Betty Modish in Colley *Cibber's The Provok'd Husband and The Careless Husband. She was the most beautiful and least vain actress of her day, but her good nature did not extend to her fellow actresses and she was constantly at odds with Kitty *Clive, Theophilus *Cibber's wife Susanna, and George Anne *Bellamy. For some years she was the mistress of *Garrick, who wrote for her the charming song 'My Lovely Peggy', and played opposite him both in London and in Dublin, where from 1752 to 1754 she appeared triumphantly in many of her most famous roles. She returned to Covent Garden at the end of 1754 as Maria in Cibber's The Non-Juror with her usual success, though she suffered somewhat from the rising popularity of Mrs Bellamy. The last male part she played was Lothario in *Rowe's The Fair Penitent, and she was also seen as Lady Randolph in *Home's Douglas when it was first seen in London in 1757, with Spranger *Barry as Young Norval. She made her last appearance as Rosalind in As You Like It in 1757, being taken ill at the beginning of the epilogue. She lingered on for three years and gave herself to good works, endowing almshouses at Teddington where she died. She is the subject of the play Masks and Faces (1852) by Tom *Taylor and Charles *Reade.

Wolfenbüttel, capital city of the Duchy of Brunswick, one of whose dukes, **Heinrich Julius** (1564–1613), was a patron of the *English Comedians. They visited him some time in the

1580s and a group of them under Thomas Sackville was attached to the ducal household from 1596 until the Duke's death. They appeared in plays written by the Duke, of which 11 are extant, dealing mainly with matrimonial discord, usually with the onus on the female side unless the husband, through sheer stupidity, deserved his punishment. Some of them are written for the old-fashioned *multiple set, but others are in the new-style single set; they show considerable English influence. The *Fool figuring in all of them is endowed with sound common sense, and in *Vincentius Ladislaus* (1595) gets the better of the braggart soldier or *Capitano imported from the *commedia dell'arte*. After the death of Heinrich Julius theatrical activity ceased, and it was not until 1688 that a theatre was built near the castle by Duke Anton Ulrich, a devotee of French and Italian light opera, who also built an indoor theatre in his new château at Salzdahlum nearby, with an open-air theatre in its garden.

Wolfit [Woolfitt], **Sir Donald** (1902–68), English actor-manager. He made his first appearance on the stage in 1920 and four years later was seen in London as Phirous in Temple Thurston's *The Wandering Jew*. In 1929–30 he was at the *Old Vic, then toured Canada as Robert Browning in Besier's *The Barretts of Wimpole Street*, and was seen in London again as Thomas Mowbray in Gordon Daviot's *Richard of Bordeaux* (1933). He was at *Stratford-upon-Avon in 1936, and the following year formed his own company, touring extensively in a Shakespearian repertory in which he also appeared for a season at the *Kingsway Theatre in 1940. During the Battle of Britain he gave over 100 lunch-time performances of scenes from Shakespeare, and he later continued to tour in Shakespeare and other classics, two of his best performances being the title-role in *Jonson's *Volpone* and Sir Giles Overreach in *Massinger's *A New Way to Pay Old Debts*. He was also excellent in the title-role of *Marlowe's *Tamburlaine* at the Old Vic in 1951, and as Lord Ogleby in *Colman and *Garrick's *The Clandestine Marriage*. In 1957 he appeared at the *Lyric Theatre, Hammersmith, in two plays by *Montherlant, *The Master of Santiago* and *Malatesta*.

His wife, **Rosalind Iden** (1911–90), was an excellent actress who appeared with him in many of his productions.

Wood, Charles Gerald (1932–), English playwright, noted for his inventive dialogue and for the military background of many of his plays—he served for five years in the regular army. His first work for the theatre was *Cockade* (1963), a triple bill, *Prisoner and Escort*, *John Thomas*, and *Spare*, featuring different aspects of army life. The farcical fantasy *Meals on Wheels* (1965) and *Fill the Stage with Happy Hours* (1966), which has a background of old-time weekly *repertory, were followed by two war plays: *Dingo* (*Royal Court, 1967), set partly in a North African prisoner-of-war camp; and *H, or Monologues at Front of Burning Cities* (*National Theatre, 1969), a savage indictment of Havelock's march to the relief of Lucknow during the Indian Mutiny. Another triple bill *Welfare* (1970) deals with the relationship between youth and age pre- and post-Second World War. *Veterans* (Royal Court, 1972), which starred John *Gielgud and John Mills as veteran actors, is set on a film location in Turkey, and arose out of Wood's experiences as script writer for the film *The Charge of the Light Brigade*. *Jingo* (*RSC, 1975) is an attack on the complacency of the British in Singapore in 1941. He returned to a film-making background with *Has 'Washington' Legs?* (NT, 1978); *Red Star* (RSC, 1984), about film-making in the Soviet Union; and *Across from the Garden of Allah* (1986), a comedy about an English screenwriter and his wife in Hollywood. His adaptation of *Pirandello's *Man, Beast and Virtue* was staged at the National Theatre in 1989.

Wood, John (date of birth undisclosed), English actor, who was on the stage for some years after leaving *Oxford, where he was President of OUDS, before his excellence was recognized. He was with the *Old Vic company, 1954–6, and made his West End début as Don Quixote in Tennessee *Williams's *Camino Real* (1957). He made his first appearance in New York in 1967 as Guildenstern in *Stoppard's *Rosencrantz and Guildenstern are Dead* (playing the Player King there in a further production in 1987). His outstanding performance in James Joyce's *Exiles* (1970), in which he uncovered the layers of meaning hidden within the text, first revealed his true quality. He then joined the *RSC, 1971–2, playing Yakov in *Gorky's *Enemies*, Sir Fopling Flutter in *Etherege's *The Man of Mode*, Brutus in *Julius Caesar*, and a psychotic Saturninus in *Titus Andronicus*. After his spindly, lecherous, and slightly manic husband in John *Mortimer's *Collaborators* (1973) he returned to the RSC in the title-role of William *Gillette's *Sherlock Holmes* and as Henry Carr in Stoppard's *Travesties* (both 1974), repeating the roles in New York in 1974 and 1975 respectively. In 1976 he was seen with the RSC in *Shaw's *The Devil's Disciple* and in the title-role of *Chekhov's *Ivanov*, and in 1978 he had an enormous success in New York in Ira Levin's thriller *Deathtrap*. Back in London in 1979, he gave an electrifying performance as Richard III with the *National Theatre company, for which he also appeared in *Schnitzler's *Undiscovered Country* in Tom Stoppard's adaptation, and in 1980 in *Vanbrugh's *The Provoked Wife*. In 1981 he returned to New York to take over from Ian *McKellen in Peter *Shaffer's

Amadeus. Again with the RSC, he was superb as Prospero in *The Tempest* in 1988 and as Solness in *Ibsen's *The Master Builder* in 1989.

Wood, Mrs John [*née* Matilda Charlotte Vining] (1831–1915), English actress and manageress belonging to a well known theatrical family, being first cousin to Fanny Vining, the mother of Fanny *Davenport. She made her first appearance in Brighton as a child, and had already made a good reputation in the English provinces as an adult actress when in 1854 she went to America, where in New York in 1856 she achieved popularity as Minnehaha in Charles Melton Walcot's *burlesque *Hiawatha*. Widowed in 1863, she took over the *Olympic Theatre in New York and ran it successfully for three years, leaving it to make her first appearance on the London stage in 1866 as Miss Miggs in a dramatization of *Dickens's *Barnaby Rudge*. From 1869 to 1877 she was manageress of the *St James's Theatre in London, where she was highly respected by her company and her audiences. She made her last appearance in New York under *Daly in 1873, and thereafter was seen only in London. She was at the *Royal Court Theatre from 1883 to 1892, appearing in the first productions of *Pinero's farces *The Magistrate* (1885), *The Schoolmistress* (1886), and *Dandy Dick* (1887). A woman of liberal views, she ruled her company kindly but firmly and spared no expense in the running and equipping of her theatre and actors.

Wood, Peter Lawrence (1928–), English director. In 1956–7 he was resident director at the *Arts Theatre, and in 1958 he directed *O'Neill's *The Iceman Cometh* and *Pinter's *The Birthday Party* in London. He made his début with the *RSC with *The Winter's Tale* (1960) and John *Whiting's *The Devils* (1961) and with the *National Theatre in 1964 with *Ibsen's *The Master Builder*. For the National he demonstrated his skill with *Restoration Comedy in *Congreve's *Love for Love* (1965 and 1985) and *The Double Dealer* (1978), *Vanbrugh's *The Provoked Wife* (1980), and *Farquhar's *The Beaux' Stratagem* (1989). His work for the company also included *Schnitzler's *Undiscovered Country* (1979), *Sheridan's *The Rivals* (1983) and *The School for Scandal* (1990), and Arthur *Miller's *The American Clock* (1986). After directing *Stoppard's *Jumpers* at the National in 1972 (NY, 1974), he was involved, with Carl *Toms as designer, in many more productions of Stoppard's work. Among his other productions were two double bills by Peter *Shaffer, *The Private Ear* and *The Public Eye* (1962; NY, 1963) and *White Liars* and *Black Comedy* (1968); and *The Prime of Miss Jean Brodie* (1966), adapted from Muriel Spark's novel.

Wood, William Burke (1779–1861), American actor, born in Montreal, Canada, where his parents, English actors who had emigrated to America, had taken refuge during the War of Independence. They later returned to the USA, where their son became the first native-born North American to achieve an important place in American theatre history. In partnership with William *Warren he was for many years manager of the *Chestnut Street Theatre in Philadelphia, and a diary which he kept from 1810 to 1833 contains much interesting material on the theatre. The list of plays given there each season shows in the earlier years a preponderance of Shakespeare and English classic comedies, perhaps because of Wood's own admirable playing of polished comedy, though he was also good in the lighter parts of tragedy. In later years more new American plays were produced.

Wood's Museum, New York, see DALY'S THEATRE 1.

Woodward, Harry (1717–77), English actor, who made his first appearance on the stage as a child, playing small parts with a juvenile company in London. A year later he was engaged by John *Rich for *Covent Garden where, as an excellent dancer and mime, he was known as Lun junior. In 1738 he went to *Drury Lane where he remained for 20 years, writing a number of *pantomimes for *Garrick in which he himself played *Harlequin. Other parts in which he excelled included Mercutio in *Romeo and Juliet*, Touchstone in *As You Like It*, Petruchio in Garrick's version of *The Taming of the Shrew*, Bobadil in *Jonson's *Every Man in His Humour*, and Marplot in his own adaptation of Mrs *Centlivre's *The Busybody*. In 1758 he went to Dublin, and with *Macklin (who soon withdrew) and Spranger *Barry took over the management of the Crow Street Theatre, where they ruined themselves. Returning to London in 1763, Woodward went back to Covent Garden, where in 1775 he created the part of Captain Absolute in *Sheridan's *The Rivals*. He remained on the stage until he died and was the last of the great Harlequins, before *Grimaldi made *Clown the chief figure in the pantomime.

Woollcott, Alexander Humphreys (1887–1943), American dramatic critic, reputedly the prototype of Sheridan Whiteside in *The Man Who Came to Dinner* (1939) by *Kaufman and *Hart. He worked as a police reporter on the *New York Times*, graduating to the drama department and becoming drama critic at the age of 27. In 1922 he went to the *New York Herald* and the following year was transferred to the *Sun*. From 1925 to 1928 he was dramatic critic for the *New York World*; he then retired to devote himself to broadcasting, lecture tours, and writing magazine articles. As a dramatic critic his judgements were capricious. He was more interested in players than in plays and missed or misjudged many important works of his time, including the plays of *O'Neill which he dismissed as completely worthless. He himself turned playwright on two occasions, collaborating with Kaufman on *The Channel Road* (1929) and *The Dark Tower* (1932), and appeared on the stage three times—in *Behrman's *Brief Moment* (1932) and *Wine of Choice* (1938), and in the touring company of *The Man Who Came to Dinner*. Though his criticism was ephemeral, he wrote engagingly, and his volumes of fugitive pieces were popular.

Worcester's Men, see QUEEN ANNE'S MEN.

World Theatre Season, annual festival of foreign plays in the original, performed by foreign companies invited to London by Peter *Daubeny and housed in the *Aldwych Theatre during the absence of the *RSC. It ran from 1964 to 1973, with a final season in 1975, the first season being part of the celebrations for Shakespeare's Quatercentenary. It was not confined to European theatre but included visits by companies from Turkey, India, and Israel, and evenings devoted to the Japanese *nō and *kabuki.

Worth, Irene (1916–), American actress, formerly a teacher, who was not seen on the stage until 1942, when she toured with Elisabeth *Bergner in Margaret Kennedy's *Escape Me Never*. A year later she made her New York début in Martin Vale's *The Two Mrs Carrolls*, and in 1944 she was in London, where she spent most of the next 30 years, her first appearance being in *Saroyan's *The Time of Your Life* (1946). She created the role of Celia Coplestone in T. S. *Eliot's *The Cocktail Party* (1949) at the *Edinburgh Festival, repeating it in New York and London, and then joined the *Old Vic company, where her Helena in *A Midsummer Night's Dream* (1951) showed a strong talent for comedy and her other roles included Desdemona in *Othello* (also 1951) and Portia in *The Merchant of Venice* (1953). After appearing in the first season of the *Stratford (Ontario) Festival in 1953 she was seen in London in N. C. *Hunter's *A Day by the Sea* (also 1953) and *Betti's *The Queen and the Rebels* (1955). Two contrasting roles—Marcelle in *Feydeau's *Hotel Paradiso* (1956) and the title-role in *Schiller's *Mary Stuart* (NY, 1957)—were followed by Sara Callifer in Graham *Greene's *The Potting Shed* (1958) before she again played Mary at the Old Vic (also 1958). After her Rosalind in *As You Like It* at Stratford, Ontario, in 1959, she starred in New York in Lillian *Hellman's *Toys in the Attic* (1960), and then joined the *RSC in London, playing Lady Macbeth, Goneril in Peter *Brook's production of *King Lear* (both 1962), and Dr Zahnd in *Dürrenmatt's *The Physicists* (1963). During another visit to New York she starred in *Albee's *Tiny Alice* (1964; RSC, 1970) and she reappeared in the West End in three roles in *Coward's *Suite in Three

Keys (1966). After an excellent performance as Hesione Hushabye in *Shaw's *Heartbreak House* in 1967, she showed great emotional force as Jocasta in *Sophocles' *Oedipus* at the Old Vic in 1968, and returned to Stratford, Ontario, in the title-role in *Ibsen's *Hedda Gabler* in 1970. Her gift for comedy gave an unusual piquancy and charm to her Arkadina in *Chekhov's *The Seagull* at *Chichester in 1973. In 1975 she returned to New York, where she was seen in Tennessee *Williams's *Sweet Bird of Youth*, as Ranevskaya in Chekhov's *The Cherry Orchard* (1977), as Winnie in *Beckett's *Happy Days* (1979), and as Ella in Ibsen's *John Gabriel Borkman* (1980). She later appeared in London as Volumnia in *Coriolanus* (*National Theatre, 1984) and in Shaw's *You Never Can Tell* (1987). She again played Volumnia in New York in 1988.

Wycherley, Sir William (1640–1716), Restoration dramatist, whose comedies, though coarse and often frankly indecent, show so much strength and savagery in their attack on the vices of the day that their author has been labelled 'a moralist at heart'. He was to some extent influenced by *Molière, but transmuted his borrowings by his own particular genius, and his style has an individuality seldom found in other writers of his time. His first play, *Love in a Wood; or, St James's Park* (1671), was followed by the somewhat uncharacteristic *The Gentleman Dancing-Master* (1672), based on *Calderón's *El maestro de danzar*. The best of his plays was *The Country Wife* (1675), in which the comedy hinges on the efforts of a jealous husband to keep his young but naturally wanton country wife from the temptations of London. It was a great success, and was revived many times up to 1748. It received a new lease of life when *Garrick adapted it as *The Country Girl* (1766) and produced it at *Drury Lane with Dorothy *Jordan in the title-role. The original version has been seen in London several times since the *Phoenix Society first revived it in 1924. Wycherley's last play, *The Plain-Dealer* (1676), was successful enough for its author to be nicknamed 'Manly' Wycherley, after the name of its hero, with whose outspoken and misanthropic bent he doubtless had much sympathy. It was occasionally revived, and in 1765 was revised by Isaac *Bickerstaffe for Drury Lane; his version held the stage until 1796.

Wyndham [Culverwell], **Sir Charles** (1837–1919), English actor-manager. He trained as a doctor, but his success in several amateur productions, under the name of Charles Wyndham which he later adopted legally, caused him to abandon medicine for the theatre, and he made his first appearance on the professional stage at the *Royalty Theatre in 1862. He then went to the USA, where the Civil War was at its height, and enlisted in the Federal army as a surgeon. He twice resigned in 1863 to appear on stage, playing Osric to John Wilkes *Booth's Hamlet in Washington and Thomas Brown to Mrs John *Wood's Pocahontas in a revival of *Barker's The Indian Princess in New York. After playing Charles Surface in *Sheridan's The School for Scandal at *Wallack's Theatre in 1869 he soon established a reputation as a light comedian, and from 1871 to 1873 led his own comedy company on an extended tour of the Middle West. Back in London in 1874 he made a great success in Bronson *Howard's Saratoga (1870), in which he had appeared in America, renaming it Brighton. It was moved to the *Criterion Theatre, with which Wyndham was to have a lifelong association, another success there being *Albery's farce The Pink Dominoes (1877). He made it one of the foremost play-houses of London, and later built and managed the New Theatre (now the *Albery) and *Wyndham's Theatre with equal success. In 1883 he took his London actors on a tour of America, the first completely English company to visit California and the Far West. On his return to England he was responsible for the production of many interesting new plays. A tall, handsome man, with a mobile, expressive face, he was at his best in high comedy but could play serious roles with conviction. One of his finest roles was David Garrick in T. W. *Robertson's play of that name, which he first revived in 1886 and made his own. He was also outstanding in Henry Arthur *Jones's The Case of Rebellious Susan (1894), The Liars (1897), and Mrs Dane's Defence (1900), Louis N. *Parker and Murray Carson's Rosemary (1896), and Hubert Davies's The Mollusc (1907).

In this last play his leading lady was a fine actress, **Mary Moore** (1869–1931), the widow of James *Albery, whom he married as his second wife in 1912. She appeared in many of his productions, and after his death continued to manage his theatres, leaving them on her death to the care of her son Bronson *Albery and his stepbrother Howard Wyndham, Wyndham's son by his first wife.

Wyndham's Theatre, London, in Charing Cross Road. Built for Charles *Wyndham, it seats 770 and has the last complete picture-frame surround in London. It opened in 1899 with a revival of T. W. *Robertson's David Garrick, in which Wyndham and his wife Mary Moore had already appeared. The first new play was a translation of *Rostand's Cyrano de Bergerac in 1900. The early years were notable for productions of Henry Arthur *Jones's Mrs Dane's Defence (1900); Charles Marlowe's When Knights Were Bold (1907); and the controversial war play An Englishman's Home (1909), written by Guy Du Maurier. Frank Curzon became manager in 1903, and in 1910 Gerald *Du

Maurier became joint manager, appearing in 1913 in a revival of *Sardou's Diplomacy. He then played Dearth in *Barrie's Dear Brutus (1917), and in 1921 scored an outstanding success as Bulldog Drummond in the play of that name by 'Sapper' (H. C. McNeile). In 1926 the first of six plays by Edgar *Wallace was staged. The stepbrothers Howard Wyndham and Bronson *Albery took over in 1931, their productions including Savory's George and Margaret (1937), Esther McCracken's Quiet Wedding (1938), and its sequel Quiet Week-End (1941), the last being the first play at this theatre to have more than 1,000 performances. The first post-war success was *Bridie's Daphne Laureola (1948), with Edith *Evans; Peter *Ustinov had a personal triumph in his own play The Love of Four Colonels (1951), as did Dorothy *Tutin in Graham *Greene's The Living Room (1953). Sandy Wilson's musical The Boy Friend began a long run in 1954. It closed in 1959 and was succeeded by Shelagh Delaney's A Taste of Honey, the first of four transfers of productions by *Theatre Workshop, which was followed by Brendan *Behan's The Hostage, Stephen Lewis's Sparrers Can't Sing in 1961, and Oh, What a Lovely War! in 1963. Joe *Orton's Entertaining Mr Sloane was seen in 1964; John *Osborne's Inadmissible Evidence and an adaptation of Muriel Spark's novel The Prime of Miss Jean Brodie, with Vanessa *Redgrave, both followed in 1965. Ronald *Millar's Abelard and Heloïse (1970) had a long run, as did Godspell (1972), a musical based on the life of Christ. In 1975 *Pinter's No Man's Land transferred from the *Old Vic, and the musical compilation Side by Side by Sondheim (1976) and Mary O'Malley's Once a Catholic (1977) both did well. The next decade brought Dario *Fo's Accidental Death of an Anarchist (1980), Peter *Nichols's Passion Play (1984), and Caryl Churchill's Serious Money (1987).

Wynn, Ed, see REVUE and VAUDEVILLE, AMERICAN.

Wynyard, Diana [Dorothy Isobel Cox] (1906–64), English actress of great beauty and distinction, who made her first appearance in London in 1925 and was a member of the *Liverpool Playhouse company from 1927 to 1929. She played small parts in London before making her New York début in 1932. Back in London she made a great success as Charlotte Brontë in Clemence *Dane's Wild Decembers (1933) and in Joyce Carey's Sweet Aloes (1934). She played Gilda in *Coward's Design for Living (1939), Linda Easterbrook in S. N. *Behrman's No Time for Comedy (1941), and Sara Muller in Lillian *Hellman's Watch on the Rhine (1942) before touring for ENSA. In 1948 and 1949 she was at the *Shakespeare Memorial Theatre, where her work in Shakespeare's leading women's roles, particularly Lady Macbeth, was much admired. She starred opposite *Gielgud at the *Phoenix Theatre as Hermione in *The Winter's Tale* (1951) and Beatrice in *Much Ado about Nothing* (1952) and appeared in Moscow as Gertrude to the Hamlet of Paul *Scofield in 1955. She was always ready to appear in new plays, however, and was seen in *Whiting's *Marching Song* (1954), Maxwell *Anderson's *The Bad Seed* (1955), Tennessee *Williams's *Camino Real* (1957), and Lillian Hellman's *Toys in the Attic* (1960). Her last appearances were in the *National Theatre company's 1963–4 season in *Hamlet* and *Frisch's *Andorra*.

Wyspiański, Stanisław (1869–1907), Polish poet, playwright, director, and scene designer. Considered the successor of *Mickiewicz, he was the first to co-ordinate and stage the latter's vast poetic drama *Forefathers' Eve* in 1901. In that and many of his other monumental productions he made innovations in stagecraft, envisaging a form of 'total theatre', which were to have a strong influence on his successors, particularly Leon *Schiller. Of his own plays, which are based on national themes and classical symbolism, the best is usually considered *The Wedding* (1901), a revival of which was seen in London during the *World Theatre Season of 1966. Other important works are *Akropolis* (1903), in which *Grotowski's company made its first appearance outside Poland in the same year; and *November Night* (1904), about the Polish insurrection of 1830, seen in London in 1975 and revived in Warsaw in 1979. Three other plays, *Liberation* (1903), *The Curse* (1905), and *The Judges* (1907), have also been revived in recent years. Wyspiański had a wide knowledge of Shakespeare, and in an introduction to a Polish version of *Hamlet* put forward a new view of the play, making the Ghost the central character, which resulted in an interesting production, entitled *The Tragical History of Hamlet*, first seen in 1901.

Y

Yablochkina, Alexandra Alexandrovna (1868–1964), famous Russian actress, who was on the stage as a child in St Petersburg (Leningrad), where her father was a stage manager. She made her adult début in 1888 at the *Maly Theatre in Moscow where she remained until her death, playing Miss Crawley in a dramatization of Thackeray's *Vanity Fair* at the age of 94. During her long career she appeared in a vast range of parts that included Ophelia, Cordelia, and Desdemona, the heroines of Oscar *Wilde and *Galsworthy, and leading roles in Russian plays from *Griboyedov and *Gorky to *Romashov and *Korneichuk. She devoted much of her time under the Tsarist regime to theatrical charities. She was chairman of the All-Russian Theatrical Society—founded in 1883 to help needy members of the theatrical profession, but after the Revolution of 1917 concentrating on artistic standards—from 1916 until her death.

Yacine, Kateb (1929–89), Algerian novelist and dramatist, whose plays are inspired by the struggle of the Algerian people against French colonial oppression and against their own internecine difficulties. His first play *Le Cadavre encerclé* (1958) had to be performed in Brussels because of the political situation in France. Together with a comic parable, *La Poudre d'intelligence*, and its sequel, *Les Ancêtres redoublent de férocité*—presented under the title *La Femme sauvage* (1967)—it forms a trilogy published as *Le Cercle des représailles*. *L'Homme aux sandales de caoutchouc* (*The Man with Rubber Sandals*, 1971), an epic account of the life of Ho-Chi-Minh, caused a political storm in Lyons, where the Théâtre du VIIIe had its subsidy cut by the town council. Although Yacine represents strikingly the massacres and tortures perpetrated by the French army, his plays are pitched on a less political and more personal level than the attitude of the authorities might suggest. *Mohamed, prends ta valise* (1972), a comic satire on the situation of immigrant workers in France, was successfully presented to Algerian communities in Paris by an Algerian company. Like *La Guerre de 2,000 ans* (1975), it is in Arab dialect. In 1988 the *Avignon Festival staged his last work, *Le Bourgeois sans culotte; ou, Le Spectre du parc Monceau*, about Robespierre.

Yakovlev, Alexei Semenovich (1773–1817), Russian actor, a member of the St Petersburg Imperial Theatre company. He made his first appearance on the stage in 1794. Helped by his excellent presence and powerful voice, and by the counsels of *Dmitrevsky, he was soon playing leading roles. His performances were very unequal, since he relied on intuition, scorned technique, and drank heavily, but he had some perceptions of greatness, and tried to initiate reforms later carried out by *Shchepkin. He died soon after collapsing on stage while playing Othello.

Yale University and **Repertory Theatre,** New Haven, Conn. There were theatrical performances annually at this American university from 1771, among the first productions being *Steele's *The Conscious Lovers*, *Farquhar's *The Beaux' Stratagem*, and in 1785 an original play, *The Mercenary Match* by 'Mr Bidwell', a student in his senior year. There has long been an amateur dramatic society for the students, the 'Yale Dramat', and in 1925 a Drama Department was inaugurated, George *Baker being its first director, with an excellent little experimental theatre, and a curriculum comprising a wide-ranging schedule of instruction in all branches of professional theatre, including design, direction, administration, playwriting, and dramatic criticism. The library, which houses the personal papers of Eugene *O'Neill, has a theatre collection which includes a vast dossier of photographs of theatrical material collected from all over Europe, begun under Allardyce *Nicoll during his term of office as head of the Drama Department and constantly being extended. In 1966 the then Dean of the School of Drama, Robert Brustein, founded the Yale Repertory Theatre. Originally intended as an adjunct to the School, it left the university premises in 1968 and moved into a church converted into a flexible theatre. In 1975 it was reconstructed to provide four auditoriums, of which the two smaller ones are used exclusively by the School, which also shares a third with the repertory company. The fourth, seating 491, with a thrust stage, is for the sole use of the repertory theatre company, among whose more adventurous productions have been an adaptation of *Aeschylus' *Prometheus Bound* by Robert Lowell, Kenneth Cavender's adaptation of *Euripides' *Bacchae*, *Aristophanes' *Frogs* (1974) with music by *Sondheim, presented in the Yale swimming pool, and *A Midsummer Night's Dream* (1975) incorporating music from Purcell's *The Fairy Queen* for the first time. Modern plays have included the world premières of Eric *Bentley's *Are You Now or Have You Ever Been . . .?* (1972), Arthur *Kopit's *Wings* (1978), and many plays by *Fugard and August *Wilson, as well as the American premières of several plays by Edward *Bond. An annual Winterfest of new plays, in addition to the regular repertoire, was inaugurated in 1981.

Yanshin, Mikhail Mikhailovich (1902–76), Soviet actor and director, who became a member of the *Moscow Art Theatre company in 1924. One of the most successful roles of his early career was Lariossik in *Bulgakov's *The Days of the Turbins* (1926), in which he showed warmth, intimacy, and a subtle sense of humour. Among his later outstanding classic parts were Gradoboyev in *Ostrovsky's *Warm Heart*, Telyegin in *Chekhov's *Uncle Vanya*, and Sir Peter Teazle in *Sheridan's *The School for Scandal*. He was Artistic Director of the Gypsy Theatre, 1937–41, and in 1950 became chief producer at the Stanislavsky Theatre in Moscow, where under his direction many new Soviet plays were produced.

Yates, Frederick Henry (1795–1842), English actor, friend of the elder Charles *Mathews, who persuaded him to go on the stage. He was first seen in Edinburgh, and in 1818 went to London, where he played Iago to the Othello of Charles Mayne *Young. In the following year he was seen as Falstaff in *Henry IV, Part One* to the Hotspur of *Macready. Having no pretensions to be anything more than a useful actor he played everything that was offered to him, but eventually concentrated on comedy where his real talent lay. In 1825 he became joint manager of the *Adelphi Theatre, and produced there a series of melodramas, farces, and burlesques, being joined on his partner's death in 1829 by Mathews. Some of the best authors of the day wrote for him, and he himself was excellent in such Dickensian parts as Fagin in *Oliver Twist*, Mr Mantalini in *Nicholas Nickleby*, and the grotesque Miss Miggs in *Barnaby Rudge*.

Yates, Richard (1706–96), English comedian, considered almost as good a *Harlequin as *Woodward. He joined the company at the second *Goodman's Fields Theatre when it opened in 1732, and remained until it was closed on the passing of the Licensing Act of 1737. He then went to *Drury Lane, where he remained for many years, playing Sir Oliver Surface in the first production of *Sheridan's *The School for Scandal* (1777). He was also good as Shakespeare's fools, and was considered the only actor of the time to have a just notion of how they should be played. His style in comedy was modelled on that of *Doggett; fine gentlemen and serious comedy lay outside his range, and he never appeared in tragedy.

Yeats, William Butler (1865–1939), Irish poet and dramatist, awarded the *Nobel Prize for literature in 1923. With Lady *Gregory he was the founder of the modern Irish dramatic movement, and as a director of the *Abbey Theatre in Dublin, where all except his earliest plays were first produced, from its foundation in 1904 until his death, he did much to ensure the integrity of the emerging national drama in spite of political and financial pressures. He also encouraged new playwrights, among them *Synge and *O'Casey, even when the theatre's repertoire moved away from the aims demonstrated in his own plays. The earliest of these were poetic dramas in the style of *Maeterlinck—*The Countess Cathleen* (1892), *The Land of Heart's Desire* (1894), and *Cathleen ni Houlihan* (1902); the last-named was written in collaboration with Lady Gregory, who was also part-author of *The Pot of Broth* (1902) and of two poetic plays dramatically more powerful than the earlier ones—*The King's Threshold* (1904) and *The Unicorn from the Stars* (1908). In the first, the function of poetry and of the poet is the major theme; the second is a bold piece of speculative thinking, based on a mystical experience. Other plays of this period were on subjects taken from the heroic legends of Ireland—*On Baile's Strand* (1904), *Deirdre* (1907), and *The Green Helmet* (1908). *The Shadowy Waters* (1911) is again a poetic play in which, as in his earlier works, the theme is remote and the experience it presents, though universal, is not revealed in terms of actual contemporary life. It was followed by *The Hour-Glass* (1914), written in verse and so performed at the Abbey; for later productions elsewhere Yeats rewrote it in prose, owing to the difficulty of finding verse-speakers who met with his approval. There are also two versions of *The Only Jealousy of Emer* (1919), the second, in prose, being entitled *Fighting the Waves*. Together with *At the Hawk's Well* (1917), *The Dreaming of the Bones* (1919), and *Calvary* (1920), the first version formed part of the 'four plays for dancers' in which extreme simplicity of design and setting is matched by a brevity of expression akin to that of the *nō play of Japan. In his last plays this plainness and severity, characteristic also of his poetry after 1916, reaches fulfilment, and the underlying thought makes stricter demands than ever upon the intelligence and imagination of the audience. This is most clearly shown in *The Player Queen* (1922), *The Words upon the Window Pane* (1934), and *Purgatory* (1938). Yeats also made new versions of *Sophocles' *Oedipus the King* (1928) and *Oedipus at Colonus* (1934). In his last play, *The Death of Cuchulain* (1939), he reverted once more to the legends of Celtic Ireland, always one of his main preoccupations.

Yefremov [Efremov], Oleg, see MOSCOW ART THEATRE and SOVREMENNIK THEATRE.

Yermolova, Maria Nikolaievna (1853–1928), Russian tragic actress, considered by *Stanislavsky the finest he had ever seen. She made her début in 1870 at the *Maly Theatre in Moscow as Emilia Galotti in *Lessing's play of that name, and soon proved herself an outstanding interpreter of such roles as Lady Macbeth, Joan of Arc in *Schiller's *Die Jungfrau von Orleans*, and *Racine's *Phèdre*. In contrast to the somewhat passive and unreal playing of such characters at the time, she made her heroines active, independent members of society. Having weathered the storms of the Russian Revolution, which she welcomed, she celebrated her stage jubilee in 1920 with full Soviet ceremonial and the honour of being named People's Artist of the Republic. A studio named after her in 1930 later became the *Yermolova Theatre.

Yermolova Theatre, Moscow, studio named in honour of the actress (above) in 1930, which acquired its present status after its fusion with the Khmelev Studio in 1937 when *Khmelev himself became its director. Its first production was *Gorky's *Children of the Sun*, *Bill-Belotserkovsky's *Storm* and *Ostrovsky's *The Poor Bride* were seen in 1938, and in 1939 Khmelev scored a triumph with an outstanding production of *As You Like It*. The theatre later saw the first production of several important Soviet plays, an adaptation of Dostoevsky's *Crime and Punishment* (1956), and modern foreign plays such as *De Filippo's *Saturday, Sunday, Monday* and *Dürrenmatt's *Play Strindberg*.

Yevreinov, Nicolai, see EVREINOV.

Yiddish Art Theatre, New York, see GARDEN THEATRE.

Yiddish Theatre, see JEWISH DRAMA.

York had its first permanent theatre building, adapted from a *tennis court in Minster Yard, in 1734. A second theatre was erected on the site of the present Theatre Royal in 1744, a larger theatre, erected on the same site in 1765, obtaining a royal patent in 1769 and thereafter being known correctly as the Theatre Royal. In 1766 the provincial theatre manager Tate *Wilkinson took over the York *circuit, which he modified to comprise York, Hull, Leeds, Doncaster, Wakefield, and Pontefract. The York theatre was remodelled in 1822, and further alterations were made to it later, so that little of the original remains. It became a theatre for touring companies in 1877 and in the 1900s was rebuilt to counteract the appeal of the new Grand Opera House (later the Empire) in New Clifford Street. In 1922 a repertory company was installed for several weeks during the summer, and from 1925 to 1929 a summer season, extending to 10 weeks, was given by a company led by Lena *Ashwell. From 1930 to 1932 the theatre was used for weekly rep. during several months each year, and in 1935 a permanent company was introduced with E.

Martin *Browne as producer. In 1951 he was also responsible for the revival of the York Cycle of *mystery plays in the ruins of St Mary's Abbey, since performed regularly at four-year intervals as part of the York Festival. The Theatre Royal, now owned by the local authority and operated by a Trust, was extensively altered and redecorated in 1967 and still functions as a repertory theatre, though some touring companies visit it. It runs a Young People's Theatre Company which works with schools and youth clubs. Seating 899, it has a foyer wing suitable for the presentation of theatrical and other exhibitions.

York Cycle, see MYSTERY PLAY.

Young, Charles, see DUFF.

Young, Charles Mayne (1777–1856), English actor, who in 1798 made his début in Liverpool as Mr Green. From there he went to Manchester and Edinburgh, where he became an intimate friend of Sir Walter Scott. He was also friendly with the elder Charles *Mathews, and with his help appeared at the *Haymarket Theatre in London in 1807 as Hamlet, Mathews playing Polonius. As one of the finest disciples of the *Kemble school of acting, Young proved himself a tower of strength in support of such players as John Philip Kemble, Mrs *Siddons, *Macready, and Miss *O'Neill, and in 1822 was with Edmund *Kean at *Drury Lane. The only new part of any importance which he created was the title-role in Miss Mitford's *Rienzi* (1828), but he was good and reliable in both comedy and tragedy, particularly as Hamlet, Macbeth, Iago in *Othello*, Falkland in *Sheridan's *The Rivals*, Sir Pertinax McSycophant in *Macklin's *The Man of the World*, Macheath in *Gay's *The Beggar's Opera*, and the Stranger in an adaptation (as *The Stranger*, 1797) of *Kotzebue's *Menschenhass und Reue*. He retired in 1832, appearing for the last time as Hamlet, again with Mathews as Polonius, Macready playing the Ghost.

Young Roscius, see BETTY.

Young Theatre Association, see CHILDREN'S THEATRE.

Young Vic Theatre, London. The first theatre to be given this name was founded in 1945 as part of the new *Old Vic Drama School, with the idea of presenting adult actors in plays suitable for young people. Under George *Devine, the company began in 1946 with *Gozzi's *King Stag* (*Il re cervo*), and did important work, as did the drama school, developing a pioneering relationship between actor training and professional performance; lack of influential support and financial stringency caused both projects to be abandoned in 1951.

The second Young Vic was founded in 1970 by Frank *Dunlop as part of the *National Theatre, becoming an independent body in

1974. Erected on a bomb-site near the Old Vic, the theatre stages good productions at low prices, with emphasis on the text and the acting rather than on elaborate staging. The first production was *Scapino*, an adaptation of *Molière's *Les Fourberies de Scapin* by Dunlop and Jim Dale, who played the title-role. Others included *Beckett's *Waiting for Godot* (1970), *Jonson's *The Alchemist*, the musical *Joseph and the Amazing Technicolor Dreamcoat* (both 1972), and *Stoppard's *Rosencrantz and Guildenstern are Dead* (several productions beginning in 1973). Frank Dunlop remained Director until 1983, except 1978–80, when Michael *Bogdanov took over. Refurbished in 1984, the theatre can now be converted from thrust (seating 450) to theatre-in-the-round (525). Two of the theatre's biggest successes, *Ibsen's *Ghosts* (1986) and *O'Neill's *A Touch of the Poet* (1988), starred Vanessa *Redgrave, the second having its British première there. Both moved to the West End, as did the British première of Arthur *Miller's adaptation of Ibsen's *An Enemy of the People* (1988).

A studio theatre seating 110 opened in 1984.

The Young Vic Youth Theatre, founded in 1988, mounts full-scale productions, holds workshops, and works in schools.

Youth Theatre, see NATIONAL YOUTH THEATRE OF GREAT BRITAIN.

Yurka [Jurka], **Blanche** (1893–1974), American actress, who made her first appearance in New York in 1907 and was later with stock companies in Buffalo, Dallas, and Philadelphia. She already had many leading parts to her credit when in 1922 she played Gertrude to the Hamlet of John *Barrymore. She later played roles in several plays by *Ibsen, in which she was outstanding: Gina in *The Wild Duck* (1928), Hedda in *Hedda Gabler*, Ellida in *The Lady from the Sea* (both 1929), and Hiordis in *The Vikings* (1930). She was also much admired as *Aristophanes' Lysistrata later in 1930 and as *Sophocles' Electra in 1932. From 1936 to 1938 she toured the USA in a solo programme of scenes from great plays. She made her only appearance in London in 1969 in *Giraudoux's *The Madwoman of Chaillot*, and was part-author of *Spring in Autumn* (1933), adapted from a play by *Martínez Sierra.

Z

Zacconi, Ermete (1857–1948), Italian actor, child of strolling players, who made his first appearance on the stage at the age of 7. While still in his twenties he was recognized as an outstanding actor of the naturalistic school—*verismo*—equalled only by *Novelli. In 1894 he formed his own company and, with *Duse, was instrumental in bringing *Ibsen to the Italian public, being the first to play Oswald Alving in *Ghosts*, which he directed himself. He produced a number of translations of plays by *Tolstoy, *Hauptmann, and *Strindberg, and played most of the important roles in Shakespeare, going as far afield as Germany, Austria, and Russia. A most versatile actor, he was considered to be at his best as Hamlet and in the title-role of Testoni's *Il Cardinale Lambertini* (1905), a comedy of intrigue based on the life of a 17th-century cardinal of Bologna.

Zanni, term used to describe collectively the servant masks of the *commedia dell'arte*; the equivalent English term is zany. Among them are *Arlecchino, *Pantalone, and (later) Columbina, the prototypes of the English *pan-

tomime players *Harlequin, Pantaloon, and *Columbine. The English *Punch, known in France as *Polichinelle, also derives from a zany, *Pulcinella, as do three famous French comic characters, Scapin, Scaramouche, and Pasquin—from *Scapino, *Scaramuccia, and *Pasquino respectively. In France, *Pedrolino became a sad, solitary *Pierrot, and in England a gregarious member of a seaside concert party. Other minor *zanni* included *Brighella and *Mezzetino. In Elizabethan times the word zany conveyed the idea of a 'clumsy imitator'.

Zarzuela, Spanish musical play or operetta, which takes its name from the Palacio de la Zarzuela, a hunting-lodge in the woods not far from Madrid. Although many early Spanish plays contained music, the first true *zarzuela* was by *Calderón—*El laurel de Apolo* (1658). The music for this and other early *zarzuelas* has been lost, but in 1933 that for the first act of Calderón's *Celos aún del aire matan* (*Jealousy, Even of the Air, Kills*, 1660) was found and published, followed in 1945 by that for the second act. The *zarzuela* flourished as a courtly

entertainment from 1660 onwards, its mythological or heroic plot being subtly designed to
flatter its royal audience, but in the 18th century
it suffered from the popularity of Italian opera.
It was revitalized by **Ramón de la Cruz** (1731–
94), who with *Las segadoras de Vallecas* (*The
Reapers of Vallecas*) introduced a new-style *zarzuela* with a plot drawn from daily life. His work
was immensely successful during his lifetime but
after his death was again eclipsed by Italian
opera, until in the second half of the 19th
century a new era began with Barbieri's *Jugar
con fuego* (*Playing with Fire*, 1851). This led to
the opening in 1856 of a Teatro de la Zarzuela
in Madrid. The modern *zarzuela*, exemplified
by the ever-popular *La verbena de la Paloma*
(*The Festival of la Paloma*), by Ricardo de la
Vega and Bretón, still exists in a form strongly
influenced by foreign revues and musicals.

Zavadsky, Yuri Alexeivich (1894–1977),
Soviet actor and director, who began his career
in 1915 as a pupil of *Vakhtangov, appearing
in his last production, *Gozzi's *Turandot*, in
1922. In 1924 he opened his own studio, and
was later appointed Director of the Gorky State
Theatre in Rostov. His first productions showed
the influence of both Vakhtangov and *Stanislavsky, but he soon evolved a personal style,
combining lyricism with excellent stagecraft,
and giving a larger place than usual to music.
Many outstanding Soviet actors were his pupils,
and *Grotowski attended his classes in play
production at the Moscow Theatre Institute.
During the Second World War, after going
briefly to the *Mossoviet Theatre in 1940, he
returned to Rostov, where he directed *Trenev's
Lyubov Yarovaya and *On the Banks of the Neva*,
*Gorky's *Enemies* and *The Philistines*, and
Othello. Back at the Mossoviet after the war,
he contributed not a little to the forward
movement of the Soviet theatre during the
1950s.

Zeamī, see NŌ PLAY.

Zeffirelli, Franco (1923–   ), Italian director
and scene designer, whose work is known for
its visual splendour. He first attracted attention
in his own country with his sets for *Visconti's
productions of *Troilus and Cressida* in Florence
in 1949 and *Chekhov's *Three Sisters* in Rome
in 1951, and he was already well known in the
world of opera when in 1960 he directed and
designed *Romeo and Juliet* at the *Old Vic (NY,
1962). He then directed and designed *Othello* at
the *Royal Shakespeare Theatre (1961); *Dumas
fils's *The Lady of the Camellias* (1963) in New
York; *Hamlet* (*Amleto*, 1964), performed by an
Italian company at the Old Vic; and *Much Ado
about Nothing* (1965) for the *National Theatre
company at the same theatre. His other work
included *Miller's *After the Fall* (Rome, 1964)
and *Albee's *Who's Afraid of Virginia Woolf?*
(Paris, 1964; Milan, 1965) and *A Delicate Balance*

(Rome, 1967). He brought the work of Eduardo
*De Filippo to the notice of a wide audience
with his production and décor for the National
Theatre in 1973 (NY, 1974) of *Saturday, Sunday,
Monday*—a piece ideally suited to his distinctive
brand of scrupulous but highly decorative
realism. His staging of the same author's *Filumena* was seen in the West End in 1977 (NY,
1980). He is also a well-known film director.

Zellerbach Theatre, see PHILADELPHIA.

Ziegfeld, Florenz (1867–1932), American
theatre manager, producer, and showman, who
perfected the American form of *revue in the
Follies, which began in 1907 and continued
almost annually until his death, being seen
intermittently thereafter until 1957; 'Ziegfeld'
was added to the title in 1911. Ziegfeld based
his show on that of the *Folies-Bergère in Paris,
with the emphasis on scenic splendour, comic
sketches, *vaudeville specialities, and beautiful
girls, many of whom later became film stars.
He was also responsible for fostering the talents
of such light entertainers as Fanny Brice,
W. C. *Fields, Eddie Cantor, and Bert *Williams, and many famous foreign artistes appeared
in New York under his management, including
Maurice *Chevalier. Among the many musicals
he presented the best was Kern and *Hammerstein's *Show Boat* (1927 and 1932); others
were *Sally* (1920), *Rio Rita* (1927), *Whoopee*
(1928), and *Coward's *Bitter Sweet* (1929). He
also presented straight plays. His shows were
seen mainly at the *New Amsterdam and
*Ziegfeld theatres.

Ziegfeld Theatre, New York, on the north-
west corner of 6th Avenue and 54th Street,
designed for Florenz *Ziegfeld by the scene
designer Joseph *Urban. With a handsome
egg-shaped auditorium, a fine, stark exterior,
and lavish backstage space, it opened in 1927
with the musical *Rio Rita*, which was followed
by Kern and *Hammerstein's *Show Boat*, *Coward's *Bitter Sweet* (1929), and the *Ziegfeld Follies*
of 1931. After Ziegfeld's death in 1932 his
theatre became a cinema, but in 1944 it was
reopened by Billy Rose, under its old name,
with the revue *Seven Lively Arts* starring Beatrice
*Lillie. Carol Channing appeared there in the
musical version of *Gentlemen Prefer Blondes*
(1949), and in 1953 the theatre housed a revival
of the *Gershwins' opera *Porgy and Bess*. From
1955 to 1963 it was used for television, but
returned to live theatre again in 1963 with an
evening of songs and sketches starring Maurice
*Chevalier, followed by Bert *Lahr in *Foxy*
(1964), a musical based on *Jonson's *Volpone*.
The building was demolished in 1967, to be
replaced in 1969 by a cinema.

Zola,   Émile-Edouard-Charles-Antoine
(1840–1902), French novelist and dramatist,
who, though mainly remembered for his novels,

in particular the 20 volumes of *Les Rougon-Macquart* (1871–93), had an important influence on the development of the French, and so of the European, theatre of his day. As the leader of the naturalistic school of literature, he much disliked the facile, optimistic works of such dramatists as *Scribe, and thought that a play should be a 'slice of life', thrown on the stage without embellishment or artifice. He exemplified his theory in his own plays, particularly *Thérèse Raquin* (1873), which he based on an earlier novel of the same name. In an English translation, this was seen in London in 1891 and has been several times revived, notably in 1938 as *Thou Shalt Not* . . . In 1945 it was produced in New York as *Therese*. Zola's other plays were less successful. The only other one to have been translated into English is a comedy, *Les Héritiers Rabourdin* (1874), which as *The Heirs of Rabourdin* was produced in 1894 by Grein's Independent Theatre Club. One of Zola's novels, *L'Assommoir*, was dramatized and in an English version, *Drink*, by Charles *Reade and Charles *Warner, had a long run in London in 1879, Warner playing the chief part.

Zorrilla, Francisco de Rojas, see ROJAS ZORRILLA.

Zorrilla y Moral, José (1817–93), Spanish poet and dramatist of the Romantic movement. His best known play is *Don Juan Tenorio* (1844), which added yet another version to the numerous interpretations of the *Don Juan legend. It is still performed in Spanish theatres, traditionally on All Saints' Day (1 Nov.). Zorrilla also wrote comedies, tragedies, and historical plays. Of the last the best known are *El zapatero*

y el rey (*The Cobbler and the King*, 1840–1), in two parts, *El puñal del Godo* (1842), and *Traidor, inconfeso y mártir* (1849), based on the life of a pretender to the throne of Portugal.

Zuckmayer, Carl (1896–1977), German dramatist, who first wrote the unsuccessful *Kreuzweg* (*Crossroads*, 1920) and *Pankraz erwacht* (*Pankraz Wakes Up*, 1925). When he turned to *realism with a comedy, *Der fröhliche Weinberg* (*The Merry Vineyard*, 1926), he had more success. This was followed by *Schinderhannes* (1927), based on a popular folk-figure of the Napoleonic Wars; *Katharina Knie* (1929), set in a travelling circus; and by *Der Hauptmann von Köpenick* (1931). This comedy of a Berlin ex-convict shoemaker who, disguised as an officer, enforces obedience to his orders merely because he is wearing military uniform was staged in London in 1971 by the *National Theatre company with Paul *Scofield. Zuckmayer left Germany in 1933, when because of his outspoken opposition to National Socialism his plays were banned by the authorities, and while in Switzerland wrote *Der Schelm von Bergen* (1934) and *Bellman* (1938). The first play he wrote after settling in the USA in 1939 was *Des Teufels General*, about Nazi Germany and the problem of resistance, first seen in Zürich in 1946. As *The Devil's General* it was produced in New York and London in 1953. Among Zuckmayer's later plays, none of which achieved a comparable success, were *Der Gesang im Feuerofen* (*The Song in the Fiery Furnace*, 1950), set in occupied France, *Das kalte Licht* (1956), based on the case of the traitorous nuclear physicist Dr Fuchs, *Die Uhr schlägt eins* (1961), and *Der Rattenfänger* (*The Pied Piper*, 1974).

A Guide to Further Reading

Publishers are located in Great Britain unless otherwise stated.

Reference Works

Encyclopedias

The Cambridge Guide to World Theatre (Cambridge University Press, 1988).

Encyclopaedia of World Drama, 5 vols. (New York: McGraw-Hill, 1984).

ESSLIN, MARTIN (ed.), *Illustrated Encyclopaedia of World Theatre* (Thames & Hudson, 1977).

GASSNER, JOHN, and QUINN EDWARD (eds.), *The Reader's Encyclopaedia of World Drama* New York: Crowell, 1969; Methuen, 1975.

GRIFFITHS, TREVOR R., and WODDIS, CAROLE, *Bloomsbury Theatre Guide* (new edn. Bloomsbury, 1989).

HARTNOLL, PHYLLIS (ed.), *The Oxford Companion to the Theatre* (4th edn. Oxford University Press, 1983).

TAYLOR, JOHN RUSSELL, *The Penguin Dictionary of the Theatre* (Methuen, 1967; New York: Barnes & Noble).

THOMSON, P., *The Everyman Companion to the Theatre* (Dent, 1987).

VAUGHN, J. A., *Drama A to Z* (New York: Ungar, 1979).

General Bibliographies

ARNOTT, J. F., and ROBINSON, J. W., *English Theatrical Literature, 1559–1900* (Society for Theatre Research, 1970).

BAKER, BLANCH M., *Theatre and Allied Arts* (New York: Blom, 1967).

CAVANAGH, JOHN, *A Bibliography of the British Theatre, 1901 to 1985* (Motley Books, 1987).

LOEWENBERG, A., *The Theatre of the British Isles, Excluding London* (Society for Theatre Research, 1950).

STRATMAN, C. J., *A Bibliography of American Theatre, Excluding New York City* (Chicago: Loyola University Press, 1965).

—— *American Theatrical Periodicals, 1798–1967: A Bibliographical Guide* (Durham, NC: Duke University Press, 1970).

—— *Britain's Theatrical Periodicals, 1720–1967: A Bibliography* (2nd edn. New York: New York Public Library, 1972).

Dictionaries and Glossaries

BOWMAN, WALTER, P., and BALL, ROBERT HAMILTON, *Theatre Language: A Dictionary of Terms in English of the Drama and Stage from Medieval to Modern Times* (New York: Theatre Arts, 1961).

GRANVILLE, WILFRED, *A Dictionary of Theatrical Terms* (Deutsch, 1952; New York: Oxford University Press, as *Theatre Language*).

Biographical Dictionaries

HERBERT, IAN (ed.), *Who's Who in the Theatre*, 17th edn., 2 vols: i: *Biographies*; ii: *Playbills* (Detroit: Gale, 1981; earlier edns., 1912–77, pub. by Pitman). Now published as *Contemporary Theatre, Film and Television* (Detroit: Gale, 1989).

NUNGEZER, EDWIN, *A Dictionary of Actors and Other Persons Associated with the Public Representation of Plays in England before 1642* (New Haven, Conn.: Yale University Press, 1929; Oxford University Press).

RIGDON, WALTER (ed.), *The Bibliographical Encyclopedia and Who's Who of the American Theatre* (New York: J. H. Heinemann, 1966).

Who Was Who in the Theatre: 1912–1976, 4 vols. (Detroit: Gale, 1978).

General Histories

ALTMAN, GEORGE, et al., *Theater Pictorial* (Berkeley, Calif.: California University Press, 1953).

BERTHOLD, MARGOT, *A History of World Theater* (New York: Ungar, 1972).

BROCKETT, OSCAR G., *History of the Theatre* (3rd edn., Boston: Allyn & Bacon, 1987).

—— *The Theatre: An Introduction* (4th edn., New York: Holt, Rinehart, 1979).

FINDLATER, RICHARD, *Banned* (on censorship) (MacGibbon & Kee, 1967).

FREEDLEY, GEORGE, and REEVES, JOHN A., *A History of the Theatre* (3rd edn., New York: Crown, 1968).

GASCOIGNE, BAMBER, *World Theatre* (Ebury Press, 1968; Boston: Little, Brown).

HARTNOLL, PHYLLIS, *A Concise History of the Theatre* (Thames & Hudson, 1968; New York: Abrams).

HARWOOD, R., *All the World's a Stage* (Secker, 1984; Boston: Little, Brown).

NICOLL, ALLARDYCE, *The Development of the Theatre* (5th edn., Harrap, 1966).

—— *World Drama from Aeschylus to Anouilh* (2nd edn., Harrap, 1976).

SOUTHERN, RICHARD, *Seven Ages of the Theatre* (2nd edn., Faber, 1964).

The Modern Drama and Theatre

BENTLEY, ERIC, *The Playwright as Thinker* (New York: Meridian, 1955; Methuen).

BRUSTEIN, ROBERT, *The Theatre of Revolt* (Boston: Little, Brown, 1964; Methuen, 1965).

CLARK, BARRETT H., and FREEDLEY, GEORGE (eds.), *A History of Modern Drama* (New York: Appleton, 1947).

ESSLIN, MARTIN, *The Theatre of the Absurd* (3rd edn., Methuen, 1974).

GASSNER, JOHN, *Directions in Modern Theatre and Drama* (rev. edn., New York: Holt, Rinehart, 1965).

GORELIK, MORDECAI, *New Theatres for Old* (Dobson, 1947; New York: Dutton, 1962).

HAYMAN, RONALD, *Theatre and Anti-theatre: New Movements since Beckett* (Secker, 1979; New York: Oxford University Press).

ROOSE-EVANS, JAMES, *Experimental Theatre from Stanislavsky to Brecht* (Studio Vista, 1970).

STYAN, J. L., *Modern Drama in Theory and Practice*, 3 vols. (Cambridge University Press, 1981).

VALENCY, M., *The End of the World: An Introduction to Contemporary Drama* (Oxford University Press, 1980).

WILLIAMS, RAYMOND, *Drama from Ibsen to Brecht* (2nd rev. edn., Penguin, 1976).

Classical Drama and Theatre

ARNOTT, PETER, *Public and Performance in Greek Theatre* (Routledge, 1989).

BEARE, W., *The Roman Stage: A Short History of Latin Drama in the Time of the Republic* (3rd edn., Methuen, 1978; Totowa, NJ: Rowman & Littlefield).

BIEBER, MARGARETE, *History of the Greek and Roman Theatre* (2nd edn., Oxford University Press, 1961; Princeton, NJ: Princeton University Press).

DUCKWORTH, G. E., *The Nature of Roman Comedy* (Princeton, NJ: Princeton University Press, 1952; Oxford University Press).

WEBSTER, T. B. L., *Greek Theatre Production* (2nd edn., Methuen, 1970; New York: Barnes & Noble).

—— *Greek Tragedy* (Oxford University Press, 1971).

Great Britain

BARNES, PHILIP, *Companion to Post-War British Theatre* (Croom Helm, 1986).

BEAUMAN, SALLY, *The Royal Shakespeare Company: A History of Ten Decades* (Oxford University Press, 1982).

BENTLEY, G. E., *The Jacobean and Caroline Stage*, 7 vols. (Oxford University Press, 1941–68).

BOOTH, MICHAEL, *English Melodrama* (Jenkins, 1965).

BRADBROOK, M. C., *History of Elizabethan Drama*, 6 vols.: i: *Themes and Conventions of Elizabethan Tragedy*; ii: *The Growth and Structure of Elizabethan Comedy*; iii: *The Rise of the Common Player: A Study of Actor and Society in Shakespeare's England*; iv: *Shakespeare and the Elizabethan Poetry: A Study of His Earlier Work in Relation to the Poetry of the Time*; v: *Shakespeare the Craftsman*; vi: *The Living Monument: Shakespeare and the Theatre of His Time* (Cambridge University Press, 1979).

—— *Shakespeare in His Context* (Harvester Wheatsheaf, 1989).

ELSOM, J., and TOMALIN, N., *The History of the National Theatre* (Cape, 1978).

HALLIDAY, F. E., *A Shakespeare Companion* (rev. edn., Duckworth, 1977).

HAPGOOD, ROBERT, *Shakespeare the Theatre-Poet* (Oxford University Press, 1988).

HAYMAN, RONALD, *British Theatre since 1955* (Oxford University Press, 1979).

HOBSON, HAROLD, *Theatre in Britain* (Phaidon, 1984).

KOTT, JAN, *Shakespeare, Our Contemporary* (2nd edn., Methuen, 1967).

LOFTIS, JOHN, *The Politics of Drama in Augustan England* (Oxford University Press, 1963).

—— *Sheridan and the Drama of Georgian England* (Blackwell, 1976).

MANDER, RAYMOND, and MITCHENSON, JOE, *Picture History of the British Theatre* (Hulton, 1957).

—— *British Music Hall* (rev. edn., Gentry Books, 1974).

MAYER, DAVID, *Harlequin in His Element: The English Pantomime, 1806–1836* (Oxford University Press, 1969).

NICOLL, ALLARDYCE, *History of English Drama, 1600–1900*, 6 vols. (Cambridge University Press, 1952–9).

—— *English Drama, 1900–1930* (Cambridge University Press, 1973).

—— *British Drama* (6th edn., rev. J. C. Trewin, Harrap, 1978; New York: Harper & Row, 1979).

—— *The Garrick Stage: Theatres and Audience in the Eighteenth Century*, ed. Sybil Rosenfeld (Manchester University Press, 1980).

ONIONS, C. T., *A Shakespeare Glossary* (3rd edn., Oxford University Press, 1986).

ROTHSTEIN, E., *Restoration Tragedy* (Westport, Conn.: Greenwood Press, 1978).

ROWELL, GEORGE, *The Victorian Theatre, 1792–1914* (2nd edn., Cambridge University Press, 1978).

ROWELL, GEORGE, *Queen Victoria Goes to the Theatre* (Elek, 1978).
—— *Theatre in the Age of Irving* (Blackwell, 1981).
SOUTHERN, RICHARD, *The Staging of Plays before Shakespeare* (Faber, 1973; New York: Theatre Arts).
—— *The Medieval Theatre in the Round* (2nd edn., Faber, 1975).
SPEAIGHT, GEORGE, *Punch and Judy: A History* (Studio Vista, 1970).
TAYLOR, JOHN RUSSELL, *The Rise and Fall of the Well-Made Play* (Methuen, 1967; New York: Hill & Wang).
—— *Anger and After: A Guide to the New British Drama* (2nd edn., rev., Eyre Methuen, 1977; New York: Hill & Wang, 1969, as *The Angry Theatre*).
—— *The Second Wave: British Drama of the Sixties* (Eyre Methuen, 1978).
TREWIN, J. C., *The Theatre since 1900* (Dakers, 1951).
—— *The Edwardian Theatre* (Blackwell, 1976).
TYNAN, KENNETH, *A View of the English Stage, 1944–63* (Davis-Poynter, 1975).
WAIN, JOHN, *The Living World of Shakespeare: A Playgoer's Guide* (Macmillan, 1980; New York: St Martin's).
WELLS, STANLEY (ed.), *Shakespeare: A Bibliographical Guide* (2nd edn., Oxford University Press, 1990).
—— and TAYLOR, GARY (eds.), *William Shakespeare: A Textual Companion* (Oxford University Press, 1988).
WICKHAM, GLYNNE, *Early English Stages 1300 to 1660: A History of the Development of Dramatic Spectacle and Stage Convention in England*, 3 vols. (Routledge; New York: Columbia University Press): i: *1300–1576* (2nd edn., 1980); ii: *1576–1660* (part 1, 1963; part 2, 1972); iii (1980).
—— *The Medieval Theatre* (Weidenfeld, 1980).
WILSON, F. P., *The English Drama, 1485–1585* (Oxford University Press, 1969).

Ireland

BELL, S. H., *The Theatre in Ulster* (Dublin: Gill & Macmillan, 1972).
CLARK, WILLIAM SMITH, *The Early Irish Stage* (Oxford University Press, 1955).
—— *The Irish Stage in the County Towns, 1720–1800* (Oxford University Press, 1965).
ELLIS-FERMOR, UNA, *The Irish Dramatic Movement* (2nd edn., Methuen, 1954; New York: Barnes & Noble).
FITZ-SIMON, C., *The Irish Theatre* (Thames & Hudson, 1983).
HOGAN, ROBERT, and KILROY, JAMES, *The Modern Irish Drama: A Documentary History*, 4 vols. (Dublin: Dolmen Press, 1975–9).

HUNT, HUGH, *The Abbey: Ireland's National Theatre, 1904–1978* (Dublin: Gill & Macmillan, 1979).

Europe

ARNOTT, PETER, *An Introduction to the French Theatre* (Macmillan, 1977).
BRUFORD, W. H., *Theatre, Drama and Audience in Goethe's Germany* (New York: Hillary, 1950; Routledge).
GORCHAKOV, NIKOLAI A., *The Theater in Soviet Russia* (New York: Columbia University Press, 1957).
HOBSON, HAROLD, *The French Theatre of Today: An English View* (Harrap, 1953; New York: Blom).
—— *French Theatre since 1830* (Calder, 1978).
INNES, CHRISTOPHER D., *Modern German Drama* (Cambridge University Press, 1979).
KENNARD, JOSEPH S., *The Italian Theatre*, 2 vols. (New York: Blom, 1964).
LOUGH, JOHN, *Paris Theatre Audiences in the Seventeenth and Eighteenth Centuries* (Oxford University Press, 1957).
—— *Seventeenth-Century French Drama: The Background* (Oxford University Press, 1979).
LUCAS, F. L., *The Drama of Ibsen and Strindberg* (New York: Macmillan, 1962; Cassell, 1963).
MARKER, F. J. and L.-L., *The Scandinavian Theatre* (Blackwell, 1975).
NICOLL, ALLARDYCE, *The World of Harlequin: A Critical Study of the Commedia dell'arte* (Cambridge University Press, 1976).
PEAK, J. HUNTER, *Social Drama in Nineteenth-Century Spain* (Chapel Hill, NC: North Carolina University Press, 1965).
SHERGOLD, N. D., *A History of the Spanish Stage, from Medieval Times until the End of the Seventeenth Century* (Oxford University Press, 1967).
SLONIM, MARC, *Russian Theater from the Empire to the Soviets* (Cleveland: World, 1961; Methuen, 1963).

United States of America

ATKINSON, BROOKS, *Broadway, 1900–1970* (rev. edn., New York: Macmillan, 1974).
BIGSBY, C. W. E., *Confrontation and Commitment: A Study of Contemporary American Drama, 1959–1966* (Macgibbon & Kee, 1967; Columbia, Mo.: University of Missouri Press, 1969).
BLUM, DANIEL, and WILLIS, JOHN, *A Pictorial History of the American Theatre, 1860–1976* (4th edn., New York: Crown, 1978).
BORDMAN, GERALD, *The American Musical Theatre* (Oxford University Press, 1978).
—— *American Musical Comedy* (Oxford University Press, 1982).

BORDMAN, GERALD, *The Oxford Companion to the American Theatre* (Oxford University Press, 1985).

—— *American Musical Revue* (Oxford University Press, 1986).

—— *The Concise Oxford Companion to the American Theatre* (Oxford University Press, 1987).

CLURMAN, HAROLD, *The Fervent Years* (about the Group Theatre) (2nd edn., New York: Hill & Wang, 1957; Macgibbon & Kee).

GILBERT, DOUGLAS, *American Vaudeville* (New York: Dover, 1963; Constable).

GREEN, STANLEY, *The World of Musical Comedy* (3rd edn. rev. and enlarged, Cranbury, NJ: Barnes, 1974; Yoseloff).

—— *Encyclopaedia of the Musical* (Cassell, 1977; New York: Dodd, Mead).

GUERNSEY, OTIS L. (ed.), *The Best Plays of . . .: the Burns Mantle Theater Year Book* (New York: Dodd, Mead). (1st pub. 1920; see under Mantle.)

—— *Directory of the American Theater 1894–1971* (New York: Dodd, Mead, 1971).

HUGHES, GLENN, *History of the American Theatre, 1700–1950* (New York: French, 1951).

KRUTCH, JOSEPH WOOD, *The American Drama since 1918* (2nd edn., New York: Random House, 1957).

LITTLE, STUART W., *Off-Broadway: The Prophetic Theater* (New York: Coward, 1972).

MANTLE, ROBERT BURNS (ed.), *The Best Plays of . . . and the Year Book of the Drama in America* (Boston: Small, 1920–5; New York: Dodd, Mead, 1926–   ).

—— *The Best Plays of 1909–19* (New York: Dodd, Mead, 1933).

—— *The Best Plays of 1899–1909* (New York: Dodd, Mead, 1944).

—— *The Best Plays of 1894–99* (New York: Dodd, Mead, 1955).

MORDDEN, ETHAN, *The American Theatre* (Oxford University Press, 1981).

ODELL, G. C. D., *Annals of the New York Stage*, 15 vols. (New York: Columbia University Press, 1927–49; AMS, 1970).

SHATTUCK, C. H., *Shakespeare on the American Stage: From the Hallams to Edwin Booth* (Washington, DC: Folger, 1976).

TAUBMAN, HOWARD, *Making of the American Theatre* (rev. edn., New York: Coward, 1967; Longmans).

TOLL, ROBERT C., *Blacking Up: The Minstrel Show in Nineteenth-Century America* (Oxford University Press, 1974).

—— *On with the Show: The First Century of Show Business in America* (Oxford University Press, 1976).

WILLS, JOHN (ed.), *Theatre World* (New York: Crown (annual)). (1st pub. 1945 by Theatre World.)

WILSON, GARFF B., *A History of American Acting* (Bloomington, Ind.: Indiana University Press, 1966).

Dramatic Forms and Theories of Drama

ARTAUD, ANTONIN, *The Theatre and Its Double* (Calder, 1970).

BENTLEY, ERIC, *The Life of the Drama* (Methuen, 1965).

BERGSON, HENRI, *Laughter* (Garden City, NY: Doubleday, 1956).

BRECHT, BERTOLT, *Brecht on Theatre* (Methuen, 1964; New York: Hill & Wang).

BROOK, PETER, *The Empty Space* (Macgibbon & Kee, 1968; New York: Atheneum).

CHARNEY, M. M., *Comedy High and Low* (Oxford University Press, 1978).

CORRIGAN, ROBERT W. (ed.), *Comedy: Meaning and Form* (San Francisco: Chandler, 1965).

—— *Tragedy: Vision and Form* (San Francisco: Chandler, 1965).

CRAIG, EDWARD GORDON, *On the Art of the Theatre* (2nd edn., Heinemann, 1955; New York: Theatre Arts).

ELLIS-FERMOR, UNA, *The Frontiers of Drama* (2nd edn., Methuen, 1964).

GROTOWSKI, JERZY, *Towards a Poor Theatre* (Methuen, 1969; New York: Simon & Schuster).

LAUTER, PAUL (ed.), *Theories of Comedy* (Garden City, NY: Doubleday, 1964).

LUCAS, F. L., *Tragedy: Serious Drama in Relation to Aristotle's Poetics* (Hogarth Press, 1927; New York: Macmillan, 1958).

MEYERHOLD, VSEVOLOD, *Meyerhold on Theatre* (Methuen, 1969; New York: Hill & Wang).

MILLER, JONATHAN, *Subsequent Performances* (Faber, 1986; New York: Viking).

NICOLL, ALLARDYCE, *The Theatre and Dramatic Theory* (Harrap, 1962; New York: Barnes & Noble).

SEYLER, ATHENE, and HAGGARD, STEPHEN, *The Craft of Comedy* (2nd edn., New York: Theatre Arts, 1957; Muller, 1958).

STEINER, GEORGE, *The Death of Tragedy* (Faber, 1961; New York: Oxford University Press, 1980).

STYAN, J. L., *The Dark Comedy: The Development of Modern Comic Tragedy* (Cambridge University Press, 1962).

WILLIAMS, RAYMOND, *Modern Tragedy* (rev. edn., Verso Editions, 1979).

Techniques of Stagecraft

ARCHER, WILLIAM, *Play-Making: A Manual of Craftsmanship* (Chapman & Hall, 1913; New York: Dover, 1960).

BENTHAM, FREDERICK, *The Art of Stage Lighting* (2nd edn., Pitman, 1976).

BURRIS-MEYER, HAROLD, and MALLORY, V., *Sound in the Theatre* (New York: Theatre Arts, 1959).

COTES, PETER, *A Handbook for the Amateur Theatre* (Oldbourne, 1957).

COURTNEY, RICHARD, *Drama for Youth* (Pitman, 1964).

FERNALD, JOHN, *Sense of Direction: The Director and His Actors* (Secker, 1968; New York: Stein & Day, 1969).

GASSNER, JOHN, *Producing the Play*, including *The New Scene Technician's Handbook* by Philip Barber (rev. edn., New York: Dryden, 1953).

GRUVER, BERT, *The Stage Manager's Handbook* (New York: DBS Publications, 1961).

HAINAUX, RENÉ (ed.), *Stage Design throughout the World since 1935* (Harrap, 1957; New York: Theatre Arts, 1965).

—— *Stage Design throughout the World since 1950* (Harrap, 1964; New York: Theatre Arts).

—— *Stage Design throughout the World since 1960* (Harrap, 1973; New York: Theatre Arts).

—— *Stage Design throughout the World 1970–75* (Harrap, 1976; New York: Theatre Arts).

HEFFNER, HERBERT C., *et al.*, *Modern Theatre Practice: A Handbook of Play Production with an Appendix on Costume and Make-up* (4th edn., New York: Appleton, 1959).

JOSEPH, STEPHEN, *Scene Painting and Design* (Pitman, 1964).

LAVER, JAMES, *Costume in the Theatre* (Harrap, 1964; New York: Hill & Wang).

MARSHALL, NORMAN, *The Producer and the Play* (3rd edn., Davis-Poynter, 1975).

PILBROW, RICHARD, *Stage Lighting* (rev. edn., Studio Vista, 1979; New York: DBS Publications).

PRIESTLEY, J. B., *The Art of the Dramatist* (Heinemann; Boston: The Writer, 1957).

ROSENFELD, SYBIL, *A Short History of Scene Design in Great Britain* (Blackwell, 1973).

—— *Georgian Scene Painters and Scene Painting* (Cambridge University Press, 1981).

WELKER, D., *Theatrical Set Design: The Basic Techniques* (2nd edn., Boston: Allyn & Bacon, 1979).

Acting

ALBRIGHT, H. and A., *Acting, the Creative Process* (3rd edn., Belmont, Calif.: Wadsworth, 1980).

BRADBROOK, M. C., *The Rise of the Common Player: A Study of Actor and Society in Shakespeare's England* (Cambridge University Press, 1979).

CALLOW, SIMON, *Being an Actor* (new edn., Penguin, 1985; New York: St Martin's Press, 1986).

CORSON, RICHARD, *Stage Make-up* (5th edn., Englewood Cliffs, NJ: Prentice-Hall, 1975).

DARLINGTON, W. A., *The Actor and His Audience* (Phoenix, 1949).

GIELGUD, JOHN, *Stage Directions* (Heinemann Educational, 1979; New York: Random House).

HODGSON, JOHN, and RICHARDS, ERNEST, *Improvisation* (Methuen, 1966).

MAROWITZ, CHARLES, *The Act of Being* (Secker, 1978; New York: Taplinger).

OLIVIER, LAURENCE, *On Acting* (new edn., Sceptre, 1987).

REDGRAVE, MICHAEL, *Mask or Face: Reflections in an Actor's Mirror* (Heinemann, 1958; New York: Theatre Arts).

—— *The Actor's Ways and Means* (Heinemann Educational, 1979; New York: Theatre Arts).

SANDERSON, M., *From Irving to Olivier* (Athlone Press, 1984; New York: St Martin's Press).

SHER, ANTONY, *Year of the King* (Chatto, 1985).

STANISLAVSKI, CONSTANTIN, *An Actor Prepares* (New York: Theatre Arts, 1936; Eyre Methuen, 1980).

—— *Building a Character* (New York: Theatre Arts, 1949; Eyre Methuen, 1979).

—— *Creating a Role* (New York: Theatre Arts, 1961; Eyre Methuen, 1981).

WILSON, GARFF B., *A History of American Acting* (Bloomington, Ind.: Indiana University Press, 1966).

Theatre Criticism

AGATE, JAMES (ed.), *The English Dramatic Critics* (Barker, 1932; New York: Hill & Wang, 1958).

HOBSON, HAROLD, *Verdict at Midnight* (first-night notices put to the test of time) (Longmans, 1952).

Individual Critics

AGATE, JAMES, *Brief Chronicles* (Cape, 1943).

—— *Red Letter Nights* (Cape, 1944; New York: Blom, 1969).

—— *Immoment Toys* (Cape, 1945; New York: Blom, 1969).

ARCHER, WILLIAM, *The Theatrical 'World'*, 5 vols. (1893–7; Walter Scott, 1894–8; New York: Blom, 1969).

BEERBOHM, MAX, *Around Theatres* (Hart-Davis, 1953; New York: Taplinger, 1969).

—— *More Theatres, 1898–1903* (Hart-Davis, 1969; New York: Taplinger).

—— *Last Theatres, 1904–1910* (Hart-Davis, 1970).

BENTLEY, ERIC, *What is Theatre?* (New York: Atheneum, 1968; Methuen, 1969).

BROWN, JOHN MASON, *Dramatis Personae* (New York: Viking, 1963; Hamilton).

CLURMAN, HAROLD, *Lies like Truth* (New York: Macmillan, 1968).

HAZLITT, WILLIAM, *Hazlitt on Theatre* (Walter Scott, 1895; New York: Hill & Wang, 1957).

HOBSON, HAROLD, *Theatre* (Longmans, 1948).
—— *Theatre 2* (Longmans, 1950).

KERR, WALTER, *Pieces at Eight* (Reinhardt, 1958).
—— *Thirty Plays hath November* (New York: Simon & Schuster, 1969).
—— *Journey to the Center of the Theater* (New York: Knopf, 1979).

NATHAN, GEORGE JEAN, *Passing Judgments* (New York: Johnson Reprint, 1969).

TYNAN, KENNETH, *He that Plays the King* (Longmans, 1950).
—— *Curtains* (Longmans, 1961; New York: Atheneum).
—— *The Sound of Two Hands Clapping* (Cape, 1975).

Autobiographies

BANKHEAD, TALLULAH, *Tallulah* (Gollancz, 1952; New York: Harper & Brothers).

BARRYMORE, LIONEL, *We Barrymores* (Davies, 1951; New York: Grosset & Dunlap).

BEHAN, BRENDAN, *Borstal Boy* (Hutchinson, 1958; New York: Knopf).

BLOOM, CLAIRE, *Limelight and After* (Penguin, 1983).

CORNELL, KATHARINE, *I Wanted to be an Actress* (New York: Random House, 1939).

COWARD, NOËL, *Present Indicative* (Heinemann, 1938; Garden City, NY: Doubleday).
—— *Future Indefinite* (Heinemann, 1954; Garden City, NY: Doubleday).

GIELGUD, JOHN, *Early Stages* (Macmillan, 1939).
—— *Distinguished Company* (Heinemann Educational, 1972).

GRENFELL, JOYCE, *Joyce Grenfell Requests the Pleasure* (Macmillan, 1976).

GUINNESS, ALEC, *Blessings in Disguise* (Hamish Hamilton, 1985; New York: Knopf).

HALL, PETER, *Peter Hall's Diaries* (Hamish Hamilton, 1984; New York: Harper & Row).

HARRISON, REX, *Rex* (Macmillan, 1974; New York: Morrow).

HART, MOSS, *Act One* (new edn., New York: St Martin's Press, 1989; Secker).

LAWRENCE, GERTRUDE, *A Star Danced* (W. H. Allen, 1945; Garden City, NY: Doubleday).

LE GALLIENNE, EVA, *At 33* (Longmans, 1940).
—— *With a Quiet Heart* (New York: Viking, 1953).

MACLIAMMÓIR, *Each Actor on His Ass* (Routledge, 1961).

MILLER, ARTHUR, *Timebends* (New York: Grove Press, 1987; Minerva).

MORTIMER, JOHN, *Clinging to the Wreckage* (Penguin, 1984).

NICHOLS, PETER, *Feeling you're Behind* (Weidenfeld, 1984).

OLIVIER, LAURENCE, *Confessions of an Actor* (Penguin, 1984).

OSBORNE, JOHN, *A Better Class of Person* (Faber, 1981; New York: Elsevier-Dutton).

PIAF, EDITH, *The Wheel of Fortune* (Peter Owen, 1965; Philadelphia: Chilton).

REDGRAVE, MICHAEL, *In my Mind's Eye* (Weidenfeld, 1983; New York: Viking).

REDGRAVE, VANESSA, *Vanessa* (Hutchinson, 1991).

RIX, BRIAN, *My Farce from my Elbow* (Secker, 1975).

USTINOV, PETER, *Dear Me* (Heinemann, 1977; Boston: Little, Brown).

WILLIAMS, EMLYN, *George* (Hamish Hamilton, 1961; New York: Random House).
—— *Emlyn* (Bodley Head, 1973).

WILLIAMS, TENNESSEE, *Memoirs* (Garden City, NY: Doubleday, 1975).

WOLFIT, DONALD, *First Interval* (Odhams, 1955).

ZEFFIRELLI, FRANCO, *Zeffirelli* (Weidenfeld, 1986).